Computing and Systems Technology Division 2015

Core Programming Area at the 2015 AIChE Annual Meeting

Salt Lake City, Utah, USA
8-13 November 2015

Volume 1 of 2

ISBN: 978-1-5108-1856-9

Printed from e-media with permission by:

Curran Associates, Inc.
57 Morehouse Lane
Red Hook, NY 12571

proceedings.com

Some format issues inherent in the e-media version may also appear in this print version.

Printed by Curran Associates, Inc. (2016)

AIChE
120 Wall Street, FL 23
New York, NY 10005-4020

Phone: (800) 242-4363
Fax: (203) 775-5177

www.aiche.org

Additional copies of this publication are available from:

Curran Associates, Inc.
57 Morehouse Lane
Red Hook, NY 12571 USA
Phone: 845-758-0400
Fax: 845-758-2634
Email: curran@proceedings.com
Web: www.proceedings.com

TABLE OF CONTENTS

VOLUME 1

VOLUME 2

437552 Optimization Models and Algorithms for Water Supply Chain Network Design and Operations in Shale Gas Production

Abstract:

In this work, we address the optimal design and operations of water supply chain networks for shale gas production [1, 2]. We develop a mixed-integer linear fractional programming (MILFP) model, the objective of which is to maximize the economic efficiency of freshwater in this shale water supply chain [2]. This model simultaneously takes into account the design and operational decisions, including freshwater source selection, multiple transportation modes, and different water management options [3-7]. Water management options include underground disposal, commercial centralized wastewater treatment (CWT), and onsite treatment (filtration, lime softening, thermal distillation) [8]. To globally optimize the resulting MILFP problem efficiently, we present three tailored solution algorithms: a parametric approach [9], a reformulation-linearization method [10], and a novel Branch-and-Bound & Charnes-Cooper transformation method. The proposed models and algorithms are illustrated through two case studies based on Marcellus shale play, in which onsite treatment shows its superiority in improving freshwater conservancy, maintaining a stable water flow, and reducing transportation burden.

References

[1] J. Gao and F. You, "Optimal design and operations of supply chain networks for water management in shale gas production: MILFP model and algorithms for the water-energy nexus," *AIChE Journal,* vol. 61, pp. 1184-1208, 2015.

[2] J. Gao and F. You, "Shale Gas Supply Chain Design and Operations towards Better Economic and Life Cycle Environmental Performance: MINLP Model and Global Optimization Algorithm," *ACS Sustainable Chemistry & Engineering*, 2015, 10.1021/acssuschemeng.5b00122.

[3] J.-P. Nicot and B. R. Scanlon, "Water Use for Shale-Gas Production in Texas, U.S," *Environmental Science & Technology,* vol. 46, pp. 3580-3586, 2012.

[4] C. E. Clark, R. M. Horner, and C. B. Harto, "Life Cycle Water Consumption for Shale Gas and Conventional Natural Gas," *Environmental Science & Technology,* vol. 47, pp. 11829-11836, 2013.

[5] M. Jiang, C. T. Hendrickson, and J. M. VanBriesen, "Life Cycle Water Consumption and Wastewater Generation Impacts of a Marcellus Shale Gas Well," *Environmental Science & Technology,* vol. 48, pp. 1911-1920, 2013.

[6] I. J. Laurenzi and G. R. Jersey, "Life Cycle Greenhouse Gas Emissions and Freshwater Consumption of Marcellus Shale Gas," *Environmental Science & Technology,* vol. 47, pp. 4896-4903, 2013.

[7] A. Vengosh, R. B. Jackson, N. Warner, T. H. Darrah, and A. Kondash, "A Critical Review of the Risks to Water Resources from Unconventional Shale Gas Development and Hydraulic Fracturing in the United States," *Environmental Science & Technology,* vol. 48, pp. 8334-8348, 2014.

[8] J. M. Wilson and J. M. VanBriesen, "Oil and Gas Produced Water Management and Surface Drinking Water Sources in Pennsylvania," *Environmental Practice,* vol. 14, pp. 288-300, 2012.

[9] Z. Zhong and F. You, "Globally convergent exact and inexact parametric algorithms for solving large-scale mixed-integer fractional programs and applications in process systems engineering," *Computers & Chemical Engineering,* vol. 61, pp. 90-101, 2014.

[10] D. J. Yue, G. Guillen-Gosalbez, and F. Q. You, "Global Optimization of Large-Scale Mixed-Integer Linear Fractional Programming Problems: A Reformulation-Linearization Method and Process Scheduling Applications," *AIChE Journal,* vol. 59, pp. 4255-4272, 2013.

436630 Metrics Driven Systems Analysis for Water & Energy Sustainability

Abstract:

This presentation focusses on showcasing how sustainability metrics can be utilized to drive energy and water sustainability outcomes in communities. Communities and local governments play an important role in developing people centric sustainable solutions as they have broad influence over activities that result in energy and water consumption within their boundary and jurisdiction. But most often their decisions are characterized by lack of quantitaitve information regarding the complex and often conflicting interactions between different aspects of the problem including the so called water-energy nexus. The authors will present how carbon footprint and water footprint metrics can be utilized within the framework of scenario modeling to help policy makers evaluate sustainability outcomes and tradeoffs associated with different policy choices. Several interesting results related to the sustainability analysis of San Luis Basin in south-central Colorado will be presented.

428508 An Industrial Ecology Approach for Managing Wastewater from Shale Gas Production

Abstract:

The rapid development of the unconventional shale gas industry offers tremendous economic benefits while simultaneously posing a critical challenge of managing vast quantities of produced and flowback water. Approximately 10-25% of the water injected during hydraulic fracturing returns to the surface as highly contaminated water with total dissolved solids reaching as high as 350,000 mg/Liter with potential detrimental impacts on human health and ecosystems. Sustainable development of the shale gas industry will require economical management of wastewater while minimizing associated environmental impacts. Membrane distillation (MD) is an emerging technology for treatment of high salinity wastewater and offers promise owing to its lower energy requirements and capital investment when compared with existing competing technologies. By utilizing the concept of industrial ecology, the economic and environmental sustainability of the MD technology can be further enhanced by a systems-level integration of MD process with waste heat sources. One such source is the waste heat available at existing natural gas compressor stations. The U.S. Natural Gas (NG) pipeline network is a highly integrated transmission grid consisting of more than 300,000 miles of transmission pipeline and 1,400 Compressor Stations (CS). In most NG CS, a portion of NG is burned to meet the energy demands of the compression process. Concurrently, about two-thirds of the fuel energy is rejected as waste heat mainly in the form of high temperature flue gas.

We present a framework to evaluate the synergies and potential of utilizing waste heat available from existing natural gas compressor stations for a novel MD technology for treatment of saline wastewater in the Marcellus shale gas play. We determine the quantity and quality of waste heat available at existing NG CS based on capacity, thermodynamic constraints, and the inlet feed temperature for MD. We also estimate the amount of wastewater that could be treated at each compressor station in Marcellus shale gas play based on experimentally measured flux rates. A Geographic Information System (GIS) based optimization framework is developed for informing decision regarding optimal wastewater routing from existing shale gas wells to hypothetical MD plants collocated with compressor stations. The results and insights from this work will be helpful in characterizing regional opportunities and constraints for management of wastewater using MD technology and waste heat sources. The implications of the work for informing sustainable management of water-energy nexus for the shale gas industry will be discussed along with a comparison with existing treatment options.

437566 Integrated Optimization of Water Supply Systems Scheduling for Energy Efficient Operations

Abstract:

In this work, we address the simultaneous optimization for the integrated water supply systems scheduling and pump operation to minimize the cost and energy consumption of pumps. The water supply systems are operated to transport water from sources to consumers [1-3]. The operation of the water supply systems is expensive because of large amount of water demands in the industry process and the daily life. Optimal scheduling of the water supply systems helps obtain remarkable savings by coordinating the pumping with time varying electricity tariffs [4]. The characteristics of the pumps can be described by a set of differential equations [5]. A great amount of energy can be saved if the operation of the pumps can be optimized. We integrate and optimize the water supply system scheduling and the pump operation simultaneously. We first formulate the integrated problem into a mixed-integer dynamic optimization (MIDO) model, which is then reformulated into a mixed-integer nonlinear programming (MINLP) full space model by discretizing the differential equations [6-8]. We propose an efficient flexible recipe method which can reduce the computational complexity significantly to solve the integrated problem [9, 10]. The flexible recipe method uses a set of discrete points to approximate the detailed dynamic optimization problems. In case study, we use a real water supply network system to demonstrate the applicability of the proposed full space modeling framework and the efficiency of the flexible recipe solution method.

References

[1] M. M. Eusuff and K. E. Lansey, "Optimization of water distribution network design using the shuffled frog leaping algorithm," *Journal of Water Resources Planning and Management,* vol. 129, pp. 210-225, 2003.

[2] A. Vasan and S. P. Simonovic, "Optimization of water distribution network design using differential evolution," *Journal of Water Resources Planning and Management,* vol. 136, pp. 279-287, 2010.

[3] D. L. Boccelli, M. E. Tryby, J. G. Uber, L. A. Rossman, M. L. Zierolf, and M. M. Polycarpou, "Optimal scheduling of booster disinfection in water distribution systems," *Journal of Water Resources Planning and Management,* vol. 124, pp. 99-111, 1998.

[4] B. Ulanicki, J. Kahler, and H. See, "Dynamic optimization approach for solving an optimal scheduling problem in water distribution systems," *Journal of Water Resources Planning and Management,* vol. 133, pp. 23-32, 2007.

[5] K. E. Lansey and K. Awumah, "Optimal pump operations considering pump switches," *Journal of Water Resources Planning and Management,* vol. 120, pp. 17-35, 1994.

[6] Y. Chu and F. You, "Integration of scheduling and control with online closed-loop implementation: Fast computational strategy and large-scale global optimization algorithm," *Computers & Chemical Engineering,* vol. 47, pp. 248-268, Dec 20 2012.

[7] H. Shi, Y. Chu, and F. You, "Novel Optimization Model and Efficient Solution Method for Integrating Dynamic Optimization with Process Operations of Continuous Manufacturing Processes," *Industrial & Engineering Chemistry Research,* vol. 54, pp. 2167-2187, 2015.

[8] H. Shi and F. You, "A Novel Adaptive Surrogate Modeling based Algorithm for Simultaneous Optimization of Sequential Batch Process Scheduling and Dynamic Operations," *Submitted,* 2015.

[9] Y. Chu and F. You, "Integrated Scheduling and Dynamic Optimization of Sequential Batch Processes with Online Implementation," *AIChE Journal,* vol. 59, pp. 2379-2406, Jul 2013.

[10] Y. Chu and F. You, "Integrated Scheduling and Dynamic Optimization of Complex Batch Processes with General Network Structure Using a Generalized Benders Decomposition Approach," *Industrial & Engineering Chemistry Research,* vol. 52, pp. 7867-7885, Jun 12 2013.

Optimization of Integrated Water and Multi-Regenerator Membrane Systems

Musah Abass and Thokozani Majozi[*]

School of Chemical and Metallurgical Engineering, University of Witwatersrand, 1 Jan Smuts Avenue, Braamfontein, Johannesburg, 2000, South Africa

Abstract (ID # 408816)

Water and energy are key resources in the process industry. The depletion of freshwater and energy resources coupled with stringent environmental regulations on effluent discharge limits have called for innovative designs for sustainable use of water and energy. This can be achieved through process integration techniques that are environmentally benign and economically feasible. This work proposes a robust water network superstructure optimisation approach for the synthesis of a multi-regenerator network for simultaneous water and energy minimisation. Two types of membrane regenerators are considered for this work, namely, electrodialysis and reverse osmosis. In each of the membrane regenerators, a detailed design model is developed and incorporated into the water network model. This presents a rigorous design and accurate cost representation as compared to the "blackbox" approach. The presence of continuous and integer variables, as well as nonlinear constraints renders the problem a MINLP. The developed model is applied to a pulp and paper case study to demonstrate its applicability, assuming a single contaminant scenario. The model was solved in GAMS using a solver BARON and the results indicate a 43.7% freshwater reduction, 50.9% decrease in wastewater generation and 47% savings in total annualised cost.

Keywords: Sustainable, Synthesis, Optimization, Reverse-osmosis, Electrodialysis.

1. Introduction

Water and energy are important resources for the development and wellbeing of humanity. They are key components in the process industry and great amounts of each resource are consumed to produce the other. Process integration is often employed in order to achieve a holistic water network for water and energy minimisation. This is done through an integrated water network that is open for direct reuse, recycle and regeneration reuse/recycle for sustainable cost-effective water and energy usage.

Water pinch techniques and mathematically based optimization techniques are the two main approaches used for optimal synthesis of water networks in process industry. Wang and Smith (1994a, b) presented the seminal work in water pinch analysis which sets to target the reuse and regeneration reuse/recycle of wastewater in order to reduce freshwater consumption. The work of Wang and Smith (1994a, b) has since developed into advanced graphical, tabular and heuristic methods for water network synthesis.

Takama *et al.* (1980) presented a superstructure optimisation approach for water minimisation in a petrochemical refinery based on a fixed mass load. The work of Takama *et al.* (1980) was later extended by several other researchers (Doyle and Smith 1997, Ahmetovic and Grossmann 2010) in order to obtain optimal freshwater usage in their respective water network superstructure. Notable among them are the work of Quesada and Grossmann (1995), Karrupiah and Grossmann (2006) and Faria and Bagajewicz (2011) who present algorithms to obtain feasible solutions in cases of the more complex nonlinear programming (NLP) and mixed integer

[*] Corresponding author: Tel: +27 11 717 7384; Fax: +27 82 456 1500
Email address: Thokozani.Majozi@wits.ac.za;

nonlinear programming (MINLP) problems that are in many cases a challenge to solve. Mathematical optimization approaches are however computationally intensive.

Tan *et al.* (2009) presented a water network superstructure with a single membrane partitioning regenerator which allows for possible reuse/recycle. The work of Tan *et al.* (2009) considered the "blackbox" approach which uses linear cost functions for the membrane regenerators. This does not give an accurate cost representation of the water network. Khor *et al.* (2011) addressed the gap on the work of Tan *et al.* (2009) by developing a detailed model representation for water network regeneration synthesis using a MINLP optimization framework. The work of Khor *et al.* (2011) was, however, limited to a single regenerator with a fixed design. In a more recent development, Yang *et al.* (2014) proposed a unifying approach by combining multiple water treatment technologies capable of treating all major contaminants. The work focused on unit specific short cut cost functions in order to gain detailed understanding of trade-offs between removal efficiency of treatment units and the cost of the units, as well as the impact on the unit design.

The membrane technologies adopted in this work involve electrodialysis (ED) and reverses osmosis (RO). Reverse osmosis (RO) is a pressure driven membrane separation process that selectively allows the passage of one or more species through the membrane unit. The application of membrane systems as regenerators generally requires enormous amount of energy. Most published work, however, uses linear cost functions or "blackbox" representation for membrane partitioning regenerators (Tan *et al.*, 2009, Khor *et al.*, 2012). This does not give an accurate cost representation of the membrane systems. There is, therefore, a need for a technique that would cater for energy minimization through detailed synthesis of membrane regeneration systems in order to obtain optimal variables that affect the operation and economics of the regenerator unit.

The main objective of this work is to develop a systematic technique that would simultaneously optimize energy and water consumption within a water network superstructure. The technique is based on an integrated water and membrane regeneration network superstructure. The membrane regenerators in this representation are ED and RO. The choice of regenerators is motivated by the increasing demand and use of membrane technology for water treatment in the process industry and also the varying potential of different membrane systems to treat specific waste ranges. Detailed synthesis of the membrane regeneration systems is conducted to determine optimal operating conditions for efficient energy usage in terms of costs. The detailed model of the regenerators is incorporated in the overall water network objective function in order to minimize freshwater and energy consumption and also give a true cost representation as compared to the "blackbox" method. It is also demonstrated that treating the removal ratio as an additional degree of freedom allowed for better performance of the integrated system. This is contrary to what is presented in literature where the removal ratio is generally treated as a parameter.

2. Problem statement

The problem statement in this work can be stated as follows:

Given:

(i) A set of water sources, J, with known flowrates and contaminant concentrations.
(ii) A set of water sinks, I, with known flowrates and known maximum allowable contaminant concentrations.
(iii) Membrane regeneration units with the potential for parallel/series connection for partial treatment of wastewater from sources for reuse/recycle.
(iv) A freshwater source, FW, with known concentration and variable and unlimited flowrate.

(v) A wastewater sink, *WW*, with maximum allowable contaminant concentration and variable and unlimited flowrate.

Determine:

(i) The minimum freshwater intake, wastewater generation, the energy consumed in the ED and RO units, and the total annualised costs for ED (TAC_e) and RO (TAC_r).
(ii) Optimal water network configuration.
(iii) Optimum design variables of the regenerators.

3. Model formulation

Based on the problem statement, water network superstructure in Figure 1 is developed. The superstructure representation is an extension of the work by Khor *et al.* (2011). The superstructure in this work incorporates multiple regenerators which are open for parallel and series connection as well as recycle and reuse of both permeate and reject streams from regenerators.

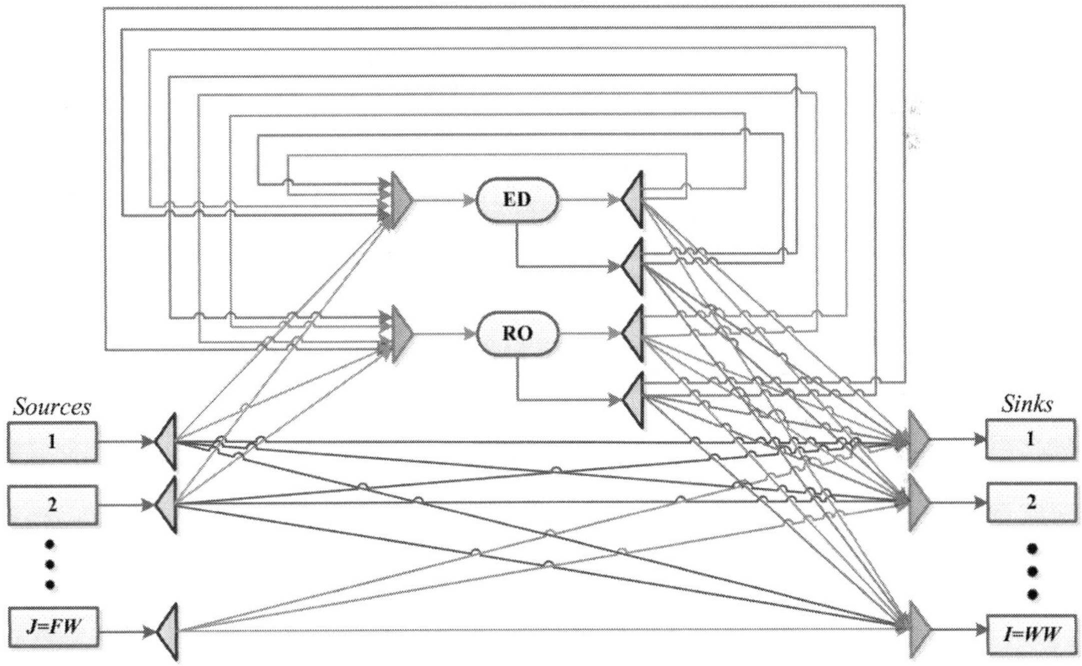

Figure 1: General water network superstructure with multiple membrane regenerators

3.1 *The objective function*

The objective function minimises the overall annualised cost of the water network. This includes total annualised costs of ED and RO units, freshwater cost, wastewater treatment cost, as well as capital and operating costs of piping interconnections. Constraint (1) represents the objective function of the overall water network superstructure. A 1-norm Manhattan distance is considered for all piping interconnections. All pipes are assumed to be of the same material properties and as a result the carbon steel pipes parameters of p and q are adopted for the piping costs.

$$Obj = \min \left(\begin{array}{l} (C_{water} FW + C_{waste} WW) AOT + TAC_e + TAC_r + \\ AF \left[\begin{array}{l} \sum_{j \in J} \sum_{i \in I} D^a_{(j,i)} \left(\frac{pQ^a_{(i,j)}}{3600\,v} + qY^a_{(j,i)} \right) + \sum_{j \in J} D^b_j \left(\frac{pQ^{ro}_j}{3600\,v} + qY^{ro}_j \right) \\ \sum_{j \in J} D^c_j \left(\frac{pQ^{ed}_j}{3600\,v} + qY^{ed}_j \right) + \sum_{i \in I} D^{ds}_i \left(\frac{pQ^{ds}_i}{3600\,v} + qY^{ds}_i \right) + D^{dr} \left(\frac{pQ^{dr}}{3600\,v} + qY^{dr} \right) \\ + D^{de} \left(\frac{pQ^{de}}{3600\,v} + qY^{de} \right) + \sum_{i \in I} D^{es}_i \left(\frac{pQ^{es}_i}{3600\,v} + qY^{es}_i \right) + D^{er} \left(\frac{pQ^{er}}{3600\,v} + qY^{er} \right) \\ + D^{ee} \left(\frac{pQ^{ee}}{3600\,v} + qY^{ee} \right) + \sum_{i \in I} D^{fs}_i \left(\frac{pQ^{fs}_i}{3600\,v} + qY^{fs}_i \right) + D^{fr} \left(\frac{pQ^{fr}}{3600\,v} + qY^{fr} \right) \\ + D^{fe} \left(\frac{pQ^{fe}}{3600v} + qY^{fe} \right) + \sum_{i \in I} D^{gs}_i \left(\frac{pQ^{gs}_i}{3600\,v} + qY^{gs}_i \right) + D^{gr} \left(\frac{pQ^{gr}}{3600\,v} + qY^{gr} \right) \\ + D^{ge} \left(\frac{pQ^{ge}}{3600\,v} + qY^{ge} \right) \end{array} \right] \end{array} \right) \quad (1)$$

The resulting mathematical model is an MINLP. The nonlinear terms are due to the presence of bilinear terms in mass balance equations and power terms in the cost functions of regeneration units.

4. Illustrative Example

The developed mathematical model is verified and applied to a pulp and paper case study adopted from Chew *et al.* (2008). The MINLP model was implemented in GAMS 24.2 using the general purpose global optimisation solver BARON which obtains a solution by using a branch-and-reduce algorithm. Table 1 shows the basic data for the plant water network which comprises five water sources including the freshwater source, and five water sinks including the wastewater sink.

Table 1: Basic data for water sources and sink

Sources, *j*			Sinks, *i*		
j	Flowrate (t/h)	Concentration (mg/L)	*i*	Flowrate (t/h)	Max. concentration (mg/L)
1	2.07	0.0002	1	3.26	0.00057
2	0.34	0.0051	2	0.34	0
3	0.024	0	3	1.34	6.16×10^{-6}
4	7.22	0.0083	4	7.22	0
FW	—	0	*WW*	—	0.01

The case study was applied to three different scenarios in order to ascertain the benefits of incorporating detailed models of the membrane regeneration units in the constraints of the water network.

(i) Case 1 considered a model of the water network without regeneration.
(ii) Case 2 considered a model of water network with regeneration units based on the "blackbox" approach.
(iii) Case 3 considered detailed synthesis of the regeneration units incorporated into the overall water network objective function.

The optimal results for all cases are presented in Table 2. The results showed significant reduction in freshwater consumption, wastewater generation as well as the total cost of the water network for case 2 as compared to the based case.

Table 2: Optimal results of water network based on the case study

	Scenario1	**Scenario 2**		**Scenario3**	
		Fixed RR	Variable RR	Fixed RR	Variable RR
Freshwater use (t/h)	18.30	11.42	9.83	11.64	10.3
% of freshwater savings		37.5	46.3	36.4	43.7
Wastewater generated (t/h)	15.79	8.89	7.3	9.11	7.75
% of wastewater saved		43.7	53.7	42.3	50.9
Total cost of water network($)	$1.170 x 10^6$	$5.97 x 10^5$	$5.08 x^5$	$6.34 x 10^5$	$6.20 x 10^5$
CPU time	0.06	687.50	2764.20	865	16709.78

The results of case 3 are displayed in Figure 2. Case 3 showed significant reduction in water network cost as well as freshwater consumption and wastewater reduction as compared to the direct water network model without regeneration. However, there was an increase in the total water network cost compared to the case 2. This is as a result of case 3 being a true representation of the total water network as it incorporates a detailed design of the membrane regeneration units and gives an accurate expression of the regeneration cost. From the results in Table 2 it is evident that variable removal ratio in case 3 presents the optimal water network configuration as presented in Figure 2, since the model is allowed to choose the performance parameters of the membrane regenerators. This resulted in 43.7 % reduction in freshwater consumption, 50.9% reduction in wastewater generation and 45 % savings in the total annualised water network cost as compared to scenario 1. The model also showed 34% and 41% savings in energy for both ED and RO units respectively.

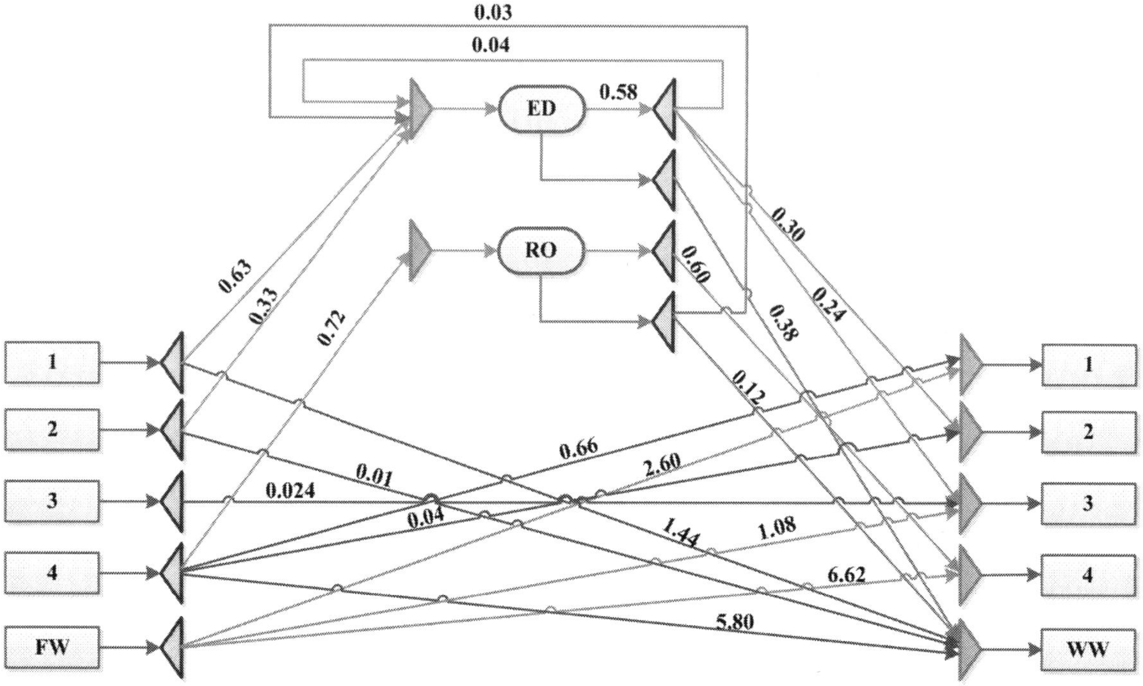

Figure 2: Optimum water network configurations for scenario 3 (Variable RR)

5. Conclusion

This work has addressed the synthesis of a multi-membrane regeneration water network by proposing a MINLP optimisation model that incorporates detailed mathematical models of ED and RO units within a water network. The developed water network allows for direct reuse/recycle and regeneration reuse/recycle. The developed MINLP model is applied to a pulp and paper industry case study and solved using GAMS/BARON. The results showed that setting the removal ratio as variable yields optimal configuration as compared to the fixed removal ratio. The model also showed that incorporating a detailed model of the regeneration unit into the overall model yields accurate expression of the regeneration cost as compared to the ''blackbox'' model. It also gives optimal operating variables of the regeneration units for minimal energy usage.

6. Nomenclature

6.1 Sets

$J = \{j|j = \text{water source}\}$

$I = \{i|i = \text{water sink}\}$

6.2 Parameters

AOT	Annual operating time
C_{waste}	Unit cost for waste treatment
C_{water}	Unit cost for freshwater
p, q	Parameter for carbon

	steel piping
v	Velocity of fluid in pipes
$D^{a}_{j,i}$	Manhattan distance from source j to sink i
D^{b}_{j}	Manhattan distance from source j to RO unit
D^{c}_{j}	Manhattan distance from source j to ED unit
D^{ds}_{i}	Manhattan distance from permeate of ED unit to sink i
D^{dr}	Manhattan distance from permeate stream of ED unit into ED unit
D^{es}_{i}	manhattan distance from reject stream of ED unit into sink i
D^{fs}_{i}	Manhattan distance from permeate stream of RO unit into sink i
D^{gs}_{i}	Manhattan distance from reject stream of RO unit into sink i

6.3 Continues variable

FW	Freshwater flowrate
WW	Wastewater flowrate
$Q^{a}_{j,i}$	Flowrate from source j to sink i
Q^{b}_{j}	Flowrate from source j to RO unit
Q^{c}_{j}	Flowrate from source j to ED unit
Q^{ds}_{i}	Flowrate from permeate stream of ED into sink i
Q^{es}_{i}	Flowrate from reject stream of ED unit into sink i
Q^{fs}_{i}	Flowrate from permeate stream of RO unit into sink i
Q^{gs}_{i}	Flowrate from reject stream of RO unit into sink i
TAC_e	Total annual cost of RO unit
TAC_r	Total annual cost of ED unit

6.4 Integer Variables

N_m	Number of reverse osmosis modules
N	Number Electrodialysis cell pairs

6.5 Binary Variables

$$y^{a}_{j,i} = \begin{cases} 1 \leftarrow \text{ Existence of piping interconnection between source and sink,} \\ 0 \leftarrow \text{Otherwise} \end{cases}$$

$$y^{b}_{j} = \begin{cases} 1 \leftarrow \text{ Existence of piping interconnection between source and RO unit,} \\ 0 \leftarrow \text{Otherwise} \end{cases}$$

$$y^{c}_{j} = \begin{cases} 1 \leftarrow \text{ Existence of piping interconnection between source and ED unit,} \\ 0 \leftarrow \text{ Otherwise} \end{cases}$$

$$y^{ds}_{i} = \begin{cases} 1 \leftarrow \text{ Existence of piping interconnection between permeate of ED unit and sink,} \\ 0 \leftarrow \text{ Otherwise} \end{cases}$$

$$y^{ds}_{i} = \begin{cases} 1 \leftarrow \text{ Existence of piping interconnection between permeate of ED unit and ED unit ,} \\ 0 \leftarrow \text{ Otherwise} \end{cases}$$

$$y^{es}_{i} = \begin{cases} 1 \leftarrow \text{ Existence of piping interconnection between reject of ED unit and sink ,} \\ 0 \leftarrow \text{ Otherwise} \end{cases}$$

$$y_i^{fs} = \begin{cases} 1 \leftarrow & \text{Existence of piping} \\ & \text{interconnection between} \\ & \text{permeate of RO unit} \\ & \text{and sink ,} \\ 0 \leftarrow & \text{Otherwise} \end{cases} \qquad y_i^{gs} = \begin{cases} 1 \leftarrow & \text{Existence of piping} \\ & \text{interconnection between} \\ & \text{reject of RO unit} \\ & \text{and sink ,} \\ 0 \leftarrow & \text{Otherwise} \end{cases}$$

7. Acknowledgement

The authors would like to thank the National Research Foundation (NRF) for funding this work under the NRF/DST Chair in Sustainable Process Engineering at the University of the Witwatersrand, South Africa.

8. References

Ahmetovic, E., & Grossmann, I. E., (2010). Strategies for Global Optimization of Integrated Process Water Networks. *ESCAPE* 20.

Chew, I. M. L., Tan, R., Ng, D. K. S., Foo, D. C. Y., Majozi, T., Gouws, J. (2008). Synthesis of Direct and Indirect and Interplant Water and Network. *Industrial Engineering Chemical Research*. 47(2008) 9485 – 9495.

Doyle, S. J & Smith, R. (1997). Targeting Water Reuse with Multiple Contaminants. *Process Safety and Environmental Protection,* 75(3) 181 -189.

Faria, D. C., & Bagajewicz, M. J., (2010). On the Appropriate Modelling of Process Plant Water Systems. American Institute of Chemical Engineers Journal, 53(3) 666 - 689.

Karrupiah, R., & Grossmann, I. E., (2006). Global Optimization for the Synthesis of Integrated Water and Systems in Chemical Processes. Computers & Chemical Engineering, 30: 650 – 673.

Khor, C. S., Foo, D. Y., El-Halwagi, M. M., Tan, R. R., & Shah, N. (2011). A Superstructure Optimization Approach for Membrane Separation –Based Water Regeneration Network Synthesis with Detailed Nonlinear Mechanistic Reverse Osmosis Model. Industrial & Engineering Chemistry Research, 50(11), 13444 – 13456.

Quesada, I. & Grossmann, I. E., (1995). Global Optimization of Bilinear Process Networks with Multicomponent Flows. Computers and Chemical Engineering, 19(12), 1219-12421.

Takama, N., Kuriyama, T., Shiroko, K. & Umeda, T. (1980). Optimal Planning of Water Allocation in Industry. Computers and Chemical Engineering, 4(80), 251 – 258.

Tan, R. R., Ng, D. K. S., Foo, D. C. Y., Aviso, K. B. (2009). A Superstructure Model for the Synthesis of Single-Contaminant Water Networks with Partitioning Regenerators. *Process Safety and Environmental Protection*, 87(9), 197 – 205.

Yang, L., Salcedo-Diaz, R., Grossmann. E. I. (2014). Water Network Optimization with Wastewater Regeneration Models. *Industrial &Engineering Chemistry Research.* 53(2014)17680 – 17695.

Wang, Y. P. & Smith, R. (1994a) Wastewater Minimization. *Chemical Engineering Science*, 49(94), 981 – 1006.

Wang, Y. P. & Smith, R. (1994b) Design of Distributed Effluent Treatment Systems. *Chemical Engineering Science,* 49(94), 3127– 3145.

424452 An MINLP Model for the Water Management in Shale Gas Operations

Abstract:

This work presents a mathematical programing approach for the optimal management of flowback water in shale gas wells in order to meet water demands in hydraulic fracturing operations considering a fixed scheduling for the completion phases of the wells. The objective function is aimed to determine the minimum total costs for the water network constituted by capital and operational costs. Capital costs are associated to the equipment acquisition of storage units, treatment technologies and disposals; whereas the operating costs considers the fresh water cost, operating cost for water treatment and the transportation costs for all transport trajectories required. Additionally, the total amount of fresh water used (environmental concern) is quantified and minimized; in this context, the methodology accounts for the flowback water reuse once it is treated and stored with the purpose to reduce the fresh water consumption, the wastewater streams and the costs. The model involves the seasonal limitations for the fresh water availability, which is an important challenge for some of the existing shale plays.

The proposed design considers that all the flowback water is treated and the outlet streams satisfy the restrictions for the pollutant concentrations in order to be reused and/or the environmental regulations to be disposed. To carry out this task, only treatment units with tested effectiveness in flowback water coming from shale gas wells have been considered. Also, the given scheduling for the fracking operations of each well taken into account can include the simultaneous operation for several fracking crews. The mathematical formulation corresponds to an MINLP multi-period model. Finally, the example problem uses updated information taken of technical reports from some of the most important existing shale regions (Marcellus and Barnette).

423943 Overview of CAST Activities and Programming

Abstract:

An overview of activities of the Computing and Systems Technology (CAST) division, including conference technical programming, will be presented. The CAST Executive Committee consists of Marianthhi Ierapetritou (Past Chair), Ray Adomaitis (Chair), Karl Schnelle (1st Vice-Chair), Stratos Pistikopoulos (2nd Vice-Chair), Martha A. Grover (Secretary-Treasurer), and the following Directors: John Richards (2013-2015), Carl Laird (2013-2015), Andreas Linninger (2014-2016), Leo Chiang (2014-2016), Michael Nikolaou (2015-2017), Mario Eden (2015-2017). The CAST Division Programming Chair is Nick Sahinidis (2014-2017).

415708 Systematic Computer-Aided Framework for Sustainable Chemical Product Design

Abstract:

Computer-aided product design (CAPD) is a method for the generation and selection of novel pure, mixed and blended chemical products [1]. In CAPD, the chemical product is generated and selected based on defined structure, property and process/application constraints. Several challenges exist for CAPD, including the accuracy of the property estimation, molecular structure generation, inclusion of sustainability, process and application targets and needs in the problem formulation. CAPD has been widely utilized for the synthesis of several types of products, such as solvents, polymers, fuels and formulated products [2]. However, for product design problems where the process needs and sustainability possess demanding constraints, the complexity of the problem is increased as the relation between product property, sustainability and process criteria is difficult to mathematically define.

In this work, a generic computer-aided framework for chemical product design is presented through a systematic framework. A CAPD problem for the generation of novel pure, mixed and blended chemical products is formulated and solved through the application of four sequential steps. In step (1), the product design problem is defined together with the process and/or application boundaries. In step (2), the CAPD problem is formulated through property constraints for pure, mixed and blended products, process/application constraints and objective function. The property constraints are carefully selected for the thermo-physical property needs and the process/application needs. Process/application and property needs are connected through an analysis of the property influence on the process/application models and thermodynamic relations. The sustainability is considered through product and process/application performance, economics and environmental impact. In step (3), the CAPD formulation is converted into a mixed-integer nonlinear program (MINLP) by set-up of constraints, objective and boundaries defined in step (2). In step (4), the MINLP is solved through a decomposed approach [3]. The decomposed approach breaks down the MINLP problem into a sub-set of programs to manage the complexity: mixed-integer linear program (MILP) for molecular generation, linear program (LP) for property constraints, non-linear program (NLP) for mixture/blend property constraints, and NLP for process constraints and objective function. This approach ensures that the optimal chemical product can be found through systematic generation and screening of alternatives based on the problem definition. The framework is implemented into a GAMS-based tool.

The application of the framework is demonstrated through sustainable working fluid design for a heat pump cycle. Working fluids used in heat pumps are continuously regulated due to the environmental issues, such as ozone depletion or global warming potential. Many of the currently used working fluids are soon to be phased out [4]. The design of novel working fluids is a challenging task as retro-fitting as well as re-designing demand increased sustainability and minimal trade-off with system performance. In the CAPD formulation, the product properties are related to the needs of heat pump cycle and its components through sensitivity analysis of the thermodynamic models and energy balances of the system. Furthermore, simple models are included for efficient assessment of the sustainability and design criteria of both the cycle and its components. It will be demonstrated that the working fluid product designed is optimal with respect to the sustainability and the heat pump cycle performance.

References:

[1] Yunus, N. A., Gernaey K. V., Woodley, J. M., Gani, R. (2014). A systematic methodology for design of tailor-made blended products, Computers & Chemical Engineering, 66, 201–213.

[2] Gani, R. (2004). Chemical product design: challenges and opportunities. Computers & Chemical Engineering, 28(12), 2441-2457.

[3] Cignitti, S., Zhang, L., & Gani, R. (2015). Computer-aided framework for design of pure, mixed and blended products. Computer Aided Chemical Engineering, 37, 2093-2098.

[4] Mota-Babiloni, A., Navarro-Esbrí, J., Barragán-Cervera, Á., Molés, F., Peris, B. (2015). Analysis based on EU Regulation No 517/2014 of new HFC/HFO mixtures as alternatives of high GWP refrigerants in refrigeration and HVAC systems, International Journal of Refrigeration, 52, 21–31.

424379 Multiparametric Dynamic Programming – Recent Advances and Future Challenges

Abstract:

Dynamic programming is a well-known and powerful paradigm for the solution of multi-stage processes. It relies on Bellman's optimality principle, which states that the optimal solution to a multi-stage optimization problem can be obtained by solving each stage recursively to optimality [1]. In particular, dynamic programming has been combined with multiparametric programming (mp-P) and was employed in explicit model-predictive control (MPC), where the MPC problem is solved explicitly offline as a function of the states.

In this work, we examine the current state-of-the-art of mp-DP algorithms. In particular, we show how all approaches presented in the literature can be categorized into direct and indirect mp-DP algorithms [2, 3]. In indirect mp-DP algorithms, one mp-P problem is solved per stage, however at each stage an additional parameter is added to the problem formulation in order to deal with the presence of nonconvexity. Hence when considering larger horizons, this procedure results in a computationally prohibitive effort. Conversely, direct mp-DP algorithms do not increase the problem size as a function of the horizon, as they directly use the principles of dynamic programming in mp-P. Using benchmark test sets, we show that current direct mp-DP algorithms are computationally attractive for hybrid systems, but are significantly outperformed by concatenation-based approaches for continuous systems. The key reasons for this are that (a) n mp-P problems need to be solved per stage, where n is the number of critical regions of the previous stage, and (b) each stage requires the comparison of n parametric profiles, which results in a significant computational burden.

Based on these considerations we propose a novel mp-DP algorithm for continuous linear state-space systems which does not require a comparison procedure while maintaining a constant problem size regardless of the horizon considered. It is based on the concept of direct mp-DP algorithms, which directly substitute the solution of the previous stage. Specifically, the optimal objective function of the previous stage is added to the objective function of currently considered stage. As this function is different for different critical regions of the parameter space, direct mp-DP algorithms require the solution of n mp-P problems.

However, it is well-known in multiparametric programming, that the optimal objective function value of explicit control problems with linear or convex quadratic cost function is convex [4]. Additionally, it was shown in the context of robust explicit control that it is possible to formulate convex piecewise functions as a single variable subject to constraints based on these function [5]. Hence, in this contribution we formulate the stage-wise mp-P using this property, which eliminates the need to formulate and solve n mp-P problems and compare their solutions. While this is indeed applicable to any convex, piecewise function, we specifically consider convex, piecewise affine objective functions as they result in a set of linear constraints which can be handled with existing solvers. Numerical examples are used to highlight the computational improvements over existing mp-DP algorithms, and its performance is compared to concatenation-based approaches.

Based on this result, we investigate new research directions in mp-DP including (a) applicability of mp-DP algorithms to the control of hybrid systems with long horizons, (b) combination of mp-DP algorithms with robust optimization and (c) computational gains in certain classes of problems involving specific characteristics such as path constraints which can be handled well using mp-DP algorithms

As an illustration, we compare the capabilities of mp-DP algorithm with concatenation-based approaches for the case of a cogeneration of heat and power (CHP) system [6]. The CHP is characterised by collaborative interactions between a power generation subsystem and a heat recovery subsystem. The control principles for the former are reduced to the manipulation of the air and gas intake in order to cover the electrical power demand. The feasible operation of the system is taken into account by formulating a model predictive controller with a single input and a single output system. This results in the presence of path constraints within the control problem formulation, which can be dealt with seamlessly using mp-DP. Numerical results will be shown and their implications will be discussed.

References

1. Bellman, R., *Dynamic programming*. Dover ed ed. 2003, Mineola, N.Y: Dover Publications.

2. Faísca, N.P., et al., *A multi-parametric programming approach for constrained dynamic programming problems*. Optimization Letters, 2008. **2**(2): p. 267–280.

3. Borrelli, F., et al., *Dynamic programming for constrained optimal control of discrete-time linear hybrid systems*. Automatica, 2005. **41**(10): p. 1709–1721.

4. Bemporad, A., et al., *The explicit linear quadratic regulator for constrained systems*. Automatica, 2002. **38**(1): p. 3–20.

5. Bemporad, A., F. Borrelli, and M. Morari, *Robust Model Predictive Control: Piecewise Linear Explicit Solution*, in *European Control Conference*. 2001: Porto, Portugal. p. 939–944.

6. Diangelakis, N.A., A.M. Manthanwar, and E.N. Pistikopoulos, *A Framework for Design and Control Optimisation: Application on a CHP System*, in *Proceedings of the 8th International Conference on Foundations of Computer-Aided Process Design*, 2014, Elsevier. p. 765–770.

428648 Mixed-Integer Model Predictive Control for Online Scheduling of HVAC Equipment in Commercial Buildings

Abstract:

With increased attention toward the negative impact of carbon and other greenhouse gas emissions on global climate, it is clear that broad changes need to be made to primary energy consumption habits. Within commercial buildings, heating, ventilation, and air conditioning (HVAC) systems are a prime target for improved supervisory control, as they are active with time-varying intensity throughout the day and combined to account for 51% of commercial building energy use in 2010. While across-the-board reduction would be ideal, existing architecture often precludes any sweeping cuts without sacrificing performance. Furthermore, due to cyclic occupation, resource consumption in buildings is highly time-varying, and specifically for electricity, the onus of meeting transient demand profiles is placed directly on suppliers who in turn pass extra costs on to consumers. Unfortunately, complicated and possibly uncertain pricing structures leave central plant operators largely adrift with only minimal feedback in the form of a monthly utility bill. Thus, existing heuristics for large-scale HVAC operation result in high cost and low efficiency, and so we propose a scheduling and control formulation to make equipment use and energy purchase decisions in a systematic manner without sacrificing occupant comfort.

In campuses or large buildings, major HVAC equipment is typically grouped in a "central plant," with thermal demand being met by producing hot or chilled water centrally (in the "waterside system") and piping it to local air-handlers where it ultimately affects air temperatures (the "airside system"). By utilizing both active and passive storage, this equipment can temporally shift energy consumption to take advantage of time-varying utility price structures and significantly reduce costs. Thus, we formulate an optimization- and model-based control system to operate central plant equipment at the lowest energy cost while meeting relevant constraints. This approach represents integration of model predictive control (MPC) and traditional production scheduling. MPC has enjoyed great success in the chemical process industries, but the binary on/off choices needed to pick from among the available equipment mean existing techniques using only continuous inputs are not directly applicable. By contrast, although many of the decisions made in scheduling problems are binary, yes-or-no choices, they are typically made on a much slower time scale, and thus care is required to balance complexity and online tractability.

Our proposed model applies to two similar cases: (1) when a model for building temperature is available to determine online how much heating or cooling is required to meet temperature constraints; and (2) when a forecast of heating and cooling demand is provided and thus a building model is not needed. In the first case, models must be able to predict with reasonable accuracy the rate of cooling that is required to maintain temperature within established comfort bounds. Since actual temperature regulation is handled by local zone controllers, model accuracy is not critical to maintaining comfort, but better airside models result in better waterside decisions. A benefit to this methodology is that pre-cooling and other passive storage techniques appear implicitly in optimal solutions insofar as they are economically optimal. For the second case, following a known demand forecast eliminates the need for an airside model, thus reducing problem size. This means that more accurate equipment models can be used, leading to better performance. However, the formulation need not change beyond removing variables and constraints that are no longer necessary. By addressing both cases with the same formulation, we ensure that our approach is modular and thus can be applied to the widest variety of buildings.

In either case, the objective is to minimize economic cost while meeting relevant production constraints (i.e., maintaining comfort levels in the building and/or meeting heating and cooling forecasts). This optimization requires making on/off and setpoint decisions for central plant equipment such as chillers, heat-recovery chillers, pumps, etc. Furthermore, the optimization must decide how to make use of active thermal storage tanks to temporally shift production to more favorable times. Thus, each optimization must consider a sufficiently long horizon to capture available cost savings. However, because of uncertainty inherent in forecasts of demand and ambient conditions, the controller must also be able to quickly re-optimize when provided with new data. This fast, closed-loop implementation avoids an overly conservative policy that would be required by slow, intermittent optimization, but it also imposes restrictions on model complexity. To this end, we formulate our optimization as a mixed-integer linear program (MILP) for which sophisticated solvers are available, allowing large instances to be solved in real time. A key consideration is the nonlinear dependence of heating/cooling production on resource consumption in process equipment. To ensure that the controller accurately predicts economic cost, nonlinear equipment models are approximated as piecewise-linear to enable use in the MILP formulation without introducing significant error to the cost calculation. This approximation scheme provides the flexibility to trade accuracy for computational speed.

After describing the mathematical formulation of the optimization problem, we present a case study for the proposed control architecture. For a system with four chillers and two temperature zones, we show that computation is sufficiently fast to find provably optimal solutions. For a larger number of temperature zones, we show that, although provably optimal solutions are much harder to find, it is still relatively easy to find quality sub-optimal solutions suitable for feedback control. For a system with a forecast for simultaneous chilled and hot water demand, we show that our model can include heat-recovery chillers with no modifications. This allows rejecting heat to process streams where it can be reused instead of simply venting to the ambient. Using a prediction horizon of 48 h with a 1 h timestep, these optimizations can be solved to optimality within 5 min, which is suitably fast to implement online. We also show improved performance relative to existing heuristic methods. A key advantage of this approach is that it makes few assumptions about the central plant equipment and can thus be applied to existing buildings without expensive retrofit. Whereas existing methods make scheduling decisions using heuristics on a slow time scale, our approach, by augmenting a scheduling model with feedback control techniques, allows equipment utilization schedules to be created and revised online and in real time, resulting in improved economic performance.

426822 Dynamic Modeling of the Nitrogen Cycle in Biochar Amended Soils: A New Approach for Rapid Screening of Designer Biochars

Abstract:

Dynamic Modeling of the Nitrogen Cycle in Biochar Amended Soils: A New Approach for Rapid Screening of Designer Biochars

Hao Sun[1,3], Caroline A. Masiello[2] and Kyriacos Zygourakis[1]

1. Department of Chemical and Biomolecular Engineering, Rice University, 6100 Main Street, Houston, TX 77005, USA

2. Departments of Earth Science and Chemistry, Rice University, 6100 Main Street, Houston, TX 77005, USA

3. Currently with Chevron Energy Technology Company, 3901 Briarpark Dr., Houston, TX 77042

ABSTRACT

Maintaining a sustainable food supply in the future without excessive environmental degradation will require novel engineering approaches for remediating the human disruption of the nitrogen cycle. Over the past several decades, food production has been able to keep pace with human population growth thanks to the development of new high-yielding crop varieties optimally grown with the help of fertilizers.

Unfortunately, more than half of the N fertilizer applied to a field is not available for plant growth due to losses caused by surface runoff, leaching into surface and ground water, or volatilization. As a result, the increased use of N as fertilizer has been linked to a variety of water pollution problems ranging from the expansion of the hypoxic zone in the Gulf of Mexico caused by eutrophication to the contamination of wells and groundwater with nitrate N. More importantly, denitrification reactions occurring in saturated soils convert nitrate to nitrous oxide, a major greenhouse gas and significant contributor to the depletion of atmospheric ozone

Several studies have reported that the addition of biochars to weathered tropical or temperate region soils decreased leaching losses of nutrients. In other cases, however, biochar amendments did not have the expected beneficial effects. Such contradictory results generated a lot of interest in the production of *designer biochars*, which are engineered to provide specific services. Although the idea of designer biochars seems reasonable, their production becomes a daunting problem. There are dozens of feedstocks from which biochars can be produced in multiple types of reactors under varying temperature and oxygen conditions, quickly leading to thousands of potential biochars with widely varying properties. Then, each of these biochars must be added to a wide range of soil types. For each biochar-soil type pair to be tested in a field trial, farmland must be allocated in statistical replicates for multiple cropping seasons (ideally for 3-5 years). Clearly, this is a very time-consuming process that may take decades to implement.

In an effort to speed up this process, we have begun to develop a computational framework that will allow us to rapidly assess the environmental performance of biochars and guide their application as soil amendments. The amendment of soils with biochar introduces a new stationary phase that necessitates an extension to the classical dual porosity models that considered pore diffusion but only one adsorbent (stationary) phase. Biochars are highly porous materials with interconnected networks of pores that span multiple length scales: from sub-nanometer micropores to macropores with sizes of the order of 10 μm or larger. Moreover, the size and adsorption capacity of biochar particles may be significantly different than the corresponding properties of soil particles.

To properly account for all these factors, we developed a dynamic convection-dispersion-adsorption model that considers nutrient transport through beds with two porous adsorbent phases: soil and biochar. Intraparticle diffusion of the nutrient in both stationary phases was also considered and Langmuir isotherms were employed to describe the local equilibria between the solute diffusing in liquid-filled pores and the solute adsorbing on the pore surfaces of the biochar and soil particles. The resulting system of PDEs was solved numerically using the method of lines.

Simulation results will be presented to demonstrate that addition of biochar can effectively slow nutrient transport through the soil if the biochar/soil ratio and crucial biochar properties (like its adsorption capacity and affinity to the sorbate) are carefully matched to the soil properties (water velocity, soil type) and the amount of rainfall or irrigation. Simulations can also track the spatial and temporal evolution of nutrient concentration profiles and identify the factors that modulate nutrient partitioning between the soil and biochar phases.

By allowing us to isolate and analyze the interactions of key components of our system, the mathematical model provides a systematic way for adjusting the amount and properties of a biochar in order to maximize its agricultural benefits. The amount of biochar applied to a field, for example, could be matched to the hydrodynamic characteristics (permeability, pore water velocity, soil type) of a specific field or to the amount of expected rainfall in order to achieve the desired nutrient retention effects. Or, we could select a feedstock and production method that would yield a biochar with optimal adsorption capacity and sorbate affinity. At the same time, the model can tell us when the application of a biochar will not work. For example, simulations have identified several cases where the addition of biochar would not provide significant benefits either because the biochar has low adsorption capacity (e.g. cation exchange capacity in the case of ammonium fertilizer), or weak affinity to the nutrient/sorbate or because it has been applied to a highly permeable soil characterized by high pore water velocities.

431315 Smart Grid: Opportunities for the Process Industries

Abstract:

The transition to more intelligent electrical grids, enabling remote and distributed generation, poses a challenge to the process industry [1]. Due to the increasing amount of uncontrollable renewable generation, e.g. wind and solar, the volatility of the availability and price of electricity has significantly increased. At the same time, each company needs to strive for better efficiency. The smart grid provides an opportunity for traditional electricity consumers to increase their awareness of their electricity consumption, as well as, also become flexible market participants as producers (prosumers) through the concept of virtual power plants, mainly enabled by industrial demand-side management [2].

There are several ways how to master the change in the dynamics between electricity producers and consumers, where smart grid is just one enabler of on-line both-way communication between the parties. These ways comprise hardware applications such as metering, visualization, analysis and monitoring and finally optimization. An ARC study shows that very few companies are utilizing this potential. In this presentation we will show some recent research results, existing software approaches and example solutions that are already today running in various industries [3]. We also briefly discuss what the main conditions are that needs to be fulfilled in order to be able to participate in the volatile energy markets. Here, the process flexibility plays a key role. Being able to adapt the production – even partially – according to the changing electricity markets can in best cases offer significant annual savings (MUSD). The key word to identify the best win-win-win scenarios is collaboration!

1. Harjunkoski, I., Scholtz, E. and Feng, X. (2014). Industry Meets the Smart Grid, Chemical Engineering Progress (CEP),August 2014, pp. 45-50
2. Hadera, H., Harjunkoski, I., Sand, G., Grossmann, I.E. and Engell, S. (2015). Optimization of Steel Production Scheduling with Complex Time-Sensitive Electricity Cost. Computers and Chemical Engineering, 76, pp. 117-136
3. Merkert, L., Harjunkoski, I., Isaksson, A., Säynevirta, S., Saarela, A. and Sand, G. (2015). Scheduling and energy – Industrial challenges and opportunities. Computers and Chemical Engineering, 72, pp. 183-198

420133 Novel Processes for the Production of Olefins and Aromatics from Methane Via Methanol: A Comprehensive Process Synthesis and Global Optimization Approach

Abstract:

Recently discovered sources of shale gas, along with the technological improvements within the shale gas industry, have greatly increased the supply of natural gas in the United States. However, several challenges have emerged due to these developments, a critical one being the proper utilization of this resource into more valuable and highly demanded products via routes that are both economical and profitable. Since natural gas is the cleanest fossil fuel, these challenges present tremendous opportunities for the proper utilization of methane and have significantly increased interest in C1 chemistry across industries and within academia and government agencies.[1] To this extent, we propose novel methods for the production of high-value aromatics and olefins from natural gas via methanol. In the mid-2000s, almost all of the approximately 4.5 million metric tons per year of methanol produced in the United States utilized the synthesis gas route.

Conversion of natural gas into methanol as an intermediate for chemicals production is exceptionally promising and remains the most probable route for the widespread commercialization of natural gas based refineries in the United States. Olefins and aromatics are valuable petrochemical products and in high demand. In 2007, the worldwide consumption of ethylene was 115 million tons and the global demand of propylene was 73.5 million tons.[2] Between 2005 and 2008, the global demand for benzene, para-xylene, ortho-xylene, and meta-xylene was approximately 40, 26, 6, and 0.4 million metric tons per year, respectively.[2] Thus, opportunities exist for widespread penetration of natural-gas based chemical refineries, albeit with challenges associated with which types of chemicals to produce and which technologies to invest in.

In this work, we propose a process synthesis superstructure that contains several direct and indirect commercialized natural gas conversion technologies, including autothermal reforming and steam reforming, among others.[3,4] To address the economic barriers often associated with the utilization of alternative feedstocks, we develop a novel branch-and-bound global optimization framework that is capable of determining the optimal technologies to produce aromatics and olefins from natural gas at the highest profit.[3-5] Several commercialized and promising olefins and aromatics production, upgrading, and separation alternatives are included within this process synthesis framework.[6-9] This approach provides an adequate baseline for comparing competing technologies and identifying bottlenecks for natural gas conversion based technologies. Several case studies are presented that explore the effect of refinery capacity and product output. The key topological decisions and analysis of uncertain parameters will be discussed.

1. Mokrani T., Scurrel M. Gas conversion to liquid fuels and chemicals: the methanol route-catalysis and processes development. Catalysis Reviews 2009, 51:1-145

2. de Klerk, A. Fischer-Tropsch Refining; Wiley-VCH Verlag & Co. KgaA: Weinheim 2011

3. Onel, O.; Niziolek, A. M.; Elia, J. A.; Baliban, R. C.; Floudas, C. A. Biomass and natural gas to liquid transportation fuels and olefins (BGTL+C2_C4): Process Synthesis and Global Optimization. Industrial & Engineering Chemistry Research 2014, 54, 359-385.

4. Niziolek, A. M.; Onel, O.; Elia, J. A.; Baliban R. C.; Floudas, C. A. Coproduction of Liquid Transportation Fuels and C6_C8 Aromatics from Biomass and Natural Gas. AIChE Journal 2015, 61, 831-856.

5. Baliban, R. C.; Elia, J. A.; Misener, R.; Floudas, C. A. Global Optimization of a MINLP Process Synthesis Model for Thermochemical Based Conversion of Hybrid Coal, Biomass, and Natural Gas to Liquid Fuels. Computers and Chemical Engineering 2012, 42, 64-86.

6. Niziolek, A. M.; Onel, O.; Hasan, F. M.; Floudas, C. A. Municipal solid waste to liquid transportation fuels – Part II: Process synthesis and global optimization strategies. Computers and Chemical Engineering 2014, 74, 184-203.

7. Niziolek, A. M.; Onel, O.; Elia, J. A.; Baliban, R. C.; Xiao, X.; Floudas, C. A. Coal and Biomass to Liquid Transportation Fuels: Process Synthesis and Global Optimization Strategies. Industrial Engineering & Chemistry Research 2014, 53, 17002-17025.

8. Baliban, R. C.; Elia, J. A.; Floudas, C. A. Novel natural gas to liquids (GTL) technologies: Process synthesis and global optimization strategies. AIChE Journal 2013, 59, 505–531.

9. Baliban, R. C.; Elia, J. A.; Floudas, C. A. Biomass and Natural Gas to Liquid Transportation Fuels: Process Synthesis, Global Optimization, and Topology Analysis. Industrial & Engineering Chem Research 2013, 52, 3381–3406.

422590 Process Design and Optimization of Chemicals' Production from Biomass Feedstocks

Abstract:

Process Design and Optimization of Chemicals' Production from Biomass Feedstocks

Zhaojia Lin[1], Abhay Athaley[1], Vladimiros Nikolakis[2], Marianthi Ierapetritou[1]

[1] Department of Chemical and Biochemical Engineering, Rutgers - The State University of New Jersey

[2.] Catalysis Center for Energy Innovation & Department of Chemical & Biomolecular Engineering, University of Delaware

This work focuses on the process design and optimization of processes for the production of chemicals from lignocellulosic biomass feedstocks. The acceptance of these products in the market place depends on their competitiveness compared with petroleum-based products not only in terms of economics but also in terms of process sustainability.[6] The development of efficient and economically sustainable routes for the production of bio-based platform chemicals such as 5-hydroxymethylfurfural (HMF), furfural, levulinic acid, etc. has attracted a lot of attention.[1,2] We investigate paths based on the non-enzymatic hydrolysis of the cellulosic and hemicellulosic fractions to produce sugar mixtures followed by "one-pot" reactive extraction to convert sugar mixtures to furan derivatives. The aforementioned chemicals can be produced from glucose or xylose solutions using reactive extraction. However, the requirement for the use of organic solvents as extractants and salts that promote the partition coefficients of HMF and furfural have a negative impact on the economics and sustainability. Moreover, xylose and glucose solutions from hydrolysis steps usually contain small amounts of glucose or xylose respectively, unless further purification process is implemented.[3] Therefore, it appears more economically and environmentally attractive to simultaneously manufacture HMF and furfural from C5 and C6 mixtures that are produced from concentrated-acid biomass hydrolysis.[4,5]

Previous studies evaluated the economics and environmental impacts of these bio-products independently;[7-9] however only few studies have explored the optimization of biobased production involving both reaction and separation steps.[10] It is essential to implement flowsheet optimization to obtain the optimal operating conditions that minimizes production costs and environmental impacts both at the production but also at the separation stage. Surrogate-based optimization is a useful tool for the cases that 1) original models are expensive to evaluate; 2) the original gradient information is inaccessible for those commercial simulation tools or the noise exists so that the gradient-based algorithms cannot be accurately used. [11,12]

This work uses techno-economic analysis and life cycle assessment to design and evaluate the production alternatives. The detailed kinetics models are included in the reaction step and the separation units are simulated using Aspen Plus. At the next stage a surrogate-based optimization approach is used to achieve the optimal alternative in terms of economic and environmental criteria. More specifically, the surrogate-based models are generated to represent one or groups of original reaction and separation units; next the surrogate-based models are connected and reformulated into multi-criteria nonlinear programing problem (NLP); and finally NLP problem is solved by a commercial solver to present the pareto optimal set of solutions.

References:

1. Bozell JJ, Petersen GR. Technology development for the production of biobased products from biorefinery carbohydrates-the US Department of Energy's "Top 10" revisited. *Green Chemistry.* 2010;12(4):539-554.

2. Dutta S, De S, Saha B, Alam MI. Advances in conversion of hemicellulosic biomass to furfural and upgrading to biofuels. *Catalysis Science & Technology.* 2012;2(10):2025-2036.

3. Harris JF, Laboratory FP. *Two-stage, dilute sulfuric acid hydrolysis of wood: an investigation of fundamentals*: U.S. Dept. of Agriculture, Forest Service, Forest Products Laboratory; 1985.

4. Chheda JN, Roman-Leshkov Y, Dumesic JA. Production of 5-hydroxymethylfurfural and furfural by dehydration of biomass-derived mono- and poly-saccharides. *Green Chemistry.* 2007;9(4):342-350.

5. Zhang J, Lin L, Liu S. Efficient Production of Furan Derivatives from a Sugar Mixture by Catalytic Process. *Energy & Fuels.* 2012/07/19 2012;26(7):4560-4567.

6. Hawkins ACG, David A.; Buelter, Thomas; Wade, James; Meinhold, Peter; Peters, Matthew W.; Gruber, Patrick R.; Evanko, William A.; Aristidou, Aristos A., Assignee, Inventors; Gevo Inc. USA, assignee. METHODS FOR THE ECONOMICAL PRODUCTION OF BIOFUEL PRECURSOR THAT IS ALSO A BIOFUEL FROM BIOMASS 2009.

7. Lin Z, Ierapetritou M, Nikolakis V. Aromatics from Lignocellulosic Biomass: Economic Analysis of the Production of p-Xylene from 5-Hydroxymethylfurfural. *AIChE Journal.* 2013;59(6):2079-2087.

8. Lin Z, Nikolakis V, Ierapetritou M. Alternative Approaches for p-Xylene Production from Starch: Techno-Economic Analysis. *Industrial & Engineering Chemistry Research.* 2014/07/02 2014;53(26):10688-10699.

9. Lin Z, Nikolakis V, Ierapetritou M. Life Cycle Assessment of Biobased p-Xylene Production. *Industrial & Engineering Chemistry Research.* 2015/03/04 2015;54(8):2366-2378.

10. Torres AI, Daoutidis P, Tsapatsis M. Continuous production of 5-hydroxymethylfurfural from fructose: a design case study. *Energy & Environmental Science.* 2010;3(10):1560-1572.

11. Caballero JA, Grossmann IE. An algorithm for the use of surrogate models in modular flowsheet optimization. *AIChE Journal.* 2008;54(10):2633-2650.

12. Henao CA, Maravelias CT. Surrogate-based superstructure optimization framework. *AIChE Journal.* 2011;57(5):1216-1232.

(91c) A Review on Life Cycle Assessments of Prioritized Renewable Chemicals and Their GHG Reduction Potential

Montazeri, M., Northeastern University

Zaimes, G. G., University of Pittsburgh

Khanna, V., University of Pittsburgh

Eckelman, M. J., Northeastern University

Originally presented on: 11/9/2015 9:10:00 - 9:30:00

Abstract:

The estimated biomass production in the world is 100 petagrams (100 Gt) of carbon per year, about half in the ocean and half on land.[1] Biomass has always been a major source of energy for humankind and is presently estimated to contribute on the order of 10?14% of the world's energy supply.[2] In today's fossil fuel-driven industrial economy, production of bio-based chemicals and fuels may have numerous benefits but should also aim to decrease life cycle environmental impacts (such as GHG emissions) to be considered as favorable alternatives compared to their conventional counterparts. The RFS2 (renewable fuel standard) program administered by the USEPA regulates minimum content of renewable fuels in regular blend and sets a threshold of greenhouse gas emissions reduction for renewable fuels of least 20% for corn ethanol and a 50-60% reduction for fuels from non-corn feedstocks, cellulosic and agricultural wastes, and biodiesel.[3] While fuels have been a focus of public policy, nearly 50 million tons of bio-based chemicals are also produced annually worldwide, but there is no commensurate threshold for GHG emissions reduction of bio-based chemicals in order to define what can be considered as a renewable chemical. This presentation will review available bio-based chemical pathways and associated non-renewable energy use and life cycle GHG emissions reductions.

The US Department of Energy has conducted two separate screening analyses [4] to identify priority bio-based chemicals based on available technologies and the market demands. These chemicals can be produced from both sugar and non-sugar components of biomass resources. In this study, we conducted a thorough literature review on life cycle assessments of 11 sugar-based and 12 lignin-based building blocks reported by USDOE. 11 other bio-based chemicals that are currently in active research and development but not listed by USDOE, are also studied. This additional group can either be produced from sugar or non-sugar components of biomass. Selected studies are comprised of cultivation, preparation, intermediate and final production processes for each chemical; environmental burdens from the use and end of life phases are excluded from this study and system boundary is set to be cradle to gate. Various sources of biomass (e.g, seeds, agricultural and woody waste, pulp and paper waste streams, and algae); technology pathways (e.g., fermentation, catalytic conversion, pyrolysis, and gasification); national or regional differences; and finally life cycle assessment allocation methods (e.g., economic and mass allocation and system expansion) were considered. Tables 1a-c show the chemicals under consideration.

Table 1. a) Sugar-based and b) lignin-based building blocks from DOE reports, and c) other bio-based building blocks of interest

Sugar-based Building Blocks
1,4-Succinic, Fumaric and Malic Acids
2,5-Furan Dicarboxylic Acid
3-Hydroxy Propionic Acid
Aspartic Acid
Glucaric Acid
Glutamic Acid
Itaconic Acid
3-Hydroxybutyrolactone
Glycerol
Sorbitol
Xylitol/Arabinitol

Lignin-based Building Blocks
Methanol/Dimethyl ether (DME)
Styrenes
Biphenyls
Phenol
Catchols, Cresols, Resorcinols
Vanillin
Vanilic Acid
Cyclohexane
Aromatic Acids
Aliphatic Acids
Wood Preservatives
Formaldehyde Free Adhesives and Binders

Other Building Blocks
Adipic Acid
Polylactic Acid (PLA)
Polyhydroxyalkanoate (PHA)
Low/ High Density Polyethylene
Propylene Glycol
Propanediol
Acrylic Acid
Butanol
Acetic Acid

Life cycle GHG emission results from each of these studies are collected and compared against common petrochemical counterparts. For the comparative LCAs, GHG changes were extracted from the articles but in cases where GHG emissions of the petrochemical counterpart was not reported, Eco-Invent unit processes were used for the estimation. Other results including non-renewable energy use was also included for most of the studied scenarios.

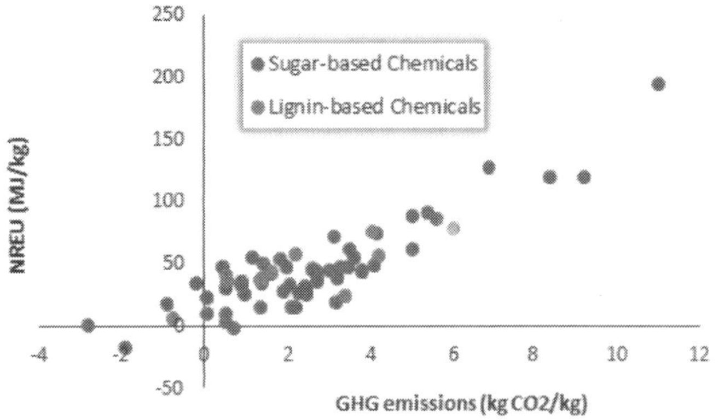

Figure 1. Non-renewable energy use versus GHG emission for sugar-based and lignin-based chemicals

Figure 1 shows the trend of non-renewable energy consumption versus greenhouse gas emissions in kg CO_2 equivalent for a functional unit of 1 kg of final product. As expected, there is a linear relationship between these parameters, as an increase in process energy consumption supplied by fossil sources will increase the emissions.

This study shows the gaps in LCA of bio-based chemicals that require more investigation, especially for lignin-based chemicals due to the complexity in primary structure of lignin and treatment methods; aspartic acid, glucaric acid, sorbitol and arabinitol are examples of sugar-based chemicals while styrene, biphenyl and cresols are examples of lignin-based chemicals included in the list but have not yet been studied on life cycle basis. Moreover, current results are dependent on processing pathways and display significant variability. The GHG results vary from >100% increase for acetic acid production from corn starch to >100% decrease for succinic acid production from corn compared to their counterparts. Even for a single chemical such as succinic acid, GHG emissions reductions range from 49% up to 100% depending on the choice of primary source, conversion and allocation methods. Bio-based chemicals that show reductions in GHG emissions also show lower fossil energy use in most cases, as shown in Figure 1.

Applying RFS2 requirements of at least 50% reduction in GHG emissions, succinic acid, PHB (polyhydroxybutyrate), xylitol, PLA (polylactic acid), polyethylene, propylene glycol and butanol show lower average values for sugar-based chemicals while methanol, PLA, PDO (propanediol), and butanol are preferable lignin-based products. Figure 2 shows these results. Average values are indicated with the full range of reported results also indicated. Blue dots show the average values while red triangles represent chemicals with no LCA studies found, indicating opportunities for further research.

Results of this study can be useful for policy makers in defining a threshold for environmental benefits should be met by renewable chemicals. There are limited data available right now to conduct a harmonization study, though. As shown in Figure 2-b, for most of the lignin-based chemicals one study was found, represented as a single point without error bars, and for the rest there are few life cycle assessment studies. On the other hand, above results are just GHG emission change, a full life cycle assessment accounting for other environmental impacts will assist finding overall benefits of renewable products. Some of the collected studies looked into other impact categories as well, but since most of the regulations are based on GHG emissions, this is the main impact category studied so far. Development of a consistent and comprehensive database for prioritized chemicals, including different pathways, sources and environmental burdens, is required for appropriate decision making.

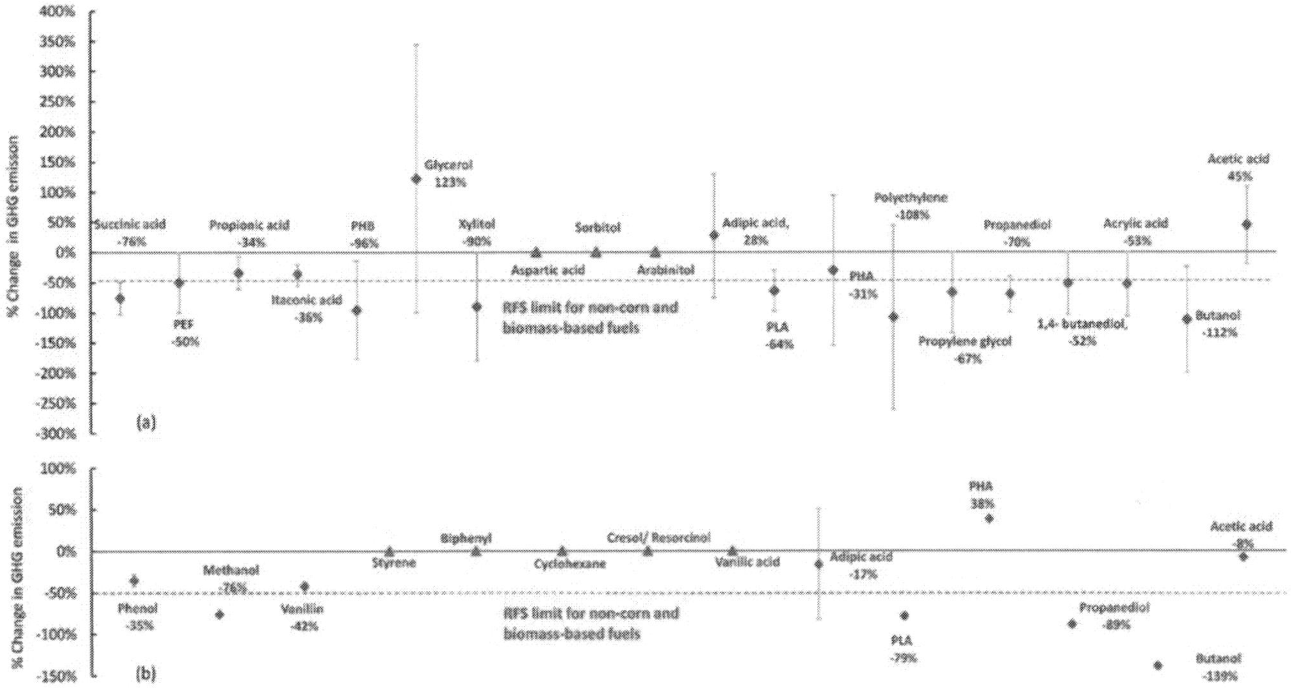

Figure 2. a) Percent change in life cycle GHG emissions of sugar-based chemicals and b) Percent change in life cycle GHG emissions of lignin-based chemicals

References:

1. Field, C. B.; Behrenfeld, M. J.; Randerson, J. T.; Falkowski, P., Primary production of the biosphere: integrating terrestrial and oceanic components. *Science* **1998,** *281* (5374), 237-240.

2. McKendry, P., Energy production from biomass (part 1): overview of biomass. *Bioresource Technology* **2002,** *83* (1), 37-46.

3. Schnepf, R.; Yacobucci, B. D. In *Renewable fuel standard (RFS): overview and issues*, CRS Report for Congress, 2010.

4. Werpy, T.; Petersen, G.; Aden, A.; Bozell, J.; Holladay, J.; White, J.; Manheim, A.; Eliot, D.; Lasure, L.; Jones, S. *Top value added chemicals from biomass. Volume 1-Results of screening for potential candidates from sugars and synthesis gas*; DTIC Document: 2004;

5. Holladay, J.; Bozell, J.; White, J.; Johnson, D., Top value-added chemicals from biomass. *Volume II?Results of Screening for Potential Candidates from Biorefinery Lignin, Report prepared by members of NREL, PNNL and University of Tennessee* **2007**.

428006 Process Simulation Models Parameterizing Multi-Sector Technoeconomic Framework to Test Viability of Emerging Bioprocesses

Abstract:

With recent advances in the development of bioprocesses, several new production schemes utilizing renewable feedstock have emerged for a wide range of chemicals. Many of these have demonstrated promising outcomes in a lab/small scale set up. Hence, lately many efforts were devoted to validate the feasibility of these processes using technoeconomic framework which intended to be used by researchers as one of the major tools to quantify the economic viability and test the potential market vulnerability.

While several exist, one of the major challenges in optimizing such a multifaceted framework is to obtain accurate process parameters, corresponding variable costs and emission estimates for the production of biochemicals. This contribution will demonstrate the manner by which these parameters are generated and subsequently linked to a Multi-Sector technoeconomic framework. Wholly, these models have the capability to capture various attributes of fundamental research and link those with the higher level economic sector, guiding the monetary investment towards the selection of robust candidate biochemicals. The resulting decision support tool can be used to study the existing market and test the viability of bioprocesses while accounting for inter and intra sector interactions in the existing value chain. The traits of this tool, bolstered with detailed process models, make it an apt venue to test various renewable process technologies, provide insights towards future production strategies and guide the research to produce bio-based products.

434302 Biolubricant: Lignin in Choline/Amino Acid Ionic Liquids

Abstract:

Lignocellulosic biomass is the most abundantly available raw resource on the earth for the production of valuable products. Utilization of cellulose in versatile promising applications has been intensely investigated in last a few decades. However, the utilization of lignin is greatly restricted due to the its resistant to degradation, being held together with strong chemical bonds and also appears to have a lot of internal hydrogen bonds. Lignin fractionation from biomass has been demonstrated in choline based ionic liquids(ILs), however, the high cost of the ILs necessitates its recycle in industrial process which is difficult due to the large energy consumption. In this work, we present a new application for lignin dissolved in Choline based amino acid ILs as a green bio-lubricant. The addition of lignin could improve choline based ILs' anti-wear properties and friction stability. Various characterization techniques including differential scanning calorimetry, thermogravimetric analysis, and rheology are used to study the effect of lignin on the thermal, physical and tribological properties of lignin/choline based ILs. The results show that the lignin could enhance the physiochemical properties simultaneously.

433512 An Exposure Reconstruction MODEL for Environmental and Consumer Product Chemicals: Application on Bisphenol

Abstract:

The estimation of external exposure and intake from biomonitoring data utilising Physiologically Based BioKinetic (PBBK) models is an ongoing research field known as exposure reconstruction or reverse dosimetry. The major benefit of this retrospective interpretation is that it can be used to quantify both aggregate and cumulative exposure. Although PBBK models are able to translate external exposure from multiple routes into internal exposure metrics addressing the effects of exposure route in the overall bioavailability, uncertainties of their use for exposure reconstruction rely on the validity of the PBBK models, the analytical error, the fitness-for-purpose of the biomonitoring data, as well as the exposure reconstruction algorithms used. Assimilation of human biomonitoring data and their translation into intake distribution accounts to a computational inversion problem, assessed by a distribution of exposures possibly associated with specific blood or urine levels.

The current study aims at the development and evaluation of an integrated framework for exposure reconstruction of environmental and consumer product chemicals. The developed framework consist of 3 basic steps. The first step aims at specifying and modifying the prior parameter distribution, the joint probability distribution, the population model and the determination of the measurement model. At the second step, the biomonitoring data is imported to the procedure and the exposure reconstruction algorithms is employeed; two different algorithms have been applied. Finally, Monte Carlo simulation is realized on the results applying a comparison of prior and posterior distribution as well as parameter independence. In principle, based on the identification of a global minimum of the error metric, the input parameters and the model setup the computational inversion techniques for exposure reconstruction can be classified as deterministic or stochastic. In this framework a stochastic exposure reconstruction method was applied and the selected algorithms were the Markov Chain Monte Carlo (MCMC) and the Differential Evolution Monte Carlo (DEMC). DEMC is the integration of the Differential Evolution (DE) and MCMC and it is a population MCMC algorithm characterized by (a) the parallel run of multiple chains and (b) the ability to choose the appropriate scale and orientation for the jumping distribution during the simulation. DE is a genetic algorithm for numerical optimization in real parameter spaces. In addition, MCMC is a stochastic process in which future states are independent of past states, given the present state. During iterative simulations, the random variables of interest are drawn from a chain of distribution that eventually converge to a stable posterior distribution. The generic PBBK model captures multi-route exposure (including inhalation, skin absorption and gastro-intestinal absorption), and incorporates detailed compartmental description, plasma protein/red cell binding and estimates the concentration of parent compounds and metabolites in several biological fluids including breast milk. The generic model was applied for the assessment of a highly controversial industrial chemical with widespread applicability in consumer goods, namely bisphenol-A (BPA). BPA is chemical of great interest cause it is an estrogen characterized by endocrine disrupting activities that are mediated via multiple molecular mechanisms as well as it is associated to neurodevelopmental defects. Parameterization of exposure scenarios were based on an extensive literature review of BPA exposure data and the generic PBBK model was properly parameterized. BPA is considered to be rapidly conjugated into BPA glucuronide (BPA-Glu), resulting in rapid elimination from the human body due to the water solubility of this metabolite. Glucuronidated bisphenol-A (BPA-Glu) is a metabolite product of BPA that can be detected in human urine samples. Exposure to BPA was reconstructed based on real-life HBM European data. Average urinary glucuronidated bisphenol-A (BPA-Glu) was 2.8 μg/L across Europe, covering different age groups. In particular the algorithm has been tested under the assumption that the average amount of 2.8 μg/L BPA-Glu to human' s urine is the results of an ordinary adult dietary schedule that includes 3 different meals. The prior knowledge of the distribution of the exposure time is based on the actual dietary schedule of the generic population. The time of the three basic dietary meals follows normal distribution and in particular, the breakfast is at 7:00 AM and the dinner is at 7:00 PM with standard deviation of 30 minutes for both and the lunch is at 2:00 PM with standard deviation of 30 minutes.

The selected exposure reconstruction algorithms (MCMC as well as DEMC) were tested for 1000 iterations. All the prior distributions converged to the actual exposure dose. The results of the exposure reconstruction illustrated that the posterior distributions include the actual exposure doses. Moreover, the posterior distributions had a reduced standard deviation and a mean value closer to the real one. However, MCMC model using the prior knowledge could not achieve a posterior distribution with a sufficient confidence interval (CI) towards the actual exposure value. The exposure dose of second and third meal were more accuratelly predicted. In practice, the results indicated that the overall daily intake is very low, approximately 1 μg/kg_bw/d, which is far below the Tolerable Daily Intake (TDI) of 50 μg/kg_bw/d proposed by the European Food Safety Authority (EFSA). Comparing the computation needs of the algorithms, the DEMC is 3 to 4 time slower than MCMC but the predictions converged faster when the algorithm was tested against synthetic data for validation purpose. The high frequency of exposure interval representing the real life exposure scenario of this study increases the computation needs. In a statistical context, the combination of the genetic algorithm and the stochastic process of MCMC that are coupled by DEMC results in more accurate predictions reducing the overall uncertainty.

This study demonstrated how an integrated framework based on a PBBK model could effectively combine biomonitoring data in order to estimate exposure, absorbed, metabolized and extracted dose on a chemical compound, since biomonitoring data reflect an internal dose associated with external exposures from all potential sources and routes. However, the exposure routes and sources (natural or synthetic; endogenous or exogenous) have to be appropriately identified and characterized before the procedure. A key issue on reverse dosimetry is the bio-persistence of a compound, since the measured biomonitoring data might reflect a recent or an older exposure regime.

The algorithm of DEMC is used for first time in exposure reconstruction models providing clear benefits in terms of biomonitoring data assimilation; however, additional optimization of the code will result in reduced computational time. Further work is expected on developing and using sufficient genetic algorithms appropriate for the exposure reconstruction models in order to reduce the computational time as well as to reduce the uncertainty of the estimated results. A generic and reliable generic (covering a large chemical space) exposure reconstruction scheme, could be a very useful tool for modern risk assessment of chemicals, utilizing the large amount of existing human biomonitoring data. In addition, higher frequency of biomonitoring sampling is needed for rapidly metabolized compounds. Overall, exposure reconstruction has the potential to provide reliable estimates of total exposure by interpreting biomonitoring data. This gives us the capability to re-run forward our model to estimate the actual biologically effective dose in the target tissue and to compare this to toxicity results obtained from in vitro testing, significantly advancing environmental health science and chemical safety regulation.

434708 Synthesis of Networks of Reactors Featuring Multiple Residence Time Distribution

Abstract:

In this work, we present a methodology to synthesize globally minimum volume reactor networks, which are allowed to employ multiple reactor types featuring a number of normalized residence time density (NRTd) functions. The synthesis problem is formulated within the Infinite DimEnsionAl State-space (IDEAS) conceptual framework, which is shown to be applicable to the problem at hand. Within the IDEAS framework, network synthesis is formulated as an infinite dimensional linear optimization problem. The IDEAS conceptual framework is realized through solution of a series of finite dimensional linear programs whose optimum values converge to the infinite program's infimum. Global optimality is established by demonstrating that the input–output information maps of the reactors with general RTd/RTD models satisfy all properties required for the application of the infinite dimensional state-space (IDEAS) approach to the RTd/RTD reactor network synthesis problem. The resulting IDEAS formulation is shown to possess a number of novel properties, which can be used to facilitate its solution. The power of the proposed methodology is demonstrated on a case study featuring a variety of reactor types, including PFR, CSTR, and SLFR (Segregated Laminar Flow Reactor).

420240 Small-Scale Gtl Processes for Utilizing Stranded Gas: Model Identification, Process Synthesis, and Global Optimization

Abstract:

Advances in the shale gas industry have generated abundant and inexpensive sources of natural gas in the United States, [1] with almost 500,000 active natural wells operating across the nation [2]. There are also many sources of "stranded gas" that are uneconomic to produce due to their low energy content, impurities, small volume of production, or lack of infrastructure. Therefore, this stranded gas is typically vented or flared. Gas-to-Liquids (GTL) processes have been considered as an alternative to monetize the stranded gas [3,4] and mitigate the negative environmental impacts. A recent review outlined that many GTL technologies are commercially and technologically developed [5]. A superstructure based global optimization framework is introduced for a GTL refinery that considers multiple conversion pathways and upgrading technologies [6]. The objective is to find the best processes and topological alternatives across different scales, products, and feedstock compositions [6].

As a novel natural gas conversion alternative, an annular microchannel reactor (AMR) is designed for the catalytic steam reforming of methane for small-scale applications [7]. Through the use of a rigorous computational fluid dynamics (CFD) model, the AMR is simulated and validated against experimental data [8]. Although the technology is promising for small scales and CFD simulations provide an accurate representation, the model [8] is computationally too expensive to consider in a superstructure-based approach [6]. Therefore, a grey-box model identification approach is developed to build a surrogate model that will accurately represent AMR behavior. A non-linear non-convex parameter estimation model is built and optimized for this motive to find a mathematical model that predicts the syngas effluent within less than 1% of the CFD simulation results.

The surrogate model is then implemented to a GTL process superstructure [6] for small-scale GTL applications. A set of case studies along with sensitivity analysis to observe the performance of the AMR at these scales is studied. The results suggest that the break-even oil prices can be as much as $20/bbl lower for technologies that utilize AMR compared to traditional reforming technologies.

References:

[1]: Energy Information Administration, Annual Energy Outlook, 2014

[2]: Energy Information Administration, Natural Gas Monthly, April 2015

[3]: Lichun, D.; Shun'an, W.; Shiyu, T.; Hongjing Z.; GTL or LNG: Which is the best way to monetize "stranded natural gas?, 2008, Petroleum Science, 5(4), 388-394.

[4]: Wood, D. A.; Nwaoha, C.; Towler, B. F; Gas-to-Liquids (GTL): A review of an industry offering several routes for monetizing natural gas, 2012, Journal of Natural Gas Science and Engineering, 9, 196-208.

[5]: Floudas, C. A.; Elia, J. A.; Baliban, R. C.; Hybrid and single feedstock energy processes for liquid transportation fuels: A critical review, 2012, Computers and Chemical Engineering, 41, 24-51

[6]: Baliban, R. C.; Elia, J. A.; Floudas, C. A.; Novel Natural Gas to Liquids Processes: Process Synthesis and Global Optimization Strategies, 2013, AIChE Journal, 59(2), 505-531.

[7]: Wilhite, B. A.; Breziner, L.; Mettes, J; Bossard, P.; Radial Microchannel Reactors (RMRs) for Efficient and Compact Steam Reforming of Methane: Experimental Demonstration and Design Simulations, 2013, Energy & Fuels, 27, 4403-4410.

[8]: Butcher, H.; Quenzel, C. J. E.; Breziner, L.; Mettes, J.; Wilhite, B. A.; Bossard, P.; Design of an annular microchannel reactor (AMR) for hydrogen and/or syngas production via methane steam reforming, 2014, International Journal of Hydrogen Energy, 39, 18046-18057.

420542 Transient Analysis of Gaseous Contamination Dispersion on Offshore Facilities

Abstract:

Many recently discovered deepwater reservoirs contain high levels of gaseous contaminants, particularly CO_2, H_2S, and SO_2. As a result, specific plans are required because the accidental leakage of gases on offshore exploration and production facilities poses significant risks to personnel health and the environment. Traditional approaches to plume simulations focus on gas distribution in the steady state, which can be achieved only after a sufficiently long time from the beginning of leakage. This work describes a computational fluid dynamics (CFD) propagation analysis of the released gas around potential sources on an offshore rig under different scenarios. In particular, a scenario of gaseous blowout is considered, and the dynamics of hazardous gas spreading through the working area are analyzed. The numerical approach used in this study is similar to the approach applied to weather forecasting and air quality prediction.

CFD simulations were performed to study CO_2 dispersion from various sources containing 80% CO_2. CO_2 is used in this study as an example, but the same method can be applied to predict dispersion of other gases, including H_2S or SO_2. The Realizable k-^{o} model for turbulence in a commercially available CFD tool was used for the multi-component gas flow, while the convective diffusion equation for species was solved for the gas dispersion calculation. A variety of dispersion scenarios were considered, including leaks in three surface well test (SWT) equipment areas, relief lines (overboard), and flare booms, with different wind directions, speeds, and gas-leak rates. From these three areas, several "worst-case" scenarios and time-dependent gas contaminant distributions were studied in detail.

Analysis of the results indicates that (1) high leak rate and low wind speed conditions result in a larger pollution area; (2) at low wind speed, the contaminant gas propagates in all directions, including upstream of the leak point; and (3) the layout of the rig on the drill ship affects the local CO_2 distribution. Detailed transient analyses show that the gaseous contaminants cover the entire rig within approximately 30 seconds, even with a low leak rate. Meanwhile, the plume size increases, and the plume can change direction during expansion.

These results help (1) determine the risk of moderate (M) and non-tolerable (NT) CO_2 levels profiles; (2) optimize the location and quantity of CO_2 sensors on the SWT plant and drill floor accordingly; and (3) assess the risks of CO_2 disposal by the flare booms and the overboard relief lines.

This work can be very helpful in analyzing gas leakage-related risks in many recently discovered reservoirs in deepwater operations. The CFD approach developed in this study can be used to forecast and monitor gas leakage on offshore vessels and rigs, where accounting for health, safety, and environment factors is required.

420803 Multi-Domain Modeling and Operation of a Steam Methane Reforming Furnace

Abstract:

In current hydrogen manufacturing plants, steam methane reforming (SMR) processes play a key role in converting methane into hydrogen, carbon monoxide and carbon dioxide by a sequence of net endothermic reactions. For industrial-level SMR processes, the furnace condition monitoring is difficult due to the large spatial scale of the furnace and of the high operating temperature. One main task of furnace operation is to avoid hot spots on reforming tube walls. Therefore, SMR furnaces are usually operated slightly below designed values to keep tube wall temperature at a safe margin away from carbon formation temperature and avoid potential tube rupture [1].

The goal of this work is to develop an industrial-level SMR computational fluid dynamics (CFD) model that can provide the temperature distributions in both furnace chamber and reforming tubes for a given set of process inputs. The model is designed to provide acceptable simulation result by certain simplification, so that the model can be used to monitor tube wall temperatures in a real-time mode. The tube wall temperature profiles will help plant operators mitigate the risk of tube rupture due to hot spots. To develop the entire SMR furnace CFD model, a full-size computer-aided design (CAD) model based on an industrial SMR plant was first designed. The CAD model includes all the components of the SMR plant, i.e., reforming tubes, burners and flue gas tunnels. Computationally efficient structured mesh with robust convergence performance was generated for the CFD model design by using ANSYS ICEM CFD [2]. We realize the CFD model of full furnace in the commercial CFD software, ANSYS Fluent [3], based on the generated mesh. Specifically, suitable approximation methodology was adopted to model the reforming tubes and the methane reforming reactions inside the tubes using the reacting channel model. The adoption of reacting channel model greatly reduces the computational burden of the real-time CFD simulation and guarantees the model accuracy at the same time.

Specifically, to realize the methane reforming reaction, we used a complete surface reaction mechanism [4] in the CFD model by reaction mechanism package, CHEMKIN [5] and also an intrinsic reaction mechanism [6] by a user defined function design. Detailed turbulent model and radiation models were used to describe the momentum and heat transport phenomena in the furnace chamber. The above CFD model design strategy has been successfully tested on a small scale model and model accuracy has been validated. The second stage of the work focusing on simulating the full-scale model and calculating maximum tube wall temperatures that ensure safety limit satisfaction is currently being conducted and the results show its promise for implementation in an industrial setting and the calculation of reforming tube-safe operation strategies.

[1] F. Beyer, J. Brightling, P. Famell, and C. Foster. Steam reforming 50 years of development and the challenges for the next 50 years. AIChE 50th Annual Safety in Ammonia Plants and Related Facilities Symposium, Toronto, Canada, Sep. 26-29, 2015.

[2] ANSYS ICEM CFD Tutorial Manual, Version 15.0, 2013.

[3] ANSYS FLUENT Theory Guide, Version 15.0, 2013.

[4] L. Maier, B. Schadel, and K. Delgado. Steam reforming of methane over nickel: development of a multi-step surface reaction mechanism. Topics in Catalysis. 2011; 54: 845-858

[5] Chemkin Theory Manual, CK-THE-10131-1307-UG-1, 2013.

[6] J. Xu and G. Froment. Methane steam reforming, methanation and water-gas shift: I. Intrinsic kinetics. AIChE Journal. 1989; 35: 88-96.

422300 Integration of High-Fidelity CO2 Sorbent Models at the Process Scale Using Dynamic Discrepancy

Abstract:

A high-fidelity model of a mesoporous silica supported, polyethylenimine (PEI)-impregnated solid sorbent for carbon capture has been incorporated into a model of a bubbling fluidized bed adsorber using Dynamic Discrepancy Reduced Modeling (DDRM). The sorbent model includes a detailed treatment of transport and amine-CO_2-H_2O interactions. Using a Bayesian approach, we calibrate the sorbent model to the Thermogravimetric (TGA) data. Discrepancy functions are included within the diffusion coefficients for diffusive species within the PEI bulk, enabling a 100-fold reduction in model order. The discrepancy functions are based on a Gaussian process in the Bayesian Smoothing Splines ANOVA framework, which provides a convenient parametric form for calibration and upscaling. The dynamic discrepancy method for scale-bridging produces probabilistic predictions at larger scales, quantifying uncertainty due to model reduction and the extrapolation inherent in model upscaling. The dynamic discrepancy method is demonstrated using TGA data for a PEI-based sorbent and an Aspen Custom Modeler-based model of a bubbling fluidized bed adsorber.

A Model-based Approach for Predicting Gas Well Leaks from Cemented Annulus and Identifying Well Construction Factors Responsible for Such Leaks

Misra Shobhit and Nikolaou Michael

Gas leaks from natural gas wells can create severe environmental and safety problems, as the leaked gas may end up in the air or water. Poor cementing job and other well completion operations of gas wells may allow gas from one rock formation zone to migrate into another and eventually leave the well. Therefore, it is important to ensure that natural gas well construction prevents such unwanted events. The design objective is known as zonal isolation and has been the subject of intense study recently.

In well construction jobs that do not ensure zonal isolation, natural gas may move through the small channels created in the cement sheath in the annulus between the well casing and well wall, and may eventually reach the well head. To detect whether a continuous flow of gas reaches the wellhead a simple procedure is followed, namely a pressure gauge followed by a needle valve are installed, and the needle valve is temporarily opened to lower pressure by bleeding off a small amount of gas, if pressure increase is detected by the pressure gauge. If gas pressure builds up again after the gas bleed off and closing of needle valve, there is clear indication that there is communication between the well head and producing sections of the well through the cemented well wall. The resulting pressure is called *sustained casing pressure* (SCP) and is undesirable. The objective of this work is to determine what factors contribute toward SCP, and prevent such factors from creating measureable SCP values. Because the effect of all relevant factors on SCP is complicated, the proposed approach relies on statistical analysis of a large volume of data. Through such analysis, a model is built that captures quantitatively the effect of related factors on SCP. These factors are associated with well drilling, cementing and hydraulic fracturing parameters (Nelson 2006).

A statistical technique called partial least squares (PLS) regression was used to create the model. The model structure involves a number of latent variables, which are linear combinations of the original variables, and which are determined through leave-one-out cross-validation in such a way, that correlation between model inputs and outputs is maximized.

Variable Importance in Projection (VIP) variable selection method was used to exclude the less relevant variables. As the term suggests, VIP score of a variable indicates the contribution of a particular X variable to the predicted value of Y (SCP in this case). In our model we excluded the variables with VIP scores less than 0.5.

The contribution of various variables to SCP is shown in the bar chart (Figure 1).

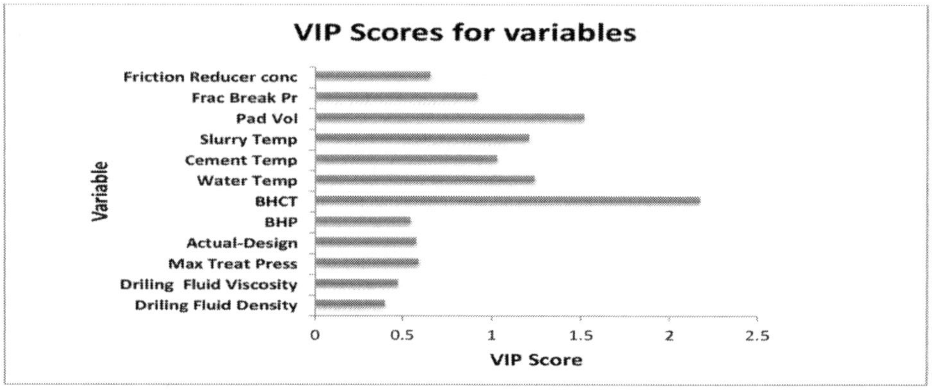

Figure1. VIP scores for variables

References:

Nelson, E. B. (2006). Well Cementing. Texas, USA.

ADDIN EN.REFLIST

418830 Offset-Free Model Predictive Control with Explicit Performance Specification

Abstract:

This work considers the problem of designing a control law to yield a desired closed-loop response subject to input constraints and plant-model mismatch. To this end, a two tiered model predictive controller is designed that first computes the best response of a certain desired form, and then computes the control action to achieve the prescribed closed--loop response. The key idea is first illustrated in the absence of plant-model mismatch for a prescribed first and second order response of the controlled variables, and illustrated using a simulation example. Subsequently, a formulation is presented to handle plant model mismatch. Finally, the proposed formulation is applied to a chemical process example.

420809 Stiction Compensation through Economic Model Predictive Control

Abstract:

Chemical process control typically involves extensive use of control valves as the final control element of a control system. Though control valve dynamics are often assumed to be very fast compared to the process dynamics or to behave linearly, significant issues, such as control loop oscillations, are often experienced in industry due to nonlinear valve dynamics [1]-[2]. One major source of valve nonlinearity is stiction, the sticking of a control valve upon a direction change of the control signal that results in stickband and slipjump [1]. Because stiction prevents control actions from being implemented when requested, there has been an interest in the development of mathematical models and compensation techniques for stiction [2]. Stiction has not yet been addressed, however, in the context of economic model predictive control (EMPC) [3]-[5]. Because EMPC is able to predict future process state trajectories and adjust the control actions accordingly to obtain economically optimal process operation under the given process model and constraints, EMPC is an alternative to other stiction compensation strategies [2].

This work extends the efforts in [6] that examined the effect of the inclusion of linear actuator dynamics in EMPC by accounting for valve stiction and its effect in the closed-loop system under EMPC. A Lyapunov-based economic model predictive control (LEMPC) strategy is developed that explicitly accounts for stiction dynamics that can be described by a nonlinear state-space model. This EMPC uses an economics-based cost function and adopts suitable constraints on the magnitude and rate of change of the calculated control actions to mitigate the impact of stiction in the closed-loop system. Closed-loop stability of nonlinear processes with valve dynamics affected by stiction is examined under the proposed LEMPC and stability conditions are derived. A chemical process example illustrates the importance of including stiction dynamics in EMPC for optimal process operation and for constraint satisfaction.

[1] Choudhury MAAS, Thornhill NF, Shah SL. Modelling valve stiction. Control Engineering Practice. 2005;13:641-658.

[2] Brásio ASR, Romanenko A, Fernandes NCP. Modeling, detection and quantification, and compensation of stiction in control loops: The state of the art. Industrial & Engineering Chemistry Research. 2014;53:15020-15040.

[3] Rawlings JB, Angeli D, Bates CN. Fundamentals of economic model predictive control. In: Proceedings of the 2012 IEEE 51st Annual Conference on Decision and Control. Maui, Hawaii, 2012:3851-3861.

[4] Heidarinejad M, Liu J, Christofides PD. Economic model predictive control of nonlinear process systems using Lyapunov techniques. AIChE Journal. 2012;58:855-870.

[5] Ellis M, Durand H, Christofides PD. A tutorial review of economic model predictive control methods. Journal of Process Control. 2014;24:1156-1178.

[6] Durand H, Ellis M, Christofides PD. Integrated design of control actuator layer and economic model predictive control for nonlinear processes. Industrial & Engineering Chemistry Research. 2014;53:20000-20012.

424125 Reformulating Nonrobust Nonlinear Model Predictive Control

Abstract:

Reformulating Nonrobust Nonlienar Model Predictive Control, D. Griffith and L.T. Biegler

Model predictive control (MPC) is an optimization based form of control that functions by solving successive optimization problems online, and it is particularly suited to multiple-input-multiple-output systems with inequality constraints due to the mathematic programming formulation of the problem. Nonlinear model predictive control (NMPC) has the added benefit of being able to take advantage of a fully nonlinear dynamic model in order to provide accuracy across a wider range of states. These traits make NMPC very attractive to chemical engineers, since chemical processes with many states, many controls, operational constraints, and significant nonlinearities are common.

Furthermore, applying NMPC to very large problems with significant nonlinear programming (NLP) problem solve time is possible thanks to advances in NLP sensitivity techniques and the development of advanced step NMPC (asNMPC) [4] and advanced multi-step NMPC (amsNMPC) [3]. These methods rely on solving the NLP problems one or more time steps in advance and updating their solutions with sensitivity information. In fact, the sensitivity updates are nearly instantaneous, so systems of arbitrary size may be controlled efficiently if enough processors are used.

Much academic research has centered on showing robust stability of the various forms of NMPC. This is typically done by using the optimal objective function value of the controller as a Lyapunov function and subsequently applying a discrete-time (DT) Lyapunov theorem to show that NMPC is input-to-state stable (ISS). The main assumptions of this analysis are on the regularity conditions of the NLP problems. If the NLP problems satisfy the Mangasarian-Fromovitz constraint qualification (MFCQ), and the general strong second order sufficient condition (GSSOSC), then it can be shown that the Lyapunov function is uniformly continuous in the state, which is the main feature necessary for showing ISS. Similar results also exist for asNMPC and amsNMPC.

However, examples exist in the literature where the necessary conditions on the NLP are violated and NMPC becomes nonrobust. This can occur even if the NLP problems are always feasible and the disturbances are arbitrarily small, so such instances are large obstacles to the implementation of NMPC. In order to resolve this, we use a reformulation [1] of the NLP problems utilizing soft constraints penalized in the objective function in place of hard inequality constraints. It can be shown that this reformulation satisfies the necessary conditions of the NLP problems, and the system therefore becomes robustly stable.

In this work, we consider three pathological examples from literature [2]. We explicitly show the traits of the NLP that lead to nonrobustness, reformulate the problem with soft inequality constraints and objective function penalties, and show that the reformulation satisfies the necessary assumptions on the NLP problems. Smooth reformulations and approximations are necessary in some cases to make the problems amenable to NLP. Furthermore, simulation results are shown for each example under multiple conditions, including asNMPC and amsNMPC. These results include comparisons of control horizon length and slack variable weight.

[1] L.T. Biegler, X. Yang, and G. Fischer. Advances in sensitivity-based nonlinear model predictive control. Journal of Process Control, Forthcoming 2015.

[2] S. Tuna A. Teel G. Grimm, M. Messina. Examples when nonlinear model predictive control is nonrobust. Automatica, 40:1729–1738, 2004.

[3] X. Yang and L.T. Biegler. Advanced-multi-step nonlinear model predictive control. Journal of Process Control, 23(8):1116 – 1128, 2013.

[4] V. Zavala and L.T. Biegler. The advanced-step nmpc controller: Optimality, stability, and robustness. Automatica, 45:86–93, 2009.

414421 Bridging the Gap Between Multigrid, Hierarchical, and Receding-Horizon Control

Abstract:

We analyze the structure of the Euler-Lagrange conditions of a lifted long-horizon optimal control problem. Here, lifting partitions the horizon into stages that are coupled using linear transition constraints (as is done in multiple shooting). The analysis reveals that the Euler-Lagrange conditions can be solved by using block Gauss-Seidel schemes and we prove that such schemes can be implemented by solving sequences of short-horizon optimal control problems. The analysis also reveals that a receding-horizon control scheme is equivalent to performing a single Gauss-Seidel sweep. Moreover, the analysis reveals that having optimal adjoint information at the transition constraints is sufficient for the receding-horizon scheme to recover the solution of the long-horizon problem. We use this observation to derive a strategy that computes approximate adjoints from a coarse long-horizon problem and we use this to correct the receding-horizon scheme. We observe that this strategy can be interpreted as a hierarchical control architecture in which a high-level controller transfers long-horizon information to a low-level, short-horizon controller. The information transfered in this hierarchy, however, is dual (adjoints) as opposed to primal (states); consequently, the proposed scheme differs from existing hierarchical control schemes. We demonstrate that the hierarchical scheme provides high quality approximate solutions of the original long-horizon optimal control problem. Moreover, this scheme can be fully implemented using off-the-shelf optimization solvers. We discuss how these new concepts are tightly connected to multigrid concepts arising in the control of distributed systems. We demonstrate our developments using optimal control problems arising in storage management for microgrids.

A FORMULATION OF ADVANCED-STEP CARLEMAN LINEARIZATION-BASED NONLINEAR MODEL PREDICTIVE CONTROL

Yizhou Fang and Antonios Armaou, the Pennsylvania State University, University Park, PA

The need of tight operating conditions in chemical, pharmaceutical, and petroleum industries has given rise to the development of advanced control methods. Model Predictive Control (MPC) started gaining attention three decades ago for optimal transitions between operating modes. The application of Dynamic Optimization (DO) enables MPC many advantages over traditional control strategies. Nonlinear MPC (NMPC) converts a constrained control problem of a nonlinear system into an optimization problem. This basic architecture makes NMPC capable of handling large state-space multi-variable systems with constraints, and dealing with model-mismatches and disturbances readily.

The computation time of control policy is required to be less than one sampling time for online operation. However, this requirement is most of the times impossible to meet when the system has high nonlinearity. That becomes one of the most significant reasons holding back the application of NMPC. As a result, there is strong motivation to develop an advanced formulation of NMPC that demands less computational effort and thus decides the control actions faster.

A Carleman linearization-based formulation of NMPC has been proposed to address this issue in (1) (2). Nonlinear systems are approached with polynomial expressions and then represented with extended bilinear forms. This enables analytical anticipation of future states and analytically providing the sensitivity of the cost function to the control signals as the searching gradient. Consequently, the computation of optimal control policy is accelerated and the computational delay is removed.

In this paper, we propose an advanced reformulation of Carleman linearization-based NMPC with a combination of Advanced-step NMPC methodology published in (3) (4) (5). Using NLP sensitivity analysis to find approximations and update solutions on-line enables this reformulation of NMPC increased capability of dealing with measurement noises and model mismatches. It is suitable to apply Advanced-step NMPC methodology to the method proposed in (1) (2) due to its bilinear formulation of the control problem. A case study example illustrating the proposed method uses a perturbed unstable nonlinear system to show the proposed NMPC controller removes computational delay while it achieves the same stability properties as traditional NMPC controllers do.

Reference

1. A. Armaou and A. Ataei, "Piece-wise constant predictive feedback control of nonlinear systems," *J. Process Control*, vol. 24, no. 4, pp. 326–335, Apr. 2014.
2. Y. Fang and A. Armaou, "Nonlinear Model Predictive Control Using a Bilinear Carleman linearization-based Formulation for Chemical Processes," *2015 Am. Control Conf.*, Jul. 2015.
3. V. M. Zavala and L. T. Biegler, "The advanced-step NMPC controller: Optimality, stability and robustness," *Automatica*, vol. 45, no. 1, pp. 86-93, Jan. 2009.
4. X. Yang and L. T. Biegler, "Advanced-multi-step nonlinear model predictive control,"*J. Process Control*, vol. 23, no. 8, pp. 1116-1128, Sep. 2013.
5. R. Huang, V. M. Zavala and L. T. Biegler, "Advanced step nonlinear model predictive control for air separation units," *J. Process Control*, vol. 19, no. 4, pp. 678-685, Apr. 2009.

425164 Approximation of Closed-Loop Dynamics for Dynamic Real-Time Optimization Calculations

Abstract:

Competitive global market conditions and demand-driven production are characteristics of the present day environment that industries face. In order to operate in a cost-optimal fashion, industrial plant economic optimization has conventionally been addressed via a multilevel process automation architecture with a time-scale separation. At the upper level, a real-time optimization (RTO) system performs economic optimization in response to demand changes or slow-varying disturbances affecting profit (Marlin and Hrymak, 1997; Darby et al., 2011). RTO interacts with the lower level process automation activities in a cascade fashion by providing economically optimal set-points for the underlying regulatory control system, which is typically model predictive control (Qin and Badgwell, 2003; Darby and Nikolaou, 2012).

The conventional RTO strategy is designed based on a steady-state model, which suffers from a limited execution frequency because economic optimization can only be executed once the process is deemed to have reached steady state. Such an approach generally does not account for the economic performance of the plant during its transition from one steady-state to another, and thus is suboptimal for processes that exhibit slow dynamics and those that experience frequent transitions. Recent advances have transformed the steady-state RTO to dynamic RTO (DRTO) based on a dynamic prediction model, thus substantially increasing the frequency at which economic optimization can be performed. Proposed DRTO strategies perform economic optimization in an open-loop fashion without taking into account the presence of the plant control system (Tosukhowong et al., 2004, Wurth et al., 2011), denoted here as an open-loop DRTO approach. In this approach, the set-points prescribed to the underlying control system are constructed from the optimal DRTO open-loop trajectories under an expectation that the closed-loop response dynamics at the plant level will follow the trajectories obtained at the DRTO level. An alternative to the multilevel configuration is a single-level, economic model predictive control (EMPC) approach that optimizes the plant economics at the controller sampling frequency. Such a strategy aims to address the issues of model inconsistencies and conflicting objectives between the traditional RTO system and the MPC control layer. In the single-level approach, the objective function could be based purely on economics (Amrit et al., 2013), or a hybrid between cost and control performance (Ellis and Christofides, 2014).

In previous work (Jamaludin and Swartz, 2014), we proposed a closed-loop DRTO strategy that fits well within the two-level industrial process automation architecture. The original closed-loop DRTO problem structure is in the form of a multilevel programming formulation, as it contains a sequence of embedded MPC controller optimization subproblems to be solved in order to generate the closed-loop response dynamics at the DRTO level. The closed-loop DRTO formulation computes the controller set-point trajectories that determine the best economics of the predicted closed-loop response over a finite optimization horizon. The multilevel problem is subsequently reformulated as a single-level mathematical program with complementarity constraints (MPCC) by transforming the MPC subproblems to constraint sets using their first-order, Karush-Kuhn-Tucker (KKT) optimality conditions. For a constrained MPC problem formulated as a convex quadratic programming (QP) problem as considered in this study, such a transformation is valid as the KKT conditions are necessary and sufficient for optimality. Details of this construct can be found in Baker and Swartz (2008). Also in our work, the complementarity constraints arising from the KKT conditions of the MPC subproblems are handled using an exact penalty formulation (Ralph and Wright, 2004) by imposing them as an additional weighted penalty term in the DRTO economic objective. With an appropriate choice of a penalty parameter, the complementarity penalty term will be negligible at the optimum, hence recovering the solution of the original MPCC problem. We have shown that the closed-loop DRTO approach outperforms the open-loop DRTO approach in terms of economics and control performance. Despite its advantages, the multilevel programming formulation significantly increases the size and solution time of the DRTO problem. Simulation studies also show that significant computational effort is spent on handling the complementarity variables.

The aim of this work is to compare different techniques for approximating the MPC closed-loop dynamics embedded at the DRTO level to allow the economic optimization problem to remain manageable and computationally tractable. We have identified and tested possible approximation techniques, which include: (i) a bilevel programming formulation, (ii) a hybrid closed/open-loop formulation, and (iii) an unconstrained MPC approximation with input clipping. In the bilevel formulation, only a single MPC optimization subproblem is needed where the prediction and control horizons are extended to match the DRTO optimization horizon, which is significantly longer in order to generate economically viable operating policies. Input trajectories from the MPC subproblem are assumed to be fully implemented over the entire horizon (Jamaludin and Swartz, 2015). The hybrid closed/open-loop formulation involves the application of a closed-loop prediction applied over a limited DRTO optimization horizon, after which an open-loop prediction is used. This means that input trajectories needed to generate the closed-loop response dynamics are obtained from the solution of a reduced number of MPC subproblems, whereas the open-loop response dynamics are generated by a full set of input trajectories from another MPC subproblem. On the other hand, closed-loop approximation via input clipping involves use of an unconstrained MPC algorithm where an input saturation mechanism is applied over the DRTO optimization horizon using complementarity constraints. The proposed approximation techniques result in a significantly reduced problem size and faster solution time in comparison to the original multilevel DRTO formulation. We found that the number of complementarity variables for the hybrid closed/open-loop formulation moderately increases when the MPC control horizon becomes longer, while other approximation techniques are insensitive to the length of control horizon. The relative performance of these methods is demonstrated via case studies. Excellent approximation of closed-loop prediction is obtained from all methods, with the economic performance largely retained.

Future work includes application of the closed-loop DRTO approach to more complex systems, and as a higher-level, centralized supervisory controller in a distributed MPC environment where coordination of multiple local MPC controllers can be addressed while making economically optimal decisions.

References

Amrit, R., Rawlings, J. B., and Biegler, L. T. (2013). Optimizing process economics online using model predictive control. Computers & Chemical Engineering, 58, 334 – 343.

Baker, R. and Swartz, C. L. E. (2008). Interior point solution of multilevel quadratic programming problems in constrained model predictive control applications. Industrial & Engineering Chemistry Research, 47(1), 81–91.

Darby, M. L., Nikolaou, M., Jones, J., and Nicholson, D. (2011). RTO: An overview and assessment of current practice. Journal of Process Control, 21(6), 874 – 884.

Darby, M. L., Nikolaou, M. (2010). MPC: Current practice and challenges. Control Engineering Practice, 20(4), 328 – 342.

Ellis, M. and Christodes, P.D. (2014). Economic model predictive control with time-varying objective function for nonlinear process systems. AIChE Journal, 60(2), 507 – 519.

Jamaludin and Swartz (2014). Effects of closed-loop dynamics in dynamic real-time optimization. AIChE Annual Meeting. Atlanta, GA, USA.

Jamaludin and Swartz (2015). A bilevel programming formulation for dynamic real-time optimization. In International Symposium on Advanced Control of Chemical Processes (ADCHEM). Whistler, BC, Canada.

Marlin, T. E. and Hrymak, A. N. (1997). Real-time operations optimization of continuous processes. In AIChE Symposium Series – Fifth International Conference on Chemical Process Control. Tahoe City, NV, USA.

Qin, S. J. and Badgwell, T. A. (2003). A survey of industrial model predictive control technology. Control Engineering Practice, 11(7), 733 – 764.

Ralph, D. and Wright, S.J. (2004). Some properties of regularization and penalization schemes for MPECs. Optimization Methods and Software, 19(5), 527 – 556.

Tosukhowong, T., Lee, J. M., Lee, J. H., and Lu, J. (2004). An introduction to a dynamic plant-wide optimization strategy for an integrated plant. Computers & Chemical Engineering, 29(1), 199 – 208.

Würth, L., Hannemann, R., and Marquardt, W. (2011). A two-layer architecture for economically optimal process control and operation. Journal of Process Control, 21(3), 311 – 321.

427297 On the Use of Transient Information for Static Real-Time Optimization

Abstract:

On the Use of Transient Information for Static Real-Time Optimization

A. Marchetti [1], G. François [2] and D. Bonvin [1]

[1]Laboratoire d'Automatique , Ecole Polytechnique Fédérale de Lausanne

EPFL – Station 9, 1015 Lausanne, Switzerland

[2] University of Applied Sciences and Arts, Western Switzerland, 2800 Delémont, Switzerland
(contact: dominique.bonvin@epfl.ch)

Keywords: Real-time optimization, Modifier adaptation, Plant-model mismatch, Gradient estimation, State estimation.

Optimal operation of chemical processes is key for meeting productivity, quality, safety and environmental objectives. Both model-based and data-driven schemes are used to compute optimal operating conditions [1]:

- The model-based techniques are intuitive and widespread, but they suffer from the effect of plant-model mismatch. For instance, an inaccurate plant model leads to operating conditions that typically are not optimal for the plant and may violate constraints. Furthermore, even with an accurate model, the presence of disturbances generally leads to a drift of the optimal operating conditions, and adaptation based on measurements is needed to maintain plant optimality.

- The data-driven optimization techniques rely on measurements to adjust the optimal inputs in real time. Consequently, real-time measurements are typically used to help achieve plant optimality. This field, which is labeled real-time optimization (RTO), has received growing attention in recent years. RTO schemes can be of two types: explicit schemes solve a numerical optimization problem repeatedly, while implicit schemes adjust the inputs on-line in a control-inspired manner.

*Explicit RTO schemes*solve a numerical optimization problem repeatedly. For example, the two-step approach uses (i) measurements to update the model parameters, and (ii) the updated model to perform the numerical optimization [2]. It has also been proposed to update the model differently. Instead of adjusting the model parameters, input-affine correction terms can be added to the cost and constraint functions of the optimization problem so that it shares the first-order optimality condition with the plant. The main advantage of the technique, labeled modifier adaptation (MA), lies in its proven ability to converge to the plant optimum, even in the presence of structural plant-model mismatch [3]. Furthermore, MA is capable of detecting the correct set of active plant constraints without additional assumptions. As a static optimization method applicable to continuous plants, MA requires waiting for steady state before taking measurements, updating the correction terms and repeating the numerical optimization. Hence, several iterations are generally required to achieve convergence. The main difficulty lies in the estimation of the steady-state plant gradient at each iteration.

In contrast, *implicit RTO schemes*, such as extremum-seeking control [4], self-optimizing control [5] and NCO tracking [6], propose to adjust the inputs on-line in a control-inspired manner. In the absence of constraints, or when assumptions can be made regarding the set of plant constraints that are active at the optimum, implicit RTO methods reduce to gradient control, as the degrees of freedom are adjusted in real time to drive the plant cost gradient to zero. Here again, the difficulty lies in the estimation of the steady-state plant gradient, which, in addition, must be performed during transient operation. This is achieved via either low-frequency plant excitation and corresponding cost measurements (as in extremum-seeking control) or the use of transient measurements together with a model of the steady-state gradient (as in self-optimizing control and NCO tracking via neighboring extremals, where the required steady-state measurements are simply replaced by the corresponding transient measurements) [7]. Implicit RTO is much more challenging when the set of active constraints is unknown, as not only the cost gradient has to be inferred from the measurements but also the set of active constraints and the constraint gradients.

This contribution proposes a framework for using MA during the transient phase toward steady state, thereby attempting to reach optimality in a single iteration to steady state. With this approach, a modified optimization problem is solved repeatedly at each optimization instant during transient, with the input-affine correction terms, which theoretically depend on steady-state plant quantities, being estimated on the basis of transient measurements. Note that such an attempt has already been documented in the literature but, as for the aforementioned implicit methods, the "steady-state" gradients were estimated using transient information in the framework of both multiple units and neighboring extremals [8]. In contrast, this work estimates the steady-state outputs from transient outputs and then uses these estimates "correctly" in the expressions for computing the steady-state gradients. For this, we propose to use the best available dynamic model and perform state estimation using an Extended Kalman Filter (EKF) framework [9]. Since the model is typically not perfect, one key parameter related to the static gain is made adjustable for each input-output pair. This way, the EKF feeds on the transient plant outputs and estimates, at the current time t, the corresponding *steady-state* outputs, which leads to the computation of the *static*gradient. The dynamic model at hand can be seen as a surrogate model that, although not sufficiently accurate globally for process optimization, can process measurements to generate an estimate of the local gradients. The approach will be illustrated on various numerical examples and then applied to the optimization of a continuous stirred-tank reactor.

References

[1] G. François and D. Bonvin, Measurement-Based Real-Time Optimization of Chemical Processes, In S. Pushpavanam, editor, *Advances in Chemical Engineering*, Vol. 43, 1-50, Academic Press (2013).

[2] T. E. Marlin and A. N. Hrymak, Real-Time Operations Optimization of Continuous Processes, In *AIChE Symposium Series - CPC-V*, Vol. 93, 156-164 (1997).

[3] A. Marchetti, B. Chachuat and D. Bonvin, Modifier-Adaptation Methodology for Real-Time Optimization, *Industrial & Engineering Chemistry Research,* 48(13), 6022-6033 (2009).

[4] K. Ariyur and M. Krstic, *Real-Time Optimization by Extremum-Seeking Control*, John Wiley, New York (2003).

[5] S. Skogestad, Plantwide Control: The Search for the Self-Optimizing Control Structure, *J. Process Control*, 10, 487-507 (2000).

[6] G. Francois, B. Srinivasan and D. Bonvin, Use of Measurements for Enforcing the Necessary Conditions of Optimality in the Presence of Constraints and Uncertainty, *J. Process Control*, 15, 701-712 (2005).

[7] G. Francois, B. Srinivasan and D. Bonvin, Comparison of Six Implicit Real-Time Optimization Schemes, *J. Européen des Systèmes Automatisés*, 46, 291-305 (2012).

[8] G. François and D. Bonvin, Use of Transient Measurements for the Optimization of Steady-State Performance via Modifier Adaptation, *Industrial & Engineering Chemistry Research*, 53(13), 5148-5159 (2014).

[9] A. H. Jazwinski, *Stochastic Processes and Filtering*, Mathematics in Science and Engineering, Academic Press (1970).

415087 On the Integration Role of Solvents in Process Synthesis-Design-Intensification: Application to Dmc/MeOH Separation

Abstract:

On the Integration Role of Solvents in Process Synthesis-Design-Intensification: Application to DMC/MeOH separation

Deenesh K. Babi, Rafiqul Gani

Department of Chemical and Bio-chemical Engineering, Technical University of Denmark, Søltofts Plads, Building 229, DK-2800, Kgs. Lyngby Denmark.

***rag@kt.dtu.dk**

Solvents (mass separating agents) play an important role in separation-based processes. For example, consider the separation of an azeotropic mixture. If the azeotrope is not pressure dependent, then a feasible separation technique that can be employed for separation of the azeotrope is usually extractive distillation. In extractive distillation the solvent affects the relative volatility of the two key compounds to be separated. In other words, for a two column distillation sequence configuration, the lighter boiling compound is obtained as the top product of the first distillation column and the heavier boiling compound is obtained as the top product of the second distillation column where the solvent is recovered (for re-use and recycle).

Therefore, the solvent design problem can be defined as follows, *given* an azeotropic mixture to be separated into two pure streams that utilizes a mass separating agent, *find* the best (optimal or near-optimal) solvent candidate (or mixture) that can perform the separation subject to economic, environmental and thermo-physical property constraints. This design problem inherently is a mixed integer non-linear programming problem because the property-process models used can be linear, non-linear or a combination of both and, numerous solvents (or solvent mixtures) can in principle be selected (Lei et al., 2015).

In this work, the generation, screening and verification of the solvent candidate follows a three stage approach, in order to, decompose the solvent design problem into manageable sub-problems. In the first stage, a number of solvent candidates are generated based on pre-defined structural constraints, for example, acyclic, cyclic and/or aromatic compounds, etc. In the second stage, the solvent candidates are screened using property constraints, for example, temperature/non-temperature dependent properties and environmental properties. In stage 3, the selected feasible solvent candidates are verified through simulation for selection of the best (optimal) solvent candidate (mixture). In stages 1-3, property models play an integration role, service plus advice role and service role respectively (Kontogeorgis and Gani, 2004).

Application of the method is highlighted for a typical azeotropic mixture separation. Di-methyl carbonate (DMC) is an important chemical because it can be used as a fuel additive and is therefore considered to be one of the better replacements for methyl tert-butyl ether. Methanol (MeOH) is used as a common raw material in the production of DMC, for example, using phosgene with hydrochloric acid as the by-product, using carbon monoxide and oxygen with water as the by-product, using a cyclic carbonate with a glycol as the by-product, etc. Therefore, recovery/separation of MeOH/DMC is an important separation sequence encountered in generating more sustainable process alternatives for the production of DMC (Babi et al., 2015, Holtbruegge et al., 2014) using MeOH as the raw material. The objective of this presentation is to present the best separation system, with the focus on solvent generation, screening and verification for extractive distillation for the separation of MeOH and DMC. The three stage approach will be presented and it will be shown that existing solvent candidates found in the literature are already generated in the "generation" stage plus new solvent candidates. In the "screening" and "verification" stages, it will be shown that two solvent candidates (not previously reported) are selected that satisfy the structural, property and environmental constraints for the effective separation and recovery of MeOH and DMC. Finally, a design of experiments method will be presented in order to cover the design and pilot testing of the best solvent candidate.

References:

Babi, D. K., Holtbruegge, J., Lutze, P., Gorak, A., Woodley, J. M., & Gani, R. (2015). Sustainable Process Synthesis-Intensification. Computers & Chemical Engineering. doi:10.1016/j.compchemeng.2015.04.030

Gani, T. & O'Connell, J. P. (2004). Computer Aided Property Estimation for Process and Product Design. Computer Aided Chemical Engineering (Vol. 19). Elsevier. doi:10.1016/S1570-7946(04)80004-X

Holtbruegge, J., Kuhlmann, H., & Lutze, P. (2014). Process analysis and economic optimization of intensified process alternatives for simultaneous industrial scale production of dimethyl carbonate and propylene glycol. Chemical Engineering Research and Design. doi:10.1016/j.cherd.2014.05.002

Zhang, L., Cignitti, S., & Gani, R. (2015). Generic Mathematical Programming Formulation and Solution for Computer-Aided Molecular Design. Computers & Chemical Engineering. doi:10.1016/j.compchemeng.2015.04.022

415250 Equation-Oriented Optimization of Processes with Dividing-Wall Columns

Abstract:

Equation-Oriented Optimization of Processes with Dividing-Wall Columns

Richard Pattison, Akash Gupta, and Michael Baldea

McKetta Department of Chemical Engineering

The University of Texas at Austin, 1 University Station C0400, Austin, TX 78712

email: mbaldea@che.utexas.edu

Distillation is one of the largest energy consumers in chemical and petrochemical processes, and typically the primary focus for improving energy efficiency and profitability. Complex column configurations, like Petlyuk (thermally coupled) columns (Figure 1, left), can substantially reduce energy use. Dividing-wall columns (DWCs) (Figure 1, right) follow the same principle as the Petlyuk configuration, but utilize a single-shell construction with a wall that partitions the stages in the middle of the column [1].

Figure 1: Left: Petlyuk column Right: Dividing-wall column

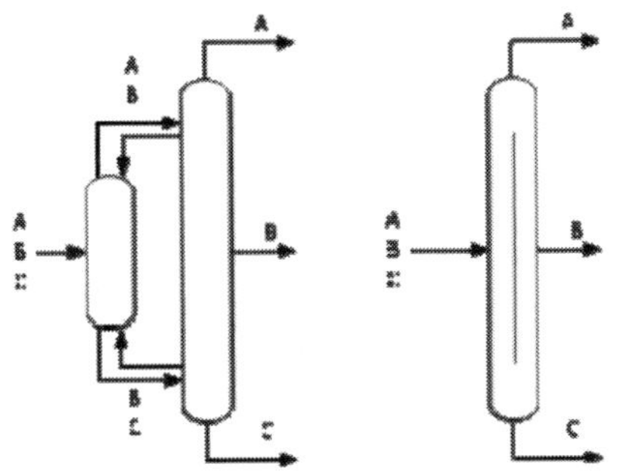

In spite of the potential economic and environmental benefits, the adoption of DWCs in industry has been slow, owing partially to the lack of a transparent and systematic method for optimal design of processes with DWCs. DWCs have more degrees of freedom than conventional distillation columns, and selecting the number stages, the feed and side draw stages, and the location of the dividing wall requires the solution of a highly nonlinear, nonconvex mixed-integer nonlinear program (MINLP).

Previous results relied on shortcut models to approximate the number of stages in each distillation cascade, thus formulating the design optimization as an NLP [2-4]. However, such approximations become inaccurate when the process mixture exhibits non-ideal behavior. On the other hand, detailed models that employ discrete decisions for activation/deactivation of stages in a column [5-7] have been employed, often reporting that initializing and solving the resulting MINLP are highly challenging tasks.

In this work, we propose a novel approach for equation-oriented modeling, simulation and optimization of DWCs and associated flowsheets. We begin with configuring distillation systems as an interconnection of units including: the reboiler, condenser, feed tray, side draw tray, and stage cascades with a variable number of stages. Specifically, a DWC has at least six stage cascades, corresponding to the stages above and below the dividing wall sections, and, within the divided section, two cascades each above and below the feed point, and above and below the side draw point. We rely on our previous results [9] to formulate the model of each distillation stage in a pseudo-transient fashion, as the combination of a mixing step and an equilibrium step that are connected via a tear stream. The formulation strategy improves the flowsheet convergence properties by decoupling the flash calculations between adjacent stages. The number of stages in each cascade is determined by defining (continuous) tray bypass efficiencies [8] to activate/deactivate stages. The proposed developments result in a pseudo-transient stage cascade model, which we demonstrate to be equivalent, at steady-state, with the conventional Material Energy Summation and Hydraulics (MESH) model of a distillation stage.

Subsequently, we show that the proposed pseudo-transient model has desirable initialization and convergence properties, and incorporate it in our previously developed pseudo-transient process flowsheet modeling and optimization framework [9].

We demonstrate the efficacy of the proposed DWC model and optimization framework on an extensive case study, focusing on the intensification of the dimethyl ether production process. Specifically, we demonstrate that the conventional two column design can be replaced with a new flowsheet structure, comprising a single DWC, which has lower capital and operating costs.

References

[1] Kaibel, G. Distillation columns with vertical partitions *Chem. Eng. Technol.* **1987**, 10, 92-98.

[2] Glinos, K.N.; Malone, M.F. Design of sidestream distillation columns. *Ind & Eng. Chem. Process Design and Development,* **1985**, 24, 822-828.

[3] Triantafyllou, C.; Smith, R. The design and optimisation of fully thermally coupled distillation columns: Process design. *Chem. Eng. Res. & Des.*, **1992**, 70, 118-132.

[4] Ramírez-Corona, N.; Jiménez-Gutiérrez, A.; Castro-Agüero, A.; Rico-Ramírez, V. Optimum design of Petlyuk and divided-wall distillation systems using a shortcut model. *Chem Eng. Res. and Des.,* **2010,** 88, 1405-1418.

[5] Dünnebier, G.; Pantelides, C.C. Optimal design of thermally coupled distillation columns. *Ind. & Eng. Chem. Res.*, **1999,** 38, 162-176.

[6] Yeomans, H.; Grossmann, I. E. Optimal design of complex distillation columns using rigorous tray-by-tray disjunctive programming models. *Ind. & Eng. Chem. Res.*, **2000**, 39, 4326-4335.

[7] Vazquez–Castillo, J.A.; Venegas–Sánchez, J.A.; Segovia–Hernández, J.G.; Hernández-Escoto, H.; Hernández, S.; Gutiérrez–Antonio, C.; Briones–Ramírez, A. Design and optimization, using genetic algorithms, of intensified distillation systems for a class of quaternary mixtures. *Comp. & Chem. Eng.*, **2009**, 33, 1841-1850.

[8] Dowling, A.W.; Biegler, L.T. A framework for efficient large scale equation-oriented flowsheet optimization. *Comp. & Chem. Eng.*, **2015**, 72, 3-20.

[9] Pattison, R.P.; Baldea, M. Equation-oriented flowsheet simulation and optimization using pseudo-transient models. *AIChE J.*, **2015**, 60, 4104-4123.

417243 Multistream Heat Exchanger Modeling and Design

Abstract:

A new model formulation and solution strategy for the design and simulation of multistream heat exchangers (MHEXs) is developed. Most current simulation-based models are over-constrained in the sense that they allow for only one unknown that can be adjusted to meet two requirements: the energy balance and second law feasibility. In many such models, a single adjustable temperature is set by the energy balance, so there is nothing left to adjust to satisfy the second law requirement; it is either satisfied or not based on the values given for the degrees of freedom in the problem, leading to temperature crossovers and other nonphysical solutions. The model proposed here allows for the specification of two additional parameters, such as the heat exchange area and/or the minimum temperature difference, as inputs to the model to free up two additional unknowns for simulation. The approach makes use of a modification of classical pinch analysis, as well as incorporating an explicit dependence of the heat transfer on the available heat exchange area to formulate a small nonsmooth equation system to model general MHEXs. Recent advances in automatic computation of derivative-like information for nonsmooth equations[1] make the method tractable, and the use of state-of-the-art nonsmooth equation solving methods[2] make the method precise and robust. Additionally, the algorithm is specifically designed such that B-subdifferential elements of the residual functions are calculated at every iteration, so the equation system is solved with a local quadratic convergence rate. A case study featuring an offshore liquefied natural gas production concept with two MHEXs, compressors and expanders is presented to highlight the flexibility and strengths of the model.

---

[1] K. A. Khan and P. I. Barton, "A vector forward mode of automatic differentiation for generalized derivative evaluation." In Press. *Optimization Methods & Software.*

[2] F. Facchinei, A. Fischer, and M. Herrich, "An LP-Newton method: nonsmooth equations, KKT systems, and nonisolated solutions," *Mathematical Programming*, vol. 146, pp. 1–36, 2014.

417701 Combined Model-Based Experimental Analysis and Process Design Leading to a Profitable Butadiene Telomerization Process

Abstract:

Most chemical products can be obtained from a limited range of platform chemicals, which are mainly produced by large scale petrochemical processes. Amongst these platform chemicals, 1-octene is an important representative serving as intermediate in the production of a broad range of end products. The production by traditional synthesis routes, such as Fischer-Tropsch synthesis or the polymerization of ethylene, covers most of the worldwide demand of 1-octene. However, both synthesis routes require expensive purification. The reason is the unselective nature of the reaction resulting in a broad product spectrum of both processes.

In contrast, in 2007 a novel synthesis route to 1-octene has been implemented by Dow Chemicals on industrial scale in Tarragona. Its key step is the homogeneously catalyzed butadiene telomerization, which allows the utilization of the C4-stream from naphta crackers as cheap and easily available butadiene source [1]. Due to its high chemoselectivity this process has the potential to replace classical processes for the production of 1-octene. However, the bottleneck of the process is the high cost for catalyst regeneration, since the catalyst becomes inactive at the end of the reaction and has to be worked-up before recycling.

Jackstell et al. [2] proposed another catalyst for butadiene telomerization, which is based on palladium modified with 1,3-dimesityl-imidazol-2-ylidene (IMes). This monocarbene-palladium complex is even more selective than the commercial catalyst utilized by Dow and can be continuously separated and recycled, since it remains active throughout the reaction.

In this work, a novel process for this catalyst is designed by applying the model-based design framework for integrated reaction-separation processes recently published by our group [3]. This hierarchical framework structures the design task by three distinctive steps: In the first step, different flowsheet alternatives are generated. In the second step, these alternative flowsheets are evaluated with shortcut methods to identify the most promising process alternatives, which are finally modeled rigorously and optimized in the third step to calculate operating point and sizing information and finally to determine the cost-optimal process structure.

Based on the insights gained through a model-based kinetic study, a tailored flowsheet for the Pd/IMes catalyst is developed and evaluated with shortcut methods. Compared to the process with the Dow catalyst, the operating cost for the PD/IMes process are significantly lower. At the optimal operating point the cost for catalyst and base is decreased by 80% and, additionaly, the process requires 70% less energy for separation. Therefore, the PD/IMes process has the potential to replace classical processes for the production of 1-octene, such as Fischer Tropsch synthesis or the oligomerization of ethylene.

Acknowledgement

The research leading to these results has received funding from the European Community's Seventh Framework Programme [FP7/2007-2013] for *SYNFLOW* under grant agreement n° 246461. Financial support for the second author by the Excellence Initiative of the German Federal and state governments through the Cluster of Excellence *Tailor-Made Fuels from Biomass* are also gratefully acknowledged.

References

[1] Bohley, R.C., Jacobsen, G.B., Pelt, H.L., Schaart, B.J., Schenk, M., van Oeffelen, D.A.G., 1992. Process for producing 1-octene. WO 92/10450.

[2] Jackstell, R., Gómez Andreu, M. Frisch, A. Selvakumar, K., Zapf, A. Klein, H., Spannenberg, A. Röttger, D., Briel, O., Karch, R., Beller, M., 2002. A highly efficient catalyst for the telomerization of 1,3-dienes with alcohols: First synthesis of a monocarbenepalladium(0)-olefin complex. Angewandte Chemie International Edition, 41(6), 986-989.

[3] Recker, S., Skiborowski, M., Redepenning, C., Marquardt, W., 2015. A unifying framework for optimization-based design of integrated reaction-separation processes. Computers & Chemical Engineering, doi:10.1016/j.compchemeng.2015.03.014.

417875 Optimal Production of DME from Switchgrass Based Syngas Via Direct Synthesis

Abstract:

Biofuels have become one of the major alternatives to reduce our dependency on fossil fuels. Typically, researchers have focused their efforts on the production of bioethanol, Fischer Tropsch liquids (FT) and biodiesel from biomass because of their easy implementation in the transportation system. However, there are a number of alternative fuels that can be produced out of the same raw materials such 2,5-dimethylfuran (DMF) or dimethyl ether (DME) from lignocelluloses and/or algae . In particular, DME is a multi-purpose fuel. It has similar properties as propane and butane, so that it can replace Liquefied Petroleum Gas, LPG. It can also be used as diesel substitute, reducing NO_x, SO_x, and particulate emissions. Furthermore, DME can also replace chlorofluorocarbons as aerosol propellant. The advantage of producing DME is that the cetane number is closer to that of crude based oil and in fact, most of the production process is similar to the production of bioethanol or FT fuelsm since DME production is based on catalytic synthesis from syngas.

In this paper, a conceptual optimal design for the production of DME and/or power from switchgrass is proposed. It is possible to produce DME using a two step reaction process comprising methanol synthesis followed by its dehydration. In the literature, a number of simulation studies on the production of DME from coal and natural gas using process simulators, ASPEN Plus, are available.[1-6] Lately biomass has also been included as raw material [7-8] Alternatively, a single step reaction to synthesize DME directly can be used. In this case, a hybrid catalyst is used avoiding the thermodynamic limitations of the methanol synthesis. The single step reaction leads to high CO conversion. Besides that, the investment is relatively lower than the two step reaction process.[9] To systematically design such a process, a superstructure is formulated embedding two different gasification technologies, direct Renugas gasification using oxygen and steam or indirect Ferco gasification, using two chambers, a gasifier fed with olivine, steam and biomass and a combustor, where the char is burned to reheat up the sand providing the energy for the gasification. Next, two reforming modes are considered, partial oxidation, which is exothermic but with a lower yield to hydrogen or steam reforming, endothermic but with a higher yield to hydrogen. Subsequently, the raw syngas is cleaned from solids, ammonia and traces of hydrocarbons. Sour gases are removed next and the gas composition is adjusted in terms of the H_2 to CO ratio. Next, DME is produced following direct synthesis using a one step technology. The process is governed by a series of equilibria. The unreacted gas can be either recycled or used within a Brayton cycle for the simultaneous production of DME and power. The liquid phase is separated using a sequence of distillation columns where $CO_2$is also desorbed and send back to the sour gaes removal system. The problem is formulated as a mixed integer nonlinear programming problem (MINLP) to determine the topology of the system, the use of the syngas to power or DME production and the operating conditions of the different units. Next, we design a heat exchanger network. Subsequently, a sensitivity analysis is performed to determine the limiting prices for DME and power to produce either or both of them. Finally, a detailed economic evaluation is carried out to determine the production and investment costs.

The optimal topology involves indirect gasification followed by steam reforming. DME production is favored over power for current electricity prices. The optimal $H_2$to CO ratio to be fed to the reactor is 1, an interesting result since it is similar to the one needed for second generation bioethanol production. Thus, the investment cost for the production of 197 kt/yr of DME adds up to 133 M$, at a price of $0.31 /kg. Apart from DME, hydrogen is also obtained (9.6 kt/yr of hydrogen) providing an interesting asset. The production of electricity is favored for electricity prices above 0.09 $/kWh. In case of producing electricity alone, up to 105 MW are produced with an investment cost 40% higher than that required for chemicals production and no hydrogen is produced. Even though, the electricity is generated at a competitive price of 0.07$/kWh.

References

[1] Lee, C-J.; Lim, Y.; Kim, H-S.; Han, C. Optimal Gas-To-Liquid Product Selection from Natural Gas under Uncertain Price Scenarios *Ind. Eng. Chem. Res.* **2009**, 48, 794–800

[2] Zhou, L.; Hu, S.; Chen, D.; Li, Y.; Zhu, B.; Jin Y. Study on Systems Based on Coal and Natural Gas for Producing Dimethyl Ether *Ind. Eng. Chem. Res.* **2009**, *48,* 4101–4108

[3] Clausen, L.R.; Elmegaard, B.; Ahrenfeldt, J.; Henriksen U. Thermodynamic analysis of small-scale dimethyl ether (DME) and methanol plants based on the efficient two-stage gasifier *Energy* **2011**, 36, 5805-5814

[4] Zhou, L.; Hu, S.; Li, Y.; Zhou, Q. Study on co-feed and co-production system based on coal and natural gas for producing DME and electricity. *Chem. Eng. J.* **2008,**136, 31–40

[5] Li, F.; Zeng, L.; Fan, L.S. Techno-Economic Analysis of Coal-Based Hydrogen and Electricity Cogeneration Processes with CO_2 Capture *Ind. Eng. Chem. Res*. **2010**, 49 (21), 11018–11028

[6] Haro, P.; Ollero, P.; Villanueva Perales, A.L.; Reyes Valle , C. Technoeconomic assessment of lignocellulosic ethanol production via DME (dimethyl ether) hydrocarbonylation *Energy* **2012**, 44 , 891-901

[7] Haro, P.; Ollero, P.; Villanueva Perales, A.L.; Gómez-Barea, A. Thermochemical biorefinery based on dimethyl ether as intermediate: Technoeconomic assessment Applied *Energy* **2013**, 102 , 950–961

[8] Pascall, A.; Adams, T. A. Semicontinuous separation of dimethyl ether (DME) produced from Biomass *Can. J. Chem. Eng.* **2013**, 91, 1001–1021

[9] Lee,S.; Gogate, M.RE.; Kulik, C.J. A novel single step Dimethyl ether (DME) synthesis in a three phase slurry reactor from CO-Ric Syngas. *Chem. Eng. Sci.* **1992**, 47, 13/14, 3769-3776

419552 Comparative Analysis of Different Retrofit Methods for Heat Exchanger Networks

Abstract:

Improving the energy efficiency of process plants through cost effective methodologies for heat exchanger networks (HENs) has received interest in recent years. To attain this, the heat transfer duties of heat exchangers within the network might be changed. Retrofit of HEN can either be performed by fixing the network structure or performing structural changes.

For a fixed network structure, there are two general ways by which retrofit can be conducted. Consider the basic equation governing heat transfer:

$$Q = UA\Delta T_{LM}F_T \qquad (1)$$

If the heat transfer duty Q is required to be increased at constant log mean temperature difference, ΔT_{LM} and correction factor, F_T, then either the value of the overall heat transfer coefficient, U or the heat transfer area, A needs to be increased.

Considering a shell-and-tube heat exchanger, for small additional heat transfer area requirement, a new tube bundle can be installed into the existing shell (or shells). If a significant amount of additional area is required, a new shell or more than one shell can be installed. However, increasing the heat transfer area can present a complex and capital intensive retrofit process.

In place of installing additional heat transfer area, the overall heat transfer coefficient can be increased i.e. heat transfer enhancement. This can be attained by changing the number of tube passes, baffle arrangements, applying tube inserts or replacing the existing plain tubes with twisted tubes on extended surfaces. Although there have been some research on the use of heat transfer enhancement for HEN retrofit[1-5], there still remains some key questions that needs to be answered. These are:

1) What are the best heat exchangers to enhance in a HEN and in what order?

2) Considering a shell-and-tube heat exchanger, what is the best enhancement technique to apply i.e. shell-side, tube-side, both side or replacing existing plain tubes with twisted tubes?

3) What is the augmentation level of each enhancement based on the exchangers' geometry?

4) How to deal with downstream effects after enhancement?

Structural changes can be used in improving the energy performance of an existing HEN by overcoming the network pinch, which restricts energy savings. However, the application of too many structural modifications i.e. resequencing and stream splitting and, additional area in the form of a new heat exchanger are not ideal in industrial practice. This is as a result of the high costs of civil and piping engineering work that needs to be carried out. There might also be production losses associated with the prolonged shutdown periods required to carry out the stipulated modifications. Adding a new heat exchanger might be very difficult to implement or not practical as a result of the constraint imposed by the existing HEN in terms of congestion. Therefore it is vital that the retrofit process based on structural changes should be kept to a minimum.

Another option in retrofit is to consider structural modifications and additional area with enhancement. This is because, when modifications are made to the existing network, there might be a reduction in the driving force of existing heat exchangers. Therefore, in addition to the modifications carried out, additional heat transfer area might be required in existing heat exchangers. If the increased Q, the new ΔT_{LM} and F_T obtained as a result of modifications carried out in the network are fixed, the relationship between the heat duty and the additional heat transfer area (ΔA) is given by:

$$Q = U(A_{existing} + \Delta A)\, \Delta T_{LM} F_T \qquad (2)$$

A benefit of heat transfer enhancement is that an enhanced heat exchanger has a higher heat transfer coefficient to exchange the same duty under smaller heat transfer area requirement. In other words, an enhanced heat exchanger with an overall heat transfer coefficient (U_E) at the same duty might be able to maintain an existing heat transfer area requirement. This relationship is given by:

$$Q = U_E A_{existing}\, \Delta T_{LM} F_T \qquad (3)$$

Therefore, depending on the additional area requirement, heat transfer enhancement can either be used in eliminating or reducing the additional heat transfer area required after structural modifications and additional area are performed.

The first part of this work presents a methodology for the application of heat transfer enhancement in HEN retrofit. The proposed methodology is a combination of a set of heuristic rules for the identification of the best candidate heat exchanger and location to apply enhancement and; a non-linear optimisation based methodology which tackles the issue of downstream effects after enhancement. The optimisation strategy is based on utility path analysis to ensure the feasibility and cost effectiveness of the retrofit process.

The second part of this work presents better insights into the identification of the best structural changes that should be applied to an existing heat exchanger network to obtain decrease in energy consumption, while ensuring the network constraints are met. This is then followed by the application of enhancement with structural modifications and additional area.

The proposed retrofit methods have been applied to a simplified crude-oil preheat train with an objective of maximising the retrofit profit i.e. difference between the profit from energy saving and total cost of retrofit. Results obtained showed that the use of structural modifications and additional area alone provided a decrease in energy consumption approximately twice that of using only heat transfer enhancement. In terms of retrofit profit, the use of only enhancement had a higher retrofit profit compared to other retrofit options. However, when enhancement was considered together with structural changes and additional area, there was a sharp increase in retrofit profit.

References

1. Pan, M., Bulatov, I., Smith, R., and Kim, J.-K. (2011a). Improving energy recovery in heat exchanger network with intensified tube-side heat transfer. Chemical Engineering, 25.
2. Pan, M., Bulatov, I., Smith, R., and Kim, J. K. (2011b). Novel optimization method for retrofitting heat exchanger networks with intensified heat transfer. Computer Aided Chemical Engineering, 29, 1864-1868.
3. Pan, M., Bulatov, I., Smith, R., and Kim, J. K. (2012). Novel MILP-based iterative method for the retrofit of heat exchanger networks with intensified heat transfer. Computers and Chemical Engineering, 42, 263-276.
4. Wang, Y., Pan, M., Bulatov, I., Smith, R., and Kim, J.-K. (2012). Application of intensified heat transfer for the retrofit of heat exchanger network. Applied Energy, 89(1), 45-59.
5. Jiang, N., Shelley, J. D., Doyle, S., and Smith, R. (2014). Heat exchanger network retrofit with a fixed network structure. Applied Energy, 127, 25-33.

431765 An Algorithm for Simultaneous Process Design and Control Under Process Disturbances and Parameter Uncertainty Using PSE Approximations

Abstract:

A new methodology for simultaneous process design and control is presented using power series expansion (PSE) functions. The proposed approach uses the idea of "back-off approach" to ensure dynamic feasibility under specific realizations in the process disturbances and parametric uncertainty. The PSE functions approximate, around a nominal operating point in the optimization variables, the cost function and the process constrains to evaluate the economics and dynamic feasibility of the system, respectively. The objective in this method is to determine the magnitude of the back-off required from the optimal steady-state design under uncertainty to make the dynamic operation feasible and economically attractive under disturbances and parametric uncertainty. In the work presented, an optimal steady state design and corresponding process parameters are integrated with dynamic feasibility test in an iterative manner to search for the optimal design of the process that ensures dynamic feasibility and that is the closest to that obtained from steady-state calculations. The work focuses on calculating various optimal design and control parameters by solving various sets of optimization problems using mathematical expressions obtained from power series expansions. These approximations are used to determine the direction in the search of optimal design parameters and operating conditions required for an economically attractive and dynamic feasible process. Case studies have been used to test the performance of this methodology, e.g., a non-isothermal continuous stirred tank reactor (CSTR) and a waste water treatment plant. The results obtained for the case studies have been compared with the standard formal integration method and have shown that this methodology has the potential to identify optimal process designs in shorter computational times.

420211 Natural Gas to Liquids, Olefins, and Aromatics: A Systematic Approach for the Optimal Production Trade-Offs

Abstract:

Natural gas to liquid transportation fuels processes[1] have been investigated for several decades, but were considered economically unfavorable because of high natural gas costs. Due to recently discovered sources of shale gas, coupled with improvements in the technology used to extract this resource, the abundance of natural gas within the United States soared in recent years. This drastic increase in the supply of natural gas has caused prices to plummet, from $7.99/MMBTU in 2008 to $2.99/MMBTU in 2015.[2] Additionally, recent studies investigating the coproduction of liquids and valuable chemicals, such as aromatics and olefins, have shown how these coproducts can significantly increase the profitability of refineries.[3,4]However, the profitability of refineries that produce a wide spectrum of products depends intricately on the ratios of these products. Therefore, we investigate the economic effects of natural gas to liquids, olefins, and aromatics refineries to determine and assess the technological and environmental tradeoffs with multiple GTL+C2C4+C6C8 alternatives using an important production parameter.

An optimization-based comprehensive process synthesis framework for the conversion of natural gas to liquid transportation fuels, olefins, and aromatics is proposed.[3,4] Several direct and indirect natural gas conversion technologies are compared to determine the optimal processing pathway.[1] Numerous hydrocarbon production alternatives are incorporated, including Fischer-Tropsch refining and methanol synthesis. Multiple commercial and novel technologies for the production of the olefins and aromatics are included. The process synthesis superstructure includes an olefins purification section and aromatics complex for the production and upgrading of these high-value chemicals. The main olefins investigated include ethylene, propylene, butenes and butadiene. The main aromatics produced include benzene, toluene, and the xylenes. A large-scale nonconvex mixed-integer nonlinear optimization (MINLP) model is formulated to encompass each of these alternatives.5 The MINLP model is solved using a powerful deterministic global optimization branch-and-bound framework and the optimal solution obtained is mathematically guaranteed to be within a few percent of the best possible value. This process synthesis framework provides an adequate baseline for comparing competing technologies.[6-9] The key production parameter is a function of the important products output from the refinery, namely, gasoline, diesel, kerosene, aromatics, and olefins.

Several case studies are presented to investigate the effect of plant capacity and production ratios on the overall profit of the refinery. The major topological decisions as a function of the production parameter will be discussed and the economic and environmental tradeoffs of the refineries will be presented.

1. Baliban, R. C.; Elia, J. A.; Floudas, C. A. Novel natural gas to liquids (GTL) technologies: Process synthesis and global optimization strategies. AIChE Journal 2013, 59, 505–531.

2. Energy Information Administration, Available at: http://www.eia.gov/dnav/ng/hist/rngwhhdm.htm, 2015

3. Onel, O.; Niziolek, A. M.; Elia, J. A.; Baliban, R. C.; Floudas, C. A. Biomass and natural gas to liquid transportation fuels and olefins (BGTL+C2_C4): Process Synthesis and Global Optimization. Industrial & Engineering Chemistry Research 2014, 54, 359-385.

4. Niziolek, A. M.; Onel, O.; Elia, J. A.; Baliban R. C.; Floudas, C. A. Coproduction of Liquid Transportation Fuels and C6_C8 Aromatics from Biomass and Natural Gas. AIChE Journal 2015, 61, 831-856.

5. Baliban, R. C.; Elia, J. A.; Misener, R.; Floudas, C. A. Global Optimization of a MINLP Process Synthesis Model for Thermochemical Based Conversion of Hybrid Coal, Biomass, and Natural Gas to Liquid Fuels. Computers and Chemical Engineering 2012, 42, 64-86.

6. Niziolek, A. M.; Onel, O.; Hasan, F. M.; Floudas, C. A. Municipal solid waste to liquid transportation fuels – Part II: Process synthesis and global optimization strategies. Computers and Chemical Engineering 2014, 74, 184-203.

7. Niziolek, A. M.; Onel, O.; Elia, J. A.; Baliban, R. C.; Xiao, X.; Floudas, C. A. Coal and Biomass to Liquid Transportation Fuels: Process Synthesis and Global Optimization Strategies. Industrial Engineering & Chemistry Research 2014, 53, 17002-17025.

8. Baliban, R. C.; Elia, J. A.; Floudas, C. A. Novel natural gas to liquids (GTL) technologies: Process synthesis and global optimization strategies. AIChE Journal 2013, 59, 505–531.

9. Baliban, R. C.; Elia, J. A.; Floudas, C. A. Biomass and Natural Gas to Liquid Transportation Fuels: Process Synthesis, Global Optimization, and Topology Analysis. Industrial & Engineering Chem Research 2013, 52, 3381–3406.

434930 A Transformation-Centric Approach to Analyzing and Solving Structured Optimization Problems

Abstract:

Computational tools for modeling mathematical programs are widely used within both academia and industry. Available commercial and open-source modeling packages support generic modeling by separating modeling constructs from instance data through concepts like sets, parameters, and parameterized constraints. However, their model representations are limited to constructs that directly correspond to established solver inputs. In general, this implies that mathematical programs must be expressed as either linear or nonlinear algebraic models; that is, a list of variables, an algebraic objective expression, and a list of algebraic constraints. Modelers must then explicitly convert non-algebraic constructs like switching decisions in disjunctive programs into algebraic relaxations (e.g., relaxing constraints implied by the decision with Big-M terms). This approach can obfuscate or eliminate the original model structure and forces the modeler to sacrifice abstraction by injecting decisions related to how to solve a model into the model representation. In this presentation, we demonstrate how high-level non-algebraic modeling constructs can be coupled with automated model transformations to improve model clarity and abstraction. This coupling provides a more flexible workflow where the modeler can explicitly apply transformations that link the structured model to a particular solver. Additionally, this modeling approach simplifies the model specification by eliminating explicit modeling of reformulation decisions. We will highlight two key non-algebraic constructs: disjunctive programming and complementarity constraints. We present transformations that rely on both direct transcription to the Big-M or convex hull relaxation, as well as heuristic procedures leverage both relaxations. Similarly, we show how to reformulate complementarity constraints as smooth nonlinear expressions or as a disjunctive expression that subsequently applies disjunctive transformations. These constructs and transformations are available in the open source Pyomo optimization project.

418380 A Software Framework for the Global Optimization of Nonconvex Two-Stage Stochastic Programs

Abstract:

Most real-life optimization models inherently contain uncertain model parameters. Stochastic programming provides a natural way of addressing such formulations, and has been extensively used in the process systems engineering literature [1-5].

This work presents a software framework for solving two-stage stochastic programs with recourse, whose deterministic equivalent form can be written as:

$$\begin{aligned} \min_{x_1,\cdots,x_s,y,z} \quad & \sum_{h=1}^{s} p_h \left(f_h(x_h) + c_{y,h}^{\mathrm{T}} y + c_{z,h}^{\mathrm{T}} z \right) \qquad \text{(P)} \\ \text{s.t.} \quad & g_h(x_h) + B_{y,h} y + B_{z,h} z \le 0, \quad \forall h \in \{1,\cdots,s\}, \\ & x_h \in X_h, \quad \forall h \in \{1,\cdots,s\}, \\ & y \in Y,\ z \in Z, \end{aligned}$$

where $X_h = \{x_h \in \Pi_h \subset \{0,1\}^{n_{xb}} \times \mathbb{R}^{n_{xc}} : r_h(x_h) \le 0\}$, $Y = \{y \in \{0,1\}^{n_y} : A_y y \le d_y\}$, $Z = \{z \in \mathbb{R}^{n_z} : A_z z \le d_z\}$, and functions $f_h : \Pi_h \to \mathbb{R}$, $g_h : \Pi_h \to \mathbb{R}^m$, and $r_h : \Pi_h \to \mathbb{R}^{m_p}$ are continuous. The uncertainties are modeled using a finite number, s, of scenarios indexed by h, each with a positive probability of occurrence $p_h > 0$. The binary and continuous first-stage decisions y and z, respectively, are made before the realization of the uncertainties, while the mixed-integer second-stage decisions, x_h, are made after the realization of scenario h. Joint polyhedral constraints in y and z are assumed (purely for notational convenience) to be modeled by the scenario constraints in Problem (P).

decomposition is an extension of generalized Benders decomposition that rigorously handles nonconvexities in the participating functions in Problem (P). A popular decomposition technique that can handle the general form of Problem (P) is Lagrangian relaxation [9, 10]. The lower bounding problem in the conventional Lagrangian relaxation technique involves the (partial) solution of the Lagrangian dual problem, which is usually computationally intensive. Our recent work [11] proposed a modified Lagrangian relaxation technique which combines Lagrangian relaxation with NGBD and bounds tightening techniques to solve Problem (P) when continuous first-stage decisions are to be made.

In this work, we provide a computational framework for solving Problem (P) which implements versions of all of the aforementioned decomposition techniques. We demonstrate the capabilities of our framework via several process systems engineering-related case studies.

[1] Sahinidis, N. V. (2004). Optimization under uncertainty: state-of-the-art and opportunities. *Computers & Chemical Engineering*, *28*(6), 971-983.

[2] Li, X., Armagan, E., Tomasgard, A., and Barton, P. I. (2011). Stochastic pooling problem for natural gas production network design and operation under uncertainty. *AIChE Journal*, *57*(8), 2120-2135.

[3] Li, X., Chen, Y. and Barton, P. I. (2012). Nonconvex generalized Benders decomposition with piecewise convex relaxations for global optimization of integrated process design and operation problems. *Industrial & Engineering Chemistry Research*, *51*(21), 7287-7299.

[4] Li, X., Sundaramoorthy, A., and Barton, P. I. (2014). Nonconvex generalized Benders decomposition. In *Optimization in Science and Engineering* (pp. 307-331). Springer New York.

[5] Karuppiah, R. and Grossmann, I. E. (2008). Global optimization of multiscenario mixed integer nonlinear programming models arising in the synthesis of integrated water networks under uncertainty. *Computers & Chemical Engineering*, *32*(1), 145-160.

[6] Benders, J. F. (1962). Partitioning procedures for solving mixed-variables programming problems. *Numerische Mathematik*, *4*(1), 238-252.

[7] Geoffrion, A. M. (1972). Generalized Benders decomposition. *Journal of Optimization Theory and Applications*, *10*(4), 237-260.

[8] Li, X., Tomasgard, A., and Barton, P. I. (2011). Nonconvex generalized Benders decomposition for stochastic separable mixed-integer nonlinear programs. *Journal of Optimization Theory and Applications*, *151*(3), 425-454.

[9] Guignard, M. and Kim, S. (1987). Lagrangean decomposition: a model yielding stronger Lagrangean bounds. *Mathematical Programming*, *39*(2), 215-228.

[10] Carøe, C. C. and Schultz, R. (1999). Dual decomposition in stochastic integer programming. *Operations Research Letters*, *24*(1), 37-45.

[11] Kannan, R. and Barton, P. I. An improved Lagrangian relaxation algorithm for general nonconvex two-stage stochastic programs. *In preparation*.

425235 Solving MINLP with Heat Exchangers: Special Structure Detection and Large-Scale Global Optimisation

Abstract:

Optimising heat exchangers networks (HEN) may increase efficiency in industrial plants; we develop deterministic global optimisation algorithms for a mixed-integer nonlinear optimisation (MINLP) model that simultaneously incorporates utility cost, equipment area, and hot / cold stream matches [1, 4]. In this work, we automatically recognise and exploit special mathematical structures common in HEN including log mean temperature difference and Chen approximation.

We computationally demonstrate the impact on the global optimisation solver ANTIGONE [3] and benchmark large-scale test cases against heuristic approaches. We also present, as an alternative, a Pyomo [2] model which solves the heat exchangers networks problems using iterative refinement and describe tradeoffs between the two techniques.

References

1. M. Escobar and I. E. Grossmann. Mixed-integer nonlinear programming models for optimal simulta- neous synthesis of heat exchangers network, 2010. Available from CyberInfrastructure for MINLP at: www.minlp.org/library/problem/index.php?i=93.

2. W. E. Hart, J.-P. Watson, and D. L. Woodruff. Pyomo: modeling and solving mathematical programs in python. Mathematical Programming Computation, 3(3):219–260, 2011.

3. R. Misener and C. A. Floudas. ANTIGONE: Algorithms for coNTinuous Integer Global Optimization of Nonlinear Equations. J. Glob. Optim., 59(2-3):503–526, 2014.

4. I. Quesada and I. E. Grossmann. Global optimization algorithm for heat exchanger networks. Ind. Eng. Chem. Res., 32(3):487–499, 1993.

430146 Design, Implementation, and Evaluation of the Branch-and-Sandwich Algorithm for Nonlinear Nonconvex Bilevel Problems

Abstract:

Optimization has become the essential tool in process systems engineering. Many of the optimization problems relevant to chemical engineering are bilevel problems in nature [1, 2] and can thus be framed as two-person, hierarchical optimization problems having a nested inner optimization problem withing the constraints of the outer optimization problem. Examples of such problems are optimization problems in design, parameters estimation in thermodynamic models, simultaneous optimization and heat integration of chemical processes, various competitive marketing strategies.

Special cases of bilevel problems, such as problems of linear (quadratic) – linear (quadratic) type, have been studied extensively and many algorithms have been proposed in the literature [2]. However, the general nonconvex form is a very challenging problem. In [3, 4], a new deterministic global optimization algorithm, named Branch-and-Sandwich (B&S), was introduced for bilevel problems with nonconvexities in both the inner and outer problems, and applied to 34 small benchmark problems.

In current work, we discuss the design (how to choose between different approaches), C++ implementation and algorithmic extensions of the B&S algorithm within the open-source MINOTAUR framework [5]. The B&S algorithm is extended to include heuristics for branching and node selection, as well as alternative management of nodes. The computational performance of the Branch-and-Sandwich algorithm is analyzed for different algorithmic variants by solving the extended set of bilevel benchmark problems and showing its applicability to engineering problems.

1. Biegler, L.T., Grossmann, I.E.: Retrospective on optimization. Comput. Chem. Eng. 28, 1169–1192 (2004).
2. Floudas, C.A., Gounaris, C.E. A review of recent advances in global optimization. J Glob Optim, 45, 3–28 (2009)
3. Kleniati, P.-M., Adjiman, C.S.: Branch-and-Sandwich: a deterministic global optimization algorithm for optimistic bilevel programming problems. Part I: Theoretical development. J Glob Optim, 60(3), 425–458 (2014).
4. Kleniati, P.-M., Adjiman, C.S.: Branch-and-Sandwich: a deterministic global optimization algorithm for optimistic bilevel programming problems. Part II: Convergence analysis and numerical results. J Glob Optim, 60(3), 459–481 (2014).
5. Mahajan, A., Leyffer, S., Munson, T., Linderoth, J., Luedtke, J.: MINOTAUR: Toolkit for Mixed Integer Nonlinear Optimization Problems, http://wiki.mcs.anl.gov/minotaur/index.php/Main_Page.

432692 The Alamo Software for Model Building, Constrained Regression, and Intelligent Experimental Design

Abstract:

The purpose of ALAMO (Automated Learning of Algebraic Models using Optimization) is to generate algebraic surrogate models of black-box systems for which a simulator or experimental setup is available. Consider a system for which the outputs z are an unknown function of the system inputs x. The software identifies a function z=f(x), i.e., a relationship between the inputs and outputs of the system, that best matches data (pairs of x and corresponding z-values) that are collected via simulation or experimentation. ALAMO can:

- build an algebraic model of a simulation or experimental black-box system
- use previously collected data for model building
- call a user-specified (simulation) function to collect measurements
- enforce response variable bounds, physical limits, and boundary conditions
- use a preexisting data set for model validation

The problems addressed by the software have long been studied in the fields of statistics, design of experiments, and machine learning. Whereas existing techniques from this literature can be used to fit data to models, ALAMO is distinctive in that it determines where to run the simulations or experiments, what models to fit, and how to determine if a model thus produced is accurate and as simple as possible. A distinguishing feature of the software is that it provides models that are as simple as possible and still accurate. These model attributes are achieved by using a recently developed optimization methodology [1]. Moreover, ALAMO is uniquely in that it utilizes theory-driven insights alongside data in its constrained regression module [2].

Following a brief review of the algorithms implemented in the software, we describe ALAMO's software design and a number of applications that ALAMO facilitates. The ALAMO models can be used to facilitate subsequent system analysis, optimization, and decision making.

References cited

1. Cozad, A., N. V. Sahinidis and D. C. Miller, Automatic learning of algebraic models for optimization, AIChE Journal, 60, 2211-2227, 2014.
2. Cozad, A., N. V. Sahinidis and D. C. Miller, A combined first-principles and data-driven approach to model building, Computers and Chemical Engineering, 73, 116 - 127, 2015.

424372 Pop – the Parametric Optimization Toolbox

Abstract:

Recent years have seen an increased interest in multi-parametric programming (mp-P), in large part due to the ever increasing number of areas where mp-P can be applied such as bilevel programming, reactive scheduling and decentralized control. This in turn has led to significant theoretical advances in fields such as multi-parametric mixed-integer programming, multi-parametric moving horizon estimation and global mp-P [1]. For the solution of the underlying mp-P problems, currently only one solver package is openly available, namely the MPT toolbox [2]. Albeit being very complete and providing a wide array of capabilities, the MPT toolbox is computationally limiting when larger problems are considered. Additionally, as it has its own class and object definitions, software interoperability becomes a challenging process especially during the closed-loop validation of the derived controllers.

In this contribution, we describe POP, the Parametric Optimization toolbox, a new, powerful toolbox for the solution of multi-parametric programming problems. For continuous mp-P problems, POP uses a geometrical approach with a variable step-size exploration strategy combined with a novel algorithm for the removal of redundant constraints. For multi-parametric mixed-integer problems a newly developed generalized framework is employed, featuring an efficient comparison procedure that enables the handling of nonconvex critical regions. Additionally, POP allows seamless formulation and solution of explicit control or scheduling problems, either via concatenation or multi-parametric dynamic programming. These capabilities naturally incorporate the ability to handle hybrid systems through the implementation of state-of-the-art multi-parametric programming algorithms. In order to facilitate the analysis and benchmarking of the developed algorithms, POP features a powerful random problem generator for all problems under consideration. It allows for the generation of problems of arbitrary dimensions, where key aspects of the algorithm can be adjusted using several parameters.

Here we examine various benchmark test sets, featuring for each class of problem 250 randomly generated mp-P problems of diverse characteristics. The simulation results are used to highlight the solution capabilities of POP and compare its performance with the MPT toolbox. They also provide valuable insight into future research directions for the improvement of the computational performance.

The applicability and software interoperability of POP is demonstrated via the explicit optimal control of the multicolumn solvent-gradient purification (MCSGP) process, which falls in the challenging class of continuous nonlinear periodic systems [3]. Using the newly developed PAROC framework (http://www.paroc-platform.co.uk/Software/POP/) [1], a high-fidelity, validated model of the system was implemented in gPROMS® ModelBuilder v4.0 [4] and reduced to a linear state-space representation employing system identification techniques. The state-space model considers 4 states, 1 input, 3 output as well as measured disturbances with a sampling time of 6s. The corresponding explicit MPC problem for different control and output horizons is formulated and solved via the POP toolbox (see Table 1).

Table 1: Detailed characteristics of the explicit optimal control of the MCSGP system.

Output Horizon	Control Horizon	Number of Parameters	Number of constraints	Number of optimization variables
10	2	40	84	2
	4		88	4
	6		92	6
	8		96	8
12	2	46	100	2
	4		104	4
	6		108	6
	8		112	8
14	2	52	116	2
	4		120	4
	6		124	6
	8		128	8

The derived controllers are validated in-silico utilizing seamless software interoperability between gPROMS® ModelBuilder and MATLAB. The closed-loop simulation results indicate successful reference tracking of the output variables and disturbance rejection.

Providing a novel and powerful toolbox for the solution of multi-parametric programming problems, this work presents POP, the Parametric Optimization toolbox. Using a newly developed random problem generator, a large number of numerical studies shows both the versatility as well as the computational efficiency of POP. This in turn allows for the development of explicit MPC controllers for very complex systems such as the MCSGP process.

Acknowledgements

This work is the result of many contributions over the last two decades. In particular, we would like to acknowledge Nikolaos A. Bozinis and Martina Wittmann-Hohlbein, who have been pivotal in developing previous versions of POP [5,6].

References

1. Pistikopoulos, E.N., et al., *PAROC - an Integrated Framework and Software Platform for the Optimization and Advanced Model-Based Control of Process Systems.* Chemical Engineering Science, 2015, **in print**.

2. Kvasnica, M., et al., *Multi-Parametric Toolbox (MPT)*, in *Hybrid Systems: Computation and Control*, R. Alur and G.J. Pappas, Editors. 2004, Springer Berlin Heidelberg. p. 448–462.

3. Aumann, L. and M. Morbidelli, *A continuous multicolumn countercurrent solvent gradient purification (MCSGP) process.* Biotechnology and Bioengineering, 2007. **98**(5): p. 1043–1055.

4. Process Systems Enterprise, gPROMS, www.psenterprise.com/gproms, 1997-2014.

5. Bozinis, N.A., et al. POP, alpha version, 2000.

6. Wittmann-Hohlbein, M., et al. POP, beta version, 2013.

421703 A Platform Facilitating Workflow Management of Multi-Task, Multi-Scale Simulations for Distributed Computing Environments

Abstract:

The `MoDeNa` project [4] is developing a framework for performing simulations of multi-scale systems. The motivation arises from the need to design and control the material structure of polyurethane foam by performing detailed simulations that connects models across all levels of *scale*, i.e. from molecular dynamics, through CFD and ending with final product. Moreover, the project focuses on reducing computational cost by employing inexpensive surrogate models that approximate the behaviour of the corresponding complex models on the lower *scales*.

The practical challenges, from a software perspective, are threefold: connecting models written in different programming languages, facilitating communication between the models and dynamically modifying the computational workflow during the simulation. The management of the computational workflow is particularly important because there is no way to know, apriori, which models will have to be executed, when, and in what order. The solution to this problem is to employ an *orchestrator*, i.e. a program that manages the computational jobs: where, when and in what order they are executed.

The most basic *orchestrator* is a que where jobs are executed in sequence and new jobs (N) are staged at the back.

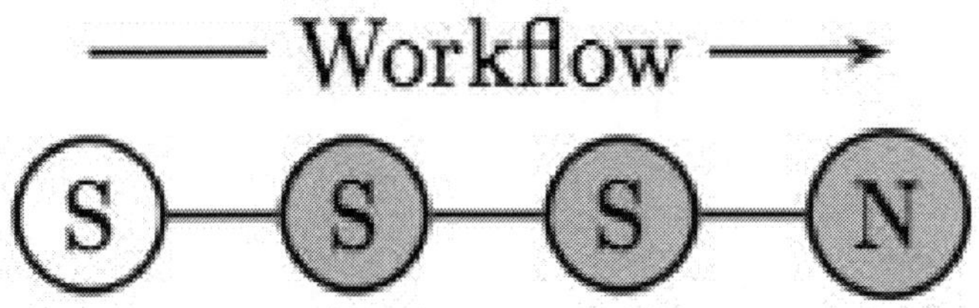

This is sufficient for systems where the sequence is known apriori, e.g. for a simple *process simulator*, but `MoDeNa` requires a more flexible *orchestrator*.

The reason is the multi-scale connection of the models, where every model requires information from models on a lower scale *before* it can be executed. Specifically, the framework must accomodate the functional dependency of models across the *scales*. The framework also allows one dependency to trigger a chain of new dependencies, some of which must be executed in sequence while others can be run in parallel. Moreover, in order to ensure the range of validity and accuracy of the surrogate models, the framework also allows design of experiments and parameter fitting procedures to be started by any model, at any time.

The following example shows how the orchestrator allows the workflow to grow with arbitrary complexity when the new job (N) and all its dependencies are added to the workflow.

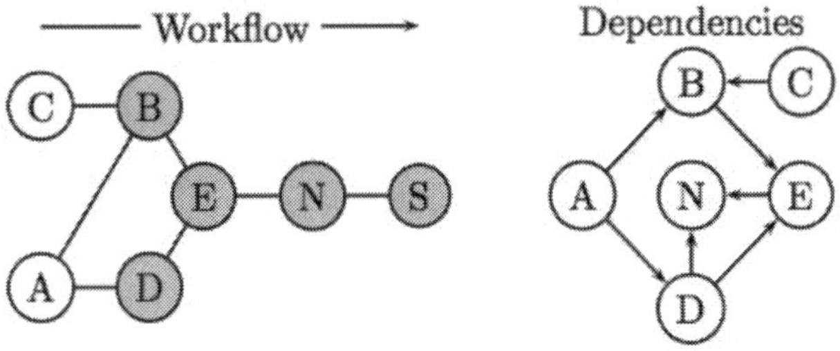

The *orchestrator* used for `MoDeNa` is the `Python`-based program `FireWorks` [2], originally developed for *the materials project* [1]. In addition to providing all the flexibility needed for dynamic management of the `MoDeNa` workflow it also suports staging of jobs on arbitrary computing resources. The use of a workflow management, such as `FireWorks`, facilitates the flexibility of connecting models from different programming languages or simulation tools. By adding a framework that facilitates communication and connection through the same backend that `FireWorks` uses (`MongoDB`[3]), `MoDeNa` makes it easier to cooperate across disciplines to create models that easily incorporate details that would otherwise be assumptions in a simplified model. References

[1] Anubhav Jain, Shyue Ping Ong, Geoffroy Hautier, Wei Chen, William Davidson Richards, Stephen Dacek, Shreyas Cholia, Dan Gunter, David Skinner, Gerbrand Ceder, and Kristin a. Persson. The Materials Project: A materials genome approach to accelerating materials innovation. *APL Materials*, 1(1):011002, 2013. [DOI | http]

[2] Fireworks: workflow software. [http]

[3] MongoDB: open-source, document database. [http]

[4] MoDeNa: The MoDeNa project. [http]

423488 Procafd: Computer Aided Tool for Synthesis-Design & Analysis of Chemical Process Flowsheets

Abstract:

In practice, chemical process synthesis-design involves identification of the processing route to reach a desired product from a specified set of raw materials, design of the operations involved in the processing route, the calculations of utility requirements, the calculations of waste and emission to the surrounding and many more. In terms of approaches to solve the synthesis-design problem three major lines of attack have emerged: (a) the knowledge based approach [1] which relies on engineering knowledge & problem insights, (b) the optimization approach [2] which relies on the use of mathematical programming techniques, (c) hybrid approach which combine two or more approaches. D'Anterroches [3] proposed a group contribution based hybrid approach to solve the synthesis-design problem where, chemical process flowsheets could be synthesized in the same way as atoms or groups of atoms are synthesized to form molecules in computer-aided molecular design (CAMD) techniques [4]. The main idea here was to apply the principle of group-contribution approach from chemical property estimation to the synthesis and design of chemical process flowsheets. That is, use process-groups representing different unit operations (reactor, distillation, flash, crystallization, etc.), bonds representing streams and/or recycles, rules for chemical feasibility also representing process flowsheet feasibility and sum of group contributions representing the performance of the flowsheet. In the next stage, the design parameters for the operations of the high ranked flowsheets are established through reverse engineering approaches based on driving forces available for each operation. In the final stage, rigorous simulation is performed to validate the synthesis-design. Note that since the flowsheet is synthesized and the operations in the flowsheet designed to match a set of design targets, there are no iterations involved as the final flowsheet is among the best, if not the best.

In this work, the flowsheet synthesis-design framework is expanded to include process-groups representing the unit operations from biochemical processes and new flowsheet property models to predict the environmental impact, process safety, product recovery and purity, which are employed to screen the generated alternatives. The extended framework implemented in the form of a computer- aided tool , ProCAFD will be highlighted through two case studies, one involving the synthesis of a chemical process flowsheet (the well-known Hydrodealkylation of toluene process) and another for a biochemical process (production of Bio-ethanol). In both cases, various process alternatives were generated quickly using the ProCAFD tool. The framework was able to generate not only the reported designs, but also various innovative designs, due to predictive nature of the methods used. The application-examples also focus on the use of SFILES notation system, developed specially for the process-group based synthesis-design method, to store and/or visualize the structural information of any process flowsheet represented by process-groups. As an extension, the SFILES notation is extended to store the process information through which representation of flowsheet alternatives for rigorous process simulation (for example, with an external process simulator) would be possible.

References:

[1] C. Jaksland, R. Gani, 1995, Chemical Engineering Science, 50, 511-530.

[2] H. Yeomans, I.E. Grossmann, 1999, Computers & Chemical Engineering, 23, 709.

[3] L. d'Anterroches, R. Gani, 2005, Fluid Phase Equilibria, 228-229, 141–146.

[4] L. Achenie, R. Gani, V. Venkatasubramanian, 2002, Computer Aided Chemical Engineering 12

437483 Designing Software for Chemical Engineers – a Product Developer's Perspective

Abstract:

Process simulators are tools widely used for the design of chemical processes and chemical products. These products are ubiquitous in process engineering work and a normal part of the engineer's work flow. These tools and associated tools like the software plug-ins designed to provide highly specialized services such as mass transfer separation modules, flare systems and advanced chemical reactors. Although the tools are widely used little is said about the process used to develop these tools, the challenges the software product developer faces and how these challenges are met. In this presentation we take an introspective look at the development of software products commonly used in process design and process simulation.

425315 Designing Chimeric Peptides Using Swarm Intelligence Optimization for the Protein Folding Problem

Abstract:

Chimeric peptides offer biological engineering methodologies adapted in materials science for wide range of applications including assembly and synthesis. Tailored chimeric functionality depends not only the sequence but also the molecular structure of the resulting molecules. Designing new chimeric peptides therefore relies on developing a better understanding of the protein folding. Here, a new *ab initio* method for finding peptide backbone angles was developed with a swarm intelligence algorithm to find peptide folds faster than using a molecular dynamics approach. Currently, many approaches to the protein folding problem have been investigated. The first approach is to experimentally determine the structures. This is the most accurate way of determining the structure, but the process is slow and requires commitments of large resources. Determining a structure experimentally without knowing that the peptide design is already successful is prohibitive. The second approach is to use molecular dynamics simulations. The main limitation of these simulations for peptide design is the low number of peptides that can be analyzed. Having larger peptide design candidate pools should help to improve chimeric peptide design. Using existing data avoids the cost of determining new structures and statistical analysis avoids the computational cost of simulating the protein folding process in the second approach. However in swarm intelligence algorithm, statistical analysis of determined folding structures may avoid the challenges of the first two approaches. In previous work[1], an antimicrobial peptide that self-assembles onto titanium implants has been developed. Here, we applied two optimization strategies to these peptide sets for finding chimeric peptide folds more effectively. We next compared their folding patterns to engineer sequence-structure function relations. The first optimization strategy uses a swarm intelligence algorithm inspired by bee hive coomunication to coordinate the selection of highly likely folds. The second approach is to match the peptide fragments with historical data of protein folds[2]. PyRosetta software[3] was used to score the likelihood of structures and to find the bond angles for the amino acid side chains. The likelihood of the structures for each approach was compared, and a time analysis was performed. Our approach offers to expand the pool of new chimeric peptide design. *The NIH-NIAMS (7R21AR062249-03) is acknowledged.*

[1] Yucesoy D, Hnilova M, Boone K, Arnold P, Snead M, Tamerler C. Chimeric Peptides as Implant Functionalization Agents for Titanium Alloy Implants with Antimicrobial Properties. JOM 2015;67:754-66.

[2] Kim DE, Chivian D, Baker D. Protein structure prediction and analysis using the Robetta server. Nucleic Acids Research 2004;32:W526-W31.

[3] Chaudhury S, Lyskov S, Gray JJ. PyRosetta: a script-based interface for implementing molecular modeling algorithms using Rosetta. Bioinformatics 2010;26:689-91.

15AIChE
Annual Meeting, Salt Lake City, UT

DTU

A New Optimization Model for Computer-Aided Molecular Design Problems

Lei Zhang, **Stefano Cignitti**, Rafiqul Gani*

PSE for SPEED Project, Department of Chemical and Biochemical Engineering, Technical University of Denmark

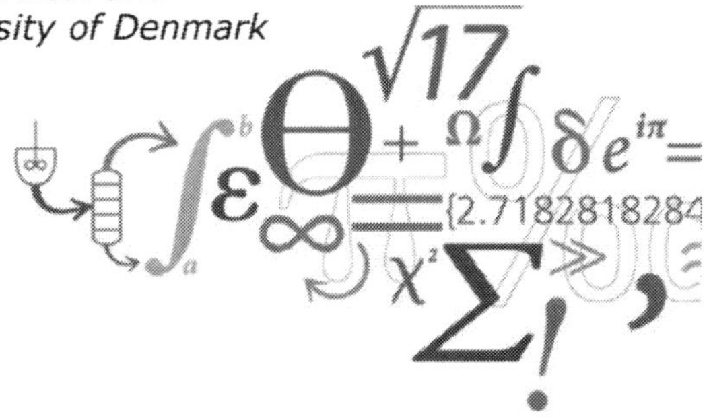

DTU Chemical Engineering
Department of Chemical and Biochemical Engineering

15AIChE
Annual Meeting, Salt Lake City, UT

DTU

Outline

- Introduction
- Computer-aided Framework for Product Design Problems
- New Optimization Model for CAMD
- Case Studies
  - Surfactant Design
  - Refrigerant Design
  - Die Attach Adhesive Design
- Conclusion

15AIChE
Annual Meeting, Salt Lake City, UT

DTU

Chemical Product Design

- The modern society needs many products for its survival.
- Which are the important products, which resources should be used to produce them and what is the role of the chemicals based manufacturing process?
- With the help of computer-aided chemical product design, the optimal, more sustainable product-process will be designed.

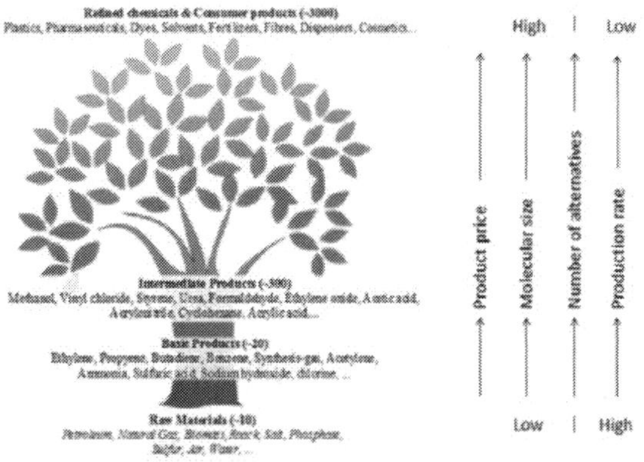

15AIChE
Annual Meeting, Salt Lake City, UT

DTU

Computer-Aided Chemical Product Design

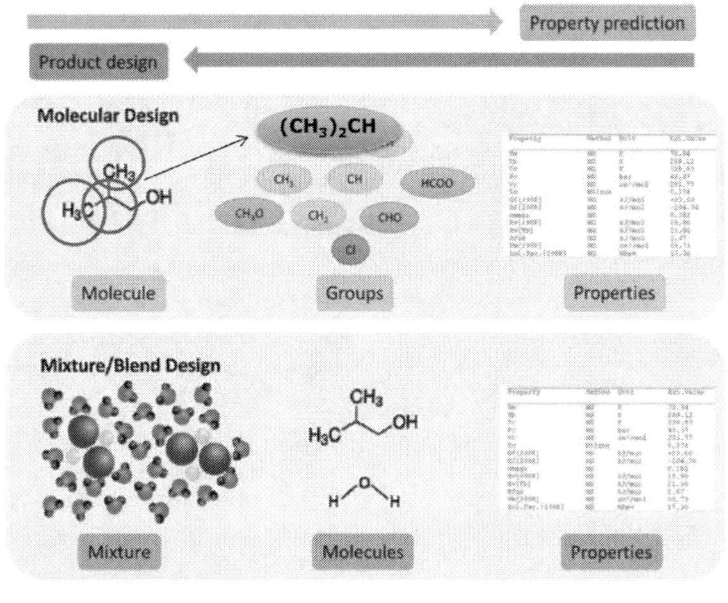

Computer-Aided Framework for Product Design Problems

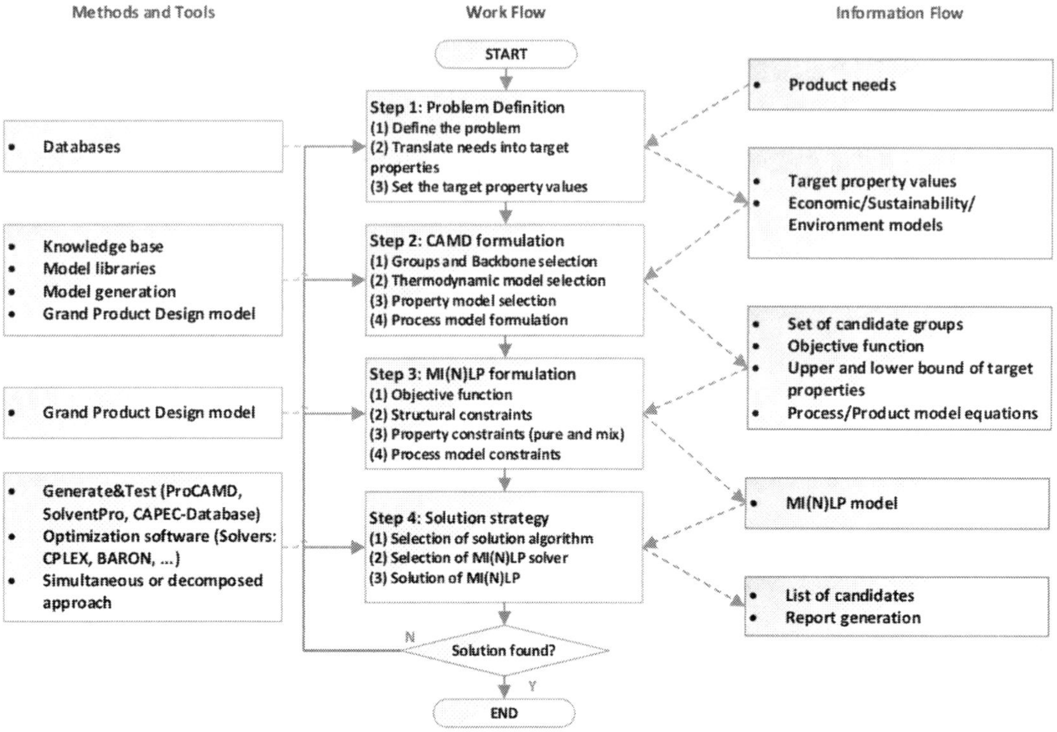

15AIChE
Annual Meeting, Salt Lake City, UT

DTU

New Optimization Model for CAMD: MINLP Formulation

$$\min / \max f_{obj}(\mathbf{X}, \mathbf{N})$$

Structural constraints: $g_1(\mathbf{N}, \mathbf{Y}) \leq 0$

Pure component property: $g_2(\mathbf{N}, \mathbf{Y}) \leq 0$

Mixture property: $g_3(\mathbf{X}, \mathbf{N}) \leq 0$

Process model and others: $g_4(\mathbf{X}, \mathbf{N}) = 0$

$$\mathbf{X} \in \mathbb{R}^n; \mathbf{N} \in \mathbb{Z}_+^m; \mathbf{Y} \in \{0, 1\}^q$$

$\mathbf{N}$: group number;
$\mathbf{Y}$: adjacency matrix;
$\mathbf{X}$: continuous variables related to the mixture and process.

- Adjacency matrix is introduced to predict the detailed structure of the molecule.
- New constraints are involved to find the number of 2nd order groups from the adjacency matrix using binary variables.

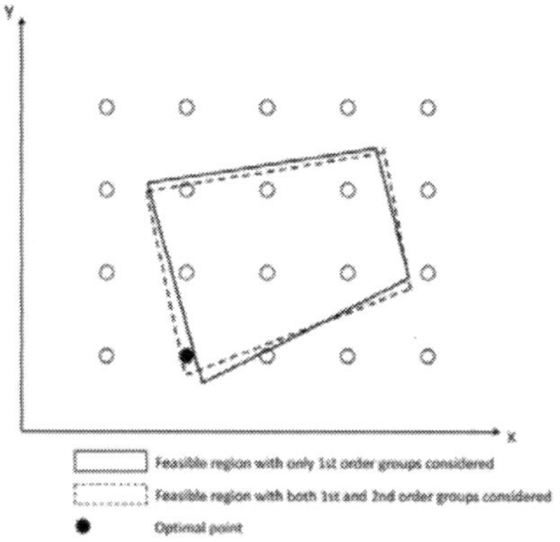

Figure: Higher order groups need to be considered simultaneously to obtain "real" optimal product

15AIChE
Annual Meeting, Salt Lake City, UT

DTU

Generic mathematical programming formulation

1. Structural constraints:

1.1 Octet rule

$$\sum_{i \in G_1} (2-\nu_i)\, n_i^{(1)} = 2q$$

$$\sum_{i_1 \neq i_2; i_1 \in G_1} n_{i_1}^{(1)} \geq n_{i_2}^{(1)} (\nu_{i_2} - 2) + 2 \quad \forall i_2 \in G_1$$

1.2 Constraints ensure only one molecule is formed (Churi and Achenie (1996)).

1.3 Group number constraints:

$$n_i^L \leq n_i^{(1)} \leq n_i^U \quad \forall i \in G_1$$

$$n^{min} \leq \sum_{i \in G_1} n_i^{(1)} \leq n^{max}$$

1.4 Adjacency matrix constraints:

$$y_{i_1,id_1,i_1,id_1} = 0 \quad \forall i_1 \in G_1, id_1 \in ID$$

$$y_{i_1,id_1,i_2,id_2} = y_{i_2,id_2,i_1,id_1} \quad \forall i \in G_1; id \in ID$$

$$\sum_{id_1 \in ID} z_{i_1,id_1} = n_{i_1}^{(1)} \quad \forall i_1 \in G_1$$

$$\sum_{i_2 \in G_1} \sum_{id_2 \in ID} y_{i_1,id_1,i_2,id_2} = \nu_{i_1} z_{i_1,id_1}$$

$$\forall i_1 \in G_1, id_1 \in ID$$

1.5 Second-order group constraints:

$$N_B^J - M\left(1 - b_{id_1,id_2,\cdots,id_{N_B}^J}\right) \leq \sum_{\text{If these two grps connect in 2nd grp. } J} y_{i_1,j_1,i_2,j_2} \leq N_B^J - 1 + M\left(b_{id_1,id_2,\cdots,id_{N_B}^J}\right) \quad \forall J \in G_2$$

$$n_j^{(2)} = \frac{1}{T} \sum_{id_1,id_2,\cdots,id_{N_B}} b_{id_1,id_2,\cdots,id_{N_B}^J} \quad \forall J \in G_2$$

2. Property constraints:

$$p_k^L \leq p_k \leq p_k^U \quad \forall k \in P$$

$$p_k = \sum_{i \in G_1} n_i^{(1)} p_k^{(1)} + \sum_{j \in G_2} n_j^{(2)} p_k^{(2)} \quad \forall k \in P$$

3. Process model, economic model and other constraints.

[1] Lei Zhang, Stefano Cignitti, Rafiqul Gani. *Computers & Chemical Engineering*, **78**, 79-84

Churi N, Achenie LE., Ind Eng Chem Res 1996;35(10):3788–94.

15AIChE
Annual Meeting, Salt Lake City, UT

DTU

Surfactant Design - Product Design (1/3)

- Step 1: Design of a UV sunscreen, in the emulsified form, with a high sun protection factor. The phenol ring is fixed as backbone structure in this case study.
- Step 2:
  - **Objective:** Toxicity should be minimized.
  - **Group selection:** CH_3 , CH_2 , CH , C , aCH , aC-OH , CH_2COO , CH_3O , CH_2O , aC-O , OCH_2CH_2OH from 220 groups (Marrero and Gani, (2001)).
  - **Backbone selection:** -C_6H_4OH

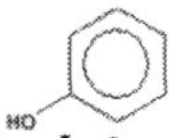

  - **Structural and property constraints:**

Need	Lower bound	Upper bound
N_G	10	15
N_F	0	8
$lc50$	3.16log(mol/L)	-
Sp	-	25MPa$^{0.5}$
clp	343.15K	-
Fp	343.15K	-
Vm	0.1cc/mol	0.3cc/mol

LC_{50}: Fathead Minnow 96-hr LC_{50}; *Sp*: Hildebrand solubility parameter; *clp*: cloud point; *Fp*: Flash point; *Vm*: Liquid molar volume at 298K.

15AIChE
Annual Meeting, Salt Lake City, UT

DTU

Surfactant Design (2/3)

- Step 3:

$$\max lc50 = lco + \sum_{i_1} n_{i_1}^{(1)} lc50_{i_1}^{(1)} + \sum_{i_2} n_{i_2}^{(2)} lc50_{i_2}^{(2)}$$

1. Structural constraints:

1.1 Octet rule and group number constraints:

$$\sum_{i} (2 - \nu_{i_1}) n_{i_1}^{(1)} = 0$$

$$\sum_{i_1, i_1 \neq i_1'} n_{i_1}^{(1)} \geq n_{i_1'}^{(1)} \left(\nu_{i_1'} - 2\right) + 2 \quad \forall i_1'$$

$$1 \leq n_{i_1}^{(1)} \leq 5 \quad \forall i$$

$$10 \leq \sum_{i_1} n_{i_1}^{(1)} \leq 15$$

1.2 Adjacency matrix constraints. (Slide 6)

1.3 Second-order group constraints. (Slide 6)

2. Pure component property constraints:

$$lco + \sum_{i_1} n_{i_1}^{(1)} lc50_{i_1}^{(1)} + \sum_{i_2} n_{i_2}^{(2)} lc50_{i_2}^{(2)} \geq 3.16$$

$$Spo + \sum_{i_1} n_{i_1}^{(1)} Sp_{i_1}^{(1)} + \sum_{i_2} n_{i_2}^{(2)} Sp_{i_2}^{(2)} \leq 25$$

$$\sum_{i_1} n_{i_1}^{(1)} clp_{i_1}^{(1)} \geq 343.15^2$$

$$Fpo + \sum_{i_1} ni_1{}^{(1)} Fp_{i_1}^{(1)} + \sum_{i_2} n_{i_2}^{(2)} Fp_{i_2}^{(2)} \geq 343.15$$

$$0.1 \leq Vmo + \sum_{i_1} ni_1{}^{(1)} Vm_{i_1}^{(1)} + \sum_{i_2} n_{i_2}^{(2)} Vm_{i_2}^{(2)} \leq 0.3$$

- #Equations: 1,324,764; #Variables: 1,286,278 (1,286,266 Integer var.).

- Step 4: This model can be solved directly using GAMS CPLEX solver.

15AIChE
Annual Meeting, Salt Lake City, UT

DTU

Surfactant Design (3/3)

Table: Optimization results of surfactant design problem

n_1	1 CH_3, 5 CH_2, 1 CH, 4 aCH, 1 aC–OH, 1 CH_2COO, 1 aC–O, 1 OCH_2CH_2OH
n_2	1 AROMRINGs1s4
$lc50$	4.124log(mol/L)
Sp	24.732MPa$^{0.5}$
clp	300.016K
Fp	552.383K
Vm	0.281cc/mol

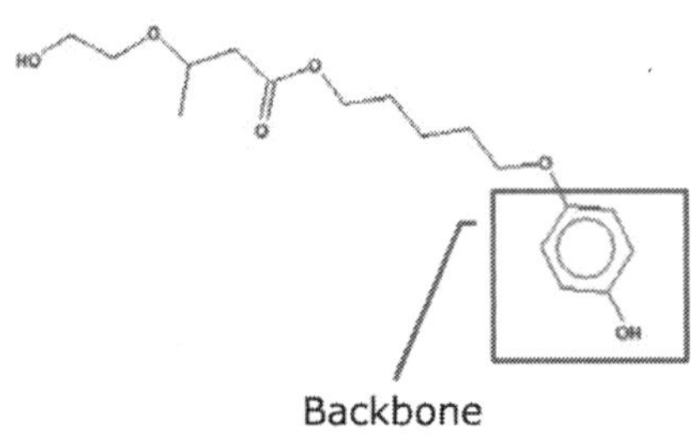

Refrigerant Design - Product-Process Design (1/3)

- Step 1: Design a new refrigerant for the replacement of R-134a, the new fluid should have (1) similar/better properties and thermodynamic behaviour; (2) reduced environmental impact.
- Step 2:
  - **Objective:** Coefficient of performance (COP) should be maximized.
  - **Group selection:** Fluids can contain C, H, F and Cl. 30 molecular groups are selected from 220 groups (Marrero and Gani, (2001)).
  - **Target properties for product and process:**

Target Property	Lower bound	Upper bound
Molar mass	-	110g/mol
Normal boiling point	-	250K
Critical temperature	350K	400K
Critical pressure	30bar	50bar
Thermal conductivity (liquid, Tb)	0.08W/m K	-
Enthalpy of Vaporization at Tb	200KJ/Kg	-
ODP	-	0
GWP	-	1400
Atmospheric Lifetime *	-	14years
Process needs	**Lower bound**	**Upper bound**
Evaporation temperature	313.15K	333.15K
Condensation temperature	265.65K	280.15K

Refrigerant Design (2/3)

$$\max COP = \frac{H_1(Z_{V1}, T_1, P1) - H_3(Z_{L3}, T_3, P_3)}{H_2(Z_{V2}, T_2, P_2) - H_1(Z_{V1}, T_1, P_1)}$$

s.t.

1. Structural constraints.
2. Property constraints.
3. Process constraints:

$$\ln \psi_{V1} = \ln \psi_{L1}$$
$$\ln \psi_{V2} = \ln \psi_{L2}$$
$$H_{L3}(T_3, P_3) = H_{L_4}(T_4, P_4)$$
$$S_{V2}(T_2, P_2) = S_{V1}(T_1, P_1)$$
$$P_1 = P4 \quad P_2 = P_3 \quad T_1 = T_4$$
$$P_2 \leq 0.8P_c \quad T_2 \leq 0.8T_c$$
$$265.65K \leq T_1 \leq 280.15K$$
$$313.15K \leq T_3 \leq 333.15K$$
$$Z_{V1}^3 - Z_{V1}^2 + Z_{V1}(-B^2 + A - B) - AB = 0$$
$$Z_{V2}^3 - Z_{V2}^2 + Z_{V2}(-B^2 + A - B) - AB = 0$$
$$Z_{V3}^3 - Z_{V3}^2 + Z_{V3}(-B^2 + A - B) - AB = 0$$

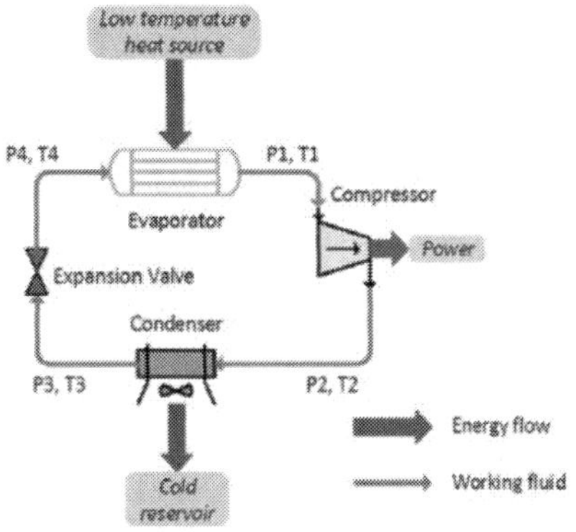

Constraints to ensure correct cubic EOS root selection (Kamath et al., 2010):

$$f(Z_{V1}) = 0 \quad f'(Z_{V1}) \geq 0 \quad f''(Z_{V1}) \geq 0$$
$$f(Z_{V2}) = 0 \quad f'(Z_{V2}) \geq 0 \quad f''(Z_{V2}) \geq 0$$
$$f(Z_{V3}) = 0 \quad f'(Z_{V3}) \geq 0 \quad f''(Z_{V3}) \geq 0$$

[1] Stefano Cignitti, et al., CAST Division Plenary

15AICHE
Annual Meeting, Salt Lake City, UT

DTU

Refrigerant Design (3/3)

Table: Optimization results for refrigerant design problem

	r134a	r152a	3,3,3-diFluoropropyne	3,3-diFluoropropene
		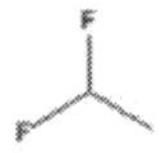		
Mw (g/mol)	66	102	94	92
Tb (K)	230	236	242	246
Tc (K)	366	374	369	392
ω	0.27	0.33	0.28	0.14
k (W/m K)	0.097	0.082	0.086	0.093
ΔH_v at Tb (J/mol)	28562	26523	18635	21312
ODP	0	0	0	0
GWP	**1370**	**133**	**134.6**	**2.4**
Atmospheric lifetime (years)	**13.4**	**1.5**	-	-
COP	**10.1**	**12.8**	**9.5**	**11.0**

15AICHE
Annual Meeting, Salt Lake City, UT

DTU

Die Attach Adhesive (DAA) Design - Product Design with Economic Analysis (1/3)

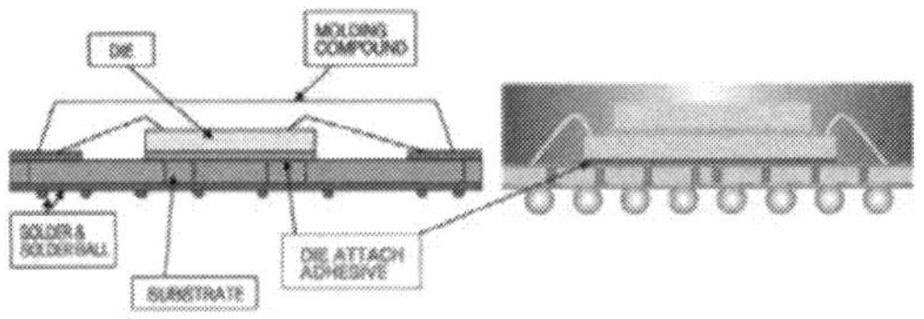

- Step 1: Die attach adhesive design used in LED lamp. It is crucial to remove the heat from the chip as fast as possible (high thermal conductivity)
- Step 2:
  - **Objective:** Net present value (NPV) should be maximized.
  - **Ingredients:** Polymer (designed), microparticle (Ag or Cu) and nanoparticle (Ag or Cu).
  - **Group selection for polymer:** -C6H4-, -O-, CH2, CH, C=O, -S-, -SO2-, from 100 groups (D.W. van Krevelen, 2009)
  - **Property constraints:** The thermal conductivity of the product should equal 30 W/m K

*Collaboration with HKUST (Kelvin Zhang and Prof. Ka M. Ng)

Die Attach Adhesive Design (2/3)

$$\max npv = \sum_{j=-n}^{m} \frac{pcf_j}{(1+R)^j}$$

s.t.

1. Structural constraints.
2. Property constraints (Group contribution method).
3. Economic model:

$$pcf_j = \begin{cases} -TDC_j - TCI_j - \Delta NWC_j & j \leq 0 \\ (p_j d_j - c_{m,j} - Adv_j - FC_j)(1-t) + tD_j + \Delta NWC_j & j > 0 \end{cases}$$

$$c_{m,j} = (w_{mp} c_{mp} + w_{np} c_{np} + c_{m,pc}) d_j \quad j \geq 1$$

$$p_j d_j = \left(\frac{\alpha_j}{\beta}\right)^{\rho} P_j^C \left(\frac{Y - p_j d_j}{P_j^C}\right)^{1-\rho} d_j^{\rho} \quad j \geq 1$$

$$Y = p_j d_j + P_j^C d_j^C \quad j \geq 1$$

Die Attach Adhesive Design (3/3)

Optimization results of die attach adhesive design problem

	Polymer			Microparticles			Nanoparticles	
Chemical	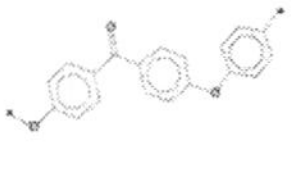			Cu			Cu	
Weight fraction	0.1187			0.7974			0.0839	
NPV	Net present value (MM$)						80.1024	
j	Years	-1	0	1	2	3	4	5
PCF_j	Project cash flow (MM$)	-1	-2.5	27.8867	27.9587	27.8819	27.8358	28.8358
p_j	Selling price ($/Kg)	-	-	1351	1351	1351	1351	1351
d_j	Demand (t)	-	-	55.4647	55.4647	55.4647	55.4647	55.4647
c_j	Product cost (MM$)	-	-	27.654	27.654	27.654	27.654	27.654
Adv_j	Advertisement cost (MM$)	-	-	0.5	0.5	0.5	0.5	0.5
$d_{c,j}$	Competitor's demand (t)	-	-	8.3561	8.3561	8.3561	8.3561	8.3561

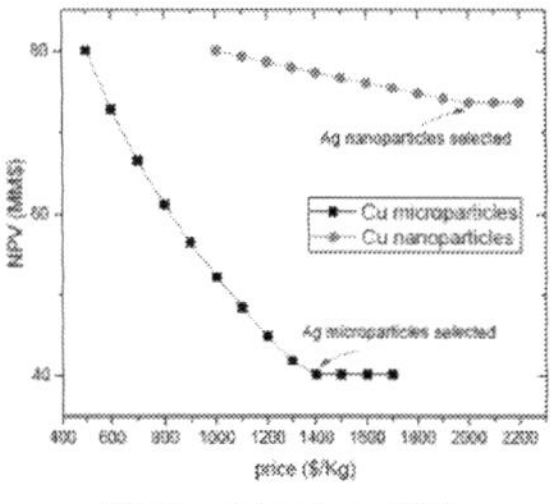

Metal particles price vs NPV

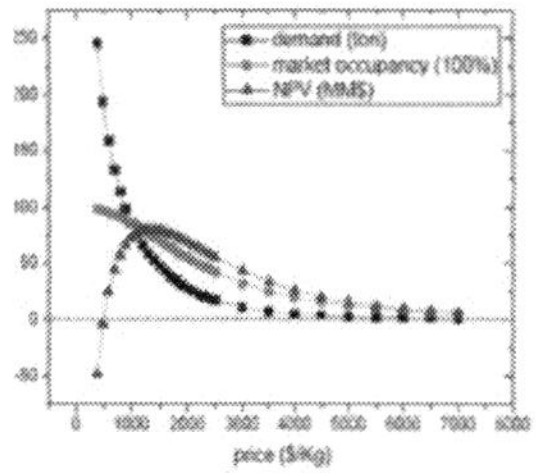

Product price vs demand, market occupancy and NPV

15AIChE
Annual Meeting, Salt Lake City, UT

DTU

Conclusion

- A computer-aided framework was developed in this work.
- This framework has been shown to systematically define and solve a large MINLP product design, product-process design, and product-process design with economic analysis problems.
- Generate and test methods can only partially solve these problems.
- Higher order description of groups and molecular structure prediction is integrated in this new MINLP mathematical programming model.
- This model is tested using a large range of product design problems (surfactants, refrigerants and polymers) and for consideration of sustainability inclusion through economic and environmental models.

17 **DTU Chemical Engineering, Technical University of Denmark** 17 December 2015

427253 Designing Optimal Mixtures Using Generalized Disjunctive Programming: Convex Hull Reformulation

Abstract:

The design of mixtures is a topic of great importance across many industrial sectors, as part of process development and of product design. Computer-Aided Mixture/blend Design (CAM^{b}D) [1] has emerged as an attractive methodology which has been successfully applied to designing various types of mixtures. Within the CAM^{b}D framework, these design problems have often been posed as Mixed-Integer Nonlinear Programming (MINLP) problems, where the optimal molecular structure is determined with respect to a set of property constraints. In most existing methodologies a reduced version of the general CAM^{b}D problem has been addressed: the number of mixture ingredients is fixed (usually a binary mixture is considered) and the identity of a compound (or of all compounds) that can participate in the mixture is chosen from a given set of candidate molecules. Solving the resulting optimization problems can be quite challenging, as these are combinatorial in nature due to the presence of binary variables and to the highly nonlinear expressions that relate composition, structure and physical properties.

In working towards a generalised CAM^{b}D problem, a novel methodology where the number, identity and compositions of mixture constituents are optimised simultaneously, was recently proposed [2,3]. In order to address the difficulties arising from the complexity of expressing the problem within a mathematical framework, a logic-based methodology, Generalized Disjunctive Programming (GDP) [4], was used to formulate the discrete choices of a mixture design problem. The design methodology was first developed for a standard formulation with a fixed number of components (restricted problem) and then generalised to allow the number of mixture ingredients to vary up to a maximum number (general problem). In this approach, the designed components in the mixture are selected from a list of candidate pure compounds. The GDP problems are formulated as MINLP problems in order to exploit the available algorithms for solving optimisation problems. The formulations are converted into MINLP problems by replacing the Boolean variables with binary ones and converting the logic propositions into linear inequalities. In this initial work [2,3], conditional constraints inside the disjunctions were formulated using the Big-M (BM) approach [4]. The BM approach is the simplest representation of a GDP problem in a mixed-integer form. However, it introduces a large parameter (Big-M) which in most cases is specified intuitively and can yield to poor relaxations. Here, we extend this work by converting the GDP models into MINLP problems using the Convex Hull (CH) reformulation [5]. The CH reformulation introduces a new set of disaggregated variables and new constraints, increasing the size of the problem. It can however be shown [6] that when the discrete domain of the CH formulation is relaxed, it gives at least as tight or tighter relaxations than BM.

The design methodology is applied to two case studies: the first one aimed to design an optimal solvent mixture to dissolve ibuprofen and the second one to design the most effective solvent mixture to extract acetic acid from water. The results of the problems studied show that the proposed formulations offer a promising approach to the problem of mixture design, as the simultaneous design of the optimal number, identity and compositions of the components that participate in a mixture, is achieved. The CH reformulation is compared to the Big-M approach and both methods appear suitable for formulating mixture problems effectively. Finally, better performance is obtained with a solvent mixture in both examples, showing the benefit of using mixtures instead of pure solvents to attain enhanced properties.

References

1. L.E.K. Achenie, R. Gani, and V. Venkatasubramanian. *Computer aided molecular design: theory and practice*. Elsevier, Amsterdam, Netherlands, 2003.
2. P.T. Akula, P.M. Kleniati, and C.S. Adjiman. On the design of optimal solvent mixtures using generalised disjunctive programming. In *22nd European Symposium on Computer Aided Process Engineering*, pages 1-5, 2012.
3. S. Jonuzaj, P.T. Akula, P.M. Kleniati, and C.S. Adjiman. The formulation of optimal mixtures using generalized disjunctive programming: choosing components from a pre-defined list. *Journal paper in preparation*, 2015.
4. R. Raman and I.E. Grossmann. Modelling and computational techniques for logic based integer programming. *Computers & Chemical Engineering*, 18(7):563-578, 1994.
5. S. Lee and I.E. Grossmann. New algorithms for nonlinear generalized disjunctive programming. *Computers & Chemical Engineering*, 24(9-10):2125-2141, 2000.
6. S. Lee and I.E. Grossmann. Generalized convex disjunctive programming: convex hull relaxation. *Computational Optimization and Applications*, 26:83-100, 2003.

432235 A COSMO-Based Approach to Computer-Aided Mixture Design

Abstract:

A relatively well-known field in the literature, Computer-Aided Molecular Design (CAMD) defines a class of problems aimed at designing a single, optimal molecular structure to best fit some purpose. While CAMD[1] has seen some success in the literature, its mixture analogue, Computer-Aided Mixture Design (CAMxD), still has some significant limitations. CAMxD is, unsurprisingly, much more complicated. This class of problems now involves the design of a blend of compounds – both every compound's molecular structure and mole fraction in the composition – and this greatly increases the number of variables in the problem and adds many more nonlinear constraints in the form phase equilibria equations, miscibility checks, and safeguards against including two molecular species which might react.

The traditional approach to CAMxD in the literature has two main shortcomings: (1) many approaches rely on enumerating every possible structure for every possible component and then checking every combination for mixture feasibility; and (2) the search space is inherently curtailed because these approaches often solve the mixture part of the problem with UNIFAC[2] method, which can only work when a regressed interaction parameter is known for every pair of molecular substructures in the blend. These issues have resulted in approaches to the CAMxD problem that can only investigate a very limited area of the molecular search space.

Our approach to the CAMxD problem circumvents these two issues by: (1) using a derivative-free optimization scheme to design for structures and test their mixture properties on-the-fly; and (2) deriving a group-contribution method to estimate COSMO-RS[3] and COSMO-SAC[4] sigma profiles, which can be used to calculate mixture properties without the need for binary interaction parameters. These two steps enable a much more efficient optimization of a much larger chemical search space. In this presentation, we will outline our approach in more detail and discuss how the COSMO group contribution methods were made. We will also demonstrate our algorithm as applied to a solvent design problem and a reaction rates optimization problem.

[1] Group-contribution based estimation of pure component properties. Jorge Marrero and Rafiqul Gani.*Fluid Phase Equilibria.* 2001. 183, 183-208.

[2] Group-contribution estimation of activity coefficients in nonideal liquid mixtures. Aage Fredenslund, Russel L. Jones, and John M. Prausnitz. *AIChE Journal.* 1975. 21(6), 1086-99.

[3] Fast solvent screening via quantum chemistry: COSMO RS approach. Frank Eckert and Andreas Klamt. " *AIChE Journal.* 2002. 48(2), 369-385.

[4] A priori phase equilibrium prediction from a segment contribution solvation model. Shiang-Tai Lin and Stanley I. Sandler. " *Industrial & engineering chemistry research.* 2002. 41(5), 899-913.

(173f) Coupling Experiments and Computational Tools to Estimate Thermal Properties for Product Design in the Fat-Based Food Industry

Teles dos Santos, M., Universidade de São Paulo - USP
Viana, I. S., Universidade de São Paulo
Ract, J. N. R., Universidade de São Paulo - USP
Le Roux, G. A. C., Polytechnic School of the University of São Paulo
Originally presented on: 11/9/2015 14:15:00 - 14:36:00

Product design has been evaluated as one of the challenges that chemical engineering has to face to help industry meet market demands for value-added products with higher performance [1]. In the food and pharmaceutical industry, the design of fat-based products is of great importance, since many foods have their desired properties related to a complex phase behavior involving lipids and many cosmetic and pharmacological products use vegetable oils as a lipidic medium in which active principles and fine chemicals are dispersed. Attempting to use Process Systems Engineering tools (namely thermodynamic modeling and computational/numerical approach), for product design, the present work coupled experimental studies and computational simulations to access the thermal properties of different scales of triacylglycerol mixtures, ranging from simple binary mixtures to ternary blends of vegetables oils.

In order to test the developed tools in raw materials used by industry, different blends formed by palm stearin (PS), fully hydrogenated soybean oil (FHSO) and canola oil were studied. The solid fat content at different temperatures, melting points and changes in heat capacity due to solid-liquid transitions are covered by the present work, both experimentally and computationally. Computational simulations and experimental results are compared and discussed. The following experiments were conducted: fatty acids composition identification, stereo distribution of fatty acids in the glycerol structure, softening point and thermal analysis (differential scanning calorimetry). The computational simulations were performed by the following steps: prediction of triacylglycerol composition of the blends by using statistical methods (random and non-random distribution of fatty acids in the glycerol structure); predictions of the distribution of molecules among solid and liquid phases by direct minimization of Gibbs free energy function, using a Generalized Reduced Gradient Method (CONOPT 3 GAMS solver). The optimization algorithm was then coupled with the main program written in FORTRAN 90, which handles the calculation of interaction parameters, melting temperature and melting enthalpy and the generation of triacylglycerols from fatty acids data [2]. The Solid-liquid equilibrium problem was then solved over a temperature range to be compared with experimental data in different temperatures and at different blend ratios (single vegetable oil, binary blends and ternary blends).

As the demand for more sophisticated fat products increases and the simple physical mixture (blending) is not sufficient for matching the desired properties, new technologies have emerged. The chemical interesterification reaction is one important technique that changes the triacylglycerol profile of the blends and, thus, modifies their physical properties. This reaction uses catalysts (e.g. sodium methoxide) and heat to promote a random distribution of fatty acids among the three positions of glycerol [3]. In order to predict how the reaction affects the final solid fat content of the blends, this work simulates the effect of the reaction by using random distribution of fatty acid in the glycerol chain and computing the melting profile of the reaction product.

REFERENCES

[1] Hill, M. (2009). "Chemical product engineering — the third paradigm." *Comput. Chem. Eng.*33, 947–953.

[2] Teles dos Santos, M.; Gerbaud, V.; le Roux, G.A.C. (2013). "Modeling and simulation of melting curves and chemical interesterification of binary blends of vegetable oils." Chem. Eng. Sci. 87, 14–22.

[3] Rodrigues-Ract, J.N.; Cotting, L.N.; Poltronieri, T.P.; da Silva, R.C.; Gioielli, L.A. (2010). "Crystallization behavior of structured lipids by chemical interesterification of milkfat and sunflower oil." Food Sci. Techn. 30, 258-267.

433479 Multi Objective Molecular Design of Reactants and Products

Abstract:

In reactive systems, computer-aided molecular design (CAMD) is performed to design solvents that enhance reaction kinetics, improve catalyst activity/lifetime as well as reactants and products. CAMD of reactants and products helps in their selection such that target properties of interest are optimized with satisfaction of property constraints. Owing to the scarcity of CAMD methodologies that design reactants and products, there has been an increased interest to address this issue. The few recently available methodologies to design reactants and products however have their shortcomings. The earliest available methodologies are restricted to cases where only structures of a single reactant and product are variable. Also, employment of linear property models and lack of a methodology to treat property constraints have been some of the other restrictions. Dev et al. [1, 2] recently developed two algorithms to design reactants and products which are unrestricted by the number of reactants and products whose structures are uknown. They also made provisions for non-linear property models and property constraints. In their first algorithm [1], only the dominant properties of the products are optimized. Also, only the products are subjected to their individual set of property constraints. In their second algorithm [2], properties that are dependent on the structures of both reactants and products are optimized. Also, all reactants and products are subjected to their respective set of property constraints. In this work, we have developed an algorithm that generates structures of reactants and products irrespective of their numbers such that each reactant and product's target property(s) is to be optimized. Also, each reactant and product is subjected to a set of property constraints. All properties have been expressed in terms of molecular structure using different types of property models. Since there exists a need to treat different types of property models on a single platform, signature descriptors have been utilized. A variety of property models can be expressed in terms of signature descriptors which are molecular building blocks. They enable the simultaneous utilization of Quantitative Structure Activity/Property Relationships (QSARs/QSPRs) and Group Contribution Models (GCMs) [3]. In the formulated problem, all property models are expressed in terms of the occurrence number of signature descriptors. Relationships have been developed that relate the occurrence number of signatures of reactants and products. This is achieved by tracking the atoms in the reactants to the products by taking into consideration the reaction mechanism. Thus the reactants and products have been structurally related. Due to this structural relation, property constraints acting on reactants and products influence the selection of both optimal reactants and products. The objective functions of each of the reactants and products display a similar influence. Since the selection of each reactant and product is dependent on the property constraints and property objective functions of the other reactants and products in addition to its own, the problem is formulated as a multi-objective optimization problem when objectives are conflicting. This problem is solved using the Augmented Epsilon-Constraint Method (AUGMECON) developed by Mavrotas [4]. AUGMECON generates pareto optimal solutions while completely avoiding weakly pareto optimal solutions. The generated solutions also include unsupported pareto optimal solutions. Specifically for multi-objective integer programming (MOIP) problems, an improved version of AUGMECON, AUGMECON2 [5] can also be utilized. AUGMECON2 is very efficient in generating the entire pareto optimal set for MOIP problems compared to AUGMECON and some earlier available methods. In this contribution, the utility and novelty of the developed methodology has been highlighted through a case study.

[1] V.A. Dev, N.G. Chemmangattuvalappil, M.R. Eden, 2014, Computer Aided Chemical Engineering, 34, 291-296

[2] V.A. Dev, N.G. Chemmangattuvalappil, M.R. Eden, 2015, Computer Aided Chemical Engineering, 37, 1445-1450

[3] N.G. Chemmangattuvalappil, C.C. Solvason, S. Bommareddy, M.R. Eden, 2010, Computers & Chemical Engineering, 34, 12, 2062-2071

[4] G. Mavrotas, 2009, Applied Mathematics and Computation, 213, 455-465

[5] G. Mavrotas, K. Florios, 2013, Applied Mathematics and Computation, 219, 9652-9669

418370 The Solar Integration National Dataset (SIND) Toolkit

Abstract:

Solar integration studies require detailed solar power output data at a large number of locations in order to perform simulations of how the power system will operate under high solar power penetration scenarios. The solar datasets that serve as inputs to these studies must realistically reflect a number of different solar power characteristics, including: ramp rates, spatial and temporal correlations, and the capacity factors of the simulated plants. The datasets must also be time synchronized with available load profiles since the same meteorological conditions that drive load patterns also drive solar power production. The SIND Toolkit presented here fulfills these requirements as a state-of-the-art national solar resource and power dataset and is made publically available to be utilized in the next generation of solar power integration studies.

The dataset consists of one minute solar irradiance and power time series at over 120,000 locations throughout the continental United States with roughly 2 km by 2km spatial resolution. The dataset covers eight years (2005 – 2012) to allow for the examination of the impacts on inter-annual variability. The one minute data is created through a statistical upsampling of 30 minute satellite-derived solar irradiation data taken from the NREL's National Solar Radiation Database (NSRDB). The upsampling utilizes the satellite data from neighboring regions to classify the timeperiod into one of six solar variability regimes, and the variability added to the time period is based on a statistical characterization of measured solar data from an extensive archive of ground truth data at locations throughout the continental United States. Finally, irradiance is converted to power through five different solar power configurations.

427878 A Study on an Integration Framework of Multiple Renewable Energy Systems

Abstract:

Increasing energy demand and globally occurring climate changes have been the key motivation for accelerating the installation of renewable energy sources as an alternative to environmentally endangering fossil fuels.
Solar and wind power generation is a well known example. A single renewable energy source may be not enough to compete against fossil fuels yet. On the other hand, the aggregation of separate renewables could have a significant impact on the overall energy system.
The integration of different kinds of energy sourcse will be an interesting topic. Multiple types of different energy sources may be much different from the original separate energy supply case. Then the issue is understanding the incorporation of multiple renewable sources. The multiple types of renewable energies may be employed in a hybrid manner. The resulting hybrid renewable energy system (HRES) should deserve further attention.
In this paper a novel framework is presented as an decision-supporting tool for designing HRES. It is essential to analyze the varying renewable energy outputs to minimize total cost of the HRES system while achieving operational stability. This paper employs a time series based scheme and actual data in the literature. This study investigated the optimal design of HRES. The energy generation using synthetic time series compare that to the previous research. Several actual numberical cases are presented to illustrate the applicability of the proposed framework.

429536 Securing Our Energy Supply: The Effect of Distributed Electricity Generation Using Natural Gas on the Interdependency of the Electric and Natural Gas Grids

Abstract:

The supply of electricity and natural gas (NG) to residential, commercial, and industrial customers relies on complex and interdependent networks. For example, much of the load following capacity of the electric generation portfolio comes from natural gas fired power plants [1]. Combined heat and power (CHP) technologies further increase the efficiency of power generation, as well as the interaction between the electricity and NG grids. As this technology can be scaled down for generation at the hundred-kW capacity and located at the neighborhood level [2], it is feasible to envision a future scenario whereby a large number of distributed CHP plants (or even smaller generation units on the order of a few kW) are deployed in a region that would at present only be serviced by the conventional electric grid.

Under the assumption of natural gas prices remaining low in the long term [3], this scenario appears as a very attractive alternative for the planning of future residential or mixed-use communities. However, the effects of the complex interactions between the electricity and natural gas grids that arise in this situation should be carefully evaluated. In this case, natural gas would be used for both large-scale power generation to supply the main electric grid and for electric generation closer to the demand sites. On the other hand, (peak) electricity demand at the local level may be met using a conventional transmission network.

In this work, we focus specifically on this pressing issue, and address the question "Does tighter integration of the electric and NG grids result in increased efficiency and environmental performance of both networks?" We develop several case studies investigating the operation and load of both the electric and NG grids when (i) CHP plants are present to service entire communities, and (ii) when smaller NG generators are located 'behind the meter' but can service several houses. Data for the electricity and NG demand of each customer are based on real-time smart-meter measurements from residences involved in the Pecan Street Project [4]. Our study also considers the impact of real-time pricing for electricity and/or NG. We formulate and solve the model for the operation of the ensemble of consumers, generators, and distribution networks as a large-scale linear program. We compare several different objectives (e.g., minimum cost to each consumer, maximum profit for the electric and NG utilities) and analyze the resulting optimal energy usage profiles in terms of the net system efficiency and environmental implications. Our results suggest that, besides offering consumers more options in terms of sourcing energy, broadening and distributing the generating portfolio leads to an increase in the efficiency of both the electric and NG grids.

[1] T. Li, M. Eremia, and M. Shahidehpour, Interdependency of Natural Gas Network and Power System Security, IEEE Trans. on Power Systems, 24(4):1817-1824, 200

[2] A. Ondeck, T. F. Edgar, and M. Baldea, Data-Driven Modeling and Optimal Operation of District-Level Combined Heat and Power and Photovoltaic Generation Systems, Submitted to Applied Energy, 2015.

[3] J. Logan, A. Lopez, T. Mai, C. Davidson, M. Bazilian, D. Arent, Natural Gas Scenarios in the US Power Sector, Energy Economics, 40:183-195, 2013.

[4] Pecan Street Inc., http://www.pecanstreet.org/energy/

422450 An Adjustable Robust Optimization Approach to Provision of Interruptible Load By Continuous Processes

Abstract:

A power grid is designed to reliably match electricity supply and demand. When real-time electricity supply falls below the demand, e.g. due to generator failures or sudden load changes, backup capacities are called upon to eliminate this supply-demand gap. One kind of such backup capacities is called operating reserve, which has to be ready upon request within minutes. Operating reserve can be provided by generators that are able to quickly increase electricity supply. Alternatively, operating reserve can be provided by electricity consumers by reducing their load and therefore reducing electricity demand; this operating reserve provided on the demand side is also referred to as interruptible load.

Providing interruptible load is attractive because one is rewarded even when no actual load reduction is required. This market incentive reflects the value of flexible loads that can react quickly to unexpected changes in the power grid. However, there is a lack of systematic methods for determining the optimal amount of interruptible load to provide. The main challenge lies in the uncertainty which stems from the fact that one does not know in advance when load reduction will actually be requested.
Recently, robust optimization has been applied to scheduling considering provision of operating reserve under uncertainty (Vujanic, et al., 2012; Zhang, et al., 2015). The applied traditional robust optimization approach guarantees feasibility for every possible realization of the uncertainty, but does not consider recourse, which makes the solution highly conservative. However, when interruptible load is provided by complex chemical processes, accounting for recourse decisions is crucial.

In this work, we use a detailed scheduling model to capture all critical operational constraints of a continuous power-intensive process, and propose an adjustable robust optimization approach (Ben-Tal, et al., 2009) that uses affine decision rules to account for recourse decisions with respect to uncertainty in load reduction demand. This results in a large-scale MILP model, to which additional redundant constraints are added to make the formulation computationally more efficient. The proposed model is applied to an industrial air separation case provided by Praxair, and the results show significant benefit from providing interruptible load. Compared to a traditional robust formulation, the adjustable robust model provides much better and more realistic solutions while guaranteeing the same level of robustness.

References

Ben-Tal, A., El Ghaoui, L. & Nemirovski, A., 2009. Robust Optimization. New Jersey: Princeton University Press.
Vujanic, R., Mariéthos, S., Goulart, P. & Morari, M., 2012. Robust Integer Optimization and Scheduling Problems for Large Electricity Consumers. 2012 American Control Conference, pp. 3108-3113.
Zhang, Q., Heuberger, C.F., Grossmann, I.E., Sundaramoorthy, A. & Pinto, J.M., 2015. Air Separation with Cryogenic Energy Storage: Optimal Scheduling Considering Electric Energy and Reserve Markets. AIChE Journal.

435044 Smart Grid Analysis of Centralized Power and Cooling for an Urban Community

Abstract:

This project aims to compare two approaches to providing building cooling and electricity to a new community of 100,000 residents. Two cases were studied and a thirty-year net present value of each case is calculated. In the baseline case, each individual building has its own HVAC equipment for cooling and is operated using electricity purchased from the grid. The second case the community purchases a natural gas combined cycle (NGCC) power plant to provide electricity for the community along with a district cooling system through the use of absorption chillers powered by waste heat generated from the power plant. Operating cost analysis utilizes the notion of real-time electricity prices to assess the benefits of the centralized power and cooling system. To provide additional the time-shifting of cooling loads, a large scale Thermal Energy Storage (TES) unit is also considered.

428662 Energy Exchange in Autonomous Isolated Networks with Renewable Energy Sources Empowered By a Multi-Agent Internet of Things (IoT) Architecture

Abstract:

This work presents a decision making methodology which is developed for the effective energy exchange between off-grid stations with Renewable Energy Sources (RES) for power generation. When multi-source energy stations (nodes) are connected, a network is formed where the decisions should be made according to the overall supply and demand of energy. Therefore, each station needs to be aware of the status of the network and to be able to share information related to the availability of its resources and its load demand. The local energy management remains a priority for each individual node whereas part of the load demand can be served when an excess of energy exists in another node of the network. Each node is able to exchange information with the monitoring and control station that supervises the operations of the entire network as part of the operations domain of the grid.

The main operation objectives are to exchange power based on demand response strategies, to maximize the usage of available stored power at network level and to utilize the total amount of available renewable energy. The desired features for the overall behaviour of the network is to ensure the security of supply, to reroute the energy based on dynamically evolving conditions and to provide an automatic decision support mechanism based on available network resources. The presented methodology facilitates the deployment of a flexible modular decision-making mechanism for networked systems that rely on a multi-agent communication structure using an Internet of Things (IoT) architectural reference model. The rules of operation are represented by propositional logic and the implementation is realized by an agent-based structure, which is executed at each node. Finally the raw data of each node are transformed into useful information which is considered by the agent and the algorithm which is responsible for the energy management within the network. The decision making framework is able to gather and act on energy and resource-related information in an automated fashion with the goal to improve the efficiency, reliability and sustainability of the production and distribution of energy. The energy sharing is based on a load balancing approach with priorities that are dynamically updated based on the status of each node and the internal handling of its energy resources and local load demand. Although this methodology is shown by a case study which is related to the energy domain, it can be easily extended to other networked systems where exchange of resources is necessary.

The decision making methodology is developed with implementation considerations for the operation of a smart grid that involves three autonomous stations with local energy storage capabilities using a lead-acid battery stack. Furthermore, there are hydrogen generation and long term hydrogen storage options. Each station has a PV array, wind generators and a diesel generator for backup. Also at one of the stations there is a Polymer Electrolyte Membrane (PEM) fuel cell and a PEM electrolyzer for the utilization and production of hydrogen respectively. In order to make this network as self-sustained as possible the main target is to minimize the use of the diesel generator. Therefore, the priority for the energy exchange is firstly to serve the local load demand and secondly to avoid the use of the generator by utilizing the available energy in the network. The protection of the subsystems within each station is also considered as respective indications for the status of the energy storage is available to the supervisory control station. This isolated grid is located at Xanthi, Greece and its operation is monitored using a Supervisory Control and Data Acquisition (SCADA) system while the Machine to Machine (M2M) communication for the decision making is implemented by the IoT enabled architecture using the agent-based middleware. The performance of the involved nodes is explored and the overall operation of the grid is presented whereas the response of each node is monitored online. An industrial grade agent-based IoT architecture can increase the resiliency and flexibility of such networks whereas the dynamic rerouting of energy can significantly improve the robustness and optimum use of the available resources.

418570 Online Optimization for Grade Transitions in Polyethylene Solution Polymerization Process

Abstract:

Various grades of linear low-density polyethylene (LLDPE) are tailored to different applications, with each grade defined by specifications of product properties such as melt index and density. Typically, several different grades are produced in the same production line, which requires frequent transitions. Given the large market of LLDPE and the current experience-based transitions, there is a need, and also room for improvement, to perform transitions and to change operation conditions in a way such that the transition time as well as the off-grade product could be minimized.

In our previous study, a mathematical model capturing the dynamics of the solution polymerization process carried out in a CSTR is developed. Besides the mass and heat balance equations, the optimization model for the entire process adopts the molecular weight moment model for the prediction of product properties and incorporates a simple, yet accurate, data-driven vapor-liquid equilibrium (VLE) model derived from rigorous calculation. Two optimization formulations, single stage and multistage optimization, are proposed to deal with single-value specification and specification bands of product properties, respectively. The results show significant improvement in transition times and off-grade production.

The offline dynamic optimization demonstrates strong potential in handling grade transition problems, but its performance can deteriorate in the presence of uncertainties, disturbances and model mismatch. In order to deal with this problem, an offline optimization strategy based on backoff constraints is taken into account to obtain robust optimal policies which can be applied to systems with different uncertainty levels.

In this talk, we extend the off-line framework described above to include online state estimation and optimal control of the large-scale grade transition problems. As described in Jung et al. (2015), two key components of the framework are state estimator and model predictive controller; both components incorporate the rigorous dynamic model we developed in our previous work. First, a shrinking horizon nonlinear model predictive control (sh-NMPC) is applied to minimize transition time and off-grade product, and to refine the control schemes. Next, an expanding horizon least squares estimation (eh-LSE) is designed to provide state estimates based on measurements in the past. At each time step, process measurements are obtained and used in eh-LSE optimization problem. The resulting state estimates serve as the initial value of an updated NMPC problem, which is then solved to reduce the overall transition time and the off-grade product. The resulting control scheme is also applied to the system at each time step so that new measurements are added. The effectiveness of the framework is demonstrated by several case studies in which uncertainties, measurement errors and disturbances are taken into account.

Reference

Tae Y. Jung, Yisu Nie, J. H. Lee, L. T. Biegler, "Model-based On-Line Optimization Framework for Semi-batch Polymerization Reactors," Proceedings of ADCHEM 2015, Whistler, BC, June, 2015

421953 Real-Time Estimation of Plasma Insulin Concentration in Patients with Type 1 Diabetes

Abstract:

Real-Time Estimation of Plasma Insulin Concentration in Patients with Type 1 Diabetes

Iman Hajizadeh, MSc[1]; Kamuran Turksoy, PhD[2]; Sediqeh Samadi, MSc[1]; Jianyuhan Feng, MSc[1]; Ali Cinar, PhD[1,2]

[1]*Department of Chemical and Biological Engineering, Illinois Institute of Technology*

[2]*Department of Biomedical Engineering, Illinois Institute of Technology*

Closed-loop control of blood glucose concentration, known as artificial pancreas (AP) system, is an alternative treatment for patients with type 1 diabetes, by automatically controlling their blood glucose levels [1]–[2]. These control systems use continuous glucose measurements to calculate optimum amount of insulin to be infused with an insulin pump in order to keep plasma glucose concentration within a safe target range to avoid health problems. Real-time estimations for plasma insulin concentration is beneficial for increasing the efficiency of AP control algorithms. This enables calculation of more realistic insulin infusion rates and prevention of hypoglycemia that would be caused by over dosing of insulin.

Our objective is to fulfill a real-time estimation of plasma insulin concentration from continuous glucose monitoring data by using a mathematical model. Hovorka's model which has been developed to describe glucose-insulin dynamic in different parts of the human body, has been incorporated into a continuous-discrete extended Kalman filter (CDEKF) to provide an estimate of the plasma insulin concentration. The CDEKF is the generalization of the Kalman filter to nonlinear systems with continuous-time state equations and discrete-time measurements. Furthermore, because of variability in system dynamics, uncertain parameters have been considered as new states in Hovorka's model to be estimated by CDEKF. This methodology has been evaluated by using clinical data from patients with Type 1 diabetes who underwent a closed-loop AP experiment. Real-time insulin estimations have been compared to plasma insulin measurements to evaluate performance and validity of the proposed methodology.

Finally, based on simulation results, it has been shown that the proposed method is able to estimate the plasma insulin concentration in real time. Furthermore, the results for estimated continuous glucose measurements and blood glucose concentration have been compared to their real values to acknowledge performance of the proposed method. This method will be beneficial for an AP system in terms of real time estimation of non-measurable variables such as plasma glucose and insulin concentrations.

[1] K. Turksoy, L. T. Quinn, E. Littlejohn, and A. Cinar, "An Integrated Multivariable Artificial Pancreas Control System," *J. Diabetes Sci. Technol.*, vol. 8, no. 3, pp. 498–507, 2014.

[2] K. Turksoy, L. Quinn, E. Littlejohn, and A. Cinar, "Multivariable Adaptive Identification and Control for Artificial Pancreas Systems," *IEEE Trans. Biomed. Eng.*, vol. 61, no. 3, pp. 883–891, 2014.

427727 Information Theory-Based Sequential Parameter Estimation of Bioreactor

Abstract:

Identification is an indispensable step in optimization and control of batch bioreactors. Bioreactor models of a practical scale system are usually nonlinear, complex, and are accompanied with a large number of parameters. Optimal Experimental Design (OED), which aims to find the most information-rich operation, is a statistical tool to guide engineers to efficiently identify a system. Despite its wide acceptance as a viable tool, current OED approaches often result in a huge optimization problem. This makes not only the computation prohibitive, but more importantly, makes the analysis of the solution infeasible.

In this work, we suggest a new scheme which estimates parameters in a sequential manner. Parameters are 'lined up' into an order and estimated by the sequence. Since the parameters are highly correlated in a typical bioreactor system, it is crucial to decide how to group and place parameters in a proper order. First, simulation is performed with a purposely inaccurate (~50%) set of parameters. The objective of this simulation is to observe qualitative behavior of the process state variables, not their exact values. Applying several criterion based on information theory such as sensitivity, we can 1) rule out unidentifiable parameters, 2) combine highly correlated parameters as a set, and 3) place those parameter sets into an order that maximizes the information gain. As a result, we can obtain the 'estimation schedule'.

In the actual implementation, optimal input is calculated at each step in a similar manner to the model predictive control. The parameter identification schedule obtained a priori is used as guidance. Maximum information gain to a certain parameter set over a predefined prediction horizon is used as an objective function. When the information gain for a certain set of parameters exceeds some predefined level, the set is considered fully identified and the objective function switches to identify the next set of parameters.

The suggested approach is illustrated with two fed-batch bioreactor models. The first is a theoretical model with five parameters. The second is a practical scale microalgal bioreactor model with more than 10 parameters. Results from both models have shown improved performance and convergence property.

The advantages of our approach are twofold. First, unlike other 'blind' approach of OED, the suggested approach can make use of the prior knowledge on the system in a quantitative sense. Second, the suggested approach can also handle other issues related to the identification that are being solved by experience or ad hoc including the number of batches required to run for identification and the amount of information that can be obtained with economical constraints.

432377 Moving Horizon Estimation Via Carleman Linearization and Stability Analysis

Abstract:

In order to control dynamic systems, knowledge of the status of the system e.g. state variables is critical. However, the online measurement of state variables is not always feasible. Consequently, there are many different approaches in the literature to estimate these unmeasurable state variables of a dynamic system. However, the estimation problem would become harder when the system is corrupted with noise signals. Moving Horizon Estimation (MHE) is one of the most successful estimation approaches in dealing with such random uncertainties [1]. The idea behind MHE design is very similar to MPC design. To handle the constraints on input and state variables, these two methods formulate and solve a constrained dynamic nonlinear optimization problem. However, solving this optimization problem is usually time-consuming and online convergence of the scheme to the optimal decision variables is sometimes infeasible. In this work, we propose a new approach that accelerates computations in MHE by exploiting Carleman linearization and prove the stability of this new design.

Extended Kalman Filter (EKF) is the alternative method for state estimation in nonlinear systems in presence of noise. In fact, EKF is a simplified version of MHE which has an analytical closed-form solution; it uses only the last measurement data and linearizes the nonlinear system around the last estimate of states. In contrast, MHE considers the entire output sequence and also takes constraints and nonlinearities into account. This work focuses on design of a new estimator using MHE's structure with a constant estimation horizon. The estimator further simplifies the associated dynamic optimization problem through use of Carleman linearization, which is more accurate than first-order Taylor expansions. Therefore, the new design of MHE gives a better balance between *accuracy* and *simplicity*. In a previous work, we employed this design method and also performed a sensitivity analysis for a system with unknown disturbance [2, 3]. In the current work, in contrast, we build on our previous work by considering system noise and also studying MHE's stability.

The stability of MHE is an important issue. This issue is particularly very thoroughly addressed for constrained linear systems. For constrained nonlinear problems, Rao *et al.* presented an algorithm to design a stable MHE by assuming boundedness of noise signals [4]. Also, they assumed the estimator is able to solve the optimization problem ideally and provide the global optimal solution. Later, some papers considered the optimization error in their MHE design [5]. Our work falls into the second category. We consider the maximum error for Carleman linearization at every time step and prove the estimation error is still bounded. Also it is shown that this error bound can be reduced to any desirable level by choosing a proper order of Carleman linearization. Finally, we employ the proposed MHE in a practical chemical engineering example. The simulation running time is reduced remarkably by employing the proposed MHE.

Reference:

[1] Haseltine, E. L., & Rawlings, J. B. (2005). Critical evaluation of extended Kalman filtering and moving-horizon estimation. *Industrial and Engineering Chemistry Research*, 44, 2451-2460.

[2] Hashemian, N., & Armaou, A. (2015). Fast moving horizon estimation of nonlinear processes

via Carleman linearization. *American Control Conference*, Chicago, IL.

[3] Hashemian, N., & Armaou, A. (2014). Moving horizon estimation using Carleman linearization and sensitivity analysis. *Annual Meeting AIChE,* Atlanta, GA.

[4] Rao, C. V., Rawlings, J. B., & Mayne, D. Q. (2003). Constrained state estimation for nonlinear discrete-time systems: stability and moving horizon approximations. *IEEE Transactions on Automatic Control,* 48, 246-258.

[5] Alessandri, A., Baglietto, M., & Battistelli, G. (2008). Moving-horizon state estimation for nonlinear discrete-time systems: New stability results and approximation schemes. *Automatica*, 44, 1753-1765.

432632 Parameter Estimation and Nonlinear Model Predictive Control Using Meshless Modelling Framework

Abstract:

Nonlinearities and complex dynamics are inherent to chemical process systems. Safe and stable operation of these processes – while maintaining profitability – in the presence of process constraints is essential to meet economic and environmental criteria. Model based control plays an important role in realizing this objective and has contributed to improved process operations. Nonlinear Model Predictive Control (NLMPC) is particularly of great interest to chemical process engineers. This is due to the ability of this controller to explicitly handle constraints, consider multivariable process interactions and intrinsically manage nonlinearities in process systems [1]. The success of NLMPC is based on accurate modelling and representation of the process. NLMPC controllers which utilize fundamental models are attractive as they can be extrapolated for a wide range of data – unlike empirical models. In this work, a new paradigm for solving the fundamental models in NLMPC controllers is presented. Artificial Neural Networks (ANN) are utilized as a meshless modelling platform to provide solution to the Ordinary Differential Equation (ODE) models of dynamic process systems [2]. Traditional numerical methods have limited differentiability, experience a significant increase in complexity with increasing number of sampling points and may be challenging to use for higher dimension problems. These limitations can be overcome by utilizing the ANN integrator to provide a solution to the ODEs. The set of system ODEs is transformed into a set of Algebraic Equations (AE) and posed as a nonlinear programming (NLP) optimization problem. The NLMPC framework, where the ODEs are integrated using the ANN, is applied to setpoint tracking, regulatory and zone control [3] studies. In process control, adequate information about the system is vital. Operational transitions could lead to changes in system parameters; and knowledge of these parameters is crucial to the operation and control of these processes. Utilizing the ANN formulation, a parameter estimation problem is formulated and solved as an NLP [2]. This offers a solution approach in which available online measurements can be utilized to compute the process parameters while simultaneously providing a solution to the system of ODEs representing the process. Case studies for parameter estimation and NLMPC are presented to demonstrate the applicability of the proposed approach.

References

[1] V. Dua, "Stability Analysis of Nonlinear Model Predictive Control : An Optimization Based Approach," no. 1995, pp. 1–6, 2005.

[2] V. Dua and P. Dua, "A Simultaneous Approach for Parameter Estimation of a System of Ordinary Differential Equations, Using Artificial Neural Network Approximation," *Ind. Eng. Chem. Res.*, vol. 51, no. 4, pp. 1809–1814, Feb. 2012.

[3] B. Grosman, E. Dassau, H. C. Zisser, L. Jovanovic, and F. J. Doyle, "Zone model predictive control: a strategy to minimize hyper- and hypoglycemic events.," *J. Diabetes Sci. Technol.*, vol. 4, no. 4, pp. 961–75, Jul. 2010.

434446 Stochastic Model Predictive Control with Bounded Inputs and Joint State Chance Constraints: Application to a Continuous Acetone-Butanol-Ethanol Fermentation Process

Abstract:

Model predictive control (MPC) is widely used for achieving high-performance operation of complex systems due to its ability to handle multivariate dynamics, system constraints, and competing sets of objectives [1,2]. To date, the majority of work on MPC of uncertain systems relies on (bounded) deterministic uncertainty descriptions. In deterministic approaches to robust MPC, the control performance is realized in terms of worst-case uncertainties and the system constraints are fulfilled with respect to all uncertainty realizations [3]. Hence, these approaches can lead to an overly conservative closed-loop control performance. Alternatively, stochastic MPC (SMPC) has recently emerged to reduce the inherent conservatism of robust control approaches by using the probabilistic information of system uncertainties in control design. SMPC enables shaping the probability distribution of system states/outputs (either in terms of the complete distribution [4], or its statistical moments [5]) and defining chance constraints, which allow the state constraints to be violated with a-priori specified level of probability.

In this talk, a full state-feedback SMPC approach is presented for linear discrete-time systems subject to independent (possibly) unbounded stochastic disturbances (aka process noise). The stochastic optimal control approach includes hard input bounds and joint state chance constraints. The key challenges in solving the SMPC problem are: 1) incorporating state feedback optimally, 2) guaranteeing hard input bounds in the presence of unbounded disturbances, 3) nonconvexity of joint state chance constraints, and 4) establishing (recursive) feasibility and stability of the control approach. A SMPC algorithm is proposed to address these challenges for the above mentioned stochastic linear setup. To obtain a tractable algorithm, the feedback control law is parameterized as an affine disturbance feedback policy, which is equivalent to affine state feedback [6]. Hard input bounds are guaranteed (for the affine disturbance feedback policy) by directly saturating the disturbances in the control law [7,8]. Joint state chance constraints are replaced with a conservative deterministic surrogate in terms of the mean and variance of the predicted states using the Cantelli-Chebyshev inequality [9]. It is demonstrated that this (and, in fact, any) SMPC problem cannot be feasible at all times due to the possibly unbounded stochastic disturbances. In addition, the inclusion of state chance constraints leads to a small region of attraction for the controller. To obviate the feasibility issue, the state chance constraints are softened using an exact penalty function method that gives the same control law as the fully-constrained SMPC problem when the problem is feasible and minimal constraint violation when the problem is infeasible. It is shown that the stochastic control algorithm will result in a stochastically stable closed-loop system by proving that the closed-loop states satisfy a geometric drift condition outside of a compact set when the system matrix is Schur stable.

The proposed SMPC approach is applied to a continuous acetone-butanol-ethanol (ABE) fermentation process [10]. The validated fermentation model consists of 12 state variables corresponding to concentration of different species in the Clostridium acetobutylicum metabolic pathway [11]. One hundred Monte Carlo simulations have been performed to assess the closed-loop performance of the SMPC approach. The simulation results indicate satisfactory closed-loop control performance in terms of keeping the process at the desired setpoints, as well as effective constraint satisfaction in the presence of stochastic disturbances.

References

[1] D. Q. Mayne, J. B. Rawlings, C. V. Rao, and P. O. Scokaert. Constrained model predictive control: Stability and optimality. *Automatica*, *36* (2000), 789-814.

[2] M. Morari and J. H. Lee. Model predictive control: Past, present and future. *Computers & Chemical Engineering*, *23* (1999): 667-682.

[3] A. Bemporad and M. Morari. Robust model predictive control: A survey. In *Robustness in Identification and Control*. Springer London, 1999, 207-226.

[4] E. A. Buehler, J. A. Paulson, A. Akhavan, and A. Mesbah. Lyapunov-based stochastic nonlinear model predictive control: Shaping the state probability density functions. In *Proceedings of the IEEE Conference on Decision and Control*, 2015, Submitted, Osaka.

[5] A. Mesbah, S. Streif, R. Findeisen, and R. D. Braatz. Stochastic nonlinear model predictive control with probabilistic constraints. In *Proceedings of the American Control Conference*, 2014: 2413-2419, Portland.

[6] P. J. Goulart, E. C. Kerrigan, and J. M. Maciejowski. Optimization over state feedback policies for robust control with constraints. *Automatica* 42 (2006): 523-533.

[7] P. Hokayem, D. Chatterjee, and J. Lygeros. On stochastic receding horizon control with bounded control inputs. In *Proceedings of IEEE Conference on Decision and Control,* 2009: 6359:6364, Shanghai.

[8] D. Chatterjee, P. Hokayem, and J. Lygeros. Stochastic receding horizon control with bounded control inputs: a vector space approach. *IEEE Transactions on Automatic Control,* 56 (2011) 2704-2710.

[9] A. W. Marshall, O. Ingram, and B. C. Arnold. Theory of Majorization and its Applications. Springer Science & Business Media, New York, 2011.

[10] S. Haus, S. Jabbari, T. Millat, H. Janssen, R. J. Fischer, H. Bahl, J. R. King, and O. A. Wolkenhauer. A systems biology approach to investigate the effect of pH-induced gene regulation on solvent production by Clostridium acetobutylicum in continuous culture. *BMC Systems Biology*, 5 (2011): 1-13.

[11] T. Lütke-Eversloh and B. Hubert. Metabolic engineering of Clostridium acetobutylicum: Recent advances to improve butanol production. *Current Opinion in Biotechnology,* 22 (2011): 634-647.

428101 A New Approach to Solving Stochastic Optimal Control Problems

Abstract:

Dynamic systems like batch processes are difficult to design and control because of time dependent nature of the system. Optimal control theory is used to establish efficient operating policies for batch unit operations. However, the time dependent nature of the system makes it difficult to deal with uncertainties. Recently, the problem of uncertainties in batch unit operation optimal control was handled by modeling the uncertainties in terms of Ito processes and Ito calculus with stochastic maximum principle methods were used to solve the resulting stochastic optimal control problem. However, stochastic maximum principle could only be used for time dependent decisions. In order to solve coupled design and control problems, a method is needed which can deal with time dependent uncertainties, static uncertainties, time dependent decisions and time independent decisions. In this paper, we are presenting a new approach based on a novel algorithm called Better Optimization of Nonlinear Uncertain Systems (BONUS). BONUS determines the search direction bypassing the need of excessive model evaluations through a reweighting scheme using Kernel Density Estimation. A simple batch reactor with uncertainties in reaction parameters is used as a study case to taste this novel approach looking obtain the optimal temperature profile that maximizes the product concentration.

422425 Integrating Lyapunov-Based Offset-Free Model Predictive Control with Subspace Identification Methods

Abstract:

Owing to their ubiquitous presence, control designs have to deal with the presence of constraints and nonlinearity. Lyapunov-Based Model Predictive Control (LMPC) is one control method that enables achieving stability from a well characterized region [1, 2]. LMPC method like other MPC methods utilizes a model of the plant. In several scenarios, a good first principles model is unavailable, however, sufficient historical data exists to possibly build a data driven model for the purpose of utilization within LMPC based approaches.

The subspace identification methods (SIM) approach uses a set of input and output data, to estimate linear time-invariant models in a state space form. These methods are exploit concepts such as geometrical projections and numerical linear algebra. Numerical robustness, fewer user parameters, MIMO systems identification, model order reduction make SIMs a good choice for industrial applications [3]. SIMs are usually non-iterative methods, which makes their application, computationally affordable, and, this simplicity would make the model update, affordable.

Motivated by the above, in this work we develop a framework to integrate LMPC with the model identified by SIMs. To this end, a data driven model is first identified and utilized within offset-free LMPC approaches. Simulation results demonstrate the effectiveness of the proposed method.

[1] Mahmood, M. and Mhaskar, P. (2012). Lyapunov-based model predictive control of stochastic nonlinear systems. Automatica, 48(9), 2271-2276.

[2] Mahmood, M. and P. Mhaskar, Constrained Control Lyapunov Function Based Model Predictive Control Design, Int. J. Rob. & Nonl. Contr., 24, 374–388, 2014.

[3] Trnka, P. (2007). Subspace identification methods. PhD thesis, Ph. D. Thesis, Czech Technical University in Prague.

422318 Computer Aided Molecular and Process Design Using Complex Process and Thermodynamic Models: A Screening Based Approach

Abstract:

The design of optimal processing materials (molecules) and optimal process variables for a given process is referred to as Computer Aided Molecular and Process Design (CAMPD). Processing materials used to achieve process goals include mass separating agents (such as solvents for absorption, extraction, leaching and adsorbents), catalysts, heat transfer fluids and reaction medium solvents. Choosing processing molecules influences the optimal process variables and vice versa. Molecular and process decision variables are linked, interacting with each other in a complex manner. Hence, neither of these decisions can be made in isolation.

The CAMPD problem involves continuous decisions such as process operating conditions (examples include pressure, temperature) and discrete decisions (examples include number of groups of a given kind in the molecule to be designed, connectivity between the groups and process topology). The objective is a process/plantwide one. The process model is a coupled nonlinear system of equations. The process model also involves numerous calls to property prediction models to obtain thermophysical properties such as density, enthalpy and chemical potentials. Property models can also be highly complex. Hence, the CAMPD problem is a MINLP problem with a large number of equality constraints.

While the evaluation of the objective is numerically challenging, an even bigger challenge is that a given molecule may be infeasible in the process. Consider a process molecule to be used as a solvent in a separation system. This molecule may or may not exist in the liquid state for a given choice of process variables. The choice of a non-liquid solvent molecule (and the corresponding process conditions) will not only result in a physically infeasible process but may also result in numerical failure when attempting to simulate or optimise the process. Hence, due to infeasible regions in the domain, failure to converge is frequently encountered in these design problems. The domain is not only discontinuous but also non convex and made more challenging from combinatorial explosion that results from the presence of discrete variables. A set of 10 functional groups in the molecular design space can generate 184755 potential "molecules" of utmost 10 groups [1]. These problems are not usually amenable to solution by general purpose MINLP solvers.

To overcome this challenge, heuristics and knowledge based screening of all molecules in the search space has been previously used [2]. This large MINLP has also been reformulated as a NLP [3, 4]. An approach based on determining a range of target properties of molecules using process requirements and then finding molecules whose properties lie within the range has been proposed [5]. A Multi objective optimisation (MOO) framework based on physical properties to identify promising molecules has also been developed [6, 7]. More recently, the HiOpt approach has been put forward in which MOO based on the performance of a simplified process has been used to identify initial guesses for the optimization of the CAMPD problem [8]. While this latter work tackles the full CAMPD problem, it does not fully address the numerical challenges posed by the simultaneous variation of molecular and process variables.

With a focus on solvent based separation systems we present an alternative screening embedded in optimisation algorithm to solve this MINLP. As mentioned before, a challenge in solving the CAMPD problem is that many molecules may be infeasible in the process based on their properties and phase behaviour in presence of the feed to be separated. Our screening methodology exploits this to reduce the solution domain and solve the problem efficiently. However, we do not generate and test molecules but choose to embed screening steps within an MINLP optimisation framework, such as previous work in the context of molecular design [9, 10]. We have proposed a series of new screening tests that help to eliminate solvents that fail the tests, but can also be used to reduce the continuous domain of process variables. These tests are implemented within the outer approximation (OA) framework [11].

A test of feed phase stability restricts the process domain to one where the feed has the desired phase (gas/liquid). This test is independent of the solvent chosen. As a solvent molecule is generated at each iteration of the OA algorithm by the master problem it goes through a series of tests. The molecule is first tested based on its pure component properties. This test is used to check if the molecule exists in the liquid state at its handling and storage temperature. Its ease and safety of usage are also tested.

A test of separation feasibility is used to checks if an infinitely long separation column with the molecule as solvent can achieve any separation of the feed. This test is equivalent to checking if the feed and the solvent can form a two phase mixture on contacting. This test is posed as an NLP optimisation problem to determine the region of the process domain where this is possible. If no feasible region is found for the problem, the solvent is rejected from the search space.

A test of purity feasibility is used to determine if the required degree of separation (or purification of the feed) may be achieved by the solvent. This test is a check on the two-phase boundary of the solvent and feed. This test is posed as an NLP optimisation problem to determine the range of the process variables where this condition is satisfied.

Molecules that fail any of the tests are eliminated from the search space. Molecules that pass are provided as input to the primal problem. Information obtained from the tests and the primal problem is progressively added to the master problem whose solution generates a new molecule.

We have implemented this algorithm in C++, with gPROMS [12] as the simulation engine and NLP solver and Gurobi [13] as the MILP solver. We demonstrate it on the design of a separation system for the absorption of carbon dioxide from natural gas. The property prediction model employed is a group contribution based version of the statistical associating fluid theory (SAFT) -γ Mie [14]. This provides a single consistent thermodynamic model across the entire fluid region. This equation of state is predictive for both pure compounds and mixtures. However, this complex and nonlinear model makes solving the CAMPD problem more challenging. The case study is solved successfully, without numerical failures, and leads to the identification of a promising solvent. The performance of the algorithm is discussed in the context of the proposed tests.

References:

1 Joback, K.G. and Stephanopoulos, G. (1995). In Advances in Chemical Engineering, Volume 21. 1995

2 Hostrup, M., Harper, P. M., and Gani, R. (1999). Computers& Chemical Engineering, 23(10):1395-1414

3 Pereira, F. E., Keskes, E., Galindo, A., Jackson, G., and Adjiman, C. S. (2011). Computers & Chemical Engineering, 35(3):474-491

4 Bardow A, Steur K, Gross J. Ind Eng Chem Res. 2010;49:2834–2840

5 Eljack, F. T., Solvason, C. C., Chemmangattuvalappil, N., and Eden, M. R. (2008). Chinese Journal of Chemical Engineering, 16(3):424-434.

6 Kim, K.-J. and Diwekar, U. M. (2002b). Industrial & Engineering Chemistry Research, 41(18):4479-4488

7 Papadopoulos, A. I. and Linke, P. (2006). Chemical Engineering Science, 61(19):6316-6336

8 Burger, J., Papaioannou, V., Gopinath S., Jackson, G., Galindo, A., and Adjiman, C. S. (2014). AIChE Journal

9 Buxton, A., Livingston, A. G., and Pistikopoulos, E. N. (1999). AIChE Journal, 45(4):817{843

10 Giovanoglou, A., Barlatier, J., Adjiman, C. S., Pistikopoulos, E. N., and Cordiner, J. L. (2003). AIChE Journal, 49(12):3095-3109

11 Duran, M. A. and Grossmann, I. E. (1986). Mathematical Programming, 36(3)

12 Process Systems Enterprise. 1997-2014. www.psenterprise.com/gproms

13 Gurobi Optimization, Inc. 2015. http://www.gurobi.com

14 Papaioannou, V., Lafitte, T., Aveñdano, C., Adjiman, C. S., Jackson, G., Müller, E. A., and Galindo, A. (2014). The Journal of Chemical Physics, 140(5)

427917 A Multi-Actor Multi-Objective Framework for the Design of Economically Optimal Processing Networks

Abstract:

The design of multiple-feedstock multiple-product processing networks involves the consideration of many different conversion technologies that link feedstocks to products through several possible intermediates. Traditional approaches aimed to find the best set of products and the most convenient processing pathways, involve the development of a conversion superstructure and the formulation and solution of a mathematical optimization problem. In recent years, such an approach has been successfully applied to the design of biorefinery networks, albeit those in which the set of target products was small and a priori known, and the technologies were well defined. However, when the set of possible products is very large or the technologies for conversion of different intermediates are still under development, the proposal and solution of a single, master, superstructure results impractical.

Motivated by the need of a more flexible approach, in our previous work we proposed a novel strategy for the design of optimal biorefinery networks. Drawing a parallel with a petrochemical network, we divided the biorefinery in (i) one supplier that produces multiple intermediates, and (ii) several consumers that upgrade these intermediates to different products. The key difference with the traditional approach is that, instead of considering a single actor process, the supplier and the consumers are considered to act independently of each other and thus seek the maximization of their own profit. Under this assumption a multiobjective optimization problem (MOP) was formulated to find the optimal set of design variables, flowrates and transfer prices for the intermediates. Using a two-level Lagrangian approach we proved that the MOP could be simplified into a separable optimization problem, to find the solution that, while attaining the same profit level as the single actor network, also maximized the profits of the individual actors. Further, we also showed that this solution was obtained when the price at which the intermediates were transferred coincides with the optimal value of the Lagrange multipliers.

In this presentation, we extend the previous framework to generic processing networks that may also be modeled by the suppliers-intermediates-consumers scheme discussed before. We begin by considering the presence of several suppliers that produce competing intermediates, and study how the earlier analysis change when a consumer can use the same processing steps to upgrade the competing intermediates and when additional processing steps are required. As a particular case, we consider the market as one of the competing suppliers. Then, we generalize the term intermediate to include the exchange of energy, and in particular, we analyze energy integration between the different consumers. Finally, the developed framework is exemplified through a case study that considers generic equations to estimate capital and operational costs for the different actors.

This work was funded by the Cooperative Agreement between the Masdar Institute of Science and Technology (Masdar Institute), Abu Dhabi, UAE and the Massachusetts Institute of Technology (MIT), Cambridge, MA, USA - Reference 02/MI/MI/CP/11/07633/GEN/G/00 for work under the Second Five Year Agreement.

430638 Incorporation of Inherent Safety and Environmental Aspects in Process Design and Supply Chain Optimization

Abstract:

ABSTRACT

The integration of inherently safer design and environmental aspects at the early phases of supply chain selection and process design provides significant benefits. It allows the highest ability to positively influence lifecycle safety, environment and cost of the project and it reduces subsequent engineering rework and modifications in the supply chain and process design. It is crucial to have a simple but effective approach to evaluate and compare the design alternatives based on the safety and environmental design aspects at the early stage of the project when available engineering information is limited.

This study proposes a framework to incorporating life-cycle safety measures for supply chain design and the use of safety metrics along with other objectives (e.g., economic, environmental) to assess and screen process technologies included in the supply chain. Top level quantitative risk assessment and environmental impact benchmarking approaches are used to evaluate designs alternatives, identify design improvement and guide decision making. It also takes into account the multiple geographic manufacturing network and its implications to transportation.

To demonstrate the approach, a case study is solved for a conceptual design of a high density polyethylene supply chain from shale gas.

Seawater Desalination

Theoretical Analysis and Economic Evaluation of Draw Solution Assisted Reverse Osmosis Process

Kiho Park, Do Yeon Kim and Dae Ryook Yang*

Process System Engineering Laboratory
Department of Chemical and Biological Engineering
Korea University

Process description

Draw solution assisted RO

- Draw Solution Assisted Reverse Osmosis (DSARO)?
  - By locating draw solution in opposite side of membrane, the osmotic pressure difference can be significantly decreased.
  - In 1st RO stage, fresh water is permeated into draw solution side, so the stage is regarded as **seawater desalination stage**.
  - After producing pure water, the concentrated draw solution is recycled in **draw solution recycling stage**.

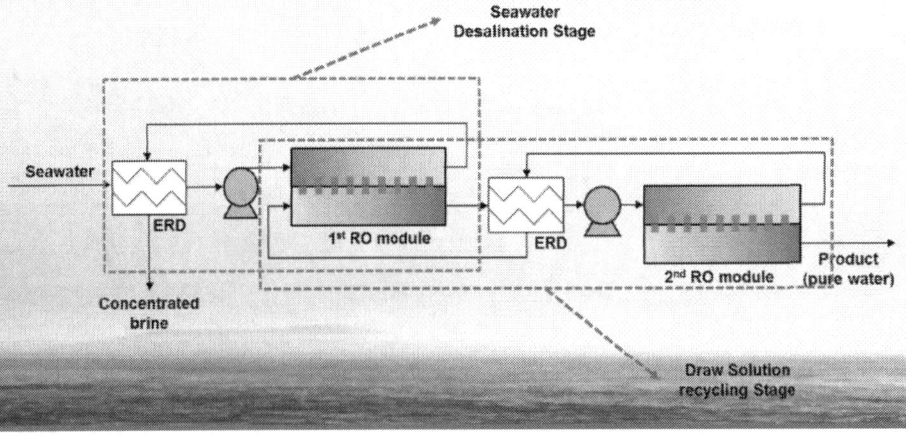

Process description

Draw solution assisted RO

- Expected advantages of the process
  - High pressure 1-staged RO process can be changed to low pressure 2-staged RO process. The energy and cost saving can be achieved.
  - By using 2-staged process, high quality water can be produced.
  - Despite high concentration seawater, seawater desalination can be easily conducted by using high concentration draw solution.

<Conventional RO Process>

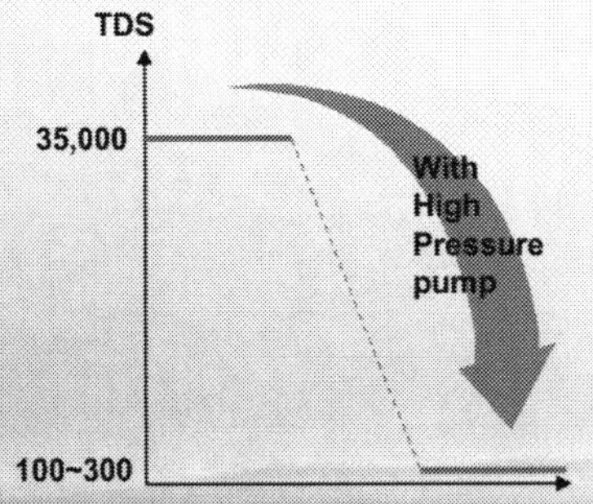

<Draw Solution Assisted RO Process>

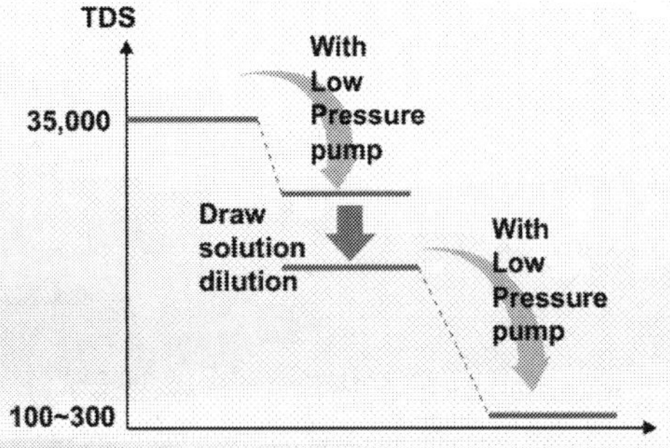

Process modeling

Draw solution assisted RO

- 1st RO and 2nd RO Process – Plate and frame module
  - Flux equations
    - Water Flux

$$J_w = A\left(\Delta P - \Delta\pi_{eff}\right)$$

where, A : water permeability
π_{eff} : effective osmotic pressure

    - Salt Flux

$$J_s = B\left(C_D - C_F\right)$$

where, B : salt permeability

    - Modified van't Hoff formula

$$\pi = \alpha C = \frac{N_{ion} R_g T C}{M_w}$$

  - External Concentration Polarization (ECP)

$$U\frac{\partial C}{\partial z} + V\frac{\partial C}{\partial y} - D\left(\frac{\partial^2 C}{\partial z^2} + \frac{\partial^2 C}{\partial y^2}\right) = 0 \longrightarrow V\frac{dC}{dy} - D\frac{d^2 C}{dy^2} = 0$$

B.C $C(0) = C_F$

$$\left.\frac{dC}{dy}\right|_{y=0} = 0$$

$$\left.D\frac{dC}{dy}\right|_{y=t} = V\left[C(t) - C_p\right]$$

$$C(t) = C_{F,m}$$

$$\frac{C_{F,m} - C_p}{C_F - C_p} = \exp\left(\frac{V}{D}t\right)$$

$$k = \frac{D}{t}$$: Film theory

$$\frac{C_{F,m} - C_p}{C_F - C_p} = \exp\left(\frac{V}{k}\right)$$

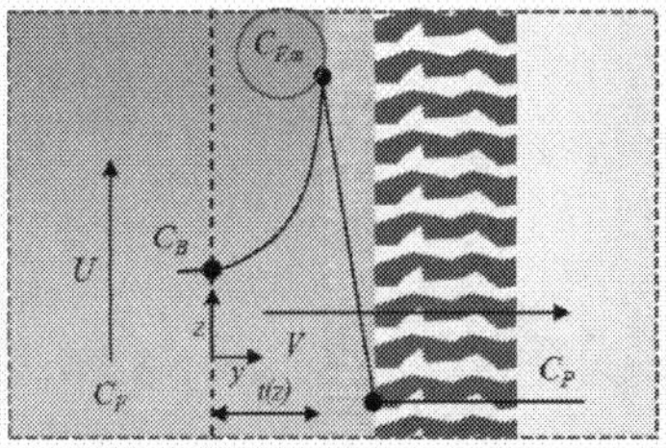

External Concentration polarization

Process modeling

Draw solution assisted RO

- 1st RO Process – Plate and frame module
  - Internal Concentration Polarization (ICP)
    - Derivation of ICP
      - Salt rejection in porous substrate consists of two components : diffusion and convection

$$\underbrace{B\left(C_{F,m}-C_{D,m}\right)}_{\text{Salt flux}}-\underbrace{D\varepsilon\frac{dC(x)}{dx}}_{\text{Diffusive flux}}+\underbrace{J_wC(x)}_{\text{Convective flux}}=0$$

$$\begin{bmatrix} \text{B.C. } C(x)=C_{D,m} \text{ at } x=0 \\ C(x)=C_{D,b} \text{ at } x=\tau t \end{bmatrix}$$

$$\frac{1}{D\varepsilon}dx=\frac{1}{B(C_F-C_D)+J_wC(x)}dC(x)$$

$$\left.\frac{1}{D\varepsilon}\right|_0^{\tau t}=\left.\frac{1}{J_w}\ln\left(B(C_F-C_D)+J_wC(x)\right)\right|_{C_{D,m}}^{C_{D,b}}$$

$$C_{D,m}=C_{F,m}-\frac{C_{F,m}-C_{D,b}\exp(-J_wK)}{1+\frac{B}{J_w}\left(\exp(-J_wK)-1\right)}$$

$K=\frac{\tau t}{D\varepsilon}$: salt resistivity in support layer

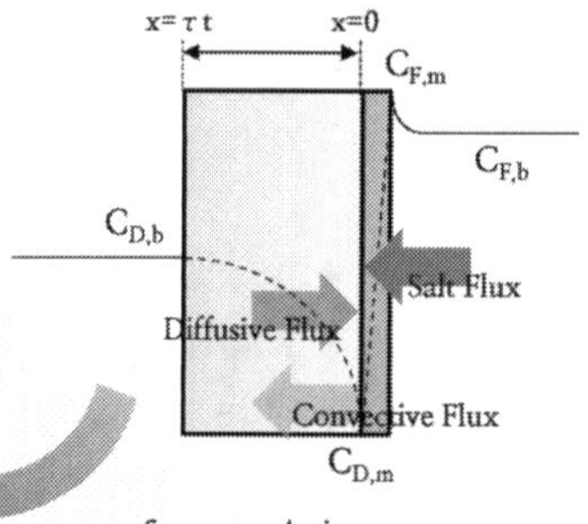

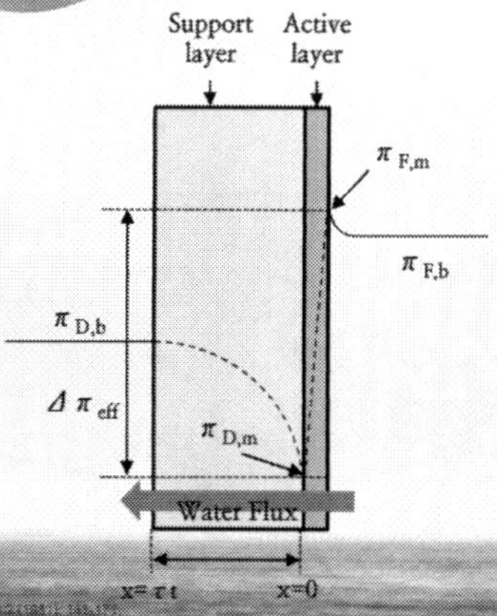

Process modeling

Draw solution assisted RO

- Comparing to FO process
  - Illustration of ICP in asymmetric membrane comparing to FO process

(a) AL-FW mode (DSARO)

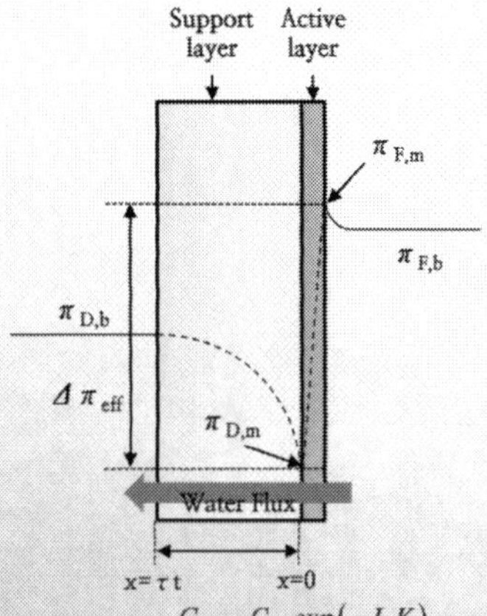

$$C_{D,m}=C_{F,m}-\frac{C_{F,m}-C_{D,b}\exp(-J_wK)}{1+\frac{B}{J_w}\left(\exp(-J_wK)-1\right)}$$

(a) AL-DS mode (FO)

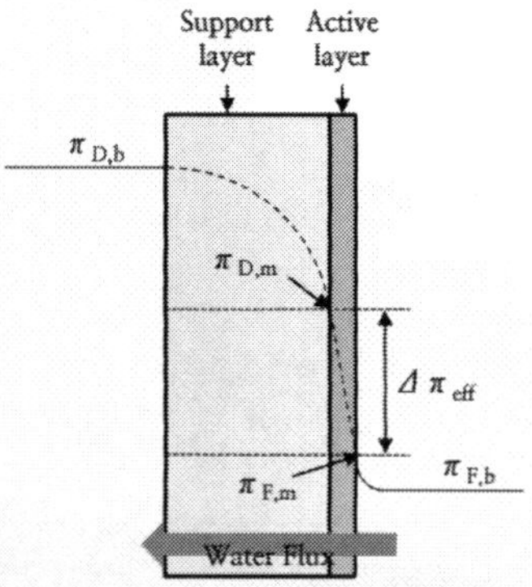

$$C_{D,m}=C_{F,m}-\frac{C_{F,m}-C_{D,b}\exp(-J_wK)}{1-\frac{B}{J_w}\left(\exp(-J_wK)-1\right)}$$

Process modeling

Draw solution assisted RO

- 1st RO Process – Plate and frame module
  - Component mass balance
    - Water
      - √ Velocity profile of y,z-direction could be negligible

$$-\frac{\partial u_{F,x}}{\partial x}-\frac{\partial u_{F,y}}{\partial y}-\frac{\partial u_{F,z}}{\partial z}+\frac{J_w}{H}=0$$

$$-\frac{\partial u_{D,x}}{\partial x}-\frac{\partial u_{D,y}}{\partial y}-\frac{\partial u_{D,z}}{\partial z}-\frac{J_w}{H}=0$$

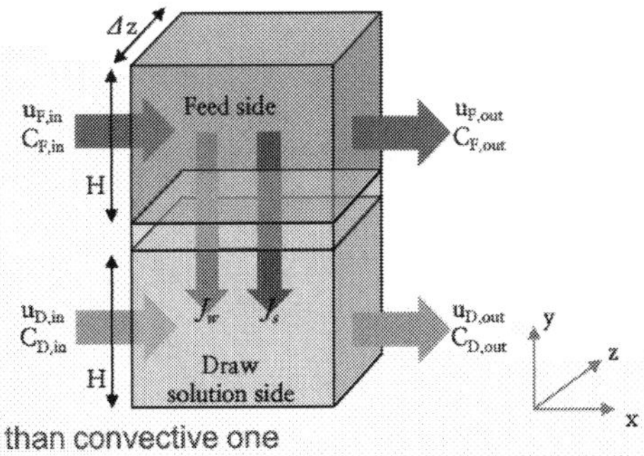

- Salt
  - √ Diffusive flux in channel is much less than convective one

$$D\left[\frac{\partial^2 C_F}{\partial x^2}+\frac{\partial^2 C_F}{\partial y^2}+\frac{\partial^2 C_F}{\partial z^2}\right]-u_{F,x}\frac{\partial C_F}{\partial x}-u_{F,y}\frac{\partial C_F}{\partial y}-u_{F,z}\frac{\partial C_F}{\partial z}+\frac{J_s}{H}=0$$

$$D\left[\frac{\partial^2 C_D}{\partial x^2}+\frac{\partial^2 C_D}{\partial y^2}+\frac{\partial^2 C_D}{\partial z^2}\right]-u_{D,x}\frac{\partial C_D}{\partial x}-u_{D,y}\frac{\partial C_D}{\partial y}-u_{D,z}\frac{\partial C_D}{\partial z}-\frac{J_s}{H}=0$$

Gu, B., et al. "Mathematical model of flat sheet membrane modules for FO process: Plate-and-frame module and spiral-wound module." Journal of Membrane Science 379.1 (2011): 403-415.

Energy consumption model

Draw solution assisted RO

- 1st RO system
  - Seawater Intake and Pretreatment

$$E_{SWIP}=\frac{P_{SWIP}F_{Feed}}{\eta_{SWIP}}$$

Ref.: Design and economics of RO seawater desalination, Desalination (1996)
P_{swip} : 5 bar, $\eta_{SWIP}=\eta_{pump}$

  - High Pressure Pump

$$E_{HPP}=\frac{\left(F_{Feed}-F_{reten}\right)}{\eta_{pump}}\left(P_{feed}-P_0\right)$$

  - Booster Pump

$$E_{BP}=\frac{F_{reten}}{\eta_{pump}}\left(P_{feed}-P_{ERD}\right)$$

* Ref.: Global optimal design of reverse osmosis networks for seawater desalination: modeling and algorithm. Desalination (2005)

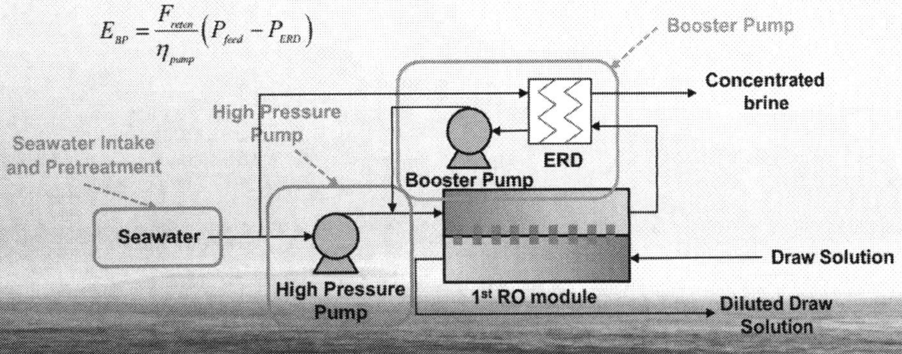

Energy consumption model

Draw solution assisted RO

- 2nd RO system
  - High Pressure Pump

$$E_{HPP} = \frac{(F_{Feed} - F_{reten})}{\eta_{pump}}(P_{feed} - P_0)$$

  - Booster Pump

$$E_{BP} = \frac{F_{reten}}{\eta_{pump}}(P_{feed} - P_{ERD})$$

* Ref.: Global optimal design of reverse osmosis networks for seawater desalination: modeling and algorithm. Desalination (2005)

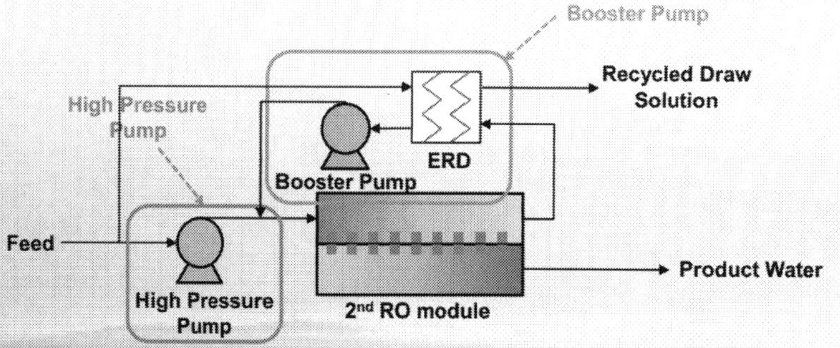

Capital cost estimation

Draw solution assisted RO

- SWRO and BWRO plant capital cost correlation equations
  - Considering real plant scale cost data, SWRO and BWRO plant capital cost can be approximated.
  - This cost equations can be applied in the cost estimation of DSARO.

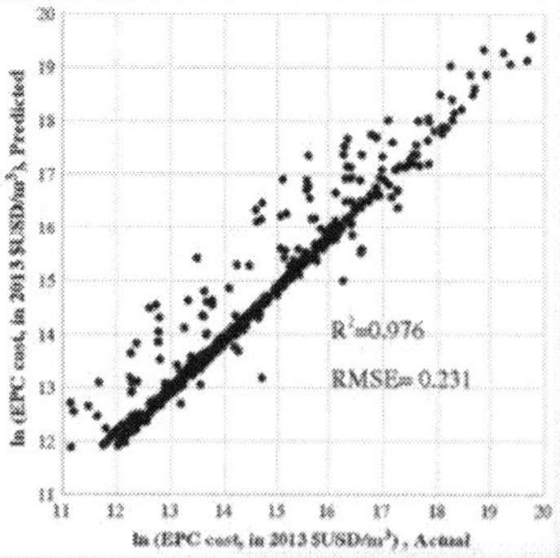

Table 1
Sub-datasets of the desalination plants inventory used in model fitting.

	Notation	Region	Observations
Dataset 1	GCC_SWRO	GCC	138
Dataset 2	GCC_BWRO		492
Dataset 3	SE_SWRO	S. Europe	129
Dataset 4	SE_BWRO		191

EPC = Engineering-Procurement-Construction Cost
CAP = Plant capacity
YEAR = EPC contract year
TYPE = Plant type
REGION = Region of the plant

$$\ln(EPC) = 58.054 + 0.939 \times \ln(CAP) - 0.028 \times YEAR + 0.830 \times TYPE - 0.074 \times REGION$$

Loutatidou, Savvina, et al. "Capital cost estimation of RO plants: GCC countries versus southern Europe." *Desalination* 347 (2014): 103-111.

Operating cost estimation

Draw solution assisted RO

- 1st RO Process
  - Cost of electricity

    $OC_{elec}[\$/\text{y}] = [W_{SWIP} + W_{HPP} + W_{BP}] \times D_{energy}[\$/\text{kWh}] \times 365[\text{day/ y}] \times L_f$

    $W[\text{kWh/day}]$
  - Membrane replacement cost
    - -20% per year (SWRO module)

      $OC_{replace}[\$/\text{y}] = 0.2[1/\text{y}]CC_{memb}[\$]$
    - -10% per year (BWRO module)

      $OC_{replace}[\$/\text{y}] = 0.1[1/\text{y}]CC_{memb}[\$]$
  - Chemical treatment cost

    $OC_{chem}[\$/\text{y}] = 0.018[\$/\text{m}^3] \times F_{feed}[\text{m}^3/\text{y}] \times L_f$
  - Annual Spares cost

    $OC_{sc}[\$/\text{y}] = 0.033[\$/\text{m}^3] \times F_{prod}[m^3/y] \times L_f$
  - Operating & maintenance cost

    $OC_{om}[\$/\text{y}] = 0.126[\$/\text{m}^3] \times F_{prod}[m^3/y] \times L_f$

[illegible] modeling and algorithm. Desalination (2005)

Operation cost estimation

Draw solution assisted RO

- 2nd RO Process
  - Cost of electricity

    $OC_{elec}[\$/\text{y}] = [W_{HPP} + W_{BP}] \times D_{energy}[\$/\text{kWh}] \times 365[\text{day/ y}] \times L_f$

    $W[\text{kWh/day}]$
  - Membrane replacement cost
    - -20% per year

      $OC_{replace}[\$/\text{y}] = 0.2[1/\text{y}]CC_{memb}[\$]$
    - -10% per year

      $OC_{replace}[\$/\text{y}] = 0.1[1/\text{y}]CC_{memb}[\$]$

* Ref : Global optimal design of reverse osmosis networks for seawater desalination: modeling and algorithm. Desalination (2005)

Simulation conditions

Draw solution assisted RO

- Operating conditions and parameters

Variables	Value	Variables	Value
Feed Concentration (ppm)	35,000	Feed flow rate (m^3/s)	0.2315
Draw Solution Concentration (ppm)	30,000	Draw solution flow rate (m^3/s)	0.0463
Feed Temperature (℃)	25	Pump efficiency (%)	74
Draw Solution Temperature (℃)	25	ERD efficiency (%)	80
Feed Pressure (bar)	45.36	Electricity unit cost ($/kWh)	0.08
Draw Pressure (bar)	1	Load factor : L_f	0.9
1st RO recovery	0.5	Plant life (year)	15
2nd RO recovery	0.7143	Interest rate (1/year)	0.03
Water Permeability (m/Pa/s)	9.39×10^{-12}	Structural parameter (m)	9.2×10^{-5}
Salt Permeability (m/s)	5.65×10^{-8}	Plant Capacity (m^3/day)	10,000

Simulation results

Draw solution assisted RO

- Process simulation results

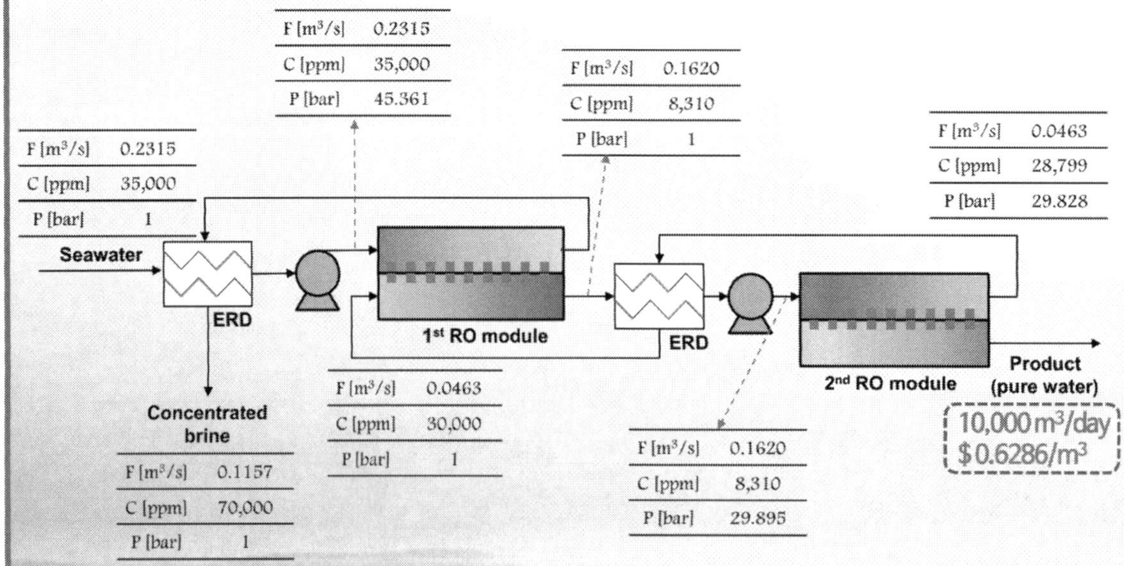

Results and discussion

Draw solution assisted RO

- Comparison with SWRO process
  - The specific water cost can be reduced more than 12%.
  - Since the pressure requirement is lower than conventional SWRO, the problems caused by high pressure can be relieved.

	SWRO					Draw Solution Assisted RO	
						1st RO	2nd RO
Series Module #	7	6	5	4	3	3	2
SWC [$]	0.7503	0.7356	0.7226	0.7144	0.7162	0.6286	
Total Capital Cost [$]	8,997,200	8,763,300	8,495,500	8,265,400	8,117,500	5,081,810	
Recovery [%]	52	52	53	54	55	50	
Pressure [bar]	62.674	63.266	65.636	68.827	74.294	45.360	29.895

Conclusions

Draw solution assisted RO

- Draw solution assisted reverse osmosis is suggested and analyzed theoretically and economically.
- The ICP equations are different comparing to FO process since the concentration of the draw solution is different.
- The process conditions are in the range of BWRO, so the BWRO capital cost correlation equation can be used.
- The cost estimation results showed that the cost of the DSARO process is 12% lower than that of the conventional SWRO process.
- If an appropriate membrane technology would be developed, the DSARO process configuration can obtain its competitiveness.

Thank You for Listening

15AIChE Annual Meeting, Salt Lake City, UT

Salt Lake Marriott Downtown
Monday 9 November 2015

NTNU

New Insight for the Design of Networks with Heat Exchange, Compression and Expansion

by **Chao Fu** and Truls Gundersen

Department of Energy and Process Engineering
Norwegian University of Science and Technology
Trondheim, Norway

NTNU

Outline of the Presentation

- Work & Heat Exchange Networks (**WHEN**s)
  - Definition, Literature and our Focus
- **Appropriate Placement** Concept
  - Brief History for Compressors & Expanders
  - New additional Insight
  - Physical Explanations
- A new **Design Procedure** for WHENs
  - Based on *4 Theorems* for ***4 Situations***:
    - Compression & Expansion above & below Ambient
  - Using the *GCC* (Grand Composite Curve)
  - With both Heat and Power, *Exergy* is our Metric
- Two Examples
  - Small Illustrative
  - Small Industrial
- The Complete Picture & the *Symmetry*
- Concluding Remarks

Definition of WHENs and Scope of Research

- Work & Heat Exchange Networks (WHENs) involve:
  - Integration of Expansion and Compression for ***Work Recovery***
  - Utilizing Heating & Cooling from Compression & Expansion for ***Heat Recovery*** – **Our Focus**
- Classical Problem Definition for Heat Exchanger Network Design has been extended:
  - *"Given a Set of Process Streams with **Supply** and **Target Temperatures and Pressures**, as well as Utilities for **Power**, **Heating** and **Cooling** → Design a Network of **Heat Exchangers**, **Expanders**, and **Compressors** in such a way that **Exergy** Consumption is minimized or **Exergy** Production is maximized".*
- Our Start: The "*ExPAnD*" Procedure (2007)

*A. Aspelund, D.O. Berstad, T. Gundersen, "An Extended Pinch Analysis and Design procedure utilizing pressure based exergy for subambient cooling", **Applied Thermal Engng.**, vol. 27, pp. 2633-2649, 2007.*

17.11.15 T. Gundersen Slide no. 3

NTNU

Previous Work on Appropriate Placement of Compressors and Expanders

- ***Homšak and Glavič (1995)*** suggested Placement of Compressors above Pinch in a discussion on Correct Integration of Reactors.
- ***Aspelund et al. (2007)*** made the following Observations:
  - "*Compressors* add Heat and should be placed *Above* Pinch"
  - "***Expanders*** add Cooling and should be placed ***Below*** Pinch"
  - This is in *contrast* to the established Design Practice
- ***Gundersen et al. (2009)*** quantified previous observations:
  - Optimal *inlet* Temperature to Compressors and Expanders that maximize Exergy Efficiency is the *Pinch* Temperature
  - But: They also observed that the Pinch may change!
- ***Wechsung et al. (2011)*** used the Insight to develop a new *Superstructure* and an *MINLP* formulation for **WHEN** Design that was successfully applied to synthesize an Offshore LNG Process

17.11.15 T. Gundersen Slide no. 4

NTNU

Our Concern and Motivation

- The *Challenges* with Compressors and Expanders:
  - Stream Temperatures and Enthalpies are changing
  - Shape of CCs and the GCC will change
  - The Pinch Point(s) and Utility Demand may (will) change
  - Even Stream Identity may (will) change
- Previous Insight may not be General?
  - The Rules have not been challenged by rigorous Thermodynamic or Mathematical Analysis
  - ***Can we increase our Insight?***

*C. Fu, T. Gundersen, "Integrating **Compressors** into heat exchanger networks **above** ambient Temperature", **AIChE Jl.**, 61(11), pp. 3770-3785, 2015.*
*C. Fu, T. Gundersen, "Integrating **Expanders** into heat exchanger networks **above** ambient temperature", **AIChE Jl.**, 61(10), pp. 3404-3422, 2015.*
*C. Fu, T. Gundersen, "**Sub-ambient** heat exchanger network design including **Compressors**", **Chem. Eng. Sci.**, 137, pp. 631-645, 2015.*
*C. Fu, T. Gundersen, "**Sub-ambient** heat exchanger network design including **Expanders**", **Chem. Eng. Sci.**, 138, pp. 712-729, 2015*

NTNU

17.11.15 T. Gundersen Slide no. 5

Related Literature

- Work Exchange Networks (**WEN**s):
  - Razib MS, Hasan MMF, Karimi IA. Preliminary synthesis of work exchange networks. *Computers & Chemical Engineering*, **36**, 262-277, 2012.
  - Liu G, Zhou H, Shen R, Feng X. A graphical method for integrating work exchange networks. *Applied Energy*, **114**, 588-599, 2014.
- Work & Heat Exchange Networks (**WHEN**s):
  - Aspelund A, Berstad DO, Gundersen T. An extended pinch analysis and design procedure utilizing pressure based exergy for subambient cooling. *Applied Thermal Engineering*, **27** (16), 2633-2649, 2007.
  - Gundersen T, Berstad DO, Aspelund A. Extending pinch analysis and process integration into pressure and fluid phase considerations. *Chemical Engineering Transactions*, **18**, 33-38, 2009.
  - Onishi VC, Ravagnani MASS, Caballero JA. Simultaneous synthesis of heat exchanger networks with pressure recovery: Optimal integration between heat and work. *AIChE Jl.*, **60** (3), 893-908, 2014.
  - Onishi VC, Ravagnani MASS, Caballero JA. Simultaneous synthesis of work exchange networks with heat integration. *Chemical Engineering Science*, **112**, 87-107, 2014.
  - Onishi VC, Ravagnani MASS, Caballero JA. Retrofit of heat exchanger networks with pressure recovery of process streams at sub-ambient conditions. *Energy Conv. & Mngmnt*, **94**, 377-393, 2015.
  - Wechsung A, Aspelund A, Gundersen T, Barton PI. Synthesis of heat exchanger networks at subambient conditions with compression and expansion of process streams. *AIChE Jl.*, **57** (8), 2090-2108, 2011.
  - Homšak M, Glavič P. Appropriate placement of energors against local pinch temperatures. *Transactions of IChemE*, **73** (A), 871-879, 1995.
- Work, Heat and **Mass** Exchange Networks:
  - Dong R, Yu Y, Zhang Z. Simultaneous optimization of integrated heat, mass and pressure exchange network using exergoeconomic method. *Applied Energy*, **136**, 1098-1109, 2014.
- Self-Heat **Recuperation** Technology
  - Kansha Y, Tsuru N, Sato K, Fushimi C, Tsutsumi A. Self-heat recuperation technology for energy saving in chemical processes. *Industrial & Engineering Chemistry Research*, **48** (16), 7682-7686, 2009.

NTNU

17.11.15 T. Gundersen Slide no. 6

Our Terminology (above Ambient)

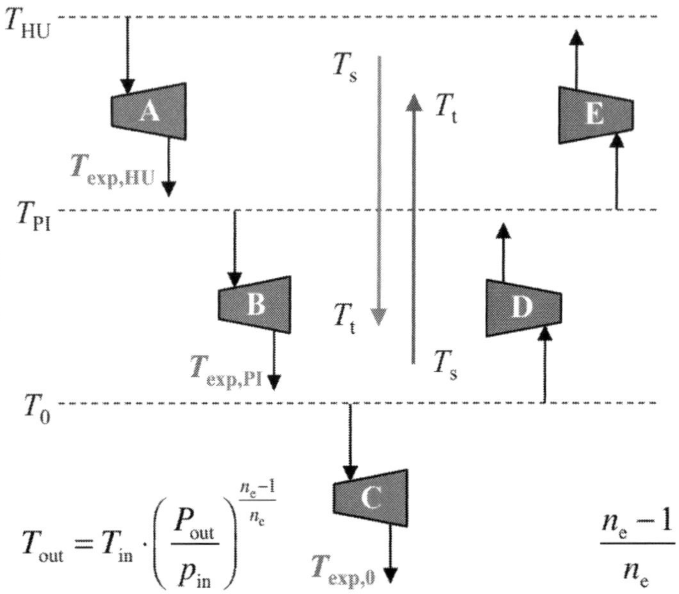

"Hot Utility Expansion" → Case A

"Pinch Expansion" → Case B

"Ambient Expansion" → Case C

"Ambient Compression" → Case D

"Pinch Compression" → Case E

$$T_{out} = T_{in} \cdot \left(\frac{P_{out}}{p_{in}} \right)^{\frac{n_e - 1}{n_e}}$$

$$\frac{n_e - 1}{n_e} = \eta_{poly} \cdot \frac{\kappa - 1}{\kappa}$$

17.11.15 T. Gundersen Slide no. 7

The Complete Picture (above & below Ambient)

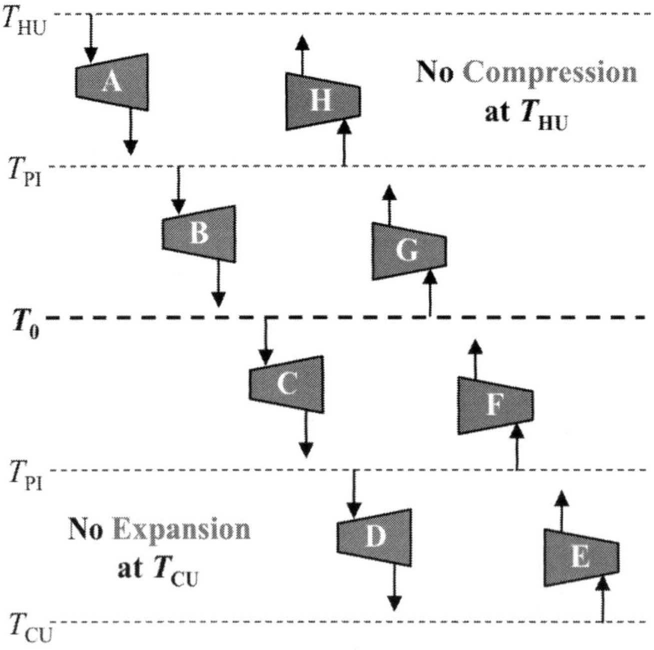

17.11.15 T. Gundersen Slide no. 8

Physical Explanation – Expanders above Ambient

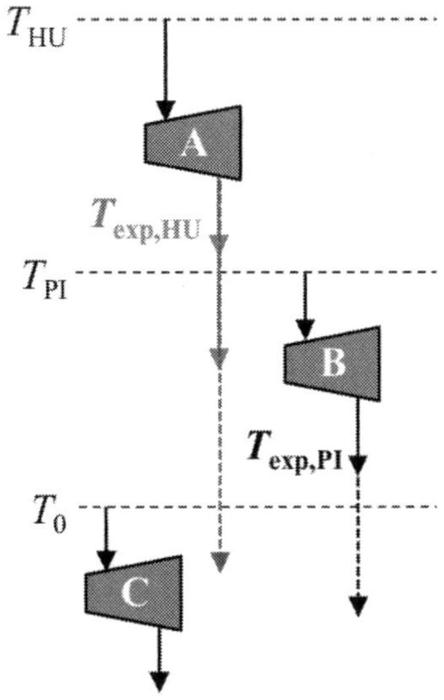

$Q_{exp,PI} = mCp \bullet (T_{PI} - T_{exp,PI})$

- 2 important Variables:
  - $Q_{exp,PI}$ and $T_{exp,HU}$
- Case I (covered by Theorem 1):
  - $Q_{exp,PI} \le Q_{C,min}$ and $T_{exp,HU} > T_0$
  - Use Pinch Expansion only
- Case II (covered by Theorem 2):
  - $Q_{exp,PI} > Q_{C,min}$ and $T_{exp,HU} \ge T_{PI}$
  - Max Expansion at T_{PI}, then at T_{HU} or T_0
- Case III (covered by Theorem 3):
  - $Q_{exp,PI} > Q_{C,min}$ and $T_0 < T_{exp,HU} < T_{PI}$
  - Increase Expansion at T_{HU}, reduce at T_{PI}
- Case IV (covered by Theorem 4):
  - $T_{exp,HU} < T_0$
  - Expansion at T_{HU}

17.11.15 T. Gundersen Slide no. 9

Graphical Procedure to find maximum Pinch Expansion (same for Compression)

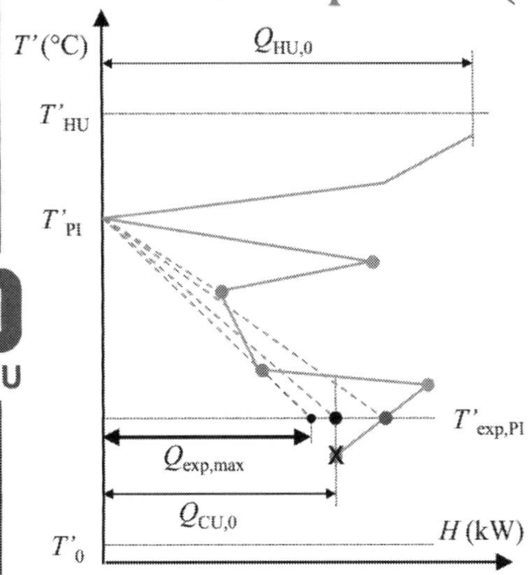

$$\left(mc_p\right)_{max} = \frac{Q_{exp,max}}{\left(T'_{PI} - T'_{exp,PI}\right)}$$

- With Pinch Expansion, the following are Potential new Pinch Points
  - Convex points between T'_{PI} and $T'_{exp,PI}$
  - The point $T'_{exp,PI}$ on the GCC
  - The lowest point on the GCC if $T'_{exp,PI}$ is not on the GCC
  - Intersection points between $T = T'_{exp,PI}$ and heat pockets
- Identification of the new Pinch
  - Draw lines between the original Pinch and the Potential new Pinch Points
  - Extend the line with the steepest slope to intersect with the line $T = T'_{exp,PI}$
  - The corresponding cooling demand is the maximum expansion work

17.11.15 T. Gundersen Slide no. 10

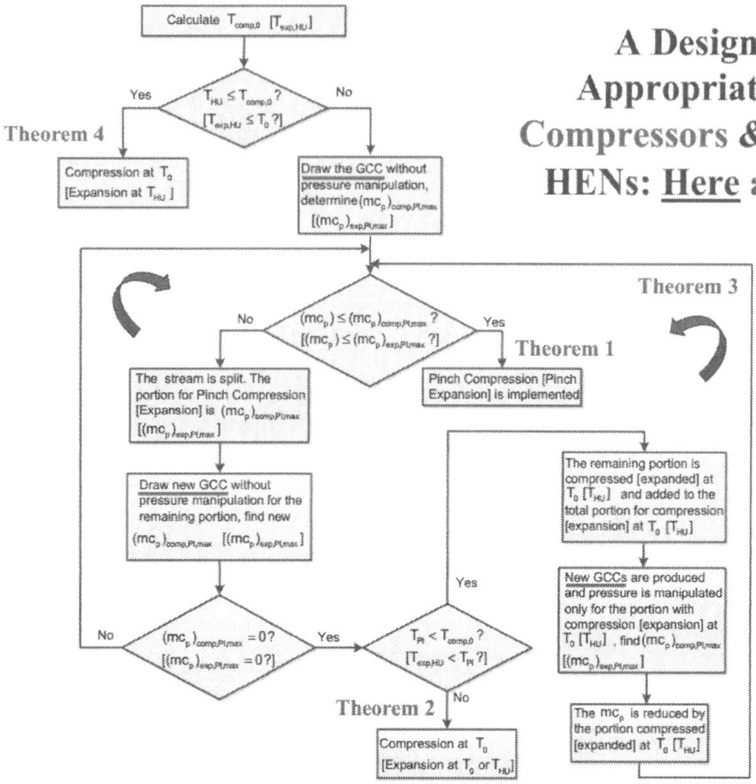

A Design Procedure for Appropriate Placement of Compressors & Expanders in HENs: <u>Here</u> above Ambient

Iterative (Loops) & Graphical (GCC) based on 4 Theorems

NTNU

17.11.15 T. Gundersen Slide no. 11

NTNU

"Experimental" Setup for Illustrative Example

- Assumptions:
  - Supply/Target States (p, T) for Streams are given
  - Compression and Expansion of only <u>one</u> Stream
  - Only <u>one</u> "hot" (288 K) and <u>one</u> cold Utility (120 K)
  - Polytropic Efficiency η_{poly} is constant
  - Ideal Gas with constant $\kappa = c_p/c_v$
- Parameters used in the Calculations
  - $\eta_{\text{poly}} = 1.0$, $\Delta T_{\min} = 4$ K, $T_0 = 288$ K and $\kappa = 1.4$
- Objective is to minimize Exergy Consumption
  - Work is assumed to be pure Exergy
  - Exergy of Hot Utility is neglected ($T_{HU} = T_0 = 288$ K)
  - Exergy of Cold Utility from Carnot Equation:

$$Ex = Q_{CU} \cdot \left(\frac{T_0}{T_{CU}} - 1 \right)$$

17.11.15 T. Gundersen Slide no. 12

Example – Illustrating the Procedure

Stream	T_s, K	T_t, K	mc_p, kW/K	ΔH, kW	p_s, bar	p_t, bar
H1	288	124	2	328	3	1
H2	252	168	4	336	-	-
C1	138	284	3	438	1	2
C2	198	235	7	259	-	-
Hot utility	288	288	-	-	-	-
Cold utility	120	120	-	-	-	-

$T_{HU} = 288$ K, $T_{CU} = 120$ K, $\Delta T_{min} = 4$ K

2 Streams subject to Pressure Manipulation:

H1 is expanded and C1 is compressed

Example – Illustrating the Procedure

Case O: Without Compression and Expansion
Case I: Compression/Expansion at Supply Temperatures
Case II: CU Compression (C1) & Ambient Expansion (H1)
Case III: Pinch Compression and Expansion
Case IV: Using the novel Design Procedure

Cases	O	I	II	III	IV
Hot utility demand, kW	145	300.2	300.2	47.6	9.6
Cold utility demand, kW	112	202.6	193.6	36.0	13.0
Pinch temperature, K	200	200	200	140	140;200;250
Compression work, kW	-	90.6	81.6	130.2	135.5
Expansion work, kW	-	155.2	155.2	108.8	99.1
Exergy consumption, kW	-	219.0	197.4	71.8	54.6

Lesson to learn: Previous "Insight" was not Rigorous

Case III – Pinch Compression and Expansion

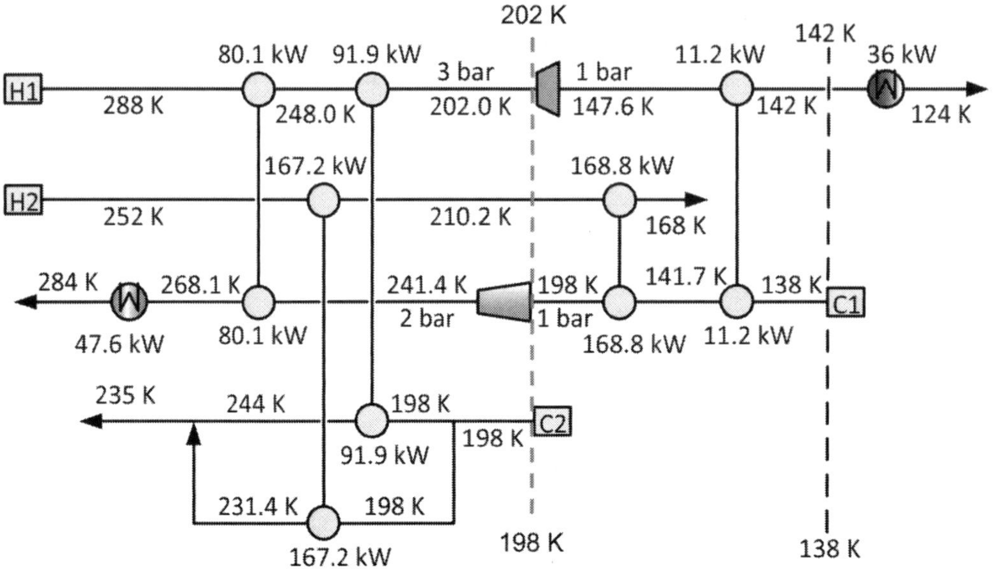

1 Compressor, 1 Expander, 7 HEs, Exergy Consumption: 71.8 kW

17.11.15 T. Gundersen Slide no. 15

Case IV – Compression and Expansion by Procedure

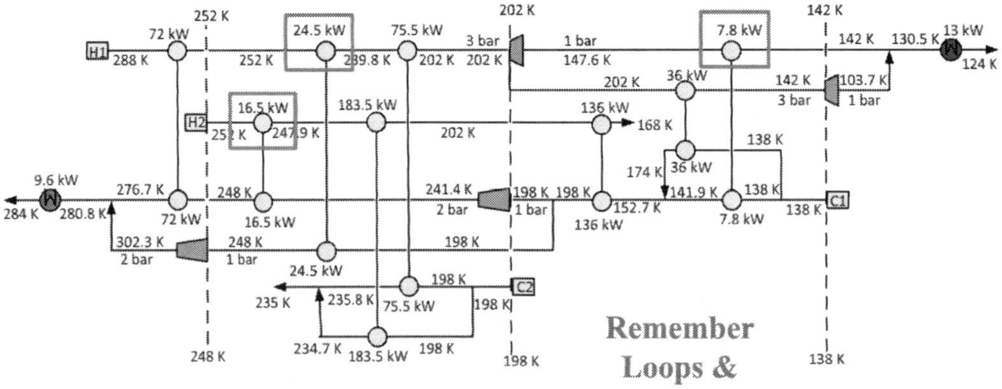

Remember Loops & Paths?

2 Compressors, 2 Expanders, 10 HEs, Exergy Consumption: 54.6 kW

17.11.15 T. Gundersen Slide no. 16

Summary for the Illustrative Example

Property	Case I	Case II	Case III	Case IV
Exergy Consumption (kW)	219.0	197.4	71.8	54.6
# of Pinch Points	1	1	1	3
# of Compressors	1	1	1	2
# of Expanders	1	1	1	2
# of Heat Exchangers	7	8	7	10
# of Stream Splits	0	0	1	4

- **Some Observations**
  - The Design Procedure should primarily be regarded as a Targeting Exercise – it aims for the "ultimate"
  - The Procedure (Case IV) results in new Pinch Points when maximizing Pinch Compression and Expansion
  - The Result is considerable Complexity (Units & Stream Splits)
  - Using only Pinch Compression and Expansion (Case III) at the original Pinch seems to be the best option in this Example

Small Industrial Example

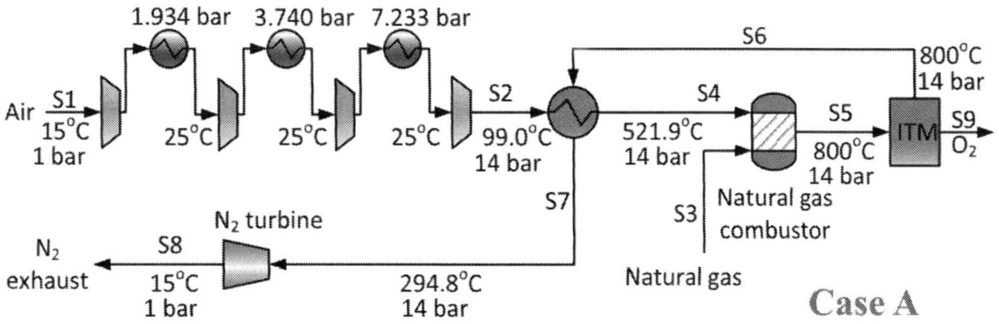

Case A

- **Briefly about this Example**
  - Membrane Separation of Air in an Oxy-Combustion Process
  - An Ion Transport Membrane (ITM) is used
  - Compressed air is preheated against N_2 rich Flue Gas
  - **Key Question**: Can Energy Consumption (Natural Gas and Power) be reduced by using the new Design Procedure?

Small Industrial Example

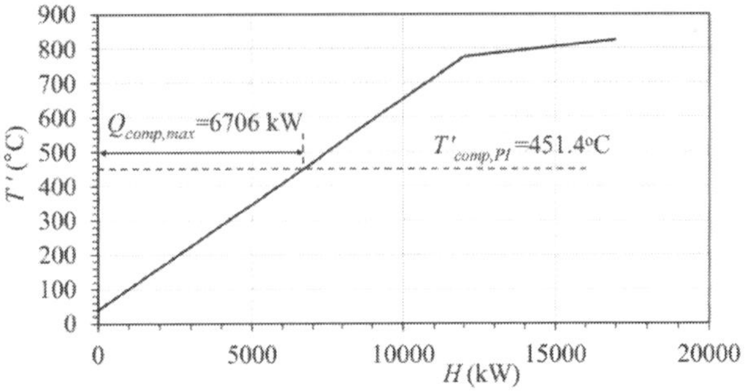

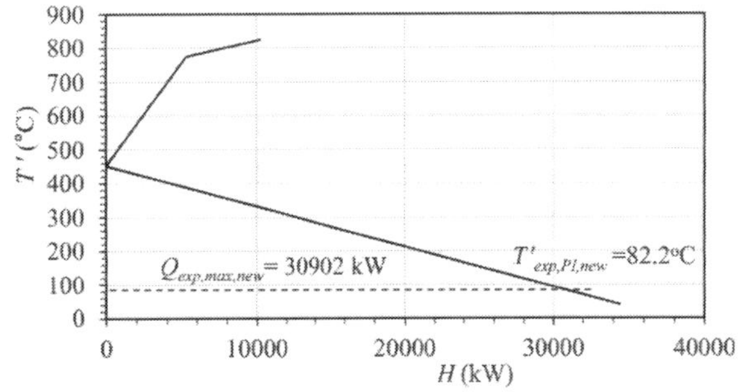

Grand Composite Curve is used as part of the Design Procedure

$\Delta T_{\min}$ **= 50°C**

Threshold Problem, but the lowest T = 15°C can be used as "Pinch" here

Maximum Heating from **Compression** is found

<u>New Pinch at 426.4°C</u>

Maximum Cooling from **Expansion** is found

17.11.15 T. Gundersen Slide no. 19

Small Industrial Example

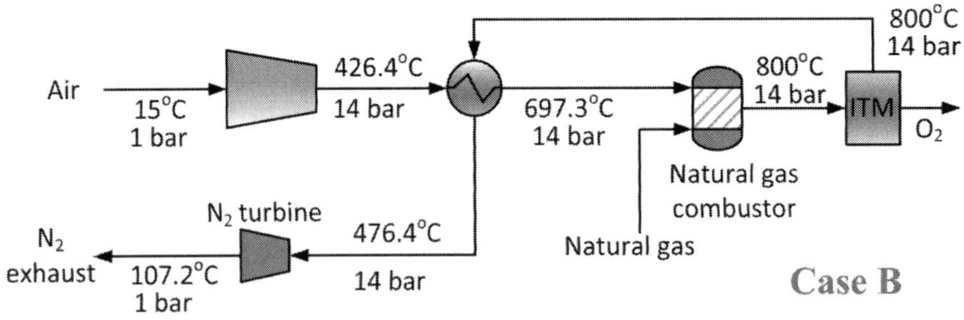

Case B

- **Some Comments about the Solution**
  - **Adiabatic Compression** is used
  - Increased Compressor Work is "invested" to reduce Fuel Consumption while also increasing Expander Work
  - Details about Performance on the next Slide

17.11.15 T. Gundersen Slide no. 20

Small Industrial Example

Energy Form (kW)	Case A	Case B	Difference
Compression Work	29,350	41,140	+ 11,790
Expansion Work	23,419	30,902	+7483
Net Work Consumption	5,931	10,238	+4307
Heat Demand (Fuel Gas)	27,815	10,275	- 17,540

- **Some Comments about the Solution**
  - The "COP" for the Net Compression Work is 4.07
  - The saved Heat (i.e. Fuel) could produce **9647 kW** in a **55%** efficient Natural Gas Combined Cycle Power Plant
  - Measured in terms of Power, the Savings are 5340 kW

The Complete Picture – Symmetric of course ☺

Theorem	Expansion Above	Compression Above	Expansion Below	Compression Below
1	$Q_{exp,PI} \le Q_{C,min}$ $T_{exp,HU} > T_0$ → **Use Pinch** Expansion	$Q_{comp,PI} \le Q_{H,min}$ $T_{comp,0} < T_{HU}$ → **Use Pinch** Compression	$Q_{exp,PI} \le Q_{C,min}$ $T_{exp,0} > T_{CU}$ → **Use Pinch** Expansion	$Q_{comp,PI} \le Q_{H,min}$ $T_{comp,CU} < T_0$ → **Use Pinch** Compression
2	$Q_{exp,PI} > Q_{C,min}$ $T_{exp,HU} \ge T_{PI}$ → Split and maximize **Pinch** Expansion, then exp. at T_{HU} or T_0	$Q_{comp,PI} > Q_{H,min}$ $T_{comp,0} \le T_{PI}$ → Split and maximize **Pinch** Compression	$Q_{exp,PI} > Q_{C,min}$ $T_{exp,0} \ge T_{PI}$ → Split and maximize **Pinch** Expansion	$Q_{comp,PI} > Q_{H,min}$ $T_{comp,CU} \le T_{PI}$ → Split and maximize **Pinch** Compression, then comp. at T_{CU} or T_0
3	$Q_{exp,PI} > Q_{C,min}$ $T_0 < T_{exp,HU} < T_{PI}$ → Increase HU Expansion & reduce **Pinch** Expansion	$Q_{comp,PI} > Q_{H,min}$ $T_{PI} < T_{comp,0} < T_{HU}$ → Increase Ambient Compression & reduce **Pinch** Compression	$Q_{exp,PI} > Q_{C,min}$ $T_{CU} < T_{exp,0} < T_{PI}$ → Increase Ambient Expansion & reduce **Pinch** Expansion	$Q_{comp,PI} > Q_{H,min}$ $T_{PI} < T_{comp,0} < T_0$ → Increase CU Compression & reduce **Pinch** Compression
4	$T_{exp,HU} \le T_0$ → Use HU Expansion	$T_{comp,0} \ge T_{HU}$ → Use Ambient Compression	$T_{exp,0} \le T_{CU}$ → Use Ambient Expansion	$T_{comp,CU} \ge T_0$ → Use CU Compression

Future Research

- Increase the Industrial Realism for our Procedure
  - Multiple Hot and Cold Utilities
  - Non-constant Utilities
  - Multi-Stage Compression and Expansion
- Our Main Objectives for Future Research
  - Establish a "Rich" but "Efficient" **Superstructure**
  - Build MINLP Models subject to Global Optimization

NTNU

17.11.15 T. Gundersen Slide no. 23

Some Final Words

- <u>Warning</u>: Relative Prices of Heat and Power do not always follow the 2nd Law of Thermodynamics! ☹
- <u>But</u>: Thermodynamics will result in good, sound and reasonable Solutions that can act as starting points for Economic Designs! ☺

NTNU

17.11.15 T. Gundersen Slide no. 24

Some extra Slides

Beyond the Heuristics in ExPAnD – Optimization has been introduced

- **The MINLP model is based on:**
  - A novel Superstructure that allows Heat Exchange, Compression and Expansion
  - New Insight on Appropriate Placement of Compressors and Expanders
    - **Compressors**: Immediately **above** Pinch
    - **Expanders**: Immediately **below** Pinch
    - In **Contrast** to current Design Practice
    - Inlet Conditions: **Pinch** Temperature
  - Minimizing Exergy Losses or Work

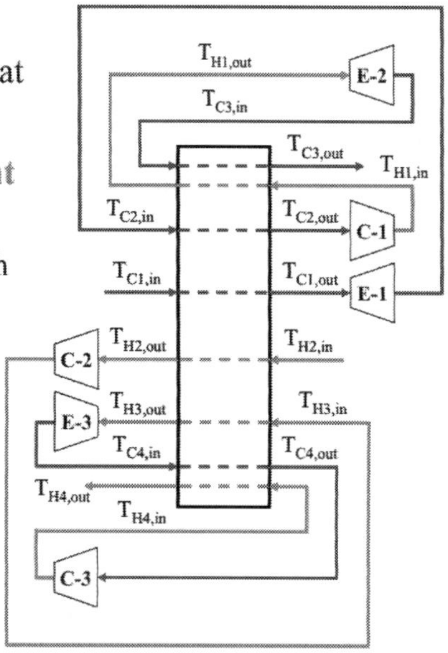

*A. Wechsung, **A. Aspelund**, T. Gundersen and P.I. Barton, "Synthesis of Heat Exchanger Networks at Subambient Conditions with Compression and Expansion of Process Streams", **AIChE Jl.**, vol. 57, no. 8, pp. 2090-2108, 2011.*

The Offshore Process in the LEC

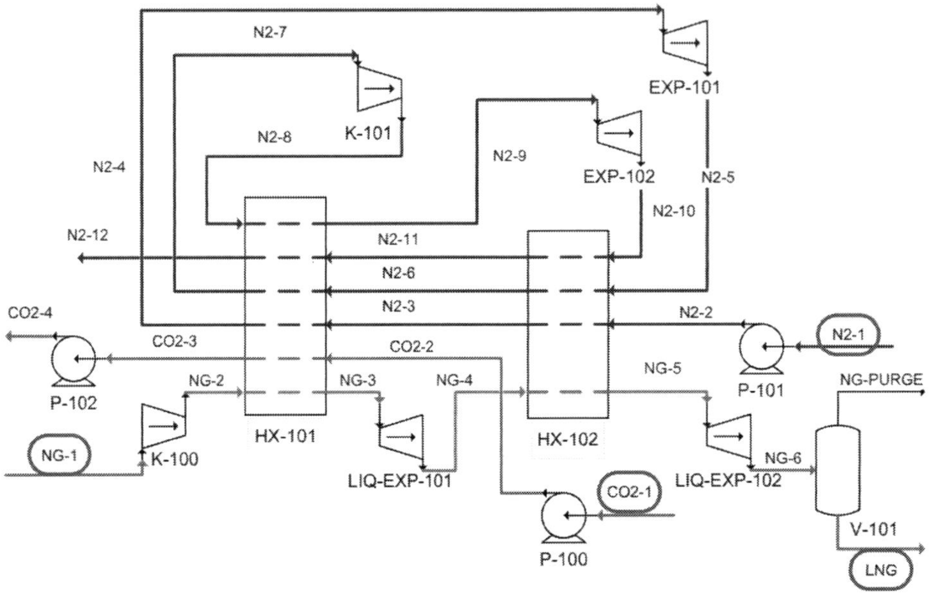

No flammable Refrigerants and no Gas Turbines !!

NTNU

Detailed Features of the Offshore Process

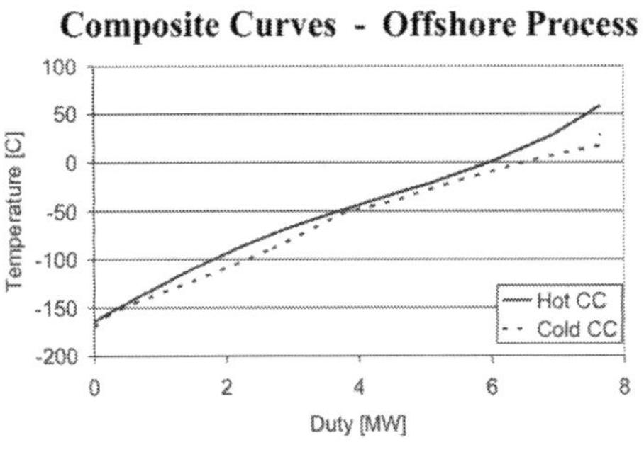

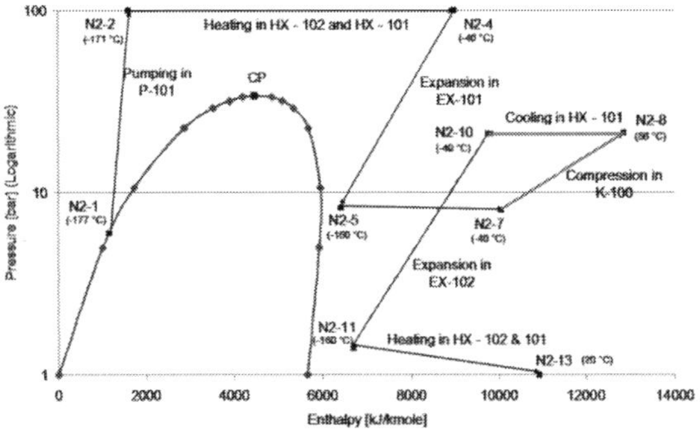

Single Components (here CO_2 and N_2) exhibit Multi-component Behavior by Compression and Expansion

Exergy Efficiency was improved from **49.7%** (evaporating LIN and LCO_2) to **85.7%** (final design from ExPAnD)

*Aspelund A. and Gundersen T. "A Liquefied Energy Chain for Transport and Utilization of Natural Gas for Power Production with CO_2 Capture and Storage – Part 2, The Offshore and the Onshore Processes", **Journal of Applied Energy**, vol. 86, pp. 793-804, 2009.*

NTNU

Applications of Pinch Analysis and Mathematical Programming Methods for Synthesizing Non-Isothermal Water Networks

Elvis Ahmetović[a,b], Zdravko Kravanja[b], François Maréchal[c], Nidret Ibrić[a], Maziar Kermani[c]

[a]University of Tuzla, Faculty of Technology, Univerzitetska 8, 75000 Tuzla, Bosnia and Herzegovina
[b]University of Maribor, Faculty of Chemistry and Chemical Engineering, Smetanova ulica 17, 2000 Maribor, Slovenia
[c]EPFL/Industrial Processes & Energy Systems Engineering Group (IPESE), Lausanne, Switzerland

15AIChE
Annual Meeting, Salt Lake City, UT

2015 AIChE Annual Meeting, November 8 – 13, 2015, Salt Lake City, UT, USA 1

Outline

- ✓ Motivation
  - ✓ Saving water and energy for resource-efficient processing
- ✓ Systematic Methods, Tools, Solution Strategies
  - ✓ Simultaneous water and energy integration <u>and synthesis of non-isothermal water networks</u>
- ✓ Applications of Pinch Analysis and Mathematical Programming Methods
  - ✓ Illustrative case studies of different complexities
- ✓ Conclusions
  - ✓ Challenges and further directions

2

Motivation

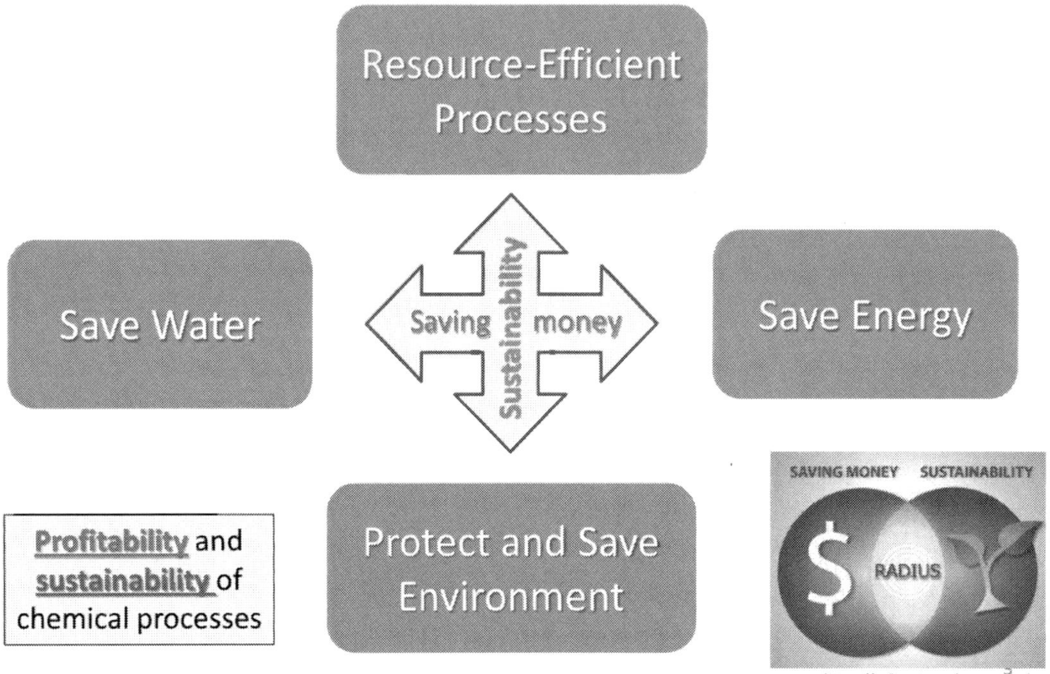

https://safaunt.wordpress.com/

Non-isothermal Water Networks

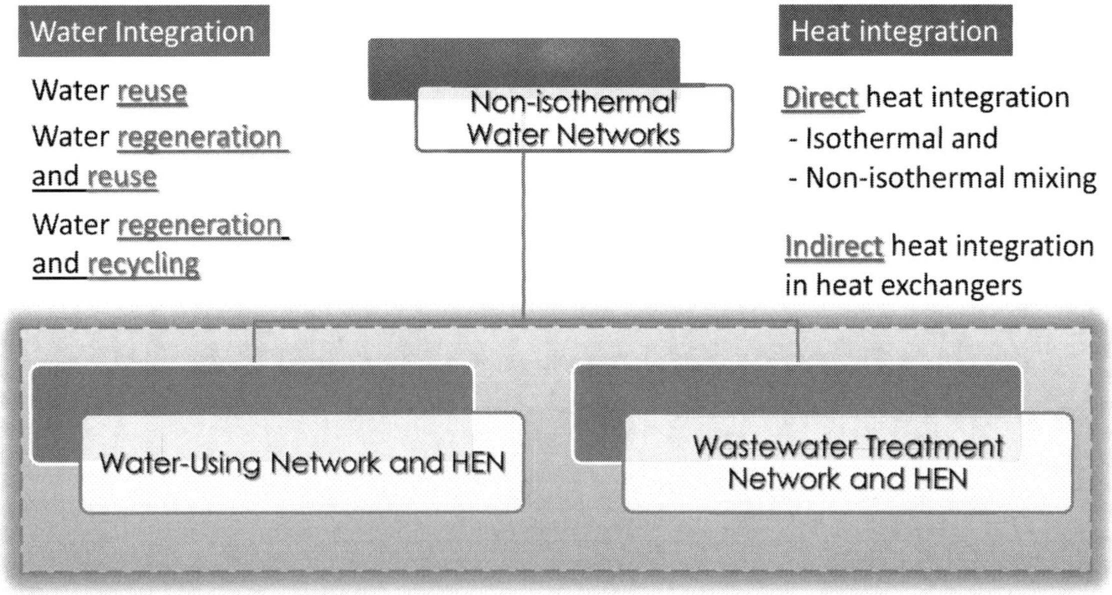

Integrated Water-Using Network, Wastewater Treatment Network and Heat Exchanger Network

4

Classifications of Methods for Non-isothermal Water Networks

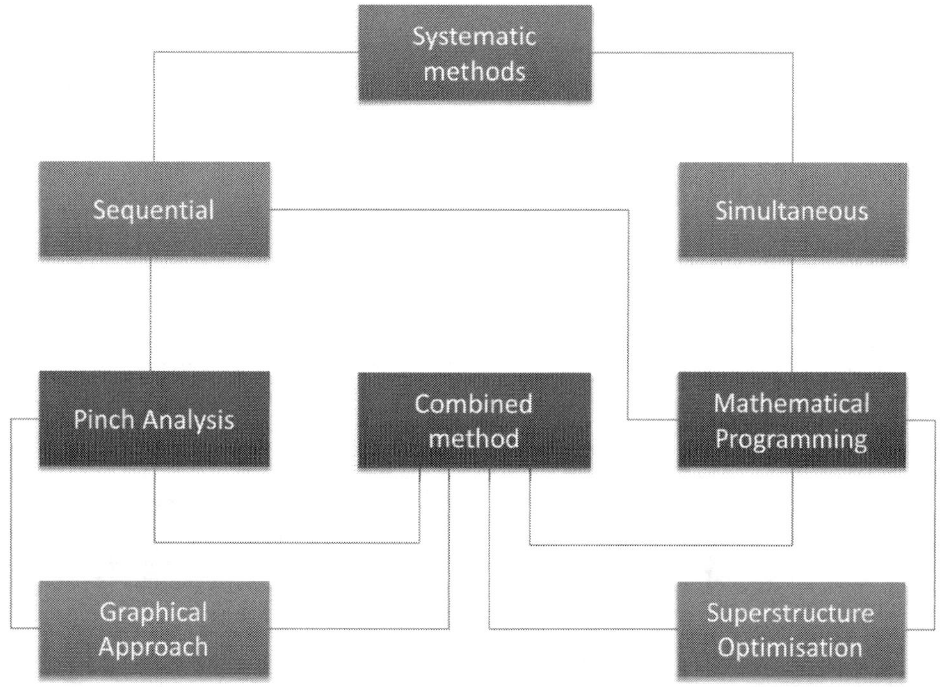

Ahmetović, E., Ibrić, N., Kravanja, Z., & Grossmann, I. E. (2015). Water and energy integration: A comprehensive literature review of non-isothermal water network synthesis. Computers & Chemical Engineering, 82, 144-171.

5

Pinch Analysis - Mathematical Programming

- Graphical solving procedure -> limitation: no restricted matches, single contaminant
- Graphical visualisation
- Heuristics
- **Targeting and Design steps**
- Sequential
- Interactions between WN-HEN can not be fully explored
- Sub-optimal solutions

Physical insights
Engineering experience

- Superstructure optimisation
  - **Superstructure formulation**
  - **Model formulation**
  - **Model solving**
  - **Results analysis**
- Superstructure incorporates all feasible options
- Easily incorporate different constraints
- Sequential and Simultaneous
- Interactions between WN-HEN can be fully explored
- Local vs Global solutions
- Non-convex and non-linear models require specialized solution strategies

Large-scale problems

Combined Approach

6

Methods, Tools, Solution Strategies

WN - Water Network; HEN - Heat Exchanger Network; WTN - Wastewater Treatment Network

Pinch analysis (PA): WN+HEN

- **Water Composite Curves** (Wang and Smith, 1994)
- **Heat Composite Curves** (Linnhoff and Hindmarsh, 1983)
- **Two-Dimensional Grid Diagram** (Savulescu and Smith, 1998)
- **Separate Systems** (Savulescu and Smith 2005)
- **Water Energy Balance Diagram** (Leewongtanawit and Kim, 2009)
- **Source and Demand Energy Composite Curves** (Savulescu et al., 2002)
- **Heat Surplus Diagram** (Manan et al., 2009)
- **Superimposed Mass and Energy Curves** (Wan Alwi et al., 2011)
- **Temperature and Concentration Order Composite Curve** (Hou et al., 2014)

Simultaneous Water and Energy Targeting and WN design → HEN Design

Combined approach: PA → MP; MP → PA

Mathematical programming (MP): WN+HEN, WN+HEN+WTN

- **Different superstructure representations**
- **Heat-integrated water-using network** Bogataj and Bagajewicz, 2008, Ahmetović and Kravanja, 2013)
- **Heat-integrated water-using and wastewater treatment network** (Bogataj and Bagajewicz, 2008, Ahmetović et al., 2014, Ibrić et al., 2014))
- **State-Space representation** (Bagajewicz, 2000, Bagajewicz et al., 2002, Dong et al., 2008)
- **Source-Demand or Source-Sink** (Renard et al., 2012; Kermani et al., 2014)
- **Mathematical models (LP, NLP, MILP, MINLP)**
- **Two-step solution strategies:** initialisation + the overall network design

WN design → HEN Design

Initialisation → HIWN or HIWTN Design

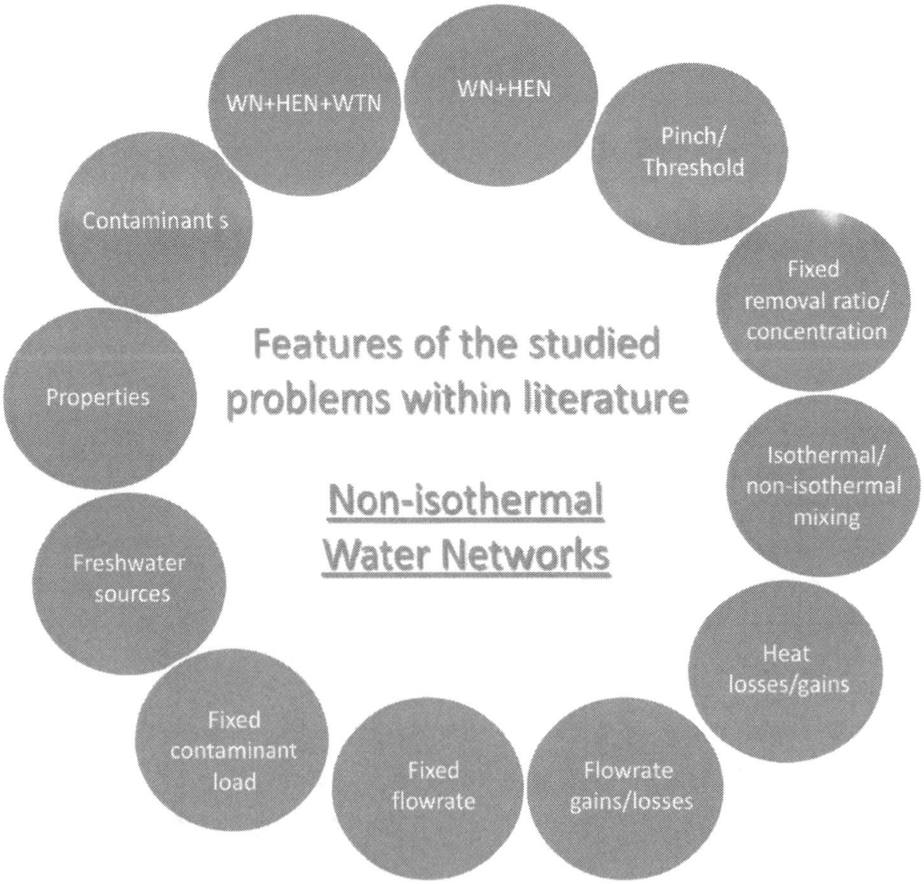

8

Examples of different complexities solved within the literature

Synthesis problem	Single contaminant		Multiple Contaminants	
	Threshold	Pinched	Threshold	Pinched
WN+HEN	3PU+HEN (13) 4PU+HEN (23)* 4PU+HEN (2)* 6PU+HEN (2) 8PU+HEN (4) 15PU+HEN (1)	4PU+HEN (2)	3PU+HEN (6)** 3PU+HEN (2)** 4PU+HEN (3)	10PU+HEN (2)
WN+HEN+WTN		3PU+TU+HEN (4)	4PU+2TU+HEN (1)	2PU+TU+HEN(4) 3PU+TU+HEN (1)*** 3PU+TU+HEN (2)*** 3PU+3TU+HEN (1)

*,**,***Different examples; Number of papers considering an example is given within parenthesis

- More than **80 papers (non-isothermal WNs)** have been published within the period **1997-2015**
- About **17 examples** of **different complexities** have been used for illustration
- **In most cases MP is used** for **multiple contaminant problems** and an overall network consisting of **WN+HEN+WTN** 9

Network designs for Case study 1: 4PU+HEN (PA)

Threshold problem! Single contaminant

C)

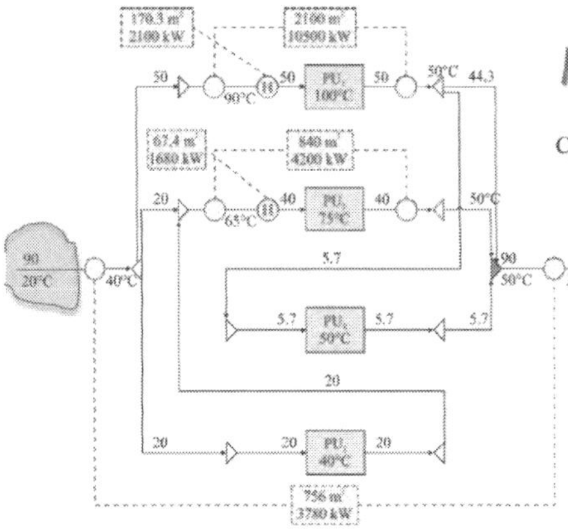

(Hou et al., 2014)
TAC: 2,728,633 $/y

D)

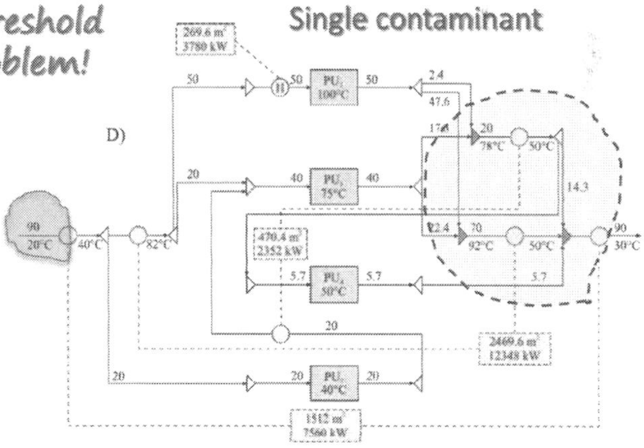

(Martínez-Patiño et al., 2012)
TAC: 2,738,979 $/y

NETWORK CHARACTERISTICS:

Freshwater consumption: 90 kg/s

Same water flowrate through the PUs

Water reused: 25.714 kg/s

5 vs. 4 heat exchangers

Non-isothermal mixing is avoided

Simpler design → Favorable to process control solutions

Network designs for Case study 1: 4PU+HEN (MP)

Single contaminant

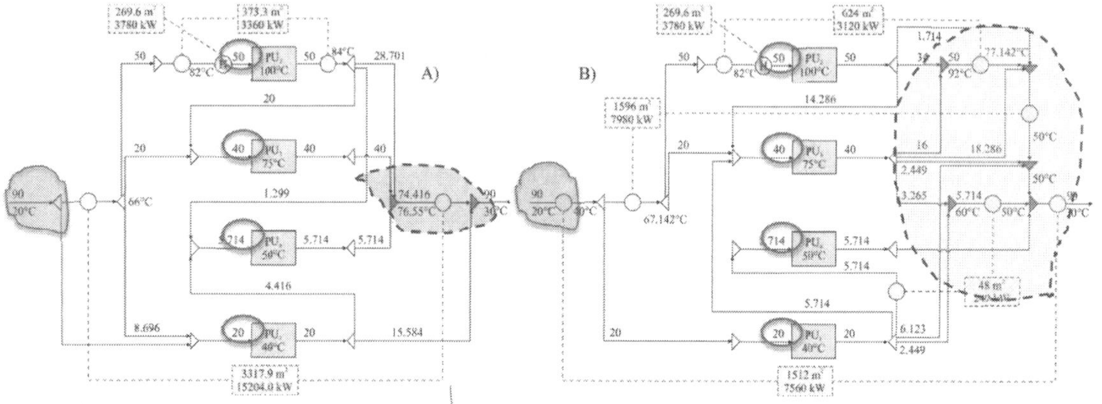

(Ahmetović & Kravanja, 2013) (Dong et al., 2008)

NETWORK CHARACTERISTICS:

Freshwater consumption: 90 kg/s

Same water flowrate through the PUs

Water reused: 25.714 kg/s

2 vs. 5 non-isothermal mixing options
3 vs. 5 heat exchangers

Beter design when comparing <u>network complexity</u> and <u>TAC</u> (2,652,959 vs. 2,738,107 $/y).

11

Network designs for Case study 2: 4PU+HEN (MP) (simultaneous approach)

Multiple contaminants

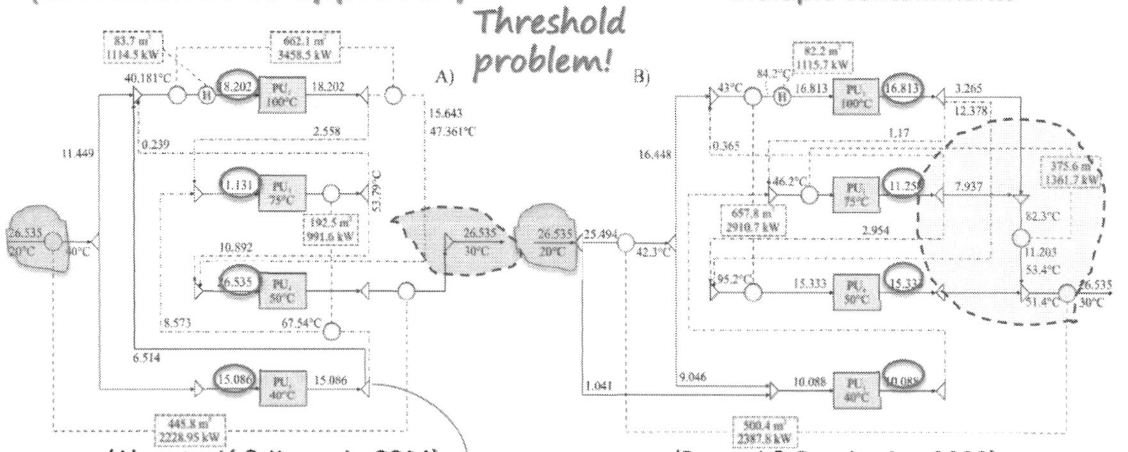

(Ahmetović & Kravanja, 2014) (Bogataj & Bagajewicz, 2008)

NETWORK CHARACTERISTICS:

Freshwater consumption: 26.535 kg/s
Different flowrates through the PUs

Similar water reuse options with different flowrates

Simpler design with minimum TAC

TAC: 889,772 vs. 907,065.7 $/y

Same freshwater and utilities consumption with different HEN designs!

Network design for Case study 2: 4PU+HEN (MP)

(sequential approach)

Multiple contaminants

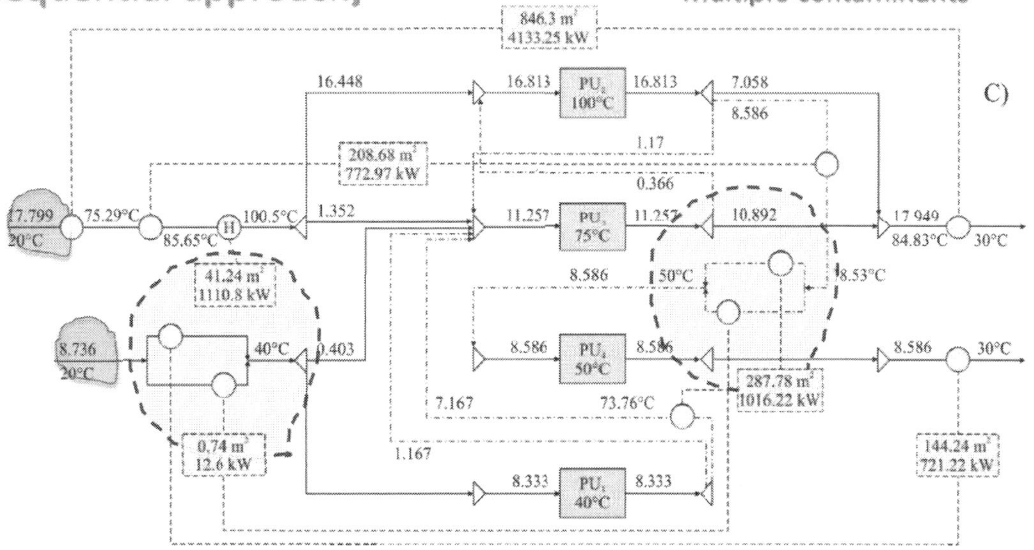

NETWORK CHARACTERISTICS:

(Ibrić et al., 2012)

Freshwater consumption: 26.535 kg/s

Streams splitting within HEN.

6 **heat exchangers** (sequential approach) **vs. 4 heat exchangers** (simultaneous approach)

TAC: 928,930 vs. 889,772 $/y

Network design for Case Study 3*: 8PU+HEN (MP)

Single contaminant

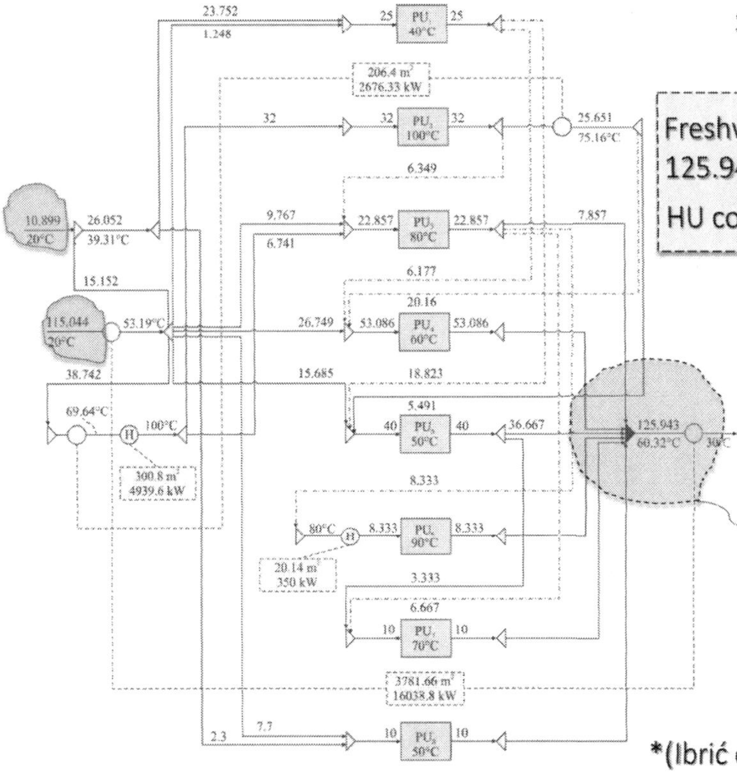

Freshwater consumption: 125.945 kg/s

HU consumption 5289.6 kW

Minimum freshwater and utility consumption. *TARGET.*

One mixing point of wastewater streams.

TAC: 3,628,022 $/y

*(Ibrić et al., 2014c)

Network design for Case study 3*: 8PU+HEN (MP)

Single contaminant

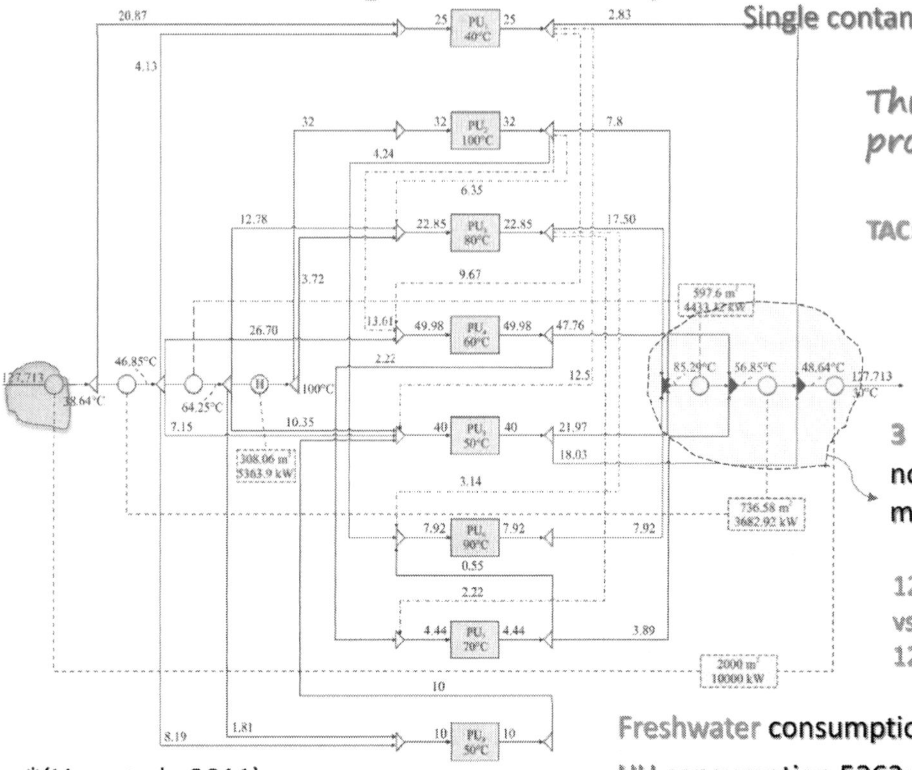

Threshold problem!

TAC: 3,628,022 **vs. 3,704,246 $/y**

3 vs. 1 non-isothermal mixing options

127.713 kg/s vs. 125.945 kg/s

Freshwater **consumption: 127.713 kg/s**

HU **consumption 5363.9 kW**

*(Liao et al., 2011)

Network designs for Case Study 4: 3PU+<u>TU</u>+HEN (MP)

Single contaminant

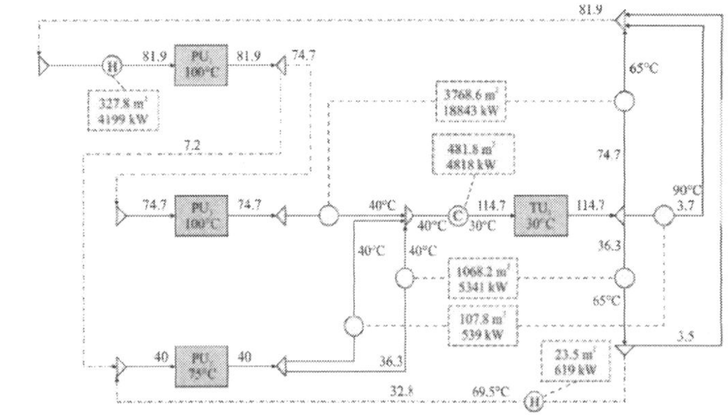

(Dong et al., 2008)

HEN investment: 410,069.6 $/y

6 heat exchangers

A)

TAC: 6,555,774 $/y

Zero Liquid Discharge!

HU and CU: **4818.95 KW**

Pinched problem!

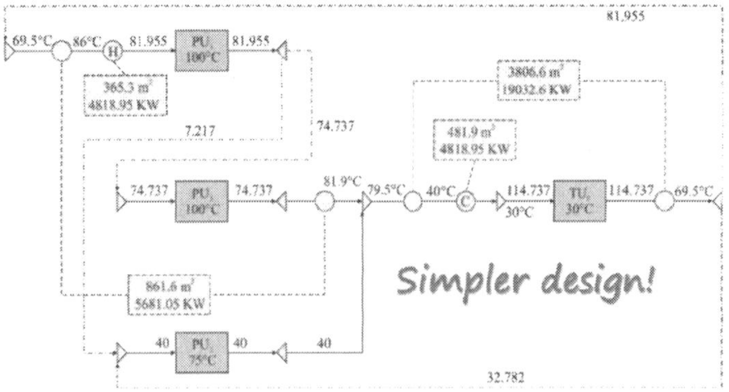

Simpler design!

HEN investment: 360,336.6 $/y

4 heat exchangers

B)

TAC: 6,506,179 $/y

(Ahmetović et al., 2014)

(Ibrić et al., 2014a)

Network design for Case study 5: 3PU+2TU+HEN (MP)

Multiple contaminants

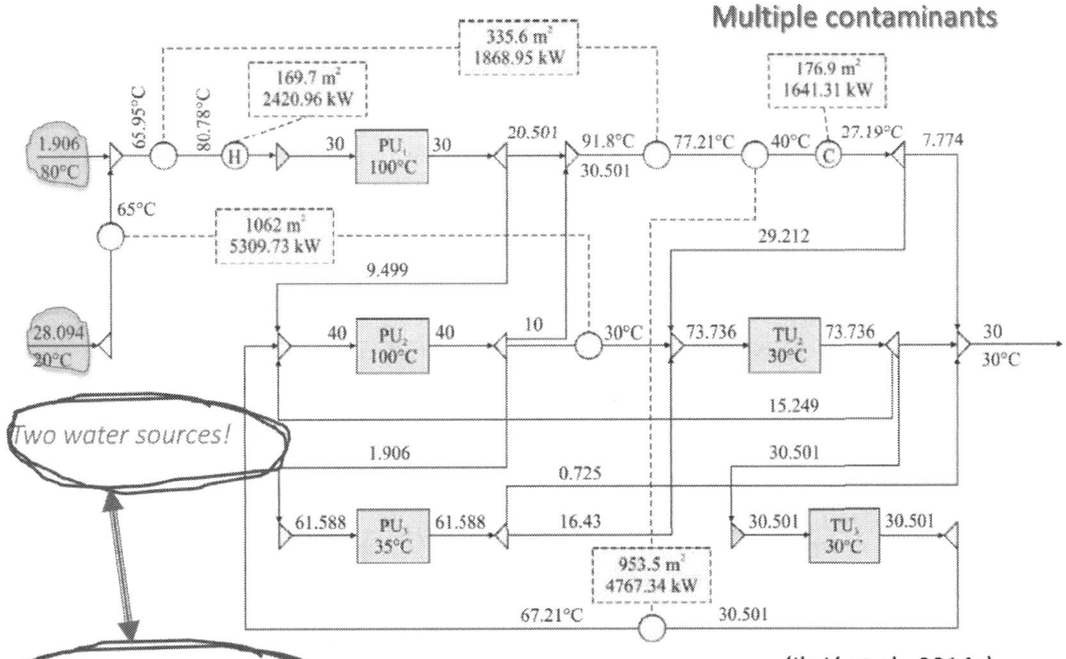

(Ibrić et al., 2014c)

HU: 2420.96 KW
CU: 1641.31 kW

Pinched problem!

TAC: 1,911,577.5 $/y

Conclusions

- ✓ Case studies of different complexities were analysed
- ✓ In most cases minimum fresh water and energy consumption (TARGET) were obtained by using both methods, PA and MP, but with different WN and HEN designs
- ✓ PA method combined with engineering insights can provide **simple** designs but obtained solutions are sub-optimal
- ✓ PA method is limited in the cases of considering the overall non-isothermal networks including multiple contaminants, water-using and wastewater treatment units.
- ✓ However, MP can succesfully address these issues, and it has been demonstrated in this presentation
- ✓ MP allows to solve the complete problem in a systematic and simultaneous way considering the a more detailed problem formulation with more detailed constraints definition.

18

Challenges and further directions

- ✓ New MP solution strategies, pinch tools or their combination are required for solving large-scale non-isothermal WNs, and it could be possible further directions within this field
- ✓ In recent years research progress has been identified considering mass, energy and property integration, and interplant non-isothermal water-using and wastewater treatment networks
- ✓ The synthesis of a combined network consisting of a process network, a water network, and a heat exchanger network could be possible further research direction in order to obtain sustainable resource utilisation, minimizing operating costs (increasing profitability), and emissions reduction (environmental protection)
- ✓ Computational efficiency is and will remain one of the main challenges as non-isothermal WN problems are NP-hard and obtaining global optimal solution is another important challenge for PSE society to master

19

Elvis Ahmetović

Zdravko Kravanja

François Maréchal

Nidret Ibrić

Maziar Kermani

Univerza v Mariboru
Fakulteta za kemijo in kemijsko tehnologijo

ÉCOLE POLYTECHNIQUE FÉDÉRALE DE LAUSANNE

Acknowledgments

FNSNF
SWISS NATIONAL SCIENCE FOUNDATION

The authors are grateful to the SCOPES 2013-2016 (Scientific Co-operation between Eastern Europe and Switzerland) joint research project (CAPE-EWWR: IZ73Z0_152622/1) for financial support as well as support from the Slovenian Research Agency (Program No. P2-0032).

20

nandri
taiku
mercé
sobodi
diolch
asante
tapadh
buznyg
aabhar
nanni
gracie
singuila
tãnan
agüyjé
ngiyabonga
meharbani
suipáy
kettu'i
spas
akpé
nagode
kurre
dankie
obrigado
mahalo
imela
merci
vinaka
welalin
tanmirt
xièxie
bayarlalaa
tanemirt
tenki
faleminderit
rahmat
chnorakaloutioun
wei
dhanyavadagalu
tumas
dhanyabaad
dziękuję
kiitos
miigwetch
ďakujem
yekeniele
akiba
enkosi
sağol
mèsi
dankewol
hvala
marahaba
barka
mauruuru
shukriya
tualumba
som
stuutiyi
dhanyavad
danke
matondo
matondi
chokrane
rahmet
gratzias
gracias
arigatô
nouari
tack
grazie
dank
motashakkeram
bedankt
choukran
tangio
dhanyavadalu
paldies
takk
salamat
thank
pilamaya
leat
sukriya
trugéré
murakoze
merkzi
děkuji
tau
mulţumesc
misaotra
tashakor
sumanga
you
fofo
wopila
sahite
danki
dankon
zikomo
manana
waita
wado

21

436315 Optimization of a Dividing Wall Batch and Semi-Batch Reactive Distillation Column for the Production of Methyl Acetate: Potential for Energy Savings

Abstract:

Batch distillation processes rely on dynamic models. Over the last three decades rigorous algorithms have been developed for different types of Batch Distillation Columns (BC), including Reactive Batch Columns, but there is still a lack of understanding about Dividing Wall Batch Reactive Distillation Columns (DWBC). In fact, no rigorous model has been reported for their solution. Reactive Batch Distillation Columns (RBDC), as Continuous Distillation Columns (CDC), are characterized by high energy consumption. An innovative solution to overcome this problem is the use of Dividing Wall Columns (DWC). The Petlyuk configuration, consisting of two coupled distillation columns [1], evolved into a more practical application by adding a vertical wall that splits the column into two separate sections. Thanks to this flexibility, DWC has found great appeal in the chemical process industry as it can separate more components in a single distillation unit thus achieving cost savings by requiring single columns instead of two and by a decrease in operation costs using a single condenser and reboiler. In fact, the use of DWC in Continuous Distillation Columns can save up to 30% in capital investment and up to 40% in operating costs [2].

In this study a model and solution strategies are proposed for the optimization of a Dividing Wall Batch and Semi-Batch Reactive Distillation Column. A profit maximization involving methanol esterification reaction for the production of methyl acetate (as the main product) is proposed. In order to determine the potential benefits, such like energy savings and reduction of batch time we propose a dynamic rigorous model that involves tray-by-tray calculations when a dividing wall and side feeds are added to the batch column for the time varying profiles. We also investigate how the distillate product composition varies when different reaction zones and different positions of the dividing wall are included in the model. The optimization variables, considered to achieve a given product specification, are batch time, vapor flowrate and reflux ratio as the control variable. The dynamic optimization problem is converted into an NLP problem by using the oriented equation method proposed by Biegler [3], finite elements and collocation points implemented in GAMS (General Algebraic Modeling System, 24.2.2). The Control Vector Parametization approach proposed by Pantelides [4,5] is solved by using gPROMS (general PROcess Modeling System, 2004) which makes use of Succesive Quadratic Programming (SQP) method.

References

[1] Slavinskii D. M. Petlyuk F. B., Platonov V. M. Thermodynamically optimal method for separating multicomponente mixtures. *Int. Chemical Engineering*, pages 555–561, 1965.

[2] Draghiciu L. Isopescu R.,Woinaroschy A. Energy reduction in a divided wall distillation. *Rev. Chim.*, 59:812–815, 2008.

[3] Biegler L. T. Solution of dynamic optimization problems by succesive quadratic programming and orthogonal collocation. *Comput. Chem. Eng.*, 8:243–248, 1984.

[4] Pantelides C. C. Vassiliadis V. S., Sargent R. W. H. Solution of a class of multistage dynamic optimization problems. Part I - Algorithmic framework. *Ind. Eng. Chem. Res.*, 33:2115–2123, 1994.

[5] Pantelides C. C. Vassiliadis V. S., Sargent R. W. H. Solution of a class of multistage dynamic optimization problems. Part II - Problems with path constraints. *Ind. Eng. Chem. Res.*, 33:2123–2133, 1994.

434619 A New Robust Scenario Approach to Supply Chain Optimization Under Uncertainty

Monday, November 9, 2015: 3:15 PM
Salon G (Salt Lake Marriott Downtown at City Creek)
Niaz Chowdhury and **Xiang Li**, Chemical Engineering, Queen's University, Kingston, ON, Canada

A new robust scenario approach to supply chain optimization under uncertainty

Niaz Chowdhury, Xiang Li*

Department of Chemical Engineering, Queen's University,

19 Division Street, Kingston, ON, K7L 3N6, Canada

The following two-stage stochastic programming models [1] are often used for supply chain optimization under uncertainty:

$$\min_{x} c^T x + E_{\xi\in\Xi}\left[Q(x,\xi)\right] \qquad (1a)$$

$$\text{s.t.} \quad Ax \le b, \qquad (1b)$$

where the first-stage variables are represented by x, the second stage cost for uncertainty realization $\xi(\in\Xi)$ is represented by,

$$Q(x,\xi) = \min\{(\alpha_0\xi+\beta_0)^T y : (\alpha_{i,1}\xi+\beta_{i,1})^T x + w_i^T y \le 0,\ i=1,\cdots,m\},$$

where α_0, β_0, $\alpha_{i,1}$, $\beta_{i,1}$, w_i are known constants. This two-stage model is often approximated by the following scenario formulation for tractable solution [1]:

$$\min_{x,y_1,\dots,y_s} c^T x + \sum_{\omega=1}^{s} p_\omega (\alpha_0\xi_\omega+\beta_0)^T y_\omega \qquad (2a)$$

$$\text{s.t.} \quad Ax \le b, \qquad (2b)$$

$$(\alpha_{i,1}\xi_\omega+\beta_{i,1})^T x + w_i^T y_\omega \le 0, \quad i=1,\cdots,m,\ \omega=1,\cdots,s, \qquad (2c)$$

where $\xi_1,\cdots,\xi_s \in \Xi$ are sampled uncertainty realizations, called scenarios, and $p_1,\cdots,p_s$ are the probabilities associated with the scenarios. The major shortcoming of the scenario formulation is that, a large number of scenarios are usually needed, to ensure that the solution is feasible for the original two-stage formulation and that the predicted expected cost is close to the true expected cost. To overcome the shortcoming of the scenario formulation, a robust scenario formulation has been proposed [2-3] based on partitioning the uncertainty region Ξ rather than sampling a finite number of uncertainty realizations from it. The resulting formulation is

$$\min_{x, U_\omega, v_\omega} c^T x + \sum_{\omega=1}^{s} p_\omega (\alpha_\omega \bar{\xi}_\omega + \beta_\omega)^T (U_\omega \bar{\xi}_\omega + v_\omega) \tag{3a}$$

$$\text{s.t.} \quad Ax \le b, \tag{3b}$$

$$\max_{\xi_\omega \in \Xi_\omega} \{(\alpha_\omega \xi_\omega + \beta_\omega)^T x + w_\omega^T (U_\omega \xi_\omega + v_\omega)\} \le 0, \quad \omega = 1, \cdots, s. \tag{3c}$$

where scenario ω is associated with uncertainty subregion Ξ_ω rather than a uncertainty realization and $\prod_{\omega=1}^{s} \Xi_\omega \supset \Xi$ is required to ensure that the solution to Formulation (3) is feasible for the original problem. The second stage decision for scenario ω is assumed to be an affine function of the uncertainty realization in Ξ_ω, i.e., $y_\omega = U_\omega \xi_\omega + v_\omega$. As a result, the coefficients of the affine function (U_ω, v_ω) rather than y_ω serve as the second-stage variables in the optimization problem.

The construction of uncertainty subregions and the reformulation of Formulation (3) into a tractable deterministic optimization problem are the two key steps for applying the robust scenario method. In previous work [2-3], the uncertainty region bounded by the infinity-norm (i.e., box uncertainty) is considered, for which the construction of uncertainty subregions is straightforward (see Figure 1(a)). In addition, in this case the uncertainty subregions are also bounded by the infinity-norm, so Formulation (3) can be equivalently transformed into a deterministic linear programming problem. However, when the original uncertainty region has an arbitrary shape, it is not clear how to effectively partition the region to yield a set of uncertainty subregions such that Formulation (3) can be transformed into a tractable deterministic optimization problem (see Figure 1(b)).

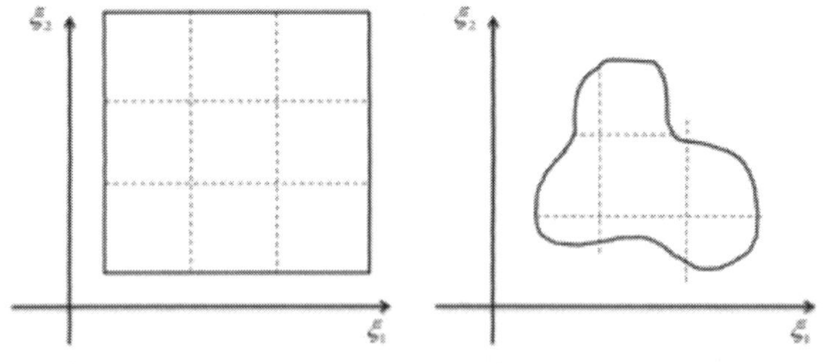

(a) Partition of a box region is straightforward. (b) It is not obvious how to effectively partition an arbitrary region.

Figure 1. Partition of uncertainty regions.

In this paper, we propose a novel robust scenario approach that constructs a sequence of box uncertainty subregions that over-estimate the original uncertainty region, and a sequence of box uncertainty subregions that under-estimate the original uncertainty region (as illustrated in Figure 2). The robust scenario formulations with the over- and under-estimation of uncertainty region can both be transformed into deterministic linear programming problems; when the number of scenarios (i.e., uncertainty subregions) increase, the over- and under-estimates of the uncertainty region become close and the optimal values of the two robust scenario formulations converge, leading to a good solution for the original two-stage stochastic programming problem. Case study of a real industrial supply chain optimization problem with ellipsoidal demand uncertainty demonstrates the advantages of the proposed robust scenario approach.

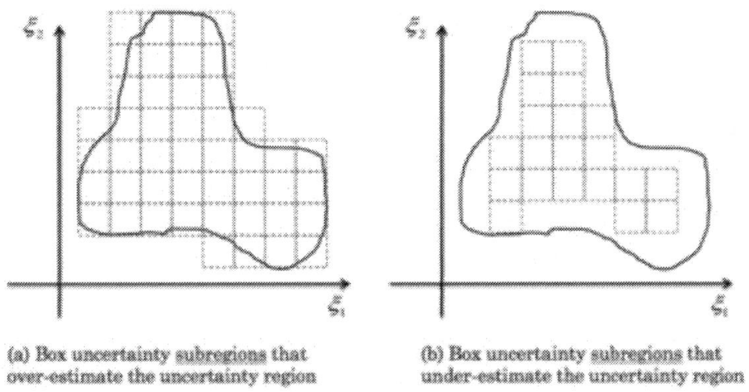

Figure 2. Over- and under-estimation of uncertainty region

References

[1] Birge, J. R.; Louveaux, F. *Introduction to stochastic programming*, Second Edition; Springer: New York, 2011.

[2] McLean, K.; Li, X. "Robust scenario formulations for strategic supply chain optimization under uncertainty", *Industrial & Engineering Chemistry Research*, 52, 5721-5734.

[3] McLean, K.; Ogbe, E.; Li, X. "Novel formulation and efficient solution strategy for strategic optimization of an industrial chemical supply chain under demand uncertainty", *Canadian Journal of Chemical Engineering*, DOI 10.1002/cjce.22173, 2015.

414106 Supply Chain Optimization Tools for the Strategic Planning of Biorefining Systems

Abstract:

A crucial element of biobased economies is the development of economically sustainable biorefining systems enabling a full exploitation of biomass (and its macrocomponents, i.e. cellulose, hemicellulose and lignin) for the production of chemicals, materials and energy. The huge variety of biomass processing paths in second generation biorefineries leads to complex product portfolios indicating the necessity of maximizing the value derived from biomass feedstock when treated in a biorefinery with the aim to create economically feasible and environmentally sustainable biobased production systems (Kokossis and Yang, 2010). One of the most important and challenging aspects in the quest of producing a set of more sustainable biobased products, is the design of a more profitable, better integrated and more sustainable biorefinery supply chain network (Akgul et al., 2011; Martín and Grossmann, 2010; Čuček et al., 2014). The biomass pretreatment, which fractionates biomass into its three macrocomponents, usually represents the highest costing part of the entire biorefining system (CIMV, 2015). Optimization tools could play a decisive role supporting the decision making process in new biorefinery systems, through the performance of sensitivity analysis, the assessment of technical or economic uncertainties as well as the identification of the major cost drivers (Kim et al., 2013).

Supply chain optimization has emphasized the integration of all the existing material and information flows between the supply chain nodes (i.e. biomass production, storage, transportation, pretreatment, conversion as well as product storage and transportation), while dealing successfully with uncertainties that usually exist in the biobased market (Sharma et al., 2013; De Meyer et al., 2014).

Complex bio-based production systems can also be handled in a powerful way with technology superstructure models. These consist of synthesis blocks that for given objective functions (e.g. profit or net present value maximization, cost or required energy or greenhouse gas emissions minimization, etc.) provide a holistic approach to the design of optimal configurations of the examined production network (Tsakalova et al., 2011; Yuan and Chen, 2012).

This work focuses on the integration of a technology superstructure with a lignocellulosic supply chain model, formulated as a mixed-integer linear programming (MILP) model and applied to an Organosolv-based biorefinery in the South-West of Hungary. The model represents a multi-echelon, multi-feedstock and multi-period supply chain model, aiming at determining the configuration of the biomass network to supply a biorefinery with the necessary biomass feedstock, maximizing the overall profit of the system and at the same time fulfilling targeted regional bioproduct demand over the entire planning horizon, including uncertainties on key parameters. Spatially-explicit features are also embedded in the model formulation, expressing the geographical availability of biomass crops and transportation links. This methodology enables a systematic design of biorefining systems as well as the evaluation of key performance indicators (KPIs) (e.g. costs, environmental impacts). GAMS® is the software tool used for this optimization task (Rosenthal, 2014).

Based on previous works (Hugo and Pistikopoulos, 2005; Zamboni et al., 2009; Giarola et al., 2011), the problem is formulated assuming as given: sets of candidate plants using known technologies (i.e. Organosolv, saccharification and co-fermentation), potential geographical sites for the plants location, available transportation modes (i.e. trucks), spatially-explicit availabilities of the raw material and a regional demand for a set of products. The planning and operational decision variables will be optimized to identify and locate the cultivation and storage sites, the production facilities, the feedstock mix supplied to the selected plants and the optimal logistics.

The biomass market features, considering 3 different types of biomass feedstock (i.e. winter wheat straw, winter barley straw and corn stover), alongside its seasonal and geographical availability, assuming 4 months of available harvest (i.e. July, August, September and October), the technical and economic characteristics of pretreatment and conversion processes as well as competitive issues are the main input elements of the proposed MILP model (BIOCORE Final Report, 2014; NOVA, 2014; Humbird et al., 2011). The examined Hungarian region, for better data accuracy, is discretised into 102 cells with a surface of 225 km^2 each, focusing on the arable land and its factors (i.e. arable land distribution and cereal straw availabilities per cell).

A single economic objective function is considered, aiming at maximizing the total profit of the whole network, subject to production, storage, transportation and demand constraints, over the entire planning horizon, which is defined to be 1 year, divided into 12 months. Subsequently, the total profit is calculated from the annual revenues and costs, where the latter term includes both annualized capital (i.e. installation investments of selected plants) and operational costs (i.e. biomass purchase, storage, transport, processing as well as product storage and transportation).In addition, the supply chain behavior is captured through the definition of both biomass and product mass balances, alongside logical constraints that have to be satisfied at each of the supply chain nodes.

The model embedding the definition of the process technology superstructure for a portfolio of selected biobased products (e.g. cellulose, hemicellulose, lignin, ethanol), will enable the analysis of the key sources of uncertainty in the planning of biorefining systems (e.g. capital and operating costs, technological yields) and their evolution over time.

Acknowledgement

The EC (Reneseng-607415 FP7-PEOPLE-2013-ITN) is gratefully acknowledged for supporting this work.

References

Akgul, O., Shah, N., & Papageorgiou, L., G. (2011). ESCAPE21 Journal, 1799-1803.

BIOCORE Final Report (2014):

http://www.biocore-europe.org/file/BIOCORE%20final%20report%281%29.pdf

CIMV (2015). http://cimv.fr/research-development/cimv-research-development.html

Čuček, L., Martín, M., Grossmann, I., E., Kravanja, Z. (2014). Computers and Chemical Engineering, 66, 57-70.

De Meyer, A., Cattrysse, D., Rasinmäki, J., & Van Orshoven, J. (2014). Renewable and Sustainable Energy Reviews, 31, 657-670.

Giarola, S., Zamboni, A., & Bezzo, F. (2011). Computers and Chemical Engineering, 35, 1782-1797.

Hugo, A., & Pistikopoulos, E. (2005). Journal of Cleaner Production, 13, 1428-1448.

Humbird, D., Davis, R., Tao, L., Kinchin, C., Hsu, D., & Aden, A. (2011). Process Design and Economics for Biochemical Conversion of Lignocellulosic Biomass to Ethanol, NREL Report, Golden Colorado, USA.

Kim, J., Sen, S., M., & Maravelias, C., T. (2013). Energy Environ. Sci., 6, 1093-1104.

Kokossis, A., C., & Yang, A. (2010). Computers and Chemical Engineering, 34, 1397-1405.

Martín, M., & Grossmann, I., E. (2010). ESCAPE20 Journal, 943-948.

NOVA (2014). NOVA paper #5 on bio-based economy 2014-11 (available at: www.bio-based.eu/markets).

Rosenthal (2014). GAMS – A User's Guide. GAMS Development Corporation, Washington, DC, USA.

Sharma, B., Ingalls, R., G., Jones, C., L., & Khanchi, A. (2013). Renewable and Sustainable Energy Reviews, 24, 608-627.

425180 Solution Approaches for Large-Scale Shale Gas Supply Chain Optimization Problems

Abstract:

Shale gas has, in recent years, emerged as a possible transition fuel that will facilitate the transformation from coal-based energy systems to less CO_2intensive energy systems [1], [2]. However, environmental impacts, especially impact on water resources, have given rise to questions and concerns and thus must be addressed in making decision related to shale development [1]–[3]. Given the cost-benefit tradeoffs , the assessment of shale gas resources has captured the attention of the research community [4]–[9] .

The planning and development of shale gas supply chains involves a number of strategic decisions that must be made to optimize the economic performance of such an investment. Three key aspects must be considered in seeking to optimize the exploitation of shale resources: (1) production schemes, which involve drilling strategy, well design, and well clustering (well-pads); (2) water management, which includes decisions such as fresh water procurement, produced water treatment (plant locations and capacities) and disposal options; and finally (3) gas processing, which encompasses gas treatment facilities decisions (plant locations and capacities), and the design of gas pipeline networks. Certainly, the simultaneous consideration of the aforementioned aspects provides substantial modeling and computation challenges. Some studies have focused mainly in the assessment of shale gas production with little or no detail concerning the gas and water transport and processing infrastructure [4]–[6], [8]. Others have primarily focused on the production schemes [7]. However, recently, Guerra et al. (2015) developed a comprehensive integrated shale gas supply chain optimization model that takes into account the different options regarding well-pad design, gas and water transport and the processing infrastructure. In its general form, the resulting optimization model is a Mixed-Integer Non-Linear Programming (MINLP) problem. However, a special case can be derived by assuming that the gas composition is constant across the shale reservoir; this simplification reduces the complexity of the model to a Mixed-Integer Linear Programming (MILP) problem. However, the general problem is an MINLP and as shown by the authors, solution of this problem using commercially available MILP/MINLP solvers can take substantial computation times or require the use of specialized solution strategies that exploit the structure and characteristics of the problem.

In this work, we explore different solution approaches for solving the large-scale shale gas supply chain optimization problem. The solution strategies employ three different devices used in the math programming literature: (1) Model reformulation, where the focus is on reducing the complexity of the model while retaining its accuracy, (2) computational enhancements, such as the use of tighter upper and lower bounds, column and row scaling, and specially structured integer cuts and finally (3) convexification and decomposition techniques. The effectiveness of these strategies are investigated using a range of case studies involving formulations constructed using data from Colombia.

Corresponding author: G.V. Reklaitis (reklaiti@purdue.edu).

References

[1] B. G. Rahm and S. J. Riha, "Evolving shale gas management: water resource risks, impacts, and lessons learned.," *Environ. Sci. Process. Impacts*, vol. 16, no. 6, pp. 1400–12, May 2014.

[2] R. D. Vidic, S. L. Brantley, J. M. Vandenbossche, D. Yoxtheimer, and J. D. Abad, "Impact of shale gas development on regional water quality.," *Science*, vol. 340, no. 6134, p. 1235009, May 2013.

[3] S. M. Olmstead, L. A. Muehlenbachs, J. Shih, Z. Chu, and A. J. Krupnick, "Shale gas development impacts on surface water quality in Pennsylvania.," *Proc. Natl. Acad. Sci. U. S. A.*, vol. 110, no. 13, pp. 4962–7, Mar. 2013.

[4] M. J. Kaiser, "Haynesville shale play economic analysis," *J. Pet. Sci. Eng.*, vol. 82–83, pp. 75–89, Feb. 2012.

[5] M. J. Kaiser, "Profitability assessment of Haynesville shale gas wells," *Energy*, vol. 38, no. 1, pp. 315–330, Feb. 2012.

[6] R. Weijermars, "Economic appraisal of shale gas plays in Continental Europe," *Appl. Energy*, vol. 106, pp. 100–115, Jun. 2013.

[7] K. C. Wilson and L. J. Durlofsky, "Optimization of shale gas field development using direct search techniques and reduced-physics models," *J. Pet. Sci. Eng.*, vol. 108, pp. 304–315, Aug. 2013.

[8] A. J. Calderón, O. J. Guerra, L. G. Papageorgiou, J. J. Siirola, and G. V. Reklaitis, "Preliminary Evaluation of Shale Gas Reservoirs : Appraisal of Different Well-pad Designs via Performance Metrics," *Ind. Eng. Chem. Res.*(submitted), 2015.

[9] O. J. Guerra, A. J. Calderón, L. G. Papageorgiou, J. J. Siirola, and G. V Reklaitis, "An Optimization Framework for the Assessment of Shale Gas Resources : Integration of Water Management and Shale Gas Supply Chain Design," *Comput. Chem. Eng.*(to be submitted), 2015.

427147 Exact Optimization Frameworks for Time-Consistent Distribution

Abstract:

In the context of optimizing distribution operations, the recent trend calls for shifting the focus from merely managing the distribution fleet towards a more explicit focus on improving service quality. Moreover, it is relatively common in practice for the distributor to provide repetitive service to customers along a multi-period time horizon. In such settings, a distributor may add value to its business by committing to consistent delivery times across the planning horizon. Doing so exposes efficiencies that reduce costs for both the distributor and the customer (e.g., by reducing the need for committing loading-dock resources) and can give the service provider significant competitive advantages. Several applications adopt such consistent service policies, including the Vendor-Managed Inventory distribution schemes that are prevalent in the chemical industries.

All previous approaches to address time-consistent vehicle routing problems [1-3] have been heuristic in nature, without providing any mathematical guarantee on solution quality. In this work, we present two exact solution approaches to address such a problem. More specifically, we study the so-called Consistent Traveling Salesman Problem (ConTSP) [4] in which one aims to design minimum-cost (or, minimum makespan) routes over a finite, multi-period horizon so as to serve a set of customers with known demands and service durations using a single vehicle. Customers may require service in only one time period (typically, residential or small customers) or over multiple periods (typically, commercial or large customers). In general, since a customer may or may not require service in a given time period, only a subset of customers need to be visited in each period. The consistency requirement applies to every customer that requires service in more than one time period, i.e., every such customer must be visited at roughly the same time in each period where service is required. This is formally imposed by requiring that the difference between the earliest and latest vehicle arrival times (along the time horizon) at each customer location does not exceed a limit. In practice, this limit may be set by either the customer or by the service provider.

Our first approach is based on a branch-and-cut framework. In this approach, we develop three alternative MILP formulations, and compare their performance when serving as the basis for a branch-and-cut solution algorithm. To this end, we develop a new family of cutting planes for the ConTSP, which we call "Inconsistent Path Elimination Constraints," and we describe an effective way to separate them in polynomial time. We also discuss how TSP cutting planes [5] can be used in this context to complement the newly proposed cuts, but we show via polyhedral analysis that, unlike the case of TSP, the TSP cutting planes do not constitute facets of the ConTSP polytope. A numerical study on a comprehensive collection of benchmark instances, which we developed by extending well-known TSPLIB datasets, shows that a formulation involving only binary arc variables and an exponential number of constraints is the computationally most efficient MILP model.

Our second approach is based on a decomposition of the multi-period problem into a sequence of single-period routing problems. More specifically, we describe a scheme to decompose the problem into single period TSPs with arrival time windows for each customer. Consistency is enforced implicitly by dynamically tightening these time windows in a branch-and-bound implementation with separable branching rules. A numerical study shows that this approach is computationally competitive to the branch-and-cut framework, while in addition, it is able to accommodate waiting times at each customer location.

Using our approaches, we have been able to solve to guaranteed optimality problems with 100 customers requiring service over a 3-period horizon, problems with 75 customers over a 5-period horizon and problems with 50 customers over a 7-period horizon. Furthermore, we estimate the "price of consistency," which is the additional cost that a distributor must incur, on average, in order to implement consistent delivery schedules. Our results show that often the additional costs are modest and, therefore, consistency of service is a value proposition that distributors should look into further.

References:

[1] C. Groër, B. Golden, and E. Wasil (2009). The Consistent Vehicle Routing Problem. *Manufacturing & Service Operations Management*, 11(4):630–643.

[2] C. D. Tarantilis, F. Stavropoulou, and P. P. Repoussis (2012). A template-based Tabu Search algorithm for the Consistent Vehicle Routing Problem. *Expert Systems with Applications*, 39(4):4233–4239.

[3] A. A. Kovacs, B. L. Golden, R. F. Hartl and S. N. Parragh (2014). Vehicle routing problems in which consistency considerations are important: A survey. *Networks*, 64(3):192–213.

[4] A. Subramanyam, C. E. Gounaris (2015). A Branch-and-Cut Framework for the Consistent Traveling Salesman Problem. Under review.

[5] D. L. Applegate, R. E. Bixby, V. Chvatal, and W. J. Cook (2007). The Traveling Salesman Problem: A Computational Study (Princeton Series in Applied Mathematics). *Princeton University Press*, USA.

427807 Multi-Criteria Decision Making Supplier Selection and Auction Based Procurement in Supply Chain Management

Abstract:

Multi-Criteria Decision Making Supplier Selection and Auction based Procurement in Supply Chain Management

Nihar Sahay and Marianthi Ierapetritou

Department of Chemical and Biochemical Engineering, Rutgers University, Piscataway, NJ

In a multi-enterprise supply chain network, the objective of each enterprise is to maximize its own objectives. Model development for such supply chain networks needs to incorporate the interactions among the different enterprises in order to assess their impact on supply chain operations. In this work, we study a multi-enterprise supply chain network where manufacturing sites and warehouses belong to a particular enterprise while raw material suppliers and retailers belong to different enterprises. In particular, the procurement of raw materials from suppliers and the distribution of products from warehouses to the retailers are studied in this work.

Over the recent years, advances in information technology have reformed the practice of supply chain operations. High-speed communication and better connectivity provide the opportunity for enterprises to participate in Internet based marketplaces. E-marketplaces offer benefits like providing a medium to the companies for developing their pricing policies, dynamically deciding the optimal mix of products, and allocating the correct capacities. Compared to the Business-to-Consumer (B2C) front, the adoption of e-marketplaces on the Business-to-Business (B2B) front has been slow and limited as B2B transactions are more complex, usually carried out based on long-term contracts, and involve multiple attributes apart from price like transportation, reliability etc.[1] However, they have gained momentum in the recent past and the advances in information technology stress the incorporation of auction-based procurement in supply chain operations. Companies can participate in auctions to distribute their products to a large number of participating retailers and take into account various factors like the number of retailers, their locations, price being offered and demand from different retailers. Peleg et al.[2] study three different procurement strategies: strategic partnership with a single supplier, online search for multiple suppliers and a combination of the two strategies. They assess the effect of the procurement process on the costs and showed that the performance of the procurement process depends on the cost and demand parameters. Jin and Yu[3]consider a two supplier, one buyer system and study the supply chain performance in a procurement setting with competition between the suppliers. They compare different auction mechanisms based on the payoffs for the buyer and the suppliers.

Supplier selection and order allocation is another problem that has been extensively studied in the supply chain literature. These are the most significant issues in the purchasing division of enterprises. It is important to have an efficient order allocation procedure that can be integrated into the ERP system of an organization.[4] Supplier selection usually involves evaluating multiple criteria like price, lead time, reliability etc. Different techniques, broadly classified into multi-criteria decision making (MCDM) techniques, mathematical programming techniques and artificial intelligence techniques, have been proposed to solve the multi-criteria supplier selection problem.[5] Due to the relative practicality of the approach, multi-criteria decision making methods are a popular class of solution techniques that are widely used to solve such problems in industrial settings.[6-8]

In this work, a multi-enterprise supply chain network is modeled where the production sites and warehouses belong to one enterprise while the raw material suppliers and retailers belong to different enterprises. The distribution of products from warehouses to retailers is based on auctions. A multi-attribute double auction is modeled as both the seller and buyer are considered to bid the in the process based on multiple attributes like price and amount of product. The supplier selection is modeled using the Analytic Hierarchy Process (AHP). The ranking procedure assigns a utility value to each supplier. The utility values are used as a basis for ranking or choice of suppliers. The AHP method requires pairwise comparisons of the different criteria and the alternatives. Such comparisons are usually imprecise. In order to model the imprecision in the pairwise comparisons required for the AHP method, fuzzy numbers are used to represent the pairwise comparisons. The relative ranking of the suppliers obtained from the solution of the fuzzy AHP is used to allocate orders to the raw material suppliers. An agent-based simulation model is developed that captures the supply chain operations including the auction based distribution of products and the allocation of orders among raw materials based on fuzzy AHP. The approach is used for small scale case studies in order to demonstrate the effects of procurement and supplier selection policies.

References:

1. Simchi-Levi D, Wu SD, Shen Z-J. *Handbook of Quantitative Supply Chain Analysis: Modeling in the E-Business Era (International Series in Operations Research \\& Management Science)*: Springer-Verlag New York, Inc.; 2004.

2. Peleg B, Lee HL, Hausman WH. Short-term e-procurement strategies versus long-term contracts. *Production and Operations Management.* Win 2002;11(4):458-479.

3. Jin MZ, Yu AJ. Procurement auctions and supply chain performance. *International Journal of Production Economics.* Apr 2015;162:192-200.

4. Lin CT, Chen CB, Ting YC. An ERP model for supplier selection in electronics industry. *Expert Systems with Applications.* Mar 2011;38(3):1760-1765.

5. Chai JY, Liu JNK, Ngai EWT. Application of decision-making techniques in supplier selection: A systematic review of literature. *Expert Systems with Applications.* Aug 2013;40(10):3872-3885.

6. Levary RR. Using the analytic hierarchy process to rank foreign suppliers based on supply risks. *Computers*

& Industrial Engineering. 2008;55(2):535-542.

7. Sevkli M. An application of the fuzzy ELECTRE method for supplier selection. *International Journal of Production Research*. 2010/06/15 2009;48(12):3393-3405.

8. Farzipoor Saen R. Developing a new data envelopment analysis methodology for supplier selection in the presence of both undesirable outputs and imprecise data. *The International Journal of Advanced Manufacturing Technology*. 2010/12/01 2010;51(9-12):1243-1250.

427906 Centralized Versus Distributed Manufacturing: A Continuous Location-Allocation Problem

Abstract:

Centralized versus Distributed Manufacturing: a Continuous Location-Allocation Problem

Advances in technology have led to the rethinking of traditional manufacturing. In recent years, the concept of Distributed Manufacturing has been discussed as a way to satisfy the new requirements of the market (e.g. customized products) and take advantage of some logistical aspects (e.g. high cost of transportation in the biomass supply chain). However, conventional large-scale centralized manufacturing can be more cost-effective due to the economies of scale.

In this paper, we address a systematic way of evaluating the profitability of centralized versus distributed manufacturing. The problem is formulated as a continuous facility location-allocation problem with limited capacity, also known as the Capacitated Multisource Weber problem. The objective of this type of problem is to generate sites in continuous space for locating new facilities in relation to a set of existing facilities located at given points in space, taking into account limited capacity and transportation costs [1].

Our generic model considers sources of raw material and consumer markets with specified locations and corresponding availabilities and demands. The problem has a number of potential facilities that can be built with specified maximum capacities. Some of them can be large-scale plants while the remaining facilities can be small/medium scale plants. The location of these plants is a variable to be determined on a two-dimensional plane. Therefore, the model has the option to choose a centralized, distributed, or hybrid configuration consisting of facilities with locations and assignments closer to the raw material or to consumers.

The problem is formulated as a mixed-integer nonlinear program (MINLP) in which the investment cost of the facilities and the transportation costs are to be minimized. The MINLP is nonconvex due to the concave investment cost functions, the Euclidean distances, and the cost dependence of distances and amounts shipped for the transportation. To deal with the nonconvexities, several reformulations are first applied to improve the computational performance of the local and global solvers. A piecewise-linear approximation is also used to approximate the concave separable terms so as to reformulate the problem as a mixed-integer quadratically-constrained quadratic program (MIQCP).

The applicability of the proposed models is illustrated with case studies for different industries, including biomass and shale gas. The results show the capability of the models to find globally optimal solutions for centralized and distributed manufacturing systems. We also compare the results for the MINLP and the approximated MIQCP models using commercial local and global optimization solvers (e.g. SBB, DICOPT, CPLEX, GUROBI, BARON, SCIP).

[1] Brimberg, Jack, Pierre Hansen, Nenad Mladenovic, and Said Salhi. "A Survey of Solution Methods for Continuous Location-Allocation Problem." *International Journal of Operations Research* 5.1 (2008): 1-12.

Non-Cooperative Games for the Optimization of Multi-Enterprise Supply Chains under Uncertainty

Kefah Hjaila, José M. Laínez-Aguirre, Luis Puigjaner
and Antonio Espuña

UNIVERSITAT POLITÈCNICA DE CATALUNYA
BARCELONATECH

15 AIChE
Annual Meeting, Salt Lake City, UT

Salt Lake City, UT, 09 Nov 2015

Motivation

❑ **Decentralized SC optimization**

- Multi-enterprises
- Individual objectives
- Conflicting goals
- Lack of Knowledge
- Uncertainty
- 3rd parties

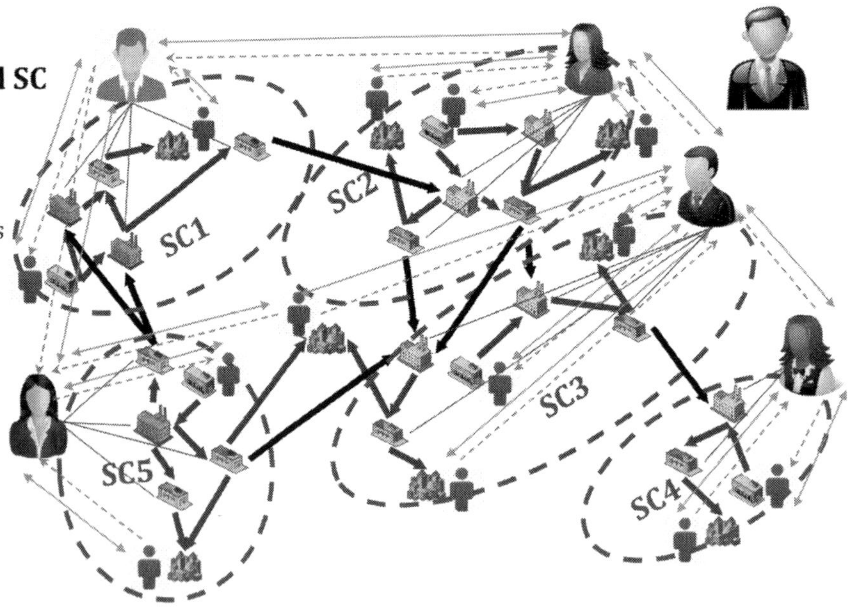

Resources flows — Ownership — Economic flows — Information flows — Conflicting resources

2/12

Literature review

Game Theory

❑ Cooperation

- Backorder costs sharing (Hennet & Arda, App. of A.Int, 2008)
- Quantity discounts (Zamarripa et al., C&CE, 2012)
- Revenue sharing (Fang & Shou, EJOR, 2015)

❑ Non-cooperation

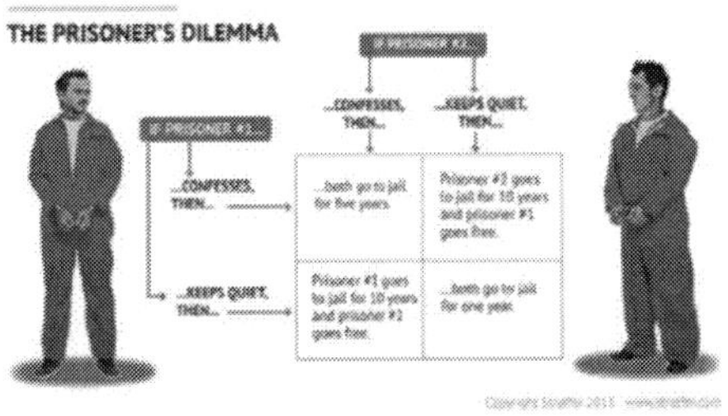

➢ **Nash Equilibrium** (Symmetrical roles)

- Fair allocation of profit
  (Ortiz-Gutiérrez et al., CACE, 2015)

➢ **Stackelberg games** (Non-symmetrical roles)

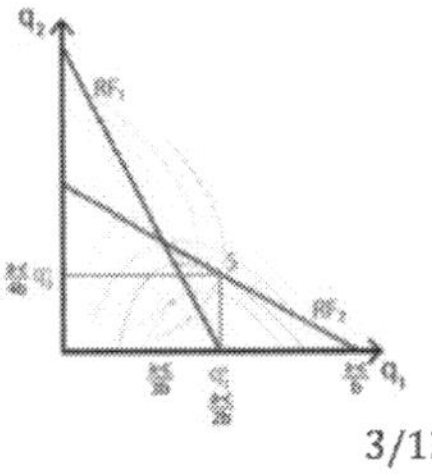

- Bi-level optimization (Yue and You, CACE, 2014)
- Scenario-Based Dynamic Negotiations (Hjaila et al., CACE,2015)

3/12

Objective

Objective: ***To set the best conditions for eventual coordination contracts between enterprises of conflicting goals.***

Methodology

❑ Stackelberg Game

- Non-cooperative
- Non-zero-sum
- Non symmetrical role

❑ Uncertainty evaluation

- Monte Carlo
- Expected Payoff matrix

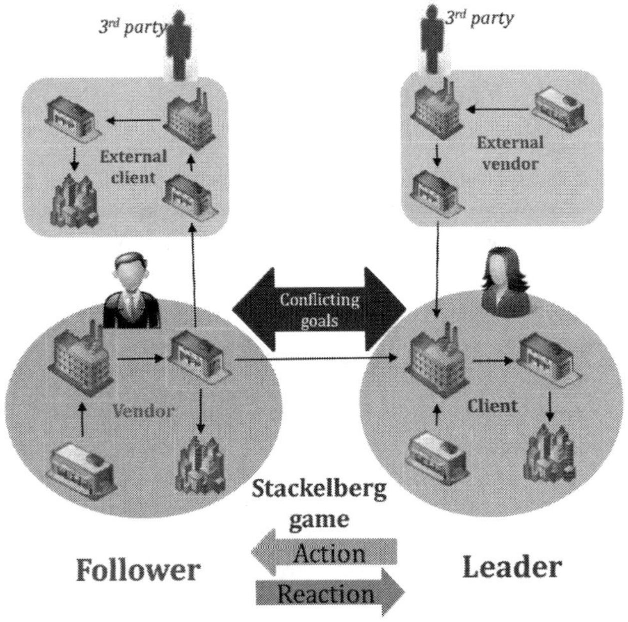

4/12

Tactical Game model

❑ Objective function

$$\max Payoff_{sc} = SALE_{sc} - COST_{sc}$$

$$SALE_{sc} = \sum_{r\in R}\sum_{m\in M}\sum_{t\in T} rp_r \cdot Q_{r,sc,m,t} + \sum_{t\in T}\sum_{\substack{r'\in R\\ r'\neq r}} pL_{r'} \cdot QF_{r',sc'\in F,t} + \sum_{t\in T}\sum_{r'*\in R} pC_{r'*} \cdot QC_{r'*,sc'\in F,t}$$

$$+\sum_{t\in T}\sum_{r'\in R} pV_{r'*} \cdot QV_{r'*,sc'\in V,t}$$

$$COST_{sc} = CRM_{sc} + CPRD_{sc} + CST_{sc} + CTR_{sc} + \sum_{t\in T}\sum_{r'\in R} pL_{r'} \cdot QF_{r',sc'\in L,t}$$

$$+\sum_{t\in T}\sum_{r'*\in R} pV_{r'*} \cdot QV_{r'*,sc'\in L,t} + \sum_{t\in T}\sum_{r'*\in R} pC_{r'*} \cdot QC_{r'*,sc'\in C,t}$$

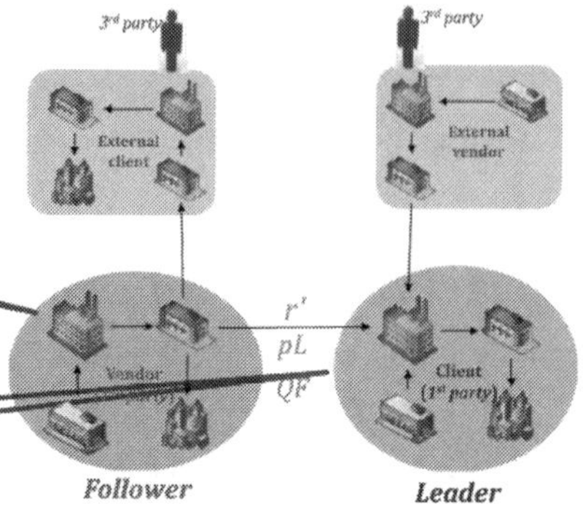

Stackelbreg-game item	r'
Stackelberg-game leader action	pL
Stackelberg-game follower response	QF

❑ Reaction function

$$\sum_{t\in T}\sum_{dc'\in DC} QF_{r',sc'\in F,dc',t} = \sum_{\substack{r\in R\\ r\neq r'}}\sum_{t\in T} prd_{r,r',sc'\in F,t} \cdot f_{r,r',sc'\in F} - \sum_{dc'\in DC}\sum_{t\in T} QC_{r'*,sc'\in F,dc',t}$$

$$= \sum_{\substack{r\in R\\ r\neq r'}}\sum_{t\in T} Manf_{r',r,sc'\in L,t} \cdot fac_{r',r,sc'\in L,t} - \sum_{dc'\in DC}\sum_{t\in T} QV_{r'*,dc',sc'\in L,t}$$

❑ Uncertainty evaluation (Monte Carlo)

$$ExPayoff_{sc'} = \frac{\sum_{n\in N} Payoff_{sc',n}}{N} \qquad prob_{sc'\in F} = \frac{No.\ of\ scenarios\ of\ improved\ profits_{sc'\in F}}{Total\ No.\ of\ scenarios_{sc'\in F}}$$

5/12

Case-study

❑ Polystyrene production SC
- 3 manufacturing sites
- 2 storage centers
- 3 markets
- 2 vendors
- 4 RM suppliers
- WWTP

❑ Energy generation SC
- 6 energy generation sites
- 3 storage centers
- 4 RMs (biomass and coal)
- 3 clients

❑ Tactical decisions (6 time periods)
- RM purchase
- Production
- Inventory
- Distribution levels

❑ Coordination contract: internal energy flows and unit price

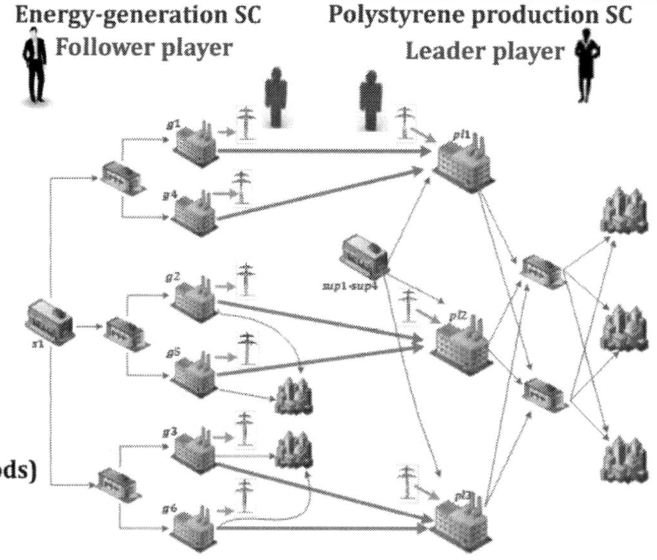

Game player SC	Model	Single equations	Single variables	Discrete variables	CPU each action (sec)
Leader	MINLP	964	1653	126	7.95
Follower	MINLP	1202	1289	180	3.85

6/12

Results- Stackelberg payoff matrix

Tradeoff: Leader expected payoff Vs. follower expected payoff

- ❑ **Higher expected standalone payoffs**

From Follower side:
- o higher expectations
- o Lower willingness to collaborate at **lower** prices

From Leader side:
- o higher expectations
- o Lower willingness to collaborate at **higher** prices

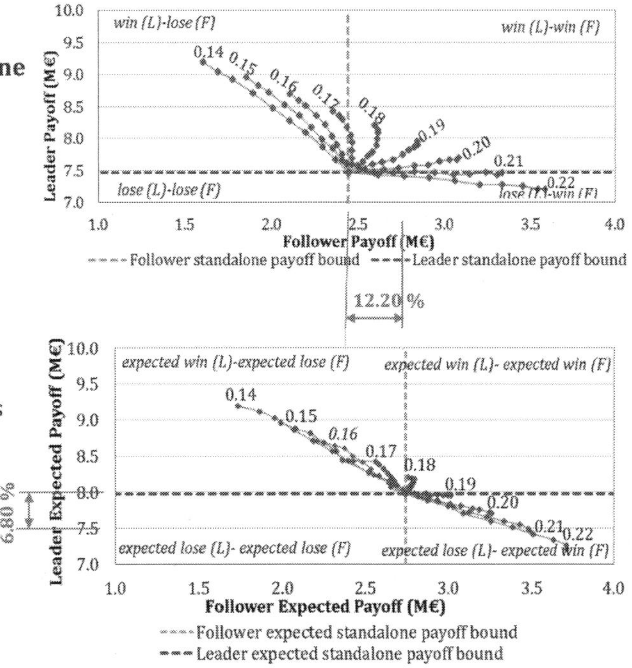

Nominal conditions

Uncertain Conditions

Results- Pareto of solutions

Nominal conditions | **Uncertain conditions**

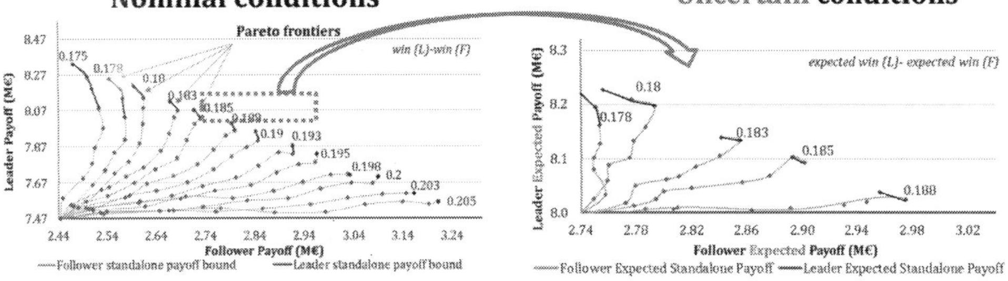

Stackelberg set of Pareto frontiers

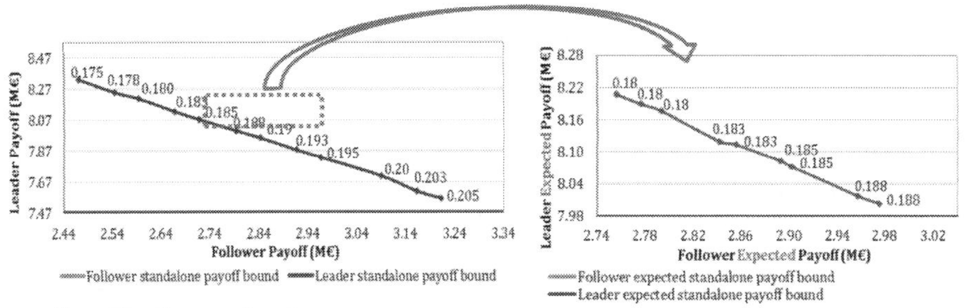

Coordination contracts
Transfer Price: 0.175-0.205 €/kWh
Internal energy: 24.71 GWh

Coordination contracts
Transfer Price: 0.18-0.188 €/kWh
Internal energy: 21.00-24.71 GWh

Results- Uncertainty reduction

- ❑ **Variance:**
  - o Lower energy amounts ➔ higher benefits variance
- ❑ **Uncertainty reduction**
  - o Follower : variance 1.03 M€^{2}
  - o Leader: variance 0.50 M€^{2}
- ❑ **Coordination leads to:**
  - o Reductions in uncertainty
  - o Higher profits expectations

Follower

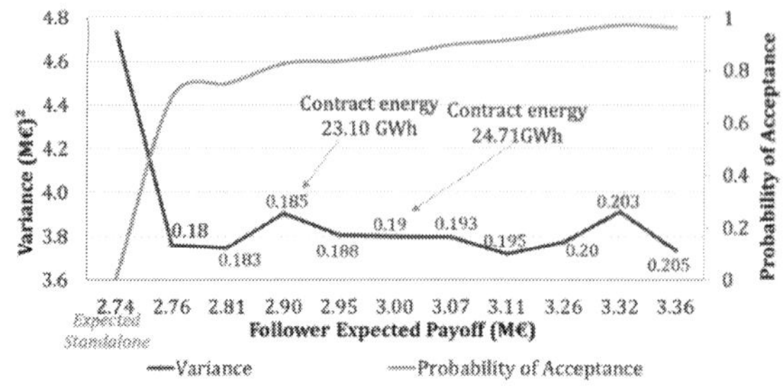

Leader

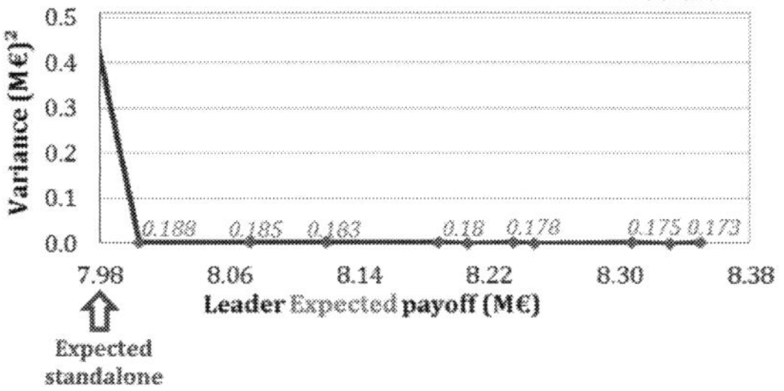

9/12

Results- tactical decisions

Follower tactical decisions: **internal energy flows**

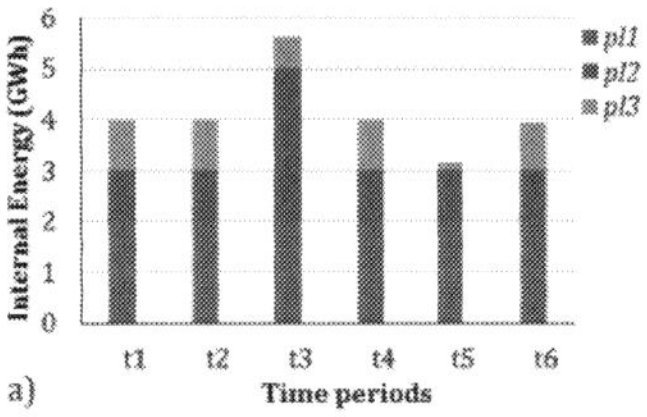

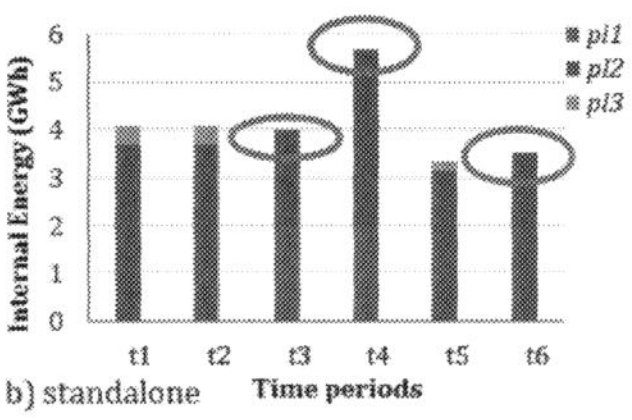

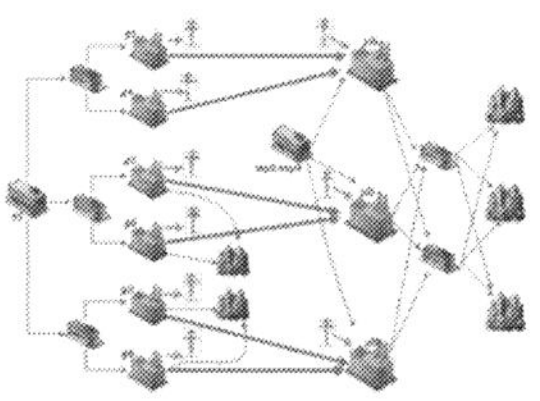

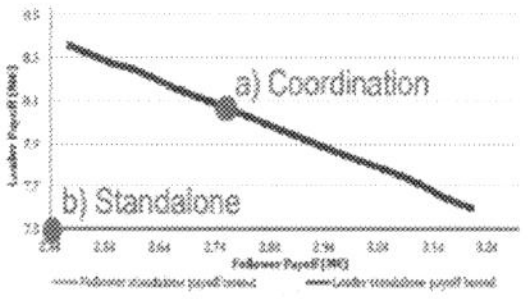

❑Leader tactical decisions: **polystyrene production**

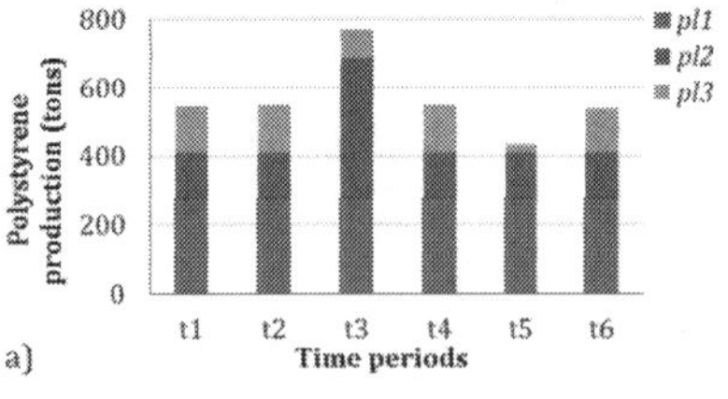

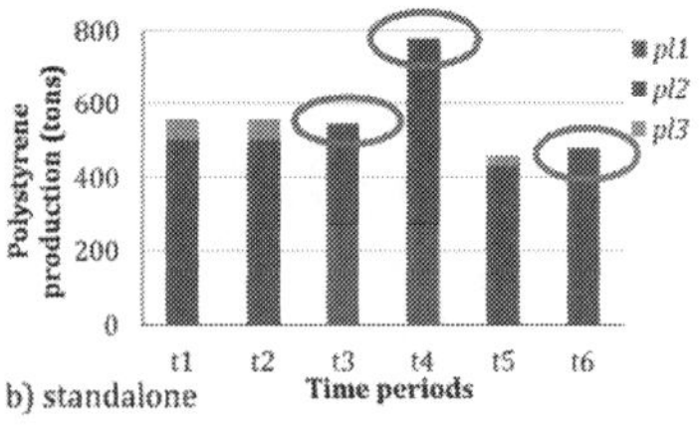

Coordination contract
- 24.71 GWh
- 0.185 €/kWh

10/12

Conclusions

- **Non-cooperative games:**
  - Set expected win-win coordination contracts
  - Solve the conflicting goals (decentralized SCs)
  - Keep expectations potential of higher individual profits
  - Lead to uncertainty reductions
  - Stress the willingness to collaborate
  - Provide different options for further negotiations

11/12

Acknowledgements

The Spanish Ministry of Economy and Competitiveness and the European Regional Development Fund, both funding the Project SIGERA (DPI2012-37154-C02-01)

GOBIERNO DE ESPAÑA MINISTERIO DE ECONOMÍA Y COMPETITIVIDAD

EUROPEAN UNION EUROPEAN REGIONAL DEVELOPMENT FUND

The Generalitat de Catalunya

- AGAUR FI program
- Grant 2014-SGR-1092-CEPEiMA.

Generalitat de Catalunya

AGAUR

12/12

430749 Supply Chain Network Design Under Outsourced Transportation and Warehousing

Abstract:

With the globalization of markets and the focus on customer satisfaction, supply chains are becoming more complex. The response of many companies to these challenges is to focus on their core business and outsource other components of their supply chain, such as the supply of raw material, transportation and warehousing.

Recognizing the importance of the topic, a large volume of research has been devoted to the optimal design of supply chain networks under different conditions. This paper considers two important elements that have usually been neglected in the literature, but are often present when the transportation and warehousing service is outsourced: discrete transportation costs and warehouse term-contracts. Efficient mixed-integer linear programming (MILP) formulations for handling these aspects are proposed for a 2-echelon, multi-commodity, multi-period problem with multiple modes of transportation in each link.

In most of the research reported on the facility location problem, transportation costs are considered to be a linear function of the transported amounts, with optional fixed costs. However, in several real-life applications the cost function is piece-wise constant, because there is a fixed cost per each transportation unit used (e.g.: trucks), and if the unit is full or half-full the transportation cost is the same. When the quantity exceeds the unit capacity, another unit is used and the cost rises to another step in the function. Five different approaches to model this situation are analyzed and compared. The first approach considers integer variables to compute the number of units of each mode of transportation used in each link at each time period. The second approach treats the cargo loading as an assignment problem in which the maximum number of units per each mode of transportation is known beforehand. The other three approaches are based on representing the stepwise nature of the cost function using SOS2 variables and General Disjunctive Programming, with its respective Big-M and Convex Hull reformulations. For a midsize problem with 3 plants, 6 potential warehouses, 8 customers, 5 products and 36 periods only the formulation with integer variables was able to solve the problem to a 0.5% gap in less than 2 minutes using the default options of the solver Gurobi 6.0.2. The other four approaches were not able to find a feasible solution after 15 minutes. The partial relaxation of the formulation with integer variables, in which the transportation units are allowed to take continuous values presents a 4% relaxation gap and solves the problem in less than 10 seconds, making it useful to obtain good lower bounds.

The second element included is warehouse contracting policies. When the service is provided by an external company there are minimum length clauses in the contracts and also if a contract is not immediately renewed, there is a minimum wait period to start a new contract. These clauses prevent the warehouse to be empty for short periods of time, making it easier for the provider to find other customers to fill the vacant periods. In order to address these requirements two extra discrete variables were added to the formulation, one to indicate when a new contract is started and another to indicate when a contract is finished. Using these variables and the original discrete variables to indicate monthly warehouse use, minimum contract length and minimum constraints were formulated. The introduction of these new variables and constraint increased the total number of discrete variables by 5% and the solution time by 20%. Finally, we report results for a large-scale industrial application.

The paper shows that important aspects present when the transportation and the warehousing is outsourced can be efficiently modeled maintaining the solution times suitable for large scale applications and to deal with uncertain parameters.

References

[1] Cordeau, Jean-François, Federico Pasin, and Marius M. Solomon. "An integrated model for logistics network design." *Annals of Operations Research* 144.1 (2006): 59-82.

NUMERICAL SIMULATION OF A DELAYED COKING REACTOR

Fabián A. Diaz[1,], Arlex Chaves[1], Maria Maradei[1], David Fuentes[1], Alexander Guzmán[2], Hector Picón[2].*

1. *Universidad Industrial de Santander, Bucaramanga, Colombia.*
2. *Instituto Colombiano del Petróleo, Bucaramanga, Colombia.*

**. First author email: fabiandiazmateus@hotmail.com*

Abstract

A numerical unsteady simulation of a pilot plant delayed coking reactor was performed, it was modelled as a three-phase system being the vacuum residue the liquid phase, the distillables the gas phase and the coke the solid phase. The dynamics of the liquid and the gaseous phase were described by continuity, momentum and energy balances; the solid phase is described only by continuity and energy balances since it is defined static and remains still where it is formed. The finite volume method was employed to carry out the discretization of the governing equations; a modified SIMPLEC algorithm achieved the coupling of the pressure-momentum equations which were solved on a 2D axisymmetric grid. A 4 hours run of the coking reactor was simulated; as this investigation is still at early stage, the only experimental parameter available is the size of the coke bed, the results of the simulation present good agreement when compared with this experimental data.

Keywords: CFD, computational fluid dynamics, vacuum residue, coke, mathematical modelling.

1. Introduction

The delayed coking process is a purely thermal process where a heavy residual oil is converted into coke, liquids and gases at temperatures above 400°C. At industrial level, the residue is heated in furnaces and rapidly transferred to a drum, the word "delayed" comes from the attempt to delay the formation of coke inside the furnace and take it only inside the drum. The process is semi-continuous because when one drum is full of coke the feed switches to a second drum while the first one is discharged and so on.

The delayed coking process is the preferred choice for refiners for the thermal conversion of petroleum residua due to the flexibility to process different types of residua (Rana *et al.*, 2007) and the superior economic benefits when compared to other processes for thermal conversion of heavy oil (Vartivariam & Andrawis, 2006).

A detailed description of the process for coke formation inside a drum has been presented by Ellis & Hardin (1993), few attempts to numerically simulate this unsteady process have been presented in the literature as most of the work focuses on the coking furnace, such as Nunhez *et al.* (2006) who performed, with their own programming code, a numerical simulation where the coke is deposited inside the pipe walls. Also, Zhang et al. (2015) performed a numerical simulation with CFD (Computational Fluid Dynamics) commercial code of a coking furnace with deposition of coke inside the pipe walls. However, numerous simulators based on correlations or statistical models have been presented in the literature, some of the most recognized are Gary & Handwerk (1975); Castiglioni (1983); Maples (2000); Zambrano (2001); Volk, Wisecarver & Sheppard (2002), Pushpalayari (2004). These models are based on the assumption that the process is in steady state and they calculate the yield of coke, liquids and gases according to the properties of the feed and the operating conditions. An interesting comparison of these correlation models for delayed coking simulation can be found in Ancheyta *et al.* (2013).

Simulation models based merely on chemical kinetics can also be found in the literature such as Bozzano & Dente (2005) and Zhou *et al.* (2007). These models also calculate the yield of coke, liquids and gases according to the properties of the feed and the operating conditions, however, the composition of the liquid is relatively more defined as it is divided into pseudo-components. An even more detailed simulator is the one presented by Tian, Shen & Liu (2012a, 2012b) based on structured oriented lumping (SOL), with this lumping the feed and the products are characterized at molecular level offering up to 7000 pseudo-components. The simulation is also based on the assumption that the process is in steady state.

In this work, a numerical unsteady simulation of a pilot plant delayed coking reactor from ICP (Colombian Institute of Petroleum) is performed; the mathematical modelling is based on unsteady continuity, momentum and energy balances for three phases (gas, liquid and solid) which are solved using CFD techniques. The results are presented throughout the run time and provide a changing coke bed which resembles the process for coke formation inside a drum presented by Ellis & Hardin (1993).

2. *The Coking reactor*

A scheme of the coking reactor from the delayed coking pilot plant at ICP is presented on Figure 1. The reactor is basically a pipe with walls that are heated by electrical resistances, given the geometry, a 2D axisymmetric grid can simulate the entire physical geometry of the reactor. The only experimental data available is the size of the coke bed, which was measured after its removal from the reactor. The complete set of operating conditions used in the experimental procedure and the simulation is on Table 1.

Table 1. Operating conditions for the delayed coking reactor

Temperature of the walls	510°C
Temperature of the feed	370°C
Feed mass rate	5g/min
Reactor diameter	2.5 NPS, SCH 40
Reactor length	569.2 mm
Operating pressure	1 atm.

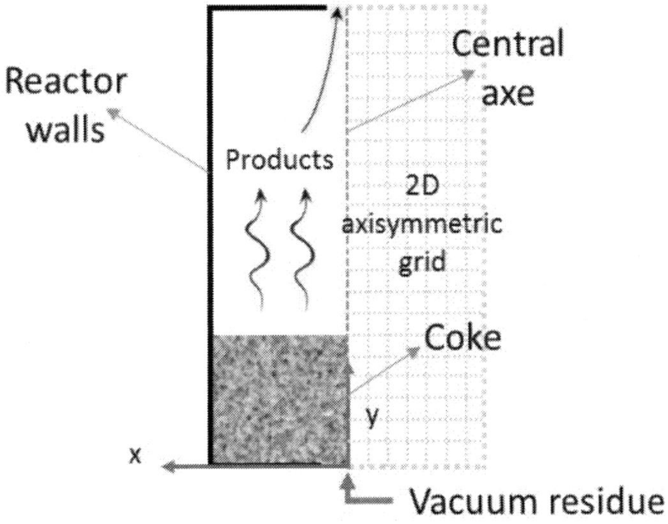

Figure 1. Scheme of the delayed coking reactor at ICP

3. *Kinetic model*

The kinetic model used in the simulation of the delayed coking reactor is taken from Diaz *et al.* (2015). It consists in a simple kinetic model where the vacuum residue decomposes into coke and distillables, in this model the reaction mechanism changes from first to second order depending on the temperature. The complete characterization of the vacuum residua used in this work can also be found on Díaz *et al.* (2015) and it corresponds to the number 3 vacuum residue.

4. *Mathematical model*

For the development of the governing equations, a series of assumptions were made in order to simplify the modelling without significantly affecting the results, these assumptions are:

- The model is 2D axisymmetric orbiting the central axe.
- The system is composed by three phases: Liquid (vacuum residue) and gas (distillables) at laminar flow and solid (coke).
- The liquid is the primary phase and the gas is the secondary.
- The chemical reaction can be model by the thermal decomposition (pyrolysis) of vacuum residue into distillables and coke.
- The coke does not have movement, it remains still at the place where it is formed.
- The momentum loss through the coke bed can be calculated as a loss through a packed bed.
- The porous medium (coke bed) is isotropic (homogenous).
- The effects of pressure and viscous heating on temperature are discarded.

4.1 Continuity equations

Continuity equation for the residue (R) and distillable (D) phase:

$$\frac{\partial(\gamma\alpha_n\rho_n)}{\partial t} + \nabla\cdot(\gamma\alpha_n\rho_n\vec{v}_n) = -\gamma\alpha_n\rho_n\Gamma_n \qquad \textit{Eq. 1}$$

Where $n = R, D$

Continuity equation for the coke phase:

$$\frac{\partial(\alpha_C \rho_C)}{\partial t} = \gamma \alpha_R \rho_R \Gamma_C \qquad \textit{Eq. 2}$$

Continuity equations can also be called volume fraction equations. The porosity of the coke bed is calculated as:

$$\gamma = 1 - \alpha_C \qquad \textit{Eq. 3}$$

4.2 Momentum equations

Momentum equation for the residue (R) and distillable (D) phase:

$$\frac{\partial(\gamma \alpha_n \rho_n \vec{v}_n)}{\partial t} + \nabla \cdot (\gamma \alpha_n \rho_n \vec{v}_n \vec{v}_n) = -\gamma \alpha_n \nabla \cdot \mathrm{P} + \nabla^2 \cdot (\gamma \alpha_n \mu_n \vec{v}_n) + \gamma \alpha_n \rho_n \vec{g} + \gamma K_{mn}(\vec{v}_m - \vec{v}_n) - \alpha_n \vec{v}_n \left(\frac{\mu_n}{\varepsilon} + \frac{C_2 \rho_n |\vec{v}_n|}{2} \right) \qquad \textit{Eq. 4}$$

Where $n, m = R, D$ and $m \neq n$

The coke phase does not have momentum equation.

4.3 Energy equations

Energy equation for the residue (R) and distillable (D) phase:

$$\frac{\partial(\gamma \alpha_n \rho_n C_{P,n} T_n)}{\partial t} + \nabla \cdot (\gamma \alpha_n \rho_n C_{P,n} \vec{v}_n T_n) = \nabla^2 \cdot (\gamma \alpha_n \kappa_n T_n) + \gamma \alpha_n \rho_n \Gamma_n \Delta \dot{H}_n + \gamma H_{mn}(T_m - T_n) + \alpha_n H_{Cn}(T_C - T_n) \qquad \textit{Eq. 5}$$

Energy equation for the coke (C) phase:

$$\frac{\partial(\alpha_C \rho_C C_{P,C} T_C)}{\partial t} = \nabla^2 \cdot (\gamma \alpha_C \kappa_C T_C) + \alpha_R H_{CR}(T_R - T_C) + \alpha_D H_{CD}(T_D - T_C) \qquad \textit{Eq. 6}$$

4.4 Parameters

The momentum exchange coefficients (K_{mn}) are calculated with the symmetric model (See ANSYS (2014)), the symmetric model is recommended where a secondary phase becomes primary phase in different places of the domain, as it happens in this simulation. The permeability (ε) and the inertial resistance factor (C_2) are calculated with the correlations from Ergun (1952). The fluid-fluid heat transfer coefficients (H_{mn}) are calculated with Tomiyama´s correlation (Tomiyama (1998)) which is appropriate for low Reynolds flow. The fluid-solid heat transfer coefficients (H_{CR}, H_{CD}) are calculated with the correlations from Wakao and Kaguei (1982) also recommended for low Reynolds flow.

5. Solution of the equations

The governing equations were solved using CFD techniques, the equations were discretized by the finite volume method and the pressure-velocity coupling was achieved by means of a SIMPLEC algorithm (Van Doormal & Raithby (1984)). The procedure proposed by Patankar & Spalding (1972), to obtain the pressure correction equation through an arrangement of the continuity and momentum equations, was used in this work.

As the momentum exchange (Drag) between the phases gas and liquid is considered in the momentum equations, the partial elimination algorithm (PEA) (Spalding & Markatos (1983)) was used to avoid convergence issues caused by the coupling of the momentum equations. In order to avoid the famous "pressure checker-board" or unrealistic pressure oscillations caused by the interpolation of the pressure, a staggered grid is used for the solution of the momentum equations.

Continuity, momentum and energy equations were solved with a Gauss-Seidel procedure. This method performs a sweeping through all the cells of the grid solving an equation of the form of equation 7 to obtain the value of the variable in the cell. This method provides an advantage in multiphase models, if the volume fraction value of certain component is zero in the cell, the value of other parameters, such as velocities, take the value of zero immediately in the cell, in this manner, the solution of the equations is more precise and faster than other solution methods.

$$\phi_P = \frac{\sum a_{NB}\phi_{NB} + \hat{S}_P}{a_P} \qquad \textit{Eq. 7}$$

The pressure correction equation was solved with a SOR (Successive over relaxation) method, this method is practically the same Gaus-Seidel method but over-relaxed with a relaxation factor between 1 – 2 that accelerates convergence. The simulator was programmed in C++ code and compiled in Linux with an Intel compiler; with a time step size of 0.01s, the simulation of the 4 hours run took about 6 days.

6. *Results*

The results of the simulation are presented according to the simulation time, phase velocities are presented as a magnitude not as a vector. Results are presented as colored contours, the colormap at the right side of the graphs represent the range of the values, volume fractions and temperatures of the phases are grouped to provide a single colormap which makes it more clear.

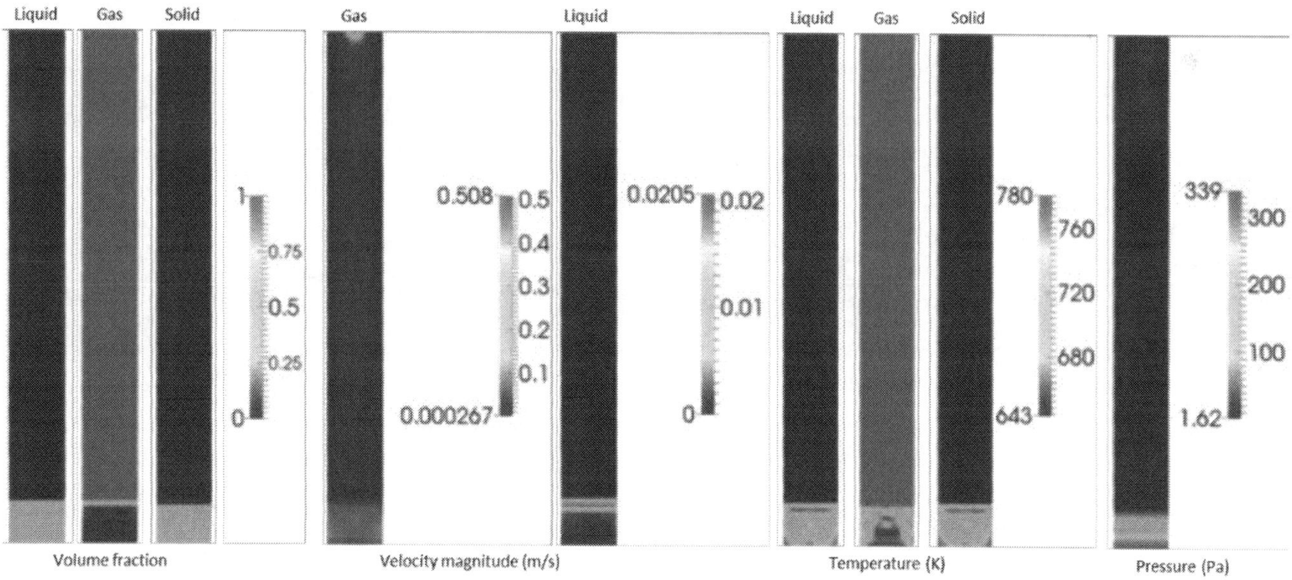

Figure 2. Results of the coking reactor simulation at a simulation time of 1h.

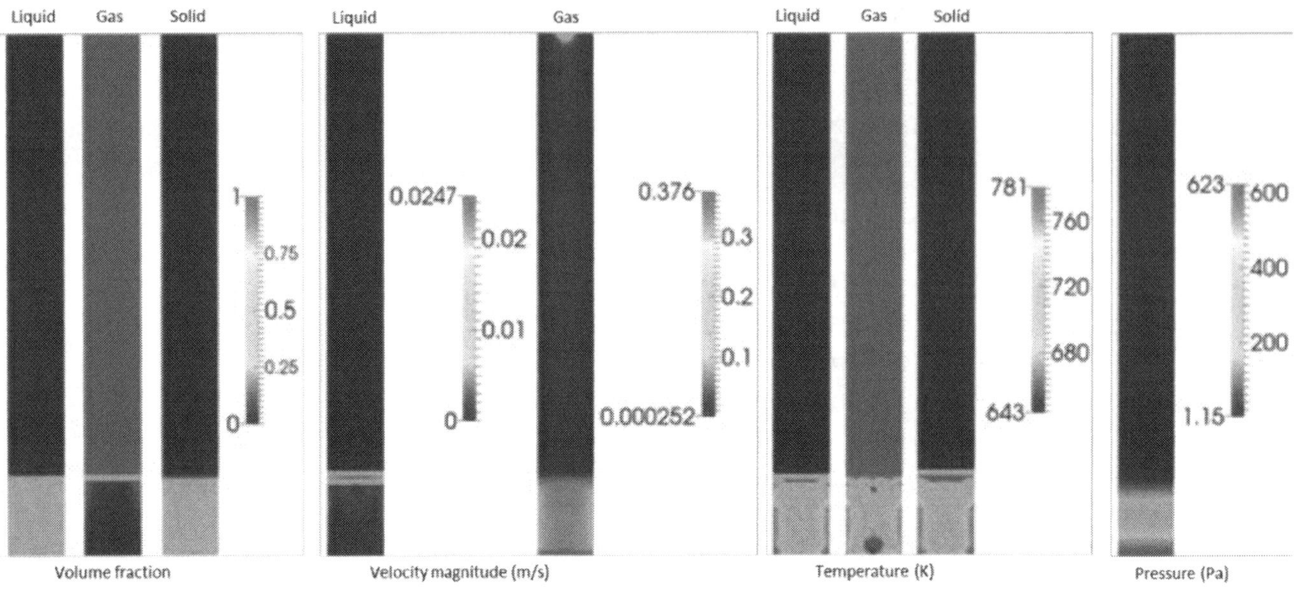

Figure 3. Results of the coking reactor simulation at a simulation time of 2h.

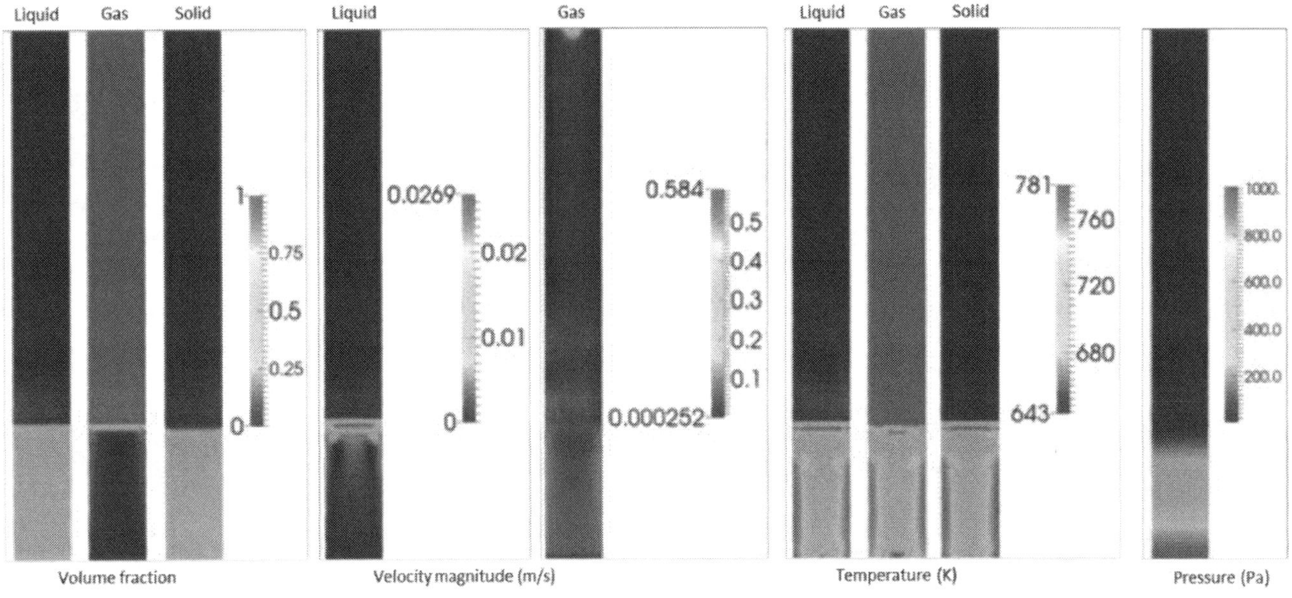

Figure 4. Results of the coking reactor simulation at a simulation time of 3h.

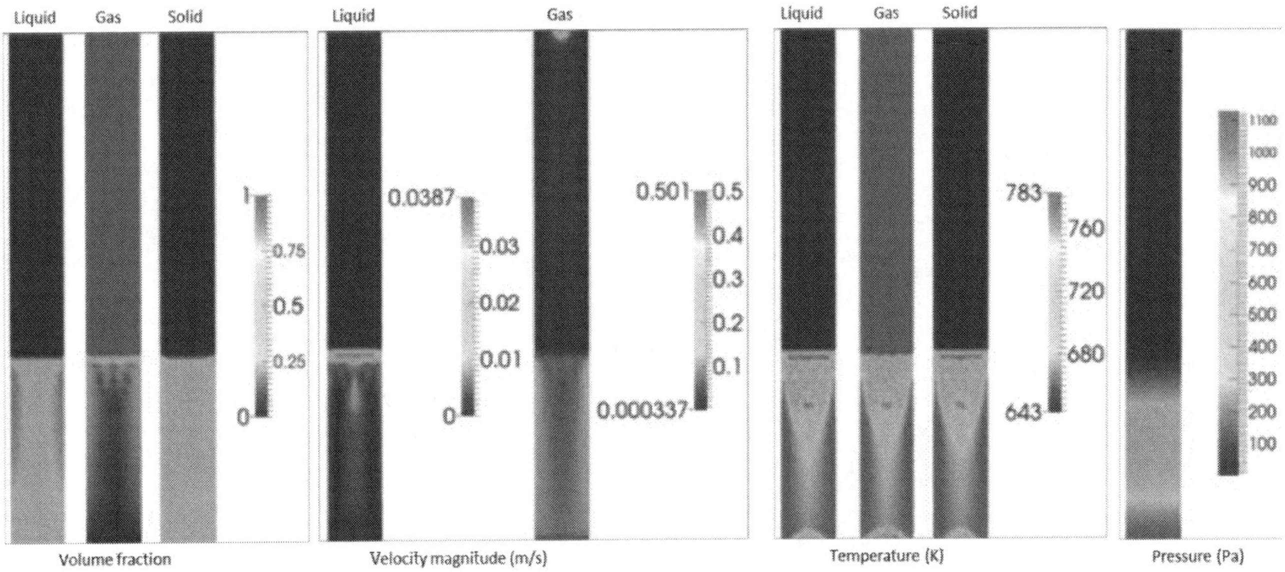

Figure 5. Results of the coking reactor simulation at a simulation time of 4h.

In figures 2 – 5 it can be seen the volume fractions, velocity magnitudes, temperatures and pressure drop calculated at 1, 2, 3 and 4 hours of run. The graphs of temperature for the liquid and solid phase in the upper part of the reactor shows a temperature of 643°C, but in the volume fraction graph it is seen that the volume fraction is zero in the same part. This occurs because the simulator is programmed to assign the lowest temperature of the system to cells with zero volume fraction of any phase. A temperature of zero degrees could not be assigned to cells with zero volume fraction because it generated an error in the simulation.

Table 2. Size of the coke bed at the end of the simulation

Size of the coke bed (calculated)	Size of the coke bed (measured)
20.78 cm	20.5 cm

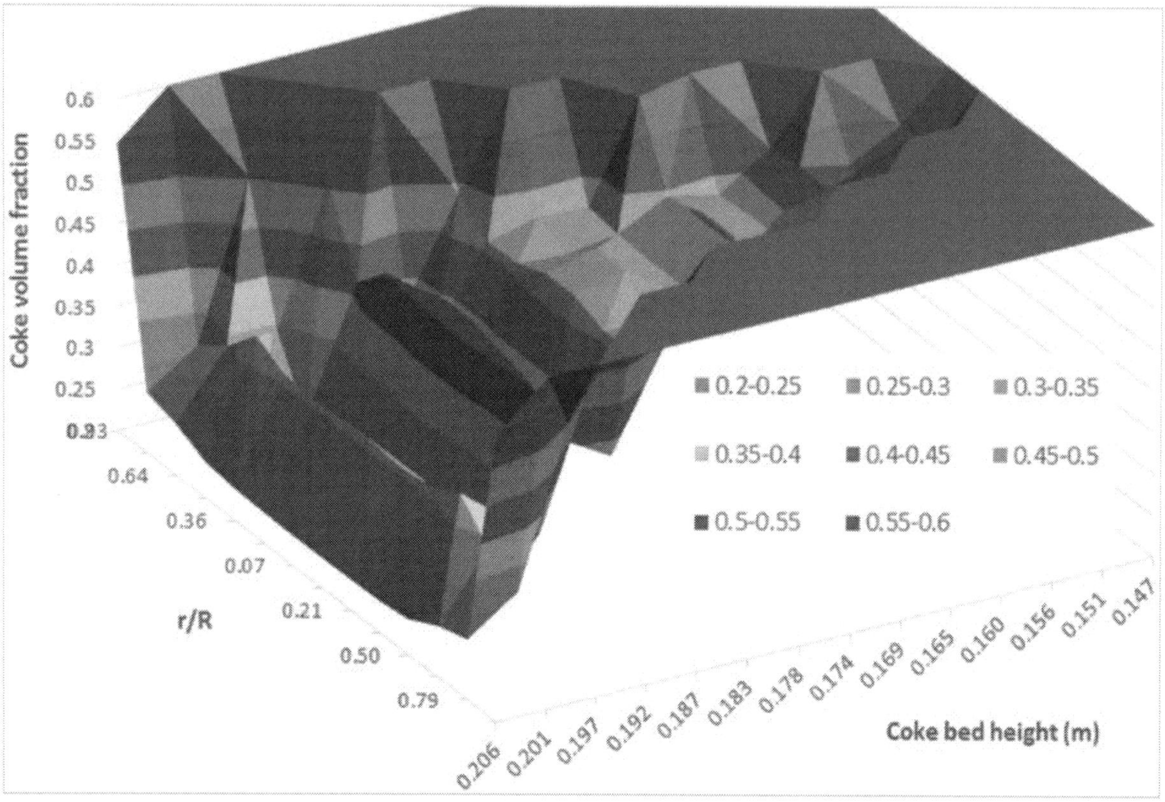

Figure 6. Shape of the upper part of the coke bed at the end of the simulation.

7. *Discussion*

The predicted and measured size of the coke bed, presented on Table 2, are practically equal. The volume fraction of coke calculated by the simulator shows a tendency to gather at the reactor walls, which means the porosity of the coke bed is lower at the walls and higher at the center of the reactor, it also provides an inverted cone shape in the upper part of the coke bed, as it can be seen on Figure 7. This observation was confirmed by visual inspection of the samples of coke extracted from the reactor. Due the small diameter of the reactor, the coke bed tends to channel towards the center.

The temperature of the three phases are practically equal (See Figures 2 – 5) which suggests that the phases are in thermal equilibrium, this is probably caused by the small diameter of the reactor. Figure 5 shows how the temperature is higher at the walls and decreases around 40°C at the top of the coke bed due to the heat absorbed by the pyrolysis reaction. Unfortunately, there is not experimental data to confirm these results yet they are consistent with what is expected in this process.

8. *Conclusions*

The size of the coke bed calculated by the numerical simulation of a delayed coking reactor presented in this work shows good agreement when compared with experimental data, even though most of the results could not be compared with experimental data, they are consistent with what is expected in this process. The results of the simulation present a porosity of the coke bed that decreases from the walls to the center of the reactor and a coke bed with the shape of an inverted cone. These observations were confirmed by visual inspection of the samples of coke extracted from the reactor, however, porosity values have not been measured to confirm the values found in the simulation.

Parameters such as temperatures, velocities and pressure drop still have to be measured to confirm the validity of the mathematical model proposed to simulate the delayed coking process.

Nomenclature

$\vec{v}$	Velocity vector	a	Convection-diffusion coefficient
t	Time	ϕ	Random variable
P	Pressure	S	Source term
T	Temperature	γ	Porosity of the coke bed
$\vec{g}$	Gravitational acceleration vector	α	Volume fraction
K	Momentum exchange coefficient	ρ	Density
C_P	Heat capacity	Γ	Reaction rate
$\Delta\dot{H}$	Heat of reaction	C_2	Inertial resistance factor
H	Heat transfer coefficient	ε	Permeability of the coke bed
μ	Viscosity	κ	Thermal conductivity

Subscripts

n	Residual phase or distillable phase	C	Coke phase
R	Residual phase	NB	Neighbored cell
D	Distillable phase	P	Core cell

References

1. Rana, Mohan S., Samano, V., Ancheyta, J. & Diaz, J.A.I. (2007). A review of recent advances on process technologies for upgrading of heavy oils and residua. *Fuel*, 86 (9), pp. 1216 – 1231.
2. Varvitariam, D. & Andrawis, H. (2006). Delayed coking schemes are most economical for heavy-oil upgrading, *Oil & Gas Journal*, 104 (6), pp. 52 – 56.
3. Ellis, P. & Hardin, E. (1993). How petroleum delayed coke forms in a drum, *Light Metals*, pp. 509 – 515.
4. Souza, B., Matos, E., Guirardello, R. & Nunhez, J. (2006). Predicting coke formation due to thermal cracking inside tubes of petrochemical fired heaters using a fast CFD formulation, *J. Pet. Sci. Eng.* 51(1-2), pp. 138 – 148.
5. Li, X., Zhang, R., Suna, Y., Jiang, B., Luo, M., Li, X. & Zhang, L. (2015). CFD modeling of phase change and coke formation in petroleum refining heaters, *Fuel Process. Technol.* 134, pp. 18 – 25.
6. Castiglioni, B. P. (1983). How to Predict Coker Yield, Hydrocarbon Processing, 62(9), pp. 77 – 79.
7. Gary, J. H. & Handwerk, G. E. (1975). Petroleum Refining – Technology and Economics, Fourth Edition, Marcel Dekker, New York, USA.
8. Maples, R. E. (2000). Petroleum Refinery Process Economics. Penn Well Publishing Co., Tulsa, USA.
9. Zambrano, T. A. (2001). Micro-correlations and data analysis, M.Sc. Thesis dissertation, Department of Chemical Engineering, University of Tulsa, Tulsa, USA.
10. Volk, M., Wisecarver, K. & Sheppard, C. (2002). Fundamentals of delayed coking joint industry project, Department of chemical engineering, University of Tulsa, Tulsa, USA.
11. Pushpalayari, A. G. (2004). Micro-Coker Continuation Studies, M.Sc. Thesis dissertation, Department of Chemical Engineering, University of Tulsa, Tulsa, USA.

12. Muñoz, J., Aguilar, R., Castañeda, L. & Ancheyta, J. (2013). Comparison of Correlations for Estimating Product Yields from Delayed Coking, *Energy & Fuels*, *27* (11), pp. 7179 – 7190.
13. Bozzano, G. & Dente, M. (2005). A Mechanistic Approach to Delayed Coking Modelling, European Symposium on Computer Aided Process Engineering – 15, 20, pp. 529–534.
14. Zhou, X. L., Chen, S. Z. & Li, C. L. (2007). A predictive kinetic model for delayed coking, *Petroleum Science and Technology,* 25 (12), pp. 1539 – 1548.
15. Tian, L., Shen, B. & Liu, J. (2012a). A delayed coking model built using the structure-oriented lumping method, *Energy and Fuels*, 26 (3), pp. 1715 – 1724.
16. Tian, L., Shen, B. & Liu, J. (2012b). Building and Application of Delayed Coking Structure-Oriented Lumping Model, *Ind. Eng. Chem. Res.*, 51 (10), pp. 3923 – 3931.
17. Diaz, F. A., Chavez, A., Fuentes, D., Maradei, M., Guzman, A. & Picon, H. (2015). Kinetic analysis of the thermal decomposition of colombian vacuum residua by thermogravimetry, *Ingenieria e investigacion.*
18. ANSYS Fluent 14.0 User Manual, in: ANSYS Inc.
19. Ergun, S. (1952). Fluid Flow through Packed Columns, *Chem. Eng. Prog.*, 48(2), pp. 89 – 94.
20. Tomiyama, A. (1998). Struggle with computational bubble dynamics. Third International Conference on Multiphase Flow, Lyon, France.
21. Wakao, N., Kaguei, S. (1982). Heat and Mass Transfer in Packed Beds, Gordon and Breach Science Publishers, New York, USA.
22. Patankar, S. & Spalding, D. (1972). A Calculation Procedure for Heat, Mass and Momentum Transfer in Three-Dimensional Parabolic Flows, *Int. J. Heat Mass Transfer*, 15(10), pp. 1787 – 1806.
23. Van Doormal, J. & Raithby, G. (1984). Enhancements of the SIMPLE Method for Predicting Incompressible Fluid Flows, *Numer. Heat Transfer*, 7, pp. 147 – 163.
24. Spalding, D. & Markatos, N. (1983). Computer Simulation of Multi-Phase Flows: A Course of Lectures and Computer Workshops, Report CFD/83/4, Mech. Eng., Imperial College Sci. Tech., London, UK.

417477 Sample Imbalance and Its Role in Understanding Drug Network Characteristics

Abstract:

Recently, several studies have shown that the targets of dangerous or ineffective compounds often have distinct topological positions in protein-protein interaction networks. These distinctions suggest that machine learning algorithms can play a valuable role in evaluating the therapeutic value of potential compounds. Here, we use compounds listed in the DrugBank databases to analyze the topological differences of the targets of clinically useful versus unsafe compounds. We show that while initially some mild distinctions are observed, these distinctions prove to be of minimal value for predicting clinical value. By applying permutation-based cross-validation, we find that the imbalances in the number of known targets between useful and unsafe compounds inhibits classifier training. We end by suggesting several methods for resolving target imbalance and discuss how to best optimize classifier training for biological networks.

422664 Using Functional Programming and GPU Parallel Computing to Improve Stochastic Optimization

Abstract:

Using Functional Programming and GPU Parallel Computing to Improve Stochastic Optimization

Peter F. Muehlebach Chemical and Petroleum Engineering Department, University of Kansas, Lawrence, KS, Andy Gil, Electrical Engineering and Computer Science, University of Kansas, Lawrence, KS, Kyle V. Camarda, Chemical and Petroleum Engineering, University of Kansas, Lawrence, KS

In chemical engineering, there is a need to isolate near-optimal solutions from large non-convex sets. This work focuses on using functional programming and parallel computing on a GPU to decrease the computation time of stochastic optimization algorithms. One way to implement optimization algorithms more efficiently is to apply parallel computing. When computing algorithms in parallel, they are limited by the number of cores on a processor. Even the highest level i7 processors can only run eight threads (two per core on four cores). However, running programs on a graphics processing unit (GPU) rather than the more traditionally utilized central processing unit (CPU) can better utilize parallel computing. Typically, GPUs have thousands of cores available to execute threads, but they cannot handle irregular calculations. Since most stochastic optimization methods require repeated simple mathematical calculations, parallel GPU computing provides an opportunity to compute solutions quickly and efficiently.

Unrecognized runtime errors and side effects can make calculations within a stochastic optimization method more challenging to make parallel than they need to be. To address these issues, this project implements Haskell, a very strongly typed, purely functional programming language. The strict typing within Haskell will eliminate the chance of unrecognized runtime errors that are common in object oriented programming languages. Since Haskell is a functional programming language, functions within Haskell do not produce side effects. Therefore, Haskell is an appropriate candidate for writing optimization algorithms to be executed on GPUs. This will improve parallelization efficiency, allowing for quicker and more comprehensive optimization solutions.

A Tabu search algorithm was written in Haskell and applied to a computational molecular design problem involving ionic liquids as mass separating agents. Tabu search is a stochastic method that uses an adaptive memory strategy. The method was first tested on a binary knapsack problem, and then adapted to fit the ionic liquid design problem. The Haskell version of the code was compared to the Microsoft VBA version of the program, which is currently implemented in Dr. Kyle Camarda's research group at the University of Kansas. In conclusion, the use of Haskell increased the efficiency of parallel computing with this algorithm. Run time curves show a substantial decrease in computation time when the algorithm is run in parallel on a GPU rather than a CPU.

427742 Optimization of Primary Separation Rate of Magnesium Electrolysis Cell Based on Thermo-Electro-Magneto-Hydrodynamics Coupling Model

Monday, November 9, 2015
Exhibit Hall 1 (Salt Palace Convention Center)
Cheng-Lin Liu[1], Ze Sun[1], Mengjie Luo[2], Gui-Min Lu[1], Xing-Fu Song[1] and Jian-Guo Yu[1,2], (1)National Engineering Research Center for Integrated Utilization of Salt Lake Resources, East China University of Science and Technology, Shanghai, China, (2)State Key laboratory of Chemical Engineering, East China University of Science and Technology, Shanghai, China

Optimization of primary separation rate of magnesium electrolysis cell based on thermo-electro-magneto-hydrodynamics coupling model

Cheng-Lin Liu[1], Ze Sun[1], Meng-Jie Luo[2], Gui-Min Lu[1], Xing-Fu Song[1], and Jian-Guo Yu[1,2,*]

[1] *National Engineering Research Center for Integrated Utilization of Salt Lake Resource, East China University of Science and Technology, Shanghai, China.*

[2] *State Key Laboratory of Chemical Engineering, East China University of Science and Technology, Shanghai, China.*

liuchenglin@ecust.edu.cn

Abstract

Magnesium has found a variety of applications due to a number of advantages including low mass density and high specific strength (Brown, 2007; Eliezer et al., 1998). Like the Hall-Herault process of aluminum production (Perron et al., 2007), the electrolysis process for magnesium is one of the most energy intensive industrial processes (Eklund et al., 2014). Thus the energy consumption and current efficiency are crucially important indexes. Due to high temperature and strong corrosive conditions involved during the industry production, it is extremely challenging to obtain the experimental data. It is also expensive and impracticable to optimizate the structure by setting up new electrolysis cells. Over the last decade, lots of research efforts have been made on the use of commercial software package to model the flow field (Jain et al., 2008), thermoelectric field (Dupuis, 2013), Electrohydrodynamic field (Rezvanpour et al., 2012), thermoelectromechanical model (Richard et al., 2001) and magnetohydrodynamic model (Severo et al., 2005). However, much attention has been paid on the aluminum reduction cell and little effort has been made on the magnesium electrolysis cell. In summary, most of the reported studies of magnesium electrolysis cell only considered the mathematical model based on one or two physical fields. Little has been found on the thermo-electro-magneto-hydrodynamics coupling model.

This paper presents a thermo-electro-magneto-hydrodynamics model for the magnesium electrolysis cells to investigate the distributions of electric filed, temperature field, magnetic field and flow field simultaneously. The model mainly need to solve the following governing equations:

$$\nabla \cdot (\sigma \nabla V) = \frac{\partial \delta}{\partial t}$$

$$\nabla \cdot (\lambda \nabla T) + Q = 0$$

$$\boldsymbol{D} = \varepsilon_0 \boldsymbol{E} + \boldsymbol{P}$$

$$\boldsymbol{B} = \mu_0 (\boldsymbol{H} + \boldsymbol{M})$$

$$\frac{\partial(\varphi_k \rho_k)}{\partial t} + \nabla \cdot (\varphi_k \rho_k \boldsymbol{u}_k) = 0, \qquad k = l, g$$

$$\frac{\partial(\varphi_k \rho_k \boldsymbol{u}_k)}{\partial t} + \nabla \cdot (\varphi_k \rho_k \boldsymbol{u}_k \boldsymbol{u}_l) = -\varphi_k \nabla p + \nabla \cdot (\varphi_k T_k^{turb}) + \varphi_k \rho_k \boldsymbol{g} + \boldsymbol{F}, \ \ k = l, g$$

As shown in Figure 1, the voltage drop of cathode, anode and electrolyte are 0.175 V, 0.549 V, and 1.007 V respectively. Electrolyte temperature is about 700 ºC. The heat generation is 2.11°Á105 W and energy dissipation is 2.13°Á105 W. The flow field of mathematical model was validated by the cold model experiments of PIV in the previous research (Liu et al., 2015). In Figure 2, three circulations, circulation A under the metal separating compartment, circulation B closed to the cathode, and circulation C closed to the free surface of the electrolysis compartment, are existed in the cell. Circulation A and C are beneficial for the separation of magnesium droplets and the magnesium droplets in the circulation B will float to the free surface of the electrolysis compartment, which will increase the contact time of magnesium and chlorine. The effects of the Lorentz force on the motion of the electrolyte is investigated in the work.

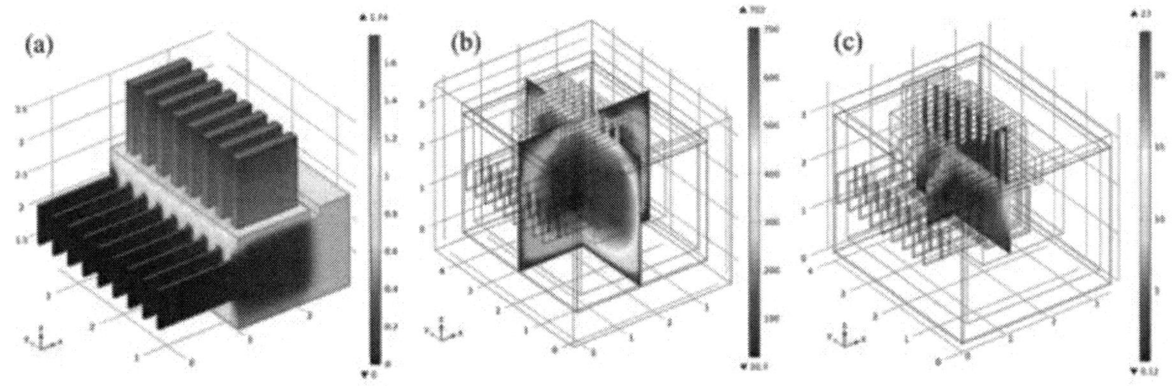

Figure 1 120 kA magnesium electrolysis cell: (a) electric field, (b) temperature field, and (c) magnetic field.

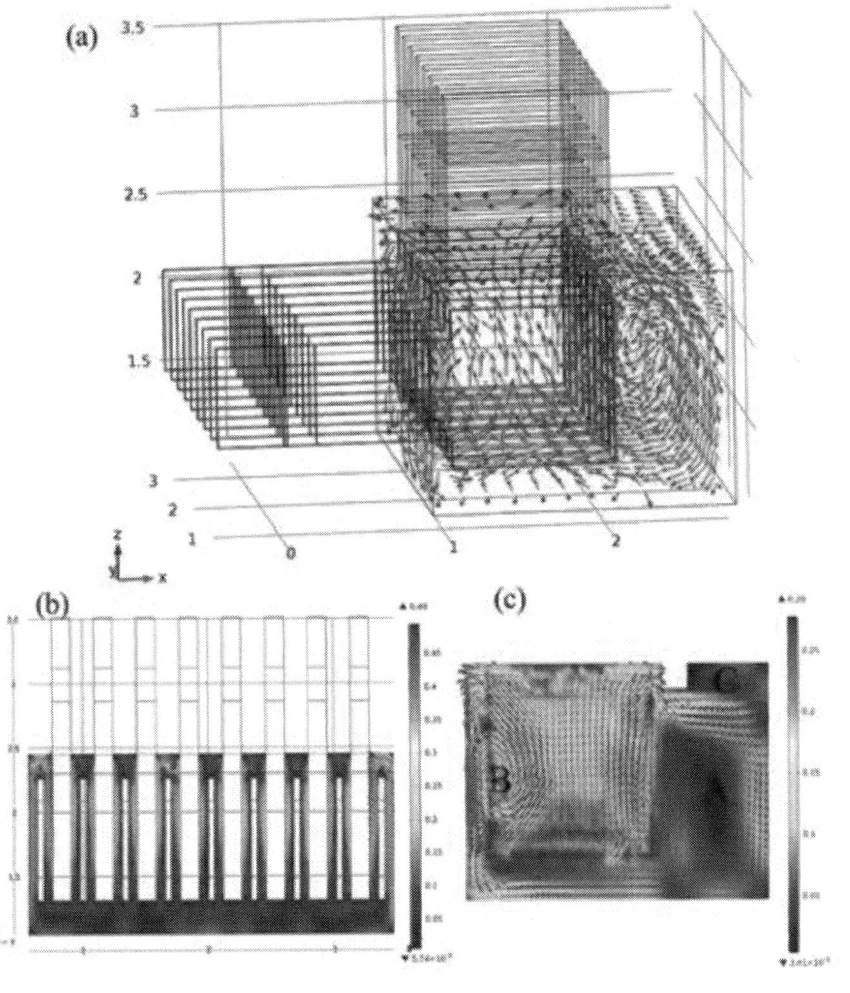

Figure 2 Flow field distribution in the 120 kA magnesium electrolysis cell (a) velocity vectors, (b) flow field in the channels, and (c) velocity vectors in the cross section.

After generated at cathode, the magnesium droplets finally gather at the free surface of the electrolysis compartment and metal separating compartment. Those magnesium droplets floating on the free surface of metal separating compartment are sucked out of the electrolysis cell; and other magnesium droplets will be drawn into the circulation of electrolyte again. Most of the magnesium droplets finally flow into the metal separating compartment after multiple cycles. Therefore, we define the process that magnesium droplets first floating on the free surface as the primary circulation of magnesium droplet. The rest of the cycles is defined as the secondary circulation of magnesium droplet. The ratio of the quantity of magnesium droplets in the metal separating compartment during the primary circulation to the whole quantity of magnesium droplets generated at the cathode is defined as the primary separation rate of magnesium droplets (PSR). In order to improve the electrolysis efficiency, the cell needs to increase its primary separation rate of magnesium droplets.

Particle tracing module of COMSOL is used to simulate the trajectory path of magnesium droplets. The density distribution of particle is proportional to the normal current density. Movement of magnesium droplets at x-z plane in the electrolyte is shown in Figure 3. At t=50 s, particles exist all over the electrolyte. After 300 s, almost all the particles float on the electrolyte and the PSR is about 16.6%, which means that the original cell has a lot of room for improvement.

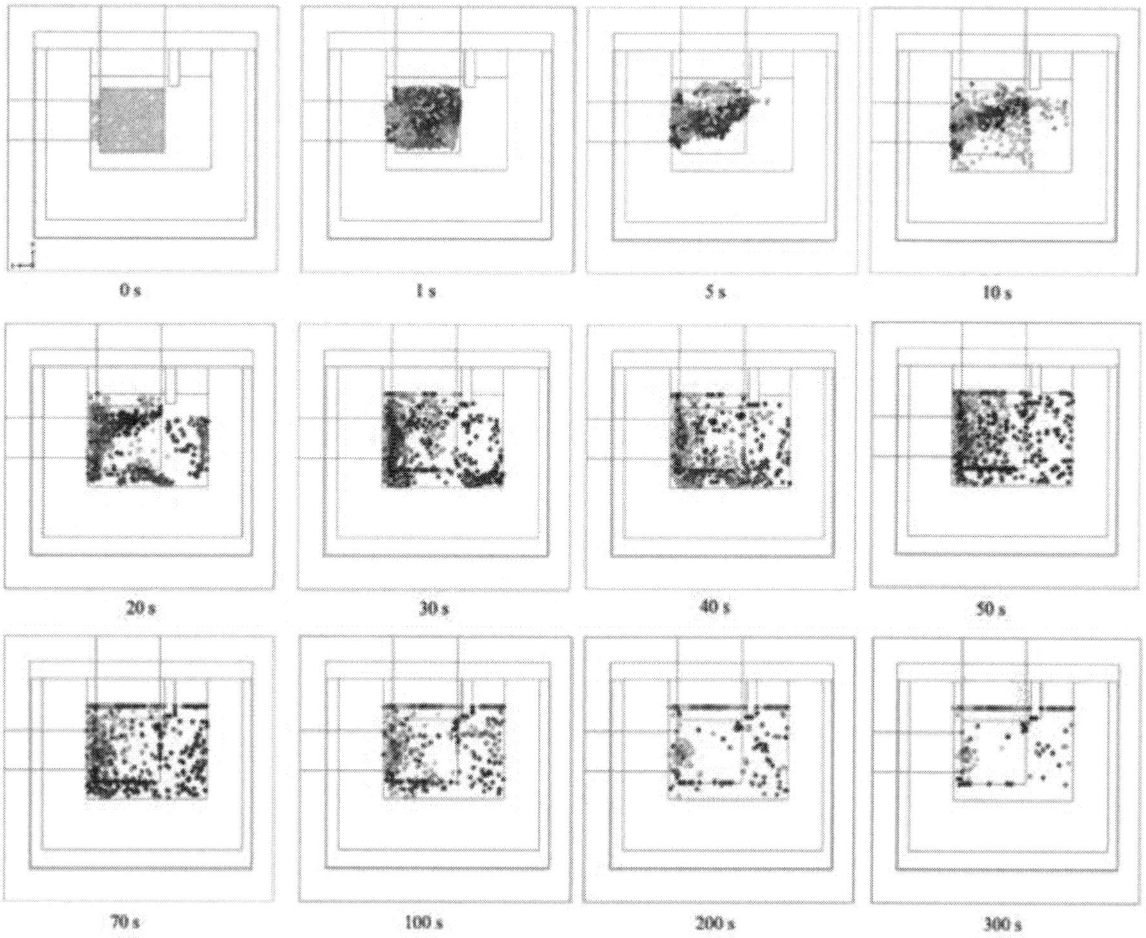

Figure 3 Movement of magnesium droplets in the electrolyte from 0 s to 300 s.

Through adjusting the structure and process parameters of magnesium electrolysis cell, the PSR has no significant improvement. Thus, two new-type of cathodes with hollow channel and separator are used in this work to farther increase the PSR. A new-type cathode with a 0.07 m hollow channel is used to gather the electrolyte and push it into the metal separating compartment. The PSR in the new type structure is about 30.1% under the same condition.Another optimized electrolysis cell, which has a separator between the electrolysis compartment and metal separating compartment, is used to guide the electrolyte flow. The separator is combined with cathode, which can increase the working area of cathode. The variation of the optimized cell with separator is 61.9%. Figure 4 shows the movement of magnesium droplets in the electrolyte from 0 s to 300 s. Unlike the original cell, most magnesium droplets in the optimized cell are involved into the circulation A at 20 s, and finally float on the free surface of electrolyte in the metal separating compartment. The separator has an obvious effect in the improvement of PSR.

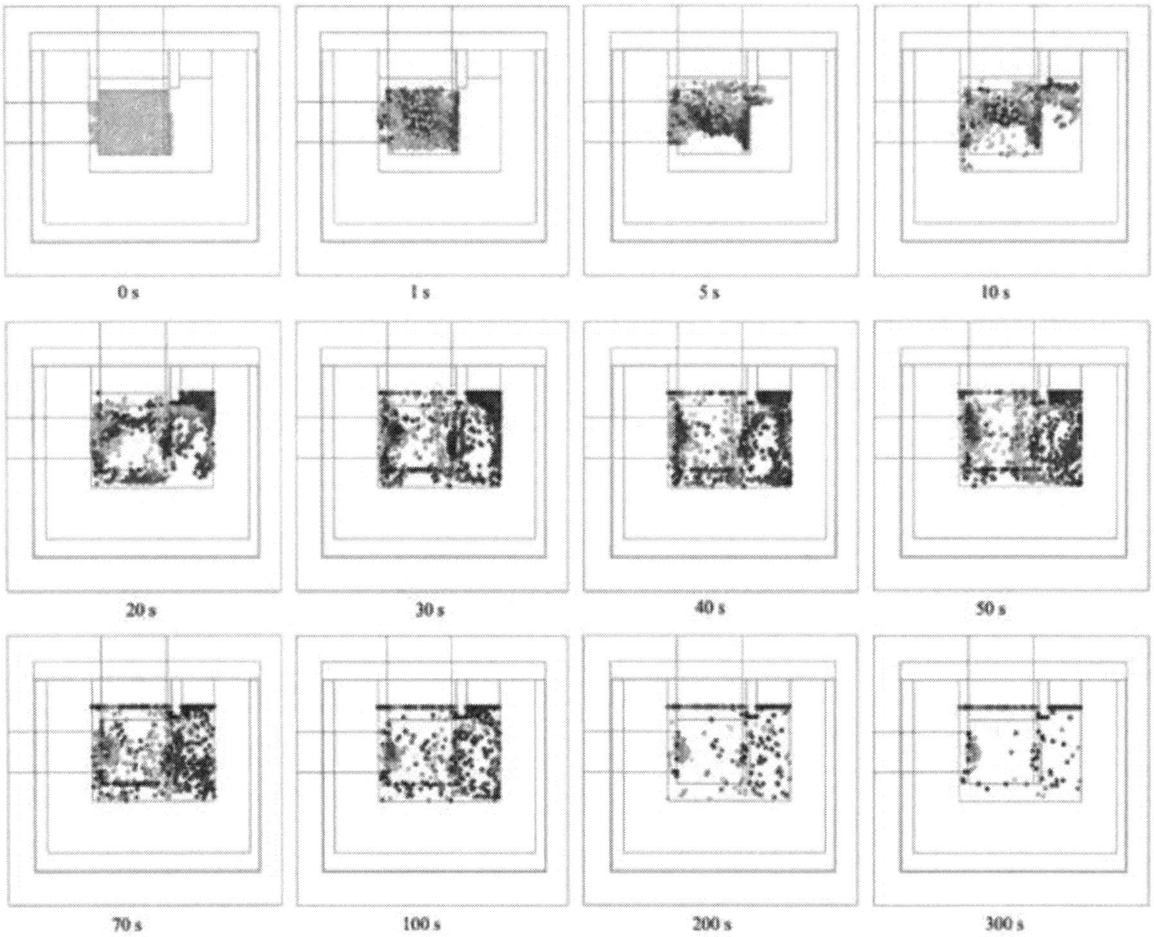

Figure 4 Movement of magnesium droplets in the electrolyte from 0 s to 300 s

Keywords: magnesium electrolysis cell, multiphysical fields, primary separation rate

References

Brown, R.E., 2007. Magnesium Industry Review 2006/2007. Light Metals 65,(1), 59.

Dupuis, M., 2013. Computation of aluminum reduction cell energy balance using ANSYS® finite element models. In: G. Bearne, M.D.A.G. (G. Bearne, M.D.A.G.)^(G. Bearne, M.D.A.G.s)|,*Essential Readings in Light Metals: Aluminum Reduction Technology. John Wiley & Sons, Inc., Hoboken, NJ, USA, pp. 409-417.

Eklund, H., Engseth, P.B., Langseth, B., Mellerud, T., Wallevik, O., 2014. An Improved Process for the Production of Magnesium. John Wiley & Sons, Inc., pp. 141-144.

Eliezer, D., Aghion, E., Froes, F.H., 1998. Magnesium Science, Technology and Applications. Advanced Performance Materials 5,(3), 201-212.

Jain, P., Biegler, L.T., Jhon, M.S., 2008. Parametric study and estimation in CFD-based PEM fuel cell models. AIChE Journal 54,(8), 2089-2100.

Liu, C., Sun, Z., Lu, G., Song, X., Yu, J., 2015. Experimental and numerical investigation of two-phase flow patterns in magnesium electrolysis cell with non-uniform current density distribution. The Canadian Journal of Chemical Engineering, 93:565¨C579.

Perron, A.L., Kiss, L.I., Poncsak, S., 2007. Mathematical model to evaluate the ohmic resistance caused by the presence of a large number of bubbles in Hall-H¨¦roult cells. Journal of applied electrochemistry 37,(3), 303-310.

Rezvanpour, A., Lim, E.W.C., Wang, C., 2012. Computational and experimental studies of electrohydrodynamic atomization for pharmaceutical particle fabrication. AIChE Journal 58,(11), 3329-3340.

Richard, D., Fafard, M., Lacroix, R., Cl¨¦ry, P., Maltais, Y., 2001. Aluminum reduction cell anode stub hole design using weakly coupled thermo-electro-mechanical finite element models. Finite elements in analysis and design 37,(4), 287-304.

Severo, D.S., Schneider, A.F., Pinto, E.C., Gusberti, V., Potocnik, V., 2005. Modeling magnetohydrodynamics of aluminum electrolysis cells with ANSYS and CFX. Light Metals, 475-480.

431588 Matrix Formulation of Unifac/Uniquac Models for the Construction of Residue Curve Maps

Abstract:

Using the matrix formulation of the UNIFAC/UNIQUAC equations, we were able to obtain explicit formulas for the evaluation of activity coefficients and their partial derivatives. These formulas lead to a direct and explicit reduction of residue curve map (RCM) models from a set of differential-algebraic equations (DAEs) of index-1 to an equivalent set of ordinary differential equation (ODEs). Thus this approach allows for a wider selection of solution methods than those limited to handling DAEs such as the mass-matrix formulations. More importantly, the standard tools can now be used for qualitative analysis of the RCM models, e.g. the explicit construction of Jacobian matrices for stability analysis. The paper also includes example applications of the methods discussed to ternary and quaternary systems that exhibit multiple azeotropes.

432023 Nonlinear Control of a Tubular Reactor with Recycle Using Modified APOD and Deim

Abstract:

In recent years, control of nonlinear distributed parameter system (DPS) has been of critical importance due to the simultaneous existence of diffusion, convection and reaction phenomena in a lot of chemical processes, such as catalytic reactors and heat exchangers. These phenomena lead to spatial variation of the process variables which significantly complicates the control and estimation problems. One approach is to construct reduced order models (ROMs) using Galerkin's method by decomposing the dependent variables into temporal part and spatial part (or basis function); the latter is obtained by analytically solving the eigenfunction problem of the linear operator of the governing equation. Since this analytical approach may not be applicable in complex systems, proper orthogonal decomposition (POD) is widely used to construct basis functions. In this algorithm, an ensemble of process operation snapshots is required. However, there is no method to guarantee the system is properly excited when collecting snapshots [1]. Another disadvantage is sampling too few snapshots may lead to inaccurate ROMs, even when the system is properly excited since their relative importance will not be captured.

One solution to this issue is adaptive proper orthogonal decomposition (APOD) [2]. In this method, basis functions are updated when new snapshots become available since the optimal basis function that captures the dominant feature of the system may change over time. However, it still has limitations. The first issue is as snapshots are discarded based on their importance; new snapshots may keep being eliminated when new snapshots show significant difference from previous ones. Another disadvantage is more basis functions might be used than necessary due to the inaccuracy of eigenvalues estimated in this algorithm.

To circumvent the above problems, we propose a refinement to APOD. First, we construct a weight matrix based on the importance vector of basis functions to keep new snapshots from being eliminated. In the basis function updating part, the residual which is norm of difference between ROM and original system is used to determine whether basis functions need to be updated and whether the size of basis needs to increase or decrease.

An issue with ROMs is that the resulting ordinary differential equations (ODEs) system may still be computationally expensive to integrate due to the way nonlinearities in the original partial differential equation(s) are handled. This can lead to delays in the computation of control action. To circumvent this issue, discrete empirical interpolation method (DEIM) [3] is adopted to reduce computational cost. In this method, the nonlinear term in governing equation in the whole domain is estimated by measurement at k points with k much less than the number of spatial grid points using nonlinear basis functions. Nonlinear basis functions are constructed off-line and the positions of those k points are determined based on DEIM algorithm.

We propose a combination of APOD and DEIM to design nonlinear controllers of reduced computational requirements to force the closed-loop DPS evolution to a desired operating point. To illustrate the performance of the proposed control scheme and investigate the stability of it, it is applied to a tubular reactor with recycle to eliminate hot spot formulation.

REFERENCES

[1] Pitchaiah, S., & Armaou, A. (2010). Output feedback control of distributed parameter systems using adaptive proper orthogonal decomposition. Industrial & Engineering Chemistry Research, 49, 10496–10509.

[2] DB Pourkargar and Antonios Armaou.(2013). Modification to adaptive model reduction for regulation of distributed parameter systems with fast transients. AIChE Journal, 59(12):4595–4611.

[3] Saifon Chaturantabut and DC Sorensen. (2010). Nonlinear model reduction via discrete empirical interpolation. SIAM Journal on Scientific Computing, 32(5):2737–2764.

STRUCTURAL SIMILARITY BETWEEN THE ORDERED PAIRS OF PRIMES AND NON-PRIMES

TAEKYOON PARK, YEONSOO KIM, AND JONG MIN LEE

Dedicated to Someone Standing on the Shoulders of Giants

ABSTRACT. There have been various approach to prove Goldbach's conjecture using analytical number theory. We go back to the starting point of this famous probelm and are able to show that the number of Goldbach partition is related to that of ordered pairs of non-primes. This proof is based on the world's first dynamic model of primes and can be a key to identify the structure of prime numbers.

1. INTRODUCTION

Although a number of studies have been conducted on prime number, many parts of primes remain unknown. Riemann hypothesis, twim prime problem and Goldbach's conjecture are typical examples of the famous unsolved problems. In 1742, Christian Goldbach suggested that every integer greater than 6 can be represented as a sum of three primes. This conjecture can be divided into two different statement: weak Goldbach's conjecture and strong Goldbach's conjecture. The weak Goldbach's conjecture states that every integer greater than 5 can be written as the sum of three primes[1]. Vinogradov showed that this conjecture holds for number larger than $3^{3^{15}}$ and this value is reduced to 3.33×10^{43000} by Chen. The strong Goldbach's conjecture states that every even integer greater than 2 can be written as the sum of two primes. Unlike the weak conjecture, much of this conjecture still remain unsolved. In this study, dynamic model for Goldbach partition is suggested and the structural similarity between the ordered pairs is shown as the first step for proving the strong Goldbach conjecture.

Date: August 7, 2014.

Key words and phrases. Goldbach's conjecture, prime number, Goldbach partition, structural similarity, number theory.

We would like to thank all who believe and support us.

2. Preliminaries: Goldbach's conjecture

Goldbach's conjecture is one of the oldest and famous unsolved problems in mathematics which states: Every even integer greater than 2 can be expressed as the sum of two primes. It can be easily expressed as follws.

$$2n = p_1 + p_2 \tag{1}$$

p_1 and p_2 are both primes for $\forall$ n $\geq$ 2, n $\in N$

This can be transformed as follows for using the geometric characteristics.

$$2n = (n - \alpha) + (n + \alpha) \tag{2}$$

$(\exists \alpha \in Z$ s.t. $(n - \alpha)$ and $(n + \alpha)$ are both primes and $0 \leq \alpha < n$ for $\forall n \geq 2, n \in N)$

This means that we have to show the existence of the integer α which makes both $(n-\alpha)$ and $(n+\alpha)$ primes. Also it can be interpreted that there are always two pair of primes which have same distance of α apart to n $\geq$ 2.

3. The main mathematical results

Let $\mathbf{G_n} = \{((n - \alpha), (n + \alpha)) \mid 0 \leq \alpha < n, \alpha \in Z\}$ be a set of given n. It is easey to classify elements in $\mathbf{G_n}$ into 4 groups: $a_n = (p,p)$, $b_n = (p,q)$, $c_n = (q,p)$ and $d_n = (q,q)$. Here p is prime nad q is not prime. Now, Goldbach's conjecture is equivalent with the following proposition.

$$n(a_n) \neq 0\, for\, \forall n \geq 2\, and\, n \in N \tag{3}$$

Since $\mathbf{G_n}$ and its 4 small groups of element varies with a value of n, now we will call $\mathbf{G_n}$ as a dynamic model of primes. It is necessary to observe a difference between $\mathbf{G_{n+1}}$ and $\mathbf{G_n}$ and 8 different flows (v_j) are introduced in Figure 1 where $\mathbf{G_n}$ is represented as a **Reactor n**. For representing the dynamic model in terms of each flow, the following function are used.

$$\pi(n) = \text{the number of prime } p \text{ which satisfies } p \leq n \tag{4}$$

$$\delta(n) = \begin{cases} 1, & \text{if n is prime} \\ 0, & \text{otherwise} \end{cases} \tag{5}$$

list numbers from 1 to 2n and fold:

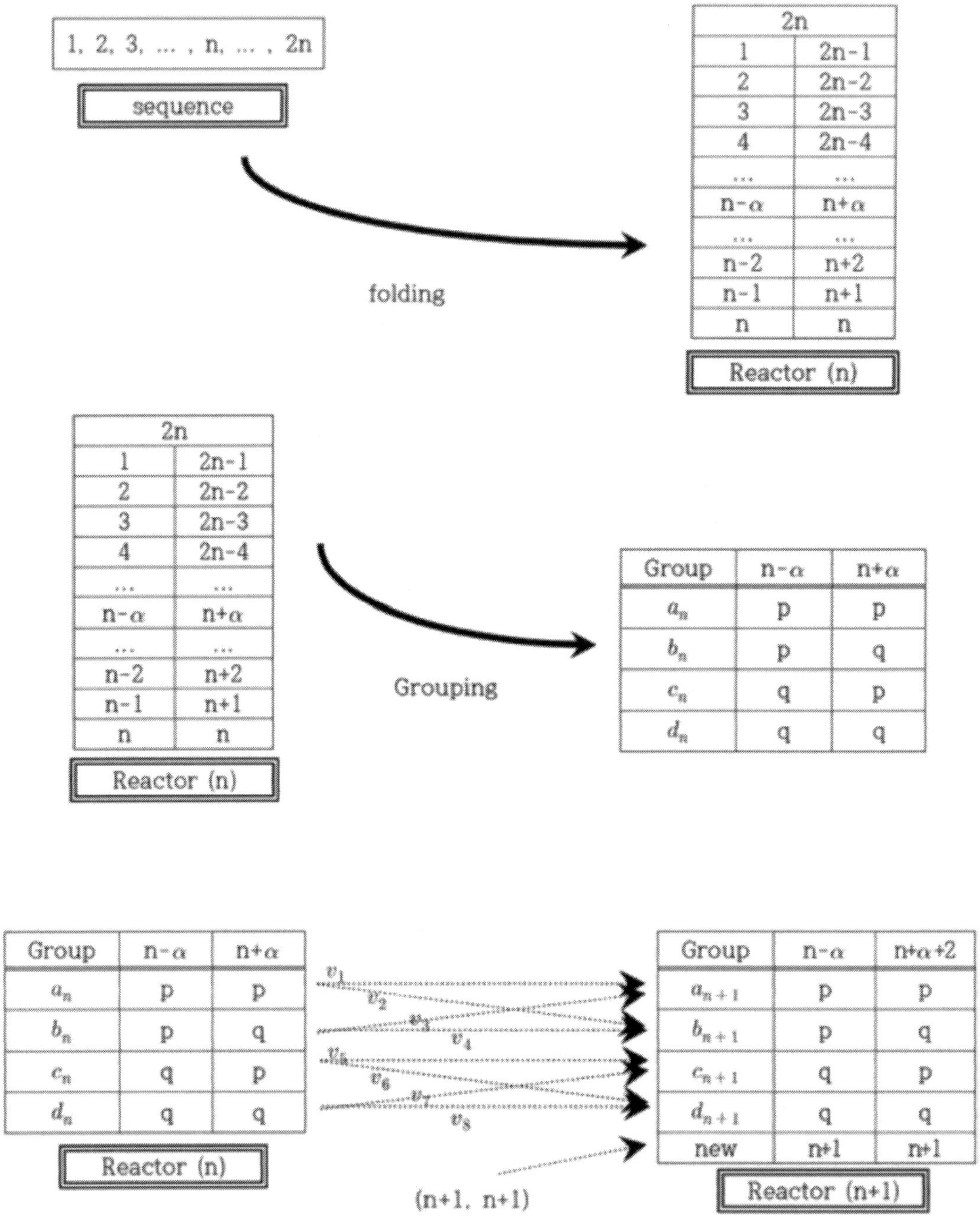

FIGURE 1. Dynamic model of ordered pairs

Now it is possible to present the dynamic model of primes in terms of v_j and given function.

First, it is obvious that n($\mathbf{G_n}$)=n, the overall balance equation can be obtained as follows.

(6)
$$n(\mathbf{G_n}) = n(a_n)+n(b_n)+n(c_n)+n(d_n) = v_1+v_2+v_3+v_4+v_5+v_6+v_7+v_8 = n$$

Second, the number of prime p such that $1 \leq p \leq n$ is only found in the left part of ordeered pairs in a_n and b_n, we can easily have the additional balance equation.

(7)
$$n(a_n) + n(b_n) = v_1 + v_2 + v_3 + v_4 = \pi(n)$$

The number of prime p such that n $\leq p \leq 2n$ is only in the right part of ordered pairs in a_n and c_n and here we get the balance equation for primes which are in right part of ordered pairs in $\mathbf{G_n}$.

(8)
$$n(a_n) + n(c_n) = v_1 + v_2 + v_5 + v_6 = \pi(2n) - \pi(n) + \delta(n)$$

Finally, when $\mathbf{G_n}$ is changed to $\mathbf{G_{n+1}}$ as n increases by one, we can observe the change of each flows and set the balance equation for primes which are in right part of ordered pairs in $\mathbf{G_{n+1}}$.

(9)
$$v_1 + v_3 + v_5 + v_7 + \delta(n+1) = \pi(2n+2) - \pi(n+1) + \delta(n+1)$$

Since we have to show $n(a_n) \neq 0$ for $n \geq 2$, it is necessary to observe Δa_n.

(10)
$$\Delta n(a_n) = n(a_{n+1}) - n(a_n) = (v_1+v_3+\delta(n+1)) - (v_1+v_2) = v_3 - v_2 + \delta(n+1)$$

Similarly, Δd_n can be obtained.

(11)
$$\Delta n(d_n) = n(d_{n+1}) - n(d_n) = (v_6+v_8+1-\delta(n+1)) - (v_7+v_8) = v_6 - v_7 + 1 - \delta(n+1)$$

If we combine the Eq.10 and Eq.11, the relation between Δa_n and Δd_n can be represented as a function of n and v_j.

(12)
$$\Delta n(a_n) = \Delta n(d_n) + v_3 - v_2 + v_7 - v_6 + 2\delta(n+1) - 1$$

To convert $(v_3 - v_2 + v_7 - v_6)$ to a function of n, we can use Eq.8 and Eq.9. Then Eq.12 can be reduced as follows.

(13) $\Delta n(a_n) = \Delta n(d_n) + \delta(2n+1) + \delta(n+1) - \delta(n) - 1 = \Delta n(d_n) + \eta(n)$

Here $\eta(n) = \delta(2n+1) + \delta(n+1) - \delta(n) - 1$.
Since both n and $(n+1)$ cannot be primes except $n = 2$, it is easy to find the value of $\eta(n)$ based on the observation of 6 cases of possible when n $\geq$ 3.

$$\eta(n) = \begin{cases} 1 & \text{if 2n+1, n+1} \in P \text{ and n} \in Q \\ 0 & \text{if 2n+1, n} \in Q \text{ and n} \in P \text{ or if 2n+1} \in P \text{ and n+1, n} \in Q \\ -1 & \text{if 2n+1, n+1, n} \in Q \text{ or if 2n+1, n} \in \text{P and n+1} \in Q \\ -2, & \text{if 2n+1, n+1} \in \text{Q and n} \in P \end{cases}$$

This relation can be seen in table I and it can be found that the similar shape in graph of $n(a_n)$ and $n(d_n)$ is due to the function $\eta(n)$. Figure 2 represents the graph of $n(a_n)$ and $n(d_n)$. Although the $n(d_n)$ has values greater than $n(a_n)$, when we focus on $\Delta n(a_n)$ and $\Delta n(d_n)$, the relation $\Delta n(a_n) - \Delta n(d_n) = \eta(n)$ holds for all n $\geq$ 3. The similar tendency of the dark band in both $n(a_n)$ and $n(d_n)$, which is generated from the bottom to the top, can be explained by this.

TABLE 1. Relation between ordered pairs of primes and non-primes

n	$\Delta n(a_n)$	$\Delta n(d_n)$	$\Theta(n)$
3	0	1	-1
4	1	1	0
5	-1	0	-1
6	1	0	1
7	0	2	-2
8	0	0	0
9	0	0	0
10	1	1	0
11	0	1	-1
12	0	0	0
13	-1	1	-2
14	1	1	0
15	-1	-1	0
16	2	2	0
17	0	2	-2
18	-2	-3	1
19	1	3	-2
20	1	1	0
21	-1	-1	0
22	1	1	0
...	...	...	...
1442	40	41	-1
1443	-50	-50	0
1444	20	21	-1
1445	14	15	-1
1446	-29	-29	0
...	...	...	...
4995	-167	-166	-1
4996	-4	-3	-1
4997	157	158	-1
4998	-156	-156	0
...	...	...	...

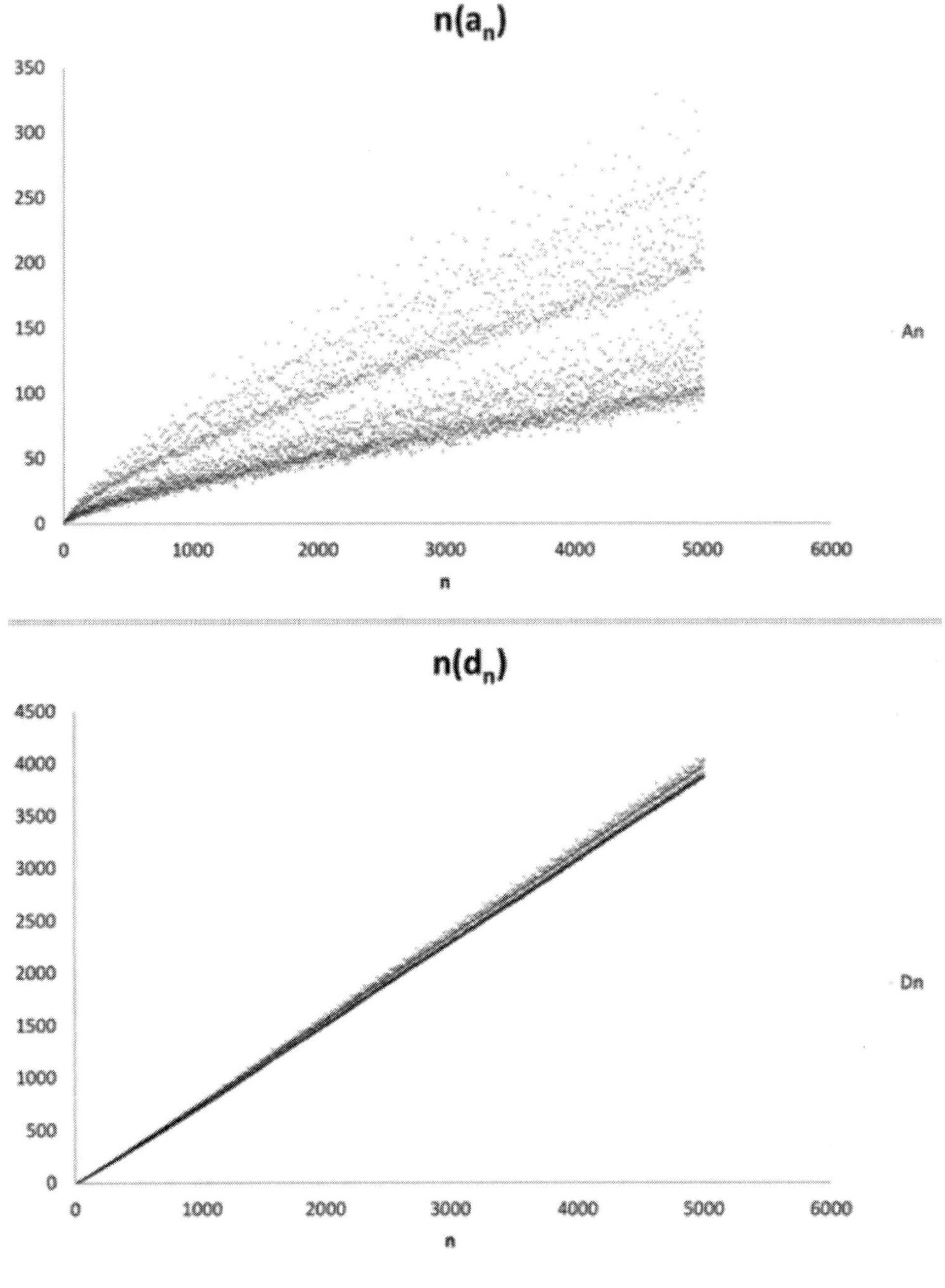

FIGURE 2. Graphs of $n(a_n)$ and $n(d_n)$

Likewise, the relation between $n(b_n)$ and $n(c_n)$ can be derived easily.

$$\Delta n(b_n) = v_3 - v_2 = -\Delta n(a_n) + \delta(n+1) \tag{14}$$

$$\Delta n(c_n) = v_7 - v_6 = -\Delta n(d_n) - \delta(n+1) - 1 \tag{15}$$

From Eq.13, Eq.14 and Eq.15,

$$\Delta n(b_n) = \Delta n(c_n) - \delta(2n+1) + \delta(n+1) + \delta(n) = \Delta n(c_n) + \theta(n) \tag{16}$$

Here $\theta(n) = -\delta(2n+1) + \delta(n+1) + \delta(n)$.
Since both n and $(n+1)$ cannot be primes except $n = 2$, it is easy to find the value of $\theta(n)$ based on the observation of 6 cases of possible when n $\geq$ 3.

$$\theta(n) = \begin{cases} 1 & \text{if 2n+1, n} \in Q \text{ and n+1} \in P \text{ or if 2n+1, n+1} \in Q \text{ and n} \in P \\ 0 & \text{if 2n+1, n, n+1} \in Q \text{ or if 2n+1, n} \in P \text{ or if 2n+1, n+1} \in P \\ -1 & \text{if 2n+1} \in P \text{ and n, n+1} \in Q \end{cases}$$

This relation can be seen in Figure 3 and it can be found that the similar shape in graph of $n(b_n)$ and $n(c_n)$ is due to the function $\theta(n)$. When we focus on $\Delta n(b_n)$ and $\Delta n(c_n)$, the relation $\Delta n(b_n) - \Delta n(c_n) = \theta(n)$ holds for all n $\geq$ 3. The similar tendency of the dark band in both $n(a_n)$ and $n(d_n)$, which is generated from the top to the bottom, can be explained by this.

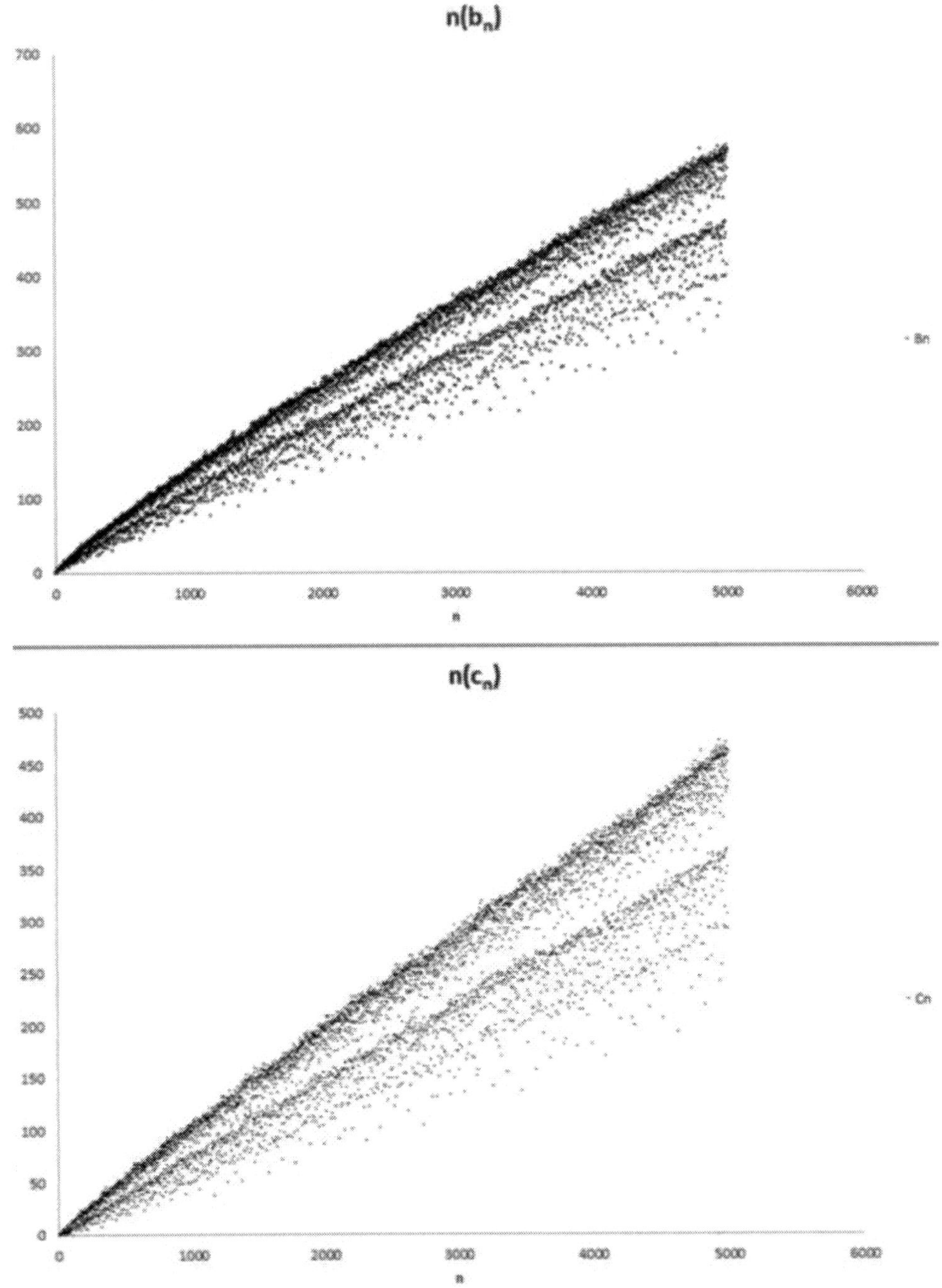

FIGURE 3. Graphs of $n(b_n)$ and $n(c_n)$

4. Conclusions

We have shown that there is a structural similarity between the ordered pairs in $\mathbf{G_n}$. Goldbach partition, $n(a_n)$, and the number of expression for even number as the sum of the two composite number, $n(d_n)$ have a similar structure which can be desribed as $\Delta n(a_n) = \Delta n(d_n) + \eta(n)$. From this, it is possible to explain the similar tendency in the graph of both $n(a_n)$ and $n(d_n)$ quantitatively. Meanwhile, it can be also found that the number of Goldbach partition has a special periodicity based on the suggested dynamic model. This will be discussed further in next paper.

References

[1] KR. Kanchu and S. Kak, *On Randomness of Goldbach Sequences.* arXiv:1207.6307

Chemical and Biological Engineering, Seoul National University, Korea
E-mail address: ptk2201@snu.ac.kr

Chemical and Biological Engineering, Seoul National University, Korea
E-mail address: younsoo333@snu.ac.kr

Chemical and Biological Engineering, Seoul National University, Korea
E-mail address: jongmin@snu.ac.kr

Modeling base to design controlled release fertilizer matching with nitrogen uptake of rice plant

Thanh H. Trinh, Universiti Teknologi PETRONAS, Seri Iskandar, Perak, MALAYSIA
, Ho Chi Minh City University of Technology, HCMC, VIETNAM

KuZilati KuShaari, Universiti Teknologi PETRONAS, Seri Iskandar, Perak, MALAYSIA

Keywords: *Fertilizer Design, Diffusion, Finite Element Method, Inverse Modeling*

Abstract

Controlled release fertilizer (CRF) has attracted a lot of interested because of its advantages on ecosystem in terms of enhancement of nutrient utilization and reducing nutrient loss from leaching, volatilization. As for supplement nutrient from fertilizer is synchronized with nutrient demand, surplus available plant nutrients in soil are avoided. Modeling, therefore, becomes a necessary tool to predict release behavior and determine good CRF design that match the nutrient requirement from the plants. The present study proposes a method to predict/design CRF parameters matching nitrogen uptake from rice plant based on multi-diffusion model. According to our results, there are several choices to manufacturer, because the design solution is not unique. However, the most stable design can be obtained as analyzed standard error (SE) variation against coating thickness for different granule sizes. In this case, the optimum urea radius and coating thickness are 1.125 mm and 115 μm, respectively. In addition, the simulated release curve can match the nitrogen uptake from rice plant according to our optimal parameters.

Introduction

Any fertilizer, whether in the natural, inorganic, or organic form, can harm the environment if misused (Chien et al., 2009). Hence, the aim of slow release or controlled release fertilizer (SRF/CRF) is to prevent nutrient loss and enhance nutrient utilization efficiency by plants (Trenkel, 1997). The amount of fertilizers recovered by plant when applied in conventional forms is only 30–50%, while CRFs make least possible losses of the fertilizer through volatilization or leaching, prevention of the seedling damage and full protection of the ecosystem in the case of biodegradable carriers (Al-Zahrani, 1999). During growth of plants, they require different quantities of nutrients for different stages. They need smaller quantities during infancy and larger amounts during the development of roots, stalk, and stem. Fertilizer demands therefore change periodically over the entire time span of plant growth (Trinh, KuShaari, Shuib, et al., 2015). It is of interest to develop a fertilizer of which the release profile can match the nutrient uptake from plants. Organic manure (natural fertilizer) has the advantage that it can provide a steady supply of nutrients over a long period of plant growth (de L'Azote, 1983). However, the rate of the release cannot be predicted. Besides, the inorganic fertilizer gives a better control of nutrient supply. The novel fertilizer, CRF, can overcome the disadvantages of both two fertilizer types and combine their advantages, i.e., steady nutrient supply over the long period, appropriate nutrient amount for each stage of plant development, and ability to predict the release behavior.

According to the report of Shaviv et al. (2003), first study on the application of controlled release technology to fertilizers was conducted in 1962 by Ortil et al. The release from a single granule of a polymer coated CRFs consists of three stages: an initial stage during which no release is observed (lag period), a stage of constant release, and finally a phase of gradual decay of the release rate. In 2015, Trinh et al. (2015) proposed a multi-diffusion model for the release of Agrium® coated urea by assistance of COMSOL Multiphysics, and they extend to the porous model (Trinh, KuShaari, and Basit, 2015) for the case of imperfect coating and soil environment. The model is robust and reliable because it was tested with a wide range of coated fertilizer from literature such as Shaviv (2003), Lu & Lee (1992), and Lu (2007).

The nutrient uptake has been investigated by numerous researchers on various plant species. Rosmarina et al. (2014) analyzed the utilization of nitrogen fertilizer from hybrid rice using ^{15}N isotopic tracer technique. To develop a good CRF for a specific plant species, the release from that fertilizer at least have to match with the nutrient uptake of the plant for the whole growth period. This study aims to set up a method, based on multi-diffusion model (Trinh, KuShaari, Shuib, et al., 2015), to predict/design CRF geometric parameters of which its release profile can match the nitrogen uptake from rice plant.

Model development

When the coating layer is saturated, urea begins to release through the coating layer by means of diffusion. Based on mass transport equation in porous medium, a transient stage for urea diffusion through the coating could be written as (Lu et al., 2007; Savithri S. and Manju M. S., 2009)

$$D_{eff}\left[\frac{\partial^2 C}{\partial r^2}+\frac{2}{r}\frac{\partial C}{\partial r}\right]=\varepsilon\frac{\partial C}{\partial t} \quad (1)$$

where C is concentration of urea in mol m^{-3}, D_{eff} is effective diffusivity of urea in porous medium in m^{2}s^{-1}, ε is porosity of the coating in percentage (%).

Since the urea pellet is motionless, it is assumed that urea flux from the interface of coating to the liquid is controlled by the diffusion of urea in liquid, and calculated by the following equation:

$$D_{urea}\frac{\partial^2 C(r,t)}{\partial r^2}+\frac{2D_{urea}}{r}\frac{\partial C(r,t)}{\partial r}=\frac{\partial C(r,t)}{\partial t} \quad (2)$$

Initial values and boundary conditions are specified for the model based on each stage of the release that can be found elsewhere (Trinh, KuShaari, Shuib, et al., 2015).

In the context of this work, it is assumed that parameters of coating material has been determined and follows Trinh et al. (Trinh and KuShaari, 2015). The model is applied to design the geometric parameters of CRF, such as radius (R_0) and coating thickness (l), and modeling scheme is illustrated in **Figure 1**. Parametric solver has been utilized to determine appropriate geometries parameters of CRF, and this information are summarized in **Table 1**. Sj rice plant has been chosen as an object in the design application, and the nitrogen uptake data are obtained from Rosmarina et al. (Rosmarina et al., 2014). By assuming that all the nutrient release can be up taken by plant, and there is no loss of nutrients during the release, a good CRF should provide a similar release profile to the nitrogen uptake curve. In other words, we can use the normalized amount of nitrogen uptake and release to compare and design CRF application. Standard error (SE) is used to calculate the difference between the nutrient release from simulation and that uptake by plant for the whole plant growth period as follows

$$SE = \sqrt{\frac{\sum\left(N_{uptake} - N_{release}\right)^2}{n}} \qquad (3)$$

where N_{uptake} and $N_{release}$ are the normalized amount of nitrogen uptake by plant and release from fertilizer at specific observation times; n is the number of observations.

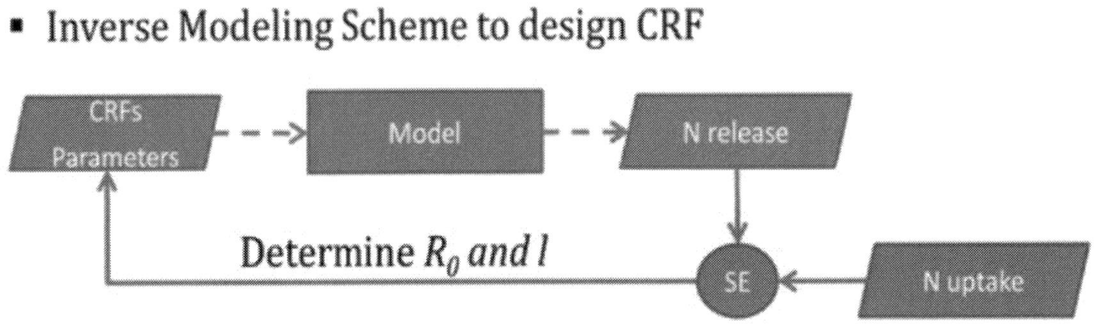

Figure 1. Modeling scheme to design CRF

Table 1. Simulation parameters are used to predict/design CRF fertilizer based on a specific coating material

Parameter	Value
Core radius (R_0), mm	0.3 – 3
Coating thickness (l), μm	10 – 150
Lag time (t_0), day	14
Effective Diffusivity (D_{eff}), m^2s^{-1}	1.72×10^{-14}

Results and Discussion

Figure 2a illustrates the relationship between coating thickness and standard error (SE) for various granule sizes (R_0). For each R_0, there will be an optimum coating thickness where the release of CRF can match the nitrogen uptake of rice plant for the whole application. These results lead to multi solutions in the design of CRF parameters. Therefore, if the real uncoated urea granule has been specified, the optimum coating thickness of the chosen material can be determined easily. However, in some circumstances, the optimum thickness may fall in an inappropriate range, i.e., quite thin in practical application or too thick comparing with the total size of the urea granule. In these cases, the designer has to consider the SE values of those optimum points.

According to **Figure 2**, there are a few solutions that can be obtained from the model. However, the urea granule of 1.125 mm gives more stable release profile as coating thickness varied in a wide range from 110 to 130 μm (presented by very small and stable SE in this range). Because coating thickness is a property that has much influence to the release profile and usually is non-uniform due to the coating process, a stable release profile in a wide range of coating thickness will be the preferable one in CRFs producing. Furthermore, in this investigation case, SE reaches minimum at 0.0308 as urea granule and coating thickness reaches 1.125 mm and 115 μm, respectively. Hence, the latter are optimum parameters which give a release profile matching with nitrogen uptake of Sj rice specie planting in Selangor.

To confirm that the release profile can match with the nutrient uptake of Sj rice, another simulation has been done. Figure 3 illustrates the profile of simulated nitrogen release at designed parameters versus that of nitrogen uptake of Sj rice in Selangor (Rosmarina et al., 2014). It clearly confirms that, with the design parameters, nutrient release profile is synchronized with that uptake requirement from Sj rice. The nutrient release profile can give best nutrient utilization efficiency for the chosen plant as the application is applied after 10–14 day of planting. Further investigation on its effective under field experiment and other rice species design will be discussed in another report.

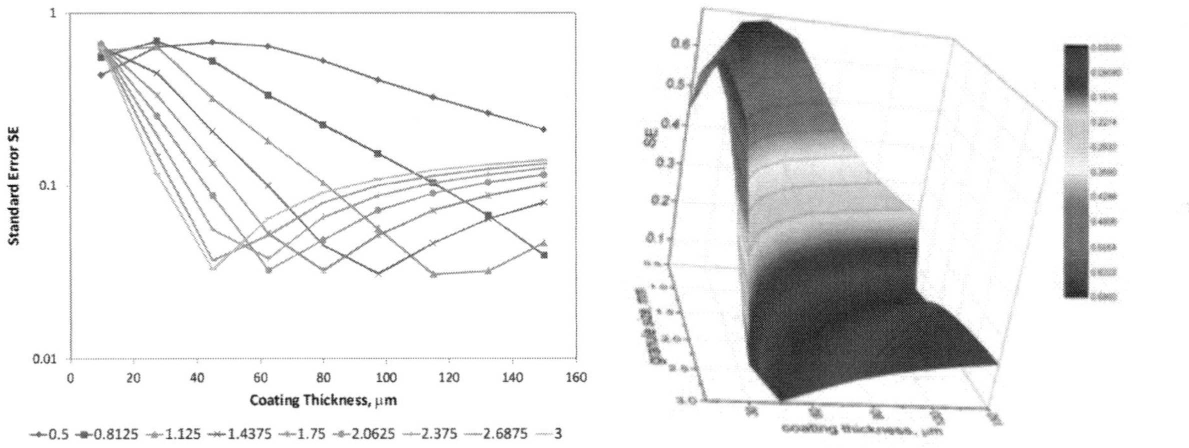

(a) (b)

Figure 2. (a) Relationship between coating thickness and SE for different granule sizes; (b) Relationship between R_0, l and (SEE) in surface plot. D_{eff} and t_0 is 1.72×10^{-14} m^2s^{-1} and 14 day, respectively.

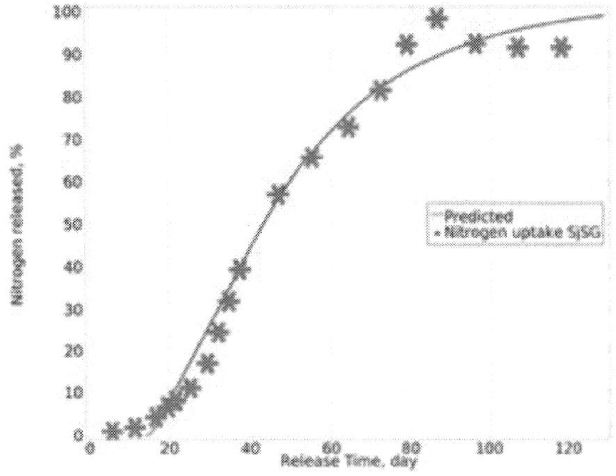

Figure 3. Simulated nitrogen release profile at designed parameters (R_0 = 1.125 mm; l = 115 μm). Start dots present nitrogen uptake from Sj rice (adapted from Rosmarina et al. **(2014, 41)**).

Conclusion

This paper had presented a solution to design CRF parameters that gives a release profile matching with nutrient requirement from plant. There are several choices to the manufacturers because the design solution is not unique. However, the most stable design can be obtained as considering the SE variation against coating thickness for different granule sizes. In the other word, the stable design should have less variation in release behavior when coating thickness varies in a wide range. Therefore, this solution is useful to those who investigate novel CRF's materials and designs.

References

1. S.H. Chien, L.I. Prochnow, H. Cantarella (2009), Chapter 8 Recent Developments of Fertilizer Production and Use to Improve Nutrient Efficiency and Minimize Environmental Impacts, In *Advances in Agronomy*, , ed. Donald L. Sparks.
2. N.C.D.É. de L'Azote (1983), Handbook on Environmental Aspects of Fertilizer Use.
3. S.M. Lu et al. (2007), Urea Release Rate from a Scoop of Coated Pure Urea Beads: Unified Extreme Analysis, *Journal of the Chinese Institute of Chemical Engineers*, 38, 3–4, pp. 295–302.
4. S.M. Lu, S.F. Lee (1992), Slow Release of Urea through Latex Film, *Journal of Controlled Release*, 18, 2, pp. 171–180.
5. A.K. Rosmarina, Y.M. Khanif, K. B. Abdul Rahim (2014), Fertilizer Nitrogen Utilization By Hybrid Rice Using ^{15}N Isotopic Tracer Technique, In *3rd OneBAJA Postgraduate Colloquium. The next Generation Green and Economical Urea.*
6. Savithri S., Manju M. S. (2009), Kinetics of the Carbothermal Reduction of Ilmenite: Grain Pellet Model, *Excerpt from the Proceedings of the COMSOL Conference 2009.*
7. Avi Shaviv, Smadar Raban, Elina Zaidel (2003), Modeling Controlled Nutrient Release from Polymer Coated Fertilizers: Diffusion Release from Single Granules, *Environ. Sci. Technol.*, 37, 10, pp. 2251–2256.
8. M.E. Trenkel (1997), Controlled-Release and Stabilized Fertilizers in Agriculture.
9. Thanh H. Trinh, KuZilati KuShaari, Anis S. Shuib, et al. (2015), Modelling the Release of Nitrogen from Controlled Release Fertiliser: Constant and Decay Release, *Biosystems Engineering*, 130, pp. 34–42.
10. Thanh H. Trinh, KuZilati KuShaari (2015), Dynamic Modelling of Nitrogen Release from Coated Urea for Prediction of Coating Material Diffusivity and Lag Time, In *ACEX2015 Abstract Book.*
11. Thanh H. Trinh, KuZilati KuShaari, Abdul Basit (2015), Modeling the Release of Nitrogen from Controlled-Release Fertilizer with Imperfect Coating in Soils and Water, *Industrial & Engineering Chemistry Research*, 54, 26, pp. 6724–6733.
12. S.M. Al-Zahrani (1999), Controlled-Release of Fertilizers: Modelling and Simulation, *International Journal of Engineering Science*, 37, 10, pp. 1299–1307.

Acknowledgment

The project is supported by OneBaja Project (Universiti Teknologi Petronas).

424577 Prediction of the Impact of Water Fraction on Emulsion Behavior Using Quantitative Structure-Property Relationship (QSPR) Modeling

Abstract:

Predicting emulsion behavior and transient properties is critical for several industrial applications, particularly in the energy and biofuels industries. The complex and dynamic nature of these systems make their computational prediction using conventional methodologies difficult. Quantitative Structure-Property Relationship (QSPR) modeling, however, is a robust framework which can be applied to predict emulsion behavior and account for the complexities that arise in these systems. This work utilizes QSPR to specifically incorporate the impact of water fraction on emulsion behavior and properties. Optical microscopy is used to quantify droplet size distributions for inclusion in the QSPR modeling framework.

(242i) Algorithm to Deal with Very Large and Very Small Numbers to Enable Utilization of Analytical Solution of Free Radical Polymerization during Gel Effect

Garg, D., Shiv Nadar University
Serra, C., University of Strasbourg
Hoarau, Y., University of Strasbourg

Algorithm to Deal With Very Large and Very Small Numbers to Enable Utilization of Analytical Solution of Free Radical Polymerization During Gel Effect

Dhiraj Kumar Garg[a], Christophe A. Serra[b], Yannick Hoarau[c]

[a]Dept. of Chemical Engineering, Shiv Nadar University, dhiraj.garg@snu.edu.in, [b]ICS (UPR 22 CNRS) & University of Strasbourg, ca.serra@unistra.fr, [c]ICUBE (UMR 7357 CNRS) & University of Strasbourg, hoarau@unistra.fr

Introduction

Numerical modelling and simulation is a great tool used extensively now-a-days. But to get correct results using it, several precautions/guidelines have been discussed by several researchers. To keep numerical simulations stable and to get correct results out of them is every researcher's priority. There are several situations which may be adverse to this requirement and everyone wants to avoid them and dealing with very large and very small numbers (inverse of very large numbers) is one of them. From machine point of view, a very large number would be some which is larger than the largest number stored and processed by that machine and the machine shows overflow for this condition. Similarly for the very small number and the machine shows underflow for the same. The other problems that are faced because of very large numbers are truncation error, round off error, subtractive cancellation and problem associated with adding a large and a small number (Chapra and Canale 2007). Exponential, power and factorial functions are some functions which are notorious in building up very large numbers rapidly.

We faced one such problem during the application of analytical solution (AS) of free radical polymerization (FRP) (Garg et al. 2014a) with gel model (Garg et al 2014b; 2014c) to simulate gel effect during polymerization.

Problem Faced

The AS for FRP worked excellently before gel effect as expected but there was some problem in getting correct results for one of the variables while modelling and simulating gel effect at lower temperatures (<90°C) using AS with different gel models. The variable was identified to be μ_2, second order moment of dead polymer chain length distribution whose AS consisted of infinite series solution for case 1 and 2 (Garg et al. 2014a) as given in eqn. (1).

$$[\mu_2]_n = [\mu_2]_{n-1}.V_{corr} + P.(1+R_{SA})^2.D_{n-1}.\left[\sum_{m=1}^{\infty}\left[\left(\frac{(C_{n-1})^m}{m.m!} - \frac{(C_{n-1}.\Delta y)^m}{m.m!}\right)\right] - ln(\Delta y)\right].V_{corr} + (2P-1).(1+R_{SA}).\{[M]_{n-1}.V_{corr} - [M]_n\} + (P-1).2f.\{[I]_{n-1}.V_{corr} - [I]_n\}$$

(1)

The important portion of this equation causing problem is given below.

$$D_{n-1}.\left[\sum_{m=1}^{\infty}\left[\left(\frac{(C_{n-1})^m}{m.m!}-\frac{(C_{n-1}\Delta y)^m}{m.m!}\right)\right]-ln(\Delta y)\right] \quad (2)$$

where $D_{n-1}=\frac{2(K_{pr}[M]_{n-1})^2}{K_t K_d}e^{-C_{n-1}}$, (3)

n represents n^{th} time step and m represents the mth term.

Considering only those terms which were causing problem, eqn. 2 reduces to eqn. 4

$$e^{-C_{n-1}}.\left[\sum_{m=1}^{\infty}\left[\left(\frac{(C_{n-1})^m}{m.m!}-\frac{(C_{n-1}\Delta y)^m}{m.m!}\right)\right]-ln(\Delta y)\right] \quad (4)$$

which, for the ease of notation, can be rewritten as

$$e^{-x}.\left[\sum_{m=1}^{\infty}\left[\left(\frac{1}{m}.\frac{x^m}{m!}-\frac{1}{m}.\frac{z^m}{m!}\right)\right]-ln(\Delta y)\right] \quad (5)$$

where $x=C_{n-1},\ z=C_{n-1}.\Delta y$ (6)

Three functions resulting in very large (or small) numbers can be noticed i.e. e^{-x}, x^m, *and* $m!$. Now, from physical aspects of the problem

$x, z >> 1$ (7)

$0<\Delta y<1$ (8)

So $x > z$ (9)

In our previous work (Garg et al. 2014) we have shown and proven that the infinite series present in the eqn. 5 are converging series for any value of x, with m^{th} term under the limit $\lim_{m\to\infty}\left(\frac{1}{m}.\frac{x^m}{m!}\right)=0$. This provides the stopping criteria to calculate the sum of the series and thus reducing the truncation error by accommodating large enough number of terms. The physical reality of the problem states that eqn. 4 should have finite small value irrespective of how large (or small) values individual functions may or can have.

Modelling and simulations are usually performed by computational software like Matlab and they have limitation on maximum and minimum numbers they can work with for a given computing machine. Just to elaborate the problem further, lets see what is the practical upper limit of values of x and m that they can have during computations against physical reality. The maximum value Matlab can process is $1.7977\times10^{+308}$. Generally such software have inbuilt library functions for mathematical functions which restricts maximum value of $m = 169$ for factorial function, $m! = 169! = 4.2691 \times 10^{+304}$ and lowest value criteria implies $m = 745$ for exponential function $e^{-745} = 4.9406 \times 10^{-324}$. Power function values corresponding to maximum value of $x = 745$ and $m = 169$ respectively are $x^m = 745^{107} = 2.0928 \times 10^{+307}$ and $x^m = 66^{169} = 3.1837 \times 10^{+307}$.

So we have practical limitations on the upper limits of values of x and m. It limits the number of terms $m \leq 169$ for $x \geq 66$ before going overflow for factorial and power function and limits the value of $x \leq 745$ for $m \geq 107$ to avoid underflow for exponential function. This limitation will have serious truncation errors during the computation of eqn. 4. The problem is further compounded by the physical aspect of FRP where, for lower temperatures < 90°C, the value of $x = C_n$ can go upto the order of 10^3 during gel effect. It is quite clear that computations will simply fail for such large values for $x = C_n$ cannot be processed. Hence the approach to evaluate various functions like exponential, power and factorial, separately and then using them for the final computation in the equation is neither an efficient nor effective strategy. Thus we need to devise some new algorithm to calculate each term in such a way that the no term or number overflows or underflows at any stage while keeping the truncation error minimum and that too without increasing the computational load by including unnecessarily large number of terms.

Solution of the problem

A new efficient and effective algorithm is proposed to avoid dealing with very large (and small) numbers during computation for any value of $x = C_n$. Following measures were taken in that:

1) Exponential, power and factorial function were not evaluated independently in eqn. 4 for each term.

2) Exponential term was clubbed with each term of infinite series instead of calculating it separately.

3) Each term of infinite series was calculated using natural log to avoid very large numbers.

4) Calculation of term of second series (RHS) is calculated from term of first series to avoid the duplication of calculations.

$$\frac{(x.\Delta y)^m}{m.m!} = \frac{(x)^m}{m.m!}.(\Delta y)^m \quad (10)$$

5) Successive terms were obtained from previous terms using a recursive relationship, thus again avoiding duplication of calculations as well as avoiding the need to use power and factorial functions.

$$m^{th}\ term = \frac{(x)^m}{m.m!} = \frac{(x)^{m-1}}{(m-1).(m-1)!}.x.\frac{m-1}{m^2} = (m-1)^{th}\ term \times x.\frac{m-1}{m^2}\ for\ m > 1 \quad (11)$$

6) Instead of calculating the sum of series and then taking the difference, sum of the series of

differences of terms were calculated with the help of a logic loop.

7) To ensure that sufficient numbers of terms are taken for the summation to keep the truncation error low as well as to avoid unnecessarily large number of terms without much gain towards summation, following criteria MUST BE USED to stop the above mentioned summation loop (step 6).

i. The relative change in the sum of the series of differences to be less than pre-specified tolerance value, AND

ii. n^{th} (current) term becomes lesser than $(n\text{-}1)^{th}$ (previous) term of the first series in eqn. 4., AND

iii. n^{th} (current) term becomes lesser than some pre-specified tolerance value.

The criteria mentioned in step 7 is the most important step to make this algorithm both efficient with running the loop only for required number of terms instead of fixed number to keep the truncation error low, and effective to accommodate large values of $x = C_n$ as well as m without ever reaching overflow or underflow.

This algorithm has been applied successfully in the AS of FRP. Just to give an idea as to how much large numbers this algorithm could handle, a successful modelling and simulation was run for step change in FRP batch reactor from 70°C to 50°C, at t = 45 min after starting the reaction. During computations at different times, the maximum value of m and C_n reached was 1408 and ≈1327 are way higher than 169 and 745 respectively as shown earlier. This gives the numbers as e^{1327}, 1327^{1408} and $1408!$ which are clearly very large numbers and whose calculations were easily bypassed by using above algorithm.

References

1. S.C. Chapra, R.P. Canale, Numerical Methods for Engineers, (2007) Tata McGraw Hill, New Delhi.

2. D.K. Garg, C.A. Serra, Y. Hoarau, D. Parida, M. Bouquey, R. Muller, Analytical Solution of Free Radical Polymerization: Derivation and Validation. Macromolecules,2014, 47 (14), 4567?4586

3. D.K. Garg, C.A. Serra, Y. Hoarau, D. Parida, M. Bouquey, R. Muller, Analytical Solution of Free Radical Polymerization: Applications- Implementing Gel Effect Using CCS Model. Macromolecules, 2014, 47 (23), 8178-8189.

4. D.K. Garg, C.A. Serra, Y. Hoarau, D. Parida, M. Bouquey, R. Muller, Analytical Solution of Free Radical Polymerization: Applications- Implementing Gel Effect Using AK Model. Macromolecules, 2014, 47 (21), 7370?7377.

409095 Elastic Net with Monte Carlo Sampling for Data-Based Modeling in Biopharmaceutical Manufacturing Facilities

Monday, November 9, 2015
Exhibit Hall 1 (Salt Palace Convention Center)

Kristen Severson[1], Jeremy G. VanAntwerp[2], Venkatesh Natarajan[3], Chris Antoniou[3], Jörg Thömmes[3] and Richard D. Braatz[4], (1)Chemical Engineering, MIT, Cambridge, MA, (2)Calvin College, Grand Rapids, MI, (3)Biogen, Cambridge, MA, (4)Department of Chemical Engineering, Massachusetts Institute of Technology, Cambridge, MA

Biopharmaceutical manufacturing involves multiple process steps that can be challenging to model using first principles. Oftentimes, operating conditions are studied in bench-scale experiments and then fixed to specific values during full-scale operations. This procedure limits the opportunity to tune process variables to correct for the effects of disturbances. Utilizing process models has the potential to increase the flexibility and controllability of the biomanufacturing processes. This work proposes a statistical modeling methodology to predict the outputs of biopharmaceutical operations. This methodology addresses two important challenging characteristics typical of data collected in the biopharmaceutical industry: limited data availability and data heterogeneity. Motivated by the final aim of control, regularization methods, specifically the elastic net, are combined with sampling techniques similar to the bootstrap to develop mathematical models that use only a small number of input variables. These techniques are of particular interest because of their ability to perform model selection and estimation simultaneously.

Process modeling techniques can be grouped into two broad categories: first-principles and data-based. This work focuses on data-based modeling, which is more often applied in biopharmaceutical manufacturing facilities. Data-based models have been applied to cell culture characterization [1] [2] [3], quality control [4] [5], process monitoring [3] [6] [7] [8], and downstream operations [3]. A drawback of current data-based methods applied in the biopharmaceutical industry is that the models that are produced are not easily interpretable because they rely on subspaces that do not have direct physical meaning.

A successful biopharmaceutical model was defined as achieving three goals: (1) model accuracy, (2) model simplicity, and (3) model interpretability. These aims have the caveat of using only a small amount of heterogeneous data, as data for biopharmaceutical manufacturing are typically both heterogeneous and relatively limited.

One way to achieve these goals is through the identification of the input variables in the process that exhibit the largest effects on the output variables. Regularization methods have been identified as possible approaches for such problems because of their ability to simultaneously handle input selection and model estimation [9]. A particular regularization method, the elastic net [10], was identified because of its ability to handle data with more measurements than observations. Elastic net is an optimization formulation for parameter estimation that is formulated as:

$$\hat{\beta}_{EN} = \arg\min_{\beta_0,\beta} \frac{1}{2N}\sum_{i=1}^{N}(y_i - \beta_0 - x_i^T\beta)^2 + \lambda P_\alpha(\beta),$$

(1)

where

$$P_\alpha(\beta) = \sum_{j=1}^{p}\frac{1-\alpha}{2}\beta_j^2 + \alpha|\beta_j|,$$

(2)

N is the number of experiments, y_i is the ith scalar response, x_i is the p-dimensional data vector at observation i, λ is a nonnegative regularization parameter, β_0 is a scalar parameter, β is a p-dimensional vector of model parameters, and α is on the interval (0,1]. Using this basis, a five-step methodology, referred to as the elastic net with Monte Carlo sampling (ENwMC) is proposed.

The first step in ENwMC is an application of the elastic net, using leave-one-out cross validation to choose the value of α. In leave-one-out cross validation, all but one of the experimental observations are used to fit the model, then the remaining experiment is used to calculate the error. This step is repeated for each possible set and then averaged. The procedure is performed for many possible combinations of the regularization parameters α and λ, where $\alpha \in (0,1]$ and λ captures the convex behavior of the error. Because α is the weighting between the ℓ_2- and ℓ_1-norm penalties and the goal is a sparse model, a value of α close to 1 is preferable. Therefore α is chosen based on a tradeoff between model dimensionality and prediction error. In some cases, this choice is trivial, as a higher value leads to a more accurate model.

Once the value of α is fixed, a test for over-fitting is performed using k-fold cross validation. Using Monte Carlo samples [11], the data are portioned into a validation set containing $1/k$ proportion of the data and a calibration set containing the rest. The elastic net, with a fixed α, is then performed and the input variables corresponding to the minimum error are recorded. This step is repeated many times to converge to the distribution of models over the possible calibration and validation sets. The frequency with which each variable is selected is then calculated. In further analysis, only the variables that were selected above a threshold frequency are considered.

The subset of selected variables is considered for inclusion in a model using best subset selection. The error of all possible ordinary least squares models of size $m \in \{1, 2, \ldots, p\}$, where p is now the dimensions that were chosen based on the threshold, is calculated. A model from this set is then selected based the tradeoff between increasing dimensionality and decreasing error. This tradeoff is easily visualized by plotting the prediction error against the model dimensions to create a Pareto curve. Plots of this type will often exhibit an "elbow." The elbow corresponds to the model dimensionality that compromises between model size and prediction error. The result of this step is the final model.

Figure 1: Simplified flowsheet of the antibody production process. The bioreactor volume was 2000L and was operated in the fed-batch mode. The column loadings were typical of an antibody purification process. [12].

The developed methodology is evaluated on an antibody manufacturing dataset (see Figure 1) and compared to well-known multivariate analysis techniques for the (bio)pharmaceutical field: principal component regression (PCR) and partial least squares (PLS). In a majority of cases, the elastic net technique out-performed PCR and PLS in terms of error and variance (see Table 1). Averaged over all of the output variables that were considered, the sum-of-squared errors decreased 27% and the variance decreased 48% using a regularized model as compared to the latent variable models. The regularized models have the added benefit of being easily interpreted in terms of the process variables.

Table 1: Comparisons of scaled error and variance for PCR, PLS, and ENwMC modeling techniques. The bold number marks the model with the best performance for each variable.

Unit Operation	Output Variable	Error using …			Variance of the prediction using…		
		PCR	PLS	ENwMC	PCR	PLS	ENwMC
Bioreactor	G0 Product Quality	3.79	4.25	**3.41**	0.146	0.148	**0.087**
	Final Titer	5.38	**3.35**	5.40	0.281	0.287	**0.178**
	DNA	7.58	6.77	**5.20**	0.209	**0.201**	0.223
	HCP	4.30	2.85	**1.67**	0.258	0.210	**0.150**
Protein A Column	DNA	4.26	4.33	**2.71**	0.151	0.143	**0.095**
	HCP	4.71	**1.60**	1.92	0.268	0.202	**0.080**
	Total Impurity	9.22	7.98	**2.40**	0.286	0.256	**0.164**
	HMW	2.08	2.54	**1.11**	0.117	0.092	**0.045**
Cation Exchange Column	HCP	**1.57**	1.99	1.96	0.226	0.132	**0.083**
	Total Impurity	7.78	**5.73**	7.18	0.323	0.348	**0.226**
	HMW	1.73	1.45	**0.32**	0.058	0.063	**0.010**
	HCP	2.63	2.59	**1.20**	0.189	0.140	**0.048**

HCP	2.63	2.59	**1.20**	0.189	0.140	**0.048**
Total Impurity	4.65	**1.56**	2.48	0.228	0.227	**0.115**
HMW	0.54	0.24	**0.23**	0.067	0.050	**0.007**

References

[1] S. M. Mercier, B. Diepenbroek, M. C. F. Dalm and R. H. Wijffels, "Mutlivariate data analysis as a PAT tool for early bioprocess development data," *Journal of Biotechnology,* vol. 167, pp. 262-270, 2013.

[2] A. Kirdar, J. Conner, J. Baclaski and A. S. Rathore, "Application of multivariate analysis toward biotech processes: Case study of a cell-culture unit operation," *Biotechnology Progress,* vol. 23, no. 1, pp. 61-67, 2007.

[3] A. S. Rathore, N. Bhushan and S. Hadpe, "Chemometrics applications in biotech processes: A review," *Biotechnology Progress,* vol. 27, no. 2, pp. 307-315, 2011.

[4] Y. Roggo, P. Chalus, L. Maurer, C. Lema-Martinez, A. Edmond and N. Jent, "A review of near infrared spectroscopy and chemometrics in pharmaceutical technologies," *Journal of Pharmaceutical and Biomedical Analysis,* vol. 44, no. 3, pp. 683-700, 2007.

[5] Z. Chen, D. Lovett and J. Morris, "Process analytical technologies and real time process control a review of some spectroscopic issues and challenges," *Journal of Process Control,* vol. 21, no. 10, pp. 1467-1482, 2011.

[6] E. Read, J. Park, R. Shah, B. S. Riley, K. A. Brorson and A. S. Rathore, "Process analytical technology (PAT) for biopharmaceutical products: Part I concepts and applications," *Biotechnology and Bioengineering,* vol. 104, no. 2, pp. 276-284, 2010.

[7] E. Read, R. Shah, B. S. Riley, J. T. Park, K. A. Brorson and A. S. Rathore, "Process analytical technology (PAT) for biopharmaceutical products: Part II concepts and applications," *Biotechnology and Bioengineering,* vol. 105, no. 2, pp. 285-295, 2010.

[8] D. Bonné, M. A. Alvarez and S. B. Jorgensen, "Data driven modeling for monitoring and control of industrial fed-batch cultivations," *Industrial & Engineering Chemistry Research,* vol. 53, pp. 7365-7381, 2013.

[9] S. Pampuri, A. Schirru, G. Fazio and G. De Nicolao, "Multilevel lasso applied to virtual metrology in semiconductor manufacturing," in *2011 IEEE International Conference on Automation Science and Engineering*, Trieste, 2001.

[10] H. Zou and T. Hastie, "Regularization and variable selection via the Elastic Net," *Journal of the Royal Statistical Society, Series B (Statistical Methodology),* vol. 67, no. 2, pp. 301-320, 2005.

[11] N. Metropolis and S. Ulam, "The Monte Carlo method," *Journal of the American Statistical Association,* vol. 44, no. 247, pp. 335-341, 1949.

[12] A. Shukla and J. Thommes, "Recent advances in large-scale production of monoclonal antibodies and related proteins," *Trends in Biotechnology,* vol. 28, no. 5, pp. 253-261, 2010.

423498 Ensemble Locally-Weighted Partial Least Squares Model and Its Application to Industrial Plants

Abstract:

In chemical plants, soft sensors are used to estimate values of process variables that are difficult to measure in real time. Values estimated by soft sensors are employed to control plants rapidly. A crucial problem is degradation of predictive ability of soft sensor models. Due to changes of process states in plants such as catalyst performance loss, changes of raw materials and fouling on pipes, errors of soft sensors increase.

To reduce the model degradation, several types of adaptive soft sensors have been developed over many years. We previously categorized the types of the model degradation and those of adaptive soft sensors, and discuss the characteristics of adaptive soft sensors to use of the right type of adaptive soft sensors for the right type of the model degradation. The three main approaches to constructing adaptive soft sensors are the moving window (MW), just-in-time (JIT), and time difference (TD) approaches. As no all-round adaptive soft sensors exist, the selective use of TD, MW and JIT models could enable the predictive ability of soft sensors to increase dramatically. However, to improve prediction accuracy of each type of adaptive soft sensors contribute the improvement of overall performance in soft sensors.

We focus JIT models in this study. Although many researches exist on similarity such as distance and correlation and ensemble learning in JIT modelling, predictive ability of JIT models depends heavily on the quality of a database. Even when data in X-variables are similar in JIT modelling, data in a y-variable are not always similar due to process changes such as catalyst performance loss and sensor and process drift, which has harmful effects to JIT models.

Data measured in similar time would have the same relationship between X and y, but old data would have a different relationship between X and y even when data in X are similar since there happen process changes such as catalyst performance loss and sensor and process drift. Meanwhile, a model constructed with only recent data cannot adapt to rapid process changes when recent data have little variation.

We therefore propose to prepare multiple sub-datasets and construct a JIT model for each sub-dataset. Locally-weighted partial least squares (LWPLS) is focused in JIT modelling in this study. By constructing LWPLS models with different sub-datasets, the models would adapt to various process states and transition between them. In addition, contamination of data in which relationships between X and y are different can be prevented for each sub-dataset by separating old data and recent data. A final predicted y-value is calculated with weighted average of y-values predicted by multiple LWPLS models. The weights are given considering predictive ability of LWPLS models. The points are the way to divide sub-datasets and a final y-value estimated using multiple LWPLS models

Through case studies using simulated and real industrial datasets, it was confirmed that the use of the proposed method enables soft sensors to operate with high predictive accuracy.

426087 PSO-Optimized BP Algorithm for Prediction of Gasoline Yield of FCC Unit

Abstract:

The on-line prediction of the gasoline yield is important for the catalytic cracking process in the refinery. The conventional algorithm based on the BP neural network can only deal with few data from the refinery and its ability in handling large amount of data is limited and large errors may be introduced. The present work suggested an advanced BP model that is optimized with PSO based on the conventional theory of the BP neural network. The present model is able to perform predictions based on the big data from a refinery through continually modifying the weights and threshold of the BP neural network in order to continually reduce the output error. The present work collected 160,000 groups of data and selected 17 factors (including the properties of the raw material and catalyst and some basic data relating to the operation) that may influence the gasoline yield of the MIP FCC unit from a refinery with a capacity of processing 1.2 million tons of oil per year. The above data was first pretreated and then normalized. The normalized data was set as the initial value and the actual data as the desired value. The conventional BP model and the PSO-optimized BP model were separately applied to predict the gasoline yield. Finally, part of the pretreated data directly collected from the unit was set as the predicted value, which was incorporated into the above two models in order to obtain the predicted value of the gasoline yield. The present work found that the conventional BP algorithm predicted the gasoline yield with a mean squared error of 5.667 while the PSO-optimized BP algorithm predicted the gasoline yield with a mean squared error of 4.644, which indicates that the PSO-optimized BP algorithm is better in predicting the gasoline yield of the catalytic cracking process.

431329 Improvement of Artificial Immune System in Fault Detection and Diagnosis of Chemical Process

Abstract:

Because of the self-adaptation and self-learning ability of artificial immune system(AIS), it is introduced to fault detection and diagnosis of chemical process. In this paper, antibodies and antigens are composed of matrices of time-sameple data of process variables, and normal antibody libraries and fault antibody libraries are later constructed based on clonal selection algorithms. Since chemical process variables are always of a great number and change frequently and irregularly, matrix of each antibody is only composed of variables of the root cause of the corresponding fault, which helps to avoid the interference of changes of other process variables, improve the prediction and accuracy of AIS and save computing time and storage. To meet the requirement of adequate historical data of different kinds of faults, antibody libraries are transplanted among chemical processes of the same kind.

425912 Sensor Modeling

Abstract:

Process model is an important computer-aided tool that has been widely used for different applications such as virtual experimentations, design of sensing and control architecture, process optimization, sensitivity analysis, operating space identification, feasibility and flexibility analysis. The measurable model outputs should ideally match with the corresponding sensors outputs that have been used for real time monitoring of any plant. In order to achieve this, process model standalone is not sufficient and sensor models are also needed. Sensor noise and specifications such as accuracy, precision, response time, operating range, resolution, sensitivity and drift can have significant effects on process monitoring and thereby on model performance [1]. Extensive work has been done in process modeling field [2]. However, no attempt till date has been made for sensor modeling. Therefore, a sensor modeling framework is highly desired to develop a sensor model to be integrated with process model. Process model integrated with sensor models should truly represents the plant and thereby the integrated model can be used more effectively for control system design, optimization and any other applications.

A process is well understood when all critical sources of variability are identified and explained and where the product quality attributes can be accurately and reliably predicted. Continuous pharmaceutical manufacturing processes often consist of a series of unit operations. To ensure acceptable and reproducible results, consideration must be given to the quality attributes of incoming materials and their processing ability for each unit operation. This can be tracked with the help of an inline/online monitoring tool. Process-sensor integrated model can be ideally used for virtual process monitoring if all source or critical process variability have been identified and included in modeling.

In this work, a systematic framework including the methods and tools has been developed for sensor modeling. The framework includes a generic methodology, a model library and an ontological knowledge-based system. The generic methodology provides step-wise guidelines to develop a sensor model. The model library consists of process model that needs to be integrated with sensor model. Examples of process models included in library are unit operations involved in continuous and batch tablet manufacturing process [2], API manufacturing process and fermentation process. Ontological knowledge base consist of commercially available sensors and corresponding performance specifications such as accuracy, precision, response time, operating range, response time, resolution, sensitivity, drift [1]. The process model provides the inputs to a sensor model while the knowledge base system provides the necessary sensor performance specifications. The applications of the framework have been demonstrated through modeling of sensors involved in continuous pharmaceutical manufacturing process.

The inline monitoring tool under consideration here is the Near-Infrared Spectrometer (NIR) [3]. NIR is the most commonly used sensor for monitoring of different variables of the continuous tablet manufacturing process such as blend uniformity, drug concentration and powder bulk density. Therefore, NIR has been considered here to demonstrate the sensor modeling approach. The temperature sensor has been also considered for modeling since it has been widely used in many processes including pharmaceutical. The NIR sensor model is proposed, developed and integrated with the existing unit operation models in the continuous tablet manufacturing process. The main motive of this paper is to simulate and match the process output results with the actual sensor outputs. Efforts have been made to take into account the sensing delay and the noise variation in the sensor models. A detailed analysis of simple sensors like thermocouples, thermistors and resistance thermometer (RTDs) are first taken into consideration. Variables like accuracy, precision, response time, operating range, resolution, sensitivity and drift are considered for the thermocouples, thermistors and the RTDs. This information is then utilized to model the more complex NIR sensor with the same variables. The performance of integrated process-sensor model for control system design has been compared with the one based on without sensor model and found to be better.

The objective of this presentation is two-fold. First to highlight the systematic framework for sensor modeling, and second to demonstrate its applications through continuous pharmaceutical manufacturing case study.

References

1. Singh, R., Gernaey, K. V., Gani, R. (2010). An ontological knowledge based system for selection of process monitoring and analysis tools. Computers & Chemical Engineering Journal, 34(7), 1137-1154.
2. Singh, R., Ierapetritou, M., Ramachandran, R. (2013). System-wide hybrid model predictive control of a continuous pharmaceutical tablet manufacturing process via direct compaction. European Journal of Pharmaceutics and Biopharmaceutics, 85(3), Part B, 1164-1182.
3. Singh, R., Sahay, A., Karry, K. M., Muzzio, F., Ierapetritou, M., Ramachandran, R. (2014). Implementation of a hybrid MPC-PID control strategy using PAT tools into a direct compaction continuous pharmaceutical tablet manufacturing pilot-plant, International Journal of Pharmaceutics, 473, 38–54.

430942 Opportunities in Spectroscopic Analysis: Chemometrics++

Abstract:

Shortly after the proliferation of spectrometers in the 1970s, multivariate analysis techniques such PLS quickly gained recognition for their ability to create predictive models from the large data sets. While these factor analysis approaches will always be an important tool for dimensionality reduction, statisticians and computer scientists continue to develop complimentary tools with ever-increasing sensitivity and specificity. Neural networks, random forests, support vector machines, lasso, and ridge regression are examples of regression algorithms that can have superior performance compared to factor analysis approaches alone.

The widespread availability of these tools, coupled with easy scripting and cloud computing setup allows for the development of workflows that can rapidly formulate and test hundreds of potential regression models and automatically identify the best balance of complexity, sensitivity, and specificity in a quantitative model. Such a workflow will be demonstrated using Python with an example set of lab NIR data starting from the raw interferograms. This presentation aims to extend the discussion of chemometric quantitation beyond conventional multivariate regression of processed absorbance spectra in order to reframe it as an application of modern applied predictive modeling.

432044 Dynamic Causal Modelling and Its Application to an Acid Gas Removal Unit

Abstract:

Dynamic Causal Modelling and Its Application to an Acid Gas Removal Unit

Temitayo Bankole, Debangsu Bhattacharyya

Department of Chemical Engineering, West Virginia University, Morgantown, WV, USA

Controlled variables are typically selected *ad hoc* without due consideration of their economic and control performances. Recently our group developed a thorough and systematic approach to controlled variable selection (Jones, et al., 2014) by considering tradeoff between economic performance and control performance. However as the process model changes during operation, the optimal controlled variables can change. Due to the large number of candidate variables and the large scale optimization problem that needs to be solved for optimal controlled variable selection, it is computationally intractable to perform the selection process dynamically for a large scale process. One option is to divide the plant sections based on their connectivity information and then perform the selection process separately for sections that are weakly connected. However, this will require determining how the connectivity information between different plant sections changes dynamically. Drawing from the neuroscience literature (Friston, 2003) that focuses on dynamic changes in the connectivity among cortical and sub-cortical areas in the human brain, a dynamic causal model is developed where the dynamic data collected from the system are used to determine how the latent connectivity and the induced connectivity change due to external and internal stimulations.

At a cortical level, the neuronal functions including their interactions can be modelled as states that dynamically evoke brain responses as a function of inputs. This is, in essence, similar to a chemical plant where the connectivity strength changes depending on the state of the system and internal and external perturbations. Thus the quantitative information about effective connectivity of the plant sections can be extracted using a bilinear approximation by using the transient data. These connectivity matrices can then be employed as a basis for selecting the controlled variables dynamically. The parameters reflecting the connectivity information are obtained by using an expectation maximization algorithm based on a Bayesian approach. Finally, the algorithm is deployed to a Selexol-based acid gas removal (AGR) unit for selective capture of CO_2 and H_2S as part of an integrated gasification combined cycle (IGCC) unit. The AGR process is highly nonlinear with strong mass and heat integration among different process sections. Our work shows that it is possible to capture the dynamic changes in the connectivity matrix by using appropriate measurement data. In this presentation, we will also discuss the computational issues and efficiency of the proposed algorithm.

Bibliography

Friston, K. J. L. H. W. P., 2003. Dynamic causal modelling. *Neuroimage,* Issue 19, pp. 1273-1302.

Jones, D., Bhattacharyya, D., Turton, R. & Zitney, &., 2014. Plant-wide control system design: Primary controlled variable selection. *Computers & Chemical Engineering,* pp. 220-234.

434029 A Multi-Scale Simulation Platform Facilitating Interactions Using Model-Based Design of Experiments and Surrogate Models

Abstract:

The development of mathematical modeling techniques and computer science over the past decades has greatly increased the possibility to perform detailed simulations of complex systems. However, detailed simulations requires significant, sometimes impractical, computational resources, even when using state of the art hardware. The conventional strategy has therefore been to focus on modeling a particular scale of a system. Consequently, it is common to assume that effects from other scales, e.g. molecular dynamics, can be neglected by invoking assumptions, such as pseudo steady state, about the behaviour or by using constitutive relations.

Multi-scale modeling, on the other hand, involves simultaneous consideration of the different scales of a system, as well as the combination of these scales into an overall model for the system. The objective is to achieve the efficiency of a higher scale model combined with the accuracy of a lower scale. In this context, the interactions between different scales play a significant role in determining the quality of the overall model. Surrogate models provide an inexpensive and efficient way of linking the different models in a multi-scale system. As long as the error of the surrogate model is kept within an acceptable range, the surrogate approximates the behaviour of the complex model on the lower scale well enough in a given design space.

The MoDeNa [1] framework aims at facilitating interactions between the models on across the different scales of the system by extensive use of surrogate models. The respective surrogate models are derived by the experts responsible for the corresponding complex models, but the surrogate models contain parameters, i.e. degrees of freedom, that are used for optimization. The parameters are determined using Design of Experiments (DoE), i.e. formulating experimental designs within a design space such that it leads to a good enough surrogate model, in combination with fitting techniques. Based on the argument that computer simulation are deterministic [3], the statistically optimal points in which the complex simulations should be performed are chosen using DoE methods such as latin hypercube sampling. Finally, the parameters of the surrogate model are fitted to results of these experiments using nonlinear least square optimisation. The procedure, as outlined in [2], is called model-based design of experiments.

However, surrogate models in MoDeNa are 'dynamic', meaning that the parameters of the model can be updated depending on where in the design space the user wants to use the model. In context of MoDeNA, the DoE and parameter fitting are defined as strategies specific to a surrogate model. This means that the model itself can employ these strategies whenever required - depending upon its structure, desired range of validity and application.

The feature is particularly useful because it allows the the user to use the surrogate model outside the design space for which the model was originally validated. MoDeNa handles this situation by expanding the design space beyond the user-specified input, in order to ensure that the input is within the range of the surrogate model, before performing new experiments in the extended space. The parameter fitting procedure is applied afterwards in order to extend the global range of the surrogate model by updating its parameters.

However, it is important to ensure that the physical boundaries of the design variables are not crossed when expanding the design space. Moreover, by studying the trajectory of the user specified inputs in previous runs of the simulation, a predictive algorithm for expanding the design space can be formulated. This helps in reducing the computational time as the surrogate model is updated in a new design space even before the user intends to compute it in that space. The accuracy of the surrogate model is ensured by validating it agains the points from the DoE that was not used in order to fit the parameters.

[1] MoDeNa. The modena project. [www.modenaproject.eu]

[2] Gaia Franceschini and Sandro Macchietto. Model-based design of experiments for parameter precision: State of the art. *Chemical Engineering Science*, 63(19):4846 - 4872, 2008. Model-Based Experimental Analysis. [DOI | http]

[3] Anthony A. Giunta, Steven F. Wojtkiewicz Jr., and Michael S. Eldred. Overview of modern design of experiments methods for computational simulations. In *41st AIAA aerospace sciences meeting and exhibit*, 2003.

434855 Gibbsian Game Theory for Competitive Decisions

Abstract:

Optimization problems involve finding parameters that minimize a given optimization function. However, in many competitive decisions involving conflict or cooperation, the optimization function for one person or group is often very different from the optimization function for another person or group, giving rise to power and political considerations. Over the past half century, competitive decision problems have been represented and solved by game theory, originated by John von Neumann. Later developments by John Nash showed that strategic games always have one or more "Nash equilibria". However, current solutions to game theory problems – here called "Nashian games" – have a number of shortcomings, and do not always conform to experimental data. For example, in the classical "Battle of the Sexes" game, three distinct equilibria arise, but no guidance is provided on how often each of the equilibria will apply. In this talk we treat strategic games as a series of chemical reactions, and so are able to use entropic considerations and minimize a type of Gibbs free energy. We solve classical games like a Prisoners' Dilemma, Battle of the Sexes, and Chicken using the "Gibbsian game theory" approach, which accounts for "entropic choices". We then show how treating strategic games as a set of chemical reactions can be used to represent situations more accurately than typical games, and solve these sets of reactions. This new way of treating social, economic, and political data in situations of conflict, power, and contested decisions could produce new possibilities for mechanistic design.

434972 Event Driven Multivariate Analysis of Eye Gaze Data for Behavior Analysis in Process Operations

Abstract:

Abstract

Accidents at onshore and offshore process facilities often lead to heavy, irrecoverable losses. Investigation reports aftermath such occurrences frequently cite human errors as a primary cause for more than 70% of accidents in the recent past (Leveson, 2004). Even at smaller scale, equipment outage, plant shutdown and various production accidents in process plants are often considered as a result of operator errors viz. slips, lapses, mistakes, violations (Kidam et al., 2010). Despite numerous interventions in the last twenty years by the governments and industry around the world to minimize the impact of human error and to improve process safety, 25% of the accidents that led to largest losses in the hydrocarbon industry over a period of 40 years have happened in the last 5 years from 2009 (Marsh,2014). Analysis suggest that some of the primary causes for human errors in process industries include increase in complexity of the processes, tight energy integration, recycles, presence of advanced control strategies, reduction in number of staffs without an effort to increase their cognitive skills in handling abnormal situations. Especially, the cognitive skills in handling plant abnormal situations play a key role. While the traditional approach has been to look at human error as the initiating event using a set of Human Error Probabilities (Munger et al., 1962), modified approaches that came later extended to cover task/environmental variables along with human-engineering-design characteristics (Swain, 1990). Further ahead, Embrey and Zaed (2010) described a set of techniques, supported by computer based tools, for predicting and preventing human errors in gas plant operations which primarily focused on Hierarchical Task Analysis and Predictive Human Error Analysis. However, none of the approaches to understand human error, till date, do not explicitly account for the cognitive behaviour of the operators under abnormal situations, an important phenomena that influences human error. To address this gap, we seek to understand the cognitive ability of a control room operator during an abnormal plant condition.

Eye movements are considered to offer a real-time window into cognition based on the "eye mind" hypothesis and therefore provide dynamic information about human's cognitive behavior (Just and Carpenter, 1976; Cooke, 2005). Therefore, researchers in various fields, including safety critical systems viz. aviation, health care, have utilized eye tracking measures to analyze different aspects pertaining to cognitive processing. Matos (2010) suggested that the eye tracker is appropriate to monitor the cognitive state of plant operators since it is non-invasive and relatively simple to install and use. In our work, we employed a Tobii TX 300 eye tracker to track and record the eye movement of participants while interacting with the HMI of a plant simulator. As the participant has to look at several dynamic components on the HMI screen (Area of Interest-AOI) while resolving an abnormality, we used AOI dwell duration, AOI visit count, AOI transitions as a measure to understand the cognitive state of the participant during the abnormality. Additionally, researchers have categorized the cognitive tasks during an operation into orientation, diagnosis and execution steps (Chu et al., 1994) and employed eye tracking measures to study them.

In this work, we demonstrate that there are distinct patterns even in the cognitive behaviors of plant operators.

This work has focused on determining the eye gaze patterns that manifest themselves while a control room operator is using a HMI to control a process abnormality. We conducted experiments where graduate students played the role of control room. From their eye gaze data, a strong correlation is evident between AOI-based measures and the orientation, diagnosis, and execution steps while rejecting a process disturbance. It was observed that while the trend pane in the HMI was the most dominant AOI during the entire course of the task, the dominance and relevance of the trend pane does vary among phases. During the initial orientation steps, the primary tag (corresponding to the variable that is directly by the disturbance) is the most used while the trend pane is ranked higher during later phases that correspond to diagnosis and execution. These insights into the cognitive behavior of control room operators provide a new approach to tackling human error in a proactive fashion.

References

Chu, Rose, et al. "Qualitative user aiding for alarm management (QUALM): an integrated demonstration of emerging technologies for aiding process control operators." *Systems, Man, and Cybernetics, 1994. Humans, Information and Technology., 1994 IEEE International Conference on*. Vol. 1. IEEE, 1994.

Cooke, L., 2005. Eye tracking: How it works and how it relates to usability. Technical Communication 52 (4), 456–463.

Embrey, D., and Zaed, S. (2010). A Set of Computer Based Tools Identifying and Preventing Human Error in Plant Operations. In *6th Global Conference on Safety, AIChE* (pp. 1–11). San Antonio.

Just, M. A., and Carpenter, P. A. (1976). Eye fixations and cognitive processes. *Cognitive Psychology*, *8*(4), 441–480. http://doi.org/10.1016/0010-0285(76)90015-3 26

Kidam, K., Hurme, M., and Hassim, M. H. (2010). Technical analysis of accident in chemical process industry and lessons learnt. *Chemical Engineering Transactions*, *19*, 451–456.

Leveson, N. (2004). A new accident model for engineering safer systems. *Safety Science*, *42*, 237–270. http://doi.org/10.1016/S0925-7535(03)00047-X

Marsh (2014). The 100 Largest Losses 1974–2013. Marsh and McLennan Companies. Marshall, S. P. (2007). Identifying cognitive state from eye metrics. *Aviation, Space, and Environmental Medicine*, *78*(5 Suppl), B165–B175.

Matos, R. (2010). *Designing eye tracking experiments to measure human behavior*. Merriënboer, J. J. G. van, and Sweller, J. (2005). *Cognitive load theory and complex learning: Recent developments and future directions. Educational Psychology Review* (Vol.17). http://doi.org/10.1007/s10648-005-3951-0

Munger, S., Smith, R., Payne, W., 1962. An index of electronic equipment operability: Data store. Tech. Rep. AIRC43-1/62-RP(l), American Institute for Research, Pittsburgh.

412235 Design of Decoupling Model-Predictive Control for Multivariable Systems

Abstract:

Most industrial processes are multivariable systems which consist of several measurement and control signals. Multivariable systems often present complicated couplings between controlled and manipulated variables so that process interactions occur. The degree of interactions increases as a result of the high demand on product quality, the attempt on process intensification and the required energy integration. A multivariable system with strong interactions between the channels can be much more difficult to control, and the decentralized control often cannot provide satisfactory decoupling performance. In this case, the decoupling control by designing a decoupler to minimize the process interactions is advised. However, the ideal decouplers are not always physically realizable. Furthermore, the main problem of this decoupling approach is the fact that the complexity of decoupler elements and apparent decoupled processes increases for high-dimensional systems, which requires model reductions.

Model predictive control (MPC) has become the method of choice in several process industries to solve difficult multivariable control problems that include inequality constraints. However, the conventional MPC design does not explicitly address the process interactions so that satisfactory decoupling performance is not necessarily attained. This study propose a novel MPC design approach that enables achieving the condition of decoupling control without adding any decouplers. To accomplish this, each control signal is conceptually decomposed into several constituent components for MPC calculation. One of the components is used to drive a controlled variable to its set-point, and the other components are used to eliminate the effects from other manipulated variables on this controlled variable. An additional term is then included in the MPC cost function to penalize the incomplete compensation of process interactions. By tuning the weighting matrices, it is possible to deal with the trade-off among the decoupling performance, the tracking performance, and the control efforts. When a large weight is given to the incomplete decoupling, the controller provides effective decoupling so that a set-point change for one controlled variable has no (or minimal) effect on the other controlled variables.

418645 Quantifying EV Battery Lifespan and Its Impact on Battery Warranty Strategy through a Battery Aging Model and a Micro-Level Transportation Network

Abstract:

The transportation sector accounts for about 70% of the total oil consumptions in the US. Conventional vehicles normally use fossil fuels as their energy sources. The depletion of fossil fuels and the threat of climate changes promote the development of transportation electrification. Several commercialized Electric Vehicles (EV) have already entered the market for a few years. However, greater deployment of EVs still faces several challenges. The concern of useful lifetime of EV batteries due to degradation is one of them. Lifespan information of populations of EV batteries is still scarce. Therefore, understanding the lifespan characteristics of EV batteries is significant for EV adoption, vehicle resale, and battery warranty strategy design.

This study quantifies how different Electric Vehicle usage patterns affect EV battery lifespans. Real world household vehicle travel information is extracted from the National Household Travel Survey (NHTS) database. A micro-level transportation model based on the Indianapolis network is built to generate realistic drive cycle data. Then whole household vehicle usage pattern information is obtained by matching the travel information with drive cycles. A physical EV energy consumption model is used to simulate the battery state-of-charge (SOC) and power profiles. With this detailed battery power data, a battery aging model derived from the literature is used to predict battery lifespan for a simulated population of vehicle usage patterns. In order to understand the effect of charging behavior on battery lifespan, two charging scenarios are examined: only charging at home, charging at home and work.

Robust Compromise Optimization Tuning Technique for MPC

André Shigueo Yamashita[(a)], *Darci Odloak*[(a)]

[(a)] *Department of Chemical Engineering, Universidade de São Paulo, Av. Prof. Luciano Gualberto, trv 3 380, 05424-970 São Paulo (SP), Brazil.*

Phone number: +551130912261

{andre.yamashita,odloak}@usp.br

Abstract

A time-domain robust tuning technique for MPC is proposed. It takes into account multi-plant uncertainty and calculates optimum tuning parameters to deal with the trade-off between robustness and performance. The robust tuning strategy in applied to tune an adapted version of the Infinite Horizon MPC, in closed-loop with a C3/C4 splitter model, identified in six operating points. A simulation illustrates how the robust tuning improves closed-loop performance in model mismatch scenarios, comparing the IHMPC tuned using both the nominal and robust version of the technique and compares the robustly tuned IHMPC to the Robust IHMPC from the literature tuned by trial and error. Even though the robust tuning strategy is not capable of ensuring robust performance, the results showed that it successfully calculated a set of parameters capable of compromising robustness and performance in multi-plant uncertainty scenarios. Moreover, the time-domain robust tuning strategy for nominal MPC yielded an alternative to the RIHMPC in model-mismatch applications.

Introduction

Model Predictive Controllers (MPC) are widely used in the oil processing and petrochemical industries because the algorithm formulation is straightforwardly applicable to both SISO and MIMO systems, moreover, it accommodates restrictions on the inputs, outputs and control actions directly in the control problem formulation (Qin & Badgwell, 2003). There is a set of parameters that affect the closed-loop performance of the MPC, which varies according to the MPC formulation. It is possible to improve the performance of the MPC by properly selecting the tuning parameters. Most of the attempts to tune MPCs in industry are based on trial and error (Liu & Wang, 2000; Al-Ghazzawi, Ali, Nouh, & Zafiriou, 2001). Nonetheless, there are tuning techniques available in the literature that take into account simplifications of the system dynamics and heuristic definition of parameters to develop analytic equations for the tuning parameters (Shridhar & Cooper, 1997); that take the sensitivity functions of the control cost function, with respect to the tuning parameters, to calculate analytical equations for the tuning parameters (K. H. Lee, Huang, & Tamayo, 2008); or that pose the tuning problem as a multi-objective optimization problem (Harris & Mellichamp, 1985; Reynoso-Meza, Garcia-Nieto, Sanchis, & Blasco, 2013; Vallerio, Van Impe, & Logist, 2014).

The classic MPC approach uses a linear model of the real process to predict the output values over the prediction horizon. Therefore, the operating range in which the liner model is valid is tight, and the operating point of the process usually changes either because of process disturbances, or due to economic optimization. It is considered in this work a multi-plant uncertainty description, in which a set of multiple linear models of the plant is available, and the plant might behave as any model throughout its operation.

The literature has proposed different approaches to deal with multi-plant uncertainty. One is to change the MPC formulation, allowing for robustness guarantees (Martins, Yamashita, Santoro, & Odloak, 2013), another is to tune the MPC considering the uncertainty model (J. H. Lee & Yu, 1994;

García-Alvarado, Ruiz-López, & Torres-Ramos, 2005; Francisco & Vega, 2010). The latter however, is mostly carried on in the frequency domain.

Robust Compromise Tuning Technique

It was presented in Yamashita & Odloak (2014) a tuning technique for MPC based in the compromise solution of a multiobjective optimization problem. The tuning goals are defined in terms of the sum of square errors between the closed-loop output responses and user-defined reference trajectories. In the original work, it was suggested first-order-plus-dead-time reference trajectories, in which its time constants reflect the degree of importance of each output, i.e. faster dynamics are assigned to more important outputs. It was shown that the proposed method is, to some degree, capable of calculating a set of tuning parameters that meet the performance criteria defined by the dynamics of the outputs.

It is enforced here the possibility to utilize the same principles to select the reference trajectories. The original method will be explained briefly and its robust extension will be detailed. Firstly, it is assumed that the controlled system is represented by a set of transfer functions $G_{i,j}(s)$, i=1,...,ny, j=1,...,nu, ny and nu being the number of outputs and inputs respectively. Moreover, multi-plant uncertainty description is assumed, meaning that a set Ω containing L models for the same systems is available. Such models have been identified at different operating points, and are linear model, thus can only reliably represent the original, non-linear system within narrow operating ranges. An incremental state-space representation (1) is derived from $G_{i,j}(s)$, which allows us to write future output predictions of the system based on current input, output and state values and the model matrices A, B and C.

$$\begin{aligned} x(k+1) &= Ax(k)+B\Delta u(k) \\ y(k) &= Cx(k) \end{aligned} \qquad (1)$$

$x(k)$ being the state vector, $y(k)$ and $\Delta u(k)$ being output and input increment vectors and A, B and C are matrices with appropriate dimensions. The tuning methodology presented here is intended for an infinite horizon Model Predictive Control (IHMPC), detailed in Santoro & Odloak (2012). Basically, control actions are calculated in order to minimize a cost function comprised of the difference between the outputs and their setpoints and the sum of control actions, subject to constraints. For brevity, a finite horizon MPC will be presented here. The MPC problem is posed as follows:

Problem 1

$$\min_{\Delta u_k} V_{1,k} = \sum_{j=0}^{p} \left\| y(k+j\,|\,k) - y_{sp} \right\|_{Q_y}^2 + \sum_{j=0}^{m-1} \left\| \Delta u(k+j\,|\,k) \right\|_R^2 \qquad (2)$$

subject to (1) and

$$u_{\min} \le u(k+j) \le u_{\max},\ j=0,\ldots,m-1 \qquad (3)$$

$$-\Delta u_{\max} \le \Delta u(k+j) \le \Delta u_{\max},\ j=0,\ldots,m-1 \qquad (4)$$

p is the prediction horizon, m is the control horizon, y_{sp} is the output setpoint, Q_y and R are positive definite weighting matrices, u_{min}, u_{max} and Δu_{max} are the lower and upper bounds on the inputs and the maximum input increment, respectively. Problem 1 defines a nominal MPC formulation, because it is assumed that the 'real system' model and the MPC model are the same ones from Ω. It is a quadratic problem with linear constraint and can be solved efficiently by QP solvers. The parameters p, m, Q_y and R are the tuning parameters of the MPC. This work puts in perspective guidelines to address the selection of Q_y and R, whereas m and p are selected following well-established guidelines in the literature (Garriga & Soroush, 2010).

Following the definition of tuning goals mentioned earlier in this work, it is possible to formally write them as in (5).

$$F_i(x)=\sum_{k=1}^{\theta_t}\left(y_i^{ref}(k)-y_i(k)\right)^2 \quad i=1,\ldots,w \tag{5}$$

$y_i^{ref}(k)$ being the discretized reference trajectory of output i at time instant k, $y_i(k)$ being the closed-loop response of output i at time instant k, θ_t being the tuning horizon and w the number of objectives. Before advancing, it is necessary to define the Utopia solution.

The solutions of a multiobjective optimization problem comprise a loci in the w-dimensional space, where w is the number of objectives. Therefore, there are infinite optimum solutions to such a problem, corresponding to the same objective function value, and the solutions are arranged in a Pareto frontier. The coordinates of the Utopia point in the w-dimensional space is found minimizing each objective individually, with respect to all the decision variables, as seen in (6).

$$F_i^{\circ}(x)=\min_x F_i(x), \quad i=1,\ldots w \tag{6}$$

subject to

$$LB\leq x\leq UB \tag{7}$$

The compromise optimization method searches for a viable solution minimizing the square of the difference between a feasible point and the Utopia point. The problem is stated as follows.

Problem 2

$$\min_x \left\|F^{\circ}-F(x)\right\|^2 \tag{8}$$

subject to (5), (6), and (7).

The extended compromise tuning strategy minimizes the cost function (8), taking into account multi-model uncertainty. Therefore, it is evaluated for all the models in Ω, deriving the extended cost function (9).

$$V_2(\omega_l)=\left\|F_{\omega_l}^{\circ}-F_{\omega_l}(x)\right\|^2, \; l=1,\ldots,L \tag{9}$$

Where ω_l indicates a model in Ω. A possible method to tackle the problem is solving the min-max optimization problem described by (10) and (7).

$$\min_x \max_{\omega_l} V_{2,b}(\omega_l), \; l=1,\ldots,L \tag{10}$$

It can be recast as a minimization problem by introducing the auxiliary variable γ, which acts as a supremum value for all admissible values of $V_{2,b}(\omega_l)$.

$$\min_x \gamma \tag{11}$$

subject to (7) and

$$V_{2,c}(\omega_l) \leq \gamma,\ l = 1,\ldots,L \tag{12}$$

Case Study

The robust compromise optimization tuning technique proposed here is illustrated in the tuning of an IHMPC in closed-loop with a 2x2 butane/propane separator system, identified at 6 operating points (Porfírio, Almeida Neto, & Odloak, 2003). The tuned IHMPC is nominal and the plant model may change throughout the simulations.

The closed-loop responses of an IHMPC tuned following the nominal version of the technique presented here, a IHMPC tuned following its robust version, and a Robust MPC from the literature (Martins et al., 2013) were compared.

The initial state of inputs and outputs are $y_0 = [1.1 \quad 50]^T$ and $u_0 = [2200 \quad 1300]^T$. The upper and lower bounds on the inputs and outputs were taken from Table 1 and $\Delta u_{\max} = [50 \quad 25]^T$. The simulation ran for 550 minutes and the set point changes are listed in Table 2 At 100 minutes, the plant model changes from model 6 to model 1, whereas the IHMPC model remains model 6. The IHMPC parameters not tuned here are T_s=1 min, m=5, S_y=diag[10^4 10^4].

Table 1: List of variables of the C3/C4 splitter system.

Tag	Variable	Range	Unit
y_1	C3 % in the butane stream	0.80 - 1.20	%
y_2	Top temperature	43 – 54	°C
u_1	Reflux flowrate	2000 - 4100	m^3/d
u_2	Hot fluid flowrate	1200 - 2200	m^3/d

Table 2: Set point changes of the C3/C4 Simulation I and tuning scenario.

Time instant (min)	$y_{1,sp}$ (%)	$y_{2,sp}$ (°C)
0	0.99	48
100	0.85	49
300	0.9	51
500	0.95	50

The tuning horizon was chosen as $\theta_t = 60$, and the pair y_1-u_2 was deemed more important than y_2-u_1. $G_{des}(s)$ was chosen based on first order approximations of the open-loop step responses of transfer functions $G_{2,1}(s)$ and $G_{1,2}(s)$ of the nominal model. The time constants of $G_{des,1,1}(s)$ and $G_{des,2,2}(s)$ were set to 10% and to 30% of the first-order approximations. The tunng problem was solved using *fmincon* (tolfun = 10^{-10}, tolx = 10^{-10}, maxfunevals = 4×10^3, tolcon = 10^{-6}, and maxiter = 400) in MATLAB 2013®. The decision variable vector was $x = [q_{y,1} \quad q_{y,2} \quad r_1 \quad r_2 \quad \lambda]$ and the initial guess was $x0 = [10 \quad 1 \quad 100 \quad 100 \quad 2000]$, The lower and upper bounds were $LB = [10 \quad 0.5 \quad 10^{-2} \quad 10^{-2} \quad -100]$ and $UB = [10 \quad 100 \quad 10^3 \quad 10^3 \quad 10^6]$.

The optimum tuning parameters calculated by the Robust Compromise technique were Q_y=diag[(10 8.96]) and R=diag([217.95 50]).

The RIHMPC control problem (Martins et al., 2013), was solved using the *CONOPT* nonlinear solver in GAMS® 23.6. The RIHMPC tuning parameters were $Qy = diag\left(\left[9\times10^5 \quad 8\times10^4\right]\right)$, $Q_u = diag\left(\left[0 \quad 0\right]\right)$, $R = diag\left(\left[5\times10^2 \quad 5\times10^3\right]\right)$, $S_y = diag\left(\left[8\times10^6 \quad 10^6\right]\right)$, $S_u = diag\left(\left[0 \quad 0\right]\right)$, and m=6. The IHMPC control problem was solved analytically disregarding the constraints on inputs and input increments at the tuning phase, and utilizing *quadprog* (default settings) in MATLAB® 2013 at the simulation phase.

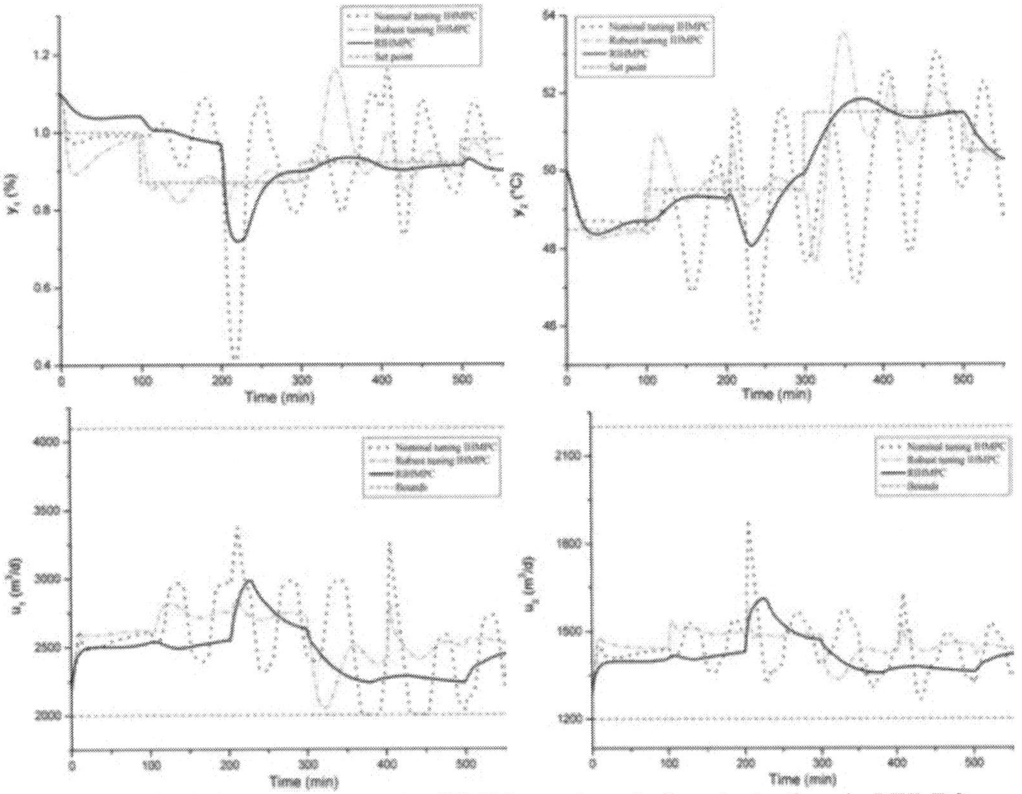

Figure 1: Simulation results, comparing IHMPCs tuned nominally and robustly and a RIHMPC.

It is observed in Figure 1 that the robust tuning techniques attenuated the closed-loop oscillatory behavior of the nominally tuned IHMPC, and improved both set point tracking and disturbance rejection capabilities. However, it yielded sluggish responses. It is observed that the RIHMPC yields sluggish performance for y_1, remarkable tracking capabilities for y_2 and non-oscillatory responses for both outputs. The RIHMPC simulation took 269 seconds while the IHMPC took only 6 seconds, both running on an Intel® Core™ i5 3.20 GHz, 4Gb RAM computer. Considering the plant was simulated over 550 time instants, the IHMPC (the values of tuning parameters do not interfere in the computing time) has a real time/time instant ratio of 0.019, while the ratio of the RIHMPC is 0.489.

Conclusions

A compromise optimization based tuning strategy for MPC under multi-plant uncertainty was developed. The tuning goals are defined with respect to the quadratic difference between user-defined reference trajectories and closed-loop output responses. The technique addresses the trade-off between performance and robustness, under plant-model mismatch, to evaluate optimum Q_y and R.

A case study illustrated the application of the proposed approach to an Infinite Horizon MPC from the literature. It was selected a butane/propane separator system, and a total of six models, identified at different operating conditions. The set point tracking performance and computational burden of nominal IHMPC tuned following nominal and robust tuning strategies and a Robust IHMPC from the literature, tuned by trial and error were compared, and it was concluded that the robust tuning technique for nominal MPC developed here is capable of reduce oscillation due to model mismatch. When compared to the RIHMPC, the robustly tuned IHMPC is not as sluggish as the former, but its responses are more oscillatory. Regarding the computational time elapsed during the simulation, the IHMPC runs for 6 seconds whereas the RIHMPC takes 269 seconds to simulate the same scenario.

Acknowledgements

The authors would like to thank the financial support by CNPq under grant 140677/2011-9, FUNDESPA and the continuous feedback and advice by Dr. Zanin from Petrobras.

References

Al-Ghazzawi, A., Ali, E., Nouh, A., & Zafiriou, E. (2001). On-line tuning strategy for model predictive controllers. *Journal of Process Control*, *11*(3), 265–284.

Francisco, M., & Vega, P. (2010). Automatic tuning of model predictive controllers based on multiobjective optimization. *Latin American Applied Research*, *40*, 255–265.

García-Alvarado, M. a., Ruiz-López, I. I., & Torres-Ramos, T. (2005). Tuning of multivariate PID controllers based on characteristic matrix eigenvalues, Lyapunov functions and robustness criteria. *Chemical Engineering Science*, *60*(4), 897–905.

Garriga, J. L., & Soroush, M. (2010). Model Predictive Control Tuning Methods: A Review. *Industrial & Engineering Chemistry Research*, *49*(8), 3505–3515.

Harris, S. L., & Mellichamp, D. A. (1985). Controller tuning using optimization to meet multiple closed-loop criteria. *AIChE Journal*, *31*(3), 484–487.

Lee, J. H., & Yu, Z. H. (1994). Tuning of model predictive controllers for robust performance. *Computers & Chemical Engineering*, *18*(1), 15–37.

Lee, K. H., Huang, B., & Tamayo, E. C. (2008). Sensitivity analysis for selective constraint and variability tuning in performance assessment of industrial MPC. *Control Engineering Practice*, *16*(10), 1195–1215.

Liu, W., & Wang, G. (2000). Auto-tuning procedure for model-based predictive controller. In *SMC 2000 Conference Proceedings. 2000 IEEE International Conference on Systems, Man and Cybernetics. "Cybernetics Evolving to Systems, Humans, Organizations, and their Complex Interactions" (Cat. No.00CH37166)* (Vol. 5, pp. 3421–3426). IEEE.

Martins, M. A. F., Yamashita, A. S., Santoro, B. F., & Odloak, D. (2013). Robust model predictive control of integrating time delay processes. *Journal of Process Control*, *23*(7), 917–932.

Porfírio, C. R., Almeida Neto, E., & Odloak, D. (2003). Multi-model predictive control of an industrial C3/C4 splitter. *Control Engineering Practice*, *11*(7), 765–779.

Qin, S. J., & Badgwell, T. A. (2003). A survey of industrial model predictive control technology. *Control Engineering Practice*, *11*(7), 733–764.

Reynoso-Meza, G., Garcia-Nieto, S., Sanchis, J., & Blasco, F. X. (2013). Controller Tuning by Means of Multi-Objective Optimization Algorithms: A Global Tuning Framework. *IEEE Transactions on Control Systems Technology*, *21*(2), 445–458.

Santoro, B. F., & Odloak, D. (2012). Closed-loop stable model predictive control of integrating systems with dead time. *Journal of Process Control*, *22*(7), 1209–1218.

Shridhar, R., & Cooper, D. J. (1997). A Tuning Strategy for Unconstrained SISO Model Predictive Control. *Ind. Eng. Chem. Res.*, *36*, 729–746.

Vallerio, M., Van Impe, J., & Logist, F. (2014). Tuning of NMPC controllers via multi-objective optimisation. *Computers & Chemical Engineering*, *61*, 38–50.

Yamashita, A. S., & Odloak, D. (2014). Compromise Optimization Tuning Strategy for Model Predictive Controllers. In *14th AIChE Annual Meeting* (p. 7). Atlanta.

422354 Flexibility Issues and Controllability Analysis of a Post-Combustion CO2 Capture Plant Integrated with a Coal FIRED POWER Plant

Abstract:

Owing to the continued use of fossil fuels in the global energy system for the foreseeable future, the International Energy Agency (IEA) suggests that CCS is an essential part of the technology portfolio needed to achieve deep reductions in global CO_2 emissions, with CCS expected to contribute around 14-20 % of total reductions in emissions by 2050 (IEA, 2010), (IPCC, 2014).

Nowadays, existing and new power plants must face the challenges of the liberalised electricity market, predictability issues regarding renewable sources and the requirement to cover intermediate and peak load constraints, to be able to respond to the variation of the electricity demand. The integration of power plants with capture plants will increase the need to develop advanced model-based control systems that are able to maintain the dynamic operability of CO_2 capture plants in the presence of operational constraints and disturbances.

One of the most promising near-term options for large-scale CCS deployment is post-combustion capture of with a monoethanolamine-type solvents. The benefits of this process are high capacity of CO_2 capture, fast reaction kinetics, inexpensive and abundant solvent. On the other hand the main challenge with this process is the substantial decrease in the efficiency of the power plant owing to the energy required for solvent regeneration that would otherwise be used to produce electricity. Further, as the power plant ramps up and down, the variation in liquid-to-gas ratios in the absorption columns can impose an important effect on the efficiency of the capture plant. Therefore the dynamic modelling and controllability analysis of the complete capture process is quite important (Bui et al., 2014).

In this paper we are going to discuss the flexibility in terms of different operating procedures of the coal power plant, and present the design and application of an advanced Model Predictive Control (MPC) approach to evaluate the controllability of a post-combustion plant in the presence of disturbances associated with the dynamic operation of the power plant. The outcome from this study is essential for future research to develop a control system for the CO_2 capture process and to develop new operability policies that will enable the safe and optimum dynamic operation of the MEA absorption process.

In literature, many papers discuss about "flexible" scenarios in the integration of a coal fired power plant with a capture plant (Chalmers et al., 2009), for example bypassing the capture in times with high electricity demand. However, a continuous process such as the amine scrubbing and regeneration process commonly used in CCS cannot be shut down and turned on at will. There must be limits to flexibility constrained by design, operation and control of the CCS plant. It needs to be recognized that when flue gas bypasses the CO_2 capture plant, hydraulics conditions of the absorber and the stripper will change substantially. A column will operate normally only when the gas flow velocity and liquid gas ratio are within range (Kister, 1992). It was also suggested that rich solvent can be stored during peak load period and can be regenerated later in off-peak period. This strategy could avoid CO_2 emission penalty because CO_2 is captured all the time. However, this strategy requires huge additional tanks and solvent inventory for buffering between peak and off-peak load period (Haines et al., 2009).

The implementation of the MPC approach addressed in this work can achieve the flexible operation of the power plant with the capture plant. The reason of choosing the centralised MPC is because it provides better performance, in terms of close-loop settling time, integral square error and compliance of operational and environmental constraints than the PI control schemes when applied in a post-combustion CO_2 capture process. The model was developed in gCCS a system modelling tool for support of design and operating decisions across the CCS chain. gCCS is built on PSE's gPROMS advanced process modelling platform, and inherits many of the platform's powerful capabilities (gCCS overview, 2014). SAFT-VR was used to describe the effect of the reactions on the fluid-phase properties and phase behaviour of the MEA-H_2O-CO_2-N_2 fluid mixture in the CO_2 capture process and the thermodynamic properties of the fluids (Mac Dowell et. al, 2013). The main control objective of this plant is to maintain the CO_2 removal fixed at a specified rate and maintain a constant composition in the CO_2 product stream. Also the liquid levels in the absorber sump and reboiler are controlled in order to maintain the solvent inventory within the system. Further the MEA concentration in the lean solvent stream needs to be below 30wt % to minimise corrosion in the equipment. The lean amine flowrate, condenser heat duty, absorber valve, reboiler valve and reboiler heat duty were selected as manipulated variables respectively.

The MPC controller as presented above, was evaluated under several scenarios, including changes in the flue gas flowrate that comes from the power plant and in the percentage of CO_2 captured in order to evaluate the amount of steam extracted from the thermodynamic cycle of the power plant during periods of high electricity demand. The results showed that the MPC control strategy can quickly adapt to any disturbance applied to the capture plant, which is essential for the profitable operation of decarbonised power plants.

References:

Bui M., Gunawan I., Verheyen V., Feron P., Meuleman E., Adeloju S., 2014, Dynamic Modelling and optimisation of flexible operation in post-combustion CO_2 capture plants - A review, Computers & Chemical Engineering, 61, 245-265.

Chalmers H, Lucquiaud M, Gibbins J, Leach M. Flexible operation of coal fired power plants with post-combustion capture of carbon dioxide. Journal of Environmental Engineering. 2009; 135:449–458.

gCCS overview: http://www.pseenterprise.com/power/ccs/gccs.html.

Haines MR, Davison JE. Designing carbon capture power plants to assist in meeting peak power demand. Energy Procedia. 2009; 1:1457–1464.

International Energy Agency, 2010, Carbon Capture and Storage, Model Regulatory Framework.

IPCC, 5th Assessment Report, Working Group III, 2014.

Kister HZ. Distillation design. New York: McGraw-Hill, 1992.

Mac Dowell N., Samsatli N., Shah N., 2013, Dynamic modelling and analysis of an amine-based post-combustion CO_2 capture absorption column, International Journal of Greenhouse Gas Control, 12, 247-258.

423444 Kinetic Modelling of Secondary Cracking of Gasoline Adopting Non Selective Catalyst Deactivation Approach

Abstract:

The Secondary cracking of gasoline using Lumping scheme was mathematically modelled and simulated by adopting non selective catalyst deactivation approach. The model was used to study the impacts of coke on catalyst and time on stream catalyst deactivation functions on secondary cracking of gasoline. The effects of olefin cracking was given importance due to their impact on distribution of aromatics in product stream and coke on catalyst. The simulation results predicts that the olefins contributes for increase in aromatics content of gasoline by aromatization pathways. But increase in aromatics and olefins in the product stream rapidly deactivates the catalyst due to deposition of coke.

Generally the effects of catalyst deactivation is categorised into Selective and Non Selective Deactivation approach. The Non Selective approach is further classified into Coke on catalyst and Time on stream. Among these Time on stream preferred over Coke on catalyst due the fact that some kinds of coke may not contribute to catalyst deactivation.

The results suggest that the prediction accuracy was greatly enhanced when the Time on stream model was used instead of Coke on stream.

425169 Bioethanol Production: Design and Control of an Alternative Extractive Distillation System

Abstract:

Bioethanol is one of the most important biofuels contributing to the mitigation of negative effects on the environment and quality of life due to the exploitation of fossil fuels. An advantage of bioethanol over other energy alternatives, such as hydrogen, is that it can be mixed with gasoline without any modification on current engines (Costa & Sodré 2010). However, to reach the bioethanol purity for biofuel applications two major processing steps must to be carried out, this due to the ethanol-water azeotrope. The first step is typically a conventional distillation tower in where near-azeotrope concentrations are achieved up to the level of 92-94 wt.%. The second step is much more complex and of greater interest. Several alternatives are available to achieve high purities (>99 wt.%), among them extractive distillation, which has being commonly used for large-scale production. Conventional extractive distillation process is performed in a sequence of two columns, where the first one (extractive column) purifies the ethanol and the other one (recovery column) splits water from the solvent, which is recycled back to the first column.

Dividing wall column (DWC) technology has demonstrated to be a feasible option to separate not only multi-component mixtures using a single shell in distillation tower, but also to be used in processes involving azeotropic, extractive and reactive distillation. As a consequence, alternative configurations to the conventional extractive system have been proposed for bioethanol purification. Hernández (Hernández 2008) and Kiss & Suszwalak (Kiss & Suszwalak 2012) proposed the use of thermally coupled distillation columns using complex extractive distillation schemes, which have demonstrated to have good dynamic control properties (Tututi-Avila et al. 2014).

Modified alternative schemes to thermally coupled systems have been considered for zeotropic purification such as those proposed by Agrawal (Agrawal 2000) and analyzed by Ramírez and Jiménez (Ramirez & Jimenez 2004) at steady state conditions and by Segovia-Hernández et al. (Segovia-Hernández et al. 2005) under closed-loop operation. Such systems consist of an alteration to the thermally coupled systems with side columns by removing the vapor-liquid coupling and adjusting the system structure to generate more operable options. In this study we consider such systems and extent them towards azeotropic systems, particularly an alternative extractive distillation system for bioethanol purification is studied. The proposed extractive distillation system consist of an extractive side-stream column from which bioethanol is removed as the distillate product. Water is removed as part of a liquid side stream and fed to a second column, which produces water as distillate product. The bottom stream from the first and second column is the solvent recycled back to the first column.

In this work, the design and control of such novel distillation system and a comparison of economic and dynamic performance to those of a conventional extractive distillation sequence and an extractive divided wall system is addressed. Steady state optimal configurations for conventional and alternative systems is obtained by simulation in Aspen Plus. Sensitivity analysis is carried out to design temperature control schemes. Responses are analyzed based upon a multi-loop framework under feed disturbances in Aspen Dynamics. The results show that the novel extractive distillation system is able of maintain the bioethanol purity under large feed disturbances, being its dynamic behavior comparable to the conventional and thermally coupled distillation systems.

References

Agrawal, R., 2000. Thermally coupled distillation with reduced number of intercolumn vapor transfers. *AIChE Journal*, 46(11), pp.2198–2210.

Costa, R.C. & Sodré, J.R., 2010. Hydrous ethanol vs. gasoline-ethanol blend: Engine performance and emissions. *Fuel*, 89(2), pp.287–293.

Hernández, S., 2008. Analysis of Energy-Efficient Complex Distillation Options to Purify Bioethanol. *Chemical Engineering & Technology*, 31(4), pp.597–603.

Kiss, A.A. & Suszwalak, D.J.P.C., 2012. Enhanced bioethanol dehydration by extractive and azeotropic distillation in dividing-wall columns. *Separation and Purification Technology*, 86(0), pp.70–78.

Ramirez, N. & Jimenez, A., 2004. Two alternatives to thermally coupled distillation systems with side columns. *AIChE Journal*, 50(11), pp.2971–2975.

Segovia-Hernández, J.G. et al., 2005. Control properties and thermodynamic analysis of two alternatives to thermally coupled distillation systems with side columns. *Chemical and Biochemical Engineering Quarterly*, 19(4), pp.325–332.

Tututi-Avila, S., Jiménez-Gutiérrez, A. & Hahn, J., 2014. Control analysis of an extractive dividing-wall column used for ethanol dehydration. *Chemical Engineering and Processing: Process Intensification*, 82, pp.88–100.

427368 Efficient Decision Making Based on a Hybrid Power Distribution Strategy - Application to a Fuel Cell Electric Vehicle

Abstract:

This work presents a decision making methodology that determines the power sharing between two sources for the fulfilment of a single load demand. The field of energy management of hybrid energy systems is of increasing importance due to the necessity for optimum energy sharing between the various interconnected subsystems that either supply or demand energy and combined constitute an integrated system. The hybrid nature of such systems requires a well-designed energy management strategy (EMS) that determines how the demanded by the load power is distributed between the available energy sources while satisfying design and operation requirements.

The proposed approach is based on a hybrid timed automaton (HTA) which is used for the realization of an overall Energy Management Strategy (EMS) of the system. At hybrid powered systems discrete and continuous operating states coexist. The HTA switches between the system's operational modes and defines the set-points for the field-level control loops. More specifically the methodology implements an EMS which is applied to a fuel cell – battery system for vehicular applications. When a fuel cell is combined with a battery various decisions should be made in order for the integrated system to operate in an optimum way, while fulfilling the frequently and rapidly changing load demand. The objectives of the EMS are the fulfilment of the load demand, the optimal use of the available hydrogen, the protection of the battery's lifetime and the preservation of the fuel cell's integrity. The system consists of a lithium-ion battery stack, a Polymer Electrolyte Membrane (PEM) fuel cell and a DC load.

A set of operation scenarios are selected based on driving profiles and are used for the sensitivity analysis, the optimization problem and for the evaluation of the overall behaviour of the system. For each subsystem a model is used and a sensitivity analysis is performed that reveals the main parameters and variables that affect the response of the integrated system. An optimization problem is formulated that examines a set of operation scenarios included in the EMS. The objective of the optimization problem is to enhance the performance of the system. A number of alternatives for energy sharing will be presented that explore the effect that different priorities have for the energy supply subsystems. The potential of the proposed EMS will be demonstrated along with the response of the various subsystems and the integrated system.

430136 Noninvasive Ultrasound Measurements of Temperature Distribution and Heat Fluxes Across Containments of Extreme Environments

Abstract:

Energy conversion often takes place in processes occurring at extreme conditions of high temperatures, pressures, chemical aggressiveness, mechanical abrasion, and other factors. In extreme environments, even the most hardened insertion sensors do not perform reliably over an extended period of operation. To overcome the difficulties in obtaining reliable temperature measurements in extreme environments, we have developed a noninvasive ultrasound method for measuring spatial distribution of temperatures in solid materials and, specifically, across containments of extreme processes, such as gasification refectories.

This presentation describes a novel approach that uses noninvasive ultrasound (US) to measure the temperature distribution in solid materials. We use the ultrasound propagation path that is structured to contain engineered or naturally occurring echogenic features which redirect a fraction of the energy of each ultrasound excitation pulse back to the transducer as a train of echoes. Their time of flight is used to determine the speed of sound in different segments of the US propagation path. Then, by using the relationship between the velocity of the ultrasound propagation and the temperature, the time of flight measurements are used to estimate the temperature distribution across the sample.

During this presentation, we will describe the results of successful laboratory demonstration and pilot-scale oxy-fuel combustor testing of this new technology. The experimental validation of the developed method shows that the estimated temperature profile is correctly captures, and that the measurement accuracy is often comparable with the traditional insertion sensors. We will discuss the advantages this approach, which include its suitability for in-situ and non-destructive measurements over a broad range of temperatures, the applicability to the measurements of conductive heat fluxes in solids, and the spatially resolved characterization of their elastic properties.

References:

Y. Jia, M. Puga, A. Butterfield, D. Christensen, K. Whitty, and M. Skliar, Ultrasound Measurements of Temperature Profile Across Gasifier Refractories: Method and Initial Validation, Energy & Fuels 2013 27 (8), 4270-4277

M. Skliar, K. Whitty, and A. Butterfield, Ultrasonic temperature measurement device, US Patent 8,801,277 B2, August 12, 2014.

430298 Unknown Input Observer-Based Nonlinear Model Predictive Control

Abstract:

Model predictive control is one of the promising advanced process control techniques for industries. The control algorithm is based on modelling of the system/s and an optimization process to generate the control action. State space models are commonly used in the nonlinear model predictive control. The control action is optimized basing on the knowledge of current state information [1]. Besides, state and output feedbacks are often required to design a stabilizing control performance while satisfying the control and state constraints [2]. However, some states can't be measured directly. A state observer (filter) is therefore used to estimate the states in the current iteration such as Kalman filter or Luenberger observer[3, 4]. However, sudden disturbances or faults will introduce large estimation error from the observer; the computed control action based on the faulty estimated state will be error prone. To deal with the sudden disturbance or faults, an unknown input observer is proposed to estimate the states. The unknown input observer aims to deal with the disturbance, which is assumed to be larger than a regular measurement/process noise (white noise). In this paper, a nonlinear unknown input observer is designed based on a nonlinear process model. The state estimation from unknown input observer was applied for the optimization process. An MPC stability theorem dealing with the asymptotic stability of the observer was proposed. An example of multi-variable batch reactor was used to demonstrate the proposed approach.

[1] Tang G-Y, Guo Z-W. Model predictive control based on state predictive observers for discrete systems with time-delay. Control and Automation, 2003 ICCA'03 Proceedings 4th International Conference on: IEEE; 2003. p. 365-9.

[2] Findeisen R, Imsland L, Allgower F, Foss BA. State and output feedback nonlinear model predictive control: An overview. European journal of control. 2003;9:190-206.

[3] Prakash J, Senthil R. Design of observer based nonlinear model predictive controller for a continuous stirred tank reactor. Journal of Process Control. 2008;18:504-14.

[4] Qian J, Dufour P, Nadri M. Observer and model predictive control for on-line parameter identification in nonlinear systems. IFAC International Symposium on Dynamics and Control of Process Systems (DYCOPS)2013. p. 571-6.

431127 Exergy Analysis and Comparison of Traditional and Modified APCI Cycles Used for Natural Gas Liquefaction

Abstract:

Nowadays natural gas (NG) is one of the most important energy sources, mainly because of its reduced environmental damage, when compared to other fossil fuels. It is also an important raw material in many chemical processes. NG may be used to produce electrical energy and steam in power cycles and can be used as fuel in vehicles. Considering security, cost and other amenities that pipelines provide, they are commonly used for NG transportation between producers and final consumers. However, when very large distances need to be covered, a pipeline almost always becomes not feasible and a liquefaction process is necessary to reduce NG volume up to 600 times, so that it can be transported by special means, mostly by ships. The APCI cycle is the most used to perform the liquefaction of NG using coupled refrigeration sub-cycles that decrease the gas temperature until approximately -160 °C. These sub-cycles operate with different working fluids and are great energy consumers. Thus, the study of alternative cycles may be attractive, for example, to replace the propane pre-cooling sub-cycle by an absorption refrigeration system (ARS), in order to reduce the electric power demand of the original APCI cycle, using waste heat available in NG processing plants (exhausted hot gases from turbines and hot streams that could be cooled, for example). In this context, this study presents a comparative thermodynamic evaluation of the traditional APCI cycle and a modified one, which includes a NH_3/H_2O single-effect ARS to replace the propane sub-cycle. Simulations of both cycles were carried out using ASPEN Hysys Version 7.3 software. The thermodynamic performance evaluation is based on the First and Second Laws of Thermodynamics and through an exergy analysis, inefficiencies were computed and identified. On energy basis, the original APCI cycle is more advantageous since its global First Law coefficient of performance (COP) is significantly greater in comparison to that of the modified APCI cycle (1.690 against 0.527). Under the exergy point of view, the modified APCI presents a global efficiency slightly higher (0.872 against 0.836, a relative difference of only 4.3%). This demonstrates that, in view of fuel and product definitions adopted for both cycles (according to their purposes of purchasing and using), the exergy destruction rate is a little more pronounced in the original APCI. In this sense, for both cycles, the main cryogenic heat exchanger is the most expressive contributor to decrease the global exergy efficiency (accounting for 41.5% and 30.3% of total exergy destruction rates in original and modified cycles, respectively), immediately followed by compressors in original APCI (25.7% of total exergy destruction rate in this cycle) and other heat exchangers in modified APCI (26.0% of total). A detailed economic analysis is desirable since it can show that despite its reduced COP, the modified APCI may be a good alternative once it uses waste energy from the process, reducing primary energy sources consumption.

431360 Statistical Process Control Applied to Continuous Processes

Abstract:

In recent years, the need to maintain safe and reliable industrial operations has taken great proportions. In the continuous process industry, in particular, this issue is even more important because the operation is often carried in full capacity, with high production speed, huge impact downtimes and low production flexibility. This scenario has stimulated the development of efficient tools for fault detection. Statistical process control (SPC) techniques are widely used in the industry of manufactured products, where their basic assumptions are often valid: independent and identically distributed measurements with normal distribution. However, in the industrial environment of continuous processes these assumptions are often violated. Thus, this study focused on the adaption of SPC techniques to continuous processes. A methodology for identifying intervals under control of a time series was proposed. Furthermore, a method for updating the monitoring parameters was defined according to the current state of the process and past operational points. Natural gas compression systems of offshore platforms were chosen as case studies using real and simulated data. The results demonstrate the ability of the techniques to detect abnormalities. It is also shown that the identification of stable intervals is adequate and that the methodology of updating parameters along the monitoring is efficient.

431561 Batch Process Monitoring By Dynamic Time Warping and k-Means Clustering

Abstract:

In the production of membranes for gas separation, the raw material goes through a series of batch processes before it is converted into a finished product. Each of these batch processes is governed by a recipe, which often involves a combination of process control logic and operator interventions. This work presents a method for monitoring the quality of a batch process by detecting deviations from the recipe, through a combination of Dynamic Time Warping (DTW) and k-Means clustering.

Historical data (e.g., flows, temperatures, pressures) are gathered from a membrane manufacturing process, which consists of several steps of variable duration. Through a combination of interactions between researchers and subject matter experts, a set of "ideal" and "bad" recipes are identified which is used as a training set for the model. Historical operation is then compared with the ideal and bad recipes, and a unified distance metric is used to then cluster the operational data using k-Means clustering. Encouraging results are reported from practice.

434258 Modeling and Control of Reversible Addition-Fragmentation Chaintransfer (RAFT) Polymerization Reaction in a Batch Reactor

Abstract:

Modeling and Control of Reversible Addition-Fragmentation chain-Transfer (RAFT) Polymerization Reaction in a Batch Reactor

Ortiz-Arroyo Arturo*., Castro-Agüero Angela, Tututi-Avila, Salvadorb.
Universidad Autónoma de Tlaxcala, Av. Apizaquito S/N, Apizaco, Tlaxcala, Mexico.
*Corresponding author: arturo.ortiz@uatx.mx,a angel.castro@uatx.mx,bsalvador.tututi@uatx.mx

Abstract
Reversible addition-fragmentation chain-transfer (RAFT) is a versatile polymer synthesis technique with many potential applications because allows to obtain polymers with complex conformational structure. Additionally, this technique provides control of molecular weight and molecular weight distribution with narrow polydispersion index. An additional advantage is that RAFT process conditions are close to the usual free radical polymerization reactions. Then, RAFT polymerization can be carried out in bulk, aqueous solutions, organic solutions, suspensions, emulsions, etc., at low temperatures. Due to these characteristics and potential applications the subject has attracted attention from researchers as is reported by Moad et al, 2008 and 2009 were at the time of writing cited more than 500 related papers. The RAFT process uses free radical initiators and monomers but, additionally, includes a Chain Transfer Agent (CTA) also called RAFT agents being the Dithioesters the most commonly used agents.
Zhang & Ray, 2001 used the moment methodology to simulate the free radical polymerization of Methyl Methacrylate (MMA) using AIBN as initiator to show the effect of the reactor type. For that purpose, Batch, semi Batch, CSTR and CSTR in series reactors are simulated in isothermal steady and non-steady state. The resulting mathematical model consisted of 27 highly non-linear ordinary differential equations (ODE) which are not included for clarity.
In this work the simulation framework of Zhang & Ray, 2000 was used in a Batch reactor for which a simplified energy equation is added which includes heat effects to/from the reactor to complete the dynamic scheme. Batch reactors are devoted to the production of high quality but low volume products like polymers and are naturally time-varying, therefore their dynamics is of vital importance due to the fact that small variations in operation conditions may degrade the product quality if the problem is not detected and corrected fast enough.
A configuration of the reactor is proposed and identified and a PID controller is proposed for the reactor temperature considering that all the reactants are present in the initial time. To test the behavior of the controller the initiator and initial monomer concentrations are modified and the temperature is controlled with a split range setup in which cooling and heating media is admitted to the reactor jacket depending on the temperature level (fig 1). It is important to mention that the main problems with polymerization reactors are the heat release and the viscosity changes.
Due to the complexity of the reactor model a simple PI control strategy on open loop is implemented and tuned using several methodologies. As the Batch reactors are naturally variant it is desirable that all the information related to the reactive system would be incorporated to the control strategy.

Fig.1 Split Range control proposed for the reactor

Zhang, M., Ray, W.H. Modeling of "Living" Free-Radical Polymerization with RAFT Chemistry. Ind. Eng. Chem. Res., 2001, 40 (20), pp 4336–4352

434410 Fault Detection and Diagnosis in Batch and Fed-Batch Bioreactor System Using PCA, PCR, PLS with Glr Method

Abstract:

Fault detection and diagnosis in batch and fed-batch bioreactor system using PCA, PCR, PLS with GLR method.

ChiranjiviBotre, M. StefanyAngarita Gómez, MajdiMansouri, Mohamed N. Nounou, Hazem N. Nounouand M. Nazmul Karim

Abstract

Multivariate statistical methods are powerful tools capable to handle huge, noise and highly correlated data sets. Principal Component Analysis (PCA), Partial least squares (PLS) as well as Principal Component Regression, linear and non-linear approaches, will be used to detect and predict faults in a batch and a fed-batch bioreactor system. Generalized likelihood ration (GLR) statistical method is used to detect the fault and it has been shown to have good fault detection abilities. Due to inherent dynamics in batch processes, multiblock and multiway methods have to be applied in which only historical data sets of past successful batches are needed. Principal Components are kept to extract the majority of variance and reduce the dimensionality of the original data set, making easier the analysis and fault detection. As soon as the fault is detected, it is necessary to find, the variable that presents the major deviation from its expected value therefore, a contribution plot for diagnosis is required.Experimental data is of batch production of carotenoids via fermentation to describe glucose consumption, metabolic products formation and depletion, and the carotenoid production in the Saccharmoycescerevisiae strain mutant SM14 with 20 g/l glucose as the carbon source. Experiments are performed to validate the results obtained from models.

Keywords: PCA, PCR,GLR, fault detection, batch process, statistical process control.

References-

1. MacGregor, J.F., and T. Kourti. "Statistical Process Control of Multivariate Processes." Control Engineering Practice 3, no. 3 (March 1995): 403–14.
2. FouziHarrou, Mohamed N. Nounou, Hazem N. Nounou and MudduMadakyaru, "Statisticalfaultdetectionusing PCA-based GLR hypothesistesting" Journal of Loss Prevention in the Process Industries 26 (2013) 129e139.
3. S. Joe Qin, "Statistical process monitoring: basics and beyond" Journal ofChemometrics 2003; 17: 480–502.

435409 Steam Load Shedding System Design Using Dynamic Simulation

Abstract:

In a refinery, steam utility is one of the most important utilities, which can be used at different temperature ranges. Proper handling of the available steam is an important factor for economic aspects of the plant. Instability in steam header pressures restricts optimum use of resources and it can also create safety issues. Constantly changing demands at various steam pressures challenge proper management and control of existing steam supply capabilities. Some emergency situations, such as boiler trip, add complexity to the above problem. Steam load shedding is required due to reduction in steam supply while demand being the same. The choice of steam load shedding location can affect economy of the process. Robust and reliable controller strategy becomes critical in such situations.

Aspen Dynamics was used as dynamic process simulation tool. A typical refinery steam system was considered as base case. A controller strategy was developed to handle various plant modes viz. normal, optimum, process upsets, and emergency (boiler trips). It also keeps the steam header pressures within specified limits. Changing steam demands for high pressure, medium pressure and low pressure steam were considered to test the controller strategy. Various situations of outage in steam supply were also considered. The methodology was to maintain header pressure, minimize steam venting and minimize steam letdowns. The target was to minimize need for external power supply while maintaining safety, and stability of the plant. This also includes reduction in wastage of steam and maximizing on-site power generation. The study has shown significant savings in the utility costs due to proper steam load shedding.

424805 Optimal Pump Network Reconfiguration Scheduling Considering Time-of-Use Electricity Prices

Abstract:

The electricity demand over the day is subject to significant fluctuations, with considerably larger demand during peak hours than off-peak [1]. An increasing penetration of intermittent renewable energy generation further increases the variations, and therefore the costs for the utilities to meet the demand. One approach to deal with this challenge is Demand Response (DR) which aims at changing the electricity end-user's demand using penalties or incentives such as variable electricity prices [2, 3].

While real time electricity pricing is only implemented in few applications so far [2], a simple way of realizing a DR strategy, Time-of-Use (TOU) rates, is widely adopted. For TOU electricity pricing, the utility sets several electricity price levels per day. The prices are fixed and are only adapted in larger time scales. A higher electricity price during peak hour clips the demand and shifts part of the demand to off-peak hours [3]. TOU plans are common for commercial and industrial consumers, but they are also implemented for residential consumers and offer an opportunity for the consumer to save costs by changing the consumption pattern according to the electricity prices.

Pump networks are widely used in many industries. According to the European Commission, pumps are the largest consumers of electrical energy in the EU [4, 5]. Especially in the paper and pulp industry or in water distribution [6], where large-scale pump networks are applied, the electricity costs incurred by pump networks are significant. Therefore, cost saving potential in pump networks is large [4, 5].

An optimal configuration of pumps can lower electricity consumption significantly [5, 4]. However, the operation of the most energy efficient configuration may not always be desirable from an economic point of view, since a higher number of pumps or technologically more elaborate pumps may also entail higher maintenance costs. Therefore, the optimal configuration, which has to consider both electricity and maintenance costs, may depend on the time of day and the electricity price levels of the TOU plan and has to be found in order to minimize operational costs.

In this work, we propose to reformulate the traditional static pump network optimization problem with consideration of TOU electricity prices and maintenance cost based on the case study in [6], to find the optimal pump network scheduling during a day. The focus of the reformulated problem is on the optimal operation of an existing network superstructure, in order to eliminate the direct consideration of the capital costs. However, the maintenance cost is correlated to the capital cost. The proposed methodology is applied to a pump network superstructure consisting of three layers, each of which is occupied by a different pump type with unique pump equations. Variable speed centrifugal pumps are assumed to allow for a change of rotational speed in each configuration. The required volume flow and pressure gradient over the pump network are assumed constant. The optimum pump network configurations for a fixed interval of electricity prices are determined for varying plant sizes by solving the reformulated MINLP-problem for global optimal solutions using the Branch-and-Reduce Optimization Navigator (BARON). It is shown that switching between different configurations can lead to overall cost savings. It is also found that the electricity prices for which a transition between two configurations is economically justifiable varies greatly with the plant size. A case with specific parameters could be found for which a reconfiguration is optimal during the peak-hours of a real TOU electricity price structure.

References:

[1] M. H. Albadi and E. F. El-Saadany, "Demand Response in Electric Markets: An Overview," in *Power Engineering Society General Meeting, 2007. IEEE*, Tampa, FL, 2007.

[2] I. Harjunkoski, X. Feng and E. Scholtz, "Smart Grid: Industry Meets the Smart Grid," *CEP*, pp. 45-50, 08 2014.

[3] C. W. Gellings, "The Concept of Demand-Side Management for Electric Utilities," in *Proceedings of the IEEE, Vol 73, No 10.*, 1985.

[4] European Commission; ETSU, AEAT PLC, *Study on Improving the Energy Efficiency of Pumps*, 2001.

[5] Hydraulic Institute, "Pump Life Cycle Analysis - A Guide to LCC Analysis for Pumping Systems - Executive Summary," 2001.

[6] T. Westerlund, F. Pettersson and I. E. Grossmann, "Optimization of Pump Configurations as a MINLP Problem," *Computers & Chemical Engineering*, pp. 845-858, 09 1994.

(244r) Modeling, Optimisation and Control of Solar Absorption Air-Conditioning System

Ocheme, S., University of Leeds

Solar thermal energy can provide abenign and environmentally friendly heat source for driving a vapour absorption refrigeration system of unit capacity using R 717 (NH_3) and water as the working fluids.The last few decades witness an upsurge in energy demand, both at the domestic and industrial cooling and heating demand. Traditionally, heating, ventilation and air-conditioning (HVAC) performance and design are based on construction of prototypes in a wind tunnel. This approach is not only time consuming but expensive. In this study, a model-based approach for the modelling, optimisation and control of a solar vapour absorption air-conditioning system is presented. A model-in-loop and hardware-in-loop model was also develop to validate the control algorithm.

434712 A Data-Knowledge Hybrid Framework for Process Monitoring and Fault Detection with the Combined Use of Big Data

Abstract:

The main purpose of this study is the faster detection of faults. Advanced control knowledges has been adapted in chemical process monitoring during last decade. In this study, we attempted to apply state estimator (e.g., Kalman filter) design to non-steady state process to detect faults faster than statistical process control (SPC) only. Due to high nonlinearity in first principles of chemistry, the system is more complex and the reach time of the observer is late. Nevertheless, we expect to this framework to approximate small faults in complex systems. So, we introduce a hybrid methodology of state estimator design of nonlinear system and SPC. Then, we test the performance of the framework with comparison against the detection by check with given historical data of the process under presence of measurement noises.

435549 Condition-Based Cleaning of Refinery HEAT Exchangers Undergoing Crude OIL Fouling

Abstract:

Gradual build-up of fouling deposit on heat transfer surfaces in refinery heat exchangers leads to significant energy losses, operating difficulties and economic penalties. The undesired deposition not only reduces the heat exchange performance of these facilities due to its low conductivity, but also leads to increased pressure drops as the flow area decreases. This blockage may lead, eventually, to losses of production. Despite any mitigation actions, cleaning is required in order to restore the performance of heat exchangers. The economic trade-off between the energy losses due to fouling, the cost of cleaning and loss of production, and on the other hand the improvement after a cleaning, is ultimately the main factor determining the scheduling of cleaning actions.

Recent work [1] led to the development of a dynamic, distributed model for the deposit layer that represents it as a multi-component system undergoing chemical reactions, such as ageing. The model, implemented in a single heat exchanger tube framework, was shown to be able to simulate for the first time with a single model fouling, full cleaning, partial cleaning and fouling resumption after a cleaning, in any order. Rate models for distinct types of cleaning methods (mechanical and chemical) were proposed. The cleaning efficacy depends on the cleaning method and the state of the deposit at the time the cleaning is performed. Fresh deposit are relatively easy to remove, but old deposits, which have been exposed to high temperatures over long periods of time, may be difficult to remove or even non-removable (depending on the cleaning method). Finally, two cleaning modes can be considered: fixed time, for which the duration of the cleaning period is fixed a priori; and condition-based, for which the cleaning lasts until a termination condition is reached. In a later work [2], this improved model of fouling deposits was implemented within Hexxcell Studio^{TM}, a dynamic, distributed, first-principle thermo-hydraulic model for a shell-and-tube heat exchanger [3,4].

In this work, the framework comprising the new deposit model integrated in the advanced heat exchanger model is used to simulate fouling cleaning cycles for an industrial multi-pass shell–and-tube heat exchanger operating in typical conditions. It is shown that various combinations of mechanical and chemical cleaning (of different extent) can be seamlessly simulated. A cost model, extending previous works to account for several costs associated to cleaning, is proposed and used to assess the most economically advantageous cleaning schedules. Finally, dynamic optimization is explored to find the optimal cleaning schedule for a simple case considering a single heat exchanger.

Acknowledgments

This research was partially performed under the UNIHEAT project. EDB, LL and SM wish to acknowledge the Skolkovo Foundation and BP for financial support. The support of Hexxcell Ltd, through provision of Hexxcell Studio™, is also acknowledged.

[1] Diaz-Bejarano, E., Coletti, F. & Macchietto, S., 2015. A new dynamic model of crude oil fouling deposits and its application to the simulation of fouling-cleaning cycles. *Submitted for publication*.

[2] Diaz-Bejarano, E., Coletti, F. and Macchietto, S., 2015. Beyond fouling factors: a reaction engineering approach to crude oil fouling modelling. In *Heat Exchanger Fouling and Cleaning XI*. Enfield (Ireland).

[3] Coletti F, Macchietto S. A Dynamic, Distributed Model of Shell-and-Tube Heat Exchangers Undergoing Crude Oil Fouling. Ind Eng Chem Res. 2011;50(8):4515–4533.

[4] Hexxcell Ltd., 2015. Hexxcell Studio^{TM}. http://www.hexxcell.com.

408186 Dynamic Manipulation of a SOFC/GT Hybrid Power Generation System Subject to Low Carbon Emissions

Abstract:

The new design of a solid oxide fuel cell (SOFC)/gas turbine (GT) hybrid power generation system can effectively reduce carbon emissions to generate the clean power. A compressed fuel processing unit (CFPU), which consists of steam methane reforming (SMR), dry reforming of methane (DRM) and compressor, produces the high pressure syngas as a source of fuel for SOFC. A combined heat and power (CHP) process is connected to meet the energy demand of the CFPU and preheat the compressed air flow in order to achieve the stand-alone cogeneration system. Finally, the multi-loop feedback control strategy is implemented to control the hybrid power outputs and suppress CO_2 emissions simultaneously.

Process Design and Optimization for Etherification of Glycerol with Isobutene

Jingjun Liu[a,b], Prodromos Daoutidis[b], Bolun Yang[a]

[a] *Department of Chemical Engineering, Xi'an Jiaotong University, Xi'an, 710049, PR China*

[b] *Department of Chemical Engineering and Materials Science, University of Minnesota, MN, 55455, USA*

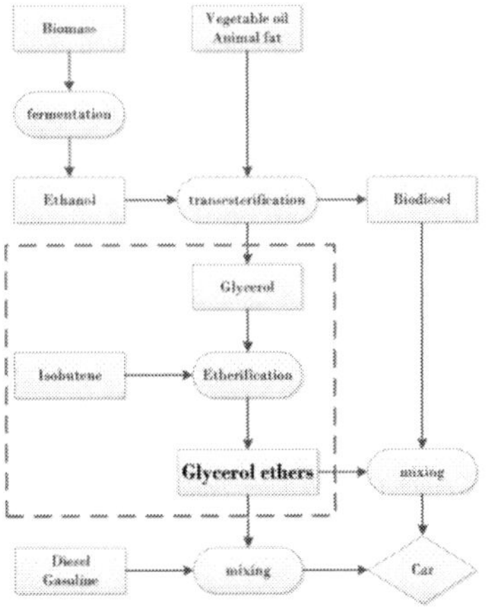

Biofuel recycling

- Biodiesel-derived glycerol is looking for outlet;
- Glycerol ethers are promising additive / component for biodiesel, diesel and gasoline;
- Glycerol ethers can reduce particulate matter emissions, and improve low-temperature properties.

Summary

- Biodiesel-derived glycerol is converted to fuel additives through etherification with isobutene.
- Di-*tert*-butyl ethers (DE) and tri-*tert*-butyl ether (TE) of glycerol are the objective product.
- A new process for the production of glycerol ethers is proposed.
- Detected dimerization of isobutene is considered in the new process.
- Latest kinetic and liquid-liquid equilibrium data are used.
- Optimization based on economic analyses is implemented.
- Isobutene cost greatly dominates the production cost of glycerol ethers.

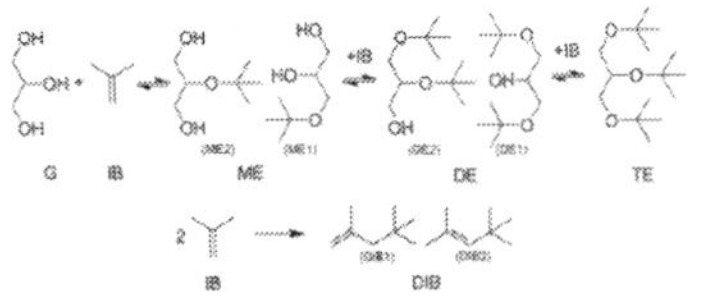

Reaction of interest

Design and optimization

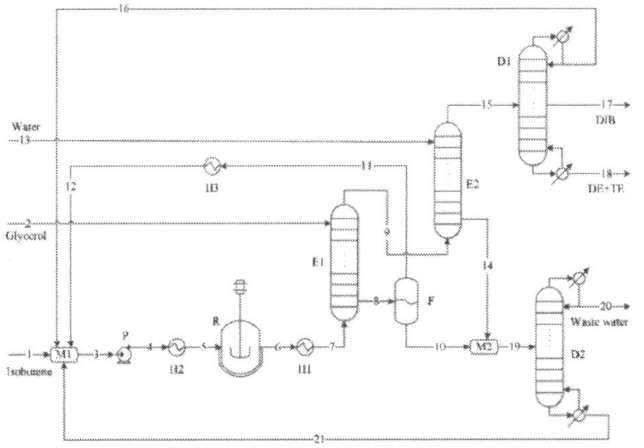

R – CSTR, E – Extraction column, D – Distillation column,
P – Pump, F – Flash tank, H – Heat exchanger, M – Mixer

- The etherification reaction carries out in a continuous stirred tank reactor (CSTR).
- Fresh glycerol is used to extract ME from the reaction product in E1 and is recycled back to the reactor.
- The extraction raffinate is washed by water to remove residual glycerol in E2, and then is fed to the distillation column D1 to separate isobutene.
- Dimers of isobutene (DIB) are taken out of the system via a side stream, and the desired di-ethers (DE) and tri-ether (TE) of glycerol are obtained at the bottom of D1.
- The glycerol in the waste water is recycled after steaming out water in the distillation column D2. The flash tank F is used to recover isobutene.

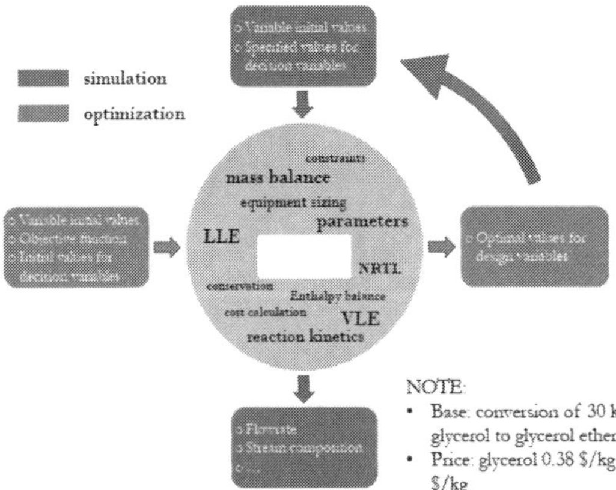

NOTE:
- Base: conversion of 30 kmol/h glycerol to glycerol ethers
- Price: glycerol 0.38 $/kg, IB 1.6 $/kg

Objective function:

Results

Values of decision variables and results

Unit	Decision variable	Initial	Optimal
R	volume, m³	25.00	28.03
	temperature, K	353	343
H1	outlet temperature, K	313	310
E1	number of trays	8	8
E2	number of trays	4	4
D1	number of trays	8	9
	feed tray	6	6
	side stream tray	4	4
	molar reflux ratio	0.50	0.06
	side stream flowrate, kmol/h	11.00	0.53
	reboiler duty, MJ/h	8500.00	9795.20
F1	temperature, K	323	391
D2	number of trays	6	6
	feed tray	5	5
	molar reflux ratio	0.40	0.83
	reboiler duty, MJ/h	2700.00	3386.20
global	IB feed rate, kmol/h	85.00	64.80
	Water feed rate, kmol/h	10.00	11.52
Main results	Glycerol conversion	1.0000	1.0000
	DE+TE yield	0.9557	0.9625
	IB utilization factor	0.7184	0.9443
	Production cost, $/kg	1.477	1.201

- The optimizing direction is to decrease the consumption of the expensive isobutene.
- The price of raw materials, especially that of isobutene largely determines the economic feasibility of this etherification process

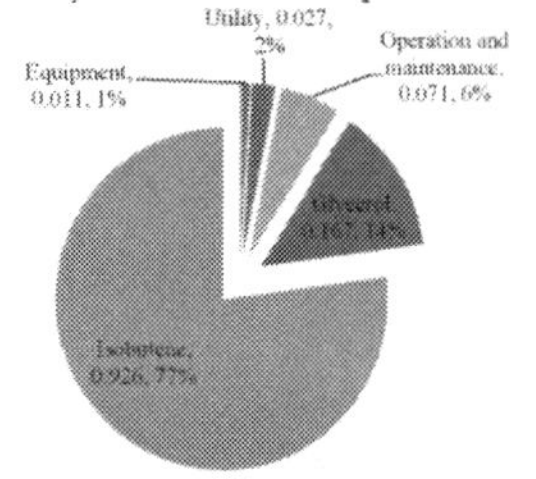

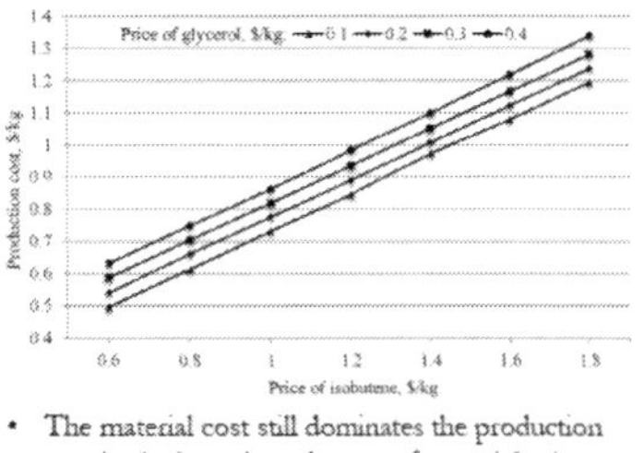

- The material cost still dominates the production cost in the investigated range of material prices.
- If the price of isobutene is less than 1.2 $/kg, the proposed process would produce glycerol ethers at a competitive cost of 1.00 $/kg.

407412 Deriving Value from Integrating Operations Control and Optimization into Engineering Design

Abstract:

The engineering design of an asset is typically executed via the front end engineering design (FEED) or front-end loading (FEL) process. FEED/FEL is a useful engineering design approach for thoroughly planning, estimating, and controlling project schedules and expenses to meet the specified technical requirements, before a project is even considered for construction. Process simulators are extremely effective in determining the viability of a project, executing a new design, or revamping existing operations. Simulators can be used to evaluate alternate scenarios—plant configurations, operating conditions, energy usage, etc. A process engineer can easily run sensitivity analyses, planning studies, operational execution runs, and on-demand decision support with the highest level of confidence and rigor.

However, to realize the true value technology provides, operations and automation systems must be built into, or on top of, detailed engineering design models. This paper will look at plant operations in a holistic view, presenting a single software solution that is able to consider safety, control, and training during the design phase. This forward thinking approach to modeling requires more work up front, but the economic benefits are enormous. Several customer applications will show the value and success of this implementation.

Use of Exergy in Design Methodologies for Energy-Efficient Sub-Ambient Processes

UNIVERSIDAD DE GUANAJUATO | NTNU

Danahe Marmolejo−Correa[a], Truls Gundersen[b]

[a]*Department of Physics Engineering, University of Guanajuato, México*

[b]*Department of Energy and Process Engineering, Norwegian University of Sciences of Technology, Norway*

Exergy Analysis

Exergy Analysis (EA) is commonly used as a ***post-design tool.***

By definition, exergy is the theoretical maximum (minimum) work obtainable (required) to bring (move) the system of interest to (from) equilibrium with an environmental state of reference

Low Temperature Processes ⟺ **Compression & Expansion**

Using Exergy in Subambient Processes?

Thermo-mechanical Exergy and Exergy of Heat

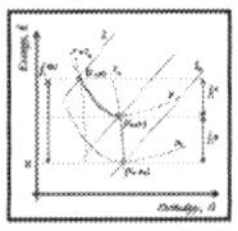

$$\dot{E}^{TM}(T,p) = \dot{E}^{T}(T,p_0) + \dot{E}^{p}(T_0,p)$$

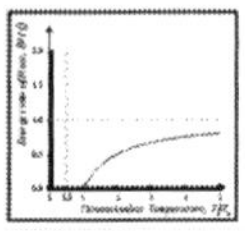

$$\dot{E}^{Q} = \begin{cases} \dot{Q} \times \eta_{Carnot} ; & T > T_0 \\ 0 ; & T = T_0 \\ -\dot{Q} \times \eta_{Carnot} ; & T < T_0 \end{cases}$$

Exergy Sources and Sinks in Heat Exchange

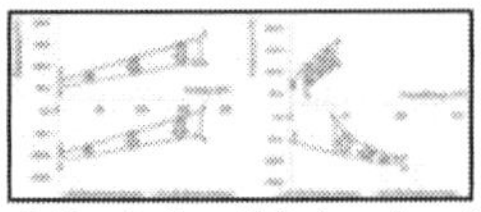

Exergy Transfer Effectiveness (ETE)

$$ETE = \frac{Exergy\ Sink}{Exergy\ Source}$$

D. Marmolejo-Correa, T. Gundersen, A new efficiency parameter for exergy analysis in low temperature processes, Int. J. Exergy. 17 (2015) 135–170.

Process Design Methodology

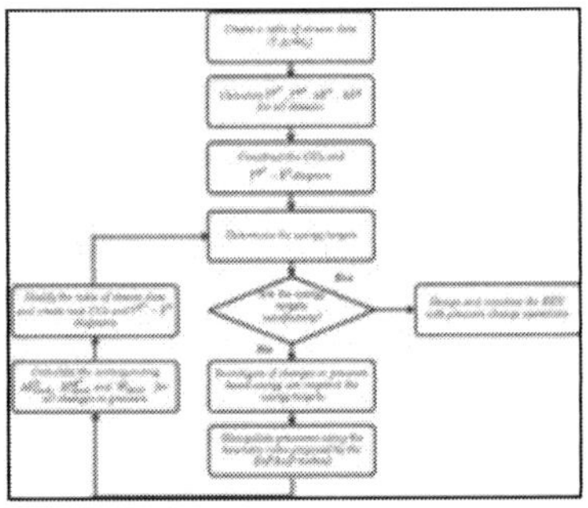

Compression and Expansion in Heat Recovery Systems

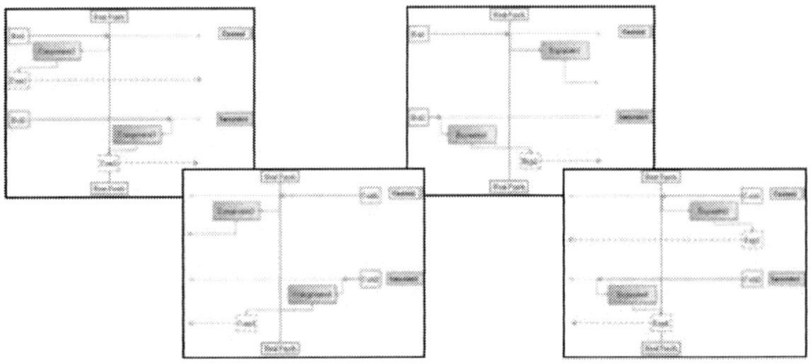

Composite Curves, a New Exergy Diagram for Heat Recovery Systems

Composite Curves (CCs)

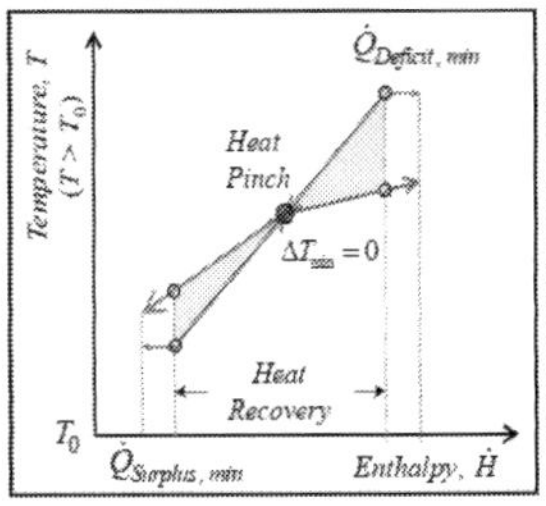

New Exergy Diagram

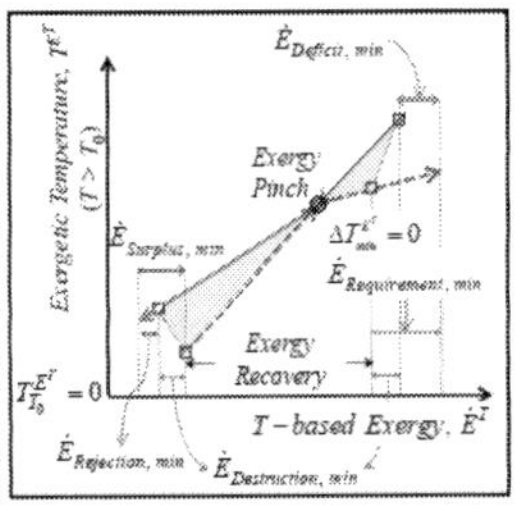

Some Characteristics:

- **Linear relation between T^{E^T} and $\dot{E}^T$.**
- **Only supply and target conditions are required.**
- **T^{E^T} is always positive.**
- **The heat and exergy pinch points are placed in corresponding enthalpy and exergy intervals.**

$$\dot{E}^{T} = \dot{m}c_p \underbrace{T_0\left[\frac{T}{T_0} - \ln\left(\frac{T}{T_0}\right) - 1\right]}_{T^{E^T}}$$

D. Marmolejo-Correa, T. Gundersen, New Graphical Representation of Exergy Applied to Low Temperature Process Design, Ind. Eng. Chem. Res. 52 (2013) 7145–7156. doi:10.1021/ie302541e.

Case Study: Before Pressure Manipulations

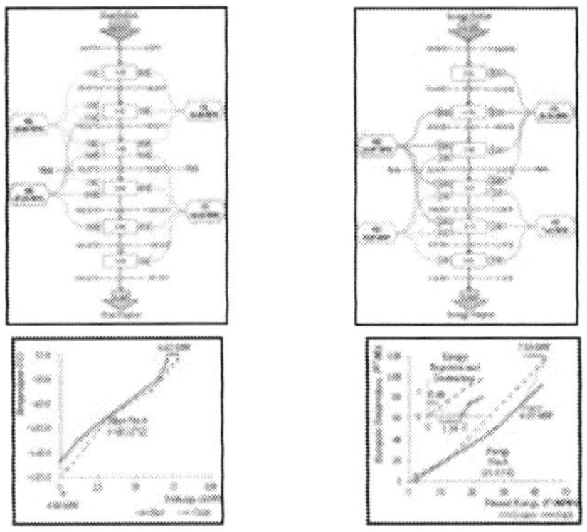

Case Study: After Pressure Manipulations (Final Design)

Ideal (ΔT_{min}=0 K, η_{Isen}=100%)

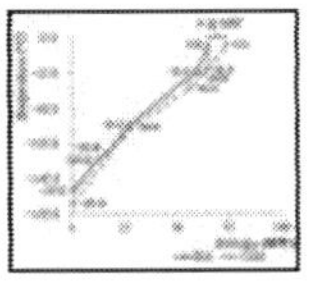

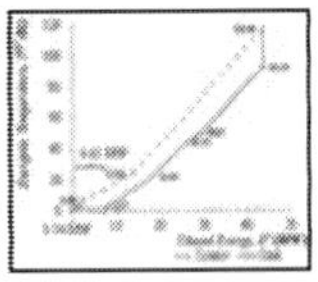

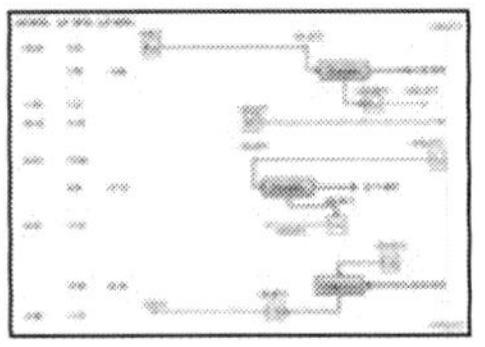

Actual (ΔT_{min}=3 K, η_{Isen}<100%)

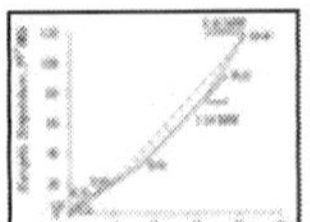

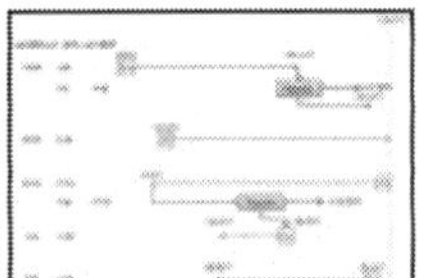

D. Marmolejo-Correa, T. Gundersen, Process Design Methodology for Energy-Efficient Processes Operating Below and Across Ambient Temperature, Submitted to AIChE J. (October 2015).

Conclusions

- **A new design procedure is presented for Heat Recovery Systems where some of the process streams are subject to pressure change (compression and expansion).**
- **The main novelty of this procedure is that exergy considerations are made at the conceptual stage of design.**
- **The heat cascade and composite curves from Pinch Analysis have been extended to an exergy cascade and a new Exergy Diagram (Exergetic Temperature).**
- **An important design tool is the novel Extended Grid Diagram, where heat exchange, compression and expansion are represented together.**
- **The process design methodology is particularly useful for processes operating below and across ambient temperature and consist of five major steps.**

Contact Information

- **Emails: d.marmolejocorrea@ugto.mx Truls.Gundersen@ntnu.no**

AIChE Annual Meeting, 2015
Salt Lake City, UT, USA, 8 - 13 November, 2015

425123 How Simulation Based Product Design in High-Pressure Polymerization Technology Leads Back to the Roots of Fundamental Science

Abstract:

Product development in high-pressure polymerization technology is a specifically demanding task. First, operation procedures from large-scale technical practice cannot be scaled down to typical lab application one to one. Second, competitive high-pressure polymerization plants operate on a scale of significant more than 100 kt/á production capacity. Both, the significant scale up as well as limitations in miniaturizing laboratory equipment are still a remarkable challenge in designing strategies for product development in high-pressure polymerization technology.

The first target is the prediction of microstructural polymer properties being determined by process conditions and strategy. During the recent years, there has been remarkable progress. Detail of predicted molecular structure has received significant improvement specifically due to the combination of deterministic modeling with Monte-Carlo techniques, so-called hybrid technology. By this, such detail is sufficient for feeding rheological models. Of course this puts significant demands on computational efforts. However, performance of computational equipment is high enough to establish such technologies in standard research. Using examples of such investigations the portability of such technologies will be discussed. The second target is to use this information to build a bridge in order to transform structural information to rheological models. By this, based on process design application properties become accessible by simulation.

With increasing insight into the polymer structure, aspects of reactor operation come into focus again that have been accepted for a long time as non-ideal behavior to be adapted. This was accepted generally as the impact of the correlation of process conditions in polymer microstructure had been the dominant problem. Now, the focus of modeling is redirected again to the non-uniform distribution of species along the radial reactor axis together with the resulting consequences. The consequences of radial species distribution in high-pressure polymerization reactors will be inspected together with their origin and impact on varying heat transfer in reactor tubes.

Systematic discrepancies in reactor modeling being observed when applying unified reactor models over a wide variety of processes and manufacturing sites indicate that it is worthwhile to search for their origin. This makes revisiting fundamentals. Both, elementary reaction kinetics and thermo-physical properties come into focus again. The access to thermo-physical properties of reaction species under process conditions such as 3000 bar and 300 °C is rare. Even inexistent is the information about their mixtures. In the major number of cases ideal mixing rules are used in modeling. However, it is of high interest to know about the impact of uncertainties of these quantities on process modeling for which reason sensitivity studies are performed and demonstrated. First attempts of a direct access of such quantities will be shown. Furthermore, diffusion limitations of elementary reactions come back into focus. Models for such mechanisms are discussed in literature. However, what remains open in many cases is the question whether their impact on the overall reaction mechanism is selective and high enough to perform clear model discrimination. First sensitivity studies with this respect will be presented.

For years the modeling of such complex systems was strictly oriented forward pushing the limits in detail in describing the polymeric microstructure. Consistent model design and multifold application in technical practice helped to learn about the potential of development and weak points. It is remarkable to note how fundamental investigations for such purposes of modeling redirect the interest back to roots of fundamental mechanisms. By this, classical disciplines experience a renaissance.

427055 Impact of Thermo-Physical Data on the Modelling of High Pressure Polymerizations

Abstract:

Nowadays modelling is a basic instrument for process optimization. Especially for technologies, which are conducted under exceptional process conditions, such as high pressure polymerizations, simulations are a welcome alternative to costly experiments in miniplants.

The synthesis of Low Density Polyethylene (LDPE) is an example for such an industrial process, which is typically operated between 100°C and 300°C and under pressures up to 3000 bar. The mathematical prediction of process parameters for the LDPE-Synthesis, for instance conversion, molecular weight distribution and branching, has been studied and refined since the early 1970s. By now the process simulation for LDPE synthesis poses a crucial tool for the LDPE-process development and optimization. But for the conduction of reliable and precise process simulations it is essential to have a very profound knowledge and understanding of the process itself, its thermodynamics and kinetics and of course the structure-properties relationship of the resulting polymer. Therefore a uniform set of parameters is essential, which can describe the polymerization reaction independently of reactor type and process. However this proves to be a highly demanding task as numerous parameters have to been known sufficiently precise within a complex reaction network and under the extreme process conditions. This is why variable parameters are still necessary in the current simulation models in order to obtain good agreement between calculation and plant data [1]. For tube reactors these parameters are e.g. the initiator efficiency and the thickness of the fouling layer inside the reactor.

The existence of these variable parameters as well as systematic discrepancies within the reactor models give reason for further investigations. In order to identify and overcome the knowledge gaps a systematic sensitivity analysis of the existing model concerning possible faulty parameters is the reasonable approach. Firstly the fundamentals, such as the thermo-physical properties, which serve as the basis of the polymerization model, should be revisited. For the simulation these are the

- density
- heat capacity
- viscosity and
- heat transfer coefficient

of ethylene, LDPE and the reaction mixture. These data have to be available in dependence on temperature, pressure and composition, respectively. For the calculation of the viscosity molecular weight and branching structure of the polymer can also be taken into account. Whereas the density, the heat capacity and the thermal conductivity primarily influence the heat balance of the model, the viscosity of the reaction mixture also influences the kinetics directly. Reaction steps that depend on polymer mobility, such as the termination reaction of the polymerization, exhibit a direct dependence on the viscosity even at low polymer concentrations.

The literature on these properties for the pure components under process conditions is scarcely available and mostly dates back to the 1950s and 60s. Furthermore it often contains extrapolations and was analyzed with outdated mathematical methods. Experimentally based data on mixture properties is virtually inexistent, thus ideal mixing rules are applied without questioning [2], [3]. Moreover it becomes evident that some of the authors obtain differing values for the thermo-physical properties at a given temperature and pressure. Thereby the discrepancies reach up to several percent.

In the next step the impact of these parameters on the process simulation of high pressure polymerizations was studied and evaluated by the example of an industrial tube reactor.

Effects could mainly be observed in the resulting temperature profile and the achieved conversion. However these effects were negligible for the heat transfer, but conspicuous for the heat capacity and striking for the viscosities. For the heat capacity a constant error was introduced and evaluated. For small errors the adjustment of the variable parameters makes it possible to diminish the deviation caused by the introduced error to a minimum. This proves the concept of the variable parameters and explains why they sometimes exhibit physically unrealistic values.

But LDPE also possesses special rheological properties due to its unique branching structure. Thus the correct prediction of the molecular mass distribution as well as the polymeric microstructure is of striking importance for this polymer as well. Therefore it is also interesting to examine whether the viscosity of the reaction mixture, which directly influences the polymerization kinetics, shows any influence on the calculated molecular mass distribution and polymeric microstructure.

[1] T. Herrmann, *Dissertation,* Darmstadt **2011.**

[2] A. Buchelli, M. L. Call, A. L. Brown, A. Bird, S. Hearn, J. Hannon, *Ind. Eng. Chem. Res.* **2005**, 44, 1474-1479.

[3] G. Luft, R. Steiner, *Chemiker-Zeitung* **1971**, 95 (1), 11-15.

429017 Production of Bio-Butanol from Wheat Straw: A Proposal for Sustainable Design

Abstract:

Production of bio-butanol from wheat straw: a proposal for sustainable design

Dorothee Luise Kurz, Randi Neerup, Anders Lodberg, Anjan Kumar Tula*

Department of Chemical and Biochemical Engineering,

Technical University of Denmark, DK-2800 Lyngby, Denmark

*antu@kt.dtu.dk

The usage of second-generation biomass is gaining more interest because it provides the opportunity to sustainably produce an important fuel using a bioprocess [1]. Butanol is an important fuel because of its similar properties to gasoline and its higher energy density than ethanol [1]. The step towards bio-based processes and bio-based fuels is both important and advantageous due to the reduction in CO_2 emissions and other greenhouse gasses that are emitted primarily from the combustion of fossil fuel.

Using a systematic, hierarchal, 12 task approach for performing (sustainable) process design, a production process for bio-butanol utilizing wheat straw as the raw material was designed and evaluated for producing 4,300 metric tons of bio-butanol per year with a yield of 0.11 kg product per kilo of raw material and a bio-butanol purity of 99.5 wt%. The reaction path considered for the production of bio-butanol is the ABE fermentation with Clostridium beijerinckii [2]. Two commercially by-products, acetone and ethanol, are produced during the reaction (carried out in a fermenter). These two by-products are of importance because acetone can be sold as a chemical solvent and ethanol can be used as biofuel [3]. The fermenter outlet consists of a multi-component mixture of butanol, acetone, ethanol and water and a minor amount of acids. First, acetone and ethanol are separated from the mixture using distillation; acetone is then further purified. Second, butanol is separated from the broth using liquid-liquid extraction. Butanol is further purified by distillation. This design is used as the base case design that is evaluated for further improvement.

Using an economic analysis, improvement targets are identified for process improvement related to heat integration and process optimization in order to reduce the process operating costs. The improvements up to this point are economic-based.

To achieve a more sustainable design, a sustainability [4] and LCA analysis [5] is performed in order to identify further improvement targets related to sustainability. These targets are met through the elimination and/or minimization of limitations in the process, identified from the sustainability analysis. The final design is a more sustainable, economic viable design. In this poster, the process design method will be presented and the output from each step related to the production of bio-butanol will be given.

References:

[1] A. Ranjan and V. S. Moholkar, "Biobutanol: science, engineering, and economics," *Int. J. Energy Res.*, vol. 36, no. 3, pp. 277-323, Mar. 2012.

[2] T. Ezeji, N. Qureshi, and H. P. Blaschek, "Butanol production from agricultural residues: Impact of degradation products on Clostridium beijerinckii growth and butanol fermentation," *Biotechnol. Bioeng.*, vol. 97, no. 6, pp. 1460-1469, 2007.

[3] E. Sklavounos, "Conditioning of SO 2 - ethanol-water (SEW) spent liquor from lignocellulosics for ABE fermentation to biofuels and chemicals," Aalto University, 2014.

[4] A. Carvalho, H. A. Matos, and R. Gani, "SustainPro-A tool for systematic process analysis, generation and evaluation of sustainable design alternatives," *Comput. Chem. Eng.*, vol. 50, pp. 8-27, Mar. 2013.

[5] S. Kalakul, P. Malakul, K. Siemanond, and R. Gani, "Integration of life cycle assessment software with tools for economic and sustainability analyses and process simulation for sustainable process design," *J. Clean. Prod.*, vol. 71, pp. 98-109, May 2014.

430219 Sustainable Design for Production of Dimethyl Carbonate from Ethylene Oxide, CO2 and Methanol

Abstract:

Sustainable Design for Production of Dimethyl Carbonate
from Ethylene Oxide, CO_2 and Methanol

Erik Haastrup[1], Rishika Chatterjee[2], Nipun Garg[3], Amata Anantpinijwatna[4]
[1]s113150@student.dtu.dk, [2]s141286@student.dtu.dk, [3]s140822@student.dtu.dk, [4]amatana@kt.dtu.dk

There is a growing awareness and research into how to mitigate the carbon dioxide (CO_2) emissions. Two attractive methods that can be applied in addressing this are, carbon capture and sequestration (CCS) (IPCC, 2005) where CO_2 is captured at emission points, and deposited under a suitable geological formation, from whence it will not escape back into the atmosphere, or CO_2 chemical utilization where CO_2 is used as a raw material for producing a chemical compound. Dimethyl Carbonate (DMC) is an important bulk chemical because it is a viable substitute for methyl-tert butyl ether (MTBE) (F.M. Mei, 2009). DMC can be produced using different reaction pathways, for example, through the reaction of phosgene with methanol, which yields DMC and HCl. This method is the traditional route, but involves hazardous chemicals, and the product has high chlorine content. Another method is the alcoholysis of urea. First urea is produced from CO_2 and ammonia. The urea then reacts with methanol to produce DMC. However, in this process, methyl carbamate is easily generated as mono-charged intermediate in the second reaction step, and this cannot be further reacted into DMC.

A more sustainable process design for the production of DMC is proposed using a systematic method that is based on a

12 task, hierarchal approach with CO_2, ethylene oxide (EO) and methanol used as raw materials and ethylene glycol (EG) as the by-product. The production target is set at 100,000 metric tons with a product and by-product purity of 99% (w/w). The reaction path synthesis chosen consists of two steps. In the first step, EO and CO_2 react (at a temperature of 110°C and pressure of 8 MPa) to produce ethylene carbonate and in the second step, ethylene carbonate and methanol react (at a temperature of 90°C and pressure of 0.3 MPa) to produce DMC and ethylene glycol. The reaction path chosen produces a by-product that, at a given purity, can be sold.

A description of the process is as follows. The first reactor outlet consists of EO, CO_2 and ethylene carbonate (EC). A flash separated as much of the CO_2 as possible, and the rest of the stream is led into reactor 2, to which is added methanol as well. The outlet stream of the second reactor contains EO, CO_2, EC, EG and DMC. First, a distillation column separates the rest of the EO and CO_2 from the mixture, and recycles this back into reactor 1. A second distillation column separates DMC and methanol from EC and EG. Both these combinations form azeotropes. The DMC/methanol azeotrope is handled using a large distillation column at high pressure, from which 99% of the DMC can be extracted at high purity. The EG/EC stream is first cleaned of leftover methanol, hence fed into a distillation column, which separates as much of the EC as possible. The azeotrope cannot be broken through distillation, so a the difference in polarity between the two is taken advantage of, and a membrane is used to separate the rest of the EC from the EG. The EC is recycled back into the second reactor, and the EG has a high enough purity to be sold commercially.

The process described here is used as the reference design, that is, the base case design which is evaluated for process improvement related to process integration (heat integration), process optimization and sustainable improvement.

First, an economic analysis is performed for identifying process limitations that are translated into design targets used for heat integration and process optimization. If these limitations are minimized and /or eliminated the design targets are met. The design is then analyzed using a sustainability (Carvalho, 2012) and LCA (Kalakul, 2014) analysis in order to further identify process limitations and other design targets related to sustainability. In this poster, a more sustainable process for the production of DMC, through the utilization of CO2 will be presented. The 12 task-based approach will be presented and the input and output of each task will be highlighted. Works Cited

Carvalho, A. M. (2012). SustainPro—A tool for systematic process analysis, generation and evaluation of sustainable design alternatives. *Computers & Chemical Engineering, 50* , 8-27.

F.M. Mei, E. a. (2009). Effective and Recoverable Homogeneous Catalysts for the Transesterification of Dimethyl Carbonate with Ethanol: Lanthanide Triflates. *Kinetics and Catalysis* , 666-670.

IPCC. (2005). *Carbon Dioxide Capture and Storage: Summary for Policymakers.*

Kalakul, S. M. (2014). Integration of life cycle assessment software with tools for economic and sustainability analyses and process simulation for sustainable process design. *Journal of Cleaner Production, 71* , 98–109.

431284 Achieving a More Sustainable Design for the Production of Di-Methyl Carbonate Via CO2 Utilization

Abstract:

Carbon dioxide (CO2) emissions are a great concern for the environment resulting in many new technologies and policies in order to mitigate these emissions. This ongoing issue has gained significant attention in recent years and poses a challenge for the 21st century. CO2 utilization can be achieved in three ways; conversion into a fuel, a feedstock for producing important chemical compounds and lastly via non-conversion utilization such as for enhanced oil recovery and as a solvent [1].

One particularly important chemical which CO2 can be converted into is dimethyl carbonate (DMC). DMC is a good replacement for toxic methylating agents and phosgene. Additionally, it can also be used as an octane booster in fuel as it significantly increases the octane number due to the high oxygen content (53.3 wt%) in comparison to other alkylcarbonates [2]. DMC can be produced from CO2 by either reacting CO2 directly with methanol or by firstly reacting ethylene oxide with CO2 to produce ethylene carbonate as an intermediate followed by a subsequent reaction with methanol [4].

In this project, a process plant was designed using the two step reactions utilising CO2 to produce DMC as well as the by-product ethylene glycol and any unconverted ethylene carbonate. A hierarchical decomposition approach was used for designing the process, by decomposing the synthesis into 12 hierarchical tasks. According to the design, the process will produce on a yearly basis around 15,000 metric tonnes of DMC (10% of the Chinese market, excluding the largest producer).

This design of the process is unique as it does not use conventional separation procedures like pressure-swing, reactive distillation or extractive distillation to break azeotropes that are commonly encountered during the separation of products, hence, avoiding the use of toxic chemicals. Instead, the separation of DMC and methanol utilises a pressure distillation column as the azeotrope is pressure sensitive and can be broken at 20 bar. Additionally, the azeotrope between ethylene glycol and ethylene carbonate is not sensitive to pressure so the possibilities of membrane separation were explored. A membrane that could separate the two compounds based on polarity was chosen, which could result in pure ethylene carbonate and almost pure ethylene glycol. Further, this membrane separation also reduces the consumption of energy in comparison to the installation of distillation columns.

Heat integration and process optimisation were then performed in attempt to reduce energy consumption and unnecessary costing according to the economic evaluation of the process. In order to meet the design targets these limitations need to be eliminated. Finally, a LCA and sustainability analysis were performed in order to reduce other limitations in the process design [3], [5].

This poster will present the 12 step hierarchical approach to synthesise a more sustainable process design for utilising CO2 to produce DMC. The results from each task and the important considerations for each stage will be highlighted.

References

[1] Michele Aresta. Carbon dioxide recovery and utilization. Springer Science & Business Media, 2003.

[2] Fredrik Bertilsson and Hans T Karlsson. Co 2 utilization options, part ii: Assessment of dimethyl carbonate production. Energy conversion and management, 37(12):1733–1739, 1996.

[3] Ana Carvalho, Henrique A Matos, and Rafiqul Gani. Sustainpro—a tool for systematic process analysis, generation and evaluation of sustainable design alternatives. Computers & Chemical Engineering, 50:8–27, 2013.

[4] Daniele Delledonne, Franco Rivetti, and Ugo Romano. Developments in the production and application of dimethylcarbonate. Applied Catalysis A: General, 221(1):241–251, 2001.

[5] Sawitree Kalakul, Pomthong Malakul, Kitipat Siemanond, and Rafiqul Gani. Integration of life cycle assessment software with tools for economic and sustainability analyses and process simulation for sustainable process design. Journal of Cleaner Production, 71:98–109, 2014.

432892 Dynamic Simulation of a Cryogenic Air Separation Unit

Abstract:

For the production of large quantity of air gases in industry, cryogenic distillation is the most economically viable process. In such a process, air is liquefied and separated in a series of distillation columns, namely, high pressure column, low pressure column, crude argon column, and pure argon column, into three pure components, oxygen, nitrogen, and argon. Over more than 100 years of process improvements, the design of such an Air Separation Unit (ASU) is typically a highly integrated process. Many of the product and intermediate streams are used as heating or cooling agents of the distillation columns, and the temperature difference is often just a few degrees. For example, the reboiler of the low pressure column and the condenser of the high pressure column are integrated together, and the liquid effluent from the high pressure column is used as the cooling agent in the condenser of the crude argon column.

For such a highly integrated process, although the steady state design has been optimized for many years, the corresponding dynamic behavior is much less studied due to its complexity, even though it is very important in operating such an ASU in a non-steady state manner. This has become increasingly more important in recent years with deregulated energy market. Because an ASU is a high energy consumer, it is desirable that it can be ramped according to the fluctuating electricity price while the product specifications are still preserved.

In our effort, a dynamic model is set up for a full ASU with first principles, with the parameter values taken from an industrial air separation plant. By solving such a model in an equation orientated simulator, the study of different dynamic scenarios can be carried out, which can be used for operation improvements, validation of new control schemes, including advanced control, controller tuning, and operator training. In this presentation, the model equations of different modules in such an ASU will be discussed, as well as the strategies used to converge such a large scale model. Simulation results will also be presented.

427707 Modeling, Simulation and Optimization of Crude Distillation Unit

Abstract:

Petroleum refinery is an industrial process plant for processing of crude oil to more useful products. Optimum functioning of the Crude Distillation unit (CDU), believed to be a vital process of petroleum refinery, is crucial for good product quality and economic benefits in the petroleum industry. The present work is aimed to investigate the general design of the CDU and to achieve optimum conditions using simulation. A heuristics based Crude Distillation Unit (CDU) is modeled in ASPEN PLUS® and used to obtain the data points which act as a substitute of working plant data to provide the process insights. A steady state model is developed in MATLAB® using the MESH equations, which predicts the temperature distribution and concentration of Naphtha over different trays. The developed model predicts the Naphtha distribution and withdrawal tray correctly, ascertaining the model accuracy. This model is then simulated for different sets of input conditions in order to obtain the specimen data points. An Artificial Neural Network (ANN) model is generated using the MATLAB neural network tool, which is further trained using sets from the specimen data points. This trained network is capable of predicting the output of a sample data point. Using this prediction capability, feed flow rate and steam flow rates are compared with the Naphtha product ratio. This provides the optimum value of feed and steam flow rates at which the Naphtha production is maximum.

Keywords: *Crude Distillation Unit (CDU), MATLAB, Artificial Neural Network (ANN), Aspen Plus, MESH equations.*

428615 Sustainable Design for Production of Styrene from Benzene and Ethylene

Abstract:

Process design – A production of Styrene from Benzene and Ethylene

Matias Falk Bjerregaard, Jonas Richard Foelsby, Michael Roland Larsen, Maria-Ona Bertran

Department of Process Engineering and department of Chemical and Biochemical Engineering, Technical University of Denmark, 2800 Kongens Lyngby, Denmark
marber@kt.dtu.dk

Styrene is an important bulk chemical because it is used for the production of a variety of products, e.g. plastic and rubber products[1].

Using a systematic method for performing sustainable process design, decomposed into 12 tasks, a more sustainable process has been developed for the production of styrene. The raw materials used are Benzene and Ethylene, and the styrene production target is set at 150.000 metric tons / year, with a 99 wt-% purity.

The raw materials, benzene and ethylene are utilized in a two-step synthesis reaction[2,3]. In the first reaction, alkylation of ethylene and benzene to form ethylbenzene and diethyl benzene as an undesired byproduct.
In the second reaction, dehydrogenization of ethylbenzene occurs to produce styrene, with benzene, ethylene, and hydrogen as the undesired by-products.

A brief description of the process is as follows. The first reactor outlet consists of ethylbenzene and diethyl benzene,

which is separated using a distillation column. Next, the second reactor outlet consists of styrene, hydrogen, ethylbenzene, and benzene. First, hydrogen is removed using a flash distillation. Next, benzene is separated using a distillation column. Lastly, ethylbenzene is separated using yet another distillation column.
The third and last reactor utilizes a recycled diethyl benzene stream for transalkylation, into ethylbenzene, which is fed into the outlet of the first reactor.
This approach is used as the base case design that is further investigated for process improvements with respect to both economic and environmental feasibility.

The base case design is analyzed using an economic analysis for identifying process improvements related to heat integration and process optimization. Improving these two aspects directly affect both the utility and operating costs. To further refine the design, a sustainability[5] and LCA analysis[4]is performed in order to identify process limitations that are translated into design targets that, if matched, eliminates process limitations.

In this poster, the task-based method for achieving a sustainable process design will be presented through the application of the design of a more sustainable process for the production of styrene. The output for each step will be highlighted and the base case design will be compared to the improved designs obtained from heat integration and process optimization and, from the sustainability analysis.

References:

1. Burridge E. Styrene. *ICIS Chemical Business*. 2006.

2. Edinburg University. Styrene process, preliminary information. Process Calculations Web site. http://www.see.ed.ac.uk/~jwp/procalcs/procalcs/study/styrene0.html. Updated 2015. Accessed 27/02, 2015.

3. Tiako Ngandjui LM, Louhibi D, Thyrion FC. **Kinetic analysis of diethylbenzene-benzene transalkylation over faujasite Y**. *CHEMICAL ENGINEERING AND PROCESSING*. 1996.

4. Carvalho, A., Matos, H. A., & Gani, R. (2013). SustainPro—A tool for systematic process analysis, generation and evaluation of sustainable design alternatives. Computers & Chemical Engineering, 50, 8–27. doi:10.1016/j.compchemeng.2012.11.007

5. Kalakul, S., Malakul, P., Siemanond, K., & Gani, R. (2014). Integration of life cycle assessment software with tools for economic and sustainability analyses and process simulation for sustainable process design. Journal of Cleaner Production, 71, 98–109. doi:10.1016/j.jclepro.2014.01.022

428926 Achieving a More Sustainable Process Design for the Production of Methanol

Monday, November 9, 2015
Exhibit Hall 1 (Salt Palace Convention Center)

Cristina Calvera Plaza and **Marta Gonzalez Garcia**, Technical University of Denmark, Copenhagen, Denmark

Achieving a More Sustainable Process Design for the Production of Methanol

Cristina Calvera Plaza, Marta Gonzalez Garcia, Ana Diez Callau, Emmanouil Papadakis*

Department of Chemical and Biochemical Engineering

Technical University of Denmark, DK-2800 Lyngby, Denmark

*empap@kt.dtu.dk

Methanol is an important chemical product because it can be used as a raw material for the production of other chemicals (1), for example dimethyl carbonate, formaldehyde and m*ethyl tert-butyl ether* and it is also one of the most produced bulk chemicals with an annual global production of 100 million metric tonnes per year (1). Methanol can be produced using different reaction paths, for example natural gas. If natural gas is used for methanol production then CO_2 is produced, utilized and can be emitted. Therefore, achieving a more sustainable design for the production of methanol is beneficial in order to reduce the process CO_2 carbon footprint.

Using a 12 step, systematic task-based approach for achieving sustainable process design, a process for the production of methanol with an annual production of 300,000 metric tonnes of methanol per year with minimum CO_2 emissions is analysed and designed. The reaction path chosen for the production of methanol utilizes CO_2 as raw material. The reactions are as follows:

$$CO_2 + 3H_2 \leftrightarrow CH_3OH + H_2O \qquad CO + 2H_2 \leftrightarrow CH_3OH \qquad CO_2 + H_2 \leftrightarrow CO + H_2O$$

The process description that uses the above reactions is explained. The first outlet stream from the reaction system is a multi-component mixture that consists of methanol, water and unreacted gases. First, the gases are separated using a flash and recycled to the reactor. Second, the methanol is purified using two distillation columns. In the first distillation, light components are separated and in the second one the water. This design is analysed for improvements and is referred to as the base case design.

The analysis consists of two parts, first an economic analysis and second, a sustainability analysis. The economic analysis provides the information in order to identify targets for improvement related to heat integration and process optimization, while the sustainability analysis (2) provides information for identification of (design) targets for achieving sustainable design, that is, for example, reduction in carbon footprint and LCA factors (global warming potential, toxicity, etc.) (3). The sustainability analysis consists of a sustainability and LCA analysis.

In this poster, the 12 task-based method for achieving (sustainable) process design will be presented and its application will be highlighted through the application of the production of methanol. The input and output related to each step will be presented.

References

1. Methanex Corporation: The Power of Agility. [on line] http://www.methanex.com.

2. *SustainPro- A tool for systematic process analysis, generation and evaluation of sustainable design alternatives.* **Carvalho, A., Matos, H.A. y Gani, R.** 8-27, s.l. : Computers & Chemical Engineering, 2013, Vol. 50.

3. *Integration of life cycle assessment software with tools for economic and sustainability analyses and process simulation for sustainable process design.* **Kalakul, S., y otros.** 98-109, s.l. : Journal of Cleaner Production, 2014, Vol. 71.

433298 A Geometric Method to Calculate the Minimum Reflux of Complex Columns Using Enthalpy-Composition Diagram

Abstract:

A Geometric Method to calculate the Minimum Reflux of Complex Columns Using Enthalpy-Composition Diagram

José D. Olguín-Ángeles , Fernando Pérez-Villaseñor, Arturo Ortiz-Arroyo, Ángel Castro-Agüero.

Facultad de Ciencias Básicas Ingeniería y Tecnología, Universidad Autónoma de Tlaxcala, Apizaco, Tlaxcala, México, 90300.

Distillation is the most common process separation used to purifying homogeneous fluid mixtures, but at the same time it is the one that consume most energy in a chemical plant. Underwood design methods have been used through many years to calculate minimum reflux of conventional (Seader et al, 2010) and complex columns (Nikolaides and Malone, 1987; Triantafyllou, C. and Smith, 1992; Ramirez et al., 2010; Muralikrishna et al., 2002). The Underwood equation is used mainly for thermodynamically ideal mixtures due the assumptions considered in this equation, therefore the designs of column distillation for separations of non-ideal mixtures are not convenient.

Doherty and Malone (1987) proposed a design method for ideal multicomponent mixtures using rigorous thermodynamic models for conventional distillation columns. This methodology can be applied to non-ideal mixtures but it must deal with azeotropic points and "tangent pinch" solutions. Doherty and Malone do not considered the energy balance, so thermal effects cannot be considered. They used the "Zero Volume Method" to find the minimum reflux ratio in a stage-to-stage solution of mass equations.

The Ponchon-Savarit method for binary mixtures is the rigorous design method when acceptable thermodynamic models are used for the specified mixture. The Ponchon-Savarit method is well known for binary separation mixtures but is not utilized for ternary or multicomponent mixtures. Reyes y col. (2000) used the Enthalpy-Concentration diagram using an iterative methodology to estimate the minimum reflux. They find the minimum reflux of conventional ideal distillation columns and compare against shortcut and stage-to-stage solutions.

In this work, it is proposed a methodology to find the minimum reflux ratio of conventional and non-conventional columns (multiple feeds and multiple side-streams). The Zero Volume Method is used to determine the minimum reflux ratio, but instead of solve stage-to-stage mass balance, an adiabatic flash solution is proposed. With the adiabatic flash, the tie lines that determine the zero volume method can be found for every column section. Besides, the zero volume method incorporates the energy balance, so that thermal effects can be considered.

The main column of the Petlyuk column is presented as a result of a non-conventional distillation column, which contains two feed streams and one liquid side stream. In a complex column, besides to find the minimum reflux for each feed, it is necessary determine which of the feeds has the maximum of all minimum reflux. With this methodology, the minimum reflux ratio of the main column of Petlyuk system was calculated, the minimum reflux ratio of each feed stream and then the minimum reflux ratio of the main Petlyuk column. The estimated minimum reflux ratio of main column was compared against to a stage-to-stage simulation with ASPEN PLUS. The simulated composition profile shows that the column operates at minimum reflux, there is a composition pinch zone, and the pinch zone is located at the correct feed stream.

References

Doherty, M. F. y Malone, M. F., "Conceptual Design of Distillation Systems", Editorial Mc Graw Hill, 1st Edition, (2001).

Muralikrishna, K., Madhavan, K. P. and Shah, S. S., "Development of Dívіding Wall Distillation Column Design Space for a Specified Separation", Trans IChemE., 80, p. 155-166, (2002).

Nikolaides, I. P. and Malone, M. F., "Approximate Design of Multiple-Feed/Side Stream Distillation Systems", Ind. Eng. Chem. Res., 26, p. 1839-1845, (1987).

Ramírez, N., Jiménez, A., Castro, A. and Rico, V., "Optimum Design of Petlyuk and Divided-Wall Distillation Systems using a Shortcut Model", Chem. Eng. Res. Des., 8, p. 1405-1418, (2010).

Reyes, J.A., Gómez, A. And Marcilla, A., "Graphical Concepts to Orient the Minimum Reflux Ratio Calculation on Ternary Mixtures Distillation", Ind. Eng. Chem. Res., 39, 3912-3919, (2000).

Seader, J.D., Henley, E.J., and Roper, D.K., "Separation Process Principles", John Wiley and Sons, 3th Edition, (2010).

Triantafyllou, C. and Smith, R. "The Design and Operation of Fully Thermally Coupled Distillation Columns", Trans. Inst. Chem. Eng., p.118-132, (1992).

433328 Olefin Furnace System Scheduling with the Consideration of Relay Cracking of Recycled Ethane/Propane

Abstract:

Ethylene is the most important product with the largest bulk productivity in the petrochemical industry. An ethylene plant employs multiple cracking furnaces in parallel to convert various hydrocarbon feed stocks to smaller hydrocarbon molecules, mostly ethylene and propylene. Among the furnaces, one furnace is normally specifically designated to crack recycled ethane and propane which are the byproducts of the cracking reaction. The continuous operational performance of cracking furnaces gradually decays because of coke formation in the reaction coils, which requires each furnace to be periodically shut down for decoking. Since some plants do not have buffers to store recycled ethane and especially recycled propane, another furnace must be picked up to relay cracking recycled ethane/propane temporary when the designated recycling furnace is in decoking. Since ethane and propane gas are easier to form coke than liquid feeds, plants normally prefer to cracking recycled ethane/propane at the beginning of the batch rather than at the ending of the batch on the temporary relay furnace to have more flexibility in adjusting production schedule. Thus it is better to let the batch starting time of the relay furnace equal to the batch ending time of the recycling furnace. When the recycling furnace finishes the previous batch and is in decoking, a relay furnace has just finished decoking and can crack recycled ethane/propane temporary. When the recycling furnace finishes decoking and starts a new batch, the recycled feed will be switched back to the designated recycling furnace and the temporary relay furnace will crack its own feed. To realize above object, a MINLP model has been developed and demonstrated by case studies.

433553 Molecular Design of Reactants and Products - a Multi-Objective Approach

Abstract:

In reactive systems, computer-aided molecular design (CAMD) is performed to design solvents that enhance reaction kinetics, improve catalyst activity/lifetime as well as reactants and products. CAMD of reactants and products helps in their selection such that target properties of interest are optimized with satisfaction of property constraints. Owing to the scarcity of CAMD methodologies that design reactants and products, there has been an increased interest to address this issue. The few recently available methodologies to design reactants and products however have their shortcomings. The earliest available methodologies are restricted to cases where only structures of a single reactant and product are variable. Also, employment of linear property models and lack of a methodology to treat property constraints have been some of the other restrictions. Dev et al. [1, 2] recently developed two algorithms to design reactants and products which are unrestricted by the number of reactants and products whose structures are uknown. They also made provisions for non-linear property models and property constraints. In their first algorithm [1], only the dominant properties of the products are optimized. Also, only the products are subjected to their individual set of property constraints. In their second algorithm [2], properties that are dependent on the structures of both reactants and products are optimized. Also, all reactants and products are subjected to their respective set of property constraints. In this work, we have developed an algorithm that generates structures of reactants and products irrespective of their numbers such that each reactant and product's target property(s) is to be optimized. Also, each reactant and product is subjected to a set of property constraints. All properties have been expressed in terms of molecular structure using different types of property models. Since there exists a need to treat different types of property models on a single platform, signature descriptors have been utilized. A variety of property models can be expressed in terms of signature descriptors which are molecular building blocks. They enable the simultaneous utilization of Quantitative Structure Activity/Property Relationships (QSARs/QSPRs) and Group Contribution Models (GCMs) [3]. In the formulated problem, all property models are expressed in terms of the occurrence number of signature descriptors. Relationships have been developed that relate the occurrence number of signatures of reactants and products. This is achieved by tracking the atoms in the reactants to the products by taking into consideration the reaction mechanism. Thus the reactants and products have been structurally related. Due to this structural relation, property constraints acting on reactants and products influence the selection of both optimal reactants and products. The objective functions of each of the reactants and products display a similar influence. Since the selection of each reactant and product is dependent on the property constraints and property objective functions of the other reactants and products in addition to its own, the problem is formulated as a multi-objective optimization problem when objectives are conflicting. This problem is solved using the Augmented Epsilon-Constraint Method (AUGMECON) developed by Mavrotas [4]. AUGMECON generates pareto optimal solutions while completely avoiding weakly pareto optimal solutions. The generated solutions also include unsupported pareto optimal solutions. Specifically for multi-objective integer programming (MOIP) problems, an improved version of AUGMECON, AUGMECON2 [5] can also be utilized. AUGMECON2 is very efficient in generating the entire pareto optimal set for MOIP problems compared to AUGMECON and some earlier available methods. In this contribution, the utility and novelty of the developed methodology has been highlighted through a case study.

[1] V.A. Dev, N.G. Chemmangattuvalappil, M.R. Eden, 2014, Computer Aided Chemical Engineering, 34, 291-296

[2] V.A. Dev, N.G. Chemmangattuvalappil, M.R. Eden, 2015, Computer Aided Chemical Engineering, 37, 1445-1450

[3] N.G. Chemmangattuvalappil, C.C. Solvason, S. Bommareddy, M.R. Eden, 2010, Computers & Chemical Engineering, 34, 12, 2062-2071

[4] G. Mavrotas, 2009, Applied Mathematics and Computation, 213, 455-465

[5] G. Mavrotas, K. Florios, 2013, Applied Mathematics and Computation, 219, 9652-9669

435033 A Semi-Parametric Independent Component Analysis Method for Multivariate Process Monitoring

Abstract:

Independent Component Analysis (ICA) is a method of decomposing a multivariate signal into statistically independent components. In contrast to its simplistic counterpart, principle component analysis, ICA is particularly useful when the signal components have non-Gaussian probability distributions. ICA is based on two major assumptions of non-Gaussian distribution and statistical independence of the components, and it involves minimizing mutual information or maximizing non-Gaussianity of the components [1].

In recent years, ICA has found applications in process monitoring and control. In multivariate process monitoring, ICA can be used to analyze sensor measurements to find statistically-independent source components that have generated the sensor data. These components can be processed further for use in monitoring and diagnosis [2] and in process control systems [3]. These applications have motivated the development of a variety of ICA methods [4, 5, 6]. In most ICA methods, probability distributions are assumed to be describable by a few leading moments of the distributions. As in practice variables can have any probability distribution, the use of a few leading moments to capture dependence structures can lead to poor predictions [7].

In this work, we propose an alternative ICA method that is based on capturing the dependence structure of variables using a semi-parametric approach. This approach combines parametric copulas with non-parametrically-estimated marginal distributions. The approach allows for modeling non-elliptical and highly nonlinear dependence structures and offers a flexible framework to represent non-Gaussian marginal distributions. The proposed ICA method offers an alternative way to efficiently calculate rotation matrices for complicated interactions that cannot be described by a few leading moments. The application and performance of the method will be shown using examples.

References

[1]. Hyvärinen, A., & Oja, E. (2000). Independent component analysis: algorithms and applications. *Neural networks*, *13*(4), 411-430.

[2]. Stefatos, G., & Hamza, A. B. (2010). Dynamic independent component analysis approach for fault detection and diagnosis. *Expert Systems with Applications*, *37*(12), 8606-8617.

[3]. Kano, M., Hasebe, S., Hashimoto, I., & Ohno, H. (2004). Evolution of multivariate statistical process control: application of independent component analysis and external analysis. *Computers & chemical engineering*, *28*(6), 1157-1166.

[4]. Favero, J. L., Silva, L. F. L., & Lage, P. L. (2014). Comparison of methods for multivariate moment inversion—Introducing the independent component analysis. *Computers & Chemical Engineering*, *60*, 41-56.

[5]. Rashid, M. M., & Yu, J. (2012). A new dissimilarity method integrating multidimensional mutual information and independent component analysis for non-Gaussian dynamic process monitoring. *Chemometrics and Intelligent Laboratory Systems*, *115*, 44-58.

[6]. Albazzaz, H., & Wang, X. Z. (2006). Multivariate statistical batch process monitoring using dynamic independent component analysis. *Computer aided chemical engineering*, *21*, 1341-1346.

[7]. Stone, J. V. (2004). *Independent Component Analysis: A Tutorial Introduction*. Cambridge, Mass.: MIT Press.

435517 Design and Development of a High Temperature PEM Fuel Cell System for Charging of Li-Ion Battery Stacks

Abstract:

This work presents the design and development of an integrated High Temperature Polymer Electrolyte Membrane (HT-PEM) fuel cell system which is used for the charging of Li-Ion battery stacks. The necessary hydrogen is provided either by a renewable hydrogen production station or from an LPG reformer system which are both located at CERTH. In this work the automated monitoring system is analyzed providing online vital information such as battery state of charge, state of health, temperature and capacity. The design is based on a novel integrated management methodology that incorporates monitoring, evaluation and control features, which are configurable, expandable and flexible according to the specific end-user requirements.

Furthermore the Battery Monitoring System (BMS) is designed and developed along with a system for the protection of each individual cell of the battery pack. More specifically, the system is developed for a battery stack that consists of 15 Li-ion battery cells of 3.2V/120Ah each, which are connected in series to produce a 48V. . In order to protect the battery and the individual cells a state of the art BMS has been designed, implemented and tested. The main requirements for this BMS are the monitoring of the state and the protection from overcharging of each individual cell and at the same time the reduction of the overall cost as much as possible. The main idea of the BMS is that while monitoring the voltage of each cell during charging, in case that a cell goes beyond a predefined limit value (set for example by the manufacturer), the charging procedure to be interrupted. This is implemented through a resistance that is connected in parallel with each specific cell such as all the charging current to go through the resistance and not the cell. This process is repeated if necessary until all cells are fully and equally charged.

When a resistance is placed in parallel to a cell its state of charge and voltage should remain constant i.e. all the charging current goes through the resistance. However, due to transient phenomena it was observed that the voltage is not remaining constant but is oscillating between two values. One of them is the value where the resistance is placed in parallel to the cell and the other one is a lower value where the resistance is removed when the cell voltage drops. It was observed that the frequency of this oscillation is different for each cell and that it is greatly depended on its state of charge and health. Moreover, it was found after several charging cycles and in case that a cell is overcharged, that its voltage continues to increase even above the limit where the resistance is placed. In this case a second safety limit is used by the BMS and when this limit is exceeded the charging process is interrupted. After all using that BMS, it is possible to balance the state of charge of all the cells and therefore to protect the state of the battery.

The flexible and modular design of the system can be applied to a wide range of traction and stationary battery applications since the involved platform is generic and can be adapted upon demand. The smart soft sensor features combined with the supervisory monitoring capabilities enable the optimum utilization of the battery pack. Finally the BMS provides informative decisions to the user that monitors the stationary system or operates the forklift and enables proactive maintenance operation to ensure that the system will provide the maximum of its capabilities.

435813 Sustainable Production of Butanol – Evaluation of a Bio-Based Process

Abstract:

Today fossil fuels take up around 80% of the global primary energy consumption, of which nearly 60 % is used by the transportation sector. Since the transportation sector is responsible for a significant fraction of the greenhouse gases emissions, substitution of oil derived fuels by biofuels, could significantly decrease environmental impacts, besides providing gains on the socio-economic levels as well. Of the renewable liquid fuel options, butanol is a promising candidate due to its higher energy density, better solubility in existing hydrocarbon fuels, lower vapor pressure and lower corrosiveness [1].

In this work, using a systematic 12 step hierarchal approach for performing (sustainable) process design, a biobutanol production process has been designed. Wheat straw has been selected as the biomass to produce 7.5 m^3 of butanol/year with a yield of 0.11 Kg of butanol per Kg of raw material. The reaction mechanism considered for the process synthesis and design is the ABE fermentation route using genetically modified strain, Clostridium beijerinckii BA101 AA3 [2], which can survive higher butanol concentrations. The fermenter outlet consists of a multi-component mixture of ABE solvents with a concentration of 13g/L of butanol together with butyric and acetic acid as main impurities. In the downstream separation process, acetone and ethanol are first recovered followed by Liquid-Liquid extraction to extract butanol from the aqueous phase. In the extraction unit mesitylene is used as the solvent due to its high distribution coefficient, butanol selectivity and low aqueous solubility [3]. Due to high boiling point difference between mesitylene and butanol 99.95% of the solvent is recovered thereby reducing the fresh solvent requirement. Finally a distillation column is used to obtain pure butanol as the bottom product while the top butanol-water Azeotrope is recycled back to the Liquid-Liquid extraction.

By using LCA, economic and sustainability analyses, hot-spots were identified to improve the base case design [4, 5].

These identified hot-spots were addressed through heat integration and process optimization. The final design generated is more sustainable and economically viable solution. In this poster, the process design method will be presented and the output from each step related to the production of bio-butanol will be given.

[1] Dürre P., Biobutanol: an attractive biofuel, *Biotechnology Journal Vol 2, 1525-1534,* December 2007.

[2] Ezeji, T.C., Qureshi, N., Blaschek, H.P. Production of acetone, butanol and ethanol by Clostridium beijerinckii BA101 and in situ recovery by gas stripping. *World Journal of Microbiology & Biotechnology, 19 595-603,* 2003.

[3] Kraemer K. Separation of butanol from acetone-butanol-ethanol fermentation by a hybrid extraction distillation process, *20th European Symposium on Computer Aided Process Engineering,* 2010.

[4] Kalakul, S., Malakul, P., Siemanond, K., & Gani, R. Integration of life cycle assessment software with tools for economic and sustainability analyses and process simulation for sustainable process design. *Journal of Cleaner Production, 71, 98–109,* 2014.

[5] Carvalho, A., Matos, H. A., & Gani, R. SustainPro—A tool for systematic process analysis, generation and evaluation of sustainable design alternatives. *Computers & Chemical Engineering, 50, 8–27,* 2013.

Characterization of Microfluidic CO_2-Water by Computational Fluid Dynamics

Chongwei Xiao, Hariganesh Bheema*
Texas A&M University-Kingsville

Abstract

Advanced eco-friendly and efficient processes are needed to decrease post-combustion carbon dioxide (CO_2) released to the atmosphere. However, low mass transfer between gas and liquid phases hinders the development of new chemical absorption process. By a means of process intensification, microfluidic reactors provide high surface to volume ratio and greatly increase gas-liquid mass transfer, which makes new processes for gas-liquid reactions possible. This intensified process also ensures instant thermal stability across the reactor and rapid heat transfer between the reactant, which greatly benefits reactions with strong heat effect. A few recent experimental and simulation works on CO_2-water system in micro-channels showed good potential of this new process. This study applied an advanced Eulerian–Multi-fluid VOF model to simulate CO_2-water two-phase flow in a microreactor with a cross-section diameter of 0.5 mm. We emphasized on simulating slug flow with the various pressure-based segregated algorithms (PSA) to determine the best algorithm to provide computation speed and accuracy. The slug formation by squeezing and dripping mechanisms were observed in the series of simulation with varied velocities and capillary numbers, which are in good agreement with published data. The effects of pressure and temperature on the slug formation and transition of the flow region were observed. This theoretic analysis provides a theoretic support for the characterization of gas-liquid flow in microchannels.

1. Introduction

Microfluidic reactor (microreactor) technology for the application in process industries has developed tremendously over the decade as a means of process intensification. The advantages associated with performing reaction in microreactors are the high surface to volume ratio ensuring thermal homogeneity across the reactor and rapid heat transfer between the reactants (DeWitt, 1999). Many investigators have observed various gas–liquid flow regimes in microchannel. The common flow regime in microreactors is slug flow, which is characterized by long gas slug separated by the liquid slug and the gas slug is separated from the wall of microchannel by a thin film. Extensive experiments have been carried out to study the flow regimes and properties of water-air two-phase flow in capillary or micro tubes, such as gas slug in a perpendicular T junction (Kashid, Renken et al. 2011), the effects of the diameter of the microchannel on the slug length (Fukano and Kariyasaki 1993, Zhao and Bi 2001), the structure of gas-liquid interfacial area (Zhang, Chen et al., 2009), and the profile of the liquid film (Seemann, Brinkmann et al., 2005).

The development and application of microreactor technology provide a potentially new solution to overcome mass transfer limitation in carbon dioxide (CO_2) absorption by aqueous solution. Yue, Chen et al. (2007) experimentally studied the physical absorption of CO_2 in water in Y- shape microreactors. It was observed that large interfacial area up to $20000 m^2/m^3$ existed in a microreactor, which exceeds conventional reactors. The overall mass transfer coefficients calculated are in the range of 0.8 to 11.01 per second, which are much higher than that of well stirred reactor, which overall volumetric mass transfer coefficient is 0.033 per second (Hill, 2006). Yue, Luo et al. (2008) studied the transition of CO_2/water in microreactors from Taylor to bubbly and from Taylor-annular to annular flow. They found that the bubble sizes did not follow a certain trend with diminishing channel size, probably because the rate of CO_2 absorption into the water that would affect the gas volume and velocity depended on channel size.

All these experiments show a tendency of process intensification and the current results are highly depending on individual experimental conditions. Limited simulations have been carried out to characterize the hydrodynamics of gas-liquid two-phase slug flow within microchannels by several interface capturing techniques. van Baten and Krishna (2004) studied the mass transfer from uprising Taylor bubbles by volume-of-fluid (VOF) model. Shirani, Ashgriz et al. (2005) developed a new method to calculate the interface pressure drop by PLIC-VOF model. Qian and Lawal (2006) studied the effects of the premixing and different hydraulic diameters on slug flow in a T-junction straight microchannel by VOF model. Fukagata, Kasagi et al. (2007) and He and Kasagi (2008) applied the Phase-Field method and the Level-Set method, respectively, to study the hydrodynamics. Both works specified void fraction and pressure drop and the flow rates were obtained after simulation, which are different from the conditions of experimental works.

With the aid of the advances in multiphase flow modelling, it is possible to characterize CO_2/water two-phase flow with advanced mathematic algorithms. In view of the need for higher accuracy and efficiency, four types of pressure-based segregated Algorithms (PSA) were applied to analyze the flow regimes. While the VOF model is widely applied to track free surface in the slug formation in microchannels (Qian and Lawal 2006, Shao, Salman et al. 2008, Shao, Gavriilidis et al. 2010), it has limitation that the velocity cannot be assigned to each phase and momentum equation is solved for a single phase. This study used Multi-fluid VOF model, which solves a set of n momentum and continuity equations for each phase and overcomes the limitation of the VOF model. The formation mechanism of the CO_2-water two-phase flow with interfacial mass transfer under various pressure and temperature conditions.

2. Numerical Model

The following assumption were considered to be reasonable to study the physical absorption of carbon dioxide in water

- Gravitational affect is negligible;
- Fluid mass diffusivity is constant;
- Both fluids are Newtonian and viscous.

The model investigated is a Y- type microreactor with a cross-section diameter of 0.5 mm consisting of two inlet channels and one co-flow channel (Fig. 1). The two inlet channels are of length 15mm and the angle between both inlets is 60°. The CO_2-water co-flow channel is of length 48mm.

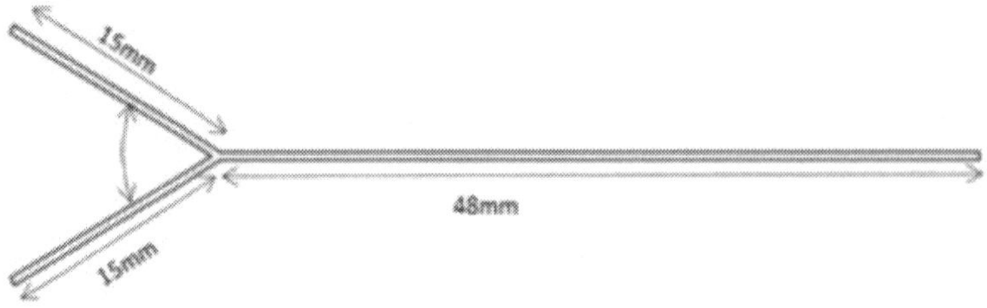

Fig. 1 Y-shape microreactor

The Eulerian model is employed for modeling multiple separate and interacting phases (liquids, gases, and solids in any combinations) and it solves n momentum equation for each phase. The Multi-fluid VOF model, a sub-model of Eulerian Model, is applied to study the interface between the two phases and in the formation of the slug and evaluate the performance of the microreactor. The hydrodynamics of the multiphase flow are described by

Continuity

$$\frac{1}{\rho_{rq}}\left(\frac{\partial}{\partial t}(\alpha_q\rho_q) + \nabla.(\alpha_q\rho_q\bar{v}_q) = \sum_{p=1}^{n}(\dot{m}_{pq} - \dot{m}_{qp})\right) \quad (1)$$

Where $\bar{v}_q$ is "the velocity of phase" and $\dot{m}$ stand for "mass transfer".

Conservation of momentum

$$\frac{\partial}{\partial t}(\alpha_q\rho_q\bar{v}_q) + \nabla.\left(\alpha_q\rho_q\bar{v}_q\bar{v}_q\right) = -\alpha_q\nabla p + \nabla.\bar{\tau}_q + \alpha_q\rho_q\bar{g}$$

$$+ \sum_{p=1}^{n}\left(K_{pq}(\bar{v}_p - \bar{v}_q) + \dot{m}_{pq}\bar{v}_{pq} - \dot{m}_{qp}\bar{v}_{qp}\right)$$

$$+ \left(\bar{F}_q + \bar{F}_{lift,q} + \bar{F}_{wl,q} + \bar{F}_{vm,q} + \bar{F}_{td,q}\right) \quad (2)$$

Where $\bar{g}$ is the gravitational acceleration.

The "continuous surface force model" (CFS) is used to reformulate surface tension into an equivalent body force. For a two-phase system, the volumetric force due to surface tension at the interface between phases 1 and 2 is given as:

$$\bar{F}_q = 2\sigma\frac{\rho k\nabla a}{\rho_1+\rho_2} \quad (3)$$

$$k = \frac{1}{|n|}\left[\left(\frac{n}{|n|}.\nabla\right)|n| - \nabla.\hat{n}\right] \quad (4)$$

$$\hat{n} = \widehat{n_w}\cos\theta + \widehat{t_w}\sin\theta \quad (5)$$

$\bar{F}_q$ in Equ. (2) is the surface tension force per volume and k is the curvature computed from the divergence of the unit surface normal.

CFD of ANSYS FLUENT provides the following segregated types of algorithms: Semi-Implicit Method for Pressure Linked Equations (SIMPLE), Semi-Implicit Method for Pressure Linked Equations-Consistent (SIMPLEC), Pressure-Implicit with Splitting of Operators (PISO) in the VOF model, and Phase coupled SIMPLE (PC-SIMPLE) in Multi-fluid VOF model. These schemes are referred to as the pressure-based segregated algorithms (PSA).

The locations of the boundaries (e.g., inlets, walls, symmetry) are very important as the poorly defined boundaries conditions have direct impacts on the convergence of the simulation. The inlet boundary conditions and wall boundary condition are set as velocity inputs and wall adhesion condition, respectively. Pressure-coupled solver is used and simulation was run with minimum time step of 1e-5. Simulations were carried out in Intel Xeon E5-1620 processor. Each simulation was run for a minimum of 4 hours to reach convergence and get a stable solution.

3. Results and Discussion

The simulations were implemented with the superficial velocities of CO_2 and water of 0.73 m/s and 0.5 m/s, respectively. Apart from inlet conditions such as gas and liquid flowrates, liquid surface tension and gas inlet size, it has been found that the degree of mixing of the two fluids at the inlet can also affect bubble size ((Qian and Lawal, 2006), (Shao, Gavriilidis et al., 2010).

3.1 Pressure-based Segregated Algorithms (PSA)

SIMPEC, PISO, and PC-SIMPLE are the variant of SIMPLE algorithm. SIMPLEC method removes the need of relaxation factor of pressure under –relaxation and SIMPLEC algorithm is seen to converge 1.2-1.3 times faster than the SIMPLE algorithm. Phase-Coupled SIMPLE (PC-SIMPLE) is an extension SIMPLE method used in multiphase flow, which uses a shared pressure-correction equation for both phases based on total volume fraction conservation and the coupling between phases is accomplished implicitly at the level of the solution matrix using a block algebraic multi-grid solver as described in (Weiss, Maruszewski et al., 1999). PC-SIMPLE algorithm is seen to converge two times faster than the SIMPLE algorithm method as this method used shared pressure correction. It is observed that the pressure drop for simulation using SIMPLE algorithm with jl=0.5 and jg=0.73 is 1432 psi, which is very low compared to the pressure drop for simulation using PC-SIMPLE algorithm of 4092 psi which is very closed to experimental result (Yue, Chen et al., 2007).

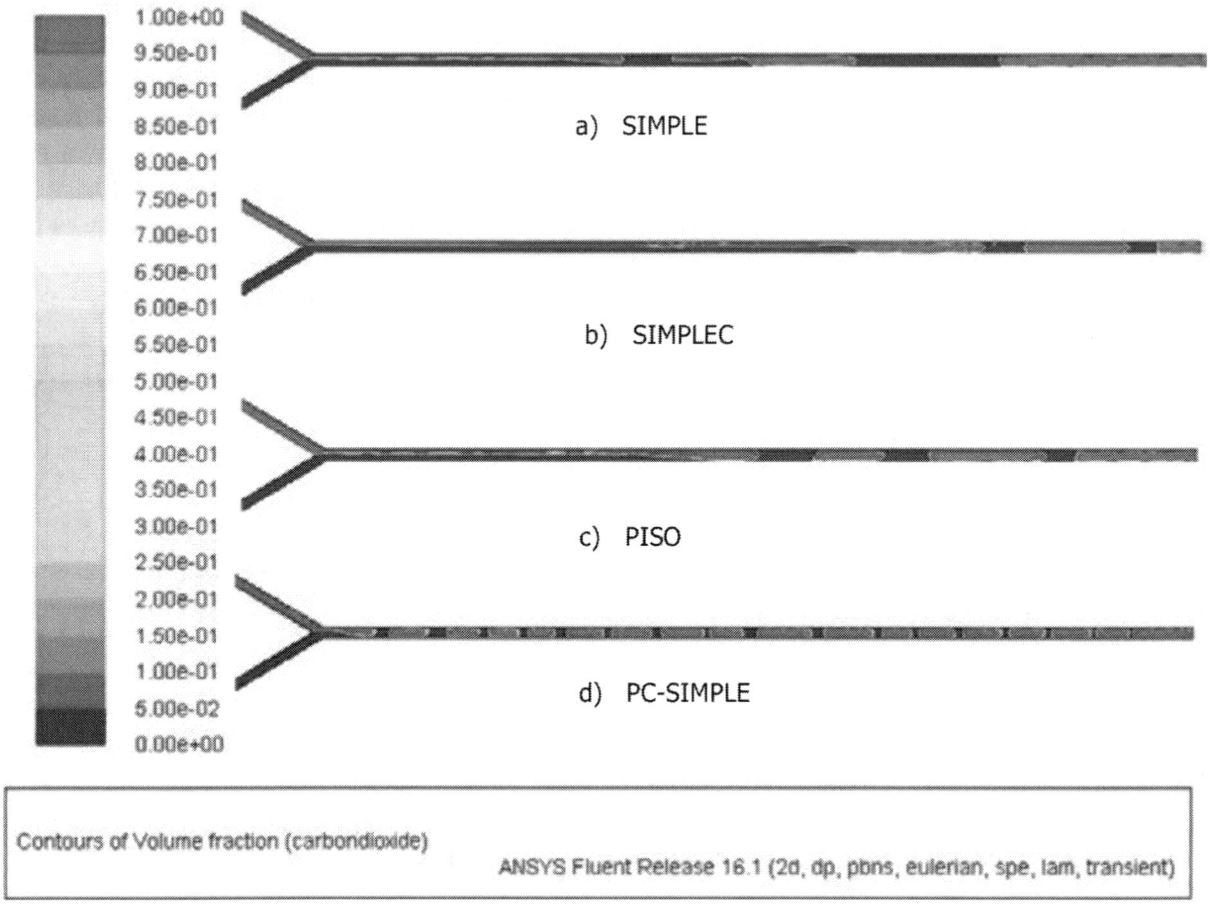

Fig. 2 Comparison of different algorithms for solving pressure-velocity linked equations (jg = 0.73 m/s, jl = 0.50 m/s);

3.2 Slug formation mechanism

The evolution of droplets and slugs in micro-fluidic T-junction geometries was explained by Garstecki et al. (2006), who discovered that break-up was not influenced by shear stresses but by the pressure drop across the generated droplet at the low capillary numbers. Through a series numerical simulation of slug generation, De Menech et al. (2008) was able to distinguish three distinct droplet formation regimes of squeezing, dripping, and jetting.

Squeezing mechanism: According to Garstecki, Fuerstman et al. (2006), Taylor slug is formed by Squeezing mechanism when the $Ca<10^{-2}$. Fig. 3 showed the numerical flow of droplet formation process, given jl=0.5m/s, jg=0.089, and Ca=0.00814.

Dripping mechanism: Taylor flow by dripping mechanism occurs at high capillary number ($Ca>10^{-2}$). The Fig. 4 showed the formation of Taylor flow, given jl=0.5m/s, jg=0.85 m/s, and Ca =0.0182.

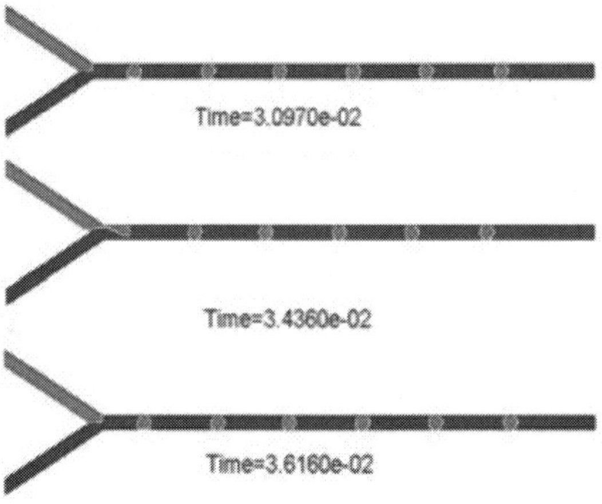

Fig. 3 Bubble formation by Squeezing mechanism

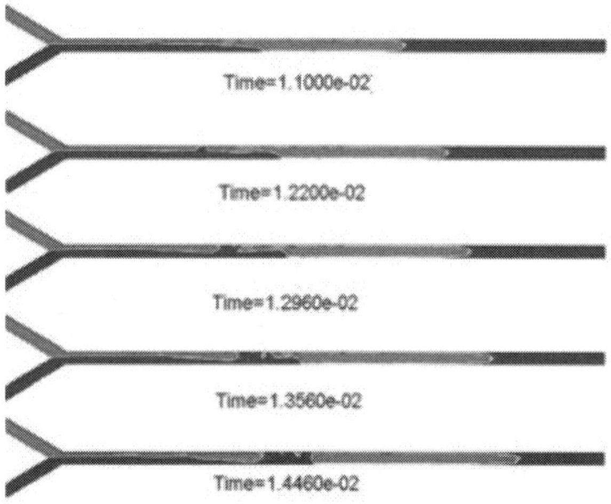

Fig. 4 Bubble formation by Dripping mechanism

3.3 Effects of Temperature and Pressure

A set of simulation with temperatures of 297.8k, 333.5K, and 343.3K were carried out with constant pressure of 1.01Mpa with jg=0.73m/s and jl=0.5m/s (Fig. 5). The surface tension plays a major role in the generation of slug. As the temperature increases, the surface tension between the fluids decreases. Here it can be observed, the slug length increased from Fig. 5a to Fig. 5b as the temperature increased. The transitions of two-phase flow regimes were observed from slug flow to churn flow and annular flow (Fig. 5). The simulation results showed that temperature affects the slug formation and the transitions of flow regimes.

A set of simulation with pressure varying from 1.01Mpa, 2 MPa, and 3Mpa were carried out with constant temperature of 297.8k with jg=0.73m/s and jl=0.5m/s. Increasing pressure increases the density of gas, which in turn changes the surface tension. When the pressure was increased, there was an increase in the slug length (Fig. 6).

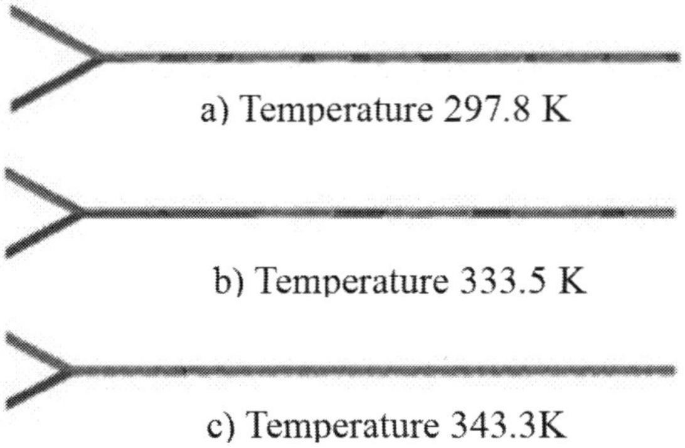

Fig. 5 The effects of temperature on flow regimes

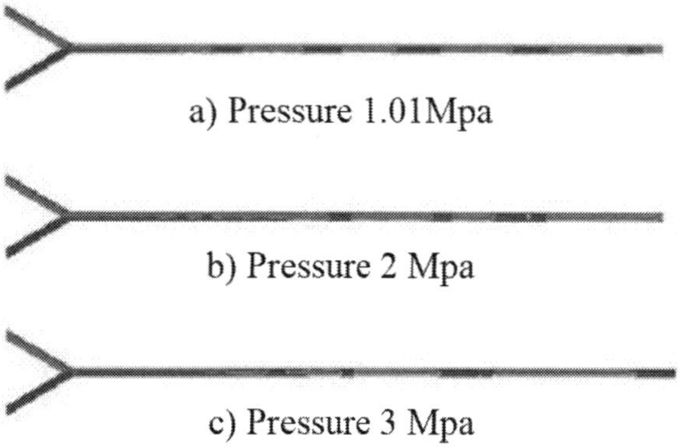

Fig. 6 The effects of pressure on flow regimes

4. Conclusion

Carbon dioxide and water two-phase were simulated using CFD in ANSYS FLUENT in a Y-shape microreactor with a cross-section diameter of 0.5 mm. Various two-phase flow regimes through the microchannel, including slug flow, slug-annular flow and churn flow, were observed and compared with (Yue, Chen et al., 2007). The slug flows simulated by the various Pressure-based Segregated Algorithms were carried out to determine the best algorithm to provide computation speed and accuracy. PC-SIMPLE algorithm was observed to converge faster than the other methods and slug formed are in a good agreement with the experimental data. The study of temperature and pressure came to the understanding that increasing pressure and temperature will change interfacial tension and increase the velocity of the bubble, which leads to the changes of bubble sizes and the transitions of flow regimes.

References

1. DeWitt, S. H. (1999). "Micro reactors for chemical synthesis." Current Opinion in Chemical Biology **3**(3): 350-356.
2. Fukagata, K., et al. (2007). "Numerical simulation of gas–liquid two-phase flow and convective heat transfer in a micro tube." International Journal of Heat and Fluid Flow **28**(1): 72-82.
3. Fukano, T. and A. Kariyasaki (1993). "Characteristics of gas-liquid two-phase flow in a capillary tube." Nuclear Engineering and Design **141**(1–2): 59-68.
4. He, Q. and N. Kasagi (2008). Numerical investigation on flow pattern and pressure drop characteristics of slug flow in a micro tube. ASME 2008 6th International Conference on Nanochannels, Microchannels, and Minichannels, Darmstadt, Germany.
5. Hill, G. A. (2006). "Measurement of Overall Volumetric Mass Transfer Coefficients for Carbon Dioxide in a Well-Mixed Reactor Using a pH Probe." Industrial & Engineering Chemistry Research **45**(16): 5796-5800.
6. Kashid, M. N., et al. (2011). "Gas–liquid and liquid–liquid mass transfer in microstructured reactors." Chemical Engineering Science **66**(17): 3876-3897.
7. Qian, D. and A. Lawal (2006). "Numerical study on gas and liquid slugs for Taylor flow in a T-junction microchannel." Chemical Engineering Science **61**(23): 7609-7625.
8. Seemann, R., et al. (2005). "Wetting morphologies at microstructured surfaces." Proceedings of the National Academy of Sciences of the United States of America **102**(6): 1848-1852.
9. Shao, N., et al. (2010). "Mass transfer during Taylor flow in microchannels with and without chemical reaction." Chemical Engineering Journal **160**(3): 873-881.
10. Shao, N., et al. (2008). "CFD simulations of the effect of inlet conditions on Taylor flow formation." International Journal of Heat and Fluid Flow **29**(6): 1603-1611.
11. Shirani, E., et al. (2005). "Interface pressure calculation based on conservation of momentum for front capturing methods." Journal of Computational Physics **203**(1): 154-175.
12. van Baten, J. M. and R. Krishna (2004). "CFD simulations of mass transfer from Taylor bubbles rising in circular capillaries." Chemical Engineering Science **59**(12): 2535-2545.
13. Weiss, J. M., et al. (1999). "Implicit Solution of Preconditioned Navier-Stokes Equations Using Algebraic Multigrid." AIAA Journal **37**(1): 29-36.

14. Yue, J., et al. (2007). "Hydrodynamics and mass transfer characteristics in gas–liquid flow through a rectangular microchannel." Chemical Engineering Science **62**(7): 2096-2108.
15. Yue, J., et al. (2008). "An experimental investigation of gas–liquid two-phase flow in single microchannel contactors." Chemical Engineering Science **63**(16): 4189-4202.
16. Zhang, H., et al. (2009). "Hydrodynamics and mass transfer of gas–liquid flow in a falling film microreactor." AIChE Journal **55**(5): 1110-1120.
17. Zhao, T. S. and Q. C. Bi (2001). "Co-current air–water two-phase flow patterns in vertical triangular microchannels." International Journal of Multiphase Flow **27**(5): 765-782.

407092 Scheduling Optimization for Regeneration of Ion Exchange Resin in a Demineralized Water Treatment Facility

Abstract:

Demineralization of water is the final treatment for production of high-quality boiler feed water in steam-dependent manufacturing as oil refineries. The process prevents metallurgical and efficiency problems in boilers by capturing dissolved ion in water until saturation or exhausting of equipment, when cleaning of ion exchange resins or reverse osmosis membranes regenerates these treatments for a new period of processing, the former within hours and the latter over months.

Used in high viscosity fluid movements, hydrogen production, and processing of raw material by promoting mass and heat transfer driving forces, steam also produces electric power in turbines in a direct way, as in powerhouse facilities, or indirectly in high to low pressure steam transformations. Due to inherent uncertainties in the demineralized water treatment facility related to ion concentration in captured freshwater, random flow distribution in parallel network, and amount and quality of condensate water from process field recycle, better scheduling predictions in the cleaning stage of saturated/exhausted equipment may prevent production bottlenecks and lower plant utilization by steam and electrical shortages as a consequence of boiler feed water unavailability.

Although optimization have been proposed for design of grassroot[1]-[4] and retrofit[5] of water treatments as well as pinch analysis for recycle, reuse, and regeneration of water[6], to the best of our knowledge, there is no work for scheduling optimization of water systems excepted for those found in irrigation and water pumping problems. In these works, the uncertainty related to inorganic salt concentration variability is neglected by fixed contaminants concentration in the inlet-flow of water consumers, treatments, etc. Fixed contaminant recovery in water treatments or contamination ratio in water-using units are also commonplace; thus, reducing these optimization problems to linear (LP) or mixed-integer linear (MILP) models, creating a gap for their applicability to industrial processes. Recently,

Yang et al.[4] considers removal efficiency variation by using shortcut models for water treatment units to exploit trade-offs between their cost and removal efficiency, where is presented a modified Lagrangean-based decomposition algorithm in order to solve the resulting nonconvex mixed-integer nonlinear programming (MINLP) problem efficiently.

We propose a scheduling optimization for short-term regeneration of ion exchange resin beds, where these treatment units are modeled as continuous-process with 2-modes of operation (processing and regenerating) using hypothetical pools for controlling accumulated processing volume until the saturation of the resins, considering pool filling during processing and drawing during regenerating. The processing volume until resin saturation is updated every each regeneration cycle by using the past data gathered continuously in the field for each ion exchange resin vessel, where measurement of electrical conductivity in the outlet-flow of the treatment alerts the necessity of regeneration. The model is an MILP problem by considering continuously the processing volume until saturation updates to reflect the uncertainties on dissolved ion concentration in the water treatment feed and random flow distribution in the parallel network. In the case of a contaminant concentration modeled as a variable, the model becomes an MINLP problem, where complicated networks and operations can be very difficult to solve, even to a feasible point.

Super dimensioned facilities capable of feeding boiler feed water constantly to avoid its unavailability is possibly the responsible for the lack of studies in scheduling optimization of regeneration of ion exchange resins. An industrial example demonstrates the approach and the resulting process-of-work improvements in the hand-operated regeneration, where a timetable for opening/closing of valves to spare time of processing before saturation of resins as well as for their regeneration can guide operator's teams in the field.

(1) Alva-Argez A, Vallianatos A, Kokossis A. A multi-contaminant transhipment model for mass exchange networks and wastewater minimisation problems. Comput Chem Eng, **1999**, 23, 1439−1453.

(2) Galan B, Grossmann IE. Optimal design of distributed wastewater treatment networks. Ind Eng Chem Res, **1998**, 37, 4036−4048.

(3) Karuppiah R, Grossmann IE. Global optimization for the synthesis of integrated water systems in chemical processes. Comput Chem Eng **2006**, 30, 650−673.

(4) Linlin Y, Salcedo-Diaz R, Grossmann IE. Global optimization for the synthesis of integrated water systems in chemical processes. Ind Eng Chem Res, **2014**, 30, 650−673.

(5) Faria DC, Bagajewicz MJ. Planning model for the design and/or retrofit of industrial water Systems. Ind Eng Chem Res, **2011**, 50, 3788−3797.

(6) Khor CS, Shah N, Mahadzir S and Elkamel A. Optimisation of petroleum refinery water network systems retrofit incorporating reuse, regeneration and recycle strategies. Can J Chem Eng. **2012**, 90, 137−143.

407398 Scheduling and Optimization for Industrial Cogeneration Plants

Abstract:

The scheduling application of cogeneration plants presented uses a first principle, steady-state, non-linear model with the following details

- Model validated continually with plant data and also used for online optimization
- Accounts for equipment that gets switched off and turned on
- Includes Process Control strategy that may use different equipment in an hierarchal manner
- Control curves and design performance curves for equipment
- Scheduler results are implemented using a model predictive controller

The scheduling application also has following capabilities

- User can input future ambient temperature
- User can switch parts of plant on or off to account for equipment contingency

The above modeling details enable the scheduler model predictions to have a close match for fuel consumption at different operating scenarios. The error from plant data is less than two percent over the entire operating range thereby allowing for efficient participation in the power scheduling market. Simpler linear models are not sufficiently accurate because they do not allow for control strategy, equipment control and performance curves, and inherent process nonlinearities.

Process Description

Scheduling of power and real-time optimization for three industrial cogeneration plants at one of Dow's Louisiana sites is presented in this paper. The three combined heat and power (CHP) plants (cogeneration plants) are capable of meeting the steam and electrical power needs of other chemical production facilities. Surplus power is produced thereby allowing Dow to participate in the day-ahead Mid-Continent Independent System Operator (MISO) power market. Power is scheduled in the MISO day-ahead market at the estimated cost to produce and adjustments to the power offer are made in real-time market thirty minutes prior to each hour.

The industrial cogeneration process at the site consists of three power plants with seven gas turbines and five steam turbines that make surplus power and four pressure levels of steam for different consumers. A steady-state model of the process that includes component material balances, energy balances and thermodynamics is developed in Aspen Plus Optimizer, Aspentech's equation oriented environment. The process model consists of equations for gas turbines, steam turbines, heat exchangers, steam headers, fuel headers, condensate system, pressures relief valves, pumps and compressors. The process model in the equation oriented environment is also appended with additional equations for process control strategy and equipment control specifications. The overall steady-state process model consists of approximately 18,000 equations.

Scheduler Implementation

The offer curve for the power schedule includes external exported power and incremental heat rate. Heat rate is the common measure of system performance in a cogeneration power plant. It is defined as the fuel energy input divided by the energy generated and has units of million BTU/MWhr. The incremental heat rate curve is able to match within two percent error from actual plant data for different ambient temperatures and operating scenarios thereby enabling better decision making for scheduling power. The close match of the power offer curve with plant data can be attributed to the following

- Fundamental nonlinear model with process control strategy and equipment control curve details
- Continuous model validation with real-time plant data

Implementation of the exported power schedule is done effectively using a model predictive controller. The power scheduling application has been in continuous use since January 2014 to enable efficient participation in day-ahead and real-time power market due to its robustness and accuracy.

Maintaining Higher Efficiency of LPG Butane Boil-Off Gas Re-Liquefaction in LPG Butane Storage Tanks Using Unsteady State Process Simulation

Ronggo Ahmad Wikanswasto[1], Yazid Bindar[2] and Hary Devianto[2],
(1)Technical Department, Badak LNG, Bontang, Indonesia,
(2)Chemical Engineering, Institute Technology Bandung, Bandung, Indonesia

Abstract

Badak LNG is a natural gas liquefaction plant, which operates LNG trains to produce LNG and refrigerated LPG. The recent annual LPG production amount is 500,000 ton/year (equivalent with 500 Million USD/year productions value). During production and storage of LPG, boil-off gas (BOG) is generated and then re-condensed in a LPG BOG Butane Condenser unit. LPG BOG Re-liquefaction system uses centrifugal compression system with total maximum electric power duty of 1.4 MW (equivalent with 2.4 Million USD/year operating cost). Cooling medium flow rate in condenser unit is controlled by using Butane Tank pressure controller or cooling medium temperature outlet controller. Normally, the pressure controller dictated the cooling medium flow rate. However, during accumulations of Non-Condensable Gas (NCG) occurrence in the condenser unit hot stream side (LPG BOG Butane) will lead to cooling medium temperature outlet controller dictation which indicates less LPG BOG Butane condensing rate. Furthermore, this less condensing rate will lead to LPG Butane tank overpressure flaring (LPG Butane product losses) accounted as LPG Butane storage efficiency for just 91.42%.

Therefore accumulation of this NCG makes LPG Butane tank pressure fail to be controlled. Hypothesis was made that by continuously venting small amount NCG, higher reliability of this tank pressure control performance could be achieved. The strategy to improve the NCG Butane BOG venting control system strategy were also formulated using unsteady state analysis of LPG Butane Tanks and Butane BOG re-liquefaction. Hysys simulator software in dynamic condition is employed to analyze unsteady state tank behavior. Activation of Fidelity feature in the simulation is critical to simulate the hydrostatic pressure of Butane BOG re-liquefaction unit in order to approach as close as possible the simulation with actual process condition during unsteady state condition.

Approach of NCG accumulation occurrence on Butane BOG re-liquefaction system can be successfully simulated in the dynamic condition simulation software. By introducing NCG disturbance into Butane BOG re-liquefaction system, the tank pressure control is failed. While tank pressure control strategy improvement by continuously venting NCG as much as 5 Nm^3/hr (0.0044 MMSCFD), LPG Butane Storage and Loading efficiency can be also dramatically improved from 91.42% to 99.96. This improvement is equivalent with annual production loss only 200,000 USD of 500 Million USD LPG Product value.

I. Introduction

I.1 LPG BOG Propane and Butane Re-liquefaction System

BADAK LNG is a natural gas liquefaction company which consists of 8 trains to produce LNG and LPG. This company reached its peak in 2001 of 1 MTPA production. Recently, BADAK LNG only operates 4 trains due to the lower feed gas delivery. Consequently, the LPG production is only around 0.5 MTPA. BADAK LNG produces mainly refrigerated LPG Propane and Butane (C3 and C4). The LPG Re-liquefaction system is utilized to recover LPG Boil-Off Gas (BOG).

LPG Re-liquefaction system is based on an auto-refrigeration process by compressing LPG Propane BOG using compressor, then condensing the LPG BOG using air cooler and finally releasing its pressure to the LPG tank pressure. LPG Butane BOG is re-liquefied by using propane cooling medium. LPG Re-liquefaction system scheme in BADAK LNG is shown in Figure-1.

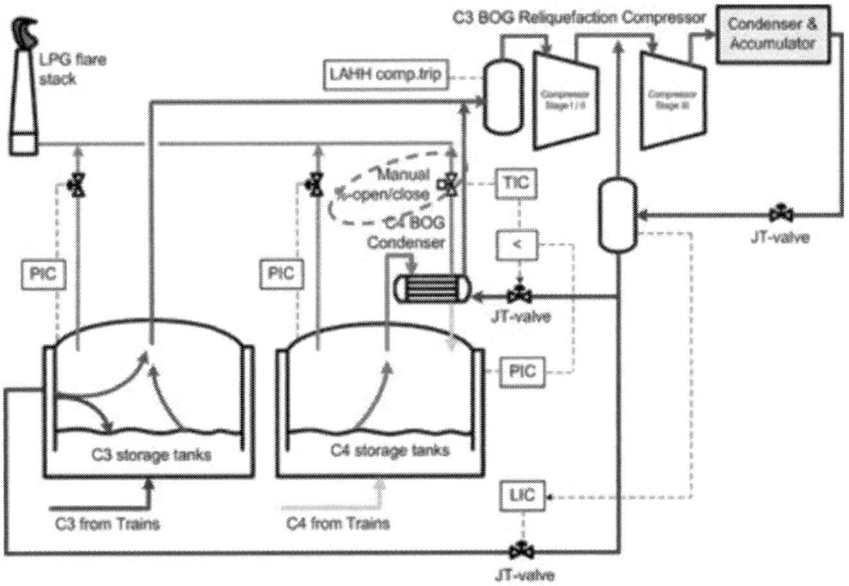

Figure-1 LPG Butane BOG re-liquefaction Unit (Butane Condenser)

I.2 LPG Butane Storage and Loading Low Yield

Table-1 BADAK LNG's LPG Butane Storage and Loading Yield

Productivity Criteria	LPG Butane
LPG Storage and Loading Yield Base Case	91.42 %
Remarks	LPG Butane product losses by non-condensable LPG BOG flaring

LPG Butane production losses in LPG Butane tanks are due to the tank overpressure flaring, related to LPG BOG accumulation in corresponding tanks. LPG BOG flaring is 8.5% (based on Table-1) equivalent to LPG BOG flaring of 1 kNm^3/hr, which is approximately equivalent to LPG losses more than 2,500 USD/hour (by assuming LPG price 1000 USD/ton, LPG density 0.6 ton/m3, LPG C3 BOG conversion 218 Nm^3/m^3 C4 liquid).

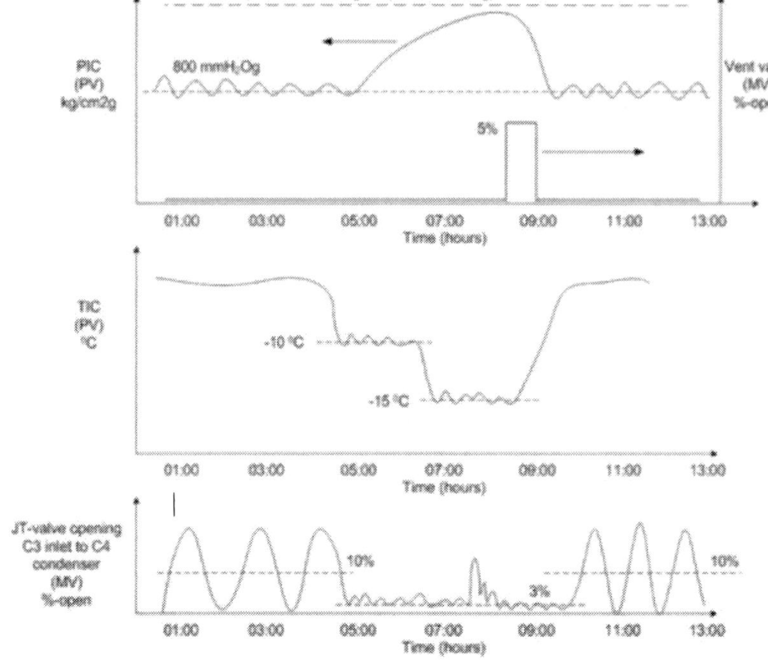

Tank overpressure to flaring SP (1200 mmH$_2$Og)
PIC
(PV)
kg/cm2g
800 mmH$_2$Og
Vent valve
(MV)
%-open
5%
01:00
03:00
05:00
07:00
09:00
11:00
13:00
Time (hours)
TIC
(PV)
°C
-10 °C
-15 °C
JT-valve opening
C3 inlet to C4
condenser
(MV)
%-open
10%
3%

Table-2 Laboratory Analysis of LPG Butane product and non-condensable LPG Butane BOG Composition

Component	LPG Butane Product (%-mol)	Non-condensable LPG Butane BOG (%-mol)	Remarks
Ethane	0	0.01	Non-condensable LPG Butane BOG is a lighter component of LPG Butane product
Propane	0.97	3.44	
i-Butane	42.24	49.71	
n-Butane	56.69	46.82	
Pentane (mix)	0.10	0.02	

I.3.1 Vapor Hydrocarbon Condensation in Shell and Tube Heat Exchanger

LPG Butane BOG is actually a form of vapor hydrocarbon, which will be re-liquefied in Butane Condenser. The condenser uses shell and tube heat exchanger E type of TEMA shell. VLE calculation inside the shell and tube on condensation side will govern the temperature and composition profile inside condenser as explained in Kern (1950), Butterworth (1988), Jensen (1988).

I.3.2 Non-Condensable Gas (NCG) Accumulation in Hydrocarbon Mixtures Condenser

NCG could be accumulated in hydrocarbon mixtures condenser which is similar to the studied Butane Condenser. The accumulation mechanism explained in Jensen (1988) in unforced condensation condenser in which buoyancy force could govern the lighter hydrocarbon component to accumulate in the top of condensation side (shell side) condenser. Binary mixture of Methane-Ethylene and water-NH3 are some of the example of NCG accumulation in the condenser which was studied for failure of total condensation which lead to partial condensation, by Stephan (2010).

Accumulation of NCG in condensation side condenser will make condensation film diminished as mentioned by Jensen (1988). The NCG layer at the interface will act as resistance to both heat and mass transfer in the condensation film interface. Another author, Som (2006) explained that heat transfer rate during NCG presence in condensation film will reduce the Nusselt number ($Nu = \frac{h.D}{k}$).

II. Methodology

II.1 LPG Storage and Loading System Unsteady State Simulation Methodology for Non-condensable Butane BOG Venting Control System Improvement

In recent condition, the non-condensable LPG Butane BOG venting to flare is operated in manual mode. LPG Tank Operating Manual (1988) stated that during period of non-condensable LPG Butane BOG accumulation, the non-condensable Butane BOG venting line to flare may be opened to release these non-condensable. However, opening this valve by DCS operator person intervention may result inconsistent action since the opening valve action is not standardized (e.g. how much %-opening of the valve should be open and for how many time duration it should be kept open). This non-condensable Butane BOG is considered as LPG Butane product losses to the flare if it vented to the flare. Therefore, an improvement for LPG Butane Tank control system for the non-condensable Butane BOG venting line standardizes action is needed.

Steady state simulation for this control system improvement is impossible since the control strategy must be tested / simulated during unsteady state condition. By conducting unsteady state/dynamic simulation, it can be demonstrated that the proposed improvement control system strategy could be applied on the actual butane tank control system.

Non-condensable Butane BOG venting control system improvement would be conducted by conducting dynamic simulation using Hysys simulator. The dynamic simulations were focused on the LPG butane tank and LPG butane BOG condenser (butane condenser). The NCG injection to the Butane Condenser is defined as disturbance for Butane Condenser operational. LPG Butane BOG enrichment of light HC component (Propane) due to buoyancy effect during LPG Butane BOG condensation is defined as disturbance in the dynamic simulation, as summarized in Table-3.

Base case development is based on Non-Condensable Gas (NCG) valve venting full closed position. LPG Butane BOG generation inside the tank will be maintained by giving heat-in leak through LPG Butane tank wall. The dynamic simulation is validated by comparing LPG Butane tank pressure ramp-up profile in the simulation with its actual pressure profile during Butane Condenser offline condition. The main parameter of tank dynamic behavior will be observed such as tank pressure and overall heat transfer coefficient in Butane Condenser. The case-1 then will be developed based on NCG valve venting manual open position. Summary for the dynamic simulation case development is shown in Table-4.

Table-3 Disturbance for Butane Condenser Performance in Dynamic Simulation

No.	Model Approach	Actual	Simulation
1	Light HC (Propane) accumulation due to LPG Butane vaporization	Possible	Not simulated
2	Fuel gas (Methane > 90%) ingress through LPG tank vacuum prevention line	Possible	Not simulated
3	LPG Butane BOG enrichment of light HC component (Propane) due to buoyancy effect during LPG Butane BOG condensation	Possible	Defined as disturbance in simulation. However simulator software is NOT capable to simulate the actual Propane enrichment mechanism.

Table-4 Case Development for LPG Butane Tank Dynamic Simulation

Case No.	Case Improvement	Model Approach
Base	NCG valve kept full close	Introduce NCG disturbance to C4 Condenser
1	NCG valve opening mode improvement (kept open to vent NCG 5 Nm^3/hr, based on plant data)	NCG disturbance is not introduced to C4 Condenser

II.2 Unsteady State LPG Butane Tank and Butane Condenser Analogy Model

Model approach analogy was developed in Figure-3 to show the relationship between butane condenser overall heat transfer coefficient (U),$\frac{dP}{dt}$, and $\frac{dm_{C4\ cond}}{dt}$. As shown in equation IV-1 and IV-2, the reduction of U (caused by presence of NCG), result in the increment of tank pressure since the butane condensation in butane condenser was not occur. Non-Condensable

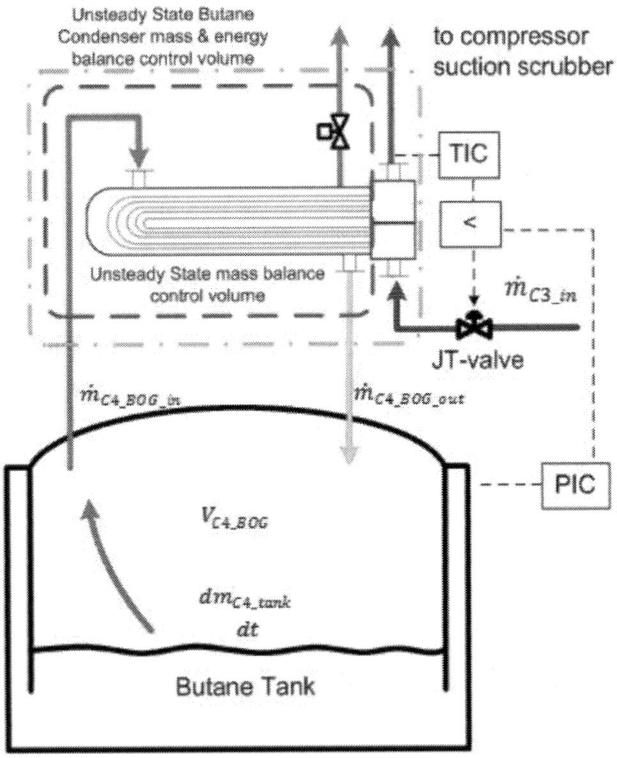

state mass balance in and out Butane Conden...

$$\frac{dm_{C4\_tank}}{dt} = \dot m_{C4\_BOG\_in} - \dot m_{C4\_BOG\_out}$$

$$\frac{d(\rho . V_{C4\_BOG})}{dt} = .\rho_{C4\_BOG} . \dot Q_{C4\_BOG} - \rho_{C4\_BOG\_cond} . \dot Q_{C4\_BOG\_cond}$$

$$\frac{d\left(\frac{P_{tank} Mw_{C4\,BOG}}{R.T} . V_{C4\_BOG}\right)}{dt} = \frac{P_{tank} Mw_{C4\,BOG}}{R.T} . \dot Q_{C4\_BOG} - \rho_{C4\_BOG\_cond} . \dot Q_{C4\_BOG\_cond}$$

$$\frac{V_{C4\_BOG} . Mw_{C4\,BOG}}{R.T} . \frac{dP_{tank}}{dt} = \frac{P_{tank} Mw_{C4\,BOG}}{R.T} . \dot Q_{C4\_BOG} - \rho_{C4\_BOG\_cond} . \dot Q_{C4\_BOG\_cond}$$

$$\frac{dP_{tank}}{dt} = \frac{P_{tank}}{V_{C4\_BOG}} . Q_{C4\,BOG} - \dot m_{C4\_BOG\_cond} . \frac{R.T}{V_{C4\_BOG} . Mw_{C4\,BOG}} \quad (I$$

steady state mass and energy balance in Condenser (as functio

$$(H_{C4\_BOG} - H_{C4\_cond}) \frac{d\dot m_{C4\_BOG\_cond}}{dt} = U.A.\Delta T_{lmtd}$$

$$(H_{C4\_BOG} - H_{C4\_cond}) \frac{d\dot m_{C4\_BOG\_cond}}{dt} = \frac{1}{\frac{1}{h_i} + \frac{\Delta x}{k} + \frac{1}{h_o}} A.\Delta T_{lmtd}$$

$$\left(H_{C4\_BOG} - H_{C4\_cond}\right)\frac{d\dot{m}_{C4\_BOG\_cond}}{dt} = \left(H_{C3,in} - H_{C3,out}\right).\dot{m}_{C3\_in} \quad \text{(IV-3)}$$

III Results and Discussions

III.1 Dynamic Simulation of LPG Butane Storage and Loading System Validation

LPG Butane Storage and Loading system was simulated in Hysys simulator by using one Butane tank with its Butane Condenser. Gravity/static head contribution in the Butane Condenser simulation was governing the flow of Butane BOG in inlet and outlet condenser. All nozzle position above the butane tank is set to 50 m to create enough static head for the butane condenser operational in the simulation. The simplification and model approach for this dynamic simulation is based on model approach shown in Figure-4 and the model development in the simulation shown in Figure-5. The dynamic simulation then validated using its overpressure profile with offline butane condenser compared to actual profile on Figure-6. The dynamic simulation was fit to purpose for its increment of pressure profile trajectory and was similar to the actual data (regardless the parabolic shapes profile on the actual pressure profile data). Butane tank reached its overpressure flaring set point for about 9 hours after butane condenser was offline. Heat-in leak to generate LPG BOG Butane was regulated by setting the overall heat transfer coefficient (U) in Butane tank, resulting in U = 20 kJ/h.m^{2}C. The LPG Butane BOG composition resulted from the simulation which justified having lighter composition than LPG Butane product composition is shown in Table-5.

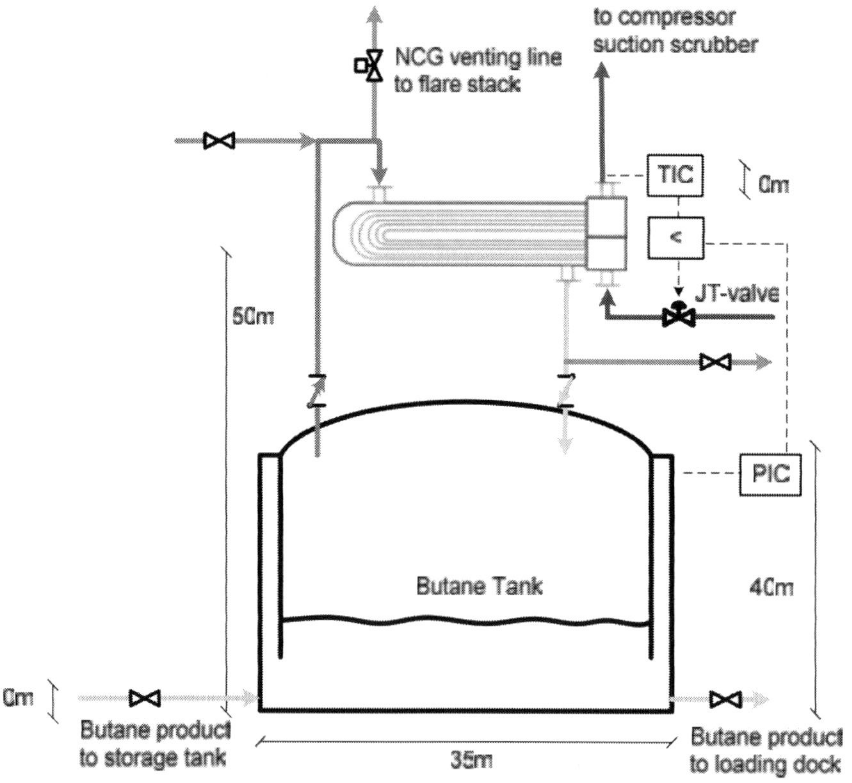

Figure-4 Simplification and Model Approach for Butane Re-liquefaction System Dynamic Simulation

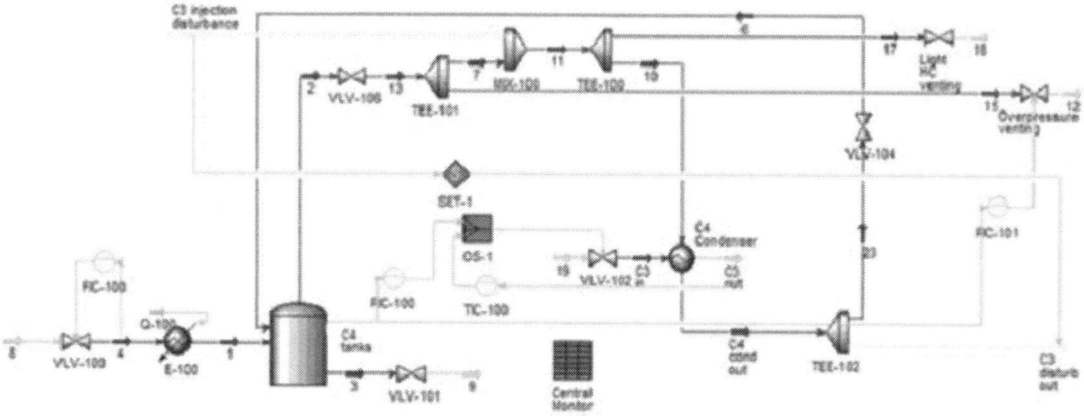

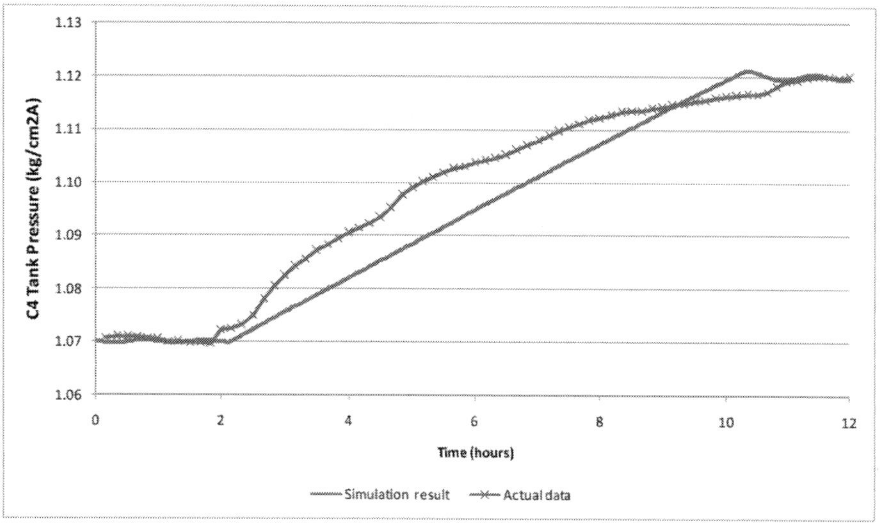

Component	Actual data of LPG Butane Product (%-mol)	Dynamic Simulation LPG Butane Product (%-mol)	Actual data of LPG Butane BOG (%-mol)	Dynamic Simulation of LPG Butane BOG(%-mol)
Ethane	0	0	0.01	0
Propane	0.97	1	3.44	7
i-Butane	42.24	42	49.71	56
n-Butane	56.69	57	46.82	37
Pentane	0.10	0	0.02	0

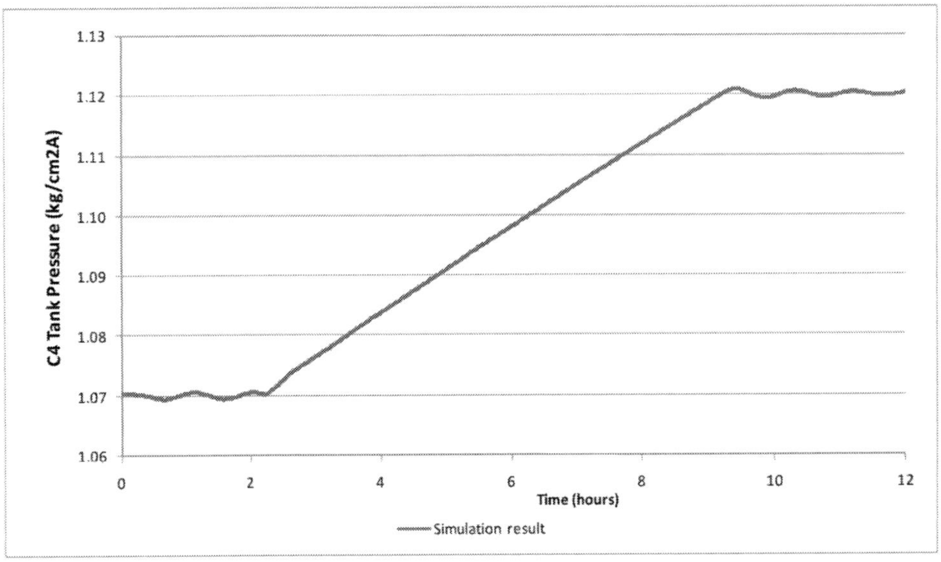
C4 Tank Pressure (kg/cm2A)
Time (hours)
Simulation result

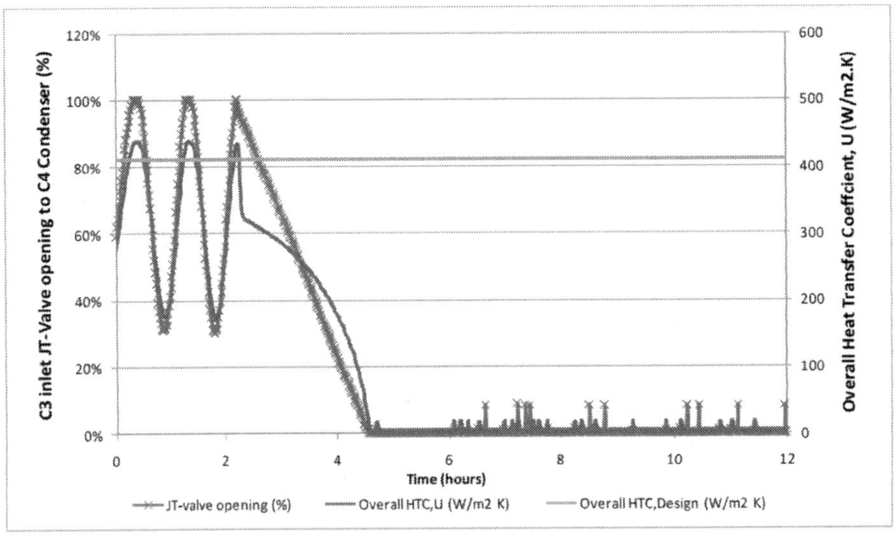
C3 inlet JT-Valve opening to C4 Condenser (%)
Overall Heat Transfer Coeffcient, U (W/m2.K)
Time (hours)
JT-valve opening (%)
Overall HTC,U (W/m2 K)
Overall HTC,Design (W/m2 K)

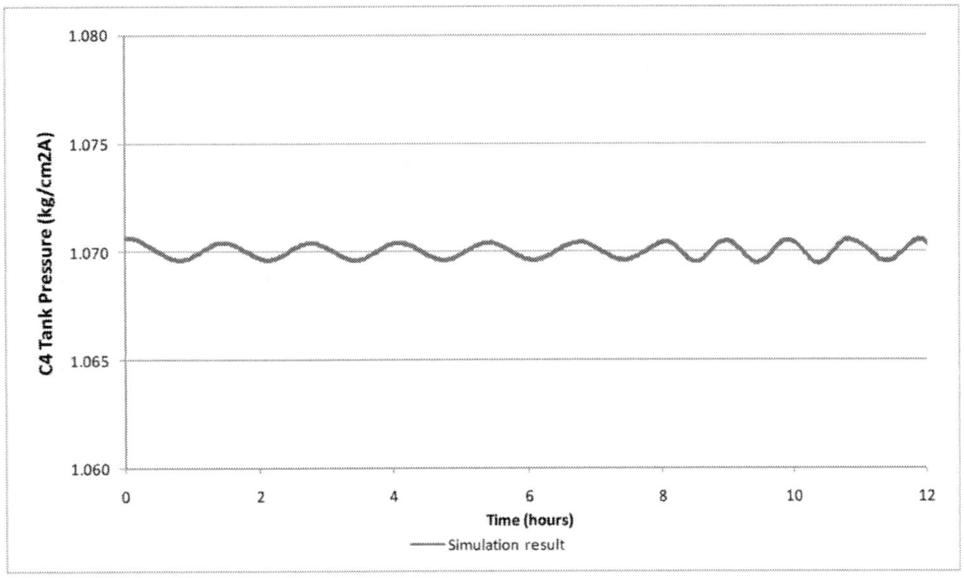
C4 Tank Pressure (kg/cm2A)
1.080
1.075
1.070
1.065
1.060
0
2
4
6
8
10
12
Time (hours)
Simulation result

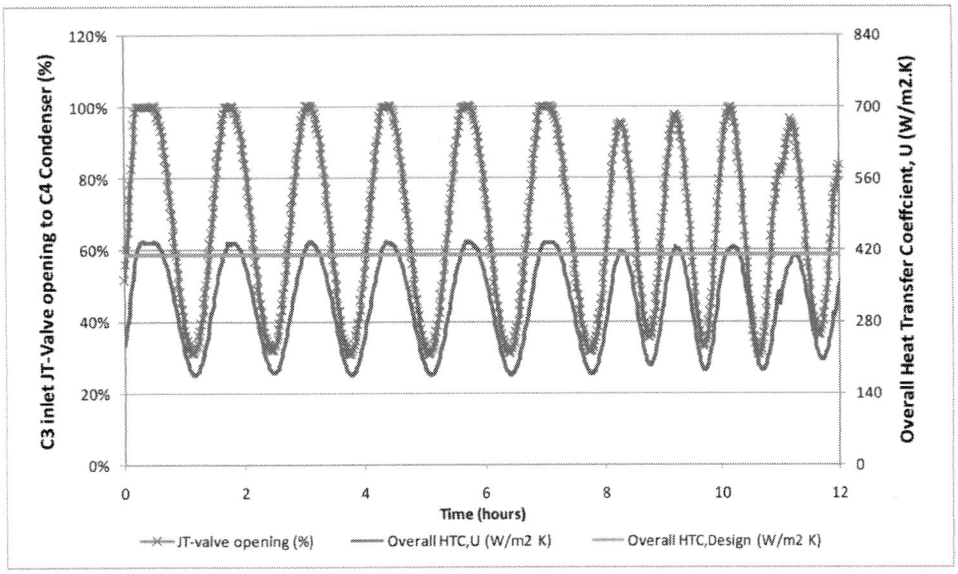
C3 inlet JT-Valve opening to C4 Condenser (%)
120%
100%
80%
60%
40%
20%
0%
Overall Heat Transfer Coeffcient, U (W/m2.K)
840
700
560
420
280
140
0
0
2
4
6
8
10
12
Time (hours)
JT-valve opening (%)
Overall HTC,U (W/m2 K)
Overall HTC,Design (W/m2 K)

IV. Conclusions

The NCG accumulation phenomenon in Butane Condenser is successfully simulated dynamically in a process simulator. NCG accumulation in Butane Condenser decrease the overall heat transfer which led to non-condensing Butane BOG furthermore it would increase the Butane tank pressure. Based on dynamic simulation results, the NCG accumulation on Butane BOG re-liquefaction system can be eliminated by continuously venting NCG as much as 5 Nm^3/hr, while maintained the Butane Condenser overall heat transfer coefficient. By conducting continuously venting NCG, the LPG Butane Storage and Loading Yield could be improved significantly from 91.42% to 99.96%.

V. Acknowledgement

This publication is supported by Master Degree cooperation program between PT Badak NGL, Bontang, East Kalimantan and Institute Technology of Bandung (ITB) in 2012 – 2013.

VI. References

1. Butterworth D. (1988) : "Condensers and Their Design", *Journal of Two Phase Flow Exchanger*, p.779-828
2. Jensen Michael K. (1988) : "Condensation With Non-Condensables and In Multicomponent Mixtures", *Journal of Two Phase Flow Exchanger*, p.779-828
3. Jorge Contreras, SENER-Spain and Jose Maria Ferrer, Aspentech Spain, (May 2005) : "LNG Receiving Terminal and Regasification Dynamic Simulation", Hydrocarbon Processing Magazine
4. Kern Donald Q. (1950) : "Process Heat Transfer", McGraw-Hill Companies, New York
5. Namjin Jang and En Sup Yoon (2011) : "Dynamic Simulation and Oprimization of LNG BOG compressor Operation in LNG Receiving Terminal", Seoul National University and Incheon Terminal Korea Gas Corp, *Korean Journal Chem.Eng.*
6. Som S.K. and Chakraborty Suman (2006) : "Film Condensation in Presence of Non-Condensable Gases Over Horizontal Tubes", *International Journal of Heat and Mass Transfer* 49, p.594-600
7. Stephan K. (2006) : "Interface Temperature and Heat Transfer in Forced Convection Laminar Film Condensation of Binary Mixtures", *International Journal of Heat and Mass Transfer* 49, p.805-809
8. Yamamoto T., Fujiwara Y., and Kitagaki S. (2011) : "Challenges of Advanced Utilization of LNG Cold in Osaka Gas Senboku LNG Terminals", *Proceedings of EcoDesign* 2011, p.148-153

424775 Optimal Design of Process Systems Under Uncertainty

Monday, November 9, 2015
Exhibit Hall 1 (Salt Palace Convention Center)
Guennadi Ostrovsky, Karpov Institute of Physical Chemistry, Moscow, Russia, **Nadir Ziyatdinov**, Process System Engineering, Kazan National Research Technological University, Kazan, Russia and Lapteva Lapteva, Process system engineering, Kazan National Reserch Technological University, Kazan, Russia

Optimal design of process systems under uncertainty

G. M. Ostrovsky[1], N. N. Ziyatdinov[2], T. V. Lapteva[2]

[1]*Karpov Institute of Physical Chemistry, Vorontsovo Pole, 10, Moscow, 103064, Russia*

Email: ostralex@yandex.ru, *Phone: 8-499-4792991*

[2]*Kazan State Technological University, Karl Marx str., 68, Kazan, 420111, Russia*

Email: nnziat@yandex.ru ; tanlapteva@yandex.ru

We will consider the problem of the optimal design of a process systems and chemical process (CP) for the case when inexact mathematical models are used. The inexactness of mathematical models arises because of the original uncertainty of chemical, physical, and economic data which are used during the CP design. The problem of the CP optimal design under uncertainty can be formulated as follows: it is necessary to create an optimal CP that would guarantee the satisfaction (exactly or with some probability) of all design specifications in the case when inexact mathematical models are used and internal and external factors change during the CP operation stage.

Usually, the following two formulations of this problem are used

1) The *formulation of the two-stage optimization problem* (TSOP) takes into account possibility of the control variables change at the operation stage. Here we suppose that at each time instant during the operation stage (a) values of all or some of the uncertain parameters can be either measured or calculated using the experimental data and (b) during the operation stage the control variables are adjusted depending on a chemical process state. This formulation can be used if it is possible to accurately estimate all or some of the uncertain parameters at the operation stage of CP.

2) The *formulation of the one-stage optimization problem* (OSOP) supposes that the control variables are constant at the operation stage. We will consider the two-stage optimization problem (TSOP) with joint chance constraints has the following form [1]

$$f^1 = \min_{d,z} E[f(d,z(\theta),\theta)] \quad (1)$$

$$\Pr\{g_j(d,z(\theta),\theta) \le 0,\ j=1,\dots,m\} \ge \alpha \qquad 0 \le \alpha < 1, \quad (2)$$

where $f(d,z,\theta)$ is a goal function, $E_\theta[f(d,z(\theta),\theta)]$ is the expected value of the goal function $f(d,z(\theta),\theta)$, d, z are vectors of design and control variables, respectively, q is a vector of uncertain parameters, $\Pr\{g_j(d,z(\theta),\theta) \le 0, j=1,\dots,m\}$ is the probability of joint satisfaction of all the constraints, i.e. it is the probability measure of the region $\Omega = \{\theta: g_j(d,z(\theta),\theta) \le 0,\ j=1,\dots,m\}$

$$\Pr\{g_j(d,z(\theta),\theta) \le 0, j=1,\dots,m\} = \int_\Omega \rho(\theta)d\theta,$$

$$E_\theta\{f(d,z(\theta),\theta)\} = \int_T f(d,z(\theta),\theta)\rho(\theta)d\theta,$$

where $T = \{-\infty \le \theta \le \infty\}$, $\rho(\theta)$ is the probability density function. We will assume that the uncertain parameters are normally distributed, random variables. The main issue in solving optimization problems under uncertainty is the computation of multiple integrals for calculation of the expected value of the objective function and probabilities of the satisfaction of the constraints. The known methods of nonlinear programming (e.g., SQP) require the calculation of multiple integrals at each iteration. This operation is very intensive computationally even when the dimensionality of the uncertain parameters vector θ is low.

We have developed a new approach for solving the TSOP. This approach is based on the following two operations. The first operation is the approximate transformation of the chance constraints into deterministic constraints (based on the approximation of the region Ω by a union of multidimensial rectangles). The second operation is based on the approximation of the goal function by a piece wise linear function. This permits to exclude operations of multiple integration when solving the OSOP. An example is given that illustrates efficiency of this approach. Removal of the operations of multiple integration permits to decrease significantly computational time of solving the OSOP.

References

[1] Ostrovsky GM, Ziyatdinov NN, Lapteva TV, Zaitsev IV. Two-stage optimization problem with chance constraints. Chem Eng Sci. 2011; 66: 3815–3828

427009 Model Based Optimization of Microalgae Cultivation in a Raceway Pond

Abstract:

In this research, a new model is proposed to predict microalgae cultivation in a raceway pond (RWP). Based on the proposed model, optimal operating condition for maximizing the operating profit is searched. Recently, many researchers have studied microalgae as potential raw material for biofuels and chemicals. However, the high cost of cultivation stands as the main obstacle to successful commercialization [1]. RWP with its relatively low cultivation costs is considered as the most probable means for the mass production of microalgae. Despite this, there exist few published research works on the modeling of microalgae growth in a RWP as it encompasses many complicated phenomena including mass hydrodynamics, light penetration, and microalgal growth. Furthermore, previously proposed RWP models, which are based on computational fluid dynamics models, are too complex to be suitable for important applications such as optimization and control [2-3]. In this research, a new RWP model is proposed. It is based on dynamics of microalgae growth, light penetration, sediment and temperature, which are important for microalgae cultivation in a RWP. Since the model is relatively simple, we can use it to optimize the operating condition. The optimization result shows that current operating practice leaves much room for improvement.

Reference

[1] J.N. Rogers, J.N. Rosenberg, B.J. Guzman, V.H. Oh, L.E. Mimbela, A. Ghassemi, M.J. Betenbaugh, G.A. Oyler and M.D. Donohue, A critical analysis of paddlewheel-driven raceway ponds for algal biofuel production at commercial scales, Algal Research 4 (2014) 76-88.

[2] M. Prussi, M. Buffi, D. Casini, D. Chiaramonti, F. Martelli, M. Carnevale, M.R. Tredici, L. Rodolfi, Experimental and numerical investigations of mixing in raceway ponds for algae cultivation, Biomass and Bioenergy 67 (2014) 390-400.

[3] O. Bernard, A.C. Boulanger, M.O. Bristeau and J. Sainte-Marie, A 2D model for hydrodynamics and biology coupling applied to algae growth simulations, ESAIM: Mathematical Modelling and Numerical Analysis 47 (2013) 1387-1412.

427249 Compact MILP Formulation for Scheduling of Multipurpose Batch Plants: Rigorous Sequencing of Tasks

Abstract:

In recent years, batch processes have been getting more attention due to their suitability for the production of small volume, high value added products. The flexibility of batch plants allows the production of different products within the same facility which mandates sharing of different types of resources (raw materials, equipment, utilities and manpower). Despite the advantage of batch plants being flexible, they also pose a challenging task to production scheduling. The profitability of these batch plants is highly dependent on the way the production schedule is optimized. Considerable scheduling models have been proposed from the academic research in the last two decades. The research direction has been to improve the existing scheduling models in terms of model size, computational time and optimal objective value in order to effectively apply and solve industrial scale problem. However, there is still a gap for improvement and new efficient model.

This study also has the same objective of developing a better scheduling model that is compact and computationally efficient. The model is rigorous in handling the different intermediate storage policies and unit wait time. Almost the binary variables, continuous variables, constraints and computational time required are reduced by 41%, 15%, 43% and 42% respectively, when compared to the recent rigorous unit-specific event based model for short-term scheduling of batch plants.

Keywords: Scheduling; Multipurpose batch plant; Optimization; MILP

414199 Multi-Agent Optimization Framework (MAOP) for Large Scale Process System Engineering Optimization Problems

Abstract:

Multi-agent Optimization Framework (MAOP) for Large Scale Process System Engineering Optimization Problems

Berhane H. Gebreslassie and Urmila M. Diwekar

Center for Uncertain Systems, Tools for Optimization and Management (CUSTOM):Vishwamitra Research Institute, Crystal Lake, IL 60012 – USA

berhane@vri-custom.org; urmila@vri-custom.org

Abstract

The multi-agent optimization (MAOP) framework provides a way of combining various algorithms in one platform and exploits the strength possessed by each algorithm. Most of large scale process system engineering problems include nonlinear and non-convex problems. Contrary to the standalone optimization algorithm framework, the MAOP strategy avoids the problem of getting stuck in local optima as well as improving the computational efficiency. In this work, we propose a multi-agent optimization framework for solving complex large scale process system engineering problems. The framework uses a variety of different algorithmic agents which include the gradient based local optimizers and metaheuristic algorithms (efficient simulated annealing, efficient genetic algorithm and efficient ant colony optimization algorithms). Each agent encapsulates a particular problem-solving procedure. We investigate the effect of cooperation among agents of the multi-agent system working in parallel and combined into a framework designed to solve large scale combinatorial optimization problems. Computational experiments are carried out using benchmark problems and real world case study. The proposed methodology enables to improve the quality of solutions and the computational efficiency in comparison with non-cooperative multi-agent framework and standalone agents. We have also examined solving the optimization problems with cooperative homogenous and heterogeneous agent systems and the results indicate that the computational efficiencies are improved when the agents are heterogeneous. Moreover, the analysis of the intra- and inter agent cooperation shows that depending on the complexity of the problem the inter- and intra-agent collaboration has a significant impact on system performance. The MAOP framework which includes the five major parts of the algorithm (figure 1) is given in the figure 2.

Fig. 1. The major parts of the MAOP algorithm and the information flow direction.

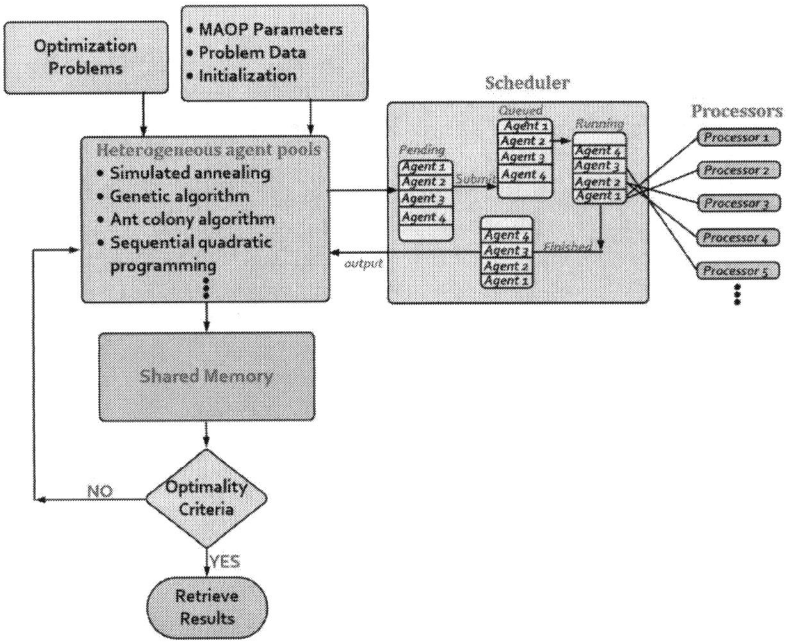

Fig. 2. The basic flow diagram of the heterogeneous MAOP algorithm

References

1. Siirola JD, Steinar Hauan S, and Westerberg AW, 2003. Toward agent-based process systems engineering: proposed framework and application to non-convex optimization. *Computers and Chemical Engineering* 27: 1801-1811.

2. Gebreslassie BH, and Diwekar UM, 2015. Efficient ant colony optimization for computer aided molecular design: Case study solvent selection problem. *Computers and Chemical Engineering* 78: 1-9.

3. Diwekar UM, Xu W, 2005. Improved genetic algorithms for deterministic optimization and optimization under uncertainty. part i. algorithms development. *Industrial and Engineering Chemistry Research* 44(18) 7132-7137.

429510 Title: An Agent-Oriented Software Framework for Real Time Optimization Solutions Implementation

Abstract:

For the past decades the increasing competition in the global market has lead the chemical process industry to look for the application of tools to optimize processes in order to enhance profitability. Real Time Optimization (RTO) is a Computer Aided Process Engineering (CAPE) technique that continuously reevaluates the process operating conditions to maximize economic benefits under operational constrains. By adjusting selected optimization variables it seeks to improve process performance iteratively using measurement data to drive the operating point toward the actual plant optimum [1].

RTO covers several concerns: real-time process data sampling through interfaces with measurement instruments, Distributed Control Systems and solutions like Plant Information System from OSI Software (www.osisoft.com); gross error detection, data reconciliation and steady-state identification; model parameters estimation and optimization. Once an optimal operational point is calculated for the plant, subjected to operational constraints, its values are sent to the process control to update set-points. Advanced control techniques like model based predictive control (MPC) are responsible for take the plant to the suggested state. Several approaches have been proposed for RTO [1].

Transforming multicomponent CAPE methods like RTO into useful tools requires appropriate software architecture and design [2]. An Agent Oriented Software Architecture (AOA) for the implementation of RTO solutions has been conceived and is presented here.

AOA and Agent Oriented Programming (AOP) are relatively new software paradigms that bring concepts from the Artificial Intelligence field into the domain of distributed systems. According to the AOA approach a software application should be modeled and built as a collection of interacting components called Agents. A definition for the term “software agent” has been addressed in several works [3, 4, 5, 6]. In general, all the definitions converged to the idea that an agent is essentially a special software component situated within and a part of an environment that senses that environment and acts on it over time, has autonomy or behaves like a human agent, working for some clients in pursuit of its own agenda with social or communication abilities.

A variety of application fields are being benefited with the use of multi-agent systems. Examples of industrial domains where this kind of systems have been fruitfully employed include process control [7], system diagnostics [8], manufacturing [9], transportation logistics [10] and network management [11]. The Foundation for Intelligent Physical Agents (FIPA) is an IEEE Computer Society standards organization that promotes agent-based technology and the interoperability of its standards with other technologies. Its specifications contain a collection of standards which are intended to promote the interoperation of heterogeneous agents and the services that they can represent (www.fipa.org).

A software framework following the proposed AOA and using some well known software design patterns and the Object Oriented Programming paradigm (OOP) has been developed also and is presented as well. In the development process, a free an open source implementation of the FIPA standards, named JADE (www.jade.tilab.com), has been used as base platform. Using the agents framework, a prototype was configured to apply RTO to the Williams-Otto study case [12] using a Model Parameter Adaptation approach [13]. For steady-state identification the methodology proposed by Cao and Rhinehart [14] was used. The process model and optimization problem were defined using the modeling language of EMSO [15] and the Ipopt algorithm [16] was used as optimization solver. Another RTO prototype for a PETROBRAS C3 Splitter distillation unit was set as well, also based on an EMSO model [17]. In both cases the framework's worth was proved.

Although proven commercial software products exist for RTO and can be parameterized they usually feature a black box style regarding the way operations are carried out. In addition, the inclusion of new RTO methods is almost always not a feasible option. That motivated the development of the framework as a technically open workbench that can be used for test, research and innovation purposes allowing the implementation of several RTO approaches. It provides a platform for the study of RTO issues, steady state identification methods, gross error detection and data reconciliation techniques. It also creates chances for the interaction with already existing modeling and simulation tools and to test the efficacy of different optimization solvers. Spots for the implementation of different control systems interfaces are also provided. It furnishes a useful environment to study in practice unexplored techniques combination and helps to open new horizons for more sophisticated RTO applications.

(1) B. Chachuat, B. Srinivasan, D. Bonvin, 2009, Adaptation Strategies for Real-Time Optimization, Comput. Chem. Eng., 33, 1557-1567.

(2) B. Braunschweig, R. Gani, 2002, Software Architectures and Tools for Computer Aided Process Engineering, Comput. Aided Chem. Eng., 11.

(3) M.R. Genesereth, S.P. Ketchpel, 1994, Software Agents. Communications of the ACM, 37(7), 48–53.

(4) M.J. Wooldridge, N.R. Jennings, 1995, Intelligent Agents: Theory and Practice. Knowledge Engineering Review, 10(2), 115–152 .

(5) S.J Russell, P. Norvig, 2003, Artificial Intelligence: a Modern Approach, 2nd ed. Prentice Hall.

(6) P.T. Tosic, G.A Agha, 2004, Towards a hierarchical taxonomy of autonomous agents, Proceedings of IEEE International Conference on Systems, Man and Cybernetics, The Hague, Netherlands, 3421-3426.

(7) N. Jennings, 1994, The Archon System and its Applications, Proceedings of the 2nd International Working Conference on Cooperating Knowledge Based Systems (CKBS-94), Dake Centre, University of Keele, UK, 13-29.

(8) M. Albert, T. Laengle, H. Woern, M. Capobianco, A. Brighenti, 2003, Multi-agent Systems for Industrial Diagnostics, Proceedings of 5th IFAC Symposium on Fault Detection, Supervision and Safety of Technical

Processes, Washington, DC , 483-488.

(9) H. Parunak, 1987, Manufacturing Experience with the Contract Net. In Huhns, M. (ed.), Distributed Artificial Intelligence, Pitman, London, 285-310.

(10) N. Neagu, K. Dorer, D. Greenwood, M. Calisti, 2006, LS/ATN: Reporting on a Successful Agent-Based Solution for Transport Logistics Optimization. Proceedings of the IEEE Workshop on Distributed Intelligent Systems (WDIS'06), Prague.

(11) D. Greenwood, G. Vitaglione, L. Keller, M. Calisti, 2006, Service Level Agreement Management with Adaptive Coordination. Proceedings of the International Conference on Networking and Services (ICNS'06), Silicon Valley, USA.

(12) T.J. Williams, R.E Otto, 1960, A Generalized Chemical Processing Model for the Investigation of Computer Control, American Institute of Electrical Engineers, Part I: Communication and Electronics, Transactions, Volume 79, Issue 5, 458 – 473.

(13) C. Y., Chen, B. Joseph, 1987, On-line optimization using a two-phase approach: An application study. Industrial and Engineering Chemistry Research, 26, 1924–1930.

(14) S. Cao, R.R. Rhinehart, An Efficient Method for On-line Identification of Steady State. Journal of Process Control 1995, 5, 363.

(15) R.D.P. Soares, A.R. Sechhi, 2003, EMSO: A new Environment for Modelling, Simulation and Optimization, Comput. Aided Chem. Eng., 14, 947-952.

(16) A. Wachter, L. Biegler, 2006, On the implementation of an interior-point filter line-search algorithm for large-scale nonlinear programming, Mathematical Programming, Volume 106, Issue 1, 25-57.

(17) D.F.M. Muñoz, L.M.P. Garcia, J.E.A. Graciano, C.A.R. Martinez, A.S. Vianna, G.A.C. Le Roux, 2013, Real-Time Optimization of an Industrial-Scale Vapor Recompression Distillation Process. Model Validation and Analysis. Ind. Eng. Chem. Res., 52, 5735-5746.

429814 Constrained Zonotopes: A New Tool for Set-Based Computations

Abstract:

Many modern control and optimization algorithms make use of sets (e.g., intervals, ellipsoids, polytopes) as basic computational objects, with the aim of characterizing some sets of interest. Sets of interest may include the range of a given function, reachable or invariant sets of dynamical systems, or sets of states or parameters consistent with a bounded-error model [1,2,3]. Since the true set of interest is often impossible to represent exactly with finite data, an enclosure of this set by an element of some class of simpler sets is typically sought instead. The choice of a set class for a given application is critical and is based on a trade-off between (i) the accuracy with which a member of the class can represent the set of interest, and (ii) the complexity of the required computations. An important consideration in (ii) is the convenience of the approximating set for its end use, which may involve checking for the inclusion of given points, as in model invalidation and fault diagnosis [2], checking for intersection with another set, as in system verification and safety analysis [1], or using the set as a constraint in an optimization problem, as in robust optimization and control [3].

This presentation introduces a new class of sets called constrained zonotopes and demonstrates that they provide a better trade-off between accuracy and efficiency than existing classes for some representative problems. While we believe that these sets will find application in a variety of optimization and control algorithms, their performance will be demonstrated here through the classical set-based state estimation problem for discrete-time linear systems with bounded noise [4], and its application to set-based fault diagnosis [5].

As with many problems in linear control and estimation, the essential set-operations required by these applications are Minkowski sums, linear mappings, and intersections. Before describing constrained zonotopes, it is informative to take stock of the capabilities of commonly used set classes with regard to these operations. The class of convex polytopes is closed under all three operations (i.e., performing any of them on members of the class results in another member of the class). Moreover, using the typical halfspace representation (H-rep), both intersections and invertible linear mappings can be computed efficiently. However, the complexity of the Minkowski sum is exponential in the dimension of the set, as is the worst-case number of halfspaces describing the result [6]. The same is true of non-invertible linear mappings (e.g., polytope projections) [7]. In contrast, if the polytope is stored as a set of vertices (V-rep), then linear mappings and Minkowski sums are much simpler, but intersection becomes NP-hard [6]. Consequently, working with convex polytopes is very costly and numerically unstable when the dimension exceeds about five, or the number of halfspaces or vertices is large.

At the other extreme, intervals, parallelotopes, and ellipsoids all provide low-complexity set representations and relatively low-cost set operations. However, intervals are not closed under linear mappings, and parallelotopes and ellipsoids are not closed under Minkowski sums or intersections. Instead, the results of these operations must be conservatively enclosed. Optimal interval enclosures are easily computed, but are often very weak. For ellipsoids, cheap heuristic enclosures are given in [4] and optimal enclosures in [8], but the intersection requires the solution of a convex optimization problem. Cheap heuristic enclosures for parallelotopes are given in [9], but optimal enclosures are not available.

Recently, zonotopes have received much attention within the control community, particularly because linear mappings and Minkowski sums can be computed exactly and efficiently [10]. Zonotopes are convex polytopes that are centrally symmetric, which makes it possible to store them in terms of their center and a set of vectors called generators. This generator representation (G-rep) is often more compact than the equivalent H- or V-rep. Moreover, linear mappings and Minkowski sums of zonotopes in G-rep can be computed through trivial identities; the first requires a single matrix-matrix product, and the second requires only memory moves. Both operations can increase the complexity of the set representation. However, in contrast to the worst-case exponential increase of the H-rep, the increase in the number of generators under these operations is only linear in the input data. Moreover, conservative order reduction techniques are available that enclose a given zonotope within another with fewer generators [11,1], providing a tunable mechanism for balancing accuracy and complexity. The main disadvantage of zonotopes is that they are not closed under intersection, and tight enclosures are difficult to compute. This leads to serious complications in, e.g., state estimation and hybrid systems verification [12].

Motivated by this discussion, the constrained zonotopes are defined by generalizing the following equivalent definition of the zonotopes: every zonotope is the image of the unit hypercube in some higher-dimensional space under an affine mapping. The constant part of this mapping is the center of the zonotope, while the linear term is comprised of its generators. To generalize this definition, we impose linear equality constraints on the unit hypercube, and define the constrained zonotopes as the affine images of these constrained hypercubes. These sets are fully described by the storing the constraint matrix and right-hand side vector in addition to the center and generator matrix. This is called the constrained generator representation (CG-rep).

With this definition, it will be shown that the constrained zonotopes are closed under Minkowsi sums and linear mappings, and that both operations can be implemented via simple identities that are hardly more complex than the corresponding identities for zonotopes. Further, we show that the constrained zonotopes are also closed under intersections, and these can be trivially computed with only memory moves. At the same time, the constrained zonotopes are also considerably more flexible than zonotopes. A simplex, for example, is a constrained zonotope that is not a zonotope (it is clearly not centrally symmetric). Remarkably, it turns out that every convex polytope is a constrained zonotope, and vice-versa. Thus, the novelty of the constrained zonotopes can be stated in two ways:

(1) When the numbers of generators and constraints are not limited, the CG-rep provides a new representation of convex polytopes that confers some significant computational advantages of zonotopes to this larger class of sets.

(2) When the numbers of generators and constraints are limited, the CG-rep describes a new class of sets that significantly extends the zonotopes while maintaining computational efficiency.

As with zonotopes, set operations on constrained zonotopes can increase the complexity of the CG-rep. This is more serious than with zonotopes because the number of generators and constraints may increase, but is considerably less serious than the growth of the H-rep discussed above. Nonetheless, accurate and efficient methods for conservatively reducing the complexity of the CG-rep are essential to computing with constrained zonotopes. Thus, a comprehensive set of generator and constraint reduction techniques are proposed and demonstrated in this work.

Finally, the advantages of constrained zonotopes are demonstrated through extensive numerical case studies. First, comparisons with several existing approaches [9,11,12] for linear discrete-time estimation are presented, showing that constrained zonotopic estimators can provide estimates with substantially smaller volumes at only modest additional cost. These estimators are further shown to scale well to high-dimensional systems. Next, these estimators are evaluated in the context of set-based fault diagnosis, where estimation accuracy is critical. Our results show that constrained zonotopic estimators are capable of reducing the time required for diagnosis by nearly half as compared to existing estimators applied to randomized examples.

References Cited

1. Althoff, M.; Stursberg, O. & Buss, M.
   Computing reachable sets of hybrid systems using a combination of zonotopes and polytopes
   Nonlinear Analysis-Hybrid Systems, 2010, 4, 233-249
2. Rosa, P.; Silvestre, C.; Shamma, J. & Athans, M.
   Fault Detection and Isolation of LTV systems using Set-Valued Observers
   Proc. of the 49th IEEE Conference on Decision and Control, 2010, 768-773
3. Mayne, D.; Rakovic, S.; Findeisen, R. & Allgower, F.
   Robust output feedback model predictive control of constrained linear systems
   Automatica , 2006, 42, 1217-1222
4. Schweppe, F.
   Recursive state estimation: Unknown but bounded errors and system inputs
   IEEE Trans. Automat. Contr., 1968, 13, 22-28
5. Scott, J. K.; Findeisen, R.; Braatz, R. D. & Raimondo, D. M.
   Input design for guaranteed fault diagnosis using zonotopes
   Automatica , 2014, 50, 1580-1589
6. Tiwary, H. R.
   On the Hardness of Computing Intersection, Union and Minkowski Sum of Polytopes
   Discrete & Computational Geometry, Springer-Verlag, 2008, 40, 469-479
7. Jones, C.; Kerrigan, E. & Maciejowski, J.
   On Polyhedral Projection and Parametric Programming
   J. Optim. Theor. Appl., Springer US, 2008, 138, 207-220
8. Chernousko, F.
   Optimal Guaranteed Estimates of Indeterminacies with the Aid of Ellipsoids. Parts I-II
   Engineering Cybernetics, 1980, 18, 1-9
9. Chisci, L.; Garulli, A. & Zappa, G.
   Recursive state bounding by parallelotopes
   Automatica , 1996, 32, 1049-1055

10. Kuhn, W.
Rigorously computed orbits of dynamical systems without the wrapping effect
Computing, 1998, 61, 47-67

11. Combastel, C.
A state bounding observer based on zonotopes
European Control Conference, 2003

12. Bravo, J.; Alamo, T. & Camacho, E.
Bounded error identification of systems with time-varying parameters
IEEE Trans. Automat. Contr., 2006, 51, 1144-1150

431306 Vcap: A Software Platform for the Assessment of Biorefinery Value Chains

Abstract:

In this work, we present VCAP (Value Chain Assessment Platform), a platform that facilitates the assessment of biorefinery value chains. VCAP combines a user friendly interface and an optimisation algorithm into a web service that is easy to use.

Current implementation facilitates the consideration of softwood and agricultural residues as the biomass input. It also facilitates consideration of a range of products including cellulose, hemicellulose in the form of sugars, and lignin. Current implementation only considers a sonication technology. That technology can be described in the following steps:

1. the feedstock is mixed with an organic acid solution to form a slurry
2. organic solvents are added to form an organosolv slurry prior to sonication
3. after sonication the solid and liquid are separated and the dried solid contributes toward the cellulose fraction
4. the liquid is further separated into the organic and aqueous fraction via water addition
5. the organic fraction contains lignin and some organic solvent, and the aqueous fraction contains the hemicellulosic sugars, the remainder of the organic solvent, and the water and acid.
6. lignin and hemicellulose are isolated via membrane or distillation technology and the organic acidic water and solvents are recycled

The optimisation algorithm employed by the platform takes into account various steps of the value chain from the collection of biomass and its transportation to different biorefining technologies and end products and assesses the economic potential of different types of biomass and/or biorefinery technologies in various geo-economic environments. As at present, VCAP can be adapted to account for any number and type of lignocellulosic biomass (e.g. hardwood, softwood, wheat straw, grasses etc.), and optimises under three different scenarios (supply, demand and biorefinery facility).

In an effort to increase the flexibility of the platform, the technology (biorefinery) module is substitutable, allowing the consideration of different technology models. VCAP also features links to external databases extracting information about biomass availability, composition and prices. Also, a mapping solution is employed to allow for a more dynamic and friendly interaction with users, regardless of the level of their expertise. The platform, developed in Java and using GLPK solver for the Mixed Integer Linear Programming problem, has been validated with a case study in Scotland.

More specifically, three scenarios have been taken into account for the case study: i) Supply-driven, ii) Demand-driven and iii) Technology-driven. The biomass type considered in all three scenarios was softwood in the form of logs and chips (sawmill by-products). The moisture contents of the selected biomass types range between 30% - 60%. Finally, candidate points for technologies were located close to power stations in order to utilise the waste heat energy.

428920 Simultaneous Scheduling of Front End Crude Transfer and Refinery Processing

Abstract:

Petroleumrefinery is a leading segment of the entire petrochemical industry. Scheduling and planning of crude oil is very important to a petroleum refinery due to the potential realization of large cost savings. It is also the one of most challenging problems for refinery operations. However, it is not easy to simulate and optimize the front-end crude-oil and refinery operations at the same time. And in reality, the crude scheduling activities are highly vulnerable to disruptions brought by various uncertainties.

In this paper, a large-scale continuous scheduling model has been employed to simulate and optimize the front-end and refinery crude-oil operations. The scope of the scheduling problem includes crude oil unloading from vessels to the crude storage tanks at onshore berths, transfer of crude from these tanks to the charging tanks, charging crude distillation units with crude mixes from the charging tanks and further processing of crude inside the refinery, which includes crude distillation, cracking, coking, reforming, hydrotreating, product blending and component recovery.

The objective of the proposed crude scheduling problem is to maximize the total operational profit; meanwhile, it should meet inventory and production demands, and try to maintain refinery plant's normal operations under different circumstances of uncertainties. Typical uncertainties come from volatile crude market, product demand changes, shipping delay, increasingly strict environmental regulations, etc. The scheduling model is a large-scale mixed integer nonlinear programming problem (MINLP). An outer-approximation based decomposition method is implemented to obtain the optimum solution of the developed MINLP scheduling model. The efficacy of the proposed rescheduling model has been demonstrated by different case studies.

433494 Dynamic Model and Estimator Development for a Smart Refractory Brick with Embedded Sensors for Gasifier Applications

Abstract:

Dynamic Model and Estimator Development for a Smart Refractory Brick with Embedded Sensors for Gasifier Applications

Qiao Huang, Prokash Paul, Debangsu Bhattacharyya

Department of Chemical Engineering, West Virginia University, Morgantown, WV 26506, USA

Edward M. Sabolsky

Department of Mechanical & Aerospace Engineering, West Virginia University, Morgantown,

WV 26506, USA

Operating temperature of a gasifier is extremely important. A lower temperature can lead to lower carbon conversion, but a higher temperature can reduce the life of the refractory resulting in a significant capital cost. Therefore, the operating temperature must be strictly monitored. Another important issue in the operation of the entrained-flow gasifiers is penetration of the molten slag into the entrained flow gasifiers causing physical and chemical changes. Thus a measure of the extent of slag penetration can be very useful not only for process monitoring, but also for developing effective control strategies. However due to the harsh environment inside the gasifier, it is practically impossible to measure any variable using current state of the art in measurement technology. If the traditional measuring devices, such as thermocouples, are inserted into gasifier through open ports within the refractory, they survive for only several weeks. Furthermore, the defects introduced by the open access ports provide paths for slag penetration reducing the refractory life further. In order to address these issues, a novel smart refractory with embedded sensors is being developed. Since the sensors are embedded within the brick and the materials for embedded sensors are carefully selected, these smart bricks results in minimal defects. While these smart bricks can measure a number of desired variables, the penetrating slag changes the thermal and electrical properties of the refractory and the layout of the sensor. Therefore, typical correlations developed by only considering the data from the pristine bricks would not result in satisfactory estimation of the variables of interest. A model-based approach is employed to circumvent this challenge. This approach can be useful to estimate not only the temperature profile but also other characteristics including stress profile, change of the material properties, and extent of slag penetration. It could be noted that due to the inaccuracy of the process model, only the model itself is not able to provide a good estimate of the variables. However, using the noisy measurement data and the model together, the accuracy of the estimated variables can be significantly improved. Therefore, an algorithm is needed in this application to get the best estimates of the variables.

To be used in the estimator algorithm, rigorous, first-principles, dynamic models of the refractory brick and the embedded sensors are developed. Both the pristine smart brick and the slag-infiltrated brick are considered in building the thermal model. The properties models are based on the data from the in-house experiment and the open literature. Models for the embedded sensors, such as thermocouple, interdigital capacitor, strain gauge, thermistor, are also developed with the consideration of the location as well as the geometries. Finally, the nonlinear unknown input filter (UIF) using the unscented Kalman filter (UKF) approach is developed. The raw measurements from the embedded sensors are used in this algorithm to obtain the desired estimates of the variables.

433509 An Ant Colony Optimization Based Methodology for the Design of Vacuum Transfer Line

Abstract:

The vacuum transfer line is an essential part of the crude oil distillation process. A well designed transfer line can not only help to increase the evaporation rate, but also improve the product's quality. An Ant Colony Optimization Based Methodology was obtained to the design of vacuum transfer line.

In the vacuum transfer line, both gas and liquid existed and mass transferred between two phases. The pressure and temperature distribution in inclined pipes was modeled by the Beggs-Brill correlation and the vaporization process was modeled by a flashing flow model. When routing the vacuum transfer line with the ACO algorithm, the selection probability was calculated by combining relative importance of pheromone deposited by ants from different inlets, the pressure drop caused by the next pipeline section and the total length of the pipeline. This design methodology is total automatic and one only need to input relevant coordinates of pipelines' inlets and the outlet. Then the iteration process will begin and the calculation result will output after the iteration. Case study was presented to demonstrate the availability and efficiency of the proposed algorithm, and computational time of this algorithm is within a couple of seconds which is far smaller than that of other algorithms.

434506 Hierarchic Cyclic Hoist Scheduling for Multi-Stage Material Handling Processes

Abstract:

In multi-stage material handling processes (MSMH), such as electroplating process and polymeric coating, the minimization of cycle duration under a customized production requirement, which is equivalent to the maximization of productivity, is the ultimate goal. Cyclic hoist scheduling (CHS) can provide a complete hoist movement plan, including pick-up and drop-off time points as well as the transportation sequence of each job. This CHS plan will be very practical for large amount of jobs in continuous production. In recent studies, CHS has been coupled with perspectives of customized productivity-ratio (CPR) and production line arrangement (PLA) to simultaneously address optimal scheduling, operation, and design issues.

In this paper, we developed a new hierarchic CHS methodology based on the previous studies. The idea is that we allow the outer CHS will contain several pseudo inner job processing cycles. The inner pseudo cycles make the cyclic hoist movement independent from cyclic job movement, so that the job processing and hoist movement can be virtually separated. In the outer CHS, job processing and hoist movement will be finally synchronized. Such arrangement will help identify better opportunities for scheduling the load of multi-job processing tanks and thus bring better solutions of CHS. Based on the proposed idea, a new integer linear programming (MILP) model has also been developed. A CHS case with considerations of CPR and PLA has been studied. The efficacy of the proposed methodology is demonstrated by compared with the previous studies.

434530 Reactive Scheduling for Optimal Decoking Operation of Ethylene Cracking Furnace System

Abstract:

Thermal cracking furnace system is the most critical section of an ethylene plant, where multiple furnaces in parallel are used to break various hydrocarbon feedstock (e.g. naphtha, ethane, and propane) into smaller-molecular hydrocarbon products such as ethylene, propylene and methane. Based on the provided inputs including product yield coefficients with respect to multiple types of feedstock, furnace performance constraints, product sale prices, and feedstock/manufacturing costs, as well as designated operational constraints, an optimal decoking schedule for the entire ethylene cracking furnace system has to be determined in order to achieve the maximized economic performance. Due to its performance-decaying operational characteristic, the cracking furnace system has to experience periodic hot shutdowns, which consequently leads to decreased throughput for the downstream section and unwanted upsets. Apart from the hot shutdown, a longer outage is required to conduct major maintenance, which is called as cold maintenance. In order for a cold maintenance to occur, the project team will provide the estimated starting/ending times to the furnace operation team, according to which a feasible schedule will be conceived. Also, in practice, because the feed supply conditions change constantly, it is necessary to implement a reactive scheduling strategy which can smartly reschedule the furnace operations with respect to any new feed delivery.

The previously published works are either focused on simulation and control aspects, or simply scheduling the furnaces for maximum average daily profits without incorporating some realistic operational constraints. In the proposed scheduling model, most of the major scheduling issues concerning an ethylene cracking furnace system have been addressed, such as multiple feedstock, multiple furnaces, hot shutdowns for decoking operations, cold shutdowns for mechanical maintenance, decaying cracking performance, non-simultaneous shutdowns, split cracking and recycled C2/C3 cracking. Specially, to mitigate the throughput upsets caused by furnace shutdowns to the downstream section, the developed scheduling model considers different measures commonly applied in industry, such as additional furnace capacity makeup. Furthermore, the scheduling model is constructed in a reactive and iterative way to seamlessly make transitions from current schedule to reschedule under any delivery of new feeds and on a rolling forward scheduling horizon. The efficacy of the developed scheduling model is demonstrated by some case studies.

434903 Ethylene Splitter Startup Strategy Development through the Integration of Dynamic Simulation and DCS Control System

Abstract:

The startup of a distillation column is composed of a set of transitional operations. Not only the dynamic behaviors during the abnormal period should be studied and analyzed, but also the control strategy of the column needs to be tested and improved to help the column start up smoothly and quickly. The integration of dynamic simulation and DCS control system provides a more reliable platform for enhancing column operating procedures and control strategy. The automation built in DCS control system can guarantee each operation conducted successfully. Finally, operation time and energy savings have an appreciable economic significance.

In this study, rigorous dynamic simulations are employed to investigate the startup of ethylene splitter in an ethylene plant. First, in the base case, the dynamic model for the column is built with normal control strategy. Second, dynamic model for ethylene splitter is built with improved control strategy. Split range control and low select control are applied in this improved control strategy. Then, the optimal startup procedures are identified. The control system built in DCS control system is integrated with dynamic model. In summary, the integration of dynamic simulation and DCS control system in this study will help identify operating procedures and control strategies to start up a distillation column quickly and safely, so as to save plant raw material and energy, reduce operational risks, and thus enhance the industrial sustainability.

435024 Optimal Synthesis of Natural Gas Reforming Based Hydrogen Production

Abstract:

In this work, we present a methodology to synthesize globally minimum energy consumption hydrogen production networks, which are allowed to employ multiple reformers, separators, and heat exchangers. The synthesis problem is formulated within the Infinite DimEnsionAl State-space (IDEAS) conceptual framework, which is shown to be applicable to the problem at hand. Within the IDEAS framework, network synthesis is formulated as an infinite dimensional linear optimization problem. The IDEAS conceptual framework is realized through solution of a series of finite dimensional linear programs whose optimum values converge to the infinite program's infimum. Global optimality is established by demonstrating that the input–output information maps of the reformers, separators, and heat exchangers satisfy all properties required for the application of the infinite dimensional state-space (IDEAS) approach to the hydrogen production network synthesis problem. The power of the proposed methodology is demonstrated on a case study featuring a variety of reactors and separators, including Gibbs equilibrium reactors, and equilibrium flash separators.

436513 Modeling of Batch Process for Tributyl Citrate Production

Abstract:

Demand of free-phthalate plasticizers has increased due to the concern about the effect that could have phthalate compounds in the human health and the trend in market to make bio-based products. Taking into account that the tributyl citrate is a potential compound to replace the phthalate plasticizers, mainly in applications that are in contact with the people like toys, medical devices and food packing, one industrial batch process for it production has been modeled.

The tributyl citrate production process involves three consecutive reactions in equilibrium, where mono and dibutyl citrate like intermediate products and water like sub-product are obtained. The process is performed in a reactor coupled to a distillation column and a decanter, the reaction take place in the liquid phase and currently is doing at constant temperature. During the operation, the vapor phase generated in the reactor that is composed by water and butanol is carried to distillation column. In the bottom of this column, butanol is obtained and recirculated to the reactor, whereas in the top, the result is a mixture water-butanol with a composition near to the binary azeotropic (heterogeneous) point. This stream is cooled and sent to the decanter, where, due to the liquid-liquid equilibrium effects, two liquid phases are obtained. Organic phase that is rich in butanol is used like reflux for the column and the aqueous phase is removed of the system.

Considering the potential that have the tributyl citrate like substitute of some phthalate plasticizers due to it is a bio-based and no-toxic product, and the trends in the plasticizers market for the next years, we have done a model for simulate this production process, with the objective to improve your productivity and thus increasing the competitiveness of this compound with respect to phthalates in terms to production costs. With the implementation of the model, we have reduced the energy consumption by kilogram of tributyl citrate and the production time by each batch.

435064 A Strategy for Improving the Competitiveness of Small Scale Biofuel Supply Chain

Abstract:

Biofuels are emerging as an important alternative to environmentally hazardous fossil fuels as a transportation fuel. To accelerate the introduction of biofuels in the current transportation fuel market dominated by fossil fuels, it is very important to supply as many customers as possible in the most economical way. The resulting biofuel supply chain problems have been actively addressed in recent years. However most previous works were concerned with designing the overall supply chain. The operational issue of the supply chain has been relatively neglected in spite of the importance. In this paper, a biofuel supply chain problem is transformed into a distribution problem and its computation method based on Ant Colony Optimization (ACO) algorithm is proposed. Some numerical examples are presented to illustrate the applicability of the proposed work with some remarks.

436707 Planning and Scheduling of Nanogrids with Multiple Generataion and Multiple Storage

Abstract:

We will present summary of our recent works including a book on the planning and scheduling of nanogrids with multiple generations and multiple storages [to complete].

414995 Numerical Simulation of Dense Gas-Solid Flow in a Loop Seal Based on an Improved EMMS Bubbling Model

Abstract:

A loop seal, as one of the most widely used non-mechanical valves in circulating fluidized beds (CFB), serves as a recycle device allowing the transfer of the solids from the high pressure standpipe to the low pressure riser, but preventing the inverse flow of the gas from the riser bottom to the standpipe. A typical loop seal consists of a supply chamber and a recycle chamber as well as a bottom connection opening or slit in between. In general, the supply chamber is operated in the minimum fluidization state, while the recycle chamber is fluidized at low gas velocities and run in the bubbling fluidization regime.

Traditional two-fluid-model (TFM) approach, as the most frequently used simulation technique for gas-solid flow, generally assumes homogeneous conditions inside a control volume, hence giving rise to the calculation deviation of interphase drag force and further to the numerical inaccuracy of multi-phase simulation. Thereby, so-called energy-minimization multi-scale (EMMS) scheme was used to simulate local heterogeneous characteristics in the CFB system with a loop seal, in which a structure-dependent drag coefficient is calculated from the EMMS model and then integrated into the TFM approach by utilizing a user defined function (UDF). However, in the loop seal the particles represent the continuous dense emulsion phase with the formation of discrete bubble-like voids, so the EMMS scheme is still open to discussion since it is based on the concept of particle clusters.

In low-velocity fluidization, gas bubbles generally begin to form as a discrete phase. A gas-solid bubbling reactor can thus be resolved into three sub-systems including the dense emulsion phase, the dilute bubble phase and the inter-phase in between. By considering both the normal pressure stress and tangential drag force between the emulsion and bubble phases, the authors proposed a stability condition for gas-solid bubbling fluidization and formulated an EMMS bubbling model. This model has been proved to be able to reasonably simulate dense gas-solid flow without introducing bubble-specific empirical correlations.

In this article, a heterogeneity index is calculated from an improved EMMS bubbling model to measure the interphase drag coefficient and further integrated into the TFM approach to simulate the loop seal for a CFB system. In bubbling fluidization this index is dependent on superficial gas velocity, so a specific drag correction scheme based on the regional division of computational domain is proposed to allow the application of various heterogeneity index correlations to different regions of the loop seal, since superficial gas velocity may differ much from the recycle to supply chamber because of the variation of superficial gas velocity itself as well as the resultant effect of the circulated solids on the actual gas flow in the loop seal. According to the uncertainty of gas flow pattern in the loop seal, several typical cases of drag correction are comparatively investigated by applying various gas velocities and/or drag models in the recycle and supply chambers. The simulation results indicate that the drag coefficients in the recycle and supply chambers should be calculated from the EMMS bubbling drag model at their respective superficial gas velocities. By incorporating with the TFM approach, the preceding optimal scheme is proved to be able to reasonably predict the parametric effects on gas inverse flow in the loop seal with an inclined internal baffle.

420134 A Transient Thermal Near-Wellbore and Completion Hydraulics Simulator for Completion Design

Abstract:

In thermal recovery, reservoir and wellbore temperature distributions are important parameters that must be calculated precisely. Most drilling and completion simulators use an overall constant heat transfer rate for the reservoir-wellbore heat balance. This approach is not accurate for thermal applications, as temperature profiles and thermal properties of the fluids change both in space and time. This article describes the development of a combined wellbore-near well region simulator capable of fast and accurate modeling of temperature and flow profiles in the stated region without the difficulties of building a three-dimensional (3D) reservoir model.

This work discusses the development of a coupled combined wellbore-near well region thermal simulator for completion design in thermal recovery fields. It consists of two parts: a two-dimensional (2D) wellbore simulator and a thermal hydraulic near-wellbore reservoir model assuming elliptical symmetry of the temperature and flow profiles around the wellbore. The simulation domain is fully discretized along the longitudinal and radial directions. Heat and mass transfer profiles in the wellbore and formation are explicitly coupled at each time step.

Reservoir models usually use rectangular grids, while the wellbore proximity (3 to 10 m) suggests radial topology is more optimal for thermal recovery modeling. The developed thermal simulator accounts for the dependence of fluid properties and thermal conductivities on both pressure and temperature. Wellbore heat transfer and production varies along the well length and with time and should consider the completion type. For testing and validation of the developed thermal solver, a 3D finite volume computational fluid dynamics (CFD) simulator was used, and good agreement was demonstrated in tested cases. For one of the validation examples, an injection of hot water into the formation was modeled. The simulator accurately predicted both the propagation of the water front and temperature field variation with the time. In contrast to many traditional wellbore simulators that account only for thermal conductivity in the reservoir, the developed simulator also considers convective heat transfer (by the movement of the fluids), making it applicable for lower completion zones.

In completion design simulations, the local near-wellbore scale of thermal recovery methods often does not justify the application of a full-field 3D reservoir simulator. Because of the 2D topology used, the proposed simulator exhibits excellent stability and speed, which are usually the weak points of thermal reservoir simulators, and provides engineers with an easy tool for modeling fluid and heat exchange between the reservoir and a wellbore over time.

(264c) Numerical Modelling and Simulation of Semi-Batch Free Radical Polymerization Reactor Operation Using Analytical Solution with AK Gel Model

Garg, D., Shiv Nadar University
Serra, C., University of Strasbourg
Hoarau, Y., University of Strasbourg
Originally presented on: 11/10/2015 9:06:00 - 9:24:00

Numerical Modelling and Simulation of Semi-Batch Free Radical Polymerization Reactor Operation Using Analytical Solution with AK Gel Model

Dhiraj Kumar Garg[a], Christophe A. Serra[b], Yannick Hoarau[c]

[a]Dept. of Chemical Engineering, Shiv Nadar University, dhiraj.garg@snu.edu.in, [b]ICS (UPR 22 CNRS) & University of Strasbourg, ca.serra@unistra.fr, [c]ICUBE (UMR 7357 CNRS) & University of Strasbourg, hoarau@unistra.fr

Introduction

There is a tremendous and growing need of improving numerical modelling and simulation of various physical and chemical phenomena to save precious resources like time, material, man-power and money required for performing costly physical experiments. Polymerization is one of them.

Batch and continuous reactors are the two most common types of reactors that the industry employs (Levenspiel, 1999). Batch reactor is transient in nature whereas continuous reactors are generally steady-state reactors. Batch reactor can undergo the variation of temperature and/or pressure along with the variation of concentration of chemical species during transient operation. But there is another category of reactors i.e. semi-continuous or semi-batch reactor, which becomes important in certain conditions. Semi-batch operation might be required to control certain aspects of chemical reactions and/or physical properties of chemical products as in case of polymerization etc. Thus semi-batch provides certain type of freedom to control the chemical reaction which otherwise is not available in batch and continuous reactors.

Modelling and simulation of ideal batch and continuous reactors is well known. But in real operation, various types of non-ideality may arise e.g. concentration gradient of various chemical species in bulk in batch and continuous stirred tank reactor (CSTR) or radially in plug flow reactor (PFR), temperature gradients etc. The modelling of the chemical process may itself be difficult or restrictive in nature. This generally happens due to lack of proper theoretical understanding, lack of adequate computational resources etc. Free radical polymerization is one such example. Thus a good modelling with a sound theoretical base is always highly desirable which could be simulated easily and whose predictions are near to the experimental data. In absence of such adequate theoretical model, semi-empirical or empirical models are devised to cater the need. Semi-batch reactor presents various and unique difficulties not only in terms of physical experiment and prediction of its output but also in terms of its modelling and simulation.

In this work, we have studied the numerical modelling and simulation of semi-batch operation of free radical polymerization (FRP) using analytical solution of FRP with AK gel model (Achilias and Kiparissides, 1992). The semi-batch operations studied are step change in reactor temperature and step addition of initiator at a certain time during reaction.

Work done so far

An analytical solution (AS) was obtained for FRP in our previous work (Garg et al. 2014a). The mathematical model was based on the method of moments which was well proven method to predict monomer conversion (XM), number average molecular weight (MWn) and weight average molecular weight (MWw) and thus polydispersity index (PDI). The kinetic scheme for FRP included the elementary steps of initiator dissociation and initiation, propagation, termination by combination, termination by disproportionation, transfer to monomer, transfer to solvent and transfer to chain transfer agent (CTA). AS was obtained for isothermal, homogeneous, variable volume batch reactor for solution homopolymerization before gel, glass and cage effect. A lot of simplification in the AS derivation was obtained using a useful assumption of quasi steady state (QSSA) for the live radical polymer chains production. AS was thoroughly validated against numerical solution as well as published experimental data for several monomer-polymer systems with different initiators, solvent fractions and temperature variations before gel effect sets-in.

AS was then extended to complete range of conversion using two gel models (Garg et al 2014b; 2014c) namely ? CCS model (Chiu, Carratt and Soong, 1983) and AK model. CCS is semi-empirical gel model based on free volume theory given by Vrentas and Duda (1977). In CCS model, various terms are clubbed into one term whose value or expression is obtained using curve fitting against experimental data. On the other hand, AK model retains those physical terms and whose values are obtained experimentally or theoretically for a given chemical species. AS was separately integrated successfully with CCS and AK models. The simulation results thus obtained matched quite well with numerical solution as well as experimental data for isothermal batch reactor.

A very less experimental data has been obtained for semi-batch reactor. Researchers have obtained the appropriate curve predicting the conversion, MWn and MWw using semi-empirical models like CCS models whose parameters were tuned by best curve fitting methods (Ray et al., 1995; Srinivas et al., 1996; Dua et al. 1996). Hence the utilization of these models was quite limited to the operating conditions and specific operations for which they were obtained. There were two types of semi-batch operations that were carried out ? one with the step change in temperature at a certain time (Srinivas et al., 1996) and other with step addition of initiator at a certain time (Dua et al. 1996). The step change in temperature is both sides, i.e. increase as well as decrease of reactor temperature. Carrying out these operations at different magnitudes as well as different times and predicting the changes in conversion and polymer properties is really a difficult task. Previous researchers were successful in obtaining the curves using the above mentioned model successfully for one type of task only, i.e. they obtained different model parameters values for temperature and initiator concentration step change separately. But as already mentioned, it had little applicability for practical applications.

Current Work

We had applied AS integrated with AK model for modelling and simulating semi-batch reactor for the same operations. Only two parameters need to be adjusted to match the experimental data and once fixed, the AS can predict quite successfully the change in conversion, MWn and MWw for both types of operations of step change in temperature as well as in initiator. The preliminary results are presented in Fig. 1 and 2. The blue and green lines represent the numerical solution for batch reactor in absence of any change in initiator concentration (as in Fig. 1) or in temperature (as in Fig. 2). The red line represents the curve by AS using AK model. The red circles represent the experimental data. It can be seen that AS predictions match quite well with the experimental data. This is a great simplification as it implies the sound theoretical basis and validity of the model and thus allows greater applicability to practical purposes like process control, process optimization etc.

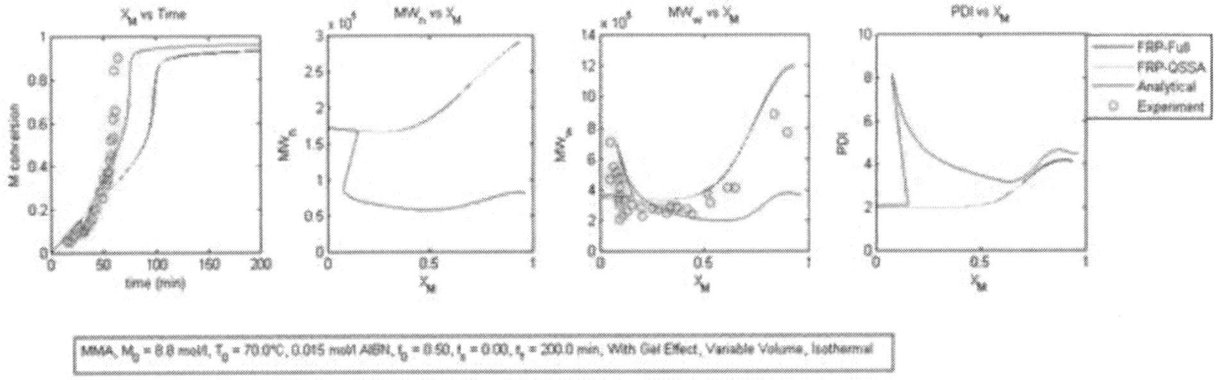

Fig. 1 ? Results for sudden addition of initiator to existing reacting solution at t = 28 min.

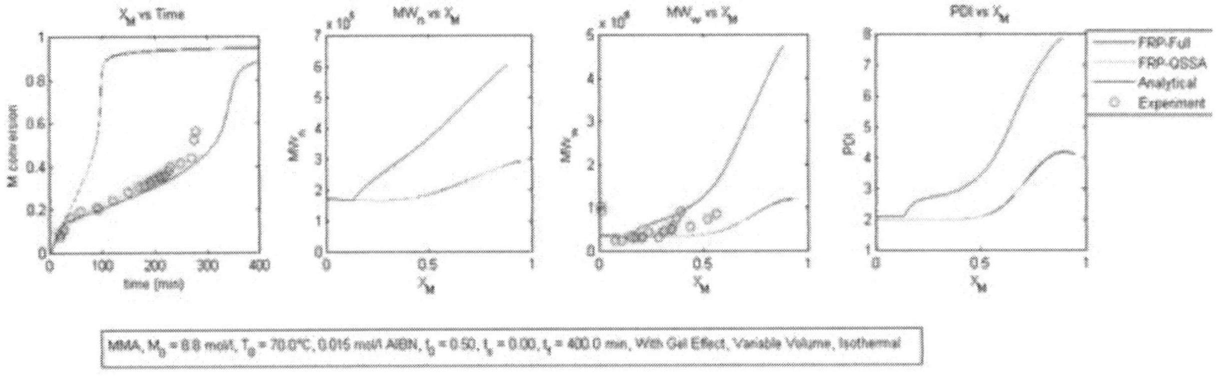

Fig. 2 ? Results for sudden drop of temperature from 70°C to 50°C at t = 25 min.

References

1) Levenspiel O (1999) Chemical Reaction Engineering. John Wiley & Sons, New York

2) Achilias, D. S.; Kiparissides, C., Development of a General Mathematical Framework for Modeling Diffusion-Controlled Free-Radical Polymerization Reactions. Macromolecules 1992,

25 (14), 3739-3750.

3) Garg DK, Serra CA, Hoarau Y, Parida D, Bouquey M, Muller R (2014a) Macromolecules 47(14):4567-4586

4) Chiu, W. Y.; Carratt, G. M.; Soong, D. S., A Computer-Model for the Gel Effect in Free-Radical Polymerization. *Macromolecules* **1983,** *16* (3), 348-357.

5) Garg DK, Serra CA, Hoarau Y, Parida D, Bouquey M, Muller R (2014b) Macromolecules 47(23):8178-8189.

6) Garg DK, Serra CA, Hoarau Y, Parida D, Bouquey M, Muller R (2014c) Macromolecules 47(21):7370?7377.

7) Vrentas JS, Duda JL (1977) J Polym Sci Polym Phys 15(3):403-416

8) Ray AB, Saraf DN, Gupta SK (1995) Polymer Engg and Science 35(16):1290-1299

9) Srinivas T, Sivakumar S, Gupta SK, Saraf DN (1996) Polymer Engg and Science 36(3):311-321

10) Dua V, Saraf DN, Gupta SK (1996) Journal of Applied Polymer Science 59:749-758.

431146 Efficient Computation of Cyclic Steady States in Periodic Adsorption Processes Using the Jfnk Method

Abstract:

Efficient Computation of Cyclic Steady States in Periodic Adsorption Processes Using the JFNK Method

Richard Pattison[a], Pieter Schmal[b], and Costas Pantelides[b]

[a]McKetta Department of Chemical Engineering, The University of Texas at Austin

[b]Process Systems Enterprise Ltd.

email: rcpattison@utexas.edu

Periodic adsorption processes (PAPs) are attractive alternatives to traditional distillation and absorption separation systems due to their relatively low capital costs and energy requirements [1]. PAPs have found many industrial applications [2], and recently have gained attention as an option for hydrogen purification for electricity generation and other purposes [3-4]. Optimizing the design of these processes is critical to realizing the potential cost and energy savings.

The modeling, simulation and optimization of PAPs is challenging for several reasons. The mathematical models of these systems are described by nonlinear partial differential algebraic equations (PDAEs), with variation in both temporal and spatial domains, and the boundary conditions change several times during each cycle simulation. A particularly challenging aspect is the need to compute the cyclic steady state (CSS) behavior of a given PAP, i.e., the state in which the variable trajectories are identical in successive cycles. This is of the most practical interest for industrial purposes.

The conventional method for calculating the CSS is to simply integrate the process model until there is no change in the state trajectories between iterations. While this is an effective and proven method, it can often require the integration of many cycles (e.g., up to 4000 have been reported [5]) to reach the CSS. Other methods have proposed formulating the problem as a system of nonlinear equations which require the values of the state variables at the beginning and end of each cycle to be identical. However, the relation between these two sets of variable values is determined by the temporal integration of a PDAE system over a cycle; calculating the Jacobian that is required for each Newton iteration is computationally expensive, and for large systems it may be practically intractable [6]. An alternative approach is to simultaneously discretize both the temporal and spatial variations within the underlying PDAEs, thereby obtaining a very large set of nonlinear algebraic relations which can then be solved using a sparse solver [7]. However, this imposes severe demands on both storage and computation; moreover, using a fixed discretization scheme may fail to result in the required accuracy of solution.

In this work, we propose an algorithm for computing the CSS in PAPs based on the Jacobian-Free Newton-Krylov (JFNK) [8] method. The latter provides the benefit of super-linear convergence without the need to form or store the actual Jacobian. The linear system required for the Newton update is solved by an iterative linear solver which uses accurate approximations of Jacobian-vector products to minimize the linear residual. The Jacobian-vector product required at each iteration is obtained via the simulation of a single cycle of the PAP.

Whilst JFNK methods are conceptually simple, their efficient operation is predicated on obtaining sufficiently fast convergence of the iterative linear solver. This is typically achieved via the choice of an appropriate preconditioner. We propose a novel preconditioner that can result in significant improvements in performance both for single "one-off" CSS determination and in the context of parametric studies where a sequence of CSSs need to be calculated for different values of process design and operation parameters.

We illustrate the benefits of the proposed method in the simulation of PAPs using a rapid pressure swing adsorption (RPSA) process and a multi-bed PSA process as case studies.

References

[1] Nikolic, D., Giovanoglou, A., Georgiadis, M.C., and Kikkinides, E.S. Generic modeling framework for gas separations using multibed pressure swing adsorption processes. *IECR*, **47**, 3156-3169, **2008**.

[2] Ruthven, D.M., Farooq, S., and Knaebel, K.M. *Pressure Swing Adsorption*. VCH Publishers: New York, **1994**.

[3] Bastos-Neto, M., Moeller, A., Staudt, R., Böhm, J., and Gläser, R. Dynamic bed measurements of CO adsorption on microporous adsorbents at high pressures for hydrogen purification processes. *Separation and Purification Technology*, **77**, 251-260, **2011**.

[4] Delgado, J.A., Águeda, V.I., Uguina, M.Á., Sotelo, J.L., Brea, P., and Grande, C. Adsorption and diffusion of H2, CO, CH4 and CO2 in BPL activated carbon and 13x zeolite: evaluation of performance in PSA hydrogen purification by simulation. *IECR,* **53,** 15414-15426, **2014**.

[5] Siettos, C.I., Pantelides, C.C., and Kevrekidis, I.G. Enabling dynamic process simulators to perform alternative tasks: A time-stepper-based toolkit for computer-aided analysis. *IECR*, **42**, 6795-6801, **2003**.

[6] Vetukuri, S.R.R., Biegler, L.T., and Walther, A. An inexact trust-region algorithm for the optimization of periodic adsorption processes. *IECR*, **49**, 12004-12013, **2010**.

[7] Nilchan, S. and Pantelides, C.C. On the optimisation of periodic adsorption processes. *Adsorption*, **4**, 113-147, **1998**.

[8] Knoll, D.A., and Keyes, D.E. Jacobian-free Newton–Krylov methods: a survey of approaches and applications. *Journal of Computational Physics*, **193**, 357-397, **2004**.

431148 3D Simulations of Fracture Dissolution Using the Openfoam Toolkit

Tuesday, November 10, 2015: 9:42 AM
Salon D (Salt Lake Marriott Downtown at City Creek)

Vitaliy Starchenko and Anthony J.C. Ladd, Chemical Engineering, University of Florida, Gainesville, FL

Although fractures take up only a small fraction (1-2%) of the available pore space in rocks they account for the majority of the subsurface transport of minerals and contaminants. Water percolating through carbonate formations will tend to dissolve the surrounding rock, due to the acidity of the dissolved CO_2. The feedback between the flow of reactant and the increase in permeability due to dissolution of the fracture faces gives rise to a number of interesting and important questions, relating to oil and gas recovery, sequestration, dam stability and the development of caves.

Numerical models of fracture dissolution typically take averages of the flow and reactant concentration across the fracture aperture, giving rise to simplified field equations that can be readily solved on a one or two-dimensional grid. However, fracture dissolution is inherently unstable and highly localized flow paths or wormholes develop [1]. Figure 1 shows a developing wormhole from an initially flat fracture, with a small region of enhanced aperture near the center of the inlet. As the wormhole grows the aperture-averaging approximation breaks down. To study the later development of wormholes, the effects of inertia and turbulence, and the competition between different flow paths a fully three-dimensional model is desirable. Here we report some preliminary developments using the OpenFOAM [2] toolkit.

The simulations solve the Stokes equation for flow in the fracture (although it is straightforward to include fluid inertia) and the convection-diffusion equation for the reactant transport. Reactions at the fracture surfaces are modeled by linear or power-law kinetics and the rates of dissolution computed. The dissolution flux is then used to modify the positions of the surfaces. The additional libraries and modifications to the OpenFOAM source needed to perform these simulations will be outlined. The major difficulty is preserving sufficient mesh quality as the fracture opens; to accomplish this we have implemented a customized source code to relax the mesh on the fracture surfaces.

Results for uniformly dissolving fractures show an excellent agreement between 2D and 3D models. At the same time first we find that the quality of the mesh and the implementation of the boundary conditions are extremely important for the resulting solution. At the meeting we will present the results of ongoing simulations of more complex fracture systems.

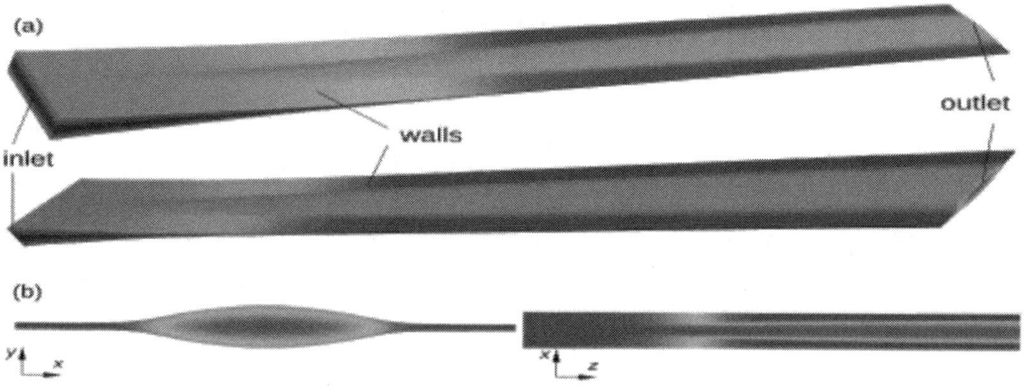

Fig 1. (a) A three-dimensional simulation of fracture dissolution based on the OpenFOAM libraries. A flat fracture was seeded with a small region of enhanced aperture at the center of the inlet. The geometry and concentration are shown after breakthrough, when the reactant has reached the outlet in significant quantities. The blue color indicates regions of low concentration and red the regions of high concentration. (b) 2D slices through the fracture, perpendicular (left) and parallel (right) to the flow.

[1] Szymczak and A. J. C. Ladd. *The initial stages of cave formation: Beyond the one-dimensional paradigm, Earth Planet. Sci. Lett.,* ***2011****, 301, 424-432.*

431633 Scale Bridging and Uncertainty Propagation in Chemical Process Modeling with Bayesian Nonparametric Regression

Abstract:

Bayesian nonparametric regression is a powerful method for building reduced models of chemical reactions and distributed systems. Motivated by Takens' theorem in dynamic systems topology, Gaussian process-based stochastic functions are insinuated into chemical system models, leading to a stochastic dynamic system of drastically reduced order. Karhunen-Loeve decomposition of the GP function kernels leads to calibration of the models to training data sets using standard approaches. Low-order, calibrated stochastic models then serve as vehicles for propagation of uncertainty across modeling length scales. This methodology can be applied to problems in design of bench and pilot-scale experiments for process design, machine learning in chemical process control, or in other settings where reduced modeling is required. Benchmarking examples derived from amine-based carbon capture and steam reformation of methane will be presented.

432337 Nonlinear Resonances and Anti-Resonances in Engineered Granular Crystals

Abstract:

Engineered granular crystals (EGCs) are closely packed assemblies of particles, contained inside a matrix, interacting through nonlinear, tensionless potentials. They are known as a flexible computational and experimental testbed for nonlinear dynamics [1] with a large range of applications, including energy absorbing layers [2], acoustic lenses [3], and rectifiers [4]. The flexibility of EGCs comes from the large number of tunable parameters in the system. The building blocks of an EGC are macroscopic particles, usually of spherical or cylindrical shape, and by changing the material properties, arrangement and shape of these building blocks, we can modify the dynamic response of the system, and therefore the energy transfer through the system. By applying external forcing to the granular crystal, we can observe periodic, quasi-periodic and chaotic responses.

EGCs belong to a broader category of media that bear a crystalline lattice structure, where the understanding of the underlying linear spectrum is very important for their linear wave transmission properties. However, recently there has been an increased interest in systems that do not possess a linear spectrum and EGCs are a perfect example since their response can be tuned from essentially linear, to weakly and highly nonlinear, depending on the external precompression applied on the crystal.

In this highly nonlinear setting we conduct a bifurcation study of a finite granular crystal with external periodic forcing, using techniques from iterative linear algebra to compute fixed points of the stroboscopic map that correspond to periodic solutions. Combining this method with continuation techniques [5] we compute one parameter branches of periodic solutions as well as two parameter branches of the bifurcations that occur. Despite the absence of a linear spectrum we identify resonant periodic propagation whereby the crystal responds at multiples of the forcing period and corresponding to local maxima of transmitted force at its right fixed boundary. In addition, we identify anti-resonances in-between resonances corresponding to minima of transmitted force. We identify the complex bifurcation diagram involving period doublings, as well as isolas (of period 3-, 4-, 5- solutions), and study the stability of such states upon frequency and amplitude variations of the external forcing. Furthermore, computing the unstable manifolds of saddle fixed points we are able to determine parameter regions with multiple attractors. This study is a stepping stone in understanding how the construction of an EGC affects the energy transfer through the system, which will allow us subsequently to design crystals that transmit/block signals with specific frequency ranges.

References:

[1] Nesterenko V. F., Dynamics of Heterogeneous Materials (Springer-Verlag, New York) 2001.

[2] C. Daraio, V. F. Nesterenko, E. Herbold, and S. Jin, "Energy Trapping and Shock Disintegration in a Composite Granular Medium," Physical Review Letters, vol. 96, p. 058002, Feb. 2006.

[3] A. Spadoni and C. Daraio, "Generation and control of sound bullets with a nonlinear acoustic lens," Proceedings of the National Academy of Sciences, vol. 107, pp. 7230–7234, Apr. 2010.

[4] N. Boechler, G. Theocharis, and C. Daraio, "Bifurcation-based acoustic switching and rectification," Nature materials, vol. 10, pp. 665-668, Sept. 2011.

[5] E. Doedel, H. B. Keller, and J. P. Kernevez, "Numerical analysis and control of bifurcation problems (I): Bifurcation in finite dimensions," International Journal of Bifurcation and Chaos, vol. 1, no. 03, pp. 493–520, 1991.

432930 A Rejection-Free Markov Chain Monte Carlo to Simulate the Dynamics of Molecular Systems

Abstract:

Monte Carlo methods are today well established as numerical tools for unraveling the complex behavior of a large variety of systems.[1-4] The broad class of MC algorithms can be divided into two main groups: methods built-up on the Metropolis scheme (MMC) and kinetic Monte Carlo methods (KMC).

The latter apply to systems evolving dynamically from state to state. As such, they can be used when the number of states is discrete and the rate constants for the state-to-state transitions are known.[5] They are therefore not suitable for molecular systems evolving in a continuous space characterized by an infinite and unknown number of states. In such cases one typically has to revert to MMC.

In MMC the system evolves through randomly sampled configurations which are accepted or rejected according to specific rules which ensure that the distribution of sampled configurations satisfies the Boltzmann equiprobability principle. The possibility to let the system evolve through unphysical trajectories makes MMC algorithms particularly efficient for calculating equilibrium properties. On the other hand, the price to pay in MMC is the lack of a physically meaningful time scale. To date, in fact, with the exception of a few cases (e.g. Brownian and spin dynamics), MMC methods do not provide information about the true time evolution of molecular systems.[3]

In this contribution a new MC algorithm for molecular systems is proposed that overcomes these limitations and allows simulating the correct dynamics of molecular systems too.

It is first shown that the equations of motion can be written equivalently as a set of equations describing a chemical reaction network, i.e., one can identify the current state of the system with the paths traveled by the atoms. This identification allows to use a KMC scheme, namely the stochastic simulation algorithm,[6] where the state-to state transitions are determined on-the-fly and a direct link between the MC time step and the true physical time is established by using propensities proportional to the velocities. The method is applied to the classical problem of hard disks and validated against event driven Molecular Dynamics. The position probabilities as well as the relaxation time of the global orientational order parameter and of the velocities converge to the correct results for decreasing step size. Additionally, the Maxwell velocity distribution is recovered directly without any need of independent samplings, as instead in traditional MMC.[7]

In perspective, the new algorithm can provide a unified numerical framework to simulate the dynamics of systems characterized by a competition between conformational relaxation and chemical reactions, as, for example, in organic light emitting diodes and enzyme catalysis.[8,9]

References

[1] Potestio et al., *Physical Review Letters*, 111, 060601 (2013).

[2] Meimaroglou D. and Kiparissides C., *Industrial & Engineering Chemistry Research*, 53, 8963-8979 (2014).

[3] Landau D.P. and Binder K., *Monte-Carlo Simulations in Statistical Physics*, Cambdridge University Press (2009).

[4] Zuniga et al., *Physical Review Letters*, 114, 155301 (2015).

[5] Voter A.F., *Introduction to Kinetic Monte Carlo Method*, in Radiation Effects in Solids, Springer (2005).

[6] Gillespie D.T., *Journal of Computational Physics*, 22, 403-434 (1976).

[7] Costa L.I., *manuscript in preparation*.

[8] Daub J. et al., *The Journal of Physical Chemistry A*, 105, 5655-5665 (2001).

[9] Wolf-Watz M. et al., *Nature Structural & Molecular Biology*, 11, 945-949 (2004).

420112 Improved *a Priori* Probabilistic Guarantees for Robust Counterpart Optimization

Abstract:
The parameters of applied mathematical models often have more than one possible value which they can achieve due to limited information or measurement error. The solutions and objective values produced from optimizing models with uncertain parameters can vary greatly based on which values the uncertain parameters realize. One method of handling uncertain parameters is to guarantee feasibility of the constraints for all possible parameter sets contained within deterministically defined parameter spaces, called uncertainty sets; the corresponding optimization model is known as a robust counterpart. If every uncertain parameter has a bounded range of possible values, the so-called "worst-case" robust counterpart can be solved [1], where each constraint's uncertainty set is defined to include all possible parameter sets; the probability that a parameter set would realize values rendering an optimal solution infeasible is zero. Less conservative solutions can also be obtained for models with bounded or unbounded uncertain parameters, where the uncertainty sets can be defined with a corresponding probability of constraint violation greater than zero [2-6]. The quality of an optimal solution of the robust counterpart relies heavily on methods that define uncertainty sets which are guaranteed to satisfy a particular upper bound on the probability of constraint violation; a tighter probabilistic bound would allow the imposition of a smaller uncertainty set with the same guarantee of feasibility and can drastically improve objective values.

We present new *a priori* bounds on the probability of constraint violation for defining uncertainty sets. The methods apply and are specially tailored to different types of uncertainty sets currently in use [6-9]. Key to the derivation of each bound is a novel theorem on fitting objects of alterable size for geometric tailoring. Situations for which the bounds are applicable include (i) bounded uncertain parameters with unknown probability distributions [6-8,10], (ii) bounded and unbounded uncertain parameters with known probability distributions [10-12], and (iii) bounded, possibly asymmetrically distributed uncertain parameters with limited information on their means, thus enabling usage of parameters matching the latter scenario for the first time. The novel bounds yield smaller uncertainty sets with the same probability of constraint violation when compared with existing methods. These methods can provide greatly improved objective values and also improve the performance of iterative algorithms which rely on *a priori* and *a posteriori* bounds to obtain less conservative solutions [13].

References:

1. Soyster, A. L. (1973). Convex programming with set-inclusive constraints and applications to inexact linear programming. *Operations research*, *21*(5), 1154-1157.

2. El Ghaoui, L., & Lebret, H. (1997). Robust solutions to least-squares problems with uncertain data. *SIAM Journal on Matrix Analysis and Applications*, *18*(4), 1035-1064.

3. El Ghaoui, L., Oustry, F., & Lebret, H. (1998). Robust solutions to uncertain semidefinite programs. *SIAM Journal on Optimization*, *9*(1), 33-52.

4. Ben-Tal, A., & Nemirovski, A. (1998). Robust convex optimization. *Mathematics of Operations Research*, *23*(4), 769-805.

5. Ben-Tal, A., & Nemirovski, A. (1999). Robust solutions of uncertain linear programs. *Operations research letters*, *25*(1), 1-13.

6. Ben-Tal, A., & Nemirovski, A. (2000). Robust solutions of linear programming problems contaminated with uncertain data. *Mathematical programming*, *88*(3), 411-424.

7. Bertsimas, D., & Sim, M. (2004). The price of robustness. *Operations research*, *52*(1), 35-53.

8. Chen, X., Sim, M., & Sun, P. (2007). A robust optimization perspective on stochastic programming. *Operations Research*, *55*(6), 1058-1071.

9. Li, Z., Ding, R., & Floudas, C. A. (2011). A comparative theoretical and computational study on robust counterpart optimization: I. Robust linear optimization and robust mixed integer linear optimization. *Industrial & engineering chemistry research*, *50*(18), 10567-10603.

10. Li, Z., Tang, Q., & Floudas, C. A. (2012). A comparative theoretical and computational study on robust counterpart optimization: II. Probabilistic guarantees on constraint satisfaction. *Industrial & engineering chemistry research*, *51*(19), 6769-6788.

11. Paschalidis, I., Kang, S., & Li, K. (2008). Distribution-dependent robust linear optimization with asymmetric uncertainty and application to optimal control. In *Proceedings of the 17th World Congress IFAC*.

12. Kang, S. C., Brisimi, T. S., & Paschalidis, I. C. (2013). Distribution-dependent robust linear optimization with applications to inventory control. *Annals of Operations Research*, 1-35.

13. Li, Z., & Floudas, C. A. (2014). A Comparative Theoretical and Computational Study on Robust Counterpart Optimization: III. Improving the Quality of Robust Solutions. *Industrial & Engineering Chemistry Research*, *53*(33), 13112-13124.

427522 Integrated Process Design and Control: Novel Applications in Energy and Environment

Abstract:

Integrated Process Design and Control: novel applications in energy and environment

Sharifzadeh Mahdi[1], Thornhill Nina F., Shah Nilay.

Centre for process Systems Engineering (CPSE), Imperial College London.

Process design and control share important decisions. When the process specifications are decided there is little room left to improve control performances. Therefore, it is highly recommended that process design and control design should be considered simultaneously. A systematic framework is needed to automatically generate alternative design and control solutions and screen them based on various competing/conflicting economic and control objectives. Such a systematic framework conforms to mixed integer optimization programming under uncertainty.

The challenge is that the integrated approach often leads to large-scale highly-combinatorial mathematical formulations which pose significant challenge for current optimization technology. This presentation provides an overview of our recent endeavours [1-8] at Centre for Process Systems Engineering (CPSE), Imperial College London, for research into integrated design and control of industrial processes. The research objective is to reduce the problem complexity and apply the methodology to challenging problems with energy and environmental applications.

We introduce a methodology for complexity reduction from Integrated Process Design and Control (IPDC) based on the notion of perfect control and process model inversion. Here, the treatment is based on the property that the inverse solution of process model can be used for evaluating the best achievable control performance. We argue that the resulting optimization framework benefits from several desirable properties. Firstly, the complexities associated with control parameterization are disentangled from the problem formulation. Secondly, simultaneous optimization of the control structure and process economy results is self-optimizing properties. Finally and most importantly, the steady-state and dynamic formulations of the so-called *inversely controlled process model (ICPM)* ensure steady-state operability and functional controllability, respectively.

We demonstrate the application of the proposed methodology using several novel applications including integrated design and control of two-series heat-integrated reactors [2], an ETBE reactive distillation [3], retrofitting an existing NGCC power plant using a solvent-based capture process [5], a liquefied natural gas (LNG) production plant [6], a micro-reactor for autothermal reforming of methanol [7] and integrated operation of renewable energy systems [8]. The features of interest include flexible operation of aforementioned processes under uncertain scenarios such as external disturbances or load reduction. Furthermore, the applied optimization solution algorithms to implement the proposed methodology are elaborated and discussed.

References

[1] **Sharifzadeh M***, (2013). Integration of process design and control: a review, *Chemical Engineering Research and Design*, 91 (12), 2515–2549 (Link).

[2] **Sharifzadeh M***, Thornhill NF, (2013). Integrated design and control using a dynamic inversely controlled process model. *Computers & Chemical Engineering,* 48, 121–134. (Link)

[3] **Sharifzadeh M***, (2013). Implementation of an inversely controlled process model for integrated design and control of an ETBE reactive distillation, *Chemical Engineering Science*. 92, 21–39. (Link)

[4] **Sharifzadeh M***, Thornhill NF, (2012).Optimal selection of control structures using a steady-state inversely controlled process model. *Computers & Chemical Engineering*, 38 (5), 126–138. (Link)

[5] **Sharifzadeh M***, Shah, N, (2015). Integrated retrofit and operation of solvent-based capture plants and natural gas combined cycle (NGCC) power plants: solvent performance comparison at industrial scale (invited paper, under review at AIChE Journal)

[6] **Sharifzadeh M***, Park J, Shah N, (2015). Integrated design and operation of liquefied natural gas plants: a case study, (in preparation).

[7] **Sharifzadeh M***, Sempou A, Shah N, (2015). Integrated design and operation of intensified processes: autothermal reforming of methanol using micro-reactors, (in preparation).

[8] **Sharifzadeh M***, Lubiano H, Shah N, (2015). Integrated design and operation of renewable energy systems under variable electricity load scenarios, (in preparation).

---

1 Corresponding author, Email: mahdi@imperial.ac.uk , **Address:** Room C603, Roderic Hill Building, Centre for Process Systems Engineering (CPSE), Department of Chemical Engineering, Imperial College London, South Kensington, London SW7 2AZ, UK.

420236 Natural Gas to Liquid Transportation Fuels: Process Synthesis Under Feedstock and Product Pricing Uncertainty Using Robust Optimization

Abstract:

Process synthesis of natural gas to liquid transportation fuels (GTL) has demonstrated the impact that the growing natural gas supply can have on the national energy landscape [1, 2]. In the search for domestic feedstocks to replace petroleum, natural gas is appealing due to low costs, growing supply, and the ability to operate at equal, or lower, levels of greenhouse gas (GHG) emissions as petroleum operations [1, 3, 4]. However, due to constantly changing energy markets, the pricing of natural gas, gasoline, diesel, and kerosene, can be unpredictable; design of refineries with uncertainty in mind is vital in order to ensure profitability. Robust optimization provides a key methodology for incorporating uncertainty into process synthesis of GTL plants. The robust counterpart framework incorporates the uncertainty of model parameters into model constraints in order to ensure that each uncertain constraint is not violated for all possible parameter values in the given uncertainty set [5,6,7].

As seen in previous studies, the primary methods for GTL conversion involve reforming to synthesis gas via autothermal reforming or steam reforming, before further upgrading to fuels via methanol synthesis and conversion or Fischer-Tropsch processes [1]. These conversion methods, in addition to other important refinery components including hydrogen and oxygen production, light gas handling, and wastewater treatment, are rigorously modeled for inclusion in a mixed-integer non-linear program (MINLP). A branch-and-bound algorithm is used to solve the MINLP to global optimality [8]. The superstructure also includes simultaneous heat and power integration [9]. The inclusion of uncertainty through robust optimization directly impacts the feedstock and product cost constraints, which must be converted from equality to inequality constraints for robust optimization theory to apply [10].

The resulting robust counterpart formulation allows for uncertainty to be handled with box, ellipsoidal, polyhedral, interval+ellipsoidal, or interval+polyhedral uncertainty sets [5]. The uncertainty of natural gas, butane, liquefied petroleum gas, electricity, gasoline, diesel, and kerosene pricing was characterized using historical data from the past five years, in order to derive nominal values and the maximum perturbation values to consider. Ninety-six case studies were conducted on a 50,000 barrel per day plant in order to compare the impact of each uncertainty set at probabilities of constraint violation ranging from 5% to 95% [6]. Using *a priori* and *a posteriori* bounds for the probability of constraint violation, an iterative method is utilized to guarantee the quality of the robust solutions [7]. The distinct impact of uncertainty on the overall profit and plant topology is discussed, with direct comparison to the nominal case study. Other key considerations in plant design, including GHG emissions and investment costs, are also presented.

[1] R. C. Baliban, J. A. Elia, and C. A. Floudas, "Novel natural gas to liquids processes: Process synthesis and global optimization strategies," AIChE Journal, vol. 59, no. 2, pp. 505–531, 2013.

[2] J. A. Elia, R. C. Baliban, and C. A. Floudas, "Nationwide, regional, and statewide energy supply chain optimization for natural gas to liquid transportation fuel (GTL) systems," Industrial & Engineering Chemistry Research, vol. 53, no. 13, pp. 5366–5397, 2014.

[3] U.S. Energy Information Administration, "United States natural gas industrial price," October 2014.

[4] U.S. Energy Information Administration, "Monthly biodiesel production report," April 2015.

[5] Z. Li, R. Ding, and C. A. Floudas, "A comparative theoretical and computational study on robust counterpart optimization: I. Robust linear optimization and robust mixed integer linear optimization," Industrial & Engineering Chemistry Research, vol. 50, no. 18, pp. 10567–10603, 2011.

[6] Z. Li, Q. Tang, and C. A. Floudas, "A comparative theoretical and computational study on robust counterpart optimization: II. probabilistic guarantees on constraint satisfaction," Industrial & Engineering Chemistry Research, vol. 51, no. 19, pp. 6769–6788, 2012.

[7] Z. Li and C. A. Floudas, "A comparative theoretical and computational study on robust counterpart optimization: III. improving the quality of robust solutions," Industrial & Engineering Chemistry Research, vol. 53, no. 33, pp. 13112–13124, 2014.

[8] R. C. Baliban, J. A. Elia, R. Misener, and C. A. Floudas, "Global optimization of a MINLP process synthesis model for thermochemical based conversion of hybrid coal, biomass, and natural gas to liquid fuels," Computers & Chemical Engineering, vol. 42, no. 0, pp. 64 – 86, 2012. European Symposium of Computer Aided Process Engineering.

[9] R. C. Baliban, J. A. Elia, and C. A. Floudas, "Simultaneous process synthesis, heat, power, and water integration of thermochemical hybrid biomass, coal, and natural gas facilities," Computers & Chemical Engineering, vol. 37, no. 0, pp. 297 – 327, 2012.

[10] X. Lin, S. L. Janak, and C. A. Floudas, "A new robust optimization approach for scheduling under uncertainty: I. Bounded uncertainty," Computers & Chemical Engineering, vol. 28, pp. 1069 – 1085, 2004. FOCAPO 2003 Special issue.

424509 Paroc - a Unified Framework Towards the Optimal Design, Operational Operation and Model-Based Control of Process Systems

Abstract:

The presence of uncertainty in process systems is one of the key reasons for deviation from set operation policies, resulting in suboptimal or even infeasible operation. As these uncertainties realize themselves on different time scales such as on a control, scheduling or design level, an integrated, comprehensive approach to consider uncertainty is required. Thus, in this contribution we demonstrate PAROC (PARametric Optimization and Control), a novel unified framework for the design, operational optimization and advanced model-based control of process systems, which decomposes this challenging problem into a series of steps, shown in the Figure below [1].

The first step comprises the formulation of a high-fidelity dynamic model of the original process, as well as its validation using various techniques such as parameter estimation and dynamic optimization. This model does not only serve as the first step in translating a real-world system into a set of equations, but also as a platform for the validation of any receding horizon policy developed.

While the high-fidelity model is in general applicable to design purposes, its complexity may render its use for the development of receding horizon policies computationally infeasible. Thus, in the second step, the validated high-fidelity model is reduced in complexity and size using system identification or advanced model reduction techniques, aiming at compromising the accuracy of the original model as little as possible [2]. This approach results in a discretized state-space model, which is used in the next step for the development of receding horizon policies such as control laws and scheduling policies [3, 4].

At this step, based on the discretized state-space model, the problem of devising a suitable receding horizon policy is formulated as a constrained optimization problem. Within our framework, this problem is solved offline employing multi-parametric programming, where the states of the system are treated as parameters and the constrained optimization problem is solved as a function thereof. Due to the parameter-dependence of the constraints, different solutions might be optimal in different parts of the parameter space. This results in a partition of the parameter space into different regions, called critical regions, and each region is associated with a corresponding optimal solution of the optimization problem as a function of the parameters. As a result we obtain the receding horizon policies explicitly as a function of the states of the system[1], and reduce the computational effort of their evaluation to a point location in the parameter space and a function evaluation.

However, when solving the receding horizon policies it is assumed that the values of the state vector are exactly known. As this might not be the case, e.g. due to noise, it is necessary to infer the state information from the available output measurements using a state estimator. While a long existing model-based technique for unconstrained state estimation is the Kalman filter, the use of constrained estimation techniques such as the moving horizon estimator (MHE) can lead to significant improvements of the estimation result by adding system knowledge [5, 6]. MHE is an estimation method that obtains the estimates by solving a constrained optimization problem given a horizon of past measurements. Thus similarly to the problem of receding horizon policies, the presented framework solves the MHE problem in a multi-parametric fashion, where the past and current measurements and inputs and the initial guess for the estimated states are the parameters of the problem [5, 7].

As a last step, the obtained receding horizon policies are validated 'in-silico' using the original high-fidelity model, thus closing the loop. This validation is of crucial importance as iterative experiments on real plants might be too costly or dangerous to run. In particular in the case of multiple objectives such as minimization of error, safe operation and economically optimal performance, the possibility of performing 'in-silico' tests of a developed control strategy allows for the fine-tuning and optimal design of the control strategy.

In order to apply the afore-described framework, we also present software solutions for the different aspects of the framework. Due to its modeling and dynamic optimization capabilities, we employ PSE's gPROMS® ModelBuilder to formulate and validate the high-fidelity model of the process. Similarly, due to its wide-spread use and numerous in-build functions, the steps of model approximation as well as formulation and solution of multiparametric programming problems is performed in MATLAB® using state-of the art software [8, 9] based on the POP® toolbox [10]. Lastly, the solution of the multi-parametric programming problem is integrated into gPROMS® using a specifically designed foreign process written in C++. This approach avoids the use of tools such as gO:MATLAB, and thus enables the use of the dynamic simulation and optimization capabilities of PSE's gPROMS®.

The applicability of this novel framework will be demonstrated on a wide range of problems such as industrial processes [11], bio-medical [5, 12, 13], cogeneration heat and power systems [14].

References

1. Pistikopoulos, E.N., et al., *PAROC—An integrated framework and software platform for the optimisation and advanced model-based control of process systems.* Chemical Engineering Science, **in print**.

2. Lambert, R.S.C., P. Rivotti, and E.N. Pistikopoulos, *A Monte-Carlo based model approximation technique for linear model predictive control of nonlinear systems.* Computers & Chemical Engineering, 2013. **54**: p. 60–67.

3. Bemporad, A., et al., *The explicit linear quadratic regulator for constrained systems.* Automatica, 2002. **38**(1): p. 3–20.

4. Kopanos, G.M. and E.N. Pistikopoulos, *Reactive Scheduling by a Multiparametric Programming Rolling Horizon Framework: A Case of a Network of Combined Heat and Power Units.* Industrial & Engineering Chemistry Research, 2014. **53**(11): p. 4366–4386.

5. Nascu, I., et al., *Simultaneous Multi-Parametric Model Predictive Control and State Estimation with Application to Distillation Column and Intravenous Anaesthesia*, in *24th European Symposium on Computer Aided Process Engineering*, 2014, Elsevier. p. 541–546.

6. Rao, C.V., *Moving Horizon Strategies for the Constrained Monitoring and Control of Nonlinear Discrete-time Systems*, in *Chemical Engineering*. 2000, University of Wisconsin-Madison.

7. Voelker, A., K. Kouramas, and E.N. Pistikopoulos, *Simultaneous state estimation and model predictive control by multi-parametric programming.* Computer Aided Chemical Engineering, 2010. **28**(C): p. 607–612.

8. Dua, V., N.A. Bozinis, and E.N. Pistikopoulos, *A multiparametric programming approach for mixed-integer quadratic engineering problems.* Computers & Chemical Engineering, 2002. **26**(4–5): p. 715–733.

9. Oberdieck, R., M. Wittmann-Hohlbein, and E.N. Pistikopoulos, *A branch and bound method for the solution of multiparametric mixed integer linear programming problems.* Journal of Global Optimization, 2014. **59**(2-3): p. 527–543.

10. ParOs, *Parametric Optimization Programming (POP)*. 2004, ParOS.

11. Papathanasiou, M.M., et al., *A control strategy for periodic systems - application to the twin-column MCSGP.* Computer Aided Chemical Engineering, 2015. **Accepted for publication**.

12. Chang, H., et al., *Robust multi-parametric model predictive control for LPV systems with application to*

anaesthesia. Journal of Process Control, 2014. **24**(10): p. 1538–1547.

13. Nascu, I., R. Oberdieck, and E.N. Pistikopoulos, *A framework for hybrid multi-parametric model-predictive control with application to intravenous anaesthesia.* Computer Aided Chemical Engineering, 2015. **Accepted for publication**.

14. Diangelakis, N.A. and E.N. Pistikopoulos, *A Decentralised Multi-parametric Model Predictive Control Study for a Domestic Heat and Power Cogeneration System.* Computer Aided Chemical Engineering, 2015. **Accepted for publication**.

---

[1] In addition, also measured disturbances and output set points are treated as parameters.

424849 On the Relation Between Flexibility Analysis and Robust Optimization

Abstract:

Optimization under uncertainty is a very active area of research, and many approaches have been proposed to systematically model uncertainty and make sound decisions using such models. One approach that has gained much popularity in recent years is robust optimization, which poses an optimization problem in a way such that it is feasible for any realization of the uncertainty within a given uncertainty set. It turns out that in the process systems engineering community, the same idea was considered in the work on process flexibility analysis (Swaney & Grossmann, 1985) a decade before the field of robust optimization emerged (Ben-Tal & Nemirovski, 1998). In this work, we establish a link between flexibility analysis and robust optimization, and show how the two approaches can potentially benefit from each other.

In the context of flexibility analysis, flexibility is defined as the capability of a system to ensure feasibility for all possible realizations of the uncertainty by proper adjustment of control variables, which correspond to the so-called recourse actions in two-stage stochastic programming. The traditional robust optimization approach does not consider recourse and therefore cannot be applied to address flexibility analysis problems. However, in recent works, researchers have proposed accounting for recourse in robust optimization by applying decision rules that are functions of the uncertain parameters (Ben-Tal, et al., 2004). The decision rule approach allows recourse; however, compared to flexibility analysis, it is conservative because we are typically restricted to affine decision rules in order to achieve tractable formulations.

Our analysis is limited to the case of linear inequality constraints and box uncertainty sets. Here, we consider the three classical problems in flexibility analysis: the flexibility test problem, the flexibility index problem, and design under flexibility constraints. For each problem, the flexibility analysis approach is compared with the robust optimization approach applying affine decision rules. Furthermore, for each problem, a new formulation is developed using both robust optimization techniques and insights from flexibility analysis. We show the connections between the three different approaches. In particular, we demonstrate that compared to flexibility analysis, the decision rule approach results in efficient formulations but is restrictive in terms of the allowed recourse actions. To overcome this limitation, the new proposed robust formulations allow full recourse by leveraging insights from flexibility analysis. Several examples are presented to illustrate the major concepts covered in this presentation.

References

Ben-Tal, A., Goryashko, A., Guslitzer, E. & Nemirovski, A., 2004. Adjustable robust solutions of uncertain linear programs. Mathematical Programming, 99(2), pp. 351-376.

Ben-Tal, A. & Nemirovski, A., 1998. Robust Convex Optimization. Mathematics of Operations Research, 23(4), pp. 769-805.

Swaney, R. E. & Grossmann, I. E., 1985. An Index for Operational Flexibility in Chemical Process Design - Part I: Formulation and Theory. AIChE Journal, 31(4), pp. 621-630.

428727 A Method of Evaluating Plant Models to Predict Safety-System Failure Probabilities

Abstract:

To estimate failure probabilities of chemical plant safety systems using Bayesian analysis, *prior* probability distributions are used in conjunction with *likelihood* probability distributions to calculate *posterior* probability distributions. The prior distributions are classically calculated from expert knowledge and/or plant data, and the likelihood distributions from plant data. When safety systems of a plant are rarely activated, the plant alarm data are sparse, which leads to Bayesian-analysis-calculated posterior distributions that depend more on prior distributions.

Moskowitz et al. (2015) introduced a method of repeated simulation to construct *informed prior distributions* for the failure probabilities of alarm and safety interlock systems. The resulting posterior distributions were shown to be more reliable than those obtained using a few alarm occurrences over extended operating periods.

In this paper, we present a method of evaluating plant models in terms of their ability to predict safety-system failure probabilities. The application and performance of the method are shown by developing and evaluating four mathematical models of an industrial steam-methane reformer (Moskowitz et al., 2015). The quality of these models is evaluated using plant measurements of effluent temperatures from reformer tubes and the surrounding furnace.

Informed prior distributions are generated using each model. When using plant feed stream data, dynamic simulations yield a close alignment between model quality and informed prior distribution quality. Steady-state simulations yield a weaker alignment.

When modeling the operator safety system, skill and attentiveness are difficult to estimate. Herein, operator responses are modeled as a function of the number of active alarms and past operator successes. Typical operator response times and response accuracies are estimated as functions of alarm count intervals and operator skill types. Resulting informed prior distributions show the effect of increasing operator skill level and process over-alarming.

Reference

Moskowitz, I. H., W. D. Seider, M. Soroush, U. G. Oktem, and J. E. Arbogast, "Chemical Process Simulation for Dynamic Risk Analysis: A Steam-Methane Reformer Case Study", *Ind. Eng. Chem. Res.*, **54**, 4347-4359 (2015).

427459 Restriction-Based Algorithms and Kkt-Based Bounding Procedures for Semi-Infinite Programming

Abstract:

Semi-infinite programs (SIPs) are mathematical programs with a finite number of variables and an infinite number of constraints. This is commonly expressed by a parameterized constraint, which has to hold for all possible values of a parameter from a set of infinite cardinality. One solution approach for SIPs is a finite discretization of this parameter set, which results in a relaxation of the original program [2]. The relaxed program is a nonlinear program (NLP) and can be solved by existing solvers to provide a lower bound to the solution of the SIP. Given an adaptive discretization scheme, a series of these lower bounding NLPs can be solved such that the lower bound converges to the solution of the SIP in the limit. This basic algorithm, however, cannot guarantee feasible iterates in finite time. Recently, several methods were proposed that solve SIPs to a feasible ε-optimal point in finite time [1,3,4,6].

In the first part of this presentation, we analyze two discretization-based bounding algorithms [4,6] that introduce (implicitly or explicitly) a restriction of the semi-infinite constraint to achieve feasible iterates. We show that for certain problems, these algorithms tend to generate an overpopulated discretization and therefore require many iterations to converge. Motivated by the comparison, we derive a hybrid method relying on modified procedures from both algorithms to generate an adaptive restriction of the semi-infinite constraint and thereby avoid overpopulation. The new method performs better on problems where overpopulation would be a problem while remaining competitive on established SIP benchmark problems [7].

In the second part of this presentation, we present numerical experiments concerning the efficacy of alternative bounding problems proposed in [5]. The bounding problems used in [4] are augmented with the KKT conditions of the lower-level problem of the SIP to achieve tighter bounds. The resulting problems are mathematical programs with equilibrium constraints (MPECs). A direct solution approach and a mixed integer reformulation are compared to the original bounding problems. Especially for upper bounding, the KKT-based approach promises tighter bounds, resulting in fewer iterations and reduced overall computational effort.

[1] B. Bhattacharjee, P. Lemonidis, W. H. Green Jr, and P. I. Barton. Global solution of semi-infinite programs. Mathematical Programming, 103(2):283–307, 2005.

[2] J. W. Blankenship and J. E. Falk. Infinitely constrained optimization problems. Journal of Optimization Theory and Applications, 19(2):261–281, 1976.

[3] C. A. Floudas and O. Stein. The adaptive convexification algorithm: a feasible point method for semi-infinite programming. SIAM Journal on Optimization 18(4):1187–1208, 2007.

[4] A. Mitsos. Global optimization of semi-infinite programs via restriction of the right-hand side. Optimization, 60(10-11):1291–1308, 2011.

[5] A. Mitsos, P. Lemonidis, C. K. Lee, and P. I. Barton. Relaxation-based bounds for semi-infinite programs. SIAM Journal on Optimization, 19(1):77–113, 2008.

[6] A. Tsoukalas and B. Rustem. A feasible point adaptation of the Blankenship and Falk algorithm for semi-infinite programming. Optimization Letters, 5(1):705-716, 2011.

[7] G. A. Watson. Numerical experiments with globally convergent methods for semi-infinite programming problems. Lecture Notes in Economics and Mathematical Systems, 215(1):193-205, 1983.

422847 Control-Relevant Decomposition of Process Networks Using Hierarchical Clustering Methods

Abstract:

Tight integration is the rule rather than the exception in chemical and energy processes. Despite obvious economic advantages, the networked nature of these processes makes their operation and control more challenging. Specifically, strong coupling between different process units limits the effectiveness of fully decentralized control approaches traditionally followed in plant-wide control.

Extensive research activity has been devoted to develop control methods that account for the networked nature of tightly integrated plants, e.g. distributed or coordinated control approaches. Key to the development and implementation of such methods is the identification of constituent sub-networks within the network that can be effectively controlled and coordinated.

The identification of such sub-networks can be viewed as a community detection problem, in which one seeks communities whose members interact strongly among them, yet are weakly coupled to the rest of the network members. Among different traditional approaches available to address such a problem, hierarchical clustering has a certain advantage, as it does not require a priori knowledge on the structure and the number of communities, which is typically the case in the problem at hand.

To this end, in this work, we propose hierarchical clustering methods to systematically decompose process networks in constituent sub-networks. Specifically, we approach this problem as a problem of clustering input and output variables, in groups of varying modularity, using an appropriate coupling measure. We postulate that relative degree is a natural such measure as it captures the directness of the effect of an input on an output, or the physical closeness between the two variables. First, an integer programming problem is formulated to obtain the input/output pairs which minimize the structural coupling in the network, captured by relative degrees. Then, an *agglomerative* hierarchical clustering method is developed, in which two input/output clusters are successively merged to identify the hierarchy of block decentralized control configurations. A notion of distance between the clusters is defined to determine the clusters to be merged. Finally, a more general method is developed within the framework of *divisive* hierarchical clustering, in which clusters are split into smaller clusters recursively. Each step of clustering is formulated as a nonlinear integer programming problem, whose objective is to maximize a *decentrality* measure of the resulting clusters, which quantifies the *compactness* of the clusters as well as the *closeness* between the clusters, also expressed in terms of relative degrees. The developed methods naturally identify the groups of process units that form compact and weakly coupled sub-networks at each level of the hierarchy.

The developed methods are scalable as they can be easily automated within a graph theoretic framework. Specifically, the relative degree, which is the key element of the methods, can be computed from equation graphs, which capture structural interdependencies among the process variables, and whose construction requires only the structural forms of the process dynamic models. Also, the computation of relative degrees can be easily implemented using e.g. the Floyd-Warshall algorithm.

The application of the developed methods is illustrated using example chemical plants. Specifically, an energy integrated solid oxide fuel cell (SOFC) system proposed in [1], and the toluene hydrodealkylation (HDA) process are considered. The results from both agglomerative and divisive hierarchical clustering methods are compared, and the discussions on the subnetwork structures are provided.

[1] Dimitrios Georgis, Sujit S. Jogwar, Ali S. Almansoori, and Prodromos Daoutidis, "Design and control of energy integrated SOFC systems for *in situ* hydrogen production and power generation", *Comput. Chem. Eng.*, 35(9):1691-1704, 2011

426697 Event-Triggered Fault-Tolerant Control of Networked Process Systems

Abstract:

Chemical plants are large-scale networked systems with tightly interconnected component subsystems coupled through material, energy and information flows and recycle. The ability of these systems to operate profitably and reliably requires the design and implementation of advanced control systems that can manage the complex dynamics and interactions between the component subsystems. This realization has motivated significant research work on the control and supervision of large-scale process networks, leading to a host of control design methodologies ranging from centralized to decentralized and distributed control architectures (e.g., see [2], [3], [6], [7] for some recent results and references).

In addition to handling complex network dynamics, another important consideration in the design of control systems for large-scale process networks is the integration of computing, communication and control at different levels of plant operations. With the emergence of the Smart Plant paradigm in recent years [1] and the increased calls for the deployment of wireless sensor/actuator networks and other advanced communication technologies in process monitoring and control systems, there is a need to address the fundamental challenges posed by the integration of communication and control. The intrinsic limitations on the computational and communication resources of wireless devices, for example, favor keeping the information exchange over the network to a minimum so as to conserve network resources, while maximizing process performance favors increased communication levels between the plant subsystems.

An approach to resolve this conflict was proposed in [4] using a quasi-decentralized networked control structure designed to balance plant performance requirements with communication costs. A key idea was to embed, within each local control system, a set of predictive models that generate estimates of the states of the neighboring units when communication is suspended, and to update the models when communication is restored. Using a fixed communication rate policy, an explicit characterization of the minimum allowable communication rate was obtained. In a subsequent study [5], a quasi-decentralized control structure using a state-dependent communication logic was developed. The key idea there was to have each unit monitor the evolution of the local state and prompt the other units to broadcast their measurements to update their models only when the unit is on the verge of instability. Compared with the fixed communication rate approach, this approach allows the plant to respond to changes in operating conditions by varying the communication rate; however, it can also lead to increased network utilization and the possibility of delays and data losses when all plant units attempt to access the network at the same time.

An alternative approach for dealing with communication constraints is the use of event triggered control strategies (e.g., [8]-[10]). In these strategies, communication over the network is typically suspended and restored in response to certain events tied to the desired closed-loop stability and performance properties. These strategies are appealing in the context of sensor/actuator networks where reducing network utilization can also reduce the energy expenditures of battery-powered wireless devices. In these studies, however, the problems of detecting and handling faults in the networked control system design were not addressed. These are important problems as fault-handling is essential to maintaining the desired closed-loop stability and performance properties. Furthermore, the presence of faults, if not appropriately dealt with, can lead to unnecessary increases in communication levels.

In this work, we present an event-triggered fault-tolerant control framework for networked process systems controlled over a resource-constrained communication medium. The objective is to maintain closed-loop stability in the presence of faults while minimizing unnecessary network utilization. To meet this objective, a set of stabilizing distributed controllers are initially designed on the basis of an approximate plant model. The local controllers utilize the available model to generate the local control action during periods of communication suspension. Updates of the model states using real-time measurements are triggered whenever the local model estimation error breaches a certain stability threshold. This threshold is designed using Lyapunov techniques and is explicitly parameterized in terms of the fault size, the plant-model mismatch and the various controller design parameters. This parameterization provides the basis for selecting a suitable fault accommodation strategy that enforces closed-loop stability, while simultaneously optimizing network resource utilization. To aid fault accommodation, fault detection and estimation are carried out locally within each subsystem using a moving-horizon optimization scheme. Based on the fault identification result, several fault accommodation measures are assessed, including adjusting the local model or controller parameters, tightening the local model estimation error threshold to increase communication frequency, and system reconfiguration at the supervisory level. The choice of a suitable strategy is made on the basis of the desired balance between control and communication requirements. Finally, the developed methodology is illustrated using a reactor-separator process network.

References:

[1] P. D. Christofides, J. Davis, N. H. El-Farra, D. Clark, K. R. D. Harris and J. N. Gipson, ``Smart Plant Operations: Vision, Progress and Challenges," *AIChE J.*, 53, 2734-2741, 2007.

[2] K. R. Jillson and B. E. Ydstie, "Process networks with decentralized inventory and flow control," *J. Process Control*, 17, 399-413, 2007.

[3] S. Jogwar, M. Baldea and P. Daoutidis, "Dynamics and control of process networks with large energy recycle," *Ind. Eng. Chem. Res.*, 48, 6087-6097, 2009.

[4] Y. Sun and N. H. El-Farra, "Quasi-decentralized model-based networked control of process systems," *Comp. Chem. Eng.*, 32, 2016-2029, 2008.

[5] Y. Sun and N. H. El-Farra, "Quasi-decentralized networked process control using an adaptive communication policy," *Proceedings of American Control Conference*, pp. 2841–2846, 2010.

[6] B. T. Stewart, S. Wright and J. B. Rawlings, "Cooperative distributed model predictive control for nonlinear systems," *J. Process Control*, 21, 698-704, 2011.

[7] P. D. Christofides, J. Liu and D. M. de la Pena. *Networked and Distributed Predictive Control: Methods and Nonlinear Process Network Applications*. Springer-Verlag, London, 2011.

[8] M. Mazo and P. Tabuada, "Decentralized event-triggered control over wireless sensor/actuator networks," *IEEE Transactions on Automatic Control*, 56, 2456–2461, 2011.

[9] X. Wang and M. D. Lemmon, "Event-triggering in distributed networked control systems," *IEEE Transactions on Automatic Control,* 56, 586–601, 2011.

[10] Y. Hu and N. H. El-Farra, ``Quasi-decentralized output feedback model predictive control of networked process systems with forecast-triggered communication," *Proceedings of American Control Conference*, pp. 2612-2617, Washington, DC, 2013.

421054 Simultaneous Control of Safety Constraint Sets and Process Economics Using Distributed Economic Model Predictive Control

Abstract:

Maintaining safe operation of chemical processes while simultaneously maximizing economic profit and meeting environmental constraints constitute issues of paramount importance in the area of process systems and control engineering. It has long been argued that process safety is fundamentally a process control problem, yet, few research efforts (e.g., [1]) have been carried out towards the direction of integrating the rather disparate domains of process safety and process control. Economic model predictive control (EMPC) [2] has attracted significant attention recently due to its ability to optimize process operation accounting directly for process economics considerations. However, there is very limited work on the problem of integrating safety considerations in EMPC to ensure simultaneously safe operation and maximization of process economic profit.

Motivated by the above considerations, this work will present a distributed economic model predictive control methodology for simultaneously coordinating in real-time the size of the safety sets in which the process state should reside at all times in order to ensure safe process operation and feedback control of the process state to optimize economics via time-varying process operation. Initially, the safeness of a region of operation with respect to a potential failure (i.e., determine if stability of the process can be maintained for any point in the region of operation if a failure occurs in a sensor or an actuator or a software component) is characterized and a safety logic is created that determines what are the safe regions of process operation under both normal and faulty conditions. Using data on probability of the potential failures, measurement feedback of the process state and future process state trajectory information as determined by the EMPC over its prediction horizon, a control system is built to determine and modify in real-time the region of safe process operation. This information is in turn used as a state constraint in the EMPC that regulates the process state to optimize economics such that the two control systems coordinate their actions. Through a rigorous stability analysis, conditions for recursive feasibility and closed-loop stability are established. The proposed method, which effectively integrates feedback control, process economics and safety considerations, is demonstrated with a chemical process example.

[1] Leveson NG, Stephanopoulos G. A system-theoretic, control-inspired view and approach to process safety. AIChE Journal. 2013;60:2-14.
[2] Ellis, M, Durand H, Christofides PD. A Tutorial Review of Economic Model Predictive Control Methods. J. Proc. Contr. 2014;24:1156-1178.

Distributed Moving Horizon State Estimation of Two-time-scale Nonlinear Systems

Xunyuan Yin and Jinfeng Liu

Department of Chemical & Materials Engineering
University of Alberta

2015 AIChE Annual Meeting

November 8-13, 2015

Salt Lake City, Utah

UNIVERSITY OF ALBERTA

Introduction

- Time-scale multiplicity
  - Arise due to physicochemical phenomena
- Common in chemical processes
  - Catalytic cracking units (Han et al., CES, 2001)
  - Reactor-distillation process networks (Chen et al., JPC, 2011)
  - Waste water treatment processes (Nagy-Kiss et al., CDC, 2010)

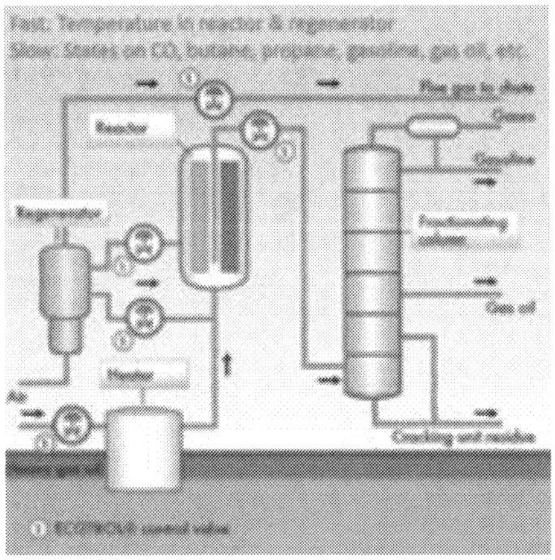

- Direct application of standard control/state estimation methods is not favourable (Kokotovic et al., Academic Press, 1986; Christofides et al., IJC, 1996)
  - Lead to ill conditioning of the system
  - Deteriorate the stability

Previous results

- Control of two-time-scale processes
  - Output feedback control of linear systems (O'Reilly, IJC, 1980; Khalil, TAC, 1987)
  - Nonlinear state feedback control (Christofides et al., IJC, 1996)
  - Robust feedback control (Christofides et al., JPC, 1997)
  - Optimal control (Kokotovic et al., Academic Press, 1986)
  - Model predictive control (Chen et al., JPC, 2011)
- State estimation of two-time-scale systems
  - Essentially state estimation on linearized models (Kiss et al., CEP, 2011)
- Systematic approaches for state estimation of two-time-scale processes need to be developed

3 of 17

Preliminaries

- State estimation of nonlinear systems
  - Extensions of linear solutions via successive linearization
    - ▷ Extended Kalman filters and extended Luenberger observers (Zeitz, SCL, 1987; Reif et al., TAC, 1999)
  - Designs that consider nonlinearity explicitly
    - ▷ Deterministic approaches: High-gain observers etc. (Gauthier et al., TAC, 1992; Ciccarella et al., IJC, 1993; Kazantizis and Kravaris, SCL, 1998)
    - ▷ Stochastic approaches: Moving horizon estimation (Rao et al., Automatica, 2001; TAC, 2003; Michalska and Mayne, TAC, 1995; Liu, CES, 2013)
- To develop state estimation schemes for two-time-scale processes in the framework of moving horizon estimation

4 of 17

Moving horizon estimation (Rao et al., TAC, 2003)

- Moving horizon estimation (MHE)

$$\min_{\tilde{X}(t_k)} \sum_{i=k-N}^{k-1} |w(t_i)|^2_{Q^{-1}} + \sum_{i=k-N}^{k} |v(t_i)|^2_{R^{-1}} + V(\tilde{x}(t_{k-N}))$$

s.t. $\dot{\tilde{x}}(t) = f(\tilde{x}(t), w(t_i)),\ t \in [t_i, t_{i+1}]$

$v(t_i) = y(t_i) - h(\tilde{x}(t_i))$

$w(t_i) \in W,\ v(t_i) \in V,\ \tilde{x}(t) \in X$

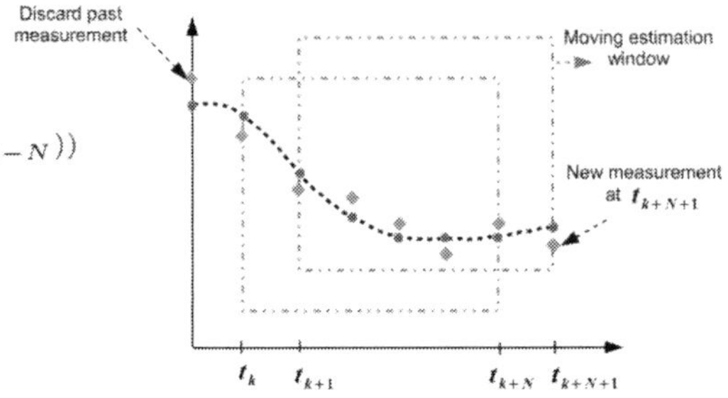

- Online optimization based approach
  - Explicit use of distribution/boundedness information on w, v, x
  - The most recent N measurements are considered
  - Past measurements are accounted for in the arrival cost $V(\tilde{x}(t_{k-N}))$
- Properties of this MHE
  - Convergence can be established if the arrival cost is well estimated
  - Constraints on disturbances and measurement noise can be handled

Observer-enhanced MHE (Liu, CES, 2013)

- Design
  - A deterministic observer used to calculate a confidence region
  - MHE gives an optimal estimate within the region

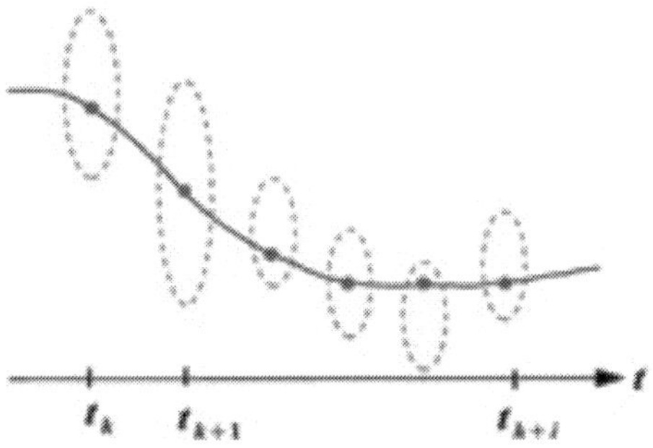

- Properties
  - Bounded estimation error for bounded uncertainties
  - Tunable convergence rate
  - Less sensitive to the accuracy of the arrival cost

System description

- Two-time-scale nonlinear system

$$
\begin{aligned}
\dot{x}_s(t) &= f(x_s(t), w_s(t), \epsilon) + \tilde{f}(x_s(t), x_f(t), \epsilon) \\
\epsilon\dot{x}_f(t) &= g(x_f(t), w_f(t), \epsilon) + \tilde{g}(x_s(t), x_f(t), \epsilon) \\
y_s(t) &= h_s(x_s(t)) + v_s(t) \\
y_f(t) &= h_f(x_f(t)) + v_f(t)
\end{aligned}
$$

  - x_s, x_f: system state variables
  - y_s, y_f: measured output variables
  - w_s, w_f: process disturbances
  - v_s, v_f: measurement noise
- Time-scale decomposition
  - Reduced-order fast system

$$
\begin{aligned}
\frac{de_f(\tau)}{d\tau} &= G(e_f, x_s, w_f) \\
e_f &:= x_f - \hat{g}(x_s, w_f)
\end{aligned}
$$

  - Reduced-order slow system

$$
\dot{\bar{x}}_s(t) = f(\bar{x}_s(t), w_s(t), 0) + \tilde{f}(\bar{x}_s(t), \hat{g}(\bar{x}_s(t), w_f(t)), 0)
$$

- It is desirable to design separate estimators for the reduced-order fast and slow systems

Distril

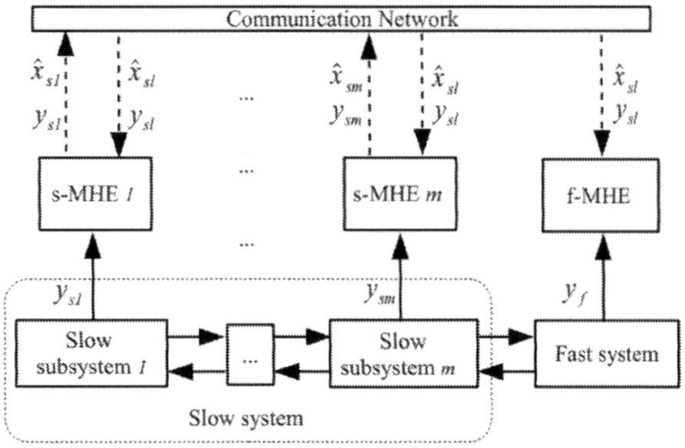

- The reduced-order slow/fast system can be divided further to slow/fast subsystems when necessary
- Subsystem estimators communicate to exchange information
- Fast system does not send information to slow subsystems

Nonlinear observers and sampling times

- Existence of nonlinear observers

$$\begin{aligned}\dot{z}_{si}(t) &= F_{si}(z_{si}(t), h_{si}(\bar{x}_{si}(t)),\ i \in \mathbb{I} \\ \dot{z}_f(\tau) &= F_f\left(z_f(\tau), x_s(\tau), h_f(e_f(\tau) + \hat{g}(x_s(\tau), 0))\right)\end{aligned}$$

  - Interactions between subsystems are not considered
- Different sampling periods for slow and fast systems

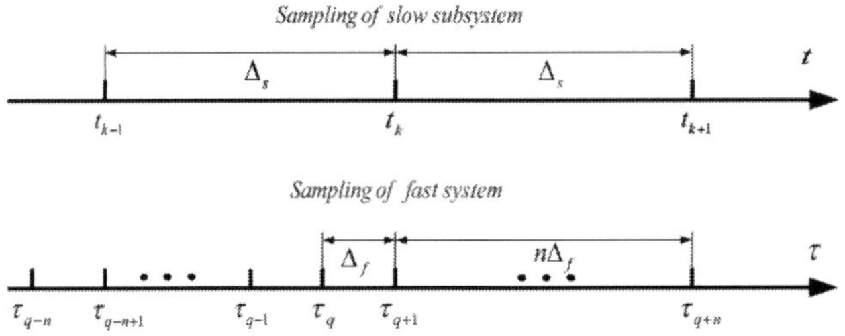

  - Sampling period Δ_s for s-MHEs is n times of Δ_f for f-MHE
  - Communication resource is saved
  - Computational burden is significantly reduced

Distributed MHE design

- Reference state estimate for s-MHEs
  - An augmented observer to generate reference state estimates $z_{si}(t_k)$
  - Interactions are considered (Zhang et al., JPC, 2013)
- Design of s-MHEs

$$\begin{aligned}\min_{\tilde{x}_{si}(t_{k-N_s}),\ldots,\tilde{x}_{si}(t_k)} & \sum_{j=k-N_s}^{k-1} |w_{si}(t_j)|^2_{Q_{si}^{-1}} + \sum_{j=k-N_s}^{k} |v_{si}(t_j)|^2_{R_{si}^{-1}} \\ \text{s.t.} \quad & \dot{\tilde{x}}_{si}(t) = f_i(\tilde{x}_{si}(t), w_{si}(t_j), 0) + \tilde{f}_i(\hat{x}_s(t_j), \hat{g}(\hat{x}_s(t_j), 0), 0), \\ & \quad t \in [t_j, t_{j+1}],\ j = k - N_s, \ldots, k-1 \\ & v_{si}(t_j) = y_{si}(t_j) - h_{si}(\tilde{x}_{si}(t_j)),\ j = k - N_s, \ldots, k \\ & w_{si}(t_j) \in \mathbb{W}_{si},\ v_{si}(t_j) \in \mathbb{V}_{si},\ \tilde{x}_{si}(t_j) \in \mathbb{X}_{si},\ j = k - N_s, \ldots, k-1 \\ & |\tilde{x}_{si}(t_k) - z_{si}(t_k)| \le \kappa_{si} |y_{si}(t_k) - h_{si}(z_{si}(t_k))|\end{aligned}$$

  - Reduced-order slow system model is used in the design
  - The design is independent on fast system state
  - Observer-enhanced design ensures a bounded estimation error

Distributed MHE design

- A reference state estimation is generated each sampling time by the auxiliary nonlinear observer
- Design of f-MHE

$$\min_{\tilde{e}_f(\tau_{q-N_f}),\ldots,\tilde{e}_f(\tau_q)} \sum_{j=q-N_f}^{q-1} |w_f(\tau_j)|^2_{Q_f^{-1}} + \sum_{j=q-N_f}^{q} |v_f(\tau_j)|^2_{R_f^{-1}}$$
$$\text{s.t.} \quad \dot{\tilde{e}}_f(\tau) = G(\tilde{e}_f(\tau), \hat{x}_s(\tau_j), w_f(\tau_j)),\ \tau \in [\tau_j, \tau_{j+1}],\ j = q-N_f, \ldots, q-1$$
$$v_f(\tau_j) = y_f(\tau_j) - h_f\left(\tilde{e}_f(\tau_j) + \hat{g}(\hat{x}_s(\tau_j), w_f(\tau_j))\right),\ j = q-N_f, \ldots, q$$
$$w_f(\tau_j) \in \mathbb{W}_f,\ v_f(\tau_j) \in \mathbb{V}_f,\ j = q-N_f, \ldots, q-1$$
$$|\tilde{e}_f(\tau_q) - z_f(\tau_q)| \leq \kappa_f |y_f(\tau_q) - h_f(z_f(\tau_q) + \hat{g}(\hat{x}_s(\tau_q), w_f(\tau_{q-1})))|$$

  - Optimal estimate of the original fast system state is calculated as

$$\hat{x}_f(\tau_q) = \hat{e}_f(\tau_q) + \hat{g}(\hat{x}_s(\tau_q), w_f^*(\tau_{q-1}))$$

  - The design has dependence on the slow system state estimate $\hat{x}_s$
  - The design is based on the reduced-order fast system
  - Observer-enhanced MHE design

Stability of the proposed design

Theorem $|e_{si}| = |\hat{x}_{si} - x_{si}|$ is a decreasing sequence and is ultimately bounded as $\lim_{t\to\infty} \sup |e_{si}| = \lim_{t\to\infty} \sup |e_{si}^{\overline{x}} + e_{si}^{x}| \leq \rho_{si,1}^{min} + \rho_{si,2}^{min}$.
$|\theta_f| = |\hat{x}_f - x_f|$ is decreasing and ultimately bounded as $\lim_{\tau\to\infty} \sup |\theta_f(\tau)| \leq \rho_f^{min} + \sum_{l\in\mathbb{I}} L_{\hat{g}}^{x_s} \left(\rho_{sl,1}^{min} + \rho_{sl,2}^{min}\right)$.

- Estimation error of entire system is bounded
- Factors affecting the convergence
  - Sampling time of f-MHE and s-MHEs
  - Magnitude of ϵ
  - Magnitude of external disturbances and measurement noise
  - Properties of the nonlinear observers for s-MHEs and f-MHE, etc.

Application to a chemical process example

- Reactor-separator process

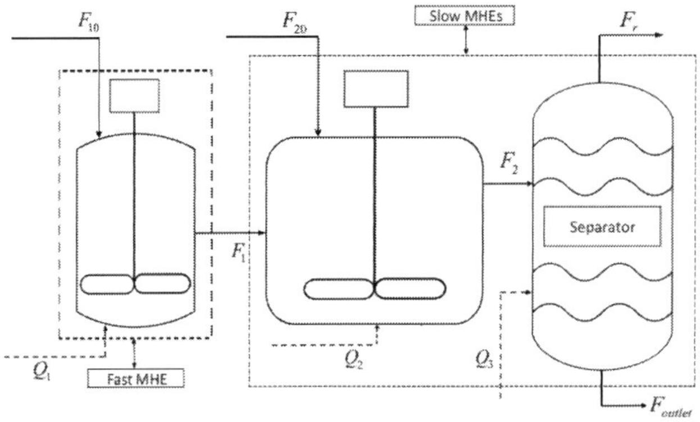

- States: x_f^A, x_f^B, T_f, x_{si}^A, x_{si}^B, T_{si}, $i = 1,\ 2$
- Outputs: T_f, T_{si}, $i = 1,\ 2$ Small parameter: $\epsilon = V_1/V_2 = 0.2$
- Divided into one reduced-order fast system and one slow system
- The slow system is further decomposed into two subsystems

Simulation results

- Black lines: Actual states   Red lines: State estimates

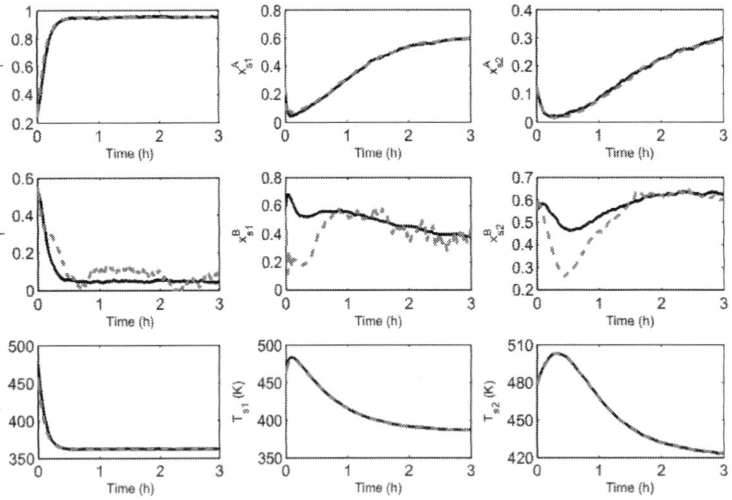

Performance comparison

- Different sampling schemes
  - s-MHEs and f-MHE evaluated at $\Delta_s = 36s$ and $\Delta_f = 18s$, respectively (Case I) v.s. all MHEs evaluated at Δ_f (Case II)

	Maximum error	Mean error	Computation time
Case I	0.6642	2.1753	382s
Case II	0.5416	2.0737	692s

  - The estimation performance in terms of the maximum and mean error values increases by about 18% and 4% in Case II
  - Computation time increases by 81% in Case II
  - Communication burden of s-MHEs dramatically increases in Case II

Performance comparison

- One directional and bi-directional communication strategies
  - One-directional as in proposed (Case I) v.s. bi-directional communication (Case III)

	Maximum error	Mean error
Case I	0.6642	2.1753
Case III	0.6683	2.1089

  - The performance for the two strategies is similar
  - Communication burden for f-MHE significantly increases in Case III

Conclusions

- Singular perturbation theory is used for subsystem decomposition considering time-scale multiplicity of nonlinear processes
- Decomposed reduced-order models are used for state estimation design
- One-directional communication strategy is employed
- A chemical process example is used to illustrate the effectiveness of the proposed DMHE

Acknowledgement

- Financial support from NSERC is gratefully acknowledged

424240 A Centralized/Decentralized Control Approach for the Multicolumn Countercurrent Solvent Gradient Purification (MCSGP) Process

Abstract:

A centralized/decentralized control approach for the Multicolumn Countercurrent Solvent Gradient Purification (MCSGP) process

Maria M. Papathanasiou[a,d], Muxin Sun[a], Fabian Steinebach[b], Thomas Müller-Späth[c], Massimo Morbidelli[b], Athanasios Mantalaris[a], Efstratios N. Pistikopoulos[d*]

[a]Dept. of Chemical Engineering, Centre for Process Systems Engineering (CPSE), Imperial College London SW7 2AZ, Lodnon, U.K

[b] Institute for Chemical and Bioengineering, ETH Zurich,

Wolfgang-Pauli-Str. 10/HCI F 129, CH-8093 Zurich, Switzerland [c] *ChromaCon AG, Technoparkstr. 1, CH-8005 Zurich, Switzerland* [d]*Artie McFerrin Department of Chemical Engineering, Texas A&M University, College Station TX 77843 *stratos@tamu.edu*

Keywords: periodic processes, multi-parametric, control, stability

Industrial applications, such as separation processes and pressure swing adsorption systems (PSA) are characterized by cyclic operation profiles and they often involve tight constraints on product purity. In order to maintain the process under optimal operation, advanced computational tools need to be developed. The latter include optimization strategies and optimal control policies that are designed to meet the process constraints. During the past few years, several research groups have been focusing on the development of such tools, aiming to tackle issues related to system periodicity and nonlinearity [1-6].

In this work we present a centralized/decentralized control approach based on the principles of the Multicolumn Countercurrent Solvent Gradient Purification (MCSGP) process. MCSGP is an industrial, semi-continuous chromatographic separation process used for the purification of various biomolecules [7]. The system is described by a highly nonlinear Partial Differential and Algebraic Equation (PDAE) model that operates under a cyclic profile. Following our previously presented work [8, 9] we demonstrate a seamless procedure for the development of advanced decentralized multi-parametric control strategies for the system at hand.

The examined system shifts periodically between two distinct configurations (Figure 1). Figure 1a illustrates the continuous operation mode, where the two columns operate connected to each other. This is usually followed by batch operation as shown in Figure 1b. For further details regarding the complete system operation the reader is referred to Krättli, et al. [10]. In this study we develop independent control strategies for each column that are then linked during the interconnected operation, tracking the integral of the outlet concentrations. In particular, the outlet of column A is treated as a disturbance by the controller operating on column B during continuous operation (Figure 1a).

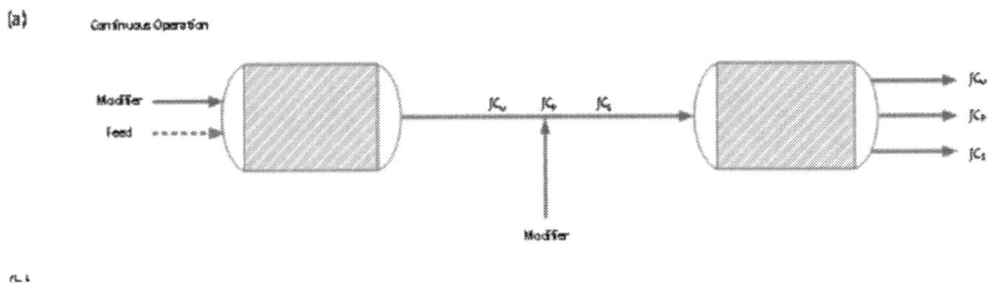

Figure 1 Chromatographic system based on the principles of MCSGP for (a) continuous and (b) batch operation mode, considering the modifier concentration as input, the integral of the three outlet concentrations as outputs and the feed composition as measured disturbance.

The periodic operation profile and the disturbances occurring from the feed stream render the preservation of the optimal operating conditions a challenging task. Therefore, to ensure optimal operation throughout the process cycle, the controller stability needs to be guaranteed. Following the framework presented by Pistikopoulos, et al. [8] the controller development is based on the derivation of a linear state space model that describes the system dynamics. The optimal periodic steady state and input strategy for the tracked process intervals are calculated off-line, minimizing the cost function of the optimization problem [11]. The online control scheme can then be applied by driving the system to the pre-computed optimal periodic steady state by using multi-parametric Model Predictive Control (mp-MPC) (Pistikopoulos, 2009). To prove the stability of the cyclic process we introduce a transformed system as suggested by to Huang, et al. [12].

The designed controllers are then tested in-silico, in a 'closed-loop' fashion, using the original process model [8]. Due to the system dynamics a repetitive output time delay is observed that corresponds to the time that the components require to reach the column exit [13]. This can be either pre-computed or related to the output and/or control horizon during the design of the control policies. Here we examine both cases and we demonstrate the evolution of the time delay with respect to both the output and the control horizon. It can be observed that both strategies are equally effective, however the second case allows greater freedom in the controller design as the setpoints are tracked independently from the time delay. The suggested strategy can efficiently track the desired outputs and retain the process under optimal operating conditions. The designed controllers guarantee stable, cyclic operation and the system time delay is successfully eliminated.

Acknowledgements

Financial support from the European Commission (OPTICO/G.A. No.280813) is gratefully acknowledged.

References

[1] S. Nilchan and C. Pantelides, "On the optimisation of periodic adsorption processes," *Adsorption,* vol. 4, pp. 113-147, 1998.

[2] Y. Kawajiri and L. T. Biegler, "Optimization strategies for simulated moving bed and PowerFeed processes," *AIChE Journal,* vol. 52, pp. 1343-1350, 2006.

[3] C. Grossmann, G. Ströhlein, M. Morari, and M. Morbidelli, "Optimizing model predictive control of the chromatographic multi-column solvent gradient purification (MCSGP) process," *Journal of Process Control,* vol. 20, pp. 618-629, 2010.

[4] P. Suvarov, A. Kienle, C. Nobre, G. De Weireld, and A. Vande Wouwer, "Cycle to cycle adaptive control of simulated moving bed chromatographic separation processes," *Journal of Process Control,* vol. 24, pp. 357-367, 2// 2014.

[5] S. Li, L. Feng, P. Benner, and A. Seidel-Morgenstern, "Using surrogate models for efficient optimization of simulated moving bed chromatography," *Computers & Chemical Engineering,* vol. 67, pp. 121-132, 8/4/ 2014.

[6] S. Liu, A. S. Simaria, S. S. Farid, and L. G. Papageorgiou, "Optimising chromatography strategies of antibody purification processes by mixed integer fractional programming techniques," *Computers & Chemical Engineering,* vol. 68, pp. 151-164, 9/4/ 2014.

[7] L. Aumann and M. Morbidelli, "A continuous multicolumn countercurrent solvent gradient purification (MCSGP) process," *Biotechnology and Bioengineering,* vol. 98, pp. 1043-1055, 2007.

[8] E. N. Pistikopoulos, N. A. Diangelakis, R. Oberdieck, M. M. Papathanasiou, I. Nascu, and M. Sun, "PAROC —An integrated framework and software platform for the optimisation and advanced model-based control of process systems," *Chemical Engineering Science,* p. In press.

[9] M. M. Papathanasiou, F. Steinebach, G. Stroehlein, T. Müller-Späth, I. Nascu, R. Oberdieck, *et al.*, "A control strategy for periodic systems - application to the twin-column MCSGP," in *Computer Aided Chemical Engineering*, ed, 2015, p. accepted for publication.

[10] M. Krättli, F. Steinebach, and M. Morbidelli, "Online control of the twin-column countercurrent solvent gradient process for biochromatography," *Journal of Chromatography A,* vol. 1293, pp. 51-59, 6/7/ 2013.

[11] D. Angeli, R. Amrit, and J. B. Rawlings, "On average performance and stability of economic model predictive control," *Automatic Control, IEEE Transactions on,* vol. 57, pp. 1615-1626, 2012.

[12] R. Huang, E. Harinath, and L. T. Biegler, "Lyapunov stability of economically oriented NMPC for cyclic processes," *Journal of Process Control,* vol. 21, pp. 501-509, 4// 2011.

[13] G. Guiochon, "Basic principles of chromatography," *Ullmann's Encyclopedia of Industrial Chemistry,* 2012.

420067 Flexible Residential Energy: Making the Case for Home Energy Management

Abstract:

Peak power loads in the United States are a costly problem. Of the total electricity generating capacity in the United States, 20% is dedicated to meet peak loads. In other words, 20% of U.S. energy generators are used only operated 5% of the time [1]. It is expensive to install additional generating capacity to guard against large fluctuations in demand. Strategies to mitigate volatility in energy consumption have the potential to reduce the need for surplus capacity. Additionally, even if strategies marginally increase electricity loads, so long as they level the overall load they can increase grid stability and allow for the use of more efficient base power plants.

Residential consumers are a significant contributor to the peak power problem because of their highly variable nature and dependence on weather. In this work, we investigate the potential for residential consumers to lower peak demand. We also show that home energy management systems with optimal controllers can reduce peak energy consumption for a community of homes while still respecting the preferences of individual users.

In our methodology we will consider the control of the largest loads contributing to peak power consumption: air-conditioning (A/C) and domestic appliances. A/C energy use correlates strongly with weather changes and is the largest load in residential homes. Appliances, such as dishwashers, individually are much smaller than A/C loads. However, the compounding effect of appliances operating simultaneously has the potential to create higher or even new power demand peaks.

First, we present a technique to derive simplified linear models of A/C energy as a function of outdoor dry-bulb temperature and indoor thermostat set points. The development of these models is based on smart meter and smart thermostat data. A non-intrusive load monitoring technique is used to disaggregate A/C electricity consumption from whole-house electricity data reported by the smart meters. A centralized model predictive control (MPC) scheme minimizes peak energy use at the neighborhood level by altering the thermostat set-points in individual homes, while respecting thermal preferences.

After the thermostat settings are planned for the day, we use a novel approach to schedule the operation of appliances, with the goal of minimizing peak load for the community. Dishwashers and washing machines/dryers were identified as potential time-shiftable appliances and are modeled from the probability distributions of start times and required completion times, as determined from consumer survey data. The scheduling problem is formulated as a mixed-integer linear program (MILP) aimed at minimizing peak load under constraints that reflect the start times and allowed delays of individual appliances in each house.

Using sample data collected from forty residential homes located in Austin, TX, we show that the proposed scheduling/control approach can minimize the peak load for the neighborhood by leveraging the physical differences and individual preferences between houses. On average, the centralized controller is able to reduce the daily peak load for the group of houses by 25% (15 kW) when compared with the original settings of the thermostat and appliance start times. In addition, the centralized controller does not significantly increase the total energy consumption and leads to a much flatter consumption profile.

1. H. Farhangi, "The path of the smart grid," *IEEE Power Energy Mag.*, vol. 8, no. 1, pp. 18–28, Jan. 2010.

(308g) Model Based Distributed Model Predictive Control of Hydrogen Standalone Combined Heat Cooling and Power System

Ocheme, S., University of Leeds

Originally presented on: 11/10/2015 10:30:00 - 10:50:00

Distributed model predictive control has witness renewed interest as a control techniques capable of inheriting the characteristics of centralised predictive control. The aim of this paper is to present a model based design approach to design and implement in a hydrogen standalone combined cooling heat and power system(HSCCHP) for a distributed generation system in a remote island. The challenges with the (HSCCHP) disturbance input to the system characterised with uncertainty. Wind power fluctuates within minutes whilst solar and river-hydro power fluctuates on longer terms, hours and days respectively. The model based predictive control schemes is robust in tracking the uncertainty that characterised the (HSCCHP).

409446 Integration of Wind and Solar Energy within Continental Biorefinery Supply Network

Abstract:

Global primary energy consumption and production is increasing for almost all fuel types especially in the emerging economies (BP, 2014). However, humankind is gradually exhausting fossil fuels even when some new discoveries are reported. Thus, people are going to have to depend on non-fossil energy sources in the future. A recent forecast including conventional and unconventional sources of oil and gas (Mohr et al., 2015) predicts that peaks of fossil fuels will occur during twenty-first century. Energy conservation, energy efficiency and renewable energy are amongst opportunities to deal with the issues of depleting non-renewable sources and progressing towards sustainability (Čuček et al., 2015). There are several incentives, plans and targets regarding substitution of fossil fuels over the next decades in various regions and countries. Among promising technologies there are those using solar radiance, wind and biomass (Davis and Martín, 2014).

In this contribution, the biorefinery supply network applied to the European continent involving lignocellulosic biomass, algae, cooking oil and grains for the production of bioethanol, biodiesel, (FAEE and FAME), FT liquids, hydrogen and food (Čuček et al., 2014) is extended to integrate wind and solar energy. Several technologies are integrated such as renewable power production from photovoltaic panels (Vujanović et al., 2014) and concentrated solar energy plants (Martín and Martín, 2013), and synthetic methane from exhaust $CO_2$and water electrolysis using renewable power (Davis and Martín, 2014). Continental biorefinery network thus becomes continental renewable energy-based supply network. Several scenarios are considered regarding targets of renewables in the energy mix based on the European perspectives over the following decades and regarding available area for renewables. The obtained results show where to exploit specific energy sources and to allocate the appropriate technologies within the regions, countries and/or continents.

References

BP (British Petroleum), 2014. BP Statistical Review of World Energy June 2014. <www.bp.com/content/dam/bp/pdf/Energy-economics/statistical-review-2014/BP-statistical-review-of-world-energy-2014-full-report.pdf> Accessed: 18.3.2015.

Čuček, L., Martín, M., Grossmann, I.E., Kravanja, Z., 2014. Large-Scale Biorefinery Supply Network – Case Study of the European Union, in: Klemeš, J.J., Varbanov, P.S., Liew, P.Y. (Eds.), Computer Aided Chemical Engineering. Elsevier, Volume 33, 319-324.

Čuček, L., Klemeš J. J., Varbanov P. S., Kravanja, Z., 2015. Significance of Environmental Footprints for Evaluating Sustainability and Security of Development, Clean Technologies and Environmental Policy, doi: 10.1007/s10098-015-0972-3.

Davis, W., Martín, M., 2014. Optimal year-round operation for methane production from $CO_2$and water using wind and/or solar energy. Journal of Cleaner Production 80, 252-261.

Martín, L., Martín, M., 2013. Optimal year-round operation of a concentrated solar energy plant in the south of Europe. Applied Thermal Engineering 59, 627-633.

Mohr, S.H., Wang, J., Ellem, G., Ward, J., Giurco, D., 2015. Projection of world fossil fuels by country. Fuel 141, 120-135.

Vujanović, A., Čuček, L., Novak Pintarič, Z., Pahor, B., Kravanja, Z., 2015. Synthesis of environmentally-benign energy self-sufficient processes under uncertainty. Journal of Cleaner Production 88, 90-104.

419580 Multi-Scale Exploration of the Technical, Economic, and Environmental Dimensions of Bio-Based Chemical Production

Abstract:

In recent years, bio-based chemicals have gained traction as a sustainable alternative to petrochemicals. However, despite rapid advances in metabolic engineering and synthetic biology, there remain significant economic and environmental challenges. In order to maximize the impact of research investment in a new bio-based chemical industry, there is a need for assessing the technological, economic, and environmental potentials of combinations of biomass feedstocks, biochemical products, bioprocess technologies, and metabolic engineering approaches in the early phase of development of cell factories.

To address this issue, we have developed a comprehensive **Mu**lti-scale framework for modeling **S**ustainable **I**ndustrial **C**hemicals production (MuSIC), which integrates modeling approaches for cellular metabolism, bioreactor design, upstream/downstream processes, as well as economic and life-cycle impact assessments. We demonstrate the use of the MuSIC framework in a case study where two major polymer precursors (1,3-propanediol and 3-hydroxypropionic acid) are produced from two biomass feedstocks (corn-based glucose and soy-based glycerol) through 66 proposed biosynthetic pathways in two host organisms (*E. coli* and *S. cerevisiae*). The MuSIC framework allows exploration of tradeoffs and interactions between economy-scale optimization objectives (*e.g.* profit maximization, emission minimization), constraints (*e.g.* land-use constraints) and process- and cell-scale technology choices (*e.g.* strain design or oxygenation conditions). We demonstrate that economy-scale assessment can be used to guide specific strain design decisions in metabolic engineering, and that these design decisions can be affected by non-intuitive dependencies across multiple scales.

422565 The Impact of Natural Gas and Natural Gas Liquids Supplies on the United States Chemical Manufacturing Industry

Abstract:

Recent advancements in horizontal drilling and hydraulic fracturing have led to an increase in domestic natural gas and natural gas liquids (NGL) production. This increase in availability, at low cost, of natural gas and associated liquids has spurred changes in both feedstock and energy use throughout the chemical manufacturing industry. Because of their low cost and high domestic availability, there is an incentive for manufacturers to use natural gas and NGL as a feedstock where possible, replacing heavy liquids such as naphtha. The increased use of NGL as a feedstock can impact the production costs of downstream materials and therefore the competitiveness of some process technologies.

Because of the complex material connections in the chemical manufacturing industries, as more NGL are used as feedstocks, many different supply chains will be impacted. Structural changes will not be restricted to the direct supply chains of NGL use (ethylene, propylene, etc.) but will also propagate throughout the entire network of chemical manufacturing operations because of the importance of byproducts in many reactions. For example, butadiene, a byproduct of ethylene cracking, is used in synthetic elastomer production, so changes in ethylene cracking technology could impact supply and cost of raw materials for rubber production.

To quantify the effects that changing NGL price and supply have on downstream chemicals, a network model of the U.S. petrochemical industry was constructed. The linear program uses stoichiometric relationships between manufacturing processes to form the structure of the network. Due to the interdependent nature of the industry, changes in feedstock availability and price can have impacts that propagate throughout the entire network, influencing production costs and the feasibility of specific processing pathways. Previously developed network models of the chemical processing industries have sought to discern the optimal industry structure (technologies chosen that meet all end product demand constraints) [1, 2]. This work extends that modeling framework to determine not only optimal technologies, but also the impact on the production costs of all downstream materials as natural gas and NGL prices change. Understanding which downstream materials are impacted by natural gas/NGL prices and the magnitude of that cost effect is important because the relationship between upstream raw material price and production costs for farther downstream materials is not always apparent. For example, a reduction in ethane feedstock price for an ethylene cracker does not mean that every product from the cracking operation will become cheaper (as evidenced by the increase in U.S. butadiene spot price from 2008 to 2012 even as ethane prices decreased) [3]. Through a series of pricing and supply scenarios, the relationship between upstream primary raw materials and 276 downstream intermediate/end product production costs is explored.

With an exogenous change in natural gas prices, 32 downstream chemicals show a change in production cost. As the selection of optimal technologies adapts to changing natural gas prices, acetaldehyde is identified as a potential bottleneck intermediate. As NGL prices are changed, 65 downstream chemicals show a change in production cost. Changes in the ratio of NGL prices did not substantially impact the industry's structure. It was also found that while availability of natural gas and NGLs and quality of crude oil do impact industry structure, raw material price more than supply availability will influence technology choices and utilization levels [4].

The abundance of these lighter feedstocks also allows for the possibility of designing new processing pathways to take advantage of their availability. This work involves using the shadow prices of materials in the dual linear program to calculate the characteristics that a new process must have to be accepted as part of the optimal solution. These results can be used to direct development of new processes. For example, considerable catalyst design work has been conducted to convert methane into a variety of aromatics [5, 6, 7]. Each different catalyst design has a different aromatic selectivity. Analyzing chemical shadow prices in the dual linear program enables calculation of the optimal aromatic selectivity that will allow for inclusion of the new process as part of the industry structure. Understanding the behavior of the target process operating in the complete industry network also allows for a robust calculation of the process cost that would make the new process competitive. This process cost can act as a starting point during further catalyst and process development.

[1] Rudd, D; Fathi-Afshar, S; Trevino, A; Stadtherr, M. Petrochemical Technology Assessment; Wiley Series in Chemical Engineering; Wiley: New York, (**1981**).

[2] Stadtherr, M; Rudd, D "Systems study of the petrochemical industry," Chem. Eng. Sci., 31, 1019–1028 (**1976**).

[3] Process Economics Program Yearbook, 2012. IHS. http://chemical.ihs.com/PEP/yearbook.htm (accessed January 2015).

[4] DeRosa, S; Allen, D "Impact of Natural Gas and Natural Gas Liquids Supplies on the United States Chemical Manufacturing Industry: Production Cost Effects and Identification of Bottleneck Intermediates," *ACS Sustainable Chem. Eng.*, *3* (3), 451–459 (**2015**).

[5] Ma, S; Guo, X; Zhao, L; Scott, S; Bao, X "Recent Progress in Methane Dehydroaromatization: From Laboratory Curiosities to Promising Technology," *J Energy Chem*, *22*, 1-20 (**2013**).

[6] Guo, X; Fang, G; Li, G; Ma, H; Fan, H; Yu, L; Ma, C; Wu, X; Deng, D; Wei, M; Tan, D; Si, R; Zhang, S; Li, J; Sun, L; Tang, Z; Pan, X; Bao, X "Direct, Nonoxidative Conversion of Methane to Ethylene, Aromatics and Hydrogen," *Science, 344*, 616-619 (**2014**).

[7] Holmen, A "Direct methane conversion to fuels & chemicals," *Catal Today*, 142, 2-8 (**2009**).

428195 Multi-Objective Optimization Applied to Sustainable Rain-Fed and Irrigated Crop Production: A Case Study of Wheat Production in Spain

Abstract:

Water scarcity is a major environmental problem worldwide with a large impact on the degradation of ecosystems and resources. Particularly, in arid and semiarid countries water consumption causes significant environmental impacts on natural resources, ecosystems and even human health. Most production systems create water shortages which affect natural systems and, in particular, the agricultural sector represents more than two-thirds of the global water use. In this backdrop, appropriate environmental strategies and policies are needed in the agricultural sector in order to achieve a more sustainable agricultural production system.

Previous works have concentrated on quantifying the environmental impact of water consumption. To achieve a more sustainable agricultural production system, however, it is necessary to integrate these descriptive metrics with tailored decision-making tools.

Optimization tools can assist decision-making in this area. In particular, multi-objective optimization is a well suited approach to address environmental issues when production objectives must be optimized without neglecting the environmental concerns. Particularly, the combined use of multi-objective optimization techniques and life cycle assessment principles is a useful tool for guiding the search towards more sustainable agriculture patterns.

This paper presents a systematic multi-objective optimization tool for optimizing rain-fed and irrigated areas considering economic and environmental criteria. Our tool is based on a linear programming model that integrates water footprint data and life cycle assessment principles. The water footprint methodology estimates the crop water requirements for green, blue and grey water in a specific location, while the life cycle assessment methodology determines the potential environmental impact on ecosystems and resources. The solution of the multi-objective linear programming model developed is given by a set of optimal Pareto alternatives, each showing a unique combination of objectives values and entailing specific optimal rain-fed and irrigated cropping areas.

The capabilities of the proposed approach have been demonstrated through its application to a real case study of wheat (*Triticum aestivum* L.) production in Spain. Results show that significant reductions in the damage to ecosystems and resources can be attained while still maintaining or even increasing the wheat production by an adequate distribution of the rain-fed and irrigated wheat areas in the Spanish watersheds. Moreover, some of the Pareto solutions found, each one entailing specific optimal crop areas in each watershed, improve the current situation simultaneously in the two objectives considered. From the set of Pareto solutions, decision and policy makers should select the best one according to their economic and environmental preferences and considering the views of the farmers and stakeholders (social concerns). Therefore, these optimal solutions represent a useful guide to define appropriate strategies and policies that will ensure a more sustainable agriculture system, minimize the environmental damage and ensure food security.

431394 3E (Engineering-Environmental-Economic) Triangle Model for Optimization of High-Gravity Carbonation Process: Establishment of Waste-to-Resource Supply Chain

Abstract:

An integrated portfolio of multi-waste treatment (i.e., steelmaking slag and wastewater) combined with CO_2 capture in the steelmaking industry can be achieved by the high-gravity carbonation (i.e., HiGCarb) process using a rotating packed bed (RPB). It was noted that a high CO_2 capture efficiency (i.e., >98%) can be achieved by the HiGCarb process with a relatively short reaction time at ambient temperature and pressure. It also has been proven that the process performance, i.e., rates of metal ion leaching and CO_2 dissolution, can be further promoted by introducing wastewater from the steelmaking industries. Furthermore, since the reacted product (e.g., carbonated solid wastes) can be used as supplementary cementitious materials, CO_2 emissions from the cement industry can be avoided if a green supply chain between the steelmaking and cement industries is established. However, several critical issues such as energy consumption, net CO_2 emission reduction, indirect CO_2 emission avoidance, and cost-benefit analysis of the HiGCarb process have still not been comprehensively addressed. Therefore, the HiGCarb process should be critically assessed through a 3E (engineering, environment, and economy) triangle model.

Since the complex relationships among different aspects can be easily visualized on a ternary plot among different scenarios, the triangle graphical presentation has been extensively used for evaluating key factors that are related but also complementary. In this study, the 3E triangle model considers the aspects of life-cycle environmental impact (LCEI) on the X axis, engineering performance (EP) on the Y axis, and life-cycle cost (LCC) on the Z axis. The objectives of this study were to (1) evaluate the energy consumption and net CO_2 capture amount by the HiGCarb process (engineering aspect); (2) quantify the environmental impacts of the HiGCarb process through LCA from the cradle-to-gate point of view (environmental aspect); (3) estimate the operating costs and revenue gained including carbon credits and product sale profits (economic aspect); and (4) determine the optimal operating modulus according to the results of the comprehensive performance evaluation using the 3E triangle model.

The results indicated that an increase in CO_2 capture performance should simultaneously reduce the potential costs and environmental impacts, which make integration of the HiGCarb process into the steelmaking industry more economically viable and environmentally friendly. It was concluded that an integrated approach to the proper treatment of alkaline wastes (i.e., wastewater and steelmaking slags) that permanently fixes CO_2 from the steelmaking industry while producing valuable supplementary cementitious materials from the cement industry can be achieved via the HiGCarb process.

436567 Using Sustainability Footprint in Optimization Framework for Prospective Design of Chemical Processes

Abstract:

Using sustainability footprint in optimization framework for prospective design of chemical processes

AIChE Annual Meeting, Salt Lake City, UT, November 8-13, 2015

Debalina Sengupta[1], Rajib Mukherje2[1], Subhas K. Sikdar[3], Mahmoud M. El-Halwagi[1]

1. *Texas A &M University, College Station, TX*
2. *Vishwamitra Research Institute, Crystal Lake, IL*
3. *NRMRL, US EPA, Cincinnati, OH,*

Abstract

A chemical product can be obtained by different process pathways involving varying amounts and types of resources, utilities, product, and byproduct formation. Evidently, the sustainability in terms of economic, environmental and social impacts of these processes will vary relatively. When such competing process options are presented, and compared to an existing process pathway, it is necessary to evaluate the most sustainable process among these competing options. Traditional process design stands on process synthesis and process analysis as the two pillars determining the technical and economic viability of the process options. Process integration techniques make use of these syntheses and analyses methods to arrive at the selection of the best process.

Sustainability of a chemical process is generally evaluated with indicators that need process and chemical properties data. These indicators individually reflect the impacts of the process on the respective areas of sustainability. For example, traditional economic indicators represent the economic dimension, environmental indicators the environmental dimension, and human health indicators represent the societal dimension. Process flow and economic data were used to compute the indicators in these categories. In order to choose among several alternative processes an overall comparative analysis is essential. To make a decision on the most sustainable process, the indicators need to be rationally aggregated into a single index. We use the sustainability footprint method (D_e) to evaluate the relative sustainability of each of the competing processes.

In this paper, we will present the results of this ongoing research on using the sustainability footprint method as a measure of overall sustainability of a process. We will also report on the research challenges faced during the use of the sustainability footprint method in mixed integer optimization framework for prospective sustainability analyses.

A Pseudo-Equilibrium Approach for Process-to-Planet Design Under Environmental Tax Policies

Rebecca J. Hanes and Bhavik R. Bakshi

William G. Lowrie Department of Chemical and Biomolecular Engineering

THE OHIO STATE UNIVERSITY

AIChE 2015 Annual Meeting
Tuesday, November 10, 2015

Motivation

Environmental policy decisions are made according to their expected effect on the economy as a whole.

The effects of new policies are also of interest at the plant scale - which *cannot* be captured with standard economic models.

Relevant questions include

1. How will the policy affect the profitability and environmental performance of an existing process?
2. How can the process be adapted to maintain profitability?
3. How will potential future policies affect a process currently in the design stage?
4. Can processes be designed to be resilient to anticipated future policies?

Methodologies

Objective: Model effects of environmental policy implementation on process design

Equipment scale

Effects of economic policy are modeled for the plant or process of interest only.

- Material, fuel or water taxes increase the cost of those inputs to the plant
- Emissions taxes add a cost of emitting to the plant's operating costs

Economy scale

Effects of economic policy are modeled for the entire economy.

- Material, fuel or water taxes increase prices economy-wide
- Emissions taxes increase operating costs of all industrial and manufacturing processes

Lam, Varbanov and Klemes. *Resour. Conserv. Recy.*, 54(5), 2010.
Ghanbari, Saxen and Grossmann. *AIChE J.*, 59(10), 2013.

Methodologies: Limitations

Equipment scale

- Ignoring large-scale effects of policy implementation risks wrong, suboptimal design decisions
- Engineering models are generally limited to smaller scales, cannot capture economy-wide effects

Economy scale

- Economic models lack information at the scale at which engineering decisions are made

Methodologies: Limitations

Equipment scale

- Ignoring large-scale effects of policy implementation risks wrong, suboptimal design decisions
- Engineering models are generally limited to smaller scales, cannot capture economy-wide effects

Economy scale

- Economic models lack information at the scale at which engineering decisions are made

- Model the effects of policy implementation at the economy scale
- Transfer those effects to the equipment scale for engineering design applications

Process to planet (P2P) modeling framework

- Integrate engineering, life cycle and economic models
- Small scale engineering models are detailed but limited
- Large scale economic models are comprehensive but coarse
- Multi-scale P2P model offers the necessary link between policy implementation at the economy scale and design decisions at the equipment scale

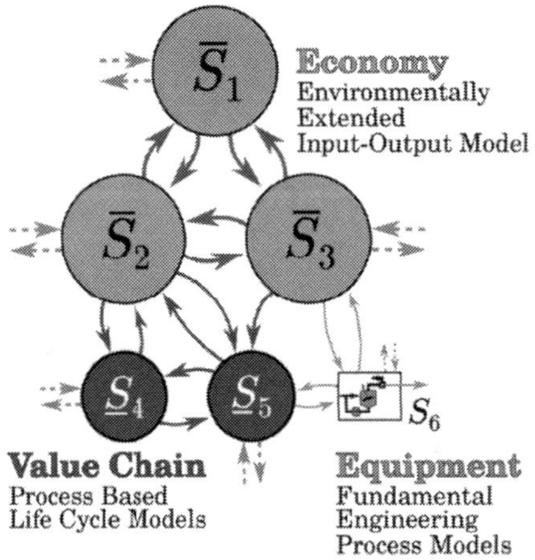

Hanes and Bakshi, *AIChE J.*, 61(10), 2015(a)(b)

Adapting P2P for design under economic policies

$$\underline{\overline{\mathbf{X}}}(\{\mathbf{z}\}) = \begin{bmatrix} \overline{\mathbf{I}} - \overline{\mathbf{A}}^{*}(\{\mathbf{z}\}) & -\underline{\mathbf{X}}_u(\{\mathbf{z}\}) & -\mathbf{X}_u^E(\{\mathbf{z}\}) \\ -\underline{\mathbf{A}}_d(\{\mathbf{z}\}) & \underline{\mathbf{X}}^{*}(\{\mathbf{z}\}) & -\mathbf{X}_u^V(\{\mathbf{z}\}) \\ -\mathbf{A}_d^E(\{\mathbf{z}\}) & -\mathbf{X}_d^V(\{\mathbf{z}\}) & \mathbf{X}(\{\mathbf{z}\}) \end{bmatrix}$$

- P2P framework uses an input-output model at the economy scale
- Commodity prices are fixed
- Require a different type of model at the economy scale in order to model and analyze economic policies

Adapting P2P for design under economic policies

$$\underline{\overline{\mathbf{X}}}(\{\mathbf{z}\}) = \begin{bmatrix} \overline{\mathbf{I}} - \overline{\mathbf{A}}^{*}(\{\mathbf{z}\}) & -\underline{\mathbf{X}}_u(\{\mathbf{z}\}) & -\mathbf{X}_u^E(\{\mathbf{z}\}) \\ -\underline{\mathbf{A}}_d(\{\mathbf{z}\}) & \underline{\mathbf{X}}^{*}(\{\mathbf{z}\}) & -\mathbf{X}_u^V(\{\mathbf{z}\}) \\ -\mathbf{A}_d^E(\{\mathbf{z}\}) & -\mathbf{X}_d^V(\{\mathbf{z}\}) & \mathbf{X}(\{\mathbf{z}\}) \end{bmatrix}$$

- P2P framework uses an input-output model at the economy scale
- Commodity prices are fixed
- Require a different type of model at the economy scale in order to model and analyze economic policies

Strategy: Replace the P2P framework's current EEIO model with a more flexible economic model

Pseudo-equilibrium (PsE) model

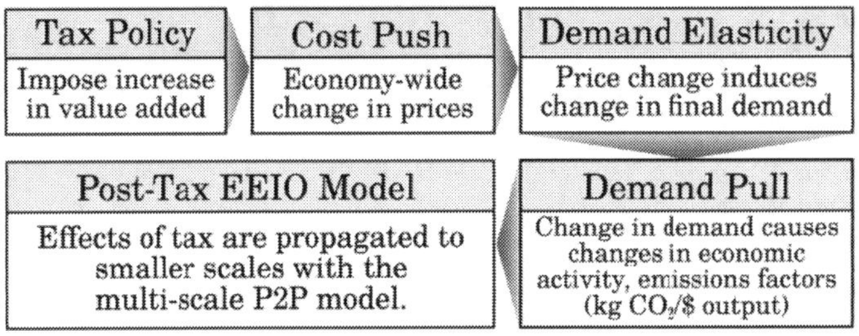

- Input-output based modeling approach
- Increase in value added represents environmental tax policy
- Cost-push model: Calculate increase in commodity prices
- Commodity price change and demand elasticity data: Calculate change in final demand
- Demand-pull model: Calculate post-policy economic activity levels, sectoral emissions factors

Choi, Bakshi and Haab, *Energy Policy*, 38, 2010.

P2P/PsE model

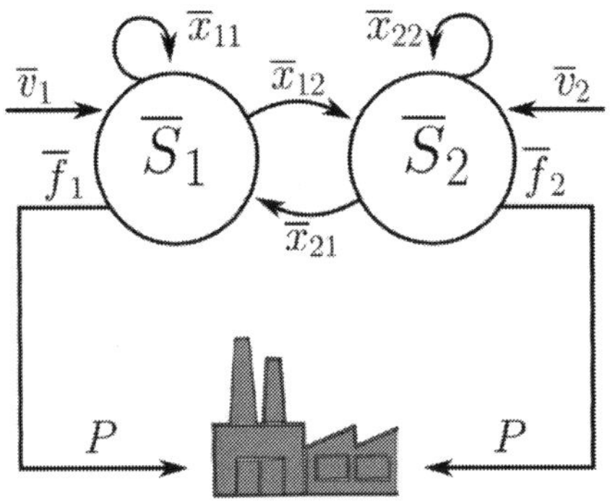

P2P/PsE model

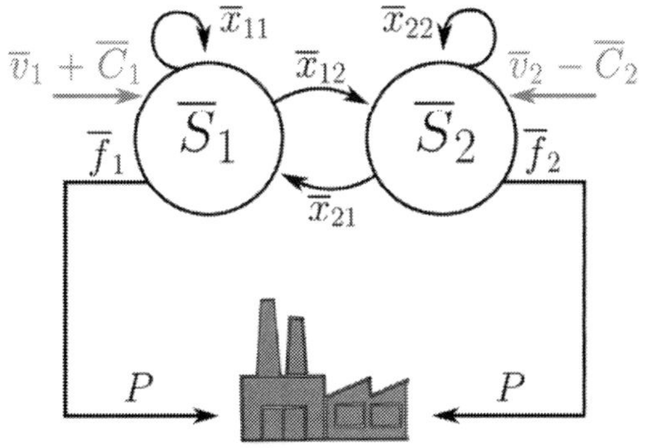

– Policy increases value added at the economy level

P2P/PsE model

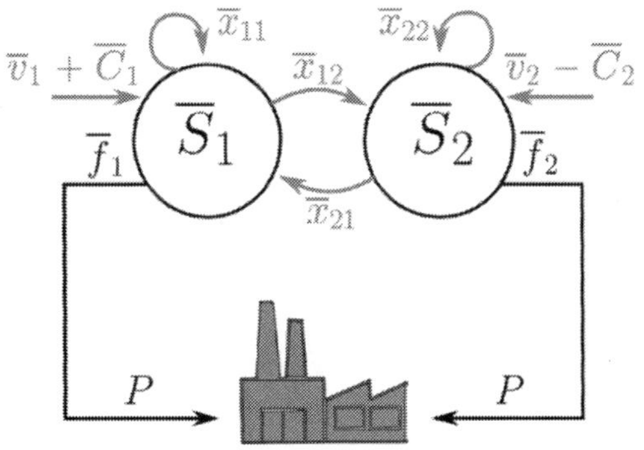

– Prices change throughout the economy

P2P/PsE model

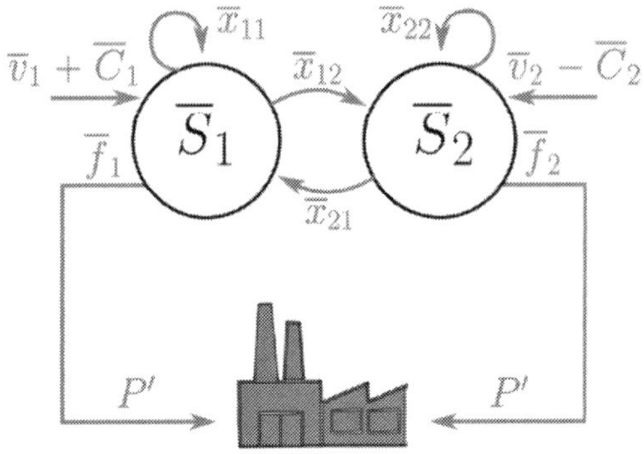

- Post-tax prices are passed down to the value chain and then to the equipment scales

P2P/PsE model

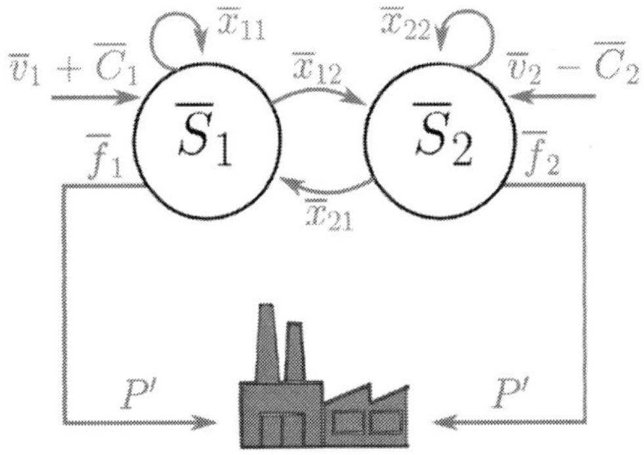

- Post-tax prices are passed down to the value chain and then to the equipment scales
- Process design is re-optimized under post-tax economic objective function

P2P/PsE model

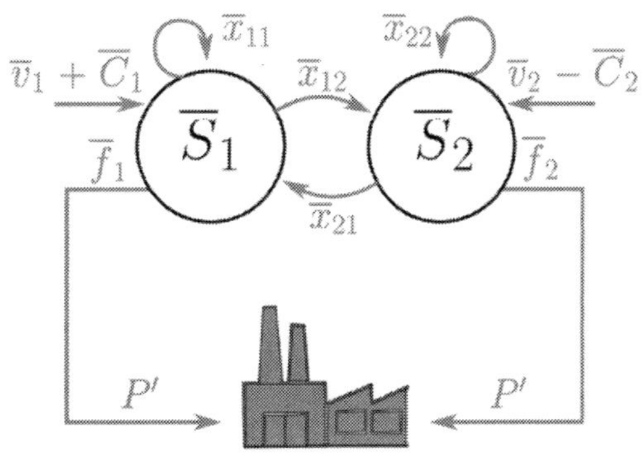

- Post-tax prices are passed down to the value chain and then to the equipment scales
- Process design is re-optimized under post-tax economic objective function
- Analyze policy's effect on emissions generated by the process and its life cycle

Demonstration of P2P/PsE model

- Compare economically optimal plant designs under three environmental tax scenarios
  1. Baseline case: No tax policy
  2. Tax modeled at equipment scale only
  3. Tax modeled at economy scale: Full P2P/PsE model

Demonstration of P2P/PsE model

- Compare economically optimal plant designs under three environmental tax scenarios
  1. Baseline case: No tax policy
  2. Tax modeled at equipment scale only
  3. Tax modeled at economy scale: Full P2P/PsE model
- P2P/PsE model consists of
  - 2007 US EEIO model
  - Demand elasticity data for 2007 from GTAP database
  - Value chain models from GREET 2015
  - Toy equipment scale model

Demonstration of P2P/PsE model

- Compare economically optimal plant designs under three environmental tax scenarios
  1. Baseline case: No tax policy
  2. Tax modeled at equipment scale only
  3. Tax modeled at economy scale: Full P2P/PsE model
- P2P/PsE model consists of
  - 2007 US EEIO model
  - Demand elasticity data for 2007 from GTAP database
  - Value chain models from GREET 2015
  - Toy equipment scale model
- Evaluate tax on CO_2 emissions at three levels: \$100, \$150 and \$200 per short ton CO_2

Demonstration of P2P/PsE model

- Compare economically optimal plant designs under three environmental tax scenarios
  1. Baseline case: No tax policy
  2. Tax modeled at equipment scale only
  3. Tax modeled at economy scale: Full P2P/PsE model
- P2P/PsE model consists of
  - 2007 US EEIO model
  - Demand elasticity data for 2007 from GTAP database
  - Value chain models from GREET 2015
  - Toy equipment scale model
- Evaluate tax on CO_2 emissions at three levels: \$100, \$150 and \$200 per short ton CO_2
- Hypothesis: Equipment scale tax model is insufficient to capture effects of environmental policy, will lead to adverse designs with higher emissions

Results: Baseline vs. equipment scale tax model

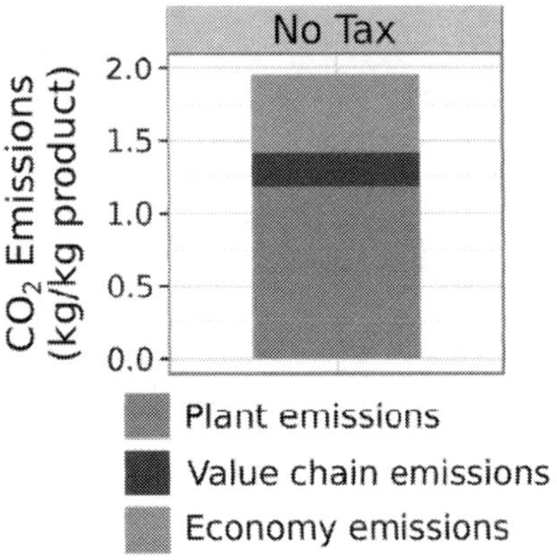

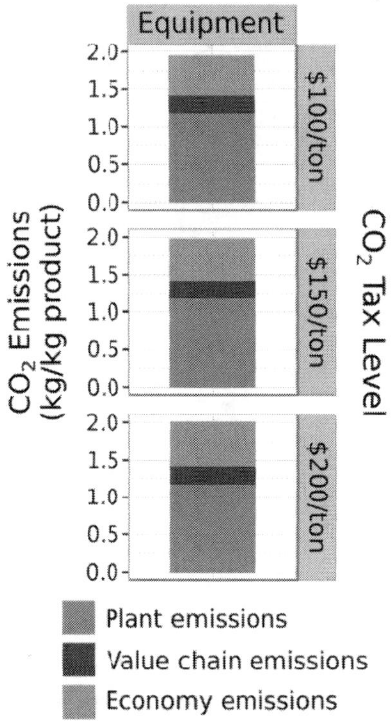

- Insignificant changes in economically optimal plant designs
- Similarly insignificant changes in CO_2 intensity

Results: Economy scale tax model

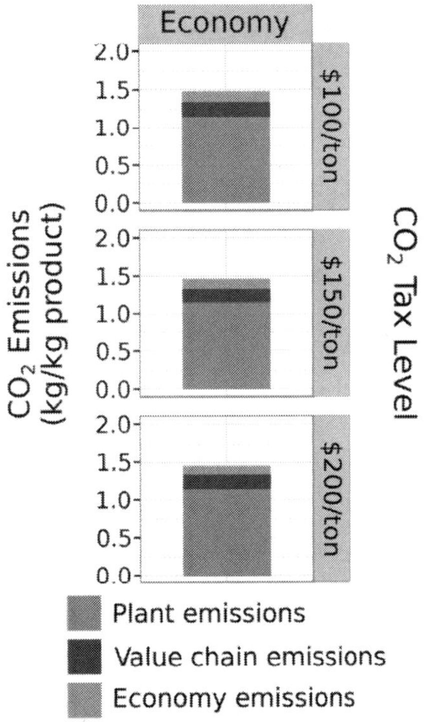

- Shift in process design to consume inputs with lower post-tax prices: less CO_2 intensive inputs
- Reduction in energy consumption, increase in water consumption
- Net result is lower CO_2 per unit product, as expected

Results: Economy scale tax model

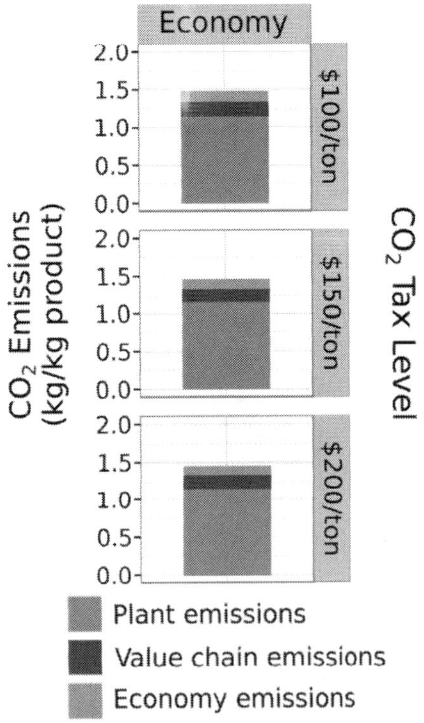

- Shift in process design to consume inputs with lower post-tax prices: less CO_2 intensive inputs
- Reduction in energy consumption, increase in water consumption
- Net result is lower CO_2 per unit product, as expected

- Next: Plant design obtained under the equipment scale tax model is evaluated under the economy scale tax model

Results: Same design, different tax models

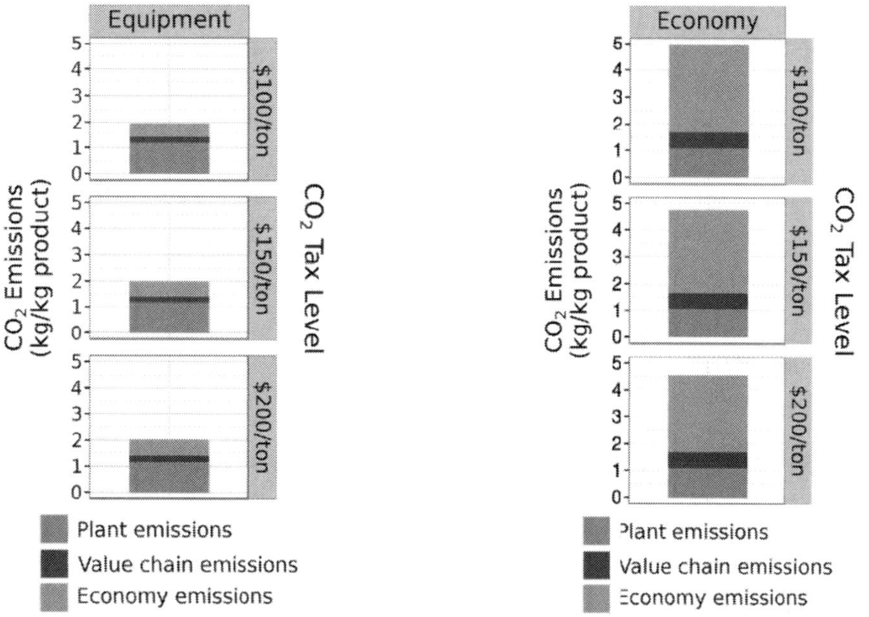

Results: Same design, different tax models

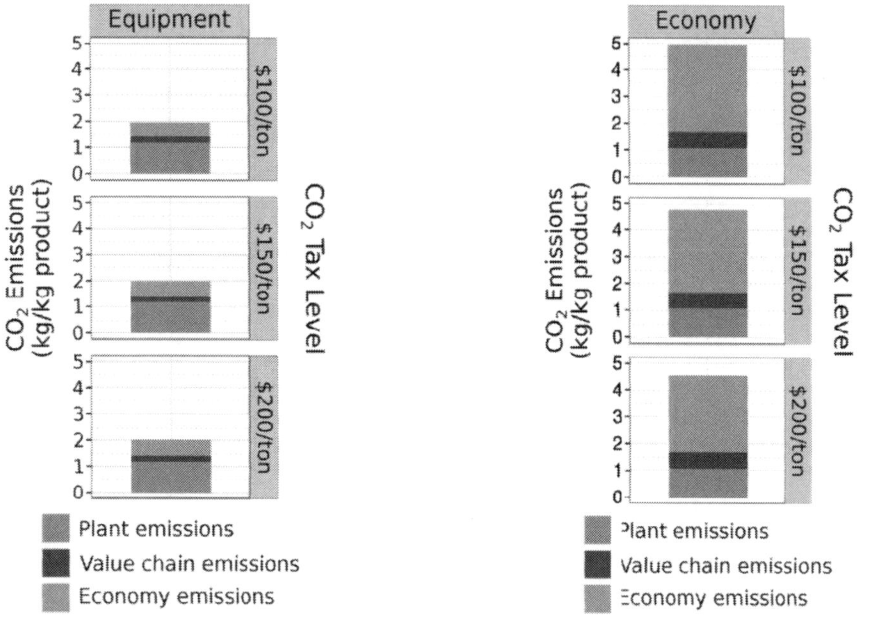

Plant design optimized under equipment scale tax model has emissions $> 100\%$ higher than design optimized under economy scale tax model

Discussion: Applicability of P2P/PsE model

- Analyze design decisions such as:
  - Utilize reagent A or reagent B?
  - What's the best fuel source for on-site steam generation?
  - Better to be energy-efficient, material-efficient, or water-efficient?
- Approach is particularly useful in early stages of process and supply chain design
- Build adaptability into process for resilience to future environmental policies

Discussion: Limitations of P2P/PsE model

- Simplified version of events before and after policy implementation: Producers cannot anticipate or prepare for policy
- Producers can adapt to the policy only by changing prices at the economy, value chain scales: All prices increase, albeit by varying amounts
- Resulting price changes are modeled either at the economy scale (71 sectors) or at the value chain scale: No regional detail

Conclusions

Pseudo-equilibrium economic policy model was incorporated into the process to planet modeling framework, successfully connecting the effects of economy scale environmental policies to engineering decisions made at the equipment scale.

Conclusions

Pseudo-equilibrium economic policy model was incorporated into the process to planet modeling framework, successfully connecting the effects of economy scale environmental policies to engineering decisions made at the equipment scale.

- Demonstrated the necessity of including economy-wide effects when modeling economic policies
- Approach is useful for early stages of design, can be used to gauge process' resilience to future policies
- Limitations of approach include simplicity of policy model, lack of temporal dynamics, lack of regional variability

Conclusions

Pseudo-equilibrium economic policy model was incorporated into the process to planet modeling framework, successfully connecting the effects of economy scale environmental policies to engineering decisions made at the equipment scale.

- Demonstrated the necessity of including economy-wide effects when modeling economic policies
- Approach is useful for early stages of design, can be used to gauge process' resilience to future policies
- Limitations of approach include simplicity of policy model, lack of temporal dynamics, lack of regional variability

Next steps

- Apply P2P/PsE model to real-world case studies
- Evaluate use of partial, general equilibrium models in place of pseudo-equilibrium model

425318 Dynamic Simulation and Multi-Objective Optimization of Flavor and Processing Time in Beer Fermentation

Abstract:

Fermentation is an essential step in beer brewing: when yeast is added to hoppened wort, sugars released from the grain during germination are fermented into ethanol and higher alcohols. In order to study, simulate and optimise the beer fermentation process, it is necessary to formulate an accurate kinetic model of the chemical system for dynamic simulation of key component concentrations. Since the entire beer production process is a highly complex series of chemical reactions with the presence of over 600 species (Vanderhaegen et al., 2006), many of the specific interactions are not quantitatively understood, rendering the construction of a complete deterministic dynamic model extremely cumbersome. In the fermentation process. A reduced-order dynamic beer fermentation model considering only the key chemical reaction pathways can be used for numerical optimisation of suitable manipulated variable profiles, providing accurate results which can be directly used for industrial implementation to an existing process. A typical optimisation goal in a beer fermentation process is achieving minimum brewing time, while also ensuring several by-product concentrations are monitored in order to maintain high product quality.

Two of the published differential algebraic equation (DAE) models have been successfully validated and indeed provide a detailed representation of the beer fermentation process (Gee and Ramirez, 1988, 1994; de Andrés-Toro, 1998). While they differ in terms of mathematical structure, chemical species considered and parameter values used, both have been demonstrated to predict the primary species profiles in good agreement with real plant data. The latter model has formed the basis of several subsequent optimisation studies which considered various systematic protocols to obtain dynamic temperature manipulation profiles capable of achieving higher production efficiency and improving industrial practice (Carrillo-Ureta et al., 2001; Xiao et al., 2003).

This paper presents a computational implementation of a beer fermentation model which enables the prediction of the dynamic profiles of all species considered: the de Andrés-Toro (1998) model consists of seven ODEs, considering a single sugar, ethanol, three distinct biomass forms and two undesirable by-products (ethyl acetate and diacetyl molecules). Our code has been validated using temperature profile inputs published by several authors (Carrillo-Ureta et al., 1999; Xiao et al., 2003) and indeed predicts species concentration profiles in excellent agreement with published results.

Dynamic optimisation of the beer fermentation process on the basis of our validated model implementation is based on an optimisation algorithm which uses an appropriate objective function (minimisation of fermentation time) and determines a temperature manipulation protocol which improves process operation by minimising fermentation time while also adhering to the prescribed composition quality constraints. The simultaneous consideration of four explicit parameters (fermentation time and concentrations of three key flavour attributes: ethanol, ethyl acetate, diacetyl compounds) enables the exploration of a vast solution space, rendering a truly multi-objective optimisation problem. A focal point of our research (in contrast to certain prior studies on beer fermentation process optimisation) is the use of real industrial plant data in order to validate our model implementation and evaluate the projected process operation improvement against the established temperature manipulation profile of current industrial practice.

LITERATURE REFERENCES

1. Carrillo-Ureta, G., P. Roberts, and V. Becerra, 2001. Genetic algorithms for optimal control of beer fermentation. *Proc. IEEE Int. Symp. Intell. Control*, 391-396.
2. de Andrés-Toro, B., et al., 1998. A kinetic model for beer production under industrial operational conditions. *Math. Comput. Simulat.* **48**(1): 65-74.
3. Gee, D.A. and W.F. Ramirez, 1988. Optimal temperature control for batch beer fermentation. *Biotechnol. Bioeng.* **31**(3): 224-234.
4. Gee, D.A. and W.F. Ramirez, 1994. A flavour model for beer fermentation. *J. Inst. Brewing* **100**(5): 321-329.
5. Vanderhaegen, B., et al., 2006. The chemistry of beer aging – a critical review. *Food Chem.* **95**(3): 357-381.
6. Xiao, J., Zhou, Z., Zhang, G., 2003. Ant colony system algorithm for the optimization of beer fermentation control. *J. Zhejiang Univ.* **5**(12): 1597-1603.

430730 A System Identification Enhanced Phenotype Phase Plane Analysis

Tuesday, November 10, 2015: 12:48 PM
Salon E (Salt Lake Marriott Downtown at City Creek)

Kyle Stone, Chemical Engineering, Auburn University, Auburn, AL, Matthew Hilliard, Chemical Engineering, Auburn Unviversity, Auburn, AL, Q. Peter He, Chemical Engineering, Tuskegee University, Tuskegee, AL and Jin Wang, Auburn University, Auburn, AL

A System Identification Enhanced Phenotype Phase Plane Analysis

Kyle Stone[1], Matthew Hilliard[1], Q. Peter He[2], and Jin Wang[1]

[1]Department of Chemical Engineering, Auburn University, Auburn, AL, 36849, USA

[2]Department of Chemical Engineering, Tuskegee University, Tuskegee, AL, 36088, USA

<>Abstract:

Genotype-phenotype relationship is fundamental to biology, and predicting different phenotypes based on the sequenced genome is one of the main goals for genome-scale metabolic model development. These models provide a holistic view of the organism's metabolism, and constraint-based metabolic flux analysis methods have been used extensively to study genome-wide cellular metabolic networks (Balagurunathan et al., 2012; Caspeta et al., 2012; Orth et al., 2010).

A prevalent method to dissect genome-scale metabolic models is phenotype phase plane (PhPP) analysis, which breaks down metabolic behavior in distinct phases through shadow prices analysis (Edwards et al., 2001). The shadow price is defined as the effect of the metabolite concentration on the objective function. Two independent variables, typically the carbon and oxygen source, are varied, and the shadow price is calculated for each metabolite. Shadow price analysis has the capability to determine if a substrate was in limited or in excess supply, and whether a product was expelled from the network. However, due to the scale and complexity involved in genome-scale models, many times only knowing whether a metabolite is limited or excess and whether it is expelled from the network is not sufficient in characterizing different phenotypes. In order to fully characterize different phenotypes predicted by a model, it is highly desirable to figure out how different pathways interact with each other for a given phenotype, and how such interactions differ from different phenotypes.

To address this challenge, the system identification based (SID) framework that we developed previously for genome-scale model validation and refinement is extended to enhance PhPP. In the SID framework, we first perturb the network through designed input sequences, i.e., designed *in silico* experiments; then apply multivariate statistical analysis tools to analyze the *in silico* results in order to extract information on how such perturbations propagate through the network; finally, we visualize the extracted knowledge against the network map to provide easy accessibility. The SID framework has been successfully applied to genome-scale model evaluation, as well as guiding the development of an improved genome-scale model of *Scheffersomyces stipitis* (Damiani et al., 2015).

Figure 1 provides an overview of the SID enhanced PhPP. In the proposed approach, *in silico* experiments can be designed to obtain the same findings obtained through shadow price analysis with less computation. More importantly, the SID framework allows us to obtain further information to help characterize the different phenotypes and identify the key differences among them. Visualization tools provided in Raven toolbox has been modified to visualize the information extracted from the SID enhanced PhPP (Agren et al., 2013). The core metabolic network model of *E. coli* were used to demonstrate the effectiveness of the proposed SID-PhPP approach (Schellenberger et al., 2011).

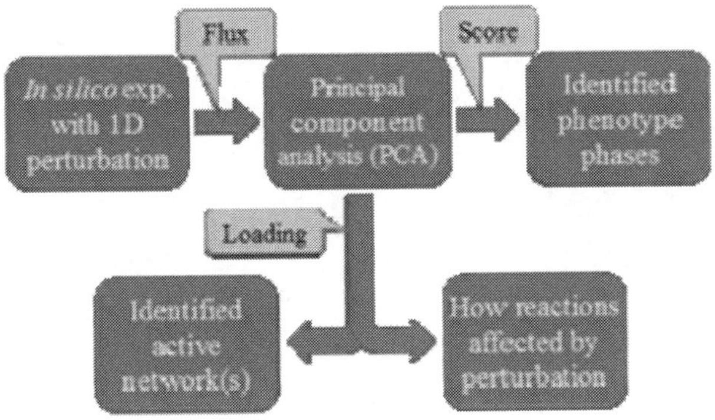

Figure 1 Overview of SID enhanced PhPP

References:

Agren R, Liu L, Shoaie S, Vongsangnak W, Nookaew I, Nielsen J. 2013. The RAVEN toolbox and its use for generating a genome-scale metabolic model for Penicillium chrysogenum. *PLoS Comput. Biol.* **9**. http://dx.plos.org/10.1371/journal.pcbi.1002980.

Balagurunathan B, Jonnalagadda S, Tan L, Srinivasan R. 2012. Reconstruction and analysis of a genome-scale metabolic model for Scheffersomyces stipitis. *Microb Cell Fact* **11**. http://www.biomedcentral.com/content/pdf/1475-2859-11-27.pdf.

Caspeta L, Shoaie S, Agren R, Nookaew I, Nielsen J. 2012. Genome-scale metabolic reconstructions of Pichia stipitis and Pichia pastoris and in silico evaluation of their potentials. *BMC Syst. Biol.* **6**. http://www.biomedcentral.com/1752-0509/6/24/.

Damiani AL, He QP, Jeffries TW, Wang J. 2015. Comprehensive Evaluation of Two Genome-scale Metabolic Network Models for Scheffersomyces stipitis. *Biotechnol. Bioeng.* http://onlinelibrary.wiley.com/doi/10.1002/bit.25535/abstract.

Edwards JS, Ramakrishna R, Palsson BO. 2001. Characterizing the metabolic phenotype: A phenotype phase plane analysis. *Biotechnol. Bioeng.* **77**:27–36. http://dx.doi.org/10.1002/bit.10047.

Orth JD, Thiele I, Palsson BØ. 2010. What is flux balance analysis? *Nat Biotechnol* **28**:245–248. http://dx.doi.org/10.1038/nbt.1614.

Schellenberger J, Que R, Fleming RM, Thiele I, Orth JD, Feist AM, Zielinski DC, Bordbar A, Lewis NE, Rahmanian S. 2011. Quantitative prediction of cellular metabolism with constraint-based models: the COBRA Toolbox v2. 0. *Nat. Protoc.* **6**:1290–1307. http://www.nature.com/nprot/journal/v6/n9/abs/nprot.2011.308.html.

432771 Nonlinear Data Mining and Parameter Reduction in Complex Reaction Networks

Abstract:

While there has been extensive work in reducing the number of variables in large, complex dynamic models, leading to models of smaller dimension, work on concomitant reduction in the number of parameters is a relatively recent endeavor; "sloppiness" and active subspaces are two promising current directions. In this paper we will explore the link between (nonlinear) data mining (via Diffusion Maps) and parameter reduction by analyzing "good" parameter sets, and determining their dimension and possible parameterizations. Combining equation-free principles with machine learning algorithms, we propose a data-based approach to resolve "sloppy" parameter directions in both singularly- and regularly-perturbed complex kinetic models, and employ it to speed up and hopefully improve parameter-fitting routines.

430225 Simulations of Stochastic Chemical Reaction Networks

Abstract:

Biological systems are wonderfully complex. In order to gain a better understanding on how the complexity dictates the biological functions and gives rise to different cellular phenotypes, it is important to investigate the underlying dynamic interactions of the biomolecular components. These interactions are governed by random events and thus stochastic models are needed to gain fundamental insight.

We have developed a closure scheme method that solves the master equation of stochastic chemical reaction models [1]. The method postulates that only a finite number of probability moments is necessary to capture all of the system's information, which can be achieved by maximizing the information entropy of the system.

Using the closure scheme, we examine the behavior of the stochastic Schlögl model, a toy, bistable chemical reaction network. We examine the behavior of the system with changes in the kinetic constants and focus on when the bifurcation to bistability occurs. Besides, we present an analysis of how the stochastic model behaves as the system approaches the thermodynamic limit.

Finally, we discuss the implementation of more advanced numerical methods in order to improve the convergence of the algorithm.

1. Smadbeck P, Kaznessis YN, "A Closure Scheme for Chemical Master Equations", Proc Natl Acad Sci U S A. 2013 Aug 27; 110(35): 14261-5 10.1073/pnas.1306481110

426403 Sensitivity Analysis for Nonsmooth Hybrid Systems

Abstract:

Dynamic chemical process systems exhibiting qualitative changes in behavior may be modeled as *hybrid discrete/continuous systems*, which are systems of differential equations that are punctuated by discrete events. At these events, state variables are permitted to jump, and the functions governing the state variables' continuous evolution may be switched. These events can be used to describe transitions in multi-stage processes, for example, or changes in flow regime or thermodynamic phase. This presentation describes new numerical methods for sensitivity analysis of hybrid systems, thereby broadening the class of hybrid systems for which sensitivity analysis is possible.

Firstly, a numerical method is presented for computing Nesterov's lexicographic derivatives [2] for implicit functions that are described in terms of nonsmooth residual functions. Lexicographic derivatives are analogs of classical derivatives of smooth systems, and provide local sensitivity information to methods for nonsmooth equation-solving and optimization.

Secondly, a numerical method is presented for computing lexicographic derivatives for *nonsmooth hybrid systems*. In these systems, the functions describing continuous dynamics, event timing, and post-event jumping are not required to be differentiable everywhere. This method combines the sensitivity analysis of nonsmooth implicit functions with our recent treatment [1] of lexicographic derivatives for nonsmooth dynamic systems. The desired sensitivities are described and computed as the state variables of an auxiliary hybrid system. This method permits sensitivity analysis of certain hybrid systems for which classical sensitivity analysis methods fail, including certain systems in which the order of discrete events encountered depends on system parameters. Examples are presented for illustration.

[1] K. Khan and P. Barton, Generalized derivatives for solutions of parametric ordinary differential equations with non-differentiable right-hand sides, *J. Optim. Theory App.*, 163 (2014), pp. 355-386.

[2] Y. Nesterov, Lexicographic differentiation of nonsmooth functions, *Math. Program. B*, 104 (2005), pp. 669-700.

421808 Time Scale Decomposition in Complex Reaction Systems: A Graph Theoretic Analysis

Abstract:

Complex reaction systems are ubiquitous. They are present in numerous chemical and biochemical systems, such as combustion, pyrolysis, nanoparticle synthesis, catalytic conversion of hydrocarbons, and cell metabolism. These systems involve a large number of reactions generating a large set of intermediate species. Automatic network generators allow generation of the exhaustive reaction network; however, the kinetic modeling of these systems is limited by their size and stiffness. Model reduction techniques involving lumping, sensitivity analysis and time-scale analysis can be used to address these challenges [1]. Model reduction using time scale analysis categorizes reactions as fast and slow, imposes pseudo-steady state hypothesis (PSSH) and/or quasi-equilibrium (QE) assumptions for the fast reactions, and generates pseudo-species which evolve in the slow time scale alone. A rigorous framework for performing these tasks for homogeneous reaction systems using singular perturbations was proposed in [2]; for the case of isothermal systems, a low-dimensional model of the slow dynamics can be expressed in terms of a set of pseudo-species obtained as a linear combination of the original species where the transformation matrix belongs to the left null space of the stoichiometric matrix of the fast reactions.

The present study proposes a graph theoretic analysis framework for identification of the pseudo-species evolving in a slow time scale. A directed bipartite graph is used to represent the species and reactions as two disjoint sets for a general chemical reaction network. An edge connecting the two sets represents species participation in a reaction as a reactant or product depending on the edge direction. Cycles consisting of a pair of species participating in fast reactions are identified using a backtracking algorithm [3]. These cycles are closed walks of two species nodes and two reaction nodes. Further, an algorithm that combines the identified cycles to form the pseudo-species, such that the pseudo-species interact with the remaining reaction network through slow reactions only, is developed. The proposed framework is scalable and applicable to heterogeneous systems as well. Three example reaction systems [2, 4] varying from $O(10^1)$ to $O(10^4)$ reactions are considered to show the efficacy of the proposed framework.

[1] Okino, M. S., and Mavrovouniotis, M.L. Simplification of Mathematical Models of Chemical Reaction Systems. *Chemical Reviews*, 98(2), 391-408, 1998.

[2] Vora, N., Daoutidis, P. Nonlinear Model Reduction of Chemical Reaction Systems. *AIChE Journal*, 47(10), 2320-2332, 2001.

[3] Tarjan, R. Enumeration of the elementary circuits of a directed graph. *SIAM J. Comput.*, 2(3), 211-216, 1973.

[4] Gerdtzen, Z.P., Daoutidis, P. and Hu, Wei-Shou. Non-linear reduction for kinetic models of metabolic reaction networks. *Metabolic Engineering*, 6(2), 140-154, 2004.

430874 Sensor Placement for Fault Diagnosis in Large-Scale Networks

Abstract:

Fault diagnosis (FD) is the process of detecting and diagnosing the cause(s) for abnormalities in a large process. This typically involves analysis of a large number of mathematical relationships among all variables. A concomitant problem is one of selecting the measured variables that should participate in the analysis from a very large number of possibilities. This sensor placement problem can be described as an optimization problem where an objective function such as the cost of the sensor network is optimized while constrained to satisfying the mathematical relationships and some FD metrics. Major difficulties that sensor placement techniques face are that these large processes are described in terms of a large number of partial differential equations (PDEs) that renders the analysis computationally complex and further, the FD metrics are generally ambiguous. Therefore, different techniques have been developed to address such difficulties in sensor placement. Cause and effect modelling is an interesting technique that simplifies the large process as a graph with process variables as nodes and arcs representing the relationship between the variables. In these representations, faults are assumed as root nodes (nodes with only outputs) that are connected to the affected variables through arcs and meaningful FD metrics can be derived. The graph-based approaches provide a qualitative information of the faults. For quantitative information of the faults (such as extent of catalyst deactivation in a packed-bed reactor), a model- based approach is used for sensor placement where usually involves state estimation of a large system of linear or nonlinear partial differential algebraic equations (PDAEs) that adds significant computational expense. For large-scale systems or networks where both qualitative and quantitative information of the faults is desired, the sensor placement problem is computationally cumbersome mainly due to nested operations involving a large number of nodes. In addition, the sensors should be placed synergistically so that the measurement information obtained at one level due to the placed sensors can be leveraged at the other level thus maximizing the efficiency of the measurement network.

In this work, we present a graph partitioning technique that decomposes a large process into subsystems and integrates the solutions derived from these subsystems to obtain the solution for the original large system. In general, naturally partitionable systems will decompose into individual non-interacting subsystems. However, many large-scale networks will not possess such natural partitions because of recycles and feedback loops. In such situations, artificial partitions that serve certain diagnosis objectives need to be created. We will describe various approaches for generation of efficient partitions of the large-scale system. Each partition has its own level of granularity which could include PDAE models, graph models and so on. Apart from the set of faults (a subset of the original fault set) present in these partitions, we propose that the interactions from the other partitions be handled as pseudo-faults. Based on such decompositions, we derive sensor placement algorithms that provide a solution to the large-scale networked process by assimilating the solutions from the partitioned systems in multiple ways. The importance of these ideas for various areas such as smart grids, water networks and large-scale fault diagnosis will be described.

432024 Impact of Spatial Segregation in Coupled Reactors with Catalytic and Autocatalytic Reactions

Abstract:

The effect of spatial segregation in catalytic reactors is mimicked by considering two well-mixed reactors with one-way and two-way coupling. The reactors in the two-reactor system are identical only in terms of feed composition and reactor space time and need not have identical volumes and volumetric feed, effluent, and exchange rates. A comprehensive analysis of static and dynamic behavior of autocatalytic reaction and a catalytic reaction exhibiting inhibition kinetics with respect to a reactant occurring in the two-reactor system is presented. Enzyme-catalyzed reactions subject to substrate inhibition belong to the latter class of reactions. A single well-mixed reactor may operate at up to three steady states. A two-reactor system may admit up to nine steady states, due to emergence of six additional steady states, at least at very low interaction rates. Three of these steady states, symmetric steady states, correspond to identical composition in the two reactors. The remaining steady states, the asymmetric steady states, correspond to different composition in the two reactors and are admissible over a range of interaction rates. Three different types of interaction are considered: (1) equal interaction, (2) unequal interaction, and (3) one-way interaction (series reactors). The six asymmetric steady states exhibit four different pairing sets with two or more limit points. When the interaction rates are equal, the asymmetric steady states are comprised of sets of steady state counterparts, each set admissible over the same range of interaction rate. In terms of merges at limit points, the asymmetric steady states exhibit two pairing sets in the operating parameter space. A switching in steady state pairs is caused through emergence of up to four additional steady states, which are admissible over the same range of interaction rate. When the interaction rates are not equal, three pairs of asymmetric steady states are admissible over different ranges of interaction rates. In terms of merges at limit points, the asymmetric steady states exhibit three pairing sets in the operating parameter space, two inherited from the equal interaction rate case. The additional pairing set corresponds to the unique pairing set for one-way interaction between the two reactors. Two such pairing sets are possible depending on which of the two interaction rates is higher. The two switches in steady state pairs are caused through emergence of four additional steady states, which are admissible over two different ranges of interaction rates. The additional pairing set is admissible over a broader region of the operating parameter space as the inequity between the interaction rates is increased. The limiting case of this inequity corresponds to two reactors in series, where there is only pairing set of the asymmetric steady states and no additional steady states emerge. Numerical illustrations reveal the rich steady state structure of the reaction schemes in coupled reactors. The steady states are tracked systematically as the interaction rate is varied. A catalytic reaction exhibiting inhibition kinetics or an autocatalytic reaction without autocatalyst decay does not lead to periodic states in two-reactor systems. The two-reactor system is operationally more flexible and more robust vis-a-vis single reactor. Emergence of additional steady states at low and intermediate values of interaction rates reveals that the two-reactor systems are examples of a complex system.

420041 Planning of Industrial-Scale Integrated Petrochemical Plants Using a Big-Data Analysis and Global Optimization Framework

Abstract:

Planning of the production of petrochemical products involves solving large-scale optimization models incorporating information such as: flowsheet topology, process capacity, operating constraints, intermediates and final product quality constraints, and market demand and costs [1-2]. Optimization of planning operations becomes significantly challenging when multiple refinery and chemical processing plants are integrated, creating complex networks and interactions through exchange of streams across plants [3]. Despite the existence of rigorous mathematical models for specific unit operations involved in petrochemical operations, the models which are suitable for entreprise-wide planning optimization must maintain a balance between accuracy and complexity in order to create overall tractable planning optimization formulations. In the past, linear formulations have predominantly been utilized, however, it has been verified that linear models lead to suboptimal solutions reducing the profitability and product quality [1-2]. Undoubtedly, planning operations have non-linear behavior in terms of the yields and properties of the intermediate and product streams, due to nonlinearities in the operation of individual unit operations as well as blending of streams. Commercially available software developed for planning operations either use linear approximation models, while in cases where non-linear models are used identification of the global optimum is not guaranteed. The need for accurate non-linear models and their integration within industrial-scale custom-made superstructures to form Mixed-Integer Non-Linear Optimization (MINLP) problems coupled with global optimal optimization methods is one of the posed future challenges in the field [1-2].

Hybrid or data-based models have been used within smaller planning problems in the literature, due to their simplicity, low computational expense and ability to accurately capture the behavior of real industrial data which is available in abundance [4-5]. One of the disadvantages of data-based modeling is their limited extrapolating abilities, requiring updating in case the operation or raw material properties have changed significantly. In this work we present the development of an industrial-scale flowsheet superstructure comprised of a hybrid set of non-linear unit operation models (first-principles and data-based). We use this grey-box framework to accurately represent an integrated petrochemical plant comprised of one refinery and two ethylene cracking plants. We show the performance of our fully automated framework which performs: (a) automated updating of unit operation model parameters based on real plant data, (b) integration of unit operation models with blending and pooling operations and mass-balance and capacity constraints, and (c) global optimization of the formulated planning problem using ANTIGONE [6]. Using the developed model, we solve multiple scenarios by modifying several uncertain parameters such as product demands, pricing, quality specifications and crude oil availability and properties and perform analysis on the obtained solutions. We also compare the obtained solutions with existing planning practices for the specific multisite plant and show that we have obtained significantly increased the profit while satisfying quality specifications.

References:

1. N.K. Shah, Z. Li, M.G. Ierapetritou, Petroleum Refining Operations: Key Issues, Advances and Opportunities, Industrial Engineering & Chemistry Research, 2011, 50, 1161-1170.

2. I. E. Grossmann, Challenges in the Application of Mathematical Programming in the Enterprise-wide Optimization of Process Industries, Theoretical Foundations of Chemical Engineering, 2014, 48(5), 555-573.

3. K. Al-Qahtani and A. Elkamel, Multisite Refinery and Petrochemical Network Design: Optimal Integration and Coordination, Industrial Engineering & Chemistry Research, 2009, 48, 814-826.

4. V. Mahalec, Y. Sanchez, Inferential monitoring and optimization of crude separation units via hybrid models, Computers and Chemical Engineering, 2012, 45, 15-26.

5. W. Li, C. Hui, A. Li, Integrating CDU, FCC and product blending models into refinery planning, Computers and Chemical Engineering, 2005, 29, 2010-2028.

6. R. Misener and C.A. Floudas, ANTIGONE: Algorithms for Continuous/ Integer Global Optimization of Nonlinear Equations, Journal of Global Optimization, 2014, 59(2-3), 503-526.

427680 Early Warning System Design: The Combination of Expert Knowledge and Historical Data

Abstract:

Early warning system design: the combination of expert knowledge and historical data

Hangzhou Wang, Faisal Khan*, Salim Ahmed

Safety and Risk Engineering Group, Faculty of Engineering and Applied Science,

Memorial University of Newfoundland, St. John's, NL, A1B 3X5, Canada

Since process industries deal with hazardous materials in daily operations, it is important to monitor the state of a process in real time to identify any vulnerable condition before it leads to a more severe event. Alarm systems are introduced to monitor these key variables states and give early warnings before severe event occur. Despite efforts to improve alarm systems, alarm flooding remains a significant problem in the process industries. Besides, these warnings only indicate these monitored variables deviations states, it is difficult to find the root cause for these warnings, or to predict what will happen with these changes. To solve this problem, an early warning system design methodology, combining expert knowledge and historical data, is proposed. The method comprises three steps as below:

1. Step 1: Identify events and scenarios in the process

1) Identify events by HAZOP analysis

2) Allocate variable to these events

3) Identify corresponded scenario and links with scenario

2. Step 2: Develop Bayesian Network based early warning system

1) Learn structure of monitored variables Bayesian Network from historical data

2) Learn parameters for monitored variables Bayesian Network with the given data set

3) Appending expert knowledge of events nodes to monitored variables Bayesian Network to construct the early warning system

3. Step 3: Implement the early warning system

1) Gather evidences of monitored variables states updating from real time variable state

2) Analyze the root causes of the current warning

3) Predict the occurrences of these events in real time

4) Take desired actions according to these occurrences probability

In this proposed methodology, with these continuously observed state changes of variables as evidence, the Bayesian Network forward inference method is used to update these events' probabilities. The Bayesian Network backward diagnosis method is used to find the root causes of the state change quickly. Once the probability of a certain event exceeds the acceptable upper threshold, an event warning message will be generated and conveyed to operators, together with the root causes if available. Compared to a conventional DCS alarm record, more information can be exploited from the same sequence of variables states' change with this proposed method; also more meaningful event warning messages are conveyed to operators, and desired action can be applied by operators in time to reduce the occurrence probability of this unfavorable event.

Keywords: Bayesian Networks, early warning system, structure learning, expert knowledge, historical data

---

* Corresponding author. Tel.: +1 709 864 8939

E-mail address: fikhan@mun.ca (Faisal Khan)

THE UNIVERSITY OF TEXAS AT AUSTIN

Data-Driven Modeling of Sequential Batch-Continuous Process

Jungup Park, Michael Baldea, Thomas Edgar

The McKetta Department of Chemical Engineering
The University of Texas at Austin

Outline

- Introduction
  - Industrial Discharge System
- Batch Monitoring
- Batch Data Alignment
- Continuous Process Modeling
- Conclusion

Introduction

- Sequential Batch-Continuous Process
  - Batch preparation followed by continuous process
    - Pharmaceutical
    - Semiconductor
    - Food industry
    - Industrial discharge system

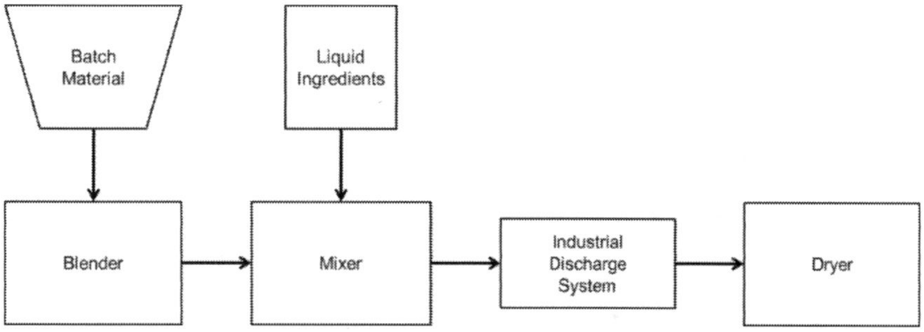

Motivation

- Final product quality is affected by
  - Batch Process
  - Continuous Process
- Detect and monitor the source of variations/disturbances in the upstream processes
- Make appropriate adjustments in the downstream process to negate the effect of the upstream disturbances

Discharge System

- Industrial Discharge System
  - Raw material preparation and blending in batch
  - Wet mixing process
  - Feeding to continuous discharge system
  - Drying and quality inspection

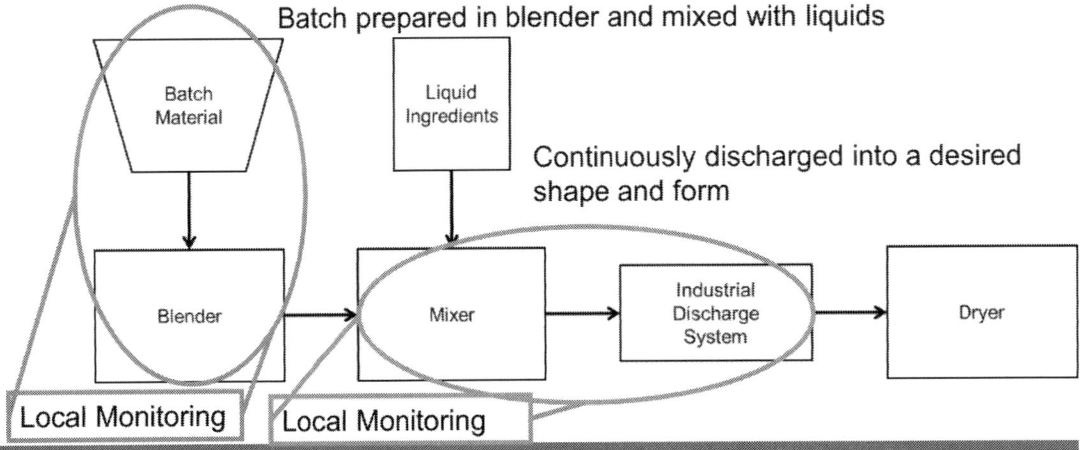

Batch Process

- Distribution of raw materials
  - Molecular weight distribution
  - Particle size distribution
  - Temperature distribution

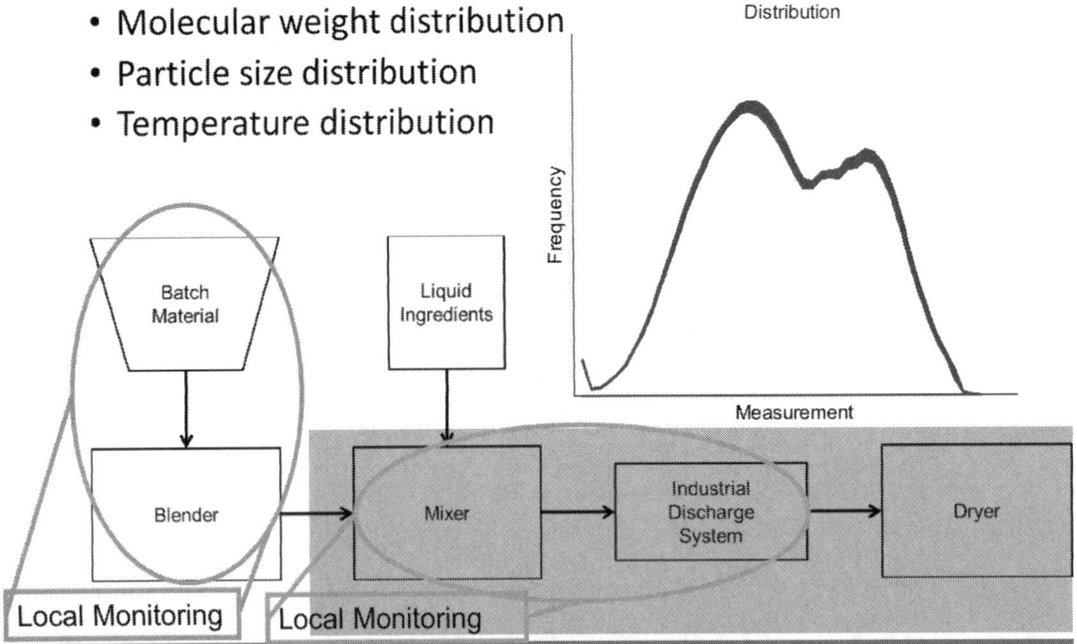

Batch Process

- Distribution measurements over time
  - Tracking individual bin does not reveal much about the overall distribution
  - Principal Component Analysis (PCA) reduces the number of variables to monitor

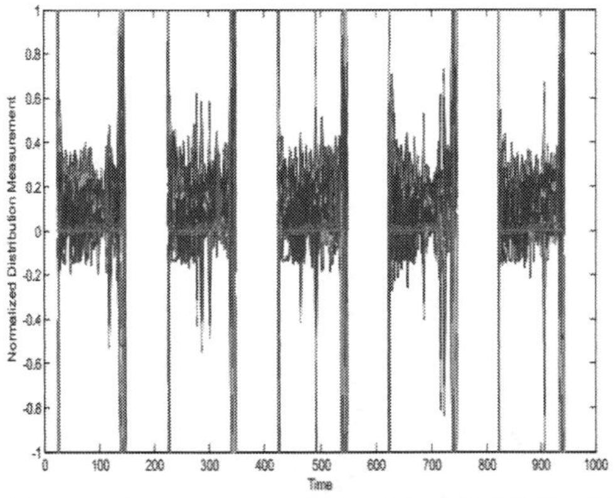

Batch Process

- PCA
  - Two PCs capture 79.5% variability in raw material distribution
  - PCs used for monitoring and modeling instead of bin measurements

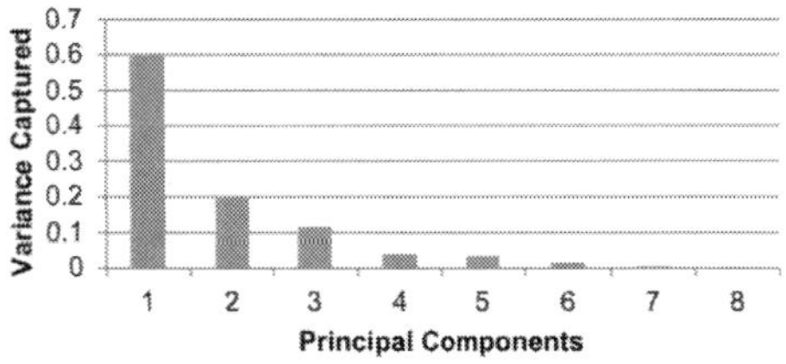

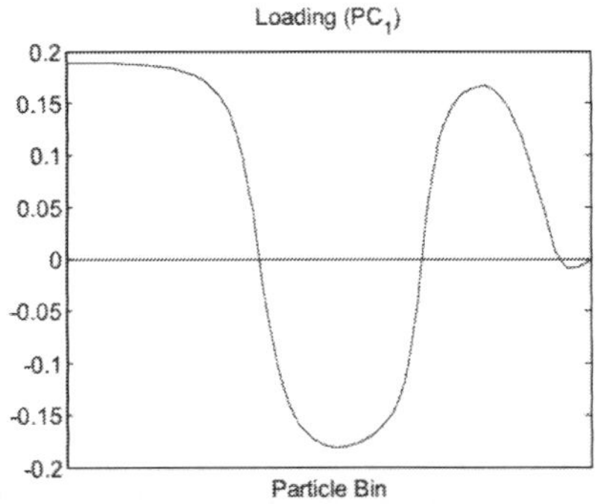

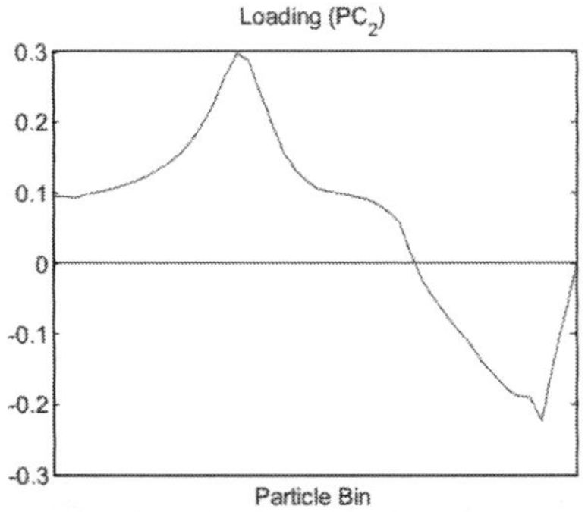

Batch Process

- Process monitoring using PCA
  - Hotelling's T^2 control chart
    - Captures variation within the model

$$T_i^2 = t_i^T \Lambda t_i$$

  - SPE control chart
    - Captures error outside the model

$$SPE_i = e_i^T e_i = x_i^T \left(I - PP^T\right) x_i$$

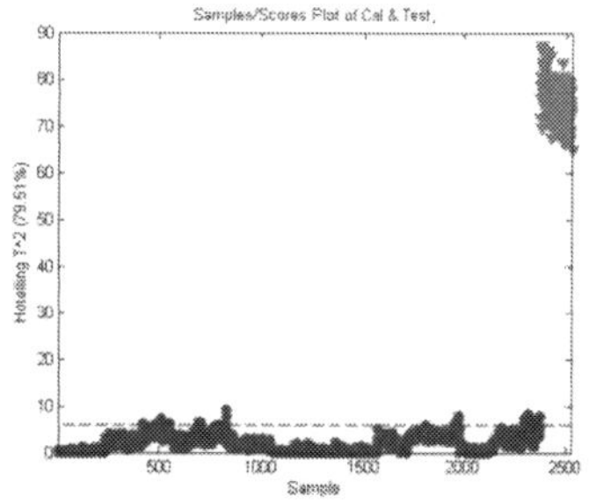

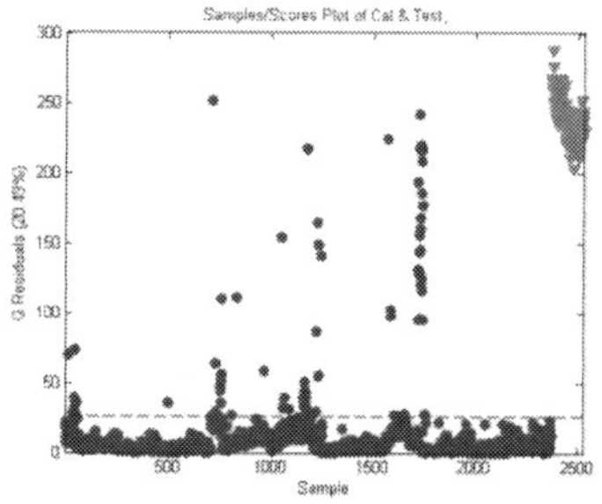

Batch Process

- Align batch dataset to continuous dataset

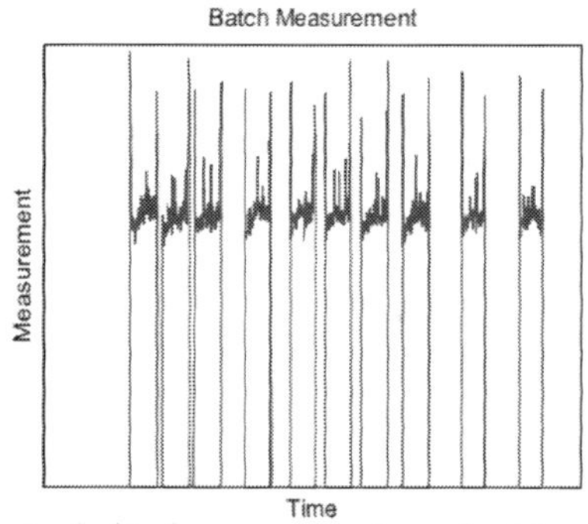

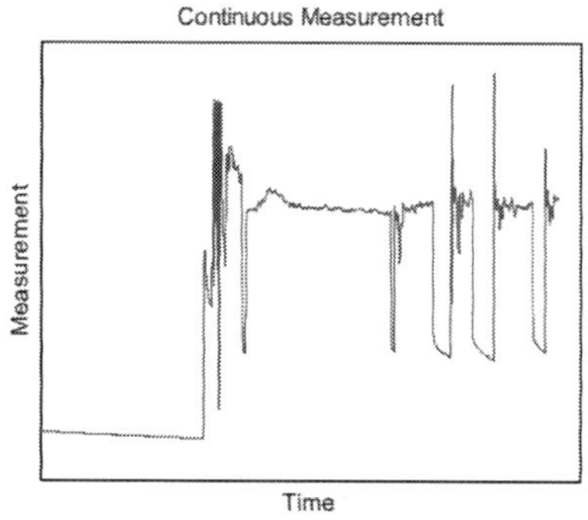

- Establish causal relationship
  - Relating upstream batch variables to downstream continuous process variables

Batch Process

- Batch data alignment
  - Filter out noise and remove outliers
  - Find nominal (average) value of the batch distribution
  - Assign the nominal value to the gaps between batches
  - Batch to batch variability is maintained

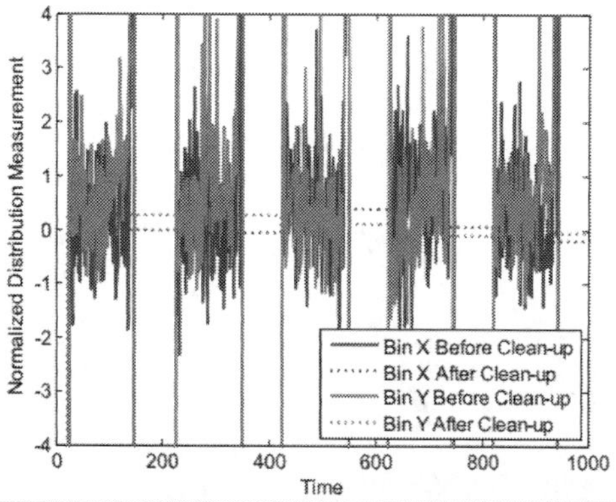

Batch Process

- Modeling batch and continuous process
  - Focused on large dataset (10+ separate production runs)
    - Effects of time delays between batch to continuous and within-batch dynamics ignored
    - Variability in production run to run is pronounced

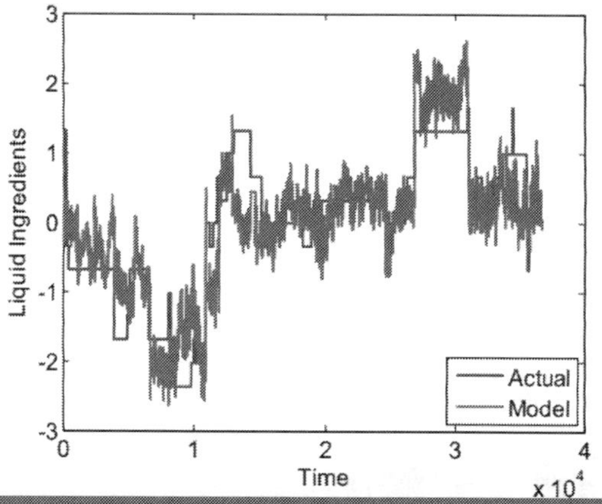

Continuous Process

- Key process variables (online continuous measurement)
  - Pressure
  - Temperature
  - Etc.

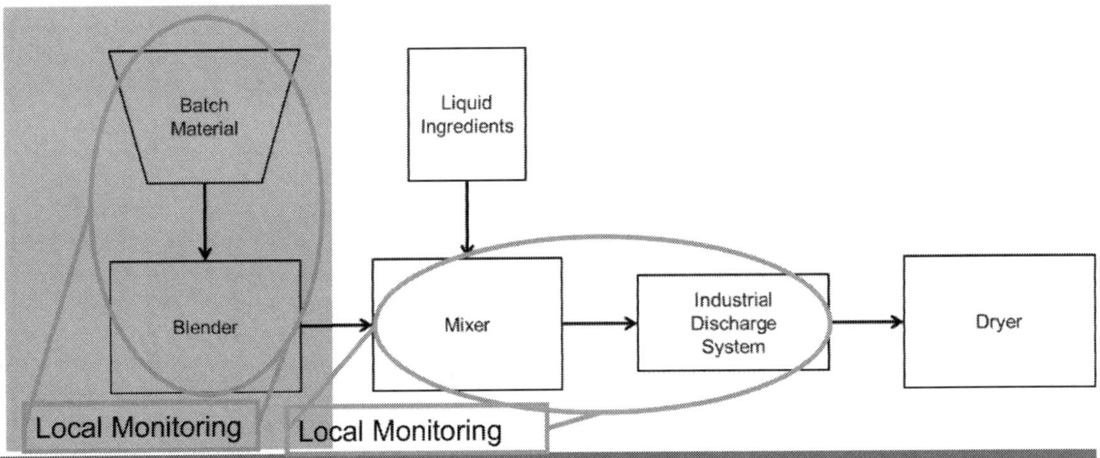

Batch and Continuous Processes

- Predicting process variable at the end of discharge system
  - Estimate of final product quality
  - Features different steady-state values
    - Similar to having multiple operating regions

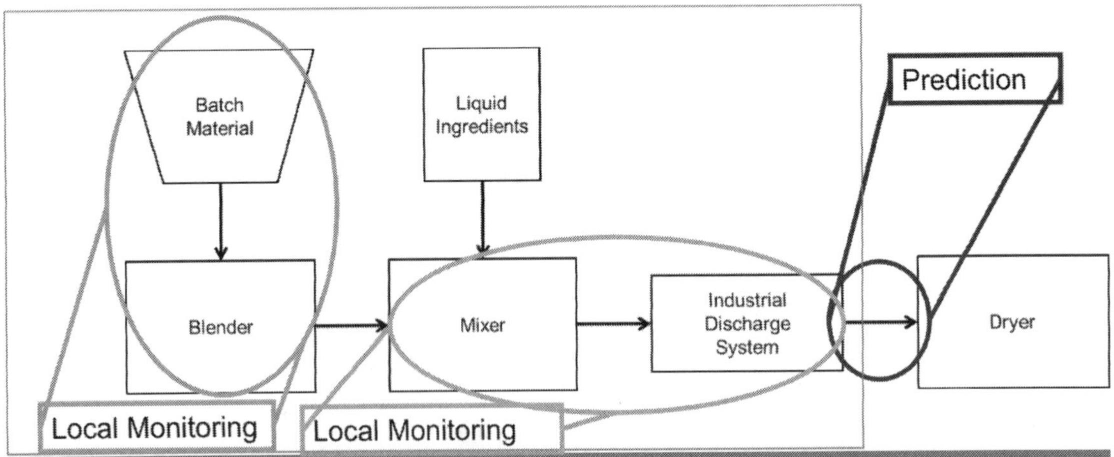

Batch and Continuous Processes

- Modeling using different length of training dataset
  - One model per one production run
  - One global model
  - One model per one cluster
    - Dataset divided into sub-groups by k-means clustering

One Model per One Production Run

- 10+ separate production runs
  - 10+ different PLS models
  - Good prediction in training dataset
  - Variability within each production run is captured

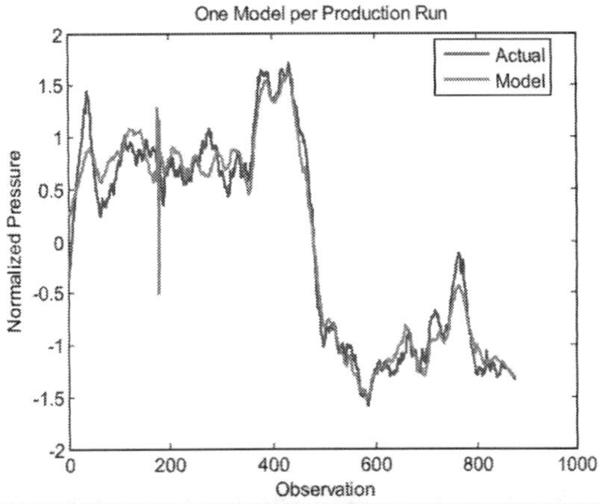

One Model per One Production Run

- 10+ separate production runs
  - Too many models
  - Bad prediction in testing dataset
    - Unable to predict different steady-states

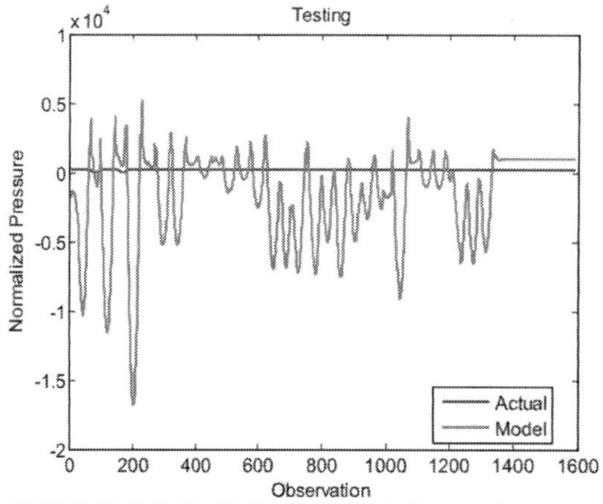

McKetta Department of Chemical Engineering

One Global Model

- 10+ separate production runs
  - 1 PLS model
  - Good prediction in training dataset
  - Variability from production run to run is captured

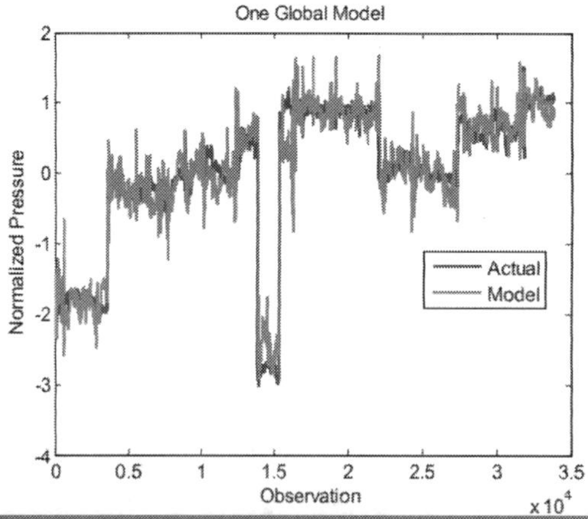

McKetta Department of Chemical Engineering

One Global Model

- 10+ separate production runs
  - Only one model
  - Bad prediction in testing dataset
    - Unable to accurately predict the variability within the production run

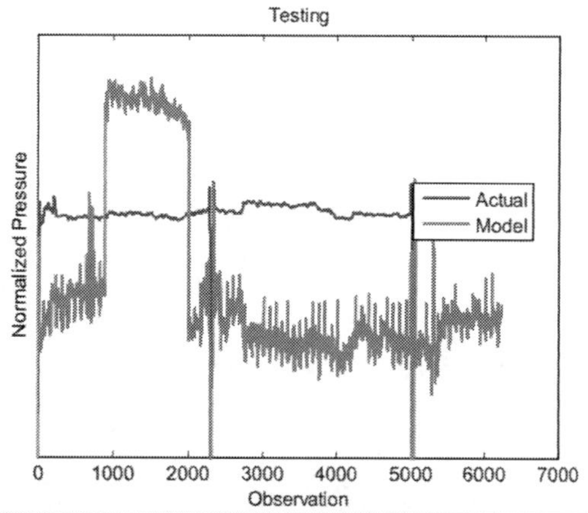

McKetta Department of Chemical Engineering

19

One Model per One Cluster

- 10+ separate production runs
  - 8 clusters (<< Number of production runs)
  - Good prediction in training dataset

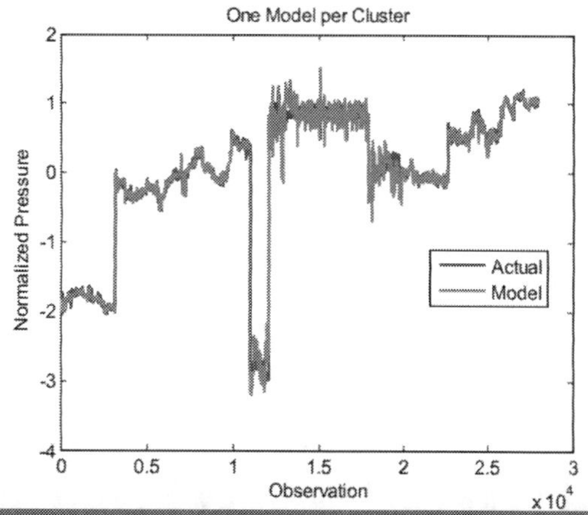

McKetta Department of Chemical Engineering

20

One Model per One Cluster

- 10+ separate production runs
  - 8 models
  - Good prediction in testing dataset
    - Able to predict the steady-state value and the variability within the production run

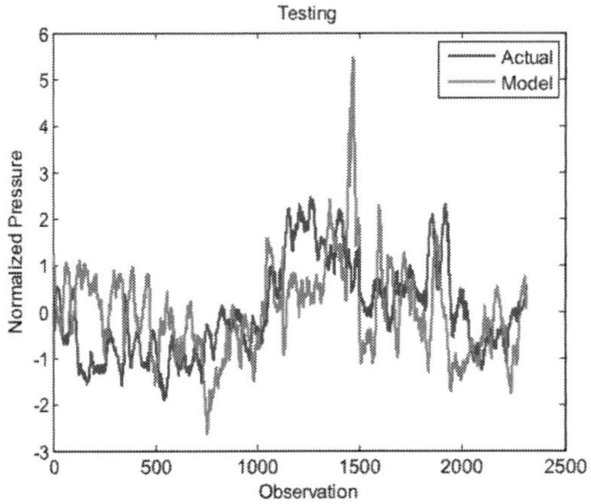

Conclusion

- Batch distribution was used to monitor the status of the batch materials.
- The batch dataset was aligned and correlated to the continuous dataset.
- Different sets of training data were used to improve the predictability in testing data.

Q&A

Determining Number of Clusters

- k-means cluster
  - Elbow method to determine the number of clusters needed

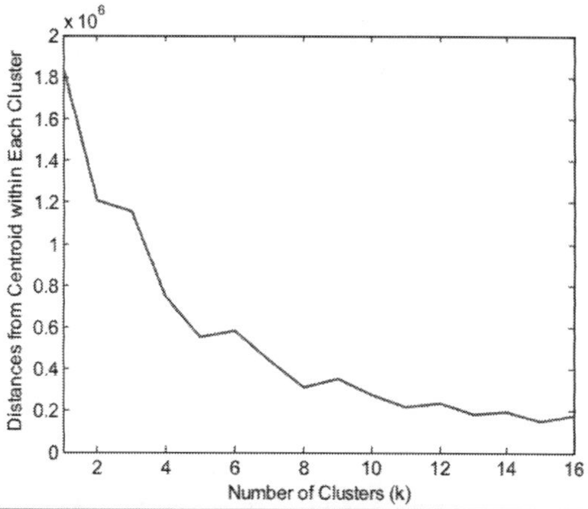

Clusters (Training Dataset)

- Each cluster falls in one cluster
  - One exception
- Production runs close to each other form the same cluster

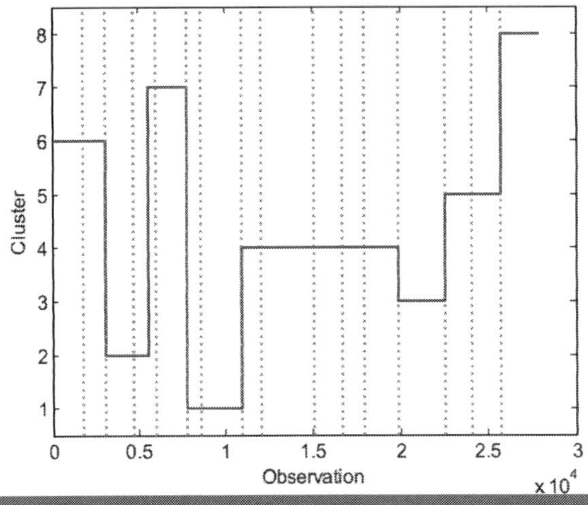

McKetta Department of Chemical Engineering 25

435364 Best Practices on Creating Soft Sensor Models for Batch Processes Utilizing Multi-Way Partial Least Squares (MPLS)

Abstract:

Soft sensors predict product quality at the end of the batch while the batch is still evolving. As a result, accurate soft sensors would allow operators to respond quicker to process deviations that impact product quality and improve process capability and yield. Data-driven methods, such as multi-way PLS (MPLS), are commonly used in deriving these batch soft sensors. Normally, a batch data set is comprised of inlet raw material conditions (sample once per batch), in-batch process measurements (sampled in real-time during the batch), and end of batch quality measurements (quality measurements such as conversion, impurity concentration or product purity). The dimensionality of a typical batch data set is very high and requires pruning prior to modeling and analysis. In addition, variable selection not only improves model performance, it also helps with increasing process interpretability of the resulting models. Nevertheless, due to the dimensionality of batch data set, identifying the most relevant variables (for estimating end of batch quality variables) can be a challenging task. Relevant variables are often observed to change depending upon the way the batch data is structured or organized. The preprocessing steps (critical for variable selection) can be summarized as follows: (1) a proper division of the phases/stages, which are typically defined in the batch recipe and followed by plant operations; (2) restructuring the data in multi-block or single-block arrangements; (3) the trajectory alignment to ensure consistent batch trajectory and identical batch length; and (4) the batch data unfolding, which converts the three-way process data $\mathbf{X}(I \times J \times K)$ into a two-way matrix, where J indicates the number of process variables at K sampling times in I batches. The combination of these four steps will have a profound impact on the interpretation and identification of the variables relevant for prediction or monitoring. Specifically, from practical modeling experiences, relying on operating phases defined in the batch recipe do not necessarily result in the most optimal phase division for modeling. This paper will demonstrate that proper phase division and data arrangement are important preprocessing steps in developing a soft sensor model for batch processes, generating a consistent variable selection and acceptable end of batch quality prediction. An industrial case study will be utilized to validate these best practices for generating a batch soft sensor model.

419054 Applying Predictive Analytics to Detect and Diagnose Impending Problems in Electric Submersible Pumps Used for Lifting Oil from Wellbores

Abstract:

<>Title: Applying Predictive Analytics to Detect and Diagnose Impending Problems in Electric Submersible Pumps

1. Objectives/Scope

The electrical submersible pump (ESP) is currently the fastest growing artificial-lift pumping technology. Deployed across 15 to 20 percent of oil-wells worldwide, ESPs are an efficient and reliable option at high production volumes and greater depths. However, ESP performance is often observed to decline gradually and reach the point of service interruption due to factors like high gas volumes, high temperature, and corrosion. The financial impact of ESP failure is substantial, from both lost production and replacement costs. Therefore, ESP performance in extensively monitored, and numerous workflows exist to suggest actions in case of breakdowns. However, such workflows are *reactive* in nature, i.e. action is taken *after* tripping or failure. Furthermore, the emerging trend in the E&P industry of using downhole sensors for real-time surveillance of parameters impacting ESP performance there is an opportunity for predicting and preventing ESP shutdowns using data analytics. Therefore, a data-driven analytical framework is proposed to advance towards a *proactive* approach to ESP health monitoring based on predictive analysis to detect impending problems, diagnose their cause, and prescribe preventive action.

2. Methods/Procedure and Process

Data-driven techniques employing multi-variate statistics are applied to detect and diagnose impending problems with ESP operation. To reach that goal, a three-stage workflow was designed and mathematical models were built for each stage.

The first stage of the workflow involves prediction of failures in advance. Historical data is first used to determine and evaluate key operational variables (decision variables) affecting ESP performance. A hybrid approach (combination of first principles and close to 20 decision variables) is then used to develop a robust principal component analysis (Robust PCA) model. The model is then used to monitor subsequent operation. An alarm is issued if a corresponding index exceeds the normal operating range established by the model, and diagnosis is offered based on the nature and value of the index or indices exceeding their normal operating range.

The second stage involved diagnostics of the potential cause leading to the failure. A mathematical model was built to determine the contribution of the various decision variables towards failures and ranking them according to these contributions. This ranking will help the operator understand the root cause and take appropriate action.

The third stage involved prescription of preventive action. It was modeled keeping in mind the operators need for a remedial solution to fix the issue. Stable operating ranges were determined for each decision variable. Operating within the stable range would not lead to an alarming event. The values of the parameters during the failures were compared against the stable operating ranges to understand which all parameters were behaving outside the acceptable limits. Those parameters were reset to values within the acceptable range.

3. Results, Observations and Conclusion

In building the predictive model based on normal operation data from an oilfield, it was demonstrated that four principal components obtained from the model output were sufficient to capture more than 80% of observed variance with 95% confidence limit.

Trends or patterns during normal operation were identified and correlated to either satisfactory operation or impending problems. This could be observed in the plot of Scores of Principal component 1 and Principal Component 2 as seen in figure 1. In this plot, the stable operating zone is clustered together as a green zone along the (0,0). The failure data sets move away from the stable zone with increasing time steps and fail at a point farther away as shown below. The increasing numbers represent the increasing time steps. The Trip 1 happened at 1518 minute, therefore the ESP was functioning abnormally for more than 24 hrs before it actually failed at the 1518th minute. With this kind of pattern plot, the failure could have been detected much in advance and could have been prevented.

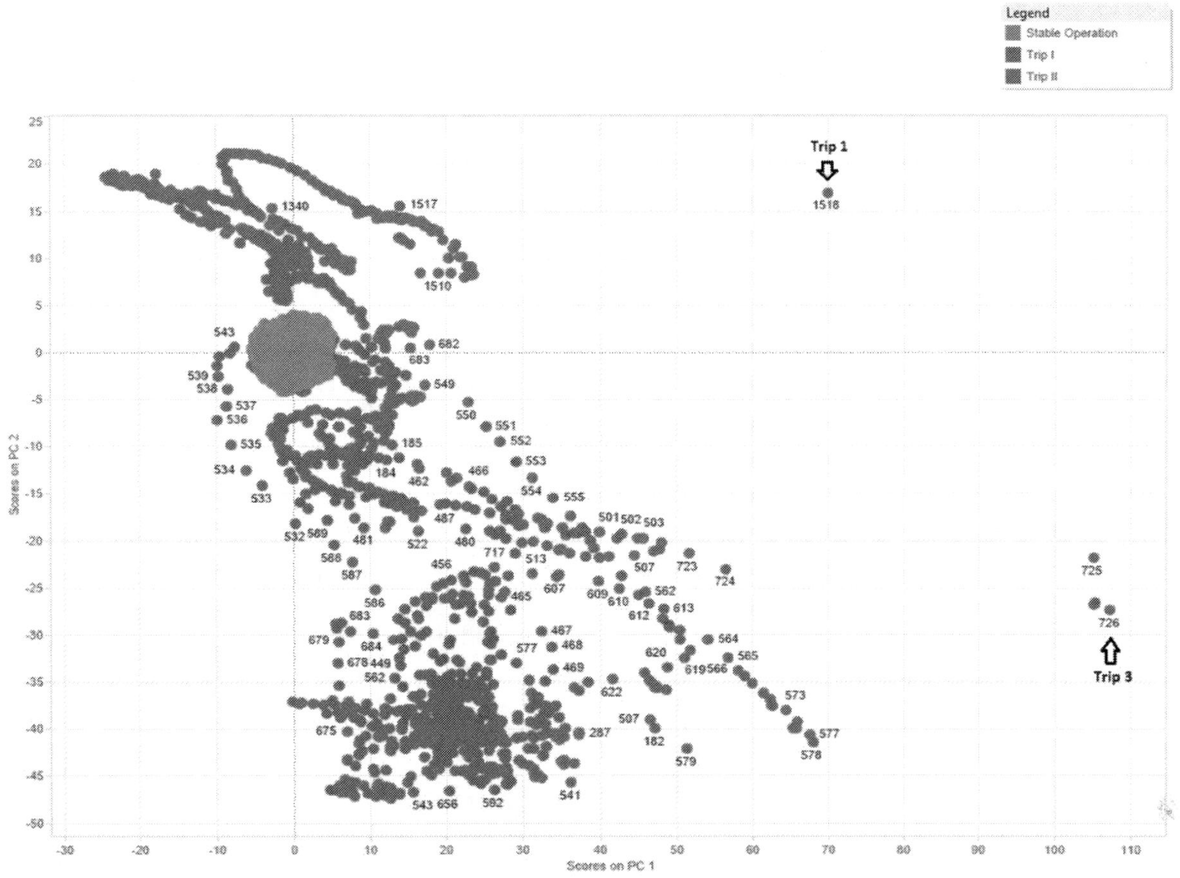

Figure 1: ESP Health Monitoring Plot

It was observed that the Hotelling T-square distribution statistic monitored over time stays within the 95% confidence limit for stable operation and exceeds the limit during unstable (transient) operation, long before an ESP trips or fails. The actual value of this statistic during a trip or failure shows an increase of more than two orders of magnitude compared to its normal value. The statistic successfully identified all impending trips or failures a few hours before they actually occurred. Had this tool been used in the field, there would have been ample time for preventive action that could mitigate or altogether avoid each problem.

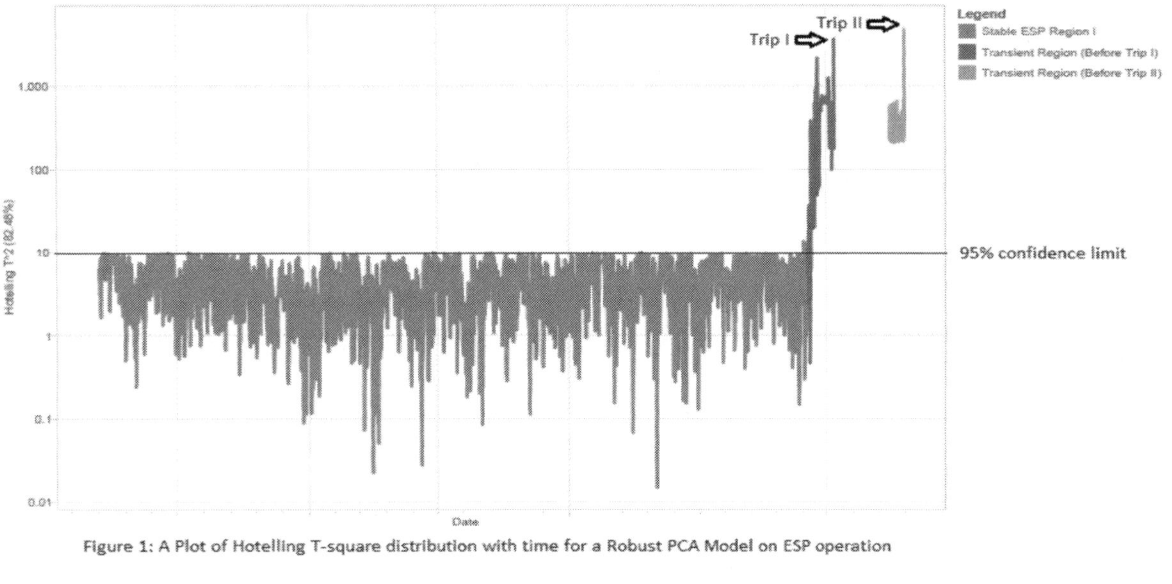

Figure 2: A plot of Hotelling T-square distribution for a Robust PCA Model

Prescription Plot in Figure 3 shows the stable ranges in green bars and the behavior of the parameters for two different events. It can be observed that some of the parameters are away from the stable zone. They can be reset to the stable zone and a failure event can be predicted in this manner.

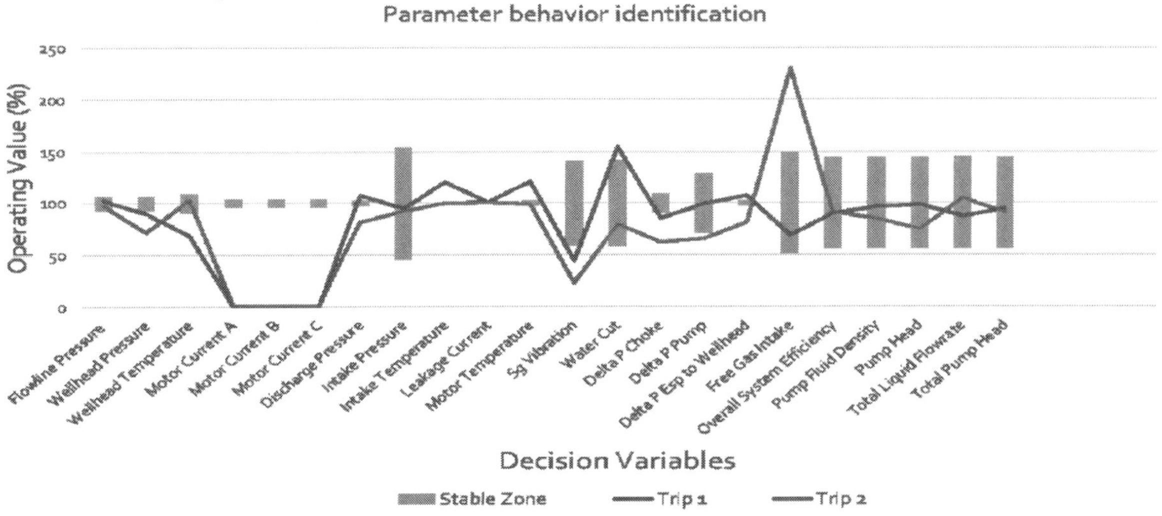

Figure 3: Prescription Plot for parameter behavior identification

4. Significance of the Subject Matter

This real-time analytical framework enables a shift towards proactive ESP monitoring to identify impending problems long before they occur thereby safeguarding ESP operation, reducing intervention costs and optimizing production. This approach creates opportunities to increase pump uptime, extend the life expectancy of ESPs and improve oilfield economics.

DATA MINING AND MONITORING FOR INDUSTRIAL SCALE CHEMICAL PROCESSES

Michael C. Thomas and Jose A. Romagnoli, Louisiana State University Cain Department of Chemical Engineering, Baton Rouge, LA

Introduction

Since the beginning of the information age, chemical plants have stored large amounts of sensor data from chemical processes. Too often, these databases become "graveyards" where the data is stored and seldom used again[1]. Process engineers and operators face the problem of information overload, where knowledge about a system is buried within a large volume of data. Recent advances in artificial intelligence and machine learning offer a way to use this data for abnormal event detection or soft sensing. Abnormal event detection, also known as fault detection and diagnosis, identifies patterns in data that give engineers insight into locating and solving process fault events.

This research presents process monitoring case studies on two industrial scale units: a separation tower and a batch polymerization reactor. Recently, a large, damaging fault occurred on the separation tower that eluded detection by process operators because it resembled a common problem with a sensor. The challenge was to first model expected the operation of the tower, then to create a process monitoring strategy that could detect the larger fault and distinguish it from the more common sensor fault. On the reactor, we focused on the problem of detecting bad product in advance of the plant's laboratory measurement abilities. Modeling the reactor is more difficult because of non-linear, time dependent batch dynamics which distinguish its operation, and data from the batches required additional processing before the product quality could be characterized.

Towards the goals of detecting novel operating conditions on the tower and analyzing the reactor product quality specification, we use the Self-Organizing Map (SOM), a classic machine learning technique used to visualize high-dimensional data. An SOM is made up of an organized array of interconnected map nodes that exist in the state space, and this structure allows us to create visualizations that summarize high-dimensional data which leads to extracting knowledge from the data. A key challenge in applying data based algorithms in an industrial environment is the scalability and general applicability of the technique. Here we implement an SOM based data mining and monitoring system on an industrial historical data set to see how it performs at detecting a previously unseen fault on the tower and characterizing the product quality of a batch reactor.

The Self Organizing Map

The Self-Organizing Map is a nonlinear, ordered, smooth mapping of high dimensional input data manifolds onto the elements of a regular low-dimensional array[2]. The makings of an SOM resemble vector quantization, but an SOM is an ordered array with a distribution that resembles that of the data. We use it in modeling chemical process data because of the way it performs vector quantization and topology preservation on training data as well as the interesting visualizations into the structure of the data it provides.

An SOM is a connected array of nodes forming a regular rectangular or hexagonal grid structure. The prototype vectors making up each map model vectors within the training data. The prototype vectors can be represented in both the data space and in the dimensionality reduced space within the map structure. A data point's representation on the map is found by locating its Best Matching Unit (BMU). The BMU of ith data vector v_i is found by finding the closest prototype vector m_k (with minimum distance $\partial(m_k, v_i)$) according to:

$$BMU_i = \arg\min\left(\partial(v_i, m_k)\right), \forall k \quad (1)$$

During training, the map vector m_k's position is updated using each data point from a training set according to the formula:

$$m_k(t+i) = m_k(t) + \alpha(t) h_{k,BMU_i}(v_i(t) - m_k(t)) \quad (2)$$

Where t is the mapping time step and α is the monotonically decreasing learning rate, h_{k,BMU_i} denotes a neighborhood kernel function centered at the BMU. The training algorithm matches data vectors to prototype vectors using Euclidean distance. The neighborhood kernel h_{k,BMU_i} centered at $m_k(t)$ is usually chosen in the Gaussian form:

$$h_{m_k,BMU_i} = exp\left(\frac{\|m_k - BMU_i\|^2}{2\sigma^2}\right) \quad (3)$$

σ(t) is the monotonically decreasing kernel width that allows for a smoothing of the prototypes composing the map[2].

SOM Visualization Tools for Process Analysis

The major advantage of SOM in process monitoring comes from the relative ease with which high dimensional structures within the data are can be visualized on the two dimensional map. For instance, the displays allow the shape of the data distribution, cluster borders, projection directions and possible dependencies between the variables to be viewed. Common visualization discussed here are component places, distance matrices, and projections as proposed by Kaski[3], Vesanto[4,5], and Himberg[6].

Distance Matrices

Distance matrices show the pairwise distances between nodes within the map structure. Using the lower dimensional structure of the SOM, colors are used to show high or low distances, which provides insight into the dominant clustering patterns present in the data. The most common visualization of the distance matrix is the U-matrix[7] which visualizes all pairwise distances between nodes in the map. In a U-matrix, the dominant clustering structure can be seen as areas with small distances within and separated by large distances. As in the basic U-matrix, visualization of the clusters can be greatly improved by augmenting the distance matrix with an additional entry between each prototype vector and each of its neighbors.

Projections

Finally, the second visualization tool considered here for process monitoring involves projections onto the map which allow us to track the position of the process relative to the map features at given points in time. The two key visualizations derived from projection are hit diagrams and trajectories. Hit diagrams show the distribution of any number of data points on a map, which can illuminate the meaning of regions observed in distance matrix plots. Trajectories project a time series of data points onto a map and connect them with lines which allow us to observe the sequential path taken by the state of the process through the map.

Industrial Case Studies

Separation Tower Description

The performance of our process monitoring scheme was tested on historical measurements from an industrial scale separation tower. The feed to the system is produced by a reactor upstream and varies depending on the petroleum grade fed. Before entering the Tower, the feed is passed through Tank 1 to remove water. Feed then enters the Tower at a high temperature and pressure, and a mixture of solvent and product leaves out the bottoms with more solvent and water leaving out of the top of the column. Tank 2 removes any remaining water before recycling the solvent.

The specific gravity sensor (SG) at the bottom of the column is used to evaluate the quality of the separation and guide the adjustment of the steam flow to Tank 2. The specific gravity is controlled by manipulating the amount of steam fed to Tank 1.

In this particular unit, a common "fault" lies in the SG and T7 sensors. These sensors work through the use of a small side stream that occasionally gets clogged with polymer. Detection of this sensor fault is fairly straightforward, but recently a larger fault occurred at the same time the sensor fault was occurring. This bigger fault, subsequently referred to as BIG FAULT, had not occurred before and resulted in significant downtime and expensive damage to equipment. Our goal was to model the sensor fault and normal operations and simultaneously diagnose the sensor fault and detect the bigger fault.

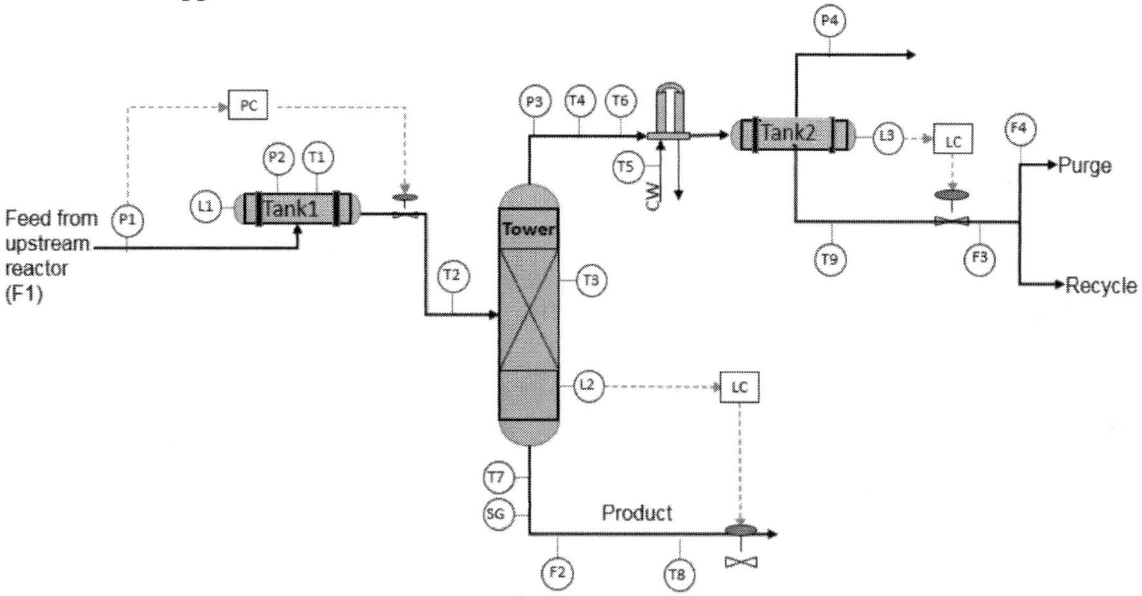

Figure 1. Separation process flow sheet

Polymerization Reactor Description

The reactor system analyzed converts a hydrocarbon feedstock into a polymer product in a batch process. Each batch proceeds through six discrete phases corresponding to steps in reactor operation: feed reactants, raise reactor temperature, run reaction, and remove products. The progression of each of these steps can be crucial to the quality of the products. Attached around the reactor are a couple of tanks that provide reflux to the reactor as well as a pump around system to increase circulation of reactants. At the end of the batch, the product is sometimes tested for quality through a lab test that takes time offline. A range of quality measurements are acceptable, but when the quality specification is above or below certain points, it creates expenses in a finishing plant including increased utility use and lost product. Our goal is to characterize the product quality specification in terms of the trajectory each sensor records as the batch progresses to make more consistent products.

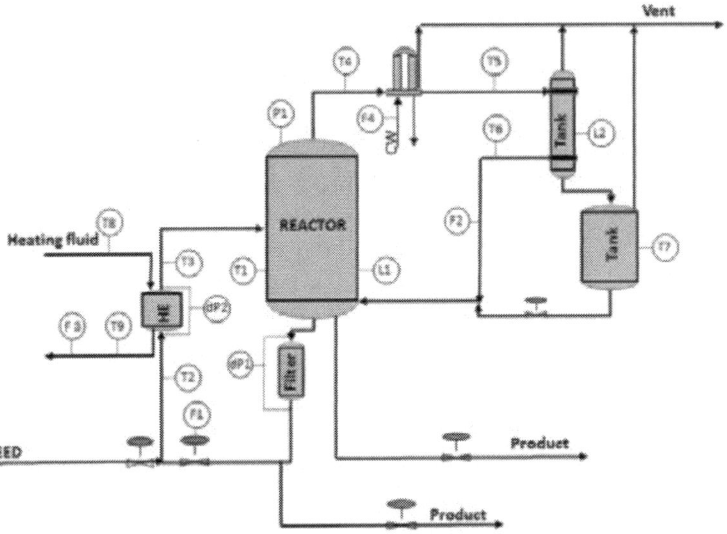

Figure 2. Reactor flow sheet

Results

Separation Tower Results

The analysis of the tower was an opportunity to test our data mining strategies in the detection of a truly novel event. The BIG FAULT had never occurred before therefore the challenge was to form a model based on historical data and check how it performed in distinguishing the familiar sensor fault from the new BIG FAULT.

To distinguish the sensor fault from normal data, three pools of data were extracted from two years of historical data: a larger data set containing all normal data for a particular grade, a smaller set with data from startup conditions, and a much smaller data set holding measurements from operations with a sensor fault. Key variables around the unit of interest were selected for monitoring, in addition to transformed variables derived from differences in temperature and pressure between different parts of the process, such as the temperature difference between the middle of the column and the top of the column, or the pressure difference between the feed pressure and the pressure in the first tank.

To achieve an even representation of the normal and fault conditions and prevent normal data from dominating the SOM representation, a simple sampling method ensured that each group is well represented during training. 1000 samples were drawn randomly with replacement from each pool of data for training, except for normal, from which 2000 data points were taken to make better use of the large repository of normal data.

Figure 3a-c gives various visualizations of the map formed after training. Figures 3a and 3b show the maps colored according to clusters of map nodes identified by k-means clustering to the SOM map nodes. Figure 3b includes a hit diagram of the sampled groups of training data which helps interpret the meaning of different regions of the SOM map. The k-means clustering separated the map into four clusters, but two are essentially a subdivision of the sensor fault group. Using the hit diagram, the two sensor fault clusters are merged, creating the final map labeling given in Figure 3c. The hit diagram also showed that points initially labelled as faulty or startup were more similar to normal operations. The k-means clustering adjusts for this noise in the training data.

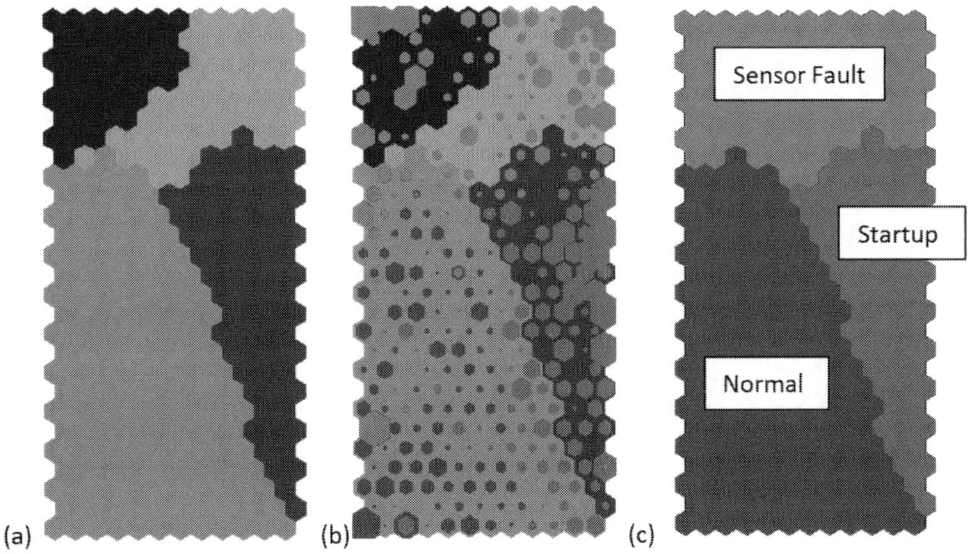

Figure 3. A k-means clustered SOM map (a) shows distinct groups in the separation tower data. The hit diagram in (b) explains the meaning of the different regions of the map. The final map used to monitor the status of the tower is given in (c).

The clustered map created can now visually monitor the state of the process visually using an SOM trajectory plot. Each time point analyzed is indicated by a black dot and connected by a line with other visualized time points. This shows us the maps units touched by the process state, and tells us whether the process is normal, in startup, or experiencing a sensor fault.

Figure 4 gives three examples of tracking the state of the process using SOM trajectory plots: 4a (startup sequence), 4b (regular sensor fault), and 4c (the novel BIG FAULT of interest). As seen in the lower graph of Figure 5, the QE is high for all points in the trajectory of the BIG FAULT and the trajectory is concentrated near the edge, indicating that while the system operates with a faulty sensor, a novel operating condition has been reached – simultaneously diagnosing a fault and detecting a novel operating condition.

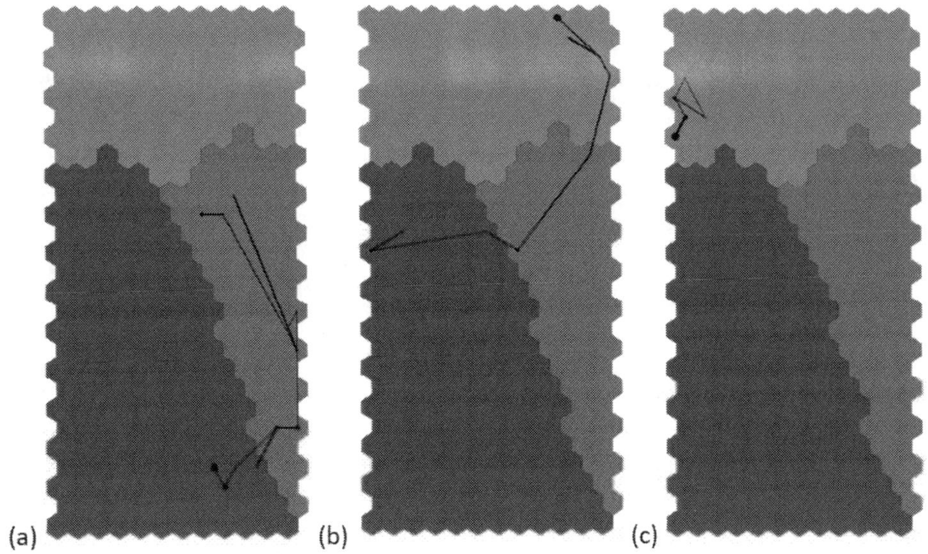

Figure 4. SOM trajectories show changes during tower operations for three events: (a) normal startup (b) startup during a temperature sensor fault (c) BIG FAULT.

Finally, a variable contribution analysis tests the map's ability to assist in detecting and diagnosing a novel event. Data from the year of the major fault was projected to the map and used to calculate a fault detection statistic based on the SOM QE of the new data vector on the process map. The sensors responsible for a novel observation can be analyzed using the variable contribution analysis used by Robertson[8].

The top of Figure 5 gives the variable contribution analysis. The BIG FAULT is characterized by a more intense difference between T3 and T7 and a highly unusual SG reading. The bottom of Figure 5 gives a plot of the QE statistic with time. This plot contains many uninteresting peaks from startups or feed grade switches, but the faulty condition is clearly indicated by an extended period of high QE around and after the time indicated by the red circle.

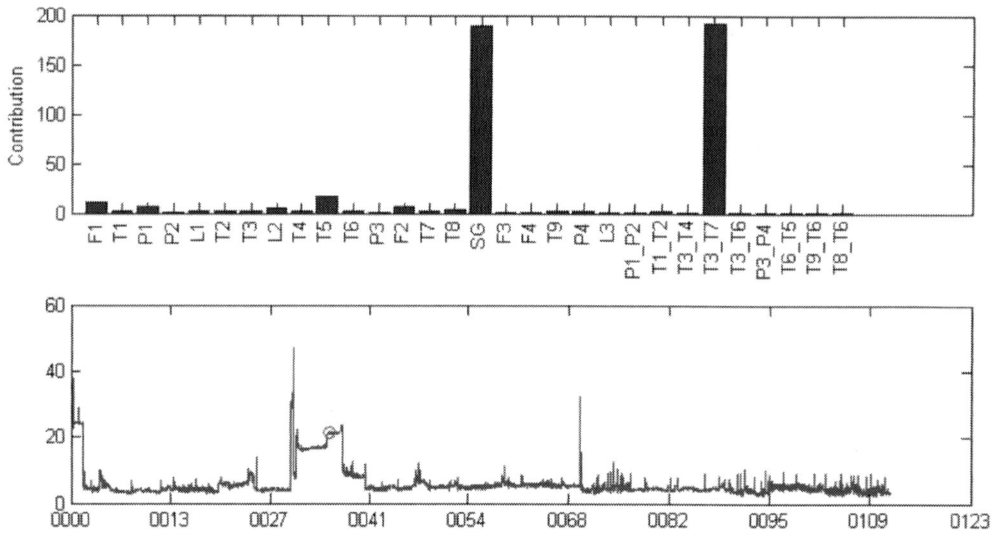

Figure 5. The upper chart indicates the variables contributing the most to the elevated fault detection statistic. The lower chart indicates the point in the historical data used to calculate the contribution plot.

Polymerization Reactor Analysis

Applying the techniques of the previous section to the polymerization reaction requires surmounting additional challenges. In contrast to the separations tower where operations are largely at steady state, a reactor where time varying nonlinear variations are a part of normal operations requires additional characterization effort. Here, the method of dynamic time warping introduced by Kassidas[9] is used to synchronize each batch before training a SOM. Several variables were added in addition to the sensor variables to incorporate needed information: lengths of the different phases programmed into batch operations and an indicator variable to assist in separating the data (-1 for low softening points, 0 for normal, and 1 for high, similar to the method of Corona[10]. Further, to compensate for the high dimensionality of the unfolded batch data (250 variables) the data are reduced in dimensionality using principal component analysis (PCA).

Figure 6 shows the maps trained to the processed data described above. We can see from the hit histogram in Figure 6b that while k-means identifies one region of the data well, others contain considerable noise. In spite of that, the three data groups separate into three contiguous regions of the map corresponding to the quality measurement analyzed: high (green) at the top, normal (blue) in the middle, and low (red) at the bottom. This structure in the map could predict the quality measurement before the lab measurement is known, giving needed knowledge to plant operators.

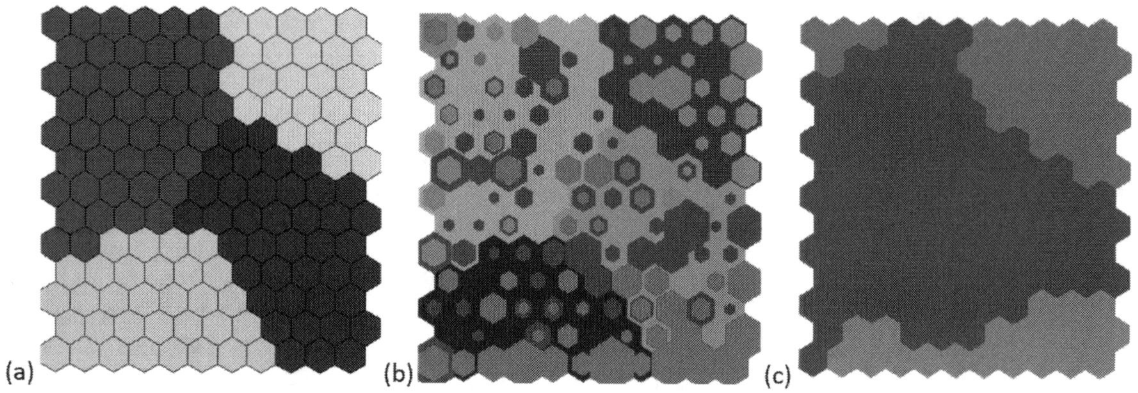

Figure 6. A k-means clustered map (a) and SOM hit diagram (b) show groups in the reactor data. The map in (c) is labelled by data voting at each map vector.

Conclusions

This research studied how self-organizing maps visualize high dimensional data and how to use this capability to mine process databases. The data was drawn from two industrial units: a separations tower operating at steady state and a polymerization reactor. On the separations tower, SOM formed the basis of a fault detection/diagnosis system that simultaneously performs novelty detection to detect new fault events. Contribution plots offer additional guidance in diagnosing newly detected faults. On the reactor, we created a preliminary characterization of batch operations with respect to the product quality measure which show that SOM is capable finding patterns in the batch data that can lead to favorable products. Future work will focus on making these techniques simpler and easily scalable to different equipment to expand their use.

References

1. Venkatasubramanian V. Drowning in Data: Informatics Modeling Challenges in a Data-Rich Networked World. *AIChE Journal*. 2009;55:2-8.
2. Kohonen T. *Self-Organizing Maps*. Berlin:Springer-Verlag; 2001.
3. Kaski S. *Data exploration using self-organizing maps.* Doctoral Dissertation, Helsinki University of Technology, Finland; 1997.
4. Vesanto J, Ahola J. Hunting for correlations in data using the self-organizing map. *Proceedings of the International ICSC Congress on Computational Intelligence Methods and Applications.* 1999;279–285.
5. Vesanto J. *Data exploration process based on the Self-Organizing Map*. Doctoral Dissertation, Helsinki University of Technology, Finland; 2002.
6. Himberg J, *From insights to innovations: Data Mining, visualization, and user interfaces.* Doctoral Dissertation, Helsinki University of Technology, Finland; 2001.
7. Ultsch A. Self-organizing neural networks for visualization and classification. *Information and Classification.*1993;307–313.
8. Robertson G, Thomas MC, Romagnoli, JA. Topological preservation techniques for nonlinear process monitoring. *Computers and Chemical Engineering*. 2015;76:1-16.
9. Kassidas A, MacGregor JF, Taylor PA. Synchronization of Batch trajectories Using Dynamic Time Warping. *AIChE Journal*. 1998;44:864-875.
10. Corona F, Mulas M, Baratti R, Romagnoli, JA. Data-Derived Analysis and Inference for an Industrail Deethanizer. *Industrial & Engineering Chemistry Research.* 2012;51:13732-13742.

Appendix

Table 1. Process measurements and derived measurements

Variable Name	Description	Derived Var. Name	Derived Var. Description
F1	Feed to upsteam reactor	P1_P2	Tank1 Feed pressure change
T1	Tank1 Temperature	T1_T2	Tank1 to Feed temperature change
P1	Tank 1 feed pressure	T3_T4	Mid-tower to top tower temperature change
P2	Tank1 pressure	T3_T7	Tower to sensor temperature change
L1	Tank1 level	T3_T6	Tower to overhead temperature change
T2	Tower feed temperature	P3_P4	Tower overhead to Tank2 temperature change
T3	Middle Tower temperature	T6_T5	Exchanger temperature gradient
L2	Tower level	T9_T6	Overhead to product temperature change
T4	Tower overhead temperature	T8_T6	Overhead and product temperature change
T5	Cooling water temperature		
T6	Tower overhead temperature		
P3	Overhead Pressure 3		
F2	Bottoms flow		
T7	Bottoms temperature		
T8	Bottoms Product temperature		
SG	Specific gravity sensor		
F3	Total Overhead product flow		
F4	Non-recycle overhead product flow		
T9	Overhead Product temperature		
P4	Vapor Purge		
L3	Tank2 Level		

430076 Learning Models of Unspecified Functional Form through Symbolic Regression

Abstract:

We address the problem of learning simple algebraic models from data obtained from simulations or experiments. Standard regression techniques seek to develop models with a pre-determined model structure or set of alternative model structures. However, in a practical setting, data is often acquired from a number of sources without clear understanding of the system at hand. Algebraic models could be used to make the system more amenable to the tasks of optimization, prediction, and control; however, a lack of insightful functional forms to use in a regression is problematic. Symbolic regression addresses this problem by learning an algebraic model of unspecified functional form from exogenous data [1].

Symbolic regression is traditionally approached with genetic programming and other heuristic algorithms. These stochastic approaches to model identification offer no guarantee of either local or global optimality, and often perform poorly in practice [2]. We show that symbolic regression can be formulated as a nonlinear nonconvex disjunctive program and can be solved to global optimality. Our approach includes steps to avoid redundant solutions and ensure that the optimal solution can be efficiently found using a branch-and-bound framework. We present extensive computational results comparing our symbolic regression approach to approaches in the existing literature.

References

[1] J.R. Koza. *Genetic Programming: On the Programming of Computers by Means of Natural Selection.* MIT Press, 1992.

[2] M.F. Korns, Accuracy in symbolic Regression. In *Genetic Programming Theory and Practice IX,* pages 129-151. Springer, 2011.

433406 Use of Fenceline Monitoring for Root Cause Analysis and Corrective Action

Abstract:

During 2010-2011, continuous emissions data was collected and analyzed for a delayed coker at a petroleum refinery in the Houston, Texas, area. The measurement system utilized multiple open-path Fourier Transform Infrared instruments and retro-reflector mirror arrays along four measurement planes. Emissions were calculated using the Vertical Radial Plume Mapping, or VRPM, method as detailed in EPA Other Test Method 10. While providing detailed information on long-term, speciated hydrocarbon emissions, lag times between data collection, analysis and availability limited usefulness for root cause analysis and corrective action. Since that time, improvements in data analysis algorithms now allow for fully-validated data, free of false positives that are presented in near-real time (approximately 2 second delay between measurement and availability). This presentation will provide an overview of state-of-the art fence-line air quality monitoring systems and how this data is being used for a variety of purposes including air emission reductions, improved community relations, and improved operations.

THE UNIVERSITY OF TEXAS AT AUSTIN

An Integrated Scheme for Oscillation Detection and Diagnosis from Industrial Data

Shu Xu[1], Willy Wojsznis[2], Mark Nixon[2], Michael Baldea[1] and Thomas F. Edgar[1]
(1)McKetta Department of Chemical Engineering, The University of Texas at Austin, Austin, TX,
(2)Innovation Center, Emerson Process Management, Austin, TX

Motivation

- In large-scale industrial processes and other complex systems, oscillations generated in a single unit can negatively affect the overall control performance of the process
- The interference of noise and the need for oscillation propagation routes pose challenges for process engineers

Methodology for oscillation detection and diagnosis

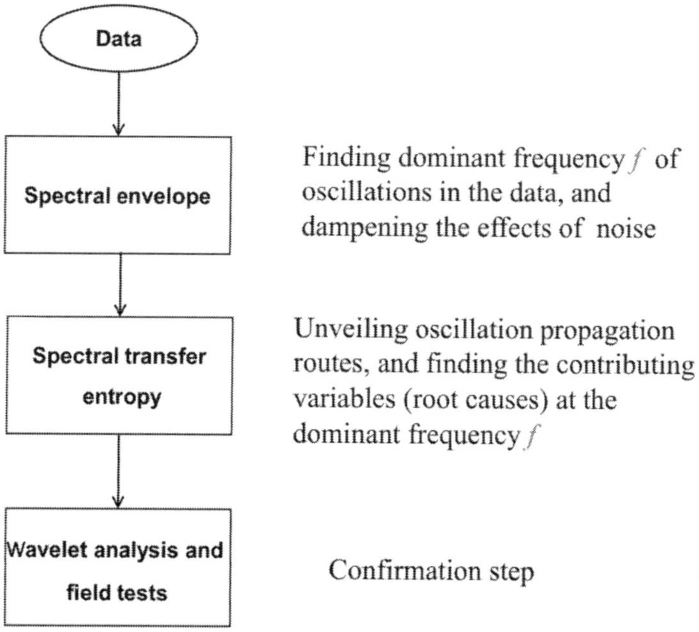

Figure 1. Methodology for oscillation detection and diagnosis

Spectral envelope method[1]

- Assume $x(t) = [x_1, x_2, \ldots, x_n]^T$ to be a time series, and each variable $x_i(t)$ has been auto-scaled to have identical power. Define the spectral envelope to be:

$$\lambda(\omega) = \sup_{\beta \neq 0} \left\{ \frac{\beta^* \mathbf{P_x}(\omega) \beta}{\beta^* \mathbf{V} \beta} \right\} \quad (1)$$

where β is an n-dimensional real or complex vector, * represents the conjugate transpose, V_x and $P_x(\omega)$ denote the covariance matrix and the power spectral density (PSD) matrix of x respectively, and ω is the normalized frequency satisfying $-1/2 \leq \omega \leq 1/2$.

- $\lambda(\omega)$ is the largest eigenvalue of $P_X(\omega)$, and $\beta(\omega)$ is the corresponding eigenvector

1. Jiang, H., Shoukat Choudhury, M. A. A., & Shah, S. L. (2007). *Journal of Process Control, 17*(2), 143-155.

Oscillation contribution index[1]

- The oscillation contribution index of $x_j(t)$ is defined to be:

$$OCI_j(\omega) = \frac{\left|\hat{\beta}_j(\omega)\right|}{2\sigma_{\hat{\beta}}(\omega)} \qquad (2)$$

Where $\sigma_\beta(\omega)$ is the standard deviation of the magnitude of the optimal scaling of all the oscillating variables.

- Commonly, variables having $OCI(\omega) > 1$ are considered as the root cause variables at frequency ω with the larger contributions at the spectral envelope peak.
- Although the OCI is convenient to calculate, it fails to give us oscillation propagation routes.

1. Jiang, H., Shoukat Choudhury, M. A. A., & Shah, S. L. (2007). *Journal of Process Control, 17*(2), 143-155.

McKetta Department of Chemical Engineering 5

Spectral Granger causality[2]

- Assume X(t) and Y(t) are two time series describing stationary stochastic processes, a bivariate autoregressive (AR) model can be built as follows:

$$\begin{bmatrix} X(t) \\ Y(t) \end{bmatrix} = \sum_{k=1}^{p} \begin{bmatrix} a_{xx,k} & a_{xy,k} \\ a_{yx,k} & a_{yy,k} \end{bmatrix} \begin{bmatrix} X(t-k) \\ Y(t-k) \end{bmatrix} + \begin{bmatrix} \varepsilon_{x,t} \\ \varepsilon_{y,t} \end{bmatrix} \qquad (3)$$

- If we perform Fourier transform on above equations, we obtain:

$$\begin{bmatrix} X(\omega) \\ Y(\omega) \end{bmatrix} = \begin{bmatrix} H_{xx}(\omega) & H_{xy}(\omega) \\ H_{yx}(\omega) & H_{yy}(\omega) \end{bmatrix} \begin{bmatrix} E_x(\omega) \\ E_y(\omega) \end{bmatrix} \qquad (4)$$

- The spectral density matrix $S(\omega)$ is derived as:

$$S(\omega) = \langle X(\omega) X^*(\omega) \rangle = H(\omega) \Sigma H^*(\omega)$$

- Where Σ is the covariance matrix of the full model residuals, and $H^*(\omega)$ is the Hermitian transpose

2. Yuan, T., & Qin, S. J. (2014). Journal of Process Control, 24(2), 450-459.

McKetta Department of Chemical Engineering 6

Spectral Granger causality and transfer entropy[3]

- The spectral Granger causality from Y to X at frequency f is:

$$I_{Y\to X}(\omega) = \ln\left(\frac{|S_X|}{|S_{X|Y}|}\right) \qquad (5)$$

- For Gaussian variables, the spectral transfer entropy $T_{Y\to X}$ can be calculated based on the following equation:

$$T_{Y\to X} = H(X(\omega)) - H(X(\omega)|Y(\omega)) = \frac{1}{2}\ln\left(\frac{|S_X|}{|S_{X|Y}|}\right) = \frac{1}{2}I_{Y\to X}(\omega) \qquad (6)$$

where $H(X)$ is the Shannon entropy of X.

- **The transfer entropy can be seen as the reduction of uncertainty of X by knowing Y**

3. Schreiber, T. (2000). Physical Review Letters, 85(2), 461-464.

Simulated case study

➢Problem formulation

$$\begin{aligned}
X_{1,n} &= 0.95\sqrt{2}X_{1,n-1} - 0.9025X_{1,n-2} + w_{1,n}\\
X_{2,n} &= 0.5X_{1,n-2} + w_{2,n}\\
X_{3,n} &= -0.4X_{1,n-3} + w_{3,n}\\
X_{4,n} &= -0.5X_{1,n-2} + 0.25\sqrt{2}X_{4,n-1} + 0.25\sqrt{2}X_{5,n-1} + w_{4,n}\\
X_{5,n} &= -0.25\sqrt{2}X_{4,n-1} + 0.25\sqrt{2}X_{5,n-1} + w_{5,n}
\end{aligned} \qquad (7)$$

➢Where $w_{i,n}$, $z_{i,n}$ are drawn from Gaussian noise with zero mean and unit variance.

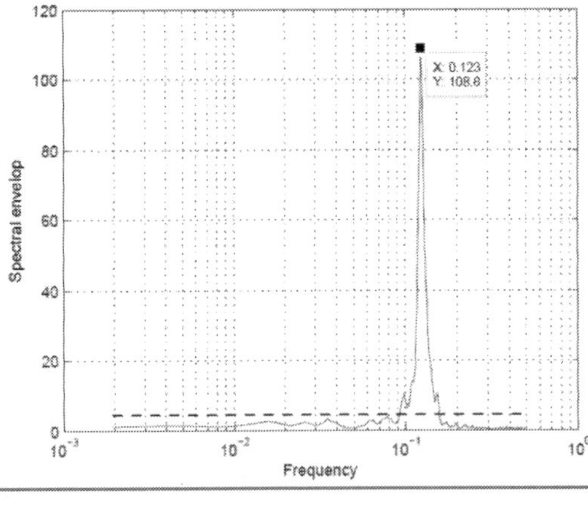

Figure 2.Spectral envelope of the simulated data

Simulated case study results

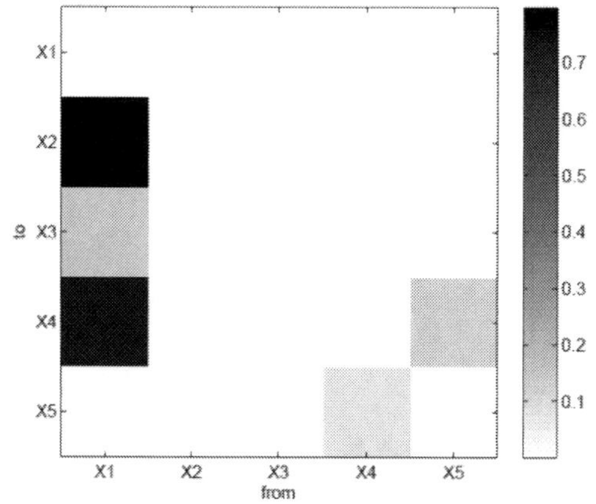

Figure 3. Spectral transfer entropy at $f = 0.123$

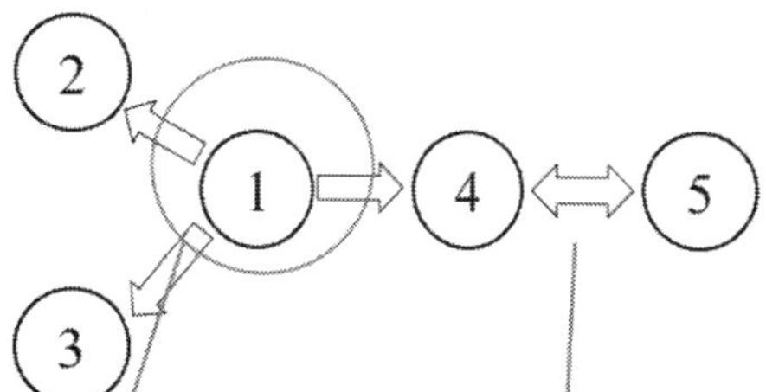

Figure 4. Process topology obtained by spectral transfer entropy

"Root cause"

Can show mutual relation

Industrial case study– Tennessee Eastman Process[4]

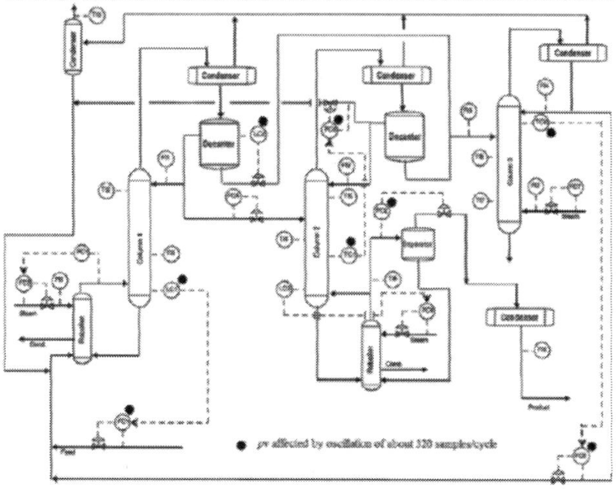

Figure 5. Process schematic. The oscillation variables are marked by circle symbols

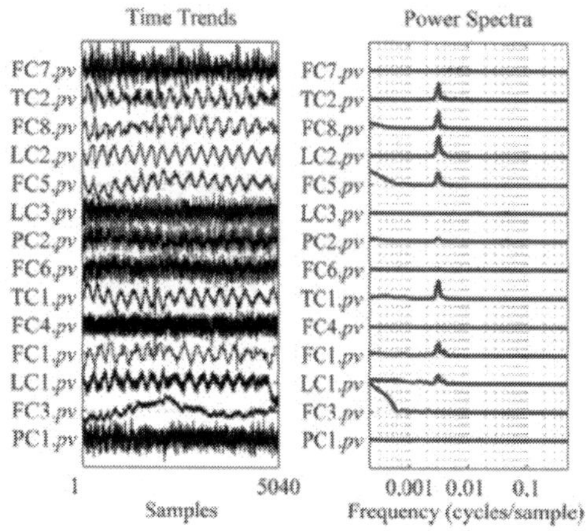

Figure 6. Time trends and power spectra of measurements of process variables (pv's)

4. Ping, D., Fan, Y., Tongwen, C., & Shah, S. L. (2013). Control Systems Technology, IEEE Transactions on, 21(6), 2052-2066.

Oscillation detection and diagnosis results(1)

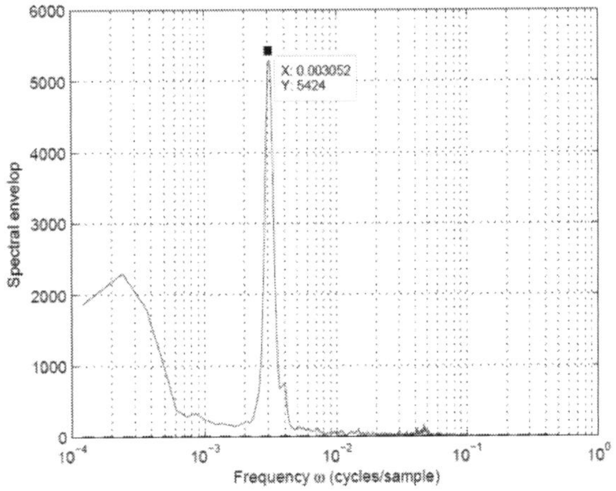

Figure7. Spectral envelope of the Eastman Chemical process data (oscillation period = 320 samples/cycle)

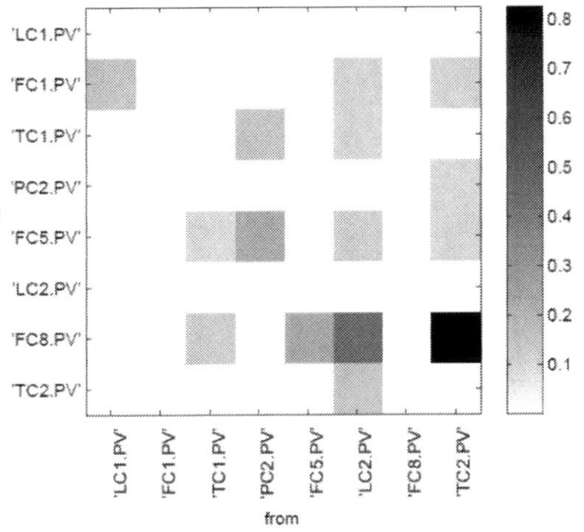

Figure 8. Spectral transfer entropy of the Eastman Chemical process data

Oscillation detection and diagnosis results(2)

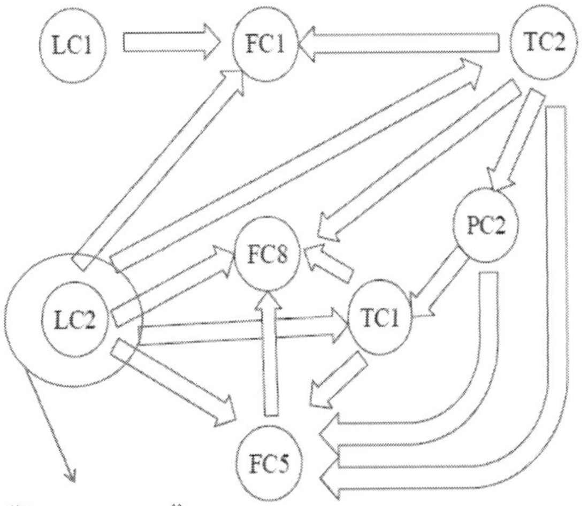

"Root cause"

Figure 9 Process topology of the Eastman Chemical process based on spectral transfer entropy

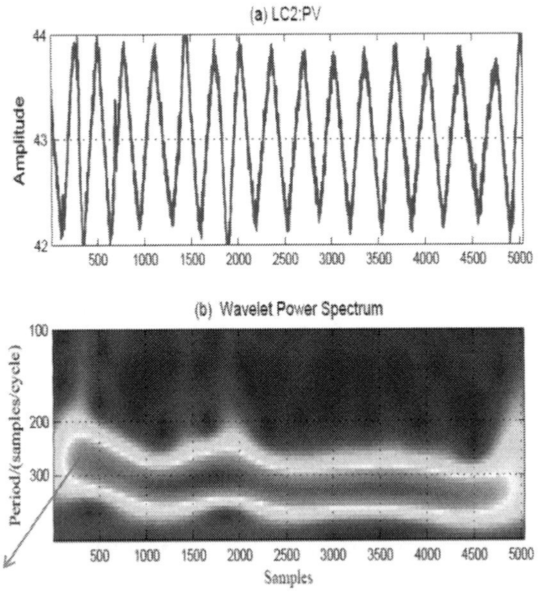

An oscillation of 320samples/cycle happens around 1000th sample point

Figure 10 Wavelet power spectrum of variable LC2.PV

Discussion

- The spectral envelope method provides a convenient and intuitive way to find the common frequency corresponding to abnormality.
- The spectral transfer entropy provides an intuitive process topology to visualize and explain the causal-effect relationship in the frequency domain, and the effects from the noise will be mitigated by the frequency selection step in the spectral envelope figure.
- The wavelet analysis works well as a confirmation step in oscillation diagnosis.

Questions?

430666 Process Data, Is More Better? a Case Study to Improve the Chemical Manufacturing Operation

Abstract:

Traditionally at Dow Chemical, multivariate analysis is often applied on process data for process troubleshooting, quality improvement and design of inferential sensors, to name a few. The case study involves multiple batch operation units with one quality measurement for each batch and the goal is to find the key process variables (x variables) that are highly correlated with the quality measurement (y variable). The ultimate purpose is to reduce variability in the y variable. The historical data available for analysis was sampled at every minute from the last two years (with data matrix of 4.7 million columns and 200 rows) but without any success in drawing any statistical conclusions. A big dataset does not guarantee the ability of extracting useful information successfully. The reason is because of lack of structure in the data due to sampling and process variability, which is a common problem of a big dataset. To overcome the challenge, a statistical design of experiment is performed in the plant with multiple measurements associated with each batch. Combination of both statistical linear mixed model and multivariate batch analysis reveal the major contributors to the variation in the y which eventually help the plant to achieve the goal.

429475 A Linear Programming Based Scenario Reduction Approach Using Transportation Distance

Abstract:

Uncertainty is pervasive in various process design and operations problems. In the corresponding optimization problems, uncertainty is often represented using samples/scenarios (e.g., stochastic programming with recourse, sample average approximation for chance constrained optimization). For the solution of the scenario based optimization problem, the quality of the solution is greatly affected by the scenarios used. In general, using a large number of scenarios can lead to higher solution reliability, but result in higher computational complexity. For many optimization problems under uncertainty, the scenario based model even becomes intractable by including a large number of scenarios into the model. Scenario reduction, an important topic of scenario based decision making, is induced with the aim of selecting a few representative scenarios from a large number of scenarios. The selected scenarios will then be used in the scenario based optimization model, so as to reduce the computational complexity.

Although it is an important topic, scenario reduction has received limited attention in literature [1-3]. Among the existing methods, the transportation distance [4] based scenario reduction was shown to be one of the most effective methods. Along this direction, a heuristic based scenario reduction method for stochastic programming was proposed in [5] and a corresponding tool SCENRED2 is available in GAMS. Li and Floudas [6] proposed a mixed integer linear optimization (MILP) based scenario reduction method, which rigorously minimizes the transportation distance (i.e., Kantorovich distance) to find the optimal subset to represent the initial superset of scenarios. This method also considers the performance of input and output space simultaneously. Although the MILP based scenario method can provide the optimal reduced subset of scenarios, the method is limited by the size of the problem. With the state-of-the-art mixed integer linear optimization solver, the method can only address moderate number of scenarios (i.e., up to 5000). Li and Floudas [7] recently proposed a method for the reduction of the huge number of scenarios (e.g., 5^{30})

generated from the factorial combination. A sequential reduction framework was proposed which significantly reduce the computational complexity. Some criteria for quantifying the quality of scenario reduction were also proposed considering that it is impractical to evaluate huge number of scenarios.

In this work, a linear programming (LP) based scenario reduction method is proposed. The objective is to address the reduction from a given large set of scenarios (much more than 5000, and not necessarily generated from the factorial combination). In the proposed algorithm, the selected scenarios and the corresponding probabilities are updated in two successive steps in an iterative fashion. The first step calculates the probabilities based on a fixed selection of scenarios and the second step updates the subset of selected scenarios. The first step is solved based on a simple linear programming problem and the second step only relies on simple cost calculations. Hence, the computational complexity of the proposed method is much less than the existing MILP based method. The proposed method works well for those scenario sets with a large size (i.e., 50000). The proposed algorithm will be illustrated with various case studies. We will also demonstrate the effectiveness of the proposed method through comparison studies with SCENRED2 [5] and the MILP based scenario reduction method [6].

References:

[1]. Dupačová J, Gröwe-Kuska N, Römisch W. Scenario reduction in stochastic programming[J]. Mathematical Programming. 2003,95(3):493-511.

[2]. Heitsch H, Römisch W, Strugarek C. Stability of multistage stochastic programs[J]. SIAM Journal on Optimization. 2006,17(2):511-25.

[3]. Karuppiah R, Martín M, Grossmann IE. A simple heuristic for reducing the number of scenarios in two-stage stochastic programming[J]. Computers & Chemical Engineering. 2010,34(8):1246-55.

[4]. Rachev ST, Rüschendorf L. Mass Transportation Problems: Volume I: Theory: Springer Science & Business Media; 1998.

[5]. Heitsch H, Römisch W. Scenario Reduction Algorithms in Stochastic Programming[J]. Computational Optimization and Applications. 2003,24(2-3):187-206.

[6]. Li Z, Floudas CA. Optimal scenario reduction framework based on distance of uncertainty distribution and output performance: I. Single reduction via mixed integer linear optimization[J]. Computers & Chemical Engineering. 2014,70:50-66.

[7]. Li Z, Floudas CA. Optimal scenario reduction framework based on distance of uncertainty distribution and output performance: II. Sequential reduction. Computers & Chemical Engineering. 2015, in publication.

432566 Bayesian Inference of Complex Chemical Process Using Reduced Order Modeling

Abstract:

Bayesian inference of parametric uncertainties is an attractive uncertainty quantification technique with its ability to seamlessly incorporate data as well as prior knowledge to update probability distributions over uncertain quantities. For complex systems with arbitrary posterior probabilities of parameters, Markov Chain Monte Carlo (MCMC) is the most commonly used method in Bayesian inference to evaluate the posterior parametric distribution. The main bottleneck, however, in applying Bayesian inference for complex systems, is that MCMC simulation involves evaluating complex models for 10^3 -10^5 realizations which becomes prohibitively expensive. Parallelization, a natural choice to reduce computational limitations, cannot be utilized as Markov chain Monte Carlo is a serial algorithm for which each new sample requires the knowledge of the previous sample in the parametric space.

Another way to reduce computational cost is to simplify the likelihood evaluation in posterior probability estimation by approximating the underlying physical models to make each step evaluation cheaper. With that motivation, there have been many different approaches studied in the past, such as as using low fidelity models applying a coarse spatial discretization [1], model approximation using global polynomials (polynomial chaos expansions) [2], reduced order models using proper orthogonal decomposition [3] and Gaussian approximation of the process [4], each with its own advantages and limitations in applicability.

In this work, we investigate the use of an artificial neural network (ANN) to develop a reduced order model in approximating the model output to apply in posterior evaluation of probabilities Using ANN as a reduced order model, the computational cost can by reduced drastically. Empirical models, in general, often fail to capture the behavior of the system over a wide range of parametric space. ANN, on the other hand can potentially represent wide regions of parametric variations as it has a learning capability which can be tuned for a large parametric space. We demonstrate the approach by applying it to estimate the impact of uncertainties of adsorption equilibrium parameters on the performance of a CO_2 adsorption process using hollow fiber sorbents.

References

1. Arridge SR, Kaipio JP, Kolehmainen V, Schweiger M, Somersalo E, Tarvainen T, Vauhkonen M. Approximation errors and model reduction with an application in optical diffusion tomography. Inverse Problems 2006; 22: 175–195
2. Marzouk, Y. M., H. N. Najm, and L. A. Rahn. Stochastic spectral methods for efficient Bayesian solution of inverse problems. Journal of Computational Physics 2007; 224 (2), 560–586.
3. Galbally D, Fidkowski K, Willcox K, Ghattas, O., Non-linear model reduction for uncertainty quantification in large-scale inverse problems. Int. J. Numer. Meth. Engng 2010; 81:1581–1608
4. Isaac T, Petra N, Stadler G, Ghattas O., Scalable and efficient algorithms for the propagation of uncertainty from data through inference to prediction for large-scale problems, with application to flow of the Antarctic ice sheet, J. Comput. Phys. (2015), *In press*

434829 On the Use of Multistage Stochastic Programming for the Design of Grid Scale Energy Storage Systems

Abstract:

It is widely recognized that a major concern with renewable energy is the fact that wind and solar sources are non-dispatchable. That is, the power produced from renewable sources is dependent on environmental conditions and is likely uncorrelated with the power demand from load centers. While fossil based sources are dispatchable and currently have the ability to respond to the full range of consumer loads, the additional range imposed by renewable sources is expected to exceed the dispatch capability of these fossil plants at the point of 20% renewable power. Thus, many have advocated the use of massive energy storage systems to provide the additional level of dispatch capability required to maintain grid solvency.

Due to the uncertainty of consumer demand as well as that of renewable generation, the problem of optimal placement of these storage units within the grid along with the selection of equipment sizes must be formulated as a stochastic program. However, rather than being a fairly simple two-stage stochastic program, the dynamics imposed by the storage devices requires the formulation to be of the far more challenging multistage class.

In this work, we propose a novel approximate solution procedure for this class of multistage stochastic programming problems. The approach utilizes the computational efficiency of the recently developed method of Economic Linear Optimal Control (ELOC) and its extension Constrained ELOC. The proposed method occurs in two stages. The first is a global search over the here-and-now variables as well as the parameters of the ELOC policy. However, this first search assumes a statistical constraint enforcement mechanism (similar to chance constrained optimization). In the second step is a gradient based search over the here-and-now variables, but used the Constrained ELOC policy to enforce point-wise-in-time constraints. To address the possible occurrence of an infeasible solution from the Constrained ELOC policy a fairly simple barrier type approach is used to improve convergence properties.

425978 A Graph-Theory Approach for Efficient Non-Anticipativity-Constraint Reduction in Multistage Stochastic Programming Problems Involving Endogenous and Exogenous Uncertainties

Abstract:

A graph-theory approach for efficient non-anticipativity-constraint reduction in multistage stochastic programming problems involving endogenous and exogenous uncertainties

Robert M. Apap & Ignacio E. Grossmann

In multistage stochastic programming problems, the number of non-anticipativity constraints (NACs) grows rapidly as the number of time periods, uncertain parameters, or realizations increases (Apap and Grossmann, 2015). It is typically desirable to eliminate as many redundant NACs as possible. In the case that the uncertainty is purely exogenous, this process is fairly straightforward since the scenario tree is fixed. For the case of purely endogenous uncertainty, however, the scenario tree is decision dependent; hence, NAC elimination is considerably more involved. Previous works in the literature, namely Goel and Grossmann (2006) and Gupta and Grossmann (2011), have proposed effective theoretical reduction properties for this class of problems. One primary assumption in these existing properties, though, is that the set of scenarios corresponds to all possible combinations of realizations of the uncertain parameters (i.e., a Cartesian product over all sets of realizations). In large, real-world problems with many uncertain parameters, the Cartesian-product approach can easily lead to instances with several thousands of scenarios or more. These problems are generally intractable. Furthermore, if some of the uncertain parameters are correlated, many of the scenarios in these large problems may in fact refer to impossible outcomes.

In this paper, we solve mixed-integer linear multistage stochastic programming problems involving endogenous and exogenous uncertainties, where the scenario sets are pre-specified and do not correspond to Cartesian products. These arbitrary scenario sets can each be viewed as the result of sampling from a joint probability distribution. Since the previous reduction properties in the literature do not apply here, efficient generation of the non-anticipativity constraints becomes a challenge. Khaligh and MirHassani (2015) recently addressed this issue for multistage stochastic programming problems under endogenous uncertainty and proposed a novel graph-theory approach for generating the minimum number of NACs in polynomial time. In the current work, we extend their approach to problems with both endogenous and exogenous uncertainties. We present several illustrative examples related to process networks and oilfield development planning in order to demonstrate the implementation of the proposed reduction algorithm. These problems are then solved with a combination of a fast heuristic and Lagrangean decomposition, as described in Apap and Grossmann (2015).

References:

Apap, R. M. & Grossmann, I. E. (2015). Models and computational strategies for multistage stochastic programming under endogenous and exogenous uncertainties. Manuscript in preparation.

Goel, V. & Grossmann, I. E. (2006). A class of stochastic programs with decision dependent uncertainty. Mathematical Programming, 108 (2-3, Ser. B), 355-394.

Gupta, V. & Grossmann, I. E. (2011). Solution strategies for multistage stochastic programming with endogenous uncertainties. Computers and Chemical Engineering, 35, 2235-2247.

Khaligh, F. H. & MirHassani, S. A. (2015). Efficient constraint reduction in multistage stochastic programming problems with endogenous uncertainty. To appear in Optimization Methods and Software.

429491 MINLP Combined with the Analytic Hierarchy Process for the Design of Chemical Processes Under Uncertainty: Application to the Synthesis of Heat Exchanger Networks

Abstract:

MINLP combined with the analytic hierarchy process for the design of chemical processes under uncertainty: Application to the synthesis of heat exchanger networks

Dauda Ibrahim, Gonzalo Guillen-Gosalbez*, Megan Jobson

Centre for Process Integration, School of Chemical Engineering and Analytical Science, The University of Manchester, The Mill, Sackville Street, Manchester M13 9PL, United Kingdom

*Corresponding author Tel: +44(0) 1613068755, E-mail address: gonzalo.guillen-gosalbez@manchester.ac.uk

Chemical processes are traditionally designed under the assumption of fixed operating parameters (temperature, pressure, composition, flow rate, etc.), which reflect nominal conditions. During plant operation, however, it is likely that the process will undergo changes in the operating parameters. The question that arises then is whether the plant will be able to cope with these new conditions that may differ (sometimes significantly) from the nominal ones.

This paper introduces a new framework for the synthesis under uncertainty of chemical processes that combines mathematical programming with the analytic hierarchy process (AHP). At the design stage, our approach takes into account explicitly the most critical uncertainties that can affect the calculations. The final goal is to identify a robust design showing good performance in the space of uncertain parameters.

The framework proposed comprises four main steps. First, the main uncertainty sources are identified and described via probability functions. Monte Carlo sampling is then applied on these distributions in order to generate representative scenarios that will be used in the calculations. In step 2, an optimization model is solved for every such scenario in order to generate a set of potential designs behaving in different ways in the uncertain parameters space. In step 3, these designs are next evaluated against all the scenarios in terms of feasibility and profitability. A set of stochastic metrics are also calculated in this step, including the expected performance, worst case and level of feasibility in different operative constraints. In step 4, the analytic hierarchy process (AHP) is applied in order to identify the most economic, feasible and flexible process design among those generated in step 2. To this end, pairwise comparisons between the metrics calculated in step 3 are first carried out using the Saaty scale. The eigenvalues of the matrix of pairwise comparisons will then be used to calculate the weights to be attached to every metric. The designs generated in step 2 will then be sorted according to these weights.

The capabilities of our approach are illustrated through its application to the design of several heat exchanger networks (HENs), in which uncertain supply temperatures and heat capacity flow rates are considered. The problem is modelled as a multi-scenario MINLP based on the original formulation of Yee and Grossmann (1990). This MINLP is solved for every scenario, which reduces the complexity associated with the original model and avoids in turn potential convergence and feasibility problems resulting from the large number of variables and non-linear nature of the problem.

Keywords: Process design, Uncertainty, Flexibility, Optimisation, HEN, MINLP, AHP

434580 Solution of Large Scale Stochastic Programming Problems for Optimal Placement of Booster Stations in Water Networks

Abstract:

Public water distribution networks are large complex systems with many access points, leading to the potential for accidental or deliberate contamination. A fixed grid of sensors can serve as an early-warning detection system to flag a contamination in the network. The problem of optimal sensor placement within these drinking water distribution systems has been extensively studied (Berry et al. (2005b); Ostfeld and Salomons (2004a); Murray et al. (2010). After a contamination has been detected by the early-warning detection system, one probable response is to collect manual water sample for laboratory analysis. Once a contamination has been confirmed by this analysis, a no-drink order could be given. However, the laboratory analysis can take several hours or more, while the contaminant continues to spread through the network. Disinfectant booster stations can help mitigate the effect of potential contamination by injecting additional (yet safe) amounts of disinfectant immediately following the initial warning (Parks and VanBriesen, 2009). Moreover, optimal placement of these booster stations can help in efficiently reducing the impact of a contamination incident.

In this work, we present a stochastic programming formulation for optimal placement of disinfectant booster stations. The formulation considers uncertainty in both location and time of a contamination scenario. For realistic large-scale water distribution networks, with hundreds of thousands of possible contamination scenarios, the formulation results in an intractably large Mixed Integer Linear Programming (MILP) problem. Fortunately, there is tremendous symmetry in the problem structure, which can be exploited to reduce the problem size by almost 5 orders of magnitude (for a 3,000 node network). The results highlight the effectiveness of booster stations in reducing the overall impact on the population, which is measured using two different metrics – mass of contaminant consumed, and population dosed above a cumulative mass threshold. Additionally, we also study the importance of various factors that influence the performance of disinfectant booster stations. These include the number and locations of sensors, the injected contaminant and its potential to react with chlorine, the location where the contaminant is introduced, the time it takes to detect the contamination, and the time delay between detection and initiation of booster disinfection.

References

Berry, J. W., Fleischer, L., Hart, W. E., Phillips, C. A., and Watson, J. P. (2005b). *Sensor Placement in Municipal Water Networks*. Journal of Water Resources Planning and Management, 131:237.

Ostfeld, A. and Salomons, E. (2004a). *Optimal Layout of Early Warning Detection Stations for Water Distribution Systems Security*. Journal of Water Resources Planning and Management, 130(5):377–385

Murray, R., Haxton, T. M., Janke, R. J., Hart, W. E., Berry, J., and Phillips, C. A. (2010). *Sensor Network Design for Drinking Water Contamination Warning Systems: A Compendium of Re- search Results and Case Studies using the TEVA-SPOT-Report.* US Environmental Protection Agency, Washington, DC. Technical report, EPA/600/R-09/141.

Parks, S. and VanBriesen, J. (2009). *Booster Disinfection for Response to Contamination in a Drinking Water Distribution System*. Journal of Water Resources Planning and Management, 135:502–511.

434949 A Multistage Stochastic Program to Evaluate Feedstock/Technology Development for Chemical Process Industry

Abstract:

Top chemical companies are shifting their focus from bulk chemicals to specialty products. For example, BASF expects to increase the share of "customized products, and functionalized materials and solutions" sales to 70% of its total by year 2020. Dow Chemical Company is "carving-out" its chlorine and epoxy assets reducing its commodity chemicals foot-print while it continues to grow in downstream specialty products in integrated plastics, electronics, and agriculture. Bayer recently announced that they will become a "pure Life Science company", focusing their business entirely on products that enhance human health and nutrition. Dupont continues to focus its product offerings on agriculture and nutrition developed using the latest advancements in biotechnology. These trends suggest that the feedstock and product portfolios along with utilized technologies of chemical process industry (CPI) may grow and be quite different compared to today's in the near future. As such, there is tremendous opportunity for investigating the impacts of these new technologies/feedstocks/products on chemical process industry.

Incorporation of new feedstocks and technologies into the existing CPI infrastructure may require significant amounts of investment. Determining the investment decisions, i.e., how much to invest, which technologies to invest in, and when to invest in each technology for research and development and for capacity expansion, is a challenging task, as there are often many emerging technologies and the future performance of these technologies is uncertain. In this work, we developed a multistage stochastic programming (MSSP) formulation accounting for both endogenous and exogenous uncertain parameters associated with the new technology investment planning (NTIP) problem.

The NTIP problem is characterized by a set of undeveloped or partially developed feedstock processing technologies and the existing CPI infrastructure system. The objective is to develop an optimal investment strategy that minimizes the total investment cost over a discrete planning horizon. The investment strategy identifies the capacity expansion timing and levels for each of the developing technologies. Each technology is assumed to go through three discrete maturity levels, (1) laboratory, (2) pilot plant, and (3) commercial. Developing technologies are unable to meet production demands unless they have reached the commercial stage. At the time of investment, the yield, the effect of capacity expansion, and the effect of research and development spending on the cost of capacity expansion are not known with certainty. These parameters are endogenous uncertain parameters because the investment decisions affect the realizations of these uncertain parameters. The demand for products is also not known with certainty, and it is represented as an exogenous uncertain parameter. In order to meet demands, the products can be produced from technologies that have been developed to commercial level or the product can be purchased.

The presented formulation models the material flow throughout the technology network using a state-task network where the states are the feedstocks, intermediates, and products, and the tasks are the technologies. Each technology has an initial maturity level. In order to represent the stages of maturity, a stage-gate framework is used. Technologies can move from one stage to the next through capacity expansion. The cost of capacity expansion is modeled using a two-factor learning curve. Capacity expansion costs can be reduced either through capacity expansion or through direct R&D spending. The rate at which each factor affects the capacity expansion cost is uncertain at the time of investment. By expanding capacity and investing in R&D, these factors are revealed. Intermediates and products can be stored, however the amount that can be stored is limited by maximum storage capacities. To incorporate the uncertainty, each uncertain parameter is modeled using a discrete uniform distribution with two possible realizations, a high and a low value. Scenarios are generated using the combinations of realizations for each uncertain parameter. For this formulation, the decisions only impact the realizations (not the distributions) of the endogenous uncertain parameters, which allows for complete enumeration of the realizations of the endogenous parameters *a priori*.

The MSSP is used to determine the investment-decision plan for the conversion of biomass and naphtha (feedstocks) to ethylene (product) with two intermediates, syngas and ethanol. A total of four different technologies are available, (1) gasification of biomass to syngas, (2) catalytic conversion of syngas to ethanol, (3) catalytic dehydration of ethanol to ethylene, and (4) the cracking of naphtha to ethylene. We assume that technologies (3) and (4) are mature meaning that they are already commercially available, and as such their yields and capacity expansion costs are known. This yields a problem with eight uncertain parameters; six of the uncertain parameters correspond to uncertain parameters relating to each technology, these parameters include the two parameters for the two factor learning curve, and the yield of each technology. Additionally, two parameters represent the demand uncertainty for each product during the planning horizon. The MSSP formulation has 1,376,256 variables and 10,289,152 .

This talk discusses in detail the size of the problem and solution time. The size of this problem grows quickly as the length of the planning horizon increases and the number of undeveloped technologies increases. Based on our preliminary results, linear growth in the number of non-mature technologies caused the size of the problem (number of constraints and variables) grew non-linearly which directly contributes to solution times. The growth in problem size is caused by an increase in the number of possible realizations of uncertain parameters which is multiplicative in nature. Even for the simple presented case, where each parameter has two possible outcomes, the number of scenarios is 256. By increasing the number of non-mature technologies by 1, the number of scenarios in the problem in increased by a factor of 8. This talk will further discuss the impact of extending the planning horizon, the relative ratio of raw material costs, demand, yield, and capacity expansion cost uncertainties on the total cost along with installed and utilized capacity of each technology throughout the planning horizon. Despite the computational challenges associated with implementing the formulation we believe that the formulation provides a flexible approach for the evaluation of developing technologies.

417309 Generalized Derivatives of Dynamic Systems with Lexicographic Linear Programs Embedded

Abstract:

Bioprocesses involving microbial communities have widespread applications in the pharmaceutical, food and biofuels industries. Despite the complexity of these systems, their modeling relies heavily on unstructured models, which are inaccurate when modeling systems involving cyclic steady states, symbiotic and competitive relationships, and multiple nutrient limitations. These shortcomings can be addressed by dynamic flux balance analysis (DFBA) [1], [2], which combines genome-scale metabolic network reconstructions with dynamic process models to model accurately the behavior of bioprocesses involving microbial communities. DFBA models result in dynamic systems with linear programs (LPs) embedded [3]. These LPs are embedded because their right-hand side depends on the dynamic states and the dynamic states depend on the solution vector of the LP. Until recently, the simulation of DFBA systems was challenging, but newly available simulators [3], [4] have made the reliable and efficient implementation of DFBA possible. These simulators transform the LP embedded into a lexicographic LP (LLP) to address complications associated with nonunique and infeasible LP solutions. In order to use DFBA to design optimal bioprocesses involving microbial communities, sensitivity information needs to be computed. Therefore, this paper presents the theory behind the computation of sensitivities for dynamic systems with LLPs embedded.

The objectives of a LLP in standard form as a function of its right-hand side are piecewise linear functions [5], and therefore, nonsmooth. This source of nonsmoothness can be propagated to the parametric dependence of the final states of the dynamic system. Therefore, there exist some parameter values for which the Jacobian of the dynamic system may not exist. Computing elements of Clarke's generalized Jacobian for complex nonsmooth functions is challenging [6], but can be done efficiently for piecewise differentiable functions, such as LLPs parameterized by their right-hand side, with lexicographic-directional (LD) derivatives [7]. LD-derivatives of dynamic systems can be computed efficiently to obtain elements of the plenary hull of the generalized Jacobian, and in some instances, elements of the generalized Jacobian [8]. This information can be provided to bundle methods to optimize DFBA systems.

This paper derives first the LD-derivatives of LLPs, which are shown to be obtained by solving related LLPs. Next, LD-derivatives of dynamic systems with LLPs embedded are computed using event detection to integrate efficiently the sensitivity ODEs, which can have discontinuous right-hand sides. Finally, the theory developed is implemented to optimize a DFBA case study.

Keywords: Linear programming, generalized Jacobian, lexicographic differentiation, LD-derivative, lexicographic optimization, nonsmooth sensitivities, nonsmooth equation solving, flux balance analysis, dynamic flux balance analysis.

References

[1] J. D. Orth, I. Thiele and B. Ø. Palsson, "What is flux balance analysis?," *Nature Biotechnology,* vol. 28, pp. 245-248, 2010.

[2] R. Mahadevan, J. Edwards and F. I. Doyle, "Dynamic flux balance analysis of diauxic growth in Escherichia coli.," *Biophysical Journal,* vol. 83, no. 3, pp. 1331-40, 2002.

[3] K. Höffner, S. M. Harwood and P. I. Barton, "A reliable simulator for dynamic flux balance analysis," *Biotechnology and Bioengineering,* vol. 110, no. 3, pp. 792-802, 2013.

[4] J. A. Gomez, K. Höffner and P. I. Barton, "DFBAlab: A fast and reliable MATLAB code for Dynamic Flux Balance Analysis," *BMC Bioinformatics,* vol. 15, p. 409, 2014.

[5] D. Bertsimas and J. N. Tsitsiklis, Introduction to Linear Optimization, Nashua, NH: Athena Scientific, 1997.

[6] F. H. Clarke, Optimization and Nonsmooth Analysis, Philadelphia: Society for Industrial and Applied Mathematics, 1990.

[7] K. A. Khan and P. Barton, "A vector forward mode of automatic differentiation for generalized derivative evaluation," *Optimization Methods & Software,* p. In Press., 2015.

[8] K. A. Khan and P. I. Barton, "Generalized Derivatives for Solutions of Parametric Ordinary Differential Equations with Non-differentiable Right-Hand Sides," *Journal of Optimization Theory and Applications,* vol. 163, no. 2, pp. 355-386, 2014.

417697 Local Optimization of Dynamic Systems with Guaranteed Feasibility of Path Constraints

Abstract:

Inequality path constraints in optimal control problems often represent safety and product quality limits and their strict satisfaction is crucial. A control vector parameterization is applied to reduce the infinite-dimensional optimal control problem to an approximate problem with finitely many decision variables. The path constraints still have to be enforced along the whole time horizon leading to a semi-infinite program (SIP). We present an algorithm that terminates in finitely many steps under relatively mild assumptions returning a guaranteed feasible solution.

The solution satisfies the first-order Karush-Kuhn-Tucker (KKT) optimality conditions to a user-specified tolerance. The algorithm is based on [1] and is an adaption of the global optimization algorithm for SIP proposed in [2]. The semi-infinite constraints are approximated by simultaneous restriction and relaxation: While the right-hand-sides of the path constraints are restricted by ε, the constraints are relaxed by enforcing them at finitely many time points only.

This approximated optimal control problem can be solved by existing dynamic optimization codes. We use the single shooting approach implemented in DyOS [3]. The output is checked for constraint violations and the time sets of the path constraints are populated iteratively by adding time points where the constraint violation is maximal. The restriction ε is reduced if no violation can be determined.

Numerical case studies with models from chemical engineering are presented and the performance is compared with existing algorithms, among others [4].

References

[1] Fu J, Faust, JMM, Chachuat B, Mitsos A. Local optimization of dynamic programs with guaranteed satisfaction of path constraints. August 2014;submitted.

[2] Mitsos A. Global optimization of semi-infinite programs via restriction of the right-hand side. Optimization 2011;60(10-11):1291–308.

[3] Schlegel M, Stockmann K, Binder T, Marquardt W. Dynamic optimization using adaptive control vector parameterization. Computers & Chemical Engineering 2005;29(8):1731–51.

[4] Chen TW, Vassiliadis VS. Inequality path constraints in optimal control: a finite iteration ε-convergent scheme based on pointwise discretization. Journal of Process Control 2005;15(3):353–62.

429360 A Parallel Multiperiod Optimization Approach for Large-Scale Dynamic Systems Under Uncertainty

Abstract:

Higher operating costs and shrinking profit margins in the chemical and petro-chemical industries is driving greater applications of advanced control techniques, as well as consideration of control at the process design stage. These applications are often model-based and require the solution of optimization formulations comprising very large systems of variables and equations that can be computationally demanding, requiring considerable computer memory and solution times. Accordingly, these applications are motivating the development and implementation of solution approaches, numerical techniques and algorithms capable of efficiently exploiting modern computational resources in terms utilizing multiprocessor systems and/or acceleration devices, all while limiting memory usage. One particular direction that we focus on in this work is a solution strategy for multiperiod dynamic optimization formulations which incorporates process models described by differential-algebraic equations (DAEs) with uncertain parameters and/or disturbance inputs. To this end, we are investigating a direct multiperiod discretization approach of the uncertainty space combined with a multiple-shooting discretization of the temporal domain, which requires the solution of an embedded dynamic model within an overall nonlinear programming algorithm.

Over the last few decades interior-point optimization approaches have been shown to be superior to other methods in terms of efficiently handling ever increasing large-scale optimization formulations. For stochastic optimization problems, multiperiod (or multiscenario) discretization is often applied where numerous scenario realizations are generated by sampling from a particular distribution or within a defined interval. The resulting formulations are posed using a sample-average objective function and constraints for all realizations [1]. Accordingly, this introduces a partially separable block structure with design variables representing coupling or complicating variables between each scenario realization. These particular formulations have led to the development of structure exploiting solution strategies that utilize efficient decomposition approaches, which either aim to reduce the dimensionality of the original formulation allowing standard dense linear algebra techniques or exploit the separability of the full-space structure allowing a portion of the algorithm to be simultaneously solved in parallel. In this latter direction, one such approach has been the determination of the Newton search direction within linear, quadratic or nonlinear programs via a Schur-complement decomposition where certain aspects of approach can be performed in parallel while the Schur-complement itself can be formed and then factored directly [2] or perhaps more efficiently, for very large linear systems with numerous complicating variables, solved iteratively using preconditioned conjugate gradient methods [3].

Our proposed solution approach extends our previous work [4] and uses a sequential-quadratic programming (SQP) algorithm where the QP subproblems are solved using an interior-point method (IPM) via the OOPS solver which employs a direct Schur-complement decomposition [2]. Additionally, a partially separable quasi-Newton strategy is used to iteratively approximate and update second-order information used in formulating the QP. Novel aspects of the work include parallelization of the DAE solution through a multiple shooting approach, coupled with parallelization of the uncertainty scenarios, and employment of an SQP-IPM large-scale nonlinear programming algorithm that exploits the structure of discretized dynamic optimization formulations via decomposition.

The focus of our presentation will be on describing our proposed strategy for efficiently decomposing and solving large-scale multiperiod dynamic optimization formulations. More specifically, we will highlight the following aspects: (1) our approach for decomposing and parallelizing embedded DAE model representations within a multiple-shooting dynamic optimization framework; (2) an assessment of the potential computation improvements of the parallel algorithm for solving embedded DAEs utilizing an adaptive DAE solver with simultaneous forward sensitivity generation; (3) an assessment of our SQP-IPM nonlinear programming approach which utilizes a structure exploiting QP solver. Our implementation makes use of a c/c++ computing environment that uses (1) OpenMP loop parallelization constructs to evaluate the embedded DAE model via the Sundials IDAS solver; (2) the QP solver OOPS for each subproblem which has the potential to be parallelized using a distributed computing MPI implementation. Case studies involving integrated design and control and/or robust control will be presented and numerical results investigated, which aim to demonstrate the potential computing speedup and efficiency of the parallelized multiperiod algorithm with increasing processors and load. Additionally, avenues for future research will be identified and some perspective provided on the application of the proposed techniques to industrial large-scale DAE models.

References

[1] A. Shapiro, D. Dentcheva and A. Ruszczynski, Lectures on Stochastic Programming, SIAM (2009).

[2] J. Gondzio and A. Grothey, Exploiting structure in parallel implementation of interior point methods for optimization, Computational Management Science 6 pp. 135-160 (2009).

[3] J. Kang, Y. Cao, D.P. Word and C.D. Laird, An interior-point method for efficient solution of block-structured NLP problems using an implicit Schur-complement decomposition, Computers & Chemical Engineering 71, pp. 563-573 (2014).

[4] I.D. Washington, C.L.E. Swartz, Design under uncertainty using parallel multiperiod dynamic optimization, AIChE Journal 60, pp. 3151-3168 (2014).

429105 Solving Large-Scale Dynamic Optimization Problems in Parallel By Exploiting Structure

Abstract:

Dynamic optimization problems are constrained by differential and algebraic equations and are found everywhere in science and engineering. A well-established method to solve these types of problems is direct transcription where the differential equations are replaced with algebraic approximations using some numerical method such as a finite-difference or Runge-Kutta (collocation) scheme. However, for problems with thousands of state variables and discretization points, direct transcription may result in nonlinear optimization problems which are too large for general-purpose optimization solvers to handle. In the case of applying an interior point solver, the dominant computational cost is solving the linear systems resulting from the Newton step. For large-scale nonlinear programming (NLP) problems, these linear systems become prohibitively expensive to solve. Furthermore, the systems quickly become too large to formulate and store in memory on a standard computer. Despite these challenges, direct transcription also has the nice property of imposing sparsity and structure on the linear systems. In this talk we discuss ways to take advantage of this property. We compare several matrix decomposition strategies which can be parallelized in order to solve large-scale dynamic optimization problems. In addition, we describe our implementation framework which can easily be applied to arbitrary dynamic optimization problems. Finally, we demonstrate the effectiveness of these techniques on some classical chemical engineering applications.

425183 Improved Interval Bounds on the Solutions of General Nonlinear ODEs Using Lifted Models with Manufactured Solution Invariants

Abstract:

We present a new technique for computing tight, pointwise-in-time interval enclosures of the solutions of nonlinear ordinary differential equations (ODEs) subject to a range of parameters and initial conditions. The computation of these state bounds is a central step in algorithms for solving dynamic optimization problems with nonlinear ODEs embedded to guaranteed global optimality. Moreover, these bounds can be used for propagating uncertainty through nonlinear dynamic models, for nonlinear state-estimation for process control, and for robust dynamic optimization problems arising in, e.g., robust model-predictive control, fault-detection, and safety verification problems.

Although state bounding methods have been extensively developed in the past several years, there remains for many problems an unsatisfactory trade-off between the computational cost of the methods and the accuracy (i.e., tightness) of the resulting bounds. This is especially true for highly nonlinear models over large parameter ranges, and is the predominant factor limiting the effectiveness of global dynamic optimization algorithms for problems of practical size.

In this work, we build upon a recently published state bounding method that exploits affine solution invariants present in many models of interest (e.g., chemical reaction kinetics). In this method, solution invariants are used to define a bounds-tightening procedure that is then incorporated into a very inexpensive state bounding method based on differential inequalities and interval arithmetic. For many problems, this scheme provides bounds that are substantially tighter than those produced by differential inequalities alone, while maintaining the low cost of the original method. The drawback, however, is that this method applies only to models with affine solution invariants, and even for such models, no further improvements to the bounds can be made after all existing invariants have been exploited.

To overcome this limitation, we present a somewhat surprising result: Significant and often dramatic improvements in state bounds can be obtained for arbitrary models by applying the aforementioned bounding technique to a lifted (i.e., higher-dimensional) system constructed by defining one or more new state variables in terms of the existing ones. These definitions constitute solution invariants for the lifted model, while the right-hand side functions of the new variables are derived by direct differentiation. While these new equations are redundant with the original model in real number arithmetic, they provide alternative enclosures in interval arithmetic that can effectively contribute to the bounding method. The use of such lifted systems provides new and unexplored degrees-of-freedom for the design of effective bounding methods, with the potential to profoundly impact global dynamic optimization capabilities. We present some preliminary insights into the optimal design of such systems with illustrative examples.

Using Dynamic Simulation to Drive Process Design, Control, and Optimization

Dynamic Simulation and Optimization I

Tuesday, November 10, 2015 2:35 - 3:00 PM

Livia Wiley

Product Marketing Manager

Schneider Electric

Agenda

> Industry Challenges

> Suggested Strategy

> Customer Applications

> Role of Academia

> Conclusion

Process Industry Challenges

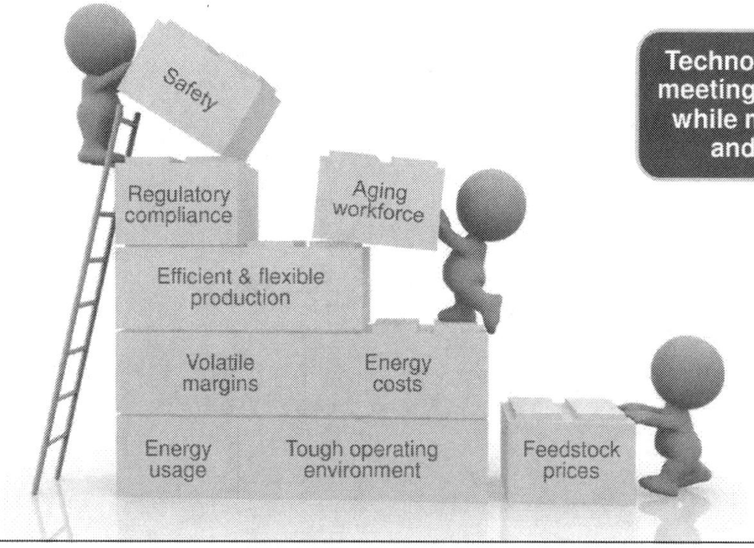

Technology is the key to meeting these challenges while maximizing profit and performance

Schneider Electric

Suggested Strategy

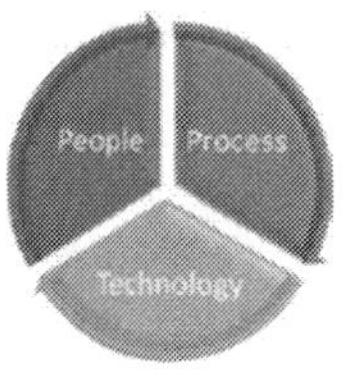

Common Modeling Approach

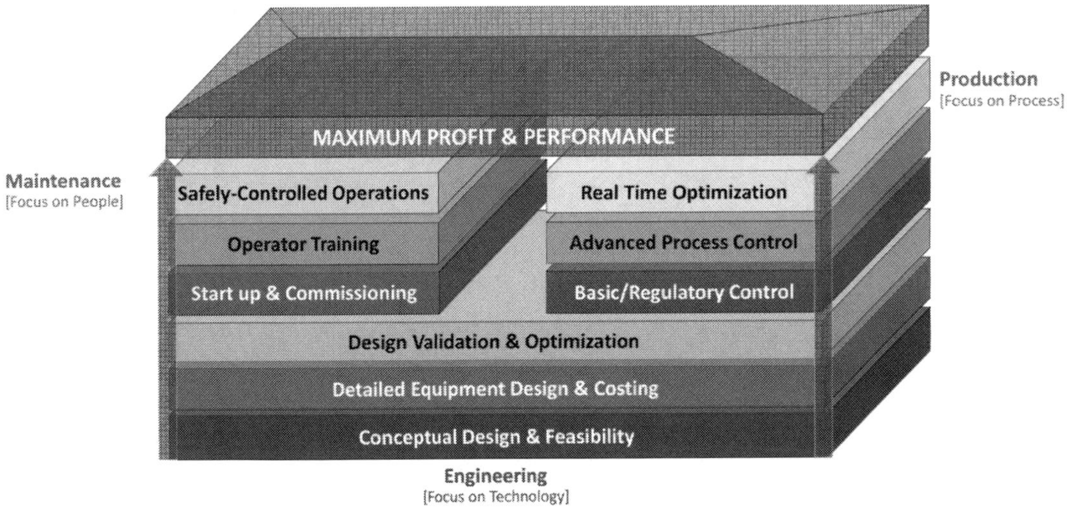

Engineering Focus on Technology

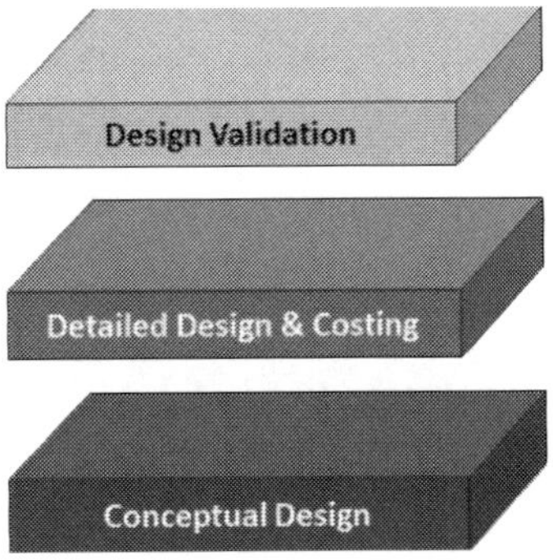

- Use dynamic simulation to validate process designs and optimize CAPEX by avoiding costly over design
  - Basis for controls checkout, control system design, OTS
- Rigorously design new processes or revamp existing facilities
  - Process modeling
  - Performance analysis
  - Energy optimization
  - Control and safety studies
  - Detailed facility design

Maintenance Focus on People

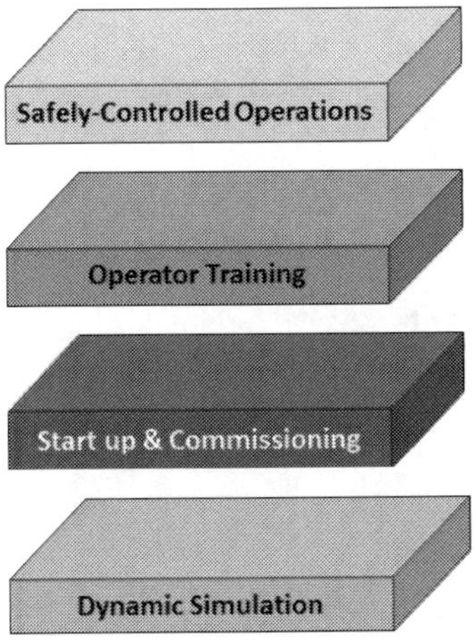

- Test common and extreme scenarios in a risk-free environment
  - Impacts asset availability, safety, and profitability
- Operators need to know how to operate the DCS and the plant via an operator training simulator (OTS)
  - Facilitate knowledge transfer while reducing training time
- Design, test, and perform system check-out prior to start-up
  - Can reduce commissioning time by several days or weeks

Production Focus on Process

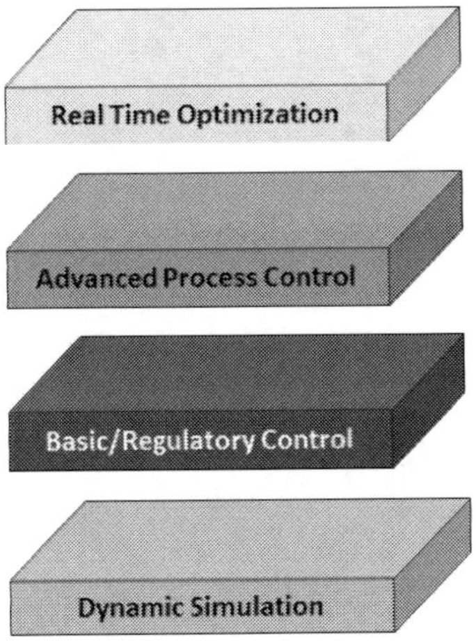

- Optimize a process in real-time given objective function(s) with specific process constraints
  - Use with process models, APCs, and data historians
- Build an APC strategy to operate the process closer to its limits by reducing variations in its operation
  - Improve product quality, throughput; reduce energy usage
- Integrate controls (DCS/PLC/SIS) to design and test basic operation, control & automation requirements
  - Provide operator training

Common Modeling Toolset

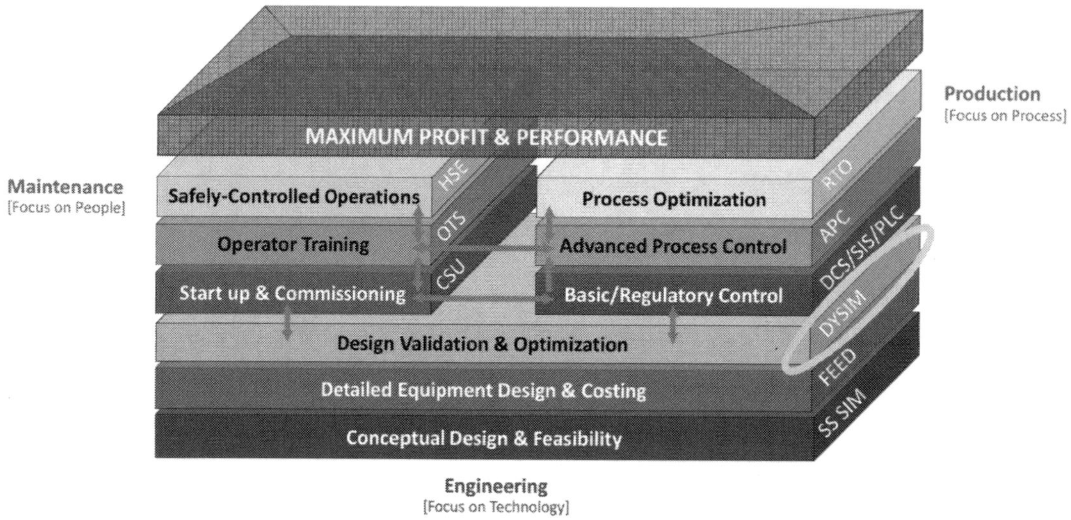

Schneider Electric

Customer Applications

Customer Value

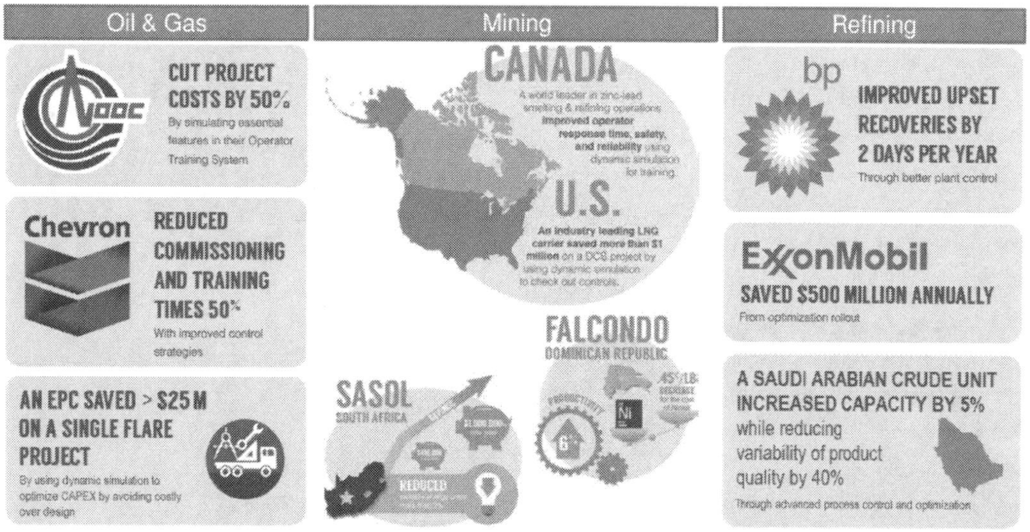

Oil & Gas Application: Chevron

- Challenges
  - Controls upgrade with safety improvements, no lost production, and shorter commissioning times
- Solution
  - DYSIM, OTS, and APC
- Benefits
  - Cut commissioning time in half, saving about $840K
  - Cut training time from months to weeks, saving four weeks over the course of the project
  - Reduced field commissioning time by 20% for instrument technicians & system integrators
  - Increased production capacity by 20% through testing of new control strategies

Mining Application: Codelco

CODELCO
Orgullo de Todos

> Challenges
  > Reduce time for controls checkout at startup
  > Provide efficient operator training
> Solution
  > 14 APC and 3 OTS projects in smelter processes
> Benefits
  > Very rapid controls checkout ROI justified cost of OTS
    - 6.5 days reduced time for controls checkout at startup
    - $2.9 MM actual project savings
  > Reduced operator training time for 36 operators
  > Justification to apply this technology at all sites
    - $370+ MM worth of opportunities identified; $50+ MM already realized

Refining Application: Reliance

Reliance
Industries Limited

> Challenges
  > Misaligned cogeneration plant & refining objectives
  > Energy costs above plan; plant has no responsibly for energy cost
> Solution
  > SS, DYSIM, APC, RTO, and OTS
> Benefits
  > $12.2M in annual energy savings achieved
    - $7.2M via performance measurement
    - $5M via advanced technology solutions
  > Payback = 2 months

Schneider Electric

Role of Academia

Addressing a Key Industry Challenge

- As operators retire, new younger engineers often take their place
  - New engineers have far less experience and learn in very different ways than the former generation
  - Need for capturing this knowledge and its transfer
- Academia is tasked with producing process engineers qualified to contribute on day one
  - Most use a steady-state process simulator (i.e., PRO/II) to teach engineering design
  - Few use dynamic simulation to teach transient behavior, control systems, and/or operator training → a sense of how a real plant operates

- Aging workforce
  - 71% of oil & gas workers are over 50
  - 50% of geophysicists and engineers will retire by 2018
  - Petroleum & engineering jobs will increase by 21% by 2022

Today's Plant View

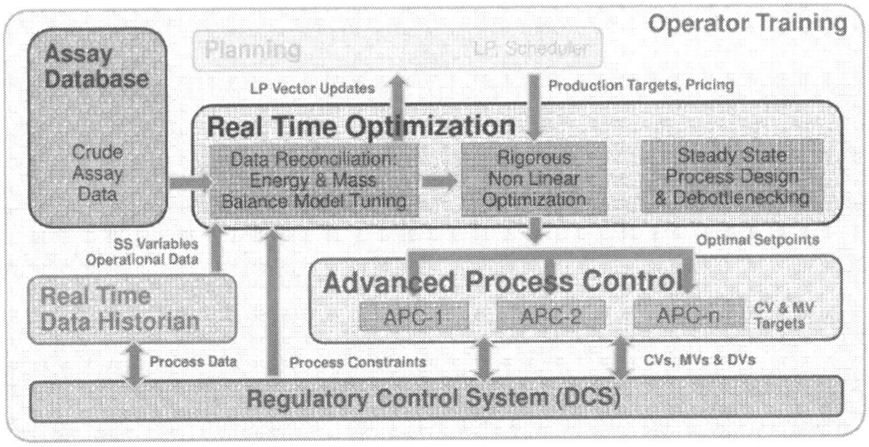

- Equipping tomorrow's engineers mean exposing students to as much hands-on experience as possible in a short period of time

Academic Application: West Virginia University

- Uses OTS as a hands-on teaching tool in their Chemical Engineering curriculum
  - IGCC power plant with 20 process areas, 200 process malfunctions, 30 accident investigations, 200 remote functions
  - Based on a dynamic simulation with an emulated control system
- Numerous benefits for students:
  - Demonstrate the real life application of controls theory and the role that control systems play
  - Show students where their current course work fits into the real plant
  - Reveal interactions between different equipment/systems/variables
  - Provide valuable operations experience to young engineers through limited but direct training on the OTS

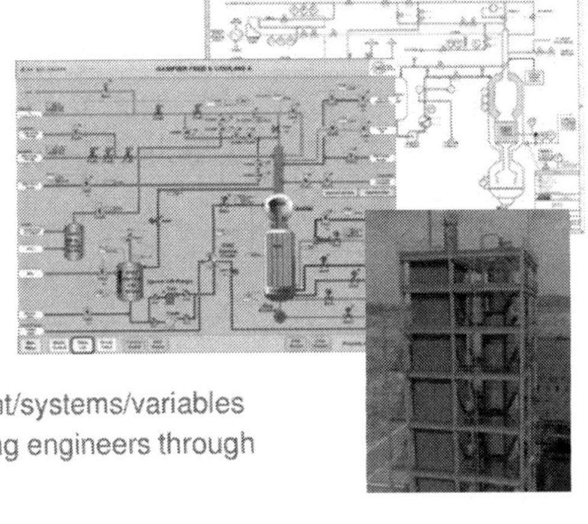

Schneider Electric

Conclusions

In Summary…

> Dynamic simulation is key to driving process design, plant control, and optimization

- ✓ Validated process designs
- ✓ Shorter plant commissioning schedules
- ✓ Lower capital expenditures
- ✓ Faster, equivalent on-the-job training
- ✓ Better operator effectiveness & agility
- ✓ Increased operational safety
- ✓ Lower operating costs
- ✓ Reduced equipment damage
- ✓ Superior control system designs
- ✓ Optimized control during transients
- ✓ Higher plant performance efficiency
- ✓ Enhanced quality
- ✓ Increased throughput
- ✓ Reduced energy usage
- ✓ Maximum sustainable plant performance
- ✓ Improved process profitability

> Applications are numerous and the value self evident

Acknowledgements

> Papers
  > Turton, Bhattacharyya, "Experience with an Operator Training Simulator in the Chemical Engineering Undergraduate Curriculum", AIChE, 2012
  > Wiley, "Accelerate Mining Operations Excellence by Investing in Added Resources – Simulation & Optimization", SME, 2015
  > Wiley, "Creating a Competitive Advantage Through Integrated Upstream Asset Management", URTeC, 2015
  > Wiley, "Deriving Value from Integrating Engineering Design and Automation with Offshore Operations", OTC, 2015
> Customer References
  > Chevron
  > Codelco
  > Reliance
  > West Virginia University

424552 Solving Multicomponent Reaction-Transport with Coupled Cellular Trajectories and Data-Driven Cellular Activation Models

Abstract:

Solving multicomponent reaction-transport with coupled cellular trajectories and data-driven cellular activation models

Yichen Lu, Mei Yan Lee, Talid Sinno, Scott L. Diamond

Institute for Medicine and Engineering, Department of Chemical and Biomolecular Engineering, University of Pennsylvania, Philadelphia PA 19104

Introduction: The release of autocrinic factors by cells often triggers autocatalytic surface reactions, all of which in the presence of diffusive and transport may further strengthen the signal and showcase interesting phenomena of cellular growth. Such situations include: blood clotting/bleeding under flow, biofilm formation, and angiogenesis/growth factor/tumor interactions. The problem of blood coagulation involves a series of autocatalytic agonists. When blood flows over disrupted vessel surfaces, activating platelets release ADP and thromboxane characterized by mobilized calcium level, while their surface becomes catalytic for thrombin production via phosphatidylserine exposure. Thrombin further facilitates clot growth by triggering fibrin production which strengthens clot mass.

Materials and Methods: Platelet calcium mobilization was measured by pairwise agonist scanning (PAS) of responses to single and pairwise delivery of low, medium, and high doses of: ADP, thromboxane mimetic (U46619), collagen mimetic (convulxin) and thrombin. A neural network model was trained (Lee et al. PLoS Comput Biol, 2015) with the PAS data for use in multiscale simulations. A multiscale model of clotting under flow (Flamm et al. Blood, 2012) was expanded for enhanced speed and accuracy. An aggregate remodeling was applied to growing clots to achieve more realistic shapes. A particle event-driven triangular element remeshing module was used for efficient and adaptive finite element method (FEM) solution of soluble concentrations. Thrombin was generated at the injury site and its transport was calculated with FEM. An activation/aggregation model involving ADP, thromboxane and thrombin was developed utilizing the calcium mobilization data from an averaged 120 neural network response.

Results and Discussion: The remodeling scheme was implemented to allow for instantaneous local translocation and rolling of single platelets and was shown to make more physiologically consistent prediction of the clot morphology. The adaptive meshing scheme was applied to provide unstructured triangular meshes that vary along with the modulating clot contour so that better resolution and efficiency were achieved for calculation of ADP/TXA2 release and thrombin transport. The activation/aggregation model characterized by the experimentally trained neural network response was able to predict platelet deposition dynamics for whole blood flowing over collagen at 200 s^{-1} wall shear rate consistent with in-silico measurements. This approach allows efficient multiscale calculations, from the single cell to the tissue level, when the cellular ensemble has evolving structure driven by coupled species transport and prevailing hemodynamics.

Conclusions: The computer-aided multiscale approach has proved useful for mimicking the growth of thrombus under flow condition and making reasonable prediction on platelet morphology and activation. More involved biological complexity can be added to the current model to study various aspects of hemostasis under different conditions.

References:

1. Lee MY, Diamond SL (2015) A Human Platelet Calcium Calculator Trained by Pairwise Agonist Scanning. PLoS Comput Biol 11(2): e1004118.
2. Flamm M, et al. Multiscale prediction of patient-specific platelet function under flow. Blood. 2012;120:190–198

Dynamics of cell induction in *Enterococcus faecalis*

Vu Tran[1], Arpan Bandyopadhyaya[2], Gary Dunny[3], Wei-Shou Hu[2], and Doraiswami Ramkrishna[1]

1. Department of Chemical Engineering, Purdue University
2. Department of Chemical Engineering and Material Science, University of Minnesota
3. Department of Microbiology and Immunology, University of Minnesota

PURDUE UNIVERSITY

NATIONAL INSTITUTES OF HEALTH

Motivations

- ***Enterococcus faecalis***: one of the 3 most common causes of hospital-acquired infections[1]
- **Endocarditis** infection by *E. faecalis* can cause death within several weeks
- Antibiotic drug resistance
- *E. faecalis* can exist in either:
  - Planktonic form (free floating)
  - **Biofilm**

1. Olawale et al., *Afr J Infect Dis*, 2011

2

Enhanced resistance of cells in Biofilms

- 60% of bacterial infections involved biofilm formation[1]
- Bacteria that adhere to medical devices or damaged tissues cause persistent infection
- A higher survival rate of bacteria in biofilms has been shown, even at a high antibiotic conc.
- In vivo, antibiotics might ***suppress*** symptoms of infection ***but fail to eradicate*** those bacterial cells still embedded in the biofilm

1. Mohamed et al., *J. of Med Mic Biol*, 2007

3

Significance of this research

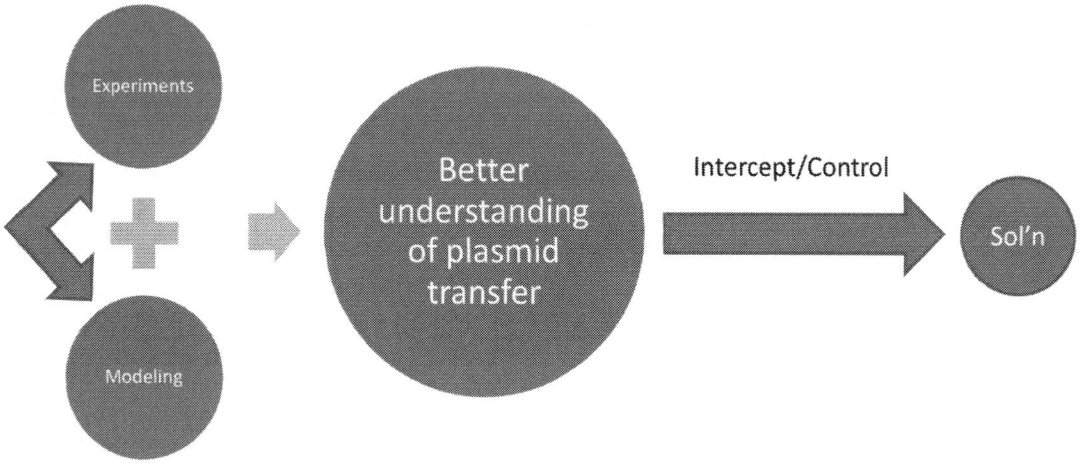

Signaling Mechanisms

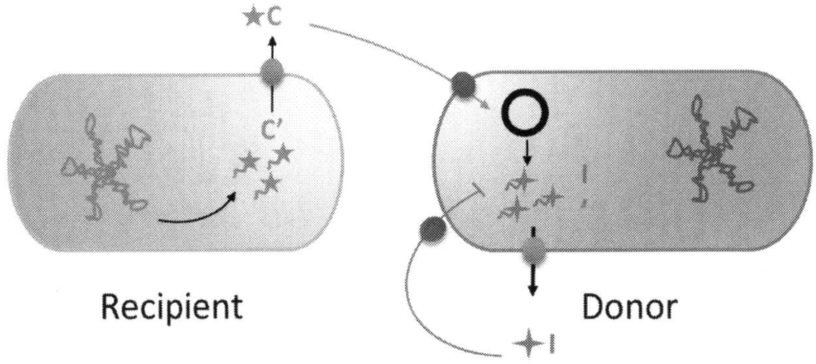

C: Mate Sensing signal (cCF10)

I: Self Sensing signal (iCF10)

Conjugation

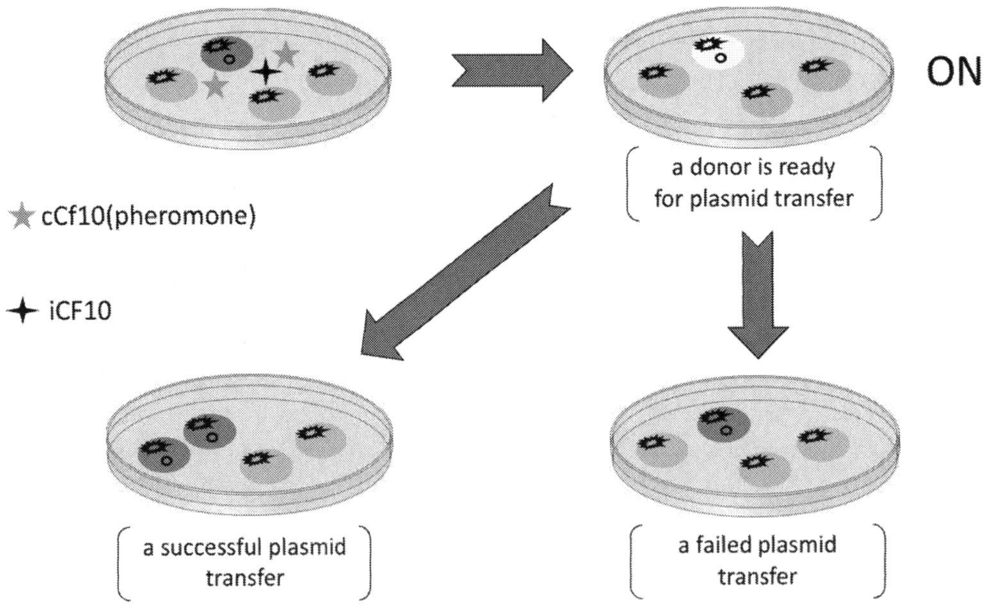

Schematic network

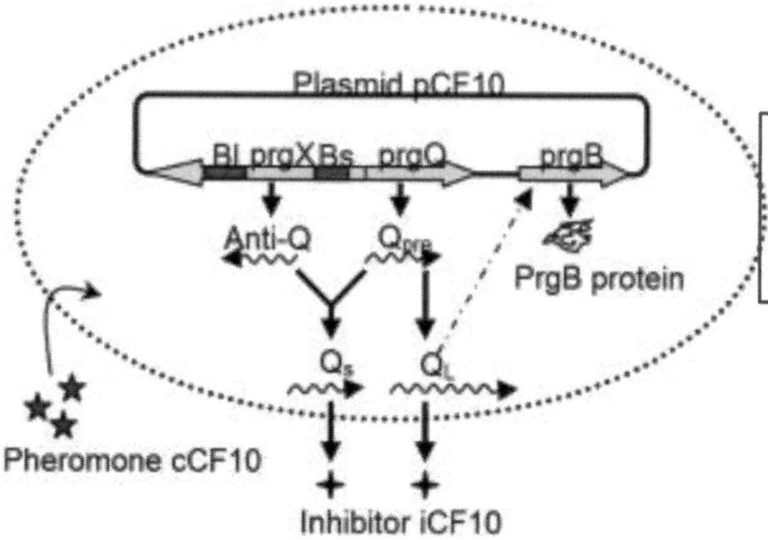

PrgB: protein that indicates conjugation
Ql: level that indicates the level of conjugation

The gene regulatory network of pCF10 conjugative system[1]

1. Shu et al., *Chem Eng Sci*, 2012

7

Descriptions

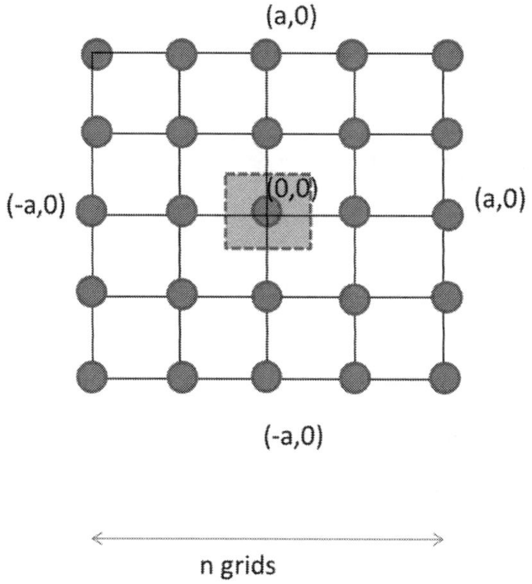

- Initially the donor is located at the center
- Only one cell can be located at each mesh pt
- The concentration within the green box is homogeneous
- Both diffusion and reactions will occur

Stochastic and dynamic models

- Biofilms are highly heterogeneous
- Concentration of particles inside the cell is low
- Dynamic behaviors of cells in biofilm are random
- Deterministic models work fairly well for the planktonic case (well-mixed) but not for biofilms
- **Stochastic modeling is required!**

Solution techniques

SDE Langevin's equation[1]

- An approximated method
- Continuous version
- Explicit version
- Difficult to deal with stiff behavior
- Less accurate

Master equation

- Discrete version
- Difficult to solve analytically
- Exact[2,3] and tau-leap methods (Monte Carlo)
- Obtain distribution of solution from different sample paths

1. Shu et al., *PLoS*, 2011
2. Gillespie, D. *J. Chem. Phys*, 1977
3. Shah et al., *AIChe*, 1977

Our approach

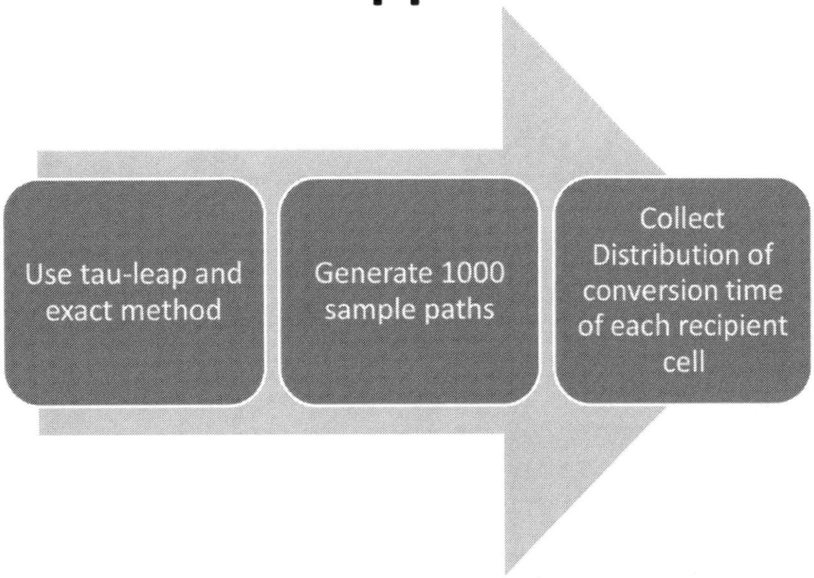

Samples to be discussed

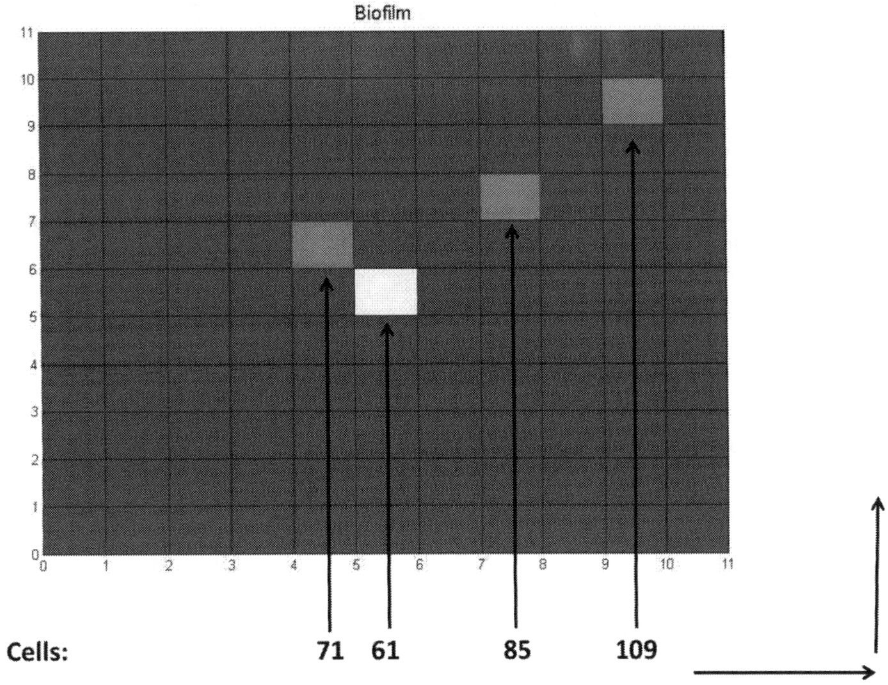

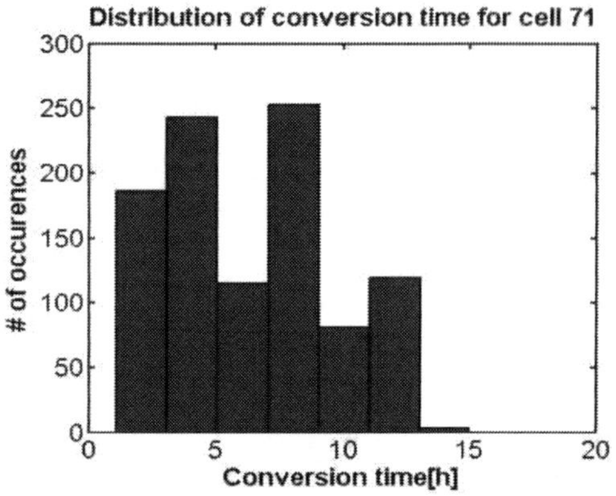

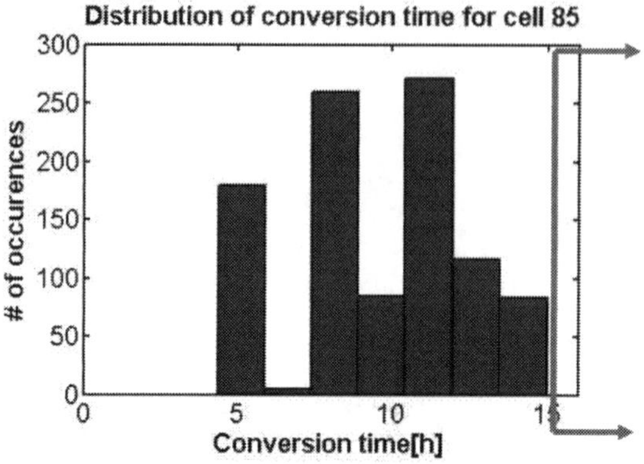

- Distribution of conversion time shifts from left to right
- Represents the process of recipient cells converting into donor cells occurs first at the center and moving outward
- Discontinuity in the histogram plots suggests the need of an increase in the number of sample paths

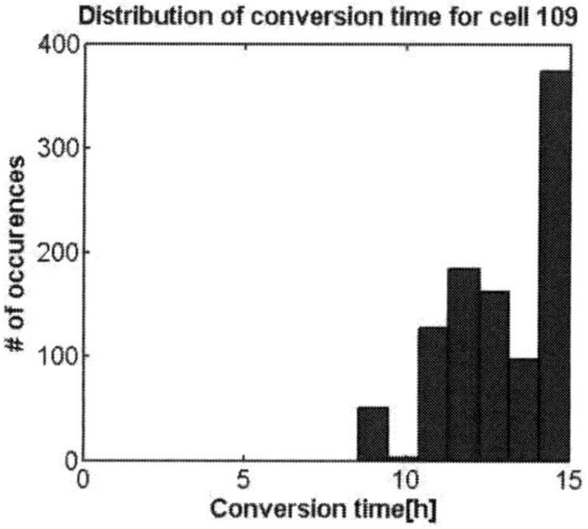

What about the rate of spreading?

- As the process progresses, cCF10 will decrease and iCF10 will increase
- We expect to see :
  1. Reduction of QI level within each donor cell
  2. Reduction of QI's highest value
     - Depletion of cCF10
     - Generation of iCF10
  3. Fluctuation due to transport effect

Averaged QI of two different cells

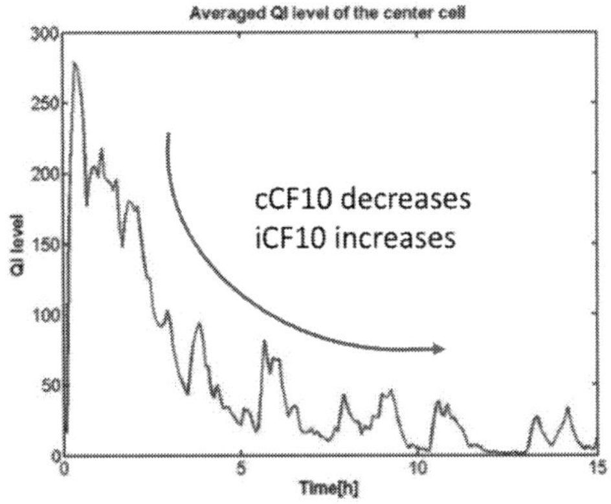

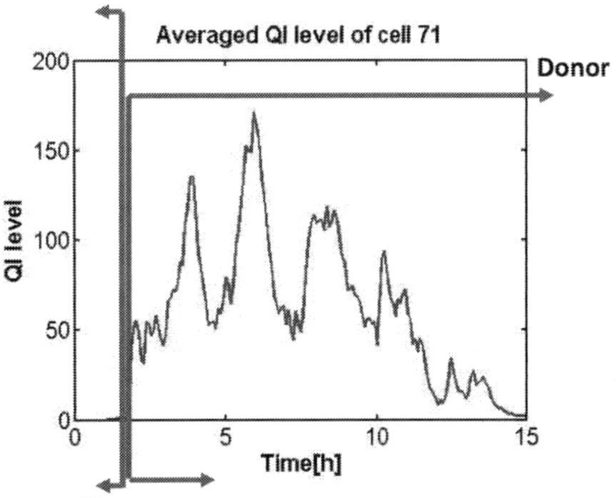

1. Reduction of QI level within each donor cell

Averaged QI of two different cells

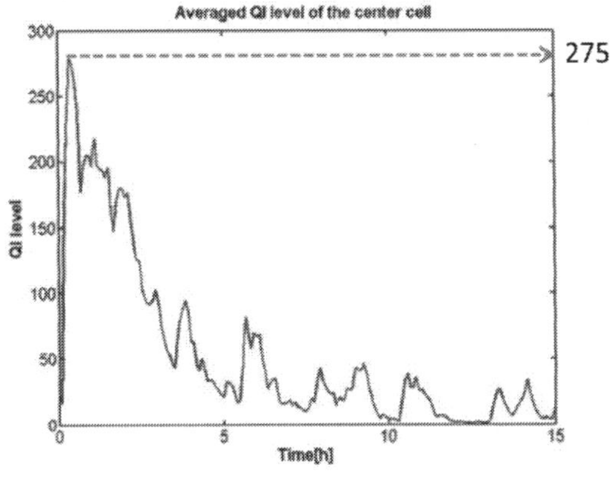

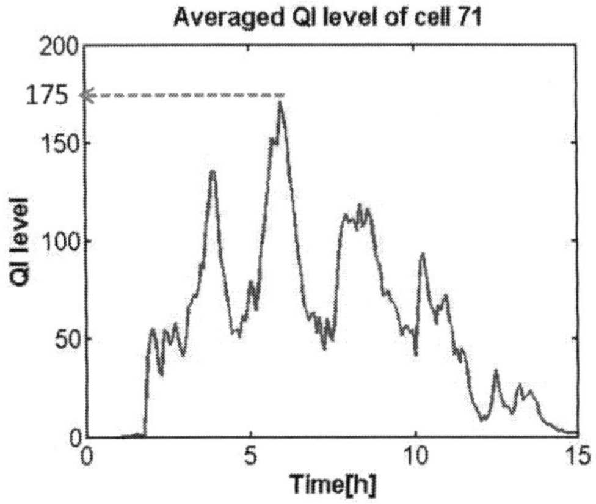

2. QI_{max} (cell 61) > QI_{max} (cell 71)
Depletion of total cCF10
Generation of iCF10

3. Fluctuation is due to the transport of signaling molecules

Ql level at different cells

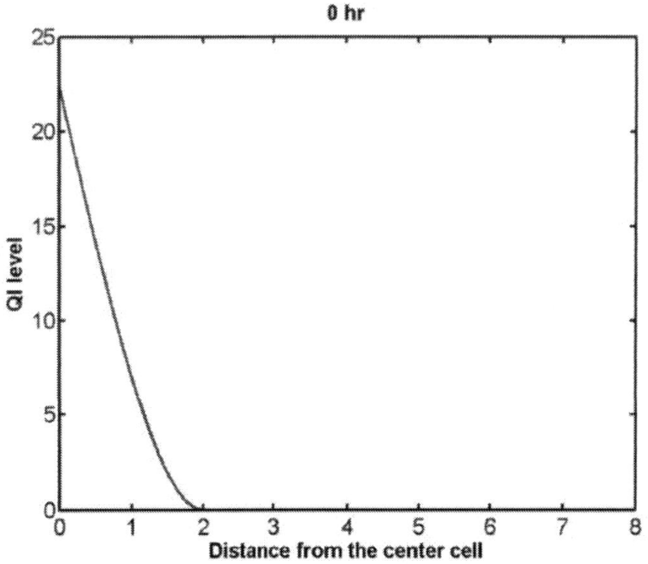

- **At the beginning:**
Ql is higher close to the center

- **As time progresses**:
More outer cells converted into donors
Ql then starts go up at those locations

- **At the end:**
Ql of the center cell reduces
More cells at the boundary are converted

Conclusions

- Capture the dynamics of the biofilm
- Capture qualitatively the spreading process
- Obtain a general picture of conjugation
- Can be used both as a confirmative and a predictive measure for experimental results
- The model is easily updated if needed

Future work

- Place more donor cells in the biofilm
- Produce more sample paths to get more accurate results
- Generalize the code to capture more phenomena such as cell division and movement
- Customize the code to speed up simulations

Acknowledgements

Dr. Doraiswami Ramkrishna
Harry Creighton Peffer
Distinguished Professor

PURDUE
UNIVERSITY

NATIONAL INSTITUTES OF HEALTH

Group Members:
Frank DeVilbiss
Conor Parks
Akancha Pandey
Parul Verma

Former Group Members:
Che-Chi Shu
Meenesh Sing
Jayachandran Devaraj
Aravinda Mandli

University of Minnesota Collaborators:
Arpan Bandyopadhyaya
Yuqing Chen, PhD
Prof. Gary Dunny
Prof. Wei-Shou Hu

428207 A Kinetic Model-Based Approach to Identify Malfunctioned Components in Signal Transduction Pathways from Data

Abstract:

Detection of malfunctioned reactions or molecules from clinical data is essential for disease treatments. Mathematical modeling approaches provide a thorough understanding of the behavior of systems, including biological networks. Therefore, mathematical models can be used to facilitate the inference of malfunctioned reactions that lead to human diseases from clinical data. For example, approaches based on the Boolean model have been utilized to determine the critical components whose dysfunction are most likely responsible for a significant impact on the output of signaling pathways [1-4]. These approaches also have the potential to directly localize faulty genes. In a study of oxidative stress response[5], Boolean modeling simulated gene sequences, and the faulty gene was detected according to the mismatch between the dysfunctional and normal sequences. Apart from model-based approaches, model-free approaches, or data-driven approaches, are commonly used for the detection of pathological states from "-omics" data based on pattern recognition techniques for patients with conditions like a gastrointestinal stromal tumor [6, 7] and various neurological diseases [8, 9]. While these existing approaches yield preliminary avenues for fault detection in signaling pathways involved in human diseases, they either disregard the transient dynamics or have insufficient details of biological systems. In addition, methods based on the Boolean model are incapable of providing accurate and unique solutions for fault diagnosis, because the use of mathematical approximations (either 0 or 1) can cause a coincidence of identical calculation results to occur, although from different combinations of Boolean variables[5].

In contrast, kinetic models, which can determine the time profiles of molecules in a signaling pathway, will provide a better platform for a more accurate and exhaustive detection of malfunctioned reactions. Significantly different from the Boolean model, a pure logical-based discrete model which considers a molecule either active or inactive and is unable to capture important intermediate states, an ordinary differential equation (ODE) model is a kinetic model which can generate continuous and quantitative profiles of molecule concentrations [10]. Kinetic modeling is therefore a more practical method of investigating malfunction pathways in diseases because the severity of diseases varies from patient to patient. In a study about the EGFR-ERK signaling pathway, based on an ODE model, the reaction rate of c-Cbl and EGFR binding was reduced by a factor of 100 or 1000 to mimic the effects of the endocytosis impairment, and active ERK concentration at steady state was measured after treatment with different concentrations of regulator to reflect the functions of selected regulators [11]. Besides, the influence of dysfunction also affects the oscillatory behavior as well [12-14]. It is reported that a biological network can exhibit diverse dynamic behaviors caused by negative feedback loops, such as long-lasting oscillations [15].

This paper presents a kinetic model based approach we developed in order to infer malfunctioned reactions in signaling pathways by quantifying the similarity between the profiles shown in the clinical data and the output profiles predicted from the model in which certain reactions/molecules are malfunctioned. The developed approach was tested in two well-studied signaling pathways, i.e., Interleukin 6 (IL-6) signal transduction[16, 17] and TNF-α/NFκB signaling pathway[18], for four abnormal clinical conditions that include the up/down regulation of single reaction rate constants and up/down regulation of single molecules. Since limited quantitative clinical data were available for these two signaling pathways, the IL-6 model was used to generate artificial clinical data for the abnormal steady-state value shown in the two key molecules (i.e., nuclear STAT3 and SOCS3). Similarly, the TNF-α/NF-κB model was used to obtain the data in which abnormal oscillation dynamic was shown in the profile of NF-κB. Different levels of random noises were added into the simulation results to represent the noisy properties shown in real clinical data. The developed approach was used to identify the malfunctioned reactions/molecules from data. The results show that the developed approach was able to successfully identify the malfunctioned reactions/preexisting molecules from the clinical data. It was found that the developed approach was noise-robust and that it was able to reveal unique solution for the faulty components in a network.

418174 Stochastic Optimal Control for Prediction of Robust Drug Dosing Policies in Superovulation Stage of in-Vitro Fertilization

Abstract:

Stochastic optimal control for prediction of robust drug dosing policies in superovulation stage of in-vitro fertilization

Kirti M. Yenkie [b] and Urmila Diwekar[a,b*]

[a]Department of Bioengineering, University of Illinois, Chicago, IL

[b]Center for Uncertain Systems: Tools for Optimization & Management (CUSTOM), Vishwamitra Research Institute, Crystal Lake, IL

***urmila@vri-custom.org**

Keywords: infertility, multiple ovulation, uncertainty, optimal control

1. Introduction and Motivation

The WHO has estimated that about 8-10% couples experience some kind of infertility problems. Medical science has come up with Assisted Reproductive Technologies (ARTs) like IVF for treatment. IVF is a four stage medical procedure and the success of superovulation, its first stage, is crucial to proceed with the next stages. Also, superovulation requires maximum medical attention, investment of time and money. The major hormone responsible for follicle growth is follicle stimulating hormone (FSH). In our previous work [1], a deterministic model for prediction of multiple follicle growth was built by drawing similarities from batch crystallization [2]. This model was validated with clinical data from 50 IVF cycles. The model fitted very well and almost 60% predictions had very low error [3]. However, some results had deviations from the expected behavior. Thus, the objective of this work is to develop a robust model which can provide a better fit for the patients with deviations in the deterministic model and then come up with a method for predicting robust hormone dosing policy.

2. Methodology 2.1. Determinstic Model

The concept of moment model in batch crystallization was used for modeling superovulation in IVF [1, 3]. The growth term in batch crystallization is temperature dependent and hence temperature is controlling variable. Similarly in IVF, follicle growth is dependent upon hormonal doses. The follicle size is converted into mathematical moments [2] by assuming them to be spherical in shape. The eq. (1) is used for converting follicle size to moments.

$$\mu_i = \sum n_i(r,t) r_i^i \Delta r_i \quad (1)$$

Here, μ_i - i[th] moment, $n_j(r, t)$ - number of follicles in j^{th} bin with mean radius as r at time t, r_j - mean radius of jth bin and Δr - range of radii variation in bins. Each moment corresponds to a feature of the follicles, like the zeroth moment represents the number, first the size, etc. The follicle growth term (G) is dependent on the amount of FSH injected (ΔC_{fsh}) to the patient at time (t) and is represented in eq. (2). Here, k and α are kinetic constants.

$$G(t) = k\Delta C_{fsh}(t)^{\alpha} \quad (2)$$

It was suggested that number of follicles activated for growth during a superovulation cycle are always constant for a patient [4], hence zeroth moment is a constant. The moment equations for the follicle dynamics can be written as in eqs. (3) and (4).

$$\mu_0 = constant \quad (3)$$

$$\frac{d\mu_i}{dt} = iG(t)\mu_{i-1}(t);\ (i = 1,2,\dots 6) \quad (4)$$

2.2. The Stochastic Model

The stochastic model for superovulation is developed in terms of Ito processes. Depending upon the behavior of the system it can modeled as one of the suitable Ito processes [5]. It was found that the uncertainties could be represented using the simple brownian motion type of Ito processes, resulting in stochastic differential equations (SDEs), i varies from 1 to 6 (eq. 5).

$$d\mu_i = iG(t)\mu_{(i-1)}dt + \sigma_i \varepsilon_t \sqrt{dt} \quad (5)$$

Here, μ_i - ith moment, σ_i - standard deviation terms for each moment equation corresponding to noise in the clinical data, ϵ_t - random numbers from unit normal distribution, dt - time interval.

2.3. Stochastic Optimal Control

Superovulation objective is to obtain eggs or follicles in high number within a specific size range (18-22mm diameter) with the aid of external hormones. Mathematically, it can be interpreted as 'minimization of the variation'. The follicle size data follows normal distribution hence, the expression for coefficient of variation (CV) is (eq. 6),

$$CV = \sqrt{\frac{\mu_0 \mu_2}{\mu_1^2} - 1} \qquad (6)$$

Thus, objective of superovulation can is to minimize the coefficient of variation at the final time ($CV(t_f)$), the control variable is FSH dose with time (C_{fsh} (t)), subjected to stochastic superovulation model. This is solved using stochastic maximum principle.

3. Results

Table 1. The results for Patient-A from stochastic optimal control

Patient A	FSH Used	FSH Optimal	No. of Follicles (9 ≤ Mean size ≥ 12)		% reduction in FSH
Follicle count - 18	1050	885	9 (used)	11 (opt)	15.7

4. Conclusions

Stochastic model for superovulation stage is used for predicting robust hormone dosing policies using stochastic control methods. Stochastic maximum principle strategy provides dosing policies which can result in better follicle size distribution on the final FSH dosing day.

5. References

[1] Yenkie, K. M., Diwekar, U., Bhalerao, V., 2013, "Modelling the superovulation stage in in-vitro fertilization", IEEE Trans. Biomed. Engg., 60(11): 3003-3008.

[2] Randolph, A. D., and M. A. Larson., 1988, Theory of Particulate Processes: Analysis and Techniques of Continuous Crystallization. San Diego, CA, Academic Press.

[3] Yenkie K.M., Diwekar U., Bhalerao V., 2014, "Modeling and prediction of outcome for the superovulation stage in In-Vitro Fertilization (IVF)", JFIV Reprod Med Genet 2: 122.

[4] Baird, D. T., 1987, "A model for follicular selection and ovulation: Lessons from Superovulation", J. steroid Biochem., 27, 15-23.

[5] Diwekar, U. 2008. Introduction to Applied Optimization, 2nd ed.; Springer NY.

429758 Modeling Signaling Interactions Controlling Heterogeneity and Fate Choice of Human Embryonic Stem Cells

Abstract:

Motivation: Human embryonic stem cells (hESCs) are an attractive raw material for regenerative medicine due to their potential to deliver a variety of clinically important mature lineages. Methods to induce and monitor self-renewal and differentiation of hESCs are now well established. However generation of pure and well-defined lineages is still a major challenge due to the heterogeneity of stem cell cultures. Recent research from diverse stem cell systems has shown that heterogeneity results from a complex interaction between variations in protein expression levels, signaling reaction rate/mechanism and the cell cycle stage [1,2]. To control heterogeneity, it is necessary to understand how above features influence network level signal propagation. **In this project, we employed a systems level approach to evaluate network dynamics and variability in differentiating hESCs by integrating computational models with experimental techniques (measuring single cell and population dynamics). The system of interest is the TGF-β/SMAD2,3 pathway and its crosstalk with PI3K/AKT pathway existing during regular endoderm differentiation of hESCs [3].**

Methods:

Experimental: For self-renewal, H1 hESCs were maintained on matrigel-coated plates in mTeSR1. Endoderm differentiation was then performed using 100 ng/ml Activin A (to activate TGF-β/SMAD2,3) with or without modulation of PI3K/AKT pathway using PI3K inhibitor, Wortmannin. Phosphorylation dynamics of participating signaling molecules at the cell population level were measured using MagPix® xMAP technology. The initial selection of key molecules was based on the study by Singh *et al.* [4]. These mainly included SMAD2,3,4,7 molecules, TGF-β

Receptors, AKT and ERK. In order to facilitate measurement of protein dynamics at the single cell level, we generated hESC cell lines expressing GFP-fused SMAD proteins. These transformed cells facilitated the measurement of nucleo-cytoplasmic shuttling kinetics of molecules from single cells using Fluorescence Recovery after Photo-bleaching (FRAP) analysis. Mathematical and computational tools described in the next section were used to guide generation of data by in-house experiments.

Mathematical Analysis: Dynamic Bayesian Network Analysis (DBN) was used to identify key sensitive molecules, hypothesize interactions and identify most informative time points. Detailed mechanistic Ordinary Differential Equation (ODE) model for the TGF-β/SMAD2,3 pathway with crosstalk interactions with PI3K/AKT was developed for a systems level analysis. The model was calibrated using Replica Exchange Ensemble Modeling (EM) and sensitive reactions were identified using computationally efficient Global Sensitivity Approach (GSA) called Random Sampling High Dimensional Model Representation. Compartmental ODE models in combination with quantitative image analysis were used to simulate photo-bleaching experiments and estimate the nucleo-cytoplasmic shuttling rates of SMAD molecules.

Results and Discussion:

We have developed a detailed ODE model representing signaling events in the PI3K/AKT pathway under self-renewal status of hESCs. Here we applied GSA to identify sensitive nodes from the pathway that modulated a critical self-renewal molecule, p-AKT. The computational cost of GSA was significantly reduced by application of a meta-model framework to capture behavior of the ODE model. Using this approach, we recently showed that a negative feedback loop in the pathway controlled the levels of p-AKT [2]. This feedback loop constrained the cell-to-cell variability of p-AKT within the population as well as between consecutive *in vitro* passages.

In the current project, we are expanding this approach to model the dominant signaling events regulating initial endoderm differentiation of hESCs. First, to identify possible crosstalk interactions between the signaling molecules, DBN analysis was performed on a detailed time series from the average population data. DBN generated for high and low PI3K conditions showed that multiple crosstalk points exist during endoderm differentiation and temporal inhibition of PI3K at the early phase of signaling was sufficient to remove crosstalk. Further, the receptor RII levels influenced the downstream molecules during the entire phase of the signaling dynamics. HESCs further showed divergence in the dynamics of regulatory SMADs, a characteristic not observed in other reported cell systems. The reason for this divergence was evaluated by analyzing parametric ensembles in the high dimensional parameter space. The nucleo-cytoplasmic shuttling kinetics of SMAD molecules with and without crosstalk with AKT at the single cell level was used to inform the ODE model. The resulting mixture of single cell ODE models was integrated and compared to the mean population dynamics for validation as well as constraining free parameters. This integration of single cell and population model enabled us to evaluate the specific reactions and network motifs that contributed most to the variation in nuclear SMAD levels, which are the determinants of endoderm differentiation. We are in the process of relating the variations in nuclear SMADs to downstream transcriptional heterogeneity.

Conclusions:

Our systems analysis results show that during signal transduction in the self-renewal state and during differentiation, network motifs have an important role in controlling the levels and heterogeneity of stem cells. Through this work, we have demonstrated that application of efficient computational tools and workflows that explore single cell and population dynamics will enable extraction of mechanistic information for systems with high heterogeneity. This will offer new avenues to control heterogeneity and improve signal propagation by targeted modulation of current culture conditions.

References:

[1] Singh, AM (2015) *Stem Cells International*, Article ID 219514. doi:10.1155/2015/219514

[2] Mathew *et al.* (2014) *Bioinformatics*, 30(16): 2334-2342

[3] Mathew *et al.* (2015) *MDPI Processes*, 3(2): 286-308

[4] Singh *et al.* (2012) *Cell Stem Cell*, 10(3), 312-326

436195 A Study on the Effects of Mechanoelectric Feedback and Mechanical Perturbation Approach on the Electrical Properties Using a Bidomain Model of Cardiac Tissue

Abstract:

A study on the effects of mechanoelectric feedback and mechanical perturbation approach on the electrical properties *using a bidomain model* of cardiac tissue

Electrical alternations in cardiac action potential duration have been shown to be a precursor to arrhythmias and sudden cardiac death (Makarov L *et al.,* 2010). Through the mechanism of excitation-contraction coupling, the presence of electrical alternans induces alternations in the heart muscle contractile activity. Also, contraction of cardiac tissue affects the process of cardiac electric wave propagation through the mechanism of so-called mechanoelectrical feedback (MEF). Electrical excitation and contraction of cardiac tissue can be linked by an electromechanical model such as the Nash-Panfilov (NP) model (Nash M P *et al.,*2004; Panfilov A V *et al.,*2005).

The electrical activity of the heart is usually modelled using the bidomain or monodomain equations. The bidomain model (Keener J *et al.,*2008) consists of a system of coupled parabolic and elliptic PDEs for two potentials in the cardiac muscle, coupled with a nonlinear system of ODEs describing the ionic currents flowing across the cardiac membrane. This model, which is computationally very expensive, is more accurate description of macroscopic electrical activity of the heart than the monodomain model and it is also the preferred model for simulating cardiac excitation, such as during cardiac pacing, when the main emphasis is how a voltage develops across the membrane. The hyperelastic Mooney-Rivlin (MR) material model has been used to model the nonlinear mechanical behavior of the myocardium (Nash M P *et al.,*2004; Panfilov A V *et al.,*2005).

In this work a strongly coupled electromechanical model, that account for the effects of electrical activity on the cardiac mechanics and the effect of mechanical deformation on electrical properties in a bidomain setting, will be utilized to study the basic electrical and mechanical properties of the heart, including action potential propagation, defibrillation, electrocardiographic (ECG) measurement, and mechanical stretch and shortening. In addition the effects of mechanical deformation via MEF and mechanical perturbation approach (Yapari *et al.,* 2014) on wave propagation will be illustrated. This will help to explore the possibility of suppressing cardiac alternans by applied mechanical perturbation approach.

References:

Makarov L, KomoliatovaV. Microvolt T-Wave Alternans during Holter Monitoring in Children and Adolescents. Annals of Noninvasive Electrocardiology 15, 138–144, 2010

Nash M P, Panfilov A V. Electromechanical model of excitable tissue to study re-entrant cardiac arrhythmias. Prog. Biophys. Mol. Biol. 85, 501–522, 2004.

Panfilov A V, Keldermann R H, Nash M. P. Self-organized pacemakers in a coupled reaction-diffusion-mechanics system, Phys. Rev. Lett. 95, 528104, 2005.

Keener J, Sneyd J. Mathematical Physiology II: Systems Physiology, Springer Science+Business Media, New York, 2008

Yapari F, Deshpande D, Belhamadia Y, Dubljevic S, Control of cardiac alternans by mechanical and electrical feedback, Phys. Rev. E 90, 012706, 2014.

416131 A Probabilistic Approach to Robust Experiment Design for Simultaneous Model Discrimination and Parameter Estimation: A Wnt Signaling Case Study

Abstract:

Dynamical models play a crucial role in the systematic analysis of complex biological systems, which are characterized by nonlinearities and uncertainties. The usual existence of several hypotheses to describe the dynamics of biological systems, along with the consideration that models of such systems typically have several unknown parameters, have motivated the use of model-based optimal experiment design (OED) approaches for model discrimination (Fedorov, 1972) and parameter estimation (Pronzato, 2008). The primary notion of OED for model discrimination and parameter estimation is to systematically design input signals aiming at, respectively, separating predictions of the rival models and maximizing the information content of experiments for accurate parameter estimation. However, a major limitation of standard OED approaches is their reliance on point estimates of the system parameters, which can be largely different from their true values, leading to (possibly) ineffective input designs. In addition, an optimal input for model discrimination is unlikely to be adequate for parameter estimation and vice versa.

This work addresses the problem of OED for simultaneous model discrimination and parameter estimation in polynomial nonlinear systems with probabilistic time-invariant uncertainties. The proposed OED approach obviates the conservatism of the deterministic robust OED formulations by explicitly accounting for the probabilistic information of system uncertainties. Chance constraints are included to tolerate predetermined levels of risk in the designed experiments in terms of state constraint violations (Mesbah and Streif, 2015). This study uses the generalized polynomial chaos framework (Xiu and Karnadiakis, 2002) in conjunction with Galerkin projection (Ghanem et al. 1991, Streif et al., 2014) for efficient propagation of the probabilistic uncertainties. The similarity between probability distributions of states/outputs is quantified in terms of the *Wasserstein distance* (Dobrushin, 1970, Vallender, 1972).

The *D-optimality* criterion is adopted as a quantitative measure for information content of the experiments. The to-be-minimized objective function used in the optimization is a geometric mean of the model discrimination and parameter estimation criteria.

The proposed optimal experiment design approach is demonstrated for the ubiquitous canonical Wnt signaling pathway, which regulates cell growth and fate determination at the molecular level (Li et al., 2010). Activation of the pathway by the wild type ligand Wnt results in accumulation of the transcriptional coactivator β-catenin in the cytoplasm. The proposed OED approach is used to discriminate between two alternative β-catenin phosphorylation mechanisms performed by two components of the pathway, i.e. CK1α and GSK3 (Li et al., 2010, Hernandez et al., 2012). In the so-called *processive* mechanism, β-catenin binds to the destruction complex and does not detach until it is fully phosphorylated and flagged for degradation. Conversely, in the *distributive* mechanism β-catenin binds and unbinds before and after each phosphorylation event, and thus not fully assembled destruction complexes (only with GSK3 or with CK1α) could also participate in the phosphorylation of β-catenin (Hernandez et al. 2012). Two dynamical models are developed to appropriately describe the distinctive events of each mechanism (Jensen et al., 2010, Hernandez et al., 2012). The input of the system is chosen to be the rate of degradation of Axin once it is sequestered by LRP on the cell membrane, the magnitude of which is related to the concentration of pathway activator Wnt (Jensen et al. 2010). Concentration of free β-catenin is assumed to be the noise-corrupted output of the system.

To evaluate the performance of the robust OED approach for simultaneous model discrimination and parameter estimation, Monte Carlo runs are performed using the optimized input profile and a nominal input profile. The simulation results reveal that the OED approach is able to effectively separate the probability distributions of model outputs over critical transient behavior of the biological system in the presence of uncertainties. In addition, the designed input leads to smaller average estimation errors for the kinetic parameters of the biological system. A key feature of the robust OED approach is the ability to ensure constraint satisfaction with a prespecified probability level in a stochastic setting.

References:

Dobrushin, R.L. (1970). Prescribing a System of Random Variables by Conditional Distributions. *SIAM Journal of Scientific Computation,* 3, 458-486.

Fedorov, V. (1972). Theory of Optimal Experiments. *Academic Press,* New York.

Ghanem R.G. and Spanos P.D. (1991). Stochastic Finite Elements: A Spectral Approach. *Springer-Verlag,* New York.

Hernandez, A., Klein, A., Kirschner, M. (2012). Kinetic Responses of β-catenin specify the sites of Wnt Control, *Science,* 338, 1337-1340.

Jensen, P.B., Pedersen, L., Krishna S., Jensen M. (2010). Wnt signaling in oncogenesis and embryogenesis: a look outside the nucleus. *Biophysical Journal,* 98, 943-950.

Li, V., Ng, S., Boersema, P., Low, T., Karthaus, W., Gerlach, J., Mohammed, S., Heck, A., Maurice, M. M., Mahmoudi, T., Clevers, S. (2010). Wnt Signaling through Inhibition of β-catenin Degradation in an Intact Axin1 Complex. *Cell,* 149, 1245-1256.

Mesbah A., Streif, S (2015). A Probabilistic Approach for Robust Optimal Experiment Designs with Chance Constraints. *ADCHEM,* In Press.

Pronzato, L. (2008). Optimal Experimental Design and Some Related Control Problems. *Automatica,* 44, 303-325.

Streif, S., Petzke, F., Mesbah, A., Findeisen, R., Braatz, R.D. (2014). Optimal Experiment Design for Probabilistic Model Discrimination using Polynomial Chaos. In *Proceedings of the IFAC World Congress,* pages 4103-4109. Cape Town, South Africa.

Vallender, S.S. (1972). Calculation of the Wasserstein Distance between Probability Distributions on the Line. *Theory of Probability and its Applications*, 18(4), 784–786.

Xiu, D., Karniadakis, G. (2002). The Wiener-Askey Polynomial Chaos for Stochastic Differential Equations. *SIAM Journal of Scientific Computation,* 24, 619-644.

434602 Redesigning the Response of T Cell Signaling Networks Using in silico Evolution

Abstract:

Cells have evolved an array of signal transduction pathways that allow them to detect and respond to their environments. T lymphocytes orchestrate adaptive immunity and act as cellular detectors of infection, using the T cell receptor (TCR) to identify molecular signatures of pathogens. The TCR signaling pathway exhibits a sharp activation threshold as a function of the binding strength of the TCR with peptide-MHC (pMHC) ligands presented on other cells. We are interested in the following question: Given a fixed network topology such as the TCR signaling network, can one find sets of parameters (kinetic rates) that give desired network responses to input? Answering such questions can help to shed insight into T cell signaling and may provide a means for designing artificial networks with desired signaling properties. Here we consider a deterministic, well-mixed model of the TCR signaling network and utilize an *in silico* evolutionary algorithm to search for sets of kinetic parameters that give altered output responses as a function of the off-rate between pMHC and TCR. We begin by shifting the activation threshold by orders of magnitude and find that the algorithm converges to solutions by altering kinetic rates in the network. Interestingly, we find that the TCR signaling network topology can exhibit even more dramatic output profiles, such as an inversion of the activation pattern in which weak TCR-pMHC binding induces activation and strong binding does not. By running multiple independent instances of the evolutionary algorithm for each desired output, we find many different sets of parameters consistent with the desired output, with distinct patterns of solutions become apparent when analyzing the resulting data. Overall, through moderate adjustments of kinetic parameters, the TCR signal transduction network has the potential to produce a wide array of input-output relations.

431047 Integrated Scheduling and Dynamic Optimization of a Cryogenic Air Separation Unit Subject to Time-Varying Electricity Prices

Abstract:

Cryogenic air separation units (ASUs) are an essential utility supplier to a number of industries, including chemicals, metal processing, electronics and water treatment. Electricity to operate compressors is the main utility required for the process and accounts for a large portion of the operating costs [1]. Optimal production scheduling is crucial for minimizing operating costs in an environment where the process is subject to time-varying electricity prices. Generally, production levels are increased at times when prices are low; excess production of nitrogen and oxygen is liquefied and stored so that the products can be used to satisfy demand when grid demand and electricity prices are at a peak [2]. This operating paradigm effectively amounts to storing energy in the form of the latent heat removed from the cryogenic products [3]. ASUs also typically supply their gas products to customers through a pipeline system, and it is important to ensure that the product quality satisfies the customer requirements during production rate transitions. Equivalently, an optimal transition pathway between two operating points, that does not violate any constraints in product quality (e.g., the impurity level) or operation (e.g., flooding in the distillation column), must be established [4, 5].

Motivated by the above, we introduce a novel integrated scheduling and dynamic optimization framework for air separation processes. Our approach consists of including a (low order) dynamic model of the ASU in the scheduling calculation as a means to avoid quality and process constraint violations. In particular, we proceed by (i) identifying a low-order model from historical operating data, that captures the dynamics of variables relevant to the scheduling calculation and (ii) developing of an integrated scheduling and control formulation for the ASU process using this model. This is a significant departure from the traditional scheduling calculation which contains limited (and typically static) information about the dynamic performance of the process.

We illustrate our framework using a detailed model of an ASU for the production of nitrogen. We discuss several continuous- and discrete-time data-driven modeling options and their implementation in the integrated scheduling and optimization framework, as well as sequential and simultaneous strategies for solving the resulting mixed-integer dynamic optimization problem. Our results show that accounting for process dynamics at the scheduling level allows for a better exploitation of real-time energy prices, and leads to improvements in ASU operating profit.

[1] R. C. Pattison and M. Baldea, Optimal Design of Air Separation Plants with Variable Electricity Pricing. In Proceedings of the 8th International Conference on Foundations of Computer-Aided Process Design, Computer Aided Chemical Engineering, 34:393-398, 2014.

[2] M. G. Ierapetritou, D. Wu, J. Vin, P. Sweeney, and M. Chigirinskiy, Cost Minimization in an Energy- Intensive Plant Using Mathematical Programming Epproaches. Industrial & Engineering Chemistry Research, 41(21):5262-5277, 2002.

[3] Q. Zhang, I.E. Grossmann, C.F. Heuberger, A. Sundaramoorthy, J.M. Pinto, Air Separation with Cryogenic Energy Storage: Optimal Scheduling Considering Electric Energy and Reserve Markets, AIChE Journal, 61(5):1547-1558, 2015.

[4] Y. Cao, Design for Dynamic Performance: Application to an Air Separation Unit, MSc Thesis, McMaster University, 2011.

[5] Y. Cao, C. Swartz, M. Baldea, Design for Dynamic Performance: Application to an Air Separation Unit, In Proceedings of the American Control Conference, 2683-2688, 2011.

429630 Optimal Campaign Continuous Manufacturing

Abstract:

Campaign continuous manufacturing (CM), characterized by relatively short operational windows such as a few weeks [1], is being explored as an alternative to traditional batch-wise manufacturing in the pharmaceutical industry. However, optimal operation in campaign CM can be challenging due to the significance of startup and shutdown phases, which can negatively affect on-specification production and plant economy. In this work, the effectiveness and computational tractability of several known optimization approaches when applied to campaign CM are investigated. Inspired by the turnpike property [2,3] in optimal control, a new approach is then proposed, which aims to maximize on-specification production directly rather than minimizing the startup/shutdown times as commonly adopted in high-volume industries. A main contribution in the new approach is that the resulting optimization formulation is guaranteed to be differentiable, despite the underlying hybrid dynamic system [4]. Thus, it can be solved reliably using gradient-based algorithms. Case studies are presented to demonstrate the effectiveness of the proposed approach.

[1] Ali M. Sahlodin, Paul Barton. Optimal Campaign Continuous Manufacturing, submitted.

[2] McKenzie, L. (1976). Turnpike theory. *Econometrica*, *44*, 841–865.

[3] Zaslavski, A. J. (2006). Turnpike Properties in the Calculus of Variations and Optimal Control. volume 80 of Nonconvex Optimization and Its Applications. Springer.

[4] Galán, S., Feehery, W. F., & Barton, P. I. (1999). Parametric sensitivity functions for hybrid discrete/continuous systems. Applied Numerical Mathematics, 31, 17 – 47.

434615 Flare Minimization for Cracking Gas Replacing Process of Ethylene Plant Start-up Via Dynamic Simulation

Abstract:

The start-up operations of ethylene plants, which are usually conducted every 3 to 6 years, generate large quantities of off-spec products that have to be flared for safety. It has been estimated that an ethylene plant with a capacity of 1.1 billion pounds of ethylene production per year will easily flare about 5.0 million pounds of ethylene during one single start-up. Flare minimization has become one of the focuses of many enterprises and researchers.

Currently, one of the flare minimization strategies is to start compressors and precool the refrigeration system before feeding cracking furnaces with start-up working mediums (SWM) such as mixed gas. The start-up operation with mixed gas is called Mixed Gas Run (MGR). Different from traditional start-up operation with cracking gas as the SWM, it is quite difficult for even experienced engineers to make a detailed start-up plan for MGR with the objective of minimum emission and energy based on their experiences. Unsuitable schemes and operations will cause considerable emission and energy loss or even severe accidents. In order to avoid above situations, a methodology of flare minimization by dynamic simulation was proposed by Xu et al to decide and validate start-up plans. However, the coupling relationships among the compression system, the refrigeration system, and the separation system have not been considered comprehensively in previous studies.

Generally MGR includes three steps: nitrogen starting-up and preliminary precooling process, mixed gas replacing and further precooling process, and cracking gas replacing and starting separation process (CGRP). In this paper, the cracking gas replacing process was quantitatively calculated and analyzed via dynamic simulation. Compression system models with performance curves and refrigeration system models with heat capacity data were developed for observing dynamic changes during the CGRP. Effects of energy saving and emission reduction of total reflux method for different distillation columns were compared and the start-up time of each key column was given. The operation sequences of key devices, the settings and interactions of critical operating parameters during the CGRP were discussed. This paper, taking a real ethylene plant as an example, demonstrates the method for determining each device operation scheme, the gantt chart of start-up operating procedures, and the quantified replacing time and emission reduction effect of the whole process.

Reference:

1. Xu, Q.; Yang, X.; Liu, C.; Li, K.; Lou, H. H.; Gossage, J. L., Chemical plant flare minimization via plantwide dynamic simulation. *Industrial & Engineering Chemistry Research* **2009**,48, (7), 3505-3512.

2. Song, G.; Qiu, T.; Zhao, J., Flare Minimization Model for Ethylene Splitter System's Shutdown. *Industrial & Engineering Chemistry Research* **2013**,52, (26), 9180-9188.

3. Yang, X.; Xu, Q.; Li, K.; Sagar, C. D., Dynamic Simulation and Optimization for the Start-up Operation of An Ethylene Oxide Plant. *Industrial & Engineering Chemistry Research* **2010**,49, (9), 4360-4371.

4. Zhao, Y.; Zhang, J.; Qiu, T.; Zhao, J.; Xu, Q., Flare Minimization during Start-Ups of an Integrated Cryogenic Separation System via Dynamic Simulation. *Industrial & Engineering Chemistry Research* **2014**, 53, (4), 1553-1562.

411963 The SEM Study and Molecular Dynamic Simulations of Nanosilica Grafted By Copolymer

Abstract:

The SEM Study and Molecular Dynamic Simulations of Nanosilica grafted by Copolymer

Shanshan Dai[a,b,*], Jinghan Zhang[a], Xiaowen Zhao[c]

a. College of Chemistry and Chemical Engineering, Southwest Petroleum University, Chengdu 610500, Sichuan, China; b. Oil & gas Field Applied Chemistry Key Laboratory of Sichuan Province, Chengdu 610500, Sichuan, China; c. State Key Laboratory of Polymer Materials Engineering, Polymer Research Institute of Sichuan University, 610065 Chengdu, China.

Acryl amide (AM) acrylic acid (AA) copolymer grafting nanosilica were prepared by inverse microemulsion polymerization to make functional materials where nano-SiO_2 provides the rigidity and strength, and the copolymer the elasticity and flexibility. The resulting microspheres are able to enter, expand, shut off and remove in the deep-layers of oilfield. SEM photographs (Fig.1) proved the structure of core-shell, but lots of quantitative relationships are still missing. To solve this problem, molecular dynamic (MD) simulation is applied in this study and the modification of inorganic particle nanosilica becomes the first key point. There are 4 kinds of coupling agents. H1, H2, H3, and H4 represent the system with methyl-, octyl-, vinyl- and ¦Ã-methacryloxy propyl-triethoxysilane respectively. As shown in Fig. 2-4, H3 is the system with the highest binding energy between nanosilica and copolymer; Radius distribution function (RDF) supplies a measure of the probability that gives the presence of an atom at the origin of an arbitrary reference frame, and the higher value illustrates the bigger grafting density; Small slope value of the best-fit line of mean square displacement (MSD) curve indicates the bad mobility of copolymer chain, namely, the copolymer chains are fixed by modified SiO_2 surface.

*Correspondence to: Shanshan DAI (E-mail: ashanscu@163.com).

Grant Sponsor: National Natural Science Foundation of China (Grant Number: 51304163); China Postdoctoral Science Foundation (Grant Number: 2014M562507XB).

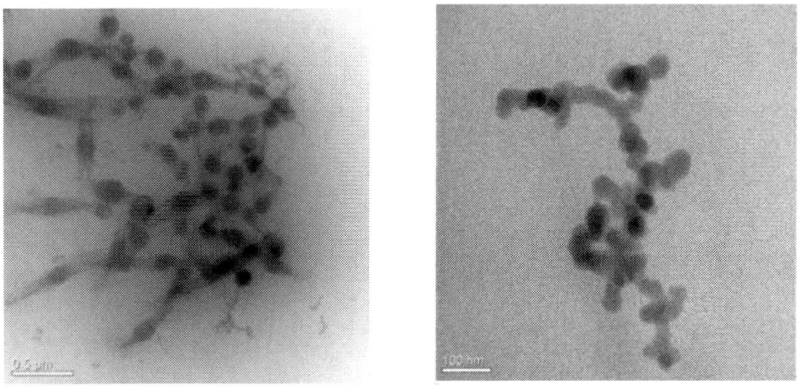

Fig. 1 The SEM of the core-shell microspheres

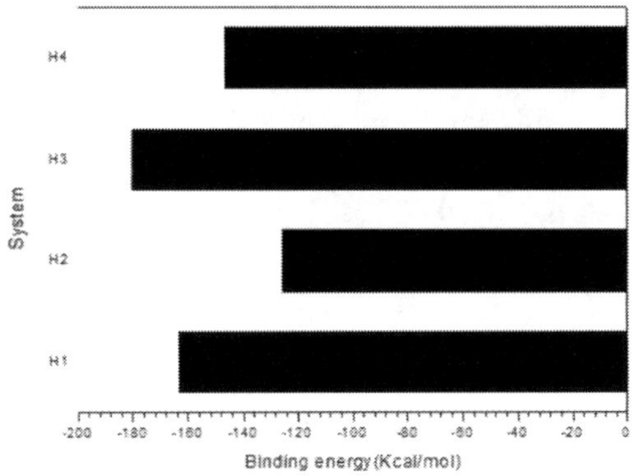

Fig. 2 The binding energy of systems with different coupling agent

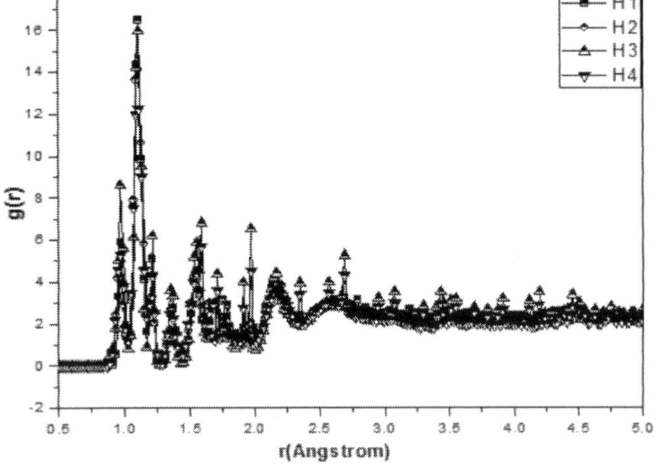

Fig. 3 The RDF of systems with different coupling agent.

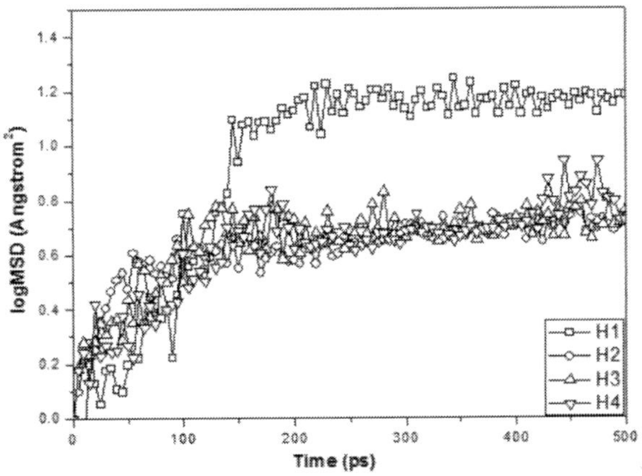

Fig. 4 The logMSDs of copolymer versus time in systems with different coupling agent

419328 Plantwide Model-Based Optimization of a Large Scale Second Generation Biorefinery

Abstract:

Second generation biorefineries reached commercial reality in 2012 [1] leading to the first commission of a large scale plant in October 2013 by Beta Renewables in Crescentino, Italy [2]. Other commercial size second generation bioethanol plants followed and are already operating or under construction: project Liberty by POET-DSM (USA), Abengoa Bioenergy (USA), DuPont (USA), Måbjerg Energy Concept (Denmark) etc. [3].

Second generation biorefineries transform lignocellulosic agricultural wastes into products with higher added value. In biorefineries employing biochemical routes, this is achieved by the following four major conversion and separation technologies: biomass pretreatment, enzymatic hydrolysis or liquefaction, fermentation and purification [4]. Lignocellulosic biomass consists of cellulose, hemicellulose (xylan and arabinan), lignin, ash and other residues in negligible amount [5]. Cellulose is protected by layers of hemicellulose and lignin, and the scope of the pretreatment phase is to expose the cellulosic fibers by relocating lignin and partially hydrolyzing the hemicellulose. Hemicellulose should only be partially hydrolyzed because most of the pretreatment by-products inhibit the downstream processes. E.g. acetic acid influences the pH of medium, which affects the enzymatic activity [6], furfural is a fermentation inhibitor, and xylose and xylooligomers strongly inhibit the enzymatic activity in the hydrolysis phase [7]. Therefore there is a trade-off between biomass pretreatment and the efficiency of the subsequent enzymatic hydrolysis and fermentation processes.

In the enzymatic hydrolysis phase, enzymes hydrolyze both cellulosic and the remaining hemicellulosic fibers. The conversion route of fibers to sugars is a competitive product inhibition process thoroughly described and analyzed in [8]. Xylose and xylooligomers are the strongest inhibitors of liquefaction, even greater than cellobiose and glucose [7]. Too little biomass pretreatment would increase the amount of hemicellulose for hydrolysis, which would eventually decrease the glucose yield due to xylose and xylooligomers inhibition. On the other hand, too much biomass pretreatment would increase the amount of fermentation inhibitors. Increasing pretreatment temperature is positive for the performance of enzymatic hydrolysis while negative for fermentation yield of ethanol.

Optimal operation depends on pretreatment temperature, enzyme dosage, and yeast seed in fermentation. The first scope of this study is to formulate optimal operation of large-scale biorefineries as a mathematical programming problem. In this formulation, the performance trade-offs between pretreatment, enzymatic hydrolysis and fermentation are formulated as objective functions subject to system dynamics taking into account the above mentioned phenomena using previously developed and validated first principles dynamic models [8, 9, 10]. E.g., too much xylan in the liquefaction phase could be overcome by significantly increasing the enzyme dosage. However, enzymes are very expensive and operation costs would increase. Therefore, price tags need to be attached to raw materials, such as: feedstock, steam for pretreatment, enzymes for hydrolysis, and yeast in the fermentation phase. Economic cost functions are formulated for maximizing the profit of the biorefinery. The final outcome would be to use the economic cost functions in order to find the best possible trade-off between the optimal operation of pretreatment, liquefaction and fermentation processes in integrated manner.

The study also includes a comprehensive sensitivity and uncertainty analysis of the optimization problem with respect to feedstock composition and kinetic parameters. A Monte Carlo technique with Latin Hypercube Sampling and correlation control is used for the uncertainty analysis [11, 12]. Uncertainties in kinetics and yield of pretreatment, hydrolysis and fermentation are found to be negligible on the economic objective function. On the other hand, inflow feed composition influences significantly the objective function.

References:

[1] J. Larsen, M. Ø. Haven, and L. Thirup. "Inbicon makes lignocellulosic ethanol a commercial reality". In: Biomass and Bioenergy 46 (Nov. 2012), pp. 36–45. DOI: 10.1016/j.biombioe.2012.03.033.

[2] Novozymes A/S. "Beta Renewables opens biofuels plant in Italy". In: Focus on Catalysts 2013.12 (Dec. 2013), p. 6.DOI: 10.1016/S1351-4180(13)70461-6.

[3] AEC. Cellulosic biofuels. Tech. rep. Advanced Ethanol Council, 2012.

[4] J. Larsen, M. Østergaard Petersen, L. Thirup, H. Wen Li, and F. Krogh Iversen. "The IBUS Process – Lignocellulosic Bioethanol Close to a Commercial Reality". In: Chemical Engineering & Technology 31.5 (May 2008), pp. 765–772. DOI: 10.1002/ceat.200800048.

[5] J. B. Kristensen, L. G. Thygesen, C. Felby, H. Jørgensen, and T. Elder. "Cell-wall structural changes in wheat straw pretreated for bioethanol production." In: Biotechnology for biofuels 1.1 (Jan. 2008), p. 5. DOI: 10.1186/1754-6834-1-5.

[6] R. M. Prunescu, M. Blanke, and G. Sin. "Modelling and L1 adaptive control of pH in bioethanol enzymatic process". In: Proceedings of the 2013 American Control Conference. 2013, pp. 2–9.

[7] Q. Qing, B. Yang, and C. E. Wyman. "Xylooligomers are strong inhibitors of cellulose hydrolysis by enzymes." In: Bioresource technology 101.24 (Dec. 2010), pp. 9624–9630. DOI: 10.1016/j.biortech.2010.06.137.

[8] R. M. Prunescu and G. Sin. "Dynamic modeling and validation of a lignocellulosic enzymatic hydrolysis process–a demonstration scale study." In: Bioresource technology 150 (Dec. 2013), pp. 393–403. DOI: 10.1016/j.biortech.2013.10.029.

[9] B. Lavarack, G. Griffin, and D. Rodman. "The acid hydrolysis of sugarcane bagasse hemicellulose to produce xylose, arabinose, glucose and other products". In: Biomass and Bioenergy 23 (2002), pp. 367–380.

[10] M. Krishnan, N. Ho, and G. Tsao. "Fermentation kinetics of ethanol production from glucose and xylose by recombinant Saccharomyces 1400 (pLNH33)". In: Applied Biochemistry and Biotechnology 77-79 (1999), pp. 373–388.

[11] R. Morales-Rodriguez, A. S. Meyer, K. V. Gernaey, and G. Sin. "Dynamic model-based evaluation of process configurations for integrated operation of hydrolysis and co-fermentation for bioethanol production from lignocellulose". In: Bioresource technology. 2011; 102.2:1174–84. DOI: 10.1016/j.biortech.2010.09.045.

[12] G. Sin, K. V. Gernaey, M. B. Neumann, M. C. M. van Loosdrecht, and W. Gujer. "Global sensitivity analysis in wastewater treatment plant model applications: prioritizing sources of uncertainty". In: Water research. 2011; 45.2:639–51. DOI: 10.1016/j.watres.2010.08.025.

425602 Forward-Backard Contractor in Verified Simulation Using Interval Analysis for Dynamic Optimisation

Abstract:

Many practically important processes are modelled by ordinary differential equations (ODEs). There is an interest in obtaining the global optima for these problems since they often offer substantial improvement compared to locally optimal solutions. Furthermore, in safety critical applications it is important to ensure that the behaviour of the process is within prescribed safe limits.

The deterministic methods for the solution of dynamic systems to global optimality are only able to address low dimensional problems. This is because in the sequential approach the integration has to be stopped early due to the generation of the overestimation in the verified simulation method.

Verified simulation seeks to enclose the trajectory of a dynamic system defined by ordinary differential equations (ODEs) within upper and lower bounds. This task is challenging when uncertainty in the parameters or the initial conditions needs to be accounted for because the method is subject to overestimation due to the dependency and wrapping effect problems (Jackson, 1975). Interval analysis is a key component in the verified simulation method since real amounts can be replaced for interval amounts and thus allowing the representation of uncertainty (Moore et al., 2009).

This work focuses on the reduction of the overestimation by means of the use of a forward-backward contractor (Jaulin et al. 2001) (also known as constraint propagation or HC4 (Benhamou et al. 1999)) in collaboration with a verified method using the Krawczyk and Newton contractors (Perez-Galvan and Bogle, 2014). The contractor is implemented in an interval Taylor series method in a stepwise fashion.

Numerical experiments using the method demonstrate the effectiveness of the method in the reduction of the overestimation. In the examples uncertainty is accounted by adding interval values in the parameters and initial conditions. Significant reduction of the overestimation and longer simulation times were observed after using the forward-backward contractor in the verified simulation method.

References

F. Benhamou, F. Goualard, L. Granvilliers, and J. F. Puget. Revisiting hull and box consistency. In International Conference on Logic Programming. Citeseer, 1999.

L. Jackson. Interval arithmetic error-bounding algorithms. SIAM Journal on Numerical Analysis, 12(2):223-238, 1975.

L. Jaulin, M Kieffer, O. Didrit, and E. Walter. Applied Interval Analysis. Springer, London, 2001.

R. E. Moore, R. B. Kearfort, and M. J. Cloud. Introduction to Interval Analysis. SIAM, Philadelphia, 2009.

C. Perez-Galvan and I. D. L. Bogle. Comparison between interval methods to solve initial value problems in chemical process design. In J. J. Klemes, P. S. Varbanov, and P. Y. Liew, editors, 24th European Symposium on Computer Aided Process Engineering, 33:1405-1410. Elsevier, 2014.

420317 Sensitivity Analysis of Molecular Design Problem for the Development of Novel Working Fluids for Power Cycles

Abstract:

In recent years there is a large availability of low-temperature heat sources in different applications such as waste heat in chemical industries and refrigeration plants as well as renewable energy sources such as biomass combustion, geothermal and solar heat sources. Power cycles are an important technology to convert such waste heat sources into usable energy. So far the low-temperature heat is not utilized efficiently for electricity generation.

To optimize the heat transfer process and the power generation, the influence of the working fluid, the cycle designs and the operating conditions is vital. Multi-criteria database search and Computer Aided Molecular Design (CAMD) can be applied to generate, test and evaluate promising pure component/mixture candidate as process fluids to help optimize cycle design and performance [1]. The problem formulation for the development of novel working fluids is a CAMD challenge both in terms of data and computational demand, because includes process related as well as property related equations.

In CAMD problems the identification of target properties is often based on expert knowledge. However, sophisticated information concerning the influence of for example a certain working fluid property parameter on the performance of the power cycle, i.e. the net power output, can facilitate the identification key properties for working fluids. In that sense a sensitivity analysis of the different parameters is suggested in this work as a systematic method to efficiently identify the target properties of the CAMD problem for working fluids.

In this study the CAMD problem for the development of novel working fluids for organic Rankine cycles (ORC) is formulated mathematically. It integrates both a system model for the ORC and property models including the Peng-Robinson equation of state [2], for the working fluids (enthalpy, entropy, etc) and group contribution (GC) models for pure component property predictions including critical properties as well as environmental and safety related properties. In addition to these property models, the system consists of coupled mass and energy balances for a pump, a condenser, a turbine and an evaporator.

As regards sensitivity analysis method, a global sensitivity analysis is performed based on Morris screening to determine which change of input parameters have important effects on the net power output. The screening is composed of individually randomized one-factor-at-a-time parameter changes over the whole parameter space [3]. This allows users to identify which properties are important (as well optimal region of property values) hence can be candidate as target properties for a working fluid design and/or search. In addition, environmental and safety properties such as the ozone depletion potential (ODP), the global warming potential (GWP) as well as flammability limits are taken into account to give a multi-criteria framework.

Furthermore, a process optimization with respect to the properties is performed to identify the optimal property values for a given process set-up [4]. This in combination with the sensitivity analysis to specify systematically the boundaries for the cycle related target properties. This facilitates an efficient screening for the optimal pure components or mixture working fluids.

The methodology will be applied in a case study of an Organic Rankine Cycle (ORC) with a low-temperature heat source. The heat source is a hot water stream from waste heat of a chemical site. Giving this pre-exquisite the method allows to identify the most favorable working fluid along with the corresponding optimal process conditions in order to get the highest possible power output.

The study presents a new approach for the identification of target properties of CAMD problems based on sensitivity analysis and shows its application for the development of novel working fluids of organic Rankine cycles for low temperature heat sources.

[1] A. I. Papadopoulos, M. Stijepovic, and P. Linke, "On the systematic design and selection of optimal working fluids for Organic Rankine Cycles," *Appl. Therm. Eng.*, vol. 30, no. 6–7, pp. 760–769, May 2010.

[2] W. Liu, D. Meinel, C. Wieland, and H. Spliethoff, "Investigation of hydrofluoroolefins as potential working fluids in organic Rankine cycle for geothermal power generation," *Energy*, vol. 67, pp. 106–116, 2014.

[3] G. Sin, K. V. Gernaey, and A. Eliasson Lantz, "Good Modeling Practice for PAT Applications: Propagation of Input Uncertainty and Sensitivity Analysis," *Biotechnol. Prog.*, vol. 25, no. 4, pp. 1043–1053, 2009.

[4] M. Lampe, M. Stavrou, J. Schilling, E. Sauer, J. Gross, and A. Bardow, "Computer-aided Molecular Design in the Continuous-Molecular Targeting Framework using Group-Contribution PC-SAFT," *Comput. Chem. Eng.*, 2015.

430643 A Framework for the Preliminary Screening of Mixtures As Post-Combustion CO2 Capture Solvent Candidates

Abstract:

The consideration of different amine solvent mixtures is receiving increased attention in post combustion CO_2 capture research to improve the chemical absorption/desorption process efficiency and costs. Binary amine mixtures are increasingly considered in aqueous solutions as CO_2 capture solvents due to their simultaneously high performance in several important properties (e.g. reaction rate, CO_2 capacity etc.). This is necessary to address shortcomings of aqueous solutions of single amines which present favorable behavior in few properties at the expense of strongly undesired effects in numerous others. The selection of the mixture composition (i.e. chemical and physical characteristics of the amines) and concentration (i.e. amount of each amine in the mixture) is clearly a very challenging problem due to a) the highly non-ideal mixture-water-CO_2 interactions, b) the availability of countless combinations of amine molecules as potential mixture components and c) the need for combined consideration of several properties as criteria for the selection of mixtures with improved CO_2 capture performance. The use of computer-aided tools can help address these challenges through models that enable accurate predictions of the desired mixture properties and systematic procedures to account for the combinatorial complexity. Although such tools have been successfully applied in the investigation of mixtures for non-CO_2 capture applications, they have been considered to a very limited extent in the field of CO_2 capture. The very few available published works focus on the development of rigorous mixture property prediction models which are indispensable for the accurate representation of mixture-water-CO_2 interactions. However, the need to consider multiple mixture components and properties as mixture selection criteria requires intense computational effort and results in complexities that limit the selection to the consideration of very few options.

In this work we propose a systematic framework for the preliminary screening of binary amine mixtures as CO_2 capture candidates considering several important properties as selection criteria. The proposed approach consists of several decision making stages which account for solvent-solvent, solvent-solvent-CO_2 and solvent-solvent-water interactions using standard group contribution models as well as equations of state and activity coefficient models to account for the vapor and liquid phase non-idealities in several relevant properties. The properties considered as selection criteria capture the molecular chemistry effects on the absorption/desorption process and include mixture solubility, vapor pressure, viscosity, density, bubble point temperature, and melting point temperature. Reactivity with CO_2 is also taken into account considering important empirical guidelines. The aim of the proposed approach is to be fast and sufficiently accurate in order to identify few useful amine combinations which may then be evaluated using rigorous prediction models or experiments, while quickly avoiding amine options of poor performance. Mixtures meeting specific performance criteria are considered in successive decision-making stages, hence a set of mixtures gradually emerges which consists of fewer but more effective CO_2 capture candidates. The desired property criteria are calculated using predictive models based a) on their potential to reflect on important capture characteristics, b) on the availability of appropriate models for their calculation and c) on the availability of sufficient data so that these models may be applied in a wide range of molecular structures. A multi-criteria assessment methodology is combined with a systematic uncertainty quantification approach to unveil important trade-offs among several important properties and to propose the mixtures that appear to be promising as CO_2 capture candidates. Uncertainty is addressed by considering multiple different models to calculate each mixture property employed as a selection criterion. Experimental data retrieved from literature are used to validate the employed property predictions models. The proposed method is applied in 200 mixtures resulting from numerous binary combinations of amines which have been previously investigated in their pure aqueous form as CO_2 capture solvents. The obtained results indicate that the proposed framework unveils useful and valid insights regarding the interactions of different amine combinations in the presence of CO_2 or water, despite the highly non-ideal mixture characteristics.

Computational Molecular Design of Water Compatible Dentin Adhesive System

Farhana Abedin,[1,2] Qiang Ye,[2] Paulette Spencer,[2,3] Kyle Camarda[4]

1 Bioengineering Graduate Program, The University of Kansas, Lawrence KS

2 Bioengineering Research Center, The University of Kansas, Lawrence, KS

3 Department of Mechanical Engineering, The University of Kansas, Lawrence, KS

4 Department of Chemical and Petroleum Engineering, The University of Kansas, Lawrence, KS

AIChE Annual Meeting, Salt Lake City, UT

November 10, 2015

Background

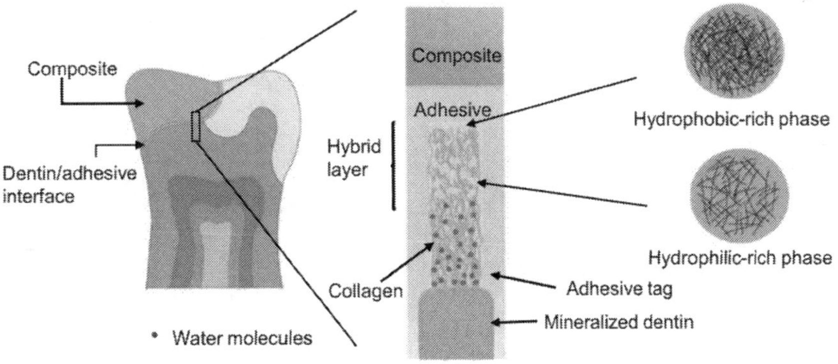

- Dental adhesive undergoes phase separation into hydrophobic & hydrophilic-rich phases
- Hydrophilic-rich phase forms a vulnerable region prone to failure
  - Poor polymerization due to lack of photoinitiators
  - Poor crosslinking due to lack of crosslinker

Objective

- To design water compatible visible light photosensitizer suitable for use in dental adhesive
- Computer-aided molecular design will be employed
- Computer-aided molecular design:
    - Reduces trial & error method
    - Rational design methodology
    - Inexpensive & efficient
    - Large search space
    - Evaluation of numerous alternatives

KU THE UNIVERSITY OF KANSAS

Computer-aided Molecular Design

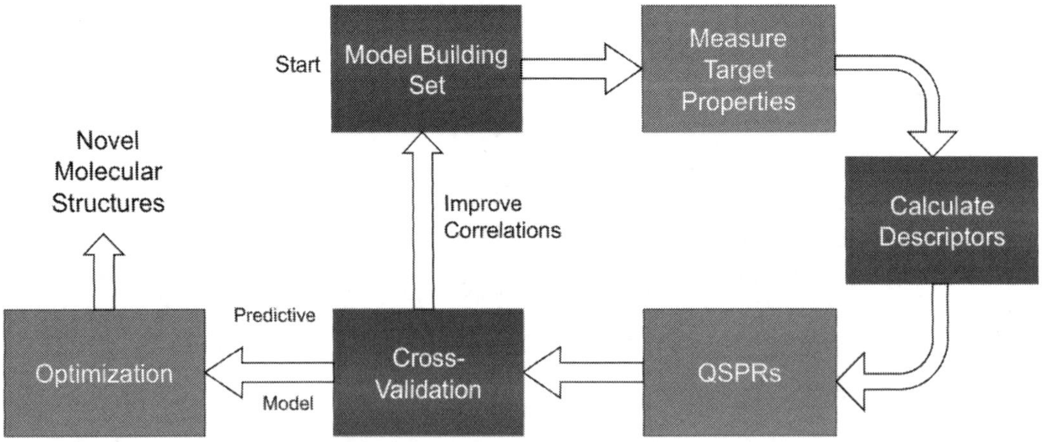

KU THE UNIVERSITY OF KANSAS

Model Building Set of Photosensitizer Molecules

- Absorption maxima within the visible wavelength
- Wide range of functional moieties
- Exhibit similar structure to some extent
- Hydrophilic in nature

Model Building Set of Photosensitizer Molecules

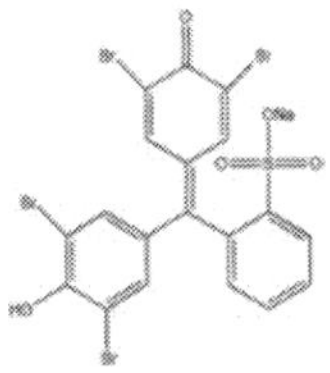

Bromophenol blue sodium salt

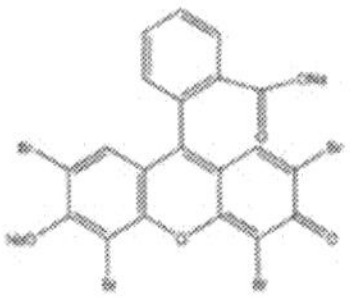

Eosin Y disodium salt

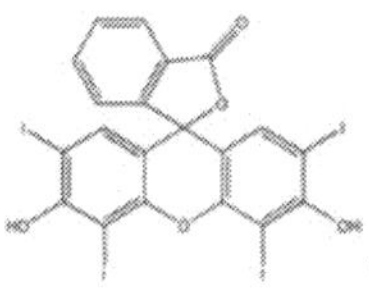

Erythrosin B

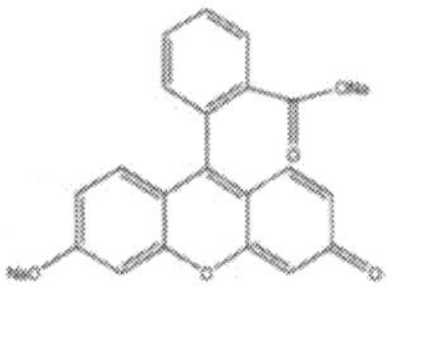

Fluorescein sodium salt

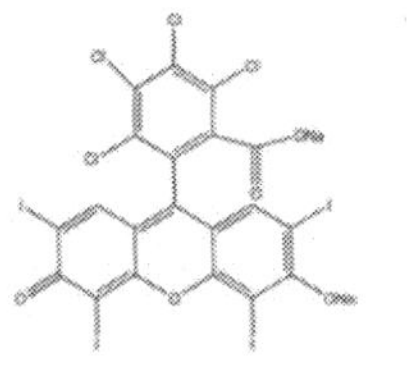

Rosebengal sodium salt

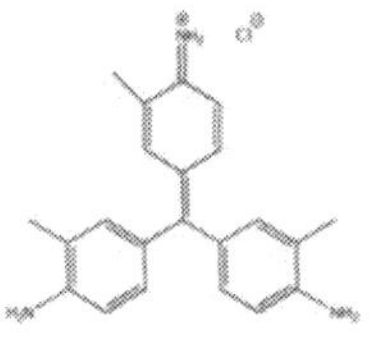

New Fuchsin

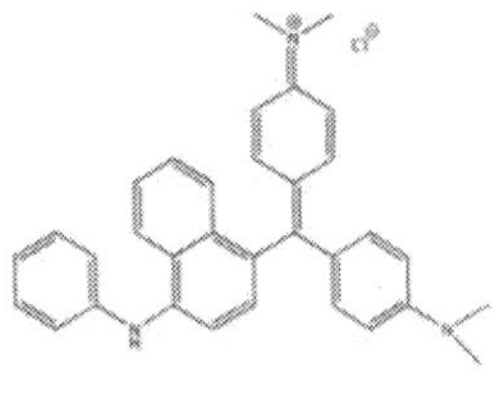

Victoria blue B

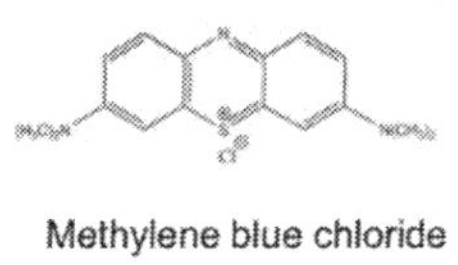

Methylene blue chloride

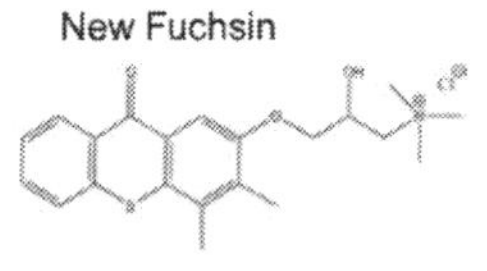

QTX

Target Properties

- **Octanol/water partition coefficient:**
  - Ratio of the concentration in octanol phase to that in the aqueous phase
  - Measure of hydrophilicity of the molecule
- **Molar extinction coefficient (ξ):**
  - Ability of a species to absorb light at a given wavelength
  - Impacts the initiation efficiency of photo-polymerization
- **Relative photon absorption efficiency:**
  - Measure of the efficiency of the photosensitizer and light curing unit (LCU) combination

KU THE UNIVERSITY OF KANSAS

Octanol/water Partition Coefficient

Obtained from literature[3, 4, 5, 6, 9]

Molecule	Experimental log P
Bromophenol blue	-2.35
Eosin Y	0.18
Erythrosin B	0.46
Fluorescein	-0.32
Methylene blue chloride	-0.96
Rosebengal	0.59
Victoria blue B	2.8
Victoria blue R	1.59

KU THE UNIVERSITY OF KANSAS

Molar Extinction Coefficient

Molecule	Experimental ξ (L mol⁻¹cm⁻¹)	ln(ξ)
Bromophenol blue sodium salt	14142 ± 510	9.56 ± 0.04
Eosin Y disodium salt	51471 ±1550	10.85 ± 0.03
Erythrosin B	13750 ± 1674	9.52 ± 0.12
Fluorescein sodium salt	46208 ± 1617	10.74 ± 0.04
Methylene blue chloride	1594 ± 232	7.37 ± 0.15
Rosebengal sodium salt	8001 ± 1792	8.97 ± 0.21
Victoria blue B	5269 ± 134	8.57 ± 0.03
New Fuchsin	76883 ± 683	11.25 ± 0.01
QTX	235 ± 154	5.32 ± 0.65

- Measured by UV-vis spectrophotometer at 480 nm wavelength
- 480 nm is approximately the peak wavelength of halogen dental curing light unit

THE UNIVERSITY OF KANSAS

Relative Photon Absorption Efficiency (PAE)

Molecule	Experimental relative PAE
Bromophenol blue sodium salt	3.14 ± 0.08
Eosin Y disodium salt	6.30 ± 0.36
Erythrosin B	2.19 ± 0.07
Fluorescein sodium salt	4.42 ± 0.07
Methylene blue chloride	0.29 ± 0.04
Rosebengal sodium salt	1.46 ± 0.03
New Fuchsin	10.65 ± 0.07
Victoria blue B	0.85 ± 0.02
QTX	0.36 ± 0.03

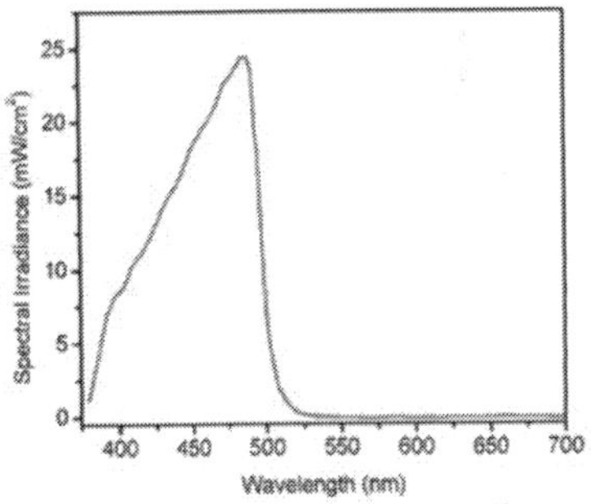

Emission spectrum of halogen LCU[10]

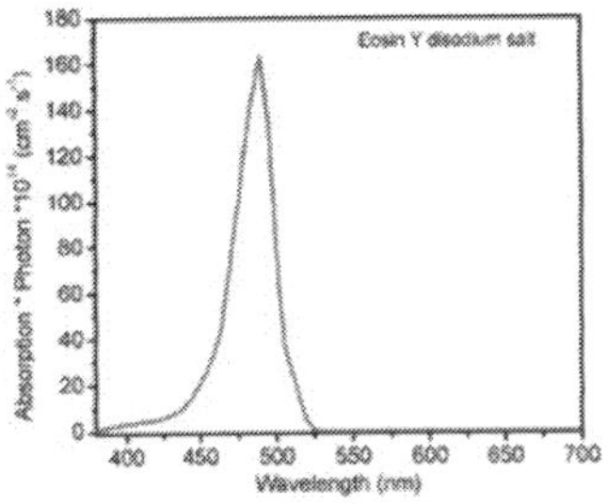

THE UNIVERSITY OF KANSAS

Molecular Descriptors

- Connectivity indices were used to develop the quantitative structure property relationships (QSPRs)
- Connectivity indices ($^{n}\chi$) describe 2D atomic & bonding features
- Valence connectivity indices ($^{n}\chi^{v}$) describe electronic configuration
    - δ = No. of non-hydrogen atoms bonded to a non-hydrogen atom
    - $\delta^{v} = \frac{z_v - N_H}{z - z_v - 1}$
    - $^{n}\chi = \sum_{i=1}^{N_s} (\prod_{k=1}^{n+1} \frac{1}{\delta_k})_i^{\frac{1}{2}}$
    - $^{n}\chi^{v} = \sum_{i=1}^{N_s} (\prod_{k=1}^{n+1} \frac{1}{\delta_k^{v}})_i^{\frac{1}{2}}$

KU THE UNIVERSITY OF KANSAS

QSPRs

- Mathematical relationships linking physicochemical properties with the structure of a molecule
- Capable of accurately predicting the physicochemical properties of a molecule
- Statistical software R was used to develop the QSPRs by linear regression
- Leaps package was employed to select descriptors

KU THE UNIVERSITY OF KANSAS

Cross-validation

- Models with higher correlation coefficient, R^2 & lowest Mallow's C_p statistic were selected
- Mallow's C_p statistic determines the optimum number of descriptors
- QSPRs models were cross-validated by leave-one-out method by using DAAG package in R
- Predictive squared correlation coefficient (Q^2) evaluates the predictive ability of the models
- $Q^2 \approx R^2$ for a good model

QSPRs for the Target Properties

QSPRs	Adjusted R^2	Q^2
$\ln(\xi) = 32.22723 + (3.95277 \,.\, {}^{0}\chi^{v}) - (15.07572 \,.\, {}^{5}\chi^{v}) - (4.60663 \,.\, {}^{0}\chi) + (0.63374 \,.\, {}^{1}\chi) - (0.62536 \,.\, {}^{3}\chi) + (4.81887 \,.\, {}^{4}\chi)$	0.99999	0.99272
$\log P = -9.08869 + (1.61343 \,.\, {}^{1}\chi) - (7.86666 \,.\, {}^{4}\chi) + (6.86586 \,.\, {}^{5}\chi) + (0.98095 \,.\, {}^{0}\chi^{v}) - (1.48700 \,.\, {}^{2}\chi^{v})$	0.99996	0.99813

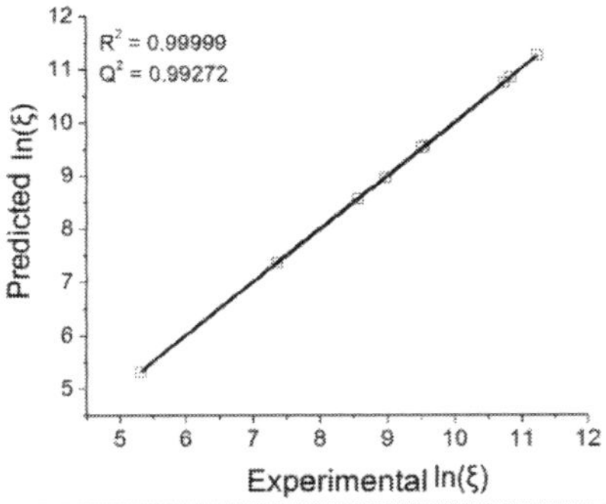

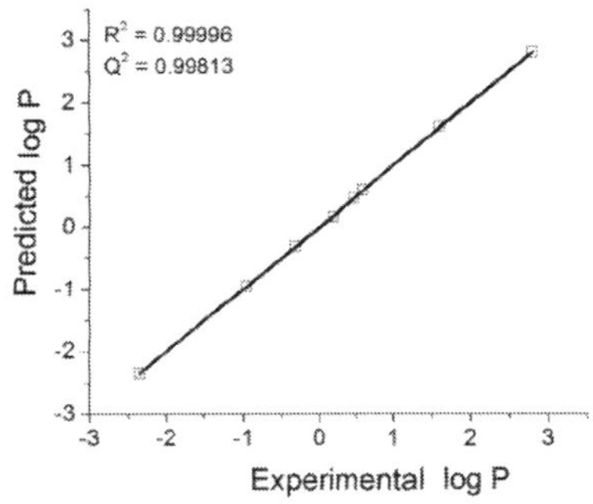

QSPRs for the Target Properties

Relative PAE = 34.3226 - (1.7485 . $^0\chi$) – (8.0407 . $^2\chi$) + (6.1803 . $^3\chi$) + (5.108 . $^5\chi$) + (9.9007 . $^2\chi^v$) - (26.9838 . $^5\chi^v$)

Adjusted R^2 = 0.9996, Q^2 = 0.9861

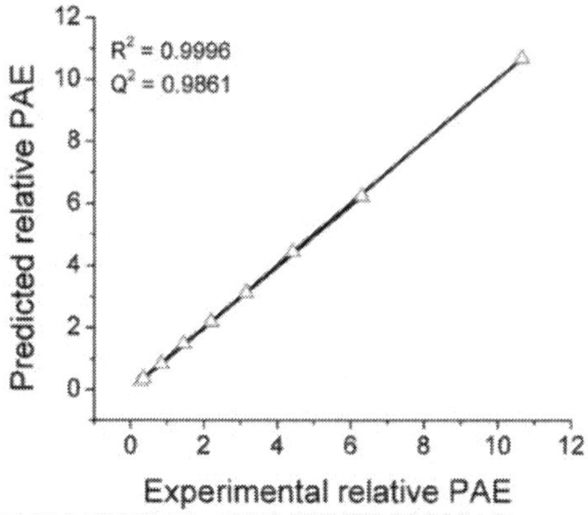

Property	Number of descriptors used	R^2	Mallow's C_p Statistic
ln(ξ)	6	0.99999	6.00845
logP	5	0.99996	5.07619
Relative PAE	6	0.9996	6.02064

KU THE UNIVERSITY OF KANSAS

Computer-aided Molecular Design Method: Inverse Problem

- Based on the molecular structures in the model building set 42 groups were created
- 5 groups could be selected from 42 at maximum
- There are approximately 100 million possibilities
- Molecular structures for solutions were generated by making local moves:
  - Swap
  - Insert
  - Delete
  - Replace

KU THE UNIVERSITY OF KANSAS

Example of Groups

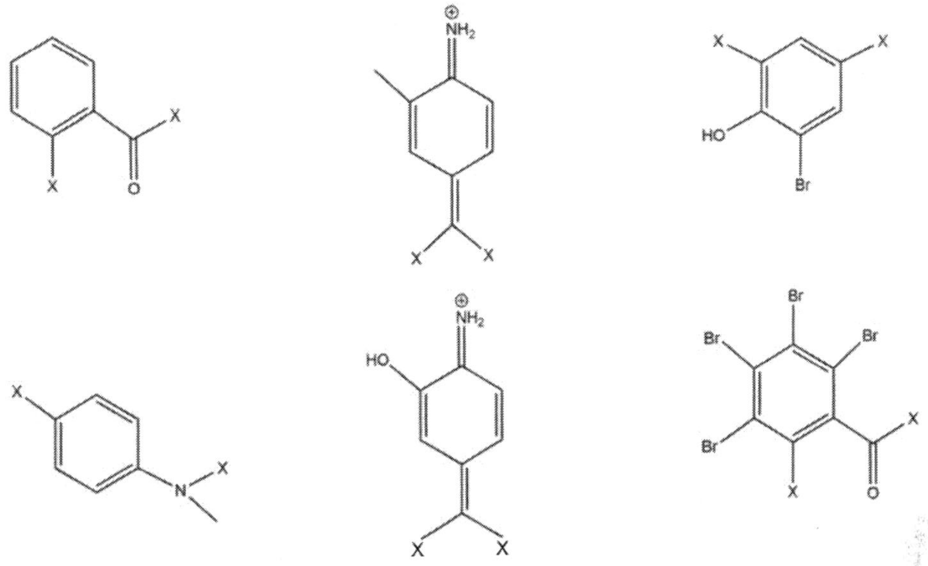

Optimization Formulation

$$\text{Minimize } s = \sum_M \frac{1}{P_m^{scale}} \left| P_m - P_m^{Target} \right|$$

$$P_m = f_m(\chi)$$

$$\chi = g(a_{ij}, w_i)$$

$$h_c(a_{ij}, w_i) \geq 0$$

S is the objective function
P is the value of property *m*
f_m is the QSPR function for property *m*
g is the function to calculate connectivity indices, χ by adjacency matrix a_{ij} & identity matrix *w*
h_c are molecular structural constraints

Target values for properties:

- log P = -0.55 (log P value for hydrophilic dental monomer)
- ln (ξ) = 9 (Approximately close to the maximum values for known candidates)
- Relative PAE = 7 (Approximately close to the maximum for known candidates)

Tabu Search

- The mixed integer non-linear problem (MINLP) was solved via Tabu Search
- Tabu Search is a stochastic method
- Stores previous solutions in a Tabu list which prevents oscillation of the solution around local optimum
- Provides locally near-optimal solutions

Candidate Optimal Photosensitizer Molecules

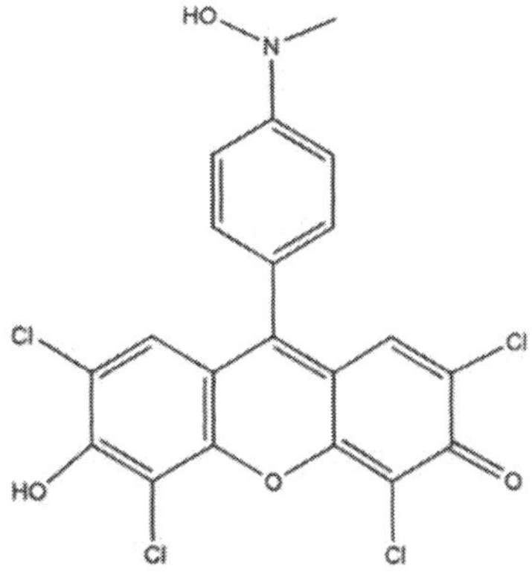

Objective = 0.2832
ln(ξ) = 9.078
log P = -0.6202
Relative PAE = 8.0548

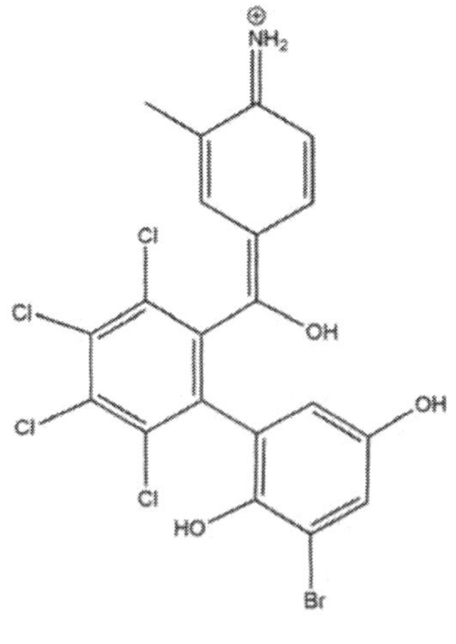

Objective = 0.1521
ln(ξ) = 9.105
log P = -0.5725
Relative PAE = 7.7199

Conclusions

- A framework to design water compatible visible light photosensitizer for dental adhesive has been demonstrated
- Two examples of possible candidate photosensitizer molecules have been proposed
- In future target properties such as degree of conversion and rate of polymerization will be incorporated

KU THE UNIVERSITY OF KANSAS

Acknowledgement

- Brock Roughton
- Xueping Ge
- Leon Song
- Md. Rajib Anwar

This investigation was supported by the research grant R01 DE022054, National Institute of Dental and Craniofacial Research, National Institutes of Health, Bethesda, MD 20892

KU THE UNIVERSITY OF KANSAS

QUESTIONS?

References

1. B. C. Roughhton, E. M. Topp, K. Camarda, 2012, Use of glass transitions in carbohydrate excipient design for lyophilized protein formulations, Computers and Chemical Engineering, 36, 208-216
2. R-Development-Core_Team, 2010, A language and environment for statistical computing. http://www.R-project.org
3. D. S. Pellosi, B. M. Estevao, J. Semensato, D. Severino, M. S. Baptista, M. J. Politi, N. Hioka, W. Caetano, 2012, Photophysical properties and interactions of xanthene dyes in aqueous micelles, Journal of photochemistry and photobiology A: Chemistry, 247, 8-15
4. M. I. Sheikh, 1976, Renal handling of phenol red II. the mechanism of substituted phenolsulphophthalein (PSP) dye transport in rabbit kidney tubules *in vitro*, Journal of Physiology, 256, 175-95
5. S. J. Wagner, A. Skripchenko, D. Robinette, J. W. Foley, L. Cincotta, 1998, Factors affecting virus photoinactivation by a series of phenothiazine dyes, Photochemistry and Photobiology, 67(3), 343-349
6. M. Wainwright, S. M. Burrow, S. G. R. Guinot, D. A. Phoenix, J. Waring, 1999, Uptake and cell-killing activities of a series of victoria blue derivatives in a mouse mammary tumour cell line, Cytotechnology, 29, 35-43
7. S. Chavali, B. Lin, D. C. Miller, K. V. Camarda, 2004, Environmentally-benign transition metal catalyst design using optimization techniques, Computers and Chemical Engineering, 28, 605-611
8. J. C. Eslick, Q. Ye, J. Park, E. M. Topp, P. Spencer, K. Camarda, 2009, A computational molecular design framework for cross-linked polymer networks, Computers and Chemical Engineering, 33, 954-963
9. S. Fujisawa and E. Masuhara, 1981, Determination of partition coefficients of acrylates, methacrylates, and vinyl monomers using high performance liquid chromatography (HPLC), Journal of Biomedical Materials Research, 15, 787-793.
10. M. Perez , F. Perez-ocon, C. Lucena-Martin, P. Rosa, 2008, Stability and reproducibility of radiometric properties of light curing units (LCUs). Part I:QTH LCUs, Dental Materials Journal, 27(2), 284-291.

AUTHOR INDEX

AUTHOR INDEX

AUTHOR INDEX

AUTHOR INDEX

AUTHOR INDEX

AUTHOR INDEX

AUTHOR INDEX

AUTHOR INDEX

AUTHOR INDEX

AIChE
120 Wall Street, FL 23
New York, NY 10005-4020

ISBN 978-1-5108-1856-9

CURRAN ASSOCIATES INC.
proceedings.com

Computing and Systems Technology Division 2015

Core Programming Area at the 2015 AIChE Annual Meeting

Salt Lake City, Utah, USA
8-13 November 2015

Volume 2 of 2

ISBN: 978-1-5108-1856-9

Printed from e-media with permission by:

Curran Associates, Inc.
57 Morehouse Lane
Red Hook, NY 12571

Printed by Curran Associates, Inc. (2016)

For permission requests, please contact AIChE
at the address below.

AIChE
120 Wall Street, FL 23
New York, NY 10005-4020

Phone: (800) 242-4363
Fax: (203) 775-5177

www.aiche.org

Additional copies of this publication are available from:

Curran Associates, Inc.
57 Morehouse Lane
Red Hook, NY 12571 USA
Phone: 845-758-0400
Fax: 845-758-2634
Email: curran@proceedings.com
Web: www.proceedings.com

TABLE OF CONTENTS

VOLUME 1

VOLUME 2

428710 Ontology Engineering Approach to Support Process Model Integration

Tuesday, November 10, 2015: 4:55 PM
Salon G (Salt Lake Marriott Downtown at City Creek)

Linsey Koo[1], Franjo Cecelja[2] and Nikolaos Trokanas[2], (1)University of Surrey, Guildford, United Kingdom, (2)Process and Information Systems Engineering Research Centre, University of Surrey, Guildford, United Kingdom

Ontology engineering approach to support process model integration

Linsey Koo, Nikolaos Trokanas, Franjo Cecelja

PRISE Group, Chemical and Process System Engineering, Faculty of Engineering and Physical Sciences, University of Surrey, Guildford, Surrey, U. K.

Process modelling and simulation is a vital tool to plan, evaluate, assess, and develop different alternatives for the design of products and processes. The complexity of problems as well as heterogeneity of modelling methods make process modelling and simulation challenging, time consuming and often tedious process requiring a wide range of expertise. Inconsistencies in model development are the main cause for redundant work. Models remain implicit to the engineers who have built them, which further limits the potential of reusability.

The only model integration framework in use, the CAPE-OPEN, addresses the issue of standardisation of interfaces to enable interoperability between simulator software components from different sources. It is the framework built around a middleware, the Common Object Request Broker Architecture (CORBA) that hosts communication between unit operations defined for a specific function and the process modelling environments. The standard specification is defined as a property package which is needed for a thermodynamic or physical property calculation. The interoperability of models, such as model selection, parameter identification, and experimental work is enabled through the connection related to the unit operations and physical properties. It is not necessary to match all parameters in order to facilitate Input-Output (I-O) matching. However, the shortcoming of the CAPE-OPEN is in the need for identifying key variables for each unit operation.

A new approach for model integration however builds upon the CAPE-OPEN framework proposes the use of ontology and replaces the CORBA object bus with more flexible semantic repository (Koo and Cecelja, 2015). Models are described by Semantic Web Services (SWS) using Ontology Web Service Description (OWL-S) as an enabler of web services through service discovery, selection, composition, and execution stages, as shown in Figure 1. The discovery stage allows formation of an integrated model through matching requests from a public repository(ies). The best match that satisfy the requestor's functionality is selected in the model selection stage. The model composition stage then formulates the chain of integrated models and execution of integrated model takes place during the execution stage.

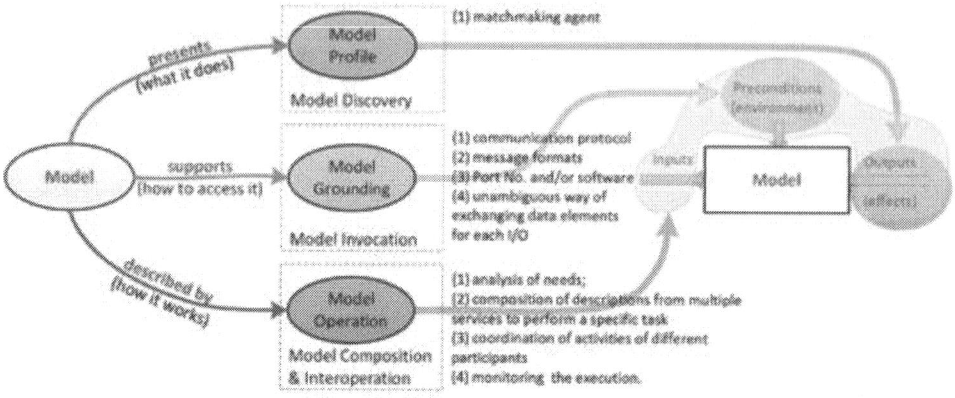

Figure 1. Semantically described model

This paper focuses on the matching parameters related to the domain of process system engineering, with emphasis placed on the role of physical properties and unit operation. We propose an extension of classification of variables defined by CAPE-OPEN with adaptation to ontology. To this end, each model representing a device (e.g. unit operations, flowsheets, subflowsheets etc.) is semantically described in domain ontology including domain assumptions and descriptions of the functionality of the model. The domain ontology guides the process of registering models and instantiation of ontology through ontology parsing. The domain ontology makes the model and data explicit and sharing terminology improves consistency. The devices in a process are connected via streams that transmit information through multiple inlets and outlets from one unit to the other. The connection between devices are described in ontology by introducing the concept of 'ports' and 'connections' (CO-LaN, 2014). The 'ports' generally describe inlets and outlets of devices and three different types of streams are distinguished: material, energy, and information, which are further described by object properties. The 'connections' are the object that is responsible for establishing a link between two ports, which contains information regarding methods, types, quantities, and units of streams.

The structural concept and classification of matching parameters defined by CAPE-OPEN have been employed in ontology. It intends to guide through the complexity of modelling methods and to facilitate I-O matching, while demonstrating the flexibility of model integration. To highlight the main feature of the ontology approach model integration, manageable scale models related to bio-refining process are used to verify the performance of the proposed approach.

References:

CO-LaN, 2014, Conceptual Design Document for CAPE-Open Project, www.co-lan.org [Accessed on 20/04/2015]

Koo, L., Cecelja, F., 2015, Model Integration Using Ontology Input-Output Matching, 25th European Symposium on Computer Aided Process Engineering Proceeding Book, Copenhagen, Denmark, pp. 2567

428762 Constructing Ontologies for Controlled Plants

Abstract:

Constructing Ontologies for Controlled Plants

Heinz A Preisig
Dept of Chemical Engineering
NTNU
Trondheim, Norway

The computer-aided modelling project [1], which has become the hallmark of our group's activity, aims at generating a component-based toolbox so we can build a zoo of computational objects [2][3], suitable to solve most of the computational engineering task we meet in our profession and beyond.

We want to construct the tools such the they enable inter-disciplinary work [4], thus impose minimal limitations on the application domain. We aim at increased productivity by making things simple on the basis from which we can assemble gradually complex problems. The user shall be in charge not the technology should impose; no repetitions, proper, and focus on the modelling and not on the technology to generate the code.

Domains we aiming at are process industry: conceptional design, simulation, flow-sheeting, detailed design, mechanical design, planning, scheduling and control [1]. Materials engineering: materials design across the scales, chemistry, crystallography, metallurgy, polymer sciences ... Mechanical engineering: computer-aided design and computer-aided manufacturing.

Currently we are working on a multi-scale modelling problem that attempts to reconstruct the mechanical behaviour of polyurethane foams. On the lowest level of the Framework 7 project we have quantum mechanic models, extending to molecular dynamics, meso-scale models to macroscopic flow models and mechanical properties models [4].

The ontology for which we have constructed a set of tools, is currently used for the description of controlled physical/chemical/biological processes and we are in the process of extending the domain to multi-scale processes of the Framework 7 project MoDeNa. Objective in this domain is to generate a standard for the model representation, which is a subset of definitions and equations extracted from the ontology and augmented with the instantiation information and possibly input data, surrogate models and solver information.

We use the term ontology for a mathematical representation of model components, which include containment information, information about what is in the containment and characteristics of all components. The first layer defines the basis build on the foundation of containment with tokens living in it all in a frame like time and space.

The presentation will focus on the foundation of the ontology, the underlying ideas and a representation of one of the ontologies we generated in our ontology-construction tool.

1. Preisig, H. A. Constructing and maintaining proper process models Comp & Chem Eng, 2010, 34(9), 1543-1555
2. Elve, Arne Tobias; Ontology Design for Representation of mathematical models MSc Thesis, NTNU, Trondheim Feb 2015
3. Preisig, H. A. A multi-layered ontology for physical-chemical-biological processes ESCAPE 21 (2011), Porto Carras, Chalkidiki, Greece, 2011, 1, ISBN 978-0-444-53711-9, 101-105
4. MoDeNa project website. Retrieved May 11, 2015 from http://www.modenaproject.eu/

419151 Optimal Operation of an Austenitization Furnace

Abstract:

Optimal Operation of an Austenitization Furnace

Vincent Heng and Michael Baldea

McKetta Department of Chemical Engineering

The University of Texas at Austin, 1 University Station C0400, Austin, TX 78712

Email: mbaldea@che.utexas.edu

The production and primary processing of metals are energy-intensive operations, accounting for about 2 quadrillion BTU (quads) of energy consumption in the United States every year. Of this, a significant percentage (up to 30% in some cases) is lost as waste heat[1], owing to both inherent inefficiencies and ineffective control strategies.

In this presentation, we examine the operation of such a metal processing system, and propose a novel approach for improving its energy efficiency. In particular, we focus on the real time optimization of an austenitization furnace currently in operation at an industrial partner. Austenitization is a heat treating operation that consists of heating metal parts, followed by rapid quenching in an oil bath, with the goal of inducing specific metal properties (e.g., tensile strength, hardness). Heating occurs in high-temperature furnaces that are operated in a continuous manner. The metal parts travel on rollers through the furnace, which is heated by ceiling and floor radiant tube burners spanning the length of the furnace. In the system under consideration, the overall heat input is on the order of 10 million BTU per hour to achieve part temperatures of 1400F, with a part residence time of four hours.

The operational objective of the process is to heat the metal to a minimum temperature threshold prior to quenching, while maintaining a relatively uniform temperature profile within the parts. During operation, part temperatures are not measured and thus cannot be explicitly controlled. Instead, furnace temperature is used as an implicit metric for the temperature of the steel parts, and is controlled by means of manipulating the flow rate of fuel (natural gas) to the burners. Operator experience and heuristics are used to determine the proper furnace temperature set points (defined for individual zones along the furnace) to achieve the desired metal exit temperatures. This empirical operation typically leads to excess energy consumption due to an "overcompensation" effect, where the furnace temperature is adjusted such that the part exit temperatures exceed (often by a fair margin) the desired temperature threshold [2].

In this work, we develop a model-based approach for improving energy efficiency in austenitization furnaces. We first describe a first-principles model of the austenitization furnace. We then develop simplified response-surface models that capture the evolution of the part temperature and energy consumption of the process as a function of the temperature set points, and utilize these models to formulate an optimization problem aimed at finding the temperature set points that minimize energy usage while obtaining the desired part exit conditions. Our results show an optimal longitudinal temperature profile with high-temperature ends and a cooler middle region. This represents a paradigm shift from the current operating prescription, which is based on increasing the temperature lengthwise from the inlet to the outlet of the furnace. We validate our findings by simulating the detailed furnace model; the simulation results confirm that significant energy savings can indeed be derived from the proposed optimal strategy. We also show that the optimal temperature profile can be easily recomputed for different operating conditions (e.g., changes in part shapes or production rate), and our strategy therefore lends itself very well to real-time implementation. We finalize our presentation with a discussion of future work concerning the implementation of the proposed strategy in the plant.

[1]V Viswanathan, R Davies, and J Holbery. Opportunity analysis for recovering energy from industrial waste heat and emissions. Pacific Northwest National Laboratory, 2005.

[2]A Thekdi. Energy efficiency improvement opportunities in process heating for the forging industry. E3M, 2010.

420769 Run-to-Run Control of PECVD of Thin Film Solar Cells

Abstract:

Thin film solar cells are receiving a growing amount of attention because of lower prices for solar equipment, higher efficiency, as well as tax and rebate incentives from the government, and provide a promising alternative to conventional energy sources. Among many solar cell systems, amorphous silicon (a-Si:H) thin film solar cells is one of the most widely used solar cell systems and plasma enhanced chemical vapor deposition (PECVD) plays a key role for the manufacturing of thin film solar cells. However, handling the varying nature of the operation of the PECVD process which persists from run-to-run remains a challenging problem due to the lack of appropriate in situ measurements for the product quality [1]. For example, a batch-to-batch change in the operating conditions or impurities in the feedstock container may significantly alter the quality of the thin film product. This is because the process model employed for in-batch control may significantly deviate from the actual process behavior, and thus, control actions derived using a nominal model are more likely to increase the variability of a process.

Motivated by this, a run-to-run (R2R) control scheme is developed and applied to a PECVD process using post-batch measurements. The major benefit of using R2R control is in its unique ability to compensate for process drift through correlations developed to describe process model parameter variations as a function of changes in the process environment [2,3]. Specifically, the present work initially focuses on the development of a multiscale modeling and operation framework for PECVD of thin film silicon solar cells with uniform thickness and desired film surface microstructure that improves light trapping efficiency. While the modeling of chemical reactions and transport phenomena in the gas-phase is based on two-dimensional in space partial differential equations, a novel microscopic model is developed for the a-Si:H thin film surface evolution, which directly accounts for four microscopic processes: physisorption, surface migration, hydrogen abstraction, and chemisorption. Then, a hybrid kinetic Monte Carlo (kMC) model is used to improve the computational efficiency without compromising the accuracy of the model [4], which is then calibrated with experimentally obtained growth rates and surface morphology data from the literature. After the wafer surface is separated into four concentric zones, the dependence of film growth rate on substrate temperature is modeled where the substrate temperature of each zone is adjusted with an independent heating element to compensate for a radially non-uniform deposition rate caused by depletion in the radical concentration in the gas-phase along the wafer [5]. Then, to deal with batch-to-batch variability, after each batch is over, the post-batch measurements including surface roughness and height-height correlation length are used to estimate off-line the process model (used in the feedback controller) parameters drifted from nominal values via a multivariable optimization problem. Additionally, the exponentially-weighted-moving-average scheme is used to deal with the remaining offset in the desired product qualities and thereby to compute a set of optimal jacket temperatures for each concentric substrate region [6]. Extensive simulations demonstrate that the use of appropriate sinusoidal wafer grating and the regulation of substrate temperature via the proposed R2R-based feedback control scheme provide a viable and effective way for the PECVD of thin film silicon solar cells with uniform thickness and film surface microstructure that optimizes light trapping.

[1] Campbell WJ, Firth SK, Toprac AJ, Edgar TF. A comparison of run-to-run control algorithms. In Proceedings of the American Control Conference, Anchorage, AK, 2002:2150-2155.

[2] Sachs E, Guo RS, Ha A. On-line process optimization and control using the sequential design of experiments. Symposium on VLSI Technology, Honolulu, HI, 1990:99-100.

[3] Wang Y, Gao F, Doyle FJ. Survey on iterative learning control, repetitive control, and run-to-run control. J. of Process Control. 2009;19:1589-1600.

[4] Tsalikis D, Baig C, Mavrantzas V, Amanatides E, Mataras D. A hybrid kinetic Monte Carlo method for simulating silicon films growth by plasma-enhanced chemical vapor deposition. J. Chem. Phys. 2013;139:204706.

[5] Crose M, Kwon JS, Nayhouse M, Ni D, Christofides PD. Multiscale modeling and operation of PECVD of thin film solar cells. Chem. Eng. Sci. 2015, in press.

[6] Kwon JS, Nayhouse M, Orkoulas G, Ni D, Christofides PD. Run-to-Run-based model predictive control of protein crystal shape in batch crystallization. Ind. & Eng. Chem. Res. 2015;54:4293-4302.

421119 Nonlinear Model Predictive Control of a Bubbling Fluidized Bed Adsorber for Post-Combustion Carbon Capture

Abstract:

When operating a post-combustion carbon capture system for power plants, specific requirements on CO_2 capture fraction should be satisfied. But the operation is subject to disturbances caused by the fast load changing of power plants. Additionally, economic issue such as high energy consumption is a major bottleneck for the commercial application of carbon capture systems. Therefore, flexible operation and control strategies are required to improve the economic performance of carbon capture systems. In this study, we develop efficient control strategies for a solid sorbent based post-combustion carbon capture system using bubbling fluidized bed (BFB) adsorbers. Computationally efficient reduced dynamic models are used to reduce the computational cost for model-based control problems. A fast nonlinear model predictive control (NMPC) algorithm is utilized to enable the online control of the carbon capture system and state estimation is required to handle the prediction error given by the reduced model.

The model for the BFB adsorber is a one-dimensional, three-region, pressure-driven model developed by national energy technology laboratory (NETL), which considers axial variations in the solid phase and the bed hydrodynamics. The reactor's hydrodynamic behavior is described by partial differential and algebraic equations (PDAEs), constructed from mass and heat conservation and hydrodynamic correlations. Detailed equations for steady-state and dynamic BFB model can be found in [1, 2]. After spatial discretization, the BFB model becomes a highly nonlinear and large scale differential and algebraic equation (DAE) system with over 10,000 equations. Considering the strong nonlinearity of the process, NMPC is used to efficiently control the carbon capture system.

Detailed first-principles BFB model can be applied for off-line optimization applications and give accurate description of the process behavior. For online control problem like NMPC, however, it requires a fast solution of dynamic optimization problems. In practice, solving NMPC problem with the original full scale BFB model can be time-consuming, which may lead to computational delay that will degenerate control performance and even destabilize the process. To reduce the computational cost for NMPC, dynamic reduced models of the BFB adsorber are developed using temporal and spatial reduction methods. From a temporal aspect, dynamic reduced models are generated using nullspace projection and eigenvalue analysis method to remove the fast dynamics of the system. From a spatial aspect, dynamic reduced models are developed using orthogonal collocation to reduce the size of the full scale model. The simulation time of the reduced dynamic models can be reduced by 70% while maintaining good prediction accuracy.

To enable online control and avoid computational delay, asNMPC is applied using the fast dynamic reduced models for the BFB adsorber, which is a sensitivity-based fast NMPC algorithm [3]. The basic idea of asNMPC is to separate optimization problems into background and on-line calculations; the open loop NMPC optimization problem is solved within the sampling time using the predicted state as initial conditions and then a fast update of NLP solution is made based on NLP sensitivity once the actual state of the system is obtained. The online update only requires a single backsolve with fixed KKT matrix and thus it greatly reduces the online computational cost. To handle the uncertainty due to reduced model error and measurement noise, a formulation of moving horizon estimation (MHE) with output and state disturbances is used to reject the disturbances and achieve offset-free control behavior [4]. After solving the modified MHE problem, both the optimal estimated states and the state and output disturbances are obtained. The predicted disturbances are then fed into NMPC to compensate for the uncertainty brought by model-plant mismatch and other disturbances. The proposed strategy can achieve good control performance in set point tracking under disturbances. Economic NMPC with reduced models will be studied to further improve the overall economic performance of the carbon capture system.

References

[1] Lee, Andrew, and David C. Miller. "A one-dimensional (1-d) three-region model for a bubbling fluidized-bed adsorber." *Industrial & Engineering Chemistry Research* 52.1 (2012): 469-484.

[2] Modekurti, Srinivasarao, Debangsu Bhattacharyya, and Stephen E. Zitney. "Dynamic Modeling and Control Studies of a Two-Stage Bubbling Fluidized Bed Adsorber-Reactor for Solid–Sorbent CO2 Capture." *Industrial & Engineering Chemistry Research* 52.30 (2013): 10250-10260.

[3] Zavala, Victor M., and Lorenz T. Biegler. "The advanced-step NMPC controller: Optimality, stability and robustness." *Automatica* 45.1 (2009): 86-93.

[4] Huang, Rui, Lorenz T. Biegler, and Sachin C. Patwardhan. "Fast offset-free nonlinear model predictive control based on moving horizon estimation." *Industrial & Engineering Chemistry Research* 49.17 (2010): 7882-7890.

424202 Simultaneous Multi-Parametric Hybrid Model Predictive Control and Estimation with Application to the Intravenous Anaesthesia

Abstract:

Simultaneous Multi-Parametric Hybrid Model Predictive Control and Estimation with Application to the Intravenous Anaesthesia

Ioana Nascu[a,b], Richard Oberdieck[a,b], Efstratios N. Pistikopoulos[b]*

[a]Dept. of Chemical Engineering, Centre for Process Systems Engineering (CPSE), Imperial College London SW7 2AZ, Lodnon, U.K

[b]Artie McFerrin Department of Chemical Engineering, Texas A&M, College Station TX (*stratos@tamu.edu)

Keywords: Anaesthesia, Explicit/Multi-Parametric Model Predictive Control, Hybrid Systems, Moving Horizon Estimation

The control of biomedical systems has become a very important and challenging research area during the last decades. One of the important challenges in controlling such systems is the presence of strong nonlinearities. These nonlinearities are included in the pharmacodynamic model, part of the mathematical model of the system which is usually represented by the Hill curve and it describes the relation between the concentration of the drug and the effect observed on the patient. Due to the S – shape characteristic of the Hill curve, a piecewise linearization around three points can be performed, resulting in a piece-wise affine formulation. The control of such a system translates in a hybrid model predictive control (hMPC) problem (Bemporad and Morari 1999) and the solution of a mixed-integer quadratic programming (MIQP) problem formulation. Since the online implementation of hMPC needs the online solution of an MIQP problem, this will introduce a high computational burden. When dealing with biomedical systems a fast control action is of great importance since the lack of it can affect the patient's safety. A way to deal with this issue is by solving the hMPC problem explicitly offline via the solution of a multi-parametric mixed-integer quadratic programming (mp-MIQP) problem where the initial states are treated as parameters and the optimization problem is solved as a function thereof (Pistikopoulos 2009). For the solution of such problems, recently novel solution procedures have been developed (Oberdieck and Pistikopoulos 2014).

The implementation of hybrid explicit/multi-parametric MPC, and in general, MPC is that it is based on the assumption that accurate, noise-free information is available for all states of the system. However, in reality, state information needs to be inferred from the available, likely noisy, output measurements using a state estimator. Dealing with system constraints entails the simultaneous use of moving horizon estimation (MHE) which can be implemented in a multi-parametric fashion (Darby and Nikolaou 2007, Voelker, Kouramas et al. 2010). By using mp-MHE we will be able to estimate the state of each individual patient when dealing with biomedical processes. This is very important since it will be able to overcome the challenge of inter- and intra- patient variability that is found in such systems. Since most biomedical process can be treated as a hybrid systems due to the piece affine formulation of the Hill function, the mp-MHE will be reformulated into a novel hybrid mp-MHE formulation that will be implemented simultaneously with hybrid multi parametric model based predictive control.

This new development is integrated in PAROC, a comprehensive step-by-step framework and software solution for the general design, operational optimization and control of process systems (Pistikopoulos, Diangelakis et al. 2014). The framework includes a high-fidelity model of the system that is approximated if the model is too complicated using discrete time models in state space form via model order reduction techniques or system identification techniques. The system model in the state space form is used to formulate an optimization problem subject to the state space model and constraints. The state space model used by the hybrid mp-MPC controller is estimated using the newly developed hybrid mp-MHE. The resulting hybrid multi-parametric programming problem is solved with state-of-the-art techniques and the solution is validated against the original high fidelity model.

In this paper the presented framework is applied on the intravenous anaesthesia process. Intravenous is an important part of surgery and intensive care unit and it represents the reversed pharmacological state of the patient where hypnosis, analgesia and muscle relaxation are guaranteed (Bailey and Haddad 2005). During the last decades, there has been a lot of research on the development of PID tuning techniques, model based strategies using predictive, (Ionescu, Machado et al. 2014), (Hodrea, Morar et al. 2012), robust (Caiado Daniela, Lemos João et al. 2013), adaptive (Haddad, Hayakawa et al. 2003), (Nascu, Nascu et al. 2012) and multi-parametric MPC (Nascu, Lambert et al. 2014), (Krieger and Pistikopoulos 2014) for the control of depth of anaesthesia. Nevertheless, control of anaesthesia, is still not a trivial task and poses a variety of challenges that need to be addressed; from the presence of nonlinearities, inter- and intra- patient variability, multivariable characteristics, variable time delays, model analysis variability.

In order to control the depth of anaesthesia and for the first step of the framework, a model that captures the dynamical response of the patient is required. The PK-PD models most commonly used for Propofol are the 4th order compartmental model described by (Schnider, Minto et al. 1998), (Minto, Schnider et al. 1997). The pharmacokinetic (PK, the linear part) – pharmacodynamic (PD, the nonlinear part) blocks denote compartmental models, used to represent the distribution of drugs in the body, i.e. mass balance. In each compartment the drug concentration is assumed to be uniform, assuming perfect and instantaneous mixing. The transport rate of the drug leaving the compartment is assumed to be proportional to the drug concentration. Once we have the state space of the model, the next step of the framework will be focused in designing the simultaneous hybrid control and estimation and their solution via mp MIQP. For this study, the simulations are performed on a set of 12 virtual generated patients under constraints and the presence of noise in the induction and maintenance phase. The performances of the controller are evaluated for the whole set of patients and under the effect disturbances and the existence of constraints. The implementation of this framework on the intravenous anaesthesia process represents an important step in the control of DOA since is able to deal with two major challenges in the process of anaesthesia: the nonlinearity by using the piece-wise linearization of the Hill curve and the inter and intra patient variability by using estimation techniques to obtain the states of each individual patients. Another advantage of the simultaneous hybrid mp-MPC and hybrid mp-MHE is that is able to overcome the noise that can corrupt the required data.

References

Bailey, J. M. and W. M. Haddad (2005). "Drug dosing control in clinical pharmacology." IEEE Control System Magazine **25**: 35-31.

Bemporad, A. and M. Morari (1999). "Control of systems integrating logic, dynamics, and constraints." Automatica **35**(3): 407-427.

Caiado Daniela, V., M. Lemos João and A. Costa Bertinho (2013). Robust control of depth of anesthesia based on H∞ design. Archives of Control Sciences. **23:** 41.

Darby, M. L. and M. Nikolaou (2007). "A parametric programming approach to moving-horizon state estimation." Automatica **43**(5): 885-891.

Haddad, W. M., T. Hayakawa and J. M. Bailey (2003). Nonlinear Adaptive Control for Intensive Care Unit Sedation and Operating Room Hypnosis.

Hodrea, R., R. Morar and I. Nascu (2012). "Predictive Control of Neuromuscular Blockade." Automation Computers Applied Mathematics **21**(4).

Ionescu, C., J. T. Machado, R. De Keyser, J. Decruyenaere and M. M. R. F. Struys (2014). "Nonlinear dynamics of the patient's response to drug effect during general anesthesia." Communications in Nonlinear Science and Numerical Simulation(0).

Krieger, A. and E. N. Pistikopoulos (2014). "Model predictive control of anesthesia under uncertainty." Computers and Chemical Engineering **71**: 699-707.

Minto, C. F., T. W. Schnider, T. D. Egan, E. J. Youngs, H. J. Lemmens, P. L. Gambus, V. Billard, J. F. Hoke, K. H. Moore, D. J. Hermann, K. T. Muir, J. W. Mandema and S. L. Shafer (1997). "Influence of age and gender on the pharmacokinetics and pharmacodynamics of reminfetanil. I. Model development." Anesthesiology **86**: 10-23.

Nascu, I., R. S. C. Lambert, A. Krieger and E. N. Pistikopoulos (2014). Simultaneous multi-parametric model predictive control and state estimation with application to distillation column and intravenous anaesthesia. Computer Aided Chemical Engineering. **33:** 541-546.

Nascu, I., I. Nascu, C. M. Ionescu and R. De Keyser (2012). Adaptive EPSAC predictive control of the hypnotic component in anesthesia.

Oberdieck, R. and E. Pistikopoulos (2014). "Explicit hybrid model-predictive control: The exact solution." Automatica **submitted**.

Pistikopoulos, E., N. A. Diangelakis, R. Oberdieck, M. M. Papathanasiou, I. Nascu and M. Sun (2014). "PAROC - an Integrated Framework and Software Platform for the Design, Operational Optimization and Advanced Model-Based Control of Process Systems." Chemical Engineering Science **submitted**.

Pistikopoulos, E. N. (2009). "Perspectives in multiparametric programming and explicit model predictive control." AIChE Journal **55**: 1918-1925.

Schnider, T. W., C. F. Minto, P. L. Gambus, C. Andresen, D. B. Goodale, S. L. Shafer and E. J. Youngs (1998). "The influence of method of administration and covariates on the pharmacokinetics of propofol in adult volunteers." Anesthesiology **88**(5): 1170-1182.

Voelker, A., K. Kouramas and E. N. Pistikopoulos (2010). Simultaneous constrained moving horizon state estimation and model predictive control by multi-parametric programming. Proceedings of the IEEE Conference on Decision and Control.

427702 Model-Based Optimal Feedback Control of Colloidal Self-Assembly

Abstract:

The ability to control the structure of a material at micro- and nano-meter scales could enable the design of new material properties. For example, when micron-scale colloidal particles are arranged in a regular crystalline structure, they interact with light in novel ways, for use in optoelectronic devices [1]. However, the rapid manufacture of defect-free materials over large length scales is challenging. The crystalline lattice is the low energy thermodynamic ground state, but in a finite time assembly process, defects are created stochastically and can be locked in. This tradeoff between thermodynamics and kinetics could potentially be circumvented using dynamic processing inputs and feedback control. Defects could be corrected in their early stages as they begin to form, by temporarily lowering the attractive potential that drives the assembly. Once the defect is healed, the voltage can be raised again to further drive the assembly.

Here we demonstrate feedback control in a colloidal assembly process consisting of about 300 micron-size particles. The assembly is driven by an applied electric field, which creates dielectrophoretic forces on the particles. Real-time optical microscopy provides the information for feedback, by tracking the individual particle locations throughout the process [2]. A Brownian dynamics model and simulation describes the forces on each particle, and is used to generate a reduced-order Markov state model for a select number of discrete voltage inputs. This Markov state model is then used to calculate the optimal feedback policy via dynamic programming, specifically through the framework of the Markov decision process [3]. This policy was implemented experimentally, and in a period of 1000 s, it was able to achieve defect-free assembly in 100 out of 100 cases, compared to 70/100 with a constant voltage input.

[1] K. A. Arpin, A. Mihi, H. T. Johnson, A. J. Baca, J. A. Rogers, J. A. Lewis, and P. V. Braun, *Multidimensional Architectures for Functional Optical Devices.* Adv. Mater., 2010. **22**(10): p. 1084-1101.

[2] J. J. Juarez and M. A. Bevan, *Feedback Controlled Colloidal Self-Assembly.* Adv. Funct. Mater., 2012. **22**(18): p. 3833-3839.

[3] Y. Xue, D. J. Beltran-Villegas, X. Tang, M. A. Bevan, and M. A. Grover, *Optimal Design of a Colloidal Self-Assembly Process,* IEEE Trans. on Cont. Sys. Tech., 2014. **22**(5): p. 1956-1963.

427798 Optimal Control of a Fed-Batch Bioreactor for Maximized Carotenoids Productivity

Abstract:

Carotenes such as β-carotene have a positive impact in human health as a precursor of vitamin A. Structural complexity of these compounds makes their chemical synthesis a difficult endeavor, facilitating the need for their biological production. Fed-batch operation is commonly used in these processes to allow for increased cell viability through the mitigation of substrate and product inhibition, thereby increasing the productivity of the carotenoid product[1,2]. This work looks to examine the operating conditions necessary to achieve the highest productivity of the desired carotenoid products when produced via fed-batch operation. First, a two-level parameter estimation method is utilized to develop a reliable kinetic model for the batch production of carotenoids via fermentation to describe the glucose consumption, metabolic product formation and depletion, and the carotenoid production in the *Saccharomyces cerevisiae* strain mutant SM14 with 20 g/L glucose as the carbon source. These models are then extended to the study of a novel fed-batch system giving the ability to use flowrate and concentration of the feed stream to act as control variables. Dynamic optimization techniques are used to determine the optimal control actions governing the dilution rate and glucose concentration in order to minimize the time necessary to maximize the carotenoid productivity in the fed-batch operation. Additional constraints on the feed flowrate and glucose concentration are considered for this finite-horizon optimal control problem.

1. Colet, R., Luccio, M. D. & Valduga, E. Fed-batch production of carotenoids by *Sporidiobolus salmonicolor* (CBS 2636): kinetic and stoichiometric parameters. *Eur Food Res Technol* **240,**173–182 (2014).
2. Ho, K., Tam, C. & Zhou, B. Growth and carotenoid production of Phaffia Rhodozyma in fed-batch cultures with different feeding methods. *Biotechnology Letters* **21,** 175–178 (1999)
3. Biegler, L. T., Cervantes, A. M. & Wächter, A. Advances in simultaneous strategies for dynamic process optimization. *Chemical Engineering Science* **57,** 575–593 (2002).
4. Senger, R. S. & Karim, M. N. Optimization of fed-batch parameters and harvest time of CHO cell cultures for a glycosylated product with multiple mechanisms of inactivation. *Biotechnol. Bioeng.* **98,** 378–390 (2007).

434008 Glucose Control in the Intensive Care Unit: Controller Design and Alarm Layer Performance Evaluation *in silico*

Tuesday, November 10, 2015: 5:03 PM
Salon D (Salt Lake Marriott Downtown at City Creek)
Timothy Knab[1], Gilles Clermont[2] and Robert S. Parker[1], (1)Chemical and Petroleum Engineering, University of Pittsburgh, Pittsburgh, PA, (2)University of Pittsburgh

As a result of serious injury or disease, patients admitted to critical care units often present with elevated blood glucose levels, a condition commonly referred to as stress hyperglycemia[1,2,3]. It has been shown[3] that a reduction in hyperglycemia through the use of intensive insulin administration to achieve tight glucose control (80-130 mg/dL) may have the potential to significantly reduce morbidity and mortality. However, special care must be taken to avoid hypogylcemia, as even a single hypoglycemic event has the capacity to eliminate any benefit of targeted glucose control (TGC)[4,5]. The careful balance between hyperglycemia reduction and hypoglycemia prevention that must be struck in a clinical setting requires careful and frequent measurement of blood glucose levels at a rate that quickly becomes burdensome on clinical staff and is untenable in a long-term, large-scale deployment.

Commercially available continuous glucose monitors (CGMs) can provide high-frequency measurements of glucose levels allowing for automated insulin and glucose delivery via a model-based control scheme, thereby alleviating the clinical burden associated with TGC. Furthermore, model-based methods can address issues of inter- and intra-patient variability using state and parameter estimation to provide superior blood glucose management and improved patient outcomes. Presented here is the development of error models for testing a proposed control algorithm[6] acting on virtual patients with simulated sensors exhibiting noise characteristics statistically similar to those observed in critical care CGM use. Additionally, the formulation of a system of alarms and fail-safes designed to assess sensor and controller performance and tighten or loosen controller constraints according to a rule-based scheme is detailed and evaluated.

We conducted an IRB-approved observational study at the University of Pittsburgh Medical Center (UPMC) where data was collected in duplicate via two subcutaneous Dexcom ®; G4 PLATINUM™ continuous glucose monitors in a cohort of 24 patients during their stay in the intensive care unit. This data provides the basis for the development of CGM error models and the fitting of virtual patients for in silico controller testing.

Simulated sensor error profiles are generated through sampling and integration of changes in CGM error, $\frac{dE}{dt}$. Stochastic trajectories of change in error are generated through sampling from representative distributions of $\frac{dE}{dt}$ compiled from the aggregation of trajectories across all patients. Integration of error derivatives is used here to generate simulated sensor error profiles as integration provides a source of continuity in the error profile that we believe to be more characteristic of CGM error and is able to recapitulate drift phenomena that random sampling of absolute errors are unable to accurately capture.

Patient-specific trajectories of $\frac{dE}{dt}$ are generated by first reconstructing a "true" blood glucose profile from which, the derivative of error can be calculated. From the CGM data in duplicate, the reconstruction is synthesized as a linear combination of the CGM signals with variable weights. Weights are regularized to control variance and ensure that the composite reconstruction approximates a simple mean. Regularization parameters are chosen such that the composite reconstruction lies within the published error bands of blood glucose measurements taken using a capillary finger-stick glucose meter.

An alternative method of error reconstruction uses a high-pass filter to extract sensor noise from physiologic changes in glucose. Physiologic glucose dynamics are slow compared to the sampling period (5 minutes), so high frequency changes in the CGM signal are likely due to CGM noise and error processes. High frequency components are extracted from the differenced CGM signal using a high-pass filter and taken as $\frac{dE}{dt}$, providing a single-sensor alternative to the reconstruction method. Error profiles using either this high-pass filtering methodology or the previously discussed reconstruction method will be generated and sampled at points corresponding to the availability of capillary finger-stick measurements in the collected patient CGM data to evaluate statistical consistency of the proposed models with clinical observation.

Virtual patients are created by fitting the Intensive Control Insulin-Nutrition-Glucose (ICING) model[7], coupled to a subcutaneous insulin model[8], to composite reconstructions of blood glucose, constructed using the weighted linear combination of CGM signals discussed above. Virtual patient fitting seeks to minimize sum squared error (SSE) between model predictions and the composite signal using insulin sensitivity (S_I) as a bounded time-varying regression parameter. Regularization is used to ensure bounded, smooth and relatively slowly changing S_I profiles consistent with physiological response.

Virtual patients are placed under the care of a previously developed[6] dual-input (insulin and glucose) zone-control MPC control algorithm, which uses moving horizon estimation (MHE) to provide an estimate of the current state of the controller model. Feedback to the controller is blood glucose as measured in duplicate by two CGMs. In these in silico trials, the simulated blood glucose signal is corrupted by the previously discussed models of sensor error. Using simulated patients and sensor noise, the controller and state estimator are tuned to filter noise and minimize performance loss compared to the unrealistic situation in which CGMs produce perfect measurements.

DSS alarms and fail-safes are developed at multiple points within the proposed control algorithm. At the sensing level, deviations both between individual CGMs and between CGMs and MHE estimated glucose are used to evaluate sensor performance, sensor faults and modify constraints according to a set of rules. A voting system is used to identify the location of single-sensor faults, and faulty measurements are ignored until either the malfunctioning sensor comes back into agreement with the remaining sensor and the estimator or an alarm is sounded. In cases of acceptable CGM-CGM and CGM-estimator consistency, sensor confidence will be assigned a rule-based performance classifier, such as excellent, good or acceptable, based on a critical mean absolute relative deviation over past data of 13% – the maximum error at which the benefits of treating using a continuously available glucose measurement continues to outweigh the potentially detrimental effects of treatment based on an erroneous signal[9]. Control action will be allowed at the maximum extent of the constraints during conditions of excellent performance and tightened otherwise.

At every time point the estimator a posteriori covariance matrix is updated according to the traditional Kalman update formula and the solution to the discrete algebraic Riccati equation[10] and can be used to provide evaluation criteria of estimator performance. If the range of blood glucose is divided into hyper-, hypo- and normoglycemia regimes and a single standard deviation in blood glucose, taken from the covariance matrix, spans multiple regimes, controller constraints are tightened to mitigate unnecessary or dangerous infusions of glucose or insulin potentially resulting in otherwise avoidable hypo- or hyperglycemia. In cases where estimates differ significantly from self-consistent CGM measurements, an alarm to clinicians is raised.

These limitations on control action are proposed to ensure patient safety even in the face of uncertain glucose measurements or state estimates, as large control actions calculated from an inaccurate state-space pose significant risk to a patient. Care is taken to choose thresholds, regimes and alarm criteria such that the control algorithm can cope with uncertainty and single-sensor failure, and only in cases where clinical intervention is expected to be required by the DSS must an alarm be raised. This design is an attempt to build-in robustness and fault tolerance such that alarms requiring immediate attention are minimized in an effort to mitigate control fatigue among clinicians. Alarm layers are tested and tuned using virtual patients with modeled CGM error, as well as unmeasured disturbances such as unannounced meals, edema caused by sensor implantation, and pressure-induced sensor fluctuation, in order to demonstrate that patient safety can be enhanced through the use of the DSS system. References

1. McCowen, K. C., Malhotra, A., and Bistrian, B. R. Critical Care Clinics 17(1), 107–124 January (2001).

2. Capes, S. E., Hunt, D., Malmberg, K., and Gerstein, H. C. The Lancet 355(9206), 773–778 March (2000).

3. Van den Berghe, G., Wouters, P., Weekers, F., Verwaest, C., Bruyninckx, F., Schetz, M., Vlasselaers, D., Ferdinande, P., Lauwers, P., and Bouillon, R. New England Journal of Medicine 345(19), 1359–1367 November (2001).

4. Finfer, S., Chittock, D. R., Su, S. Y.-S., Blair, D., Foster, D., Dhingra, V., Bellomo, R., Cook, D., Dodek, P., Henderson, W. R., Hébert, P. C., Heritier, S., Heyland, D. K.,

McArthur, C., McDonald, E., Mitchell, I., Myburgh, J. A., Norton, R., Potter, J., Robinson, B. G., and Ronco, J. J. The New England journal of medicine 360(13), 1283–97 March (2009).

5. Brunkhorst, F. M., Engel, C., Bloos, F., Meier-Hellmann, A., Ragaller, M., Weiler, N., Moerer, O., Gruendling, M., Oppert, M., Grond, S., Olthoff, D., Jaschinski, U., John, S., Rossaint, R., Welte, T., Schaefer, M., Kern, P., Kuhnt, E., Kiehntopf, M., Hartog, C., Natanson, C., Loeffler, M., and Reinhart, K. New England Journal of Medicine 358(2), 125–139 January (2008).

6. Knab, T., Clermont, G., and Parker, R. S. In Proceedings of ADCHEM 2015 (ADCHEM 2015, Whistler, BC, 2015).

7. Lin, J., Razak, N. N., Pretty, C. G., Le Compte, A., Docherty, P., Parente, J. D., Shaw, G. M., Hann, C. E., and Geoffrey Chase, J. Computer methods and programs in biomedicine 102(2), 192–205 May (2011).

8. Vilkhovoy, M., Pritchard-Bell, A., Clermont, G., and Parker, R. 10988–10993, August (2014).

9. Boyd, J. C. and Bruns, D. E. Clinical Chemistry 60(12), 1463–1465 December (2014).

10. Anderson, B. D. O. and Moore, J. B. Optimal Filtering. Dover Publications, Mineola, NY, January (2005).

434744 Nonlinear Model Predictive Control of Managed Pressure Drilling Based on Hammerstein-Wiener Piecewise Linear Models

Abstract:

Automation systems are increasingly important in many industries, especially those in competitive scenarios that encourage improved economic margins and reduced risk. Automation systems not only improve safety and convenience but also, perhaps most beneficially, enable optimization strategies that are less consistent with manual operation. The oil well drilling industry is transitioning to automation systems like the refining and petrochemical industry. The rapidly changing crude oil market combined with developing alternative energy sources creates renewed interest in drilling automation. Recent outstanding breakthroughs in the drilling industry give many opportunities to apply a variety of automation and optimization strategies. Most of all, managed pressure drilling (MPD) opens the way to control the drilling process in a closed loop fashion. However, the inherent nonlinearity and semi-batch nature of the drilling process gives many challenges to apply an automation strategy.

Typically, two approaches are used to develop a nonlinear model predictive control (NMPC) schemes. The first approach is based on first principles models that describe the fundamental physics of the process. The second approach employs a derivative form of empirical models, termed "black box models", identified from the existing process data. While first principles models describe the process dynamics in more detail, they normally require more computational time to solve the complicated fundamental equations [1]. Computation time of solving first principles models might not be adequate for control purposes in real time applications. Many previous research studies in drilling automation have established low order first principles models by capturing the main dynamics of the drilling process [2-5]. Although using low order models reduces the computational time significantly, those models use several variables that are not measured and must be estimated. In contrast with first principles models, black box models retain the all advantages of linear MPC that is widely used in many industries [6]. Hammerstein-Wiener models are the most widely implemented method of black box nonlinear models in industry [7].

This work discusses the development of Hammerstein-Wiener based NMPC for the drilling process. The Hammerstein-Wiener models employ a piecewise linear function to process the input value. These Hammerstein-Wiener models apply a nonlinear static element and linear dynamic element, allowing a gain-scheduling concept for nonlinear control [7]. The control performance and computation time of the Hammerstein-Wiener based NMPC is compared to the typical lower order first principles NMPC in various scenarios that frequently occur in drilling operations, such as unexpected gas influx from the formation due to a drill pipe connection procedure. The case study is verified with a high fidelity drilling simulator under certain well conditions [8] shown in Table 1.

Table1. Well conditions of the WeMod (high-fidelity hydraulic simulator)

Parameter	Value
Well depth	11,800 ft
Riser inner diameter	19''
Water depth	590 ft
Casing inner diameter	9''
Casing depth	7,100 ft
Drill string average outer diameter	4.5''
BHA length	150 ft
BHA average outer diameter	6.7''
Open hole/bit size	8.5''
Reservoir depth	9840 ft
Reservoir Pore Pressure	401.0 bar/1.364 s.g.
Initial mud density	1.24 s.g.

This investigation also details strategies to estimate downhole pressure. Previous research studies consider the automation strategy assuming that reliable downhole measurements from wired drill pipe (WDP) technology are available [9]. However, the majority of drill rigs are still operating without WDP. In order to apply the control system to a majority of drilling rigs in current operation, it is necessary to establish the framework to estimate the downhole properties instead of directly measuring the value. In order to estimate the downhole pressure, a moving horizon estimator (MHE) and a lower order model are employed.

References

1. Rohit S. Patwardhan, S.L.a.S.L.S., *Nonlinear Model Predictive Contol using PLS based Hamerstein and Wiener Models.* AICHE, 1997.

2. Nygaard, G. and G. Nvdal, *Nonlinear model predictive control scheme for stabilizing annulus pressure during oil well drilling.* Journal of Process Control, 2006. **16**(7): p. 719-732.

3. Siahaan, H.B. and G. Nygaard. *On modeling1 and observer design of fluid flow dynamics for petroleum drilling operations*. in *47th IEEE Conference on Decision and Control, CDC 2008, December 9, 2008 - December 11, 2008.* 2008. Cancun, Mexico: Institute of Electrical and Electronics Engineers Inc.

4. Nygaard, G. and G. Nvdal. *Modelling two-phase flow for control design in oil well drilling*. in *2005 IEEE International Conference on Control Applications, CCA, August 28, 2005 - August 31, 2005*. 2005. Toronto, ON, Canada: Institute of Electrical and Electronics Engineers Inc.

5. Kaasa, G.-O., *A simple dynamic model of drilling for control*, in *StatoilHydro Research Centre* 2007: Porsgrunn.

6. Ania Lusso´ n Cervantes, O.E.A., *, Jose´ L. Figueroa, *A nonlinear model predictive control system based on Wiener piecewise linear models.* Journal of Process Control, 2002. **13**: p. 655-666.

7. G. Shafiee, M.M.A., M.R. Jahed-Motlagh, A.A. Jalali, *Nonlinear predictive control of a polymerization reactor based on piecewise linear Wiener model.* Chemical Engineering Journal, 2008. **143**.

8. Gravdal, J.E., *Real-Time Evaluatioon of Kick During Managed Pressure Drilling Based on Wired Drill Pipe Teemetry.* SPE Journal, 2009. **IPTC 13959**.

9. Asgharzadeh Shishavan, R., Hubbell, C., Perez, H.D., Hedengren, J.D., and Pixton, D.S., *Combined Rate of Penetration and Pressure Regulation for Drilling Optimization Using High Speed Telemetry.* SPE Drilling and Completions 2015. **30**(1): p. 17-26.

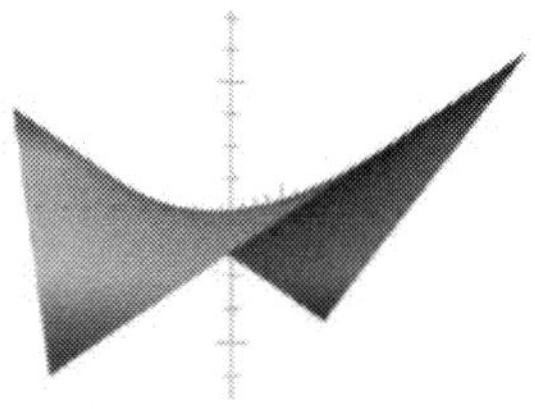

Closing the Gap Between Multiparametric Disaggregation and Piecewise McCormick Relaxations for MIQCPs

Pedro M. Castro (pmcastro@fc.ul.pt)
http://www.researcherid.com/rid/C-3642-2008

Associate Researcher
CMAFCIO
Faculdade de Ciências
Universidade de Lisboa

IF INVESTIGADOR FCT

U LISBOA UNIVERSIDADE DE LISBOA

Problem definition (MINLP)

- Mixed-integer bilinear program (MIQCP)

$$\min f_0(\boldsymbol{x},\boldsymbol{y})$$

$$f_q(\boldsymbol{x},\boldsymbol{y}) \leq 0 \quad q \in Q\backslash\{0\}$$

$$f_q(\boldsymbol{x},\boldsymbol{y}) = \sum_{(i,j)\in BL} a_{ijq}x_ix_j + B_q\boldsymbol{x} + C_q\boldsymbol{y} + d_q \qquad q \in Q \qquad \textbf{(P)}$$

$$0 \leq \boldsymbol{x}^L \leq \boldsymbol{x} \leq \boldsymbol{x}^U$$

$$\boldsymbol{x} \in \mathbb{R}^m, \boldsymbol{y} \in \{0,1\}^r$$

Relaxation gives lower bound

$\boldsymbol{y} = \emptyset \Longrightarrow$ QCP (NLP)

$\boldsymbol{x}$ – continuous variables
$\boldsymbol{y}$ – binary variables
BL - (i,j) index set
$i \neq j$ – strictly bilinear problems
a_{ijq}, B_q, C_q, d_q - parameters

Motivation

- Problem (**P**) frequently encountered in engineering
  - Process networks, generalized pooling
    - Quesada & Grossmann 95
    - Floudas & co-workers 06,11
- Wastewater treatment networks
  - Galan & Grossmann 98
    - Design for steady state (NLP)

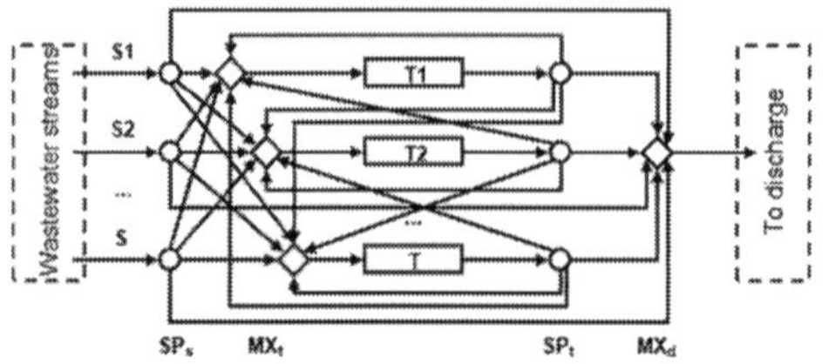

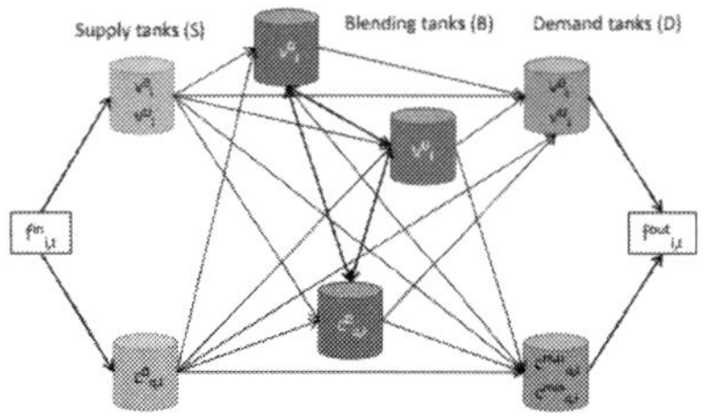

- Multiperiod blending in oil refineries
  - Kolodziej et al. 13
    - Scheduling component (MINLP)
- Main issue
  - Multiple input streams to a mixing node or tank
    - Characterized by flowrate & set of properties
  - Bilinear terms appear in mass balances
    - Need to estimate properties of output stream or tank blend

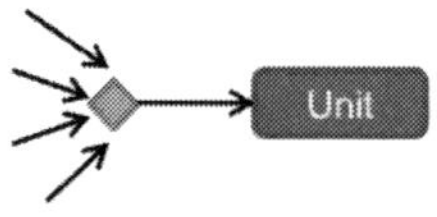

Introduction

- Problem (**P**) is non-convex ⇒ multiple local solutions
  - Gradient based solvers unable to certify optimality
- Need for global optimization algorithms
  - What do they have in common?
    - LP or mixed-integer linear (MILP) relaxation of (**P**) → LB
      - McCormick (76) envelopes (LP)
      - Piecewise McCormick envelopes (Grossmann & co-workers 05-06)
      - Multiparametric disaggregation (Castro & co-workers 13)
    - Solution of (**P**) with fast local solver → UB
      - Using LB as starting point

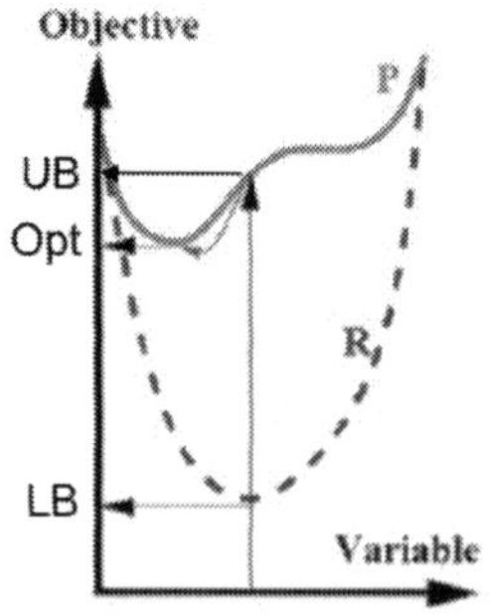

- Tight relaxation critical to ensure convergence
  - Relative optimality gap (UB-LB)/LB<ε

Piecewise McCormick relaxation $w_{ij} = x_i x_j$

- Domain of x_j divided into N uniform partitions

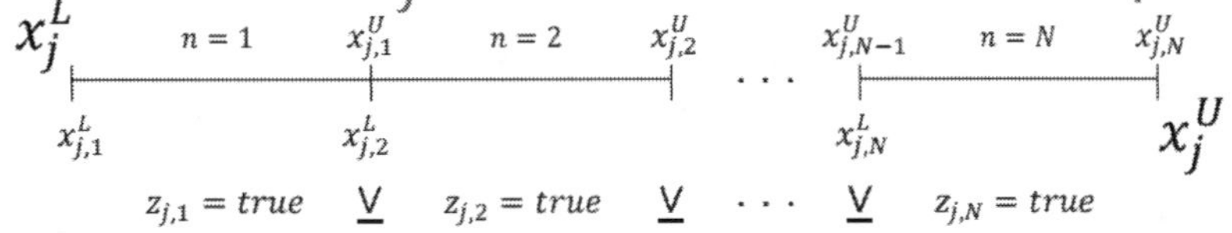

Optimization will pick single partition

Generalized Disjunctive Program (Balas 79; Raman & Grossmann 94)

$$\underline{\bigvee_n}\begin{bmatrix} Z_{jn} \\ w_{ij} \geq x_i \cdot x_{jn}^L + x_i^L \cdot x_j - x_i^L \cdot x_{jn}^L \\ w_{ij} \geq x_i \cdot x_{jn}^U + x_i^U \cdot x_j - x_i^U \cdot x_{jn}^U \\ w_{ij} \leq x_i \cdot x_{jn}^L + x_i^U \cdot x_j - x_i^U \cdot x_{jn}^L \\ w_{ij} \leq x_i \cdot x_{jn}^U + x_i^L \cdot x_j - x_i^L \cdot x_{jn}^U \\ x_{jn}^L \leq x_j \leq x_{jn}^U \end{bmatrix}$$

$$x_i^L \leq x_i \leq x_i^U$$
$$x_{jn}^L = x_j^L + (x_j^U - x_j^L) \cdot (n-1)/N$$
$$x_{jn}^U = x_j^L + (x_j^U - x_j^L) \cdot n/N$$

Convex hull

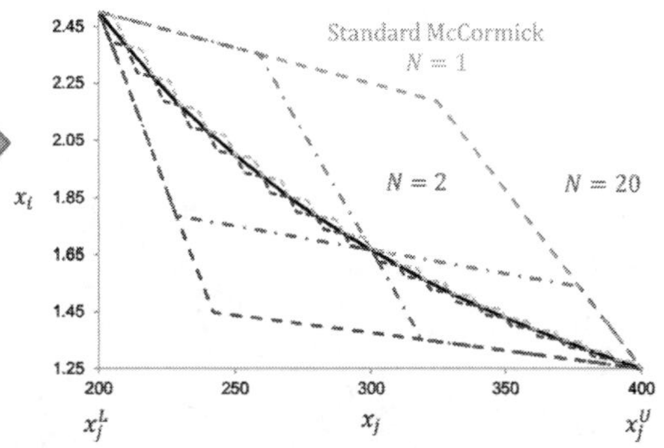

Normalized multiparametric disaggregation

- Main concept of new approach (NMDT)
  - Compute x_j as linear combination of its lower & upper bounds
    - $x_j = x_j^L + \lambda_j (x_j^U - x_j^L) \ \forall j$
  - Discretize $\lambda_j \in [0,1]$ using decimal representation system
    - Continuous domain (accuracy level p): $\lambda_j = \sum_{l=p}^{-1} \lambda_{jl} + \Delta\lambda_j \ \forall j$
      $0 \leq \Delta\lambda_j \leq 10^p \ \forall j$
  - Pick one digit $k \in \{0,1,\dots,9\}$ for every power l (Teles et al. 13)
    - $\bigvee_{k=0}^{9} \begin{bmatrix} z_{jkl} \\ \lambda_{jl} = 10^l \cdot k \end{bmatrix} \forall j, l \in \{p, \dots, -1\}$
- How to obtain $\lambda_j = 0.4385$ for $p = -2$?
  - $z_{j,4,-1} = 1, z_{j,3,-2} = 1, \Delta\lambda_j = 0.0085 \leq 0.01 = 10^{-2}$

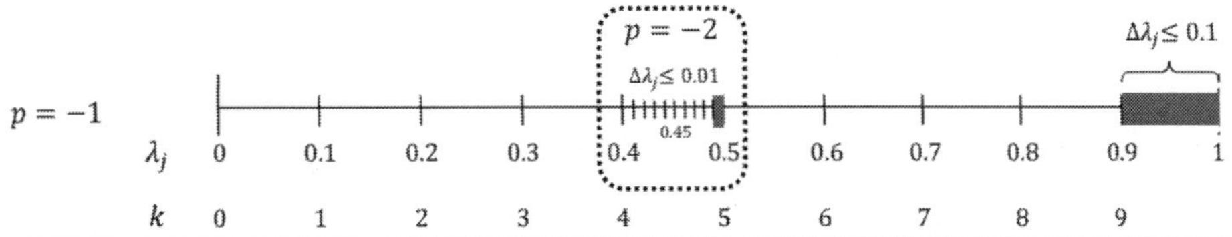

Remaining sets of constraints (MILP)

- Derivation of relaxation problem (**PR**)
  - Convex hull reformulation of disjunction (Balas 85)
    - $\lambda_j = \sum_{l=p}^{-1} \sum_{k=0}^{9} 10^l \cdot k \cdot z_{jkl} + \Delta\lambda_j \ \forall j$
    - $\sum_{k=0}^{9} z_{jkl} = 1 \ \forall j, l \in \{p, \dots, -1\}$
  - Relaxation of bilinear term
    - $w_{ij} = x_i x_j^L + v_{ij}(x_j^U - x_j^L) \ \forall (i,j)$
  - After exact linearization (Oral & Kettani 92)
    - $v_{ij} = \sum_{l=p}^{-1} \sum_{k=0}^{9} 10^l \cdot k \cdot \hat{x}_{ijkl} + \Delta v_{ij} \ \forall (i,j)$
    - $z_{jkl} x_i^L \le \hat{x}_{ijkl} \le z_{jkl} x_i^U \ \forall (i,j), k \in \{0, \dots, 9\}, l \in \{p, \dots, -1\}$
    - $x_i = \sum_{k=0}^{9} \hat{x}_{ijkl} \ \forall (i,j), l \in \{p, \dots, -1\}$
  - McCormick relaxation of new bilinear terms
    - $x_i^L \cdot \Delta\lambda_j \le \Delta v_{ij} \le x_i^U \cdot \Delta\lambda_j \ \forall (i,j)$
    - $(x_i - x_i^U) \cdot 10^p + x_i^U \cdot \Delta\lambda_j \le \Delta v_{ij} \le (x_i - x_i^L) \cdot 10^p + x_i^L \cdot \Delta\lambda_j \ \forall (i,j)$

$$\bigvee_{k=0}^{9} \begin{bmatrix} z_{jkl} \\ \lambda_{jl} = 10^l \cdot k \end{bmatrix}$$

$$x_j = x_j^L + \lambda_j (x_j^U - x_j^L) \ \forall j$$

$$w_{ij} = x_i x_j$$

$$v_{ij} = x_i \lambda_j$$

$$\Delta v_{ij} = x_i \cdot \Delta\lambda_j$$

Important properties

- Solution of (**PR**) yields a lower bound for (**P**)
  - $f_0^R = \min_{x,y} f_0^R(x,y) \le f_0 = \min_{x,y} f_0(x,y)$
- (**PR**) can be made as tight as desired
  - As p approaches $-\infty$, f_0^R approaches f_0
    - Outline of proof:
      - $\lim_{p \to -\infty} \Delta\lambda_j = \lim_{p \to -\infty} 10^p = 0$
      - $\lim_{p \to -\infty} \Delta v_{ij} = \lim_{p \to -\infty} (x_i^L \cdot \Delta\lambda_j) = \lim_{p \to -\infty} (x_i^U \cdot \Delta\lambda_j) = x_i^U \cdot \lim_{p \to -\infty} \Delta\lambda_j = x_i^U \cdot 0 = 0$
      - $l \in \{-\infty, \dots, -1\} = \mathbb{Z}^-$
      - (**PR**) becomes equivalent to (**P**)
- Optimal solution from (**PR**) provides good initialization point for the solution of (**P**)

Global optimization algorithm

- Solve (**PR**) and (**P**) in sequence until gap<ε
  - Step 0. $p := -1$, let upper bound $f_0^* = +\infty$.
  - Step 1. Solve (**PR**) to obtain f_0^R and point (x^R, y^R).
  - Step 2. Add constraint $y = y^R$, reducing (**P**) to an NLP. Using x^R as starting point, solve NLP with local solver to obtain upper bound f_0^* (if feasible).
  - Step 3. If $(f_0^* - f_0^R)/f_0^R \leq \varepsilon$, STOP, solution (x^*, y^*) is globally optimal. Else, $p := p - 1$, return to step 1.

Advantages of normalized MDT

- Double improvement in relaxation quality after bound contraction
  - Tighter variable bounds
  - More partitions for the same domain
    - Larger problem size compared to MDT
      - Tradeoff involved

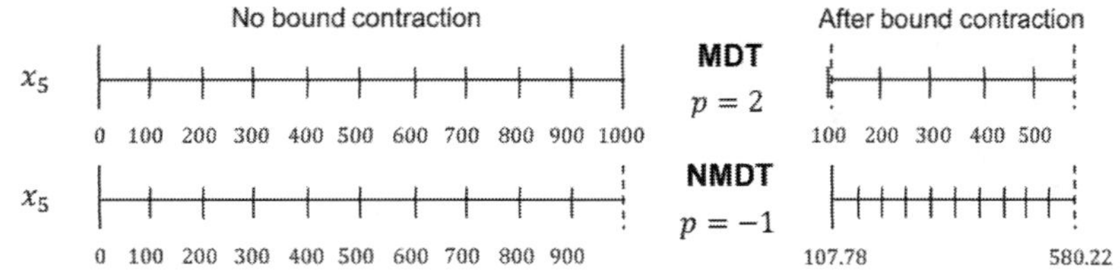

Preceded by bound contraction?		No	Yes	No	Yes	No	Yes
Approach	*p*	Optimality gap (%)		Binary variables		CPUs	
MDT	2	56.1	55.1	60	26	0.38	0.25
NMDT	-1	56.1	28.1	50	50	0.38	0.37
MDT	1	6.95	3.39	110	76	1.13	0.73
NMDT	-2	6.95	1.66	100	100	1.15	1.82
MDT	-2	0.0071	0.0026	260	226	1172	756
NMDT	-5	0.0071	0.0016	250	250	626	2736

Advantages of normalized MDT (2)

- Direct comparison with piecewise McCormick (PCM)
  - Relation between accuracy parameters: $N = 10^{-p}$
  - # binary variables per x_j
    - PCM: N
    - NMDT: $10 \log N$
  - Scales much better with problem size!
    - Relaxation not as tight for quadratic terms
- Which is the best for N=10?

Performance profiles (Dolan & Moré 02)

KPI: CPUs

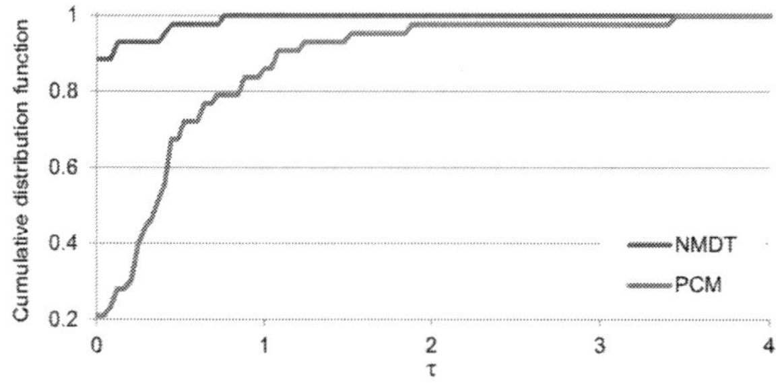

Numerical experiments

- 43 benchmark problems with strictly bilinear terms
  - NLPs
    - 18 water-using network design & 15 wastewater treatment design
  - MINLPs (www.minlp.org)
    - 7 multiperiod blending & 3 hydroelectric scheduling
- GAMS 24.3, Intel i7-4790 (3.6 GHz), 8 GB RAM
  - NMDT & PCM algorithms
    - CPLEX 12.6 (8 threads) (MILP)
    - CONOPT 3.16C (NLP)
  - Global optimization solvers
    - BARON 14.0.3 & GloMIQO 2.3
- Key performance indicators

Maximum resource limit=3600 CPUs (18,000 for hydro problems)
Relative optimality tolerance $\varepsilon=10^{-6}$

	Optimal solutions				
Algorithm	Proven	Found	Suboptimal solutions	No solutions	Best performer
NMDT	21	39	4	0	21
GloMIQO	23	39	4	0	18
BARON	17	41	2	0	2
PCM	12	34	8	1	2
				Total	43

Comparison to commercial solvers

- Performance profiles, optimality gap
  - NMDT obtains consistently lower gaps
    - Maximum=3.57% vs. 41.4% GloMIQO, 72.1% BARON
  - Clear ranking: NMDT>GloMIQO>BARON>PCM
    - Remains constant throughout x-axis

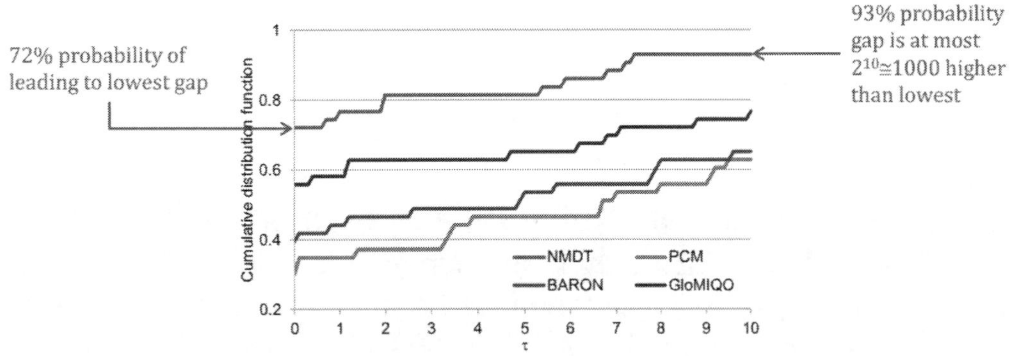

Integrating NMDT with spatial B&B

- Proven optimal solutions go from 21 to 33!
  - Better performance in NLPs than MINLPs
  - GloMIQO fastest at proving optimality
  - BARON excels in multiperiod blending problems

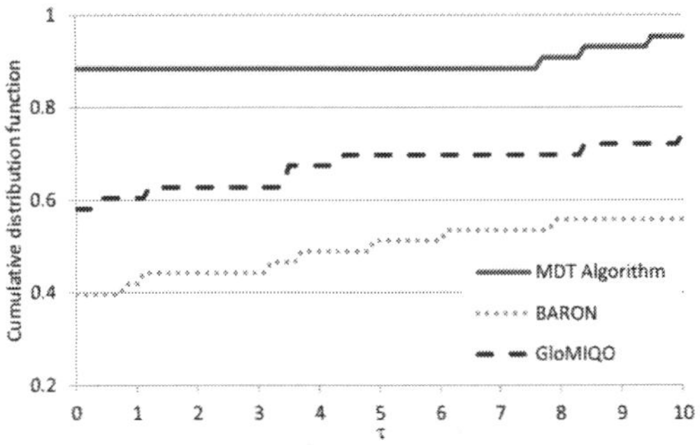

Optimality gap

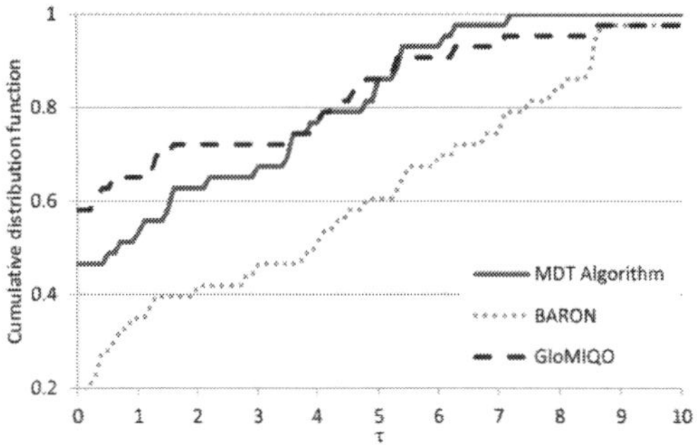

Computational time

Conclusions

- New MILP relaxation model for a class of MINLPs
  - Upgrade of multiparametric disaggregation technique
    - Discretizes variables up to chosen accuracy level
    - Better suited for integration with bound contracting procedures that are part of spatial B&B algorithms
  - As tight as piecewise McCormick for strictly bilinear terms
    - Significantly faster with increase in # partitions due to logarithmic vs. linear increase in problem size
  - Used to build rigorous global optimization algorithm
    - Performs significantly better than state-of-the-art solvers

- Further details
  - J. Glob. Optim. doi: 10.1007/s10898-015-0342-z
- Acknowledgments
  - Fundação para a Ciência e Tecnologia

429580 A New Portfolio of Relaxations for Global Optimization of NLPs and Minlps

Abstract:

Factorable programming techniques have been in the past in order to form convex relaxations of general global optimization problems [1, 2, 3]. Often, convex relaxations are further relaxed with polyhedral relaxations, thus facilitating outer approximation through efficient and reliable linear programming (LP) techniques [4, 5, 6]. In addition to LP relaxations, a non-linear programming (NLP) relaxation can be used for bounding. The use of hybrid LP and NLP relaxations was found beneficial for global optimization, especially in cases for which a problem becomes convex after branching at any given node of a branch-and-bound tree search [7].

In recent work, we incorporated mixed-integer linear programming (MILP) in the portfolio of relaxations of the BARON software [8]. MILP relaxations often result in much better lower bounds and feasible integer solutions, thus significantly reducing the number of nodes of the search tree for MINLPs. In this paper, we further extend BARON's portfolio of relaxations to include piecewise approximations of NLPs and MINLPs. Nonconvex functions can be approximated with piecewise linear outer-estimators, a process that leads to MILP relaxations even for purely continuous NLPs [9]. The main challenge in this context is how to balance complexity and tightness of a relaxation. On the one hand, MILP and NLP relaxations are tighter than their LP counterparts. However, this increased tightness comes with increased, often prohibitively expensive relaxation times. We present techniques for the automatic incorporation of piecewise linear relaxations as part of the portfolio of relaxations in BARON [4, 5, 6]. Key in our approach is the incorporation of learning mechanisms that balance accuracy and costs of relaxations. This process is dynamic, thus resulting in different types of relaxations being used for different problems, with the corresponding decision taking place during run time. We find that this approach is advantageous for NLPs and MINLPs with difficult continuous nonconvex components. We report extensive computational results with these new portfolios of relaxations in BARON on problems from a collection of publicly available test sets.

References

1. G. P. McCormick. Computability of global solutions to factorable nonconvex programs: Part I-Convex underestimating problems. Mathematical Programming, 10:147-175, 1976.
2. H. S. Ryoo and N. V. Sahinidis. Global optimization of nonconvex NLPs and MINLPs with applications in process design. Computers & Chemical Engineering, 19:551-566, 1995.
3. E. M. B. Smith and C. C. Pantelides. A symbolic reformulation/spatial branch-and-bound algorithm for the global optimisation of nonconvex MINLPs. Computers and Chemical Engineering, 23:457-478, 1999.
4. M. Tawarmalani and N. V. Sahinidis. Convexification and Global Optimization in Continuous and Mixed-Integer Nonlinear Programming: Theory, Algorithms, Software, and Applications. Kluwer Academic Publishers, Boston MA, 2002.
5. M. Tawarmalani and N. V. Sahinidis. Global optimization of mixed integer nonlinear programs: A theoretical and computational study. Mathematical Programming, 99:563-591, 2004.
6. M. Tawarmalani and N. V. Sahinidis. A polyhedral branch-and-cut approach to global optimization. Mathematical Programming, 103:225-249, 2005.
7. A. Khajavirad and N. V. Sahinidis. A hybrid LP/NLP paradigm for global optimization relaxations. Mathematical Programming Computation. Under revision.
8. M. Kilinc and N. V. Sahinidis. Global Optimization of Mixed-Integer Nonlinear Optimization Problems. AIChE meeting, 2014.
9. M. L. Bergamini, P. Aguirre, and I. E. Grossmann. Logic-based outer approximation for globally optimal synthesis of process networks. Computers & chemical engineering, 29(9):1914-1933, 2005.

420194 Convex Envelopes of Product-Separable Edge-Concave Functions for Global Optimization

Abstract:

Convex underestimators are utilized by deterministic global optimization algorithms to relax nonconvex functions. The convergence of these algorithms relies heavily on the tightness of underestimators relative to the original nonconvex functions. We focus on a broad class of functions which have a vertex polyhedral convex envelope over a polytope domain and thus have exploitable special structure with a finite generating set [1,2]. A quite general sufficient condition for a function to have a vertex polyhedral convex envelope is for it to be edge-concave, that is, for it to be concave along the edge directions of the domain [3,4]. This class of functions includes many functional forms prevalent in chemical engineering applications, such as the set of all concave functions, the set of all multilinear functions, and many signomial functional forms over a box domain [5,6].

We present the explicit convex envelopes of edge-concave functions over a box which are product-separable, a condition which has been conjectured to apply to all non-concave edge-concave functions [3]. We defined conditions applicable to component univariate functions to qualify and classify product-separable edge-concave functions. The complexity of this diverse class of functions was reduced by grouping functions by the characteristics of their component univariate functions. The facets of the convex envelopes of every group are then determined by using the appropriate triangulation of the domain into simplices [7,8]. This procedure results in the explicit convex envelope represented by facets which are easily utilized in an optimization modeling platform or global optimization algorithm.

References:

1. Falk, J. E., & Hoffman, K. R. (1976). A successive underestimation method for concave minimization problems. *Mathematics of operations research, 1*(3), 251-259.

2. Rikun, A. D. (1997). A convex envelope formula for multilinear functions. *Journal of Global Optimization, 10*(4), 425-437.

3. Tardella, F. (2004). On the existence of polyhedral convex envelopes. In *Frontiers in Global Optimization*(pp. 563-573). Springer US.

4. Tardella, F. (2008). Existence and sum decomposition of vertex polyhedral convex envelopes. *Optimization Letters, 2*(3), 363-375.

5. Meyer, C. A., & Floudas, C. A. (2005). Convex envelopes for edge-concave functions. *Mathematical programming, 103*(2), 207-224.

6. Misener, R., & Floudas, C. A. (2012). Global optimization of mixed-integer quadratically-constrained quadratic programs (MIQCQP) through piecewise-linear and edge-concave relaxations. *Mathematical programming, 136*(1), 155-182.

7. Meyer, C. A., & Floudas, C. A. (2004). Trilinear monomials with positive or negative domains: Facets of the convex and concave envelopes. In *Frontiers in global optimization*(pp. 327-352). Springer US.

8. Meyer, C. A., & Floudas, C. A. (2004). Trilinear monomials with mixed sign domains: Facets of the convex and concave envelopes. *Journal of Global Optimization, 29*(2), 125-155.

421381 Mccormick Relaxations: Extension to Multivariate Outer Function and Analysis of Convergence Rates

Abstract:

Converging convex/concave relaxations is of primal significance for the global solution of nonconvex programs. High convergence rates of the relaxations of the nonconvex functions assure finite termination of methods relying on the estimators in the limits, e.g., [8].
The technique for constructing convex and concave relaxations for factorable functions was originally presented by McCormick [4]. Factorable functions are constructed by using a finite recursive composition of binary sums, binary products and a known library of univariate intrinsic functions. Propagation of the relaxations of functions of the form $F_1{^\circ}f_1 + F_2{^\circ}f_2 \cdot F_3{^\circ}f_3$ is possible with two main theorems. Tsoukalas and Mitsos [9] extended the McCormick univariate to multivariate outer functions, i.e., functions of the form $g=F(f_1,...,f_m)$. It was also shown in [9] that this extension provides at least as tight relaxations as the relaxations obtained by the original McCormick rules and in special cases, such as the binary product, minimum and maximum functions and fractional terms, strictly tighter relaxations.
The well-known branch-and-bound (B&B) algorithm performs better if fast convergence of the relaxations can be achieved, since nodes with almost optimal objective value can only be fathomed if the relaxations used in the computations are sufficiently tight. The discussions in [3, 6, 10] on the cluster effect and additionally the recent article about convergence of geometric B&B methods [7] give a deeper insight on the problem.
Bompadre and Mitsos [2] discussed the concept of convergence order of McCormick relaxations. The analysis is based on interval analysis concepts [1, 5], where the convergence orders of the relaxations is considered by the comparison of the estimation error to the decrease of the range of the function.
In this work we establish the convergence order in the McCormick relaxations developed in [9]. We consider the convergence orders in the Hausdorff metric by comparing the range of the original functions to the range of its estimators and pointwise convergence orders by comparing the maximum distance between the original functions and its estimators over a given domain. Similar to the results in [2] the convergence orders are resolved by the orders of the inclusion functions of the inner functions and the convergence order of the multivariate outer function. We show supported by examples that in the case of the binary sum with the usage of results in [9], the convergence rate remains the same, while for the binary product and the minimum function higher convergence orders are possible.

References

[1] G. Alefeld and G. Mayer. Interval analysis: Theory and applications. Journal of Computational and Applied Mathematics, 121(12):421-464, 2000.
[2] A. Bompadre and A. Mitsos. Convergence rate of McCormick relaxations. Journal of Global Optimization, 52(1):1-28, 2012.
[3] K. Du and R. B. Kearfott. The cluster problem in multivariate global optimization. Journal of Global Optimization, 5(3):253-265, 1994.
[4] G. P. McCormick. Computability of global solutions to factorable nonconvex programs: Part I-Convex underestimating problems. Mathematical Programming, 10(1):147-175, 1976.
[5] R. E. Moore and F. Bierbaum. Methods and applications of interval analysis (SIAM Studies in Applied and Numerical Mathematics). Society for Industrial & Applied Math, 1979.
[6] A. Neumaier. Complete search in continuous global optimization and constraint satisfaction. Acta Numerica, 13:271-369, 5 2004.
[7] A. Schöbel and D. Scholz. The theoretical and empirical rate of convergence for geometric branch-and-bound methods. Journal of Global Optimization, 48(3):473-495, 2010.
[8] M. Tawarmalani and N. V. Sahinidis. A polyhedral branch-and-cut approach to global optimization. Mathematical Programming, 103(2):225-249, 2005.
[9] A. Tsoukalas and A. Mitsos. Multivariate McCormick relaxations. Journal of Global Optimization, 59:633-662, 2014.
[10] A. Wechsung, S. D. Schaber, and P. I. Barton. The cluster problem revisited. Journal of Global Optimization, 58(3):429-438, 2014.

418710 Convergence-Order Analysis of Branch-and-Bound Algorithms for Constrained Problems

Abstract:

Global optimization has found widespread applications in chemical engineering [1]. Deterministic global optimization techniques attempt to determine an approximate optimal solution within a specified tolerance and terminate with a certificate of its optimality in finite time [2]. The performance of branch-and-bound algorithms for deterministic global optimization is strongly dependent on the ability to construct tight and rapidly convergent convex relaxations of nonconvex functions.

Recently, Bompadre and coworkers [3, 4] analyzed the convergence order of convex relaxations based on McCormick, Taylor, and McCormick-Taylor models. They showed that if a function is in C^2, the scheme of relaxations corresponding to its convex and concave envelopes is at least second-order pointwise convergent which, in turn, implies Hausdorff convergence of at least second-order. Establishing that a scheme of relaxations is at least second-order Hausdorff convergent is important from many viewpoints, notably in mitigating the so-called cluster problem in unconstrained problems [5-7]. A similar analysis of clustering and convergence order for constrained optimization algorithms is lacking.

In this work, we investigate the convergence order of branch-and-bound algorithms in global optimization by extending the convergence analysis of Bompadre and coworkers to constrained problems. Specifically, we propose a definition of the convergence order for constrained optimization problems, and analyze the convergence order of commonly used branch-and-bound schemes.

[1] Floudas, C. A. and Gounaris, C. E. (2009). A review of recent advances in global optimization. *Journal of Global Optimization*, *45*(1), 3-38.

[2] Horst, R. and Tuy, H. (1996). *Global optimization: Deterministic approaches*. Springer Science & Business Media.

[3] Bompadre, A. and Mitsos, A. (2012). Convergence rate of McCormick relaxations. *Journal of Global Optimization*, *52*(1), 1-28.

[4] Bompadre, A., Mitsos, A., and Chachuat, B. (2013). Convergence analysis of Taylor models and McCormick-Taylor models. *Journal of Global Optimization*, *57*(1), 75-114.

[5] Du, K. and Kearfott, R. B. (1994). The cluster problem in multivariate global optimization. *Journal of Global Optimization*, *5*(3), 253-265.

[6] Wechsung, A., Schaber, S. D., and Barton, P. I. (2014). The cluster problem revisited. *Journal of Global Optimization*, *58*(3), 429-438.

[7] Goldsztejn, A., Domes, F., and Chevalier, B. (2014). First order rejection tests for multiple-objective optimization. *Journal of Global Optimization*, *58*(4), 653-672.

427736 Enhancing the Performance of the Abb Algorithm through the Use of Interval and Convexity Tests

Abstract:

As system complexity grows, the deterministic global optimization of engineering problems becomes more relevant than ever before. Solving problems to global optimality with a guarantee is essential in many classes of problems, e.g. phase equilibrium [1], where the global solution is the only physically meaningful state of the real system, and parameter estimation [2]. Deterministic global optimization is also becoming increasingly important in many disciplines, as the difference between finding the global solution or a local one may lead to increased competitiveness. Furthermore, deterministic global optimization methods can provide a guaranteed answer which is otherwise impossible to retrieve in the general case: whether a problem is feasible or not. This information can be invaluable in heavily constrained problems where even finding a feasible starting point is a challenge.

Many approaches have been proposed over the past four decades to solve problems to global optimality in a deterministic manner. One such approach for general engineering problems is the αBB algorithm, developed by Androulakis, Maranas and Floudas [3], with numerous subsequent contributions [4][5][6]. However, while this algorithm can deal with many classes of nonconvexity, it can exhibit slow convergence in some classes of problems, especially as the number of nonlinear variables increases.

We investigate improvements to the performance of the αBB algorithm for general nonlinear non-convex problems using different approaches to underestimation and problem formulation. Our strategies include the use of interval arithmetic during the bounding process, and the use of different approaches to problem formulation in order to generate the lower bounding problem. Problem formulation is well known to impact the performance of spatial branch and bound algorithms [7], therefore the lower bounding problem is reformulated such that any underlying convexity is exploited to produce tighter underestimators. We demonstrate the impact of these approaches across a set of test problems from standard libraries, using a C++ implementation of the αBB algorithm in the open source MINOTAUR [8] framework.

[1] Floudas, C.A., "Deterministic Global Optimization in Design, Control, and Computational Chemistry," *Large Scale Optimization with Applications,* **93**, pp. 129-184, (1997)

[2] Amaran, S. and Sahinidis, N.V., "Global Optimization of Nonlinear Least-squares Problems by Branch-and- bound and Optimality Constraints," *Top*, **20**, pp. 154–172 (2012)

[3] Androulakis, I.P. et al., "αBB: A Global Optimization Method for General Con- strained Nonconvex Problems," *Journal of Global Optimization,* **7**, pp. 337–363 (1995)

[4] Adjiman, C.S. et al., "A global Optimization Method, αBB, for General Twice-Differentiable Constrained NLPs–I. Theoretical Advances," *Computers & Chemical Engineering* , **22**, pp. 1137–1158 (1998)

[5] Akrotirianakis, I.G. et al., "A New Class of Improved Convex Underestimators for Twice Continuously Differentiable Constrained NLPs," *Journal of Global Optimization,* **30**, pp. 367–390 (2004)

[6] Skjal, A. et al. "A Generalization of the Classical αBB Convex Underestimation via Diagonal and Nondiagonal Quadratic Terms," *Journal of Optimization Theory and Applications,* **154**, pp. 462–490 (2012)

[7] Misener, R., and Floudas, C. A., "ANTIGONE: Algorithms for coNTinuous / Integer Global Optimization of Nonlinear Equations," *Journal of Global Optimization*, **59**, pp. 503-526 (2014)

[8] Mahajan, A. et al., "MINOTAUR: Toolkit for Mixed Integer Nonlinear Optimization Problems," Argonne National Lab, Chicago (2008-2015)

421157 Modeling Abstractions and Frameworks for Optimization Problems with Differential Equations in Pyomo

Abstract:

The first part of this talk presents pyomo.dae, an open-source tool for abstracting and discretizing optimization problems constrained by differential equations. Our tool does not require differential equations to have a particular form and is able to represent high-order differential equations as well as partial differential equations on bounded rectangular domains. This provides users a new level of modeling flexibility that is not found in most algebraic modeling languages and optimization frameworks. Furthermore, pyomo.dae provides several automatic discretization methods for converting models containing differential equations to finite-dimensional algebraic models that can be solved with standard optimization solvers.

Dynamic models are often used within more advanced methods such as moving horizon estimation (MHE) or nonlinear model predictive control (NMPC). While these methods have been widely applied in the literature, there is no standard for implementing them. The second part of this talk introduces a general framework for dynamic optimization within Pyomo. This framework provides easier implementation and use of several advanced optimization techniques by both experts and non-experts alike.

The final part of this talk demonstrates pyomo.dae and our dynamic optimization framework on several illustrative examples, showing how it can be used to tackle a variety of dynamic optimization problems. We present the finite difference and collocation discretization schemes that have been implemented and demonstrate how simple it is to combine such schemes to create new, non-intuitive solution strategies and perform discretization studies. We also show how our framework for dynamic optimization drastically simplifies the implementation of MHE and NMPC. Finally, we discuss how this package enables the exploration of new modeling paradigms such as combining pyomo.dae with other packages in Pyomo for stochastic or disjunctive programming.

437053 Thin-Film Deposition Process Dynamics on Manifolds and Graphs

Abstract:

There is a long history associated with the mathematical analysis of chemical reaction dynamics to generate reduced-order models, particularly in homogeneous reaction systems. Combustion reaction dynamics research, for example, has produced such studies as Maas and Pope (1992) and Lam and Goussis (1994); the process systems community has contributed significantly as well (e.g., Vlachos, 1996; Vora and Daoutidis, 2001). By comparison, heterogeneous reacting systems have received less attention, especially in the realm of thin-film processing.

Effective modeling of heterogeneous reaction dynamics is crucial to Atomic Layer Deposition (ALD) high-throughput system process design and optimization. In ALD, thin films are deposited in discrete (usually sub-monolayer) increments by exposing the growth surface to the gas-phase reactive precursors in an alternating manner. Unlike chemical vapor deposition, ALD does not possess steady state growth modes; instead, continuous ALD operation consists of limit-cycle behavior.

Generating accurate experimental data useful for studying intrinsic ALD reaction rates is challenging because of the dynamic nature of the process, the wide time-scale range of the dynamic processes, and the small surface concentrations of some of the reaction intermediate species. Therefore, researchers have turned to quantum chemical computations to model and compare ALD reaction mechanisms (Elliott, 2012). Because these methods primarily produce static information regarding reaction energetics and transition state configurations, the transitions between surface states can only be described in terms of equilibrium relations. However, conventional transition-state theory does provide rate expressions for the activated surface reactions, and rate equations for precursor species adsorption and desorption processes likewise can be formulated.

Given the modeling situation described, we should expect the overall ALD reaction network (RN) model to consist of a differential-algebraic equation (DAE) system; a key objective of our work in this area has been to develop rational methods to formulate ALD reaction kinetics models in the form of a well-posed DAE system. By writing all reactions in terms of their net-forward rates, including equilibrium processes by adding a fictitious time constant $\epsilon << 1$ s, the pure differential-equation system then can be factored to decouple the reaction terms (Remmers, et. al, 2015; Vora and Daoutidis, 2001). Success of this factorization procedure indicates that the outer solution to this singular perturbation problem retains the dynamics required to accurately model the deposition process. A number of reaction-independent modes also are normally produced by this process, reflecting the conserved quantities of the process and elimination of the redundant dynamic modes. Computing the span of the species space associated with the slow and conserved modes while ignoring minor surface species can further reduce the dynamic dimension of the deposition system by identifying combinations of finite-rate processes that approximate new equilibrium relationships.

The stoichiometry array that premultiplies the reaction-rate array of the model described is a adjacency matrix connecting reaction rates to the time-rate of change of the reaction species, and so can be used to create a (bipartite) species-reaction (SR) graph. The SR graph has been used previously to analyze RN for the potential of multiplicity (Craciun and Feinberg, 2006). Here we will use SR graphs to provide additional insight into our factorization procedure and to diagnose structural problems within thin-film surface RN models.

References

Craciun, G. and M. Feinberg, Multiple equilibria in complex chemical reaction networks II: The species-reaction graph, SIAM J. Appl. Math. 66 1320-1338 (2006).

Elliott, S. D., Atomic-scale simulation of ALD chemistry, Semicond. Sci. Technol. 27 074008 (2012).

Lam, S. H. and D. A. Goussis, The CSP method for simplifying kinetics, Int. J. Chem. Kinetics 26 461-486 (1994).

Maas, U. and S. B. Pope, Simplifying chemical kinetics: Intrinsic low-dimensional manifolds in composition space, Combustion and Flame 88 239-264 (1992).

Remmers, E. M., C. D. Travis, and R. A. Adomaitis, Reaction factorization for the dynamic analysis of atomic layer deposition kinetics, Chemical Engineering Science 127 374-391 (2015).

Vlachos, D. G., Reduction of detailed kinetic mechanisms for ignition and extinction of premixed hydrogen/air flames, Chem. Engng Sci. 51 3979-3993 (1996).

Vora, N. and P. Daoutidis, Nonlinear model reduction of chemical reaction systems, AIChE J. 47 2320-2332 (2001).

437055 Distributed Extremum-Seeking Control over Networks of Dynamically Coupled Unstable Dynamic Agents

Abstract:

Real-Time Optimization (RTO) is a process automation technology whose objective is to predict the economically optimal process operating policy in the near term. The application of RTO typically requires a considerable expenditure for manufacturers. It involves the development of a process model and the implementation of a suitable optimization routine that can solve the optimization problem in real-time. To circumvent these requirements, a number of alternative techniques have been proposed to solve steady-state optimization problems. One such RTO technique is extremum-seeking control (ESC). This approach, which dates back to the 1920s provides a mechanism by which a system can be driven to the optimum of a measured variable of interest. ESC can be viewed as an empirical RTO implementation in which no exact model description is required, but where the objective function of the RTO problem is available from process measurements. ESC provides an effective control system design technique that can be used to steer an unknown dynamical system to an equilibrium that optimizes a cost function.

When dealing with complex dynamical systems, it is generally recognized that overall process objectives are difficult to achieve due to the computational complexity associated with centralized approaches. Thus, a decentralized or a distributed optimization approach is usually favoured in large-scale RTO systems design. In this approach, global process objectives are achieved by solving several local RTO subproblems. The distributed optimization task is said to be non-cooperative when each local RTO achieves its local optimization objectives. Non-cooperative RTO problems have been tackled using ESC by several researchers. The distributed optimization task can also be cooperative when the local RTOs coordinate actions to optimize the sum of their assigned costs. A particular class of distributed cooperative optimization has been the subject several studies. For a class of unconstrained optimization problems, it is shown that it is possible to achieve overall system objectives by solving local problems and communicating the optimization results via the network. Few ESC techniques have been proposed to solve decentralized and distributed optimization problems. For constrained optimization problems, the Alternating Direct Method of Multipliers (ADMM) can be used to solve distributed and coordinated optimization problems. A constrained distributed optimization approach was presented where a projection operator approach is used. The approach can effectively solve distributed optimization problems subject to a known computable projection to a known convex set. Extensions of cooperative optimization techniques to more general network architectures have been proposed.

The control of networks of multi-agent systems with unknown dynamics has been treated in a number of studies have presented a consensus approach for a class of time-varying dynamical networks with unknown stable, decoupled system dynamics and first order agent dynamics with dynamic coupling. In general, existing design approaches for multi-agent systems with unknown agent dynamics are limited to multi-agent dynamics where the only dynamic coupling arises from consensus or communication protocols. This study proposes the design of a method of distributed optimization over networks of dynamic agents with unknown coupled unstable dynamics. The network dynamics are described by a large-scale unknown unstable nonlinear dynamical system operating over a local actuator and sensor communication network. Each agent has access to a local sensor measurement and a certain number of actuators. It is also able to communicate its sensor information with neighbouring agents. The local input-output dynamics of each agent are assumed to be unknown and can be affected by actuator variables and state variables from the network dynamics. To the best of our knowledge this class of multi-agent dynamical systems has not been treated in the literature. A distributed extremum-seeking controller is proposed to solve the optimization problem. The main feature of existing ESC approaches is the use of some variation of a gradient descent algorithm. This approach requires the integration of an estimate of the gradient of the unknown cost function subject to some form of filter and a dither signal. ESC problems cannot be systematically solved in the absence of time-scale separations. As a result, the stabilization of unknown nonlinear systems with the help of ESC cannot be tackled. This paper proposes a novel proportional-integral ESC (PIESC) design technique that is implemented in a distributed environment to design cooperative systems to solve a distributed optimization problem over networks of unknown unstable dynamic agents. The agent dynamics implement a PIESC controller that can solve the real-time optimization without the explicit need for a two time-scale approach. The main contribution of this paper is to show that the distributed PIESC can be effectively applied to the design of real-time optimization control systems that can stabilize the network of unstable dynamics to the unknown optimum of the total plant cost. The result offers a novel mechanism to design distributed control systems with unknown dynamic interactions. Application to large-scale economic model-predictive controller design is addressed.

Challenges and Opportunities of System Engineering in Industry

Rui Huang
Research Engineer
United technologies Research Center
huangr@utrc.utc.com
860-610-7655

AIChE Annual Meeting
Salt Lake City
November 11th 2015

United Technologies Research Center

United Technologies Corporation Business

The largest building supplier in the world

Commercial

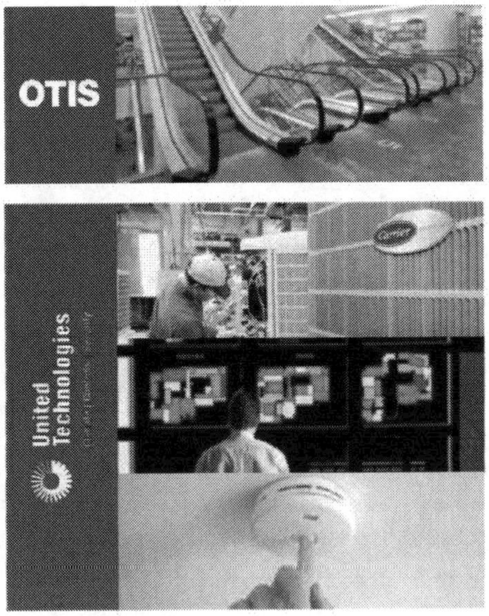

Aerospace

OTIS

United Technologies Research Center

2

Team and Acknowledgement

Johan Åkesson	Modelon
Miroslav Barić	United Technologies Research Center (former)
Lorenz Biegler	Carnegie Mellon University
John Burns	Virginia Tech
John Cassidy	United Technologies Corporation
Eugene Cliff	Virginia Tech
Clas Jacobson	United Technologies Corporation
Carl Laird	Purdue University/Modelon
Manfred Morari	ETH Zurich
Slaven Peleš	United Technologies Research Center (former)
Drăguna Vrabie	United Technologies Research Center

United Technologies Research Center

3

Key Points

Energy efficient building (situation)

Market conditions cost, labeling and regulations require more energy efficient and environmental-friendly buildings. They pose challenges and opportunities for building suppliers to rethink the design and operation workflows

Takeaway:

- ***System level engineering (not limited to component level) DOES benefit***
- ***Optimization-based approaches provide additional capabilities***
- ***Uncertainty analysis and fault handling need more work***

Open Problems:

- ***Modeling issues:*** industry-accepted, system compatibility
- ***Diagnostic for users:*** workflows for engineers to identify modeling issues
- ***Uncertainty analysis:*** industry-accepted approaches
- ***Computational issues:*** solving large optimization problems online

United Technologies Research Center

4

Agenda

Context and problem definition

The market conditions and regulations for buildings

Why is it important to revisit the design and operation?

How do we define system engineering?

System level design and operation workflows based on optimization

Providing diagnostic information for users.

The system level supervisory control or real time optimization

Coordinating system level operation for energy benefits

Uncertainty analysis and fault operation

Presenting the state-of-the-art and challenges

The challenges and open opportunities

United Technologies Research Center

Building Energy Challenges

The energy challenges need innovative design and operation

Buildings consume

- 39% of total U.S. energy
- 71% of U.S. electricity
- 54% of U.S. natural gas

Buildings produce 48% of U.S. Carbon emissions

Annual energy costs for U.S. commercial buildings & industrial facilities: $400 B

Portion of energy in buildings used inefficiently or unnecessarily: 30%

The only energy end-use sector showing growth in energy intensity

- 17% growth 1985 – 2000
- 1.7% growth projected through 2025

European Union Requirements:

Buildings:

1. From 2019 all new buildings produce as much energy as they consume
2. Member states set minimum targets for zero-energy buildings in 2020.

Residential

1. After 2018 must generate as much as consumed via solar, heat pump and conservation
2. Member states set energy targets for existing buildings by 2015.

United Technologies Research Center

Regulatory Environment Challenges

The regulation require innovative design and operation

EU Fluorinated Gas regulation

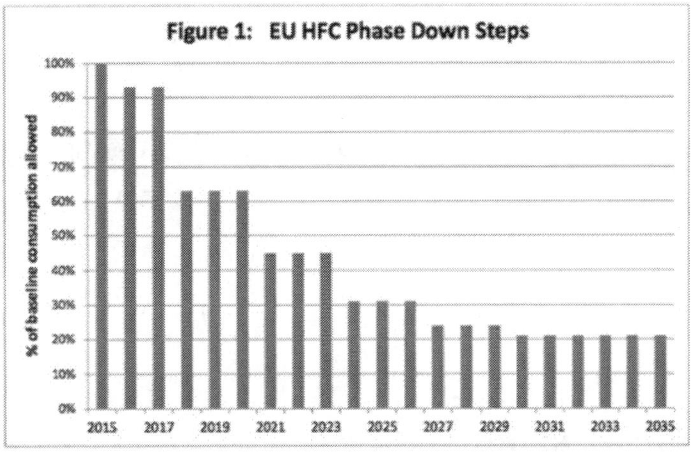

Source: fridgehub

- 2030 it will cut the EU's F-gas emissions by two-thirds compared with 2014 levels.
- Increasing number of HFC taxes throughout Europe which increases the product prices by ~10%

United Technologies Research Center

7

The Implications

The need and benefit for system engineering

Need to rethink system design and operation
Or going out of business

There is a huge potential for advancing humankind by optimizing energy systems
Unfortunately *system engineering* is only partially used in their design

United Technologies Research Center

8

SYSTEMS ENGINEERING

Definition & Scope

Systems engineering is a methodology for product system level design, *optimization* and verification that:

- Provides guarantees of performance and reliability against customer **requirements**
- Produces modular, extensible **architectures** for products
- Exploits **model-based analytical tools** and techniques
- Coordinated execution of a prescriptive, repeatable and measurable **design process**

United Technologies Research Center

The System Level View for Buildings

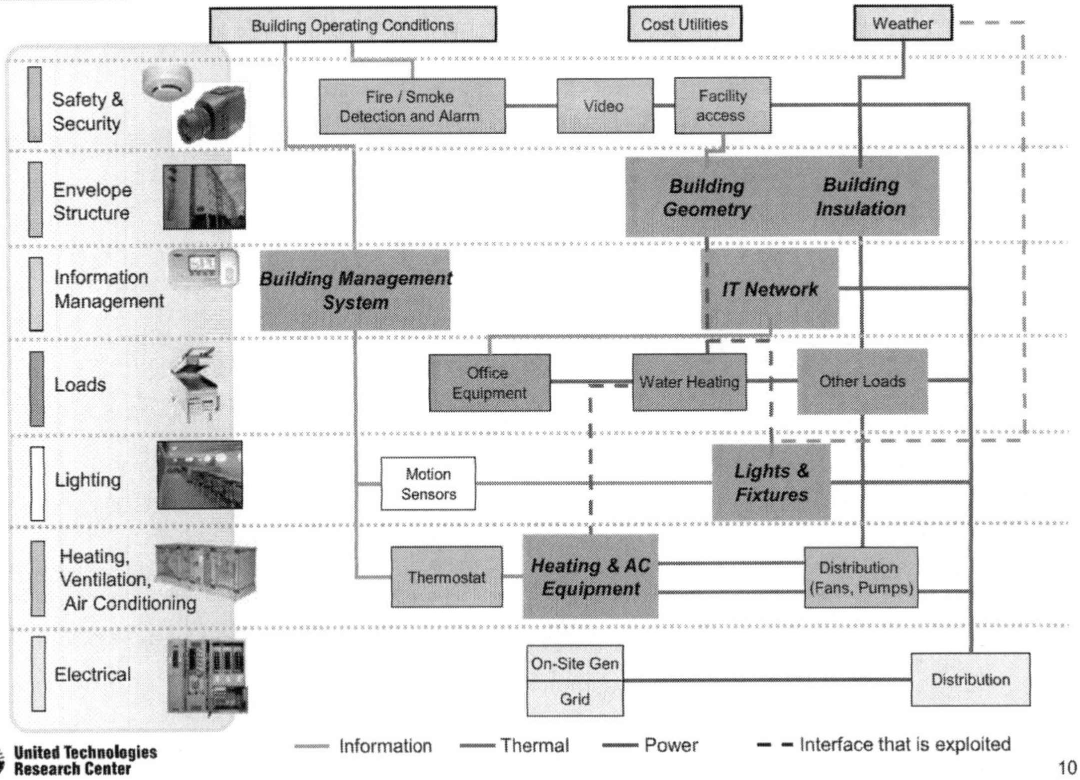

United Technologies Research Center

10

The Successful Examples

Highly efficient buildings exist

Energy Retrofit
10-30% Reduction

Cityfront Sheraton
Chicago IL
1.2M ft^2, 300 kWhr/m^2
5753 HDD, 3391 CDD
VS chiller, VFD fans, VFD pumps
Condensing boilers & DHW

- **Different types of equipment for space conditioning & ventilation**
- **Increasing design integration of subsystems & control**

LEED Design
20-50% Reduction

Tulane Lavin Bernie
New Orleans LA
150K ft^2, 150 kWhr/m^2
1513 HDD, 6910 CDD
Porous Radiant Ceiling, Humidity Control
Zoning, Efficient Lighting, Shading

Very Low Energy
>50% Reduction

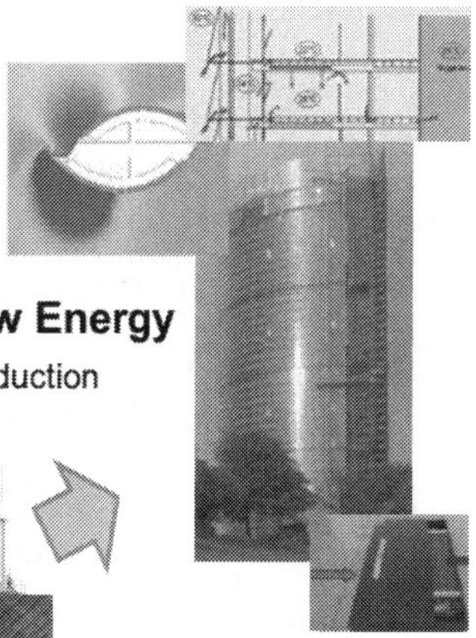

Bonn Germany
1M ft^2, 75 kWhr/m^2
6331 HDD, 1820 CDD
No fans or Ducts
Slab cooling
Façade preheat
Night cool

United Technologies Research Center

Agenda

Context and problem definition
The market conditions and regulations for buildings
Why is it important to revisit the design and operation?
How do we define system engineering?

System level design and operation workflows based on optimization
Providing diagnostic information for users.

The system level supervisory control or real time optimization
Coordinating system level operation for energy benefits

Uncertainty analysis and fault operation
Presenting the state-of-the-art and challenges

The challenges and open opportunities

United Technologies Research Center

12

Engineering Standard Work

Definition & Case Study

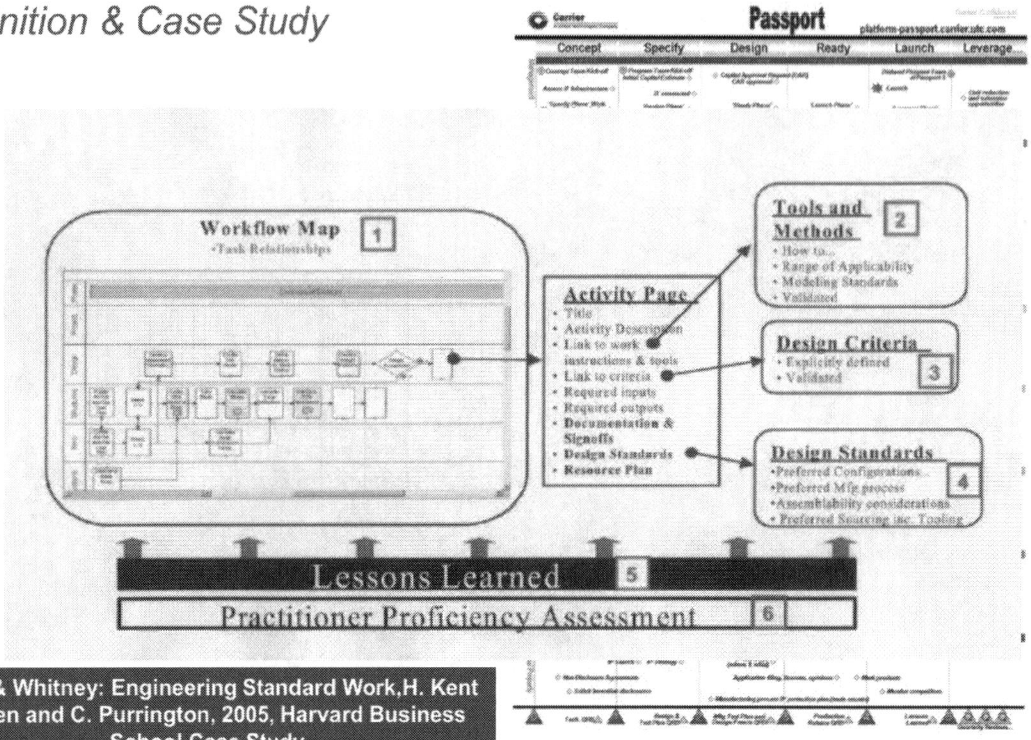

Pratt & Whitney: Engineering Standard Work, H. Kent Bowen and C. Purrington, 2005, Harvard Business School Case Study

The Current System Design Workflow

Inefficient and time consuming

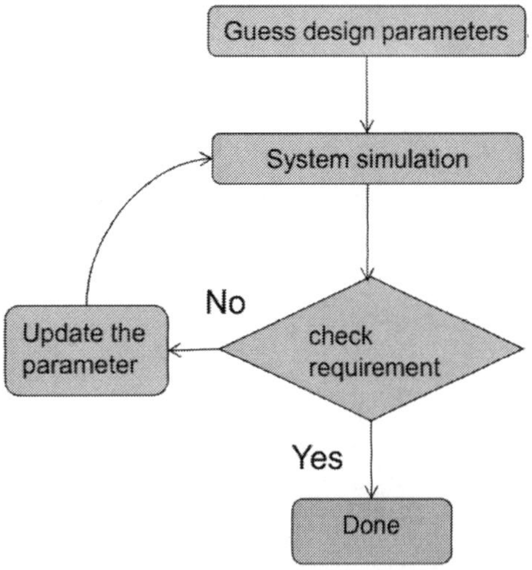

System design workflow

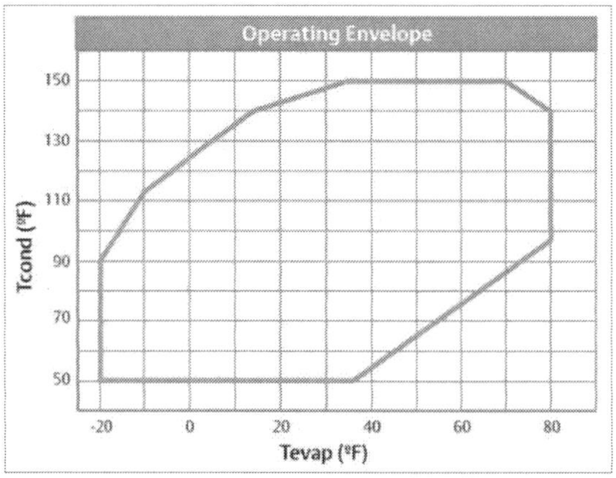

Source: Emerson climate

It limits to solving steady state simulation problems

<u>The system models are assembled from components</u>

Additional requirements further complicate the design flow

United Technologies Research Center

The Current System Design Workflow (2)

Optimal design based on multiple simulations

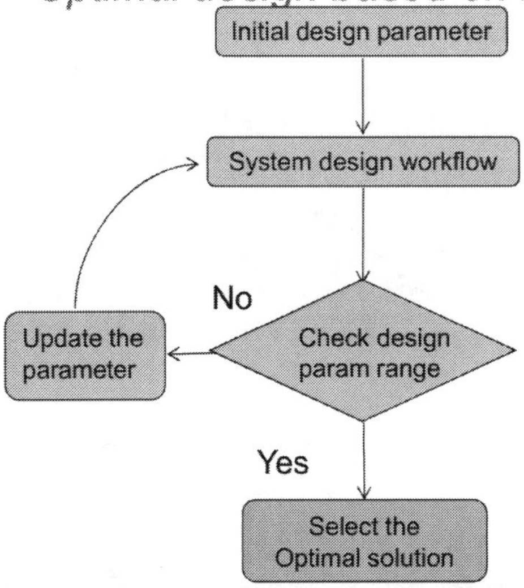

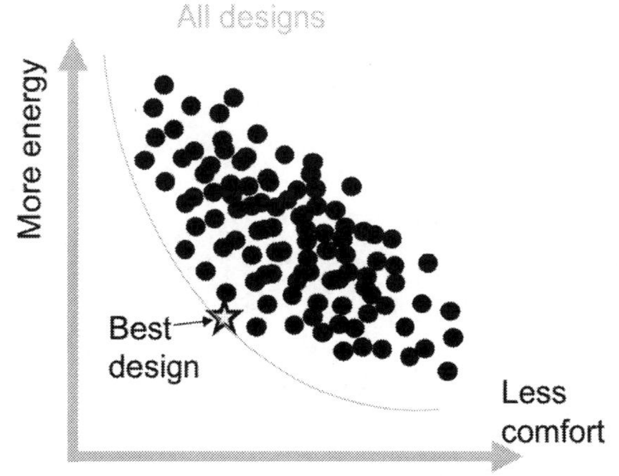

Optimal design based on multiple simulations

The optimal design workflow is based on the running multiple simulations. Additional objectives increases the difficulty (….cost, weight, noise, emissions, sales, productivity, ….)

Optimization Based Workflow – feasibility formulation

Identify the modeling inconsistency and valid operating range

$$\begin{aligned} \min_{x} \quad & \|y-\bar{y}\|_Y^2 + \sum_{i}^{n_x} w_i \left(q^{(i)} + r^{(i)}\right) \\ \text{s.t.} \quad & f(x,y,\bar{\theta}) = q - r \\ & x_L \le x \le x_U,\ y_L \le y \le y_U \\ & g(x,y,\bar{\theta}) \le 0, \quad q, r \ge 0 \end{aligned}$$

x	state variables
y	control variables (manipulated)
$\bar{y}$	reference or desired value for controls
$\bar{\theta}$	boundary conditions
$f(x,y,\bar{\theta})$	system equations
$g(x,y,\bar{\theta})$	feasibility constraints

- Define a set of constraints based on physical, computational, or operational boundaries (*g*)
- Seeks a feasible solution for system if exists
- System equations *f* are relaxed, penalized with L1-norm
- If weights are large enough and system is feasible, $q \approx r \approx 0$
- Regularization to handle multiple solutions. Find the closest to reference.

Optimization based workflow- targeting formulation

Solve the set point reachability problem

$$\begin{aligned}\min_{x} \quad & \|y-\bar{y}\|_Y^2 + \sum_{i}^{n_c} w_i \left(q^{(i)} + r^{(i)}\right) \\ \text{s.t.} \quad & f(x,y,\bar{\theta}) = 0 \\ & x_L \le x \le x_U,\ y_L \le y \le y_U \\ & g(x,y,\bar{\theta}) \le 0,\ \ q,r \ge 0 \\ & c(x,y,\bar{\theta}) = q - r\end{aligned}$$

x	state variables
y	control variables (manipulated)
$\bar{y}$	reference or desired value for controls
$\bar{\theta}$	boundary conditions
$f(x,y,\bar{\theta})$	system equations
$g(x,y,\bar{\theta})$	feasibility constraints
$c(x,y,\bar{\theta})$	targeting constraints

- Define constraints (*c*) to specify performance targets
- Seek a value of the controls for which the feasibility constraints and targets are satisfied
- Targeting constraints relaxed, penalized with L1-norm
- If weights are large enough and target is achievable, then $q \approx r \approx 0$
- There may be multiple solutions, so regularization is added. Find the value for the controls that is closest to a reference or desired value for *y*.
- It directly extends to system optimization.

United Technologies Research Center

17

Optimization based workflow

Provide diagnostic information & system verification for users

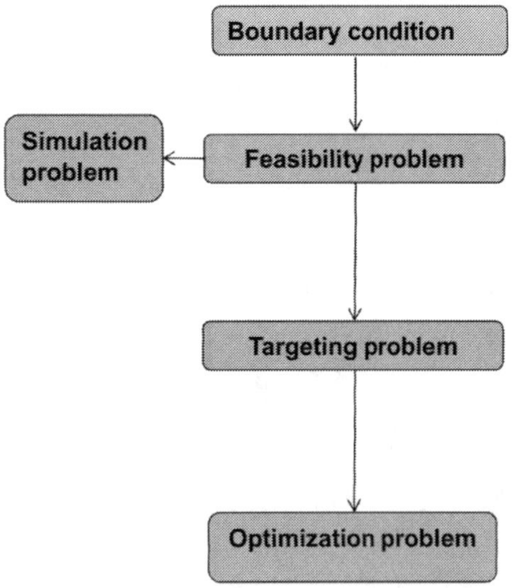

Can the system operate at current operating range? If not, identify the potential modeling error and identify the feasible region. If yes, the simulation formulation can be used to solve the problem. *Diagnostic information for users.*

The system is feasible at the given boundary condition, can it reach the specified performance target? If not, identify & modify the limiting targeting constraint. *System verification information for users.*

The system can reach the target at the given boundary condition, the process optimization finds the optimal solution w.r.t. system objective function.

United Technologies Research Center

18

Example: Vapor Compression System

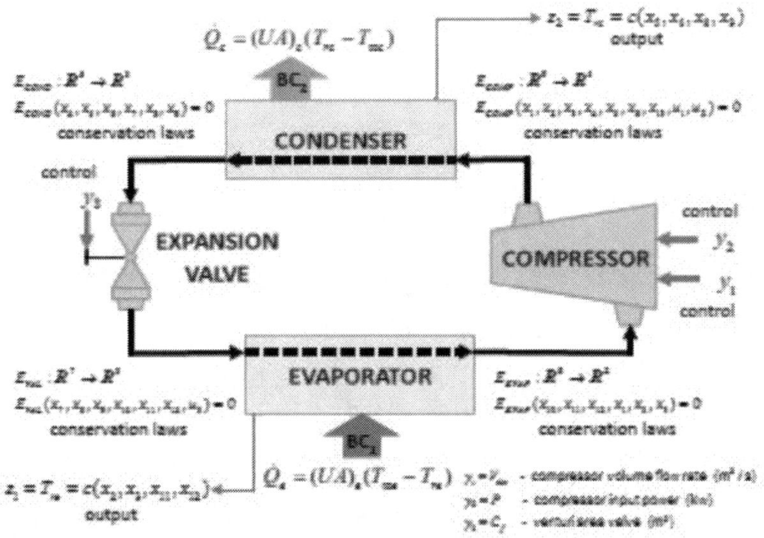

13 system variables,
3 control variables
2 boundary conditions

United Technologies Research Center

19

Example: Vapor Compression System

The targeting formulation verifies the system operating condition

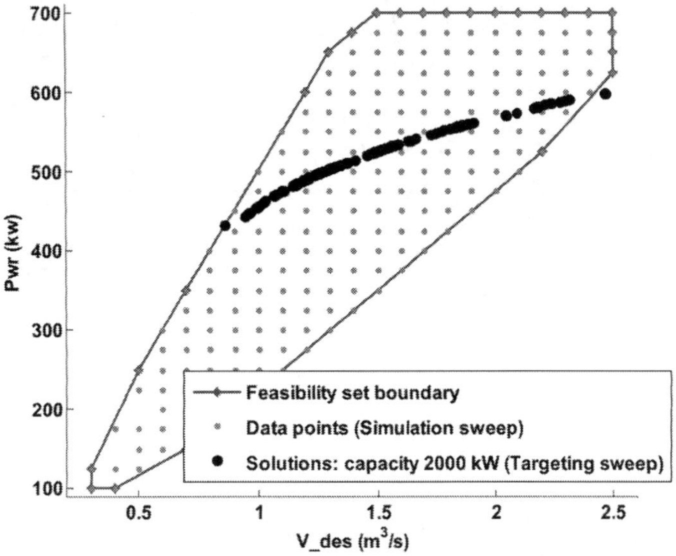

- The targeting problem finds multiple solutions for the cooling capacity equals 2000 W. It verifies the system capacity.

United Technologies Research Center

20

Agenda

Context and problem definition

The market conditions and regulations for buildings

Why is it important to revisit the design and operation?

How do we define system engineering?

System level design and operation workflows based on optimization

Providing diagnostic information for users.

The system level supervisory control or real time optimization

Coordinating system level operation for energy benefits

Uncertainty analysis and fault operation

Presenting the state-of-the-art and challenges

The challenges and open opportunities

United Technologies Research Center

21

Supervisory Control

Definition and challenges

The economic benefit of system control is from coordination of system operation.

Definition & Goal:

1. Optimization: adapt control targets for economic optimization
2. Constraint management: obey operational constraints
3. Sequence transitions, e.g. start-up, shut down, load line operation, reaction to failure

Challenges:

1. system model for plant-wide supervisory control
2. dynamics matter.
3. computational requirement (large scale and online)

United Technologies Research Center

22

Example: Supervisory control – VCS

The benefit of supervisory control

The optimization problem is able to find the point which maximize the coefficient of performance while reaching 2000 W set point.

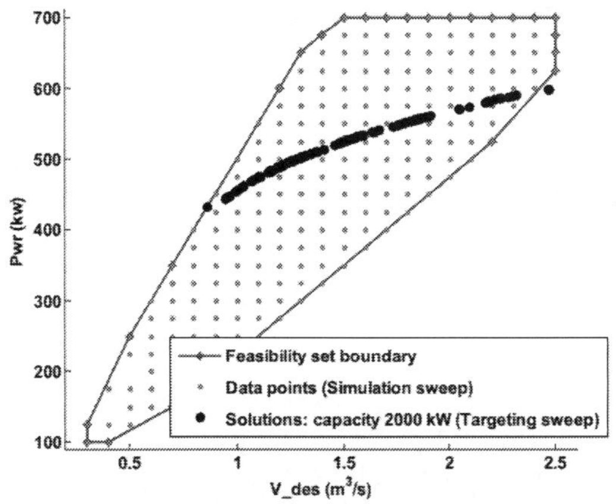

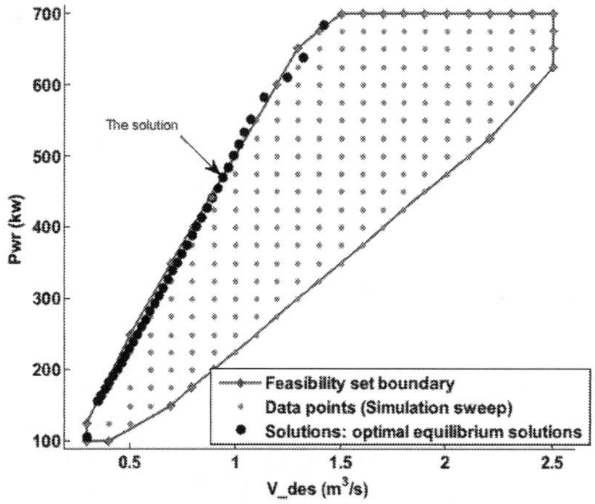

The optimal operation is on the left boundary. It is more economic to operate the VCS on the left boundary in the sequence transition.

United Technologies Research Center

Agenda

Context and problem definition

The market conditions and regulations for buildings

Why is it important to revisit the design and operation?

How do we define system engineering?

System level design and operation workflows based on optimization

Providing diagnostic information for users.

The system level supervisory control or real time optimization

Coordinating system level operation for energy benefits

Uncertainty analysis and fault operation

Presenting the state-of-the-art and challenges

The challenges and open opportunities

United Technologies Research Center

Uncertainty – State of the Art

Situation:

- In a given system design, modeling and prediction of normal operation is challenging but typically straight forward.
- For failure analysis, a different mindset is needed, hypothesizing what can break is not as straight forward.

An approach: uncertainty is (largely) a computational issue, need to attack from how to model uncertainty, how to get data, how to compute, how to exploit system structure...

How to put in a process or design flow & use control to mitigate uncertainty

Challenges: Fault handling (and detection) are a consequence of uncertainty... Critical for operations and critical to get at the complexity of software intensive systems.

United Technologies Research Center

25

Optimization formulation with uncertainty

The approach with computational challenges

$$\min_{x} \phi(x, y, p, \hat{\theta}, u_i)$$
$$\text{s.t. } f(x, y, p, \hat{\theta}, u_i) = 0$$
$$x_L \le x \le x_U, p_L \le p \le p_U,$$
$$y_L \le y \le y_U, \ u_i \in \mathcal{U}$$
$$g(x, y, p, \hat{\theta}, u_i) \le 0, \ i \in \mathcal{I}$$

x	state variable
y	control variables (manipulated)
p	design parameters
$\hat{\theta}$	boundary conditions
u_i	uncertainty parameters
$\mathcal{U}$	uncertainty set
$f(x, y, p, \hat{\theta}, u_i)$	system equations
$g(x, y, p, \hat{\theta}, u_i)$	feasibility constraints
$\phi(x, y, p, \hat{\theta}, u_i)$	system objective (maximizing COP)

- Uses multi-scenario uncertainty parameter (u_i).
- The formulation chooses the finite amount of sampling points in the uncertainty set U.
- Covering the entire uncertainty sets leads to infinite sampling points.
- It leads to large scale optimization problems, thus computational challenges.

United Technologies Research Center

26

Agenda

Context and problem definition

The market conditions and regulations for buildings

Why is it important to revisit the design and operation?

How do we define system engineering?

System level design and operation workflows based on optimization

Providing diagnostic information for users.

The system level supervisory control or real time optimization

Coordinating system level operation for energy benefits

Uncertainty analysis and fault operation

Presenting the state-of-the-art and challenges

The challenges and open opportunities

United Technologies Research Center

27

Challenges & Opportunities

Modeling

- Identification, creation, and standardization of industry-accepted (dynamic) models for system engineering
- Model compatibility at the system level

Diagnostic for users

- Standard problem formulations and workflows providing diagnostic information for users who are not experts

Constraint handling in simulation

- Standard workflows to satisfy the constraint in simulation

Uncertainty Analysis

- Improvement for sample-based approach, problem specific?

Computational challenges

- Solve large scale optimization problems online (supervisory control and uncertainty handling)

United Technologies Research Center

28

Key Points

Energy efficient building (situation)

Market conditions cost, labeling and regulations require more energy efficient and environmental-friendly buildings. They pose challenges and opportunities for building suppliers to rethink the design and operation workflows

Takeaway:

- ***System level engineering (not limited to component level) DOES benefit***
- ***Optimization-based approaches provide additional capabilities***
- ***Uncertainty analysis and fault handling need more work***

Open Problems:

- ***Modeling issues:*** industry-accepted, system compatibility
- ***Diagnostic for users:*** workflows for engineers to identify modeling issues
- ***Uncertainty analysis:*** industry-accepted approaches
- ***Computational issues:*** solving large optimization problems online

United Technologies Research Center

29

420675 On Multiscale, Multidomain Modeling and Parallel Computation: Application to Crystal Shape Prediction in Crystallization

Abstract:

Multiscale modeling provides a powerful tool for fundamental understanding and quantitative prediction of complex process behavior and has tremendous potential for technological impact in the chemical, pharmaceutical and microelectronics industries [1]. The increasing capabilities of high-performance computing have strengthened this belief and triggered a growing interest for multiscale, multidomain modeling. Moreover, chemical engineers are uniquely positioned among other scientists and engineers to make significant contributions to the field of multiscale process modeling due to our unique discipline ranging from fundamental chemistry to large-scale chemical processes. More specifically, kinetic Monte Carlo (kMC) model is receiving a growing amount of attention to carry out dynamic simulations of microscopic/mesoscopic process behavior. The standard kMC algorithm is a serial algorithm since only one event can occur at each time step. However, for some problems one needs to simulate systems with larger length and time scales than what can be simulated using a serial algorithm and available computing power. For these problems, it would be desirable to develop efficient parallel kMC algorithms so that many processors can be used simultaneously in order to carry out realistic computations over extended time and length scales.

Motivated by this, we focus on the development of a parallelized multiscale, multidomain modeling strategy that directly reduces computation requirements without compromising the accuracy of established chemical models. In general, creating a parallel computation program from a sequential one consists of three steps: 1) decomposition of the original problem into tasks or domains, 2) assignment of tasks to processors, and 3) orchestration (i.e., uses a programming mechanism) to communicate and synchronize the information flow among all processors involved [2]. In this work, we applied a parallelized multiscale modeling strategy according to the above three steps to a batch crystallization process. First, we decomposed the nucleation and crystal growth processes in a batch crystallization system into the collection of tasks where each task represents the crystal growth of a nucleated crystal [3]. Second, tasks (i.e., simulating the crystal growth of nucleated crystals) are assigned according to a modulus function: crystal number modules equal to the number of processors available [4]. This is a number-based assignment assuming all processors have identical processor speed and memory. Third, the message passing interface (MPI) settings that use the information passing between the processors is used to link the macroscopic model (e.g., mass and energy balance equations) to the microscopic models (e.g., kMC model). The coupled simulation follows a master-worker MPI computational scheme [5]: there is a processor (i.e., master) that is responsible for decomposing an original sequential problem (e.g., multiscale model to simulate batch crystallization system) into a number of tasks (e.g., crystal growth of a group of crystals) and allocating those tasks to processors (i.e., workers). Workers are responsible for solving assigned tasks and when a worker completes the simulation of an assigned task, it notifies the master. Then, the master allocates a new task to the processor. Each processor waits until it has received all the data computed at the previous iteration, coming from other processors before beginning its following iteration, allowing for a synchronization of the iterations over the processors of the system. Therefore, we are able to achieve significant decrease in time required to complete the batch simulation as the number of cores are increased. A series of results demonstrating the computational efficiency of the approach will be presented.

[1] Maroudas D. Multiscale modeling of hard materials: challenges and opportunities for chemical engineering. AIChE J. 2000;46:878-882.

[2] Culler DE, Singh JP, Gupta A. Parallel computer architecture: a hardware/software approach. Morgan Kaufmann Publishers. 1999.

[3] Kwon JS, Nayhouse M, Christofides PD, Orkoulas G. Modeling and control of protein crystal shape and size in batch crystallization. AIChE J. 2013;59:2317-2327.

[4] Kwon JS, Nayhouse M, Christofides PD, Orkoulas G. Modeling and control of shape distribution of protein crystal aggregates. Chem. Eng. Sci. 2013;104:484-497.

[5] Chandy M, Taylor S. An introduction to parallel programming. Jones and Bartlett Publishers. 1992.

437056 Recent Advances in Uncertainty Analysis and Control in Multiscale Process Systems with an Application to Thin Film Deposition

Abstract:

Thin film deposition is a key batch process in semiconductor manufacturing that is characterized by phenomena that evolve at different time and length scales. To capture the multiscale phenomena occurring in this process, the evolution of the thin film is often modeled using nonlinear partial differential equations (PDEs) embedded with lattice-based kinetic Monte Carlo (KMC) simulations. Strong dependence of the electrical and mechanical properties of thin films on their microstructure, e.g. surface roughness and film thickness, has motivated research on control and optimization in thin film growth. Despite the extensive body of research, there is still a significant discrepancy between the expected performance and the actual yield that can be accomplished employing the current control methodologies. This gap is mainly related to the complexities associated with the multiscale phenomena in the thin film deposition process, lack of practical online *in-situ* sensors at the fine-scale level, and uncertainties in the mechanisms and parameters of the system[1]. This research presents recent approaches that have been developed that make the control and optimization of multiscale processes more realistic by partially addressing these issues.

Although the recently introduced optical *in-situ* sensors have motivated the development and implementation of feedback control in the thin film deposition process, their application is still limited in practice since most of the measurements are only available at the end of the batch. This has motivated the development of suitable optimization and control approaches that do not have access to online fine-scale measurements. An approach to perform offline optimization of the deposition process will be presented based on desired product quality specification in the presence of model-plant mismatch[2-3]. The uncertainty propagation is performed employing power series expansion (PSE). To provide a computationally tractable optimization, the sensitivities in PSEs are evaluated using reduced-order lattices in the KMC models. The results from this research have demonstrated the need to develop robust strategies for this process and have shown that optimal control trajectories obtained from robust optimization formulations can produce thin films under stringent product constraints.

One key challenge in multiscale modelling analysis is the lack of closed-form models, which are essential for model-based control and optimization applications. To address this issue, closed-form models that can predict the control objectives in the presence of uncertainty have been developed[4]. The robust performance is quantified by estimates of the distribution of the controlled variables employing PSE. Since these models can predict the controlled outputs efficiently, i.e. the film's roughness and growth rate, they can either be used as estimators for feedback control purposes in the lack of sensors or as a basis of Model predictive control (MPC) framework. Thus, a multivariable robust estimator that predicts the surface roughness and growth rate under uncertainty has been designed. To ensure that the control objectives are met at the end of the batch, the robust estimator is designed such that it computes bounds on the process outputs. The estimator has been coupled with traditional feedback controllers to evaluate the performance of the system in the lack of online measurements and under uncertainty in the multiscale model parameters.

MPC is an effective advanced control framework widely used in the industry. A robust nonlinear model predictive control (NMPC) application for the thin film deposition process will be presented[5]. The NMPC makes use of closed-form models, which were identified offline and are able to predict the thin film's roughness and thickness at a predefined probability of satisfaction. The NMPC framework aims to minimize the final surface roughness while satisfying constraints on the control actions and film thickness at the end of the deposition process. Since the identification is performed at a specific probability limit, hard constraints were specified in that NMPC formulation. Accordingly, conservative control actions are predicted by the NMPC algorithm. To improve the robust performance of NMPC using soft constraints, closed-form models were developed to estimate the first and second-order statistical moments of the thin film properties under uncertainty in the multiscale model parameters. The results show that NMPC is an efficient technique that can be implemented to address controllability in thin film deposition while explicitly considering uncertainty in the analysis.

The outcomes derived from this study have shown that parameter uncertainty plays a key role in the design of efficient control and optimization techniques in thin film deposition and that must be explicitly considered in the multiscale modelling analysis of these systems to produce thin films with specific high-quality requirements.

[1] Ricardez-Sandoval, L.A., 2011. Current challenges in the design and control of multiscale systems. Can. J. Chem. Eng. 89, 1324–1341.

[2] Rasoulian, S., Ricardez-Sandoval, L.A., 2014. Uncertainty analysis and robust optimization of multiscale process systems with application to epitaxial thin film growth. Chem. Eng. Sci. 116, 590–600.

[3] Rasoulian, S., Ricardez-Sandoval, 2015. Worst-case and distributional robustness analysis of a thin film deposition process. Proceedings of the 9th International Symposium on Advanced Control of Chemical Processes, ADCHEM, Vancouver, Canada.

[4] Rasoulian, S., Ricardez-Sandoval, L.A.. Robust multivariable estimation and control in an epitaxial thin film growth process under uncertainty. Submitted to Journal of Process Control (2014).

[5] Rasoulian, S., Ricardez-Sandoval, L.A., A robust nonlinear model predictive controller for a multiscale thin film deposition process. Chem. Eng. Sci. (2015), http://dx.doi.org/10.1016/j.ces.2015.02.002

435723 Improved Performance Via the Inverted Classroom

Abstract:

This study examined student performance in an inverted thermodynamics course (lectures provided by video outside of class) compared to a traditional lecture class. Students in the inverted class performed better on their exams. Students in the bottom third of the inverted course showed the greatest improvement. These bottom third students had a C exam average compared to a D+ average of their traditional lecture counterparts. When material was presented in videos longer than 15 minutes, students did not perform as well when compared to material presented in shorter videos. Students also needed to be quizzed on video material frequently to incentivize the students to keep up with the video lectures and spend the time to comprehend the material presented outside of class.

429041 A Collection of Virtual Experiments with Tracking of Student Interraction Data

Abstract:

In the area of online pedagogy for engineers, it is difficult to find accessible interactive simulations which may be used to both educate and assess students. We have developed a collection of online simulations and a means of tracking student usage, and have used these simulations in the teaching of core chemical engineering concepts.

Multiple simulations have been created for this site and are used as the focus of problems for student homework. The simulations are written in PHP and Javascript to create dynamic, interactive environments. They are accessible on any internet-capable device, from mobile phones to desktops, without the need to install proprietary software or apps, such as *Adobe Flash*, which might limit their reach to certain operating systems. Furthermore, such an approach does not require the user to permit Java code to run on their machine. This collection of simulations covers a wide range of core chemical engineering concepts: reaction kinetics, fluid dynamics, microbial growth, and more.

Around this collection of simulations we have created a database system for tracking students and their performance. Students create a unique user profile containing information which may be used to track progress through our curriculum and the effect of our curriculum on underrepresented populations. Profile information includes names, contact information, university ID, race and ethnicity, gender, age, and educational background. Student information is stored in a protected mySQL database and their login requires an encrypted password.

Students are assigned to experiment with the simulations in order to determine various, randomly selected physical constants, such as diffusivities or reaction rate constants. Simulated data may be analyzed by students online or exported to be analyzed in, for example, *Microsoft Excel*. Once a student submits an answer they receive immediate feedback. Student performance is used to both assign credit for course assignments and to give the student "experience points" which accumulate as they progress through various simulations, thereby adding a gaming aspect to the site.

All student interactions with the simulations are recorded, such as changing of simulation settings, watching instructional screencasts, the time of day, all mouse movements, their IP address, and more. This information has been used to discover insights into student use of such simulations and improve the teaching material in general. Visualization tools have been created to aid instructors in grading and in deciphering student usage data online.

In this work we present the results of using this online system as a means to educate and assess students, and as a means to accommodate problems which require student experimentation within ever-expanding class sizes.

432473 The Energy Sustainability Remote Laboratory (ESRL)

Abstract:

Increasing world-wide energy demand coupled with finite conventional energy resources calls for a portfolio of sustainable energy alternatives and new technologies. We address the educational aspects of this challenge by, first, the development of lab modules (designed for use in 3rd and 4th year Chemical and Mechanical Engineering lab courses) which promote student understanding of sustainable energy practices and alternative energy sources; second, the provision of data from remote industrial and pilot-plant facilities that students will use with the modules. An important added benefit is that our efforts will provide faculty with some relief from the logistic nightmares now endemic in lab classes due to increasing engineering enrollments and constrained budgets.

The lab modules and educational products in renewable and advanced energy technologies are formulated in a common fashion to provide real–time data, explanatory videos and other tutorials, suggested assignments and germs of assignments accompanied by incomplete calculation (solution) templates, assessment instruments, and results of our own assessment experiences. These materials are all designed to maximize student learning while allowing faculty to focus more on the planning / engineering / design aspects of the labs, as opposed to the nuts and bolts of equipment operation and data collection. Among the modules to be discussed are: (1) Cogeneration / Data Reconciliation; (2) Cogeneration / Ideal Gas Performance; (3) Emissions Prediction and Remediation in Combustion Processes; (4) Using Photovoltaic (PV) Solar Collector Manufacturer's Data in a PV Facility; (6) Biological pH Swing Crystallization; (7) Correlating Particle Size Distribution – Milling Characteristics in Biomass Feed Preparation; (7) Data Reconciliation and Instrument Bias in Power Plants.

The primary impact is in the development and hosting of the experiment-rich, energy sustainability interactive resource website. The site's open structure will allow any student to review background information on the technologies, to access real–time data from each of the technology areas, and even to post feedback. We realize that as student enrollment increases, the pressure in our lab classes for "not the same as last round" and for "meaningful experiences" falls totally on faculty shoulders. We believe the open site structure providing faculty with suggested assignments, detailed solutions, assessment instruments, and results of our own assessment experiences will encourage use.

425981 Use of Interactive Online Videos to Enhance Student Understanding of Thermodynamic Efficiency

Abstract:

ChBE 162 is a required foundational class for undergraduate chemical engineering students. Roughly 60 sophomores take this class every year. Based on student self-evaluations of the course and performances on exams, students lack confidence in applying the concept of efficiency to thermodynamic systems. While they are able to calculate efficiency for a simple process when given the relevant thermodynamic equations, their conceptual understanding for developing such equations is lacking. This gap is apparent when students perform such analyses on analogous processes and wrongly attempt to force-fit prior equations that no longer apply.

The root of the problem is that the concept of efficiency when combined with the use of thermodynamic variables is un-relatable for many students. In order to address this disconnect, we are developing instructional analogies to illustrate the concept of efficiency using more familiar systems, such as those that involve monetary transactions as they should be more tangible. As an example, when a monetary transaction is made at a financial institution, a transaction fee may be charged of say 3%; the efficiency for this step will be 97%. When part of a larger collection of transfers, this cost affects an overall efficiency. The transaction fees in essence symbolize losses in efficiency, analogous to energy losses that may occur during specific unit operations within an overall chemical process.

The use of such analogies activate student learning prior to class. I have developed a series of web-delivered videos, roughly 10 minute each, to explore the concept of efficiency. Students are required to watch these videos for answering problem set questions. The first sets of videos introduce the concept of efficiency and its importance in energy generation systems. These videos develop the analogies for conceptually relating contributors to thermodynamic efficiency to cost-bearing financial transactions. The next set of videos focus on process units within chemical engineering systems that affect efficiency such as pumps and turbines. Here, thermodynamic limits that define efficiency are considered. The final videos relate the overall determination of efficiency to the interconnected function of multiple components together in a chemical process. An example is that of a refrigeration system that consists of a condenser, evaporator, throttle valve and a compressor. Each of these videos is embedded with 3-5 questions to provide interactive feedback.

The improvement in student learning was assessed by two main ways. Firstly, exam questions from several years ago were re-administered in Spring 2015. The performance on the exam questions is compared between 2015 and prior years. Secondly, student attitude surveys were administered to gauge student perceptions towards the videos.

431617 Scenari + Moodle = Self-Training Module for a Flipped Classroom in Distance Lifelong Learning or in Traditional Teaching

Abstract:

The Conservatoire National des Arts et Métiers (CNAM) is a public education institution dedicated to professional lifelong higher education established under the Convention in 1794 on a proposal by Abbe Henri Grégoire "to perfect national industry" and "illuminate the ignorance of he who does not know and the poverty of he who has no means of knowing". Its mission is to provide continuing education open to all, anywhere.

The CNAM chemical engineering team started to offer its distance learning program in 2003; since 2007, 90% of the credits towards the engineering degree can be obtained at a distance; presence is compulsory only for experimental labs. Over the years, many tools have been tested, but until recently only traditional teaching practices were used. The experiment of flipped classroom described here took place in fall 2014.

This paper presents feedback on the first flipped classroom experiment in chemical engineering (applied fluid mechanics, pressure drop and pumps) at the CNAM. It is part of the increasing use of Information and Communication Technologies in Teaching (ICTT) in such courses over the last ten years. This is a turning point in the pedagogical approach, and this presentation will attempt to assess the potential of this strategy.

The self-training module was developed with the publishing chain ScenariOpale. The module is inserted in the learning environment Moodle which allows the use of discussion forums and the development of rich quizzes. Virtual classrooms organized with AdobeConnect complete the system.

This first experiment was overall judged to be very positive, with more in-depth acquisition of knowledge and skills. It came at the price of a substantial investment by the teacher and the ICTT engineers in this initial year of implementation, and required a special effort by students, who were not at all familiar with this approach; all the participants unanimously said that it requires more time than traditional teaching. Most appreciated it, however, stressing in particular their feeling of having understood and assimilated the course content better, although a small number expressed a preference for traditional teaching methods.

Much work is currently underway to improve the quality of the self-training module and to increase the number of activities. To support this effort, we have received funding from Unit (Digital University for Engineering and Technology); as a result, the system will be enriched with more content until summer 2015 and tested with various audiences during september 2015. We added in the self-training module numerous quizzes, videos available on YouTube and simulations proposed by LearnChemE. We have also developed our own animations and videos on educational plants.

This paper presents not only the feedback on the flipped classroom experiment in the context of distance lifelong learning, but also the feedback through more traditional teachings at several levels and for different audiences. We will discuss in particular Russell's no significant difference.

434247 Progress on the SMART-CN Education Modules for Engineering Curriculum

Abstract:

Progress on the SMART-CN Education Modules for Engineering Curriculum

Debalina Sengupta, Yinlun Huang, Thomas F. Edgar, Cliff I. Davidson, Mario R. Eden, Mahmoud M. El-Halwagi*

* corresponding author, el-halwagi@tamu.edu

The Sustainable Manufacturing Advances in Research and Technology (SMART) project funded by National Science Foundation aims to bridge the gap between academic knowledge discovery and industrial technology innovation for sustainable manufacturing. The SMART project involves a multidisciplinary team which has created an interdisciplinary and international Research Coordination Network (RCN) through the joint effort among a number of leading academic laboratories, centers, non-government organizations, and major manufacturing industries.

Meeting the educational goals of the SMART CN has begun through the creation of research coordination among member universities and developing online modules specific to sustainable manufacturing. These modules are highly structured class room ready materials for use by instructors at undergraduate or graduate level. These modules can also be used by industrial entities aiming to further their knowledge in sustainable manufacturing through self-study.

The presentation will provide an overview of this educational component of the project with emphasis on the ongoing development of modules supported by computer tools for incorporation into engineering curriculum. These modules include topics like life cycle assessment, green design and engineering, green chemistry, process intensification, process integration, optimization, process safety etc. among the core module elements and can be adopted by chemical, mechanical, civil, environmental and other engineering disciplines. Extensive case studies demonstrating these concepts are also provided for classroom teaching, homework assignments and term projects. The modules already available through the CACHE website have been reviewed by corporate sustainability teams, as well as academic instructors. This presentation also discusses future plans for module development and dissemination.

Reference: http://cache.org/super-store

417299 Multi-Scale Optimization Methods for Oxycombustion Power System Design

Abstract:

Motivated by the threat of climate change, engineers are developing new power system technologies with increased efficiencies and reduced CO_2 emissions. Flowsheet optimization methods help reduce technology deployment times and carbon capture cost by systematically exploring different design configurations and manipulating operating conditions to improve one or more performance metrics (e.g., cost of electricity). However, applying traditional flowsheet optimization tools to power systems requires new methodologies for embedded multi-scale models of complex subsystems, such as reactors or boilers. In this paper, we present recent work on a trust region optimization algorithm and its application to oxycombustion system design, in which the fuel (e.g., coal, biomass, etc.) is combusted in a N_2 lean environment, producing a flue gas rich in CO_2(70 – 85%) for economical sequestration.

In contrast to previous systems analysis of the oxycombustion system, we are modeling the entire system in an equation-based framework. First and second derivative calculations from the algebraic model allow the use of efficient large-scale nonlinear programming algorithms. With these methods, sensitivity information is easily obtained from the multipliers at the optimum rather than having to perturb design parameters. In addition, the framework allows simultaneous optimization of the oxycombustion subsystems (e.g., air separation unit, steam cycle, CO_2 processing unit and compression train, etc.) rather than analyzing the subsystems in isolation. We have demonstrated this approach by optimizing the sub-ambient separations and their accompanying multistream heat exchangers for oxycombustion power systems[1,2].

Optimization of the steam cycle in an oxycombustion process requires special care. Flue gas recycle is used to replace N_2 in air by mainly CO_2 and H_2O to regulate the flame temperature and the heat transfer profile. As a consequence, boiler performance correlations for air fired (N_2 atmosphere) boilers are not applicable, and discretized partial differential equation based numerical models are preferred. This presents a challenge for optimization with state-of-the-art nonlinear programming algorithms, which assume that accurate derivative information is cheaply available. In this work, we use a hybrid zonal boiler model that considers detailed 3D radiation behavior[3].

A common approach for including computationally expensive black box functions/simulations in nonlinear programming algorithms is through surrogate models. However, experience has shown that no matter how much care is placed into constructing these models, the optimizer is very good at exploiting the error to artificially improve the objective. We are developing a trust region algorithm[4]that is capable of adaptively managing surrogate model error, with provable convergence to critical points of the true problem (with the full, detailed model). The trust region filter algorithm has been applied to both air-fired and oxy-fired steam cycle case studies to integrate the boiler model as well as external function calls to obtain steam table based thermodynamics. The air fired results confirm that our methodology matches with expected designs from the literature, while the oxy-fired results provide insight on key oxycombustion design decisions.

References:

1. Dowling, A. W., & Biegler, L. T. (2015). A framework for efficient large scale equation-oriented flowsheet optimization. *Computers & Chemical Engineering*, *72*, 3–20.
2. Dowling, A. W., Balwani, C., Gao, Q., Biegler, L.T. (2015), Optimization of Sub-ambient Separation Systems with Embedded Cubic Equation of State Thermodynamic Models and Complementarity Constraints. *Computers & Chemical Engineering*(in press).
3. Ma, J., Dowling, A., Eason, J., Biegler, L., & Miller, D. (2014). Development of First Principle Boiler Model and Its Reduced Order Model for the Optimization of Oxy-combustion Power Generation System. In *39th International Technical Conference on Clean Coal & Fuel Systems*.
4. Eason, J. P., and Biegler, L. T. 2015. Reduced model trust region methods for embedding complex simulations in optimization problems. In *12th International Symposium on Process Systems Engineering and 25th European Symposium on Computer Aided Process Engineering*.

417893 Wet Vs. Dry Cooling for the Operation of Concentrated Solar Plants

Abstract:

Energy consumption has increased over the last decades and more sustainable sources such as biomass, solar and wind are being evaluated. Together with it, a recent concern is water consumption since, according to recent reports, two thirds of the plant will suffer water stress in the next decade (Rosengrant et al., 2002). There is a strong connection between energy production and water consumption that is well established for fossil based sources of energy (Water in the West, 2013). While thermal power plants and biomass based facilities have the luxury of being allocated in areas with access to freshwater, solar based thermal facilities present another challenge. Typically regions where the solar incidence is high present low water availability. Therefore, the production of power from concentrated solar power, in spite of the plentiful source of energy, is limited by the cooling capabilities. There are two main systems, wet cooling using a cooling tower and dry cooling, using typically an A-frame. The first one has the drawback of the water consumed. The second one does not require water for its operation, but it consumes a fraction of the produced power to operate the fans.

In this work we present the year-round optimization of the operation of a concentrated solar power facility. Such plants comprise a solar field, a molten salts storage system, the power block consisting of a regenerative Rankine cycle and a cooling system, using either an A-frame or a cooling tower. The salts are heated up using solar energy and stored so that day operation of the facility is secured. A system of three heat exchangers is used to heat up water to saturation, evaporate and superheat it. Only a fraction of the salts are directly used to superheat the steam, that is sent to the high pressure section of the turbine. The rest is used to reheat up the steam leaving the first body of the turbine before it is fed to the second section. The entire flow of salts is then used to heat up the recycled water and evaporate it. In the second body of the turbine, part of the steam is extracted at a medium pressure and it is used to heat up the condensate. The rest of the steam is finally expanded to an exhaust pressure, condensed and recycled. The condensation of the low pressure steam is performed either using a heat exchanger, cooled with water that is next cool down using a cooling tower, or directly using an A-frame system. The aim is to analyze the tradeoffs between the savings in water using dry cooling versus the power consumed due to the operation of the fans. We locate the plant in the south of Europe, Almería (Spain), where high solar radiation is available. The operation of the plant is a function of the solar incidence as well as the climate and atmospheric conditions. The optimization of the system is formulated as a multiperiod Non-linear Programming problem (NLP) that is solved for the optimal production of electricity over a year defining the main operating variables of the thermal and cooling cycles, the operating pressures and temperatures at the different bodies of the turbine, the fraction of salts used for reheating up the steam before feeding it to the second body of the turbine, the fraction of steam extracted to reheat up the liquid, the wet cooling tower operation (air flow and water losses by evaporation) or the A-frame air cooler performance (number of fans and units needed, the fans power consumption and the air inlet and outlet temperatures). Finally, both systems are compared based on the mitigated CO_2 emissions due to the production of power from renewable sources

The operation of the wet cooled facility produces a maximum of 25 MW in summer and a minimum of 9.5 MW in winter with an annual production cost of electricity of 0.15 €/kWh consuming an average of 2.1 L_{warter}/kWh. The investment for the plant is 260 M€, close to that of a similar plant already in operation. For the dry cooled facility, the cooling system consumes an average of 5% of the energy produced, increasing the production cost of energy up to 0.16 €/kWh and the investment to 265 M€. The advantage is the negligible water consumption. In terms of CO_2 mitigation, it turns out that, for the particular allocation, using the wet system is 5% better. Finally, scale up studies reveal that the production cost can decrease by half while the investment per unit of power should become competitive with current coal based power plants if solar and coal facilities present similar production capacities.

References

Rosegrant, M. W.; Cai, X.; Cline, S. A. (2002) *Food Policy Report; Global Water Outlook to 2025 (AVerting an Impending Crisis)*; International Food Policy Research Institute: Washington, DC, 2002.

Water in the West (2013) Water and Energy Nexus: A literature Review. http://waterinthewest.stanford.edu/sites/default/files/Water-Energy_Lit_Review.pdf

418347 Stochastic Scheduling for Microgrid Power Systems with Constrained External Power Exchange

Abstract:

Microgrid power systems are of interest as a potential avenue for increasing the penetration of distributed renewables, increasing resiliency, decreasing the carbon intensity of energy supply, and facilitating the liberalization of the power market [1]. Existing literature on microgrids has covered topics such as the optimal sizing and technology selection [2-3], the effect of stochasticity on microgrid design [4], real-time control to maintain internal power quality [5], and scheduling to ensure robust power supply and minimal operating expenses [6-8].

An aspect that has received little attention in microgrid scheduling is the nature of the power exchange with the surrounding macrogrid. Due to the stochastic nature of renewable power and load, power exchange at this interface may be erratic which could lead to macrogrid instability and place undue stress on utility-scale plants [1]. In this work, we propose a novel formulation of the scheduling problem in which we ensure that power exchange with the macrogrid is well behaved. Specifically, commitments for power exchange with the macrogrid are made in a day-ahead fashion and have a limited ramp rate so that the microgrid operates as a "good citizen" of the grid. Thus, from an external perspective, the microgrid has limited stochasticity (realized power exchange must lie close to the commitments made) and mild ramping (which is conducive to tracking by large power plants).

A case study is carried out for a prototype microgrid consisting of photovoltaics, microturbines, a battery bank, and a bi-directional connection to the macrogrid. Measured residential load and rooftop PV production data over a 1-year period is used. The unit commitment problem is formulated as a chance-constrained mixed integer linear program over a 48 hour receding horizon. The chance constraints ensure that the first-stage variables (discrete unit states and day-ahead power exchange commitment) are able to adequately satisfy the power balance subject to the inherent stochasticity. Using forecasts for power demand and renewable availability, the chance-constrained power balance is transformed into a set of linear inequalities. The schedule resulting from this optimization is compared to two alternative formulations: the deterministic unit commitment (where only the nominal power balance is considered) and the naïve unit commitment (where only the nominal power balance is considered and grid exchange is not scheduled ahead of time).

The performance is analyzed with respect to the operational cost, amount of PV power curtailed, the probability of maintaining the power balance without violating power exchange commitments, and accuracy of predicted second-stage variables (e.g. unit setpoints and storage levels). Computation time is also considered since this unit commitment is intended to be one layer of an on-line hierarchical control system. Finally, additional benefits of this method are discussed such as the intelligent determination of reserve capacity via the chance constrained power balance (as opposed to determining reserve margins *a priori* as in traditional power systems scheduling).

[1] Colson, C. M., and M. H. Nehrir. "A review of challenges to real-time power management of microgrids." *Power & Energy Society General Meeting, 2009. PES'09. IEEE*. IEEE, 2009.

[2] Liu, P., M. C. Georgiadis, and E. N. Pistikopoulos. "An energy systems engineering approach for the design and operation of microgrids in residential applications." *Chemical Engineering Research and Design* 91.10 (2013): 2054-2069.

[3] Koutroulis, E., et al. "Methodology for optimal sizing of stand-alone photovoltaic/wind-generator systems using genetic algorithms." *Solar energy* 80.9 (2006): 1072-1088.

[4] Zhou, Z., et al. "A two-stage stochastic programming model for the optimal design of distributed energy systems." *Applied Energy* 103 (2013): 135-144.

[5] Piagi, P., and R. H. Lasseter. "Autonomous control of microgrids." *IEEE Power Engineering Society General Meeting, 2006.*

[6] Khodaei, A. "Resiliency-oriented microgrid optimal scheduling." *IEEE Transactions on Smart Grid* 5.4 (2014): 1584-1591.

[7] Zhang, Y., N. Gatsis, and G. B. Giannakis. "Robust energy management for microgrids with high-penetration renewables." *IEEE Transactions on Sustainable Energy* 4.4 (2013): 944-953.

[8] Trifkovic, M., et al. "Dynamic real time optimization and control of a hybrid energy system." *AIChE Journal* 60.7 (2014): 2546-2556.

418883 Optimization of Chemical-Looping Combustion Fixed-Bed Reactors in Natural Gas-Fired Combined Cycle Power Plants

Abstract:

The increasing rate of emissions of carbon dioxide (CO_2) from combustion of fossil fuels is widely accepted as the cause of global warming. The power generation sector alone is responsible for 40% of the total anthropogenic CO_2 emissions.[1] Therefore, interest exists in controling the CO_2 emissions through carbon capture and sequestration technologies. The available methods entail the separation of the CO_2 from flue gases (post-combustion), de-carbonization of the fuel (pre-combustion), or use of pure oxygen for combustion (oxy-fuel combustion).[2] One of the greatest barriers facing the implementation of power generation technologies with CO_2 capture is the high capital cost and high energy penalty associated with gas separation.[3] Thus, considerable research and investment are aimed at developing new strategies that enable CO_2 separation to be accomplished at low-cost.

Chemical-looping combustion (CLC) has been identified as one of the most promising strategies to reduce CO_2 emissions from power plants. The principle of CLC is that combustion of a hydrocarbon fuel is carried out in two stages, so that the fuel and air streams are not mixed. Thus, CLC intrinsically eliminates the need for additional CO_2 separation equipment and the associated energy penalty. Fixed-bed reactors are one possible option for CLC, because it can operate at high pressures (10-20 bar), which is a requirement for the modern combined cycle-based power plants. To implement CLC in the fixed-bed reactor, the gas flows dynamically alternate between reducing, oxidizing, and purge conditions. This concept takes advantage of the different reaction and heat fronts moving along the axial direction of the bed. After the oxidation cycle, the heat stored in the bed is used to produce a high temperature gas stream, which can be effectively converted into mechanical work in a combined cycle. The gas turbine, however, poses strict constraints to the variability of the purge gas temperature, which is challenging for CLC because of its transient, batch-type cyclic operation.

In this work, an optimal design of the CLC reactor system based on dynamically operated fixed-bed reactors is proposed. An optimization problem is formulated to maximize the efficiency of the power plant. The fluctuations of the exhaust gas stream of the CLC reactor are minimized to satisfy the standard operating conditions of commercial gas turbines. The set of manipulated variables consists of the cycle configuration (*i.e.*, time interval and mass flows), number of reactors, and properties of the CLC oxygen carrier. Additionally, the CLC reactor needs to operate at a high conversion of the fuel (<98%), high CO_2 capture efficiency (>90%), and low pressure drop (<8%).[4–6] The temperature inside the reactor must always be below the melting point of the CLC oxygen carrier. Thus, the design of the CLC reactor system presents a challenging, heavily constrained non-linear mixed integer problem (MINLP). In this work, we show that at the optimal process conditions, the fixed-bed reactor performs significantly better than the conventional configuration, with minimal transient impact on the gas turbines downstream the reactor. Furthermore, we demonstrate the feasibility of the integrated CLC power plant for continuous operation at different loads, despite the inherent batch-type operation of the CLC reactor.

Acknowledgement

This material is based upon work supported by the National Science Foundation under Grant No. 1054718.

References

(1) *Department of Energy and Environmental Protection Agency, Carbon dioxide emssions from the generation of electric power in the United States*; 2000.

(2) Hossain, M. M.; de Lasa, H. I. *Chem. Eng. Sci.* **2008**, *63* (18), 4433–4451.

(3) Adanez, J.; Abad, A.; Garcia-Labiano, F.; Gayan, P.; De Diego, L. F. *Prog. Energy Combust. Sci.* **2012**, *38* (2), 215–282.

(4) Lawal, A.; Wang, M.; Stephenson, P.; Obi, O. *Fuel* **2012**, *101*, 115–128.

(5) Woods, M. C.; Capicotto, P. J.; Haslbeck, J. L.; Kuehn, N. J.; Matuszewski, M.; Pinkerton, L. L.; Rutkowski, M. D.; Schoff, R. L.; Vaysman, V.; DOE/NETL; Klara, J. M. *Cost and Performance Baseline for Fossil Energy Plants*; DOE/NETL, 2007; Vol. 1.

(6) Erlach, B.; Schmidt, M.; Tsatsaronis, G. *Energy* **2011**, *36* (6), 3804–3815.

424259 Integrated Process Design for Efficient Solar Thermal Hydrogen and Power Production

Abstract:

For the past century, energy sector has been predominantly dependent on fossil energy resources. However, concerns regarding limited fossil fuel reserves and increasing greenhouse gas (GHG) emissions from fossil fuels accelerate advancements and implementation of renewable energy conversion processes such as wind turbines, photovoltaic devices, and concentrated solar power plants. Among these alternatives, solar energy conversion processes are particularly prominent since solar energy is the only energy source that can alone meet all human energy needs with its tremendous potential. Yet intermittencies and land availability constraints for solar energy collection are the grand challenges for solar energy conversion technologies and demand high conversion efficiency and synergistically integrated energy storage solutions. Recently, it has been shown that chemical energy storage cycles that utilizes solar heat and electricity to synthesize a carbonaceous material, provides efficient and energy dense storage options for uninterrupted large scale energy supply to the grid [1-2]. Furthermore, the potential of solar energy usage in biofuel production has been investigated [3]. The common aspect of these studies is the dependence on the production of solar hydrogen. Analysis of theoretical and achievable sun-to-hydrogen efficiency for solar thermal hydrogen production revealed that the solar thermal hydrogen production has a great potential to efficiently produce hydrogen [4].

Here, we introduce a novel integration strategy for the coproduction of hydrogen and electricity from concentrated solar thermal energy [5]. The integration of solar thermal power production and solar thermal hydrogen production techniques reduces the exergy losses associated with either of the systems and thus provide an efficient solution. Solar power production processes are evaluated based on the process *exergy efficiency* that refers to the fraction of incident solar exergy that is directly recovered as the net exergy output, which is defined as the sum of electricity and the hydrogen exergy output. Process simulations for the proposed integration are performed in an integrated Matlab and Aspen Plus platform. The operating conditions of various units and topological structure of the integrated process are determined via sensitivity analysis and optimization using genetic algorithm in Matlab.

The integrated coproduction process provides a synergistic process that minimizes exergetic losses and increases production efficiency. The novel features of the integrated process and the modeling approach are presented. Two solar thermal hydrogen production techniques: (i) single step membrane, and (ii) two-step metal oxide cycle are presented as a case study to demonstrate the benefit of the proposed integration strategy. Efficient coproduction of hydrogen and electricity presents a continuous power supply solution that can achieve high sun-to-electricity efficiencies when combined with the proper hydrogen power cycle and also creates other opportunities since hydrogen has numerous uses in chemical industry, biofuel production, and transportation sector.

[1] Al-musleh EI, Mallapragada DS, Agrawal R. Continuous power supply from a baseload renewable power plant. Applied Energy, 122, 83-93, 2014.

[2] Gençer E, Al-musleh E, Mallapragada D, Agrawal R. Uninterrupted Renewable Power through Chemical Storage Cycles. Current Opinion in Chemical Engineering, 5, 29-36, 2014.

[3] Mallapragada DS, Tawarmalani M, Agrawal R. Synthesis of augmented biofuel processes using solar energy. AIChE Journal, 60(7), 2533-2545, 2014.

[4] Mallapragada DS, Agrawal R. Limiting and achievable efficiencies for solar thermal hydrogen production. International Journal of Hydrogen Energy, 39(1), 62-75, 2014.

[5] Gençer E, Tawarmalani M, Agrawal R, Integrated solar thermal hydrogen and power coproduction process for continuous power supply and production of chemicals, Computer Aided Chemical Engineering, 37, 2291-2296, 2015.

430768 A Systematic Sensitivity Analysis Approach for the Design of Organic Rankine Cycles and the Selection of Efficient Working Fluids Under Operational Variability

Abstract:

Organic Rankine Cycles (ORCs) have received increased attention in recent years, spanning a wide range of applications including power generation from industrial waste heat recovery, geothermal energy and solar irradiation, to name but a few. Their operation is based on the extraction of heat, which is used to evaporate an appropriate organic working fluid subsequently expanded in a turbine to produce work. ORC process flowsheets involve various complex options including multiple pressure loops, expansion stages etc. to increase the energetic and exergetic efficiencies and improve economics. Their use with intermittent heat sources implies that ORCs are often required to operate under variable heat input conditions, while as in every other process system internal variability is always important due to leaks, malfunctions, fouling and so forth. Unless such variability is accounted for, the ORC ability to perform satisfactorily under conditions different from the nominal design settings, namely on the ORC operability, will be significantly jeopardized. This is because potential unexpected changes in the system operation may result in deviations from the performance for which the system was initially designed. To ensure efficient operability, ORC systems need to be designed so that they are sufficiently capable to handle operating variations. Working fluids are inherent components of ORC operation performance, hence their impact on operability should also be accounted for together with ORC design decisions for a wide range of operating conditions. Whereas some working fluids or ORC system designs and operating characteristics may be less sensitive to such changes, others may significantly deviate from their expected performance, eventually failing to meet the desired operating specifications. Published research considers such issues as an afterthought to the selection of working fluids and determination of optimum ORC features under nominal operating conditions. The employed methods are often heuristic as design or operating system parameters are selected within an arbitrarily defined range and with very limited consideration of their combined effects into the system performance. Such approaches fail to provide assurances regarding the validity of the resulting insights or that they will not be affected by the consideration of additional parameters or more extensive design and operating ranges.

In our current work, we propose a systematic method which explicitly considers the impacts of working fluid and ORC design/operating decisions on ORC operability. The proposed method is based on a systematic sensitivity analysis procedure aiming to identify working fluids and ORC operating/design characteristics that optimize the overall performance under nominal operation while simultaneously minimizing the system sensitivity under variability. The method facilitates a) the identification of design parameters with high influence on the overall ORC-working fluid performance, b) the quantification of the overall system sensitivity with respect to these parameters and c) the incorporation of sensitivity metrics to the decision-making process of optimum working fluid and ORC design characteristics. It is based on the development of a sensitivity matrix that incorporates the impact of ORC operating conditions and working fluids on a set of relevant ORC performance measures (e.g., energetic or exergetic efficiencies, irreversibility and so forth). The singular value decomposition of the sensitivity matrix enables the calculation of the major directions in the operating condition space that cause the largest variability in the performance indicators. The entries in such dominant directions determine the impact of each operational condition in these directions. It is therefore possible to explore the behavior of the ORC system along a small number of directions of variability that encompass the combined effect of multiple operating conditions and obtain clear knowledge about the system behavior under perturbed conditions. The variation is not limited to small in magnitude variations but allows a comprehensive outlook of the system attitude for large deviations from the nominal operating point. The investigation is capable of identifying critical process constraint violations and performance features for a wide range around the region where the system exhibits the greatest sensitivity. Through such a procedure a quick but efficient screening and rank-ordering of alternative ORC-working fluid systems can be achieved utilizing steady-state simulations. The proposed aggregate sensitivity metric encapsulates the system variability and acts as an additional objective function in the multicriteria assessment problem, along with all performance indices calculated under nominal operation. The proposed method is presented within a formal mathematical analysis framework which avoids arbitrary assumptions regarding the effects of the selected design and operating parameters and subjective interpretations of the obtained results. It also remains generic and independent of the proposed application.

The proposed developments are illustrated considering solar collectors connected with a storage tank to provide the necessary heat for ORC operation. Multiple working fluid mixtures [1] are evaluated together with several operating and design parameters (e.g., mass flowrates and temperatures, collector area, tank volume and so forth) considering multiple performance indices (e.g., net work output, energetic efficiency, irreversibility, volume ratio across the turbine and so forth). Mixtures are considered as working media because they facilitate increased exergy efficiency while maintaining relatively high energy efficiency due to the thermodynamic characteristics of phase change, but result in a decision making problem of increased complexity due to the need to assess different types of molecules at various concentrations [2]. The obtained results indicate that in order to avoid significant performance losses, variability should be systematically considered as part of the working fluid and ORC design and selection process.

Cited References

[1] Papadopoulos, A.I., Stijepovic, M., Linke, P., Seferlis, P., Voutetakis, S., Toward optimum working fluid mixtures for Organic Rankine Cycles using molecular design and sensitivity analysis, Industrial and Engineering Chemistry Research 52 (34) (2013) 12116-12133.

[2] P. Mavrou, A. I. Papadopoulos, M. Z. Stijepovic, P. Seferlis, P. Linke, S. Voutetakis, Novel and conventional working fluid mixtures for solar Rankine cycles: Performance assessment and multi-criteria selection, Applied Thermal Engineering 75 (2015) 384-396.

434518 Parallel Solution of the Contingency-Constrained Acopf Problem

Abstract:

Efficient operation of an interconnected, multi-source power network is challenging. The alternating current optimal power flow (ACOPF) problem can be solved to determine the optimal generator setpoints that minimize the total operating cost across an electrical transmission network under nominal conditions. However, to guarantee the electrical system reliability and resiliency to failure, many possible contingencies, such as transmission line outages, must be taken into consideration. It is possible to formulate a two-stage stochastic programming problem that solves the ACOPF problem while including all possible single-line outage scenarios.

In this paper, the N-1 contingency-constrained ACOPF problem is formulated as a nonlinear stochastic programming problem within Pyomo, a flexible, Python-based mathematical modeling language. This problem uses a variant of the rectangular current-voltage (IV) formulation within a two-stage stochastic programming formulation whose extensive form includes instances of the ACOPF for nominal operation and each valid N-1 contingency. In this formulation, each scenario is a complete two-stage ACOPF formulation with one of the transmission lines broken. While the nonlinear ACOPF problem itself is solvable to local optimality using off-the-shelf NLP solvers, when considering realistic network sizes and all N-1 contingencies, the resulting nonlinear extensive form outstrips the capabilities of existing serial solvers. For example, in the well-known example case2383wp the single-scenario ACOPF problem has over 60,000 variables, while the N-1 contingency-constrained problem has tens of millions of variables. We present the N-1 contingency-constrained problem formulation, two parallel decomposition approaches, and computational timing results that demonstrate efficient solution of this problem on realistic network sizes.

Electro-thermal Electricity Storage for decentralized power and heat generation

Dr.-Ing. Annelies Vandersickel*, Amir Aboueldahab
Institute for Energy Systems
Technical University Munich

*annelies.vandersickel@tum.de

Electro-thermal Electricity Storage for decentralized power and heat generation

- Motivation
- kW-Scale Design
- Economic Assessment
- Conclusion & Outlook

Motivation

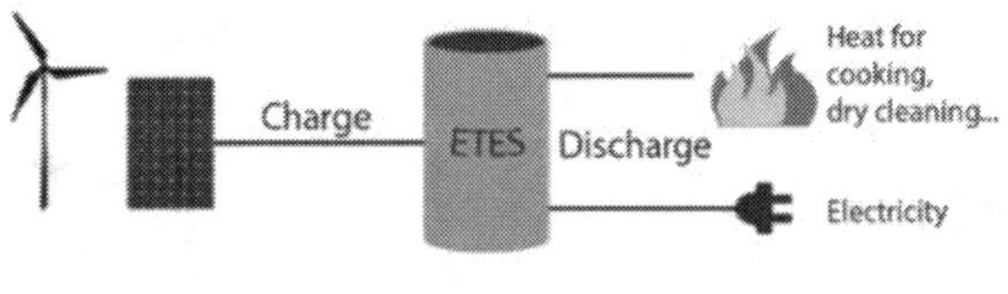

Electro-thermal Electricity Storage

- off-grid power supply in developing countries
- high lead acid pollution due to short battery life time & lack of recycling systems

- applicable in distributed or off-grid systems
- environmentally friendly, robust and possibly low-cost storage system
- combined **heat and electricity** storage for household applications

System Design for low cost

- Off-the-shelves components for kW-scale
- Water for high T thermal storage
- No cold storage
- No sub-ambient p

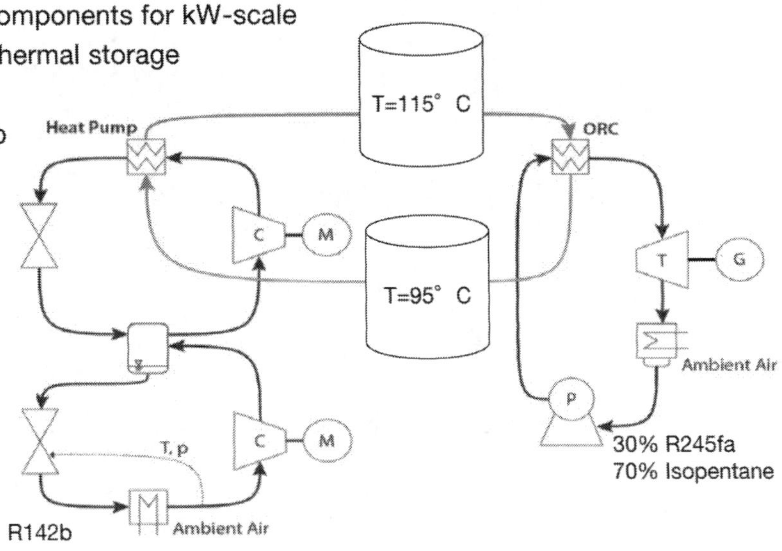

System Design – Key Results

Base case conditions:
- $p_{cond} < 32\ bar$
- $\eta_{comp} = 0{,}65$
- $\eta_{turb} = 0{,}65$
- $\Delta T_{pp,evp} = 10\ K$
- $\Delta T_{pp,cond} = 10\ K$
- $\Delta T_{oh,evp} = 5\ K$
- $T_{amb} = 15\ °C$

⇩

$\eta_{RT} = 13{,}4\%$

- Base case assumptions are based on currently available off-the-shelve system components
- Round-trip efficiency limited by
  - low compressor/turbine efficiency
  - High pinch point in the heat exchanger
  - Temperature spread between hot and cold tank

Economic Assessment for Off-Grid applications

2 Systems
- ETES based
- Battery based

2 Scenarios for Germany
- Low seasonal Variability
- High Seasonal Variability

ETES based System

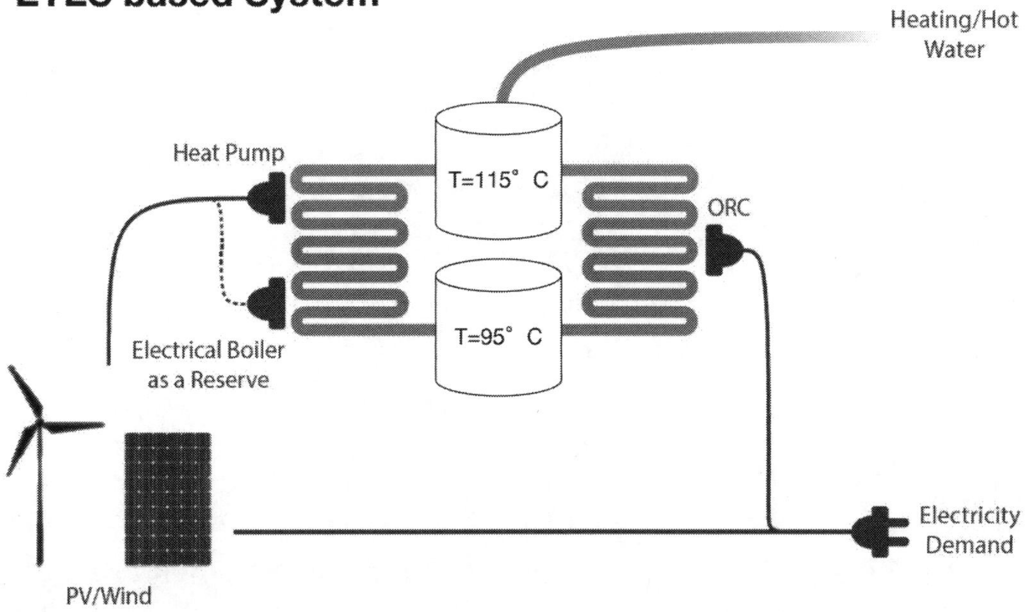

Battery based System

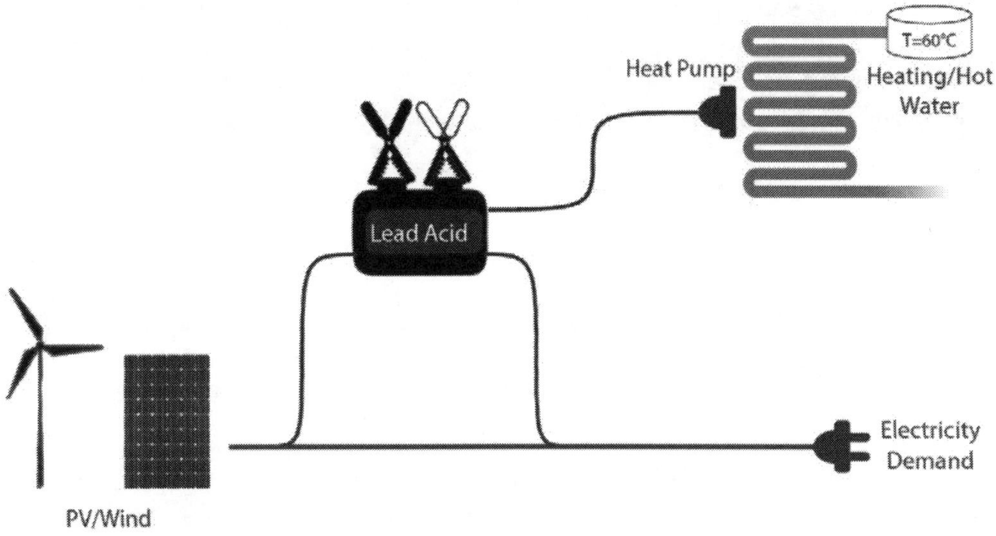

Methodology

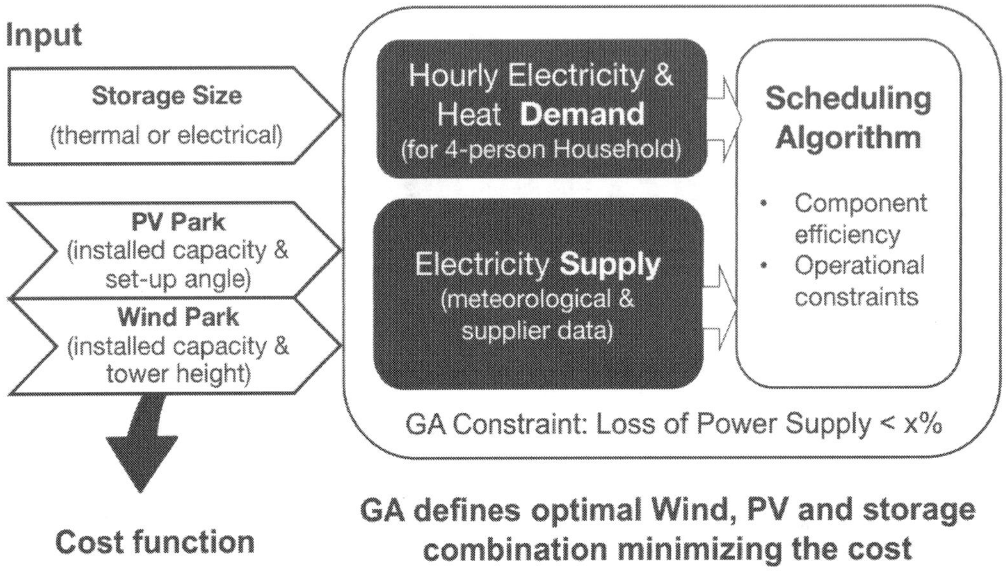

Modeling approach

- Hourly simulation over the course of a year
- Heat[1] and electricity demand of a 4-person household randomly generated using monthly and daily typical demand curves[2]
- PV / Wind yield is modeled using manufacturer information and meteorological data
- ORC & HP
  - efficiency as a function of ambient temperature ($T_{amb,min,HP} = 6°C$)
  - max load determined via parametric study: $\boldsymbol{P_{in,max} = 13\ kW} \mid \boldsymbol{P_{out,max} = 3.5\ kW}$
  - min load of compressor/turbine: $P_{C,min} = 0.33 \cdot P_{C,max} \mid P_{T,min} = 0.23 \cdot P_{T,max}$
  - no part-load behavior represented
- If the heat pump is not able to transform (all of) the excess electricity an (additional) electrical boiler is used ($\eta_{boiler} = 0{,}95$)

[1] heating and warm water demand via VDI-Guideline 2067 **[VDI Guideline 2000]**
[2] modeling approach by Paatero & Lund **[Paatero and Lund 2006]**

Results – Scenario 1 - Seasonal storage

- Munich, Germany
  - Low average wind velocity (2,6 m/s)
  - Little solar insolation during winter months

→ High seasonal variance

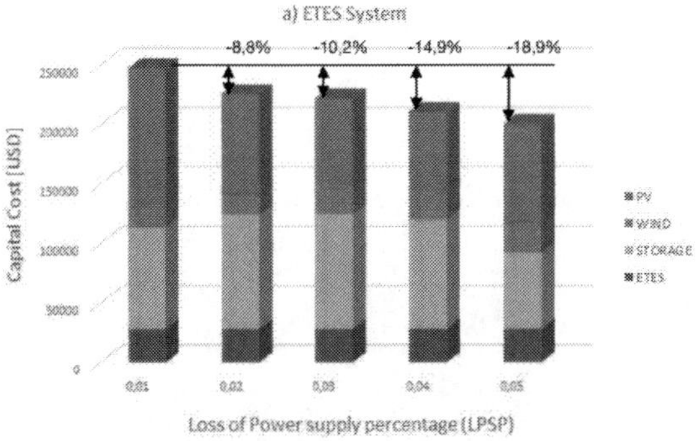

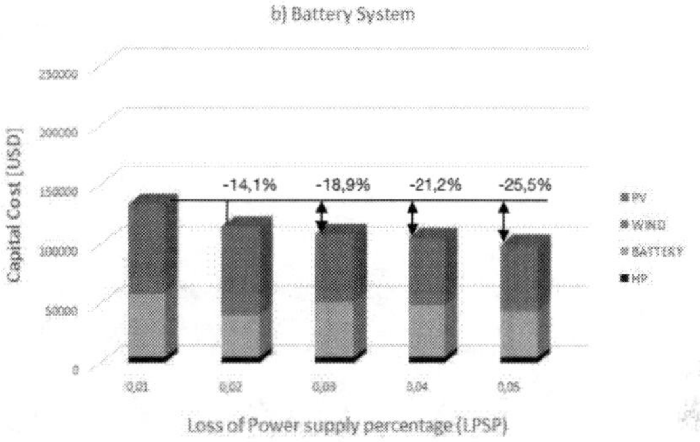

→ Battery system about 50% cheaper

Results – Scenario 2 – Short term storage

- Norderney, Germany
  - Wind velocity consistent at approx. 6 m/s
  - Little solar insolation during winter months

→ Low seasonal variance due to good wind conditions

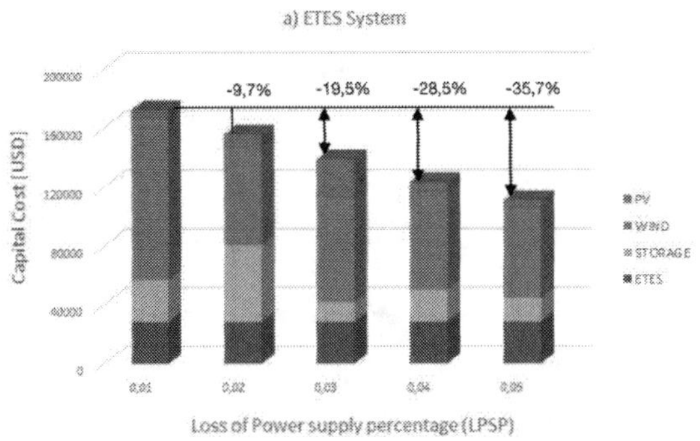

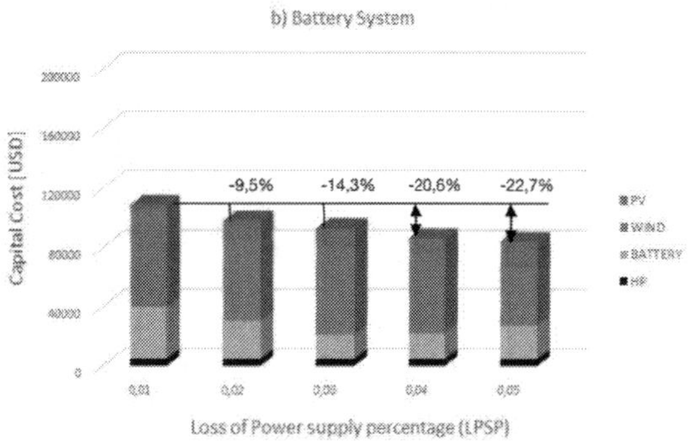

→ Compared to scenario 1 about 30% cheaper
→ Battery System still about 40% cheaper

Results – Direct Comparison of all scenario

- Loss of Power Supply Percentage set to 2%
- Common design value for developing countries

			SCENARIO 1		SCENARIO 2	
			ETES	Battery	ETES	Battery
PV	Installed Capacity	[kW]	46.8	30.9	4.3	6.7
	set-up angle	[°]	35	35	30	30
WIND	Installed Capacity	[kW]	2.2	2.3	7.7	6.2
	Tower Height	[m]	20	13	28	17
STORAGE	a) Water Tank	[kWh_{th}]	6500	-	3500	-
	b) Battery Capacity	[kWh_{el}]	-	190	-	140

Lehrstuhl für Energiesysteme
Technische Universität München

Conclusion

- Thermo-economic assessment of an Electro-thermal Electricity Storage based on currently available off-the-shelve system components

➔ For German demand scenarios & with current component costs: ETES is not economically viable compared to lead-acid batteries

➔ Main reason for high costs being the low round-trip efficiency

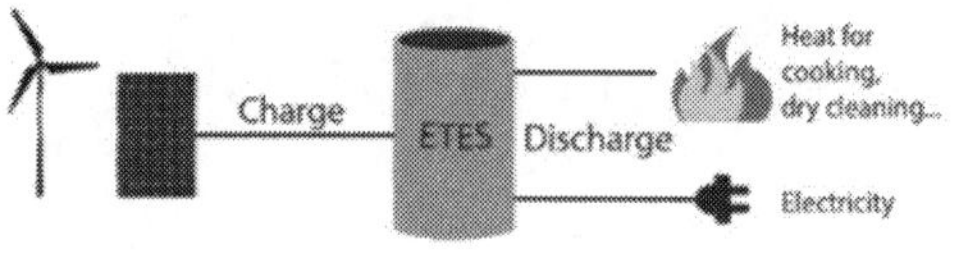

Outlook

- Extend assessment to demand profiles for developing countries
  → Higher Heat to Electricity demand ratio (steam-cooking)

- Detailed assessment of the developments required to increase round-trip efficiency at low cost
  - In particular the ORC Turbine & HP Compressor (scroll for high T)
  - Optimized heat exchanger design with smaller pinch points
  - Efficiencies of 24-30% expected

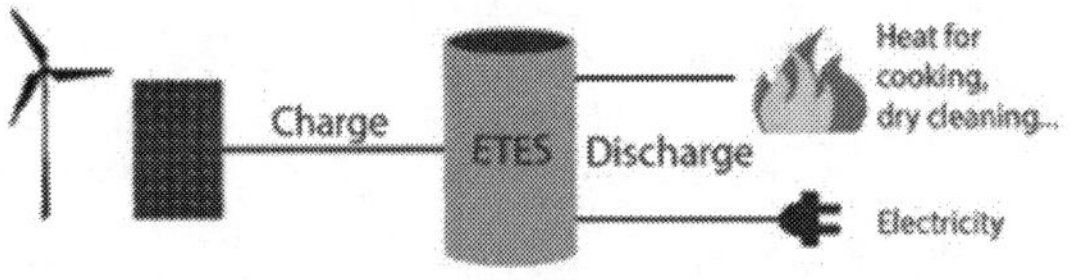

American Institute of Chemical Engineers Annual Meeting, Salt Lake City, UT, November 8-13, 2015

Optimization of Continuous Crystallization Processes to Maintain Content Uniformity

Christopher Burcham
Small Molecule Design and Development
Eli Lilly and Company

Niall Mitchell
Life Sciences, Consumer Goods and Specialty & Agrochemicals
Process Systems Enterprise Limited

PSE

Lilly

Desire End to End Understanding of Process Changes to Product Performance

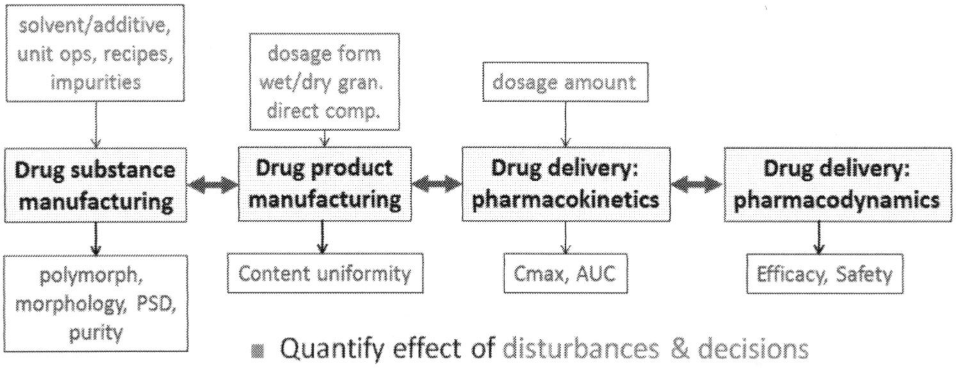

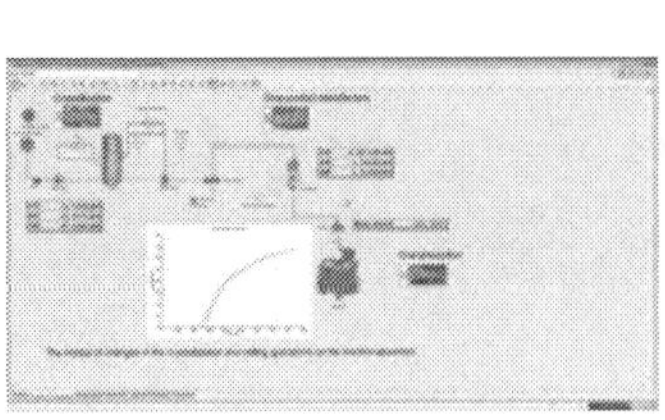

- Quantify effect of disturbances & decisions
  - uncertainty in process knowledge
  - common cause variability

  on Key Performance Indicators
  - Critical Quality Attributes
  - process economics, operability, safety

Desire End to End Understanding of Process Changes to Product Performance

solven unit o im

Drug s manu

ery: namics

poly morph p

ty

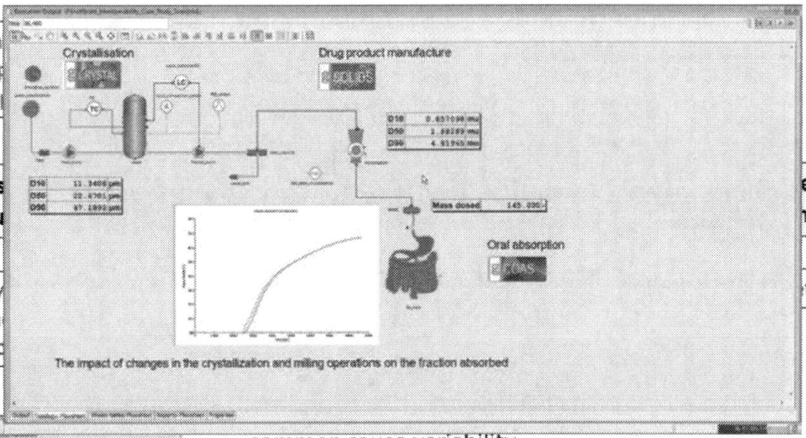

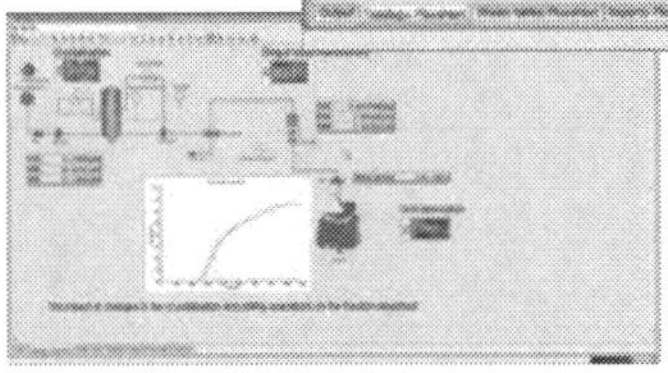

- common cause variability

on Key Performance Indicators

- Critical Quality Attributes
- process economics, operability, safety

Systems Based Pharmaceutics

- A holistic approach to the development and optimization of drug manufacture and drug delivery.
- A single model based framework for the modelling of:
  - drug substance manufacture
  - drug product manufacture
  - oral absorption and pharmacokinetics
  - *Beginning to end manufacturing process.*
- Considers the impact of changes to the crystallization process on downstream unit operations (ie drug product operations such as roller compaction, tablet compression), on quality indicators (ie dissolution testing) and performance expectations (ie absorption modeling).
- A framework to support optimization across the complete manufacturing system.

Objective

To optimize a continuous crystallization process such that:

1. Physical property control is achieved via crystallization conditions.
2. Drug product quality constraints (e.g. unit dose uniformity) are met.
3. Process meets manufacturing requirements for yield and productivity.

- Utilize a population balance model coupled with product constraints (unit dose uniformity in a drug product) and manufacturing constraints (filtration time).
  - Determine optimal processing conditions to maximize yield.
- Illustrative example:
  - Model low dose drug with crystallization kinetics for paracetamol.
    - High growth, low nucleation

Physical Property Control for Pharmaceuticals: Solid Oral Dosage Forms

- Drug substance physical properties, namely particle size can effect:
  - Content uniformity.

  - Bioavailability

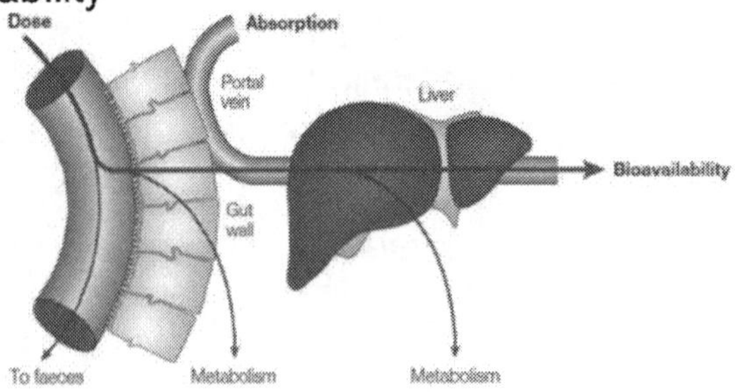

van de Waterbeemd, H, and Gifford, E. (2003). *ADMET In Silico Modelling: Towards Prediction Paradise?* Nat. Rev. Drug Disc. **2**, pp 192-204

Content Uniformity

- The USP guidance for acceptable Uniformity of Dosage Units
  - USP38-NF33 S1, <905>Uniformity of Dosage Units
  - Tier 1 (n= 10) testing
    - Provided the mean potency is within 1.5% of target, then a tablet RSD for potency not more than 6.25%
  - Tier 2 (n=30) testing
    - Provided the mean potency is within 1.5% of target, then a tablet RSD for potency not more than 12.5%
    - No tablet potency outside 75%-125% for up to n=30 tablets
- In brief, RSD < 6.25%

Content Uniformity Model

- Content uniformity models are used to predict volumetric diameter needed to meet USP Stage 1 criteria for tablets
- Model assumptions:
  - Random, dilute mixture (Poisson distribution)
  - Spherical particles
  - Representative PSD of drug substance in dosage unit
  - No segregation etc.

$$C_v = y\left(\frac{\pi\rho}{6G}\right)^{1/2}\left(\sum_i f_i D_i^3\right)^{1/2}$$

$f_i \equiv$ Mass fraction of drug existing as particles of diameter, d_i

$G \equiv$ Dose

$\rho \equiv$ True density

$y \equiv$ Mass fraction of excipient

Johnson MCR 1972. Particle size distribution of the active ingredient for solid dosage forms of low dosage. Pharm Act Helv 47A:546-559.

Content Uniformity Model

$$G = \sum_i n p_i m_i \equiv \text{Mass of drug in a sample}$$

$$\mathrm{var}(G) = \sum_i n p_i m_i^2$$

$$n p_i m_i = p_i M$$

$$\mathrm{var}(G) = M \sum_i p_i m_i$$

$$m_i = \frac{\pi}{6} d_i^3 \rho$$

$$\sigma(G) = \left(\frac{\pi \rho M}{6}\right)^{1/2} \left(\sum_i p_i d_i^3\right)^{1/2}$$

$$p_i = f_i \frac{G}{M}$$

$$C_v = \left(\frac{\pi \rho}{6G}\right)^{1/2} \left(\sum_i f_i d_i^3\right)^{1/2}$$

- Spherical particles
- Ideal mixing
- Small drug fraction
- Sampling corresponds to a Poisson distribution

$n \equiv$ Total number of particles in the sample

$m_i \equiv$ Mass of drug in a particle of diameter, d_i

$p_i \equiv$ Number fraction of drug existing in the sample as particles of diameter, d_i

$f_i \equiv$ Mass fraction of drug existing as particles of diameter, d_i

$M \equiv$ Total sample mass

Johnson MCR 1972. Particle size distribution of the active ingredient for solid dosage forms of low dosage. Pharm Act Helv 47A:546-559.

Content Uniformity Models

USP38-NF33 S1, <905> Uniformity of Dosage Units.

- Rohr's et al model:
  - Assumes log-normal PSD
  - Design tool – target x_{90}.

$$d_{50} = \sqrt[3]{\left(\frac{6Dose}{\pi\rho}\right)\exp\left(-4.5\ln^2\sigma_g\left(\frac{RSD}{100}\right)^2\right)}\ 10^3$$

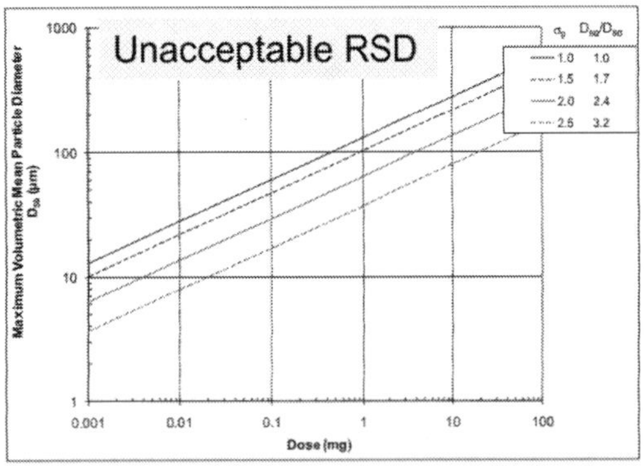

$C_v = 3.84\%$ $\rho = 1.3$ g/cm^3

Rohr BR, *et al* (2006). *J of Pharm. Sci.* 95(5) pp 1049-1059.

- D[6,3] model:
  - No assumptions about PSD shape.

$$RSD = \sqrt{\frac{\pi}{6} \cdot D[6,3]^3 \cdot \frac{\rho}{Dose}}$$

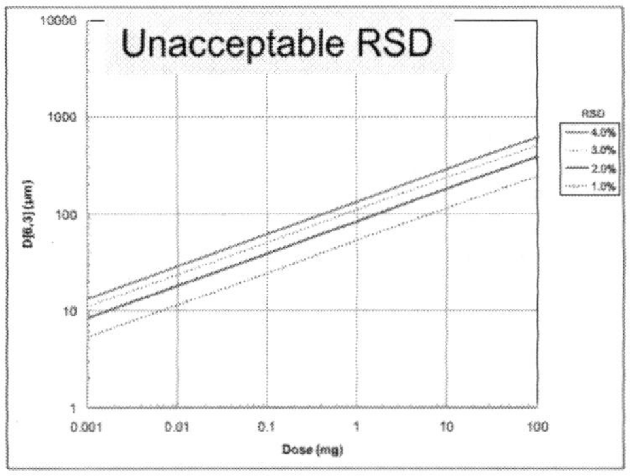

Hilden, J. *et al* (2012). *J Pharm. Sci.* 101(7) pp 2364-2371.

Absorption

- Dependent on solubility (thermodynamic), dissolution rate (kinetic) and permeability.

Physiology of Drug Absorption

Class I	Class II
• High solubility • High permeability • Generally well absorbed	• Low solubility • High permeability • Solubility / dissolution limited absorption
Class III	**Class IV**
• High solubility • Low permeability • Permeability limited absorption	• Low solubility • Low permeability • Generally poorly absorbed

BCS I if entire dose dissolves in 250ml (pH 1 to pH 8)

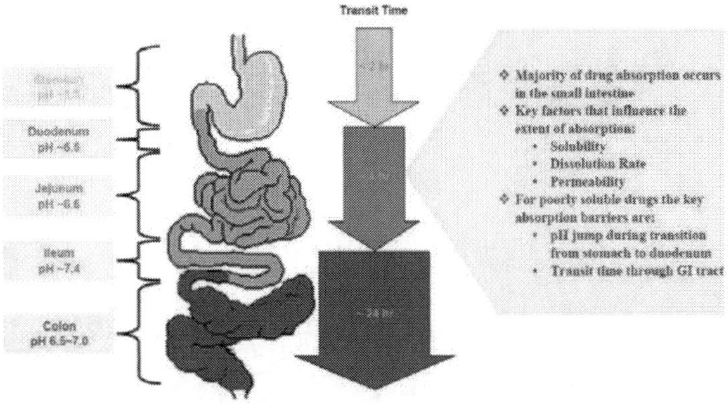

Amidon, G. L. et al. (1995). *Pharm Res*, **12**, 3, pp 413-420.

Martin, A. (1993). *Physical Pharmacy*, 4th Ed. Lea and Febinger, Philadelphia.
Martinez, M.N. and Amidon, G. J. (2002). *J. Clin. Pharmacol.*, **42**, pp 620-643.

Dissolution Rate Limited

- Drug substance physical properties, namely particle size effects:
  - Absorption/bioavailability of the drug substance can be impacted by the dissolution rate of the drug substance according to the Noyes Whitney Equation:

$$\frac{dm}{dt} = \frac{D_{AB}}{\lambda} A\left(C^{*} - C_{b}\right)$$

$$A = \left(Dose\right) SSA$$

$$SSA \propto \frac{1}{D}$$

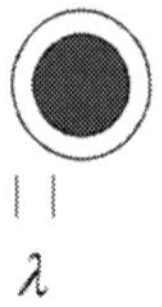

Impact of Particle Size on Content Uniformity as a Function of Dose

- True density of 1.33 g/cm^3, log-normal distribution with a standard deviation of 1.72, $X_{90}/X_{50} = 2$

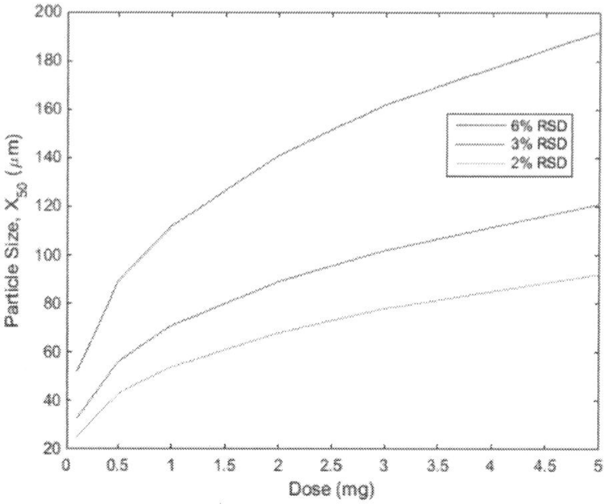

Allowable and Attainable Regions

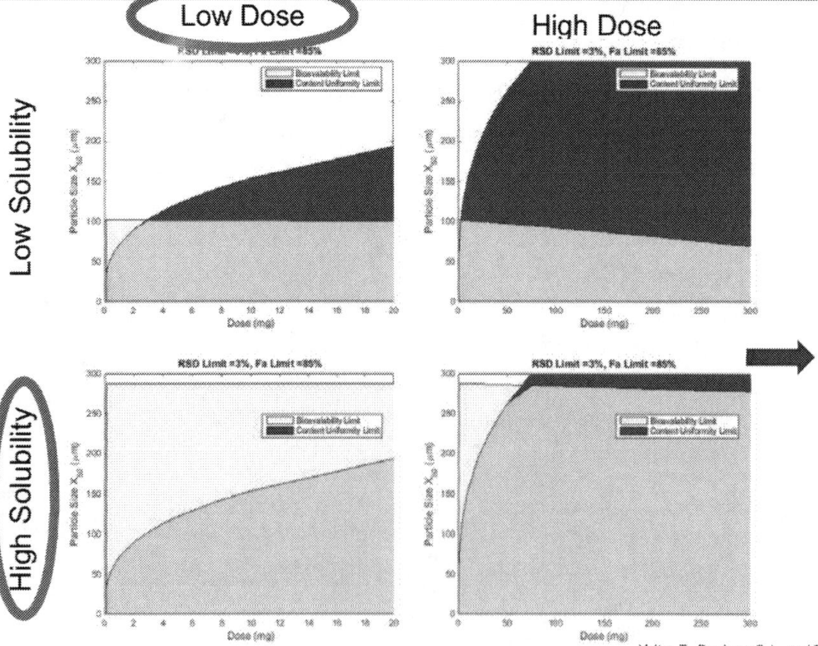

Oh D-M, Curl RL, Amidon GL 1993. Estimating the fraction dose absorbed from suspensions of poorly soluble compounds in humans: A mathematical model. Pharm Res 10(2):264-270.

Vetter, T., Burcham, C. L., and Doherty, M. F. (2014). "Regions of Attainable Particle Sizes in Continuous and Batch Crystallization Processes." *Chem. Eng. Sci.*, **106**, 167-180.

Allowable and Attainable Regions

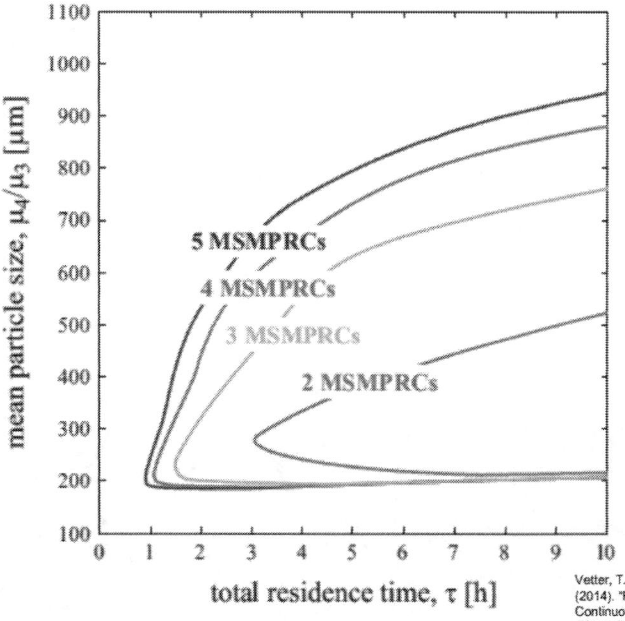

Vetter, T., Burcham, C. L., and Doherty, M. F. (2014). "Regions of Attainable Particle Sizes in Continuous and Batch Crystallization Processes." *Chem. Eng. Sci.*, **106**, 167-180.

15

Impact to Manufacturability – Filtration Time

- Filtration time[1]:

$$t = \left(\frac{\mu c \alpha}{2 A^2 \Delta p} \right) V(t)^2 + \frac{\mu R_m}{A \Delta p} V(t)$$

$$c = \frac{c_s}{1 - [m_F / m_c - 1] c_s / \rho}$$

- Cake resistance dependence upon particle size[2]:

$$\alpha = \frac{18}{\rho} \frac{K}{d_{vg}^2 \exp(4 \ln^2 \sigma_g)} \frac{\nu(\varepsilon)}{\varepsilon^2}$$

$$\nu(\varepsilon) = \frac{10(1-\varepsilon)}{\varepsilon}$$

t	time
V	Filtrate Volume
μ	Filtrate Viscosity
α	Cake Resistance
A	Filter Cross Sectional Area
Δp	Pressure Drop over the Filter
R_m	Filter Media Resistance
m_F	Wet Cake Mass
m_c	Dry Cake Mass
c_s	Concentration of Solids in the Slurry
$\rho_{filtrate}$	Density of the Filtrate
ε	Cake Porosity
ρ	True Density of the Solid Phase
d_{vg}	Volume equivalent diameter on a number basis
K	Dynamic Shape Factor
σ_g	Lognormal Standard Deviation
ν	Void Fraction

1. McCabe, W. L., Smith, J. C., and Harriott, P. (1985). *Unit Operations of Chemical Engineering*, 2nd ed., McGraw-Hill, New York, pp. 873-877.
2. Endo, Y. and Alonso, M. (2001) 'Physical meaning of specific cake resistance and effects of cake properties in compressible cake filtration', *Filtration & Separation*, pp. 42-46.

Modeling the Crystal Size

- Population Balance Model

$$\frac{\partial n(L,t)V}{\partial t} = F_{in}n_{in} - F_{out}n(L,t) - V\frac{\partial(G(L)n(L,t))}{\partial L} + (B-D)V$$

- Birth (B)
  - Nucleation
    - Primary
      - Homogeneous (from solution)
      - Heterogeneous
    - Secondary
      - From existing crystals
        - Attrition
        - Activated
- Death (D)
  - Dissolution
- Birth and death
  - Attrition, breakage and agglomeration
- Growth (G)

Kinetic Models and Parameters

- Nucleation
  - Primary

$$B_{prim} = k_n'(\Delta C)^n$$

  - Secondary nucleation:

$$B_{sec} = k_N m_2 (\Delta C)^b$$

- Growth

$$G(L) = k_g \exp\left(-\frac{E_{A,g}}{RT}\right)(\Delta C)^g$$
$$\text{if } C - C^* > 0$$

Parameter	Value	Units
k_n'	1.597*10[10]	(#/min)
n	2.276	-
k_g	9.979	(m/s)(m^3/kmol)^g
$E_{A,g}$	40.56	kJ/mol
g	1.602	-

Agitation Rate (rpm)	k_N (#/s m^2)	b
200	2.656*10[7]	2.232
250	6.397*10[7]	2.339
300	26.632*10[7]	2.314

- No agglomeration, attrition, breakage.

Frawley, P.J., Mitchell, N.A., O'Ciardha, C.T., and Hutton, K.W., (2012). "The effects of supersaturation, temperature, agitation and seed surface area on the secondary nucleation of paracetamol in ethanol solutions." *Chem Eng Sci*, **75**, pp. 183–197.

Flowsheet

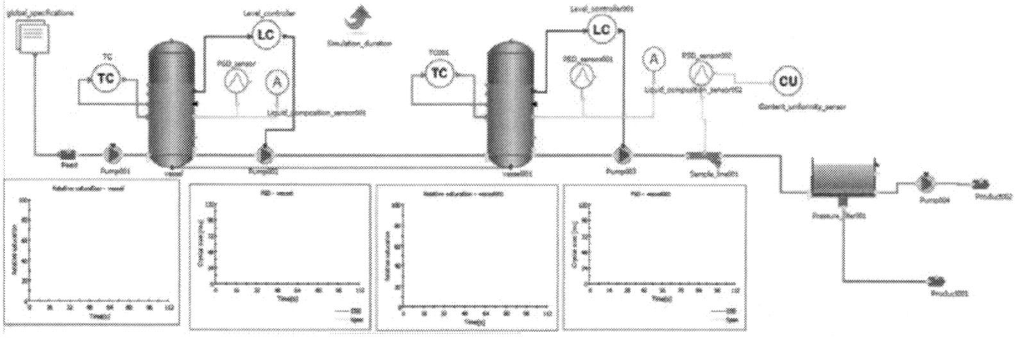

11/9/2015 2015 AIChE Annual Meeting 19

Flowsheet with Continuous Seeding

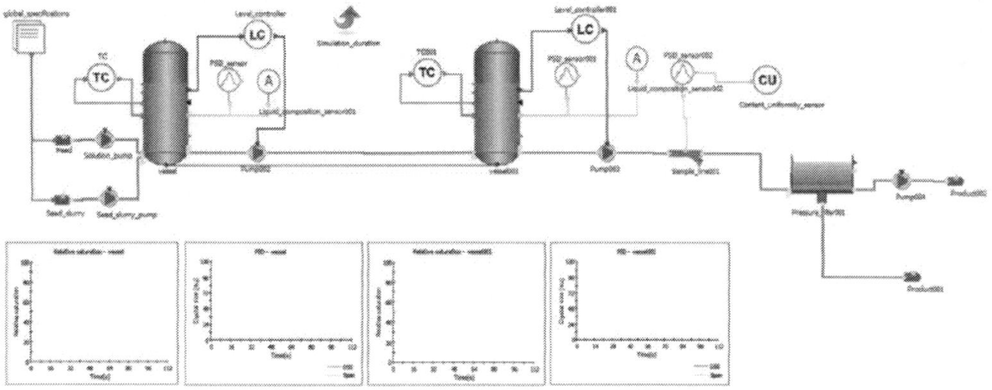

11/9/2015 2015 AIChE Annual Meeting 20

Optimization Problem

- Optimize: Maximize X_{50} of the product

- Constraints:
  - Tablet RSD < 3%
  - Cake resistance < 10^{10} m/kg
  - Relative supersaturation < 2.0
  - Yield > 90% theoretical yield
- Controls:
  - *Fraction of seed to first MSMPR (0.01 to 0.5 wt/wt)*
  - Temperature in the first MSMPR (30 to 50 °C)
  - Residence time in the first MSMPR (40 to 270 min)
  - Temperature in the second MSMPR (5 °C)
  - Residence time in the second MSMPR (40 to 270 min)
  - Solution concentration in the feed

Attainable Dose Range (Without Continuous Seeding)

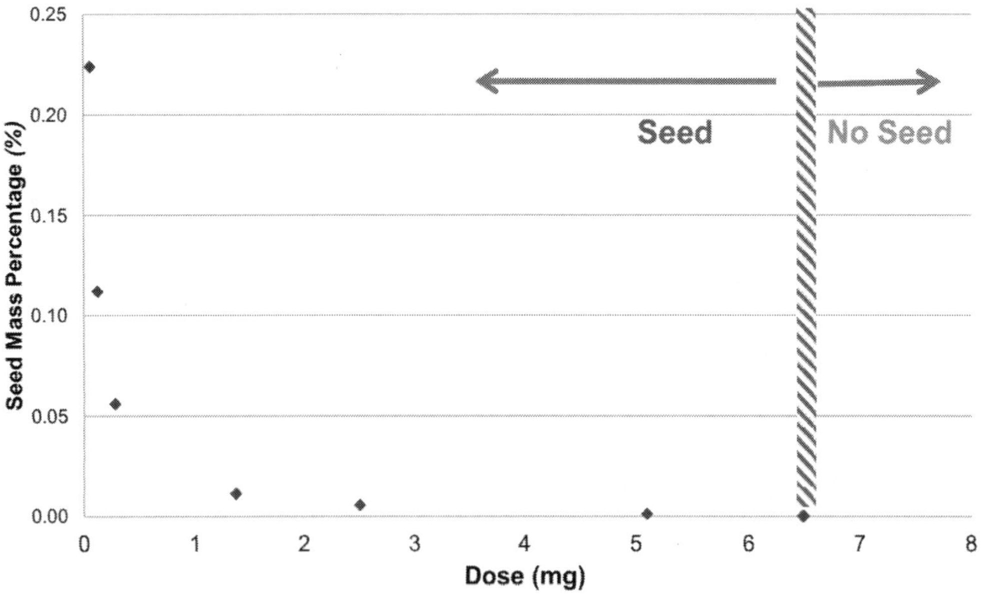

Increasing Secondary Nucleation

By Increasing Agitation Rate

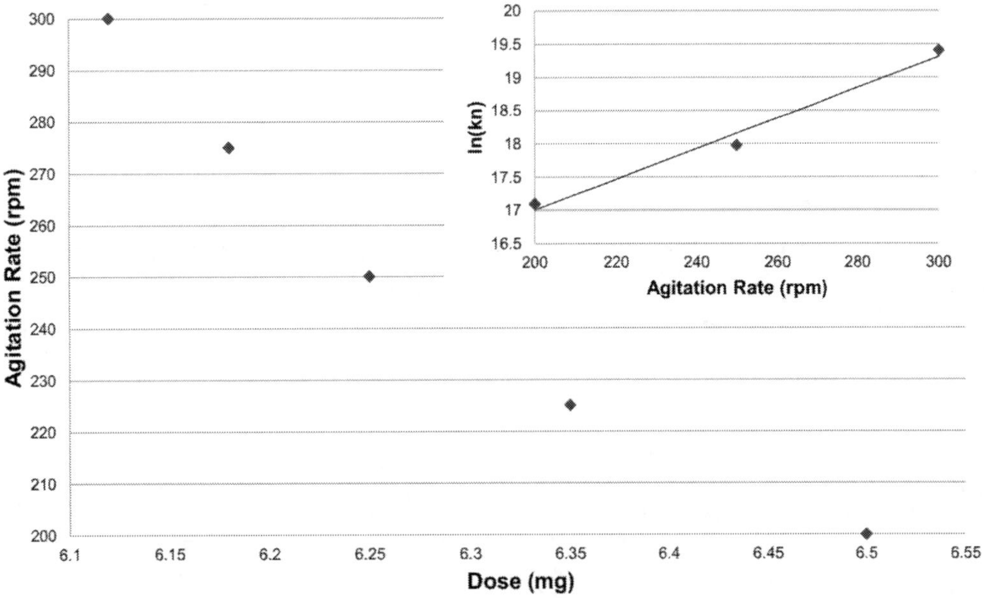

11/9/2015 2015 AIChE Annual Meeting 23

Continuously Seed MSMPR 1

0.000011 kg seed/kg feed, Fixed feed rate

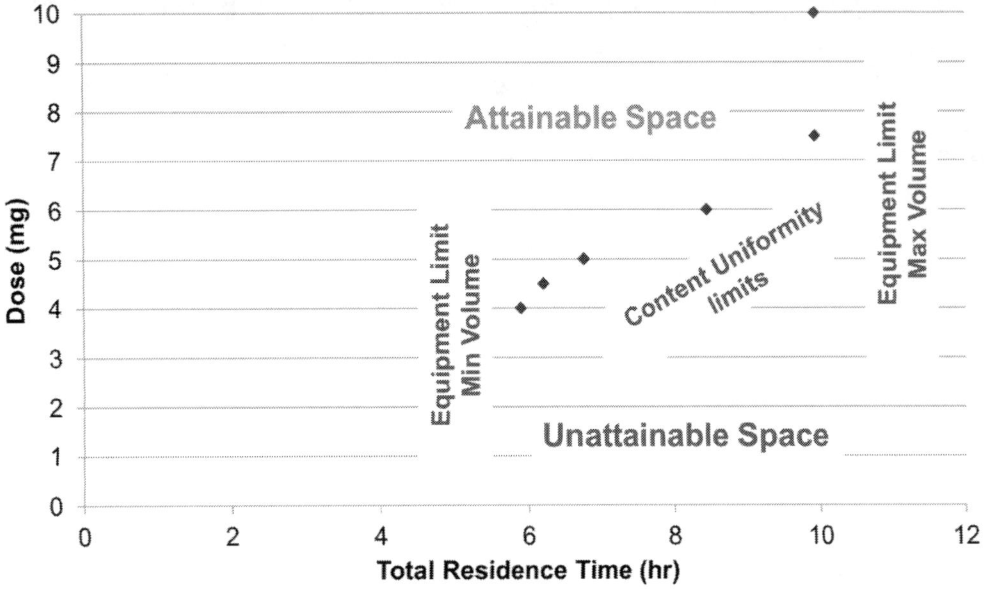

11/9/2015 2015 AIChE Annual Meeting 24

Conclusions

- Integrating product quality constraints (content uniformity) with manufacturing constraints (filtration time), economic constraints (yield) with a process model (operational parameters) allows for attainable regions to be probed.
  - Minimizes experimentation.
  - Allows for "what-if" scenarios to be considered.
- For this case, with the crystallization kinetics considered,
  - Seeding is needed to meet constraints on content uniformity RSD (3%) for doses below 7 mg due.
  - Increasing nucleation by an order of magnitude (via increasing the agitation level) is not sufficient to decrease the product particle size.

2015 AIChE Annual Meeting
November 8-13, Salt Lake City, US

Application of Raman and ATR-UV/Vis Spectroscopy for a Model-Free and Model-Based Active Polymorphic Feedback Control of Crystallization Processes

E. Simone, A. Borsos, Z.K. Nagy

PURDUE UNIVERSITY   ROYAL SOCIETY OF CHEMISTRY   erc European Research Council   Loughborough University

Overview

- Introduction (Polymorphism of OABA, Experimental rig)
- Model-free active polymorphic feedback control (APFC)
- Model-based APFC:
  - Model development
  - Parameters identification and validation
  - Optimization
- Conclusions

PURDUE UNIVERSITY   ROYAL SOCIETY OF CHEMISTRY   erc European Research Council   Loughborough University

Polymorphism of OABA

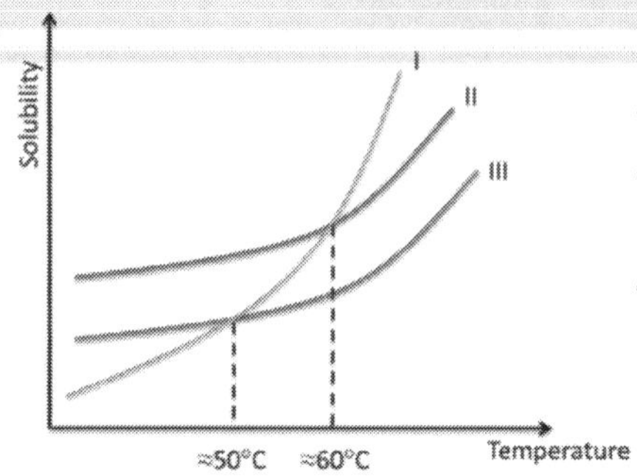

- Three known forms of OABA
- Form I and II are normally nucleated from solution
- Form III is very difficult to nucleate but can be obtained by melt crystallization or polymorphic transformation above 60° C

Form I: orthorhombic cell, **zwitterions** and neutral molecules (1:1 ratio)

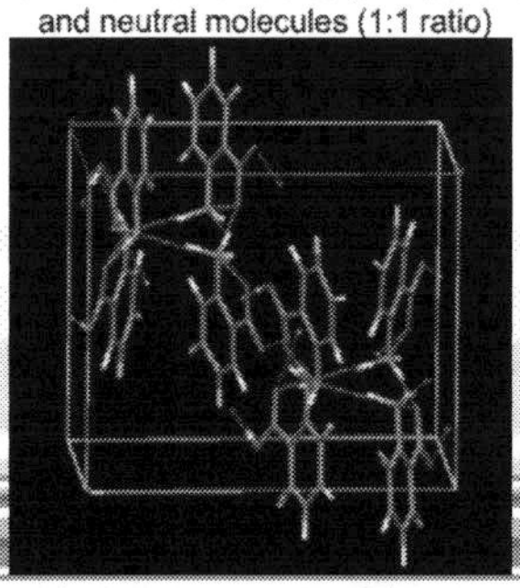

Form II: orthorhombic cell, **dimers** of neutral molecules

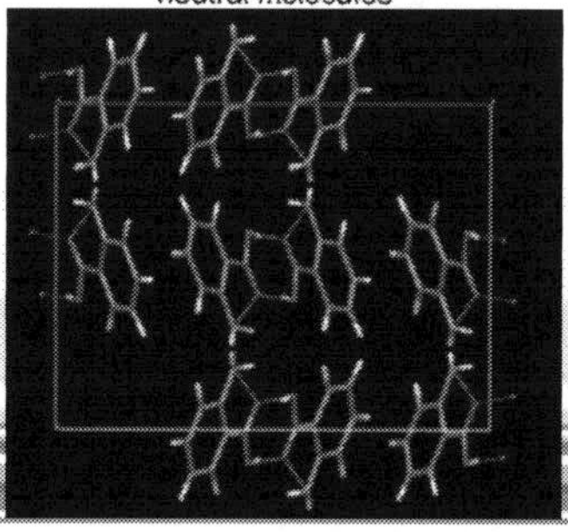

Form III: monoclinic cell, **dimers** of neutral molecules

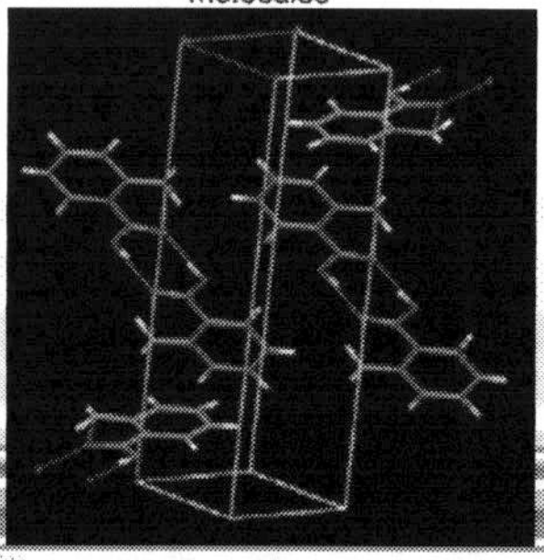

Jiang et al., *Crystal Growth and Design*, 2010

Experimental rig

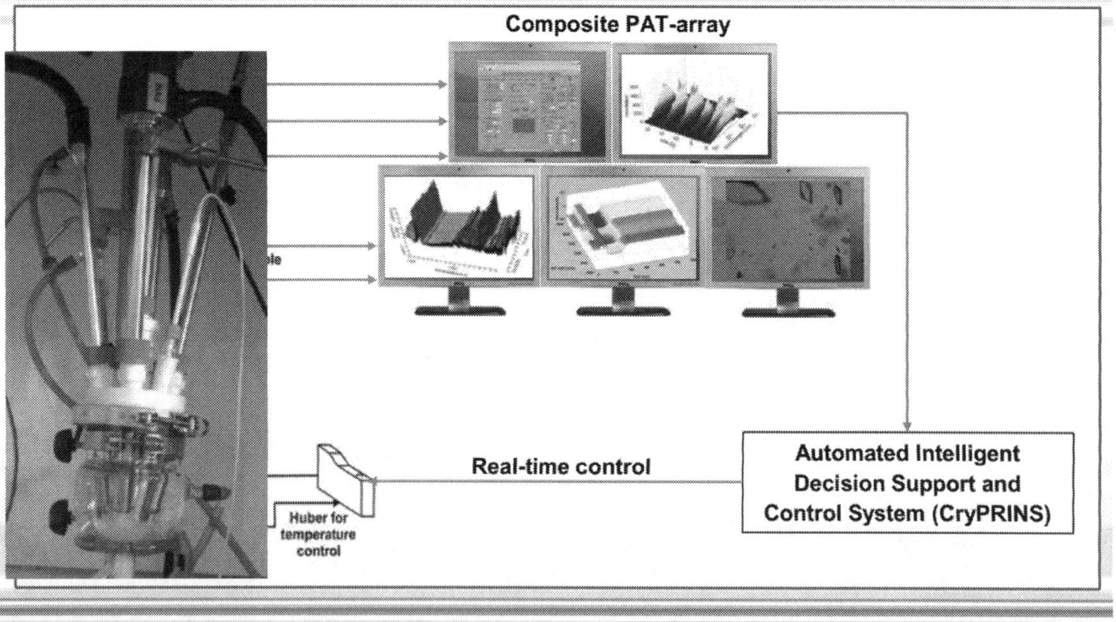

Definition of Active Polymorphic Feedback Control

- Detects the formation of a **polymorphic mixture** and eliminates the metastable form by triggering a *controlled dissolution cycle*, and allowing the growth of the stable form using *supersaturation control*
- Uses a combination of **Raman spectroscopy** (calibration-free), **ATR-UV/Vis spectroscopy** and **CryPRINS**

PURDUE UNIVERSITY | ROYAL SOCIETY OF CHEMISTRY | erc European Research Council | Loughborough University

Active polymorphic feedback control (APFC) strategy

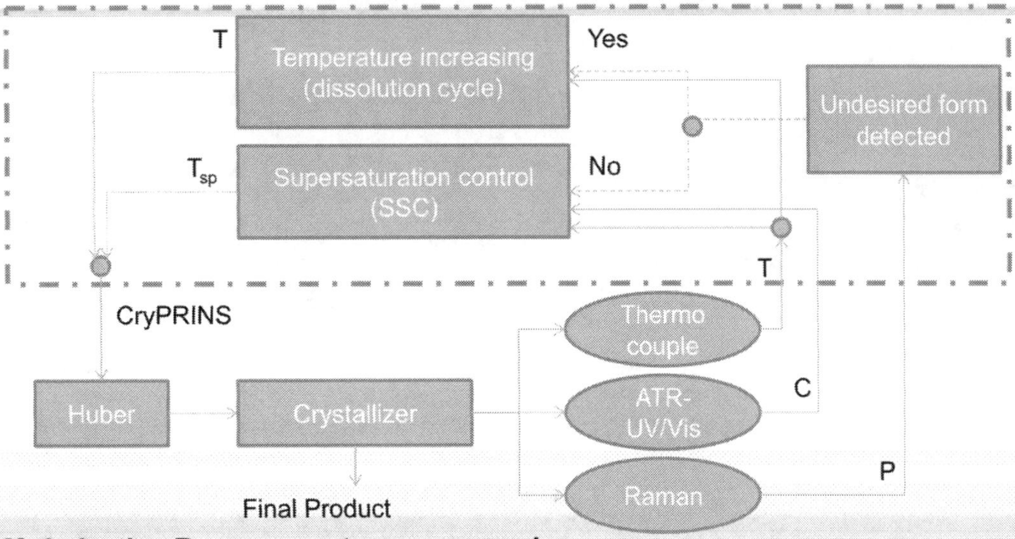

- 2° derivative Raman spectrum was used
- The Raman peak at 770 cm^{-1} was used to reveal the presence of the metastable form in the slurry

PURDUE UNIVERSITY | ROYAL SOCIETY OF CHEMISTRY | (Simone et al., *Crystal growth and design*, 2014) | erc European Research Council | Loughborough University

APFC using Raman and supersaturation control (1)

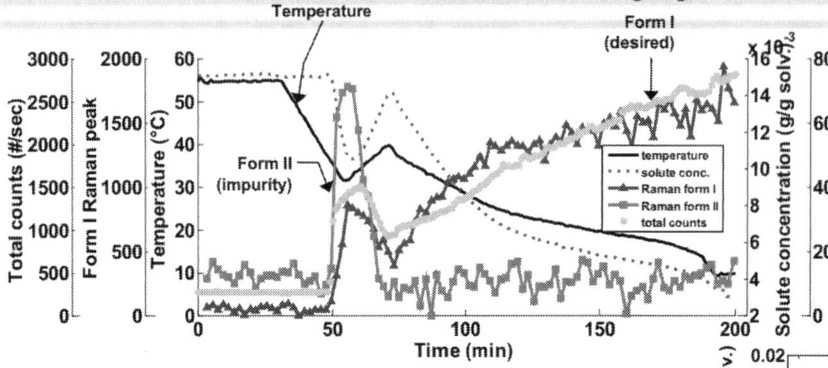

- Seeding of mixtures of OABA polymorphs
- PVM to check the purity of the final form

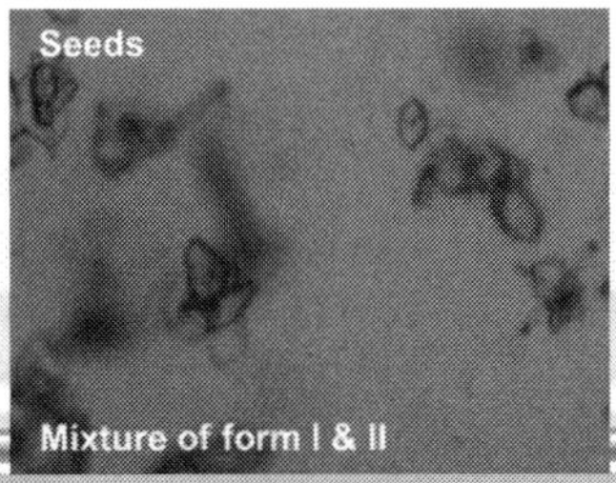

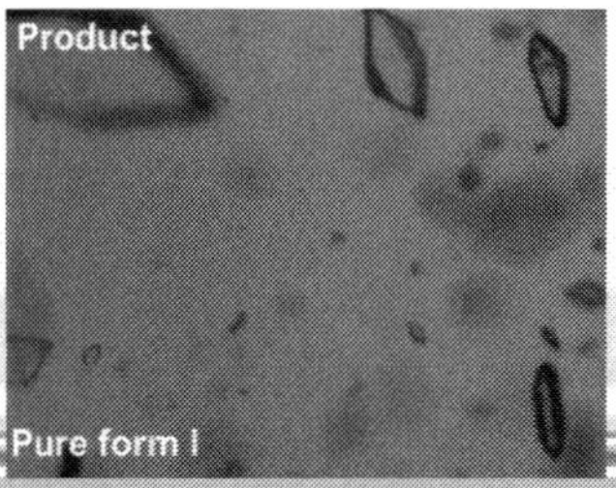

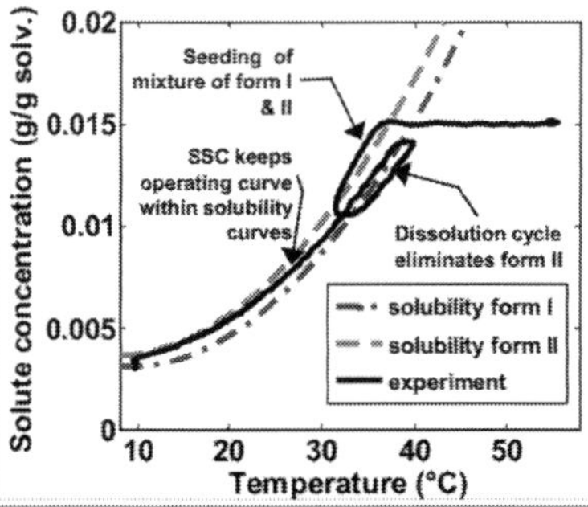

PURDUE UNIVERSITY | ROYAL SOCIETY OF CHEMISTRY | erc European Research Council | Loughborough University

(Simone et al., *Crystal growth and design,* 2014)

APFC using Raman and supersaturation control (2)

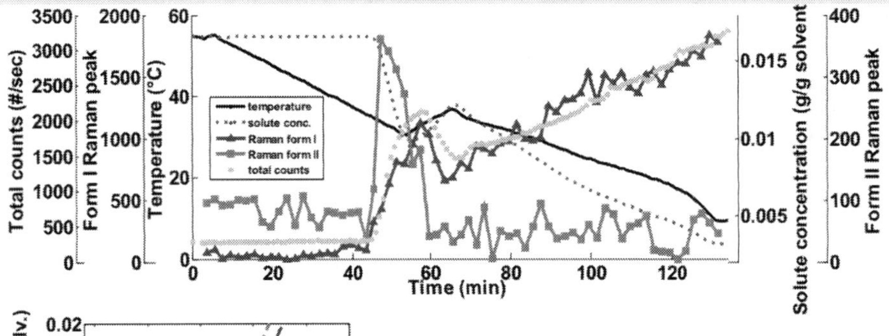

- Nucleation of mixtures of OABA polymorphs

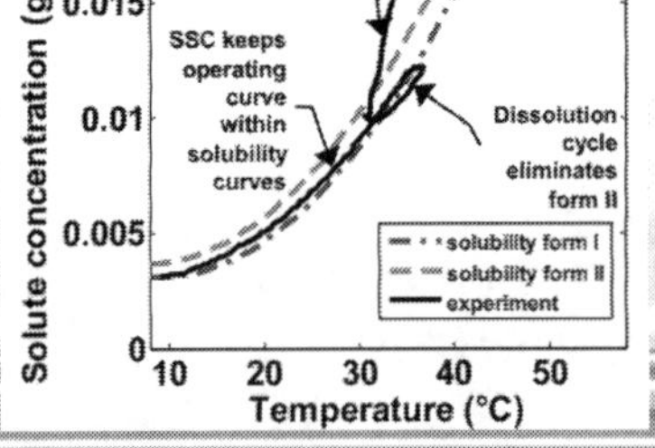

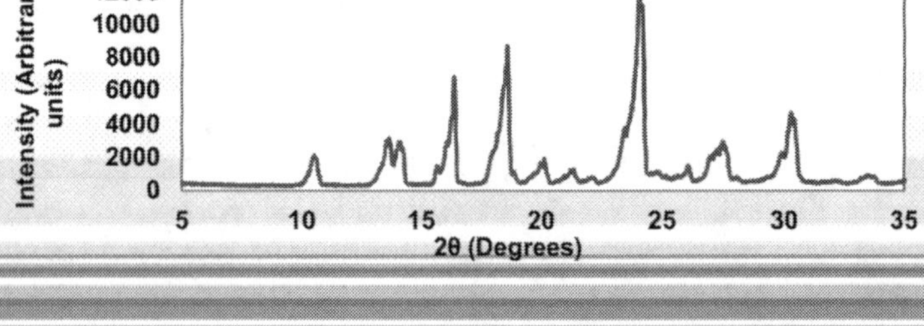

PURDUE UNIVERSITY | ROYAL SOCIETY OF CHEMISTRY | erc European Research Council | Loughborough University

(Simone et al., *Crystal growth and design,* 2014)

Kinetic parameters for the model based APFC

- Dissolution of form II
- Dissolution of form I
- Growth of form II
- Growth of form I
- Nucleation of form I (primary and secondary)

PURDUE UNIVERSITY · Royal Society of Chemistry · erc European Research Council · Loughborough University

Population balance equations

- Two sets of equations, one for each polymorph
- Method of moments used to express the PBEs

$$\frac{d\mu_{0,I}}{dt} = B_I$$

$$\frac{d\mu_{i,I}}{dt} = i(G_I - D_I)\mu_{i-1,I} \qquad i = 1,2,3$$

$$if\ \mu_{1,I} \leq 0\ then\ D_I = 0$$

$$\frac{d\mu_{0,II}}{dt} = 0$$

$$\frac{d\mu_{i,II}}{dt} = i(G_{II} - D_{II})\mu_{i-1,II} \quad i = 1,2,3$$

$$if\ \mu_{1,II} \leq 0\ then\ D_{II} = 0$$

$$\frac{dC_{liq}}{dt} = 3k_v\rho_c\left(-G_{II}\mu_{2,II} - G_I\mu_{2,I} + D_{II}\mu_{2,II}\right)$$

$$D_{I,II} = A_{dI,II}exp\left(\frac{-B_{dI,II}}{T}\right)(1 - S_{I,II}) \quad G_{I,II} = A_{gI,II}exp\left(\frac{-B_{gI,II}}{T}\right)(S_{I,II} - 1)^{a_{I,II}} \quad S_{I,II} = \frac{C}{C_{satI,II}}$$

$$B_I = A_bS_I^{a_b} + K_{b1}S_Iexp\left(\frac{-K_{b2}}{(ln\ S_I)^2}\right) \qquad B_{II} = 0$$

(Hermanto et al., *AIChE Journal*, 2007)
(Wantha & Flood, *Chemical Engineering and Technology*, 2013)

PURDUE UNIVERSITY · Royal Society of Chemistry · erc European Research Council · Loughborough University

Kinetics parameters estimation

- **Specific experimental design** to separate different mechanisms
- **Concentration only** fitted (using *fminsearch* and *fmincon* in Matlab 2013)
- Initial crystal size distribution from **Malvern Mastersizer**

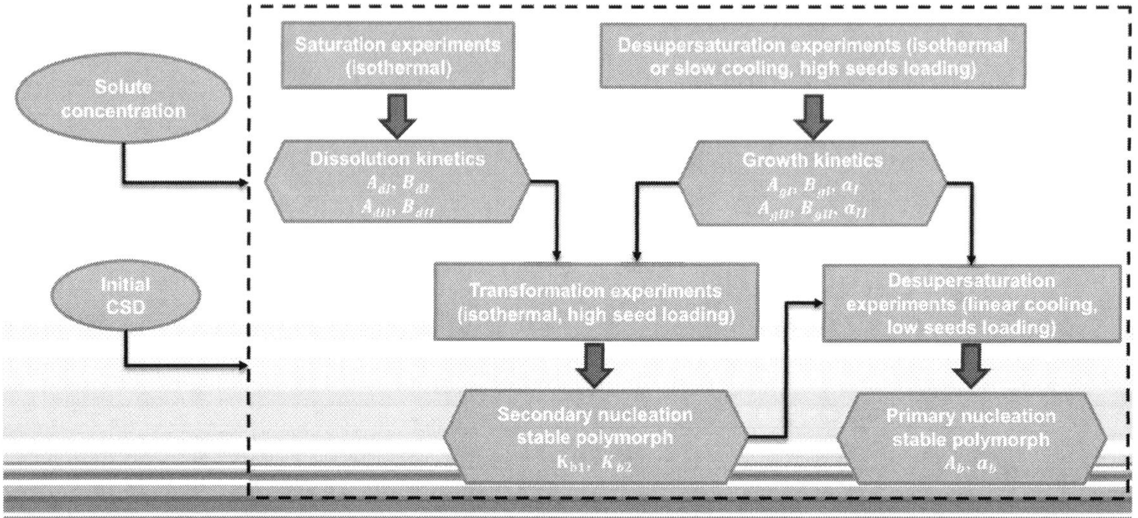

PURDUE UNIVERSITY ROYAL SOCIETY OF CHEMISTRY

(Simone et al., *manuscript in preparation*)

erc European Research Council Loughborough University

Estimated kinetic parameters

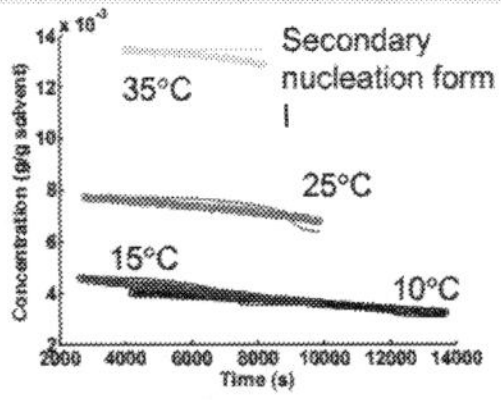

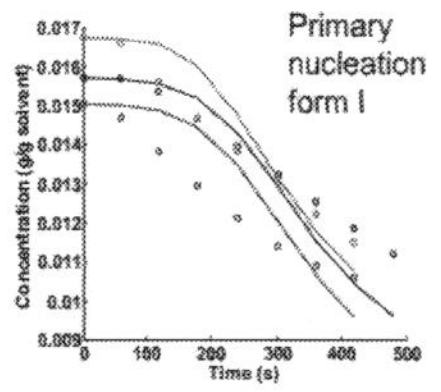

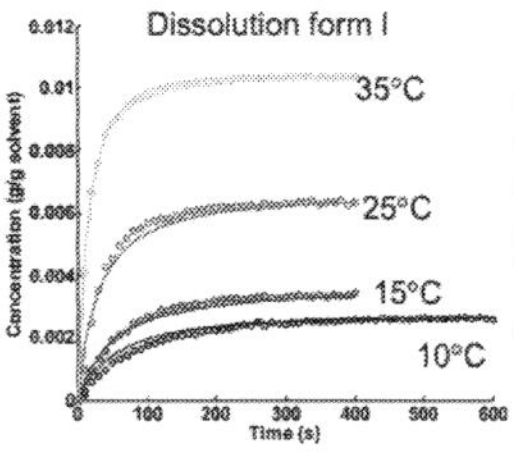

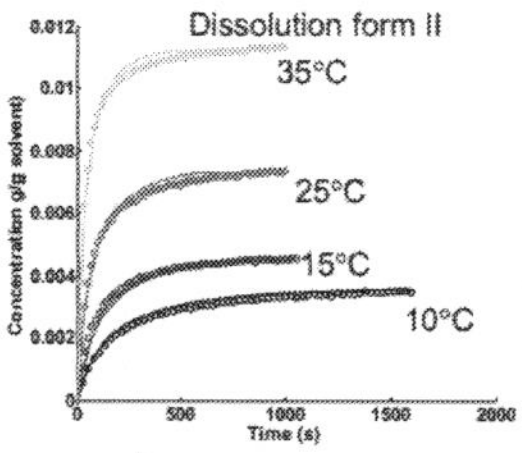

Kinetic parameter	Constant	Value	Upper limit of the interval of confidence	Lower limit of the interval of confidence
D_I	A_{dI}	4.3675	2.078	9.181
D_I	B_{dI}	4376.9	4160.1	4604.5
D_{II}	A_{dII}	1.148e-3	7.478e-4	1.764e-3
D_{II}	B_{dII}	1845.3	1719.7	1980.1
G_I	A_{gI}	2.385e8	1.149e8	4.950e8
G_I	B_{gI}	11033.6	10832.4	11238.5
G_I	a_I	1.131	1.018	1.245
G_{II}	A_{gII}	1.258e-4	6.15e-6	2.576e-3
G_{II}	B_{gII}	2032.9	1378.3	2998.3
G_{II}	a_{II}	1.003	0.766	1.239
B_I	K_{b1}	1.574e10	1.149e10	2.156e10
B_I	K_{b2}	1.483	0.380	0.409
B_I	A_b	1.656e8	7.385e6	3.710e9
B_I	a_b	2.36	-13.813	18.53

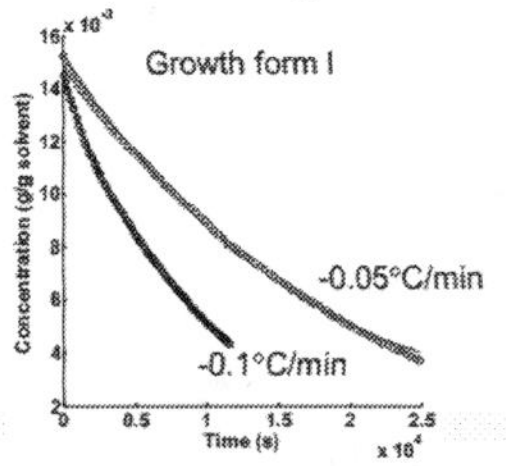

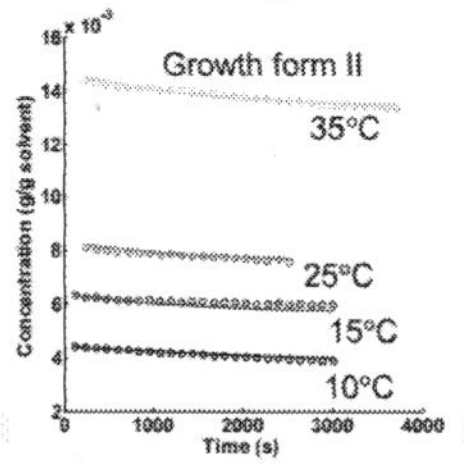

- ❑ Primary nucleation is the more uncertain parameter estimated
- ❑ The best fitting was obtained for growth and dissolution kinetics

PURDUE UNIVERSITY ROYAL SOCIETY OF CHEMISTRY

(Simone et al., *manuscript in preparation*)

erc European Research Council Loughborough University

Model validation (form II)

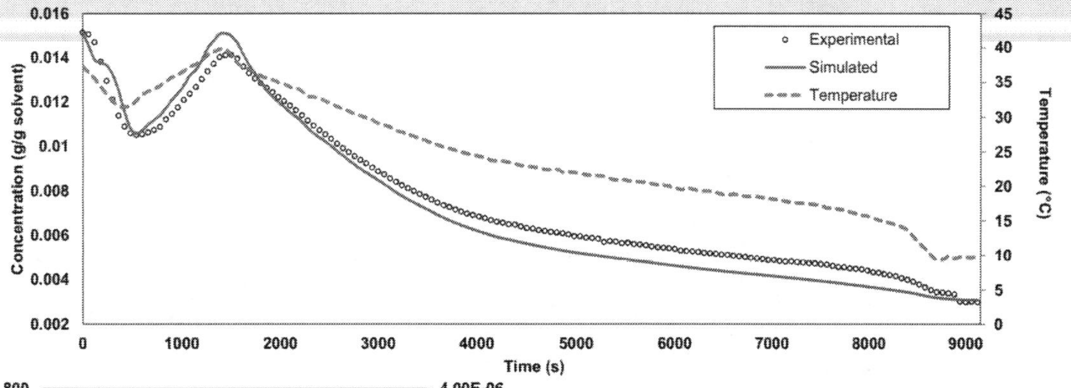

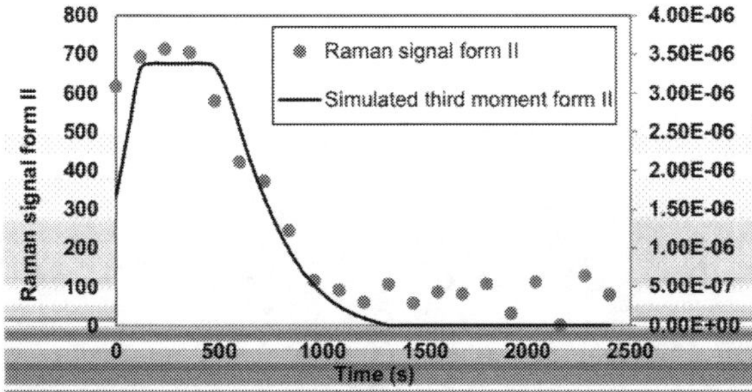

- The kinetics parameters estimated were used to calculate solute concentration during an APFC
- The simulated profile is very close to the experimental

PURDUE UNIVERSITY · ROYAL SOCIETY OF CHEMISTRY · erc European Research Council · Loughborough University

(Simone et al., *manuscript in preparation*)

Model validation (form I)

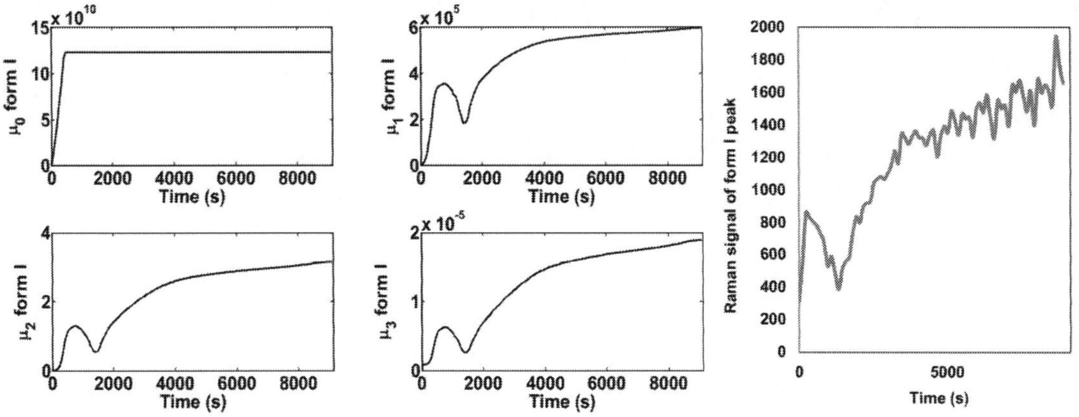

Initial conditions:

- Seeding temperature: 37.26°C
- Concentration of the solution: 0.015 g/g solvent
- Total mass of solid in the vessel: 0.6g
- Mass of solvent: 400 g

PURDUE UNIVERSITY · ROYAL SOCIETY OF CHEMISTRY · erc European Research Council · Loughborough University

(Simone et al., *manuscript in preparation*)

Optimization conditions

- Minimization performed using both the *genetic algorithm toolbox* and *fmincon* in Matlab 2013

$$\min_{T(k)}\left(-\bar{L}_{I,end}\right)\ with\ \bar{L}_{I,end}=\frac{\mu_{1,end}}{\mu_{0,end}}$$

Optimization problem

Objective function

$$-0.5\le\frac{dT}{dt}\le 0.5$$

$$11°C\le T\le 45°C$$

$$c_{end}\le c_{max,end}=0.005\text{g/g solvent}$$

$$\mu_{II1,end}=0$$

Constraints

PURDUE UNIVERSITY

ROYAL SOCIETY OF CHEMISTRY

(Simone et al., *manuscript in preparation*)

erc European Research Council

Loughborough University

Optimal temperature profile

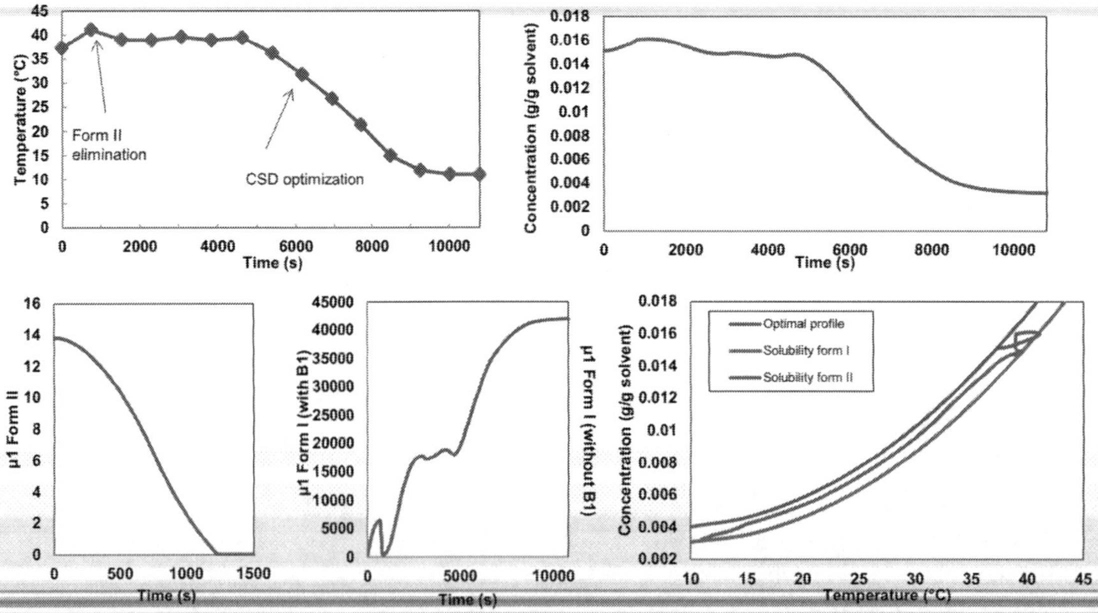

PURDUE UNIVERSITY

ROYAL SOCIETY OF CHEMISTRY

(Seki et al., *www.aidic.it/isic18/webpapers/73**Seki**.pdf*)

(Majumder & Nagy, *AIChE Journal*, 2013)

erc European Research Council

Loughborough University

Conclusions

- The **effectiveness** of the APFC in obtaining the pure desired polymorph of OABA was demonstrated **experimentally**
- The principal kinetic parameters of OABA were estimated using **sequential experiments to decouple mechanisms**
- A model based on the **method of moments** was validated and applied in a process optimization framework
- The optimal profile has two dissolution cycles to both eliminate form II and maximize the size of form I

PURDUE UNIVERSITY   ROYAL SOCIETY OF CHEMISTRY   erc European Research Council   Loughborough University

Thanks for your attention

E. Simone, Z.K. Nagy

e.simone@lboro.ac.uk

Acknowledgements:

- European Research Council for the funding (grant no. **280106-CrySys**)
- Royal Society of Chemistry for the travel grant

erc European Research Council Established by the European Commission   ROYAL SOCIETY OF CHEMISTRY

Description of Technical Importance

Batch crystallization is commonly employed in the fine-chemical and pharmaceutical industries to obtain solid products of high purity. We are exploring the use of crystallization for a different application: the separation of non-radioactive salts from nuclear waste. If successful, such an operation could significantly reduce costs and expedite waste clean up efforts.

Regardless of the application, control of the mean crystal size is often critically important; yet, establishing such control has remained an outstanding issue. In this work we leverage process analytical technology to develop a feedback control method and demonstrate—for the first time—explicit control over the mean size of crystals produced by unseeded, batch cooling crystallization.

Mass-Count Framework
for Crystal Size Control

Daniel J. Griffin, Martha A. Grover,
Yoshiaki Kawajiri, and Ronald W. Rousseau

Georgia Institute of Technology

Background and Introduction

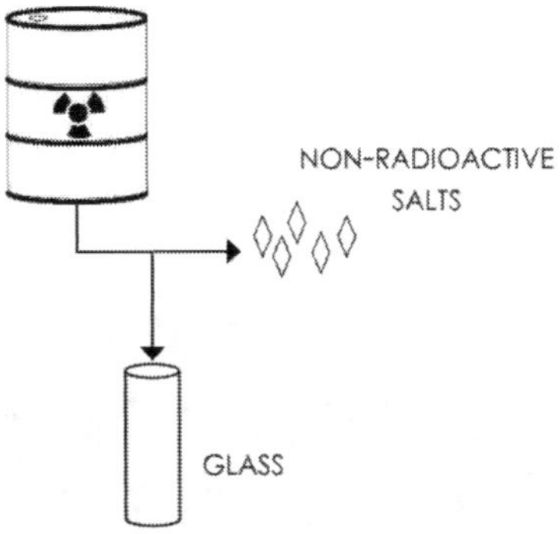

Legacy Waste Cleanup

TANK FARM C UNDER CONSTRUCTION (HANFORD SITE, WA)

56 MILLION GALLONS OF RADIOACTIVE WASTE STORED IN 177 TANKS

US DOE CLEANUP:

REMOVE, **VITRIFY (TURN INTO GLASS)**,

AND ENCASE IN STEEL CANISTERS

RESEARCH ON SEPARATION OPERATIONS

4

Crystallization of Salts from Waste

CRYSTALLIZATION

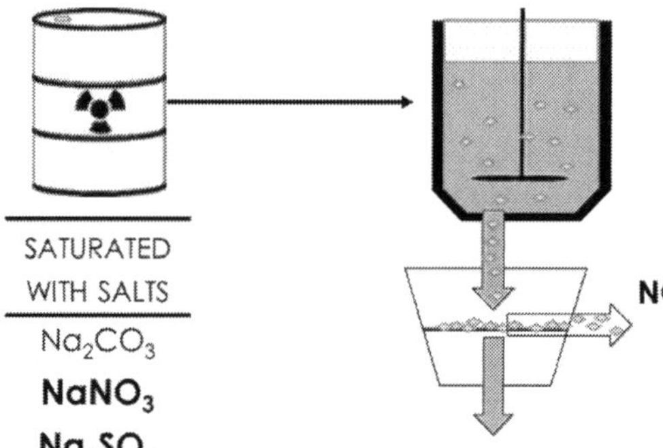

SATURATED WITH SALTS

Na_2CO_3

$NaNO_3$

Na_2SO_4

$NaNO_2$

...

NON-RADIOACTIVE SODIUM SALTS

RADIOACTIVE LIQUID **STREAM** (VITRIFIED)

EFFECTIVE SEPARATION

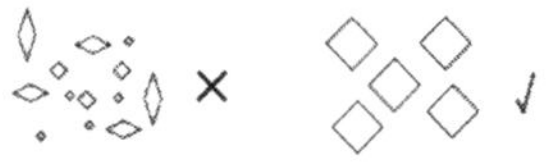

NEED TO **CONTROL CRYSTALLIZATION TO PRODUCE LARGE CRYSTALS THAT CAN BE EASILY FILTERED**

5

Research Objective and Outline

OBJECTIVE

establish a methodology for reliably controlling the **mean crystal size**

unseeded, batch cooling crystallization

OUTLINE

1. EXPERIMENTAL SYSTEM AND ONLINE MEASUREMENT TOOLS
2. MASS-COUNT FRAMEWORK
3. DATA-DRIVEN MODELING AND DYNAMICS PROGRAMMING
4. RESULTS

6

1. Experimental System and Online Measurement Tools

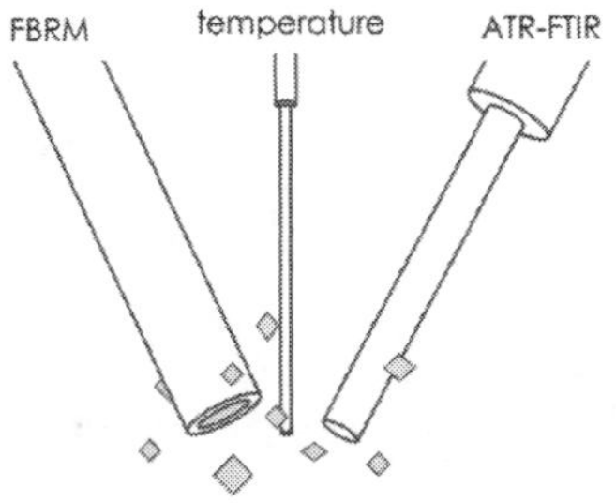

Experimental System

study crystallization from a two-component electrolytic solution

110 g **$NaNO_3$** and 7.25 g **Na_2SO_4**
per 100 g **water**

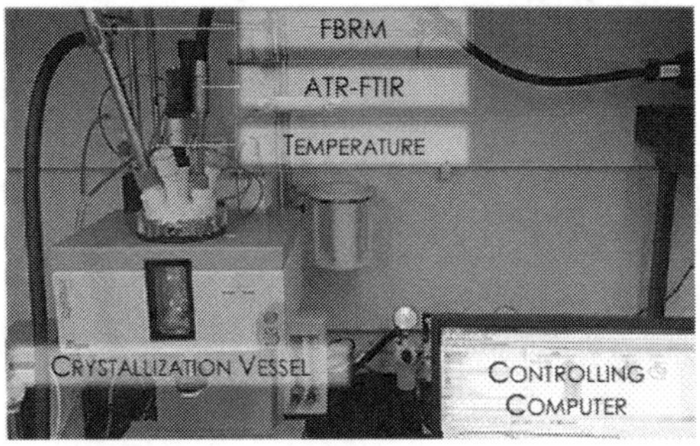

lab crystallization system

8

Online Measurement Tools

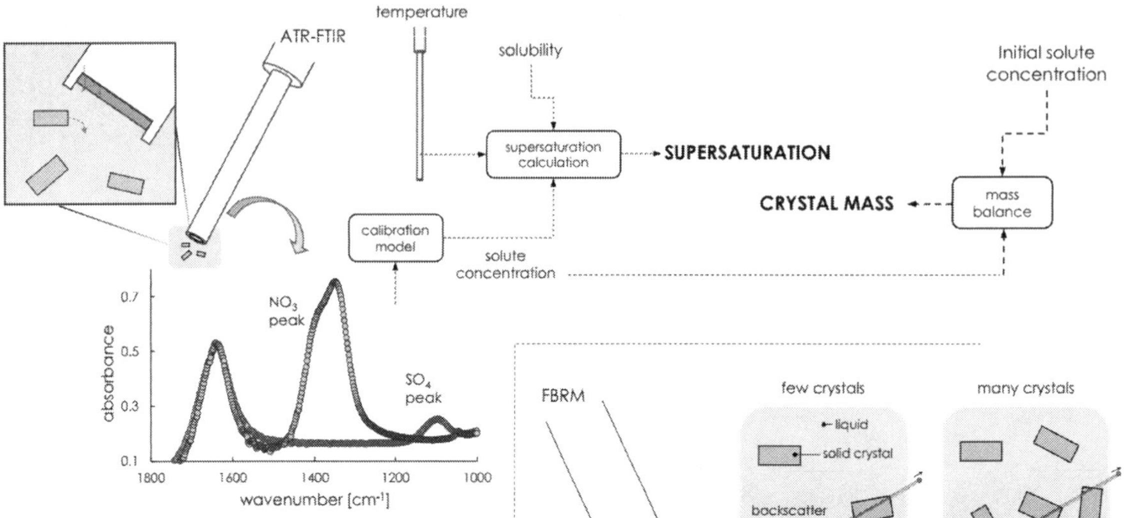

Dunuwila, D. D. and Berglund, K. A. 1997. **ATR FTIR Spectroscopy for In Situ Measurement of Supersaturation**. *Journal of Crystal Growth*.

FBRM: **CHORD COUNT** ~ CRYSTAL NUMBER

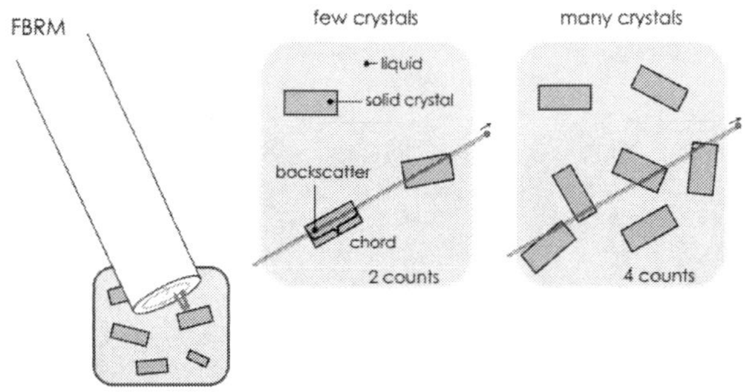

Heinrich, J. and Ulrich, J. 2012. **Application of Laser-Backscattering Instruments for In Situ Monitoring of Crystallization Processes - A Review.** *Chemical Engineering & Technology*.

9

2. Crystallization Dynamics from a Different Perspective:

Mass-Count Framework

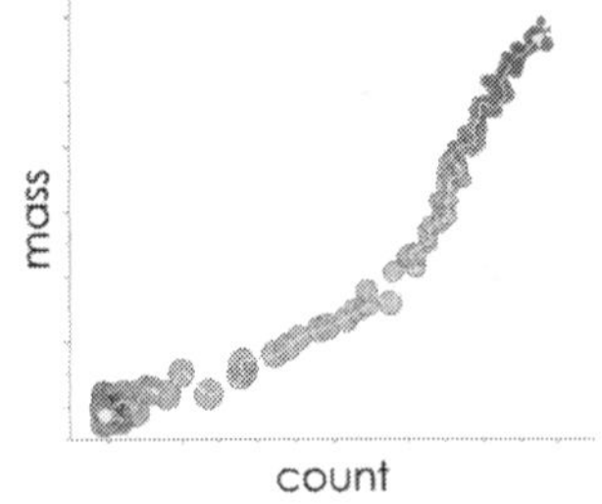

Mass-Count Plot

UNSEEDED, COOLING CRYSTALLIZATION

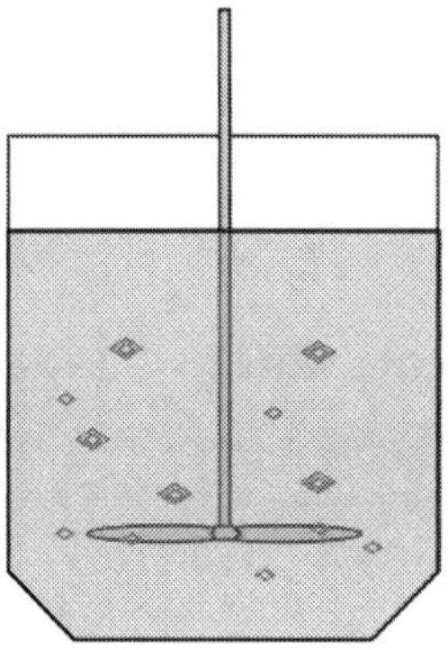

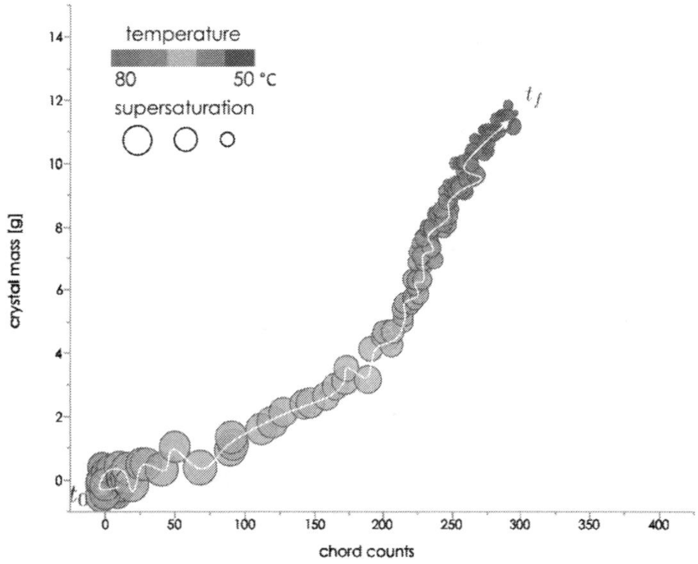

Griffin, D. J., Grover, M. A., Kawajiri, Y., and Rousseau, R. W. 2015. **Mass-Count Plots for Crystal Size Control.** *Chemical Engineering Science.*

11

Mass-per-Count Indicates Crystal Size

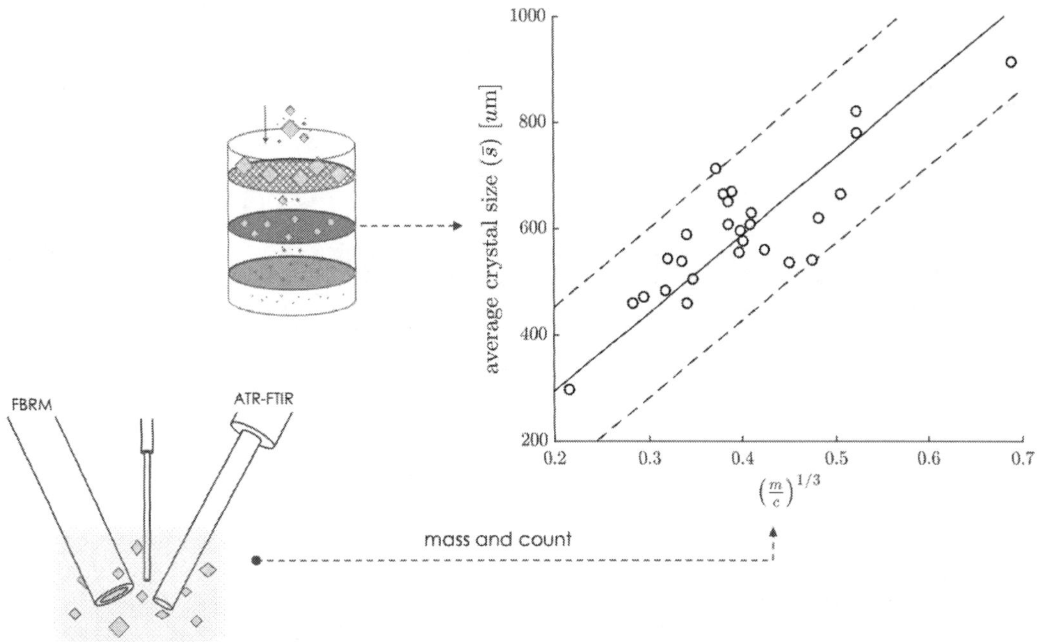

Griffin, D. J., Grover, M. A., Kawajiri, Y., and Rousseau, R. W. 2015. **Combining ATR-FTIR and FBRM for Feedback on Crystal Size.** *Proceedings of the American Control Conference*, Chicago, IL.

12

Clarified Control Objective

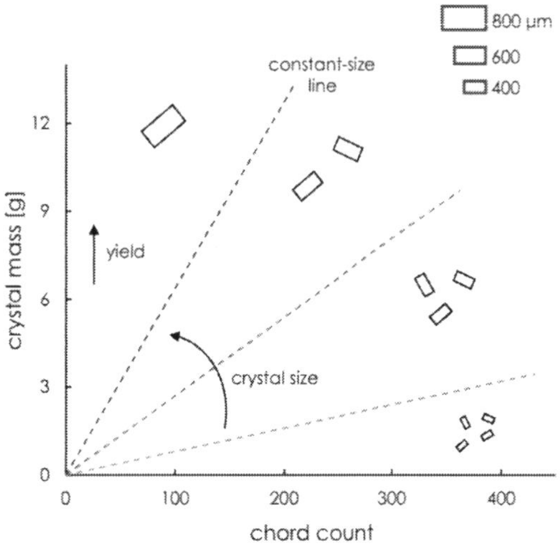

Clarified Control Objective: reach a target crystal mass and number

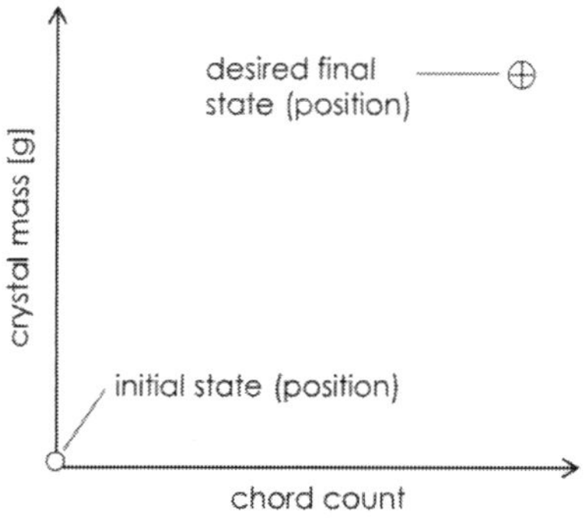

13

3. Data-Driven Modeling and Dynamic Programming

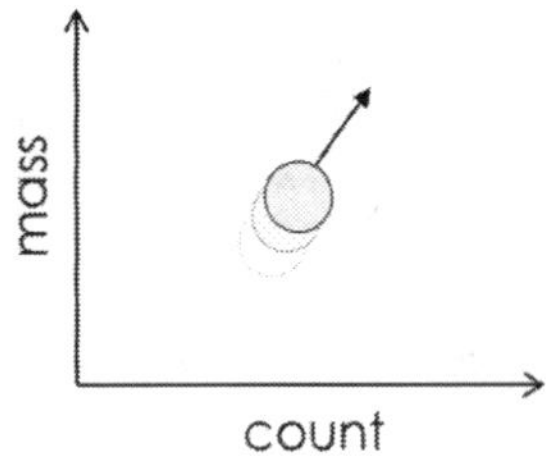

Data-Driven Modeling and Dynamic Programming

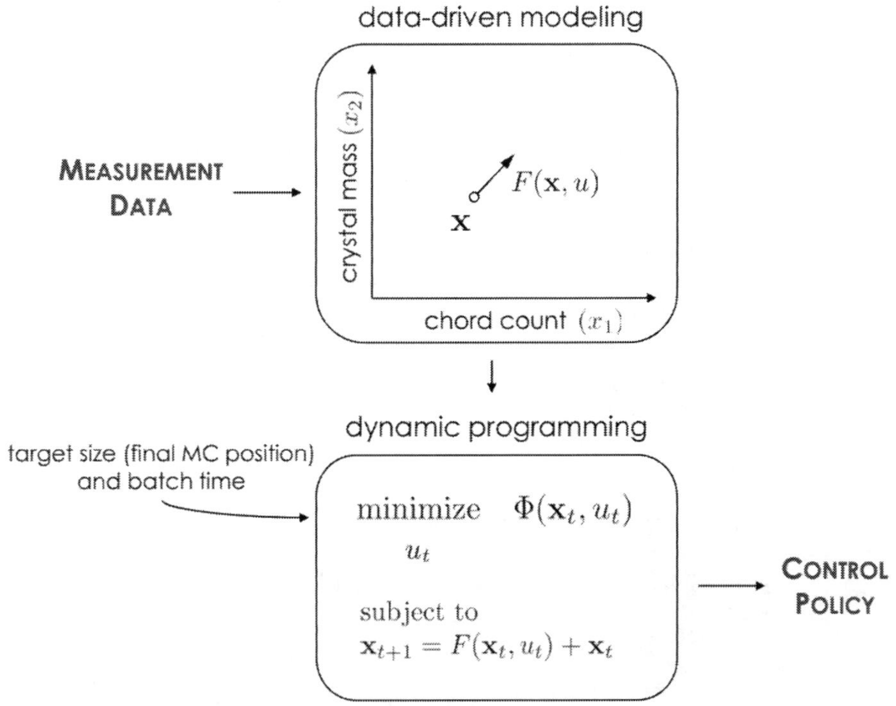

15

Model From Data

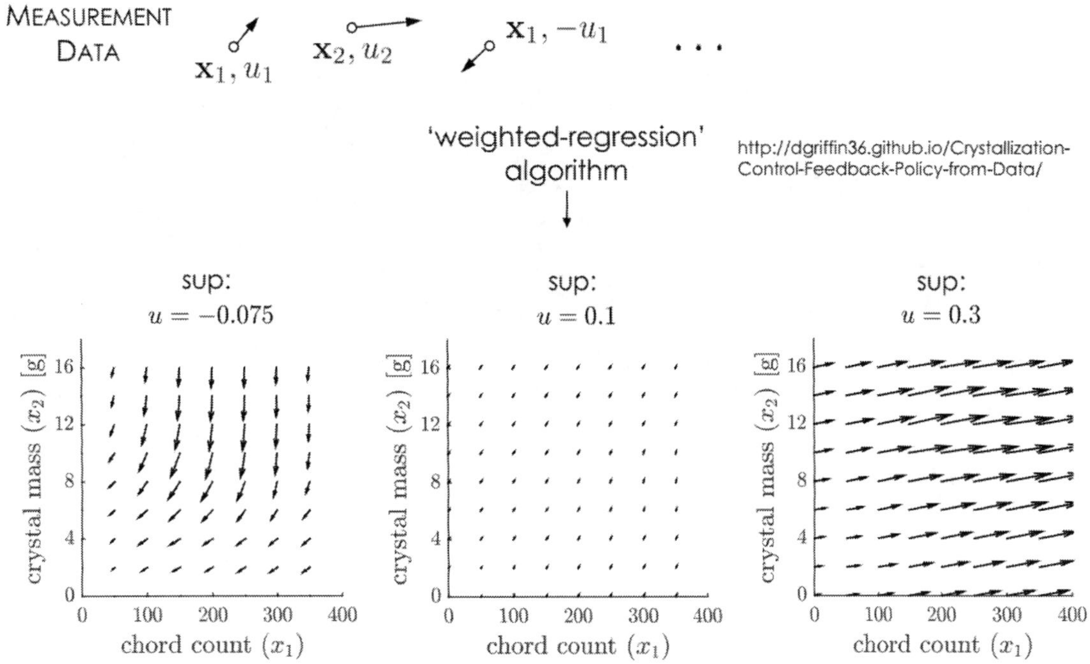

16

Dynamic Programming

Input-effort

$$\underset{\text{control policy } u_\tau = \pi(\mathbf{x}_\tau)}{\text{minimize}} \left\{ \rho \sum_{\tau=0}^{N-1} u_\tau^2 + \sum_{\tau=1}^{N} (t_\tau / t_N)^\gamma d(\mathbf{x}_\tau, \mathbf{x}^\oplus) \right\}$$

batch time

distance-to-target

$$\text{subject to} \quad \mathbf{x}_{\tau+1} = F\big(\mathbf{x}_\tau, u_\tau = \pi(\mathbf{x}_\tau)\big)\Delta t + \mathbf{x}_\tau, \ \tau = 0, \ldots, N-1;$$

$$\mathbf{x}_0 = \mathbf{x}^{\text{init}}.$$

model of dynamics

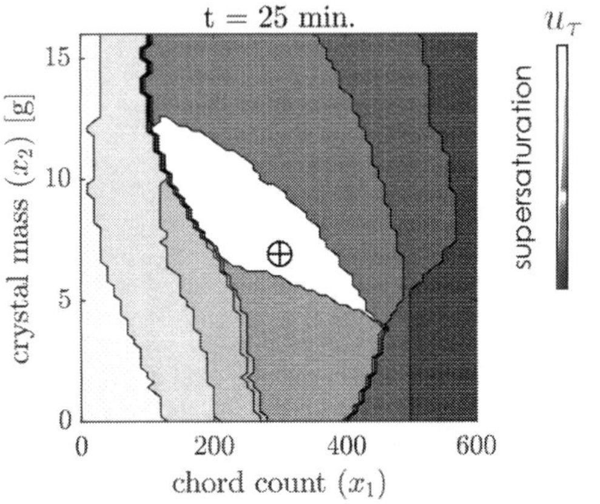

http://dgriffin36.github.io/Crystallization-Control-Feedback-Policy-from-Data/

17

Feedback Control Strategy

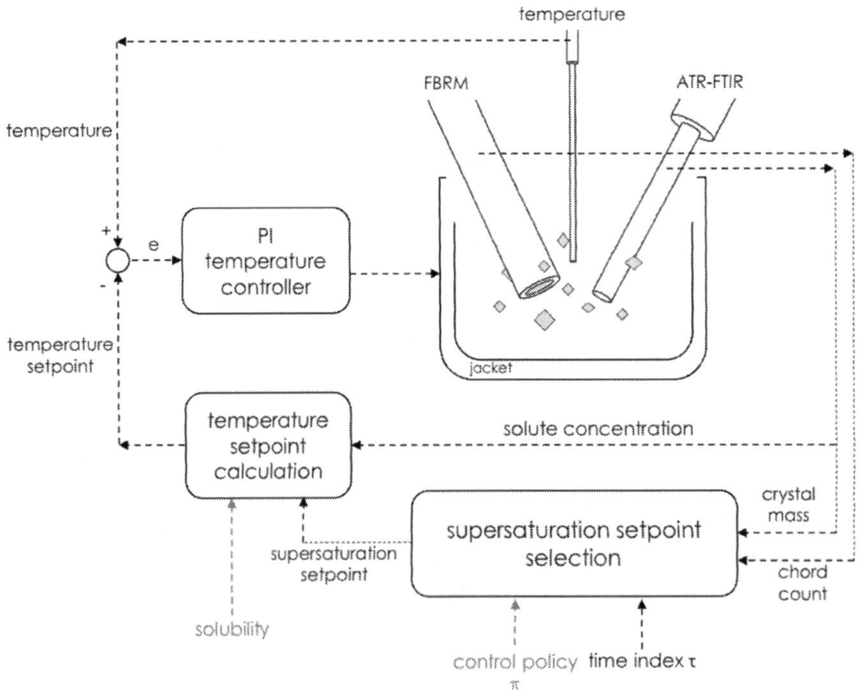

18

4. Results:

Application of strategy to control crystal size

Three Test Runs

Run number	Target final position: (count, mass (g))	Target crystal size (µm)	Set batch time (min.)
1	(300, 7)	420 (261–579)*	30
2	(200, 10)	540 (381–701)	60
3	(75, 11)	775 (612–937)	120

*95% confidence interval for the measured average crystal size if the mass-count target is hit exactly

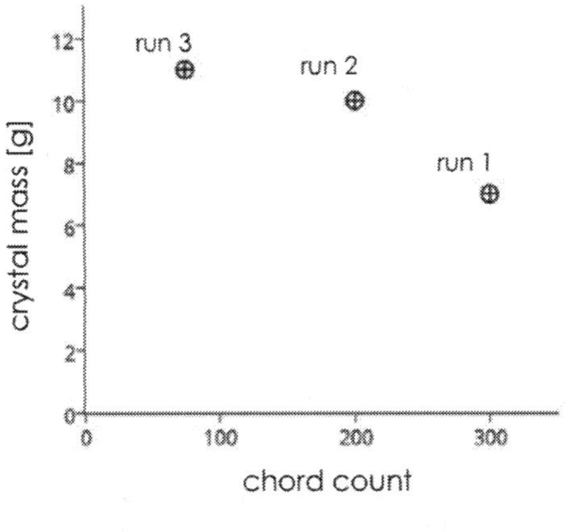

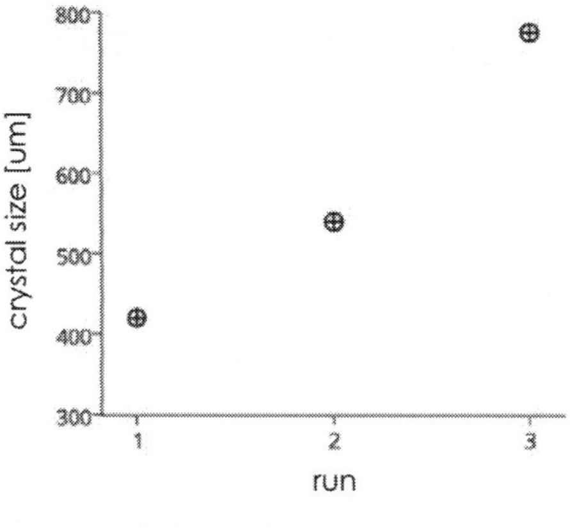

for each run we identify the appropriate control policy

$$u_\tau = \pi(\mathbf{x}_\tau)$$

20

Results

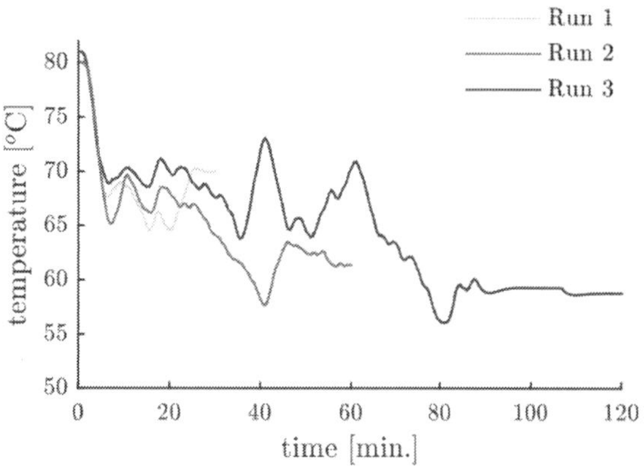

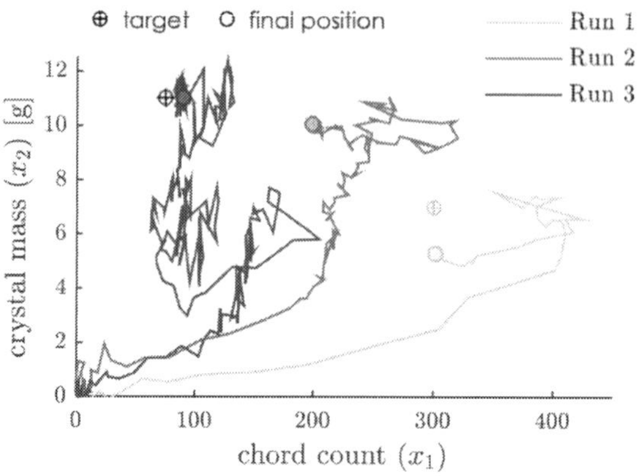

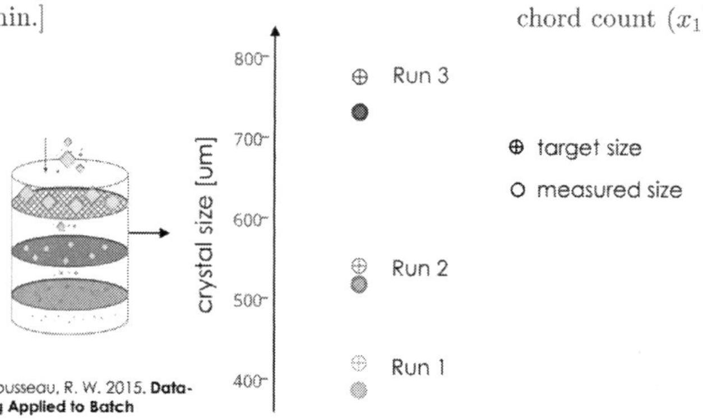

Griffin, D. J., Grover, M. A., Kawajiri, Y., and Rousseau, R. W. 2015. **Data-Driven Modeling and Dynamics Programming Applied to Batch Cooling Crystallization.** *Submitted.*

21

Summary and Conclusion

Online Monitoring of Key Solution and Solid State Properties

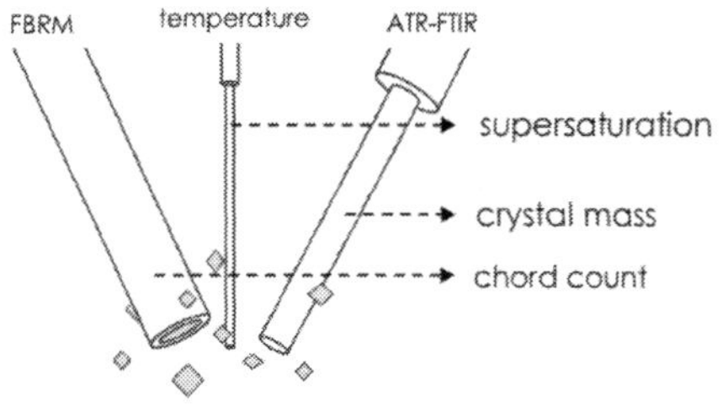

New Perspective of Dynamics

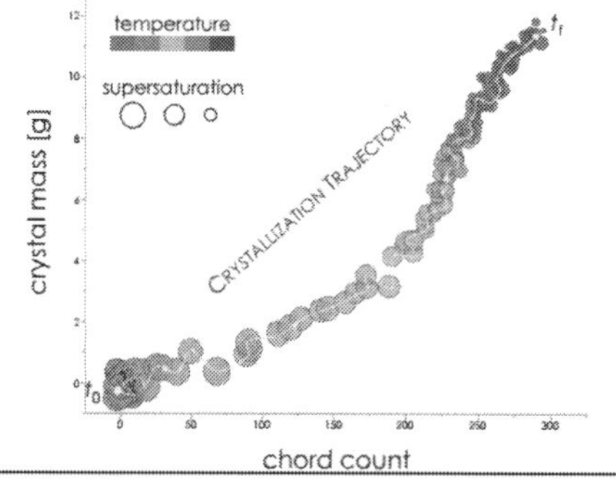

Modeling and Optimal Control Formulation

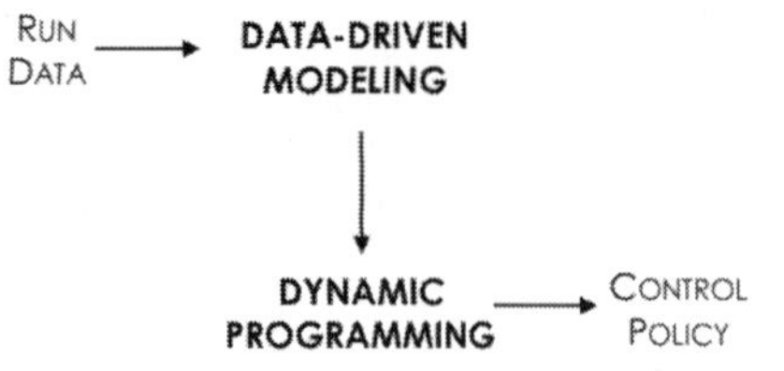

Confirm Control of Crystal Size

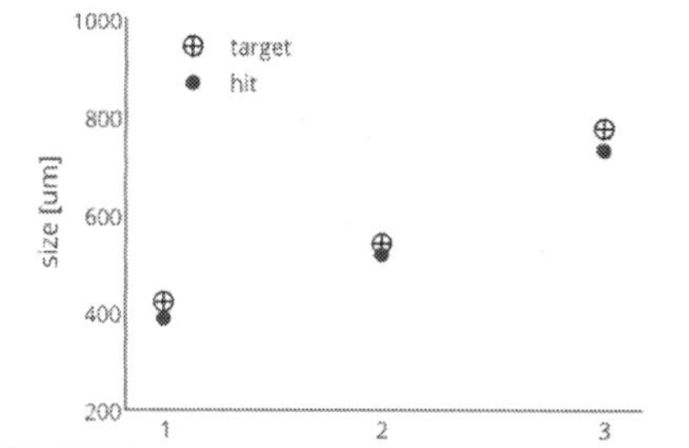

Acknowledgments

Grover, Kawajiri, and Rousseau Research Groups

Prof. Evangelos A. Theodorou (Aerospace Engineering GATECH)

- Consortium for Risk Evaluation with Stakeholder Participation (CRESP)
- Nuclear Energy University Programs (NEUP)
- Innovations in Fuel Cycle Research
- Georgia Research Alliance, and the Cecil J. "Pete" Silas Endowment

CRESP
Consortium For Risk Evaluation with Stakeholder Participation

NEUP Nuclear Energy University Programs
U.S. Department of Energy

innovations
in Fuel Cycle Research

Georgia Tech

23

434276 Nonlinear Model Predictive Control of a Batch Crystallization Process

Abstract:

Batch crystallization is crucial in the pharmaceutical industry because more than 90% of the active pharmaceutical ingredients (API) are in the form of crystals. The crystal size and shape distribution is important for both downstream processing and product quality. The size and shape distribution can be controlled by manipulating the cooling profile of the reactor, which directly affects the supersaturation ratio. The dynamic evolution of the crystal size and shape distribution has been modeled as a system of differential algebraic equations derived using the method of moments [1].

Here, we present a nonlinear model predictive control (NMPC) formulation for batch crystallization using the JModelica [2] platform. This DAE-constrained optimization problem is solved by discretizing the system using Radau collocation and optimizing the resulting algebraic nonlinear problem with Ipopt. We analyze the performance of the NMPC in terms of set point change, model mismatch, and system noise. For all cases, the performance of the NMPC is much better than the open-loop optimal control strategy.

Nevertheless, many state variables cannot be measured accurately online and some model parameters change from batch to batch. This challenge drives the need for an optimization approach like nonlinear moving horizon estimation (NMHE) to estimate unknown states and parameters prior to solving the NMPC problem. Therefore, using the same JModelica framework, we also present a NMHE formulation for the batch crystallization process. Given the batch process, and combining these two formulations, we obtain an expanding horizon estimation problem and a receding horizon model predictive control problem. We test the performance of this control strategy using a case study of a 90-minute batch crystallization process with 90 control steps and sampling steps. The combined solution time for the NMHE and the NMPC formulations is well within the sampling interval, allowing for real world application of the control strategy. Furthermore, significant performance improvements are possible with this online, model-based control strategy over the existing open-loop optimal solution.

References:

[1] Acevedo, D., Tandy, Y., and Nagy, Z. K., Multiobjective Optimization of an Unseeded Batch Cooling Crystallizer for Shape and Size Manipulation, Industrial & Engineering Chemistry Research, 54(7), pp 2156-2166,2015.

[2] Åkesson, J., Årzén, K. E., Gäfvert, M., Bergdahl, T., & Tummescheit, H., Modeling and optimization with Optimica and JModelica. org—Languages and tools for solving large-scale dynamic optimization problems, Computers & Chemical Engineering, 34(11), pp1737-1749, 2010.

423850 Modeling of Crystallization of Melt-Based Oral Dosages in a Drop-on-Demand Manufacturing System

Abstract:

Under the NSF supported Engineering Research Center for Structured Organic Particulate Systems (NSF ERC-SOPS), a dropwise additive manufacturing process for solid oral drug production has been developed as an alternative to conventional pharmaceutical manufacturing methods [1]. This mini manufacturing process for the production of pharmaceuticals utilizes drop on demand printing technology for automated and controlled deposition of melt-based drug formulations onto edible substrates [2]. A supervisory control system, including on-line monitoring, automation and closed-loop control, is implemented on the process, in order to produce individual dosage forms with the desired dosage amount and crystal morphology [3].

The temperature of the deposited drops is controlled indirectly by controlling the temperature of the substrate using a Peltier device placed beneath the substrate on the xy-stage. This allows control of the rate of solidification of melts and thus control of nucleation and crystallization phenomena. Precise control of the drop solidification process occurring on the substrate is very important since the crystallization temperature profile has a strong effect on product solid-state characteristics, influencing the dissolution properties and hence the bioavailability of the drug [3]. Recently, Içten et al. [3] reported that applying a fast cooling rate of 10 °C/min to the deposits containing naproxen results in faster dissolution profiles compared to a slow cooling rate of 1 °C/min, indicating that the crystallization temperature profiles can be used to enhance the solubility of dissolution limited drugs such as naproxen.

This paper presents a model based on the non-isothermal Avrami kinetic equation [4] for the solidification and crystallization processes of the drug deposits produced using the dropwise additive manufacturing process. Temperature dependent crystallization kinetics is monitored under different cooling profiles via an optical microscopy with a hot-stage. Using the proposed model, the temperature profiles leading to the desired solid-state parameters such as the mean feature size can be optimized, which allows achievement of the desired product quality, in this case, dissolution rate.

1. L. Hirshfield, A. Giridhar, L. S. Taylor, M. T. Harris, G. V. Reklaitis, "Dropwise Additive Manufacturing of Pharmaceutical Products for Solvent-based Dosage Forms," *Journal of Pharmaceutical Sciences*, vol. 103, issue 2, pages 496-506, 2014.
2. E. Içten, A. Giridhar, L. Taylor, Z.K. Nagy, G.V. Reklaitis, 2014, Dropwise Additive Manufacturing of Pharmaceutical Products for Melt-based Dosage Forms, Journal of Pharmaceutical Sciences, DOI 10.1002/jps.24367.
3. E. Içten, Z. K. Nagy, G.V. Reklaitis, 2014, Supervisory Control of a Drop on Demand Mini-manufacturing System for Pharmaceuticals, Comp. Aided Process Eng., 33, 535-540.
4. M. Avrami, 1939, Kinetics of phase change I General Theory, Journal of Chemical Physics, 7, 12, 1103-1112.

421342 Towards Predictive Modeling of Crystallization Fouling: Taking into Account Fouling Layer Structures

Abstract:

Towards Predictive Modeling of Crystallization Fouling: Taking into Account Fouling Layer Structures

Jie Xiao, Feng Zhang, Jian Han, Xiao Dong Chen

Suzhou Key Laboratory of Green Chemical Engineering

School of Chemical and Environmental Engineering

College of Chemistry, Chemical Engineering and Materials Science

Soochow University, Suzhou, Jiangsu Province 215123, China

Abstract

The fouling layers formed on heat exchanger surfaces exhibit complicated structures, which essentially determine flow hydrodynamics, fouling kinetics and hence the heat transfer performance. In-depth understanding of deposit layer formation and its structural effects on the fouling process could lead to effective fouling mitigation or even prevention strategies. The exploration tasks, however, are not trivial due to a lack of knowledge of the crystallization process and the complicated deposit structure.

In this talk, I will summarize our past and on-going efforts on numerical simulation of different stages of a complex crystallization fouling process, focusing on deposit layer growth and its impact on the fouling process. The developed method incorporates a pseudo-dynamic scheme where the dynamic fouling process is approximated as a set of sequential steady-state processes taken place in a continuously varying geometric domain. This unique approach allows the characterization of mass, momentum and heat conservations of a continuous flow of liquid over a growing fouling layer. Dynamic evolution of the fouling layer surface and its intricate interactions with hydrodynamics and fouling kinetics can then be rigorously taken into account. The introduced model was validated using the experimental data for a calcium sulphate fouling system. The effects of the solution chemistry and operating conditions on fouling resistance evolution were quantified through a comprehensive parametric study. Furthermore, we are developing methods that can be used to quantitatively evaluate the effects of fouling layers' *interior* structures on fouling dynamics. We have systematically investigated four representative schemes for fouling layer characterization: i.e., a homogeneous porous medium that is impermeable to water (HoIm), a heterogeneous porous medium that is impermeable to water (HeIm), a homogeneous porous medium that is permeable to water (HoPe), and a heterogeneous porous medium that is permeable to water (HePe). Under the same operational conditions, four models offer significantly different prediction results on the fluid velocity, temperature distribution and fouling resistance. It can be concluded that numerical model development should take the fouling layer structure into account, and the scheme of HePe that best resembles a real fouling layer structure should be a promising option.

Keywords: Crystallization fouling; fouling inhibition; fouling layer structure; CFD model; pseudo-dynamic scheme

434926 Modeling Crystallization in Hydrogels Under Double-Diffusion Conditions

Abstract:

In nature, the deposition of biominerals usually takes place in gel-like extracellular matrix environments where the transport of ions to growth sites is regulated. These matrices play an important role in controlling the nucleation, growth, and the final morphology of biominerals. Hydrogels can serve as excellent matrix models for biomineralization. The hydrogel matrix provides a template for crystallization, while at the same time controls the diffusion of ions, regulates the nucleation and growth rates, and directs the hierarchical organization of the minerals. In this work, we investigate the crystallization of calcium oxalate, the primary mineral constituent of kidney stones, in hydrogels. We employ a double-diffusion experimental set-up in which a hydrogel (serving as the crystallization medium) is brought in contact with calcium and oxalate ion reservoirs that serve as sources for the reactants, thus allowing for counter diffusion of the reactant ions across the hydrogel column and precipitation of calcium oxalate in the gel. We study the nucleation, growth, polymorphic outcome and morphological evolution of the calcium oxalate crystals as affected by the gel density, reservoir concentrations, molar stock solution ratios, and additives. The experimentally observed precipitation patterns (time and location within the gel) in the calcium oxalate-gel systems are in very good agreement with theoretical predictions based on the relative diffusion coefficients of both ions. Such understanding of the diffusion and precipitation processes can aid in predicting the time scales of nucleation and growth of the crystals and can facilitate further use of gel matrices as templates for biomineral formation or for development of bio-inspired materials.

422312 Distributed Renewable Fuel and Power: Challenges and Opportunities

Abstract:

Distributed renewable fuel and power: challenges and opportunities

Andrew Allman and Prodromos Daoutidis

Recent research in the development of novel energy systems has been moving towards a more decentralized, renewable based infrastructure. In particular, distributed microgrids and biorefineries have been identified as systems that can increase the resiliency of the energy infrastructure, reduce the carbon emissions associated with the production of power and fuels, and mitigate the losses, emissions, and risks inherent in long distance transport of power and fuels. Extensive research exists on the design and operation of microgrids and biorefineries, and previous studies suggest that synergies between the two systems may make a combined system more efficient than the sum of its parts [1], [2]. For example, including an electrolyzer with a renewable microgrid allows for carbon-neutral hydrogen generation which can be used for shifting syngas, generating ammonia, or reacting platform chemicals to fuels in a biorefinery. Additionally, heat, power, and produced fuel can be passed between the two systems.

Motivated by these considerations, this work considers combined power and fuel polygeneration systems. A reference architecture is proposed combining a microgrid with wind turbines, solar panels, microturbines, and hydrogen storage with a biorefinery containing both gasification-based and thermochemical fuel producing pathways. Such an architecture inherently contains multiple time scales: power and heat demands must be met instantaneously, but fuel demands can be met over daily or weekly requirements. The combined model contains a large number of variables from the microgrid (required due to its shorter time scale) and nonlinearities from the biorefinery (required for accuracy in capturing chemical transformations). As such, optimizing the combined system design is more difficult than optimizing each system individually. We address the optimal design of such a "system of systems" by decomposing and reformulating the problem in a way that isolates different time scales and nonlinearities. We also discuss and provide insights on a number of research challenges and opportunitites that emerge in the design, operation, and control of such polygeneration systems. On the enterprise level, fuel and power supply chains must be considered to determine locations where distributed polygeneration is economically viable. On a system-wide level, unit scheduling programs that are proactive and robust to uncertainties in supply and demand are needed. On a local level, novel control architectures must be developed that take into account the complex interconnections between subsystems, the multiple time scales present in the system, and the different operating states of the microgrid.

[1] R. Agrawal, N. R. Singh, F. H. Ribeiro, W. N. Delgass, D. F. Perkis, and W. E. Tyner, "Synergy in the hybrid thermochemical-biological processes for liquid fuel production", Computers and Chemical Engineering, vol. 33, pp. 2012-2017, 2009.

[2] B. Subhadra and M. Edwards, "An integrated renewable energy park approach for algal biofuel production in United States", Energy Policy, vol. 38, pp. 4897-4902, 2010.

430181 A Model-Based Framework for Fault-Tolerant Dispatch of Distributed Energy Resources

Abstract:

Distributed energy resources (DERs) are composed of modular energy generation units such as fuel cells, photovoltaic arrays, small-scale wind turbines, battery storage and other distributed generation technologies that are deployed close to the point of consumption. The modular nature of DERs allows them to be integrated with existing grid infrastructure or implemented in a stand-alone manner. DERs have aided in the integration of sustainable energy resources into the power grid, but these new resources pose fundamental challenges including fluctuations in generation from intermittent availability of renewable resources, as well as the resulting increased network communication necessary to coordinate local generation supply with load demand. These and other challenges can be dealt with through use of process control and monitoring techniques (e.g., [1]-[4]).

Process control is necessary to ensure that the load demand is met and that the economical operation of each DER is maintained. However, the stability and performance of DER smart grids have not been rigorously assessed in the presence of faults at the local and network levels. This is significant given the fact that the distributed power market is driven by the need for reliable high-quality power, and the substantial impact local disruptions in power flow can have when the DERs are integrated to support grid operations. In this context, fault-tolerant control (FTC) is an important tool for reducing the possible deterioration in power quality in the face of faults and uncertainties, resulting in increased reliability of the network.

In prior work, FTC has been studied in the context of a small-scale network of solid oxide fuel cells (SOFCs), where the focus has been on the detection and recovery from faults at the local level without supervisory oversight [5]. DER health monitoring took place through use of an alarm threshold on a properly designed observer-based output residual. Exploiting inherent actuator redundancy in SOFCs, a number of stabilizing controller configurations were designed, and a methodology for active switching between the configurations in the event of a threshold breach was developed. The main contribution of this work was the characterization of a stable region within which each controller configuration could operate. However, the faults considered were limited to total and abrupt component failures, thus negating the need for fault estimation and accommodation. Furthermore, the detection scheme was stability-based, and thus not suited for detecting performance-degrading faults that do not compromise stability. The results also did not address the supervisory actions needed in the case of irreparable local faults potentially leading to total system failure.

The objective of this contribution is to extend our previous work through the development of a model-based framework for integrated fault detection, estimation and accommodation in DER networks. The proposed framework extends our previous results in three important directions. In one direction, fault estimation capabilities are explicitly integrated into the fault detection mechanism through the use of a data-based moving-horizon parameter estimation scheme which uses past input and output data to generate an estimate of the local fault magnitude. A key feature of this scheme is that it can still be used for fault identification even after fault recovery, thus allowing for timely fault detection in the event of consecutive system faults. This is an advantage over residual generation approaches where the alarm threshold needs to be redesigned following every fault recovery event. In another direction, the proposed framework considers the use of fault accommodation, rather than reconfiguration, to compensate for local faults. Through a stability analysis, an explicit characterization of closed-loop stability for each component is obtained as a function of the fault size and the controller design parameters, and is used as a metric in determining the appropriate post-fault response. Finally, an optimization-based supervisory control system is designed to reconfigure the power demand distribution in the network by adjusting the set-points of the functioning DERs to compensate for the loss in power supply caused by irreparable local failures. The developed fault-tolerant dispatch framework is illustrated through a simulation case study.

References:

[1] S. Barsali, M. Ceraolo, P. Pelacchi, and D. Poli, "Control techniques of dispersed generators to improve the continuity of electricity supply," *Proceedings of IEEE Power Engineering Society Winter Meeting*, pp. 789–794, 2002.

[2] M. Marei, E. El-Saadany, and M. Salama, "A novel control algorithm for the DG interface to mitigate power quality problems," *IEEE Transactions on Power Delivery*, 19, 1384-1392, 2004.

[3] M. Marwali and A. Keyhani, "Control of distributed generation systems - part I: Voltages and currents control," *IEEE Transactions on Power Electronics*, 19, 1541–1550, 2004.

[4] M. Deshmukh and S. Deshmukh, "Modeling of hybrid renewable energy systems", *Renewable and Sustainable Energy Reviews*, 12, 235 –249, 2008.

[5] Y. Sun, S. Ghantasala and N. H. El-Farra, "Monitoring and Fault-Tolerant Control of Distributed Power Generation: Application to Solid Oxide Fuel Cells," *Proceedings of American Control Conference*, pp.448-453, 2010.

421000 Data-Driven Optimal Design of Combined Heat and Power for Residential Neighborhoods

Abstract:

Data-Driven Optimal Design of Combined Heat and Power for Residential Neighborhoods

Abigail D. Ondeck, Michael Baldea, and Thomas F. Edgar

McKetta Department of Chemical Engineering

The University of Texas at Austin, 1 University Station C0400, Austin, TX 78712

email: aondeck@utexas.edu

Combined heat and power (CHP) facilities are a very promising path to reducing CO_2 emissions and increasing efficiency in the power generation sector, especially when combined with residential solar photovoltaic (PV) power generation. CHP facilities rely on natural gas, a cleaner fuel than coal, to generate electricity, as well as hot and chilled water to provide heating and cooling. The ability to supply these essential residential utilities in an efficient way on a medium to large scale opens the path for combining district cooling, heating and power generation, and suggests that CHP plants are an appealing choice as an integrated utility supplier for the neighborhood of the future. Yet, there are currently no CHP plants that serve exclusively residential neighborhoods, and published works exploring this possibility are scarce.

In our previous work [1], we demonstrated that a CHP plant with PV integration can in principle meet the demands of a residential neighborhood during the summer months. However, a CHP plant must be optimally sized to maximize efficiency in producing electricity, heating, and cooling, and to lower the capital and marginal costs. High generation efficiency and low costs are typically achieved by producing the minimum amount of thermal energy necessary to meet thermal loads, while not limiting the amount of electricity generated. The majority of the approaches described in the literature for optimal CHP plant sizing (e.g. load-duration curve, multi-criteria sizing) thus assume that a connection to the grid is available to act as a profitable outlet for any electricity generated in excess.

In the paper, we describe a novel strategy for the optimal sizing and integration of a CHP plant as a utility producer for a residential neighborhood operating in island (i.e., grid-disconnected) mode, and the potential for using photovoltaics and energy storage, implemented in a centralized fashion, to alleviate fluctuations in residential demand. Utilizing data collected by Pecan Street Research Inc., a non-profit smart grid demonstration project headquartered at The University of Texas at Austin, residential heating, cooling, and electricity demand are analyzed and evaluated. These data are then used to create a time-resolved energy profile describing residential energy use, which is subsequently serves as a basis for optimizing the size of the CHP plant so that all utility demands are met, while minimizing the capital and marginal costs. Based on the optimally sized CHP plant, we compute an optimal operating schedule while accounting for PV electricity generation and energy storage.

References:

[1] Ondeck, A.D., Baldea, M., Edgar, T.F. (2015). *Data-Driven Modeling and Optimal Operation of District-level Combined Heat and Power and Photovoltaic Power Generation System*. Applied Energy, submitted.

426819 Reduction in Cycling of the Boilers By Using Large-Scale Energy Storage of Cryogenic Carbon Capture

Abstract:

With the continuing increase in electricity consumption and penetration of renewable energy, power grids will require more responsive power generation units. When electricity demand is high or when renewable energy sources have high fluctuations, reliable power sources should be used to meet the demand and compensate for the fluctuations in the renewables. This becomes especially important for some of the renewables such as wind and solar that are very intermittent. In such events, two main approaches are used to meet the power demand or compensate for the fluctuations: (1) using responsive power generation units such as gas- or oil- fired turbines (2) cycling the baseline generation units in advance. However, running gas- or oil-fired turbines (as the spinning/non-spinning reserves) is very expensive and inefficient. Cycling the baseline generation units, such as coal-fired boilers, also imposes various thermal damage to the boilers [1]. A third approach to overcome the abovementioned problems is to use large-scale energy storage systems. Energy storage offers several advantages to the power generation units. The most important function of the energy storage is the peak-shaving of the electricity load. This means that whenever power could be produced in excess, it is generated and saved in an energy storage medium. Then, during periods with high electricity demand, stored energy is used to meet the demand. When energy storage is used, the baseline power generators are not obliged to cycle to meet the peak demand. In addition, by appropriately designing an energy storage system, there is less need for the responsive but expensive generators (spinning/non-spinning reserves) because the required electricity during peak hours is met from the energy storage. Other advantages of the energy storage systems include the ability to have energy arbitrage, the possibility to smooth the variations in renewable energy sources and to increase their contribution to the grid, and power grid stability [2].

Different energy storage systems have been considered in the literature. The most widely applied technologies for energy storage are batteries, compressed air energy storage (CAES), pumped hydro-electricity energy storage (PHES), and thermal storage [3], [4]. CAES and PHES are 56-85% efficient; however, they have specific geographical requirement that limit their applicability. Batteries are very efficient (90%) and have fast response to load changes and do not have geographical limitations but they are expensive [5]. In this study, thermal energy storage, offered by Cryogenic Carbon Capture TM (CCC) [6], is used for integration with power generation units. The energy storage offered by the CCC process is +95% efficient and it does not have the geographical limitations mentioned above [5]. The cost of the energy storage system of the CCC process is also small compared to the constructional cost of the CCC process. The economic benefits of the energy storage of the CCC process is high enough that it can recover a significant fraction of the cost of construction of the CCC process [7].

Cryogenic carbon capture is a new technology for the separation of CO_2 from flue gases generated by the power plants. CO_2 is removed from the flue gas by cooling the gas down to a temperature below the desublimation point of the CO_2. Solid CO_2 is then separated from the slurry by using filters [5]. The cooling medium for the separation is provided by two refrigeration cycles. CF_4 is used as the refrigerant in one of the cycles while liquefied natural gas (LNG) is used in the other one. These refrigeration cycles consume the most energy. The energy storage, however, provides the flexibility to produce LNG in excess when electricity demand is low or when renewables are available. Overproduced LNG is then stored inside a well-insulated tank. During periods with high electricity demand, LNG is recovered from the tank and is used in the CCC process. Using stored LNG leads to significant reduction in the power consumption of the refrigeration cycles. This means that more power is available for the grid during peak hours. In addition, LNG is vaporized after removing heat from the CCC process and the vaporized LNG is used for power production in a gas turbine [8]. The power produced from the gas turbine can be used to meet the power demand of the CCC process as well as the residential demand. In addition, by producing gas power from the stored LNG, it is possible to keep the baseline power unit from cycling and avoid running the spinning/non-spinning reserves when there is a peak electricity demand or high fluctuations in renewables. The possibility of producing power in a gas turbine and the reduction in energy consumption of the CCC process during peak hours are both the features that are offered by using an energy storage system. The significance of the energy storage becomes of great importance when the regulations for the CO_2 emissions of the power plants become effective in near future.

In a recent study, integrating the CCC process and the associated energy storage system with a single load-following power generation unit was studied [7], [8]. In this study, no cycling cost was assumed for the load-following boiler but the rate of change for the boiler power output was limited to 7%/min. This study represents the recently designed power generation units with load-following capability. However, there are still many baseline power generation units. In the current study, the impact of using the CCC process and the associated energy storage on a baseline coal-fired generation unit (1800 MW) is considered. Gas turbine (fed from the vaporized LNG also recovered from the energy storage) is able to produce up to 1000 MW power while the maximum wind power output adopted in this study is 390 MW. Maximum residential demand is also 2000 MW. The performance of the hybrid system of baseline power generation unit and the CCC process shows that by using the energy storage, the power output from the boiler is kept at the baseline value. Total power demand is met by using a combination of power production from steam boiler, gas turbine, and wind. All of the available wind power is also utilized in meeting the electricity demand. As for the load-following case, reduction in the power consumption in the refrigeration cycles is another important conclusion from this investigation.

As a second case study, leveling the fluctuations in the power output from a load-following boiler is considered. In this case, the boiler can have unconstrained fluctuations in the power output; however, a cycling cost is considered for each load-following event. This case study represents a more realistic operational strategy compared to the previous study in [7]. For a case study with energy storage, it is seen that dedicating a cycling cost to the movement in boiler power output results in significantly lower fluctuations in the boiler power output. Considering an appropriate size for the energy storage as well as the cycling costs for the boiler lead to minimal cycling in the boiler power output, even in the event of a sudden increase in power demand. As mentioned before, the power provided by the gas turbine is used to meet the peak electricity demand without needing to frequently cycle the boiler. Supplying power from the gas turbine is a direct result of using energy storage. The long-term results of having less cycling in the boiler power output are the longer lifetime and lower maintenance costs of the boiler. In addition to the saving in the capital and maintenance costs of boiler, significant saving in the operating cost of the power plant with energy storage is observed. Without energy storage, the power generation unit has to meet the demand by running the spinning/non-spinning reserves. Energy storage reduces the need for running the spinning/non- spinning reserves which significantly reduces the operational costs.

[1] N. Kumar, P. Besuner, S. Lefton, D. Agan, and D. Hilleman, "Power Plant Cycling Costs," 2012.

[2] S. M. Safdarnejad, J. D. Hedengrena, and L. L. Baxter, "Effect of Cryogenic Carbon Capture (CCC) on Smart Power Grids," in *American Institute of Chemical Engineers (AIChE)*, 2015.

[3] J. Nease and T. A. A. II, "Coal-fuelled systems for peaking power with 100% {CO2} capture through integration of solid oxide fuel cells with compressed air energy storage," *J. Power Sources*, vol. 251, no. 0, pp. 92–107, 2014.

[4] D. Connolly, H. Lund, B. V Mathiesen, E. Pican, and M. Leahy, "The technical and economic implications of integrating fluctuating renewable energy using energy storage," *Renew. Energy*, vol. 43, no. 0, pp. 47–60, 2012.

[5] M. Jensen, "Energy Processes Enabled by Cryogenic Carbon Capture," Brigham Young University, 2015.

[6] "Sustainable Energy Solutions Company." [Online]. Available: http://sesinnovation.com/.

[7] S. M. Safdarnejad, J. D. Hedengrena, and L. L. Baxter, "Plant-level Dynamic Optimization of Cryogenic Carbon Capture with Conventional and Renewable Power Sources," *Appl. Energy*, 2015.

[8] S. M. Safdarnejad, L. Kennington, L. Baxter, and J. Hedengren, "Investigating the Impact of Cryogenic Carbon Capture on the Performance of Power Plants," American Control Conference (ACC), Chicago, IL, 2015.

423695 Electrochemical Impedance Spectroscopy Analysis of Polymer-Electrolyte Dye Sensitized Solar Cells Using First-Principles Macroscopic Modeling

Abstract:

Dye sensitized solar cells (DSSCs) are commonly regarded as a promising next generation solar cell technology[1] because they have good performance at ~13% and can be made relatively inexpensively without intensive purification and fabrication steps. However, the liquid electrolyte of conventional DSSCs is prone to leakage and evaporation, which hinders DSSC applications, durability, and thermal stability. Novel polymer electrolyte DSSCs can overcome the shortcomings and can enhance the cell I-V behavior. Recent experimental studies have demonstrated that polymer electrolytes are good alternatives to liquid electrolytes in DSSCs[2-4].

There have been relatively few studies that combine experimental and computational/theoretical approaches to better understand DSSC processes and optimize DSSC performance. Currently, there is a need for a better understanding of how polymer electrolyte chemistry affects the cell performance. Accurate first-principles models can provide this better understanding. By properly accounting for the physical and electrochemical processes occurring inside the cells, the models can elucidate competing effects within DSSC components, can identify factors that affect the overall cell performance, and can guide experimental work in designing cell materials.

Continuing our previous work[5, 6] on the mathematical modeling of liquid-electrolyte and polymer-electrolyte DSSCs, in this work we theoretically and experimentally investigate the effects of polymer-electrolyte chemistry on the performance of polymer-electrolyte DSSCs. Using steady-state and transient models we predict current-voltage behavior and model the frequency response of DSSCs to simulate electrochemical impedance spectroscopy (EIS). The frequency response is performed to investigate the electronic and ionic processes in DSSCs. Using the macroscopic models, we conduct parametric sensitivity studies to identify key parameters that limit cell efficiency and recognize how the parameters affect cell behavior. Furthermore, to gain a complete understanding of the photochemical processes inside the DSSCs, the first-principles macroscopic modeling is coupled with experiments involving initiated chemical vapor deposition to synthesize and incorporate the polymer electrolytes within the mesoporous TiO_2 photoanode. Using the experimental data and the model, we will elucidate the effects of polymer chemistry on the cell performance, and determine how and why the photochemical processes inside such solar cells are altered by the polymer electrolytes. Parameter estimation using the macroscopic model indicates that a shift in the conduction band of TiO_2 and a suppression of the back electron transfer at the dye-TiO_2-electrolyte interface are induced by the chemical groups of the polymers.

1. O. Mari, O. Kei, S. Ai, T. Hideyuki, H. Nozomu, E. Akira, T. Hiromitsu, K. Momoji and M. Akira, *Japanese Journal of Applied Physics*, 2010, 49, 04DP10.

2. A. F. Nogueira, J. R. Durrant and M. A. De Paoli, *Advanced Materials*, 2001, 13, 826-830.

3. P. Wang, S. M. Zakeeruddin, J. E. Moser, M. K. Nazeeruddin, T. Sekiguchi and M. Gratzel, *Nat Mater*, 2003, 2, 402-407.

4. S. Nejati and K. K. S. Lau, *Nano Letters*, 2010, 11, 419-423.

5. M. Bavarian, S. Nejati, K. K. S. Lau, D. Lee and M. Soroush, *Industrial & Engineering Chemistry Research*, 2014, 53(13), 5234-5247.

6. Y. Y. Smolin, S. Nejati, M. Bavarian, D. Lee, K. K. S. Lau and M. Soroush, *Journal of Power Sources*, 2015, 274, 156-164.

425990 Design of Advanced Controllers for Post-Combustion CO2 Capture Process Integrated with a Supercritical Pulverized Coal Plant

Abstract:

Design of Advanced Controllers for Post-Combustion CO_2 Capture Process Integrated with a Supercritical Pulverized Coal Plant

Qiang Zhang, Debangsu Bhattacharyya, and Richard Turton

Department of Chemical Engineering, West Virginia University,

Morgantown WV, 26506, USA

Focus of this presentation will be on development of advanced control strategies for an MEA-based, post-combustion CO_2 capture and compression unit integrated with a 550MWe commercial supercritical pulverized coal power plant. The steady-state process design indicated that multiple parallel trains of CO_2 capture processes are required for treating the flue gas from a commercial sized power plant. Thus the dynamic model used in this work comprises of two parallel trains of CO_2 capture processes integrated with the steam-side of a supercritical coal fed power plant. The dynamic model of the CO_2 capture unit is rigorous with user models incorporated into the Aspen Plus Dynamics® environment for improving the model accuracy under design and off-design operations.

Linear model predictive control (LMPC) is found to provide good control performance of the CO_2 capture rate at a set-point of 90% in the face of typical disturbances such as change in the flue gas flow rate and composition. However, when the integration of the power plant and the CO_2 capture processes is considered then the situation becomes more complicated. For example, for a load-following power plant with sliding pressure operation, as the flue gas flowrate changes in response to the load changes, there is a corresponding change in the pressure and temperature of the low pressure steam available for the CO_2 stripper reboiler. Since the time constants for the CO_2 capture process are much longer than for the power plant and the loading of the MEA is strongly influenced by the stripper reboiler temperature the response of the overall system becomes highly non-linear. In addition, considerable time delay in the process and strong interaction between multiple variables result in a challenging problem. The control problem becomes far more difficult when uncertainties in the input and outputs as well as gradual loss in the absorber and stripper performances are considered. Performance of the LMPC is found to be poor under these scenarios. A number of advanced controllers such as nonlinear MPC (NMPC), H_∞ robust controller, and robust MPC are designed with novel strategies for uncertainty quantifications and rigorous analysis of tradeoff between performance and stability. In addition, a number of strategies has been developed to reduce the computational expense of the control system.

430228 Stabilization of Biogas Production in a Two-Stage Reactor Process

Abstract:

Production of methane via the anaerobic digestion of biogenic material is an alternative renewable energy source. Among various methods for energy production from biomass (e.g. bioethanol, pyrolysis, biodiesel, etc.), biogas has shown to be a robust, cheap and independent process. It can run on a wide variety of substrates, contributes to carbon and methane capture, does not require sterile conditions, can be very stable under conservative operation, has minimal safety requirements, and the digestate may be used as fertilizer. Still, with the process being carried out by a large symbiotic and constantly changing microbial population (mainly due to mutations, non-sterile feeding, and low dilution rates), and with the sludge making state monitoring costly and maintenance-intensive, there are major challenges in the dynamic analysis and control of the process. This has led to conservative operation of the biogas production process, far from optimal yields and lacking the necessary flexibility to comply with the fluctuations in both, the electricity and feedstock supply grid.

In this work, we study a two-stage reactor design to enable maximization of production by stabilizing the dynamics of the process near the critical dilution rates. In this way, it is possible to achieve: 1) higher yield, 2) higher methane concentration in the output gas, and 3) a more flexible gas production.

The anaerobic digestion process involves four main biochemical steps, namely: 1) hydrolysis, 2) acidogenesis, 3) acetogenesis, and 4) methanogenesis. These steps are carried out by different microbial populations with varying characteristics, growth rates and substrate affinities. Dividing the process in different stages to exploit the characteristics of different populations has already been proposed in the literature [1]. Nevertheless, since optimal performance of the process is near wash-out, increasing the size of the stability region at such operating conditions is of great importance. In order to be able to develop a successful two-reactor biogas production technology, it is necessary to build appropriate models for each stage, as well as advanced control strategies.

A global optimization at steady state biogas production in two reactors shows that an increment of methane flow rate per volume day of 28% is possible against a single reactor concept. Additionally, the concentration of methane in the biogas can also be increased up to 40% (reaching almost 70% methane content). Unfortunately, although stable, these optimal operating points are in the vicinity of washout so that small disturbances in the process may have severe consequences.

The present work addresses problems of feedback stabilization in a two-stage reactor system operating near critical dilution rates, for maximal biogas production. These are tackled through reduced versions of a state-of-the-art anaerobic digestion model [2], specifically tailored to the operating conditions, as well as concepts of constant-yield control [3], for the design of stabilizing controllers. Using methane, carbon dioxide, and pH as measurements, the control law is designed to stabilize the system by manipulating the dilution rates in each reactor and the recycling. Lyapunov methods are used to derive estimates of the size of the stability region in open loop and in closed loop. The results show that stable operation of the two-stage reactor process is feasible, leading to increased yield and methane purity.

References:

1. Parawira, W., et al., *Energy production from agricultural residues: high methane yields in pilot-scale two-stage anaerobic digestion.* Biomass and Bioenergy, 2008. **32**(1): p. 44-50.

2. Bernard, O., et al., *Dynamical model development and parameter identification for an anaerobic wastewater treatment process.* Biotechnology and Bioengineering, 2001. **75**(4): p. 424-438.

3. Savoglidis, G. and C. Kravaris, *Constant-yield control of continuous bioreactors.* Chemical Engineering Journal, 2013. **228**: p. 1234-1247.

433247 A Generic Framework for the Robust Control of Fuel Cell Energy System

Abstract:

A Generic Framework for the Robust Control of Fuel Cell Energy System

Amit M. Manthanwar*† and Efstratios N. Pistikopoulos‡

†Department of Chemical Engineering, Imperial College London, England, UK

‡Artie McFerrin Department of Chemical Engineering, Texas A&M University, USA

Abstract: 10 - COMPUTING AND SYSTEMS TECHNOLOGY DIVISION

Targeted Session: 10B05 Modeling, Control and Optimization of Energy Systems

With hydrogen envisioned as an energy carrier, fuel cell is the sustainable energy enabling technology. Amongst all fuel cell types, a polymer electrolyte membrane fuel cell is the most widely researched fuel cell type. Operating at low temperature and having higher energy conversion efficiency compared to an internal combustion engine, it offers distinct benefits and has found widespread adoption in many areas including the transport and stationary energy sector. Regardless of application and the type of fuel cell used, balance of plant system integration brings additional challenges affecting durability and performance of the overall system, [1]. Although the fuel cell stack technology has reached a level of maturity by approaching a target cost of $30 per kW to be competitive against internal combustion engines, systemwide control automation for better electricity, water and thermal management still remains the main research priorities. The majority of fuel cell system failures and forced outages (~90%) are due to lack of system integration and are the result of balance of plant events, [2]. These statistics reinforce the need for better system integration and more efficient controller design. Therefore three important aspects for improving fuel cell systems are: (a) reducing manufacturing and operational costs; (b) improving safety, durability and reliability; (c) improving operational performance by efficient controller design. These

three aspects are intertwined and the effective modelling, optimisation and control automation strategy plays a key role in addressing operational challenges and significantly improving system performance, durability and operational costs.

The real time control of fuel cell system using a well established technology of model predictive

control requires an accurate mathematical model of the system. The dynamic phenomenon taking place inside fuel cell involves complex interaction of mass transport, energy transport and electrochemical kinetics. These uncertain dynamics determine the overall operational performance of the system. In addition, the real time control requires computing machinery to run the optimisation algorithms. This compounds the manufacturing and operational costs. After a decade long theoretical research carried out in the area of mathematical optimisation and control theory, [3], we are poised to make a significant contribution of the developed theory applied to the energy systems applications, in particular to the experimental investigation of fuel cells at multiple scales. This work presents a generic framework for the development of efficient fuel cell controllers by integrating high fidelity mathematical models of the fuel cell system and integrated subsystems with online and offline model-based predictive control algorithms. Additionally, research presents the development of state-of-the-art testing and validation facility for characterising various designs as well as benchmarking operational policies of the polymer electrolyte membrane fuel cell system operating under uncertainty. Research envisions to further adapt and deploy in real-time the advanced optimisation and control algorithms to the various energy systems applications.

References

[1] C. Ziogou, E. N. Pistikopoulos, M. C. Georgiadis, S. Voutetakis, and S. Papadopoulou, "Empowering the performance of advanced nmpc by multiparametric programming–an application to a pem fuel cell system," Ind. Eng. Chem. Res., vol. 52, pp. 4863–4873, Mar. 2013.

[2] K. Wipke, S. Sprik, J. Kurtz, T. Ramsden, C. Ainscough, and G. Saur, "Final results from u.s. fcev learning demonstration," in EVS26 International Battery, Hybrid and Fuel Cell Electric Vehicle Symposium, (Los Angeles, California), May 6–9 2012.

[3] E. N. Pistikopoulos, "From multi-parametric programming theory to mpc-on-a-chip multi-scale systems applications," Computers and Chemical Engineering, no. 0, 2012.

*The financial support from EPSRC grants (EP/I014640/1 and EP/K503381/1) and sponsorships from ARM Incorporation is gratefully acknowledged. Corresponding author: amit@imperial.ac.uk

430226 Surrogate Based Derivative Free Optimization Methodology for Supply Chain Management

Abstract:

Surrogate based Derivative Free Optimization Methodology for Supply Chain Management

Nihar Sahay and Marianthi Ierapetritou

Department of Chemical and Biochemical Engineering, Rutgers University, Piscataway, NJ

Simulation models are one of the most effective tools to study supply chains. Compared to analytical and mathematical programming techniques, they offer the capability to include a greater deal of fidelity. Different kinds of simulation models have been widely used to study various aspects of supply chain management. These models provide a very convenient approach to generate various "what-if" scenarios and find the optimal values of the discrete variables. However stand-alone simulation models cannot be used to optimize the continuous variables. It is necessary to couple the simulation model with an optimization approach in order to find the optimal values of the continuous variables. During the recent years, supply chains have evolved into global, highly complex networks and the overall supply chain operations are a result of numerous autonomous, adaptive and intelligent entities. A high fidelity simulation model is not only difficult to develop but also difficult to use in an optimization framework due to computational complexity involved in each function evaluation. In order to optimize the variables in these simulation models, deterministic optimization solvers cannot be used as the derivatives are unavailable. Also, since the simulations take long times to run, it is not possible to perform a large number of simulation runs in order to approximate the derivatives. Taking these factors into consideration, we propose a surrogate based derivative free optimization methodology to solve a supply chain planning problem.

Simulation based optimization approaches have been extensively studied in the literature. These approaches have been used to solve diverse problems. Jung et al.[1] propose a simulation based optimization framework to determine the safety level of each product at each production site in order to meet the required customer satisfaction. The objective is the minimization of the expected value of the cost of the supply chain. The optimization framework consists of an outer and an inner loop. The outer loop consists of a stochastic simulation coupled with a stochastic gradient based search module for finding the optimal safety level. The inner loop consists of the discrete event simulation model coupled with a deterministic planning and scheduling optimization model. Wan et al.[2] propose a simulation based optimization framework that includes domain reduction, use of least square support vector machine to construct the surrogate model, and incremental sampling by maximizing Bayesian information and expected improvement. The framework is used to optimize the inventory levels in a three stage supply chain. Mele et al.[3] propose a simulation based optimization framework for parameter optimization of supply chain networks. The logical rules defined for the agents in the simulation model are parametrized and then a genetic algorithm is used to optimize the parameter values. The framework is used to determine the optimal values of the parameters associated with the inventory control policy. Derivative free optimization methods have been comprehensively reviewed by Rios and Sahinidis.[4] Development of surrogate models for solving derivative free optimization problems has also received a great deal of attention in the literature.[5-8]

In this work, a surrogate based derivative-free optimization approach is proposed to optimize a supply chain planning problem. An initial global sampling is performed to construct an initial surrogate model using the expensive simulation model, which is improved by adaptive sampling. A kriging model[9] is used to develop the surrogate approximation whereas, a trust region based framework is used to optimize the surrogate model. The proposed optimization framework addresses issues related to the number of function calls required and also the limitations related to high dimensionality. A high dimensional supply chain planning problem is used to demonstrate the effectiveness of the proposed framework. The detailed agent based simulation model that is used as the black box function is able to provide a realistic representation of the supply chain operations.

References:

1. Jung JY, Blau G, Pekny JF, Reklaitis GV, Eversdyk D. A simulation based optimization approach to supply chain management under demand uncertainty. *Computers & Chemical Engineering.* 2004;28(10):2087-2106.

2. Wan X, Pekny JF, Reklaitis GV. Simulation-based optimization with surrogate models—Application to supply chain management. *Computers & Chemical Engineering.* 2005;29(6):1317-1328.

3. Mele FD, Guillén G, Espuña A, Puigjaner L. A Simulation-Based Optimization Framework for Parameter Optimization of Supply-Chain Networks. *Industrial & Engineering Chemistry Research.* 2006/04/01 2006;45(9):3133-3148.

4. Rios L, Sahinidis N. Derivative-free optimization: a review of algorithms and comparison of software implementations. *Journal of Global Optimization.* 2013/07/01 2013;56(3):1247-1293.

5. Boukouvala F, Ierapetritou M. Surrogate-Based Optimization of Expensive Flowsheet Modeling for Continuous Pharmaceutical Manufacturing. *Journal of Pharmaceutical Innovation.* 2013/06/01 2013;8(2):131-145.

6. Booker AJ, Dennis JE, Jr., Frank PD, Serafini DB, Torczon V, Trosset MW. A rigorous framework for optimization of expensive functions by surrogates. *Structural optimization.* 1999/02/01 1999;17(1):1-13.

7. Davis E, Ierapetritou M. A kriging method for the solution of nonlinear programs with black-box functions. *AIChE Journal.* 2007;53(8):2001-2012.

8. Huang D, Allen TT, Notz WI, Zeng N. Global Optimization of Stochastic Black-Box Systems via Sequential Kriging Meta-Models. *Journal of Global Optimization.* 2006/03/01 2006;34(3):441-466.

9. Sacks J, Welch WJ, Mitchell TJ, Wynn HP. Design and Analysis of Computer Experiments. 1989/11 1989(4):409-423.

420105 Constrained Grey-Box Global Optimization of High Dimensional Problems

Abstract:

The use of deterministic global optimization methods based on analytical C^2 functions is prohibitive in a large portion of engineering applications which have one or multiple of the following characteristics: (a) high computational cost which implies that the calculation of derivatives is not viable, (b) numerical noise which leads to unreliable derivative information, and (c) lack of closed-form model equations which implies lack of derivative information [1-2]. Optimization without derivatives is becoming necessary in many recent applications dealing with detailed multiscale expensive simulations (i.e., finite element method models, computational fluid dynamic models and non-linear partial differential equation systems). In this work we are interested inconstrained grey-box systems, which are defined as optimization formulations which contain a set of explicitly known equations (constraints and/or objective function) and a set of explicitly unknown equations (constraints and/or objective). In prior work, we have shown that well-designed sampling coupled with iterative fitting of surrogate models coupled with deterministic global optimization has a competitive performance in locating global solutions at a low sampling expense [3-4]. However, treatment of a large set of general constraints (number of constraints > 50) in high dimensional problems (number of dimensions > 100) still remains a challenge in the field of constrained grey-box optimization due to the computational expense associated with collecting samples in high-dimensional spaces, solving large parameter estimation problems and finally optimizing the formulated grey-box formulations globally. Existing approaches in the literature treat such problems as pure "black-boxes", disregarding any information in the form of constraints or problem structure and limiting their communication and integration with existing deterministic optimization software. Moreover, existing methods have been tested on high dimensional box-constrained problems using local optimization concepts- which are highly dependent on the existence of a good initial point [5].

The ARGONAUT framework (Algorithms for Global Optimization of constraiNed Grey-Box compUtational sysTems), is comprised of several key components which are enhanced in this work to handle higher dimensional problems and problems with a large number of constraints. The main components of the algorithm are: (a) Optimized sampling in constrained regions defined by any known information of the problem, (b) Bounds tightening when this is possible based on known information, (c) Selection of optimal surrogate functions for each individual unknown model from a larger pool of possible regression and interpolation techniques (i.e., linear, quadratic, signomial, kriging and radial-basis functions), (d) Exploration of model sparsity, (e) Sequential treatment of variables based on variable importance and (f) Parallelization of sampling, parameter estimation, model validation and global optimization procedures for computational savings. The proposed framework is highly intertwined with the deterministic global optimization solver ANTIGONE [6] which is necessary for optimizing the constrained non-linear grey-box formulations providing a diverse set of local and global solutions which are good candidates for further sampling. We provide thorough analysis of results on a large set of test problems, and propose different strategies based on the size of the problem as well as the computational expense of the simulation model. Finally, the performance of the developed algorithms are compared to existing software for constrained optimization without derivatives.

References:

1. Conn, A.R., Scheinberg, K. Vicente, L.N.:Introduction to Derivative-free Optimization. MPS-SIAM Series on Optimization, vol. 8, SIAM, Philadelphia (2009).

2. Boukouvala, F. , Misener, R., Floudas, C.A.: Global Optimization Advances in Mixed-Integer Nonlinear Programming, MINLP and Constrained Derivative-free Optimization, CDFO. European Journal of Operational Research. Under Review (2015).

3. Hasan, M.M.F., Boukouvala, F., First, E.L., Floudas, C.A.: Nationwide, Regional, and Statewide CO2 Capture, Utilization, and Sequestration Supply Chain Network Optimization. Industrial Engineering & Chemistry Research 53(18), 7489-7506 (2014).

4. Boukouvala, F., Hasan, M.M.F., Floudas, C.A.: Global Optimization of General Constrained Grey-Box Models: New Method and its Application to Constrained PDEs. Submitted to Journal of Global Optimization (2015).

5. Adjengue, L., Audet, C., Yahia, I.B.: A variance-based method to rank input variables of the Mesh Adaptive Direct Search algorithm. Optim Letters 8, 1599- 1610 (2014).

6. Misener, R., Floudas, C.A.: ANTIGONE: Algorithms for Continuous/ Integer Global Optimization of Nonlinear Equations, Journal of Global Optimization 59(2-3), 503-526 (2014).

429508 Branch-and-Model: A Model-Based Derivative-Free Global Optimization Algorithm

Abstract:

Derivative-free optimization (DFO) addresses the problem of optimizing in the absence of an algebraic model. DFO algorithms can be used to optimize over simulations and experiments, when an algebraic model is not easily available. The subject has found renewed interest in recent time as many DFO algorithms and software have reached a level of maturity that facilitates the successful solution of many applications [1]. Unfortunately, as reported in a recent comparison of 22 DFO solvers on over 500 test problems in [1], the success of current DFO software implementations is typically limited to problems with up to 10 or 20 degrees of freedom. Clearly, there is a strong need for advances in the DFO area.

In this paper, we address the optimization of a deterministic function over a box-bounded domain in the absence of an algebraic model for and its derivatives. We assume that can be calculated via simulation or experiment, although the cost of such a calculation may be significant. We present a novel model-based algorithm for this problem. This algorithm, named branch-and-model (BAM) is a DFO algorithm that builds a model around each evaluated point using available surrounding evaluated points. The algorithm performs a dense sense and employs a local search algorithm to identify local optima. We have proved that this algorithm converges to a global optimum of the problem. We present extensive computational experience with the algorithm and comparisons with other DFO algorithms on a publically available collection of 500 test problems ranging from 2 variables to 30 variables [1].

[1] Rios, L. M. and N. V. Sahinidis, Derivative-free optimization: A review of algorithms and comparison of software implementations, Journal of Global Optimization, 56, 1247-1293, 2013.

427138 Bifurcation and Stability Analysis of Nonlinear Dynamic Systems Using Complete Search

Wednesday, November 11, 2015: 1:45 PM
Salon F (Salt Lake Marriott Downtown at City Creek)

Mario E. Villanueva[1], Hugo Sant'Ana Pereira[2], Jai Rajyaguru[1], Rob Krams[2] and **Benoit Chachuat**[1], (1)Centre for Process Systems Engineering, Imperial College London, London, United Kingdom, (2)Department of Bioengineering, Imperial College London, London, United Kingdom

Dynamic systems found in chemical and biological engineering are typically governed by parametric nonlinear dynamic systems of the form

$$\dot{x}(t) = f(x(t), p)$$

where $x(t) \in \mathbb{R}^{n_x}$ and $p \in \mathbb{R}^{n_p}$ denote the state and parameter vectors, respectively. An important aspect of the asymptotic study of such systems involves computing the equilibrium manifold $\Gamma := \{(\hat{x}, \hat{p}) \in \mathbb{R}^{n_x+n_p} \,|\, f(\hat{x}, \hat{p}) = 0\}$, and analyzing its stability. Another related aspect is the computation of bifurcation points, whereby qualitative change in the dynamics occur. Typically, stability and bifurcation analysis are performed using continuation methods [1]. Such methods use numerical integration starting from a grid of initial conditions to identify stable equilibrium points. Then, parameter perturbation and local search are used to reconstruct the equilibrium manifold. Due to their local nature, the results obtained using numerical continuation may be incomplete, which may lead to missing unconnected branches for instance.

The focus of this paper is on using complete-search methods to rigorously enclose the equilibrium manifold and all the bifurcation points on a domain $Z_0 := X_0 \times P_0 \subseteq \mathbb{R}^{n_x+n_p}$. A set-inversion algorithm [3] is used in order to compute partitions $\mathbb{Z}_{\text{in}}, \mathbb{Z}_{\text{bnd}} \subseteq \mathbb{R}^{n_x+n_p}$ satisfying

$$\bigcup_{Z \in \mathbb{Z}_{\text{in}}} Z \subseteq \Gamma \subseteq \bigcup_{Z \in \mathbb{Z}_{\text{bnd}} \cup \mathbb{Z}_{\text{in}}} Z$$

Enclosures of the image of f under Z are computed using polynomial models, in particular Chebyshev model arithmetic [8], and optimization-based domain reduction is applied to accelerate the convergence of the set-inversion algorithm [5]. Furthermore, a set-valued implicit equation solver [7] is used to compute a Chebyshev model representation of the boundary of the set whenever possible. It is important to note that, in this proposed approach, bifurcation points can be computed without computing the equilibrium manifold first. This is done by appending algebraic conditions for bifurcation (steady-state or Hopf bifurcations) [2] to the equilibrium constraints. Another contribution of the paper is a validated a posteriori stability analysis of the equilibrium points. This analysis is based on an validated extension of the Neville elimination algorithm for Hurwitz matrices [6] using Chebyshev models. Bifurcation points can also be obtained as a by-product of this stability analysis.

The performance of the proposed algorithm is illustrated with several case studies of various complexity drawn from chemical engineering and systems biology. For illustration, the figure below shows the results of a bifurcation and stability analysis for a simple nutrient-resource-consumer model [4] with respect to the nutrient density parameter. Stable equilibrium points are shown in blue, unstable points in red, steady-state bifurcation points are shown as circles and Hopf bifurcations as squares.

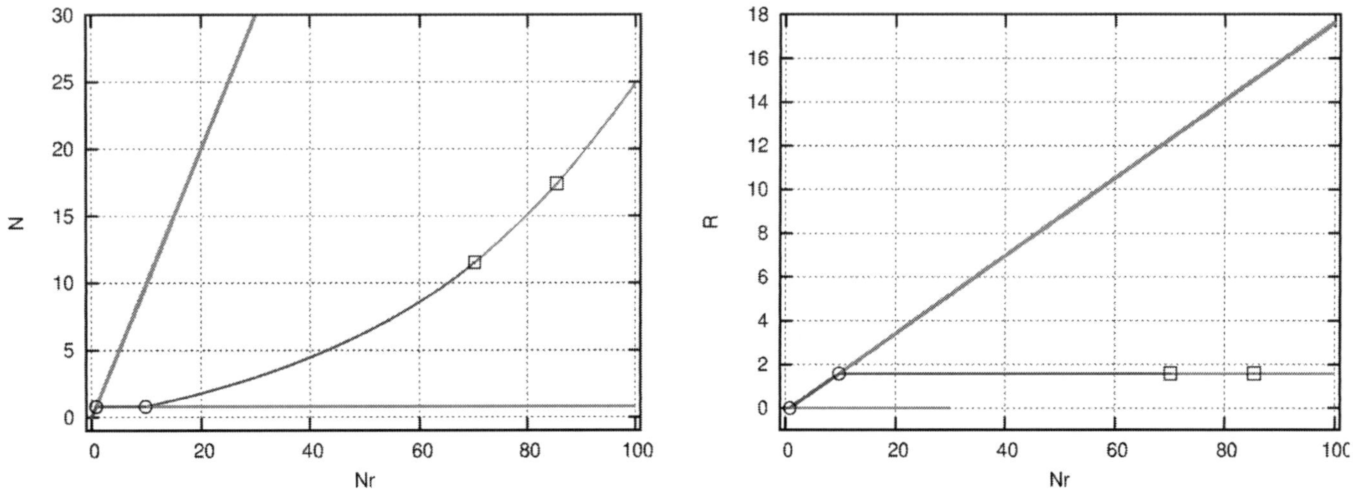

References:

1. E. L. Allgower and K. Georg. *Numerical Continuation Methods*. Springer Series in Computational Mathematics. Springer-Verlag, Berlin, 1990.
2. M. El Kahoui and A. Weber. Deciding Hopf bifurcations by quantifier elimination in a software-component architecture. *Journal of Symbolic Computation*, 30(2):161-179, 2000.
3. L. Jaulin and E. Walter. Set inversion via interval analysis for nonlinear bounded-error estimation. *Automatica*, 29(4):1053-1064, July 1993.
4. B. W. Kooi. Numerical bifurcation analysis of ecosystems in a spatially homogeneous environment. *Acta Biotheoretica*, 51(3):189-222, 2003.
5. R. Paulen, M. E. Villanueva, and B. Chachuat. Guaranteed parameter estimation of nonlinear dynamic systems using high-order bounding techniques with domain and CPU-time reduction strategies. *IMA Journal of Mathematical Control & Information*, in press (doi: 10.1093/imamci/dnu055), 2015.
6. J. M. Peña. Characterizations and stable tests for the Routh-Hurwitz conditions and for total positivity. *Linear Algebra & Its Applications*, 393:319-332, 2004.
7. J. Rajyaguru and B. Chachuat. Taylor models in deterministic global optimization for Large-Scale systems with few degrees of freedom. *Computer Aided Chemical Engineering*, 32:973-978, 2013.
8. J. Rajyaguru, M. E. Villanueva, B. Houska, and B. Chachuat. Higher-order inclusions of nonlinear systems by Chebyshev models. In *Proc. AIChE Annual Meeting 2014*, Paper 385510, 2014.

428800 Experiments with Distributed Optimization for Non-Convex Optimization Problems

Abstract:

We focus on distributed optimization for a class of non-convex optimization problems that are relevant to the process systems engineering community. Distributed optimization refers to the process of solving a centralized optimization problem in a distributed fashion where a collection of "agents" cooperatively and rigorously optimize the centralized problem. It is a branch of optimization that has links to parallel computing, problem decomposition, decentralized computing, agent-based computing, etc. We focus on the following "almost" separable optimization problem. The functions in the objective function can be non-convex and are sufficiently smooth.

min sum_{i=1}^{N} f_i(x_i)

s.t. sum_{i=1}^{N} A_i x_i = b

We will make a case for why this structure is extremely relevant to process optimization and why it is important to solve these problems using distributed optimization techniques. While most of the activity in the broader optimization community is for convex optimization problems, we will talk about our experience with solving distributed non-convex optimization problems using the alternating direction method of multipliers (ADMM) [1, 2]. We will present examples and results from pooling/blending optimization, stochastic programming, and a series of 2-variable test examples that highlight the difficulties (mostly having to do with the Augmented Lagrangian parameter) associated with ADMM as well as potential advantages of doing so. ADMM methods typically have sub-linear / linear global convergence rates for convex problems. Motivated by [3], we present an algorithmic framework for distributed optimization that has good local convergence rates. We also discuss a specific instance of this framework that employs an L1 exact penalty function as a measure of progress and highlight how this is different from the classical Augmented Lagrangian-based methods.

References:

[1] J. Eckstein and D. P. Bertsekas, "On the Douglas-Rachford splitting method and the proximal point algorithm for maximal monotone operators", Mathematical Programming, 55, pp. 293-318, 1992

[2] S. Boyd, N. Parikh, E. Chu, B. Peleato and J. Eckstein, "Distributed Optimization and Statistical Learning via the Alternating Direction Method of Multipliers", Foundations and Trends in Machine Learning, Vol. 3, No. 1, pp. 1–122, 2010

[3] B. Houska, J. Frasch, M. Diehl, "An Augmented Lagrangian Based Algorithm for Distributed Non-convex Optimization", http://www.optimization-online.org/DB_HTML/2014/07/4427.html

435180 A Cross Decomposition Method for Stochastic Nonconvex Mixed-Integer Nonlinear Programming

Wednesday, November 11, 2015: 2:35 PM
Salon F (Salt Lake Marriott Downtown at City Creek)
Emmanuel Ogbe and **Xiang Li**, Chemical Engineering, Queen's University, Kingston, ON, Canada

A cross decomposition approach to stochastic nonconvex mixed-integer nonlinear programming

Emmanuel Ogbe, Xiang Li*

Department of Chemical Engineering, Queen's University,

19 Division Street, Kingston, ON, K7L 3N6, Canada

When considering uncertainties explicitly, many process systems engineering problems (such as process design and operation problems) can be cast as a two-stage stochastic program [1]; when nonlinear models and integer decisions are involved, the stochastic program leads to a large-scale nonconvex mixed nonlinear programming (MINLP) problem in the following general form:

$$
\begin{aligned}
\min\ & \sum_{\omega=1}^{s} \left[f_{0,\omega}(x_0) + f_{1,\omega}(x_{1,\omega}) \right] \\
s.t.\ & g_{0,\omega}(x_0) + g_{1,\omega}(x_{1,\omega}) \le 0, \quad \omega = 1,\cdots,s, \\
& x_{1,\omega} \subset X_{1,\omega}, \quad \omega = 1,\cdots,s, \\
& x_0 \subset X_0.
\end{aligned} \qquad \text{(P)}
$$

Here ω indicates different scenarios, and the size of the problem depends linearly on the number of scenarios included. Problem (P) has a decomposable structure in the sense that, when x_0 (called linking variable) is fixed, the problem is naturally decomposable over the scenarios. Therefore, a decomposition approach is often computationally more efficient than a regular solution method for solving Problem (P).

Various decomposition strategies have been developed in the literature for solving Problem (P). When Problem (P) is partially convex (with respect to $x_{1,\omega}$) and some constraint qualification holds, Benders decomposition [2] or generalized Benders decomposition [3] can be used to solve Problem (P) to global optimality. Furthermore, if Problem (P) is fully convex, then Dantzig-Wolfe decomposition [4] can also be used to achieve an optimal solution. However, when set $X_{1,\omega}$ or/and functions $f_{1,\omega}$, $g_{1,\omega}$ are nonconvex, classical decomposition approaches cannot guarantee convergence to a global optimal solution due to the loss of strong duality. Recently, two rigorous decomposition-based global optimization methods have been developed for solving Problem (P) that includes nonconvexity in x_0 space. One method utilizes Lagrangian decomposition [5][6], but needing branch-and-bound search in the x_0 space to guarantee global optimality. The other is a variant of (generalized) Benders decomposition that yields values for x_0 via solving convex lower bounding problems, but the convergence to a global optimal solution is guaranteed only when x_0 are integer variables [7].

This paper presents a cross decomposition approach that guarantees finding an ε-optimal solution in finite time, without the need of explicit branch-and-bound search and the convexity assumption for the problem. This approach is motivated by a cross decomposition method [8] that was originally developed for solving mixed-integer linear programming (MILP) problems with special structures. The basic idea of cross decomposition is to use two different decomposition approaches iteratively. The one that is more computationally efficient but less rigorous is performed more frequently, so that good quality solutions or even a global optimal solution can be generated quickly. The one that is relatively inefficient but able to guarantee convergence is performed only when necessary, so that the convergence of lower and upper bounds is guaranteed. It will be shown that, a cross decomposition framework that combines a variant of Dantzig-Wolfe decomposition (also a variant of Lagrangian decomposition) and generalized Benders decomposition, can be used to solve Problem (P) to global optimality, regardless of the convexity of the problem. It will be also shown that, with extensive domain reduction computation at each iteration of the cross decomposition, the efficiency of the cross decomposition can be significantly improved. The computational advantages of the proposed cross decomposition method over state-of-the-art global optimization solvers will be demonstrated through several process network design and operation problems.

References

[1] Birge, J. R.; Louveaux, F. *Introduction to stochastic programming*, Second Edition; Springer: New York, 2011.

[2] Benders, J. F. Partitioning procedures for solving mixed-variables programming problems. *Numer. Math.* **1962**, 4, 238-252.

[3] Geoffrion, A. M. Generalized Benders decomposition. *Journal of Optimization Theory and Applications* **1972**, 10, 237–260.

[4] Dantzig, G. B.; Wolfe, P. Decomposition principle for linear programs. *Operations Research* **1960**, 8, 101-111.

[5] Karuppiah, R.; Grossmann, I. E. A Lagrangean based branch-and-cut algorithm for global optimization of nonconvex mixed-integer nonlinear programs with decomposable structures. *Journal of Global Optimization* **2008**, 41, 163-186.

[6] Carze, C. C.; Schultz, R. Dual decomposition in stochastic integer programming. *Operations Research Letters* **1999**, 24, 37–45.

[7] Li, X.; Tomasgard, A.; Barton, P. I. Nonconvex generalized Benders decomposition for stochastic separable mixed-integer nonlinear programs. *Journal of Optimization Theory and Applications* **2011**, 151, 425-454.

[8] Van Roy, T. J. Cross decomposition for mixed integer programming. *Mathematical Programming* **1983**, 25, 46-63.

436726 Computer-Aided Tools for Teaching Sustainable Product-Process Design

Abstract:

Computer-Aided Tools for Teaching Sustainable Product-Process Design

Deenesh K. Babi

Department of Chemical and Bio-chemical Engineering, Technical University of Denmark, Søltofts Plads, Building 229, DK-2800, Kgs. Lyngby Denmark

In teaching a course for performing (sustainable) process-product design, different computer-aided tools for generation, selection (and screening) and verification/simulation should be incorporated within the framework of the course. For example, consider the computer-aided process design of a chemical where mass and energy balance data (obtained from the rigorous models) are used for performing a sustainability and LCA analysis. These analyses identify process limitations that are translated into design targets that, if matched, removes and/or eliminates the process limitation, thereby improving the design. This is the approach implemented in teaching (sustainable) process design at the Technical University of Denmark, Department of Chemical and Bio-chemical Engineering. For the sustainability and LCA analysis, the model proposed by the IChemE for calculation of sustainability metrics and the cradle to the gate approach based on ISO framework (ISO 14040) are used, respectively (Carvalho et al., 2013, Kalakul et al., 2014).

In generating a feasible product-process design, it is imperative that the student understands concepts related to phase equilibria, reaction engineering and separation processes. In teaching a course at an advanced level related to product-process design, it is important to disseminate within the course current concepts applied in both R&D and industry. For example, consider the following: *given* a set of customer/market needs (constraints) for a product, *design* a prototype product that satisfies the design constraints (product design) or *given* a CO_2 rich stream as part of a power plant effluent, *design* a chemical process that has a negative net CO_2 emission (objective function, process design). The student can use a 3-stage approach for performing product-process design. In stage 1, for product, needs are translated into target properties that must be satisfied and, for process, feasible reactions are selected that utilizes CO_2. In stage 2, for product, using property models, compounds (that will constitute the final product) are selected that satisfy property/stability constraints and, for process, flowsheet alternatives are generated for the selected CO_2 reactions. In stage 3, for product, the product is verified trough experiments and /or model based calculations and, for process, the best process for CO_2 utilization is selected based on the calculation and ordering of the objective function followed by process optimization.

In each stage, computer-aided tools are utilized. In stage 1, a database is used (for example, ICAS database). In stage 2, computer-aided molecular design tools are used, for example, ProCAMD (Gani et al., 1997), property prediction tools using group contribution methods, for example, ProPred (Gani et al., 1997), simulation tools, for example, VPPD (the product simulator), PROII (process simulator), etc. In stage 3, model-based tools, for example, MoT (Gani et al., 1997).

In this presentation, a product-process design problem will be formulated and solved, presenting the various computer-aided tools utilized in courses related to (sustainable) product-process design and separation process design. The presentation will interactively highlight the input and output of each tool through examples of different design problems posed to students.

References

Carvalho, A., Matos, H. A., & Gani, R. (2013). SustainPro—A tool for systematic process analysis, generation and evaluation of sustainable design alternatives. Computers & Chemical Engineering, 50, 8–27.

Gani, R., Hytoft, G., Jaksland, C., & Jensen, A. K. (1997). An integrated computer aided system for integrated design of chemical processes. Computers & Chemical Engineering, 21(10), 1135–1146.

Kalakul, S., Malakul, P., Siemanond, K., & Gani, R. (2014). Integration of life cycle assessment software with tools for economic and sustainability analyses and process simulation for sustainable process design. Journal of Cleaner Production, 71, 98–109

Mattei, M., Yunus, N. A., Kalakul, S., Kontogeorgis, G. M., Woodley, J. M., Gernaey, K. V., Gani, R., (2014). The Virtual Product-Process Design Laboratory for Structured Chemical Product Design and Analysis, Computer Aided Chemical Engineering33, 61-66

434466 Sage Math - an Open Source Computer Algebra System

Abstract:

The choice of a programming language and software package is always a major consideration when planning a course on computational methods in chemical engineering. Feature set, ease of use, extensibility, external adoption, and cost should all play a role in the decision. In this presentation I will make the argument that when it comes to software, the best things in life are free.

434455 Developing Graphical Interfaces for Solving Distillation Problems Trough Graphical and Short-Cut Methods

Abstract:

Almost as long as there have been user interfaces, there have been special software systems and tools to help design and implement the user interface software. Many of these tools have demonstrated significant productivity gains for programmers and have become important commercial products. On the other hand, developing interfaces applications is a difficult task; people have to address various common software engineering problems and interfaces-specific issues. To help the developers, a number of patterns have been proposed by the software community. In this work, we have developed interfaces for solving distillation problems in chemical processes. The tool developed help to the professor and student to understand the construction of graphical and shortcut methods for solving distillation problems.

433862 Using Vanderbilt's Chemical Process Innovation Center for Immersive Chemical Engineering Design and Laboratory Courses

Abstract:

A state-of-the-art Chemical Process Innovation Center was recently created at Vanderbilt to allow immersive education of our chemical engineering undergraduates. Specifically, the Chemical Process Innovation Center is a facility that includes our undergraduate unit operations laboratory modules, most of which are custom designed and built at Vanderbilt, wireless network technology with Apple iMac's, Apple TV's, mobile student work stations, and ten 65 inch HDTV monitors positioned throughout the center. In addition to the unit operations labs and the design curriculum that will be discussed, a measurement lab program using any number of our over 50 sensors and interface units is also included in the center. A specific description of the integration of chemical engineering lab, design and safety courses will be provided along with a detailed description of the center. This approach will highlight the immersion experience that chemical engineering students at Vanderbilt receive through the use of this center.

433214 Population Balances in Undergraduate Education in Chemical Engineering

Abstract:

Particle breakage and agglomeration occurs in a variety of processing units in chemical processing at the macroscale, and in colloidal systems as well. Population balance equations (PBEs) are commonly used to model the evolution of particle size distributions due to breakage and/or agglomeration. The difficulty with using PBEs in an undergraduate class is that these partial integral differential equations are difficult to solve and have few analytical solutions. When introducing PBEs in an undergraduate class, the result is that students focus more on the computational techniques than the concepts of breakage and agglomeration and the effects of the parameters. To address this, a preliminary version of a computer software has been developed for use in a particle and crystallization technology elective course, and a nanotechology in chemical applications course. This preliminary version of the software will be presented and discussed.

USING TEMPLATES FOR DEMONSTRATING GOOD PROGRAMMING PRACTICES

Mordechai Shacham, Ben-Gurion University, Beer-Sheva, Israel
Michael B. Cutlip, University of Connecticut, Storrs, CT, USA

Introduction

Computer-based numerical problem solving (CBNPS) is one of the most widespread applications of the computers in chemical engineering education and practice (Shacham et al. 2009). Mathematical software packages such as Excel®, MAPLE™, MATHCAD®, MATLAB®, Mathematica®, and POLYMATH™ are currently used routinely for numerical problem solving in engineering education. In CBNPS the emphasis is usually on the preparation of the correct mathematical model of the problem at hand and the use of the proper numerical method for solution. Less information is typically given regarding the display of the results is typically presented. It is important to have a preferred format for such a results display.

Good examples for the content and the format of the results display are provided by many process simulation programs. These programs have been extensively used for the last fifty years (since the mid nineteen sixties), and this long experience has identified the essential components of the report providing the results. Consider for example the results of an Ammonia synthesis process as presented by the UniSim (www.honeywellprocess.com) simulation program (Figure 1). Observe that all the variables can be displayed at the solution upon the user's request including the variable names, variable numerical values and the corresponding units. Thus for maximal benefit of CBNPS, all this information should be displayed. In addition, it is preferable to display each result the first time it becomes available in order to assist in the debugging of the problem.

The generation of such an extensive report requires a great deal of programming effort. However, this makes the programs difficult to understand and follow and makes it challenging to separate the important problem specific information in the program.

Program "Templates" have been prepared that enable problem solution along with the results display based on the information provided by the user. This enables a user to solve the problem and display the computational results with sufficient detail. It also provides only the problem specific data and the associated mathematical models. The Templates were prepared using the MATLAB[1] software package. The mathematical models of the problems were coded using Polymath[2] and were converted to MATLAB functions automatically using the utility available in Polymath.

Templates were prepared for solving systems of nonlinear algebraic equations (NLE) and detailed in this paper, systems of ordinary differential equations (ODE), polynomial & linear regression (POLYREG), multiple linear regression (MLNREG) and nonlinear regression (NLREG).

---

[1]MATLAB is a product of MathWorks, Inc., <http://www.mathworks.com>
[2]Polymath is a product of Polymath Software, <http://www.polymath-software.com>

Templates for Nonlinear Systems of Equations (NLE)

Systems of NLE may contain constants, explicit variables whose value can be calculated by simple substitution of constants, and other explicit variables and implicit variables whose values should be calculated by simultaneous solution of implicit equations of the form

f(**x**) = expressions that contain implicit variables and may contain constants and explicit variables **f**(**x**) = **0.**

The notation used in the program for the variables associated with the NLE constants and variables is shown in Table 1. This information includes the numerical values (in a numerical array) of the variables (and constants) and their names and units (stored in structures). Names and units must be specified by the user for all the constants and variables. Numerical values must be specified only for the constants and for the implicit variables (initial guess). Additionally a title for the problem should be provided (stored as text in the variable *Pt*).

The MATLAB *fsolve* library function is used to solve the NLE system. In the data input section the user may modify the parameter *TolFun* that is used as convergence tolerance by fsolve (See more details on this issue, later, in the example.).

In addition to the data, the user must provide two functions representing the mathematical model of the problem: *function fx = MNLEfun(x, Const)* and function *Evar = Evarfun(x, Const)*. The first function obtains the constant values and the current implicit variable values as input parameters and returns the current f(x) to the *fsolve* function. The second function obtains the constant values and the current implicit variable values as input parameters, and returns the current explicit variable values.

From the user provided data and functions, the Template first displays the constant values, then the values of the explicit and lastly the implicit variables along with the function values at the initial guesses. These values can be very helpful in debugging the problem. After the solution is obtained, the same information is displayed at the solution. The use of the NLE solver template is demonstrated by solving a three stage ultrafiltration process presented by Foley (2011) and Cutlip and Shacham (2013).

Demonstration of the NLE System Template for an Ultrafiltration Process

A typical 3-stage feed-and-bleed ultrafiltration process is shown in Figure 2. Fresh feed with a volumetric flow rate of Q_0 (m^3/s) and solute concentration of c_0 (g/L) is mixed with retentate (product stream) recycle from the first stage and enters the first membrane module. Some of the solvent (and possibly some of the solute) pass through the membrane and exits the unit as permeate (filtrate). The concentrated product stream (with solid concentration of c_1 g/L) is partially recycled in order to increase the flow rate into the unit to ensure well mixed conditions. The product (bleed) stream from the first unit is fed into the second stage with a volumetric flow rate of Q_1 (m^3/s).

A detailed discussion of the multi-stage ultrafiltration process, including the associated design equations, is provided, for example, by Seader and Henley(2006). Here the assumptions suggested by Foley (2011) are used for simplification of the model. These assumptions are that 1) No solute passes the membrane (complete rejection) and 2) The gel polarization model applies, thus the membrane is operating at limiting flux.

The mathematical model of the 3-stage ultrafiltration process is displayed (using the POLYMATH notation) in Table 2. The specified constants, in this case, are Q_0 – the feed flow rate to the 1st stage, k – the mass transfer coefficient, A – the membrane area per stage, c_0 – the feed solute concentration and c_g – the gel concentration. The explicit variables are the membrane fluxes (j_1, j_2 and j_3) and the outlet flow rates (Q_1, Q_2 and Q_3) from the three stages. The implicit variables are the solid concentrations (c_1, c_2 and c_3) in the three stages, and the functions (f_1, f_2 and f_3) represent the mass balance equations on the three stages.

The section of the input data that the user need to provide is shown in detail in Table 3. It includes the variable *Pt* that contains the problem title, the desired convergence tolerance (*TolFun*) value, the names and units of all the constants and variables, the numerical values for the constants and the initial estimates for the implicit variables.

In Table 4 the user supplied function *MNLEfun* is shown. This function obtains the constant values and the current implicit variable values (rows 2 through 9) as input parameters, calculates the explicit variables (rows 10 through 15) and returns the current f(x) (calculated in rows 16 through 18).

The *Evarfun* function obtains the constant values and the current implicit variable values as input parameters and returns the current explicit variable values. Parts of this function are shown in Table 5. Rows 2 through 15 in this function are the same as in *MNLEfun.* In row 16 the explicit variable values are inserted into the *Evar* array so that they can be displayed by the program.

Introducing the user supplied data and the *MNLEfun* and *Evarfun* functions into the template enable solving the problem and display the data and results as shown in Tables 6 and 7. Initial estimates for the solid concentration at the three stages are selected in the range between c_0 = 10 (g/L) and cg = 300 (g/L) (Table 6). Because of the use of the SI units (like m^3/s, for flow rate) all the explicit variable and function values are of the order of 10^{-6}. Consequently when using the default convergence tolerance of *TolFun* = 10^{-6}, the *fsolve* function judges that "Equation solved at initial point" and does not attempt to improve the solution. In order to obtain the correct solid concentration values, the function must be normalized to the -1 to +1 range or the value of *TolFun* must be considerably reduced. Thus the display of the variable values, in this case, helps to identify a numerical problem with the problem formulation and enables modifications that yield the correct solution.

A reduction in TolFun to 10^{-15} yields the correct solution that is shown in Table 7. Observe that the solid concentrations are significantly different from the initial estimates, and the function values are reduced by seven (or more) orders of magnitude.

Conclusions

The Templates developed in this work enable presentation of complete sets of results in numerical solution of problems that require minimal programming efforts. The model function, which includes appropriate comments as provided by the user (*such as NLEfun*), provides clear and complete documentation of the mathematical model of the problem. The results printed by the template also provide details of the solution which is comparable to the reports provided by simulation programs. Thus, the use of these templates enables clear and complete documentation of the problem and its solution with minimal effort on the user's behalf.

A complete set of templates and the associated explanations and examples are available at ftp://ftp.bgu.ac.il/shacham/templates. These are designated as: NLE (Nonlinear Equations as presented in this work), ODE (Ordinary Differential Equations), POLREG (Polynomial Regression), MLNREG (Multiple Linear Regression) and NLREG (Nonlinear Regression).

The templates were used in some homework assignments given in the "Process Simulation" course at the Ben-Gurion University of the Negev. The students found them easy to use and appreciated the help they provided in collecting and organizing the required input data and in displaying the complete sets of results.

References

1. Cutlip, M. B. and M. Shacham (2013), "Continuous Feed and Bleed Ultrafiltration – a Demonstration of the Advantages of the Modular Approach for Modeling Multi-Stage Processes", Chemical Engineering Education, 47(3), pp. 170-176.
2. Foley, G. (2011), "Solution of Nonlinear Algebraic Equations in the Analysis, Design and Optimization of Continuous Ultrafiltration", Chemical Engineering Education, 45 (1), pp. 59-64.
3. Seader, J. D. and E. J. Henley (2006), *Separation Process Principles*, 2nd Ed, New York, NY, Wiley.

Table 1. The Notation Used in the Program to Represent Properties of Variable

Variable			
Type	**Value**[1]	**Name**[2]	**Units**[2]
Constant	Const(i)	Cn{1,i}	Cu{1,i}
Explicit	Evar(i)	Evarn{1,i}	Evaru{1,i}
Implicit	x(i)	Vn{1,i}	Vu{1,i}

[1] MATLAB variable type: numerical array element

[2] MATLAB variable type: element of a structure containing text

Table 2. Mathematical Model of the Ultrafiltration Problem (Using POLYMATH Notation)

No.	Equation/ # Comment
1	#Model equations
2	**#First Stage**
3	j1=k*ln(cg/c1) #Membrane Flux (m/s)
4	Q1=c0*Q0/c1 #Outlet flow-rate (m^3/s)
5	f(c1) = Q0-Q1-j1*A #Mass balance
6	**#Second Stage**
7	j2=k*ln(cg/c2) #Membrane Flux (m/s)
8	Q2=c1*Q1/c2 #Outlet flow-rate (m^3/s)
9	f(c2)=Q1-Q2-j2*A #Mass balance
10	**#Third Stage**
11	j3=k*ln(cg/c3) #Membrane Flux (m/s)
12	Q3=c2*Q2/c3 #Outlet flow-rate (m^3/s)
13	f(c3)=Q2-Q3-j3*A #Mass balance
14	**#Problem specific data**
15	Q0=1/(60*1000) #Feed flow rate (m^3/s)
16	k=3.5e-6 #Mass transfer coefficient (m/s)
17	A=0.9 # Membrane area per stage (m^2)
18	c0=10 # Feed solute concentration (g/L)
19	cg =300 # Gel concentration (g/L)
20	**# Initial estimates**
21	c1(0)=20 #1st stage solid conc. (g/L)
22	c2(0)=50 #2nd stage solid conc. (g/L)
23	c3(0)=170 #3rd stage solid conc. (g/L)

Table 3. User Input Data for the Ultrafiltration Problem

No.	Command % Comment
1	**% Start of user's data input section**
2	Pt = 'CEE 2013 - Continuous feed and bleed ultrafiltration';
3	options = optimset('TolFun',[1e-6]); % Convergence tolerance, default value
4	**% Constant's numerical values, names and units**
5	cg = 300; c0 = 10; A = .9; k = .0000035; Q0 = 1 / (60 * 1000);
6	Const(1)= cg; Const(2) = c0; Const(3) = A; Const(4) = k; Const(5) = Q0;
7	Cn{1,1}={'cg'}; Cn{1,2}={'c0'}; Cn{1,3}={'A'}; Cn{1,4}={'k'}; Cn{1,5}={'Q0'}; Cu{1,1}={'g/L'}; Cu{1,2}={'g/L'}; Cu{1,3}={'m^3'}; Cu{1,4}={'m/s'};
8	Cu{1,5}={'m^3/s'};
9	**%Explicit variable's names and units**
10	Evarn{1,1}={'j1'}; Evarn{1,2}={'j2'}; Evarn{1,3}={'j3'}; Evarn{1,4}={'Q1'};
11	Evarn{1,5}={'Q2'}; Evarn{1,6}={'Q3'}; Evaru{1,1}={'m/s'}; Evaru{1,2}={'m/s'}; Evaru{1,3}={'m/s'};
12	Evaru{1,4}={'m^3/s'};
13	Evaru{1,5}={'m^3/s'}; Evaru{1,6}={'m^3/s'};
14	**%Implicit variable's names, units and initial guess**
15	Vn{1,1}={'c1'}; Vn{1,2}={'c2'}; Vn{1,3}={'c3'};
16	Vu{1,1}={'g/L'}; Vu{1,2}={'g/L'};Vu{1,3}={'g/L'};
17	x = [40. 70. 150.]; % initial guess vector
18	**% End of user's data input section**

Table 4. User Supplied Function MNLEfun for the Ultrafiltration Problem

No.	Command % Comment
1	function fx = MNLEfun(x,Const);
2	c1 = x(1); % Solid concentration (g/L), 1st stage
3	c2 = x(2); % Solid concentration (g/L), 2nd stage
4	c3 = x(3); % Solid concentration (g/L), 3rd stage
5	cg = Const(1); % Gel concentration (g/L)
6	c0 = Const(2); % Feed solute concentration (g/L)
7	A = Const(3); % Membrane area (m^2) for a stage
8	k = Const(4); % Mass transfer coefficient (m/s)
9	Q0 = Const(5); % Feed flow rate (m^3/s)
10	j1 = k * log(cg / c1); % Membrane flux (m/s), 1st stage
11	Q1 = c0 * Q0 / c1; % Stage 1 outlet flow rate (m^3/s)
12	j2 = k * log(cg / c2); % Membrane flux (m/s), 2nd stage
13	Q2 = c1 * Q1 / c2; % Stage 2 outlet flow rate (m^3/s)
14	j3 = k * log(cg / c3); % % Membrane flux (m/s), 3rd stage
15	Q3 = c2 * Q2 / c3; % Stage 3 outlet flow rate (m^3/s)
16	fx(1,1) = Q0 - Q1 - (j1 * A); % Mass balance on stage 1
17	fx(2,1) = Q1 - Q2 - (j2 * A); % Mass balance on stage 2
18	fx(3,1) = Q2 - Q3 - (j3 * A); % Mass balance on stage 3

Table 5. Parts[1] of the User Supplied Function *Evarfun* for the Ultrafiltration Problem

No.	Command % Comment
1	function Evar=Evarfun(x,Const)
2	c1 = x(1); % Solid concentration (g/L), 1st stage
•	
•	
•	
15	Q3 = c2 * Q2 / c3; % Stage 3 outlet flow rate (m^3/s)
16	Evar(1)= j1 ; Evar(2)= j2; Evar(3)= j3; Evar(4)=Q1; Evar(5)=Q2; Evar(6)=Q3;

[1] Rows 2 through 15 are the same as in Table 4

Table 6. Constant and initial variable values for the ultrafiltration problem

CEE 2013 - Continuous feed and bleed ultrafiltration	
Constants	A = 0.9 (m^3)
cg = 300 (g/L)	k = 3.5e-006 (m/s)
c0 = 10 (g/L)	Q0 = 1.6667e-005 (m^3/s)
Variable values at the initial estimate	
Implicit Variables	Explicit variables
c1 = 40 (g/L)	j1 = 7.0522e-006 (m/s)
c2 = 70 (g/L)	j2 = 5.0935e-006 (m/s)
c3 = 150 (g/L)	j3 = 2.426e-006 (m/s)
Function values	Q1 = 4.1667e-006 (m^3/s)
f1 = 6.1531e-006	Q2 = 2.381e-006 (m^3/s)
f2 = -2.7984e-006	Q3 = 1.1111e-006 (m^3/s)
f3 = -9.1357e-007	

Table 7. Variable values at the solution for the ultrafiltration problem

Variable values at the solution	
Implicit Variables	Explicit variables
c1 = 20.3483 (g/L)	j1 = 9.4177e-006 (m/s)
c2 = 56.6691 (g/L)	j2 = 5.8329e-006 (m/s)
c3 = 163.1191 (g/L)	j3 = 2.1326e-006 (m/s)
Function values	Q1 = 8.1907e-006 (m^3/s)
f1 = -5.0958e-013	Q2 = 2.9411e-006 (m^3/s)
f2 = 4.1394e-013	Q3 = 1.0217e-006 (m^3/s)
f3 = 3.3259e-015	

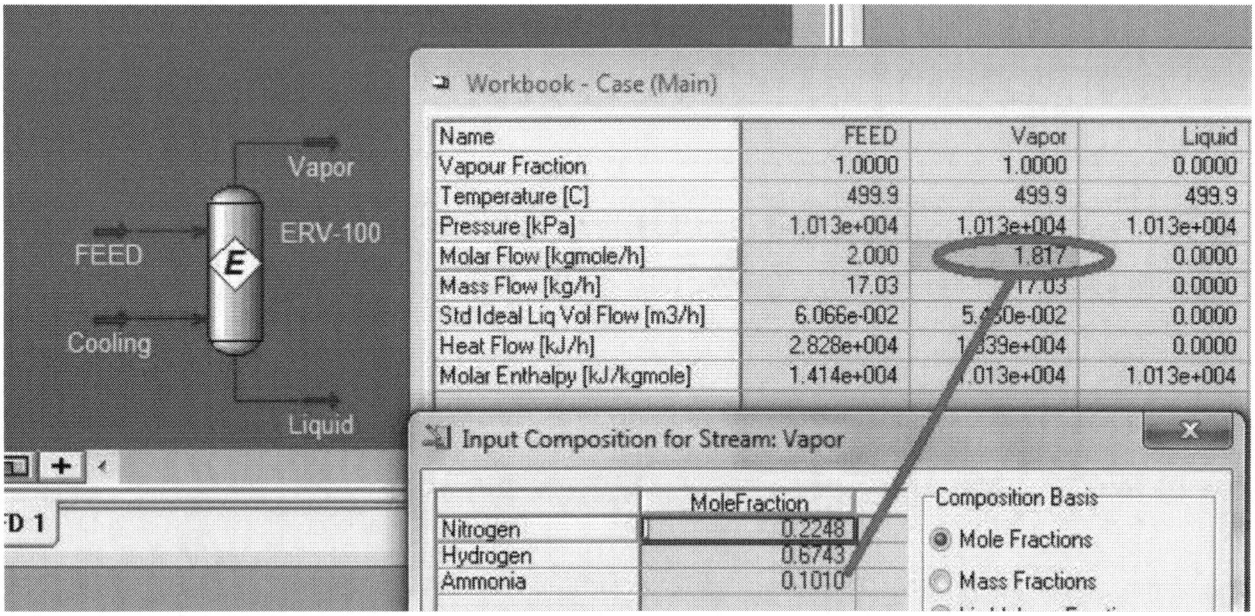

Figure 1. Partial results of an ammonia synthesis process obtained by the UNISIM program

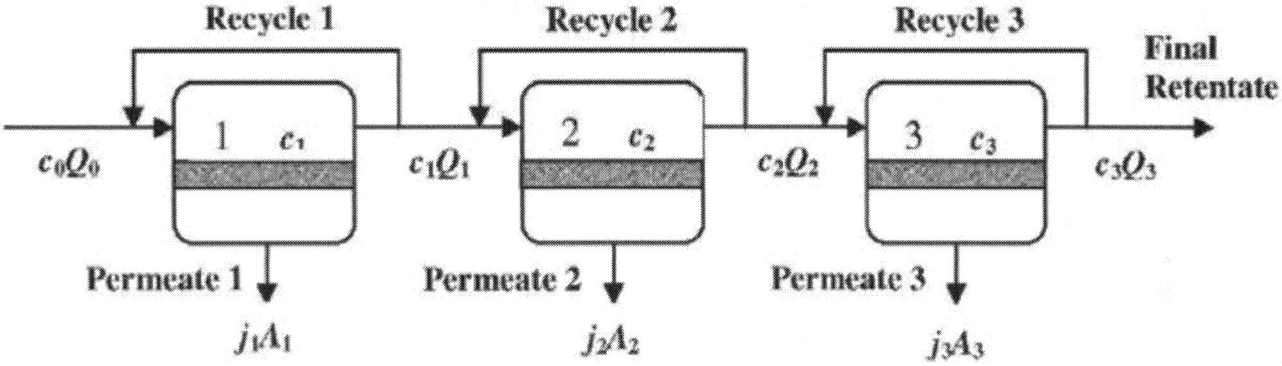

Figure 2. Schematic plot of the multi-stage ultra-filtration system

425558 Teaching Aspen Plus and Other Engineering Software through Video at the University of Delaware

Abstract:

Many students learn to use ASPEN Plus® simulation software as part of their Senior Design course. The University of Delaware (UD) Department of Chemical and Biomolecular Engineering (CBE) has recognized the importance of introducing students to advanced engineering software such as ASPEN Plus® early in the curriculum. UD freshman taking their first chemical engineering course in material and energy balances (CHEG112) are now taught how to model industrially relevant chemical processes using ASPEN Plus®. This year, freshman in CHEG112 were taught how to simulate the hydrogenation of benzene to cyclohexane using ASPEN Plus®. Often, the lack of time available in the classroom and knowledge of ASPEN Plus® prevent the inclusion of such material in the first chemical engineering course. Therefore, we have created a series of ASPEN Plus® tutorial videos that students can access on-line outside of the classroom to learn how to model a variety of unit operations including mixers, heat exchanges, reactors, separators, etc. One advantage to introducing advanced topics using ASPEN Plus® is for the students to make a connection between what they are currently learning and what they will learn in future course work.

This presentation will focus on the use of screen-shot software using Camtasia® to create a library of videos for teaching chemical engineering students how to use ASPEN Plus®. We have also used Camtasia in the Fluid Mechanics course to teach AFT Fathom and ANSYS Fluent. Other courses have used screen-capture software to introduce Matlab to chemical engineering students.

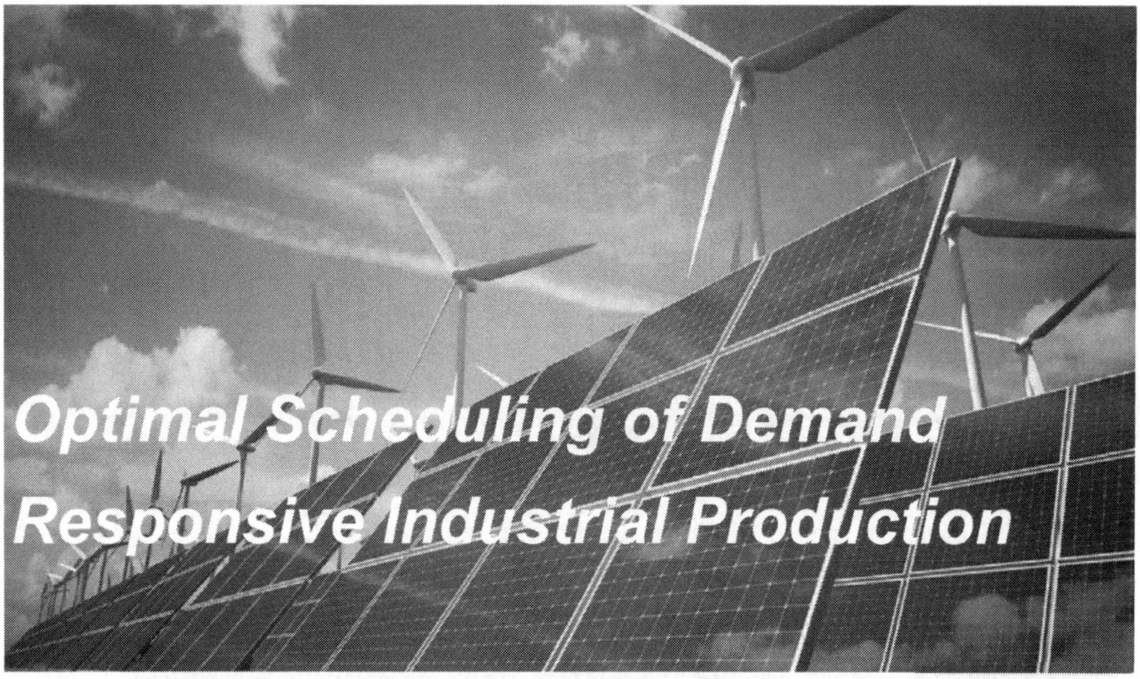

Optimal Scheduling of Demand Responsive Industrial Production

Xiaonan Wang, Ahmet Palazoglu and Nael H. El-Farra

Department of Chemical Engineering and Materials Science

2015 AIChE Annual Meeting
Salt Lake City, UT
November 11, 2015

UCDAVIS

Outline

Energy Management

Demand Responsive Industrial Process

- Introduction
- Component Modeling
  - Hybrid Renewable Energy Systems
  - Chlor-Alkali Plant
- Operational Optimization

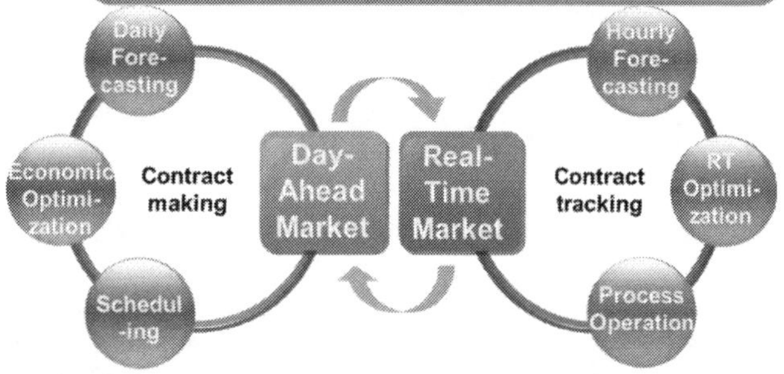

- Results, Discussion and Conclusions

2

Introduction

Motivation for research on ENERGY

➢ *Generation* Side

Goal of California Energy Commission
33% from Renewables by 2020

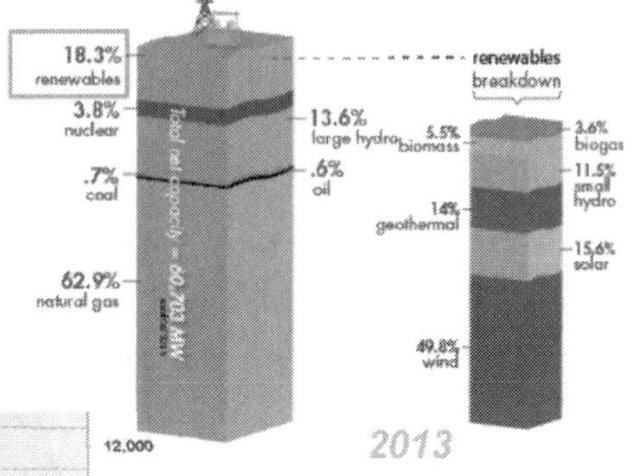

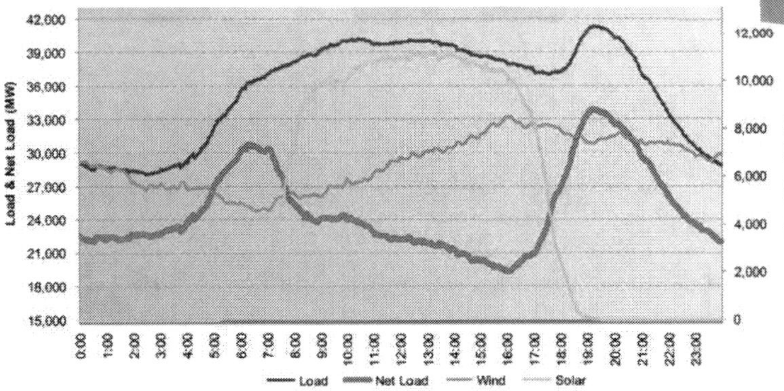

Projected net demand (load and renewable) profiles in April 2020

Hybrid Renewable Energy System (HRES)

L. Xu, D. Tretheway, Flexible ramping products, CAISO Proposal.

3

Introduction

Motivation for research on ENERGY

➢ *Demand* Side

Energy intensive process/ Demand-Responsive loads

Time-of-use (TOU)
- Different unit prices for usage during different blocks of time

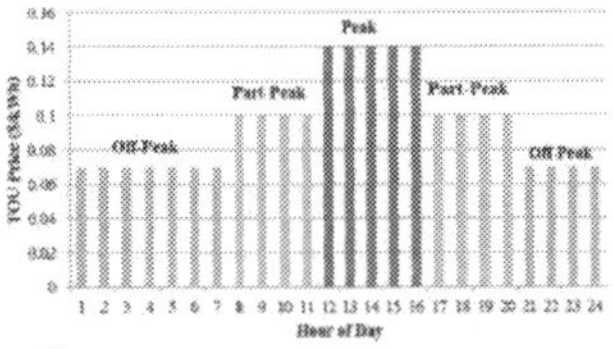

Real-time pricing (RTP)
- Price fluctuating hourly reflecting changes in the wholesale price of electricity

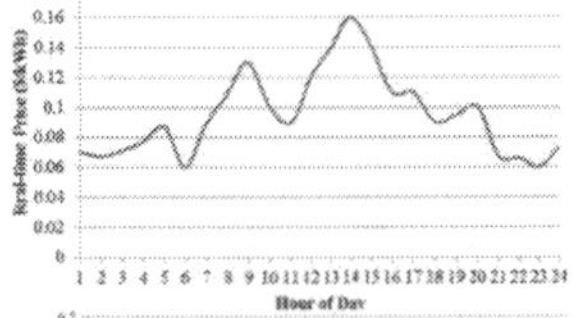

Critical Peak Pricing (CPP)
- Pre-specified high rate for usage designated by the utility to be a critical peak period; a price discount received during non-CPP periods.

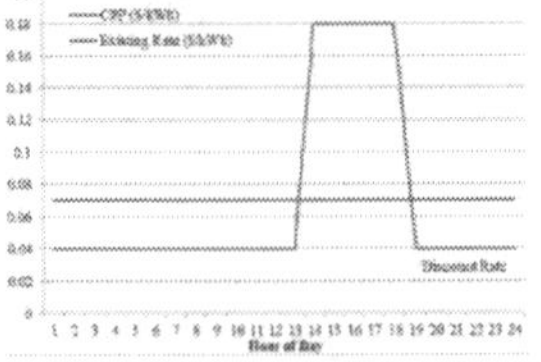

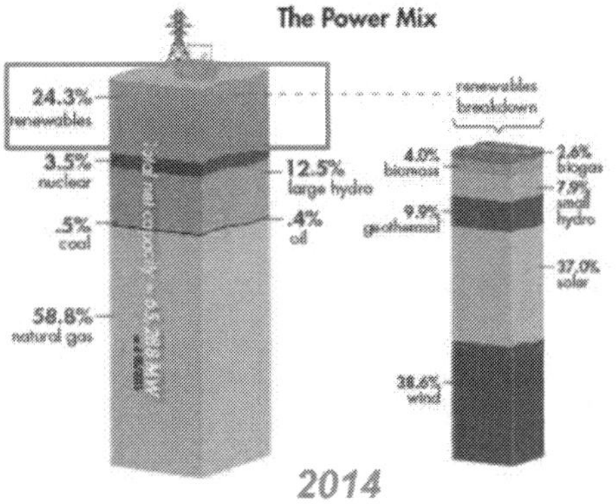

Price-based demand response options

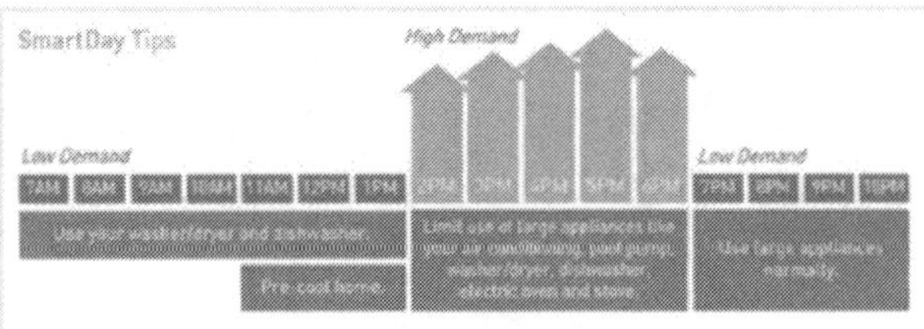

Smart Rate| PG&E

4

http://www.caiso.com/informed/Pages/CleanGrid/default.aspx

Chlor-Alkali Process

Produce chlorine, caustic soda and hydrogen

- Optimal dispatch policy for a grid connected industrial plant

Energy intensive process

- Renewable supply – wind/solar/fuel cell
- Grid connected – buying and selling power
- Demand Management

Hydrogen "Economy"

- Byproduct recovery
- Energy recovery

5

Modeling: Solar Energy

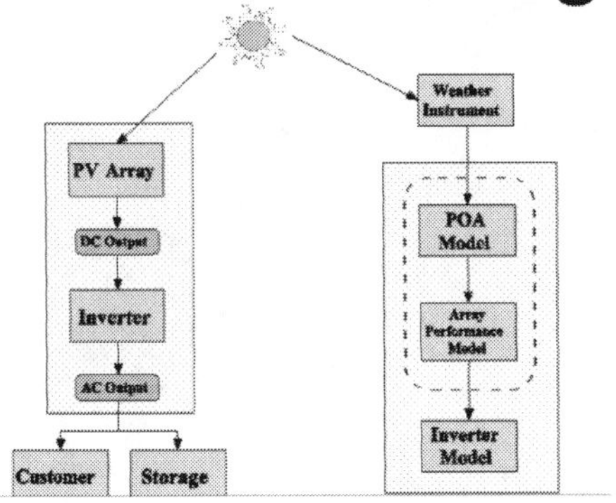

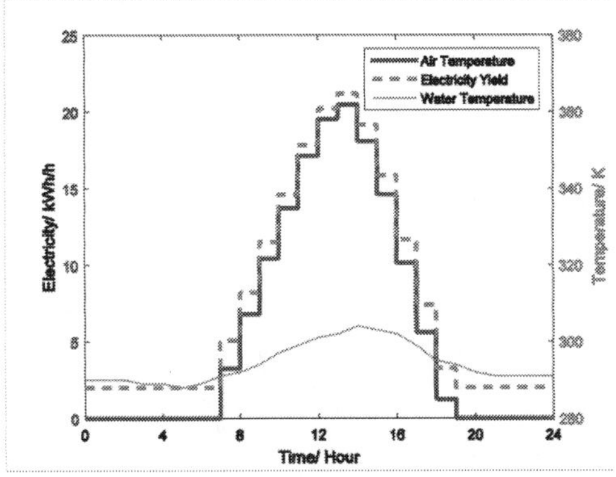

Photovoltaic/Thermal hybrid solar system (PVT system)

- **Plane-of-Array (POA) model:**
- **PV performance model**:
  - Input: global irradiance, temperature

$$I_L - I_S\{\exp[\alpha(v_{pv} + R_S i_{pv})] - 1\} - \frac{v_{pv} + R_S i_{pv}}{R_{sh}} - i_{pv} = 0$$
$$p = v_{pv} i_{pv}$$

5-parameter model (W. Soto et al., 2005)

- **Thermal collector**

$$\eta_{el} = \frac{V_{mpp} I_{mpp}}{G A_{pv}} \qquad \eta_{th} = \frac{\dot{m} C_p (T_{out} - T_{in})}{G A_{pv}}$$

$$\eta_{th} = \tau_\alpha - \tau \eta_{el}$$

6

Modeling: Wind Energy

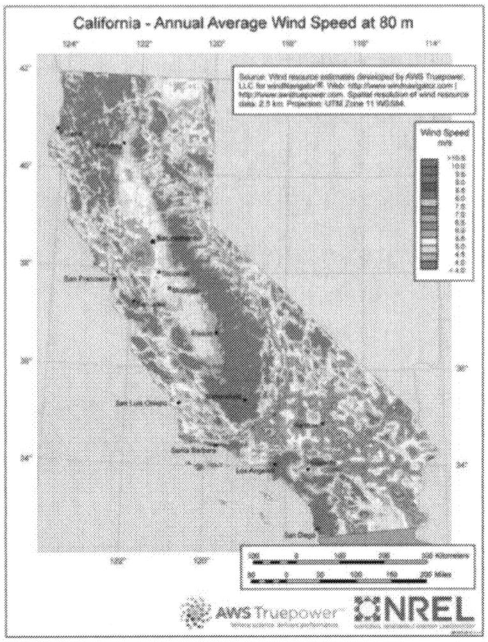

- **Efficiency:**
  - Less than **59.3%**

  Conversion of kinetic energy of the wind to mechanical energy of the turning rotor
  - Only **10%-30%**

  A complete wind turbine system, considering strength and durability
- **Power:** $P_t = \frac{1}{2} C_p(\lambda)\rho A v_w^3$

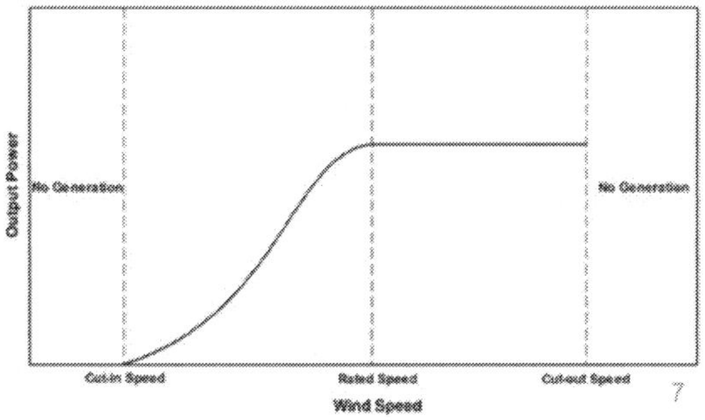

7

Modeling: Hydrogen

Electrolysis:
Chlorine production
Hydrogen Recovery
Storage:
Satisfy unmet demand due to intermittent supply
Fuel Cells:
Reverse process
Use hydrogen to produce energy

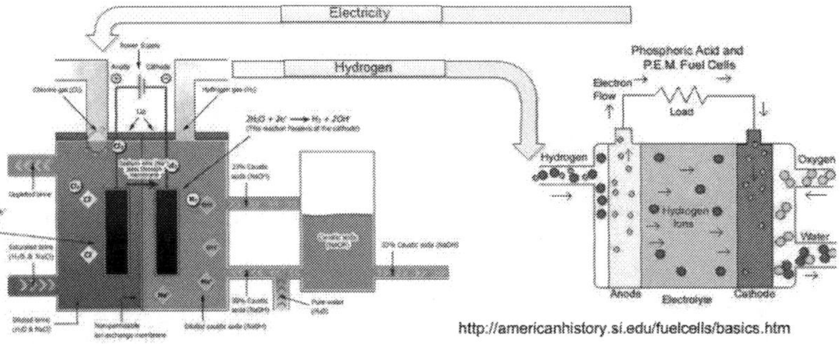

http://americanhistory.si.edu/fuelcells/basics.htm

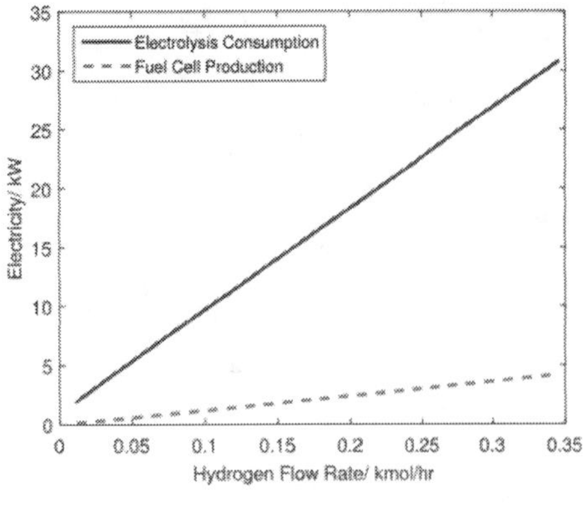

Anode

$$2Cl^-(aq) \rightarrow Cl_2(g) + 2e^-$$

Cathode

$$2Na^+(aq) + 2H_2O + 2e^- \rightarrow H_2(g) + 2Na^+(aq) + 2OH^-(aq)$$

Relations between electricity power (consumed by electrolysis and produced by fuel cells) and hydrogen flow rates

8

Modeling: Plant Diagram

PVT/Wind Energy - Chlor-Alkali Electrolysis - Hydrogen storage - Fuel cells

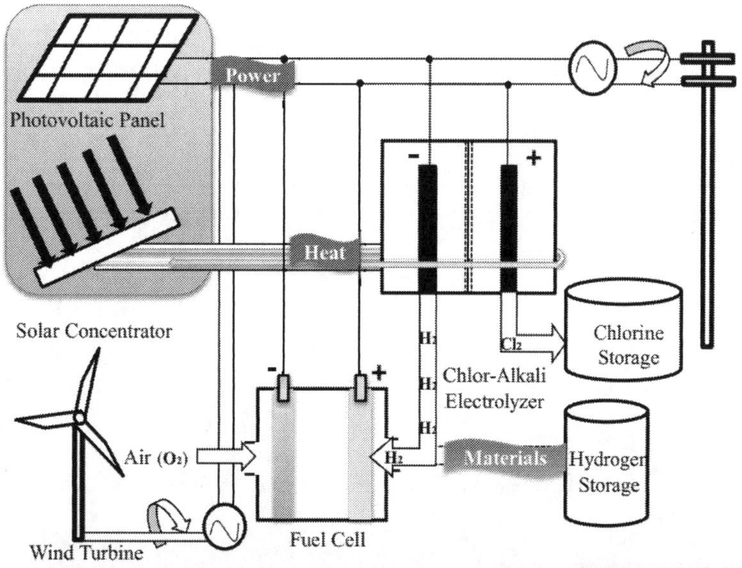

Produce chlorine, caustic soda and hydrogen through electrolysis

Wang, X., Teichgraeber, H., Palazoglu, A., & El-Farra, N. H. (2014). An economic receding horizon optimization approach for energy management in the chlor-alkali process with hybrid renewable energy generation. *Journal of Process Control*, 24(8), 1318-1327.

9

Operational Optimization

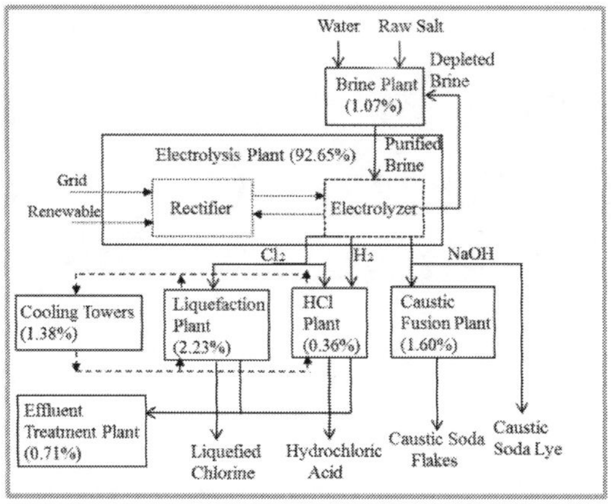

Chlor-Alkali process flow diagram with energy consumption percent (adapted from Babu and Ashok, 2008).

Step 3

Production and generation scheduling is made and real-time operation carried out

Step 2

Contract is reviewed by Smart Grid and discounted pricing scheme decided

Step 1

Electricity consumption patterns and amount are estimated and sent to the electricity providers

10

C. Babu and S. Ashok, "Peak load management in electrolytic process industries," *IEEE Transactions on Power Systems*, vol. 23, no. 2, pp. 399–405, 2008.

Day-ahead Optimization Formulation

Economic Objective (Operation & Environmental Cost)

$$\min \sum_{t=0}^{24} \alpha \left(\sum_i (P^t_{i,ref} C^t_i - P^t_{grid\_sell,ref} C^t_{grid\_sell}) \right)^2 + \beta \left(\sum_i P^t_{i,ref} Cen^t_i \right)^2$$

$$w.r.t. \; T^t_{op}, P^t_{PV,ref}, P^t_{wind,ref}, P^t_{FC_{in},ref}, P^t_{FC_{out},ref}, P^t_{grid\_buy,ref}, P^j_{grid\_sell,ref}$$

$$s.t. P^t_{required} = L_1(Q^t_{product}, T^t_{op})$$

$$T^t_{op} \le L_2(P^t_{th}, \eta^t_{th}) \quad \text{(PVT system)}$$

$$P^t_{required} \le \sum_i P^t_{i,ref} - P^t_{grid\_sell,ref}$$

$$P^t_{i,\min} \le P^t_{i,ref} \le P^t_{i,\max}$$

$$|P_{i,ref}((k+1)\triangle) - P_{i,ref}(k\triangle)| \le dP_{i,\max}, k = 0, 1, ..., 23$$

$$i = \{PV, Wind, FC_{in}, FC_{out}, grid\_buy\}$$

- Modeling of individual components
- Convex quadratic programming
- Decision variables are the power outputs/exchanges at each hour over a day
- Weighting factors emphasize priorities

Optimization over one day

11

Price Structure

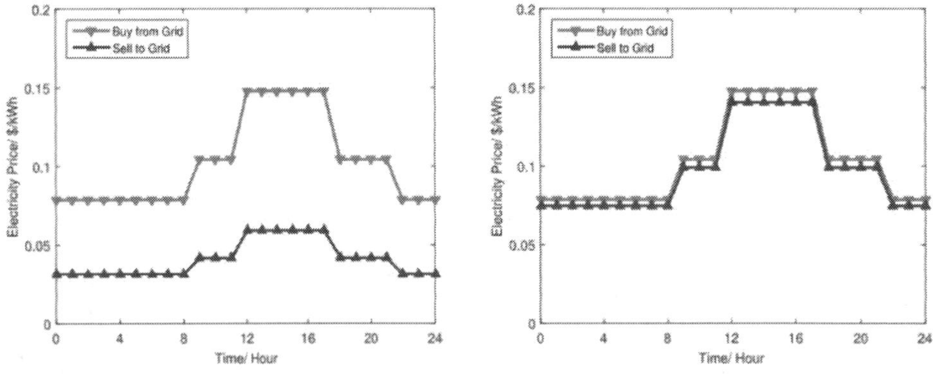

(a) (b)

Time-varying electricity price: (a) normal conditions; (b) incentives for selling electricity.

The contract between electricity providers and users becomes meaningful to provide more competitive pricing schemes to encourage load tracking:

- ✓ how much electricity to be purchased
- ✓ how much to be sold-back

12

Day-ahead Scheduling and Contract Design

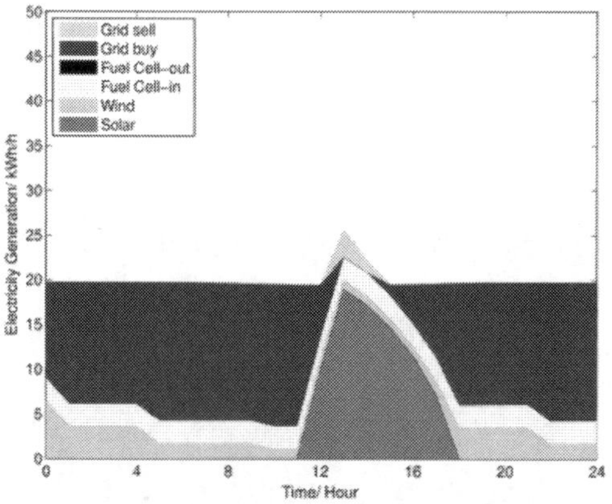

(a) Day-ahead energy dispatch strategies

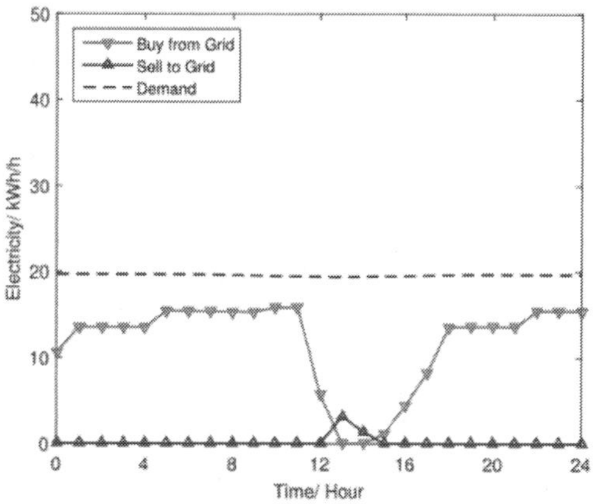

(b) Optimal contract for buying and selling electricity

- One can see the complementary relations between all resources.
- Purchased hydrogen used less due to cost.

- Electricity is bought during morning evening hours.
- Electricity is sold when generation exceeds demand.

13

Day-ahead Scheduling and Contract Design

Demand Response

$$\sum_{t=0}^{24} Q_{product}(t) \geq Q_{product}(daily)$$

- Allowing the system to change its operating level among all possible (continuous) states.
- Output satisfies required daily production level.

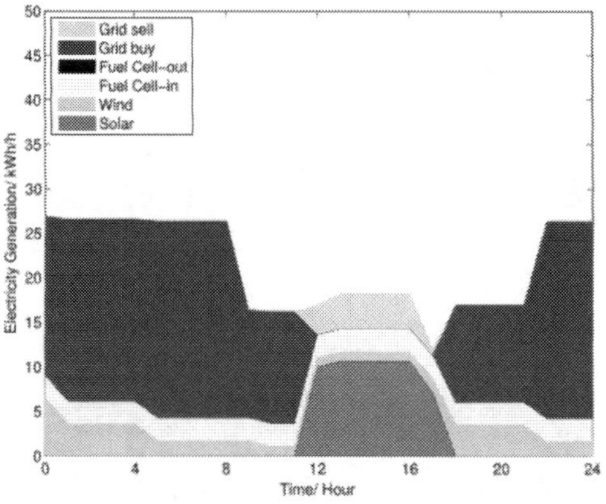

(a) Day-ahead energy dispatch strategies

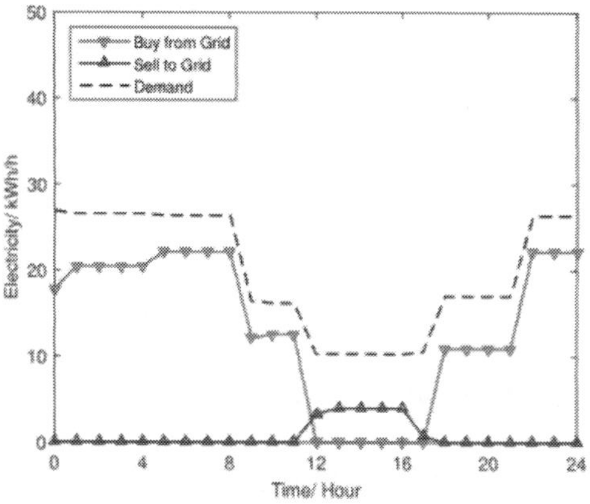

(b) Optimal contract for buying and selling electricity

14

Real-time Optimization Formulation

$$\min \sum_{t=j}^{j+N\Delta} \alpha \left(\sum_i (P_{i,ref}^t C_i^t - P_{grid\_sell,ref}^t C_{grid\_sell}^t) \right)^2 + \beta \left(\sum_i P_{i,ref}^t Cen_i^t \right)^2$$
$$+ \lambda \left(y_i^t - y_{i,r}^t \right)^2 \quad \text{(Contract tracking)}$$

$$w.r.t. T_{op}^j, P_{PV,ref}^j, P_{wind,ref}^j, P_{FC_{in},ref}^j, P_{FC_{out},ref}^j, P_{grid\_buy,ref}^j, P_{grid\_sell,ref}^j$$
$$s.t. \forall t \in [j\Delta, (j+1)\Delta)$$
$$P_{required}^t = L_1(Q_{product}^t, T_{op}^t)$$
$$T_{op}^t \le L_2(P_{th}^t, \eta_{th}^t)$$
$$P_{required}^t \le \sum_i P_{i,ref}^t - P_{grid\_sell,ref}^t$$
$$P_{i,\min}^t \le P_{i,ref}^t \le P_{i,\max}^t$$
$$|P_{i,ref}((k+1)\Delta) - P_{i,ref}(k\Delta)| \le dP_{i,\max}, k = 0, 1, ..., N-1$$
$$i = \{PV, Wind, FC_{in}, FC_{out}, grid\_buy\}$$

- Operation state is varying based on cost of electricity and the availability of renewable resources.
- Trade-off between maximizing profits and following the scheduled electricity trade.

15

Real-time Optimization and Contract Following

$$s = \begin{cases} 1, \text{when } Peak\ hours \\ 0, \text{when } Normal\ hours \\ -1, \text{when } Off-peak\ hours \end{cases}$$

Last term in the objective function:

$$\lambda \cdot s \cdot (P_{grid\_buy}^t - P_{contract\_buy}^t)$$

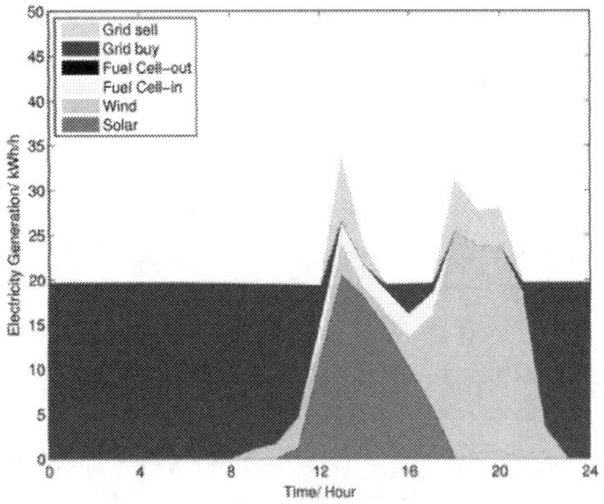

(a) Real-time energy dispatch strategies

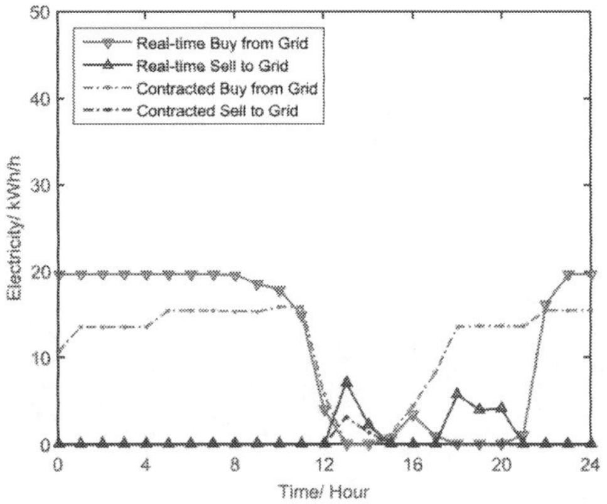

(b) Optimal profiles for buying and selling electricity compared with contracts

16

Real-time Optimization and Contract Following

Demand Response in Real-time Operation

$$\sum_{t=0}^{24} Q_{product}(t) \geq Q_{product}(daily)$$

Deviations from contracted buying and selling curve induce certain penalties based on peak or off-peak hours of the day.

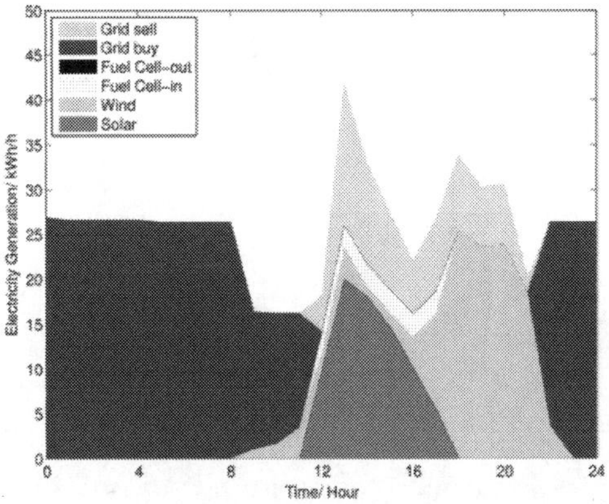

(a) Real-time energy dispatch

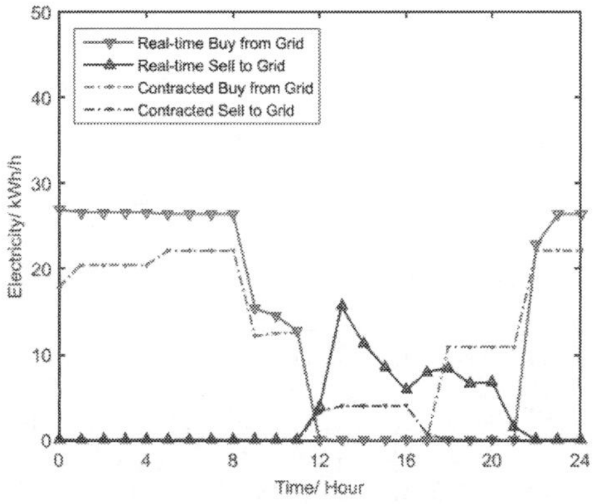

(b) Optimal profiles for buying and selling electricity compared with contracts

17

Economical Results

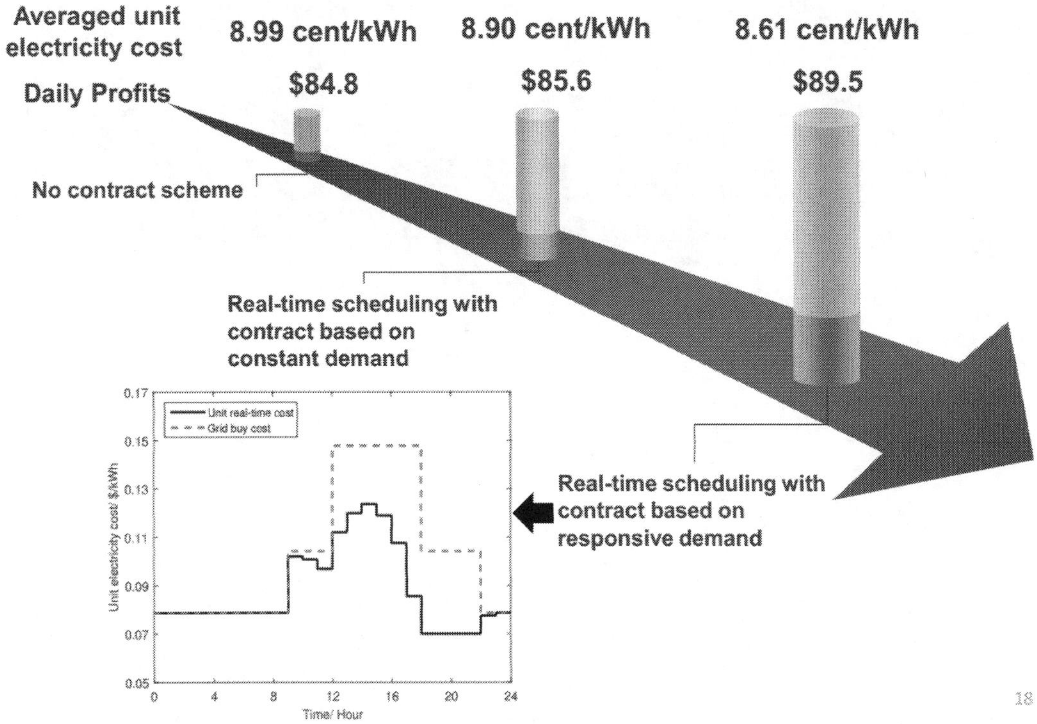

18

Summary

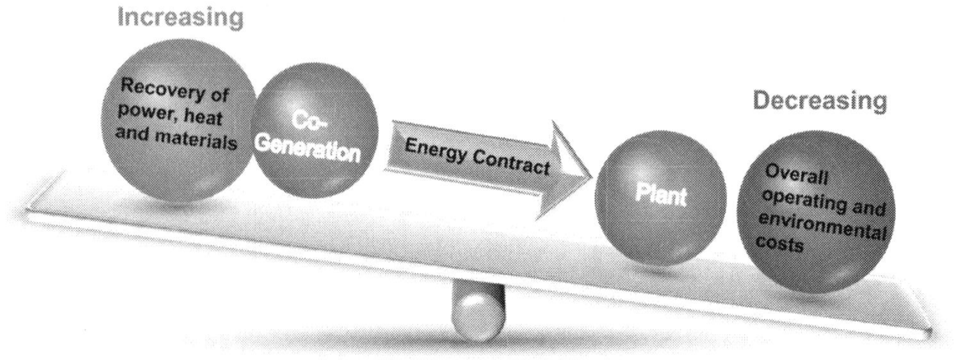

- ✓ Optimal operation of grid-connected hybrid PV-thermal, wind and fuel cells energy conversion systems to supply power to process industries (e.g., chlor-alkali plant)
- ✓ Electricity contract and incentive scheme designed for efficient operation
- ✓ Day-ahead and real-time operational strategies optimized to increase profits

19

Thank you!

Any Questions?

20

Reference

[1] T. Jin, M. Mechehoul, Ordering electricity via internet and its potentials for smart grid systems, Smart Grid, IEEE Transactions on 1 (3) (2010) 302–310.
[2] K. Nolde, M. Morari, Electrical load tracking scheduling of a steel plant, Computers & Chemical Engineering 34 (11) (2010) 1899–1903.
[3] W. Saad, Z. Han, H. V. Poor, T. Basar, Game-theoretic methods for the smart grid: An overview of microgrid systems, demand-side management, and smart grid communications, Signal Processing Magazine, IEEE 29 (5) (2012) 86–105.
[4] E. Worrell, D. Phylipsen, D. Einstein, N. Martin, Energy use and energy intensity of the US chemical industry, Tech. rep., Lawrence Berkeley National Lab., CA (US) (2000).
[5] T. F. OBrien, T. V. Bommaraju, F. Hine, Handbook of chlor-alkali technology, Vol. 1, Springer, 2005.
[6] M. Paulus, F. Borggrefe, The potential of demand-side management in energy-intensive industries for electricity markets in Germany, Applied Energy 88 (2) (2011) 432–441.
[7] M. Broeren, D. Saygin, M. Patel, Forecasting global developments in the basic chemical industry for environmental policy analysis, Energy Policy 64 (2014) 273–287.
[8] L. Xu, D. Tretheway, Flexible ramping products, CAISO Proposal.
[9] X.Wang, H.Teichgraeber, A.Palazoglu, N.H.El-Farra, Aneconomicrecedinghorizon optimization approach for energy management in the chlor-alkali process with hybrid renewable energy generation, Journal of Process Control 24 (8) (2014) 1318–1327.
[10] S. Mitra, I. E. Grossmann, J. M. Pinto, N. Arora, Optimal production planning under time-sensitive electricity prices for continuous power-intensive processes, Computers & Chemical Engineering 38 (2012) 171–184.
[11] P. Cappers, C. Goldman, D. Kathan, Demand response in US electricity markets: Empirical evidence, Energy 35 (4) (2010) 1526–1535.

UNIVERSITY OF CALGARY

Power Management of Microgrids: Multiparametric Programming Approach to Improved Handling of System Uncertainties

Evar Umeozor, Milana Trifkovic
University of Calgary

SCHULICH School of Engineering

November 11, 2015
AIChE Annual Meeting
Salt Lake City, Utah, USA

Motivation

Past/Present

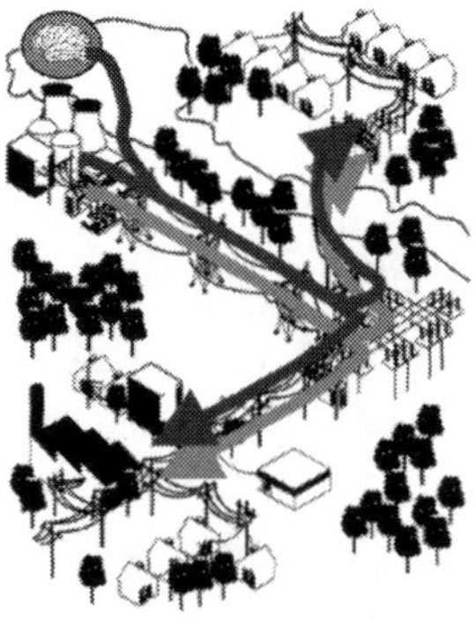

- Centralized operation and control.
- Passive transmission and distribution.
- Vulnerable

Future

- Distributed operation and control – Smart Peripherals
- Renewable Energy Sources
- Integrated communications
- More efficient loads
- More robust

Through integration of renewables into the current power system we can decrease environmental pollution and improve power reliability

UNIVERSITY OF CALGARY

2

Motivation

Present

Power Produced

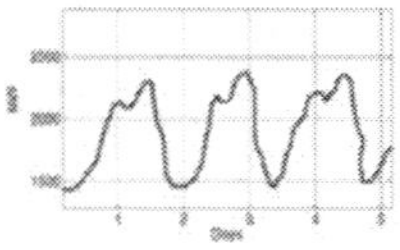

=

Power Consumed

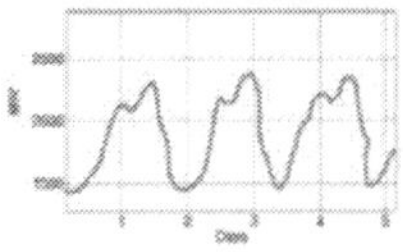

Power balance
easily achieved

≠

Power Produced

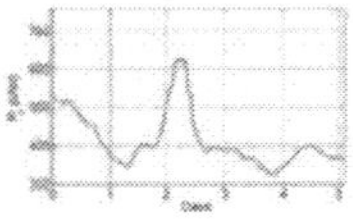

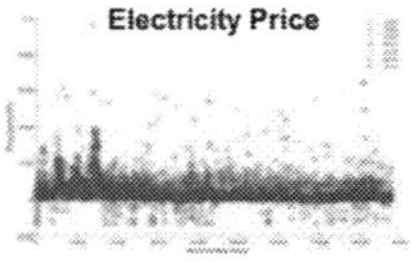

Future

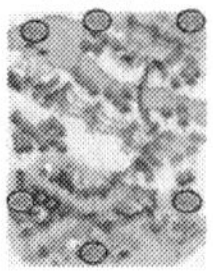

- Intermittency in renewable resources
- Dynamic demand
- Changing electricity price

We need new modeling and optimization tools to achieve efficient operation of complex microgrid systems

UNIVERSITY OF CALGARY

3

Objective

Develop optimal operational schedules for a set of microgrid technologies

Case Study (Hybrid Energy System built in Sarnia, Ontario - Renewable and Conventional Energy Sources)

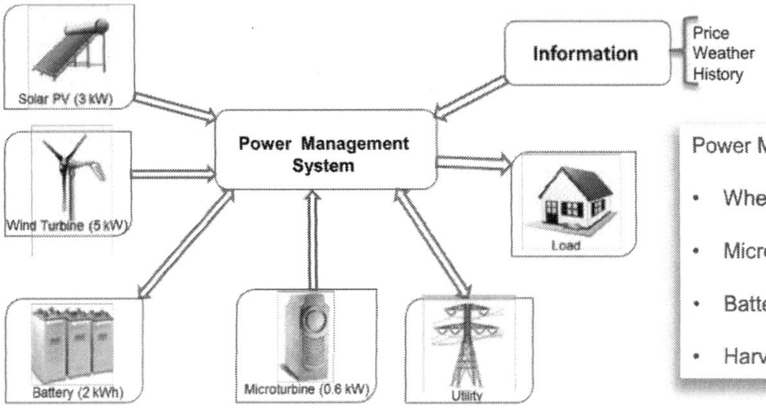

Power Management System Decisions:

- When to buy/sell electric power
- Microturbine startup/shutdown
- Batteries charging/discharging
- Harvesting maximum renewables

Operate optimally in the face of uncertainties

UNIVERSITY OF CALGARY

4

Optimization under Uncertainty

Traditional Approach – Online Optimization

- Utilize forecasts for weather to solve optimization problem over a finite horizon (e.g. 24 hours)
- Implement optimal solution in each optimization interval (e.g. 1hr)
- Re-optimize at next sampling time with the knowledge of the current states

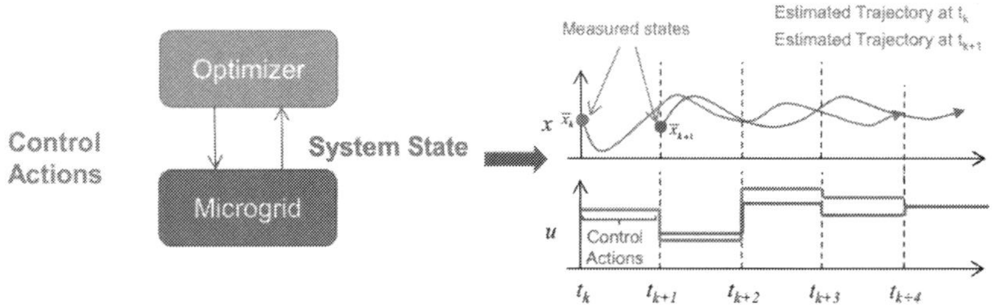

Drawbacks

Computationally Expensive
- Multiple computations
- Dimensionality

Poor Handling of Uncertainty
- Forecast errors
- Time-scale mismatch

Trifkovic, Marvin, Daoutidis, & Sheikhzadeh, (2014); Morari, Dua & Pistikopoulos (2001); Kwon, Bruckstein, & Bemporad (1983, 1989)

UNIVERSITY OF CALGARY

5

Optimization under Uncertainty

Parametric Optimization Approach

- Parameterize renewable resource variability into grid points
- Solve a bi-level optimization problem as a function of the parameters in an offline manner
- Obtain the coordinates in the space of parameters where the solutions are valid

Well-bounded systems!
Design and Operational limits

- Cut-in and cut-out wind power
- Maximum insolation on location

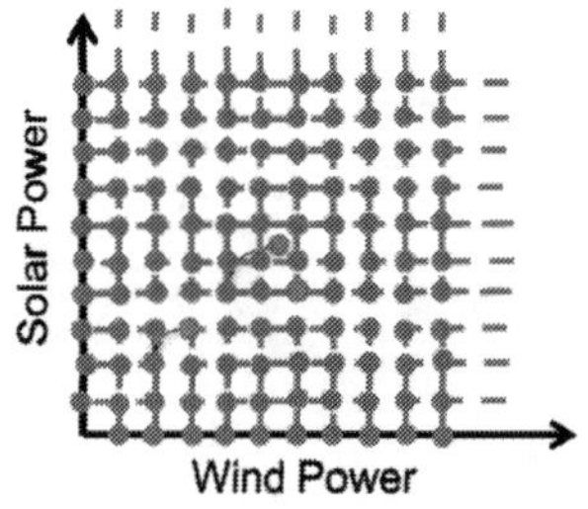

- Optimization problem solution as a function of solar and wind power
- Realized renewable power

Take actions according to the closest realizable "node" in parametric space

Dua & Pistikopoulos [2000] Ind. Eng. Chem. Res.

Finer resolution of the parametric space results in more efficient operation but increased computational cost – need to find optimal resolution factor (rf)

UNIVERSITY OF CALGARY

6

Optimization under Uncertainty

Online Optimization

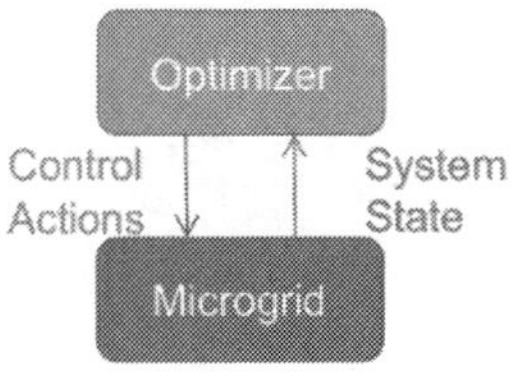

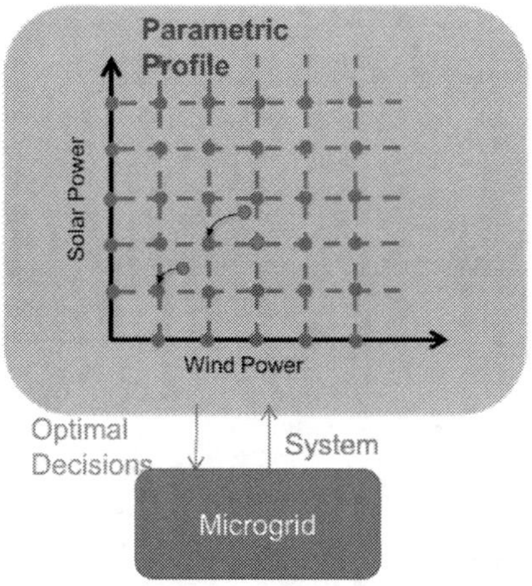

Expected outcomes

- Achieve proposed objectives while satisfying all economic, logical and units' lifetime requirements
- Reduced computational effort
- Leverage design capacities and significantly reduce dependency on the accuracy of the weather forecasts

Achieve "on-line optimization via off-line optimization"

UNIVERSITY OF CALGARY

7

Parametric Optimization Problem

$$\min_{rf} Q = \{Q^k \in \mathbb{R}: Q^k(\boldsymbol{\theta}), \boldsymbol{\theta} \in \Theta\}$$

$s.t.$

$$Q^k(\theta) = \min_{x_k \in X, y_k \in Y} M_{w,k} + M_{s,k} + M_{m,k} + M_{u,k} + M_{b,k}$$

$$M_{w,k} = y_{w,k} h_w + (c_w - c_{w,k})\theta_{w,k}$$

$$M_{s,k} = y_{s,k} h_s + (c_s - c_{s,k})\theta_{s,k}$$

$$M_{g,k} = y_{g,k}(h_g + m_{sc}) + c_g P_{g,k}$$

$$M_{u,k} = e_{p,k} P_{ub,k} - e_{p,k} P_{us,k}$$

$$M_{b,k} = h_b(y_{b1,k} + y_{b0,k}) - e_{p,k}(P_{b1,k} + P_{b0,k})$$

$$y_{g,k} P_{g,k} + \theta_{w,k} + \theta_{s,k} + P_{ub,k} + P_{b1,k} - L_{1,k} - L_{2,k} = P_{us,k} + P_{b0,k}$$

$$y_{g,k} P_{g,k} + \theta_{w,k} + \theta_{s,k} + P_{ub,k} + P_{b1,k} \geq L_{1,k} + L_{2,k}$$

$$P_{ub,k} \leq (1 - u_k)(L_{1,k} + L_{2,k})$$

$$P_{us,k} \leq u_k(\theta_{w,k} + \theta_{s,k})$$

$$P_{b1,k} \leq y_{b1,k}\left(\frac{SOC_{b,k} - SOC_{min}}{\Delta k}\right)$$

$$P_{b0,k} \leq y_{b0,k}\left(\frac{SOC_{max} - SOC_{b,k}}{\Delta k}\right)$$

$$SOC_{b,k+1} = SOC_{b,k} + E_{b0,k} - E_{b1,k}$$

$$y_{g,k} + u_k \leq 1$$

$$y_{b0,k} + y_{b1,k} \leq 1$$

$$e_{p,k} = \begin{cases} \leq e_{pc} \rightarrow (y_{b0,k} - u_k \geq 0) \\ > e_{pc} \rightarrow (y_{b1,k} - u_k \geq 0) \end{cases}$$

$$SOC_{min} \leq SOC_{b,k} \leq SOC_{max}$$

$$P_{b0,k}, P_{b1,k}, P_{g,k}, P_{ub,k}, P_{us,k}, C_{bat}, L_{1,k}, L_{2,k} \geq 0$$

$$P_{g,k} \leq y_{g,k} \times L_{1,k}$$

$$r_k - s_k = y_{m,k} - y_{m,k-1} \qquad \forall k \in T$$

$$y_{m,k} \geq r_k + \sum_{\substack{n>0 \\ n \in N^{on}}} r_{k,n} \qquad \forall k \in T: \tau^{on} \geq 4$$

$$1 - y_{m,k} \geq s_k + \sum_{\substack{n>0 \\ n \in N^{off}}} s_{k,n} \qquad \forall k \in T: \tau^{off} \geq 4$$

Objective function
- Minimize net cost of operation for the realized renewable resource scenario

Equality constraints
- Component net O&M costs
- Power and energy balances

Inequality constraints
- Renewables availability
- Equipment longevity
- Demand satisfaction

Solve an mp-MILP problem to optimally schedule a set of microgrid technologies

UNIVERSITY OF CALGARY

8

Microgrid Power Management System (MPMS)

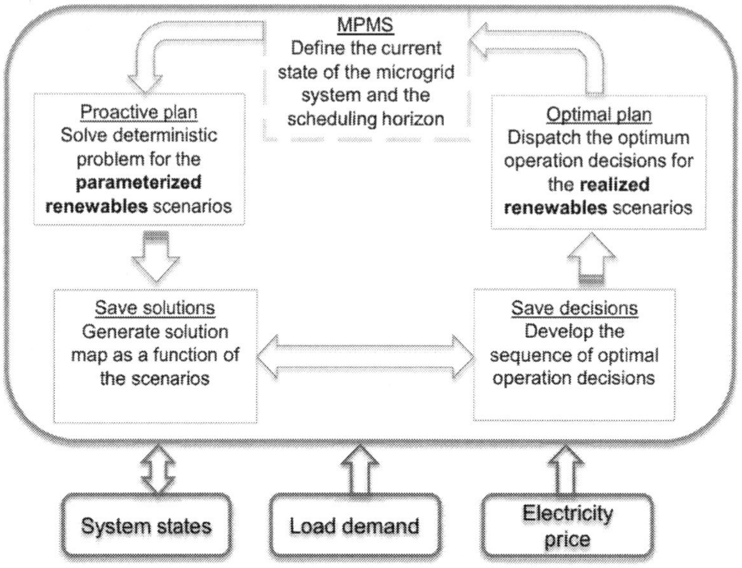

UNIVERSITY OF CALGARY

9

Case Study – Sarnia, Ontario

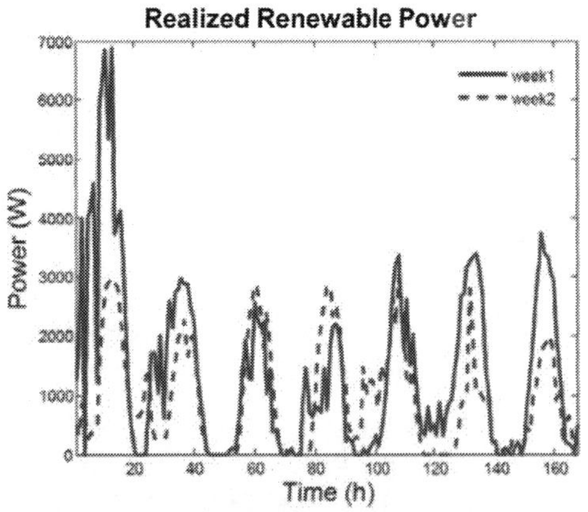

The realized variations are observed for two weeks in Ontario. Variability in the realized renewable resources can be very drastic.

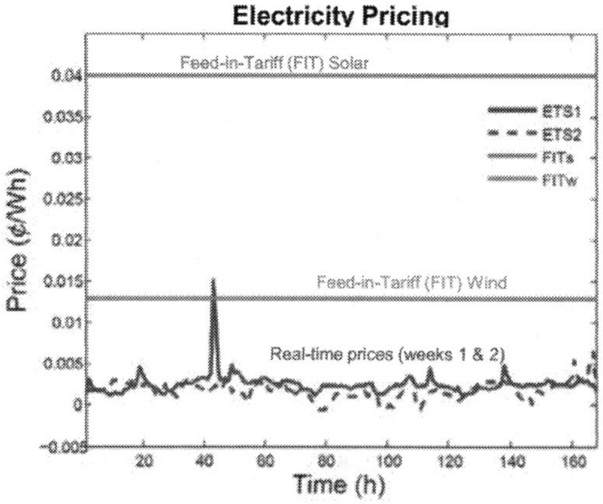

For a prosumer whose local consumption is incentivised but sells to (and buys from) the utility at real-time market price

UNIVERSITY OF CALGARY

10

Results – Battery State of Charge

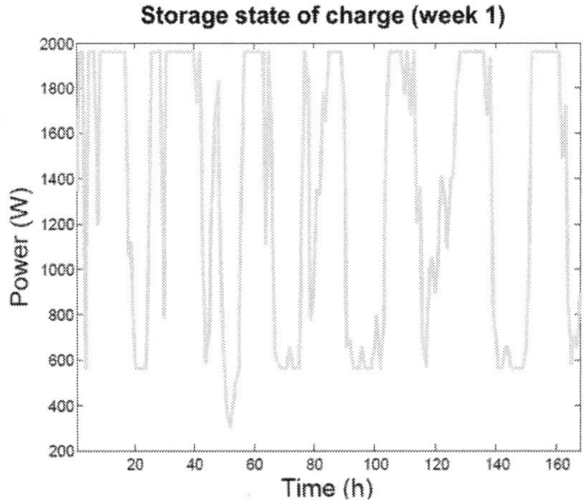

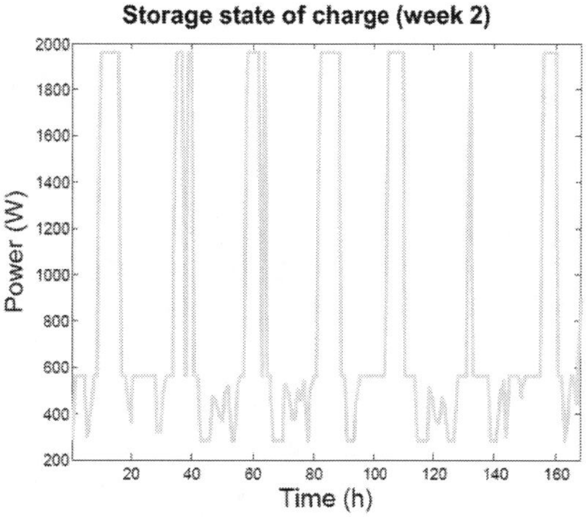

Storage serves for both power and economic buffering
Charging → more renewables and/or low price
Discharging → less renewables and/or high price

UNIVERSITY OF CALGARY

11

Results – Microturbine Scheduling

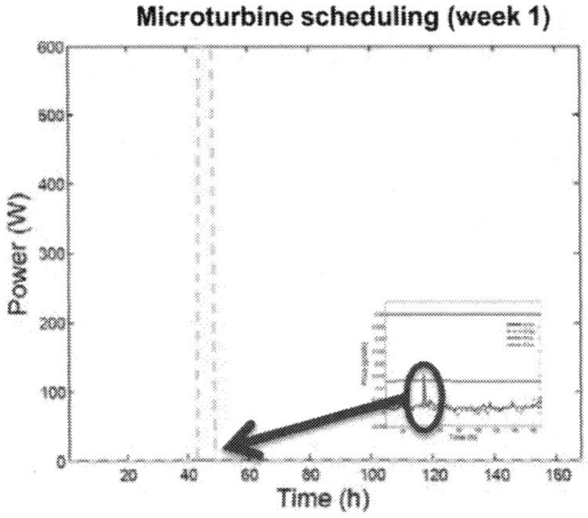

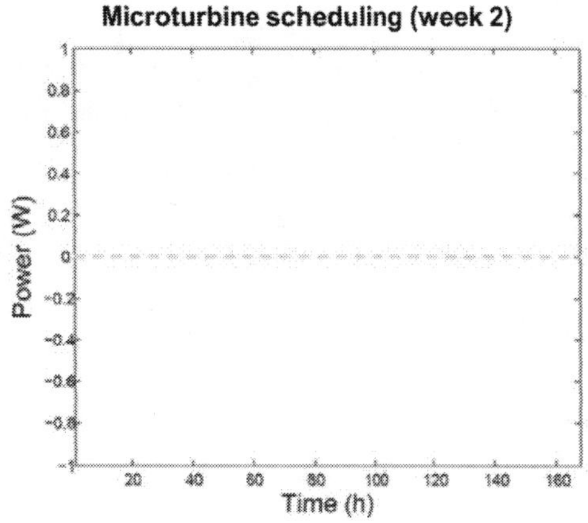

Microturbine is activated only when the electricity price is high and there is no renewable energy

UNIVERSITY OF CALGARY

12

Results – Grid Interaction

Utility interaction (week 1)

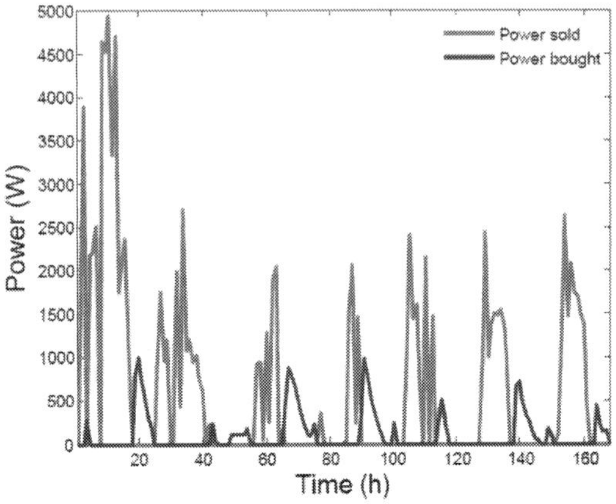

Utility interaction (week 2)

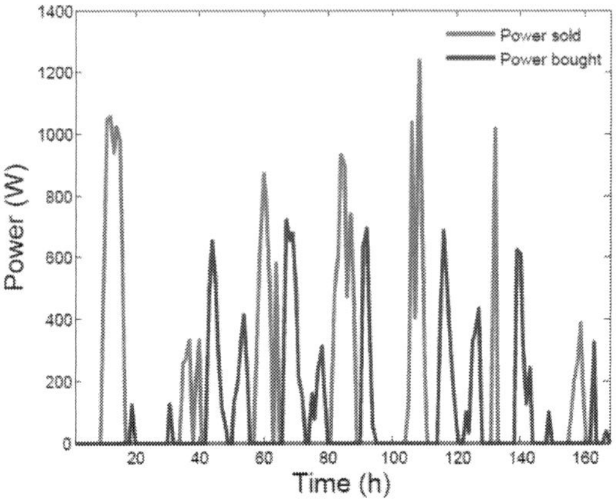

Less power bought when the electricity price is high and there is no renewable energy

Results – Power Balance

Power balance (week 1)

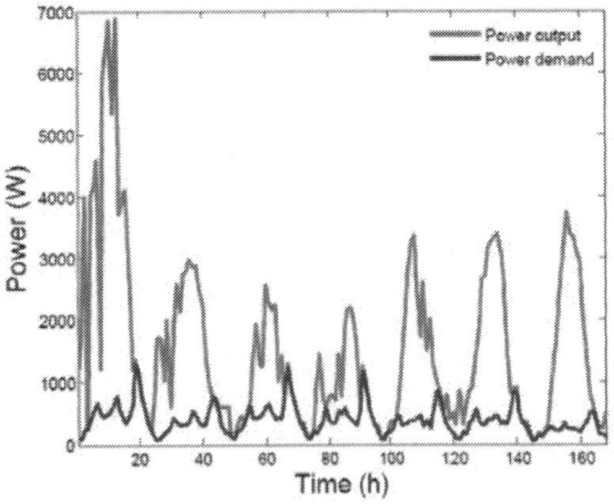

Power balance (week 2)

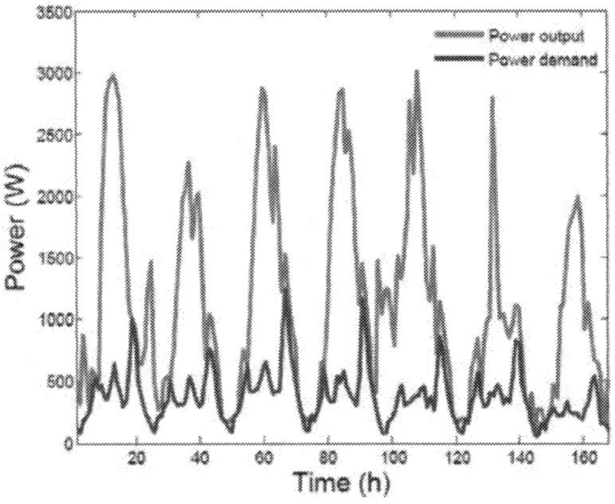

All local load demand are always satisfied

Results – Net Cost

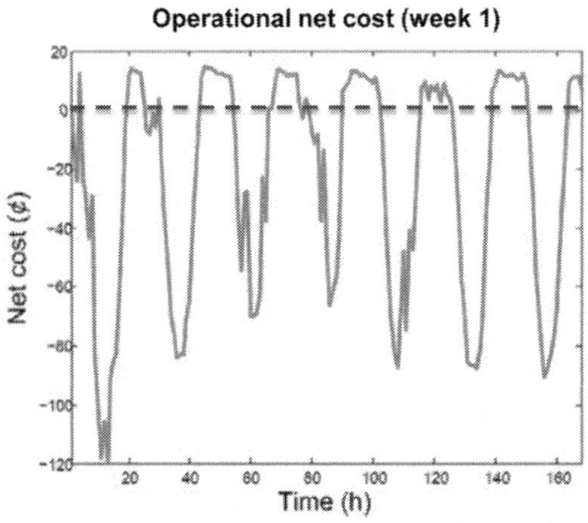

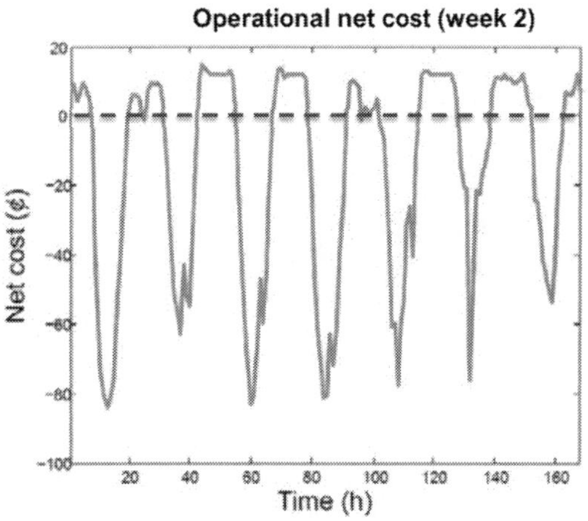

System minimizes net cost of the operational plan

UNIVERSITY OF CALGARY

Summary

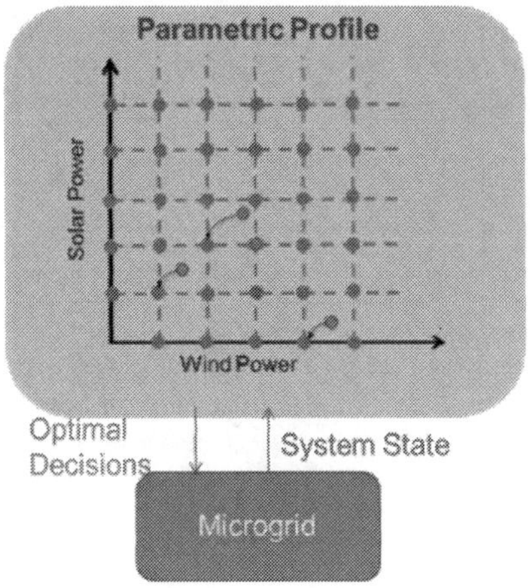

- Parameterize the uncertain renewable resources using the operational/design bounds on the wind and solar power systems
- Eliminate solution dependence on meteorological info
- Easily deployable for real-time studies
- Promising approach for microgrid energy management

Future Work

- Load programming (Demand-side management)
- Electricity price modelling
- Implement on the existing system

UNIVERSITY OF CALGARY

Acknowledgements

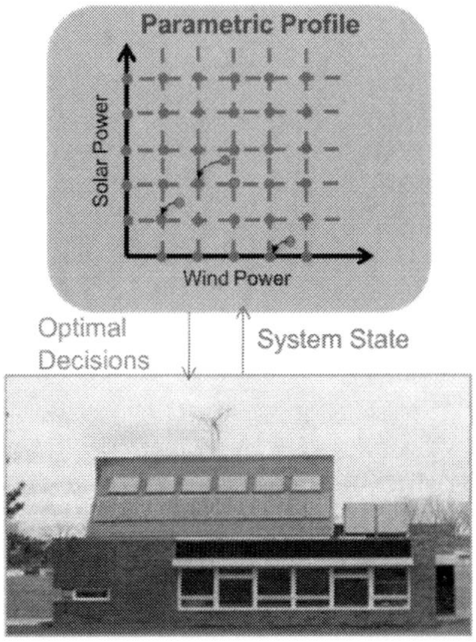

- **Bluewater Power Corporation**
- **Natural Sciences and Engineering Research Council of Canada**

UNIVERSITY OF CALGARY

17

SUPPLEMENTARY SLIDES

UNIVERSITY OF CALGARY

18

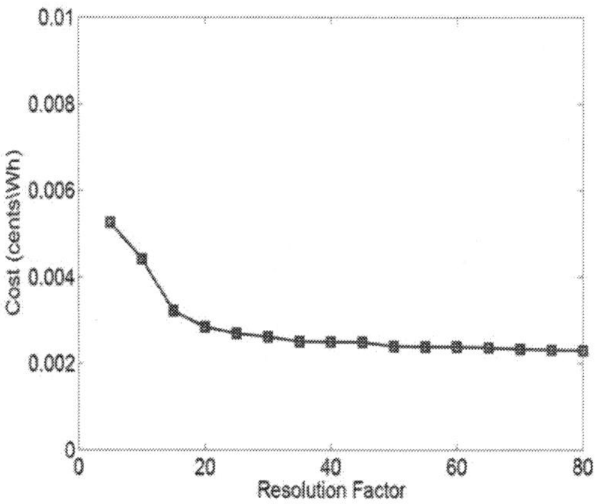

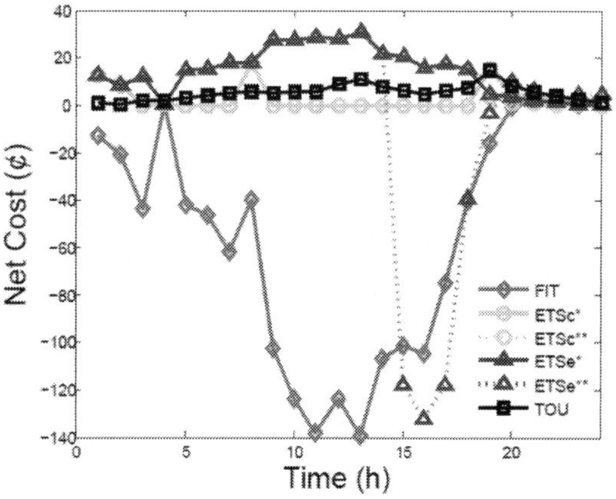

UNIVERSITY OF CALGARY

19

Table 1. Cost characteristics of the microgrid components			
Component	Depreciation ($cents/h$)	Maintenance cost ($cents/Wh$)	Lifetime (yrs)
WT	7.5	0.0008	20
MT	1.68	0.002	20
PV	18	0.0003	20
Battery	12	0.00015	1

Table 2. Simulation parameters

cut in speed	3 m/s
cut out speed	25 m/s
rated speed	12 m/s
reference height	10 m
reference insolation	1000 W/m^2
MT startup cost	20 $cents$

UNIVERSITY OF CALGARY

20

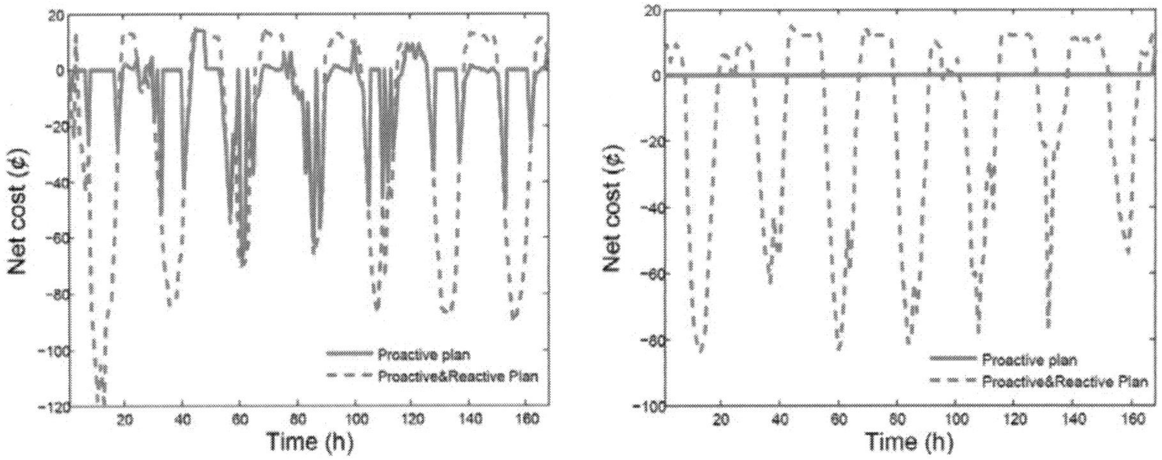

Combining proactive and reactive schema improves economic performance

UNIVERSITY OF CALGARY

21

425468 On the Optimal Sizing and Dispatch Problem for Microgrids with Stochastic Generators – When Is the Expected-Cost Differentiable?

Abstract:

This presentation concerns the problem of jointly optimizing the component sizes and dispatching rules for microgrids incorporating intermittent renewable resources such and wind and solar. The stochasticity introduced by these renewable resources, and to a lesser extent by the loads, plays a central role in this problem. First, it is well known that sizing components (e.g., generators and storage units) based on average resource availabilities can lead to designs that are highly unreliable or suboptimal in reality. Moreover, dispatching decisions must be made under large uncertainties about the future demand and resource availabilities, despite the fact that the relative costs and benefits of the competing options may change radically across alternative futures.

Thus, the optimization problem at hand involves complex trade-offs between cost and reliability arising from the interplay of both design and operational decisions over time. To capture these trade-offs, a number of detailed simulation models have recently been developed. A given design is typically evaluated through an hourly time-series simulation over the course of a year. Hourly load, irradiance, and wind speed data are supplied by the user or stochastically generated from monthly averages. In each hour, these data are used to compute the net available power and dispatch decisions are made, typically based on a fixed set of logical rules parameterized by user-defined thresholds. Finally, model equations are solved for each component to update the system state and evaluate performance over the hour. The high temporal resolution of these simulations makes it possible to incorporate accurate performance and operational cost models for the system components. Moreover, these simulations can be driven with randomized load and generation profiles to accurately evaluate the expected cost of a system, including both stochastic operational costs and reliability metrics.

In light of the preceding discussion, the optimal microgrid sizing and dispatch problem can be formulated as a nonlinear program in which the objective is defined as the expected-value of randomized time-series simulations. In fact, this simulation-optimization formulation has been used extensively in conjunction with heuristic optimization algorithms that view the stochastic simulation as a black-box. However, in the interest of enabling more efficient optimization approaches, we consider here the important question of whether or not the expected-cost is a continuously differentiable function of the decision variables (i.e., the component sizes and parameters influencing the dispatching rules). We begin by formulating time-series microgrid simulations generically as discrete-time stochastic hybrid systems (DTSHS) with associated stage and terminal costs. Alarmingly, for any fixed realization of uncertainty, we find that discrete dispatching decisions can introduce discontinuities in the system cost at every time-step along the simulation horizon. This observation carries over to finite sample averages of the cost, so that formally optimizing such an average would require the introduction of very many binary variables (there is a relation to the conventional multistage stochastic programming formulation of the problem here). Remarkably, our findings show that the true expected-value of the cost function is nonetheless continuously differentiable under very general conditions, and hence is amenable to gradient-based optimization strategies. We provide sufficient conditions for continuous differentiability and demonstrate their verification for some representative microgrid models. We also highlight particular model features and dispatching strategies that may lead to violations. Finally, we present improved optimization results using a stochastic gradient-descent algorithm.

432717 Model Predictive Control of Integrated Gasification Combined Cycle Power Plants with Membrane Reactors for Carbon Capture

Abstract:

In recent years, facing the challenge of global climate change, technologies of power plants with carbon capture have been studied worldwide. Integrated gasification combined cycle (IGCC) plants, which have potential for higher efficiencies than the conventional pulverized coal-fired power plants [1], are an emerging candidate for carbon capture. In this work, we analyze the implementation of model predictive control (MPC) strategies for an IGCC power plant equipped with a water gas shift membrane reactor (WGS-MR) unit for carbon capture. A model for this IGCC power plant represented by a system of differential algebraic equations was previously developed in MATLAB/Octave [2]. Using this model, two centralized control strategies based on linear and nonlinear MPC algorithms are presented to address different power generation scenarios.

The linear MPC control strategy is based on a dynamic matrix control (DMC) method. For this method, a multiple-input multiple-output step response matrix is obtained employing step tests in the IGCC plant. The nonlinear MPC control strategy, NLMPC [2, 3], is based on the conversion of the differential algebraic equation system into a large-scale nonlinear programming (NLP) problem, which is solved using IPOPT, an efficient interior point-based large-scale nonlinear optimization algorithm. In the proposed multivariate control strategies, the following variables are considered as controlled output variables (CVs): carbon capture rate, power generation, temperature of cooled syngas stream, temperature of cooled permeate stream, steam to CO ratio at the WGS-MR inlet and hydrogen purity in the permeate stream. Also, the selected manipulated variables (MVs) are: water flow rates for syngas and permeate cooling, steam injection flow to syngas for the facilitation of the WGS reaction, total coal/water slurry, oxygen enriched air flow for the gasifier, sweep flow for the WGS-MR and air flow for the gas turbine.

A number of control scenarios for the IGCC power plant are addressed. These scenarios include: (i) setpoint tracking associated with a power generation demand change, in which the IGCC power plant demand imposes a step increase from its original power generation steady point; (ii) disturbance rejection related to the variability in the quality of the coal feed. In this case, the power generation should be kept at the setpoint while the carbon content in the coal/slurry is reduced; and (iii) a combination of both setpoint tracking and disturbance rejection cases. In this presentation, results on the closed-loop responses for these different scenarios will be analyzed and compared considering the advanced linear and nonlinear MPC control strategies.

References

[1] Lima, F. V. P. Daoutidis, M. Tsapatsis, and J. J. Marano. Modeling and optimization of membrane reactors for carbon capture in Integrated Gasification Combined Cycle units. Ind. Eng. Chem. Res., 51(15):5480–5489, 2012.

[2] Lima, F. V., R. Amrit, M. Tsapatsis, and P. Daoutidis. Nonlinear model predictive control of IGCC plants with membrane reactors for carbon capture. In Proceedings of the American Control Conference. Washington, DC, June 17-19 2013.

[3] R. Amrit, J. B. Rawlings, and L. T. Biegler. Optimizing process economics online using model predictive control. Computers and Chemical Engineering, 58:334-343, 2013.

425064 A MULTI-Objective Optimization Approach to Optimal Sensor Location Problem in IGCC POWER Plants

Abstract:

A MULTI-OBJECTIVE OPTIMIZATION APPROACH TO OPTIMAL SENSOR LOCATION PROBLEM IN IGCC POWER PLANTS

ABSTRACT

In order to operate an IGCC power plant in a safe and stable manner, many input and output process parameters need to be monitored. However, due to economic and operational constraints it is infeasible to place sensors at locations pertaining to all of these parameters. Hence, it becomes important to select the most effective sensor locations which can lead us to gain maximum information about the plant conditions. Drawing on the work done by "Lee & Diwekar", this paper attempts to broaden the realm of the optimal sensor location problem to address simultaneous optimization of multiple objective functions. In addition to "Fisher Information", two other objective functions, viz., "thermal efficiency of the power plant" and "cost", will be considered. Practical issues present in an IGCC power plant such as harsh physical conditions and variability in process parameters make the optimal sensor location problem an especially complicated one. Advanced simulation software and data from a set of existent sensors are used to collect information about the output parameters which do not have any sensors to gauge them yet. In order to solve this real world large scale problem, we use a novel algorithm called Better Optimization of Nonlinear Uncertain Systems (BONUS). BONUS works in probability distribution space and avoids sampling for each optimization and derivative calculations iterations.

A non-linear stochastic multi-objective problem has been solved to determine the non-dominated or Pareto set which contains different options for optimal sensor network locations that can simultaneously satisfy the three objective functions, (maximizing observability, minimizing cost and maximizing the IGCC power plant thermal efficiency).

434902 Smart Grid Coordination of an IGCC Plant: Comparison of Multi-Stage Stochastic Programming and Empc

Abstract:

The Integrated Gasification Combined Cycle (IGCC) possesses a number of benefits over traditional power generation plants, ranging from increased energy conversion efficiency to flex-fuel and carbon capture opportunities. A less known benefit of the IGCC configuration is the ability to load track electricity market prices. The idea being that material storage can be used to time-shift power production away from periods of low power value to periods of high power value.

Due to the uncertainty of electricity prices, the problem of sizing material storage units within the process must be formulated as a stochastic program. However, rather than being a fairly simple two-stage stochastic program, the dynamics imposed by the storage devices requires the formulation to be of the far more challenging multistage class. Due to the computationally intense nature of multistage stochastic programming, many have advocated the use of Economic MPC. The objective of this work is to provide a head-to-head comparison of feedback policies generated from multi-stage stochastic programming and EMPC. This comparison of policies provides a validation of EMPC based equipment design procedures.

432561 Deployment of a Search Space Reduction Algorithm to an NMPC Problem - Comparative Experimental Analysis to a Fuel Cell System

Abstract:

Nowadays, the increase in energy demands, the tighter environmental regulations and various economic considerations require for the systems and processes to operate over a wider range of conditions and often near the boundaries of their operating region. Moreover, the interconnection and cooperation of various systems into integrated units is a prerequisite for efficient resource handling and allocation. In the sight of these issues, advance process control, such as Model Predictive Control (MPC) aims to facilitate the improvement of the response and overall behavior of systems and processes. In general the impact of control technology is evident in a wide range of application areas, including fuel cells, as it is the necessary facilitator for achieving desired objectives and fulfilling application-specific goals. Fuel cells represent a versatile and efficient electricity generation source that can be applied in a wide range of industries - from vehicles and primary energy systems to autonomous back-up power stations and portable consumer electronics devices. Fuel cell systems exhibit fast dynamics, nonlinearities and uncertainties that constitute challenges requiring appropriate control in order to be confronted effectively.

This work presents an alternative way of combining components from two advanced MPC methodologies for a multi-variable nonlinear control problem in order to systematically exploit their synergistic benefits. More specifically at the core of a nonlinear model predictive control (NMPC) formulation a nonlinear programming (NLP) problem is solved utilising a discretized nonlinear dynamic model. Prior to the online solution of the NLP problem a pre-processing search space reduction (SSR) algorithm is applied. The SSR algorithm adjusts the search space of selected variables, based on a pre-computed augmented low-complexity piecewise affine (PWA) approximation of the feasible space.

The behavior of the proposed synergetic formulation is illustrated through a multivariable nonlinear control problem involving the operation of a small scale fully automated Polymer Electrolyte Membrane (PEM) fuel cell unit. Overall, four different MPC-based approaches are formulated and subsequently evaluated by addressing the control issues that arise during the operation of the fuel cell system. The aim of comparative experimental study is to explore the behavior of MPC-based configurations and summarize the characteristics of each approach with respect to the specific control problem. To achieve this objective an experimental scenario is formed which is applied to the PEM fuel cell unit using each time a different control configuration. The proposed framework is developed and deployed online to an industrial-grade Supervisory Control and Data Acquisition (SCADA) system. The response of the multivariable nonlinear controllers are assessed through a set of experimental studies, illustrating that the control objectives are achieved and the fuel cell system operates at a stable environment regardless of the varying operating conditions. The performance of the deployed controllers is assessed based on their response, their qualitative response characteristics and the required energy for the heat-up and the cooling during each experiment. The comparative experimental study presents the benefits and challenges of each configuration.

423126 Addressing Control Challenges of Discontinuous Processes with Multi-Fidelity Model Predictive Control

Abstract:

Petroleum currently fulfills 32% of total national energy demand, more than any other source of energy, and is likely to continue to do so for at least the next 30 years[1]. While the need for alternative sources of energy is clear, the growth and maturity of renewable energies has been slow and unable to meet energy demands, especially in transportation. Therefore, to meet the growing demand for energy, more efficient, robust, and reliable technological advances are needed in the petroleum industry. The petroleum industry is historically divided into upstream and downstream divisions. The upstream division finds and extracts oil and gas from geologic formations, while the downstream division refines the crude oil and gas into usable products. The downstream sector has seen many technological advances in process control and optimization, but many processes in the upstream sector, such as oil well drilling, still lack any significant automation[2]. When automation is optimized, it improves safety and efficiency over manual control by responding faster to process disturbances and by operating closer to process constraints. However, a unique challenge of automating Managed Pressure Drilling (MPD) for oil and gas is the discontinuous nature of the process[2]. An oil or gas well is created by drilling into the earth for several hundred to several thousand feet, stopping to insert and cement in place a segment of casing pipe to the well bore, then repeating the process until the target depth is reached. Before the casing pipe is inserted into the well bore, the string of drill pipe must be removed from the well, set aside, and then placed back in the well to resume operations. As the well deepens, more drill pipe is connected to the drillstring. At the bottom of the drillstring, a Bottom Hole Assembly (BHA), consisting of measurement and steering equipment, is attached to the drill bit. The drill bit is cooled by the drilling fluid, or mud, which also moves the rock cuttings to the surface and maintains pressure in the well annulus (see Figure 1). The well annulus pressure consistently needs to be greater than the geologic reservoir pressure to prevent hydrocarbons from entering the well during the drilling process. If the mud pressure in the well is too high, it can damage the rock formation; if it is too low, hydrocarbons from the subsurface reservoir can come to the surface in an uncontrolled and dangerous manner. When this catastrophe happens it is known as a blowout. The well bore pressure must be maintained within a small range of pressures that balances the reservoir fluid pressure to prevent damaged formations and blowouts. Maintaining the pressure balance in the well is the goal of MPD[3].

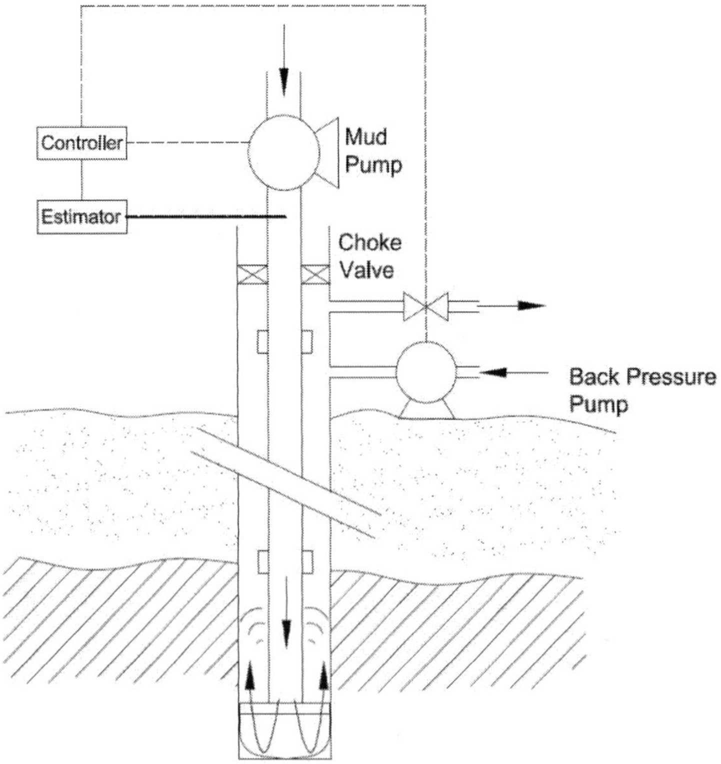

Figure 1 A simplified diagram of the MPD process

Currently, industrial MPD is completely manual. However, it has been demonstrated that an automated controller can maintain borehole pressure and reject disturbances faster and more accurately than manual control by using Nonlinear Model Predictive Control (NMPC)[4]. A controller can coordinate the mud pump flow rate and the annulus choke valve opening to reach the desired bit pressure set point[5]. One of the challenges with automated control of MPD is the necessary stopping and starting of operations. This discontinuous process can cause pressure sensors in the drillstring and BHA to lose calibration and controller models to lose tuning in the intervals. To address these challenges, this work simulates multiple control models arranged in an ensemble control structure to maintain control over various process stages: normal drilling operations, pipe reconnection procedures, and the disturbance of unwanted gas influx from the reservoir to the well bore. The novel control structure allows switching among control models of varying accuracy. A high fidelity model of the process is used when available, but the long computation time makes it unavailable for every control instance. The faster and less accurate low-order first principles model is the primary controller, yet it may not converge to an optimal solution at every time step due to the use of dynamic parameter estimation using Moving Horizon Estimation (MHE). An empirical First Order Plus Dead Time (FOPDT) model generally converges to an optimal solution at each time step, yet it is the least accurate model available. The controller makes use of these multi-fidelity models based on the availability and accuracy of the past model predictions compared to the current process state. The model solution with the least error is used to control the process at the present control instance. A limitation on the rate of change of process inputs is set to avoid sudden jumps when switching between models. An additional benefit of this control structure is the ability to tune one model while another is being used to control the process. Thus, model tuning can be accomplished without interrupting the drilling process[6]. Hence, the redundant model control structure offers benefits typically associated with redundant process measurements. This novel ensemble controller is capable of maintaining control over the discontinuous MPD process.

References

1. *Annual Energy Outlook* 2014, U. S. Energy Information Administration. p. A-1.

2. Godhavn, J.M., et al., *Drilling Seeking Automation Control Solutions*, in *18th IFAC World Congress*. 2011: Milano, Italy.

3. Breyholtz, O., G. Nygaard, and M. Nikolaou. *Automatic control of managed pressure drilling*. in *American Control Conference (ACC), 2010*. 2010.

4. Asgharzadeh Shishavan, R., et al., *Combined Rate of Penetration and Pressure Regulation for Drilling Optimization by Use of High-Speed Telemetry* SPE Drilling and Completion Journal, 2015. **30**(1).

5. Breyholtz, O., et al. *Evaluating control designs for co-ordinating pump rates and choke valve during managed pressure drilling operations*. in *Control Applications, (CCA) & Intelligent Control, (ISIC), 2009 IEEE*. 2009.

6. Sui, D., et al., *Ensemble Methods for Process Monitoring in Oil and Gas Industry Operations*. Journal Of Natural Gas Science and Engineering, 2011. **3**: p. 748-753.

429133 Mathematical Modeling of Electrohydrodynamic Flow in Tumor Cells for Tumor Treating Fields (TTF) Therapy

Abstract:

Tumor treating fields (TTF) is a new type of cancer therapy that was recently approved by the Food and Drug Administration (FDA) in March 2011 for use in glioblastoma multiforme (GBM) patients that dramatically increases quality of life, while still being just as effective as traditional chemotherapy. TTF therapy uses alternating (AC) applied electrical fields to interfere with tumor cell division by disrupting mitosis via the movement of charged molecules within the cells, altering their internal structure. The processes of electrophoresis, electroosmosis and dielectrophoresis play a key role in the effectiveness of this therapy. Thus, here we will present a mathematical model that describes electrohydrodynamic flow within a single, dividing tumor cell during anaphase and cytokinesis by using idealized geometrical structures to describe the cell. Time dependent velocity profiles as a function of electrical field, cell phase and geometrical parameters indicate situations in which flow reversals can occur within the cells. These results can aid in ultimately predicting and optimizing tumor cell response to TTF therapy.

433542 Mathematical Modeling of the Extracellular Matrix in Cancer Metastasis

Abstract:

Metastatic cancer accounts for 90% of cancer deaths in the U.S. Early in metastasis, cancer cells detach from the primary tumor site and move through the extracellular matrix (ECM) surrounding the tumor into a blood vessel from which the cancer can spread throughout the body. The ECM is a composite tissue composed primarily of a network of macromolecules including insoluble fibers and fluid-filled pores. Normal ECM serves as a physical barrier to the movement of cells. Cancer cells manipulate the ECM through cellular signaling pathways that activate enzymatic remodeling of the structure of the ECM to facilitate cell migration. ECM remodeling leads to a spatially heterogeneous ECM that changes over time. However, the mechanism of ECM remodeling and the impact of ECM remodeling on physical properties that influence metastatic cell migration are not well understood (Lu, Weaver, & Werb, 2012). Experimental approaches alone cannot decouple the complexities of the many interacting processes during ECM remodeling. Thus, there is a critical need for alternative approaches to understand how ECM composition and structure are modified during metastasis and to determine if ECM properties may be restored to non-metastatic values. Computational models considering simultaneous physiological, chemical, and biomechanical interactions can be developed to enhance quantitative understanding of ECM remodeling in cancer.

In this work, we present a computational model to of ECM remodeling in the early stages of cancer metastasis. The model includes the dynamic reaction and diffusion processes involving ECM-degrading matrix metalloproteinases (MMPs) and ECM-cross-linking lysyl-oxidase (LOX) and their impact on the collagen fibers in the ECM to predict ECM physical properties. This model improves upon existing models for the ECM in tumor metastasis by considering the transport variations due to the heterogeneous medium and the enzymatic reactions of both MMPs and LOX to both degrade and cross-link the ECM in tandem. The concentrations of the MMPs, LOX, and ECM basement membrane and interstitial collagen fibers are tracked with a mechanistic mathematical model that is solved computationally. The simulation results are assessed as to their ability to accurately predict stiffness changes due to ECM remodeling in the metastatic disease state as compared to published experimental data in the literature.

LU, P., WEAVER, V. M. & WERB, Z. 2012. The extracellular matrix: a dynamic niche in cancer progression. *J Cell Biol,* 196, 395-406.

435011 Computational Model of Single Cell Transcriptional Regulation and Cellular Networks Driving Liver Regeneration Following Surgical Resection

Abstract:

Liver resection is used clinically as a treatment for hepatocellular carcinoma and as a technique to enable live liver transplant. In both cases, the resected liver initiates a highly orchestrated regeneration program that is expected to culminate in complete restoration of liver mass. Despite decades of study, our understanding of the regulatory mechanisms underlying this repair process remains incomplete. This lack of complete understanding limits the ability to intervene in cases where a resected liver fails to regenerate to the levels required to support normal physiological function. Nevertheless, the availability of clinical data and pathology samples from regenerating and non-regenerating cases present an opportunity to employ modeling approaches that integrate data from different functional scales to better understand the mechanisms underlying clinical resection responses. Our present study seeks to address the gap in mechanistic understanding of liver regeneration by using a combination of single-cell based transcriptional analysis and computational network modeling, which can then be integrated with clinical data to improve prognostic outcomes.

Initiation and progression of regeneration relies on the coordination of the responses of multiple cell types over time. A canonical description of the regeneration process after resection includes transient activation of the resident macrophages by the injury, followed by priming of hepatocytes to re-enter cell cycle, growth factor production by hepatic stellate cells and sinusoidal endothelial cells, proliferation of hepatocytes, proliferation of non-parenchymal cells, and termination of proliferative processes via remodeling the matrix microenvironment. Disruption of this coordinated sequence can lead to deficits in liver mass recovery, as clinically observed in chronic liver disease.

We developed a computational model that accounts for the multi-scale nature of liver regeneration by integrating physiological-scale interactions; activation phenotypes of hepatocytes and liver non-parenchymal cells; and molecular signaling networks. We explored a range of model dynamics to identify parameter sets that account for experimentally observed regeneration profiles of multiple animal models of liver disease as well as the human liver transplant case. Model simulations predicted that aberrant activation dynamics of hepatocytes and hepatic stellate cells are key factors suppressing liver regeneration in multiple disease contexts.

We tested these predictions in the alcoholic liver disease case by acquiring a new experimental data set on gene regulation in hundreds of single and pooled cells obtained through laser capture microdissection of liver tissue. We analyzed the high-dimensional single cell scale gene regulation data to identify distinct cell populations, and extrapolated from the single-cell scale to cell phenotype distribution in whole tissue. Our approach characterized the transcriptional state of individual cells in the *in vivo* regenerating liver at a level of detail not previously achievable. Characterization of transcriptional profiles revealed cell-type specific, functionally distinct states for hepatocytes as well as stellate cells in our data. Surprisingly, we found the two liver cell types distributed among all identified functional states irrespective of the experimental condition (baseline or regenerating, alcohol adapted or control). However, relative proportions of the cells distributed among the identified functional states changed based on regeneration stage as well as adaptation to chronic ethanol intake, consistent with predictions of the computational model. These results suggest that chronic alcohol use impairs liver regeneration via significant alteration of cellular networks that are critical to initiation and progression of liver repair following resection. Our integrated experimental and computational analysis points towards new avenues for therapeutic intervention based on renormalizing the cellular functional states for improving surgical outcomes.

429373 Modeling the Dynamics of Neuroendocrine-Immune Interactions in Collagen-Induced Arthritis

Abstract:

Modeling the Dynamics of Neuroendocrine-Immune Interactions in Collagen-induced Arthritis

R. Rao[1], D. DuBois[2, 3], R. Almon[2, 3], W.J. Jusko[2] and I.P. Androulakis[1, 4]

[1]Department of Chemical and Biochemical Engineering, Rutgers, the State University of New Jersey, Piscataway, NJ, 08854, USA

[2]Department of Pharmaceutical Sciences, School of Pharmacy and Pharmaceutical Sciences, State University of New York at Buffalo, SUNY Buffalo, NY 14260, USA

[3]Department of Biological Sciences, State University of New York at Buffalo, SUNY Buffalo, NY 14260, USA

[4]Department of Biomedical Engineering, Rutgers, the State University of New Jersey, Piscataway, NJ, 08854, USA

Rheumatoid arthritis (RA) is a chronic inflammatory disease that affects about 1% of the population in the developed world, and is primarily characterized by synovial inflammation and cartilage deterioration, typically in the joints of the hands and feet (1). The hypothalamic-pituitary-adrenal (HPA) axis, a critical component of the neuroendocrine system, is thought to be an important regulator of disease activity. It has been hypothesized that adrenal insufficiency; the inability of the HPA axis to mount an adequate anti-inflammatory in response to ongoing chronic inflammation in RA might contribute to disease pathophysiology (2). In addition, circadian variation in clinical symptoms has been well documented in RA patients. Moreover, a disruption in the natural circadian rhythms of major mediators of the HPA axis, such as cortisol has been observed (3). A mechanistic understanding of the interplay between the circadian rhythms of the HPA axis and the immune system might provide useful insights into RA pathologies and optimization of treatment regimens.

Animal models of RA have been extensively used to study disease pathogenesis and in the discovery of anti-arthritic drugs (4). A rodent model of collagen induced arthritis (CIA), which involves immunization of the animal with an emulsion of the complete Freund's adjuvant and type II collagen, is considered to be the gold standard in vivo model for RA (5). In this work we discuss the development of a mathematical model that aims to evaluate the dynamic interactions between the HPA axis and pro-inflammatory cytokines, and their downstream effects on paw edema, a key disease end-point in CIA. In particular, we investigate the changes in circadian secretion of corticosterone (CST) and pro-inflammatory cytokines after induction of CIA, that ultimately result in circadian variability in paw edema. Moreover, we recapitulate the phenomenon of adrenal insufficiency that has been observed in both CIA as well as RA patients. These simulations will aid in the characterization of neuroendocrine-immune response in animal models of RA.

Our model builds on earlier work (6-9) and is comprised of three components; a central oscillatory compartment that simulates the circadian secretion of CST in the HPA axis, a peripheral compartment that captures the downstream effects of secreted CST on cytokine expression, and finally, a compartment that accounts for the disease endpoints, specifically, paw edema. The oscillations in the central compartment are generated due to the negative feedback between CST and corticotropin-releasing hormone (CRH)/adrenocorticotropic hormone (ACTH) using a modified Goodwin oscillator model that incorporates Michaelis-Menten kinetics to express the degradation terms in the hypothalamus, pituitary, and adrenals. The HPA axis is positively modulated by the expression of cytokines in the peripheral compartment, which in turn is balanced by the negative feedback of CST. This results in an interlinked positive and negative feedback circuit that enables the investigation of the dynamics of HPA axis-cytokine interactions upon the onset of inflammation. In this model, we assume that the severity of paw edema is in direct correlation to the expression of pro-inflammatory cytokines. Furthermore, we explicitly model the emergence of adrenal insufficiency in response to chronically elevated cytokines by postulating an indirect response mechanism that effectively results in stimulating the clearance of ACTH in the HPA axis. Disease progression is initiated by a delay in the onset of the increased pro-inflammatory cytokine expression by using a simple transduction model (6).

Model predictions demonstrate that elevated cytokine expression results in circadian variability in paw edema progression, reminiscent of the circadian rhythms of disease symptoms exhibited by RA patients. Notably, our model predicts that an increase in mean secretion of CST after the induction of the disease is accompanied by a decrease in the amplitude of the CST oscillation. This phenomenon persists even after the development of adrenal insufficiency and the consequent drop in mean CST secretion toward levels present prior to the onset of the disease. This suppression of circadian rhythm has been experimentally observed in a number of important neuroendocrine and HPA axis mediators during adjuvant induced arthritis, including; ACTH, CST and adrenocortical ornithine decarboxylase (10). Further analysis of possible alterations in the phase of oscillation of CST and the cytokines during disease progression is currently ongoing.

In conclusion, our model captures key features of chronic inflammation in animal models of RA. In doing so, it highlights the importance of accounting for circadian rhythms while studying the neuroendocrine-immune system interactions in RA.

References

1. McInnes IB, Schett G. The Pathogenesis of Rheumatoid Arthritis. N Engl J Med. 2011 Dec 8;365(23):2205–19.

2. Harbuz MS, Korendowych E, Jessop DS, Crown AL, Li pdfan S. L., Kirwan JR. Hypothalamo-pituitary-adrenal axis dysregulation in patients with rheumatoid arthritis after the dexamethasone/corticotrophin releasing factor test. J Endocrinol. 2003 Jul;178(1):55–60.

3. Cutolo M, Masi AT. Circadian Rhythms and Arthritis. Rheum Dis Clin N Am. 2005 Feb;31(1):115–29.

4. Kannan K, Ortmann RA, Kimpel D. Animal models of rheumatoid arthritis and their relevance to human disease. Pathophysiology. 2005 Oct;12(3):167–81.

5. Brand DD, Latham KA, Rosloniec EF. Collagen-induced arthritis. Nat Protoc. 2007;2(5):1269–75.

6. Earp JC, DuBois DC, Molano DS, Pyszczynski NA, Keller CE, Almon RR, et al. Modeling Corticosteroid Effects in a Rat Model of Rheumatoid Arthritis I: Mechanistic Disease Progression Model for the Time Course of Collagen-Induced Arthritis in Lewis Rats. J Pharmacol Exp Ther. 2008 Aug;326(2):532–45.

7. Mavroudis PD, Corbett SA, Calvano SE, Androulakis IP. Mathematical modeling of light-mediated HPA axis activity and downstream implications on the entrainment of peripheral clock genes. Physiol Genomics. 2014 Oct 15;46(20):766–78.

8. Sriram K, Rodriguez-Fernandez M, Doyle FJ. Modeling Cortisol Dynamics in the Neuro-endocrine Axis Distinguishes Normal, Depression, and Post-traumatic Stress Disorder (PTSD) in Humans. PLoS Comput Biol [Internet]. 2012 Feb [cited 2015 May 1];8(2). Available from: http://www.ncbi.nlm.nih.gov/pmc/articles/PMC3280965/

9. Ramakrishnan R, DuBois DC, Almon RR, Pyszczynski NA, Jusko WJ. Fifth-generation model for corticosteroid pharmacodynamics: application to steady-state receptor down-regulation and enzyme induction patterns during seven-day continuous infusion of methylprednisolone in rats. J Pharmacokinet Pharmacodyn. 2002 Feb;29(1):1–24.

10. Cardinali DP, Esquifino AI. Circadian disorganization in experimental arthritis. Neurosignals. 2003 Dec;12(6):267–82.

431123 Development of a Hypertensive Ovine Model to Study Implantation of Autologous Arteries and Veins

Abstract:

Hypertension is a leading cardiovascular risk factor worldwide, playing a significant role in ischemic heart disease, stroke and atherosclerosis. Co-existence of systemic hypertension in patients undergoing infrainguinal bypass surgery was estimated at 70% and 68% of re-intervention cases in coronary artery grafting procedures were reported to be hypertensive. Venous hypertension also leads to chronic vein insufficiency leading to superficial vein incompetence, limiting the availability of autologous vein grafts. Here, we describe development of a hypertensive ovine model that was employed to test the performace of autologous arteries or veins when transplanted under clinically realistic conditions.

Our study involved the development of an ovine renal hypertension model utilizing the 2K1C model, where nephrectomy was performed on the left kidney and renal artery stenosis was performed on the right kidney after 3 weeks. Following the onset of hypertension, carotid artery grafts (CAG) or facial vein grafts (FVG) were implanted into carotid arteries of hypertensive (HT) and normotensive sheep (NT). Autologous carotid artery (CA) was used as a control. In addition, valuable real-time hemodynamic data was collected through instrumentation that was inserted during grafting. The instrumentation included a Doppler flow probe, ultrasonic crystals and in-line pressure catheters to measure flow rate, dynamic changes of graft diameter and arterial pressure.

With the 1K1C model, we were able to achieve elevated systolic as well as diastolic pressures, leading to increased Mean Arterial Pressure (MAP). Our in-vivo hemodynamics data revealed that in hypertensive animals the pulse wave velocity increased and the dynamic radial compliance was significantly reduced, indicating increased stiffness, a hallmark of hypertensive vessels. Surprisingly, both CAG and FVG showed significantly reduced mechanical properties as well as collagen and elastin content following grafting, albeit FVG were affected to a higher extent by HT. Smooth muscle cell remodeling, as well as vascular function was also more significantly affected by HT in FVG as compared to CAG. Interestingly, transplanted FVG into HT animals were infiltrated by pro-inflammatory macrophages, while in NT animals, transplanted FVG were infiltrated mostly by macrophages of an anti-inflammatory phenotype.

We present here a pre-clinical large animal hypertensive model, which allows for real-time hemodynamic data monitoring, to evaluate remodeling and performance of autologous vessels. Our study revealed that autologous arteries lose many of the vascular properties when grafted under NT conditions, but are not affected by HT as much as the veins. On the contrary, grafted veins showed constructive remodeling under NT, while under HT they performed poorly. This data has repercussions in choosing appropriate graftable vessels harvested from patients suffering from hypertension.

Development of Modeling Approaches to Predict Effects of Facilitated Wound Closure on Scarring

Stephanie Jorgensen
Dr. Robby Sanders

Department of Chemical Engineering
Tennessee Technological University

TTU

1

Outline

- Overall Motivation
- Phases of Wound Healing and Types of Facilitated Closure
- Approach Taken Towards a Solution
- Development of a Model
- Conclusions

TTU

2

Motivation

- Millions of people are injured in a way requiring wound closure (via home remedies and/or suturing), and many have wounds that result in excessive scarring and complications.
  - ~3.1% of ER patients undergo suturing procedures each year[1].
  - Keloid and hypertrophic scars frequently occur in these patients (71% occurrence)[2].

Idea: To create a mathematical model to predict the type of wound closure that will result in minimal scarring

3

Phases of Wound Healing and Types of Facilitated Closure

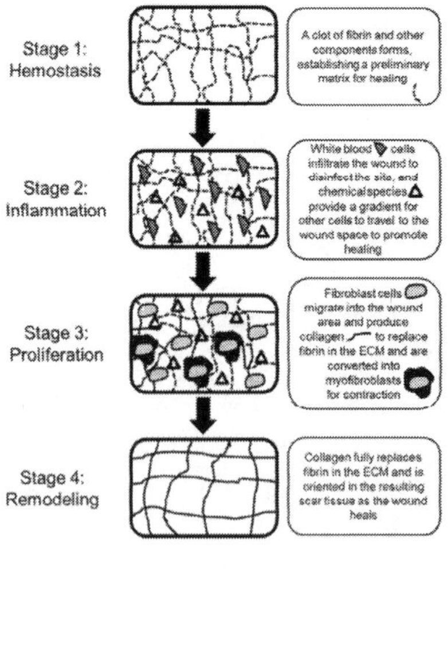

- Types of facilitated closure include
  - Sutures/stitches
  - Staples
  - Adhesives (tape, glue)

4

Approach: Collaborate with Healthcare Professionals

- Gain perspective of "needs" in healthcare through interaction with healthcare professionals
  - o Developed a collaboration with faculty in the TTU School of Nursing
    - Facilitated access to Cookeville Regional Medical Center's ER to observe suturing and gain understanding about decisions relevant to suture material properties and stitching technique
    - *Lesson learned:* Techniques for closure/facilitated healing are based on experience (e.g., actual suturing in the clinic after limited training)

TTU

5

Approach: Conduct Comprehensive Literature Review

- Identify gaps between prior work by conducting a comprehensive literature review
  - o Models included in 48 peer-reviewed articles (1981-2014) were classified using a two-tiered approach based on mathematical techniques applied and purposes of the models (including inputs and outputs).

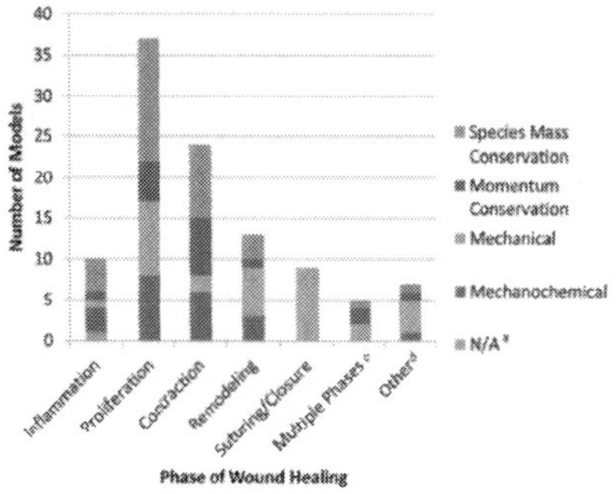

TTU

6

Approach: Conduct Comprehensive Literature Review

- A gap in the literature was identified.
  - Specifically, a lack of modeling of the events in inflammation and remodeling phases was observed, and few models were focused on effects of facilitated closure during wound healing.
- Opportunities for future mathematical modeling of wound healing exist with respect to these unexplored phases.

Goal: To create a mathematical model to predict the type of wound closure that will result in minimal scarring

7

Envisioning the Modeling Approach

- Determining the wound geometry and making assumptions for simplification (what's been done and considerations for my work)
  - Physical representations of the systems/domains utilized in typical models (from the comprehensive literature review) were identified, and the images were compared to typical wound domains that have been sutured (as illustrated in various medical resources).

8

1-D linear wound geometry

Has been used to investigate how concentration gradients affect fibroblast levels and tissue displacement during healing under normal and abnormal conditions

2-D, rectangular wound geometry

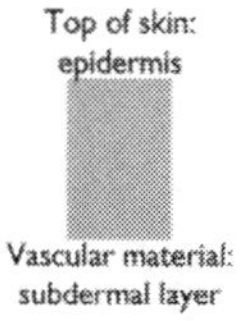

Has been used to determine how fiber alignment results from movement of cells within the wound and conclude contact guidance results in increased speed of healing

Circular wound geometry

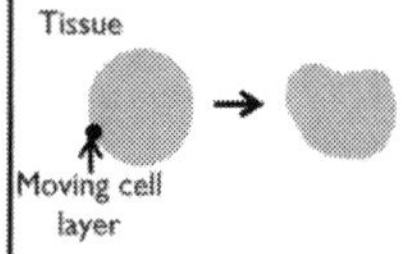

Has been used to determine how inhibition of lamellipod formation in cells and restriction of proliferation at wound edge affect wound closure

Fusiform wound geometry

Has been used to determine regions of high and low stresses induced in wounds exhibiting facilitated closure

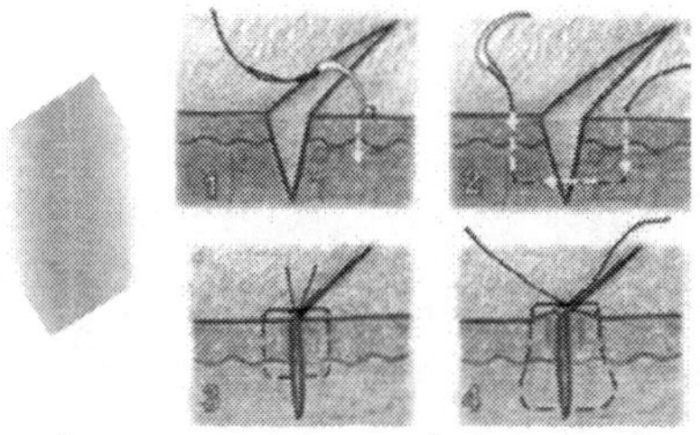

3 4

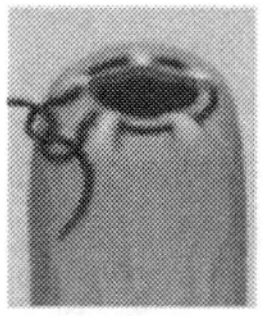

5

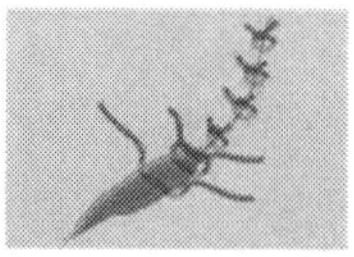

6

9

Envisioning the Modeling Approach

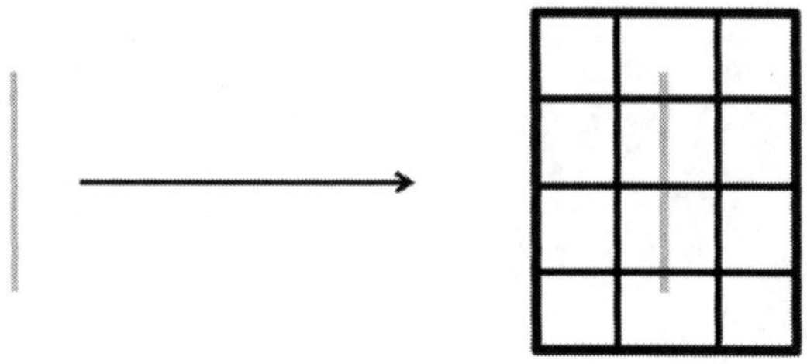

- Advantages:
  - Simple geometry for considerations of conservation equations
- Disadvantages:
  - Less physiologically relevant since wounds requiring suturing would be multi-dimensional

10

Envisioning the Modeling Approach

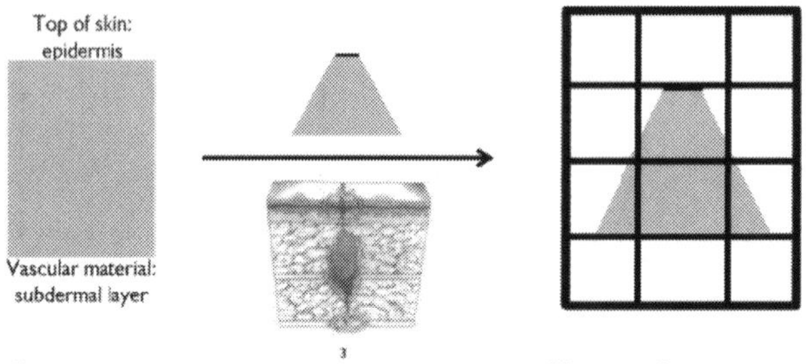

- Advantages:
  - Relatively simple geometry for considerations of conservation equations
- Disadvantages:
  - Not fully physiologically relevant as the "dead space" may not contain tissue elements associated with a typical healing response

11

Envisioning the Modeling Approach

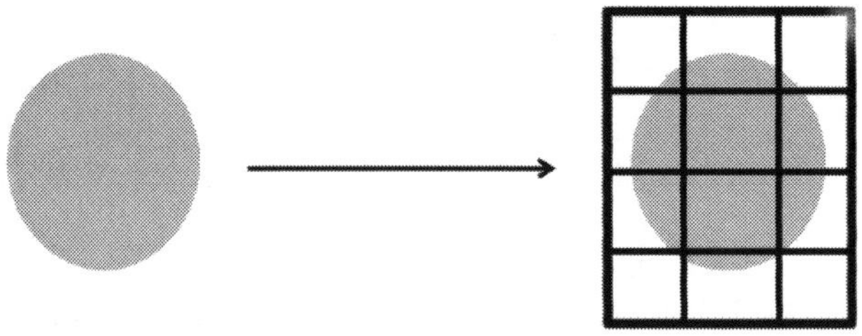

- Advantages:
  - Gives evidence of healing in different domains
- Disadvantages:
  - Geometry introduces many "trouble areas" in FEM—refinement of grid is necessary
  - Suture method presented (purse-string method) would pose a difficult boundary

12

Envisioning the Modeling Approach

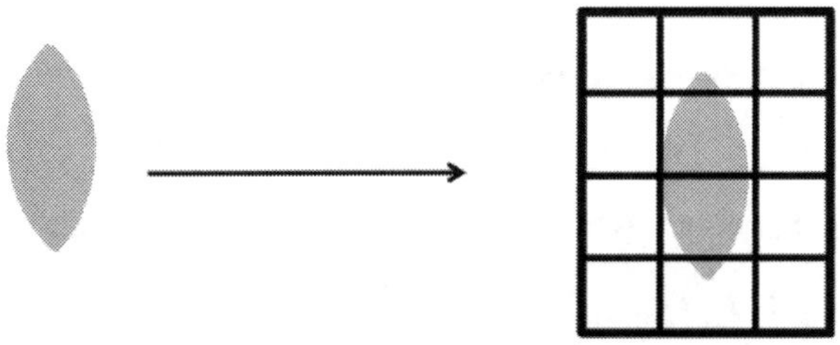

- Advantages:
  - Gives evidence of healing in different domains, and some wounds are intentionally made fusiform to achieve minimal scarring
- Disadvantages:
  - Complex for initial modeling efforts which is further complicated if consideration of subdermal forces is to be made

13

A Finite Element Approach

- Leveraging information from Larrabee and Galt[7]: we envision applying a force balance to wound boundaries using a triangular-type domain
- Determine the displacement of tissue along the wound boundary and within the healing wound as a result of facilitated closure

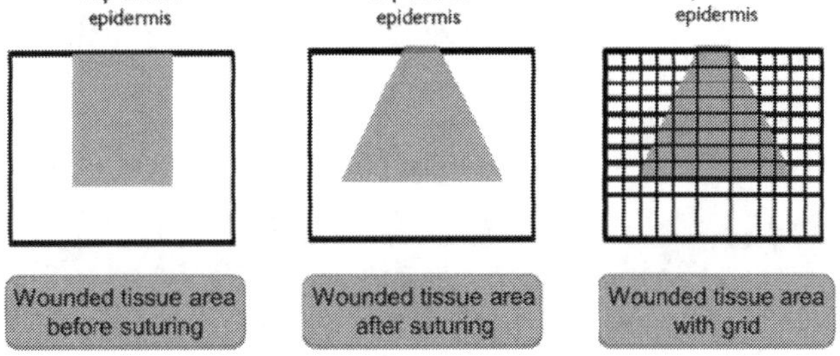

14

Conclusions

- Opportunities exist to better consider the effects of facilitated closure via suturing on the occurrences during the phases of wound healing
- Defining the system as it relates to facilitated closure/suturing is critical
- Next steps include generating a finite element mesh and/or model equations with appropriate boundary/initial conditions for the chosen domain

15

Citations

- 1 *National Hospital Ambulatory Medical Care Survey.* Centers for Disease Control and Prevention, 2010. Web. 25 Sept. 2014.
- 2 Li-Tsang, Cecilia W. P., Lau, Joy C. M., and Chan, Chetwyn C.H. "Prevalence of hypertrophic scar formation and its characteristics among the Chinese population." *Burns* 31 (2005): 610-616. Web. 15 Sept 2014.
- 3 Ethicon, Inc.,. *Wound Closure Manual.* Web. 24 Sept. 2015.
- 4 Brady, Mark. 'Optimal Suture Results: Slideshow'. 2015. Presentation.
- 5 Javed, Imran. 'Wound Management'. 2015. Presentation.
- 6 Brady, Mark. 'Optimal Suture Results: Slideshow'. 2015. Presentation.
- 7 Larrabee, W. F., and J. A. Galt. 1986. A finite element model of skin deformation, III. The finite element model. *The Laryngoscope.* **96**, 413-418.

16

Acknowledgements

- TTU Office of Research, Department of Chemical Engineering
- Research Committee: Dr. Robby Sanders, Dr. Pedro Arce, Dr. Jeffrey Rice, Dr. Jennifer Pascal, Dr. Melissa Geist, Dr. Yung-way Liu
- Sponsor: Chemical Engineers in Medicine
- Co-Sponsor: Computing Systems and Technology Division
- Chair: David Shaw
- Co-chair: Charles Reid

TTU

17

Questions?

TTU

18

A collision model of erythrocyte aggregates in shear flow

S.K. Ahuja

Abstract:

Platelet (thrombocyte) aggregation is one of the first physiological reactions to vessel wall injury. Aggregation in platelets and neutrophils (PMNs) is relevant to the progression of thrombosis, acute myocardial infarction, or unstable angina. Platelet aggregation at sites of vascular injury is central to the arrest of bleeding and for subsequent vascular repair; Excessive thrombocyte aggregation response at sites of atherosclerotic plaque rupture can lead to the development of vascular occlusive thrombi. In order to prevent it from happening, platelets move to the vessel wall and collide with either the damaged site or previously adhered platelets. Hemostasis is affected by erythrocytes when blood viscosity is increased through increase in concentration of erythrocytes or aggregation of erythrocyte takes place or when erythrocyte structure changes on deformation.

Erythrocyte (Red Blood Cell) aggregation is large at low shear rates and as shear rates increases aggregates break up. Bridging model and depletion model are used to analyze thrombus, clot formation. The adhesion of erythrocytes to each other and of erythrocyte to thrombocytes can take place. In the bridging model, it is assumed that fibrinogen or dextran molecules non-specifically adsorb onto the cell membrane and form a "bridge" to the adjacent cell. In contrast, the depletion model proposes the opposite. In this model, aggregation occurs as the concentration of macromolecules near an erythrocyte surfaces in close proximity is depleted compared to the concentration of the bulk phase, resulting in a net depletion force. Erythrocyte aggregation depends on their elastic properties. A model is proposed which considers collisions of thrombocyte with another thrombocyte or with an erythrocyte. Factors such as sizes of thrombocyte and erythrocyte, concentration of erythrocyte, shear rates in the thrombocyte to thrombocyte adhesion and thrombocyte to erythrocyte adhesion are discussed and compared with available experimental results.

Introduction:

Blood clot (thrombi) formation is a complex biological process involving an extensive system of biochemical (coagulation) reactions, platelet hydrodynamics, platelet–platelet and platelet–blood vessel wall interactions leading to ligand–receptor adhesion bond formation and platelet activation. Platelet adhesion to the vessel wall is one of the first events associated with formation of haemostatic clots and pathological thrombi. Damage or alteration of a blood vessel lining can result in activation of flowing platelets and their subsequent aggregation at sites of vascular injury. The ability of platelets to tether to and translocate on injured vascular endothelium relies on the interaction between the platelet glycoprotein receptor Ib α (GPIb α) and the A1 domain of vonWillebrand factor (vWF-A1) [1]

Platelet activation and homotypic aggregation induced by high shear stress or by chemical agonists (e.g., ADP, collagen, or thrombin) are involved in atherosclerosis and accompanying thrombosis cell aggregation and disaggregation in a linear shear flow. Platelet adhesion to collagen via blood glycoprotein vonWillebrand Factor (vWF) is important for thrombosis initiation at high shear. VWF is shear dependent that transitions from a globular to elongated conformation under high shear rate, enhancing absorption onto a collagen surface and platelet adhesion. A collision model of shear induced cell aggregation and disaggregation is governed by receptor–ligand binding. Proaggregation forces are generated by chemical affinity of cellular receptors and their ligands. Anti-aggregation forces are generated by elastic collision, electrostatic barrier and osmotic pressure It was revealed that platelet aggregation rate does not simply increase with blood-flow velocity; rather, it exhibits a maximum with a subsequent decrease due to the growing hydrodynamic forces that inhibit platelet adhesion The combination of hydrodynamic features of microvasculature and nonlinear shear-dependent platelet aggregation rate may, in principle, stop the growth of a thrombus. [2,3].

Erythrocyte form aggregates in rouleaux formation with macromolecules (fibrinogen or dextran) present in plasma. The first step is the rapid formation of unstable catch–slip bonds which slow platelets and cause platelet flipping along the damaged surface. The platelet receptor component, GPIb α, forms transient bonds with the vWF exposed at the injury site. Transient tethering and subsequent flipping (or rolling) and pausing of platelets on the vessel surface result in stable bonds slowly between platelet receptors and ligands (often integrin αIIbβ3 binding with Vwf or fibrinogen) bound to the damaged wall or the surface of the thrombus. As the blood clot grows, platelet–platelet interaction becomes one of the major factors determining clot growth rate and integrity as platelets expose GPIIbIIIa receptors which permit platelet–platelet adhesion via fibrinogen. Adhesion of platelets to the injured surface is also affected by shear rates of the flow. At high shear, platelet integrin $\alpha 2 \beta 1$ and GPVI receptors are not sufficient to initiate binding to collagen, and binding of the GPIb α receptor to vWF immobilized on collagen becomes essential in platelet adhesion[3]. *The stiffness of the platelet not only determines the shape and morphology of the clot, but also affects clot mechanical properties, as platelet stiffness determines cell shape when exposed to various flow conditions and contact interaction with other cells and blood vessel wall.* This will affect the number of receptor–ligand pairs in platelet–platelet and platelet–substrate interactions. Platelet stiffness is also an important property reflecting platelet functioning, because it reorganizes its structure during activation or as a response to physiological or pathological conditions [4].

Roll of Erythrocyte in blood clot::Thrombus structure has two distinct regions: a densely packed "core" nearest the injury site (composed of highly activated platelets that have undergone αgranule exocytosis and phosphatidylserine exposure) and a loose "shell" overlaying the core composed of less activated platelets that have retained a discoid shape The thrombus shell consists of loosely associated platelets that are subject to continual removal by flow forces. Within the thrombus core, platelet agonists may become enriched since molecular transit may be reduced by the core having restricted pores. The presence of erythrocytes in contracted blood clots has been recognized and sometimes exploited; for example, during the time of medical bloodletting, the size of the contracted clot from blood removed from the patient was used as a measure of erythrocyte mass to determine when the procedure should cease. Erythrocytes are uniquely deformable cells with shapes determined by their geometric and material properties that regularly undergo extensive, reversible shape changes under the influence of intravascular fluid forces and while passing through vessels in the microcirculation smaller than their diameter. However, the effects on erythrocytes of the forces generated by clot contraction and their impact on hemostasis have not heretofore been investigated [5].

There are possible mechanisms through which erythrocytes influence the balance of thrombosis and bleeding. Erythrocytosis can be caused by diseases affecting the bone marrow (primary erythrocytosis, such as polycythemia vera) or by disease and environmental conditions that affect blood oxygen saturation (secondary erythrocytosis). In secondary erythrocytosis, the bone marrow produces more erythrocytes to counterbalance the decrease in oxygen saturation. An increase in hematocrit results in an increase in viscosity that decreases the blood flow and can lead to the development of a thrombus. In addition, an elevated hematocrit promotes the transport of platelets and coagulation factors toward the vessel wall, thereby increasing collisions of platelets with the vasculature and with themselves. In a laminar blood flow, the highly deformable and biconcave shaped erythrocytes push platelets to the peripheral, and they themselves migrate to the center of the flow. This phenomenon is also called axial migration. The axial migration of erythrocytes causes platelets to be concentrated on the edge of the vessel wall. Erythrocytes are able to aggregate when suspended in aqueous solutions containing large plasma proteins (e.g., fibrinogen) or polymers (e.g., high-molecular weight dextran). At stasis or when fluid shear rates are sufficiently low, erythrocytes tend to form regular, linear, face-to-face structures (i.e., rouleaux), or three-dimensional aggregates that are difficult to disperse. Mechanisms involved in erythrocyte aggregation are based on two models.. The bridging model proposes that erythrocyte aggregation occurs when bridging forces generated from the adsorption of macromolecules onto adjacent cell surfaces exceed disaggregation forces generated from electrostatic repulsion, membrane strain, and mechanical shearing [6,7].

The other model is the depletion model where erythrocyte aggregation occurs because of a lower localized protein or polymer concentration near the cell surface compared with the suspending medium leading to an osmotic gradient and a depletion interaction. Although it is not yet possible to decide the superiority between these two mechanisms several studies favoring the depletion model have recently been published. In large blood vessels under low shear conditions, where blood can be regarded as a continuous fluid, enhanced erythrocyte aggregation increases blood viscosity and overall hydrodynamic resistance. In the microvasculature, however, enhanced erythrocyte aggregation increases axial migration of erythrocytes, resulting in an increased platelet concentration near the vessel wall, as well as a decreased local viscosity and reduced frictional resistance.

In addition, axial accumulation of erythrocytes also promotes plasma skimming and the Fahraeus effect, which in turn may result in decreased hematocrit values, thereby reducing blood viscosity and flow resistance. This reduction in local viscosity and consequent reduction in wall shear stress leads to a lower local nitric oxide bioavailability. Nitric oxide is an important substance in maintaining the quiescent status of endothelial cells and platelets. The decrease in local nitric oxide may promote platelet activation and tilt the hemostatic balance. Experimental studies indicate that lateral platelet movement is caused by continuous rebounding collisions between erythrocytes. These net collisions produce shear-induced diffusion (dispersion), which locally mixes the plasma and platelets near erythrocytes [8, 9, 10].. In a mesoscopic approach employing the immersed boundary method was used to simulate the motion of RBCs, modelled as deformable capsules, flowing through two-dimensional channels 20–300 mm in height. The results showed migration of RBCs perpendicular to the wall and formation of a cell-free layer[11].

In the immersed boundary method was also employed to model RBC migration in two dimensional channels of 6–12 mm in height; cell migration from the vessel wall was also observed. In the dissipative particle dynamics method was applied to blood flow in microtubes ranging from 10 to 40 mm in diameter, and cell migration away from the wall to the tube center was reported A different approach is represented by continuum models, which are able to simulate larger domains for longer time scales. An example of this approach can be found in where blood flow in vessels with diameter ranging from 40 to 100 mm was investigated using the shear-induced migration model proposed in. Particle migration away from the wall was also observed in this case. RBC distribution influences blood viscosity, and therefore the velocity profile, and also a series of biological activities such as oxygen distribution in the lumen and the scavenging of nitric oxide (NO) by haemoglobin in the RBCs[12]..

Erythrocyte aggregation: Platelets first became activated at locations where shear rates are greater than the critical shear rate (critical shear regions) of the first and second bends. Thrombus first grows along the wall in the downstream direction and then later in the upstream direction. Collision with erythrocyte with thrombus results in thrombus to roll along the wall. RBC collisions caused thrombi to continually roll along the wall and exit, The collision rates are consistent with the observation that the smaller molecules are more dynamic, and, therefore, have higher collision rates than the larger molecules. In a single layer of platelets that could contain many platelets. The colloid particle population is distributed by mass (or size) in a cluster mass distribution (CMD). The CMD evolves according to a population balance equation (PBE) which satisfies the principle of mass conservation. Smoluchowski was the first to apply statistical mechanics for aggregation of colloid particles.[13,14]. In processes where hydrodynamic forces are dominant, fragmentation may occur. Simultaneous aggregation and fragmentation by fluid shear is encountered in a number of processes. Aggregate breakup occurs by two mechanisms: erosion and splitting. Erosion is the detachment of small particles from surface, of aggregate and splitting is the breakup into pieces of size comparable to the parent aggregate. Competition between attractive van der Waals-Hamaker and repulsive double-layer forces determines the stability or instability of many colloidal systems. The interaction between colloidal particles can be modeled by the Derjaguin-Landau-Verwey-Overbeek (DLVO theory) where DLVO potential, which is the sum of the screened electrostatic repulsion between their double-layers and of the van der Waals-Hamaker attraction [15]. Shear-induced aggregation and Brownian aggregation are two mechanisms that exist simultaneously in sheared system.

Blood viscosity, is affected by various factors such as hematocrit, erythrocyte deformability and aggregation, osmolality, and plasma proteins. Rheological parameters of blood that have the potential to indicate blood circulation adequacy: yield shear stress (YSS) and disaggregating shear stress (DSS). YSS and DSS reflect the strength of red blood cell (RBC) aggregation in suspension under static and dynamic conditions, respectively. YSS, defined as the critical stress to disperse RBC aggregates under static conditions, was found to be dependent upon hematocrit, fibrinogen, and red cell deformability, but not temperature. DSS, defined as the minimum shear stress to disperse RBC aggregates under dynamic conditions, is dependent upon fibrinogen, red cell deformability, and temperature but not hematocrit[16].

Hydrodynamic stresses in elongation or shear in either laminar or turbulent regime can affect dispersion stability and breakage of colloidal aggregates in fractal or non-fractal morphology [16]. Balance between clustering (structure-formation) and breakup (structure-failure) is critical enabler in analyzing structural origin of the rheological properties of complex fluids.

$$\frac{dn_i}{dt} = \frac{1}{2}\sum_{j=1}^{i-1} k^{(a)}(i-j,j)n_{i-j}n_j - \sum_{j=1}^{\infty} k^{(a)}(i,j)n_in_j - k^{(b)}(i)n_i + \sum_{j=i+1}^{\infty} \beta(i,j)k^{(b)}(j)n_j,$$

---------------------------------(1)

where the subscripts i and j refer to different size classes, n_i is the number of aggregates having i monomers per unit volume, and $\beta(i, j)$ is the fragment size distribution[26] (here binary fragmentation). The kernels $k^{(a)}$ and $k^{(b)}$ depend on the type of flow

$$k^{(b)}(i) = \dot{\gamma}\exp\left(-\frac{2F_c}{5\pi\rho_i^2\eta\dot{\gamma}}\right)$$

On application of the analytical aggregation kernels for simple laminar shear (orthokinetic aggregation), and thermal motion (perikinetic aggregation) to model the coagulation kinetics of Brownian particles under laminar shear flow and superimposing the two processes gives the aggregation kernel as

$$k^{(a)}(i,j) = c_o\left(\frac{4}{3}\dot{\gamma}(\rho_i+\rho_j)^3\right) + c_p\left(\frac{2k_BT(\rho_i+\rho_j)}{3\eta_0}(1/\rho_i+1/\rho_j)\right),$$

-------------------(2)

where γ is the effective shear rate, ρ is the aggregate radius, c is the collision efficiency for the orthokinetic (perikinetic) aggregation, T is the temperature (here T is fixed to 300 K), and η is the viscosity of the suspending fluid. The collision efficiency terms correct the aggregation kernel, which are derived for neutral impermeable particles, which in practice overestimates the collision frequency if the particles exhibit nonneutral interaction potentials or have reduced permeabilities.

Collision take place between the two erythrocytes and the aggregation dynamics is modeled in the form of mass balance equations originally written by Smoluchowski. Fragmentation is the opposite process when an aggregate splits into smaller pieces. It may happen spontaneously, due to thermal fluctuations. The fragmentation is also possible due to collisions of aggregates. In this case the kinetic energy of their relative motion transforms into kinetic energy and surface energy of debris. The analytical aggregation kernels for simple laminar shear (orthokinetic aggregation), and thermal motion (perikinetic aggregation) are used to model the coagulation kinetics of Brownian particles under laminar shear flow. The two aggregation mechanisms are independent over a wide range of conditions [17]. Therefore, the aggregation rate J (the subscripts i j are dropped to refer to equal-sized spheres) in a system undergoing both shear-induced aggregation and Brownian aggregation can be written as

$$J = \frac{8k_B T n^2}{3\eta W_B}\left[1 + \frac{4}{3\pi}\frac{W_B}{W_S}P_e\right] \text{------(3)}$$

W_B is the stability ratio introduced into the Brownian aggregation due to the potential barrier in DLVO theory, η is the plasma viscosity, γ is the surface tension and *Pe* is referred to as Péclet number, i.e.

$$Pe = \frac{3\pi\eta a^3\gamma}{k_B T} \text{------(4)}$$

W_B is the stability ratio due to dispersion forces (van der Waals-Hamaker attraction), W_S is the stability ratio which includes hydrodynamic interaction. W_B is given in terms of maximum potential barrier and stability ratio due to dispersion forces.

$$W_B = W_\infty + 0.25\left[\exp\left(\frac{\varphi_{\max}}{k_B T}\right) - 1\right] \text{------ (5)}$$

For erythrocyte spheres of identical composition in plasma where the separation distance is much smaller than the particle radius, the van der Waals contribution to the interaction potential can be calculated, to close approximation, as

$$\varphi_{vdW} = -\frac{aA_H}{12D} \text{------(6)}$$

In the limit of constant surface charge density *Φe* is approximated as

$$\varphi_e \approx 2\pi\varepsilon\varepsilon_0 a\psi_s^{\,2} In\left[\frac{1}{1-\exp(-\kappa D)}\right] \text{------ (7)}$$

where ε is the dielectric constant of the medium, ε_0 is the permittivity of free space, ψ_S is the particle surface potential, and κ is the Debye parameter. Replacing the surface potential with the more accessible zeta potential, ζ, the total interaction potential is given by

$$\varphi = -\frac{aA_H}{12D} + 2\pi\varepsilon\varepsilon_0 a\zeta^2 In\left[\frac{1}{1-\exp(-\kappa D)}\right] \text{------(8)}$$

The bulk property yield stress, τ_y, is proportional to the inter-particle force at the separation distance times the number of particle contacts per unit area. The number of particle contacts scales as ϕ^2 / a^2 where φ is the volume fraction of the solids and a is the particle radius.

$$\tau_y \propto \frac{\phi^2}{a^2}\frac{\varphi(D_0)}{D_0} \text{--(9)}$$

Using the interaction potential from DLVO theory and electrostatic potential, Φe given in equations 4 and5 and putting in Equation 7, yield stress is given by

$$\tau_y \propto \frac{\phi^2}{a}\left(\frac{A_H}{12D_0^2} - \frac{2\pi\varepsilon\varepsilon_0}{D_0}\zeta^2 In\left[\frac{1}{1-\exp(-\kappa D_0)}\right]\right) \text{---(10)}$$

The above equation shows that at low zeta potential, van der Waals forces dominate. As the zeta potential becomes significant, yield stress increases as the square of zeta potential.
The effective volume fraction of the erythrocyte flocs is given by following equation

$$\phi_f = \left(\frac{d_f}{d_0}\right)^{3-D}\phi_0 \text{--(11)}$$

where d_f, d_0 and ϕ_0 are the diameter of a floc, diameter of a primary particle and volume fraction of primary particles. D is the fractal dimension and is in the range of 1≤ D≤3. When van der Waal forces dominate, yield stress is dependent on the diameter of the floc as

$$\tau_y = \frac{A}{9\pi d_0 h^2}\left(\frac{\phi_0}{0.52}\right)^{\frac{2}{3-D}} \text{--(12)}$$

where A is Hamaker constant and h is the minimum separation distance between particle surfaces. At the condition of maximum cubic packing, the volume fraction is 0.52. In conclusion, aggregation of erythrocytes increase as particle size decreases, hematocrit level increases, viscosity increases and shear rate decreases. Experimental results on the single cell level and theoretical analyses of the depletion model indicate the influence of the cell shape on the binding strength.

Experimental results indicate, aggregation is observed to be a two-step process: the formation of short linear stacks with few erythrocytes followed by their coalescence into long linear and branched rouleaux. As the shear rate increases the large rouleaux break up into smaller ones, and at higher values the suspension ultimately becomes one of mono-dispersed erythrocytes. This process then reverses as the shear rate is decreased [18].

Yield stress of blood is found to lie between 0.0015and 0.005 Pa at hematocrit concentratio, Ht = 0.45 i.e. a threshold stress for flow to begin often estimated by the extrapolation of measured shear stress to the zero shear rate on the basis of the Casson's equation[19]. Yield stress is power law dependent on hematocrit level and is fitted to a power law dependence and a constant that is a function of function of fibrinogen concentration. Red blood cell (RBC) agglutination depends on intermolecular attractive forces (hydrophobic bonds, Van der Wall forces, electrostatic forces and hydrogen bonds) and repulsive interactions (zeta potential). The force involved in RBC aggregation can be measured using double optical tweezers, in normal serum, in the presence of erythrocyte antibodies and associated to agglutination potentiator solutions Dextran, low ionic strength solution [LISS] and enzymes [20]. Electrostatic repulsive force extends from the surfaces of normal RBC's for at least 16 nm, preventing the aggregation of normally charged cells by the short Dextran 20. Relation between aggregating force (of fibrinogen) and disaggregating force (due to electrostatic repulsion among erythrocytes) in erythrocyte aggregation can be investigated using a rheoscope where erythrocyte aggregation increases with the increase of molecular weight of bridging macromolecules. Increase in electrostatic repulsion between cells may explain the disaggregation of normal RBC's at high concentrations of dextrans.

References:

[1] Varga-Szabo D, Pleines I, Nieswandt B. 2008. Cell adhesion mechanisms in platelets. Arterioscler Thromb Vasc Biol 28: 403–412.

[2] Doggett TA, Girdhar G, Lawshe A, Schmidtke DW, Laurenzi IJ, Diamond SL, Diacovo TG. 2002 Selectin-like kinetics and biomechanics promote rapid platelet adhesion in flow: the GPIb(alpha)-vWF tether bond. *Biophys. J.* **83**, 194–205.

[3} Tokarev, A Butylin ,A. and F. I. Ataullakhanov. 2011. Platelet adhesion from shear blood flow is controlled by near-wall rebounding collisions with erythrocytes. Biophys. J. 100:799–808.

[4] Yago T *et al.* 2008 Platelet glycoprotein Ibalpha forms catch bonds with human WT vWF but not with type 2B vonWillebrand disease vWF. *J. Clin. Invest.* **118**, 3195–3207.

[5] Ruggeri, Z. M., and G. L. Mendolicchio. 2007. Adhesion mechanisms in platelet function. Circ. Res. 100:1673–1685.

[6] Bas De Laat Clinical Advances in Hematology & Oncology Volume 12, Issue 10 October 2014.

[7] Doddi SK, Bagchi P. 2009 Three-dimensional computational modeling of multiple deformable cells flowing in microvessels. Phys. Rev. E Stat. Non lin. Soft Matter Phys. 79, 046318

[8] Zhang J, Johnson PC, Popel AS. 2007 An immersed boundary lattice Boltzmann approach to simulate deformable liquid capsules and its application to microscopic blood flows. Phys. Biol. 4, 285–295.

[9] Babler, M. U., M. Morbidelli.2007 Analysis of the aggregation fragmentation population balance equation with application to coagulation. J. Colloid Interface Sci. 316(2):428– 441,

[10] Doddi SK, Bagchi P. 2009 Three-dimensional computational modeling of multiple deformable cells flowing in microvessels. Phys. Rev. E Stat. Nonlin. Soft Matter Phys. 79, 046318. (doi:10.1103/ PhysRevE.79.046318)

[11]. A. Kumar and M. D. Graham, "Margination and seg- regation in confined flows of blood and other multicomponent suspensions," Soft Matter (2012), doi: 10.1039/C2SM25943E.

[12] O.Baskurt, B. Neu, H. Meiselman, Red Blood Cell Aggregation, CRC Press Taylor and Francis Group, 2012

[13] N. V. Brilliantov, A. S. Bodrova and P. L. Krapivsky,2009. J. Stat. Mech. P06011

[14] R. Hellmuth,1,2 M. S. Bruzzi, and N J. Quinlan, 2015,Analysis of Shear-Induced Platelet Aggregation and Breakup, Annals of Biomedical Engineering, DOI: 10.1007/s10439-015-1409-1

[15]P. Decuzzi and M. Ferrari, "Modulating cellular adhesion through nanotopography," Biomaterials, vol. 31, no. 1, pp. 173–179, 2010.

[16] J.Jung, 1, Byoung-Kwon Lee, and S Shin.2014, Yield shear stress and disaggregating shear stress of human blood, Korea-Australia Rheology Journal, Vol.26, No.2,.191-198
.[17]M Mohtaschemi, A Puisto, X Illab and M. J. Alavaa, 2014, Rheology dynamics of aggregating colloidal suspensions, Soft Matter, 2014, 10, 2971.
[18] [17] Q. Zhao, L. G. Durand, L. Allard, and G. Cloutier, Effects of a sudden flow reduction on red blood cell rouleau formation and orientation using RF backscattered power, Ultrasound Med. Biol., **24**, 503–511, 1998
[19] A. L. Copley, C. R. Huang, and R. G. King,1973, Rheogoniometric studies of whole human blood at shear rates from 1,000-0.0009 sec−1. Part I. Experimental findings, Biorheology, **10**, 17–22.
[20] H. Bäumler and E. Donath, Does dextran realy significantly increase the surface potential of human red blood cells?, Stud.Biophysica **120** (1987), 113–122.

420230 Big Data Analysis for Selecting Clinically Relevant Biomarkers: A Global Optimization Framework

Abstract:

Biomarkers are measurable indicators of biological processes that can be applied in clinical settings for disease diagnosis and prognosis, risk-factor assessment, disease staging, and as indicators of treatment efficacy. High-throughput –omics platforms enables the interface of personalized medicine with the advantages and challenges of big data. Typical studies generate expansive datasets with thousands of candidate biomarkers (i.e., data features). For a biomarker to be accepted into clinical praxis, it must be subjected to large-scale, often expensive clinical validation stages [1]; the ultimate success of a discovery-phase biomarker study lies in its ability to produce a small subset of biomarkers with the greatest probability of success in large-scale targeted studies [1,2]. This high data dimensionality per sample is almost always coupled with a comparatively low number of samples in the discovery phase, yielding a statistically difficult feature selection problem with a high probability of overfitting or of selecting data artifacts as meaningful candidates [3].

We present a novel optimization model that selects the optimal subset of candidate biomarkers. The model is self-regularized and can be solved to global optimality to obtain the best subset of biomarkers as per the objective function. We display the utility and power of the model by applying it to four well-known expansive cancer genomics datasets from the literature [4-7] which study differences between diseased and healthy patients or between different types of cancer. All datasets have an extremely low (< 0.02) samples-to-features ratio. We utilized an established evaluation protocol [8] to compare our method to the current state of the art [9]. Very small subsets of genes selected by the model are able to classify new samples from each of the disease systems with high sensitivity, specificity, and accuracy, and are selected robustly and stably in the face of random data permutations [8,10,11]. The model comprises a general methodology which applies to any scenario, biomedical or otherwise, where an optimally descriptive set of features must be extracted from a vast pool of mostly irrelevant possibilities.

References:

1. Rifai, N., Gillette, M. A., & Carr, S. A. (2006). Protein biomarker discovery and validation: the long and uncertain path to clinical utility. *Nature biotechnology*, *24*(8), 971-983.

2. Srinivas, P. R., Verma, M., Zhao, Y., & Srivastava, S. (2002). Proteomics for cancer biomarker discovery. *Clinical chemistry*, *48*(8), 1160-1169.

3. Rubingh, C. M., Bijlsma, S., Derks, E. P., Bobeldijk, I., Verheij, E. R., Kochhar, S., & Smilde, A. K. (2006). Assessing the performance of statistical validation tools for megavariate metabolomics data. *Metabolomics*, *2*(2), 53-61.

4. Alon, U., Barkai, N., Notterman, D. A., Gish, K., Ybarra, S., Mack, D., & Levine, A. J. (1999). Broad patterns of gene expression revealed by clustering analysis of tumor and normal colon tissues probed by oligonucleotide arrays. *Proceedings of the National Academy of Sciences*, *96*(12), 6745-6750.

5. Golub, T. R., Slonim, D. K., Tamayo, P., Huard, C., Gaasenbeek, M., Mesirov, J. P., ... & Lander, E. S. (1999). Molecular classification of cancer: class discovery and class prediction by gene expression monitoring. *science*, *286*(5439), 531-537.

6. Alizadeh, A. A., Eisen, M. B., Davis, R. E., Ma, C., Lossos, I. S., Rosenwald, A., ... & Staudt, L. M. (2000). Distinct types of diffuse large B-cell lymphoma identified by gene expression profiling. *Nature*, *403*(6769), 503-511.

7. Singh, D., Febbo, P. G., Ross, K., Jackson, D. G., Manola, J., Ladd, C., ... & Sellers, W. R. (2002). Gene expression correlates of clinical prostate cancer behavior. *Cancer cell*, *1*(2), 203-209.

8. Abeel, T., Helleputte, T., Van de Peer, Y., Dupont, P., & Saeys, Y. (2010). Robust biomarker identification for cancer diagnosis with ensemble feature selection methods. *Bioinformatics*, *26*(3), 392-398.

9. Guyon, I., Weston, J., Barnhill, S., & Vapnik, V. (2002). Gene selection for cancer classification using support vector machines. *Machine learning*, *46*(1-3), 389-422.

10. He, Z., & Yu, W. (2010). Stable feature selection for biomarker discovery. *Computational biology and chemistry*, *34*(4), 215-225.

11. Kuncheva, L. I. (2007, February). A stability index for feature selection. In *Artificial intelligence and applications* (pp. 421-427).

436664 Atomic Layer Deposition Processes: Understanding the Algebraic and Geometrical Structure of Dynamic Reaction Models

Abstract:

While substantial effort has been invested in understanding the fundamentals of atomic layer deposition (ALD) surface reaction mechanisms and how surface reaction models are integrated with reactor-scale precursor transport simulations (see, e.g., [1-3]), ambiguities remain in modeling the surface species dynamics during the different phases of an ALD process cycle. For example, open questions exist regarding how reactive surface species in amorphous ALD films are defined, how rate-limiting steps are identified in surface reaction networks, how equilibrium processes are distinguished from their finite-rate counterparts, and how redundant modeling equations can be pruned from the ALD surface reaction network. Therefore, the objective of this work is to take the first steps towards developing a mathematical framework that can be used to decompose ALD surface reaction dynamics according to the time-scales of the elementary reactions and to assess whether the resulting simulation problem is well-posed and so will provide physically meaningful simulation results.

In this paper, we present a recently developed reaction factorization procedure [4] that explicitly separates the slow (deposition), fast (equilibrium), and instantaneous (conserved) modes of thin- film deposition models describing the dynamics of the precursor, surface, and deposition chemical species for ALD processes. The reaction factorization procedure provides an unambiguous means of translating sequences of adsorption, equilibrium, and irreversible reactions characterizing a deposition system into a low-dimensional differential-algebraic (DAE) system when the reaction kinetics are predicted using transition-state theory, an approach that builds directly on such ALD DFT studies as those described in [1]. The factorization eliminates redundant dynamic modes and provides a wealth of other structural information about the deposition model. An implicit Euler procedure then is used to numerically solve the singular- perturbation problem describing the time-evolution of the reaction species on the manifold defined by the combination of the equilibrium relationships and conserved quantities.

Our modeling and analysis approach will be described in the context of an alumina ALD process based on the TMA/water precursor system. Even with the existing bodies of modeling research for these two systems, our approach to reaction kinetics modeling will reveal new insights into the rate-limiting processes of each deposition reaction network by quantifying the flux of ALD precursors and surface species through the nodes of an ALD reaction graph.

References

[1] Elliott, S. D., Semicond. Sci. Tech. 27, 1–10 (2012)
[2] Holmqvist, A., T. Torndahl and S. Stenstrom, Chem. Eng. Sci. 81, 260–272 (2012) [3] Travis, C. D. and R. A. Adomaitis, Processes 1, 128–152 (2013)
[4] Remmers, E., C. D. Travis, and R. A. Adomaitis, Chem. Eng. Sci., in press, (2015)

422920 Nonlinear Analysis of the Surface Morphological Evolution of Stressed Crystalline Materials and Heteroepitaxial Thin Films

Abstract:

Strained crystalline materials are found in many electronic and photonic devices, such as in integrated circuits in microelectronics, with strain resulting from the manufacturing processes, and in heteroepitaxial thin films, where strain is induced by the lattice mismatch between the film and the substrate. The resulting accumulation of elastic strain energy has a significant impact on the morphological stability of the material, triggering well-known instabilities. These include the Asaro-Tiller/Grinfeld (ATG) instability that causes surface cracking in stressed bulk materials and the Stranski-Krastanov (SK) instability that leads to formation of three-dimensional (3D) islands in heteroepitaxial films.

Numerical simulations of the surface morphological evolution of uniaxially stressed elastic crystalline solids have demonstrated that, in addition to ATG (surface cracking) instabilities, long-wavelength perturbations from the planar surface morphology can trigger a tip-splitting instability that causes formation of a pattern of secondary ripples, which cannot be explained by linear stability theory. In this presentation, we develop a weakly nonlinear stability theory, which can explain the occurrence of such secondary rippling instabilities and predict the number of secondary ripples that form on the surface as a function of perturbation wavelength. The theory shows that this type of surface pattern formation arises entirely due to the competition between surface energy and elastic strain energy, regardless of surface diffusional anisotropy or the action of externally applied fields. The origin of secondary rippling is explained through nonlinear terms included in the analysis that generate sub-harmonic ripples in the surface morphology with wave numbers that are multiples of the original surface perturbation wave number. Based on the weakly nonlinear theory, we have derived simple analytical expressions that predict the critical wavelength for the onset of secondary rippling, the increase of the number of secondary ripples with increasing perturbation wavelength, and how the onset of the secondary rippling instability and the rippled surface pattern are affected by surface diffusional anisotropy and the action of an applied electric field. The conclusions of the theory are validated by systematic comparisons with results of self-consistent dynamical simulations of surface morphological evolution. Further quantitative analysis based on a wavelength selection nonlinear theory also will be presented.

Following the same approach, we have carried out a weakly nonlinear analysis of the surface morphological evolution of a heteroepitaxial thin film on a substrate. We have found that in addition to the SK instability, long-wavelength perturbations also can trigger nonlinear secondary instabilities, resulting in the splitting of a single quantum dot (i.e., 3D island) into multiple quantum dots of smaller sizes. The theory also can predict the critical wavelength of the initial surface perturbation for the onset of the secondary nonlinear instability. A systematic protocol of self-consistent dynamical simulations based on a fully nonlinear continuum model of epitaxial film driven morphological evolution were conducted to assess the validity of the theoretical results. The theoretical predictions were found to be in good qualitative and quantitative agreement with the simulation results. Our study sets the stage for exploiting nonlinear surface phenomena toward engineering tunable-size nanoscale surface features.

420732 Investigation of Film Porosity and Hydrogen Content Dependence on Deposition Conditions in a PECVD Process Using a Multiscale Model

Abstract:

Plasma enhanced chemical vapor deposition (PECVD) is widely used in the microelectronics and solar cell industries to deposit thin films from a mixture of gas-phase species onto a solid substrate [1]. Specifically, in the solar cell industry, PECVD is broadly used in the production of thin film silicon solar cells to deposit amorphous silicon semiconductor layers due to reduced manufacturing costs via silane recycling and the possibility for extremely low operating temperatures (<250 C) while achieving optimal dielectric properties of amorphous silicon thin films. The material of interest here, hydrogenated amorphous silicon (a-Si:H), differs from crystalline silicon by the lack of long-range order and the high content of bonded hydrogen (typically around 10% in device-quality a-Si:H). Although useful for opto-electronic devices, this material suffers from degradation via the well characterized Staebler-Wronski effect; whereby, in the first six months of operation the defect density increases under intense light, accompanied by a 10-15% drop in efficiency [2,3].

Motivated by the above considerations, a multiscale modeling and operation framework is developed to explore the relationship between standard operating conditions (e.g., 150 < T < 250 C, 1 < P < 5 bar) and the quality of deposited films, in an effort to develop deposition techniques that are capable of producing high-quality materials at reasonable growth rates. In the present work, quality is a function of two conditions: the hydrogen content which affects the stability and optical band-gap, and the void fraction which plays a role in the electronic and structural properties of the film [4]. Specifically, the continuous gas phase is modeled via a system of one dimensional reaction and transport equations at discrete locations across the wafer surface. The resulting concentration of silane and hydrogen radicals is then applied to a hybrid kinetic Monte Carlo (kMC) model which captures the complex surface chemistry involved in the growth of a-Si:H films [5]. Unlike traditional solid-on-solid schemes, the triangular, two-dimensional lattice employed in this study creates the possibility for overhangs and voids to develop within the growing film. Additionally, both silicon and hydrogen radicals interact on the surface and within the lattice bulk, yielding a more accurate growth mechanism than in previous studies. After completion of the deposition process, post batch measurements allow for the hydrogen content and void fraction to be characterized. Simulations across a wide range of deposition conditions reveal that the reactor temperature and the ratio of SiH_4 to H_2in the feed are critical parameters which play a key role in determining the quality of the film microstructure and the concentration of trapped hydrogen species.

[1] Kern W. Thin Film Processes II. Academic Press. 1991

[2] Rech B, Wagner H. Potential of amorphous silicon for solar cells. Appl. Phys. A. 1999;69:155-167.

[3] Nelson J. The Physics of Solar Cells. Imperial College Press. 2003

[4] Smets AHM, Kessels WMM, Van de Sanden MCM. Vacancies and voids in hydrogenated amorphous silicon. Appl. Phys. Let. 2003;82:1547-1549.

[5] Crose M, Kwon JS, Nayhouse M, Ni D, Christofides PD. Multiscale modeling and operation of PECVD of thin film solar cells. Chem. Eng. Sci. 2015, in press.

431958 Stochastic Nonlinear Model Predictive Control Applied to an Epitaxial Thin Film Growth Process Under Uncertainty

Abstract:

The design of a nonlinear Model Predictive Controller that regulates the performance of an epitaxial thin film growth process under distributional uncertainty is presented. The epitaxial deposition process has been modeled employing nonlinear partial differential equations (PDEs) coupled with lattice-based kinetic Monte Carlo (KMC) simulations, i.e. a multi-scale model. Since the KMC simulations are computationally prohibitive for online control applications, reduced order models (ROM) are developed in this work based on data-driven simulations of the multiscale system. The key innovative feature in the ROMs is that they explicitly account for distributional uncertainty in the multiscale system's parameters. Power series expansions are employed in this work for uncertainty propagation. Therefore, the resulting ROMs quantify the key distributional characteristics in the controlled variables due to changes in the key manipulated variables, e.g. substrate temperature, and under distributional uncertainty in the parameters, e.g. the activation energy in the migration process occurring in the thin film's surface. The ROMs have been validated using open-loop simulations of the multiscale process and have been embedded within a nonlinear model predictive control (NMPC) framework to predict estimates of the control actions that comply with the process control objectives at a user-defined probability of satisfaction, i.e. a stochastic NMPC. The shrinking horizon approach is adopted to minimize the roughness of the thin film while satisfying the constraints on the applied substrate temperature at each time interval and film thickness at the end of the deposition process (at a given probability of occurrence).

429448 Development of a Multiscale Simulation Approach for Modeling Nanostructured Si-Based Anodes in Lithium Ion Batteries

Abstract:

Recently, significant interest has been focused on Lithium ion batteries containing silicon based anodes for their ability to lead to increased storage capacity due to the significant gain in Lithium uptake evidenced by the silicon anode in comparison to more traditional graphitic or carbon based anodes. However, the increased capacity seen in lithium insertion into amorphous silicon (a-Si) has a substantial drawback: The full insertion of lithium into the anode is accompanied by a roughly 300% volume increase of the anode material upon full lithiation, coupled with a substantial reduction in the mechanical properties of the lithiated silicon. As such, while the capacity of a-Si to accept lithium is prodigious, the stabilitiy of the silicon anode on repeated cycling remains problematic due to the evolution of high internal stresses from expansion coupled with the decrease in mechanical stability as the silicon bonding network is disrupted by lithium insertion and bonding.

Since the chemical process of lithium insertion is fundamental to the destabilization of the silicon anode mechanical properties, avenues apart from chemical modification of the anode material have been investigated in order to improve the cyclability and performance of the silicon anode. Some success has been seen by reinforcing the silicon phase with external coatings, or changing the physical size and shape of the silicon composing the anode phase to allow for more even lithium penetration. While such techniques are evocative of potential avenues for increasing the mechanical stability while preserving the viability of lithium storage in amorphous silicon anodes, they are not prescriptive. In order to try to enable rational design of electrodes, we have therefore embarked on an effort to develop a robust computational multi-scale model of the Li/a-Si composite system which will allow for a phenomenological understanding of the stress evolution and failure of these systems, and which may therefore serve as a basis for establishing design parameters to promote more reliable future anode designs.

This talk will focus on our efforts to build a dynamically launched, computational framework to support the necessary elements of multi-scale modeling of the Li/a-Si anode evolution upon charge/discharge cycles. While the context is the specific sets of models and equations which allow for the calculation of this particular system, we have developed a generic framework for such multi-scale approaches which allows us to couple microscopic sensitivity (in the guise of the changing mobility of the Lithium within the partially lithiated a-Si at arbitrary stress state) with macroscopic response (via a chemo-reactive material-point method (MPM) simulation) by enacting dynamic and on-demand calculation of the key elements of the MPM constitutive equations via atomistic molecular dynamics and ab initio simulations. We will discuss the construction of this framework, the necessary assumptions and relations that must hold true in order to allow for this coupling, and the statistical techniques (based upon uncertainty quantification) used to guide the launch of the on-demand sub-grid calculations, which modify the MPM material response. By providing a concrete demonstration of our multi-scale approach, we will also demonstrate the generic sets of constraints under which it is possible to enact clean transition between differing simulation scales, in order to help to begin to establish the capability for consistently tying dynamic multiple scale simulations into a cohesive overall material model which may span multiple time/length scales and dynamic processes seamlessly.

429658 A Cellular Automaton Approach for Modeling Biomass Pyrolysis

Abstract:

Lignocellulosic biomass is a chemically and morphologically heterogeneous material. Furthermore, this heterogeneity is in-part responsible for the vast number of thermal decomposition products seen in pyrolysis events. While modeling of biomass pyrolysis has been a subject of much research in past years, at the macro and microscopic scales, the majority of this work has focused on a range of continuum-based approaches. While effective at capturing global outcomes, these approaches are less tractable as frameworks for capturing microstructural effects and upscaling molecular information. This work demonstrates the use of cellular automata (CA) as an alternative platform for the modeling and simulation of biomass pyrolysis. Efficient at capturing transport and chemical processes in highly heterogeneous system, CA is effective for modeling microstructural changes that occur as a result of chemical and physical transformations. A number of benchmark trials demonstrated the convergence of the CA to global continuum outcomes. Application of the CA to images of actual biomass show promise for this platform as an intermediate length-scale tool capable of predicting char morphologies that mimic experimental outcomes.

427831 Coding and Decoding Using Microfluidic Loops: An Analytical Approach

Abstract:

Droplet microfluidics is the science and technology of systems that process and manipulate small (pico- litres) volumes of fluids, using channels of dimensions less than a millimetre. The fluid flow in the channels is in the creeping flow regime where the governing equations are linear and reversible. But experiments in the literature have shown that a network device as simple as a microfluidic loop can result in complex patterns. A droplet reaching the junction of the loop chooses one of the two branches based on the local flow rates, which depend on branch lengths and the number of drops present in the branches. This discrete decision making step makes the problem non-linear and irreversible. A train of equally spaced droplets (1 period) entering the loop device gets converted to a periodically spaced train of droplets with one, two or even chaotic periods.

This loop device can be thought of as an encoder that passively converts a signal (train of drops) into a coded signal (periodic train of drops). Then it would be interesting to understand if there could be a decoder which is a loop device or several loops in series which can essentially convert the coded signal back to the original signal. This would mean that the decoder should convert an n-period input to a single period output. We propose an analytical scheme based on continuous phase redistribution and repeatability of events to find the different possible solutions for the lengths of the branches of the decoder loop. Multiple solutions exist to a certain decoding problem, based on the sequence of occurrence of events (drops entering and exiting the loop). The number of possibilities scale up with the periodicity of the signal being decoded. However, there exists one specific sequence of events, a pattern common to all decoding solutions that can decode any periodicity to a single period in the output signal. The problem of decoding any period signal to 1 period signal now scales down to arranging for this sequence of events to occur by optimising the branch lengths. Decoding of 2 period and 3 period to 1 period using this sequence of events is demonstrated and extension to the decoding higher period signals is indicated.

437187 Temperature Control of Microchannel Reactors Using Bimetallic Thermally-Actuated Valves

Wednesday, November 11, 2015: 12:30 PM
Salon D (Salt Lake Marriott Downtown at City Creek)

Richard Pattison, McKetta Department of Chemical Engineering, University of Texas at Austin, Austin, TX, Akash Gupta, Department of Chemical Engineering, The University of Texas at Austin, Austin, TX, Melissa Donahue, James R. Fair Process Science and Technology Center, Austin, TX and Michael Baldea, McKetta Department of Chemical Engineering, The University of Texas at Austin, Austin, TX

Temperature Control of Microchannel Reactors Using Bimetallic Thermally-Actuated Valves

Richard Pattison, Akash Gupta, Melissa Donahue, and Michael Baldea

McKetta Department of Chemical Engineering

The University of Texas at Austin, 1 University Station C0400, Austin, TX 78712

email: mbaldea@che.utexas.edu

Microchannel reactors coupling exothermic and endothermic reactions are one of the most prominent results of the process intensification paradigm. The high surface-area-to-volume characteristics of CPRs result in a process with minimal transport limitations and consequently, a size and cost that are an order of magnitude less than conventional reactors of the same capacity [1].

While CPRs bring significant potential economic benefits, they also pose significant control and operational challenges. Synchronizing heat generation and consumption in the reactors is particularly challenging, and if not properly addressed, could lead to the formation of temperature hot spots that may damage the catalyst coating of the plates or the reactor structure itself [2-4]. Furthermore, in practical scenarios, the reactors are subject to disturbances in the inlet conditions and potentially unequal distribution of flow to the channels, which may further contribute to the formation of hot spots. The implementation of measurements (temperature, composition, and/or flow rate) at the channel level is practically challenging. As a consequence, thus far, the majority of feedback control systems for CPRs rely on measuring the properties (typically temperature) of the bulk outlet streams, and on boundary actuation, in the sense that the total (rather than channel-wise) flow rate of fuel is adjusted in response to changes in reactor outputs.

In this work, we present a novel approach for microchannel reactor control, that provides actuation and temperature control at the individual channel level. Specifically, we introduce a new class of thermally-actuated valves constructed from bimetallic strips [5]. Bimetallic strips consist of two different metal strips that are rigidly attached. The difference in thermal expansion properties of the two metals causes the strips to deflect upon changes in temperature, transforming temperature variations into mechanical displacement. Thus, affixing bimetallic strips to either side of the combustion channels in microchannel reactors is equivalent to implementing a thermally-actuated valve as seen in Figure 1.

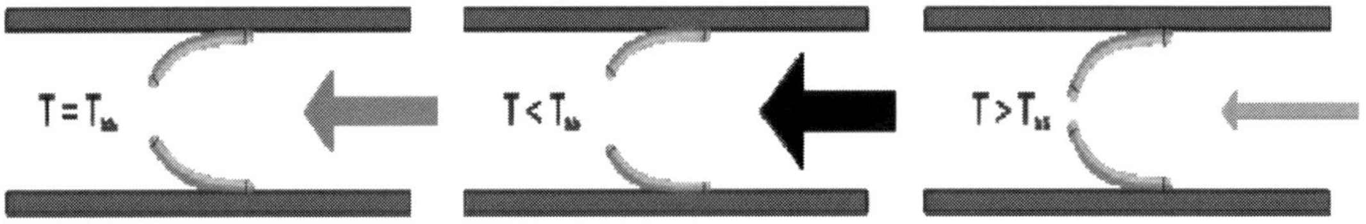

Figure 1: Response of a thermally-actuated valve to temperature changes. Tss represents the design temperature (adapted from [5]).

In this system, the flow rate through the channel is dictated by the pressure drop across the valve. If the operation deviates from the nominal conditions, temperature changes in the reactor will result in a deflection of the strips and a change in the valve position, and consequently an adjustment in the fuel flow rate. For example, in the presence of a disturbance that reduces the amount of heat consumed in the reforming channels, the temperature in the reactor will rise causing the strips to deflect toward the channel center. Consequently, the fuel flow rate will be restricted, and, in turn, the heat generated in the combustion channel will drop, returning reactor temperature towards the nominal value.

Using a detailed model of a steam methane reforming CPR as a case study, we illustrate the considerations involved in the design of thermally-actuated control valves, and demonstrate via dynamic simulations their effectiveness for temperature control when the reactor is subject to flow maldistribution and operational disturbances.

References

[1] Zanfir, M.; Gavriilidis, A. Catalytic combustion assisted methane steam reforming in a catalytic plat reactor. *Chem. Eng. Sci.* **2003,** *58*, 3947-3960.

[2] Baldea, M.; Daoutidis, P. Dynamics and Control of Autothermal Reactors for the Production of Hydrogen. *Chem. Eng. Sci.* **2007,** *62*, 3218-3230.

[3] Zanfir, M.; Gavriilidis, A. Influence of flow arrangement in catalytic plate reactors for methane steam reforming. *Chem. Eng. Res. & Des.* **2004,** *82*, 252-258.

[4] Vaccaro, S.; Malangone, L.; Ciambelli, P. Modelling of a catalytic micro-reactor coupling endothermic methane reforming and combustion. *Int. J. of Chem. Reactor Eng.* **2010**, *8*.

[5] Pattison, R.C., Donahue, M.M., Gupta, A., Baldea, M. Localized Temperature Control in Microchannel Reactors Using Bimetallic Thermally-Actuated Valves, Ind. Eng. Chem. Res., in press

424226 Advanced Control Strategies Towards the Intensification of Monoclonal Antibody Production

Abstract:

Advanced control strategies towards the intensification of monoclonal antibody production

Maria M. Papathanasiou[a,d], Ana Quiroga Campano[a], Fabian Steinebach[b], Thomas Mueller-Spaeth[c], Massimo Morbidelli[b], Athanasios Mantalaris[a], Efstratios N. Pistikopoulos[d]*

[a]*Dept. of Chemical Engineering, Centre for Process Systems Engineering (CPSE), Imperial College London SW7 2AZ, London, U.K*

[b]*Institute for Chemical and Bioengineering, ETH Zurich, Wolfgang-Pauli-Str. 10/HCI F 129, CH-8093 Zurich, Switzerland*

[c]*ChromaCon AG, Technoparkstr. 1, CH-8005 Zurich, Switzerland*

[d]*Artie McFerrin Department of Chemical Engineering, Texas A&M University, College Station TX 77843*

**stratos@tamu.edu*

Keywords: process intensification, continuous processing, monoclonal antibodies, multi-parametric control

Process intensification plays a dominant role in the improvement of sustainability and productivity of biopharmaceutical processes. In particular, during the past few years there has been increasing demand on product quality and higher titers in the production of monoclonal antibodies (mAbs) [1]. The latter consists of two main parts: (1) the upstream processing (USP), where the cells are cultured and the therapeutic agent is produced and (2) the downstream processing (DSP) that involves the isolation/purification of the targeted product. Both the upstream and the downstream processing, however, are affected by various parameters related to the design of the bioreactor, the methods chosen for separation/purification as well as the operation mode (e.g. batch or continuous).

In order to develop environmental friendly, cost effective processes, current industrial trends suggest gradual shift from batch to continuous operation. Along with process intensification efforts, this can be facilitated by the use of advanced computational tools that will derive continuous operating conditions under minimum cost. In this work we present a complete computational study on the development of advanced control strategies for the production and purification of monoclonal antibodies. We follow the PAROC framework and software platform [2, 3] for the development of advanced multi-parametric control strategies that comprises four main steps: (i) development of a high-fidelity process model, (ii) approximation of the complex, process model into a linear state space representation, (iii) design of the multi-parametric controller, (iv) 'closed-loop', in-silico validation of the controller against the process model. The controller development is based on the derivation of multi-parametric Model Predictive Control (mp-MPC) policies that allow the solution of the optimization problem offline, improving the controller performance during online operation.

The examined system is described by (1) the upstream and (2) the downstream processing. The upstream process considers fed-batch culturing of GS-NS0 cells producing a chimeric IgG4 antibody. The mathematical model describing the system is based on multiple energy metabolic pathways and considers standard Monod kinetics. The model monitors the population of viable, early apoptotic and death cells, as well as the consumption/production profile of glucose and lactate, amino acids and the ATP balance. It comprises 16 differential and 23 algebraic equations, 41 variables and 45 parameters. The model is simulated in gPROMS® ModelBuilder v4.0.0 and the results are tested against experimental data. Due to its high complexity, the model cannot be used directly for the design of advanced controllers. It has to be therefore reduced to a linear state space model used for the formulation and solution of the multi-parametric programming problem (MATLAB®).

The purification of the upstream mixture (downstream processing) is performed using the Multicolumn Countercurrent Solvent Gradient Purification (MCSGP) process [4]. Similarly, using the model described by Krättli, et al. [5] we develop an advanced control strategy to ensure maximum recovery yield, while satisfying purity constraints [3, 6]. The 'in-silico' connection of the two systems is realized using the output of the upstream processing as a disturbance for the downstream system.

The results of this study form a solid basis for advanced "in-silico" experiments to determine the optimal operation conditions and evaluate the efficiency of the designed controllers. Additionally, such experiments will highlight the process bottlenecks and will provide the user with the necessary insight on the system dynamics, thus facilitating the realization of continuous monoclonal antibody production.

Acknowledgements

Financial support from the European Commission (OPTICO/G.A. No.280813) is gratefully acknowledged.

References

[1] P. Gronemeyer, R. Ditz, and J. Strube, "Trends in Upstream and Downstream Process Development for Antibody Manufacturing," *Bioengineering,* vol. 1, pp. 188-212, 2014.

[2] E. N. Pistikopoulos, "Perspectives in multiparametric programming and explicit model predictive control," *AIChE Journal,* vol. 55, pp. 1918-1925, 2009.

[3] E. N. Pistikopoulos, N. A. Diangelakis, R. Oberdieck, M. M. Papathanasiou, I. Nascu, and M. Sun, "PAROC

—An integrated framework and software platform for the optimisation and advanced model-based control of process systems," *Chemical Engineering Science,* p. In press.

[4] L. Aumann and M. Morbidelli, "A continuous multicolumn countercurrent solvent gradient purification (MCSGP) process," *Biotechnology and Bioengineering,* vol. 98, pp. 1043-1055, 2007.

[5] M. Krättli, F. Steinebach, and M. Morbidelli, "Online control of the twin-column countercurrent solvent gradient process for biochromatography," *Journal of Chromatography A,* vol. 1293, pp. 51-59, 6/7/ 2013.

[6] M. M. Papathanasiou, F. Steinebach, G. Stroehlein, T. Müller-Späth, I. Nascu, R. Oberdieck, *et al.*, "A control strategy for periodic systems - application to the twin-column MCSGP," in *Computer Aided Chemical Engineering*, ed, 2015, p. accepted for publication.

425075 Modeling and Analysis of Heat Recirculating Microreactor for Catalytic Combustion of Propane

Abstract:

Combustion in catalytic microreactors finds several applications due to the high energy density of hydrocarbon fuels. These include portable power generation, micro-propulsion, decentralised processing of fuels in thermally integrated micro-reactors, etc. Due to high surface area to volume ratio, high rates of catalytic reactions can be obtained compared to conventional devices. However, the large surface area also makes the system susceptible to extinction due to heat losses.

Heat recirculating geometries have shown to increase the range of stable operation of micro-reactors. The working principle for heat recirculating reactors is that the incoming cold reactant stream is preheated by transferring the excess enthalpy of the hot product stream. This involves carrying out reaction in a heat exchanger like assembly, to facilitate heat exchange between hot outlet and cold reactant streams. This can be achieved in simple geometries such as counter-current reactor, or in more complex Swiss-Roll geometries. Swiss-Roll geometry is especially popular in homogeneous and catalytically stabilised combustion in micro-burners.

In this work, we present numerical simulations for a new heat recirculating geometry. Specifically, we modify the Swiss Roll geometry to develop a heat recirculating micro-reactor with a spiral geometry. The inner walls are coated with catalyst. Thorough numerical simulations are presented for understanding the role of heat recirculation on thermal management and stability of the microreactor. The proposed geometry is simpler than the Swiss-Roll, while retaining the main advantage of larger area for heat transfer between the combustion channel and the incoming cold inlet stream. Coupling the micro-reactor with energy generation is also investigated using simple heuristics. We analyse the effect of geometry on stability of propane combustion in the micro-reactor, and compare the reactor with more traditional geometries.

436943 An Integrated, Multi-Stage, Multi-Scale Framework for Achieving Sustainable Process Synthesis-Intensification-Control

Abstract:

The chemical and biochemical industry needs major reductions in energy consumption, waste generation, etc., in order to remain competitive through the design and operation of more sustainable chemical and biochemical processes. These required reductions can be addressed through process synthesis-intensification-control, that is, the efficient use of raw materials (feedstock), the use of sustainable technologies and the design (and control) of processes that directly impact and improves sustainability/LCA factors. The unit operations concept, which has been sufficient until now, is one of the most used for performing process synthesis (and intensification) because it allows the association of tasks (functions) with the processing route to be followed. At the unit operations scale (Jaksland et al., 1995) and task scale (Siirola, 1996) alternatives are limited to existing (well-known) unit operations and therefore, may not be able to generate new integrations/combinations of intensified existing equipment.

However, to find innovative processes designs, extensions of the current concepts are necessary. Here, process synthesis-intensification using a phenomena based process synthesis method (Lutze et al., 2013) can play a major role because it provides the opportunity to perform the same tasks in a more sustainable way, new/novel unit operations can be generated (Lutze et al, 2013) and more sustainable processes can be designed (Babi et al., 2014).

An integrated, multi-stage, multi-scale, computer-aided framework has been developed in order to perform process synthesis-intensification-control. The framework operates at different scales, the unit operations scale, task scale and phenomena scale. In stage 1, process synthesis is performed (at the unit operations scale) using computer-aided flowsheet design (CAFD) (Tula et al, 2015), considering a superstructure of all possible alternatives based on known technologies, in order to generate a base case design. In stage 2, the base case design is decomposed into the smallest constituent units, that is, tasks then the involved phenomena. The base case design is analysed using economic and sustainability analyses in order to identify process limitations (hot-spots) that are translated into intensification design targets. In stage 3, an integrated task-phenomena-based synthesis-intensification method is embedded and applied (Babi et al., 2015) that consists of combining the phenomena to fulfil tasks, which are then translated into unit operations that constitute the (more sustainable) flowsheet alternatives which satisfy the intensification design targets. In this way, truly predictive and innovative solutions are generated much in the same way as atoms are combined to form molecules with desired properties (that is, analogous to computer-aided molecular design). The final stage involves validation and selection. Here detailed model-based calculations and/or experiments are performed to evaluate and compare the new solutions. In stage 4, design-control integration issues are validated/fine-tuned in order to generate the more sustainable controller structure. Note that in stages 1-3 design decisions are structured and made in such a way that the designed process(s) will also have the best opportunity for control (Mansouri et al., 2015).

In this presentation, the integrated process synthesis-intensification-control framework will be presented together with the corresponding databases, computer-aided models and tools needed to achieve sustainable synthesis-intensification. Different examples, related to synthesis, intensification and control will be presented for each stage of the framework.

References

1. Babi, D. K., Holtbruegge, J., Lutze, P., Gorak, A., Woodley, J. M., & Gani, R. (2015). Sustainable Process Synthesis-Intensification. Computers & Chemical Engineering.
2. Babi, D. K., Lutze, P., Woodley, J. M., & Gani, R. (2014). A Process Synthesis-Intensification Framework for the Development of Sustainable Membrane-based Operations. Chemical Engineering and Processing: Process Intensification, 86, 173–195.
3. Jaksland, C. A., Gani, R., & Lien, K. M. (1995). Separation process design and synthesis based on thermodynamic insights. Chemical Engineering Science, 50(3), 511–530.
4. Lutze, P., Babi, D. K., Woodley, J. M., & Gani, R. (2013). Phenomena Based Methodology for Process Synthesis Incorporating Process Intensification. Industrial & Engineering Chemistry Research, 52(22), 7127–7144.
5. Mansuori S, Sales Cruz M, Huusom JK, Gani R. In: Integrated process design and control of reactive distillation processes. Submitted for the International Symposium on Advanced Control of Chemical Processes – ADCHEM 2015; 2014
6. Siirola, J. J. (1996). Strategic process synthesis: Advances in the hierarchical approach. Computers & Chemical Engineering, 20, S1637–S1643.
7. Tula, A. K., Eden, M. R., & Gani, R. (2015). Process synthesis, design and analysis using a process-group contribution method. Computers & Chemical Engineering.

435211 On Quantification of the Attainable Region for Separator Networks with Varying Network Outlet Flowrates

Abstract:

In this work the process network attainable region (AR) concept is explored for separator networks in which the network outlet flow rates are allowed to vary. For separator unit models to which the IDEAS conceptual framework is applicable, it is shown that identification of AR boundary membership is equivalent to feasibility assessment of an infinite linear program (ILP). A number of important AR properties are then theoretically established, including AR convexity for fixed network inlet flowrate ratios, and representation of the AR in a process network outlet state space of reduced dimension. Finite dimensional approximations of the aforementioned ILP are then employed in creating increasingly accurate representations of the variable-flow-rate separator AR. A case study involving construction of the AR for the vapor-liquid equilibrium based separation of a ternary azeotropic mixture will be presented.

430169 Multi-Model Operability Approach As a Tool for Process Intensification: Application to a Membrane Reactor for Direct Methane Aromatization

Abstract:

This presentation introduces a novel multi-model operability-based approach for design and intensification of processes described by high-dimensional and nonlinear models. This novel approach is applied to membrane reactors (MRs) for the direct methane aromatization (DMA) conversion to benzene and hydrogen. In the last decades, MRs have emerged as examples of intensified processes as they combine reaction and separation unit operations in one process improving the system efficiency and lowering the cost[1,2,3]. Specifically, in the DMA-MR process, higher methane conversion than in conventional reactors is enabled due to the selective hydrogen removal from the reaction side through the membrane. This removal shifts the reaction equilibrium towards the products, thus increasing methane conversion and benzene production rate. The nonlinear and multivariable nature of the DMA-MR system associated with the reaction-transport phenomena interplay make this system challenging to design and optimize.

Process operability was developed as an approach for the design and control of complex chemical processes[4]. In particular, process operability enables the verification of a design's ability to achieve the feasible region of operation associated with process specifications, considering the physical limitations of the process design. Operability describes the relationship between the input (manipulated and/or disturbance) and output variables through a linear/nonlinear mapping obtained by employing the process model[5]. This approach has been applied to chemical processes that are described by high-dimensional linear models[6]. For the design of nonlinear systems, a low-dimensional operability implementation for the DMA-MR was analyzed using a subset of the system variables, including: membrane selectivity and permeance, reactor pressure and diameter, methane conversion and benzene production rate[7,8]. The results of this previous study demonstrated the promising capabilities of the operability approach to provide guidelines for experimental research, in terms of determining the set of membrane characteristics that should be obtained to achieve desired process specifications. However, the DMA-MR process intensification analysis using process operability and the full set of process variables has not been addressed, especially due to the limitation of current operability methods for nonlinear systems in terms of the problem size that they can address.

In this research, we first explore the extension of previously formulated operability algorithms for large-scale linear systems[9] to develop a multi-model operability approach for high-dimensional and nonlinear models. This approach employs sensitivity calculations and linear programing tools to quantify the output space as a function of the input space using the minimum number of linearized models that are necessary to adequately describe the nonlinear process operating envelope. The proposed approach is then applied as a tool for process intensification of complex high-dimensional systems. The performed operability calculations consider the multivariable nature of the DMA-MR process, by including the following variables: (i) reactor: diameter, length, temperature, pressure, and input flows; (ii) membrane: area, permeability, and selectivity; and (iii) economics: membrane and catalyst costs. Preliminary results on the application of the operability method as a tool for process intensification show reduction of the DMA-MR footprint (reactor volume and membrane surface area) for an equivalent level of performance. This case study indicates that operability can be a powerful tool for process intensification of membrane reactors and other complex chemical processes.

References

1. Stankiewicz A. I., Moulin, J. A. Process intensification: transforming chemical engineering. *Chem. Eng. Prog.* **2000**, 96(1), 22-34.
2. Drioli E., Stankiewicz A. I., Macedonio F. Membrane engineering in process intensification - An overview. *J. Membr. Sci.* **2011**, 380(1–2), 1–8.
3. Brunetti A., Caravella A., Drioli E., Barbieri G. Process intensification by membrane reactors: high-temperature water gas shift reaction as single stage for syngas upgrading. *Chem. Eng. Technol.* **2012**, 35(7), 1238–1248.
4. Vinson D. R., Georgakis C. A new measure of process output controllability. *J. Proc. Cont.* **2000**, 10(2-3), 185-194.
5. Georgakis C., Uzturk, D., Subramanian S., Vinson D. R. On the operability of continuous processes. *Control Eng. Prac.* **2003**, 11 (8), 859–869.
6. Lima F. V., Georgakis C. Design of output constraints for model-based non-square controllers using interval operability. *J. Proc. Cont.* **2008**, 18(6):610-620.
7. Carrasco J. C., Lima F. V. Nonlinear operability of a membrane reactor for direct methane aromatization. *In Proceedings of the 2015 IFAC ADCHEM Symposium*, British Columbia, Canada, June **2015**.
8. Carrasco J. C., Liu D., Lima, F. V. Modeling and nonlinear operability analysis of a membrane reactor for direct methane aromatization. *Presented at 2014 AIChE Annual Meeting*, Atlanta, Georgia, November **2014**.
9. Lima F. V., Georgakis C. Input-output operability of control systems: the steady-state case. *J. Proc. Cont.* **2010**, 20(6):769-776.

425808 Synthesis of Azeotropic Reactive Distillation Networks Involving Quaternary Mixtures

Abstract:

The synthesis of an isobaric reactive distillation network involving a quaternary azeotropic mixture is investigated in this work. The network is comprised of vapor-liquid equilibrium trays, with liquid holdup that changes from tray to tray. Kinetically and/or equilibrium limited reactions are assumed to occur in the liquid phase.

The Infinite DimEnsionAl State-Space (IDEAS) conceptual framework, is first shown to be applicable to the problem, and is then employed to formulate an infinite linear program (ILP) that can rigorously assess the feasibility of the considered process synthesis problem. The formulated IDEAS ILP's infimum value is approximated by the sequence of optimal values of a number of finite dimensional linear programs (FLP) of ever-increasing size. The method is demonstrated on a case study involving the isobaric reactive distillation of an acetic acid/isopropanol/isopropyl acetate/water azeotropic mixture.

422063 Lagrangean Disjunctive Branch and Bound for Linear Generalized Disjunctive Programs

Abstract:

Many optimization problems are represented with algebraic equations, using continuous and discrete variables, giving rise to Mixed-Integer (Linear or Nonlinear) Programs (MILP/ MINLP). An alternative way to represent this type of problems is Generalized Disjunctive Programming (GDP) proposed by Raman and Grossmann [1] that involves not only algebraic equations, but also disjunctions and logic propositions. This higher level representation, which is of special importance in engineering, allows us to exploit the logic structure of the problem to obtain better formulations and improved solution methods.

GDP problems are normally reformulated as MILP/MINLP models to exploit the developments of these types of solvers. Big M (BM) and Hull Reformulation (HR) are the two traditional GDP to MILP/MINLP reformulations. The BM provides a formulation that is usually weaker than the HR (i.e. the continuous relaxation of the HR is at least as tight as that of the BM, and typically tighter). However, the BM reformulation yields an MILP/MINLP that is smaller in size than the HR. Alternatively to reformulation, specialized methods have been developed to solve generalized disjunctive programs: disjunctive branch and bound [2], and logic-based outer approximation [3].

The traditional disjunctive branch and bound requires the solution of the continuous relaxation of the MILP/MINLP reformulation of the GDP at every node, using either the BM or HR. The BM disjunctive branch and bound typically requires the evaluation of more nodes than the HR disjunctive branch and bound. However, the time required to evaluate each node is smaller in the BM disjunctive branch and bound. An important property of the disjunctive branch and bound is that the time required to evaluate each node decreases as the search tree progresses.

In this work we first present a novel Lagrangean relaxation scheme for the continuous relaxation of the HR. Normally, the Lagrangean relaxation technique requires the user to identify the "complicated constraints" in each problem. In the proposed Lagrangean decomposition, the complicating constraints are the same for any linear GDP. These complicating constraints are related to the (HR) reformulation.Therefore, the proposed decomposition is general and applicable to any linear GDP (i.e. also applicable to any MILP, since any MILP can be represented as a linear GDP). This Lagrangean relaxation has two important properties. The first one is that it is simpler to solve than the HR, as it is usually expected from the Lagrangean relaxation of a problem. The second one, and perhaps the most relevant, is that its solution (a continuous LP) always assigns 0-1 values to the binary variables. It is important to note that a solution to the Lagrangean relaxation does not necessarily satisfy all of the constraints of the HR.

We present a solution method that exploits the proposed Lagrangean relaxation of the HR (which is general to any linear GDP, is "easy to solve", and its solution always provides 0-1 values to binary variables). This solution method is a disjunctive branch and bound for linear GDPs that makes use of the Lagrangean relaxation.

The proposed method solves three LPs at each node: the continuous relaxation of the BM, the proposed Lagrangean relaxation of the HR, and an LP in which all of the discrete variables are fixed. The continuous relaxation of the BM serves the same purpose as in the traditional disjunctive branch and bound: if it is infeasible or integer feasible the node is pruned; if the relaxation is worse than a known feasible solution then it is pruned; otherwise, one branches on that node. The proposed Lagrangean relaxation has two purposes. The first one is to provide a stronger lower bound than the BM relaxation, which will occur in some nodes. The second one is to provide a value for the binary variables. The discrete decisions are then fixed at this value, and a corresponding LP is evaluated (e.g. the third LP evaluation at each node). Note that the Lagrangean relaxation requires the use of Lagrangean multipliers. These are calculated at the root node by solving the continuous relaxation of the HR, and can be updated later in the search tree. From a general point of view, the Lagrangean relaxation is used as a heuristic to find feasible solutions at every node. Considering that it only requires the solution of two "simple" LPs, and that the Lagrangean relaxation is a good approximation of the linear GDP, the method is very efficient.

We illustrate the application of this algorithm with several linear GDP problems, including strip packing problems and randomly generated instances. The results show that the proposed method performs much better than the traditional disjunctive branch and bound (either BM or HR). The method requires fewer nodes to find feasible solutions and the optimal solution. Furthermore, in most cases the search tree requires the evaluation of considerably fewer nodes. Compared to the HR disjunctive branch and bound, the method also requires less time per node.

References:

[1] Raman, Ramesh, and Ignacio E. Grossmann. "*Modelling and computational techniques for logic based integer programming*", Computers & Chemical Engineering. 1994; 18(7), 563-578.

[2] Beaumont, Nicholas. "*An algorithm for disjunctive programs*", European Journal of Operational Research.1990; 48(3), 362-371.

[3] Türkay, Metin, and Ignacio E. Grossmann. "*Logic-based MINLP algorithms for the optimal synthesis of process networks*", Computers & Chemical Engineering. 1996; 20(8), 959-978.

415525 Interior Point Regularization Strategies for Mathematical Programs with Equilibrium Constraints

Abstract:

Mathematical Programs with equilibrium constraints (MPEC) are known as difficult optimization problems because of the violation of constraint qualifications (CQ) such as Mangasarian-Fromovitz constraint qualification (MFCQ). Without a CQ, the optimum may not be a KKT point. So MPECs are challenging for most NLP solvers, which are based on the KKT conditions. However, MPECs are widely used for some classes of disjunctions in a model. In process optimization, complementarities allow for models containing disappearance of phases, flow reversal, safety valve options, and some other discrete events.

To solve MPECs with NLP solvers, one approach is to add a parameter ε to relax the complimentary constraints and make MFCQ hold in the feasible points. By solving a series of relaxed NLP problems, the optimal solution is approached as ε goes to zero. Alternatively, we can move the complementary constraints to a penalty term in the objective function. However, the relaxation approach slows down the solving process and the penalty approach has difficulty in determining the penalty terms.

As problem size and degrees of freedoms increase for these relaxed NLPs, active set methods become less efficient due to their combinatorial complexity of identifying the active constraints. Alternatively, interior point methods add a logarithmic barrier function to the objective function for the inequality constraints, resulting in cheaper iterations with Newton steps. As one of the most popular interior point NLP solvers, IPOPT is widely used because of its high performance. However, interior point methods require certain regularity conditions for the solution path to exist. Unfortunately, MPECs do not satisfy these conditions and even relaxed NLPs may be difficult to solve with interior points methods.

In this presentation, we present a new regularization algorithm for IPOPT. The new regularization algorithm uses concepts from active set methods, and identifies an independent subset of active constraints. For interior point methods, this is achieved by adding big-M terms in dependent rows that essentially eliminate dependent constraints. This results in more accurate Newton steps and faster convergence to a solution.

When IPOPT can't get the sufficient improvement in the line search, it switches to a restoration phase to improve the infeasibility, but often leads to slow convergence. Here we introduce a new penalty restoration phase, as an alternative of the restoration phase in current version IPOPT, by adding the objective function as a penalty term. A nice convergent property is shown inside the restoration phase and the algorithm works well for degenerate or infeasible problem sets. 167 MPEC problems are tested with the proposed algorithm and the improvements over current version of IPOPT are demonstrated.

Algorithmic Innovations and Software for Dual Decomposition of Stochastic Mixed Integer Programming

Kibaek Kim
Mathematics and Computer Science Division
Argonne National Laboratory

Joint work with: Victor M. Zavala

AIChE 2015 Annual Meeting

Stochastic Mixed Integer Programming (SMIP)

- General formulation of SMIP:

$$z = \min_x \left\{ c^T x + \mathbb{E}_\xi[Q(x,\xi)] : Ax = b, x \in X \right\}$$

  - Recourse function of an integer program:

$$Q(x,\xi) := \min_y \left\{ q(\xi)^T y : W(\xi) y = h(\xi) - T(\xi) x,\ y \in Y \right\}$$

  - X and Y represent integer restrictions on a subset of the decision variables x and y, respectively.

- Extensive form of SMIP:

$$\begin{aligned} z = \min_{x_s, y_s} \quad & \sum_{s \in \mathcal{S}} p_s \left(c^T x_s + q_s^T y_s \right) \\ \text{s.t.} \quad & (x_s, y_s) \in G_s,\ s \in \mathcal{S}, \\ & \sum_{s \in \mathcal{S}} H_s x_s = 0 \quad (\lambda) \end{aligned}$$

Non-anticipaticity constraints

$$G_s := \left\{ (x_s, y_s) : Ax_s = b,\ T_s x_s + W_s y_s = h_s,\ x_s \in X,\ y_s \in Y \right\}$$

Technical Challenges of SMIP

- Evaluating the scenario number of recourse functions for a given first-stage decision x:
  - (Infinitely) many first-stage decisions x
  - Often requires many scenarios for statistical significance
  - Solving many (NP-hard) mixed integer programs
- Non-convexity and discontinuity:
  - Difficult to find a global optimal solution
  - Does not fit in traditional Benders decomposition framework
- Few solution tools available
  - PIPS [1] (Argonne): continuous
  - PySP [2] (Sandia): progressive hedging for SMIP
  - ddsip [3]: dual decomposition method for SMIP
  - LINDO and FortSP (OptiRisk): Benders' decomposition

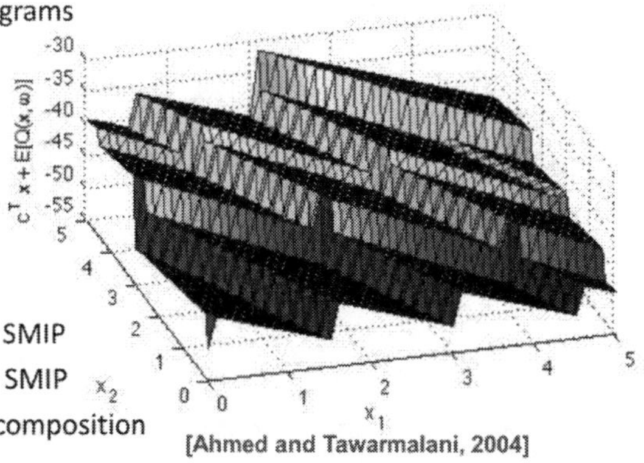

[Ahmed and Tawarmalani, 2004]

1. Petra, C.G., Anitescu, M.: A preconditioning technique for schur complement systems arising in stochastic optimization. Computational Optimization and Applications 52(2), 315–344 (2012)
2. Watson, J.P., Woodruff, D.L., Hart, W.E.: PySP: modeling and solving stochastic pro- grams in python. Mathematical Programming Computation 4(2), 109–149 (2012)
3. Märkert,A., Gollmer, R.: Users Guide to ddsip–A C Package for the Dual Decomposition of Two-Stage Stochastic Programs with Mixed-Integer Recourse (2014)

State-of-the-Art Methods for SMIP

- Benders decomposition (a.k.a. L-shaped method) [1,2]
  - Assumption: continuous recourse
    - i.e., need to relax the second-stage integrality
  - Convexification techniques for certain problem classes
  - Specialized branching technique
- Dual decomposition [3]
  - Assumption: complete recourse (e.g., not true in unit commitment)
    - Difficult to recover feasible solutions
    - Difficult to obtain upper bounds (so, don't know duality gap)
  - Can find a good lower bound
- Progressive hedging [4]
  - No theoretical convergence
  - May find an upper bound with heuristic

1. Benders, Jacques F. "Partitioning procedures for solving mixed-variables programming problems." *Numerische mathematik* 4.1 (1962): 238-252.
2. Van Slyke, Richard M., and Roger Wets. "L-shaped linear programs with applications to optimal control and stochastic programming." *SIAM Journal on Applied Mathematics* 17.4 (1969): 638-663.
3. CarøE, Claus C., and Rüdiger Schultz. "Dual decomposition in stochastic integer programming." *Operations Research Letters* 24.1 (1999): 37-45.
4. Rockafellar, R. Tyrrell, and Roger J-B. Wets. "Scenarios and policy aggregation in optimization under uncertainty." *Mathematics of operations research* 16.1 (1991): 119-147.

Dual Decomposition

- Lagrangian relaxation of *non-anticipativity* constraints:

$$D(\lambda) := \min_{x_s, y_s} \left\{ \sum_{s \in \mathcal{S}} L_s(x_s, y_s, \lambda) : (x_s, y_x) \in G_s,\ s \in \mathcal{S} \right\}$$

$$L_s(x_s, y_s, \lambda) = p_s \left(c^T x_s + q_s^T y_s \right) + \lambda^T (H_s x_s)$$

- Scenario *decomposition*:

$$D(\lambda) = \sum_{s \in \mathcal{S}} D_s(\lambda) \qquad D_s(\lambda) := \min_{x_s, y_s} \{ L_s(x_s, y_s, \lambda) : (x_s, y_x) \in G_s \}$$

- *Lower bound* of SMIP for any given λ
- Seek to obtain the best lower bound by solving:

$$z_{\text{UB}} \geq z \geq z_{\text{LD}} := \max_{\lambda} \sum_{s \in \mathcal{S}} D_s(\lambda)$$

5

Dual Variable Update 1: Subgradient Method

- Popular in non-differentiable optimization
- The dual variable is updated as:

$$\lambda^{k+1} = \lambda^k - \alpha^k \sum_{s \in \mathcal{S}} H_s x_s^k$$ *Subgradient of $D(\lambda)$ at λ^k*

- Different step-size rules are available:
  - Constant step size: $\alpha_k = h$
  - Constant step length: $\alpha_k = h / \| \sum H_s x_s^k \|_2$
  - Square summable but not summable

    $$\sum_{k=1}^{\infty} \alpha_k^2 < \infty, \ \sum_{k=1}^{\infty} \alpha_k = \infty$$

  - Practical approach [1]

    $$\alpha_k := \beta_k \frac{z_{\text{UB}} - D(\lambda^k)}{\| \sum_{s \in \mathcal{S}} H_s x_s^k \|_2^2}$$

  Determined by convergent behavior of step-size

- ***Cannot prove optimality***

1. Fisher, Marshall L. "The Lagrangian relaxation method for solving integer programming problems." *Management science* 50.12_supplement (2004): 1861-1871.

6

Dual Variable Update 2: Cutting-Plane Method

- Outer approximation to solve the Lagrangian dual problem

$$\max_{\lambda} \sum_{s \in \mathcal{S}} D_s(\lambda)$$

- The master problem is given by

$$\begin{aligned} m_k = \max_{\theta_s, \lambda} \quad & \sum_{s \in \mathcal{S}} \theta_s \\ \text{s.t.} \quad & \theta_s \leq D_s(\lambda^l) + (H_s x_s^l)^T (\lambda - \lambda^l), \\ & s \in \mathcal{S},\ l = 0, 1, \ldots, k \end{aligned}$$

- **Key Property:** Piecewise linear concave in λ

$$\begin{aligned} D_s(\lambda) = \min_{x_s, y_s} \quad & p_s \left(c^T x_s + q_s^T y_s\right) + \lambda^T \left(H_s x_s\right) \\ \text{s.t.} \quad & (x_s, y_s) \in \text{conv.hull}(G_s) \end{aligned}$$

- *Finite convergence with optimality*
- *Stabilization issue*

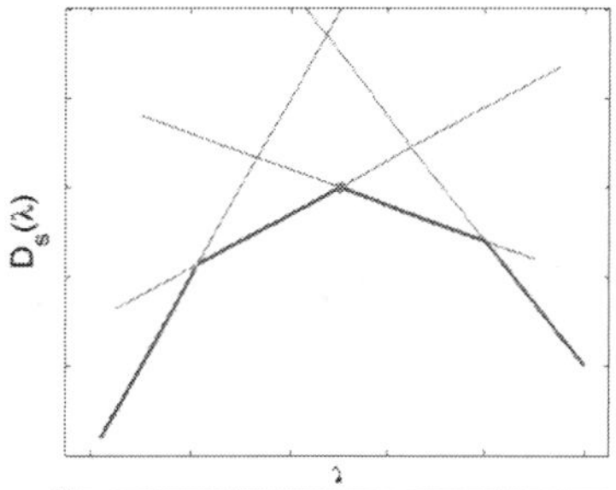

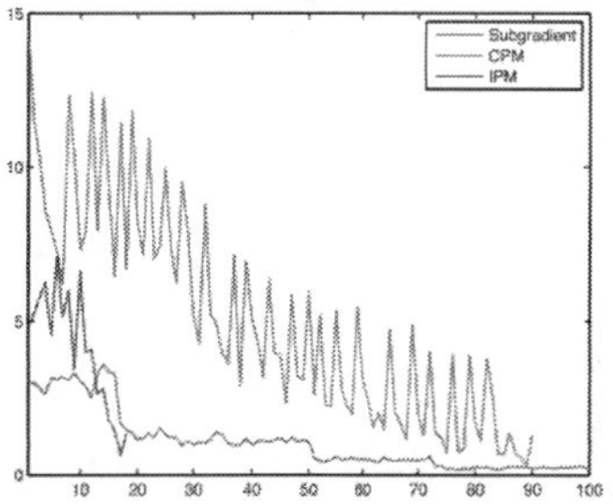

7

In This Talk,

- Difficult to recover feasible solutions
  - No upper bound
  - No duality gap

- Stabilization issue in cutting-plane method
  - Finite, but slow convergence
  - Numerical instability

8

Algorithmic Innovation 1:
Tightening Inequalities for the Subproblems

- Case 1: A feasible subproblem solution that is infeasible with respect to SMIP.

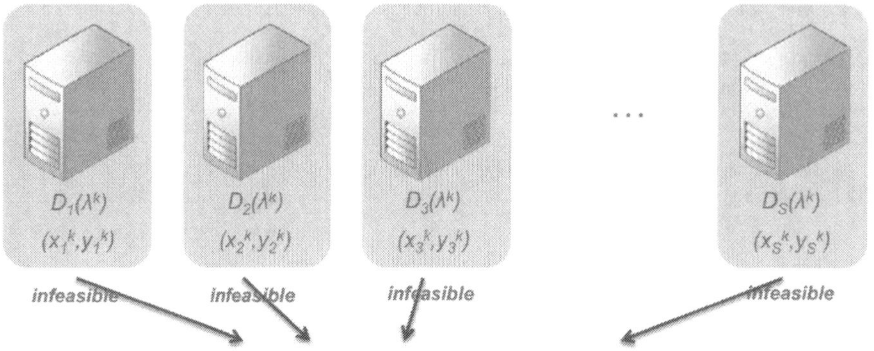

$$z_{\mathrm{UB}}(x) = c^T x + \min_{y_s} \sum_{s \in \mathcal{S}} p_s q_s^T y_s$$

NOTE: can also be decomposable in s

$$\text{s.t.} \quad W_s y_s = h_s - T_s x, \ s \in \mathcal{S},$$
$$y_s \in Y, \ s \in \mathcal{S}$$

Algorithmic Innovation 1:
Tightening Inequalities for the Subproblems

Proposition. Let $(\hat{x}, \hat{y}) \in G_s$ for some $s \in \mathcal{S}$ and assume that there exists an scenario $s' \neq s$ such that there does not exist $y \in \mathbb{R}^{n_2}$ satisfying $(\hat{x}, y) \in G_{s'}$ for fixed $\hat{x}$. Let $\mu_{s'}$ be an optimal solution of the linear programming problem

$$\max_{\mu} \left\{ \mu^T (h_{s'} - T_{s'}\hat{x}) : \mu^T W_{s'} \leq 0, \ |\mu| \leq 1 \right\}, \tag{1}$$

where the absolute value $|\cdot|$ is taken componentwise. The inequality

$$\mu_{s'}^T (h_{s'} - T_{s'}x) \leq 0 \tag{2}$$

excludes $\hat{x}$ from the set $\{x : (x, y) \in G_{s'}\}$ and is also valid for SMIP.

- **IDEA:** Finding a separating hyperplane by Farkas' lemma to exclude the infeasible solution
- **Finiteness:** There are only a finite number of optimal bases to the problem.

Algorithmic Innovation 1: Tightening Inequalities for the Subproblems

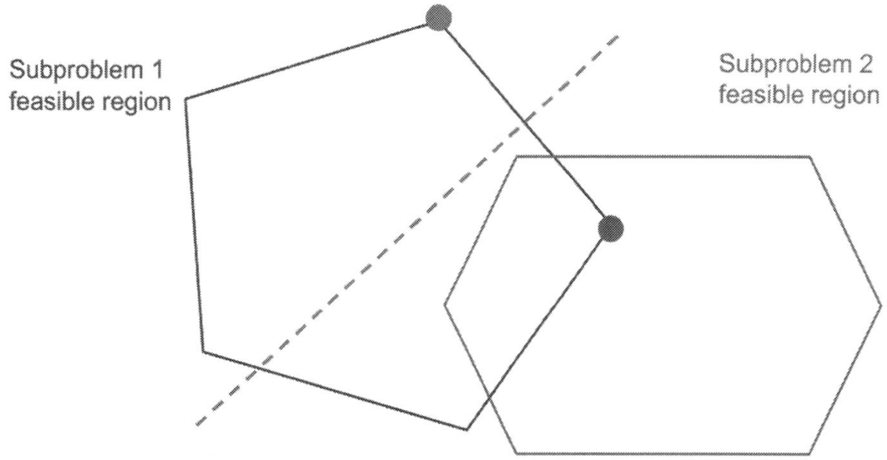

11

Algorithmic Innovation 1: Tightening Inequalities for the Subproblems

- Case 2: A feasible subproblem solution that is also feasible with respect to SMIP.

Proposition. Assume that there exists $y \in \mathbb{R}^{n_2}$ such that $(\hat{x}, y) \in G_{s'}$ for all $s' \in \mathcal{S}$ and for fixed $\hat{x}$. Let π_s be the optimal solution of the following recourse problem for each $s \in \mathcal{S}$:

$$\max_{\pi} \left\{ \pi^T (h_s - T_s\hat{x}) : \pi^T W_s \leq q_s \right\}. \tag{1}$$

The inequality

$$c^T x + \sum_{s \in \mathcal{S}} \pi_s^T (h_s - T_s x) \leq z_{\text{UB}} \tag{2}$$

is valid for SMIP.

- **IDEA:** Finding a supporting hyperplane that represents a lower bound of the objective function of SMIP
- *Parametric cut*: The inequality is parameterized by upper bound z_{UB}.
- **Finiteness:** There are only a finite number of optimal bases to *each* problem.

12

Algorithmic Innovation 1:
Tightening Inequalities for the Subproblems

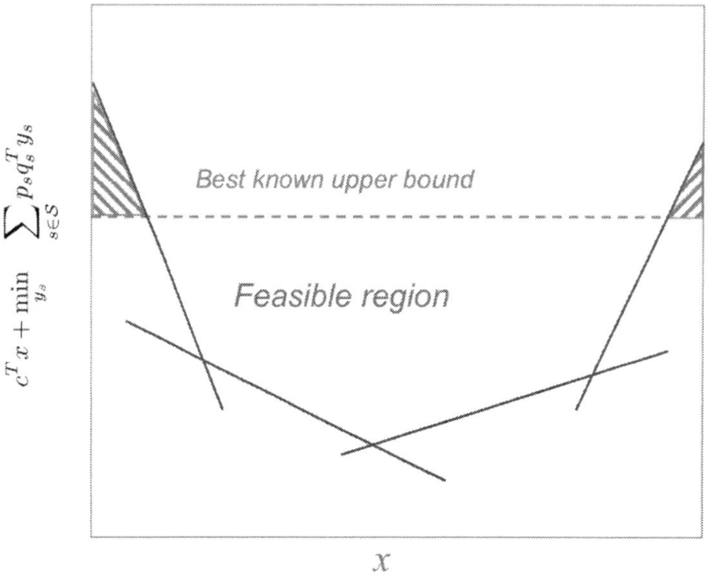

13

Algorithmic Innovation 2:
Interior-Point Cutting-Plane Method

- The cutting-plane solutions of the Master problem can
  - oscillate significantly;
  - suffer from degeneracy.

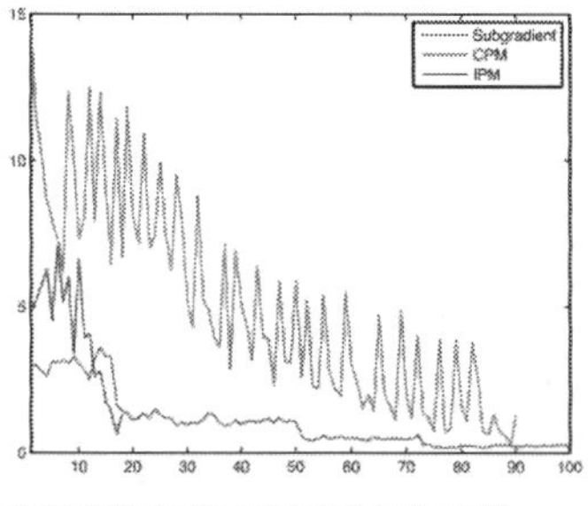

- Interior-Point Cutting-Plane Method
  - Solve the master by using interior-point method to find a feasible solution
    - as compared to an extreme optimal point
  - NOTE: The master has inequalities only.
- Prioritized Termination Criteria:
  1. $\sum_{s \in \mathcal{S}} \theta_s^k \geq z_{\text{UB}}$
  2. Duality gap $< \epsilon_{\text{IPM}}^k$

$$\epsilon_{\text{IPM}}^k := \min \left\{ \epsilon_{\text{IPM}}^{max}, \frac{g_{k-1}(\tilde{\theta}^{k-1}, \tilde{\pi}^{k-1})}{\delta} + \frac{\tilde{m}_{k-1} - \sum_{s \in \mathcal{S}} D_s(\tilde{\lambda}^{k-1})}{1 + |\tilde{m}_{k-1}|} \right\},$$

where $\tilde{m}_{k-1} := \sum_{s \in \mathcal{S}} \tilde{\theta}_s^{k-1}$ and $\delta > 1$ is the degree of optimality.

14

Algorithmic Innovation 2: Stabilization of the Master Problem

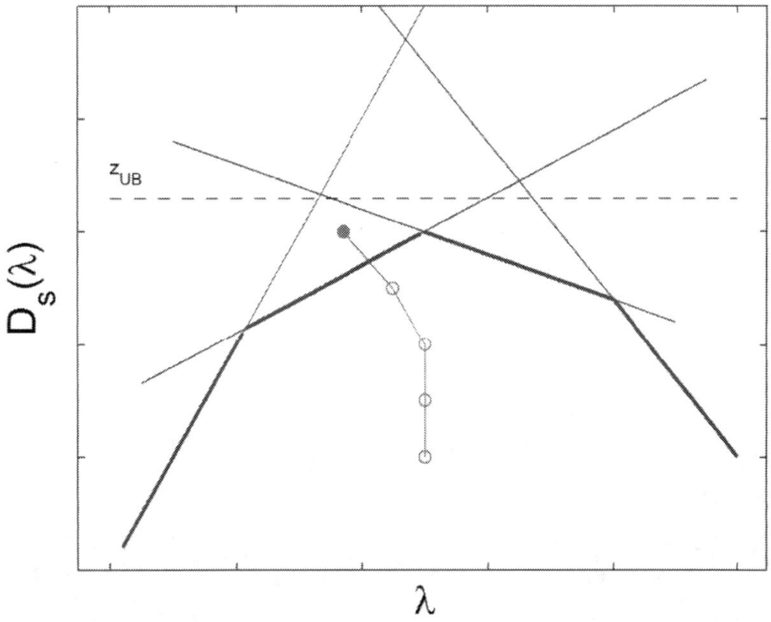

DSP: Decomposition methods for Structured Programming

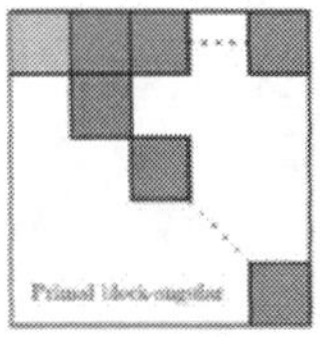

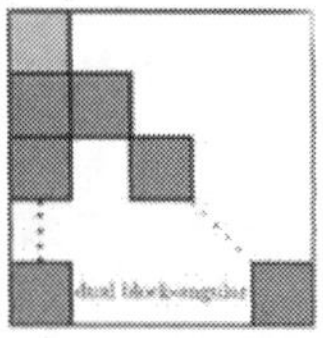

- ***Object-oriented open-source software package***
  - ***Decomposition*** methods for ***Structured Programming***
  - ***Parallel*** computing
  - Currently implemented for Stochastic Programs
- *JuMP.jl* package [1]
  - Algebraic modeling language for mathematical programming (LP, MIP, SOCP and NLP).
  - As fast as AMPL, ***faster*** than GAMS and other open-source languages [1]
  - Currently supports many open-source solvers and commercial solvers
- *StochJuMP.jl* package [2]
  - Extension of *JuMP* package for stochastic programming
  - Supports distributed memory computing via **MPI** communication – *Scalable!*
  - Users: ***PIPS*** and ***DSP***

1. **Computing in operations research using Julia**. Miles Lubin, Iain Dunning (2015), *INFORMS Journal on Computing*, 27:238–248, doi:10.1287/ijoc.2014.0623, http://dx.doi.org/10.1287/ijoc.2014.0623
2. **Parallel algebraic modeling for stochastic optimization**. Joey Huchette, Miles Lubin, Cosmin Petra (2014), *HPTCDL'14 Proceedings of the 1st Workshop on High Performance Technical Computing in Dynamic Languages*, 29–35, doi:10.1109/HPTCDL.2014.6

Example: *DSP-StochJuMP*

$$\min \left\{ -1.5x_1 - 4x_2 + \sum_{s=1}^{3} p_s Q(x_1, x_2, \xi_1^s, \xi_2^s) : x_1, x_2 \in \{0, \ldots, 5\} \right\},$$

where

$$\begin{aligned} Q(x_1, x_2, \xi_1^s, \xi_2^s) = \min_{y_1, y_2, y_3, y_4} \quad & -16y_1 + 19y_2 + 23y_3 + 28y_4 \\ \text{s.t.} \quad & 2y_1 + 3y_2 + 4y_3 + 5y_4 \le \xi_1^s - x_1 \\ & 6y_1 + y_2 + 3y_3 + 2y_4 \le \xi_2^s - x_2 \\ & y_1, y_2, y_3, y_4 \in \{0, 1\} \end{aligned}$$

and $(\xi_1^s, \xi_2^s) \in \{(7,7), (11,11), (13,13)\}$ with probability $1/3$.

Only 16 lines of Julia script!

```
using DSPsolver, StochJuMP, MPI; # Load packages
MPI.Init(); # Initialize MPI
m = StochasticModel(3); # Create a Model object with three scenarios
xi = [[7,7] [11,11] [13,13]]; # random parameter
@defVar(m, 0 <= x[i=1:2] <= 5, Int);
@setObjective(m, Min, -1.5*x[1]-4*x[2]);
for s in 1:3
  q = StochasticBlock(m, 1/3);
  @defVal(q, y[j=1:4], Bin);
  @setObjective(q, Min, -16*y[1]+19*y[2]+23*y[3]+28*y[4]);
  @addConstraint(q, 2*y[1]+3*y[2]+4*y[3]+5*y[4]<=xi[1,s]-x[1]);
  @addConstraint(q, 6*y[1]+1*y[2]+3*y[3]+2*y[4]<=xi[2,s]-x[2]);
end
DSPsolver.loadProblem(m); # Load model m to DSP
DSPsolver.solve(DSP_SOLVER_DD); # Solve problem using dual decomposition
MPI.Finalize(); # Finalize MPI
```


Computational Experiment I: SIPLIB Test Instances

- **SIPLIB:** publically available test library for stochastic integer programming
  - http://www2.isye.gatech.edu/~sahmed/siplib/
  - ***DCAP***: *complete recourse,* first-stage **mixed**-integer and second-stage **pure** integer
    - Number of scenarios: 200, 300 and 500
  - ***SSLP***: *complete recourse,* first-stage **pure** integer and second-stage **mixed**-integer
    - Number of scenarios: 5, 10, 15, 50 and 100

Name	# Rows	# Columns	# Integers
dcap233_200	3006	5412	5406
dcap233_300	4506	8112	8106
dcap233_500	7506	13512	13506
dcap243_200	3606	7212	7206
dcap243_300	5406	10812	10806
dcap243_500	9006	18012	18006
dcap332_200	2406	4812	4806
dcap332_300	3606	7212	7206
dcap332_500	6006	12012	12006
dcap342_200	2806	6412	6406
dcap342_300	4206	9612	9606
dcap342_500	7006	16012	16006

Name	# Rows	# Columns	# Integers
sslp_5_25_50	1501	6505	6255
sslp_5_25_100	3001	13005	12505
sslp_15_45_5	301	3465	3390
sslp_15_45_10	601	6915	6765
sslp_15_45_15	901	10365	10135
sslp_10_50_50	3001	25510	25010
sslp_10_50_100	6001	51010	50010

- Will compare *Subgradient*, *Cutting-Plane*, and *DSP*

Computational Experiment 1: SIPLIB Benchmark - Optimality Gap (%)

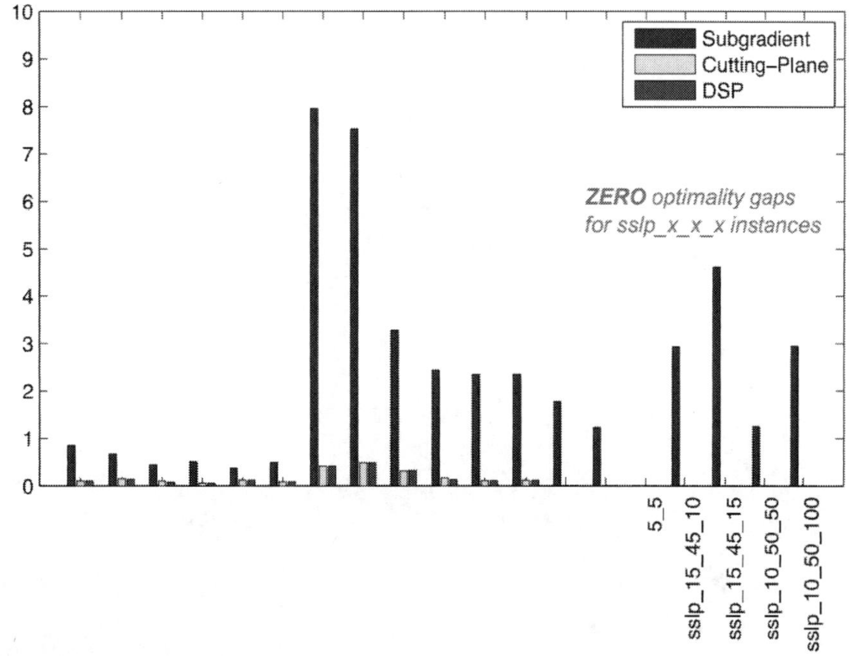

19

Computational Experiment I: SIPLIB Benchmark - # of Iterations

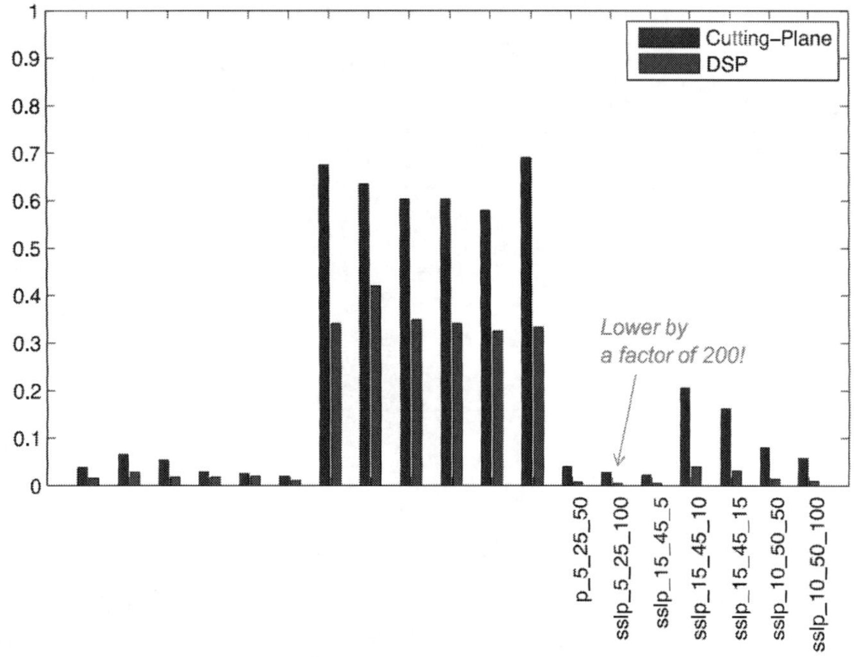

20

Computational Experiment I: SIPLIB Benchmark - Solution Time

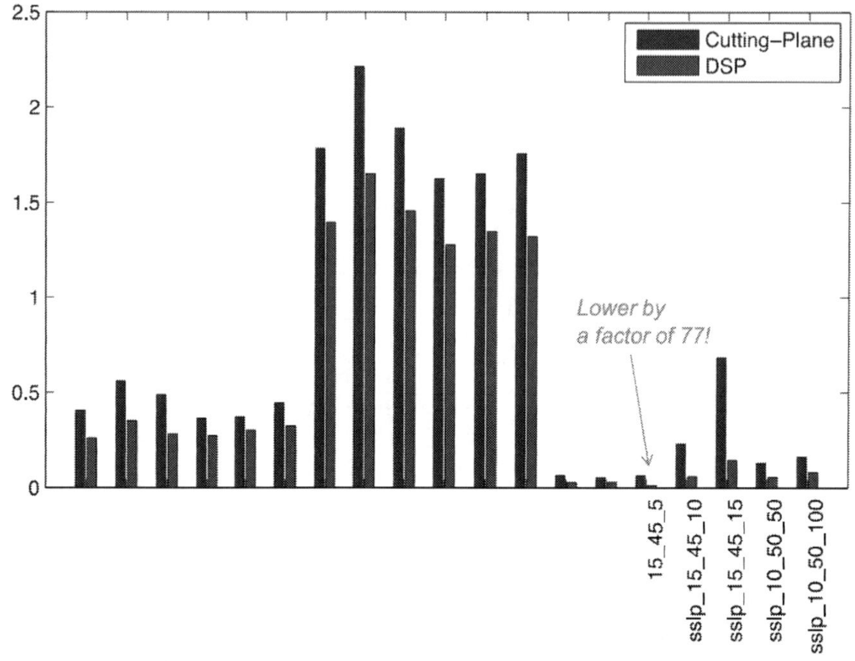

Computational Experiment I: SIPLIB (sslp_15_45_15) - Solution Progress

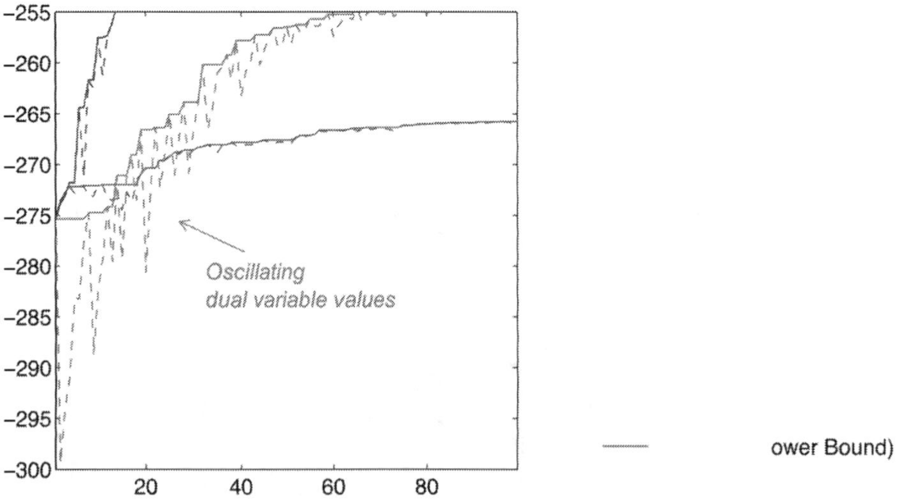

Computational Experiment: Stochastic Unit Commitment

- Unit Commitment with Uncertain Wind Power Generation [1]
  - IEEE 118-bus system (54 generators and 186 transmission lines)
  - 17 generators have quick-start capabilities on demand.
  - Demand load: 3,095 MW on average with a peak of 3,733 MW
  - 3 identical wind farms
  - Wind power: 494 MW on average with a peak of 916 MW (from 64 scenarios)
  - Constraints:
    - Network flow constraints
    - Transmission line constraints
    - Ramping constraints
    - Minimum up/down constraints
    - Spinning reserve constraints

Scenarios	# Rows	# Columns	# Integers
4	120,015	38,880	2,592
8	229,303	75,168	4,320
16	447,879	147,744	7,776
32	885,031	292,896	14,688
64	1,759,335	583,200	28,512

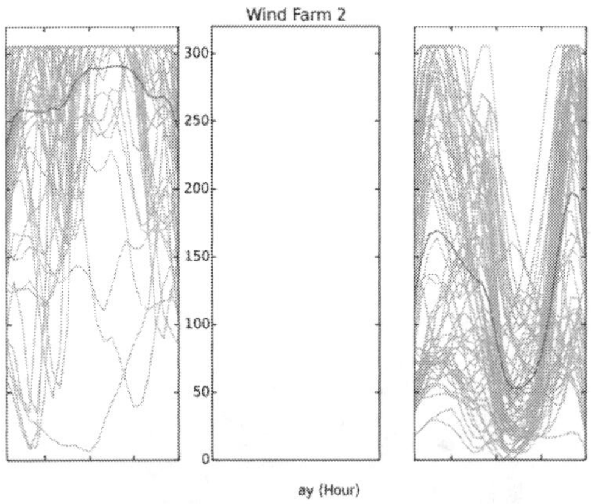

[Scenarios generated from Illinois data by Bessac]

1. Lee, C., Liu, C., Mehrotra, S., Shahidehpour, M. (2014). Modeling Transmission Line Constraints in Two-Stage Robust Unit Commitment Problem. IEEE Transactions on Power Systems, 29(3), 1221–1231. doi:10.1109/TPWRS.2013.2291498

Computational Experiment: Stochastic Unit Commitment

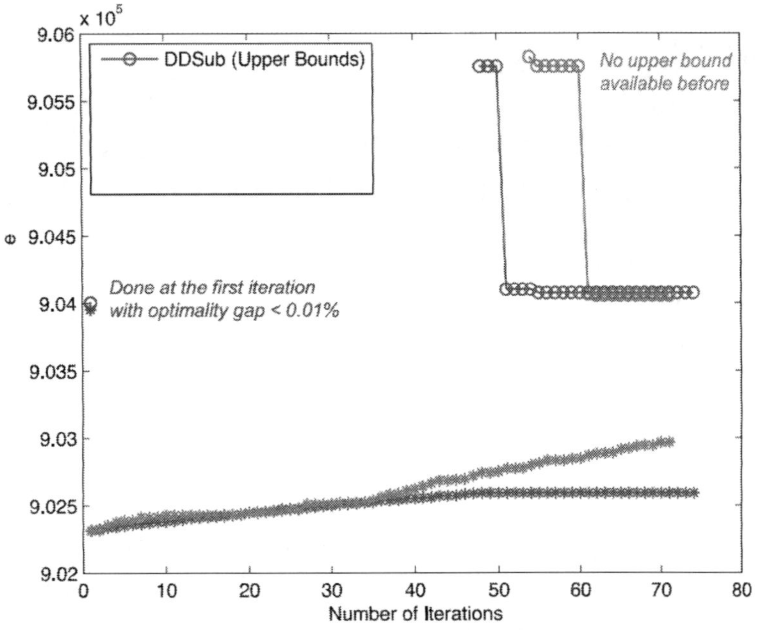

Computational Experiment 2: Stochastic Unit Commitment

Numerical results from solving Extensive forms with *SCIP-3.1.1*

Scenarios	Branch-and-Cut Nodes	Upper Bound	Lower Bound	Gap (%)	Time (sec.)
4	88831	907035.3	906089.9	0.01	6632
8	58235	904068.1	903567.8	0.05	> 21600
16	3505	900806.1	900200.3	0.07	> 21600
32	9	907536.0	901759.8	0.64	> 21600
64	1	∞	33605.4	∞	> 21600

Numerical results from *DSP*

Scenarios	Iterations	Upper Bound	Lower Bound	Gap (%)	Time/Iter. (sec.)
4	1	907046.1	906979.1	< 0.01	551
8	1	904006.6	903953.5	< 0.01	667
16	1	900706.3	900650.7	< 0.01	764
32	19	903227.7	903149.9	< 0.01	390
64	16	895118.0	895044.6	< 0.01	895

Does NOT increase in number of scenarios

Scale in number of scenarios!

25

Thank You

- DSP software package is available from
  - Website: http://www.mcs.anl.gov/research/projects/DSP/
  - Github: https://github.com/Argonne-National-Laboratory/DSP
  - Including *StochJuMP* script for stochastic unit commitment
- Research paper is available from
  - http://www.optimization-online.org/DB_FILE/2015/06/4960.pdf

26

423337 Solving General Mixed Integer Bilevel Linear Programs (MIBLP): A Decomposition Algorithm

Abstract:

In spite of the fact that a wide range of applications fit the bilevel programming framework, real-world implementations of bilevel programming are still relatively scarce. This is mainly due to the lack of efficient algorithms for solving bilevel programming problems. The complexity of bilevel optimization lies in the property that the constraint region of the upper-level program is partially determined by the solutions to another optimization problem. Therefore, bilevel optimization problems are typically formulated as a lower-level program embedded in an upper-level program. In this work, we propose a novel algorithm for solving mixed-integer bilevel linear programming (MIBLP) problems, which is a class of bilevel programming problems. MIBLP is intrinsically difficult to solve because 1) the bilevel feasible region can be non-convex and disconnected; 2) the supremum (infimum) can be unattainable; and 3) removing the integrality constraints does not necessarily provide a valid relaxation of the original MIBLP problem . Relevant literature also includes studies on mixed-integer bilevel nonlinear programming (MIBNLP) problems.

To the best of our knowledge, existing algorithms for MIBLPs are either subject to several simplifying assumptions or designated to special classes of MIBLP problems. The algorithm proposed in this work is the first practical algorithm that deals with the most general form of MIBLPs. The development of our algorithm involves two major steps: reformulation and decomposition. An iterative solution procedure is proposed that converges to the global optimal solution in finite iterations. Detailed solution procedures are presented for solving two classical literature examples. We propose 50 randomly-generated numerical instances. The computational results show that the proposed algorithm can MIBLP problems within reasonable computational times with up to 100 upper-level continuous variables, 100 upper-level binary variables, 100 lower-level continuous variables, 100 lower-level binary variables, 80 upper-level constraints, and 80 lower-level constraints. To demonstrate a real-world application, we further apply the algorithm to a hierarchical supply chain planning problem, formulated as an MIBLP. A total of 35 case instances are proposed with varying numbers of plants and products. Computational results show that the proposed algorithm can solve these MIBLP problems within reasonable computational times with up to 12 upper-level continuous variables, 12 upper-level binary variables, 144 lower-level continuous variables, 144 lower-level binary variables, 13 upper-level constraints, and 180 lower-level constraints.

432655 A Branch and Bound Algorithm for Discrete Bilevel Optimization

Abstract:

Bilevel optimization problems arise in situations where two independent decision-makers choose their actions sequentially with the goal of optimizing their own objective functions. Bilevel optimization problems are known to yield solutions corresponding to Stackelberg equilibria of extensive games with perfect information. The main characteristic of these games is that both players, leader and follower, are considered rational decision-makers that can predict and observe the actions of the opponent. In particular, the leader is aware the decision criterion and the constraints of the follower when choosing its actions, and the follower observes the actions of the leader before making its decisions.

A bilevel optimization problem can be described as a mathematical program with an optimization problem in the constraints. In general, bilevel optimization problems are very challenging because their feasible region is noncovex; however, for the case in which the lower-level problem is a linear program (LP), some reformulation methods have proved to be successful to solve them. These methods replace the lower-level problem with a set of constraints that only allows optimal responses of the follower to be feasible in the reformulation. In particular, two equivalent reformulation methods have shown to be effective: KKT reformulation and duality-based reformulation.

The reformulation methods used for bilevel optimization with LP lower level rely on their necessary and sufficient conditions for optimality. However, for the case in which the lower-level problem contains discrete variables, these reformulations are no longer valid. Some search procedures have been proposed to solve bilevel optimization problems with discrete variables in the lower level, but most of them are based on implicit enumeration schemes that are only effective for small-scale problems.

This research proposes a solution algorithm that explores the decision space of the leader using branch and bound, and identifies regions of the leader's feasible region that imply the same reaction of the follower. Using this method, large regions of the leader's decision space can be characterize by solving few Mixed-Integer Linear Programs (MILPs). The branch and bound scheme is efficient because at each node defining a subregion of the leader's decision space, an upper and lower bound can be found. These bounds can be used to fathom subregions that do not contain the bilevel optimal solution before they are completely characterized.

The branch and bound algorithm is used to solve two small examples that illustrate how the feasible regions are explored and the bilevel optimal solutions are identified. Additionally, the algorithm is used to solve randomly generated instances of different sizes. The results show that the proposed branch and bound scheme solves small and middle-sized instances in reasonable computational times.

424392 The Exact Solution of Multiparametric Mixed-Integer Quadratic Programming Problems

Abstract:

In recent years, multiparametric programming in general and multiparametric mixed-integer quadratic programming (mp-MIQP) in particular has received a growing interest due to its applicability in areas such as explicit optimal control and reactive scheduling [1]. In general, mp-MIQP problems consist of a quadratic objective function $z(\theta)$ subject to linear constraints and a polytopic parameter space. The corresponding solution is a partitioning of the feasible parameter space into a number of so-called critical regions CR_i, each of which is associated with the corresponding optimal affine solution $x_i(\theta)$, the optimal combination of binary variables y_i and quadratic objective function $z_i(\theta)$.

However, obtaining this solution is very challenging due to two key difficulties:

- The presence of the binary variables: This has been approached via (a) global optimization [2], (b) branch-and-bound algorithms [3, 4] and (c) exhaustive enumeration [5].

- The comparison of the parametric profiles for each feasible combination of binary variables: Such a comparison procedure is necessary in order to obtain a single, optimal parametric profile for the solution of the mp-MIQP. Due to the quadratic nature of the objective function $z(\theta)$, the transition from the one solution being optimal to another might be quadratic, resulting in non-polytopic partitions. Earlier approaches have avoided the formation of such regions by assigning multiple solutions $x_{i,j}(\theta), y_{i,j}, z_{i,j}(\theta)$ to one critical region CR_i, which is referred to as an envelope of solutions. The presence such envelopes requires the online comparison of all solutions assigned to a specific critical region, a procedure which conceptually undermines the idea of explicit MPC as it introduces online computational burden.

In this work, we present a novel mp-MIQP algorithm which circumvents the generation of envelopes of solutions by enabling the seamless handling of non-polytopic critical regions [6]. This strategy can be decomposed into a series of steps:

1. Determination of a candidate combination of binary variables considering the non-polytopic critical region CR.

2. Calculation of a polytope Ξ such that CR Ξ via McCormick relaxations [7].

3. The candidate combination of binary variables is fixed in the original mp-MIQP problem and the resulting multiparametric quadratic programming (mp-QP) problem is solved for θ Ξ using existing solvers [8].

4. Given the problem is feasible, the resulting, polytopic parametric profile is compared with the parametric profile of the upper bound. This comparison can be decomposed into two steps:

a. Determination whether one of the solutions is optimal over the entire range of the critical region considered

b. If this is not the case, partition the considered critical region using the difference of the optimal objective function of the mp-QP subproblem in the critical region considered and the objective function of the current best upper bound. Due to the quadratic nature of these functions this might result in quadratically constrained critical region.

5. As the comparison procedure was performed in Ξ, it is necessary to perform a post-processing step by re-introducing the quadratic constraints of CR. As this might result in empty regions, a quadratically constrained quadratic programming (QCQP) feasibility problem is solved for each newly formed region.

6. The next iteration of the algorithm begins.

This newly developed approach towards the solution of mp-MIQP problems is applied to several numerical examples, where its computational performance is compared against existing approaches. In particular, it is shown that despite the necessity of solving a QCQP feasibility problem, the computational effort remains manageable as each critical region is only associated with one objective function value.

Additionally, we apply the proposed procedure to explicit hybrid control problems, which can be formulated as mixed-integer quadratic programming (MIQP) problems [9]. In particular, we consider the well-known hybrid control problem posed in [9], and show how the proposed procedure yields the exact partitioning of the state-space, hence reducing the online computational effort for this system to a point location and a function evaluations.

References

1. Pistikopoulos, E.N., et al., *PAROC - an Integrated Framework and Software Platform for the Optimization and Advanced Model-Based Control of Process Systems.* Chemical Engineering Science, 2015, **in print**.

2. Dua, V., N.A. Bozinis, and E.N. Pistikopoulos, *A multiparametric programming approach for mixed-integer quadratic engineering problems.* Computers & Chemical Engineering, 2002. **26**(4–5): p. 715–733.

3. Oberdieck, R., M. Wittmann-Hohlbein, and E.N. Pistikopoulos, *A branch and bound method for the solution of multiparametric mixed integer linear programming problems.* Journal of Global Optimization, 2014. **59**(2-3): p. 527–543.

4. Axehill, D., et al., *A parametric branch and bound approach to suboptimal explicit hybrid MPC.* Automatica, 2014. **50**(1): p. 240–246.

5. Borrelli, F., *Constrained optimal control of linear and hybrid systems.* Lecture Notes in Control and Information Sciences. 2003, Berlin; New York: Springer. 1 online resource (xvii, 206.

6. Oberdieck, R. and E.N. Pistikopoulos, *Explicit Hybrid Model-Predictive Control: The exact solution.* Automatica, 2015. **Accepted for publication**.

7. McCormick, G.P., *Computability of global solutions to factorable nonconvex programs: Part I — Convex underestimating problems.* Mathematical Programming, 1976. **10**(1): p. 147–175.

8. ParOs, *Parametric Optimization Programming (POP)*. 2004, ParOS.

9. Bemporad, A. and M. Morari, *Control of systems integrating logic, dynamics, and constraints.* Automatica, 1999. **35**(3): p. 407–427.

428779 Robust Optimization Under Correlated Uncertainty within a Single Constraint and Across Multiple Constraints

Abstract:

In deterministic optimization problems, it is assumed that all the model parameters are known and take certain nominal values. However, many realistic parameters are uncertain and subject to random distributions. Assuming deterministic or nominal values for those uncertain parameters will lead to decisions that are infeasible or suboptimal for practical implementation. Thus, it is important to incorporate parameter uncertainty into the optimization problem. Various approaches such as robust optimization, stochastic programming with recourse and chance constraint have been proposed in the past to address uncertainty in optimization problems. Among them, robust optimization deals with the parameter uncertainty by introducing an uncertainty set which covers part or the whole region of the uncertainty space. The target is to select the best solution that is feasible for any realizations of the uncertain parameters in the uncertainty set [1].

The uncertainty set induced robust optimization framework has been extensively studied in literature and applied to various decision making problems. However, existing robust optimization methods generally assume that the uncertainties in the parameters are independent [2-4]. In practice, the uncertainties may be correlated with each other. For instance, the price and demand of crude oil are correlated. As a result, the traditional robust optimization methods that ignore the correlation may lead to a solution that is too conservative. In this work, we present novel results on robust optimization under correlated uncertainties that is either within a single constraint or across multiple constraints.

First, we propose the robust optimization framework for correlated uncertainty within a single constraint. Robust counterpart optimization formulations are derived for five different types of uncertain set. Specifically, box, polyhedral and ellipsoidal type of uncertainty set induced robust optimization formulations are derived for unbounded and correlated uncertainties within a constraint, and "interval+polyhedral" and "interval+ellipsoidal" set induced robust optimization formulations are derived for bounded uncertainties and correlated uncertainties within a constraint.

Second, we study the robust optimization method for correlated uncertainty in multiple constraints. Robust optimization formulations are derived based on the aforementioned results for correlated uncertainty in a single constraint and the recent work on joint chance constrained optimization [5]. Chance constraint models the uncertainty in an optimization problem by enforcing the probability of constraint satisfaction under parameter uncertainty. As an important application of robust optimization, chance constrained optimization problem can be solved with robust optimization approximation. Since joint chance constraint enforces several constraints to be satisfied simultaneously under uncertainty, it is very complicated and only very few works have been reported in the literature (e.g. [6]). In this work, we derive the robust optimization formulations for joint chance constraint with correlated uncertainty across multiple constraints.

Finally, numerical examples are studied to illustrate the proposed robust optimization framework for correlated uncertainty within a constraint and in multiple constraints. Different cases of the correlations are considered to demonstrate the effectiveness of the proposed robust optimization method under correlated uncertainty.

Reference:

[1] Ben-Tal, A., & Nemirovski, A. (1999). Robust solutions of uncertain linear programs. Operations research letters, 25(1), 1-13.

[2] Li, Z., Ding, R., & Floudas, C. A. (2011). A comparative theoretical and computational study on robust counterpart optimization: I. Robust linear optimization and robust mixed integer linear optimization. Industrial & engineering chemistry research, 50(18), 10567-10603.

[3] Li, Z., Tang, Q., & Floudas, C. A. (2012). A comparative theoretical and computational study on robust counterpart optimization: II. Probabilistic guarantees on constraint satisfaction. Industrial & engineering chemistry research, 51(19), 6769-6788.

[4] Li, Z., & Floudas, C. A. (2014). A Comparative Theoretical and Computational Study on Robust Counterpart Optimization: III. Improving the Quality of Robust Solutions. Industrial & Engineering Chemistry Research, 53(33), 13112-13124.

[5] Yuan, Y., Li, Z., & Huang, B. (2015). Robust Optimization Approximation for Joint Chance Constrained Optimization Problem. Journal of Global Optimization. Under review.

[6] Chen, W., Sim, M., Sun, J., & Teo, C. P. (2010). From CVaR to uncertainty set: Implications in joint chance-constrained optimization. Operations research, 58(2), 470-485.

420866 On the Role of Constraints in Economic Model Predictive Control

Abstract:

Economic model predictive control (EMPC) has recently attracted substantial attention within the chemical process control and energy systems communities because of its unique ability to control nonlinear processes/systems while reconciling process economic optimization and process control (e.g., [1]-[4]; see, also, the recent review [5] for comprehensive reference list of recent papers on EMPC). Unlike tracking model predictive control which uses a quadratic stage cost, EMPC uses a general, user-defined stage cost that reflects the process economics. Owing to the generality of the stage cost, achieving recursive feasibility of the EMPC optimization problem, closed-loop stability, and certain guarantees on closed-loop economic performance properties do not generally follow by applying EMPC unless additional conditions are satisfied (e.g., additional constraints are imposed on the EMPC optimization problem and/or a sufficiently long prediction horizon be used).

A brief overview of EMPC methods is provided with a particular focus on the role of constraints imposed in the EMPC optimization problem to address recursive feasibility, closed-loop stability, and closed-loop performance. Three main types of constraints are considered including terminal equality constraints, terminal region constraints, and constraints designed via Lyapunov-based techniques. An explanation of the motivation behind each type of constraint is provided. Additionally, the conditions that must be satisfied to yield the three closed-loop properties (feasibility, stability, and performance) are presented and explained. A well-known chemical engineering example, a non-isothermal CSTR with a second-order reaction, is used to illustrate effectiveness and theoretical performance properties of time-varying operation to improve closed-loop economic performance compared to steady-state operation and to demonstrate the impact of economically motivated constraints on optimal operation.

[1] Angeli D, Amrit R, Rawlings JB. On average performance and stability of economic model predictive control. IEEE Transactions on Automatic Control. 2012;57:1615-1626.

[2] Amrit R, Rawlings JB, Angeli D. Economic optimization using model predictive control with a terminal cost. Annual Reviews in Control. 2011;35:178-186.

[3] Huang R, Harinath E, Biegler LT. Lyapunov stability of economically oriented NMPC for cyclic processes. Journal of Process Control. 2011;21:501-509.

[4] Heidarinejad M, Liu J, Christofides PD. Economic model predictive control of nonlinear process systems using Lyapunov techniques. AIChE Journal. 2012;58:855-870.

[5] Ellis M, Durand H, Christofides PD. A tutorial review of economic model predictive control methods. Journal of Process Control, 2014;24:1156--1178.

430156 Dual Model Predictive Control (DMPC): State of the Art

Abstract:

Dual adaptive controllers find the optimal inputs for uncertain systems. Dual controllers trade off control against anticipated learning by taking into account the value of future information gained by excitation. As a result, the model error is actively minimized while the performance objective is optimized. In other words, the dual tasks of an optimal control for an unknown system are (1) directing the output toward the reference using caution while (2) investigating the system for learning ("exploration" or "probing"). The uncertainty is reduced over time and the controller gradually shifts more focus to optimizing for the classic performance objective. From a Bayesian perspective it follows that the dual controls actually are the optimal controls for control under uncertainty.

Significant progress has been made in designing algorithms that solve approximations to the dual control in the last few years. This presentation gives and overview some recent advances in dual receding-horizon control, or dual model-predictive control (DMPC). While solutions to dual control problems have been possible to obtain with dynamic programming (DP) for very small systems (one parameter), the "curse of dimensionality" prevents DP from being a practical method for dual control synthesis. Recent novel re-formulations using the MPC approach enable efficient computation using nonlinear programming.

The class of methods we review is based on reformulating the expected output error with respect to the future decision sequence. The resulting deterministic optimal-control problem can be cast as a quadratically constrained-quadratic program (QCQP), a problem class for which there exists efficient solvers capable of finding global optima. This formulation also permits efficient implementation of probabilistic (chance) output constraints. The approach is particularly relevant for industrial MPC implementations, where FIR and step-response models are common. While exact reformulation of the objective function into deterministic form is not possible for all systems classes, the results we present here give insight into how uncertain systems are optimally excited during normal operation and can aid the design of approximate and heuristic dual controllers for more general systems.

The approaches we discuss rely on least-squares estimation of the unknown parameters, a method that provides the first two statistical moments of the stochastic variables: the means and the covariances. For the class of systems we focus on here the future error covariances of the parameters are deterministic functions of the control inputs, which means it is possible to propagate the uncertainty forward in time. This enables exact uncertainty prediction in the finite-horizon optimal-control problems, and is the foundation of exact optimal dual control.

The presentation contains demonstrations of how dual control can improve closed-loop performance, as well as some examples of control problems that can be solved with dual controllers (like the admissibility problem of adaptive control). We conclude the presentation with some thoughts on future directions for dual control research and applications.

418527 Output Feedback Lyapunov-Based Stochastic Nonlinear Model Predictive Control

Abstract:

Deterministic robust model predictive control approaches have been developed to account for worst-case uncertainties [1]. These approaches can lead to conservative closed-loop performance due to the low probability of occurrence of worst-case uncertainties. Stochastic model predictive control approaches address this issue by incorporating the probabilistic description of stochastic system uncertainties into the optimal control problem (e.g., see [2] and the references therein). This work presents an output feedback stochastic nonlinear model predictive control (SNMPC) approach for a class of nonlinear systems with unbounded stochastic uncertainties. The SNMPC approach aims to shape (multivariate) probability density function (PDF) of the stochastic states in presence of input and joint state chance constraints. The output feedback control approach accounts for the fact that system states cannot be readily measured in most practical applications, which in turn necessitates the estimation of system states from available measurements.

This work uses the Fokker-Planck (FP) equation [3] to describe the dynamic evolution of the PDF of states. The FP equation characterizes the complete multivariate PDF arisen from the stochastic system uncertainties in initial states and disturbances. Characterizing the complete PDF of states allows for shaping the PDF with respect to any desired (multivariate) PDF, and computing chance constraints of any complexity directly without conservative approximations. The value function of the SNMPC approach is defined in terms of the Hellinger distance [4], which quantifies the degree of overlap between the target PDF and the predicted PDF of states. Asymptotic stability in the probabilistic sense is ensured by using a Lyapunov-based control law. Closed-loop stability is ensured by designing a stability constraint in terms of a stochastic control Lyapunov function, which explicitly characterizes stability in a probabilistic sense [5]. To facilitate output feedback control, a continuous-discrete particle filter [6] based on Bayesian estimation is used for estimating the states. In the prediction step of the particle filter, the transition PDF of the states is obtained by solving the FP equation. This PDF is then substituted in the Bayes' equation, along with the PDF of the measurements, to yield the posterior PDF of the states.

The proposed output feedback SNMPC approach is implemented on a continuous acetone-butanol-ethanol (ABE) fermentation process [7]. The simulation results indicate that the control approach is able to shape the PDF of states, while ensuring the satisfaction of the state constraints with a (least) prespecified probability level. The simulation case study demonstrates the capability of the SNMPC approach to systematically seek tradeoffs between the closed-loop performance and robustness to system stochasticities.

References

1. Bemporad A., Morari M. Robust model predictive control: A survey. *In Robustness in Identification and Control*, 1999; 207-226.
2. Mesbah A., Streif S., Findeisen R., Braatz R. D. Stochastic nonlinear model predictive control with probabilistic constraints. *Proceedings of the American Control Conference*, 2014; 2413-2419.
3. Risken H. *The Fokker-Planck Equation: Methods of Solution and Applications*. Springer-Verlag, Berlin, 1984.
4. Gibbs A.L., Su F.E. On choosing and bounding probability metrics. *International Statistical Review*, 2002; 70:419-435.
5. Deng H. and Krstic M. Output-feedback stabilization of stochastic nonlinear systems driven by noise of unknown covariance. *Systems and Control Letters*, 2000, 39:173-182.
6. Xia Y., Deng Z., Li L., Geng, X. A new continuous-discrete particle filter for continuous-discrete nonlinear systems. *Information Sciences*, 2013; 242:64-75.
7. Haus S., Jabbari S., Millat T., Janssen H., Fischer R. J., Bahl H., King J.R., Wolkenhauer O. A systems biology approach to investigate the effect of pH-induced gene regulation on solvent production by Clostridium acetobutylicum in continuous culture. *BMC Systems Biology*, 2011, 5:1-13.

435874 Constructing Constrained Control Lyapunov Functions for Control Affine Nonlinear Systems

Abstract:

Summary

Systems exhibiting nonlinearity, constraints, and uncertainty are ubiquitous in practice and can pose significant challenges. Previous results have shown that control designs which do not acknowledge these features can under-perform more advanced techniques or even result in closed-loop instability [1]. In many applications, particularly those systems with input constraints, control designs are sought that guarantee stabilization to the origin from the largest possible set of initial conditions. This set has been termed the null controllable region (NCR) [2].

Constrained Control Lyapunov Functions (CCLF) are Lyapunov functions designed so that the control laws which guarantee their decay are admissible and stabilizing for the constrained system with a region of attraction equal to the NCR [3]. Clearly, control designs using CLFs that are not CCLFs only guarantee stabilization within subsets of the NCR [4]. There currently do not exist results that identify the NCR of a general nonlinear system or give controls laws for its stabilization everywhere within the NCR.

Motivated by the above, in this work we consider the problem of determining the NCR of control-affine nonlinear unstable systems with constrained controls.

Our construction of the nonlinear NCR is a generalization of the results for linear systems [1], where it was shown that for single-input linear systems of arbitrary dimension, the NCR of the system was covered by extremal trajectories of the reverse-time system, i.e. its reachable set. It was shown that these trajectories are induced solely by all bang-bang controls with a known number of switches. To determine the NCR for nonlinear systems, we will exploit the fact that time-optimal trajectories traverse the boundary of the system's reachable set [5, 6].

We will then show that the NCR is a CCLF and that decay of the trajectory into successive concentric sub-NCRs, corresponding to proportionally smaller input requirements, is sufficient to result in stabilization to the origin. To illustrate the proposed approach, we will employ the CCLF in a model predictive controller that guarantees stability everywhere by constraining the control action to those which force the trajectory into successive sub-shells of the NCR.

References

[1] T. Hu, Z. L. and Qiu, L. (2001). Stabilization of exponentially unstable linear systems with saturating actuators. IEEE Transactions on Automatic Control, 46(6), 973-979.

[2] T. Hu, Z. L. and Qiu, L. (2002). An explicit description of null controllable regions of linear systems with saturating actuators. Systems & Control Letters, 47, 65-78.

[3] Mahmood, M. and Mhaskar, P. (2014). Constrained control Lyapunov function based model predictive control design. Int. J. Robust Nonlinear Control, 24, 374-388.

[4] Mahmood, M. and Mhaskar, P. (2008). Enhanced stability regions for model predictive control of nonlinear process systems. Proceedings of the American Control Conference,

1133-1138.

[5] Lewis, A. (2006). The Maximum Principle of Pontryagin in Control and in Optimal Control. Department of Mathematics and Statistics, Queen's University, Kingston, Canada.

[6] Shattler, H. and Ledzewicz, U. (2012). Geometric Optimal Control. Springer, New York.

[7] Sussmann, H. (1987). The structure of time optimal trajectories for single input systems in the plane. SIAM Journal of Control and Optimization, 25(2), 433-465.

432772 Passivity Based Control of Chemical Reactors Using Reaction Invariants

Abstract:

We develop an approach to deal with the nonlinear control of the CSTR using the concept of using passivity based control to control one or more reaction invariants. More precisely, the paper shows that the CSTR dynamics can be split into reaction variant and reaction invariant dynamics through the use of linear combinations of system variables (called inventories). This separation is based the stoichiometric information only. As the reaction invariants are exponentially stable, the control design for an exponential stabilization of the reactor is simply solved by stabilizing the reaction variants. In this paper we show how this can be achieved using a combination of nonlinear (adaptive) feed-forward and passivity based feedback control of the reaction variants. In some applications there may be more variants than it is possible to independently stabilize using nonlinear control. In such cases the uncontrolled variants may or may not be stable. We show how the stability of these dynamics can be analyzed using Feinberg's reaction network theory based on deficiency analysis. The theoretical results of the paper are illustrated by application of the theory to example systems. These include a first order chemical reaction with multiplicity behavior, second order and higher order systems. Non-adaptive and adaptive control systems are shown to give the same asymptotic behavior after the learning phase has been completed. During the course of the paper presentation we will reveiw briefly the historical background of the use of reaction invariants in the area of process control, starting with the seminal work of Fjeld, Asbjornsen and Astrom leading to our own contribtions and the the work of Bonvin and his students at EPFL.

428982 Design and Implementation of a Biomimetic Control Strategy for Chemical Processes Based on Efficient Ant Colony Optimization

Abstract:

In the last decades, many biological systems have been a source of inspiration for advanced control methods because of the successes of these systems in solving difficult problems encountered in nature. For example, the behavior of natural groups such as ants, bees and swarms demonstrates that self-organization and cooperation by following simple rules of interaction can result in a wide range of optimal patterns [1]. Inspired by the ant's rule of pursuit idea, the present work introduces a novel biomimetic optimal control strategy for nonlinear chemical process control. In this approach, starting from an initially feasible trajectory for the leader agent (or ant), each follower agent improves its path towards the setpoint by employing optimal control laws. As the number of agent progresses, the trajectories converge to an optimal solution. The proposed approach overcomes the computational time limitations of the previously introduced algorithm based on the modified version of the Generalized Sampled Local Pursuit (GSLP) strategy [2]. The previous strategy used the gradient-based solver *dynopt* [3] to tackle the intermediate optimal control problems for the leader-follower local interactions. To address the computational challenges of the gradient-based approach, the present biomimetic controller framework employs the computationally efficient optimal control solver entitled Efficient Ant Colony Optimization (EACO).

The EACO algorithm is a metaheuristic optimization technique inspired by the ants' foraging behavior which utilizes probabilistic and stochastic concepts for solving large-scale optimization problems. In basic Ant Colony Optimization (ACO) algorithms, artificial ants are stochastic candidate solution construction procedures that exploit a pheromone model and possibly available heuristic information of the mathematical model. The artificial pheromone trails are the sole means of communication among the artificial ants. Pheromone decay allows the artificial ants to forget the past history and focus on new promising search directions. The pheromone values are updated according to the information learned and the algorithmic procedure leads to accurate and potentially, a global optimal solution [4]. The EACO algorithm improves the performance of the conventional ACO algorithm for combinatorial, continuous and mixed-variable optimization problems by using the Hammersley sequence sampling (HSS) for probabilistic elements of ACO. The initial solution archive diversity for continuous and mixed-variable optimization problems plays an important role in the performance of the ACO algorithm. The uniformity property of the HSS technique is exploited to avoid clustering of the initial solution archive in a small region of the potential solution space. Moreover, the ACO algorithm is a probabilistic method, in which several random probability functions are involved in the algorithm procedure. The distribution of these random numbers affects the performance of the ACO algorithm. At this step, the multidimensional uniformity property of HSS is also introduced to improve the computational efficiency of the ACO algorithm. The capabilities of the proposed methodology were illustrated using benchmark problems and real world case studies [5].

The developed biomimetic controller approach consists of the novel combination of the EACO technique with the modified GSLP strategy for chemical process control. The applicability of the proposed strategy is demonstrated by its implementation for the optimal control of a fermentation process in which the objective is to track the product concentration setpoint. The challenges in this fermentation process are the steady-state multiplicity and the oscillations in the concentration profiles that affect process efficiency [6]. Closed-loop results associated with this implementation show that the biomimetic controller successfully overcomes these process challenges. In addition, the performance of the proposed approach is compared to the classical PI controller as well as to the modified GSLP-*dynopt* strategy. The results of such comparisons indicate potential improvement in the solution quality and reduction in computational time for the biomimetic controller when compared to its counterparts. Therefore, the proposed GSLP-EACO control approach provides a computationally efficient alternative to the gradient-based optimal control algorithms for optimization of large-scale dynamic systems.

References:

[1] Hristu-Varsakelis D. and Shao C. "A bio-inspired pursuit strategy for optimal control with partially constrained final state". *Automatica* **2007**;43:1265-1273.

[2] Lima F. V., Li S., Mirlekar G. V., Sridhar L. N. and Ruiz-Mercado G. J., "Modeling and advanced control for sustainable process systems". Submitted to Sustainability in the Analysis, Synthesis and Design of Chemical Engineering Processes, G. Ruiz-Mercado and H. Cabezas (eds.), Elsevier, **2015**.

[3] Cizniar M., Fikar M., Latifi M. A., "MATLAB DYNamic OPTimisation code", User's Guide, Version 4.1, **2010**.

[4] Dorigo, M., Stutzle, T., "Ant colony optimization theory". A Brandford Book, the MIT Press, Cambridge, Massachusetts, **2004**.

[5] Gebreslassie B. H. and Diwekar U. M., "Efficient ant colony optimization (EACO) for computer aided molecular design: Case study solvent selection problem". *Computers and Chemical Engineering* **2015**:78:1-9.

[6] Sridhar L. N. "Elimination of oscillations in fermentation processes". *AIChE Journal* **2011**;57(9):2397-2405.

Adaptive output feedback control of transport-reaction processes via two layer system adaptations

Davood Babaei Pourkargar and Antonios Armaou

Compared to adaptive control of lumped chemical processes which has been well studied in the last decades, the adaptive control problem of transport-reaction processes is a more challenging task that needs more advanced mathematical tools [1]. The spatiotemporal dynamic behavior of such systems which can be mathematically modeled by parabolic partial differential equations (PDEs) usually includes unknown terms due to unidentified terms or uncertainty in transport and reaction mechanisms. This is typical in the chemical and advanced material industries with examples ranging from packed and flow reactors, crystallization and polymerization process, chemical vapor deposition to microelectronic fabrication processes and it showcases the importance of system identification and adaptive control methodologies in industries.

The infinite-dimensional representation of parabolic PDEs which describes the aforesaid transport-reaction processes can in principle be partitioned into a finite-dimensional slow (and possibly unstable) and an infinite dimensional fast (and stable) subsystems [2]. The time-scale separation between the subsystems can also be identified by analytical or statistical approaches. Taking advantage of such separation we apply Galerkin projection to derive a low-dimensional reduced order model (ROM) which can be employed as the basis for adaptive controller designs. To implement such strategy we require the basis functions of the system to discretize the governing PDEs to ROMs in the form of low-dimensional ordinary differential equations (ODEs). For systems with known linear spatial differential operators defined over simple domains we can find an analytical basis functions. However we cannot identify the required basis functions analytically when the spatial differential operator is nonlinear or contains the unknown parameters, or when the process operates over irregular domains. For such cases statistical techniques such as proper orthogonal decomposition (POD) can be applied to compute the dominant empirical basis functions based on the available solution profiles of the system. The essential requirement of the POD-like approaches is the a-priori availability of solution profiles of the governing PDEs which present the global spatiotemporal dynamics of the studied system. To overcome such a limiting requirement we employ adaptive proper orthogonal decomposition (APOD) to recursively compute the set of required empirical basis functions during process evolution [3, 4].

In this work we consider the output feedback control problem of parabolic transport-reaction processes in the presence of unknown parameters. We classify the general form of the parabolic distributed chemical processes based on the required basis functions computation methods. Then to generalize the model reduction based adaptive control method we develop a two-layer system adaptation approach. In the external layer we employ APOD to recursively revise the set of empirical basis functions and reconstruct the ROM as needed during process operation. In the internal layer we derive a Lyapunov-based adaptive output feedback controller to deal with the unknown parameters of the recursively updated ROM. A Luenberger-type dynamic observer is also synthesized to estimate the state of the ROM required by the output feedback control structure. The simulation results are presented for adaptive control of thermal spatiotemporal dynamics in a tubular reactor in the presence of unknown diffusivity coefficient, axial velocity and reaction kinetics.

[1] A. SMYSHLYAEV, M. KRSTIC, Adaptive control of parabolic PDEs, Princeton Univ. Press, Princeton, NJ, 2010.

[2] P.D. CHRISTOFIDES, Nonlinear and robust control of PDE systems, Birkhauser, New York, 2000.

[3] D. BABAEI POURKARGAR, A. ARMAOU, Design of APOD-based switching dynamic observers and output feedback control for a class of nonlinear distributed parameter systems, Chem. Eng. Sci., DOI:10.1016/j.ces.2015.02.032, 2015.

[4] D. BABAEI POURKARGAR, A. ARMAOU, Modification to adaptive model reduction for regulation of distributed parameter systems with fast transients, AIChE J., 59(12): 4595–4611, 2013.

Estimation of spatially distributed processes via adaptive model reduction using mobile sensors network

Davood Babaei Pourkargar and Antonios Armaou

Even though the state estimation and monitoring problems have reached certain maturity for lumped dynamic processes described by ordinary differential equations (ODEs), they remain challenging tasks for spatially distributed processes (SDPs). SDPs can be mathematically modeled by a set of partial differential equations (PDEs), frequently arise in chemical and advanced materials industries can be exemplified by packed and fluidized bed reactors, polymerization and crystallization processes, chemical vapor deposition processes, plasma discharge systems and microelectronic fabrication processes. The infinite dimensional representation of such systems can be approximated by a finite number of ODEs because a time-scale separation can in practice be identified between the finite-dimensional slow (and possibly unstable) and infinite-dimensional fast (and stable) subsystems [1]. The resulting system approximations in the form of low-dimensional ODEs can be applied as the basis for Luenberger-type dynamic observer designs. Practically, such set of ODEs are constructed by discretizing governing PDEs using Galerkin's method. However, in general the required basis functions by Galerkin's method cannot be computed by analytical methods, specifically when the describing PDEs are nonlinear and/or the process operates over irregular domains.

To extend the applicability of model reduction approaches to general SDPs, statistical approaches such as proper orthogonal decomposition (POD) may be applied to compute an optimal set of empirical basis functions. The essential requirement of the POD-like approaches is the a-priori availability of solution profiles of the governing PDEs which present the global spatiotemporal dynamics of the studied system. To overcome such vital requirement we employ adaptive proper orthogonal decomposition (APOD) to recursively compute this set of required empirical basis functions as needed during the process evolution [2, 3]. An unavoidable assumption of previously developed methodologies for dynamic observer synthesis that employ data driven model reduction techniques like APOD is that the complete snapshots which span the whole process domain are accessible as time evolves. Obtaining such information might not be feasible owing to high sensor costs and limited availability of sensors which sample all the regions of the process domain.

We propose two methods to estimate the required snapshots as a demarcation point to develop process estimation methodologies which relax both the complete snapshot assumption and the representative ensemble requirement; (1) using a PDE-based Luenberger-type dynamic observer which employs moving sensors information, (2) a combination of scanning sensor networks and APOD methodology. To extend the applicability of APOD to situations when the availability of distributed sensing is limited, we base the proposed work on Gappy-APOD [4]. A small numbers of moving sensors that provide continuous local measurements are employed to recursively construct the approximate reduced order model.

The basic premise of the proposed method is that a set of mobile sensors achieve better estimation performance than a set of immobile sensors [5]. To enhance the performance of the state estimator, a network of sensors that are capable of moving within the spatial domain is utilized. The objective is to provide mobile sensor control policies that aim to improve the state estimate. The metric for such an estimate improvement is taken to be the expected state estimation error.

The effectiveness of the proposed estimation approaches are then successfully implemented on temperature estimation in a tubular reactor where the thermal spatiotemporal dynamics can be mathematically modeled by a PDE. Beyond the obvious savings on communication costs and sensor network size, the study of this problem provides a tool for the assessment of the robustness of the synthesized dynamic observers and will allow us to identify the fundamental limits on the monitoring structure tolerance to communication suspension.

[1] P.D. CHRISTOFIDES, Nonlinear and robust control of PDE systems, Birkhauser, New York, 2000.

[2] D. BABAEI POURKARGAR, A. ARMAOU, Design of APOD-based switching dynamic observers and output feedback control for a class of nonlinear distributed parameter systems, Chem. Eng. Sci., DOI:10.1016/j.ces.2015.02.032, 2015.

[3] D. BABAEI POURKARGAR, A. ARMAOU, Modification to adaptive model reduction for regulation of distributed parameter systems with fast transients, AIChE J., 59(12): 4595–4611, 2013.

[4] S. PITCHAIAH, A. ARMAOU, Output feedback control of dissipative PDE systems with partial sensor information based on adaptive model reduction, AIChE J., 59(3), 747–760, 2013.

[5] M.A. DEMETRIOU, I.I. HUSSEIN, Estimation of spatially distributed processes using mobile spatially distributed sensor network. SIAM J. Control Optim., 48(1):266-291, 2009.

424740 Sensor Fault Estimation and Accommodation in Networked Distributed Processes

Abstract:

The problems of fault identification and fault accommodation have been the subject of considerable interest in the control research community in response to the ever increasing requirements on the reliable and safe operation of automatic control systems. At present, however, only relatively few studies have been devoted to the development of systematic methods for the diagnosis and handling of faults in spatially distributed processes. This is an important problem given that many industrial processes are characterized by spatial variations and are modeled by Partial Differential Equations (PDEs) (e.g., [1], [2]). The focus of existing results in this direction has been on the detection and compensation of either actuator (e.g., [3]–[6]) or component faults (e.g., see [7]). By contrast, the problem of sensor fault-tolerant control has not attracted as much attention despite the common occurrence of sensor faults in practice and the detrimental effect they can have on the monitoring capability and overall control quality of the closed-loop system.

An effort to tackle this problem was made in [8] where a sensor fault detection and reconfiguration strategy was developed to reduce the performance deterioration associated with sensor faults in highly dissipative distributed processes. Fault detection was achieved by means of a residual generation approach in which the residual was compared against a suitably designed fault-free alarm threshold. In this study, however, the faults considered were assumed not to jeopardize closed-loop stability. The detection approach was stability-based and thus could leave partial faults that do not compromise closed-loop stability go undetected. Furthermore, the problems of identifying the location and the magnitude of the sensor faults were not addressed. It was assumed instead that a suitable mechanism for fault identification was already built into the system and able to determine the size and location of the fault. Finally, and similar to other existing methods for fault-tolerant control of distributed parameter systems, the fault-tolerant control structure was designed within a conventional feedback control setting with continuous and flawless sensor-controller communication.With the increased integration of communication networks in control systems, the impact of sensor-controller communication constraints (e.g., resource limitations, communication delays) on closed-loop stability can no longer be ignored and needs to be explicitly accounted for (e.g., see [9]).

Motivated by these considerations, we present in this work an integrated approach for data-based fault identification and accommodation of sensor faults in a class of spatially distributed processes with low-order dynamics, and sensor-controller communication constraints. The main idea is to synthesize a model-based networked output feedback controller based on an appropriate reduced-order system obtained via model order reduction of the highly-dissipative PDE. By analyzing the fault-free behavior of the networked closed-loop state, the closed-loop stability region is characterized in terms of the sensor-controller communication rate, the communication delay size, the model uncertainty, the control actuator placement and the controller and estimator design parameters. The fault identification mechanism is then implemented by solving a moving-horizon optimization problem, which is embedded in the sensors, to estimate the size of the faults using the historical sampled output and input data. Once the faults are identified, estimates of the fault magnitudes are sent to the controller at the next sensor transmission time. These estimates, together with the characterization of the networked closed-loop stability region, are used to determine the suitable measures for fault accommodation, which include updating the control model parameters, adjusting the control actuator placement, and/or switching to a new stabilizing feedback gain. The proposed methodology is illustrated using a simulated diffusionreaction process subject to both abrupt and gradual sensor faults.

References:

[1] P. D. Christofides, *Nonlinear and Robust Control of PDE Systems: Methods and Applications to Transport-Reaction Processes*. Boston: Birkhauser, 2001.

[2] A. Smyshlyaev and M. Krstic, *Adaptive Control of Parabolic PDEs*. Princeton University Press, 2010.

[3] M. Demetriou, "A model-based fault detection and diagnosis scheme for distributed parameter systems: A learning systems approach," *ESAIM-Control Optimisation and Calculus of Variations*, vol. 7, pp. 43–67, 2002.

[4] S. Ghantasala and N. H. El-Farra, "Robust actuator fault isolation and management in constrained uncertain parabolic PDE systems," *Automatica*, vol. 45, pp. 2368–2373, 2009.

[5] Z. Yao and N. H. El-Farra, "Robust fault detection and reconfigurable control of uncertain sampled-data distributed processes," in *Proceedings of 50th IEEE Conference on Decision and Control*, Orlando, FL, 2011, pp. 4925–4930.

[6] A. Baniamerian and K. Khorasani, "Fault detection and isolation of dissipative parabolic PDEs: Finite-dimensional geometric approach," in *Proceedings of American Control Conference*, Montreal, Canada, 2012, pp. 5894–5899.

[7] A. Armaou and M. Demetriou, "Robust detection and accommodation of incipient component faults in nonlinear distributed processes," *AIChE J.*, vol. 54, pp. 2651–2662, 2008.

[8] Z. Yao and N. H. El-Farra, "Performance-based sensor reconfiguration for fault-tolerant control of uncertain spatially distributed processes," in *Proceedings of 19th IFAC World Congress*, Cape Town, South Africa, 2014, pp. 5193–5198.

[9] Y. Sun, S. Ghantasala, and N. H. El-Farra, "Networked control of spatially distributed processes with sensor-controller communication constraints," in *Proceedings of American Control Conference*, St. Louis, MO, 2009, pp. 2489–2494.

Single-Step Formulation of Feedback Control for an Exothermic Plug-Flow Reactor

Qingqing Xu
Stevan Dubljevic

11 November 2015

Process Control
Department of Chemical and Materials Engineering
University of Alberta

Outline

- System Representaion
- Optimal Steady State
- Single-step Full State Feedback Control
- Control Simulation

System Representation

◇ A nonlinear exothermic plug-flow reactor takes the following PDE form:

$$\frac{\partial T}{\partial t} = -v\frac{\partial T}{\partial \zeta} - \frac{\Delta H}{\rho C_p}k_0Ce^{-\frac{E}{RT}} - \frac{4h}{\rho C_p d}(T - T_w)$$
$$\frac{\partial C}{\partial t} = -v\frac{\partial C}{\partial \zeta} - k_0Ce^{-\frac{E}{RT}}$$

with initial condition and boundary condition:

$$t = 0, \quad T(\zeta, 0) = T_{in}, \quad C(\zeta, 0) = C_{in}$$
$$z = 0, \quad T(0, t) = T_0, \quad C(0, t) = C_0$$

Assume initial conditions and boundary conditions as $T_{in} = 340(K)$, $C_{in} = 0.02(mol \cdot L^{-1})$, $T_0 = 340(K)$, $C_0 = 0.02(mol \cdot L^{-1})$.

System Representation

◇ By defining state and input variables as $x_1 = \frac{T - T_{in}}{T_{in}}$, $x_2 = \frac{C_{in} - C}{C_{in}}$, $u = \frac{T_w - T_{in}}{T_{in}}$ and process parameters as $\alpha = k_0e^{-E/RT_{in}}$, $\beta = \frac{4h}{\rho C_p d}$, $\delta = -\frac{\Delta H}{\rho C_p}\frac{C_{in}}{T_{in}}$, $\gamma = \frac{E}{RT_{in}}$, the system becomes:

$$\frac{\partial x_1}{\partial t} = -v\frac{\partial x_1}{\partial \zeta} + \alpha\delta(1 - x_2)e^{\gamma x_1/1+x_1} - \beta x_1 + \beta u$$
$$\frac{\partial x_2}{\partial t} = -v\frac{\partial x_2}{\partial \zeta} + \alpha(1 - x_2)e^{\gamma x_1/1+x_1}$$

with initial condition and boundary condition:

$$x_1(\zeta, 0) = 0 \quad x_2(\zeta, 0) = 0$$
$$x_1(0, t) = 0 \quad x_2(0, t) = 0$$

Process parameters are $v = 0.1(m \cdot s^{-1})$, $\alpha = 0.0581$, $\beta = 0.2$, $\delta = 0.25$, $\gamma = 16.6607$.

Optimal Steady State

- In steady state, the above partial deferential equations of the system reduce to ordinary differential equations:

$$v\frac{\partial x_{1ss}}{\partial \zeta} = \alpha\delta(1 - x_{2ss})e^{\gamma x_{1ss}/1+x_{1ss}} - \beta x_{1ss} + \beta u_{ss}, \quad x_{1ss}(0) = 0$$
$$v\frac{\partial x_{2ss}}{\partial \zeta} = \alpha(1 - x_{2ss})e^{\gamma x_{1ss}/1+x_{1ss}}, \quad x_{2ss}(0) = 0$$

- The optimal control of the system at steady state is to find control $u^*(\zeta)$ which causes the system to follow an admissible trajectory x^* while minimizing performance criterion:

$$\min_{u} \quad J = h[x(L)] + \int_0^L g[x(\zeta)]d\zeta$$

*Smets, Ilse Y and Dochain, Denis and Van Impe, Jan F, Optimal temperature control of a steady-state exothermic plug-flow reactor, AIChE Journal, Vol.48, No.2, 2002.

Optimal Steady State

- In this work, we consider a cost criterion of the following type:

$$\min_{u} \quad J = (1 - A)(1 - x_2(L)) + \frac{A}{K}\int_0^L x_1^2(\zeta)d\zeta$$

  where A is the trade-off coefficient between terminal and integral cost, and K is a weighting factor to bring the two costs in the same order of magnitude.
- According to the Minimum Principle of Pontryagin, the optimal heat exchanger temperature profile $u^*(\zeta)$ can be maximum-minimum profile or maximum-singular-minimum profile. Assume $A = 0.7$ and $K = 250,000$.

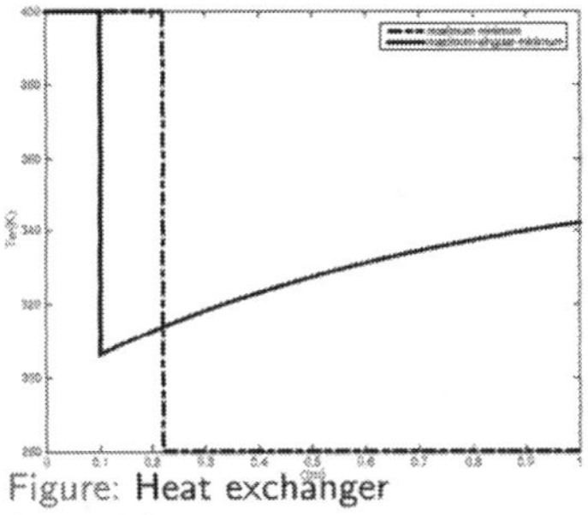

Figure: Heat exchanger temperature

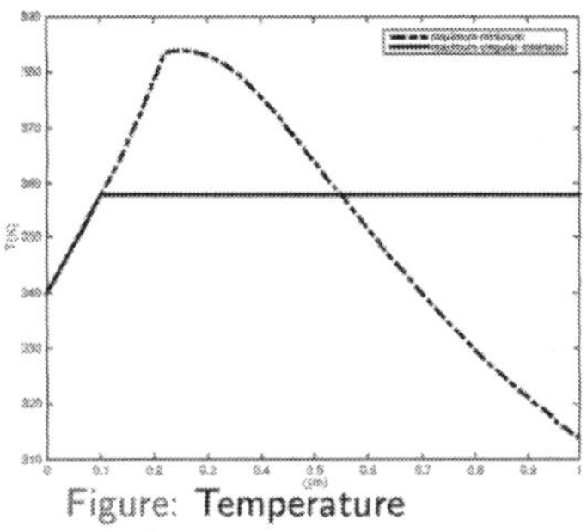

Figure: Temperature

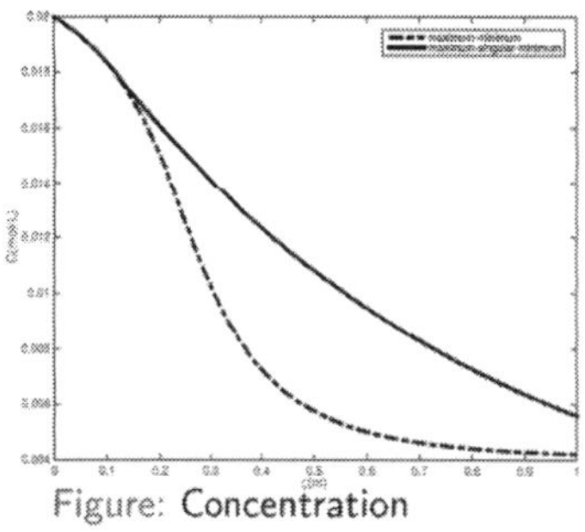

Figure: Concentration

Single-step Full State Feedback Control

- The finite dimensional state space realization of the Luenberger **single-step** formulation is given for the linear system as follows:

$$\dot{x}(t) = Ax(t) + Bu(t), \;\; x(0) = \bar{x}_0$$

 where A is a linear operator and B is an input operator.

- The system is transformed by the linear coordinate transformation $z(t) = Tx(t)$ with the linear state feedback control law $u(t) = -Kz(t)$ which yields desired closed-loop dynamics:

$$\dot{z}(t) = A_d z(t), \;\; z(0) = T^{-1}\bar{x}_0$$

 where A_d is matrix with desired "target" eigenvalues.

- The closed loop dynamics in the original coordinates with desirable eigenvalues can be written as follows:

$$\dot{x}(t) = T^{-1}A_d T x(t)$$

Single-step Full State Feedback Control

- The second-order Taylor expansion of the system at optimal steady-state of interest is as following:

$$\begin{bmatrix} \frac{\partial \tilde{x}_1}{\partial t} \\ \frac{\partial \tilde{x}_2}{\partial t} \end{bmatrix} = -v \begin{bmatrix} \frac{\partial \tilde{x}_1}{\partial \zeta} \\ \frac{\partial \tilde{x}_2}{\partial \zeta} \end{bmatrix} + \begin{bmatrix} J_{11}(\zeta) & J_{12}(\zeta) \\ J_{21}(\zeta) & J_{22}(\zeta) \end{bmatrix} \begin{bmatrix} \tilde{x}_1 \\ \tilde{x}_2 \end{bmatrix}$$
$$+ \begin{bmatrix} \frac{1}{2}H_{11}(\zeta)\tilde{x}_1^2 + H_{12}(\zeta)\tilde{x}_1\tilde{x}_2 + \frac{1}{2}H_{13}(\zeta)\tilde{x}_2^2 \\ \frac{1}{2}H_{21}(\zeta)\tilde{x}_1^2 + H_{22}(\zeta)\tilde{x}_1\tilde{x}_2 + \frac{1}{2}H_{23}(\zeta)\tilde{x}_2^2 \end{bmatrix} + \begin{bmatrix} \beta \\ 0 \end{bmatrix} \tilde{u}(t)$$

 where $\tilde{x}_1 = x_1 - x_{1ss}$, $\tilde{x}_2 = x_2 - x_{2ss}$ and $\tilde{u} = u - u_{ss}$.

Single-step Full State Feedback Control

- The controller synthesis goal is to apply the state feedback control law $\tilde{u} = -Kz(\tilde{x}(\zeta, t)$ and the nonlinear transformation $z(\tilde{x}(\zeta, t)) = W(\zeta)\tilde{x}(\zeta, t) + P(\zeta)\tilde{x}^2(\zeta, t)$ simultaneously. The state feedback control law is as following:

$$\begin{aligned} \tilde{u} &= -Kz \\ &= -\begin{bmatrix} K_1 & K_2 \end{bmatrix} \begin{bmatrix} w_{11}(\zeta)\tilde{x}_1 + w_{12}(\zeta)\tilde{x}_2 + P_{11}(\zeta)\tilde{x}_1^2 + P_{12}(\zeta)\tilde{x}_1\tilde{x}_2 + P_{13}(\zeta)\tilde{x}_2^2 \\ w_{21}(\zeta)\tilde{x}_1 + w_{22}(\zeta)\tilde{x}_2 + P_{21}(\zeta)\tilde{x}_1^2 + P_{22}(\zeta)\tilde{x}_1\tilde{x}_2 + P_{23}(\zeta)\tilde{x}_2^2 \end{bmatrix} \end{aligned}$$

- With control, we will achieve desired closed-loop target dynamics:

$$\begin{bmatrix} \frac{\partial z_1}{\partial t} \\ \frac{\partial z_2}{\partial t} \end{bmatrix} = -v \begin{bmatrix} \frac{\partial z_1}{\partial \zeta} \\ \frac{\partial z_2}{\partial \zeta} \end{bmatrix} + \begin{bmatrix} \sigma_{11}(\zeta) & \sigma_{12}(\zeta) \\ \sigma_{21}(\zeta) & \sigma_{22}(\zeta) \end{bmatrix} \begin{bmatrix} z_1 \\ z_2 \end{bmatrix}$$

In the simulation studies, we choose $K = [8 \quad 4.5]$ and $\sigma = [-0.25 \quad -0.1; 0.075 \quad -0.45]$.

Control Simulation

- Linear Control

The linear control synthesis is to consider only the first-order Taylor expansion of the system and is obtained by determining the numerical solution of the transformation $w(\zeta)$. We choose $w(1) = [0.08; 0.03; 0.005; 0.01]$.

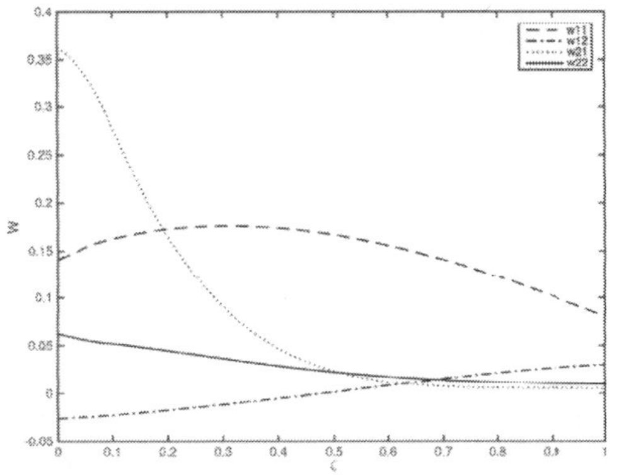

Figure: Linear transfrom operator $w_{ij}(\zeta)$, for $i, j = 1, 2$

Control Simulation

◇ Linear Control

For the initial condition given as $x_1 = x_{1ss} + 0.05e^{\zeta}$ and $x_2 = x_{2ss} - 0.1e^{\zeta}$, the linear controller is able to stabilize the state around spatially uniform unstable steady state.

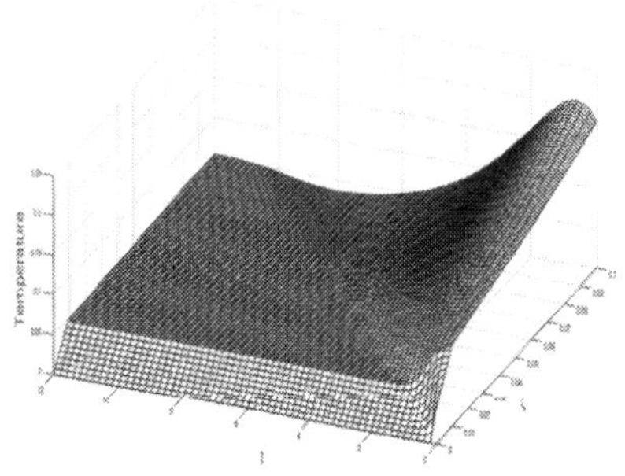

Figure: Temperature

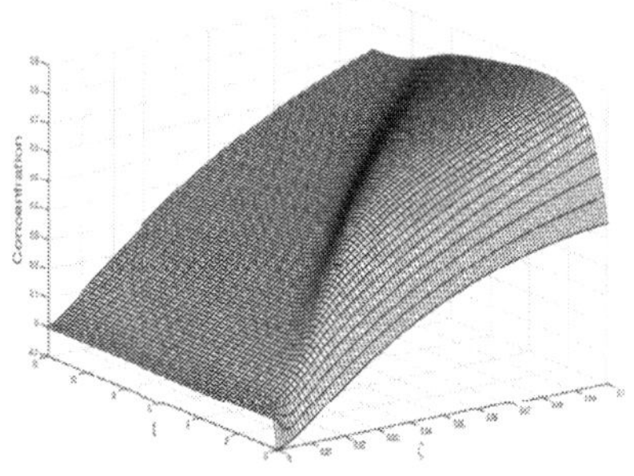

Figure: Concentration

Control Simulation

◇ Quadratic Control

The quadratic control synthesis, which contains the second-order terms of the system, is obtained by determining the numerical solution of the transformation $w(\zeta)$ and $P(\zeta)$. We choose $P(1) = [0.01; 0.01; 0.02; 0.03; 0.01; 0.02]$.

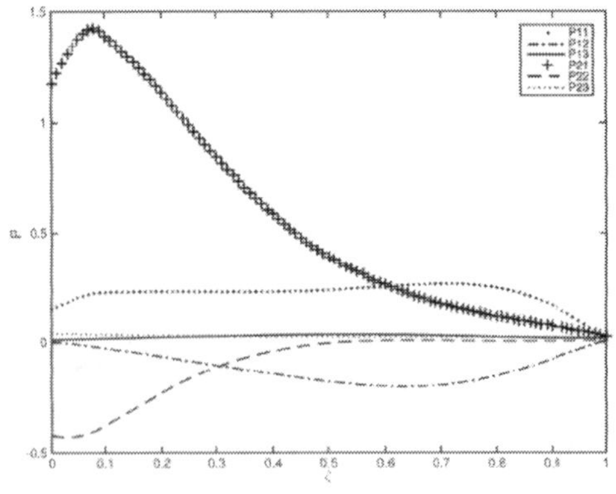

Figure: Nonlinear transfrom operator $P_{ij}(\zeta)$, for $i = 1, 2; j = 1, 2, 3$

Control Simulation

◇ Quadratic Control

For the initial condition given as $x_1 = x_{1ss} + 0.05e^{\zeta}$ and $x_2 = x_{2ss} - 0.1e^{\zeta}$, the quadratic controller is able to stabilize the state around spatially uniform unstable steady state.

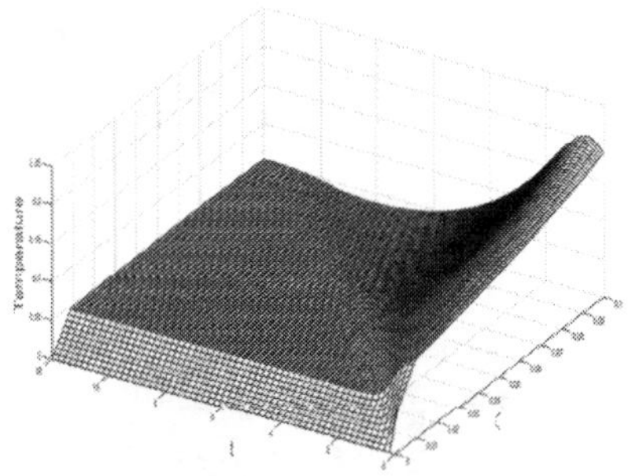

Figure: Temperature

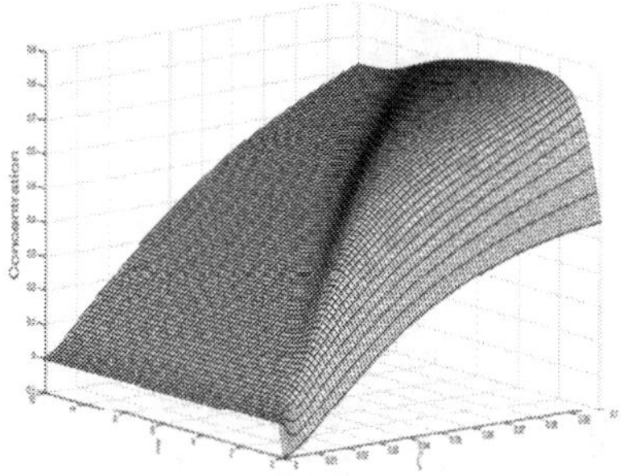

Figure: Concentration

Control Simulation

◇ Norms of Linear Control and Quadratic Control

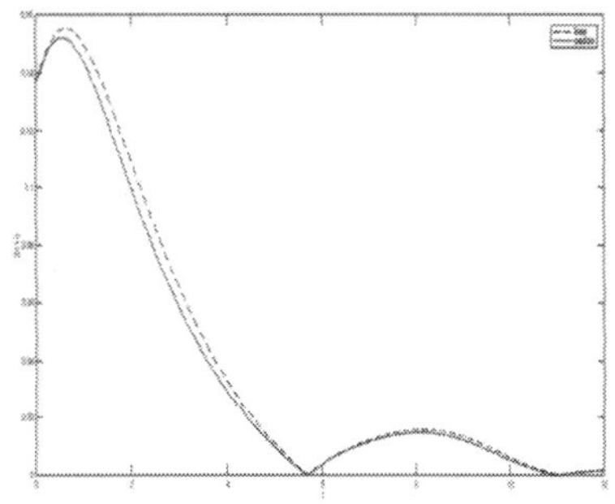

Figure: Temperature

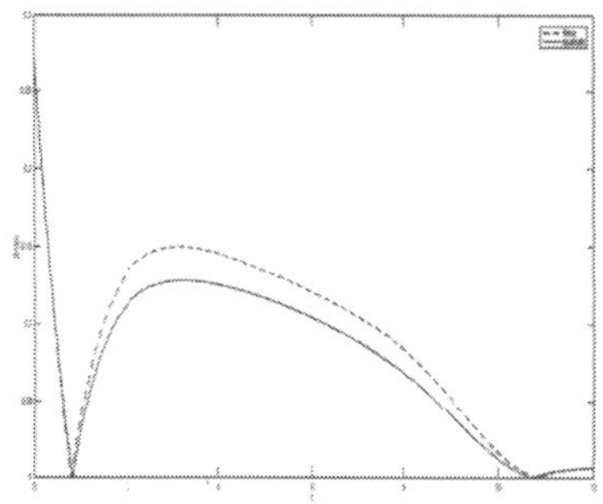

Figure: Concentration

Conclusion

◇ The optimal heat exchanger temperature profile is determined by a cost criterion between process performance and heat loss.

◇ Single-step full state feedback controller is designed in this work for the nonlinear first-order hyperbolic PDEs system of exothermic plug-flow reactor.

◇ Two controllers, linear controller and nonlinear controller, are developed and they can successfully stabilize the unstable nonlinear first-order hyperbolic PDEs system.

◇ The quadratic controller has a good control performance for stabilizing the first order hyperbolic PDEs around desired steady state.

428544 Observations and Analysis of Index Infinity Dae Systems Describing Reactors and Reacting Flows

Abstract:

Mathematical models that describe the steady-state behavior of chemical reactors and reacting flows in which volumetric and/or wall reactions occur can be generally classified as elliptic or parabolic. The behavior of elliptic models is well studied in the literature and they are known to display a variety of solutions and bifurcations (e.g. multiple homogeneous solutions and various types of patterned/asymmetric solutions). In the literature, the so called "boundary layer models (in which axial diffusion/conduction is neglected) used to describe these systems are treated as parabolic equations and having a unique solution. Such a solution is usually obtained by discretization of the transverse coordinate and integration in the axial direction. However, it was shown in our earlier work that these are index infinity differential-algebraic equation (DAE) systems, and could have an infinite number of solutions. Further, the integration of such models along the flow direction (forward) and against the flow direction (backward) could lead to different solutions. In this work, we present further observations and analytical as well as numerical methods for determining the various solutions of such index infinity DAE systems. We consider three specific systems to illustrate the main ideas and techniques : (i) a DAE system consisting of one differential equation (in axial coordinate) and one algebraic equation, having infinite number of solutions (such a system is obtained by transverse averaging of the partial differential equation models) (ii) a parabolic equation in two variables (axial and transverse) that has nonlinearity in the boundary conditions (iii) a parabolic equation in two variables with linear boundary conditions but a nonlinear source term. For these cases, we present bifurcation analysis as well as numerical methods for determining all the solutions with particular emphasis on how to eliminate the Gibbs' phenomenon (that arises due to discontinuous solutions) and numerical diffusion (that can lead to spurious solutions).

433088 Slow-Manifold Order Reduction of Reaction-Diffusion Equations with Dirichlet Boundary Conditions

Abstract:

The derivation of low-dimensional averaged models of reaction-diffusion equations greatly facilitates the analysis, optimization and control of dynamic reactive systems. Examples of systems where order reduction has been successful are catalyst pores, wall coated reactors, single phase tubular reactors, catalytic solid slabs etc. In most past studies of these systems, the boundaries are either impermeable (no-flux conditions) or admit a slow flux, such that the boundary condition may be considered as a small deviation from a no-flux condition. In the limit of very fast diffusion, these systems admit a manifold of equilibria which have a zero eigenvalue, corresponding to the null-space of the diffusion operator. This manifold of equilibria is the set of constant concentration modes without transverse variations. Application of slow (center) manifold reduction (SMR) results in a reduced model in terms of the transverse average concentration. However, if the same problem is analyzed with Dirichlet boundary conditions (homogeneous or non-homogeneous), as in cases when the concentration at the boundary is fixed, then the equilibria of zero eigenvalues do not exist. This occurs because the diffusion operator with Dirichlet boundary conditions does not have a null space. Therefore the reduction techniques used on no-flux problems cannot be applied to problems with Dirichlet boundary conditions.

In this work, we devise techniques for applying SMR to Dirichlet problems. The key idea is to embed the Dirichlet problem within a larger family of problems by the introduction of a suitable parameter (α). This embedding is carried out such that a manifold of equilibria with zero eigenvalues exists in the limit of a critical value of the parameter α. We use the model system of reaction-diffusion in a 1-D slab to demonstrate two different order reduction techniques. We examine the validity and accuracy of the reduced order models for the averaged concentration. Finally, we apply the reduced models to analyze the dynamics of some systems with Dirichlet boundary conditions commonly encountered in reaction engineering.

Keywords: center manifold reduction, averaging, reaction-diffusion, Dirichlet boundary conditions

433582 Simulation and Development of an Estimation Technique for a Bi-Component Aggregation Systems

Abstract:

In chemical engineering, material science and biology there are many systems that require modeling the evolution of particle population. Modeling efforts usually result in deriving integro-differential equations. The particles in these systems, depending on the system, are characterized by one or more properties such as type, size or composition. However, the governing equation obtained from the population balance is not easy to solve especially for multivariate systems. One of the simplest examples with a bivariate distribution function is coagulation in pharmaceutical applications. In such systems, an excipient (solvent) binds the drug particles (solute) to make larger granules. We would like to estimate the output's unmeasurable properties i.e. statistical information, for control purposes. This work focuses on two-component coagulation processes and estimates the output distribution of particles in time.

The rate of coagulation is quantified by a kernel. In literature, there are analytical solutions for population balance equations in which the kernels are constant or additive [1-4]. However, the dynamic of agglomeration is more complicated and its rate is described by a Brownian kernel. Constant-number Monte Carlo (cNMC) is an algorithm to simulate these more complicated processes [5,6]. In this work, we modified this algorithm for a continuous coagulation process. Although extremely useful, Monte Carlo simulations are computationally demanding and impractical in online applications. The computations are necessary for real-time state estimation during the evolution process. The method of moments is an approach to reduce the original integro-differential equation to a set of ordinary differential equations (ODEs). The state variables of these ODEs are the moments of particles' distribution. To ensure moment closure, we use Taylor expansion to approximate some moments with fractional index with respect to the considered ones. This approach has already been employed but only for single variable distribution functions [7]. However, to the best knowledge of authors, there is no effort in the literature for reduction of the original integro-differential equations in multivariate systems. The results show the probabilistic moments obtained by method of moments in time are in agreement with cNMC simulation results.

The measurement of number of particles in the outflow stream is possible, whereas measuring the other output properties online which specify the further moments is not easy. The goal in this work is to estimate these probabilistic moments by measuring the output population of particles. In addition, we assume the feed flow concentration is affected by a white uniformly distributed noise signal. As a result, to obtain the statistical information, we need to employ an estimator which deals with uncertainties in the system. Moving Horizon Estimation (MHE) is a successful estimation technique for unmeasurable states in presence of noise in dynamics or output measurements. Although this method has many advantages over other alternative methods, it needs to solve a dynamic optimization at every sampling time which makes it slow. To deal with this issue, we use a new design proposed in our previous work [8,9]. This approach exploits Carleman linearization to give a more accurate approximation compared to the traditional linearization. This approach not only provides the analytical solution of the system, but also derives the KKT matrix which decreases the dynamic optimization computations significantly. Simulation results show this approach reduces the simulation running time significantly while the performance is at the same level as using the traditional MHE.

References:

[1] Lushinkov AA. Evolution of coagulation systems: III. Coagulating mixtures. *J of Colloid & interface Sci.* 1976;54:94-101.

[2] Gelbard FM, Seinfeld JH. Coagulation and growth of a multicomponent aerosol. *J of Colloid & Interface Sci.* 1978;63:472-479.

[3] Krapivsky PL, Ben-Naim E. Aggregation with multiple conservation laws. *Phys. Rev. E.* 1996;53:291-298.

[4] Vigil RD, Zi RM. On the scaling theory of two-component aggregation. *Chem Eng Sci.* 1998;53:1725-1729.

[5] Matsoukas T, Lee K, & Kim T. Mixing of components in two-component aggregation, *American Institute of Chemical Engineers.* 2006;52, 3088-3099.

[6] Marshall Jr. C. L, Rajniak P, & Matsoukas T. Multi-component population balance modeling of granulation with continuous addition of binder, *Powder Technology.* 2013; 236, 211-220.

[7] Yu M, Lin J & Chan T. A new moment method for solving the coagulation equation for particles in Brownian motion. *Aerosol Sci. & Technol.* 2008;42:705-713.

[8] Hashemian N & Armaou, A. Fast moving horizon estimation of nonlinear processes via Carleman linearization. *American Control Conference*, Chicago, IL, 2015.

[9] Hashemian N & Armaou A. Moving horizon estimation using Carleman linearization and sensitivity analysis. *Annual AIChE Meeting,* Atlanta, GA, 2014.

423455 A New Computationally Efficient Data Assimilation Approach for Finite Element Models

Abstract:

Recent advancements in the field of echocardiography have introduced various methods to image blood flow in the heart [1]. Our particular interest is in the left ventricle (LV) of the heart, which pumps oxygenated blood from the lungs out through the aorta. One method for imaging blood flow is injecting FDA-approved micro-bubbles into the left ventricle, and then, using the motion of the micro-bubbles and the frame rate of the ultrasound scan (i.e., using Particle Imagining Velocimetry or echo-PIV), the blood velocity can be calculated. In addition to blood velocity, echocardiologists are also interested in calculating pressure gradients and other flow properties, but this is not currently possible because the velocity data obtained is two-dimensional and noisy. Our goal is to assimilate two-dimensional velocity data from micro-bubble ultrasound experiments into a three-dimensional computer model. In order to achieve this objective a numerical method is needed that can approximate the solution of a system of differential equation and assimilate an arbitrary number of noisy experimental data points at arbitrary locations within the domain of interests to provide a 'most probable' approximate solution that is properly influenced by the experimental data [2].

We propose a new numerical method for combing two-dimensional noisy echo-PIV data, which is more computationally efficient than previous approaches [3]. The approximate solution is calculated using continuous interior penalty finite element method (IPFEM) coupled with a least-squares finite element method (LSFEM) for integrating the noisy echo-PIV data. This framework allows the flexibility of using any existing numerical approaches for the differential equations that model the original problem, and then complement the solution using a weighted LSFEM framework for data assimilation. The choice of LSFEM approach is due the flexibility when assimilating noisy echo-PIV data since the method can strongly weight more accurate echo-PIV data and use a lower weight for less accurate data. The new numerical approach have been used to predict the 3-dimensional LV blood flow. Results from the current method show the impact of assimilating echo-PIV data combining the data with a full 3-dimensional simulation result.

REFERENCES

1. Borazjani, I., et al., Left ventricular flow analysis: recent advances in numerical methods and applications in cardiac ultrasound. Computational and mathematical methods in medicine, 2013. 2013.
2. Heys, J.J., et al., Weighted least-squares finite elements based on particle imaging velocimetry data. Journal of Computational Physics, 2010. 229(1): p. 107-118.
3. Rajaraman, P.K., et al., Echocardiographic particle imaging velocimetry data assimilation with least square finite element methods. Computers & Mathematics with Applications, 2014. 68(11): p. 1569-1580.

Nonlinear Observer Design for a Class of Hyperbolic PDE Systems

Xiaodong Xu
Stevan Dubljevic

Department of Chemical and Materials Engineering
University of Alberta, Edmonton, Canada

11 Nov. 2015

2015 AIChE
Salt Lake City, Utah, US

Motivation:

- Most chemical reaction processes can be described by semi-linear hyperbolic partial differential equations (PDEs).
- Boundary actuation and boundary observation are common in distributed parameter systems (DPSs).

Objective:

- A nonlinear form of boundary observer is synthesized for a class of semi-linear hyperbolic PDEs.
- An analytic state transformation is utilized to obtain the output injection such that the convergence of observer can be ensured.

Relevant work of others:

- Kazantzis, Nikolaos, and Costas Kravaris. Singular PDEs and the single-step formulation of feedback linearization with pole placement. Systems & Control Letters 39.2 (2000): 115-122.

  — The feedback linearization formulation is applied to finite-dimensional nonlinear systems.

- Zeitz, M. The extended Luenberger observer for nonlinear systems. Systems & Control Letters 9.2 (1987): 149-156.

  — The proposed observer consists of a copy of nonlinear plant and correction for finite-dimensional setting.

System Description:

Semi-linear hyperbolic PDE systems:

$$\partial_t x(z,t) = -\partial_z x(z,t) + f(x(z,t))$$

for $(z,t) \in (0,1) \times \mathbb{R}^+$ with the boundary condition (BC)

$$x(0,t) = u(t),\ t \in \mathbb{R}^+$$

the initial condition (IC) and the output

$$x(z,0) = x_0(z), y(t) = x(1,t),\ z \in [0,1]$$

on domain $(z,t) \in [0,1] \times \mathbb{R}^+$. $x(z,t) \in \mathcal{X}$ denotes the state variable in the Hilbert state space $\mathcal{X} = L^2(0,1)$. The nonlinear function $f(x)$ satisfies the following assumptions:

A.1 $f(x)$ is assumed to analytic in x near origin $x = 0$..

A.2 $f(x) = 0$ at the point $x = 0$ and $a_1 = \partial f(0)/\partial x \neq 0$.

$$f(x) = \sum_{k=1}^{+\infty} a_k x^k = a_1 x + \bar{f}(x) \text{ with } a_k = \frac{\partial f^k(0)}{\partial x^k}$$

$$\begin{aligned} \partial_t x(z,t) &= -v(z)\partial_z x(z,t) + \alpha(z)x(z,t) \\ &\quad + f(x(z,t)),\ z \in [0,1],\ v(z) > 0 \\ x(0,t) &= u(t),\ t \in R^+ \\ y(t) &= x(1,t),\ t \in R^+ \end{aligned}$$

by employing the change of variables $z' = \int_0^z 1/v(s)ds/D$, $t' = t/D$, and $x(z,t) = x'(z',t')\exp(p(z'))$ with $D = \int_0^1 1/v(s)ds$, $p(z') = D\int_0^{z'} \alpha(\zeta)|_{z=z'^{-1}(\zeta)}d\zeta$ and $f(\cdot)$ mapped accordingly

$$\begin{aligned} \partial_t x(z,t) &= -\partial_z x(z,t) + f(x(z,t)),\ z \in [0,1],\ v(z) > 0 \\ x(0,t) &= u(t),\ t \in R^+ \\ y(t) &= x(1,t),\ t \in R^+ \end{aligned}$$

Equivalent linear Target System:

Construct the following distributed-parameter target system

$$\begin{aligned} \partial_t w(z,t) &= \Phi\left(w(z,t), y(t)\right) \\ &= -\partial_z w(z,t) + bw(z,t) + k_2 w(1,t) - k_1 y(t) \\ w(0,t) &= \Theta(x(0,t)) = \Theta(u(t)) \\ w(z,0) &= \Theta(x(z,0)) \end{aligned}$$

for $(z,t) \in [0,1] \times \mathbb{R}^+$ on the state space $\mathcal{W} \subset L^2(0,1)$. $\Phi : \mathcal{W} \times \mathbb{R} \mapsto \mathcal{W}$. The parameters b, k_2 and k_1 are some known constants.

Nonlinear state transformation:

$$w(z,t) = \Theta(x(z,t))$$

Equivalent linear Target System:

Construct the following distributed-parameter target system

$$\begin{aligned} \partial_t w(z,t) &= \Phi\left(w(z,t), y(t)\right) \\ &= -\partial_z w(z,t) + bw(z,t) + k_2 w(1,t) - k_1 y(t) \\ w(0,t) &= \Theta(x(0,t)) = \Theta(u(t)) \\ w(z,0) &= \Theta(x(z,0)) \end{aligned}$$

for $(z,t) \in [0,1] \times \mathbb{R}^+$ on the state space $\mathcal{W} \subset L^2(0,1)$. $\Phi : \mathcal{W} \times \mathbb{R} \mapsto \mathcal{W}$. The parameters b, k_2 and k_1 are some known constants.

Nonlinear state transformation:

$$w(z,t) = \Theta(x(z,t))$$

Associated first order PDE:

$$\begin{aligned} \frac{\partial \Theta(x)}{\partial x}\left[-\partial_z x + f(x)\right] &= -\partial_z \Theta(x) + b\Theta(x) + k_2\Theta(x(1,t)) - k_1 x(1,t) \\ \Theta(0) &= 0 \end{aligned}$$

Observer for the target system:

Same structure with the target system (**Initial condition**):

$$\begin{aligned} \partial_t \hat{w}(z,t) &= \Phi\left(\hat{w}(z,t), y(t)\right) \\ &= -\partial_z \hat{w}(z,t) + b\hat{w}(z,t) + k_2 \hat{w}(1,t) - k_1 y(t) \\ \hat{w}(0,t) &= \Theta(u(t)) \\ \hat{w}(z,0) &= \hat{w}_0(z) \neq \Theta(x(z,0)) \end{aligned}$$

The observer error dynamic $\tilde{w}_e(z,t) = w(z,t) - \hat{w}(z,t)$

$$\begin{aligned} \partial_t \tilde{w}_e(z,t) &= \Phi\left(\tilde{w}_e(z,t), y(t)\right) \\ &= -\partial_z \tilde{w}_e(z,t) + b\tilde{w}_e(z,t) + k_2 \tilde{w}_e(1,t) \\ \tilde{w}_e(0,t) &= 0 \\ \tilde{w}_e(z,0) &= \Theta(x(z,0)) - \hat{w}_0(z) \end{aligned}$$

Exponentially Stable

Observer for the nonlinear plant
Construct Luenberger extended observer for the nonlinear plant:

$$
\begin{aligned}
\partial_t \hat{x}(z,t) &= -\partial_z \hat{x}(z,t) + f(\hat{x}(z,t)) \\
&\quad - \left(\frac{\partial \Theta(\hat{x})}{\partial \hat{x}}\right)^{-1} k_1 \left(y(t) - \hat{x}(1,t)\right) \\
\hat{x}(0,t) &= u(t) \\
\hat{x}(z,0) &= \hat{x}_0(z)
\end{aligned}
$$

such that

$$
\begin{aligned}
\partial_t \tilde{\theta}(z,t) &= -\partial_z \tilde{\theta}(z,t) + b\tilde{\theta}(z,t) + k_2\tilde{\theta}(1,t) \\
\tilde{\theta}(0,t) &= 0
\end{aligned}
$$

where $\tilde{\theta}(z,t) = \Theta(x(z,t)) - \Theta(\hat{x}(z,t))$.

$$\hat{w}(z,t) = \Theta(\hat{x}(z,t)), \quad \tilde{\theta}(z,t) = \tilde{w}_e(z,t)$$

Analytic solution of associated PDE

$$
\begin{aligned}
\frac{\partial \Theta(x)}{\partial x}\left[-\partial_z x + f(x)\right] &= -\partial_z \Theta(x) + b\Theta(x) + k_2\Theta(x(1,t)) - k_1 x(1,t) \\
\Theta(0) &= 0
\end{aligned}
$$

$$\frac{\partial \Theta(x)}{\partial x}\phi(w,x) = \psi(w,x)$$

Analytic solution of associated PDE

$$\frac{\partial\Theta(x)}{\partial x}\left[-\partial_z x + f(x)\right] = -\partial_z\Theta(x) + b\Theta(x) + k_2\Theta(x(1,t)) - k_1 x(1,t)$$
$$\Theta(0) = 0$$

$$\frac{\partial\Theta(x)}{\partial x}\phi(w,x) = \psi(w,x)$$

Theorem

Assume that $f : \mathcal{X} \mapsto \mathcal{X}$ is real analytic function with $f(0) = 0$. The parameters a_1 and b satisfy the above assumptions. Suppose there exists an invertible function $P_1(z)$ such that

$$\frac{dP_1(z)}{dz} + (a_1 - b)P_1(z) = 0$$
$$P_1(1) = \frac{k_1}{k_2} \neq 0$$

and

$$\Theta(x(1,t)) - P_1(1)x(1,t) = 0$$

Then there exists a unique real analytic solation $w = \Theta(x)$ to the above PDE locally in the neighborhood of $x = 0$ with $\Theta(0) = 0$ and $\frac{\partial\Theta(0)}{\partial x} = P_1(z)$.

Truncated expression of $\Theta(x)$

$$\Theta(x) = P_1(z)x + P_2(z)x^2 + \cdots + P_k(z)x^k + \cdots$$

$$\frac{d}{dz}\begin{bmatrix} P_1(z) \\ P_2(z) \\ \vdots \\ P_k(z) \\ \vdots \end{bmatrix} = \begin{bmatrix} -(a_1-b) & 0 & \cdots & 0 & \cdots \\ -a_2 & -(2a_1-b) & \cdots & 0 & \cdots \\ \vdots & \vdots & \ddots & \vdots & \vdots \\ -a_k & -2a_{k-1} & \cdots & -(ka_1-b) & \\ \vdots & \vdots & \cdots & \vdots & \ddots \end{bmatrix}\begin{bmatrix} P_1(z) \\ P_2(z) \\ \vdots \\ P_k(z) \\ \vdots \end{bmatrix}$$

with $P_1(1)= k_1/k_2,\ P_2(1) = 0,\ \cdots,\ P_k(1) = 0,\ \cdots$

Example:

Nonlinear transport process

$$\begin{aligned}
\partial_t x(z,t) &= -\partial_z x(z,t) + 1.7\sin\left(x(z,t)\right) \\
x(0,t) &= \sin(2\pi t) \\
x(z,0) &= 5\sin(2\pi z) \\
y(t) &= x(1,t)
\end{aligned}$$

The extended Luenberger observer:

$$\begin{aligned}
\partial_t \hat{x}(z,t) &= -\partial_z \hat{x}(z,t) + 1.7\sin\left(\hat{x}(z,t)\right) \\
&\quad -L(z,t)\left(y(t) - \hat{y}(t)\right) \\
\hat{x}(0,t) &= \sin(2\pi t) \\
\hat{x}(z,0) &= -7\cos(2\pi z) \\
\hat{y}(t) &= \hat{x}(1,t)
\end{aligned}$$

where $L(z,t) = \left(\frac{\partial \Theta(\hat{x})}{\partial \hat{x}}\right)^{-1} k_1$

Coefficients in the Taylor expansion of $f(x)$ are
$a_1 = 1.7$, $a_2 = 0$, $a_3 = -0.2833$, $a_4 = 0$ and $a_5 = 0.0142$.
If we select $b = 0$, $k_2 = -5$ and $k_1 = -10$.

$$\Theta(\hat{x}) \approx P_1(z)\hat{x} + P_3(z)\hat{x}^3 + P_5(z)\hat{x}^5$$

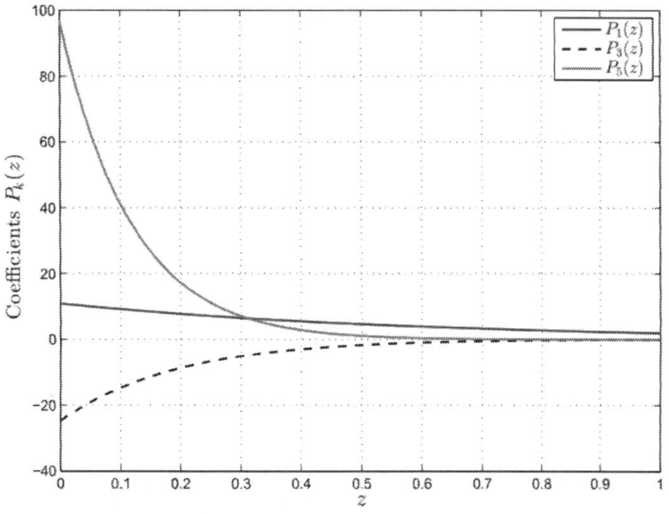

Figure: The coefficients of Taylor expansion for nonlinear transformation function $\Theta(x)$. When $b = 0$ and $k_1 = -10$, $|P_1(z)| < |P_3(z)| < |P_5(z)|$. The transformation function $\Theta(x)$ is far away from the linear part $P_1(z)x$.

This problem appears because the approximation is not good enough.

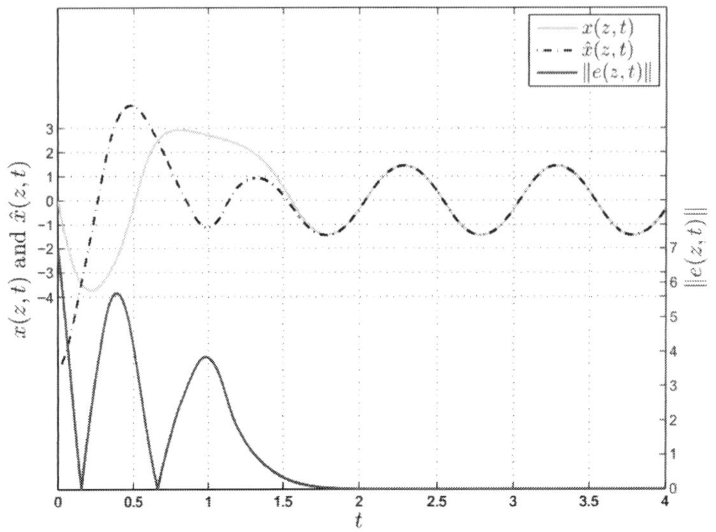

Figure: The plant trajectory $x(z,t)$ and the observer trajectory $\hat{x}(z,t)$. The observer does not converges to the plant state quickly.

$b = 7$, $k_2 = -5$ and $k_1 = -10$

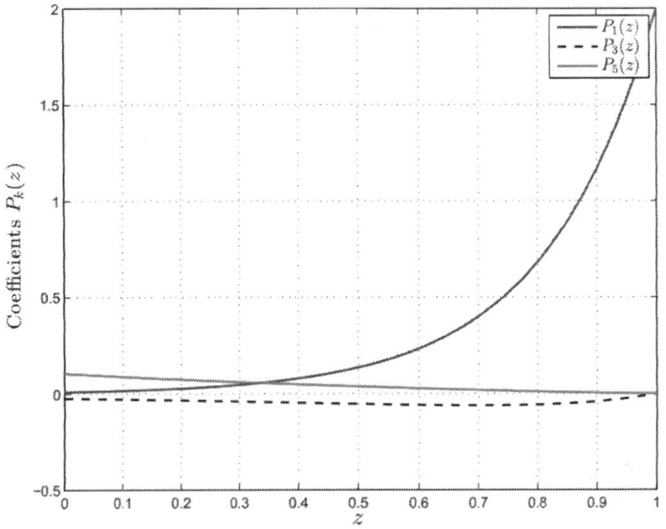

Figure: The coefficients of Taylor expansion for nonlinear transformation function $\Theta(x)$. When $b = 7$ and $k_1 = -10$, $|P_1(z)| >> |P_3(z)| > |P_5(z)|$. The transformation function $\Theta(x)$ is close to the linear part $P_1(z)x$.

Simulation Results

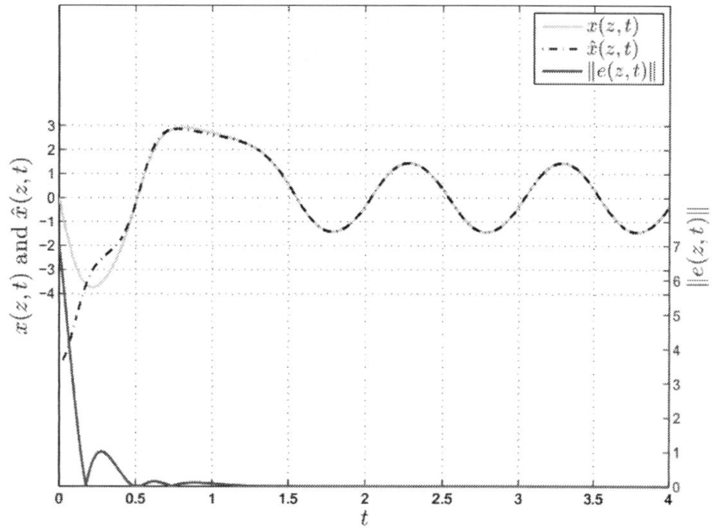

Figure: The plant trajectory $x(z,t)$ and the observer trajectory $\hat{x}(z,t)$. The observer error almost vanishes after $0.5s$.

Conclusion:

- The extended Luenberger observer is designed.
- Analytic nonlinear state transformation is applied to compute the output injection weight to ensure the convergence of the observer.
- To obtain the nonlinear state transformation function $\Theta(x)$, the associated PDE with sepcific boundary conditions is solved. (**Convergent Solution**)
- The performance of the extended Luenberger observer for the nonlinear transport process is demonstrated via a simulation study and the guidance of choice of design parameter is provided.

435459 ISA-95 Based Scheduler Using Flexible Heuristics

Abstract:

The main challenge in productizing scheduling solutions is often twofold: 1) Defining a landscape that can host the algorithmic environment, collect information necessary to formulate a realistic problem instances and communicate the results into production, 2) Providing an algorithm that works efficiently for various cases and provides - if not optimal - solutions that can be used in practice.
In this presentation we will address both challenges and present a scheduling solution that uses ISA-95 standard for importing the information based on which the problem is modeled. To ensure an efficient operation the solution uses a number of heuristic algorithms, which in many cases perform close-to-optimally. Different use cases are shown, along the discussion on how the algorithms performance can be affected both in terms of efficiency and quality.

Harjunkoski, I. and Bauer, R. (2014). Sharing data for production scheduling using the ISA-95 standard. *Frontiers in Energy Research*, 2/44, pp. 1-15

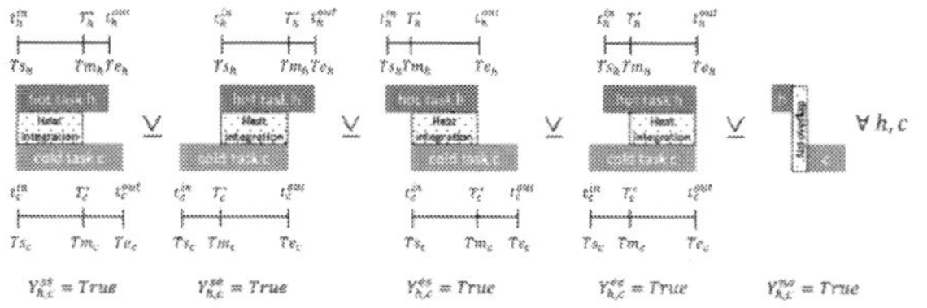

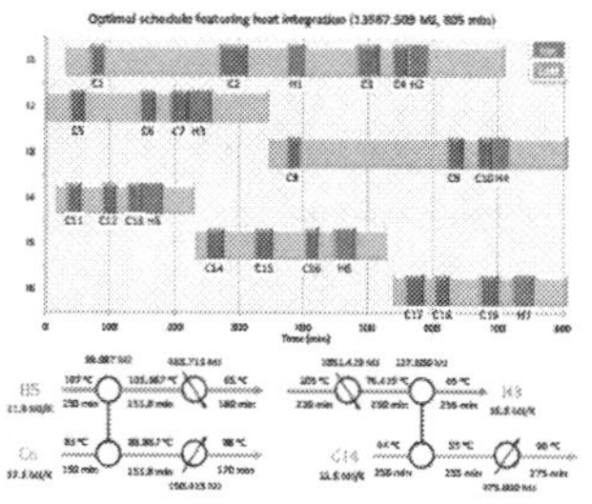

A Generalized Disjunctive Programming Model for Simultaneous Scheduling and Heat Integration of a Vegetable Oil Refinery

Pedro M. Castro

Bruno Custódio

Henrique A. Matos

INVESTIGADOR FCT

Ciências ULisboa

Unilever Jerónimo Martins

Técnico Lisboa

U Lisboa Universidade de Lisboa

Motivation

- Energy efficiency: effective way to combat climate change
- System optimization methods help identify energy saving potentials in industry
  - Strategic plan US DOE
  - EU framework program for research & innovation (Horizon 2020)
    - Savings ≥ 13% expected
- Process integration widely applied to continuous plants
  - Pinch analysis (Linnhoff et al. 1982)
  - Mathematical programming (Biegler et al. 1997)
- Not so much to batch plants due to following challenges
  - Modelling: tackle scheduling & heat integration simultaneously
  - Operational: control of dynamic heat exchanger network
  - Economical: energy savings occur over a fraction of the day, return on equipment investment more difficult

Problem statement

- Single-stage batch plant with multiple tasks per product i
- Parallel units, fixed product-unit assignments $m \in M_i$
- Subset of tasks $k \in K_i$ requires:
  - Heating $\Rightarrow$ Cold task c (stream ps)
  - Cooling $\Rightarrow$ Hot task h (stream ps)
- Characterized by:
  - Duration $p_{i,k}$; Heat capacity cp_{ps}; Initial t_{ps}^{in} & target temperatures t_{ps}^{out}
- Heat integration between tasks of different units

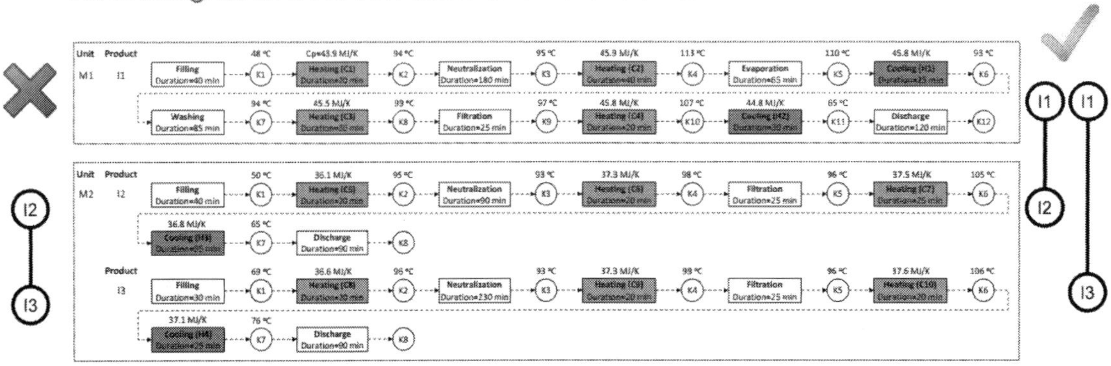

Direct heat integration model

- Adds timing element to Yee et al. (1990) model
  - Simultaneous targeting & heat exchanger network design
- Hot/cold stream goes through sequence of cooling/heating stages
  - # stages/slots specified by the user
- Parallel matches allowed in each stage
  - Process streams and external utility

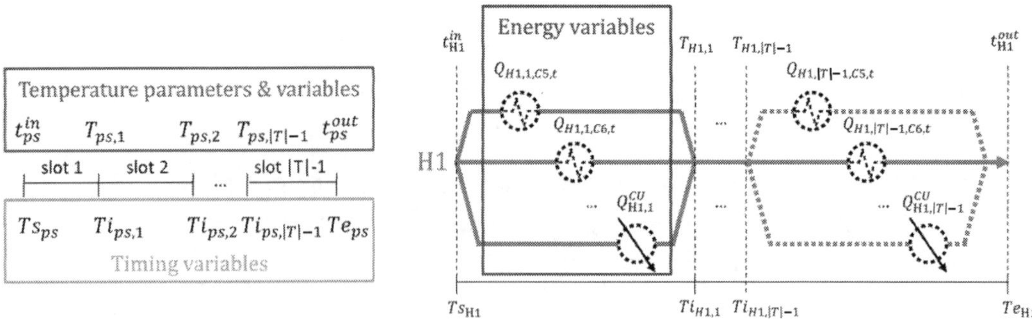

GDP heat integration model (2 stages)

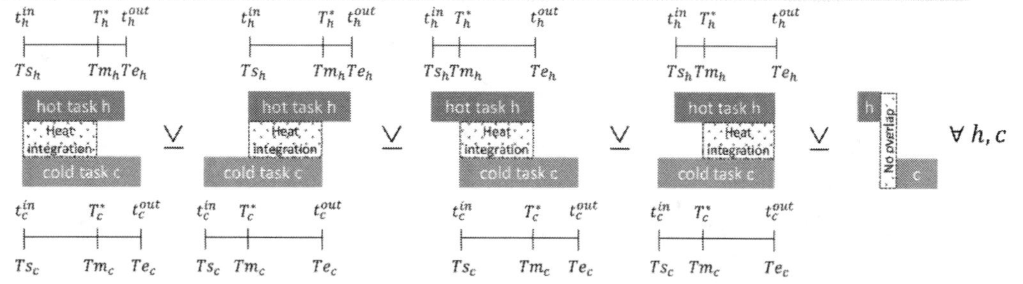

$Y_{h,c}^{ss} = True$ $\quad Y_{h,c}^{se} = True$ $\quad Y_{h,c}^{es} = True$ $\quad Y_{h,c}^{ee} = True$ $\quad Y_{h,c}^{no} = True$

- Timing, temperature driving force & energy bounds

$$\begin{bmatrix} Y_{h,c}^{ss} \\ Ts_h = Ts_c \\ Tm_h = Tm_c \\ T_h^* \geq T_c^* + \Delta t \\ Q_{h,c}^{ss} \leq q_{h,c} \\ Q_{h,c}^{se}, Q_{h,c}^{es}, Q_{h,c}^{ee} = 0 \end{bmatrix} \veebar \begin{bmatrix} Y_{h,c}^{se} \\ Ts_h = Tm_c \\ Tm_h = Te_c \\ T_h^* \geq t_c^{out} + \Delta t \\ Q_{h,c}^{se} \leq q_{h,c} \\ Q_{h,c}^{ss}, Q_{h,c}^{es}, Q_{h,c}^{ee} = 0 \end{bmatrix} \veebar \begin{bmatrix} Y_{h,c}^{es} \\ Tm_h = Ts_c \\ Te_h = Tm_c \\ t_h^{out} \geq T_c^* + \Delta t \\ Q_{h,c}^{es} \leq q_{h,c} \\ Q_{h,c}^{ss}, Q_{h,c}^{se}, Q_{h,c}^{ee} = 0 \end{bmatrix} \veebar \begin{bmatrix} Y_{h,c}^{ee} \\ Tm_h = Tm_c \\ Te_h = Te_c \\ t_h^{out} \geq t_c^{out} + \Delta t \\ Q_{h,c}^{ee} \leq q_{h,c} \\ Q_{h,c}^{ss}, Q_{h,c}^{se}, Q_{h,c}^{es} = 0 \end{bmatrix} \veebar \begin{bmatrix} Y_{h,c}^{no} \\ Ts_h \geq 0 \\ Ts_c \geq 0 \\ Q_{h,c}^{ss}, Q_{h,c}^{se}, Q_{h,c}^{es}, Q_{h,c}^{ee} = 0 \end{bmatrix} \forall h, c$$

- Energy balances

$$cp_h(t_h^{in} - T_h^*) = \sum_c (Q_{h,c}^{ss} + Q_{h,c}^{se}) + Q_h^{CUs} \; \forall h \qquad cp_c(T_c^* - t_c^{in}) = \sum_h (Q_{h,c}^{ss} + Q_{h,c}^{es}) + Q_c^{HUs} \; \forall c$$

$$cp_h(T_h^* - t_h^{out}) = \sum_c (Q_{h,c}^{es} + Q_{h,c}^{ee}) + Q_h^{CUe} \; \forall h \qquad cp_c(t_c^{out} - T_c^*) = \sum_h (Q_{h,c}^{se} + Q_{h,c}^{ee}) + Q_c^{HUe} \; \forall c$$

Utility consumption

$$UT = \sum_h (Q_h^{CUs} + Q_h^{CUe}) + \sum_c (Q_c^{HUs} + Q_c^{HUe})$$

Scheduling model (continuous-time)

- Sequence products allocated to same unit
  - General precedence variables $X_{i,i'}$ (Méndez et al. 2001)
  - Tasks of a production recipe treated as aggregate task

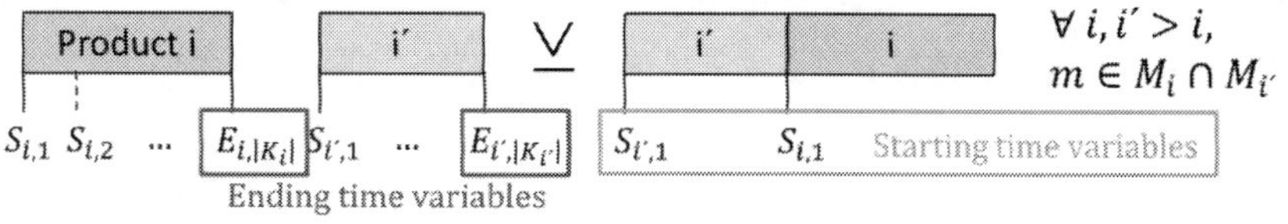

$X_{i,i'} = True$ $\qquad X_{i,i'} = False$

$$\begin{bmatrix} X_{i,i'} \\ E_{i,|K_i|} \leq S_{i',1} \end{bmatrix} \veebar \begin{bmatrix} \neg X_{i,i'} \\ E_{i',|K_{i'}|} \leq S_{i,1} \end{bmatrix} \forall i, i' > i, m \in M_i \cap M_{i'}$$

- Timing constraints

$$E_{i,k} = S_{i,k} + p_{i,k} \; \forall i, k \in K_i \qquad S_{i,k+1} = E_{i,k} \; \forall i, k \in K_i \backslash |K_i|$$

Definition of makespan

$$MK \geq E_{i,|K_i|} \; \forall i$$

Linking the two models

- Just need to relate timing variables
  - Easy to replace scheduling model (more complex environment)
    - STN/RTN-based model for multipurpose plants

Scheduling & heat integration model

Heat integration model

$$Ts_{ps} = \sum_i \sum_{k:ps \in PS_{i,k}} S_{i,k} \ \forall ps$$

$$Te_{ps} = \sum_i \sum_{k:ps \in PS_{i,k}} E_{i,k} \ \forall ps$$

Scheduling model

- Heat-integration layer chooses temperature profile

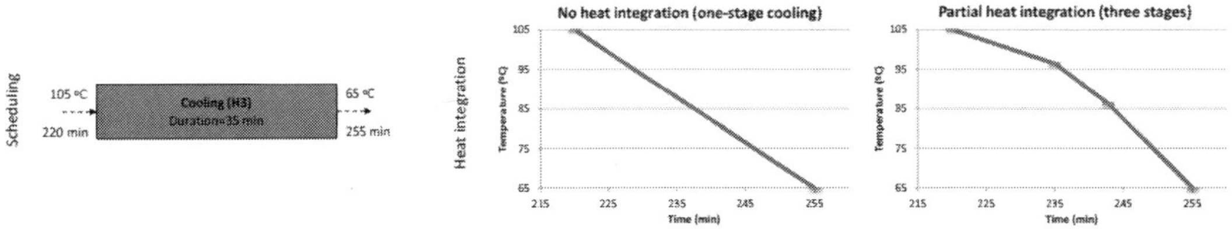

Bi-objective optimization

- Tradeoff makespan vs. utility consumption
  - If heating/cooling task waits for partner ⇒ production delays for minimum utility usage
  - Heat integration still possible for minimum makespan
- Pareto front through ε-constraint method

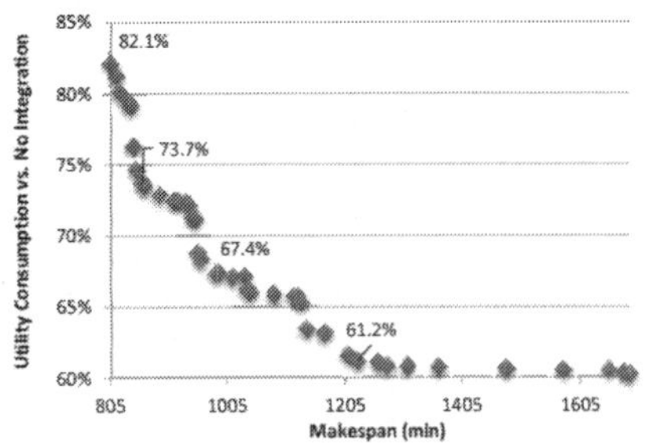

Energy savings

18 %

40 %

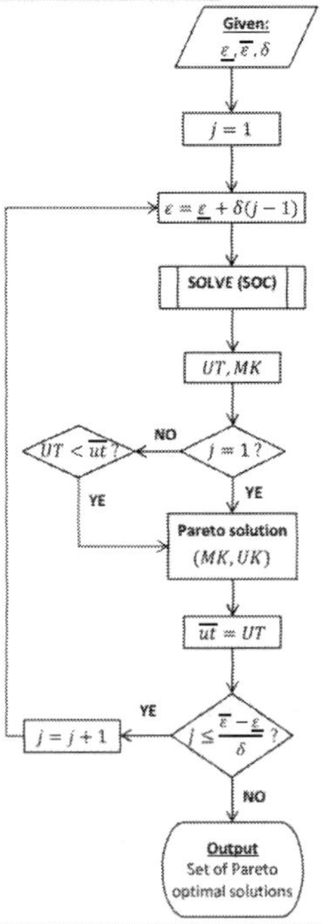

2 or 3 temperature-changing stages?

- Utility savings (& CPUs) increase with number of stages

P18S

Number of stages	1	2	3	4
Utility savings (%)	0	33.6	34.0	34.0
Computational time (CPUs)	0.12	0.73	53.0	1558

P26S

1	2	3
0	37.7	39.2
0.11	18.4	*3600**

- Optimal solution changes slightly for minimum makespan (P26S)

H5-C6 match lasts 1.9 min!

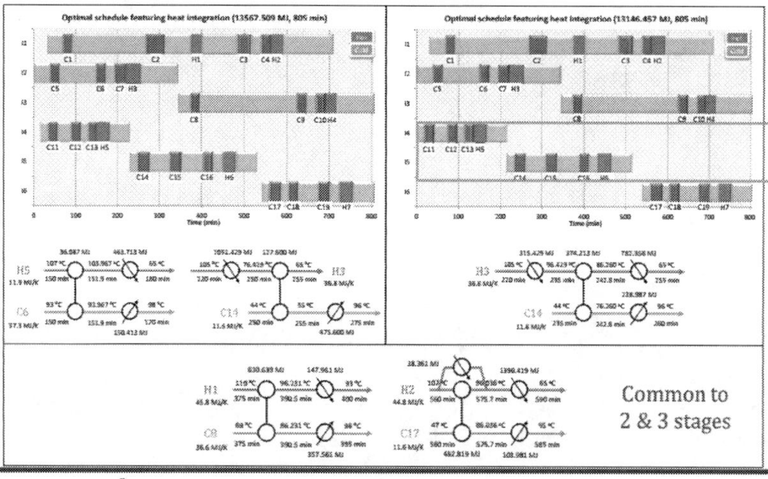

H3 only stream with 3 stages!

Pareto optimal solutions

- Smoother frontier (more points) for 3 stages

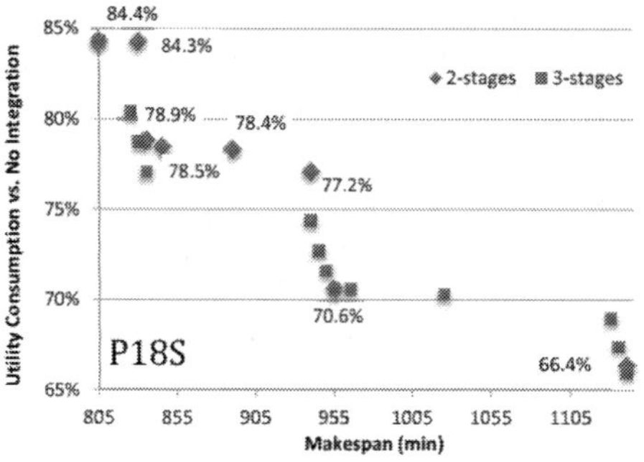

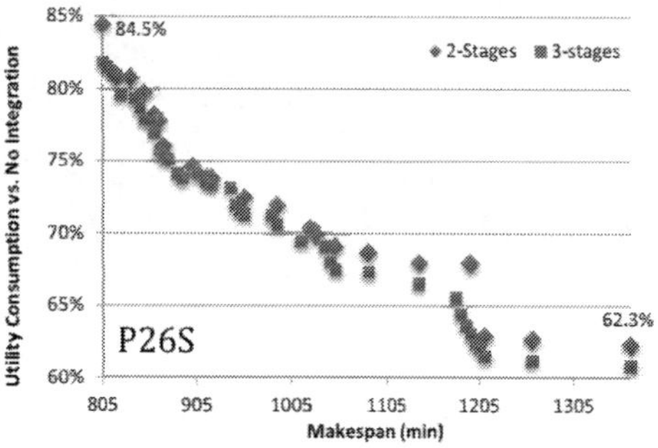

- Total computational time prohibitively high
  - 4 times longer than minimum makespan
    - Not enough time to wait for optimization!

Problem/Stages	2	3
P18S	29.5	927
P26S	463	202652
P33S	171971	-

CPLEX 12.6 parallel deterministic mode (up to 8 threads)
Relative optimality tolerance=10^{-6}
Intel i7-4790 CPU @ 3.6 GHz, 8 GB of RAM
Windows 7, 64-bit, SSD

Two interesting solutions (P26S)

(890 min, 73.8%) (1050 min ↑30%, 67.5%)

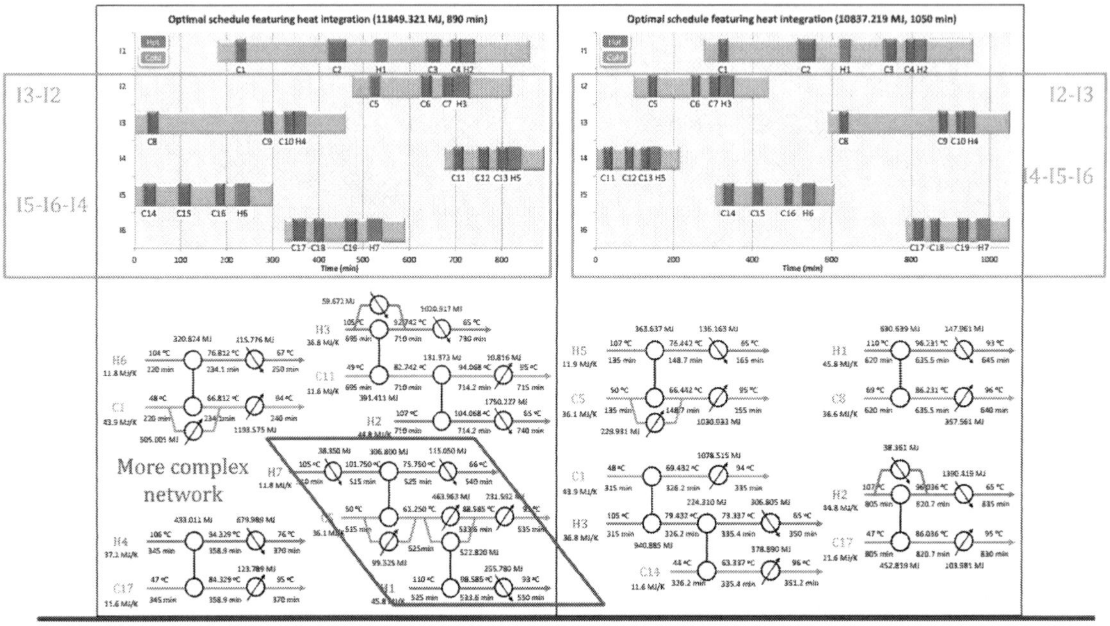

Conclusions

- New MILP model for improving energy efficiency
  - Simultaneous scheduling & heat integration in batch plants
  - Superstructure with multiple temperature-changing stages
    - Allows for parallel matches between process streams and external utilities
  - Multi-objective problem
    - Makespan vs. utility consumption
- Industrial case study of vegetable oil refinery
  - 39% potential savings compared to no heat integration
    - Major delays necessary for optimal synchronization of tasks
  - 15.6% without compromising productivity
  - Variety of Pareto optimal solutions to choose from
- Future work
  - Also include indirect heat-integration ⇒ further savings
  - Relax isothermal mixing constraint ⇒ MINLP problem
- Further details
  - CACE 82 (2015) 172-185
- Acknowledgments:

  - Fundação para a Ciência e Tecnologia
    - Investigador FCT 2013, PTDC/EQU-ESI/118253/2010, grant SFRH/BDE/51346/2011

Optimal Design of Natural Gas Dryers

Yasser Al Wahedi[a*], Arwa H. Rabie[b], Prodromos Daoutidis[c*]

[a]*Department of Chemical Engineering, Abu Dhabi Petroleum Institute, P.O. Box 2533, Abu Dhabi, United Arab Emirates*

[b]*Department of Research & Technology, Abu Dhabi Gas Industries Ltd. (GASCO), Abu Dhabi, P.O. Box 665, Abu Dhabi, United Arab Emirates*

[c] *Department of Chemical Engineering and Materials Science, University of Minnesota, Minneapolis, MN-55455, USA*

AIChE Annual Meeting

Marriot , Salon D
Wednesday, November 12th 2015

RESEARCH & DEVELOPMENT
GAS SUB-COMMITTEE

Problem Motivation

- **World Energy Demand:**
  - Increase in energy demand.
  - Gas Supply Required Increases (by 131% from 2000 till 2009)(1)
  - LNG Market is Expected to Increase in the coming future
- **Dehydration of Gas:**
  - Natural Gas used for LNG is dehydrated by Molecular Sieves.
  - Total Market Size of LNG ranges between 120 to 160 billion as of 2014(3)
- **Importance of Optimal Design**
  - Market size and expected increase in LNG demand.
  - Deficiencies in current design approaches.

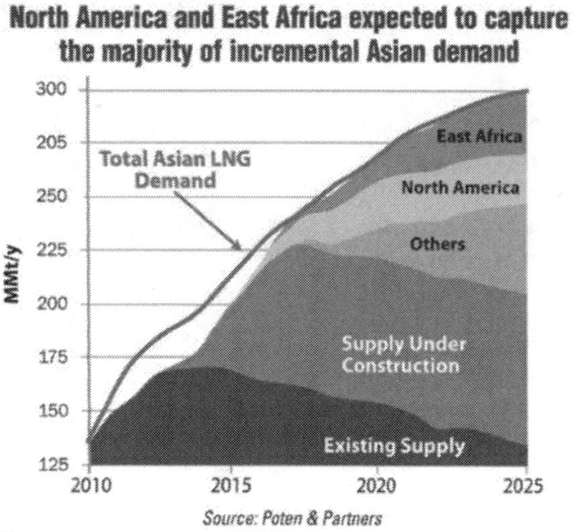

http://www.alaskajournal.com/business-and-finance/2013-12-04/lng-101-spot-market-pricing-economics

(1) http://en.wikipedia.org/wiki/World_energy_consumption
(3) http://globalchange.mit.edu/files/document/MITJPSPGC_Rpt27...

RESEARCH & DEVELOPMENT
GAS SUB-COMMITTEE 2

Typical Process Schemes for NG Dehydration

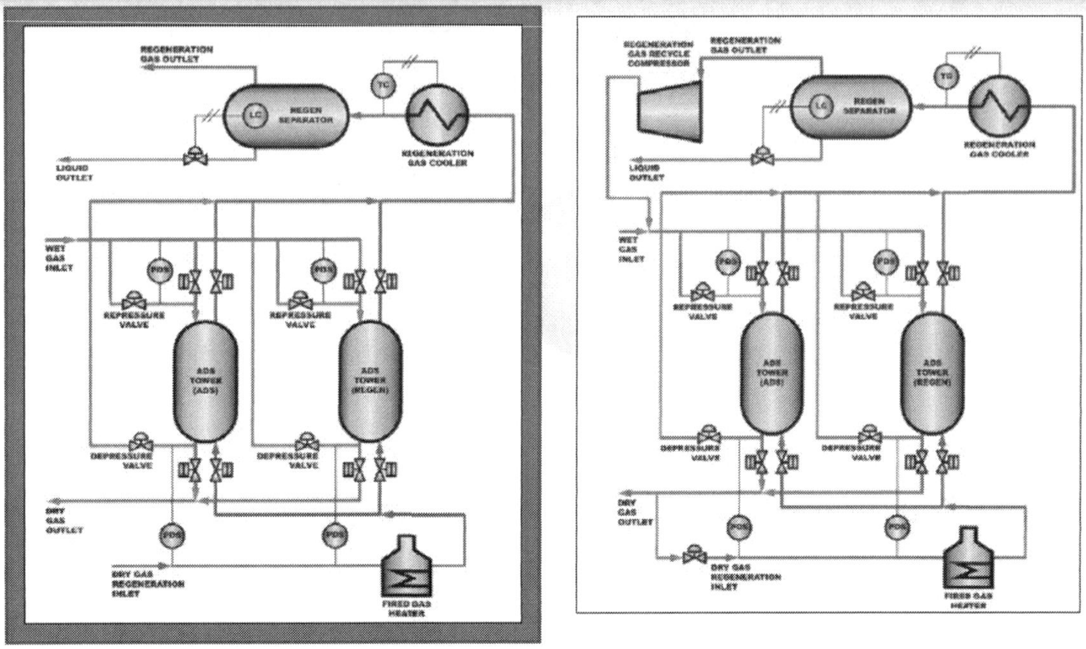

F. S. Manning and R. E. Thompson, *Oilfield Processing of Petroleum: Natural gas*: PennWell Books, 1991.

RESEARCH & DEVELOPMENT
GAS SUB-COMMITTEE

Design Approaches

Common approach used based upon experimentally determined "effective" sorbent capacity in mol/kg & Target Pressure Drop (Wunder, Oil & Gas, 1962)

Rely on assumed cycle timings and simplified model of the cycle

We Target An Optimization Design Protocol which Assures design optimality by Virtue of Lowest NPV while capturing the complexities of kinetics

Total ensued costs during project life (NPV) is not considered

Design can satisfy the requirement but no knowledge on possible cost savings

RESEARCH & DEVELOPMENT
GAS SUB-COMMITTEE

The Optimization Problem Structure

- Objective function is the Net Present Value of Costs (NPVC).
- Eight design variables.
- Four Constraints.

$$NPVC = \underbrace{6\cdot\left(P_{bed}+P_{v}+P_{FHR}+P_{comp}+P_{HxC}+P_{Sep}\right)}_{TCI}+$$
$$\underbrace{\sum_{j=1}^{J}\left(P_{W}+P_{HDR}+P_{CW}\right)\left(\frac{1+r_{\inf}}{1+r_{dis}}\right)^{j}}_{OPEX}+\underbrace{\sum_{j=n^{*}f}^{J}\left(P_{bed}\right)\left(\frac{1+r_{\inf}}{1+r_{dis}}\right)^{j}}_{BED}$$

Design Variable	Description
L	Bed Length (m)
D	Bed Diameter (m)
N_{ads}	Number of Vessels undergoing adsorp.
N_{reg}	Number of Vessels undergoing regen.
Q_{reg}	Molar flow of regeneration gas (kmol·h^{-1})
T_{regG}	Temperature of regeneration gas (°C)
t_{reg}	Allocated time for regeneration (h)
t_{ads}	Adsorption cycle time (h)

Constraint Description	Mathematical Expression
Assigned adsorption cycle time should not exceed the breakthrough time predicted via simulation given L,D and N.	$t_{ads} \le t_{ads,max}$ (2)
Total regeneration time should not exceed the assigned adsorption cycle time.	$t_{ads} \ge \frac{N_{ads}}{N_{reg}}\left(t_{reg}+t_{heating}+t_{cooling}\right)$ (3)
Assigned regeneration cycle should exceed the minimum required time for regeneration predicted via simulations given L,D, N and N_1	$t_{reg} \ge t_{reg,min}$ (4)
TSA unit pressure drop should not exceed the maximum allowable pressure drop ΔP_{max}	$\Delta_{max} \ge L\left(\frac{150\mu(1-\varepsilon_b)^2 U_f}{4R_p^2\varepsilon_b^3}+\frac{1.75\rho_f(1-\varepsilon_b)U_f^2}{2R_p\varepsilon_b^3}\right)$ (5)

RESEARCH & DEVELOPMENT
GAS SUB-COMMITTEE 5

Overall Algorithm for Optimization(1)

- **Optimization Step 1**
  - Use Rhee et al. solution for the isothermal equilibrium adsorption/regeneration case(2).
  - Use of efficiency factors might cause deviations from reality.
  - Algorithm assumes phase change like regeneration
  - Algorithm do not include temperature temporal and spatial variations.
  - Account for vessel steel heating.

- **Optimization Step 2**
  - Full Cycle Optimization Done using Gproms 4.0
  - Solve governing mass and energy PDEs.
  - Losses to vessel steel included in the model.

(1) Al Wahedi et al., CES, 2015
(2) (Rhee et al., Trans. Of Roy. Soc. Of London, 1972)

RESEARCH & DEVELOPMENT
GAS SUB-COMMITTEE 6

Implementing the Algorithm

Assumptions:
- 20-100 kPA Pressure drop
- Pellet Dia. is 1 cm
- 1000 regenerations
- Adsorbent cost 1.7 $/kg

Adsorbent:
- Zeolite 3A[1]
- Design Capacity at end of life.
- Void fraction of 0.4

Category	Feed Gas
Pressure (bar)	69.3
Temperature (°C)	28
Compressibility Factor	0.82
Feed Flow (kmol h^{-1})	31,525
Methane	8.31E-01
Ethane	8.73E-02
Propane	4.22E-02
i-Butane	8.31E-03
n-Butane	1.57E-02
i-Pentane	4.27E-03
n-Pentane	3.93E-03
n-Hexane	3.20E-03
Nitrogen	3.19E-03
H_2O	7.05E-04

(1) M. Simo et al, Ind Eng Chem Res, 2009

RESEARCH & DEVELOPMENT
GAS SUB-COMMITTEE 26

The Results

- **Total computational time is ~2 hours per case**
- **OPEX comprise the majority of the cost**

- **NPV ranges from 4.3 to 5.1 $/MMSCF**
- **OPEX ranges from 2.1 to 2.9 $/MMSCF**

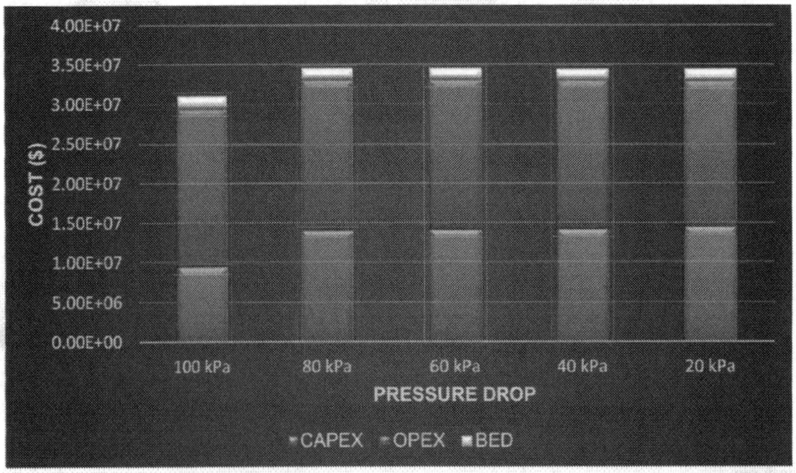

RESEARCH & DEVELOPMENT
GAS SUB-COMMITTEE 8

Conclusions

- Opportunity for optimizing long term cost in dehydration processes' design.
- Operating Costs (OPEX) comprises the majority of the NPV and hence must be considered over the course of the plant life.
- For the studied case, computed NPV ranged from 4.3 to 5.1 $/MMSCF while OPEX ranged from 2.1 to 2.9 $/MMSCF.
- The importance of considering NPV as the objective function.

Acknowledgement

- Dr. Frank Geuzebroek
- Abu Dhabi Gas Industries Ltd
- Gas Sub Committee for their kind funds.

Back Up Slide

RESEARCH & DEVELOPMENT
GAS SUB-COMMITTEE

Adsorption Systems Optimization Literature Review

- **General Comments:**
  - Literature Focuses on PSA.
  - Optimization Coupled with Kinetics and tackled simultaneously.
  - Most of studies focus on single point optimization.
  - Problem Break Down.

Optimization suffers from Long computational times (~ 100s hrs), Objective functions do not represent all costs, and absence of implementation of global search engines.

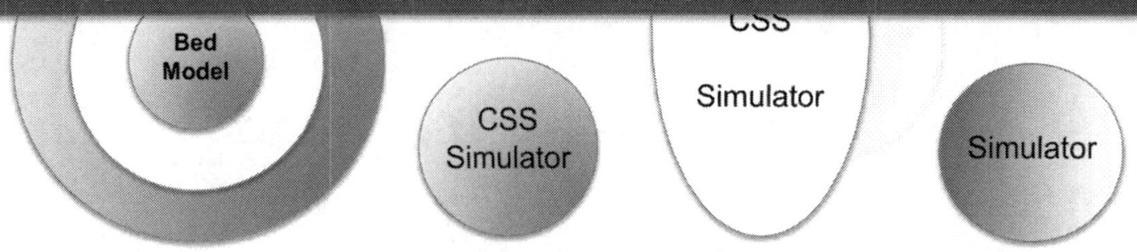

(a) (Jiang et al. ,2003) (Sankararao et al., 2007) (Jiang et al. ,2003)

CSS: Cyclic Steady State

RESEARCH & DEVELOPMENT
GAS SUB-COMMITTEE

Enabling Bioprocess Development Using Virtual Plant Technology

John D. Bomberger and Tracy L. Clarke-Pringle
DuPont Engineering Research and Technology
Dynamic Simulation, Control, and Optimization Group

AIChE Annual Meeting, Salt Lake City, UT
Nov 11, 2015

Virtual Plant Technology

- A "Virtual Plant" is a surrogate for the real process/plant
- It is an environment where the fundamental engineering equations* that describe a process can be executed repeatedly (in a dynamic simulation), to show the evolving behavior of the process as time proceeds.
- It's a "flight simulator" for a manufacturing process

*equations: material and energy balances, thermodynamics, reaction kinetics

 2

Why Dynamic Modeling for Virtual Plants?

- Processes, particularly batch processes, are often too complex to be able to think through entire scenarios or consequences of actions
- We can't run unlimited experiments
  - equipment limitations, time limitations
  - cost of experiments
  - the future organism or equipment may not exist
  - it may not be safe
- Dynamic modeling provides a safe and educational environment to test out all ideas and gain insights into how a batch will evolve.
  - A 40 hour batch can be run in minutes
- Dynamic modeling provides quantitative data and improves our ability to make knowledgeable decisions

A Competitive Advantage

- When you combine Virtual Plant Technology with good process engineers, you end up with a winning combination
- When you give a good engineer access to a Virtual Plant, they will learn from it:
  - Evaluate and test process alternatives and process operability
  - Evaluate ideas for optimizing the process
  - Control strategy design and testing
    - This includes both batch automation and regulatory controls
    - Addressing how abnormal situations are handled
    - Rapid prototyping for advanced process control
  - PAS configuration testing
  - Operator Training Simulator
- Doing this "virtually" is faster, cheaper, and safer than using the real process

DuPont Experience with Virtual Plants: TMODS

- DuPont™ TMODS was created by DuPont engineers circa 1990. It is a first-principles, engineering, dynamic simulator
  - Invented by Bjorn Tyreus and Ed Longwell (late 1980s)
  - Developed into a world-class simulator by Bob Cox and Ross Wilcox (1990 – 2013)
- Today, this DuPont-owned software is managed and supported to meet DuPont needs
  - The core development team: Tracy Clarke-Pringle (lead), John Bomberger, Jorge Castro, Yuris Fuentes, Mike Zotti
  - We are users and developers
- DuET: Dynamic Simulation, Control and Optimization (DySCO) group – world class process modeling and control experts, and power users of TMODS
- The 300+ users across DuPont (process and process control engineers)
  - A wide range of power users and casual users

Using DuPont™ TMODS

- DuPont™ TMODS model are created from a library of unit operations
- Users provide specifications for each unit op (size, # trays, UA, valve CVs, etc.)
- Models are easy to construct, and can encompass whole plants
- Interactive simulations run on business grade laptops for maximum accessibility

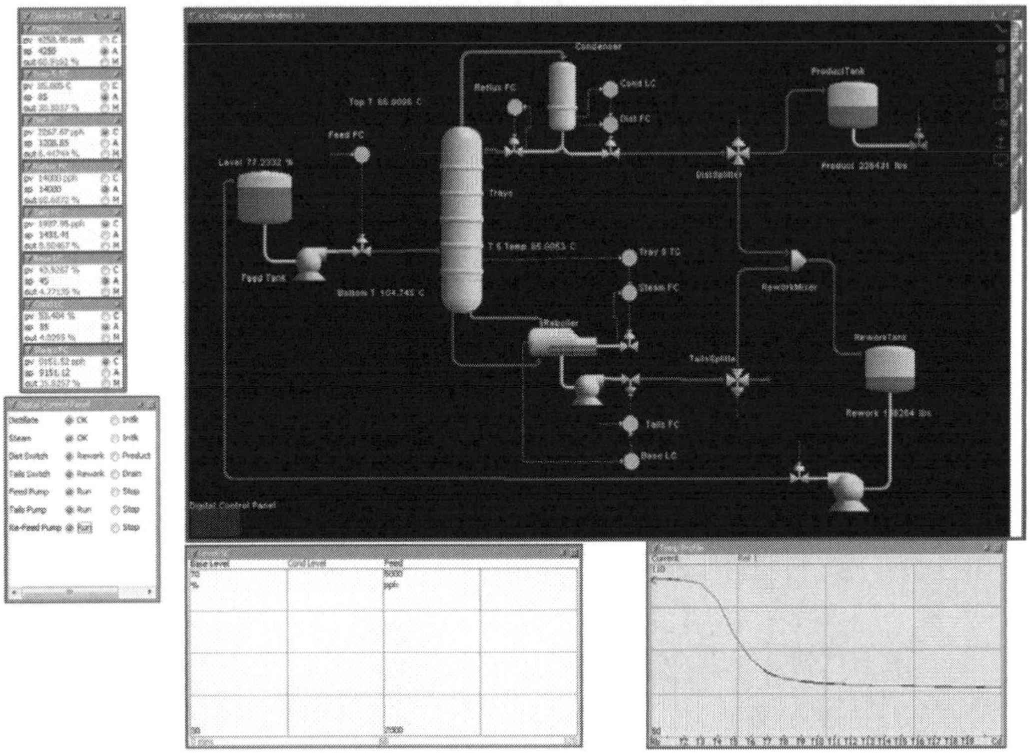

 Public

Virtual Plant Technology and Bioprocess Development

- DuPont has made a clear, long-term commitment to developing and operating bioprocesses
  - Modeling and simulation are critical in bioprocess development
    - Complicated flowsheets
    - Extensive use of batch and combined continuous / batch processing
    - Profitable in a narrow window of process conditions.
  - Dynamic models are valuable tools to develop process understanding
- Using Virtual Plant Technology in this area:
  - Accelerate growth of new businesses
  - Deliver productivity improvements

 Public

Project Life Cycle: Pre-commercialization through operation

- Example 1: Cellulosic Ethanol
  - Batch control design and functional description
  - Configuration checkout (FAT)
  - Startup procedure development and testing
  - Operator Training
- Example 2: Enzymatic Polymerization
  - Demonstration of Integrated operation
  - Source of dynamic data for individual unit operations, complementing Aspen and XLS flowsheet models
  - Plant wide modeling was started very early in the project life cycle

 9

Example 1: Cellulosic Ethanol

- Virtual Plant Technology was used for both pilot and commercial scale design and training

- Control System Functional Descriptions
  - High level control strategy narratives written from control strategies designed and tested on DuPont™ TMODS plant models
  - Critical batch automation was designed on TMODS
  - Detailed functional description for batch automation was prepared
  - BatchML export from TMODS was used to guide system integrators while programming batch logic

 10

Example 1: Cellulosic Ethanol

- DCS Configuration
  - DCS was *required* to be connected to DuPont™ TMODS models during Factory Acceptance Test of control system
  - Permitted thorough testing of all regulatory controls and batch automation
  - Potential time savings on Site Acceptance Test and configuration re-work
- Startup Procedures
  - Models of some areas were used to prepare and test startup procedures
  - Especially critical for highly interactive, energy integrated process

Example 1: Cellulosic Ethanol

- Operator Training
  - A DCS connected Operator Training Simulator was created
  - Uses DCS identical (in most aspects) to plant DCS
  - In OTS, DuPont™ TMODS replaces inputs from plant with inputs from model
  - DCS outputs (controller outputs, digital signals) drive TMODS model

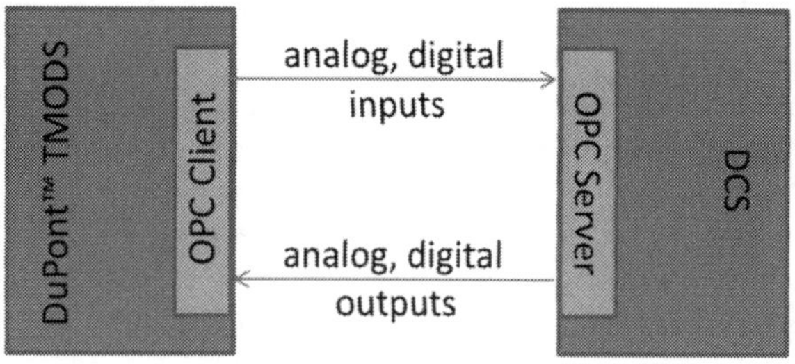

Example 2: Enzymatic Polymerization

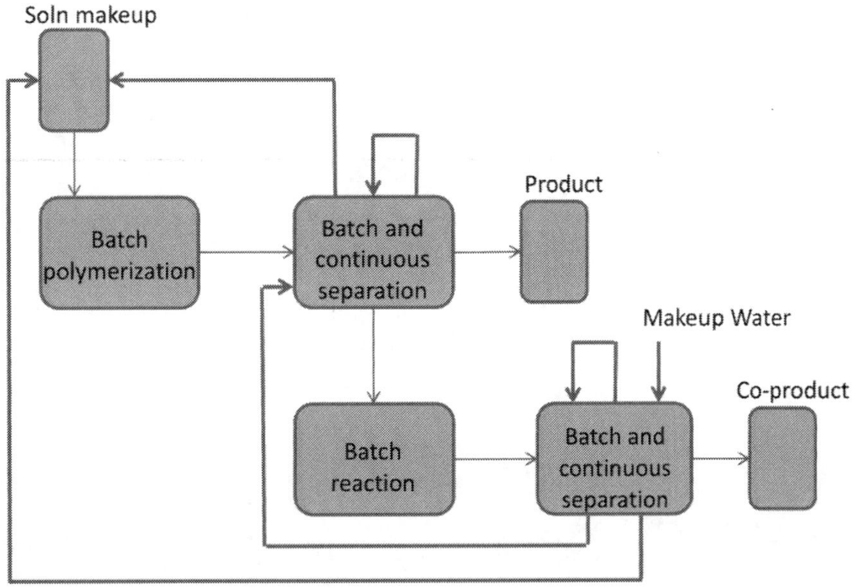

 Public

Example 2: Enzymatic Polymerization

- The process design relies heavily on water reuse, recycle, and integration within and between two co-processes
  - Water recycle is necessary for process economics
- However, the piloting strategy is to demonstrate pieces of the process at a toll facility
  - The only place the full integrated process will be demonstrated is in dynamic model

 Public

Example 2: Enzymatic Polymerization

- The process is a highly integrated combination of continuous, batch, and dynamically varying operations. Automation will be required to achieve design rates while minimizing labor costs
- DuPont™ TMODS provides a way to efficiently model the entire flowsheet dynamically. It is the best way we have of understanding plant-wide interactions to design and test the controls
- We have a preliminary automation design that optimizes operability of the integrated plant. Key elements:
  - Automation utilized to adjust batch, set rates based on need
  - Advanced regulatory control for soln tank inventory and concentration control
  - Surge tanks in between batch and continuous unit operations
  - Separation tanks watchdog to avoid tank drain or overflow

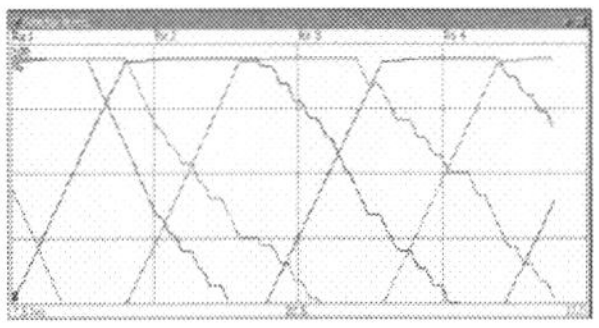
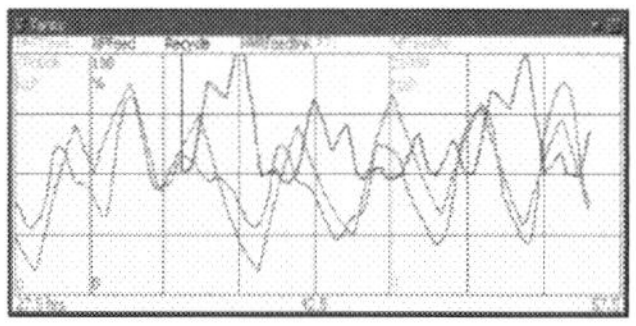
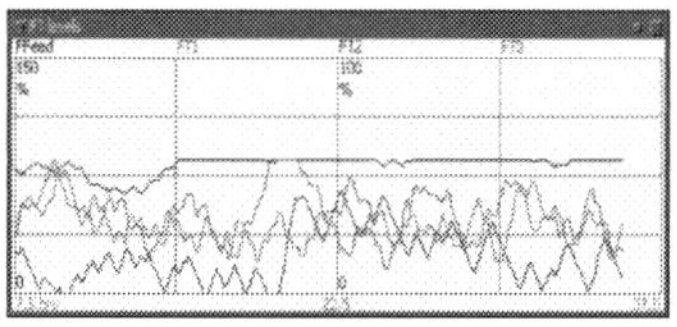

 Public 15

Summary

- Virtual Plant Technology is becoming an integral part of our commercialization of bioprocess technology
- It is used throughout our project lifecycle: Development, control system configuration and testing, training, and process improvement
- We find it a valuable way to save time and money during development, piloting, and commissioning

Future developments

- Smart manufacturing
- Virtual Plant as library of process and control system knowledge

 Public 16

421770 Dynamic Modeling, Advanced Control, Diagnosis and Optimization of Large-Scale Lignocellulosic Biorefineries

Abstract:

Second generation biorefineries transform agricultural wastes into biochemicals with higher added value, e.g. bioethanol, which is thought to become a primary component in liquid fuels [1]. Extensive endeavours have been conducted to make the production process feasible on a large scale, and recently several commercial size biorefineries became operational: Beta Renewables (Italy, 2014), Abengoa Bioenergy (USA, 2014), POET-DSM (USA, 2014), GranBio (Brazil, 2014) [2], while others are under construction, e.g. the Måbjerg Energy Consortium in Denmark, soon to be commissioned.

This paper presents the findings of a 3 years PhD project that was run by Technical University of Denmark (DTU) in collaboration with the Danish energy company DONG Energy A/S. The company owns a large second generation biorefinery in Kalundborg, Denmark, also known as the Inbicon demonstration plant [3]. The goal of the project was to utilize real-time data extracted from the large scale facility to formulate and validate first principle dynamic models of the plant. These models were then further utilized to derive model-based advanced tools for process optimization, control, diagnosis and monitoring.

The Inbicon biorefinery converts wheat straw into bioethanol utilizing steam, enzymes, and genetically modified yeast. The biomass is first pretreated in a steam pressurized and continuous thermal reactor where lignin is relocated, and hemicellulose partially hydrolyzed such that cellulose becomes more accessible to enzymes. The biorefinery is integrated with a nearby power plant following the IBUS principle for reducing steam costs [4]. During the pretreatment, by-products are also created such as organic acids, furfural, and pseudo-lignin, which act as inhibitors in downstream processes. The pretreated fibers consist of cellulose and xylan, which are then liquefied in the enzymatic hydrolysis process with the help of enzymes. High glucose and xylose yields are thus obtained for co-fermentation. Ethanol is recovered in distillation columns followed by molecular sieves for achieving a high concentration ethanol. Lignin is separated in the first column and recovered as bio-pellets in an evaporation unit. The bio-pellets are then burnt in the nearby power plant for steam generation.

The first part of the study presents a large scale dynamic model of the plant, separated in modules for pretreatment, enzymatic hydrolysis, and fermentation. The pretreatment and enzymatic hydrolysis models have been validated and analyzed in [5] and [6] together with a comprehensive sensitivity and uncertainty analysis. The models embed mass and energy balances with a complex conversion route. Computational fluid dynamics is used to model transportation in large tanks capturing tank profiles, and delays due to plug flows. As an application, the dynamic models are used to construct real-time observers that act both as measurement filters and soft sensors for variables that are not measured, e.g. biomass composition in any point of the process, or viscosity change in the enzymatic hydrolysis tanks.

In the second part of the study, besides a basic control layer for flow rates and tank levels, an L1 adaptive control layer [7] is constructed for biomass pretreatment temperature [8] and pH in enzymatic hydrolysis [9]. These controllers provide a better operation across multiple nominal points without the need of retuning. The basic control layer is characterized by a fast feedback reaction.

The next step is to build an optimization layer, which searches for optimal values regarding the pretreatment temperature, enzyme dosage, and yeast seed in fermentation. The biorefinery has to be treated in an integrated manner because there is interaction between pretreatment and the downstream processes. When biomass is pretreated, by-products are also created that act as inhibitors in enzymatic hydrolysis and fermentation. Economic cost functions are formulated to identify the best trade-offs between the refinery steps such that profit is maximized. Sensitivity and uncertainty analysis is also performed with Monte Carlo simulations and Latin Hypercube Sampling on feedstock composition and kinetic parameters following the methodology from [10, 11].

In the last step of the study, diagnosis methods are implemented to identify failures that can occur across the plant. Focus is placed on batch contamination with lactic acid bacteria in fermentation.

References:

[1] R. Datta, M. a. Maher, C. Jones, and R. W. Brinker. “Ethanol-the primary renewable liquid fuel”. In: Journal of Chemical Technology and Biotechnology 86.4 (2011), pp. 473–480. DOI: 10.1002/jctb. 2580.

[2] R. C. Brown, T. Brown, S. Capareda, B. Dale, D. Edwards, V. Estes, C. Granda, M. Holtzapple, S. Lonkar, and C. E. Wyman. “SBE Supplement: Lignocellulosic Biofuels”. In: AIChE CEP (2015), pp. 33–64.

[3] J. Larsen, M. Ø. Haven, and L. Thirup. “Inbicon makes lignocellulosic ethanol a commercial reality”. In: Biomass and Bioenergy 46 (Nov. 2012), pp. 36–45. DOI: 10.1016/j.biombioe.2012.03.033.

[4] J. Larsen, M. Ø. Petersen, L. Thirup, H. W. Li, F. Krogh Iversen, M. Ø. Petersen, L. Thirup, H. W. Li, and F. K. Iversen. “The IBUS Process – Lignocellulosic Bioethanol Close to a Commercial Reality”. In: Chemical Engineering & Technology 31.5 (May 2008), pp. 765–772. DOI: 10.1002/ceat.200800048.

[5] R. M. Prunescu, M. Blanke, J. G. Jakobsen, and G. Sin. “Dynamic modeling and validation of a hydrothermal biomass pretreatment process - A demonstration scale study”. In: AIChE Journal (Under review).

[6] R. M. Prunescu and G. Sin. “Dynamic modeling and validation of a lignocellulosic enzymatic hydrolysis process - A demonstration scale study”. In: Bioresource Technology 150 (Dec. 2013), pp. 393–403. DOI: 10.1016/j.biortech.2013.10.029.

[7] N. Hovakimyan and C. Cao. L1 Adaptive Control Theory. SIAM, 2010.

[8] R. M. Prunescu, M. Blanke, and G. Sin. “Modelling and L1 Adaptive Control of Temperature in Biomass Pretreatment”. In: Annual Conference on Decision and Control. 2013.

[9] R. M. Prunescu, M. Blanke, and G. Sin. “Modelling and L1 adaptive control of pH in bioethanol enzymatic process”. In: Proceedings of the 2013 American Control Conference. 2013, pp. 2–9.

[10] G. Sin, K. V. Gernaey, M. B. Neumann, M. C. M. van Loosdrecht, and W. Gujer. “Global sensitivity analysis in wastewater treatment plant model applications: Prioritizing sources of uncertainty”. In: Water Research 45.2 (Jan. 2011), pp. 639–651. DOI: 10.1016/j.watres.2010.08.025.

[11] R. Morales-Rodriguez, A. S. Meyer, K. V. Gernaey, and G. Sin. “A framework for model-based optimization of bioprocesses under uncertainty: Lignocellulosic ethanol production case”. In: Computers & Chemical Engineering 42 (July 2012), pp. 115–129. DOI: 10.1016/j.compchemeng.2011.12.004.

428815 Efficient Real-Time Operation of an Industrial Gas Network through Global Optimization

Abstract:

One of Air Liquide's major business activities is the separation of air into its pure components: oxygen, nitrogen and argon. The products can be in either liquid or gaseous phase. The gas products are often supplied to various customers in a region via gas pipelines which connect the plants to the customers. Each plant may contain one or more of air separation units (ASU) which are at the heart of the air separation process. The production capability of each plant for the different products is determined by the plant layout and the design of each of the air separation units. ASUs are highly energy intensive. Hence, the operating margin of an air separation plant is highly susceptible to the rapid fluctuations in electricity prices. The customers usually take gases at different conditions, and thus the gas pipelines have to be operated at different pressure levels through the operation of downstream and network compressors. Further, customer demands are not always stable but varying. Given these two major factors of fluctuating prices and consumer demands, real-time optimization techniques are necessary to determine the most profitable way of operating gas pipeline networks while satisfying a variety of process and network constraints.

Real-time optimization of a network necessitates a quality model describing the underlying processes and network. Air separation units operate in a fashion that can be described only with highly nonlinear and nonconvex functional forms. Unfortunately, detailed first-principles models can be very large, posing a significant challenge for optimization algorithms. For the pipeline in this study, Air Liquide developed a model to describe both plant and pipeline operations using mass and energy balances and, through regression, on historical plant data. By combining overall mass and energy balances with regression, we were able to maintain the accuracy of the model while preventing the model size to grow too large.

The second set of modeling challenges are the process constraints that necessitate many discrete decisions such as whether certain compressors and air separation units should be turned on or off. Modelling these discrete decisions through the use of integer variables required us to use a mixed-integer nonlinear programming model. Due to nonconvexities this model possesses multiple local optima. Global optimization techniques were found necessary in order to avoid inferior locally optimal solutions. The model must be efficiently solved in reasonable time to allow for its use in real-time optimization. Further, the size of the model and numerical characteristics introduced through regression data pose considerable challenges for all global solvers. We overcame these challenges by pursuing a model reduction and scaling strategy that facilitated reliable and robust solutions to the model. Detailed computational results are presented showcasing a highly successful industry-academia collaborative effort between Air Liquide and Carnegie Mellon University to address several modeling and numerical solution challenges that arose in the context of this challenging pipeline operational problem.

429479 Mixed-Integer Programming Solution Methods for Inventory Routing Problems

Abstract:

To reduce the distribution cost in supply chains (SCs), vendor managed inventory (VMI) policy is adopted in many chemical industries. Under this policy, the vendor can decide when and how much to serve each customer, on the condition that the inventories of customers are kept between the agreed minimum and maximum levels. To make the distribution decisions under a VMI policy, the inventory routing problem (IRP) needs to be solved, in which the vehicle routes and schedules are decided simultaneously with delivery amounts and visiting times. A wide range of constraints, which include access windows of customers, maximum working/driving time limit for drivers, and vehicle capacities, make IRP a hard problem. To address this challenge, we have developed advanced solution methods, including preprocessing and decomposition methods.

First, in the preprocessing algorithm, we eliminate some nodes and arcs in the SC network that are very unlikely to be visited or used in the current horizon. Based on the information of real-time inventory levels, estimated future consumption and customer locations, the preprocessing algorithm reduces the size of the SC network by removing some redundant nodes and arcs for the current horizon.

Second, the proposed decomposition method considers an upper level vehicle routing problem, and a lower level scheduling problem. The upper level vehicle routing problem determines the optimal routes and route-vehicle pairings to minimize the transportation cost while making sure that the minimum customer "demands" are satisfied (i.e., the inventory at the end of horizon is greater than a minimum terminal level). The upper level solution gives a lower bound on the minimization objective as it is a relaxation of the original problem. Using the solution of upper level problem as input, we solve the lower level distribution scheduling problem with customers, vehicles, drivers and all constraints. The lower level solution yields an upper bound, since it is a feasible solution for the IRP. The upper and lower subproblems are linked by an iterative algorithm. When the lower level scheduling problem is infeasible, or the lower bound is strictly less than the upper bound, integer cuts are generated and added to the upper level problem. The lower level infeasibility can be due to (1) an infeasible assignments of routes to a driver, (2) access window constraints, and (3) inventory level violations. The discrepancies between the upper and lower bounds can be due to longer working times or smaller delivery amounts found by the lower level problem compared to that of the upper level problem. The iterative algorithm terminates when the gap of the upper and lower bounds is closed.

Furthermore, we show how our methods account for various practical constraints. These constraints include the modeling of time-varying consumption profiles, truck-customer compatibility, serve-first requirements for some customers, and penalization of low inventory levels. Finally, we discuss instances with different products, which lead to distribution networks of various sizes.

417901 Optimal Integration of a Self Sustained Algae Based Facility with Solar and/or Wind Energy

Abstract:

Biodiesel is typically produced using methanol in the transesterification of the oil. While the use of methanol has been justified from the technical and economic points of view, quicker reaction times and cheaper than any other alcohol, it is mainly produced from coal or natural gas. Therefore, the second most important raw material in the production of biodiesel is currently a fossil based chemical. Ethanol can be produced from algae and can be competitively used instead for the production of biodiesel, namely fatty acid ethyl ester (FAEE), Martín & Grossmann (2013a). Alternatively, methanol can be produced from renewable sources including biodiesel main byproduct, glycerol, Martín & Grossmann (2013b). However, only two thirds of the demand of methanol could be covered in this way and the production cost was also higher.

In this work, we consider the design of an integrated facility that uses CO_2 as carbon source and solar and / or wind energy to produce methanol and algae based biodiesel. On the one hand, solar energy is used to grow algae and to produce power, solar photovoltaics (PV), for water splitting producing hydrogen and oxygen (Levene et al., 2005). Wind turbines can also be used for power production. Lately, it has been proved that CO_2 can be hydrogenated to produce methanol (Van-Dal, Bouallou 2013; Trudewind et al 2014a&b), which represents a promising use for the captured CO_2. By producing methanol in this way, it is not only renewable, but it represents a way to store energy from intermittent sources. On the other hand, the oil is produced by growing microalgae from where lipids are extracted. The transesterification step is similar to current biodiesel production (Martín & Grossmann, 2012). The optimal flowsheet is synthesized with a multiperiod MINLP model whose solution provides the optimal source of energy and the operating conditions over a year for an average production of 60 Mgal/yr of FAME. Simultaneous optimization and heat integration is implemented within the formulation to address the tradeoff related to the operation of the reactors, governed by chemical equilibria, and the preparation and separation stages. Next, a heat exchanger network is designed and finally, a water network is synthesized.

The proposed design uses only solar energy. The advantage is that the methanol and oil production rates are the same. The drawback can be related to the area available for the ponds and the solar panels allocation and the idle equipment along the year due to the variability of solar incidence. The investment costs adds up to 120 M€ for, a production cost of 0.85€/gal (0.28 €/kg). The energy consumption is 3.2 MJ/gal of FAME, capturing 4.05 kg CO_2 per kg of FAME and consuming 1L of water per littre of methanol produced, including the operation of the cooling tower and the boiler.

References

Levene, J.I., Mann, M.K. , Margolis, R. , Milbrandt, A., 2005. An Analysis of Hydrogen Production from Renewable Electricity Sources Conference Paper . NREL/CP-560-37612 September 2005

Martín, M., Grossmann, I.E. (2012) Simultaneous optimization and heat integration for biodiesel production from cooking oil and algae. Ind. Eng. Chem Res. 51 (23) 7998–8014

Martín, M., Grossmann, I.E., (2013) Optimal engineered algae composition for the integrated simultaneous production of bioethanol and biodiesel AIChE J. 59 (8) 2872–2883

Martín, M.; Grossmann, I.E. (2013) ASI: Towards the optimal integrated production of biodiesel with internal recycling of methanol produced from glycerol. Environmental Progress & Sustainable Energy 32(4) 791-801

Levene, J.I., Mann, M.K. , Margolis, R. , Milbrandt, A., 2005. An Analysis of Hydrogen Production from Renewable Electricity Sources Conference Paper . NREL/CP-560-37612 September 2005

Trudewind, C.A., Schreiber, A., Haumann, D., 2014. Photocatalytic methanol and methane production using captured CO2 from coal power plants. Part I a Life Cycle Assessment. J. Cleaner Prod. http://dx.doi.org/10.1016/j.jclepro.2014.02.014

Trudewind, C.A., Schreiber, A., Haumann, D., 2014. Photocatalytic methanol and methane production using captured CO2 from coal power plants. Part II – Well-to-Wheel analysis on fuels for passenger transportation services J. Cleaner Prod. http://dx.doi.org/10.1016/j.jclepro.2014.02.024

Van-Dal., E.S., Bouallou, C., 2013. Design and simulation of a methanol production plant from Co3 hydrogenation. J. Cleaner Prod. 57, 38-45.

421936 Biomass to Liquid Transportation Fuels Utilizing Biological and Thermochemical Conversion: Process Synthesis and Global Optimization

Abstract:

Biomass is a valuable and currently underutilized feedstock for the energy sector in the United States. Petroleum, a non-renewable resource, accounts for over 35% of the total national energy consumption and is primarily used in the transportation industry [1]. Renewable resources only contribute to 4.7% of liquid fuels production, and thus a significant impact can be made through the use of biomass as a feedstock [1]. Due to absorption of CO_2 during photosynthesis, the greenhouse gas reduction resulting from biomass feedstocks is substantial. This, in addition to the domestic feedstock availability of biomass, gives ample motivation for the synthesis and optimization of biomass to liquid transportation fuel (BTL) refineries [2]. Currently, ethanol is the primary fuel produced from biomass, yet it is a poor fuel replacement in the current transportation infrastructure [3]. Technology exists to convert biomass to gasoline, diesel, and kerosene, and process synthesis utilizing global optimization provides a mathematical approach to accurately model and optimize a BTL plant [4].

Thermochemical methods for converting biomass to fuels are well-known and have been studied using process synthesis for a variety of possible feedstocks [5, 6, 7]. The process synthesis model, which utilize gasification of biomass and fuel production via Fischer-Tropsch or methanol synthesis routes, forms a mixed-integer non-linear program (MINLP) which can be solved to global optimality using a rigorous branch-and-bound framework. These plants can have a significant impact on the liquid fuel supply chain [8]. In the refinery, CO_2 can be recycled, vented, or sequestered, and the process incorporates simultaneous heat, power, and water integration [9]. The superstructure yields the breakeven oil price required for the biorefinery to be competitive in today's energy markets. However, to this point, studies on BTL plants have yet to incorporate biological conversion of biomass into the MINLP superstructure.

The MixAlco process, in which biomass is converted to carboxylic acid salts through fermentation before upgrading to liquid fuels, provides a unique starting point for incorporation of biological routes into the MINLP superstructure [10, 11]. The upgrading of carboxylic acids requires hydrogen, which can be provided through pressure swing adsorption of synthesis gas provided by gasification of lignin residues leaving the fermenter. Gasification of lignin also allows for MixAlco integration with a synthesis gas conversion method. After input-output modeling of the MixAlco process, biological and thermochemical conversion of switchgrass are compared through case studies at capacities ranging from 1000 barrels per day (1 kBD) to 200 kBD. Important financial considerations, including the overall cost of liquid fuels production, the investment cost, greenhouse gas emissions, and product distribution, are analyzed and discussed. Topological decisions for the biorefinery, along with the synergistic impacts of incorporating biological and thermochemical conversion of biomass simultaneously, are also examined.

[1] U.S. Energy Information Administration, "Monthly energy review," March 2015.

[2] L. R. Lynd, E. Larson, N. Greene, M. Laser, J. Sheehan, B. E. Dale, S. McLaughlin, and M. Wang, "The role of biomass in America's energy future: framing the analysis," Biofuels, Bioproducts and Biorefining, vol. 3, no. 2, pp. 113–123, 2009.

[3] J. C. Serrano-Ruiz and J. A. Dumesic, "Catalytic routes for the conversion of biomass into liquid hydrocarbon transportation fuels," Energy Environ. Sci., vol. 4, pp. 83–99, 2011.

[4] R. C. Baliban, J. A. Elia, R. Misener, and C. A. Floudas, "Global optimization of a MINLP process synthesis model for thermochemical based conversion of hybrid coal, biomass, and natural gas to liquid fuels," Computers & Chemical Engineering, vol. 42, no. 0, pp. 64 – 86, 2012. European Symposium of Computer Aided Process Engineering.

[5] R. C. Baliban, J. A. Elia, and C. A. Floudas, "Biomass to liquid transportation fuels (BTL) systems: process synthesis and global optimization framework," Energy Environ. Sci., vol. 6, pp. 267–287, 2013.

[6] R. C. Baliban, J. A. Elia, C. A. Floudas, B. Gurau, M. B. Weingarten, and S. D. Klotz, "Hardwood biomass to gasoline, diesel, and jet fuel: 1. Process synthesis and global optimization of a thermochemical refinery," Energy & Fuels, vol. 27, no. 8, pp. 4302–4324, 2013.

[7] R. C. Baliban, J. A. Elia, C. A. Floudas, X. Xiao, Z. Zhang, J. Li, H. Cao, J. Ma, Y. Qiao, and X. Hu, "Thermochemical conversion of duckweed biomass to gasoline, diesel, and jet fuel: Process synthesis and global optimization," Industrial & Engineering Chemistry Research, vol. 52, no. 33, pp. 11436–11450, 2013.

[8] J. A. Elia, R. C. Baliban, C. A. Floudas, B. Gurau, M. B. Weingarten, and S. D. Klotz, "Hardwood biomass to gasoline, diesel, and jet fuel: 2. Supply chain optimization framework for a network of thermochemical refineries," Energy & Fuels, vol. 27, no. 8, pp. 4325–4352, 2013.

[9] R. C. Baliban, J. A. Elia, and C. A. Floudas, "Simultaneous process synthesis, heat, power, and water integration of thermochemical hybrid biomass, coal, and natural gas facilities," Computers & Chemical Engineering, vol. 37, no. 0, pp. 297 – 327, 2012.

[10] C. Granda, M. Holtzapple, G. Luce, K. Searcy, and D. Mamrosh, "Carboxylate platform: The MixAlco process part 2: Process economics," Applied Biochemistry and Biotechnology, vol. 156, no. 1-3, pp. 107–124, 2009.

[11] V. Pham, M. Holtzapple, and M. M. El-Halwagi, "Technoeconomic analysis of a lignocellulose-to-hydrocarbons process using a carboxylate platform," Integrated Biorefineries: Design, Analysis, and Optimization, p. 157, 2012.

MACC

McMaster University

DESIGN AND ASSESSMENT OF NOVEL LIGNOCELLULOSIC THERMOCHEMICAL BIOMASS TO BUTANOL CONFIGURATIONS

CHINEDU OKOLI*, THOMAS ADAMS II

DEPT. OF CHEMICAL ENGINEERING
MCMASTER UNIVERSITY
HAMILTON, ONTARIO

American Institute of Chemical Engineers Conference, 8th – 13th November 2015

MACC

INTRODUCTION

McMaster University

Lignocellulosic biofuels

- ❖ Address sustainable society challenges:
  - Global climate change
  - Energy security
  - Land use (food vs. fuel)

Agricultural residue

Switchgrass

- ❖ Energy content estimates of annual forest and agricultural residues produced in Canada is put at **5.1 EJ**[1]
  - Equivalent to **62%** of the country's energy derived via fossil fuel combustion

Wood chips

- ❖ U.S. congress in 2007 mandated the annual production of **21 billion gallons** of (ligno)cellulosic biofuels by 2022[2]
  - A **vast increase** from 2011 production estimates of 5 billion gallons

1 EJ (Exajoule) = 10^{15} kJ

[1] Wood and Layzell, 2003. BIOCAP Canada Foundation; 18 – 24.
[2] Congress U.S., 2007. Public Law; 2: 110 – 40.

2

MACC

INTRODUCTION

McMaster University

Why biobutanol

- Potential replacement for bioethanol as a gasoline substitute
- Advantages over bioethanol include[3,4]:
  - a higher energy content
  - lower affinity for water
  - better compatibility with existing internal combustion engines and fuel pipeline networks

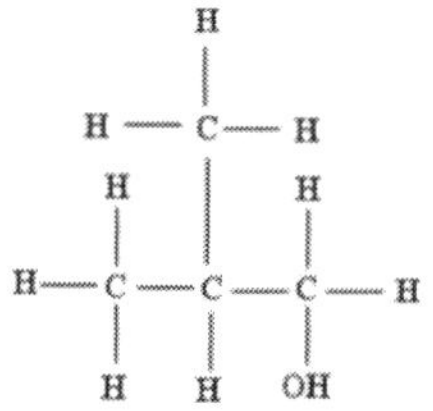

Iso-butanol molecule

Property	Ethanol	iso-Butanol	Gasoline
Density at 20 °C (g/cm^3)	0.794	0.802	0.791
Research Octane number	112 – 122	102 – 105	85 - 87
Energy content (% of gasoline)	65	**82**	100
Water solubility (wt %)	100	**8.5**	-

[3] Kumar and Gayen, 2011. App Energy; 88(6): 1999 – 2012.
[4] Ranjan and Moholkar, 2012. Int J Energ Res; 36(3): 277 – 323.

3

MACC

RESEARCH MOTIVATION

McMaster University

- Past work[5]
  - Novel thermochemical route for biobutanol production proposed
  - No external hot utilities and power required
  - Minimum butanol selling price (***MBSP**) of $3.14/ gallon
  - Mixed Alcohol Synthesis (MAS) reactor model based on yield and conversion data
    - Impact of varying reactor design parameters cannot be studied
  - Futuristic MAS catalyst conversion based on National Renewable Energy Laboratory targets

- This work
  - Detailed kinetic reactor model based on a modified high pressure methanol synthesis catalyst[6]
  - Impact of reactor design parameters and recycle structure on butanol production can be investigated
  - Additional configurations in which dividing wall columns are used for butanol recovery
  - What are the impacts on MBSP

***MBSP** - the **selling price** of the **butanol product** over the plant's **life time** at which the **NPV** is zero*

[5] Okoli and Adams., 2014. Ind. Eng. Chem. Res., 53(28): 11427 – 11441
[6] Beretta et al., 1998. Ind. Eng. Chem. Res., 37: 3896 – 3908.

4

METHODOLOGY - configurations

McMaster University

Long recycle configuration

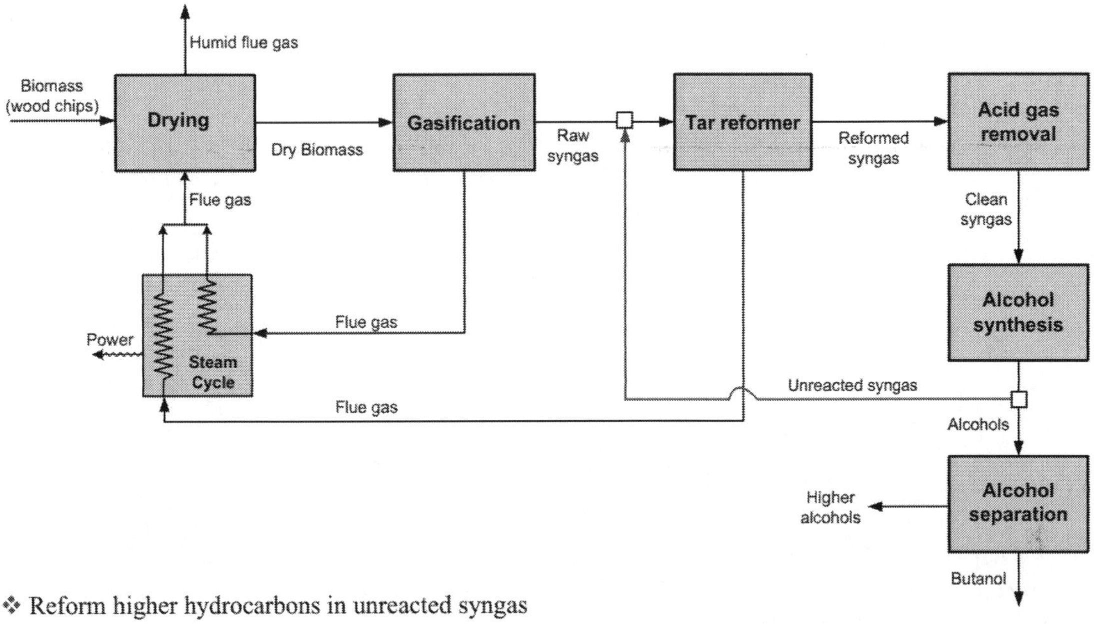

❖ Reform higher hydrocarbons in unreacted syngas

5

METHODOLOGY - configurations

McMaster University

Intermediate recycle configuration

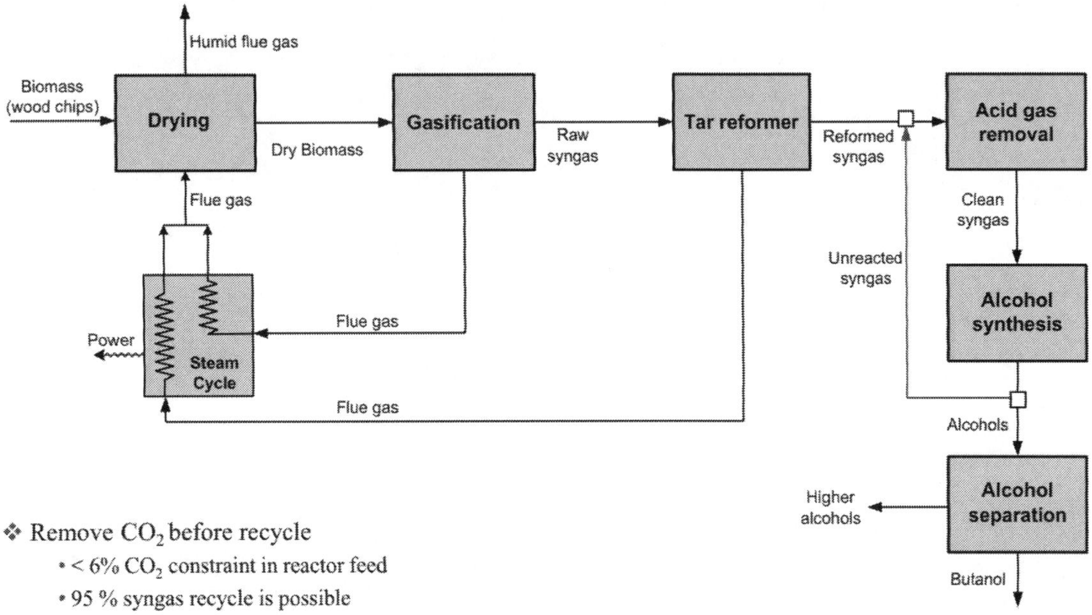

❖ Remove CO_2 before recycle
- < 6% CO_2 constraint in reactor feed
- 95 % syngas recycle is possible
- High recompression costs

6

METHODOLOGY - configurations

McMaster University

Short recycle configuration

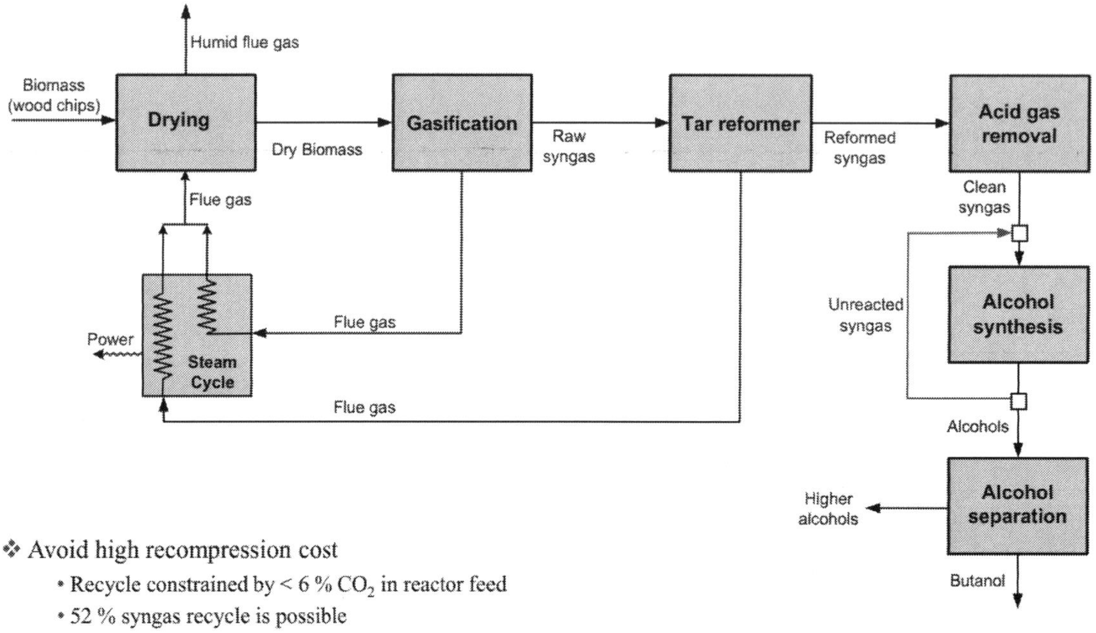

- Avoid high recompression cost
  - Recycle constrained by < 6 % CO_2 in reactor feed
  - 52 % syngas recycle is possible

7

METHODOLOGY – reactor design[6]

McMaster University

- Kinetic model of Beretta et al.[7] is implemented in Aspen Plus
  - Langmuir Hinshelwood Hougen Watson form
- Design objective - maximize butanol yield
  - Increase reactor temperature
  - Reduce GHSV (gas hourly space velocity)
  - Increase reactor pressure
  - Reduce inlet % CO_2
  - Utilize optimal H_2/CO ratio
- Reaction set

$$CO + 2H_2 \leftrightarrow CH_3OH$$
$$CO + H_2O \leftrightarrow CO_2 + H_2$$
$$2CH_3OH \rightarrow C_2H_5OH + H_2O$$
$$CH_3OH + C_2H_5OH \rightarrow C_3H_7OH + H_2O$$
$$\boxed{CH_3OH + C_3H_7OH \rightarrow iC_4H_9OH + H_2O}$$
$$2CH_3OH + C_3H_7OH \rightarrow C_5H_{11}OH + 2H_2O$$
$$2CH_3OH \rightarrow (CH_3)_2O + H_2O$$
$$CO + 3H_2 \rightarrow CH_4 + H_2O$$
$$2CO + 5H_2 \rightarrow C_2H_6 + 2H_2O$$

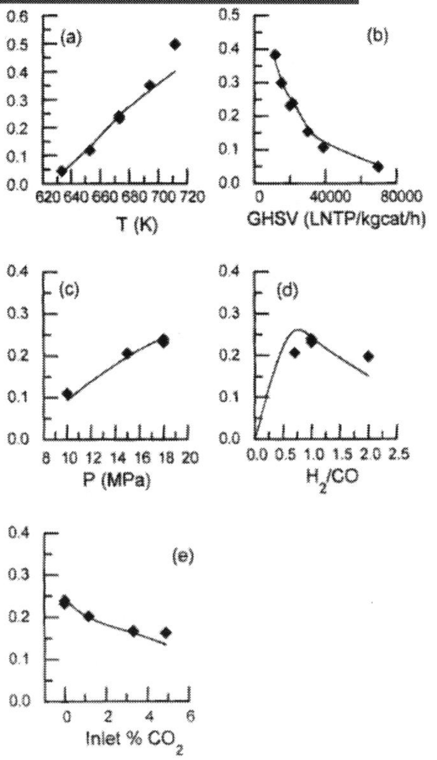

Figure 6. Experimental and calculated trends of isobutyl alcohol concentration with varying operating conditions.

[6] Beretta et al., 1998. Ind. Eng. Chem. Res., 37: 3896 – 3908.

8

MACC

METHODOLOGY – DWC design[7]

McMaster University

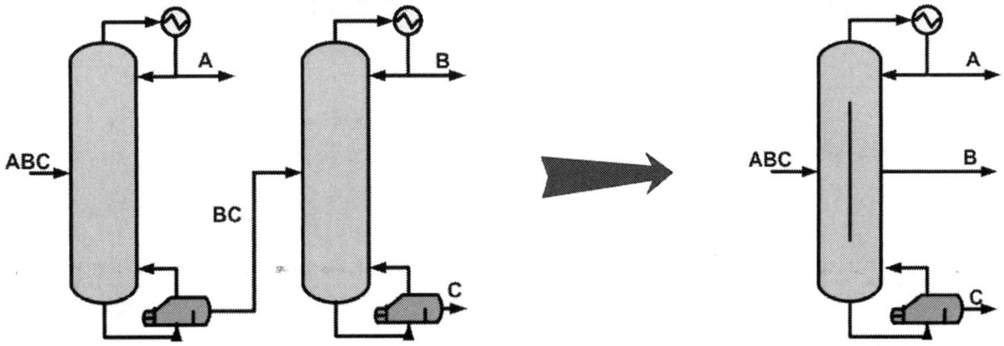

- Fully thermally coupled distillation column integrating several separation steps in a single shell
  - Only one reboiler and one condenser
  - Single or multiple partition walls
- Avoid internal remixing of streams
  - Minimizes the entropy of mixing and thus
  - Less energy required for separation
- DWCs are designed using
  - Short-cut method (V_{min} diagram[8]) to generate good initial guesses
  - Particle Swarm Optimization coupled with Aspen Plus to determine optimal design parameters

[7] Okoli and Adams., 2015. Chem. Eng. Proc: Proc Intensif., 95: 302 – 316.
[8] H alvorsen and Skogestad, 2003. Ind. Eng. Chem. Res., 42: 596 – 604.

9

MACC

METHODOLOGY

McMaster University

Base case economic assumptions

Biomass flowrate	2,000 dry tonnes per day
Wood chips price	$75 per tonne
Higher alcohol sales price	$4.11/gallon
Gasoline price	$3.44/gallon
Cooling water price	0.05 cents/L
Plant life	30 years
Plant on stream hours	7884 h/year
Plant financing by debt/equity	50 % / 50 %
Shareholders return rate	10 % after tax
Debt financing (Length/ interest rate)	10 years / 8 %
Income tax rate	35%

10

MACC

RESULTS & DISCUSSION

McMaster University

Techno-economic results

	Long recycle	Inter. recycle	Short recycle
Product yields (tonnes/hr)			
Butanol	5.3	11.2	7.5
Higher alcohols	3.1	1.1	2.8
% products yield per feed	10.1	**14.7**	12.3
Power (MW)			
Power geneneration	94.0	104.9	76.9
Power consumption	81.7	104.5	67.0
Net Power	12.3	0.4	9.9
% Plant efficency (LHV basis)	23.4	29.3	27.2

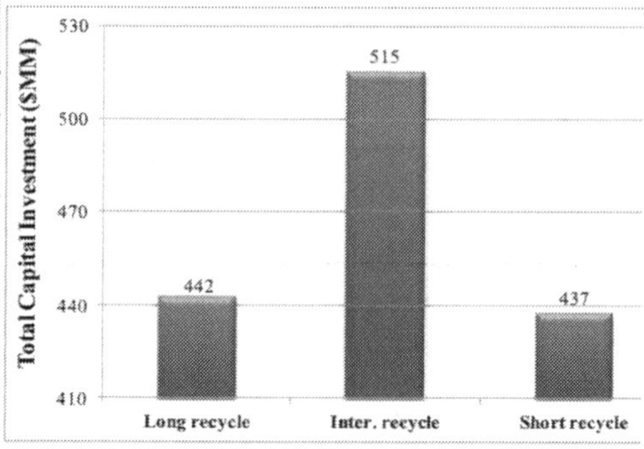

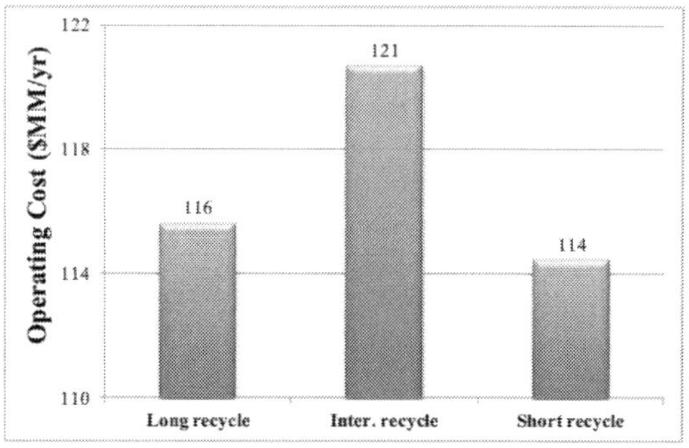

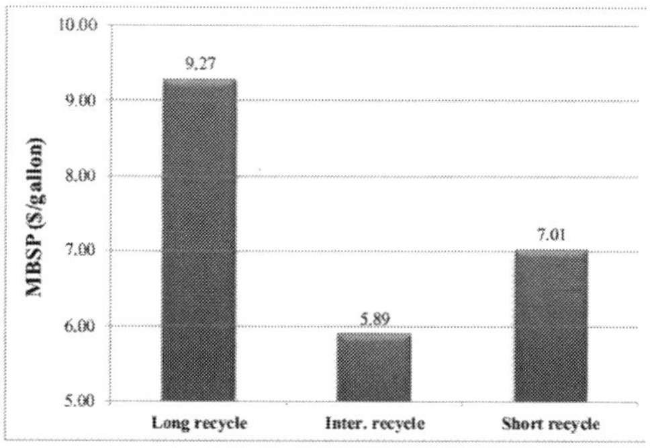

11

MACC

RESULTS & DISCUSSION

McMaster University

Techno-economic results - DWCs

	Total Cap. Invest. ($MM)	Operating Cost ($MM/yr)	MBSP ($/gallon)
Long recycle	442	116	9.27
Long recycle + DWC	428	115	9.24
Inter. recycle	515	121	5.89
Inter. recycle + DWC	498	120	**5.85**
Short recycle	437	114	7.01
Short recycle + DWC	424	114	7.00

12

MACC

RESULTS & DISCUSSION

McMaster University

Capital cost breakdown

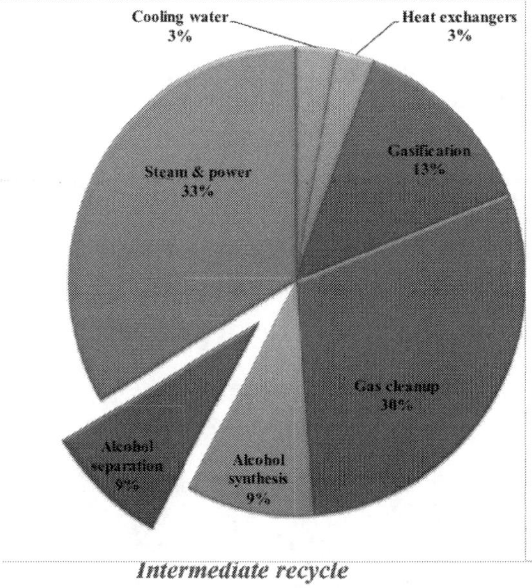

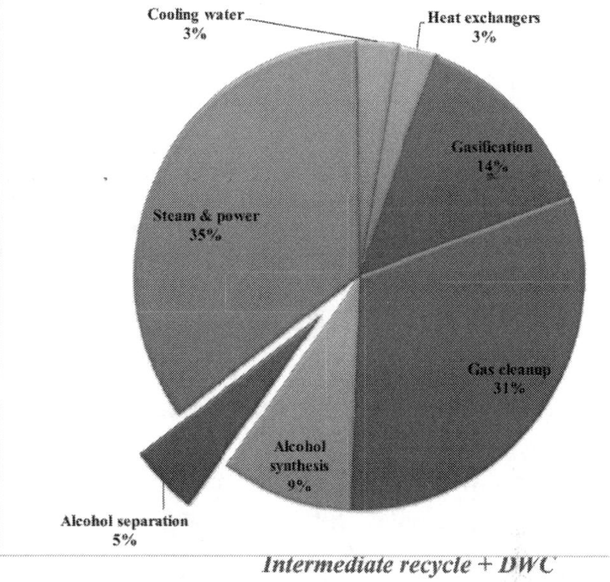

- Replacement with DWC results in $17 MM savings in Total Capital Investment
  - 4 cents/gallon reduction in MBSP

13

MACC

RESULTS & DISCUSSION

McMaster University

Capital cost breakdown

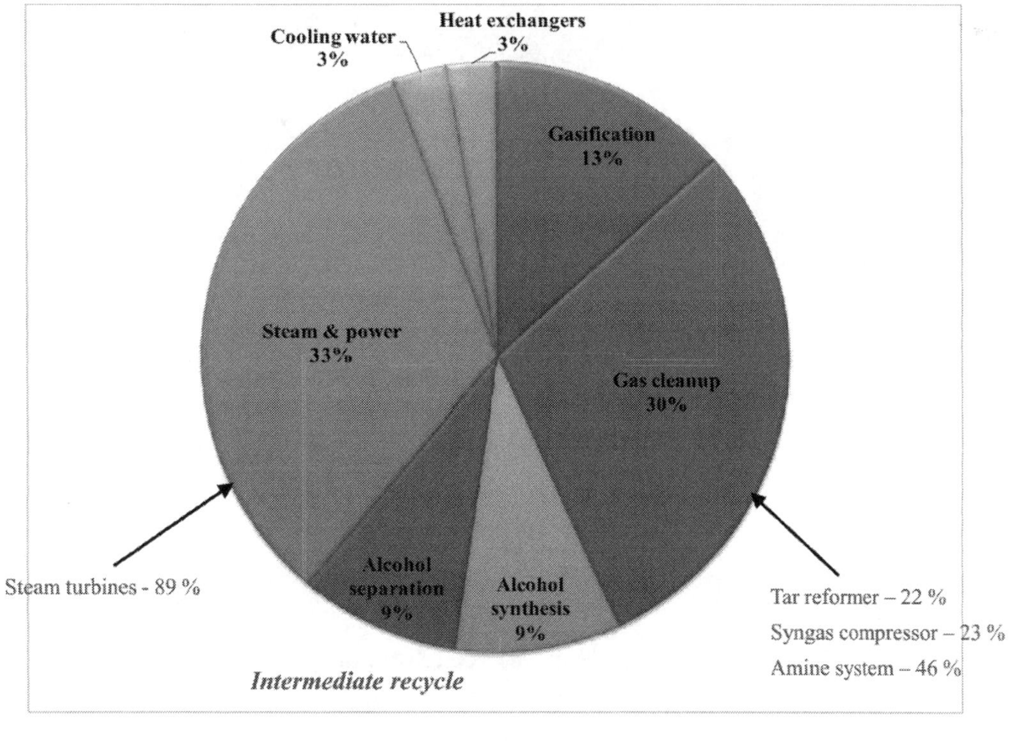

14

MACC

RESULTS & DISCUSSION

McMaster University

Operating cost – interm. recycle

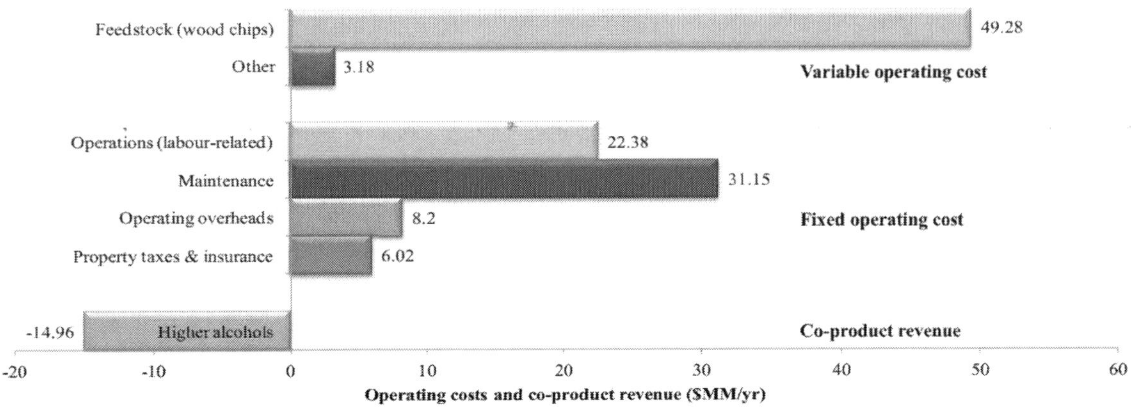

15

MACC

RESULTS & DISCUSSION

McMaster University

Sensitivity analysis – interm. recycle

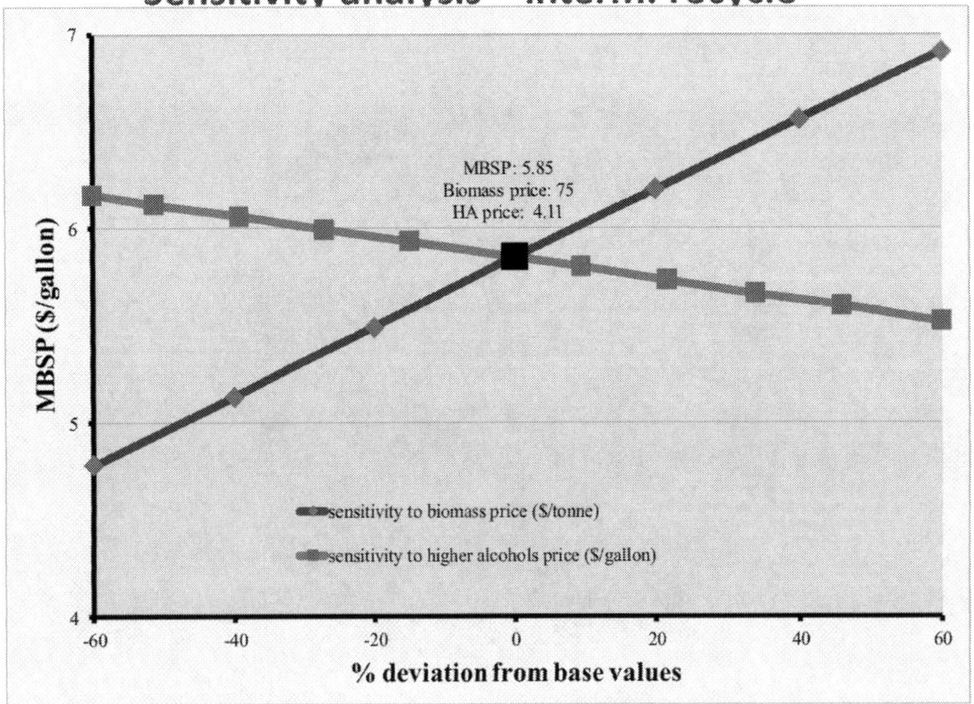

Sensitivity analysis of MBSP ($/ gallon) wrt. change in biomass price ($/dry tonne) and higher alcohols price ($/gallon)

16

MACC

RESULTS & DISCUSSION

McMaster University

Potential process improvement options

- Cheaper feedstock
  - Ultimate analysis (wt % dry basis) for a variety of lignocellulose biomass in Ontario[9]

	Switchgrass	Wheat straw	Corn stover	Hardwood
Carbon	45.5	43.4	43.7	48.3
Hydrogen	6.1	6	6.1	6
Nitrogen	0.9	0.8	0.5	0.2
Sulfur	0.1	0.1	0.1	0
Oxygen	41.7	44.5	44.6	45.1
Ash	5.7	7.7	5.1	0.4
LHV (MJ/kg)	18	18	19	19

  - Feedstock prices vary from $24 to $66 per dry tonne at farm gate
    - Prices depend on demand, location and harvest yields
    - Transportation costs, supply chain and feedstock availability also important

- Replace secondary tar reforming with primary tar reforming
  - Potential to reduce capital cost by eliminating external tar reformer
  - Primary tar reforming technology not fully understood (hasn't been demonstrated commercially)[10]

[9] Ontario ministry of agriculture, food and rural affairs: http://www.omafra.gov.on.ca/english/engineer/facts/11-033.htm#6
[10] Devi et al., 2003. biomass and bioenergy, 24(12): 125 – 140.

17

MACC

RESULTS & DISCUSSION

McMaster University

Potential process improvement options

- Combust natural gas as heat source for endothermic tar reforming
  - Increases butanol yield as syngas flowrate to alcohol synthesis reactor is increased
  - biomass syngas H_2/CO ratio of 0.5 preferable to reformed natural gas syngas ratio of 2
  - butanol fuel is less "green"

- Purchase steam and power
  - Capital cost of steam & power generation section is significantly reduced
  - Location of butanol plant to a utility supplier might not be optimal
  - Financing might be much easier

18

MACC **CONCLUSION AND FUTURE WORK** McMaster University

Conclusion

- Process systems engineering tools have been utilized to assess the techno-economics of a thermochemical route for biobutanol production
- The selected syngas recycle configuration is critical to butanol yield and MBSP
  - Intermediate recycle configuration yields MBSP of **$5.89/ gallon**
- Configurations with DWC provide improvements in MBSP
  - **4 cents/gallon** reduction in MBSP obtained for interm. + DWC config over interm. config
  - **$1 million** per year reduction in operating cost
  - **$17 million** reduction in Total Capital Investment
- Techno-economic results motivate various potential improvement options

Acknowledgement

Ontario Research Fund

Ontario Research Fund

19

430689 Conceptual Process Design with Energetic Analysis for a Bio-Based 2-Butanone Production

Abstract:

In view of a rising energy demand and a depletion of fossil raw material, the request for renewable feedstock for the energy and chemical market is growing. In a recent screening for innovative fuel candidates, 2-butanone has emerged as a promising alternative. In addition, 2-butanone is a component used mainly as solvent for paints and is primarily produced by dehydrogenation of 2-butanol, which is gained by hydration of n-butenes via petrochemical synthesis. Hence, a sustainable production of 2-butanone could open up several opportunities in bioeconomy.

However, a bio-based production route combining pretreatment, reaction and purification has not been investigated yet and thus the potential of bio-based 2-butanone production as a fuel is unknown. This contribution pursues a holistic analysis of alternative 2-butanone production routes. It relies on a screening of reaction pathways to identify promising routes, uses an incremental refinement of process synthesis steps [1] and thus provides an comparative overview of production processes including state-of-the-art heat integration and first results in terms of the economic potential for the production of 2-butanone. The design is based on sugars as feedstock for comparison to mature ethanol processes.

For the bio-based production of 2-butanone several alternative synthesis routes starting from glucose are conceivable from literature. First, mass balances for different routes are set up without detail design of separation in order to select the most promising path by means of the overall yield. A efficiency can be expected from the high titer and high productivities of 2,3-butanediol fermentation ([2],[3]) and from high selectivity's in the chemo-catalytic conversion by a solid-acid catalyst [4].The fermentation to 2,3-butanediol and dehydration to 2-butanone is thus pursued further.

In the next step, conceptual process design was performed including 2,3-butanediol fermentation, 2,3-butanediol separation, the dehydration reaction to 2-butanone and the following purification of the product. Due to multiple approaches for the separation step an incremental refinement approach is used to find a competitive production process. To this end, both well-known distillation processes as well as hybrid separation strategies seem feasible.

The process alternatives were evaluated by means of the energy demand i.e., based on the minimum energy calculations for distillation columns with reliable shortcut methods and optimistic assumptions in case of other unit operations. The most promising strategy is a heat integrated process, which shows an energy demand of 0.43 $MJ/MJ_{2\text{-butanone}}$ after pinch analysis and application of vapor recompression.

The analysis of the simulation of the processes identifies the separation of 2,3-butanediol as a high boiling constituent from the fermentation broth and the separation of mixtures with non-ideal distillation behavior as key issues of this route. If the costly water separation could be circumvented, 2-butanone unveils a potential in bioeconomy similar to that of ethanol.

In conclusion, this contribution identifies potentials and bottlenecks of bio-based 2-butanone production. Different reaction and separation routes involving alternative devices are compared. By this comparative analysis a production of 2-butanone via 2,3-butanediol is proposed. A hybrid separation concept for 2,3-butanediol separation, which consists of membrane separation, extraction and a distillation sequence, is the most promising process concept with 0.43 $MJ/MJ_{2\text{ butanone}}$. In comparison to the bio-ethanol production, this process concept seems competitive if an improvement of the separation of 2,3-butanediol from water is achieved.

Acknowledgement

This work was funded by the federal ministry of food and agriculture and was performed as part of the Cluster of Excellence "Tailor-Made Fuels from Biomass", which is funded by the Excellence Initiative by the German federal and state governments.

References

[1] Recker, S., Skiborowski, M., Redepenning, C. and Marquardt, W. (2015): A unifying framework for optimization-based design of integrated reaction-separation processes. *Computers & Chemical Engineering*.

[2] Ji, X. J. , Huang, H. and Ouyang, P. K. (2011): Microbial 2,3-butanediol production: A state-of-the-art review. *Biotechnology Advances, 29*(3), 351-364.

[3] Ma, C., Wang, A., Qin, J., Li, L.., Ai, X., Jiang, T., Tang, H. and Xu, P. (2009): Enhanced 2,3-butanediol production by *Klebsiella pneumoniae* SDM. *Appl Microbiol Biotechnol,82*(1), 49-57.

[4] Multer, A., McGraw, N., Hohn, K. and Vadlani, P. (2012): Production of methyl ethyl ketone from biomass using a hybrid biochemical/catalytic approach. *Industrial & Engineering Chemistry Research, 52*(1),56-60.

434772 Systematic Design and Synthesis of Integrated Multi-Product Biorefinery Processes

Abstract:

Depletion of petroleum sources combined with fluctuations in the global oil market, and increasingly strict environmental requirements prompt the need to discover alternative fuel/chemical-production technologies. In this presentation we systematically design and synthesize a cost-effective and environmentally-benign multi-product biorefinery process. Besides the production of traditional liquid transportation fuels, propylene production is also included as a platform chemical that has a gap between demand and supply. In the designed conversion process, biomass (hybrid poplar) is firstly directly or indirectly gasified to syngas which, after treatment and cleanup, is subsequently converted to liquid hydrocarbons and methanol through Fischer-Tropsch (FT) synthesis and methanol synthesis, respectively. Gasoline, kerosene, and diesel are obtained from upgrading of the FT raw liquids via ZSM-5 [1] and hydro-processing approaches, while propylene is obtained from the "on-purpose" methanol-to-propylene (MTP) process. Any light gas products are recycled and reformed to make additional syngas to improve the overall carbon conversion. Finally, all generated CO_2 is converted to produce acetic acid.

We initially develop a biorefinery superstructure, which considers multiple alternatives for certain conversion sections, and formulate a mixed integer nonlinear programming (MINLP) model. Global power, heat, and water networks [2-4] are contained within the formulated framework to minimize the utility requirements and wastewater discharge. Data-driven techniques [5,6] are used to generate simple surrogate models with high accuracies for the indirect/direct heated-gasifier to reduce the computational complexity. The bilinear, trilinear, quadrilinear, and concave terms are reformulated for enhancing the MINLP model to be solved to global optimality [7,8]. Through solving the optimization model, we present parametric/structural analyses to better elucidate the optimal topology under various scenarios with respects to volatile feedstock/product prices, market requirements as well as constraints on waste emissions. We then check the relationship between process profit and ratio of syngas towards the FT and methanol synthesis units, respectively. The total process profit, when the produced CO_2 is converted to acetic acid vs. when it is sequestrated, is also discussed.

References:

[1]. Bechtel. Baseline design/economics for advanced Fischer-Tropsch technology. Contract No. *DE-AC22-91PC90027*. 1992.

[2]. Duran MA, Grossmann IE. Simultaneous optimization and heat integration of chemical processes. *AIChE J*. 1986; 32: 123-138.

[3]. Bruno JC, Fernandez F, Castells F, Grossmann IE. A rigorous MINLP model for the optimal synthesis and operation of utility plants. *Trasn IChemE*, 1998; 76: 246-258.

[4]. Ahmetovic E, Grossmann IE. Global superstructure optimization for the design of integrated process networks. *AIChE J*. 2011; 57: 434-457.

[5]. Schmidt M, Lipson H. Distilling free-form natural laws from experimental data. *Science*. 2009; 324: 81-85.

[6]. Cozad A, Sahinidis NV, Miller DC. A combined first-principles and data-driven approach to model building. *Comput Chem Eng*. 2015; 73: 116-127.

[7]. Gounaris CE, Misener R, Floudas CA. Computational comparison of piecewise-linear relaxations for pooling problems. *Ind. Eng. Chem. Res*. 2009; 48: 5742-5766.

[8]. Cafieri S, Lee J, Liberti L. On convex relaxations of quadrilinear terms. *J Glob Optim*. 2010; 47: 661-685.

433729 Techno-Economic Analysis of a Direct Coal-Biomass to Liquids (CBTL) Plant with CO2 Capture and Storage (CCS)

Abstract:

Techno-economic analysis of a Direct Coal-Biomass to Liquids (CBTL) Plant with CO_2 Capture and Storage (CCS)

Yuan Jiang, Debangsu Bhattacharyya

Department of Chemical Engineering, West Virginia University, Morgantown, WV 26506, USA

Due to the insecurity of crude oil supply, various studies have been conducted for the direct liquefaction of coal. However, because of the low H/C ratio in coal, a large amount of gaseous H_2 is required for converting coal to syncrude and upgrading syncrude to on-spec gasoline and diesel. As a result, these processes are usually associated with high CO_2 emission from H_2 production plants that use technologies such as natural gas steam reforming or coal gasification. In order to reduce net CO_2 emissions, biomass can be co-fed with coal. The CCS technologies can also be applied for further reduction of the carbon footprint, but at the cost of increasing capital and operating costs. In this talk we will present a techno-economic study conducted in Aspen Process Economic Analyzer® (APEA®) environment for a direct CBTL plant with CCS using the high fidelity process models developed in Aspen Plus®and Excel.

The direct CBTL plant with CCS is designed based on the two-stage coal liquefaction process investigated by HTI, where inline hydrotreating is considered.[1] In this study, a small amount of biomass is co-feed with coal to the direct liquefaction reactor. The product from the liquefaction reactor is sent to an inline hydrotreater, and then to a hydrocarbon recovery unit being separated for further upgrading. Vacuum distillation and supercritical extraction technology are considered for recovering heavy oil that is used as the liquefaction solvent. Due to the abundant supply of shale gas in the Marcellus Shale region, H_2 required in the system is produced from the shale gas steam reforming and partial oxidization of the liquefaction residual. Both chemical and physical absorption processes are considered for CCS and optimized for the direct CBTL plant depending on the sources of the CO_2 containing stream. It can be noted that in a direct CBTL plant, the streams such as the syngas from the H_2 plant differ widely in composition and CO_2partial pressure in comparison to the streams such as the flue gas from furnaces.

The process model is validated by using the data obtained from the open literature. The economic model is developed using Aspen Process Economic Analyzer (APEA®). For the equipment items the costs of which cannot be estimated in APEA®, cost correlations are developed using the data available in the open literature. The outside battery limit (OSBL) section is designed using Aspen Analyzer Utility Modules (AUM) and then integrated with the inside battery limit (ISBL) units for economic analysis. In this presentation, we will discuss the modeling approach and the impacts of the technology choice and design parameters on the various economic matrices such as net present value (NPV), payout date, internal rate of return (IRR), and break-even price of the direct CBTL plant with CCS.

Reference

(1) Comolli, A.G., Lee, L.K., Pradhan, V.R., Stalzer, T.H., Harris, E.C., Mountainland, D.M., Karolkiewicz, W.F. and Pablacio, R.M., Direct Liquefaction Proof-of-Concept Facility, Technical Progress Report POC Run 01, Contract No. AC22-92PC92148, Hydrocarbon Research Inc., 1995.

419787 An Integrated Framework for Scheduling and Control Using Fast Model Predictive Control

Abstract:

An integrated approach that models the scheduling and control simultaneously addresses the information sharing between scheduling and control levels, leading to the integrated decision making that is overall optimal (Terrazas-Moreno, et al., 2008). Meanwhile, the integration increases the complexity of modeling and thus the computational requirements of the solution approaches (Engell and Harjunkoski, 2012). In online applications where disturbance is non-negligible, the integration of scheduling and control requires a repetitive solution of the integrated problem at real time in order to handle the state deviation caused by disturbance (Zhuge and Ierapetritou, 2012).

In this study we propose an integrated framework that involves two control loops for the online integration of scheduling and control. In the outer loop we approximate the original process dynamics using a piece-wise affine (PWA) model and incorporate it with the scheduling level. This leads to an integrated problem that is subject to linear constraints. The integrated problem at the outer loop generates both the production scheduling and the state reference for control problem. The scheduling solution and state reference are provided to the inner loop where fast Model Predictive Control (MPC) is used to track the state reference and to compute the exact control solution online. Note that these two loops correspond to different models and different time scales. The outer loop uses the integrated model and solves it in the period of days or weeks while the inner loop uses the process dynamic model and updates the solution in seconds. Essentially the inner loop is solved much more frequently than the outer loop does. Following this approach the outer loop is able to achieve an overall optimality for both scheduling and control levels efficiently, and the inner loop fast MPC is able to respond quickly to the local disturbance. Note that the proposed approach is applicable in a dynamic market environment as the outer loop incorporates the market information such as demand and price. The proposed framework solves an integrated model to guarantee the overall optimality and employes fast MPC to respond timely to process disturbances. While solving the integrated problem the latest market information such as material price and product demands are incorporated into the integrated problem, and the scheduling solutions are updated accordingly.

When disturbance is detected which means that a state deviation from the referece is measured, the state information is feedback to the inner fast MPC or outer integrated problem. A threshold is introduced to determine the feedback. If the state deviation is less than the threshold, the state is feedback to inner fast MPC and locally treated by fast MPC. Otherwise if the state deviation is large (higher than the threshold) it is feedback to the integrated problem and the scheduling solution is updated accordingly. The threshold is an empirical value and it is determined to avoid unecessary changes of the scheduling solution when disturbances are small enough that can be handled efficiently in the control level, whereas updates in the scheduling solution are considered if significant disturbance occurs. Results of case studies show the feasibility and efficiency of the proposed approach.

Reference:

Terrazas-Moreno, S.; Flores-Tlacuahuac, A.; Grossmann, I. E. Simultaneous design, scheduling, and optimal control of a methyl-methacrylate continuous polymerization reactor. AIChE Journal 2008, 54, (12), 3160-3170.

Engell, S.; Harjunkoski, I. Optimal operation: Scheduling, advanced control and their integration. Computers & Chemical Engineering 2012, 47, 121-133.

Zhuge, J.; Ierapetritou, M. G. Integration of Scheduling and Control with Closed Loop Implementation. Industrial & Engineering Chemistry Research 2012, 51, (25), 8550-8565.

434730 On the Alleviation of Inventory Creep in Process Scheduling

Abstract:

On the Alleviation of Inventory Creep in Process Scheduling

Jin Zhang and Donald J. Chmielewski

In the work of Lima et al., [1], the notion of inventory creep within a process scheduling context was introduced. In short, the inventory creep phenomenon is a gradual reduction of material inventory over time. To alleviate this myopic behavior, the authors of [1] advocate the use of larger prediction horizons, but quickly run into computational tractability issues. Recently, similar inventory creep phenomena have been observed in the context of Economic MPC, [2, 3, 4]. In [2] and [3], inventory creep was virtually eliminated by the use of a surrogate MPC objective function. This quadratic objective function was constructed to be inverse optimal with respect to an appropriately defined Economic Linear Optimal Control (ELOC) policy [5]. In [3], the original economic objective function is retained, but is appended by an ELOC derived final cost term. This final cost is an approximation of the original objective function (from the end of the prediction horizon to infinity) resulted in an approximate infinite-horizon EMPC policy.

In the current effort, the approximate infinite-horizon EMPC policy is applied to a simplified version of the long-term scheduling problem found in [1]. This example is distinct from those of [2-4] in that many of the decision variables of the scheduling problem are restricted to integer values. This fact has two implications. First, these integer restrictions will significantly increase the computational complexity of the original scheduling version of the EMPC problem, and create a much greater need for reductions in horizon size. The second issue is that ELOC is incapable of enforcing integer constraints. However, development of an ELOC policy with integer constraint relaxed, leads to a final cost term that sufficiently approximates the economic cost from the end of the horizon to infinity. The result is an approximate infinite-horizon policy that enforces integer constraints over the finite horizon, but due to the appropriately selected final cost term is able to employ much short prediction horizons (requiring much less computational effort) while observing virtually zero inventory creep.

[1] Lima, R. M.; Grossmann, I. E.; Jiao, Y. "Long-term scheduling of a single-unit multi-product continuous process to manufacture high performance glass." Comput. Chem. Eng. 2011, 35 (3), 554–574.

[2] Omell, B.P.; D.J. Chmielewski, "IGCC Power Plant Dispatch using Infinite-Horizon Economic Model Predictive Control" Ind. Eng. Chem. Res., 52(9) pp 3151-3164 (2013).

[3] Mendoza-Serrano, D.I.; D. J. Chmielewski, "Smart Grid Coordination in Building HVAC Systems: Computational Efficiency of Constrained ELOC " Sci. Tech. Built Env., in press (2015)

[4] Omell, B.P.; D.J. Chmielewski, "On the Stability of Infinite Horizon Economic MPC," Annual Meeting of the AIChE, San Francisco, CA, 2013.

[5] Peng, J.K.; A.M. Manthanwar; D.J. Chmielewski, "On the Tuning of Predictive Controllers: The Minimally Backed-off Operating Point Selection Problem," Ind. Eng. Chem. Res., 44, pp 7814-7822, (2005).

430979 Fault Detection-Based Triggers for Rescheduling of Batch Process Operations

Abstract:

The scheduling of batch process operations has been the subject of significant research in the past decades. The state task network (STN) and resource task network (RTN) formulations [1, 2] and their modifications can now be used to determine the sequence and duration of operations to execute within the production line in both sequential and more complex, flow shop setups. Handling disturbances and recomputing the production schedules in the presence of unforeseen events (e.g., arrival of urgent orders, unexpected changes in feed materials or operating conditions, and unit break-downs) is equally important. This challenge can be addressed in an "ab-initio" fashion via scheduling under uncertainty [3, 4]. Reactive approaches that correct the schedule upon the occurrence of an event (i.e., performing a rescheduling calculation) have also been proposed [5].

Rescheduling calculations must, however, be performed judiciously, as imposing frequent schedule changes leads to schedule "nervousness" [6] and possibly deleterious economic effects. This, in turn, calls for a careful choice of the rescheduling trigger, and requires the response to unforeseen events to find a balance between the economical optimum and potential upset on the process.

Rescheduling approaches discussed in the literature thus far are largely reactive, in the sense that rescheduling is only triggered once an event has affected system performance, and focus largely on events that affect directly the scheduling layer (e.g., new/rush order arrival or machine break-down). On the other hand, dealing with events that affect the process dynamics (i.e., disturbances affecting operation of one unit within the system, without leading to unit break-down) has received less attention, under the likely implicit assumption that their impact will be addressed by the local control system. However, recent studies have shown that accounting for process dynamics at the scheduling level can lead to significant benefits [7], and suggest that disturbances must be dealt with in an integrated fashion at both the scheduling and control levels.

Motivated by the above, in this work, we focus on integrating dynamics and control considerations in the scheduling by developing dynamics-based metrics and triggers for rescheduling calculations. In particular, we rely on an integrated scheduling and dynamic optimization calculation to develop reference trajectories for the process states and inputs. Subsequently, we compute the discrepancy between real-time data and these reference values. Based on this, we show that deviations of the input (rather than state or output) trajectories of the process can be used for process fault detection and as a proactive trigger for rescheduling calculations. Furthermore, we illustrate that using these metrics enables us to detect issues during production before the end of an event, significantly mitigating the impact of potential deviation in, e.g., product quality. We present a case study demonstrating these concepts using a flowshop example, and show that rescheduling of operations (whether right-shifting or complete rescheduling) in the presence of dynamic disturbances can lead to significant economic benefits.

[1] E. Kondili, C.C. Pantelides, and R.W.H. Sargent. "A general algorithm for short-term scheduling of batch operations—I. MILP formulation." *Computers & Chemical Engineering,* 17: 211-227, 2003.

[2] C.C. Pantelides, "Unified frameworks for optimal process planning and scheduling." *Proceedings on the second conference on foundations of computer aided operations*. New York: Cache Publications, 1994.

[3] Z. Li and M.G. Ierapetritou. "Process scheduling under uncertainty: Review and challenges." *Computers & Chemical Engineering* 32:715-727, 2008.

[4] P.M. Verderame, J.A. Elia, J. Li, and C.A. Floudas. "Planning and scheduling under uncertainty: a review across multiple sectors." *Industrial & engineering chemistry research, 49*(9):3993-4017, 2010.

[5] J.P. Vin, and M.G. Ierapetritou. "A new approach for efficient rescheduling of multiproduct batch plants." *Industrial & engineering chemistry research* 39:4228-4238, 2000.

[6] G.E. Vieira, J.W. Herrmann, and E. Lin. "Rescheduling manufacturing systems: a framework of strategies, policies, and methods." *Journal of scheduling* 6:39-62, 2003.

[7] Y. Nie, L.T. Biegler, J.M. Wassick, "Integrated Scheduling and Dynamic Optimization of Batch Processes Using State Equipment Networks", *AIChE Jounal*, 58(11):3416-3432, 2012.

429270 Closed-Loop Properties of a Mixed Scheduling and Control Problem

Abstract:

Closed-Loop Properties of a Mixed Scheduling and Control Problem
Douglas Allan, James B. Rawlings, and Thomas Badgwell
Integration of production scheduling and process control is a subject rich with research opportunities. The high level, low-detail models often used by schedulers can result in suboptimal process economics because they do not take into account detailed process dynamics, and sometimes the resulting schedules are found to be infeasible at the control level. In recent years, advances have been made in combining scheduling and control. One major approach formulates one large problem containing all the aspects of both scheduling and control into a mixed-integer dynamic optimization (MIDO). The MIDO is then discretized into a mixed-integer nonlinear program (MINLP), which is in turn decomposed into a mixed-integer linear program (MILP) and nonlinear program (NLP), which can be solved iteratively to a solution [1,2,3].
This approach can result in great improvements in process economics, but is fundamentally an open-loop control strategy. Process disturbances or unplanned changes in schedule can disrupt the control sequence calculated offline, which then can result in poor performance. Several feedback strategies have been proposed. Instead of calculating control moves offline, Chu and You calculated the optimal tuning parameters of many multiloop PI controllers [4]. Unfortunately, this control structure precludes more advanced control strategies, such as model predictive control (MPC) which limits the system performance. Chu and You have also proposed a method of repairing a schedule calculated offline in a structured way in order to reduce the online computational burden [5]. Finally, Zhuge and Ierapetritou (2014) implemented a control strategy to resolve the optimization problem if the process deviates farenough from the target trajectory [6].
Simply running the open-loop strategies online (when computationally tractable) does not necessarily provide good performance. It is well known in the control community that optimality doesn't necessarily guarantee even closed-loop

stability. The concept of stability is new to the scheduling community, but recent efforts to apply the methods of MPC directly to a scheduling problem have shown that stability is indeed meaningful in a scheduling context, and that when neglected, inventory can increase without bound [7]. Therefore, we propose to analyze the closed-loop behavior of a mixed scheduling and control problem. Economic MPC provides a good framework to solve this problem and analyze stability and closed-loop behavior.

In this work we develop a simplified blending process example in which one product is delivered continuously into a pipeline and another is fed to a product tank. The tank is drained periodically by customers who have signed contracts for delivery of a full tank with a given composition at a specified time. As is typical in such contracts, there is a severe economic penalty if the product tank is not full with the correct product composition when the customer arrives to take delivery. Therefore the most important goal of the scheduling/control system is to create a full tank of the product with the correct composition when the customer arrives. In addition, the scheduling/control system must be flexible enough to handle disturbances in the delivery schedule, such as when a customer signals that they will be late picking up the product. In this case the combination of a delayed order and rush order must be accounted for by the dynamic controller. We propose to present a solution and closed-loop analysis of a class of problems displaying these features.

References

[1] M. Baldea and I. Harjunkoski. Integrated production scheduling and process control: A systematicreview. Computers & Chemical Engineering 71:377-390, September 2014

[2] Y. Nie, L. T. Biegler, C. M. Villa, and J. M. Wassick. Discrete Time Formulation for the Integrationof Scheduling and Dynamic Optimization. Industrial & Engineering Chemistry Research, September 2014

[3] A. Flores-Tlacuahuac and I. E. Grossmann. Simultaneous Cyclic Scheduling and Control of a Multiproduct CSTR. Industrial & Engineering Chemistry Research 45 (20):6698–6712, 2006

[4] Y. Chu and F. You. Integration of scheduling and control with online closed-loop implementation: Fast computational strategy and large-scale global optimization algorithm. Computers & Chemical Engineering 47:248-268, December 2012

[5] Y. Chu and F. You. Moving horizon approach of integrating scheduling and control for sequential batch processes. AIChE Journal 60 (5):1654-1671,January 2014

[6] J. Zhuge and M. G. Ierapetritou. Integration of Scheduling and Control with Closed Loop Implementation. Industrial & Engineering Chemistry Research 51 (25): 8550-8565, June 2012

[7] K. Subramanian, J. B. Rawlings, C. T. Maravelias. Economic model predictive control for inventory management in supply chains. Computers & Chemical Engineering 64:71-80, May 2014

424038 Simultaneous Design, Control and Operational Optimisation of a Domestic CHP System

Abstract:

In process systems engineering, the design of economically profitable and operationally optimal processes has been an active area of research for over 25 years. Significant contributions have been made throughout the years that consider the problem through a variety of approaches, particularly the consideration of (i) operability and design under uncertainty [1] as well as (ii) the system controllability with regulatory control structure selection and its tuning [2, 3] and (iii) the design and model predictive control with stability and robustness aspects taken into account, in a unified framework [4]. During the last few years there has also been great interest in the field of optimal operation, with several studies in literature exploring the interactions between shorter term operational decisions (regulatory and model-based control) with their longer term counterparts (scheduling and planning techniques) thus attempting to bridge the gap between the two [5-8]. The aim of this approach targets the long-term optimal operation of a process while taking into account its control aspects and the inherent handling of uncertainty [9].

In this work we present a study where the interactions of all three aspects of decision making of the process -- from the design aspects of the process and control elements to the optimal scheduling and control strategies -- in a single framework are considered, through its application on a series on micro-CHP units destined for domestic use. We consider the design aspects of the process as uncertain during the design procedure of the model-based controllers as well as the rolling-horizon scheduling. The latter takes into account the dynamics of the system through the consideration of the interactions between the control and scheduling via the use of a bridging model [5]. The integrated system of control and scheduling is solved into a multi-parametric fashion, thus allowing the acquisition of the exact solution of the problem a priori, while considering both (i) the design aspects, and (ii) the operational aspects of the process as uncertain. The optimally operational process is then introduced to dynamic optimisation which targets the design optimisation.

The principles introduced in [10] form the basis for this work. More specifically, a high fidelity model of an internal combustion engine equipped, natural gas powered, domestic CHP plant has been developed and introduced into the gPROMS® simulation and optimisation environment [11]. The model is able to simulate the simultaneous production of electrical power and usable heat. Aspects of the model such as the size of the internal combustion engine as well as a hot water buffer tank are considered for the optimal design. The model is subjected into approximation techniques in order for simplified state-space models with minimal loss of accuracy to be acquired. The approximation technique involves the use of the System Identification Toolbox of MATLAB®. Two subsystems are identified which reflect into the inherent distinct operation of the CHP -- the power generation subsystem and the heat recovery subsystem. The approximation procedure takes into account the design variables of the high fidelity model. The design of the mp-MPCs is based on the afore-mentioned state space models thus resulting into a design dependent decentralized control scheme [12]. The principles described in [13] are followed for the design of the rolling-horizon scheduling approach. In this case, the design aspects of the process are also treated as uncertainty. Finally, the design dependent bridging model is developed that closes the gap between the short-term and long-term operation of the process. All three rolling-horizon policies are introduced into the gPROMS® environment via the development of a software tool described extensively in [10]. Figure 1 graphically represents the set-up of the system and its components.

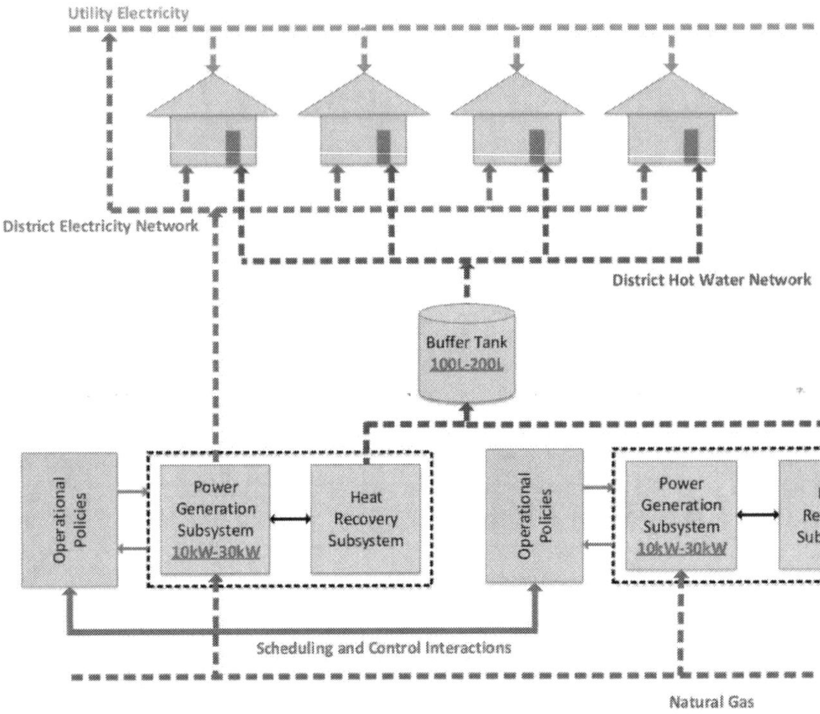

Figure 1 CHP System setup for domestic use with design uncertainty

At this point, the system can be simulated for given horizons, for a range of designs. The system is then introduced into the gOPT® optimisation functionality of gPROMS® that targets the design optimisation. At every iteration of the single vector shooting optimisation procedure, the design of the system is explored for a predefined demand profile, thus causing the uncertainty to realize itself. The operational objective of the design-dependent control is minimal mismatch to a set-point which is provided by the economically optimal solution of the design-dependent rolling-horizon scheduling and bridging model. The dynamic design optimisation targets the cost of acquisition of the system components via the transition of the cost on a daily basis, given a 10-year loan repayment.

Through this technique, we manage to target all three aspects of designing economically and operationally optimal processes with the use of a high-fidelity model and design-dependent rolling-horizon strategies. Most importantly though, we manage to provide a framework that can be applied to a large variety of processed without compromising the accuracy that a high-fidelity model provides and at the same time using advanced control and scheduling techniques in a closed-loop fashion.

References

1. van Schijndel, J.M.G. and E.N. Pistikopoulos, Towards the integration of process design, process control and process operability - Current stares and furore trends. Foundations of Computer-Aided Process Design, 2000. 96: p. 99--112.

2. Mohideen, M.J., J.D. Perkins, and E.N. Pistikopoulos, Optimal design of dynamic systems under uncertainty. AIChE Journal, 1996. 42(8): p. 2251--2272.

3. Bansal, V., et al., Simultaneous design and control optimisation under uncertainty. Computers & Chemical Engineering, 2000. 24(2 7): p. 261--266.

4. Sakizlis, V., J.D. Perkins, and E.N. Pistikopoulos, Parametric Controllers in Simultaneous Process and Control Design Optimization. Industrial & Engineering Chemistry Research, 2003. 42(20): p. 4545--4563.

5. Baldea, M. and I. Harjunkoski, Integrated production scheduling and process control: A systematic review. Computers & Chemical Engineering, 2014. 71: p. 377--390.

6. Zhuge, J. and M.G. Ierapetritou, Integration of scheduling and control for batch processes using multi-parametric model predictive control. AIChE Journal, 2014. 60(9): p. 3169--3183.

7. Subramanian, K., C.T. Maravelias, and J.B. Rawlings, A state-space model for chemical production scheduling. Computers & Chemical Engineering, 2012. 47(0): p. 97--110.

8. Engell, S. and I. Harjunkoski, Optimal operation: Scheduling, advanced control and their integration. Computers and Chemical Engineering, 2012. 47: p. 121-133.

9. Patil, B.P., E. Maia, and L.A. Ricardez-Sandoval, Integration of scheduling, design, and control of multiproduct chemical processes under uncertainty. AIChE Journal, 2015. In press

10. Pistikopoulos, E.N., et al., PAROC-An integrated framework and software platform for the optimisation and advanced model-based control of process systems. Chemical Engineering Science, 2015.

11. Diangelakis, N.A., C. Panos, and E.N. Pistikopoulos, Design optimization of an internal combustion engine powered CHP system for residential scale application. Computational Management Science, 2014. 11(3): p. 237--266.

12. Diangelakis, N.A. and E.N. Pistikopoulos, A Decentralised Multi-parametric Model Predictive Control Study for a Domestic Heat and Power Cogeneration System, in Computer Aided Chemical Engineering2015. Accepted for publication

13. Kopanos, G.M. and E.N. Pistikopoulos, Reactive scheduling by a multiparametric programming rolling horizon framework: A case of a network of combined heat and power units. Industrial and Engineering Chemistry Research, 2014. 53(11): p. 4366-4386.

432871 Integrating Scheduling and Control for a Mixed-Mode Process

Abstract:

The control of mixed-mode manufacturing operations that contain both batch units and continuous operations presents the challenge of synchronizing the continuous operations to the batch sequencing. This presentation will describe an approach that has been applied to an industrial scale process. Results will be presented and future research directions will be suggested.

430634 Data-Based Modeling and Control of Batch Emulsion Polymerization

Abstract:

Emulsion polymerization is a popular approach in which water-insoluble monomer is dispersed in water by agitation and the addition of a surfactant. Once an emulsion has been formed, initiator is added starting the polymerization reaction in the aqueous phase. As polymers react, they reach a critical length at which point they form a new phase which is distinct from the monomer and aqueous phases. The continued reaction of polymers within the polymer particle phase is the key advantage of this reaction method.

Benefits of emulsion polymerization include the ability to obtain high monomer conversion, high molecular weight, and low VOC while maintaining a low overall viscosity within the reactor. Another advantage of emulsion polymerization is that the high heat capacity of the aqueous phase is able to absorb the heat from strongly exothermic polymerization reactions. However, these benefits come at the cost of complicating the modeling and control problem.

The control objective for emulsion polymerization systems, like many other batch processes, is to reach the desired terminal quality. Specifically, the quality variable of interest is the particle size distribution (PSD). To achieve the desired PSD, the available inputs to the process include the temperature of the reactor and the addition (either in a continuous fashion or at discrete times) of ingredients such as monomer, initiator, surfactant, and water. Of particular interest is the addition of surfactant. The presence of surfactant above a critical concentration (the critical micelle concentration) facilitates the formation of new polymer particles. Therefore, addition of surfactant has a marked impact on the final particle size distribution.

In general emulsion polymerization, can be decomposed into three distinct intervals. The transitions between intervals is governed by the consumption of species present in the reactor. Namely, the transition between interval I and interval

II occurs when the concentration of surfactant falls below the critical concentration to form micelles and the transition between interval II and III occurs when the monomer droplets (monomer phase) is consumed. Each interval is governed by distinct dynamics that arise as the rate-limiting steps in the reaction change. These effects must be explicitly considered in the control problem.

In addition to the complexity introduced by multiple intervals and phases, emulsion polymerization also exhibits difficulties common to typical batch control problems. By nature, the process transits a large range of operating conditions (states). As a result, any dynamic model of the process must reflect the non-linear nature of the underlying system. Clearly, the traditional control approach of using linear models to regulate operation around a single steady-state are not applicable. Instead, the control approach for emulsion polymerization must account for the non-linear nature of the system to achieve the desired PSD.

One of the key difficulties in reaching a desired PSD is that the PSD is not easily measured online. As a result, secondary measurable variables (such as temperature and certain aqueous phase concentrations) must be used to implement feedback control. In general, there are two approaches to handle lack of quality measurements discussed in literature. These are trajectory tracking approaches and inferential modeling approaches. In the former approach, controllers are designed to tightly track predetermined set-point trajectories for measurable process variables. In the inferential approach, a model of the process is used to estimate the unmeasured quality variable from the available measurements. Contributions employing both of these approaches have been made to the emulsion polymerization literature.

Crowley et al. consider optimal input trajectories (of surfactant and initiator) for trajectory tracking control of the PSD for an emulsion polymerization system.[1] Flores-Cerrillo and MacGregor consider temperature trajectory tracking for emulsion polymerization using latent variable models.[2]

Flores-Cerrillo and MacGregor also considered direct control of a bimodal PSD for an emulsion polymerization reaction. In their approach, they use partial least squares (PLS) inferential models to predict the final PSD at a per-determined sample time partway though the batch. Then they implement a control action in the form of a discrete addition of surfactant. Results in this work are shown for an isothermal simulation of an emulsion polymerization process.[3]

Previously, we considered the quality control of batch polymerization reactors using data-based model predictive control. Specifically, we considered the modeling problem into two steps. First, we constructed a nonlinear dynamic model by combining local linear ARX models with weights depending on lagged outputs and inputs. The resulting model provided process measurement predictions which were then used in a PLS model to infer the resulting batch quality. The combined models (the output prediction model and the PLS model) provided a causal model of the process which was then implemented in a MPC scheme.[4, 5]

In this work we consider a new approach to PSD control for emulsion polymerization reactions. Specifically we develop a data-driven model of the process to relate measurable process variables and the desired properties of the PSD. The motivation behind this work is the understanding that most physical systems can be described by a (potentially unmeasured) Markov state. Therefore, the desired quality (PSD) should be a function of that state. The approach, then, is to identify a state-space model of the process and use it in a MPC framework to drive the process to the state corresponding to the desired quality. The approach can be broken into two steps: identification of a state-space model from available input-output data and appropriate application of the resulting model in an MPC framework.

In order to identify models from existing (possibly closed loop) data, we apply a well studied identification technique called subspace identification. The principal advantage of subspace methods is their ability to identify state-space models directly from input-output data. Furthermore, subspace methods are non iterative eliminating any issues with convergence. However, subspace methods are also constrained by a list of assumptions. These effectively restrict the application of the methods based on stationarity, linearity, and excitation. In this work, we show how subspace methods can be adapted to batch processes despite these restrictions.

Once a reliable model has been identified via subspace identification, it remains to implement the model in an MPC framework capable of driving the process to the desired quality (PSD). This MPC must reflect the adaptations made in the model identification phase. Specifically, the non-linearity of the system must be addressed.

The methodology described above is shown in application to the emulsion polymerization system. To this end, a highly detailed, non-isothermal, population balance model is used to generate historical training data and validate the proposed model and control design.[6] Results demonstrate the efficacy of the methodology for control of emulsion polymerization reactors. Furthermore, the applicability of the methodology to a broader class of batch and semi-batch reactors is discussed. References

[1] Crowley T, Meadows E, Kostoulas E, Doyle F. Control of particle size distribution described by a population balance model of semibatch emulsion polymerization Journal of Process Control. 2000;10:419-432. 14th IFAC World Congress, Beijing, Peoples R China, JUL 05-09, 1999.

[2] Flores-Cerrillo J, MacGregor J. Latent variable MPC for trajectory tracking in batch processes Journal of Process Control. 2005;15:651-663.

[3] Flores-Cerrillo J, MacGregor J. Control of particle size distributions in emulsion semibatch polymerization using mid-course correction policies Industrial & Engineering Chemistry Research. 2002;41:1805-1814.

[4] Corbett B, Macdonald B, Mhaskar P. Model Predictive Quality Control of Polymethyl Methacrylate IEEE Transactions on Control Systems Technology. 2015;23:687-692.

[5] Aumi S, Corbett B, Clarke-Pringle T, Mhaskar P. Data-driven model predictive quality control of batch processes AIChE Journal. 2013;59:2852-2861.

[6] Meadows E, Crowley T, Immanuel C, Doyle F. Nonisothermal modeling and sensitivity studies for batch and semibatch emulsion polymerization of styrene Industrial & Engineering Chemistry Research. 2003;42:555-567.

431736 Development and Validation of a Reduced-Index Dynamic Model of an Industrial High-Purity Column

Abstract:

This work is part of the development and implementation of a state-of-art Real Time Optimization (RTO) system to an industrial-scale depropenizer column, a vapor recompression distillation unit located at Refinaria de Paulínia, Paulínia, São Paulo, Brazil and owned by Petrobras S.A. Dynamic modeling and simulation validation using real process data of the unit are carried out in an equation-oriented environment, EMSO (Environment for Modeling, Simulation and Optimization)(Soares and Secchi, 2003).

The depropenizer is a high purity distillation column with high nonlinear behavior because of the strong interactions due to the vapor recompression system (Mendoza et al., 2013). Furthermore, the difference between internal and external material/energy flows causes a complex multi-time-scale dynamics (Jogwar and Daoutidis, 2009). Modeling such process is a challenging problem due to these characteristics.

In addition, the modeling of dynamic equilibrium processes often results in higher index DAE systems (Pantelides et al., 1988). Usually, phenomenological relationships are used to solve the index problem, but this approach gives rise to errors as a result of unknown parameters and project details that are assumed. Considering that the column's response to composition changes, in general, takes place over a timescale one or two order of magnitude lower than those of flow rate changes, an approach similar to a proportional loop with arbitrarily large gain is used as an alternative to solve the index problem (Ponton and Gawthrop, 1991). The dynamic model structure is based on the steady state model presented by Mendoza et al. (2013) and contains more than eight thousand equations. An extensive validation with real process data shows that the proposed approach is able to predict the dynamic behavior of the column properly. The application of the model in Moving Horizon Estimation is under implementation.

Jogwar, S. S., & Daoutidis, P. (2009, June). Vapor Recompression Distillation: Multi-Scale dynamics and Control. In American Control Conference, 2009. ACC'09. (pp. 647-652). IEEE.

Mendoza, D. F., Palacio, L. M., Graciano, J. E., Riascos, C. A., Vianna Jr, A. S., & Le Roux, G. C. (2013). Real-Time Optimization of an Industrial-Scale Vapor Recompression Distillation Process. Model Validation and Analysis. In Industrial & Engineering Chemistry Research, 52(16), 5735-5746.

Pantelides, C. C., Gritsis, D., Morison, K. R., & Sargent, R. W. H. (1988). The mathematical modelling of transient systems using differential-algebraic equations. In Computers & chemical engineering, 12(5), 449-454.

Ponton, J. W., & Gawthrop, P. J. (1991). Systematic construction of dynamic models for phase equilibrium processes. In Computers & chemical engineering, 15(12), 803-808.

Rincon, F.D., Le Roux, G.A.C., & Lima, F.V. (2013). Robust Calorimetric Estimation of Semi-Continuous and Batch Emulsion Polymerization Systems with Covariance Estimation. In Macromolecular Reaction Engineering, 8(6), 456-466.

Soares, R. D. P., & Secchi, A. R. (2003). EMSO: A new environment for modelling, simulation and optimisation. In Computer Aided Chemical Engineering, 14, 947-952.

426428 Dynamic Modeling with Correlated Inputs: Theory, Method and Experimental Demonstration

Abstract:

When modeling dynamic processes for several inputs with freely existing data, such as data collected with normal process operations, the ability to accurately model the output response for a given input change is impeded when inputs are cross- (i.e., pair-wised) correlated as this adversely affects accurate estimation of the causative effects of inputs on the response variable. The causative effects of the inputs can be evaluated functionally and analytically via the Jacobian Matrix which is done in this work for NARMAX and Wiener structures that are linear and nonlinear in model parameters. This analysis shows that the Wiener structure with physically-based nonlinear parameterization is superior. This conclusion is also supported in this work by a modeling study on a real distillation column consisting of eight test runs over a period of three years.

In chemical processes, output variables are determined by the values of input variables, which may or may not be measured and some input variables may not even be known. Mathematical modeling is the process of mapping measured variables to output variables using a mathematical formulation. In mathematical modeling a structure must be selected and any unknown coefficients (i.e., parameters) must be estimated. For a general model structure, let its expectation be represented as $h_i = f(\boldsymbol{X}_i;\boldsymbol{q})$, where h_i is the expected value of the response (i.e., output) at the ith sampling time, $i = 1, \ldots, n$; $\boldsymbol{X}_i$ is the vector of input values at the ith sampling time; and $\boldsymbol{q}$ is the vector of unknown model parameters with $\boldsymbol{q} = [q_1 \ldots q_q]^{\mathrm{T}}$. Therefore, the element of its Jacobian Matrix, $\mathbf{J}^{n\mathrm{x}q}$, in the ith row and j^{th} column is that is, Moreover, the jth column represents q_j and its column vector represents the change in the response space as q_j changes for the set of experimental conditions. If two columns, say j and k, are orthogonal, their correlation coefficient is zero, and the information to estimate q_j is decoupled (i.e., separate or independent) from the information to estimate q_k, and vice versa. The advantage of this is that causative relationships of inputs on the response can be obtained and standard estimation errors of parameters are minimum[1]. Correlated columns in the Jacobian Matrix arise from pairwise correlation of inputs. Thus, obtaining orthogonal columns for each parameter necessitates setting $\boldsymbol{X}_i$ according to some predetermined statistical experimental design. However, running a statistically designed experiment is not always practical or possible. The ability to accurately model the output response for given inputs is impeded when inputs are cross-correlated. Consequently, model identification often involves the use of experimental data with correlated inputs that result in columns of **J** that are not orthogonal (i.e., correlated). Notwithstanding, for a given set of experimental data, with pairwise correlated inputs, the best model (in the sense of evaluating causative relationship between inputs and response and extracting scientific knowledge from parameters) will be the one that produces columns of **J** with the smallest pair-wise correlation. Therefore, the goal of this work is to evaluate dynamic model structures based on pair-wise correlation of the columns of **J** when the inputs are pair-wise correlated. This scope is restricted to transfer function models that are linear in the time-dependent process variables. More specifically, this work compares structures that are developed from transfer functions that are applicable to Nonlinear Autoregressive Moving Average with eXogenous variables (NARMAX) and Wiener models.

The basic difference of NARMAX and Wiener transfer functions is the characteristic equation which is the same for each input for NARMAX but can be different for each input for Wiener. With discrete-time modeling, this work will evaluate linear and nonlinear regression transfer function models of NARMAX and Wiener networks. The engineering literature typically defines linear models based on the form of the time-dependent variables in the differential equations of transfer functions. However, this scope is parameter estimation with focus on **J** which is based on the behavior of the parameters. Hence, this work adopts the statistical definition of linear and nonlinear models in parameter estimation. More specifically, a linear model in this work is one that is linear in parameters and a nonlinear model is one that is nonlinear in parameters. All parameters in this work are estimated with the least squares criterion.

From a search in the process identification literature, we discovered that for NARMAX models and its subclasses, linear forms in parameters are widely, if not exclusively, used. It also revealed that the common practice of Wiener Modeling is the use of transfer functions that are linear in the estimated parameters. The only discrete-time Wiener approach that we found to use nonlinear transfer functions is the one developed by Rollins and co-workers. The nonlinear parameterized NARMAX structure will be developed in this work following this approach.

Spatiotemporal modeling and identification of CO_2 adsorption columns

Davood Babaei Pourkargar, Seyed Mehdi Kamali, Robert M. Rioux, and Antonios Armaou

Global warming and climate change have recently received widespread attention due to increasing CO_2 emissions [1] that necessitate developing new processes to capture anthropogenic greenhouse gases from the flue gas stream of coal-fired power plants for electricity generation. While a wide range of CO_2 capture technologies have been proposed for post-combustion processes in chemical industries and power plants [2], adsorption on solid-based beds is recognized as one of the most efficient capture processes which is able to handle high throughputs. Although there is a large selection of industrial adsorbents, aminosilica adsorbent packed beds have recently received increasing attention in CO_2 adsorption due to their exceptional capacity and adsorbing CO_2 selectivity over all the other major components of the gas stream being processed [3].

The dynamic behavior of a given CO_2 adsorption column is practically described by a breakthrough curve which presents the temporal profile of the CO_2 concentration at the column outlet. This curve contains essential operational properties of the adsorption process can be obtained by direct experimentation or mathematical modeling. Compared to the experimental method which is typically time-consuming and expensive, mathematical modeling is economically efficient and has recently attracted increasing interest. To apply such a mathematical model as a basis for process design and control, it should be able to accurately predict the experimental behavior employing process parameters and conditions.

We focus on a rigorous model for isothermal CO_2 adsorption columns which describes the spatiotemporal dynamics of the CO_2 concentrations in the bulk and solid bed by a set of partial differential equations. Such mathematical model which considers both dispersion and convection phenomena provides the spatiotemporal behavior of the adsorption rate and circumvents the unphysical simplifying assumption of the linear driving force rate and unified rate through the column length which has been applied in the previous proposed models [3, 4, 5]. The proposed model is the base for a characterization approach seeking to compute physical quantities originating from mass balance laws such as the adsorption rate constant and capacity from a set of experimental data without using empirical parameters assumptions which have been invoked in previous research work [3, 4, 5].

The spatiotemporal dynamics of CO_2 adsorption in an aminosilica packed bed are successfully predicted by the proposed model. The dispersion coefficient, adsorption rate constant and capacity of the bed are then identified using a set of experimental CO_2 concentration measurements at the adsorption column outlet for different temperatures using a dynamic optimization formulation which is solved using shooting methods [6]. Finally, we compute the heat of adsorption for different temperatures and identify the activation energy of adsorption based on an Arrhenius temperature-dependence assumption.

[1] S. PACALA, R. SOCOLOW, Stabilization wedges: Solving the climate problem for the next 50 years with current technologies, Science, 305:968-972, 2004.

[2] K.Z. HOUSE, C.F. HARVEY, M.J. AZIZ, D.P. SCHRAG, The energy penalty of post-combustion CO_2 capture & storage and its implications for retrofitting the U.S. installed base, Energy Environ. Sci., 2:193-205, 2009.

[3] P. BOLLINI, N.A. BRUNELLI, S.A. DIDAS, C.W. JONES, Dynamics of CO_2 adsorption on Amine adsorbents. 1. Impact of heat effects, Ind. Eng. Chem. Res., 51:15145-15152, 2012.

[4] A. HEYDARI-GORJI, A. SAYARI, CO2 capture on Polyethylenimine-impregnated hydrophobic mesoporous Dilica: Experimental and kinetic modeling. Chem. Eng. J. 173:72–79, 2011.

[5] J. KALYANARAMAN, Y. FAN, R.P. LIVELY, W.J. KOROS, C.W. JONES, M.J. REALLF, Y. KAWAJIRI, Modeling and experimental validation of carbon dioxide sorption on hollow fibers loaded with Silica-supported Poly(ethylenimine), Chem. Eng. J., 259:737-751, 2015.

[6] V.S. VASSILIADIS, R.W.H. SARGENT, C.C. PANTELIADES, Solution of a class of multistage dynamic optimization problems. 1. Problems without path constraints & 2. Problems with path constraints, Ind. Eng. Chem. Res., 33:2111-2133, 1994.

422258 Identification of a Reduced Order Data-Driven Model of a Hydrogen Production Plant

Abstract:

Hydrogen is one of the most important chemical components widely used in petroleum and chemical industries. An economic way to commercially produce hydrogen is to utilize steam methane reforming [1]. The process consists of a number of material and heat flows, resulting in an intricate network of material and energy. Natural gas (NG) and superheated steam are fed to a chemical reactor called reformer, which consists of catalyst tubes filled with nickel reforming catalyst. Here, majority of the hydrogen is produced. The gas out of the reformer is then processed through another reactor to further produce hydrogen. Finally, the hydrogen flow is purified in a pressure swing adsorber (PSA), where high purity hydrogen is produced. The reformer exit temperature is an important process variable for this process and is expected to be kept at a desired level by heating the reformer. This heat is provided by burning the off-gas from the PSA and NG fuel stream. A fan is used to supply air to the burners and another one draws the combustion products, termed as flue gas, out of the reformer box.

This process presents various constraints and challenges during the start-up as well as nominal operation of the plant such as, the reformer exit temperature should be maintained at a desirable level. Further, the firebox pressure should not breach its lower and upper limits for safety. If the pressure is too low, the fire can be extinguished. If it is too high, it may impose safety hazards to facility and personnel [2].

In this work, we design a safe and optimal start-up strategy for the unit as it goes from the shutdown conditions to the nominal operating phase. A desired characteristic is to optimize the process in such a way that it minimizes the start-up time and the cost while achieving the desired specifications of the hydrogen produced. To this end, first a high fidelity model of the entire plant is developed to be used as a test bed. Then a data driven model [3-4] utilizing simulation data from the high fidelity model is developed for the purpose of optimizing the startup. Simulation results are presented that illustrate the improvement achieved by implementation of the optimization based startup.

[1] JL Lynn and BH Bory. Kirk-othmer encyclopedia of chemical technology, 2000.

[2] Miao Du, Prashant Mhaskar, Yu Zhu, and Jesus Flores-Cerrillo (2014). Safe-Parking of a Hydrogen Production Unit. Industrial & Engineering Chemistry Research, 53 (19), 8147-8154.

[3] S. Aumi, B. Corbett, P. Mhaskar, and T. Clarke-Pringle. Data-based modeling and control of nylon-6, 6 batch polymerization. Control Systems Technology, IEEE Transactions on, 21(1):94-106, Jan 2013.

[4] Aumi, S. and P. Mhaskar. An Adaptive Data-based Modeling Approach for Predictive Control of Batch Systems. *Chem. Eng. Sci.*, 21, 94-106, 2013.

426362 Feedforward Control with Block-Oriented Modeling with Demonstration to Nonlinear Parametrized Wiener Modeling

Abstract:

This work proposes a feedforward control (FFC) approach with block-oriented modeling. It is demonstrated using a nonlinear parametrized Wiener Modeling approach in feedback feedforward (FBFF) control. The process used for this demonstration is a simulated continuous stirred tank reactor (CSTR) with four (4) measured inputs. The Wiener model is nonlinear in the physically-based dynamic parameters of the transfer functions and linear in the static parameters of the static gain function. The static gain function has a second order linear regression form with interaction and quadratic terms. The Wiener model is built under open-loop conditions using a Box-Behnken statistical experimental design consisting of 27 sequential step tests. Under a sequence of multiple input changes, the addition of feedforward control (FFC) reduced the standard deviation of the controlled variable from its set point by 70%.

Traditional feedback control (FBC) makes adjustments to some manipulated variable after the process has deviated from its desired operating condition or set point (SP) for the controlled variable. In the past few decades, more sophisticated control algorithms have been introduced that use models determined from process data to achieve tighter control. Some of the most well-known model-based control systems include FFC, internal model control (IMC), and model predictive control (MPC). However, while FFC is the only model-based control system that can totally cancel the effect of a disturbance, theoretically, it appears to have seen limited implementation in real processes as indicated by the very low number of articles in the process control literature. For effective FFC the model must consists of only inputs and the inputs have to map accurately and causatively to changes in the controlled variable which is very difficult to achieve in modeling real data for several reasons. The first one is due to shifts in the output caused by changes in the levels of unmeasured disturbances. Minimizing the number of unmeasured disturbances by measuring them and including them in the set of inputs can help to alleviate this behavior but as the number of inputs in the model increases, the complexity of the model increases and parameter estimation becomes more difficult. Furthermore, as the number of inputs increase, the longer the data collection will need to be and thus, the greater the likelihood of shifts in the output from changes in unmeasured inputs. Secondly, causative input modeling accuracy will suffer when the inputs are pairwise cross-correlated, especially for structures with linear parametrization. Finally, over the input operating space, the inputs typically map nonlinearly to the response space due to interactive and curvilinear relationships between the inputs and the output. As a consequence of these modeling challenges, FFC in practice has often been limited to a few variables, narrow input ranges, and linear static gain functionality.

Some models for nonlinear gain behavior have been proposed for model-based controllers including radial basis functions (RBF), genetic algorithms (GA), Nonlinear Auto Regressive Models And eXogenous inputs (NARMAX) models, and block-oriented models (BOMs). An important advantage of NARMAX and BOMs is that they can use transfer functions, i.e., linear dynamic equations with physically interpretable parameters. However, a limitation of the NARMAX structure is that all of its transfer functions have the same characteristic equation or denominator dynamics. BOMs use the outputs from blocks of dynamic (transfer) functions that are linear (L) differential equations as inputs to functions that can be nonlinear (N) with respect to static gain parameters. The simplest of the BOMs is the Hammerstein network (NL), which has an N block followed by an L block and the Wiener network (LN), which reverses the order of these two blocks. More complicated block-oriented structures include sandwich models such as an LNL network, which has linear dynamic blocks, followed by a nonlinear static block, followed by a second linear dynamic block. When the inputs can have different dynamic behavior, the Wiener network is the preferred choice over Hammerstein and is superior to NARMAX because the inputs can have completely different dynamic structures.

A number of researchers have studied the identification of model parameters for the Hammerstein and Wiener networks. The LNL network has not gotten as much attention, but some have proposed methods for its parameter identification. There has been much progress over the last decade in the identification of BOMs and recently, by taking a nonlinear parametrized approach for estimation of the dynamic parameters, accurate Wiener modeling was demonstrated using nine (9) inputs on a real distillation process with large variation due to unmeasured disturbances and with highly pairwise cross-correlation of the inputs. While there has been progress in the use of BOMs in model based control, progress of FFC using BOMs appears to have been limited to single input models. Thus, the objective of this work is the development of a general FFC framework for multiple-input BOMs in FFC with nonlinear static gain behavior. This framework will be demonstrated using a nonlinear parametrized Wiener modelwith a second-order static gain structure on a simulated CSTR. The Wiener model is built under open-loop conditions using a Box-Behnken statistical experimental design consisting of 27 runs or sequential step tests. Under a sequence of multiple input changes, the addition of feedforward control reduced the standard deviation of the controlled variable from its set point by nearly 70%.

430674 Development of a Recursive Time Series Model for Growth of Mammalian Cells Used in Monoclonal Antibody Production

Abstract:

Development of A Recursive Time Series Model for Growth of Mammalian Cells Used in Monoclonal Antibody Production

Jingwei Gan, Satish J. Parulekar, and Ali Cinar

Department of Chemical and Biological Engineering

Illinois Institute of Technology, Chicago, IL 60616

Abstract

Monoclonal antibodies (MAbs) are high value proteins with wide variety of applications including diagnostic testing and therapeutic treatment. MABs derived from hybridoma cell cultures are used extensively in diagnostic assays and have found increasing use in therapeutic applications, affinity production systems, and in vivo imaging. The cell lines, hybridoma, are produced by fusing B cells (lymphocytes) from immunized animals with myeloma cells. The simplest approach for producing a MAb in vitro is to grow the hybridoma cells in batch and fed-batch cultures and recover and purify the MAb from the culture medium. The large scale production of MAbs occurs via in vitro cultivation of hybridoma cell lines in bioreactors using techniques similar to those used for microbial cultivation. A cost-effective production of MAbs requires an understanding of the effects of bioreactor process variables on the physiology of hybridoma cells. Within mammalian expression platforms, Chinese hamster ovary cell-based system remain the most commonly used expression system. The development of methods for optimization of cell growth and cell productivity has become a crucial issue to enhance MAb yield in in vitro production.

Traditional operations of these culture systems at production scales are mainly based on empirical knowledge. There has been limited effort on development of first principle-based mathematical models for these cultures at laboratory scale operations. Due to complexity of the models and uncertainties associated with estimation of model parameters, the first principle models are not suitable for online prediction and control of mammalian cell cultures. In order to achieve accurate online prediction and control, simple data driven models may be more suitable for this. In this work, time series models for viable cell, glucose and glutamine concentrations in fed-batch mammalian cell cultures are developed. Results from non-recursive and recursive models are compared. The framework for developing these models is described. The key culture variables considered are concentrations of viable cells, glucose, and glutamine.

In order to generate 'experiment data', we use a first principles model proposed by Kontoravdi et al. (2005). This model structure is suggested by many researchers,while the values suggested for each parameter may vary. In order to generate random 'experiments' when using this model, random variation for each parameter is used with their mean values suggested by Kontoravdi et al. (2005) and standard deviations computed by using the reported values by all authors who used the same cell line.

Viable cell concentration in cell culture is expressed by a recursive ARMAX model with stability constrain proposed by Turksoy et al. (2014). The time series model expresses current and future viable cell concentrations as a function of past values of viable cell concentrations and multiple inputs. Model parameters are recursively estimated by using a least square estimation algorithm. Model stability is guaranteed by converting the time series model into state space representation and ensuring that all eigenvalues lie within unit circle. Glucose and glutamine concentrations are modeled by using a dual rate model proposed by Ding and Chen (2004). After modifying the time series model by using a polynomial transformation technique, the resulting dual rate model can be used to represent dual sampling rate systems where inputs and outputs have different sampling rates. Using a dual rate model combined with recursive estimation with stability constraint, we propose recursive dual rate model with stable recursive parameter estimation to estimate the glucose and glutamine concentrations in fed-batch culture.

References:

[1] Ding, Feng, and Tongwen Chen. "Combined parameter and output estimation of dual-rate systems using an auxiliary model." *Automatica* 40.10 (2004): 1739-1748.

[2] Turksoy, Kamuran, et al. "Multivariable adaptive identification and control for artificial pancreas systems." Biomedical Engineering, IEEE Transactions on61.3 (2014): 883-891.

[3] Kontoravdi, Cleo, et al. "Application of Global Sensitivity Analysis to Determine Goals for Design of Experiments: An Example Study on Antibody Producing Cell Cultures." *Biotechnology progress* 21.4 (2005): 1128-1135.

435332 Modeling & Simulation Trends and Application of Optimization in Polymerization Processes

Abstract:

Innovations in process modeling technologies and tools for the polymer processes evolved recently. Key innovations in various polymer process modeling technologies have emerged and made it possible to develop high fidelity polymer process models that can be used for process design and process optimization. The use of process modeling technology in the polymer industry has been rare till recent past. The modeling scope was limited to the expertise domains of modeling i.e., polymer reaction engineering and reactor modeling. Mathematical modeling is a powerful tool for the development of process understanding in polymerization.

The modeling functionalities to address key polymer modeling issues and challenges majorly include polymer molecular structure, polymerization reaction engineering, phase equilibrium, mass transfer, parameter estimation, structure-property relationship etc. Due to the complex nature of polymer chemistry and physics, there are still many unsolved modeling issues and opportunities for future innovations.

This work focuses on the establishment of fundamental modeling concepts relevant to catalysis & polymer processing and presents a critical review on recent modeling and simulation trends along with applications of various optimization approaches.

413744 Actuator and Sensor Fault Isolation of Nonlinear Systems Subject to Uncertainty

Abstract:

Actuator and sensor fault isolation of nonlinear systems subject to uncertainty

Automatic control techniques have been widely employed in industry to increase efficiency and profitability of the processes. However, this comes at a price of increased impact of abnormalities in major control equipment such as actuators and sensors. This realization has motivated design of frameworks for fault detection and isolation (FDI) and fault tolerant control (FTC). Important issues such as system nonlinearities and existence of uncertainties must be considered in FDI and FTC design to be able to successfully decrease or tackle the severity of the fault effects.

The FDI problem for nonlinear systems has been considered widely in the literature during the past decade (see, e.g., [1], [2], [3] and [4]). Most of the existing results, however, focused on isolation of single actuator or single sensor faults. Recently, results have enabled distinguishing between simultaneous sensor and actuator faults, where the results are derived assuming no uncertainty (see e.g., [5]).

In [6], the problem of isolation of complex actuator faults (occurrence of several actuator faults in same order of differentiation) in the presence of uncertainty is handled by explicitly characterizing the way the faults affect the nonlinear process system, and driving the system to a point that enables fault isolation. The problem of FDI in the presence of uncertainty has also been studied (see e.g., [7] and [8]) using adaptive estimation techniques. First a fault detection scheme is designed which simply uses output estimation error as residual. Then, a bank of fault isolation estimators is designed using adaptive estimation techniques. The existing results, however, consider only single fault scenarios. In summary, there is a lack of results for nonlinear systems subject to uncertainty where the problem of fault detection and isolation for simultaneous actuator and sensor faults is addressed.

Motivated by the above considerations, this work considers the problem of actuator and sensor fault isolation for simultaneous faults in the presence of uncertainty. This is achieved by building a bank of residuals, each using an appropriate subset of the available measurements (and associated state estimators), to determine the expected behavior of the system and compare with the observed evolution. To this end, at first we establish boundedness of estimation error for discretized fashion high gain observer presented in [4] in the presence of uncertainty. Next, we propose the fault detection and isolation design, which comprises a bank of fault detection and isolation filters that trigger an alarm based on appropriately defined residuals breaching their thresholds. Thresholds are defined in a way that they account for the impact of the uncertainty on the estimation error and the prediction of the expected system behavior. In this way, the fault isolation mechanism explicitly accounts for effect of the uncertainty. Then we present the detectability conditions for single and simultaneous faults. The efficacy of proposed FDI framework in presence of uncertainty and measurement noise is illustrated using a chemical reactor example.

References

[1] Prashant Mhaskar, Charles McFall, Adiwinata Gani, Panagiotis D Christofides, and James F Davis. Isolation and handling of actuator faults in nonlinear systems. Automatica, 44(1):53-62, 2008.

[2] Chudong Tong, Nael H El-Farra, Ahmet Palazoglu, and Xuefeng Yan. Fault detection and isolation in hybrid process systems using a combined data-driven and observer-design methodology. AIChE Journal, 60(8):2805-2814, 2014.

[3] Prashant Mhaskar, Jinfeng Liu, and P Christofides. Fault-tolerant process control. Springer, 2013.

[4] Miao Du and Prashant Mhaskar. Isolation and handling of sensor faults in nonlinear systems. Automatica, 50(4):1066-1074, 2014.

[5] Miao Du, James Scott, and Prashant Mhaskar. Actuator and sensor fault isolation of nonlinear process systems. Chemical Engineering Science, 104:294-303, 2013.

[6] Miao Du and Prashant Mhaskar. Active fault isolation of nonlinear process systems. AIChE Journal, 59(7):2435-2453, 2013.

[7] Xiaodong Zhang, Marios M Polycarpou, and Thomas Parisini. Fault diagnosis of a class of nonlinear uncertain systems with Lipchitz nonlinearities using adaptive estimation. Automatica, 46(2):290-299, 2010.

[8] Xiaodong Zhang. Sensor bias fault detection and isolation in a class of nonlinear uncertain systems using adaptive estimation. Automatic Control, IEEE Transactions on, 56(5):1220-1226, 2011.

420797 Detection and Isolation of Batch-to-Batch Parametric Drift in Protein Crystallization

Abstract:

Modeling and control of batch crystallization is of significant interest to pharmaceutical industry because the bioavailability of drugs highly depends on the size and shape distributions of their active pharmaceutical ingredients. Within this context, the dominant dynamic behavior of the evolution of key crystallization variables is modeled through a process model, which is used for the design of an in-batch model predictive controller (MPC) to regulate the average crystal size and shape distributions obtained at the end of the batch process. However, due to unknown batch-to-batch parametric drifts, the process model employed for in-batch control and estimation purposes may significantly deviate from the actual process behavior. More specifically, small batch-to-batch changes in the pH level, buffer concentration, or impurity concentration in the feedstock container may significantly alter the quality (e.g., size and shape) of the crystal products [1]. There are previous contributions that have considered run-to-run (R2R) control schemes that deal with batch-to-batch parametric drifts in a variety of batch processes [2-5].

In the present work, in order to further refine the R2R approach by relaxing the requirement of the post-batch measurements over multiple batch runs, we focus on the design of parametric drift detection and isolation (PDDI) scheme for the detection and isolation of the parametric drift. Using this scheme, it becomes easier to precisely calculate the magnitude of the process drift because parametric drift candidates are narrowed down from the set of all possible drift candidates. The proposed PDDI scheme uses in-batch (protein solute concentration and crystallizer temperature) and post-batch measurements (crystal product attributes) and consists of two parts: preparatory stage before batch-to-batch operation and post-batch stage during batch-to-batch operation. The goal of the preparatory stage is to compute the threshold values and signatures for each parametric drift using simulations and batch process common cause variance described by noise. During the batch-to-batch operation, the proposed PDDI system monitors closed-loop process residuals, which are computed by taking the difference between the time profiles of the states obtained through in-batch and post-batch measurements, from the time profiles of the states obtained from the drift-free simulation with noise. Then, we compare the residuals with signatures obtained in the preparatory stage for each parametric drift for isolation of a parametric drift. The PDDI system estimates the magnitude of the parametric drift via a multi-variable optimization problem and updates the parameters of the batch process model used in the in-batch MPC system to compute a set of jacket temperatures for the production of crystals with a desired shape distribution in the next batch. The performance of the MPC with the proposed PDDI scheme is demonstrated by applying it to a multiscale simulation of a batch protein crystallization process with parametric drifts.

[1] Cacioppo E, Pusey ML. The solubility of the tetragonal form of hen egg white lysozyme from pH 4.0 to 5.4. J. of Crystal Growth. 1991;114:286-292.

[2] Campbell WJ, Firth SK, Toprac AJ, Edgar TF. A comparison of run-to-run control algorithms. In Proceedings of the American Control Conference, Anchorage, AK, 2002:2150-2155.

[3] Wang Y, Gao F, Doyle FJ. Survey on iterative learning control, repetitive control, and run-to-run control. J. of Process Control. 2009;19:1589-1600.

[4] Kwon JS, Nayhouse M, Orkoulas G, Ni D, Christofides PD. Run-to-Run-based model predictive control of protein crystal shape in batch crystallization. Ind. & Eng. Chem. Res. 2015;54:4293-4302.

[5] Kwon JS, Nayhouse M, Orkoulas G, Ni D, Christofides PD. A method for handling batch-to-batch parametric drift using moving horizon estimation: application to run-to-run MPC of batch crystallization. Chem. Eng. Sci. 2015;127:210-219.

424755 Systematic Fault Detection and Classification in Semiconductor Manufacturing Process

Abstract:

Semiconductor manufacturing is composed of various processes such as photo lithography, dry etch, diffusion, ion implantation, thin film deposition, cleaning and chemical mechanical planarization. Among these processes, plasma related processes occupy more than 30% of the whole manufacturing steps. Rapidly progressed and continuous shrinkage of the chip size makes 14nm semiconductor device commercialization, which enables developing microprocessors for smart device as smartphone, medical instrument and IoTs (Internet of Things). However, the physical difficulties expected in materials and patterning technology to cross the 10nm threshold by inherent complexity of plasma, lack of plasma sensors. Cost-effective manufacturing through monitoring and control in plasma etching that many different approaches applied in equipment and process control, new sensor systems, in-situ and integrated metrology and fab automation to improve yield while simultaneously decreasing circuit geometries. Systematic fault detection and classification (FDC) is introduced to an automatic determination of abnormal equipment state, execution of alarms, and assignment of the cause of detected faults. So FDC reduces uncertainty in process operation, noise reduction, metrology accuracy and stability will be increased. In this study, we used MSPC methodology such as principal component analysis (PCA), factor analysis (FA) and mixture model of k-means clustering to process mode classification, detecting the fault, and revealing causal relationship will lead to minimized tool downtime and increased throughput within a system. The application and performance of the proposed method in fault detection and classification will be shown by lab scale CCP (Capacitance Coupled Plasma) chamber operation examples.

425513 Fault Detection and Accommodation in Sampled Data Process Systems with Measurements and Actuation Errors

Abstract:

Development of systematic methods for the detection and handling of faults in chemical processes is a fundamental problem at the interface of process control and operations. The realization that malfunctions in the measurement sensors, control actuators or process equipment can cause unrecoverable losses, and may result in unacceptable performance deterioration or even instability, has led to an extensive body of research work on the development of various process monitoring and fault-tolerant control methods over the past few decades (e.g., see [1]–[4] for some results and references in this area).

While the problem is typically formulated and addressed within the conventional feedback control setting, which assumes a direct and flawless communication link between the sensors and the controller, the increasing complexity of the process/controller interface (e.g., owing to the use of network-based communication) motivates re-examining this formulation to suit the needs of the real-time operation practices. In particular, the inherent limitations on the information transmission and processing capabilities of the measurement system and the sensor-controller communication medium can erode the diagnostic and fault-tolerance capabilities of the control system if not properly accounted for. Issues such as resource constraints, data sampling and losses, processing and communication delays, measurement quantization, and real-time scheduling constraints, challenge many of the assumptions in traditional process monitoring and control methods and need to be integrated explicitly in the fault-tolerant control system design methodology.

An effort to address some of these challenges was initiated in [5] where a model-based framework for actuator fault detection and reconfiguration using discretely-sampled and delayed measurements was developed. The main idea was to develop an integrated framework for actuator handling which consists of a family of robust output feedback controllers, observer-based fault detection filters that account for the discrete sampling and delayed availability of measurements, and a switching law that orchestrates the transition from the faulty actuator configuration to a healthy fallback following fault detection.

While control system reconfiguration may be necessary for the recovery from severe failure situations, it can also be a costly strategy, especially as the realization of reconfigurable control requires the availability and use of redundant configurations, which can raise process operating costs. In cases involving only partial faults, fault accommodation by means of updating the process model and/or control design parameters based on an estimation of the fault magnitude, represents a more efficient and less costly alternative to maintaining process operation at the desired level, even under the faulty conditions. This benefit was demonstrated in [6] where a stability-based fault accommodation scheme was devised and implemented. These results were further generalized in [7] where an optimization-based approach for fault estimation was incorporated into the design methodology to facilitate the implementation of the fault accommodation strategy.

A key consideration in the previous studies is the assumption that the measurements used for monitoring, reconfiguration and/or accommodation are exact, and the assumption that the control action computed by the controller is also transmitted accurately to the control actuators. In many practical situations, however, these assumptions need to be re-examined in light of the possible degradation in the quality and precision of the transmitted sensor and actuator data over the network. Possible sources for the presence of measurement and actuation errors include interference with other signals in the field in the case of wireless networks, errors due to noisy sensor readings which are subject to drift over time, the use of discrete sensors (e.g., binary sensors) and discrete actuators (on/off valves) in the control system which cannot be varied continuously, the use of data fusion and aggregation techniques, and the need to conserve network bandwidth by reducing data precision. The end result in all these situations is that only inexact values of the state and/or input variables will be available, and the resulting errors may cause a significant deterioration in the closed-loop control and monitoring qualities, if not handled properly in the control system design.

Motivated by these considerations, we present in this contribution a methodology for the design of fault-tolerant control systems for process systems with actuator faults, discretely sampled and delayed measurements, and bounded sensor and actuator errors. A key objective of this study is to elucidate the key design parameters that can be varied to mitigate the impact of degradation in the quality of transmitted sensor and actuator data on the monitoring and fault-tolerance capability of the closed-loop system. Initially, a model-based controller that stabilizes the fault-free closed-loop system in the absence of sensor and actuator errors is synthesized. The controller includes an inter-sample model predictor to compensate for the discrete measurement sampling and a propagation unit to compensate for the delayed availability of measurements. Lyapunov techniques are them utilized to analyze the implementation of the model-based controller in the presence of bounded measurement and actuation errors. A precise characterization of the closed-loop stability region is obtained in terms of the sensor/actuator error bounds, the sensor data transmission rate, the delay size, the magnitude of fault, the size of the process-model mismatch and the choice of control configuration. The characteristic fault-free closed-loop behavior is used as the basis for deriving a time-varying residual alarm threshold for fault detection, while the stability region characterization is used to decide a suitable fault accommodation strategy. Finally, the design and implementation of the integrated monitoring and fault-tolerant control architecture are demonstrated using a chemical process example.

References:

[1] M. Blanke, M. Kinnaert, J. Lunze, and M. Staroswiecki, *Diagnosis and Fault-Tolerant Control*. Berlin, Germany: Springer, 2003.

[2] R. Isermann, *Fault-Diagnosis Systems: An Introduction from Fault Detection to Fault Tolerance*. Berlin, Germany: Springer, 2005.

[3] Y. Zhang and J. Jiang, "Bibliographical Review on Reconfigurable Fault-Tolerant Control Systems," *Annu. Rev. in Contr.*, vol. 32, pp. 229–252, 2008.

[4] P. Mhaskar, J. Liu, and P. D. Christofides, *Fault-Tolerant Process Control: Methods and Applications*. London, England: Springer-Verlag, 2013.

[5] Y. Sun and N. H. El-Farra, "Model-Based Fault Detection and Fault-Tolerant Control of Process Systems with Sampled and Delayed Measurements," *Proceedings of 18th IFAC World Congress*, pp. 2749-2754, Milan, Italy, 2011.

[6] T. Napasindayao and N. H. El-Farra, "Fault Detection and Accommodation in Particulate Processes with Sampled and Delayed Measurements," *Ind. Eng. Chem. Res.*, 52, 12490–12499, 2013.

[7] T. Napasindayao and N. H. El-Farra, "Model-based Fault-Tolerant Control of Uncertain Particulate Processes: Integrating Fault Detection, Estimation and Accommodation," *Proceedings of 9th IFAC Symposium on Advanced Control of Chemical Processes,* to appear, Whistler, British Columbia, Canada, 2015.

429556 Data Reconciliation in Open Reaction Systems Using the Concept of Extents

Abstract:

Data Reconciliation in Open Reaction Systems using the Concept of Extents

S. Srinivasan[1], J. Billeter[1], S. Narasimhan[2] and D. Bonvin[1]

[1] *Laboratoire d'Automatique, Ecole Polytechnique Fédérale de Lausanne*

EPFL – Station 9, 1015 Lausanne, Switzerland

[2] *Department of Chemical Engineering, Indian Institute of Technology, Madras*

(contact: dominique.bonvin@epfl.ch)

Keywords: Data reconciliation, Vessel extents, Plant-model mismatch, Monotonicity constraints, Open reaction systems.

Kinetic models of chemical reaction systems are typically represented in terms of state variables, such as concentrations, temperature and partial pressures [1]. These state variables in turn depend on the underlying reactions, transfer phenomena, and transport due to the inlet and outlet flows. Kinetic models are derived from first principles – material and energy balances – and are expressed as a set of differential and algebraic equations (DAE), with the differential equations describing the evolution over time and the algebraic equations describing relationships that have to be satisfied at each time instant [2]. The identification of kinetic models along with the estimation of their parameters is carried out using measurements obtained from experiments performed under well-chosen and often ideal experimental conditions [3]. Kinetic models can then be adapted to non-ideal process operations using real-time measurements for the purpose of monitoring, control and optimization [4-5].

Measurements are inherently corrupted with noise. The quality of measurements made during the identification experiment affects kinetic identification, while the quality of measurements made during process operation affects its efficiency. Data reconciliation techniques rely on balance equations (here algebraic constraints) to improve the accuracy of these measurements, with more relationships leading to better reconciliation [6]. Hence, data reconciliation can be formulated as a constrained optimization problem in terms of measured and reconciled variables. However, since these variables are typically involved in more than one rate process, it is difficult to add additional shape constraints involving for example monotonicity.

Several alternative representations of reaction systems in terms of variants and invariants have been proposed in the literature. Asbjørnsen and co-workers [7], for example, introduced a two-way decomposition into reaction variants and reaction invariants. Unfortunately, the transformed variables are also flow variant. Srinivasan et al. [8] introduced a nonlinear decomposition into reaction variants, flow variants, and reaction and flow invariants. Note that all these representations use abstract variables that do not carry any physical meaning. Recently, a representation of reaction systems using a linear transformation has been proposed. The transformed states, called *vessel extents*, have a clear physical meaning, and, in addition, each vessel extent is associated with a single rate process [9].

Srinivasan et al. [10] have recently shown that the transformation to vessel extents allows a general formulation of process constraints under all common operating conditions, namely, batch, semi-batch and continuous mode. In addition, the extents are monotonically increasing in the absence of an outlet stream (batch and semi-batch mode). The same authors have also shown that the addition of monotonicity constraints to the data-reconciliation problem improves the accuracy of the reconciled estimates. Unfortunately, the monotonicity of extents cannot be guaranteed in the presence of an outlet stream.

In this contribution, piecewise monotonicity constraints on extents are applied for the purpose of data reconciliation in the presence of an outlet stream, that is, in continuous operation mode. A general procedure is presented to identify regions where these state variables are monotonically increasing or decreasing. The strength of these piecewise shape constraints for the task of data reconciliation will be illustrated via simulated examples.

References

[1] O. Levenspiel. *Chemical Reaction Engineering*, Wiley, 1972.

[2] C. G. Hill and W. R. Thatcher. *Introduction to Chemical Engineering Kinetics and Reactor Design*, John Wiley & Sons, 2014.

[3] A. Bardow and W. Marquardt. Incremental and simultaneous identification of reaction kinetics: Methods and comparison, *Chem. Eng. Sci.,* 59, 2673-2684, 2004.

[4] A. Marchetti, B. Chachuat and D. Bonvin. Modifier-adaptation methodology for real-time optimization, *Ind. Eng. Chem. Res.,* 48(13), 6022-6033, 2009.

[5] S. Srinivasan, J. Billeter and D. Bonvin. On the use of extents for process monitoring and fault diagnosis, 14th

Conf. on Chemometrics in Analytical Chemistry, Richmond (USA), 2014.

[6] S. Narasimhan and C. Jordache. *Data Reconciliation and Gross Error Detection*, Elsevier, 1999.

[7] O. A. Asbjørnsen. Reaction invariants in the control of continuous chemical reactors, *Chem. Eng. Sci.,* 27, 709-717, 1972.

[8] B. Srinivasan, M. Amrhein, and D. Bonvin. Reaction and flow variants/invariants in chemical reaction systems with inlet and outlet streams, *AIChE J.,* 44(8), 1858-1867, 1998.

[9] D. Rodrigues, S. Srinivasan, J. Billeter and D. Bonvin. Variant and invariant states for reaction systems, *Comput. and Chem. Eng.,* 73, 23-33, 2015.

[10] S. Srinivasan, J. Billeter, S. Narasimhan and D. Bonvin. Data reconciliation in reacting systems using the concept of extents, 25th European Symp. on Computer Aided Process Engineering, Copenhagen (Denmark), 2015.

431685 State Observer for Process Monitoring of Melt Polycondensation Reactors

Abstract:

Polycondensation is an important polymerization method for the synthesis of polyesters, polyamides, etc. A commercially important polycondensation product is polyethylene terephthalate (PET), which is mainly used in synthetic fibers, water bottles and molding materials. It is manufactured by a four-stage process which consists of transesterification, prepolymerization, melt polycondensation and solid state polycondensation. The main polycondensation reaction is reversible and thus equilibrium limited [1]. In order to achieve a high molecular weight, byproduct (i.e., ethylene glycol) needs to be removed to favor the forward reaction by applying vacuum to the reactor in the finishing stage. Besides, the product in this stage determines the quality of final product, e.g. purity, melting point, average molecular weight, polydispersity index, etc. Therefore, control and monitoring of the finishing stage is of great importance.

However, lack of fast on-line sensors makes process monitoring of polycondensation very difficult. There is a substantial need to develop soft sensors which can estimate those unmeasured state variables accurately. In this study, a state observer based on exact error linearization with eigenvalue assignment [2,3] is used to reconstruct unmeasured state variables and estimate the degree of polymerization (DP).

Although a detailed kinetic model which includes all the polymeric species can describe the degree of polymerization and molecular weight distribution (MWD) quite accurately, it requires large computational effort and may not be suitable for control problems. For this reason, standard modeling approaches for the finishing stage only account for reactions between functional end groups.

In our work, melt polycondensation of PET in continuous stirred-tank reactor (CSTR) is studied. The four-state reaction-mass transfer model developed by Rafler et al. [4] and verified experimentally, was used throughout the present study. The concentration of ethylene glycol, hydroxyl end group, carboxyl end group and ester group in the melt are the four states of the model. The hydroxyl end group in the melt can be inferred from a model using torque, temperature and stirrer speed, which should be calibrated for the reactor [5]. The concentration of carboxyl end group is measurable by on-line acidimetric titration with addition of an indicator [6,7].

In the CSTR model, the mass transfer parameter is affected by bubbles, geometry of the reactor, and operating condition. Therefore, it needs to be determined from experimental data. Based on the two measurements, an observer was built for estimating the states as well as the mass transfer parameter, which provided fast-converging accurate estimates. After obtaining the mass transfer parameter off-line, we considered the hydroxyl end group as the only measurement in on-line operation. We have found that the concentration of ethylene glycol and ester group can be estimated accurately through a fast-converging state observer. For the carboxyl end group, which is unobservable, an open loop observer was built, whose rate of convergence turned out to be reasonable, even though it is not adjustable. Simulation results indicated that DP is estimated accurately.

References:

[1] Jacobsen, Lance L., and W. Harmon Ray. "Analysis and design of melt and solution polycondensation processes." AIChE Journal 38.6 (1992): 911-925.

[2] Kazantzis, Nikolaos, and Costas Kravaris. "Nonlinear observer design using Lyapunov's auxiliary theorem." Systems & Control Letters 34.5 (1998): 241-247.

[3] Kazantzis, Nikolaos, Costas Kravaris, and Raymond A. Wright. "Nonlinear observer design for process monitoring." Industrial & Engineering Chemistry Research 39.2 (2000): 408-419.

[4] Rafler, Gerald, et al. "Kinetics of mass transfer in the melt polycondensation of poly (ethylene terephthalate)." Journal of Macromolecular Science—Chemistry 22.10 (1985): 1413-1427.

[5] Appelhaus, Paul, and Sebastian Engell. "Design and implementation of an extended observer for the polymerization of polyethylenterephthalate." Chemical Engineering Science 51.10 (1996): 1919-1926.

[6] Pohl, Herbert A. "Determination of carboxyl end groups in polyester, polyethylene terephthalate." Analytical Chemistry 26.10 (1954): 1614-1616.

[7] Kinniburgh, D. G., C. J. Milne, and P. Venema. "Design and construction of a personal-computer-based automatic titrator." Soil Science Society of America Journal 59.2 (1995): 417-422.

A DATA-DRIVEN MULTIDIMENSIONAL VISUALIZATION TECHNIQUE FOR PROCESS FAULT DETECTION AND DIAGNOSIS

Shriram Gajjar and Ahmet Palazoglu

Department of Chemical Engineering & Materials Science
University of California, Davis

2015 AIChE Annual Meeting
Nov 12, 2015

Background

(A)* Oil Refinery-A large scale process **(B)*** Operator monitoring a large scale process **(C)** Univariate Control Chart

Safety, Maintain optimal operations and increase profitability

*source: images.google.com

Outline

- Introduction to Parallel Coordinates
- Case study: Tennessee Eastman Benchmark Process
- PCA and control limits
- Fault Detection Method
- Fault Identification – Random Forests
- Results
- Conclusions and Future Work

3

Introduction: Advantages of Parallel Coordinates[1]

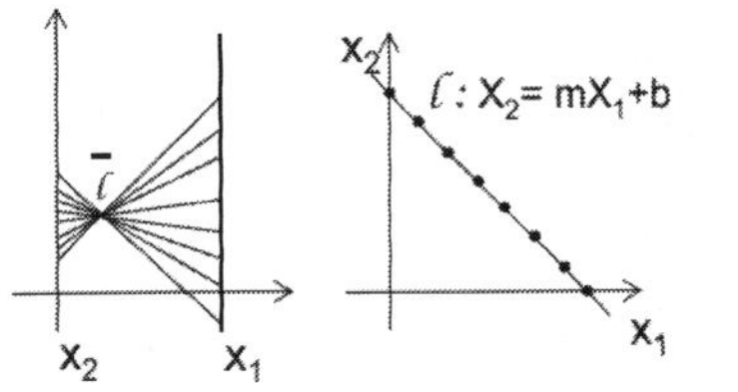

In the plane parallel coordinates induce duality . [1]

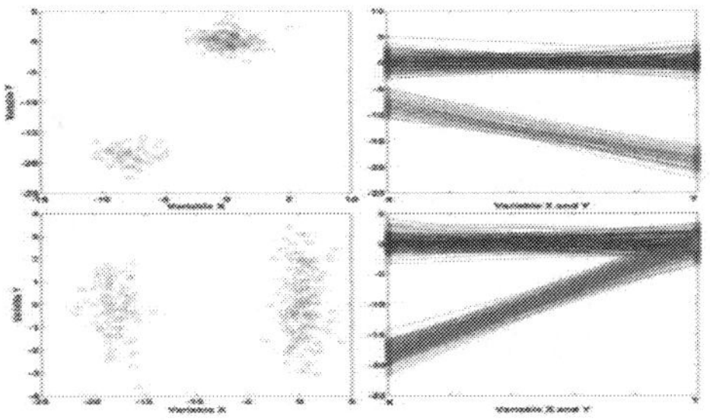

Example of cluster identification in parallel coordinates

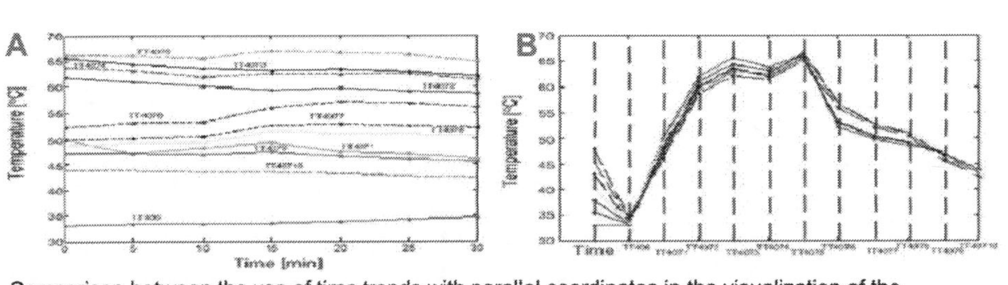

Comparison between the use of time trends with parallel coordinates in the visualization of the temperature profiles along the absorber column[2]

[1] Inselberg, The Visual Computer, 1:69-91,1985
[2] Dunia et al, AIChE J,59:445-456,2013

4

Case Study: Tennessee Eastman Process

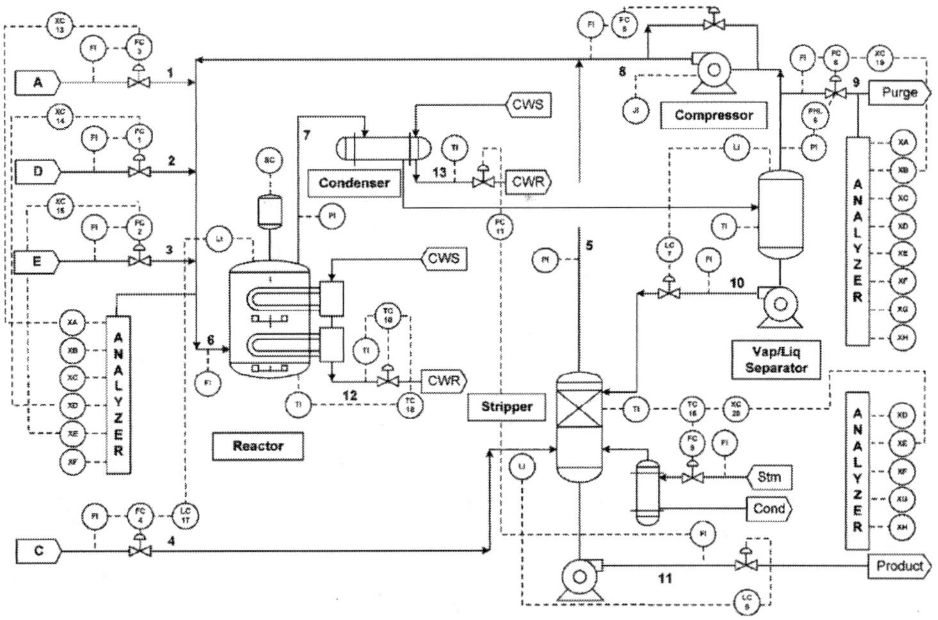

5

Monitoring Variables

NO.	Process variables	NO.	Process variables	NO	Process variables
1	A feed	12	Product separator level	23	D feed flow valve
2	D feed	13	Product separator pressure	24	E feed flow valve
3	E feed	14	Product separator under flow	25	A feed flow valve
4	Total feed	15	Stripper level	26	Total feed flow valve
5	Recycle flow	16	Stripper pressure	27	Compressor recycle valve
6	Reactor feed rate	17	Stripper under flow	28	Purge valve
7	Reactor pressure	18	Stripper temperature	29	Separator pot liquid flow valve
8	Reactor level	19	Stripper steam Flow	30	Stripper liquid product flow valve
9	Reactor temperature	20	Compressor work	31	Stripper steam valve
10	Purge rate	21	Reactor cooling water outlet temperature	32	Reactor cooling water flow
11	Product separator temperature	22	Separator cooling water outlet temperature	33	Condenser cooling water flow

Process Faults

No.	Description	Type
1	A/C feed ratio, B composition constant (stream 4)	Step
2	B composition, A/C ratio constant (stream 4)	Step
3	D feed temperature (stream 2)	Step
4	Reactor cooling water inlet temperature	Step
5	Condenser cooling water inlet temperature	Step
6	A feed loss (stream 1)	Step
7	C header pressure loss – reduced availability (stream 4)	Step
8	A,B, C feed composition (stream 4)	Random variation
9	D feed temperature (stream 2)	Random variation
10	C feed temperature (stream 4)	Random variation
11	Reactor cooling water inlet temperature	Random variation
12	Condenser cooling water inlet temperature	Random variation
13	Reaction kinetics	Slow drift
14	Reactor cooling water valve	Sticking
15	Condenser cooling water valve	Sticking
16	Unknown	Unknown
17	Unknown	Unknown
18	Unknown	Unknown
19	Unknown	Unknown
20	Unknown	Unknown
21	The valve for stream 4 was fixed at the steady-state position	Constant Position

6

Dimensionality Reduction : Principal Component Analysis (PCA)

- For a data matrix $X \in \boldsymbol{R}^{n \times m}$ scaled to *zero mean and unit variance,* the covariance matrix of *X* is defined as:

$$\mathrm{cov}(X) = \Sigma = \frac{X^T X}{n-1}$$

- The decomposition is expressed as follows:

$$X = TP^T + E$$

where $T \in \boldsymbol{R}^{n \times m}$ and $P \in \boldsymbol{R}^{m \times m}$ are the score matrix and the loading matrix, respectively.

- P and Λ are the eigenvectors and eigenvalues of Σ

A

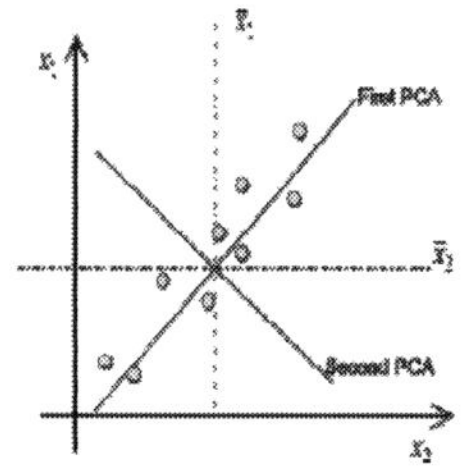

B

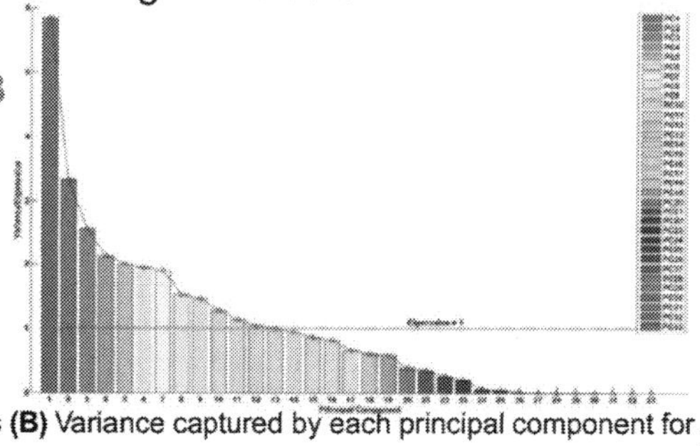

(A) Example of a PC subspace for two variables **(B)** Variance captured by each principal component for TEP with 960 samples and 33 variables

7

Fault Detection with PCA

- Hoteling's T^2 and Q (or SPE) are calculated as follows[3]:

$$T^2 = t^T \Lambda^{-1} t = x^T P_l \Lambda^{-1} P_l^T x \qquad Q = e^T e = \sum_{i=1}^{l} e_i^2$$

- Upper control limits are calculated as follows[3]:

$$T_\alpha^2 = \frac{(n-1)l}{(n-l)} F_\alpha(l, n-l) \qquad Q_\alpha = g\chi^2_{h,\alpha}; \quad g = \frac{v}{2m}, h = \frac{2m^2}{v}$$

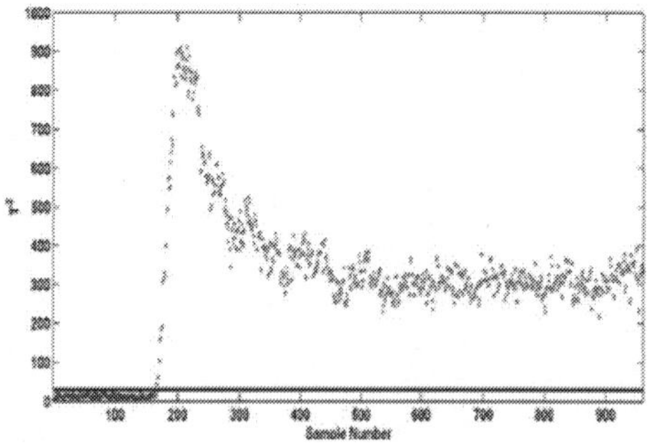

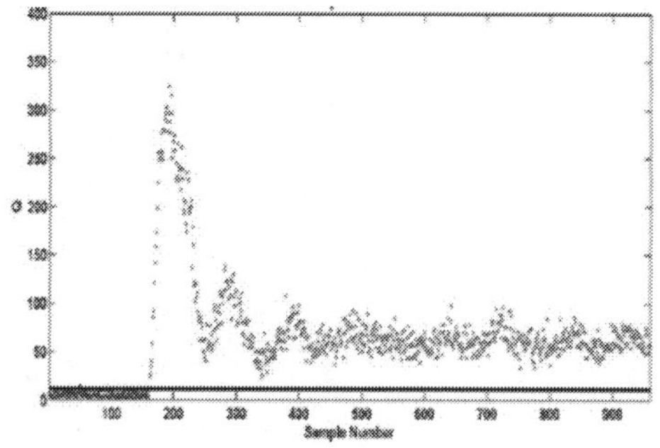

T^2 and Q statistics for a step type fault, Fault # 1.

[3] Tong et al., J. of Zhejiang University-SCIENCE A (App. Phys. & Engg.),14(7): 520-534, 2013

8

Proposed Approach : Visualization

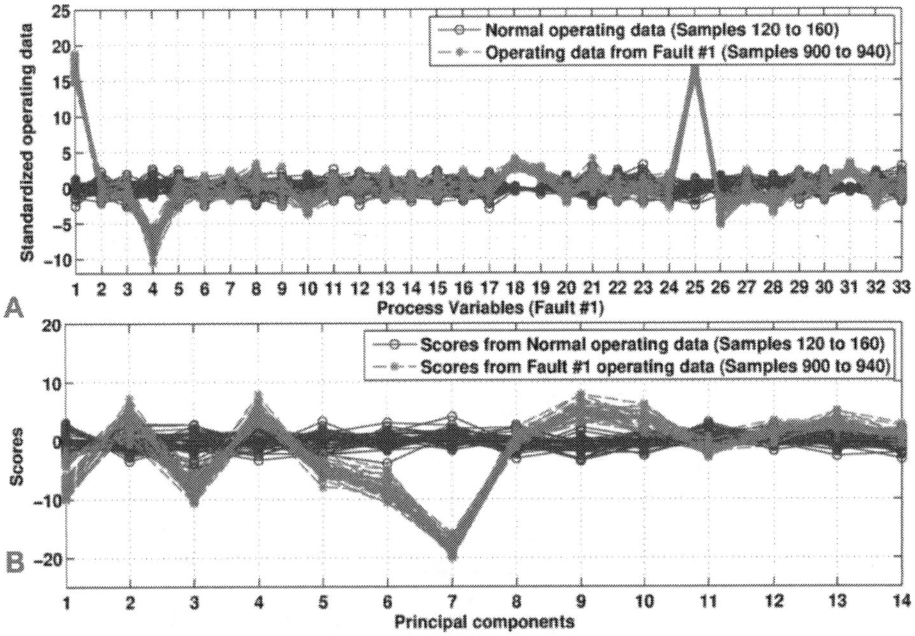

Process monitoring visualization in parallel coordinates for the 33 process variables in fault #1 operating data **(A)** Normal data visualization of all process variables **(B)** Data visualization using PCA 9

Proposed Approach : Defining *run* rules

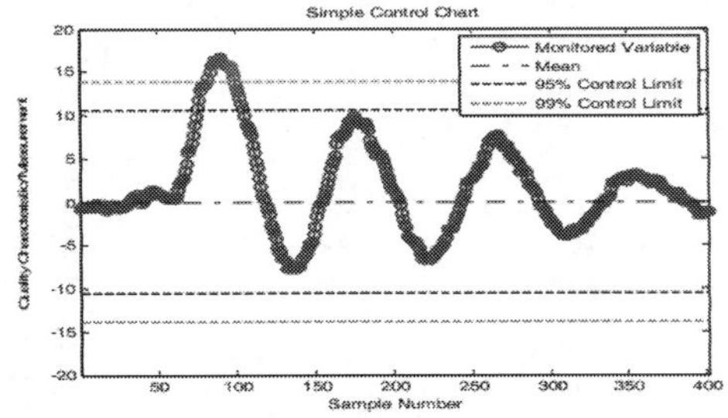

Western Electric Handbook (1956) suggests a set of decision rules for detecting non-random patterns on control charts. Specifically, it suggests concluding that the process is out of control if either of the follows[*]:

1. A single point exceeds the 3-sigma control limits;
2. Two out of three successive points exceed the 2-sigma warning limits;
3. Four out of five successive points exceed the 1-sigma limits;
4. Eight successive points fall on one side of the center line.
5. Eight successive points forming a run up or a run down
6. A nonrandom or unusual pattern in data

Fault is declared when two successive sample exceed the 2-sigma control limits for the same PC

10

* Cinar Ali, Palazoglu Ahmet, Kayihan Ferhan.(2007). "Chemical Process Performance Evaluation". FL: CRC Press

Results: Process monitoring

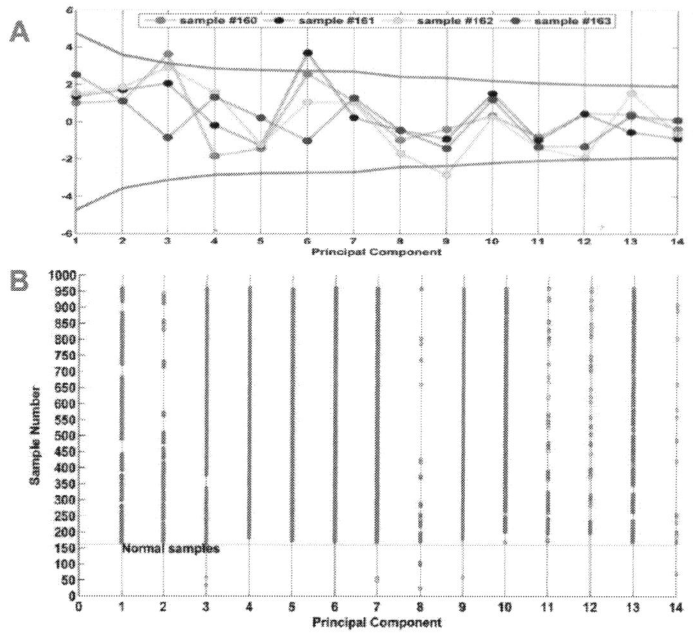

Fault # 1 data visualization **(A)** Parallel coordinate plot with control limits **(B)** Fault detection visualization

11

Results: Fault detection

Missed detection rate														
Fault	PCCL			RSCL			Either, PCCL or RSCL			Proposed Method			Russell, Chiang, and Braatz	
	Type I	True positive	Type II	Type I	True positive	Type II	Type I	True positive	Type II	PCCL	RSCL	Either	PCA T^2	PCA Q
1	9	793	7	18	800	0	26	800	0	0.009	0.000	0.000	0.008	0.003
2	8	790	10	19	789	11	26	795	5	0.013	0.014	0.006	0.020	0.014
3	22	136	664	51	168	632	70	266	534	0.830	0.790	0.668	0.998	0.991
4	10	466	334	20	799	1	27	800	0	0.418	0.001	0.000	0.956	0.038
5	10	301	499	20	799	1	27	799	1	0.624	0.001	0.001	0.775	0.746
6	4	796	4	19	799	1	23	799	1	0.005	0.001	0.001	0.011	0.000
7	7	799	1	22	799	1	28	799	1	0.001	0.001	0.001	0.085	0.000
8	8	788	12	36	770	30	43	795	5	0.015	0.038	0.006	0.034	0.024
9	49	127	673	41	153	647	77	250	550	0.841	0.809	0.688	0.994	0.981
10	7	444	356	9	724	76	15	739	61	0.445	0.095	0.076	0.666	0.659
11	8	469	331	17	718	82	24	736	64	0.414	0.103	0.080	0.794	0.356
12	38	789	11	18	795	5	52	797	3	0.014	0.006	0.004	0.029	0.025
13	11	755	45	17	771	29	27	772	28	0.056	0.036	0.035	0.060	0.045
14	11	787	13	11	799	1	20	799	1	0.016	0.001	0.001	0.158	0.000
15	14	162	638	18	189	611	31	294	506	0.798	0.764	0.633	0.988	0.973
16	61	362	438	37	726	74	80	759	41	0.548	0.093	0.051	0.834	0.755
17	7	698	102	29	786	14	35	787	13	0.128	0.018	0.016	0.259	0.108
18	6	726	74	21	734	66	27	744	56	0.093	0.083	0.070	0.113	0.101
19	13	237	563	24	581	219	36	660	140	0.704	0.274	0.175	0.996	0.873
20	8	432	368	21	646	154	29	684	116	0.460	0.193	0.145	0.701	0.550
21	23	389	411	67	594	206	78	606	194	0.514	0.258	0.243	0.736	0.570

Detection time																					
Fault	1	2	3	4	5	6	7	8	9	10	11	12	13	14	15	16	17	18	19	20	21
PCCL	161	165	162	162	162	162	162	164	160	169	160	163	182	162	160	162	160	165	162	160	162
RSCL	160	167	170	161	162	164	162	160	160	166	167	160	160	162	197	162	162	164	172	216	160
Either, PCCL or RSCL	160	165	162	161	162	162	162	160	160	166	160	160	160	162	160	162	160	164	162	160	160

12

[4] Russell et al., Chemo. and Intelligent Lab. Systems, 51 (2000) 81-93

Fault identification/diagnosis: Unique signatures

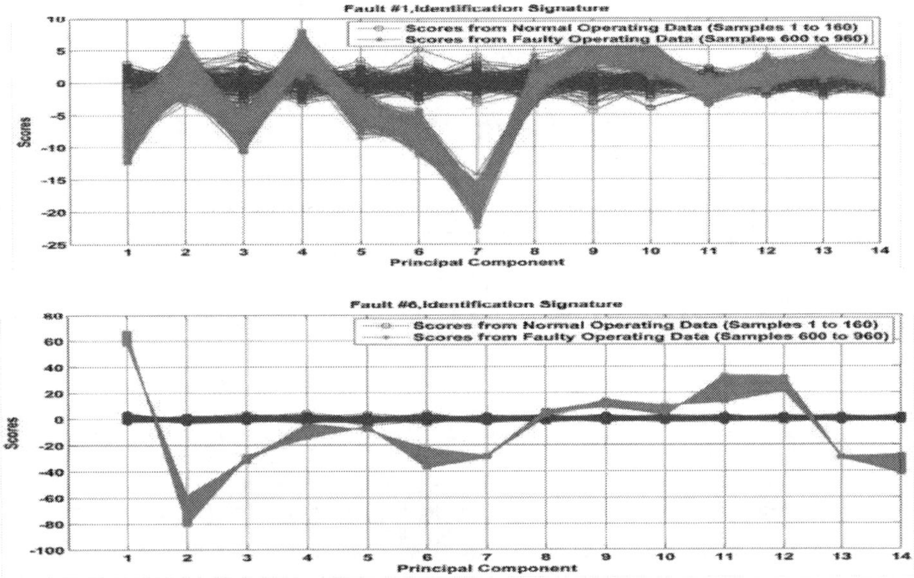

Fault # 1 and Fault # 6 signatures.

13

Fault Identification/diagnosis : *Random Forests*[5]

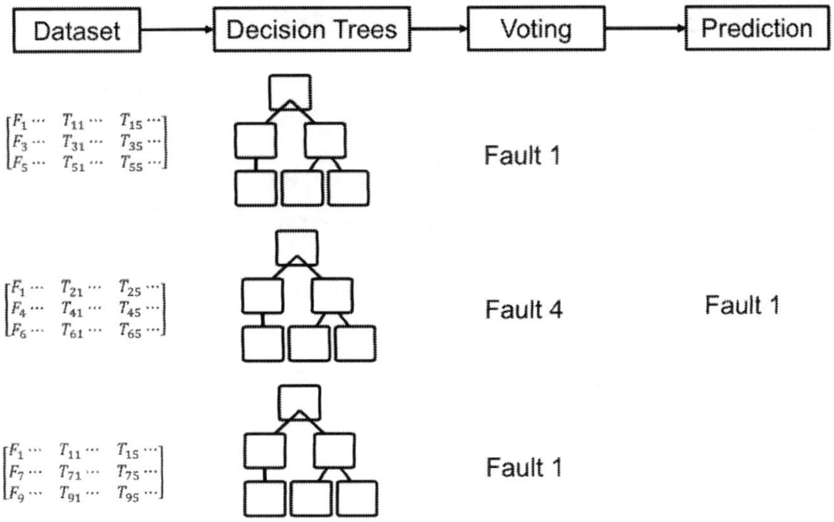

Schematic for Random Forests algorithm

[5] L. Breiman, Random Forests, Machine Learning, 45 (2001) 5-32

14

Results : Fault Identification/diagnosis

Variables included in training data	Fault diagnosis accuracy (%)	Training dataset
14 PC scores	67.86	All faults
14 PC scores and residuals	71.32	All faults

Fault	Correct diagnosis rate (%) *
1	30
2	95
3	0
4	5
5	25
6	100
7	65
8	0
9	5
10	0
11	15
12	0
13	5
14	5
15	5
16	0
17	85
18	30
19	0
20	0

Fault	Random Forests	
	Sensitivity (%)	Specificity (%)
1	98.75	100
2	97.29	100
3	19.63	95.87
4	89.14	97.11
5	41.57	97.54
6	98.93	100
7	100	100
8	95.34	99.65
9	23.56	96.34
10	51.63	97.36
11	45.97	98.97
12	85.93	99.31
13	88.88	99.8
14	92.85	99.68
15	32	94.85
16	30.33	97.65
17	75.37	99.42
18	81.17	99.91
19	51.42	97.85
20	49.7	97.43

** Li et al., J. of Adv. in sig. processing, 2011 (83)*

Fault diagnosis results using model based on training dataset that contained 14 PC scores from all faults (does not include scores of normal operating data)

15

Conclusions and Future Research

- PCA space is visualized through parallel coordinates.
- Control limits established on PCs.
- Run rules dictate fault detection performance
- Detection performance better than classical approaches.
- Unique signatures of faults in PC parallel coordinates facilitate fault identification.
- Fault propagation analysis pending.

16

THANK YOU!
Q&A

17

434732 A Method of Proactive Model-Based Alarm System Design

Abstract:

Process safety has been an active field of research for many decades. Along with the development of more sophisticated control systems to improve product quality and plant efficiency, modern safety systems have evolved to comply with tighter safety regulations, criteria, and standards required by governments, regulatory agencies, and the process industries.

Alarm systems provide a direct sensible feedback from the current status of an operating plant. This is usually done through monitoring the signal transmitted by a physical sensor connected to a process variable of interest and converting the signal to a visual or annunciator alarm whenever the signal crosses a limit. This is to alert the plant personnel to take proper safety action to prevent an incident which is about to occur as a result of the observed deviation. However, multiple problems have emerged with the advancement of the alarm systems. First, as implementation of alarm systems is becoming cheaper and more feasible, an unnecessarily large number of alarms are employed, leading to many false alarms. Second, complex interactions among variables of a process plant can cause many alarms to trigger redundantly or unnecessarily in response to a single event. Different approaches have been introduced to address these problems, including alarm signal processing [1], multivariate analysis [2], alarm threshold optimization [3], and model-based monitoring [4, 5].

In this work, we combine plant-wide operability analysis and alarm systems, and present a method of soft (model-based) alarm system design. The method is model-based as it makes use of a deterministic first-principles model to simulate the plant dynamic behavior over a receding time-horizon into the future. Such simulations provide a basis to project the plant operability status in the future and to generate soft alarms. The proposed method offers the following advantages. First, the method systematically accounts for interactions among plant variables, as it utilizes a first-principles model with an adequate level of complexity. Second, the method can provide alarms tied to unmeasurable state variables. This feature is particularly interesting as conventional alarms respond to measurable plant variables only. Third, since the soft alarm system can provide alarm signals before an incident occurs, operators can take the proper precautionary and mitigating actions proactively. Techniques that ensure the computational efficiency of the method will be presented. The application and performance of the method will be shown using a polymerization plant example.

References

[1]. Chen, T. (2010). On reducing false alarms in multivariate statistical process control. *Chemical Engineering Research and Design*, *88*(4), 430-436.

[2]. Stefatos, G., & Hamza, A. B. (2010). Dynamic independent component analysis approach for fault detection and diagnosis. *Expert Systems with Applications*, *37*(12), 8606-8617.

[3]. Jiang, R. Y. (2010). Optimization of alarm threshold and sequential inspection scheme. *Reliability Engineering & System Safety*, *95*(3), 208-215.

[4]. Mohseni Ahooyi, T., Soroush, M., Arbogast, J. E., Seider, W. D., & Oktem, U. G. (2014). Maximum likelihood maximum entropy constrained probability density function estimation for prediction of rare events. *AIChE Journal*, *60*(3), 1013-1026.

[5]. Mohseni Ahooyi, T., Arbogast, J. E., & Soroush, M. (2015). Applications of the Rolling Pin Method. 1. An Efficient Alternative to Bayesian Network Modeling and Inference. *Industrial & Engineering Chemistry Research*. *54* (16), 4316–4325.

A Methodology for a Sustainable CO_2 Capture and Utilization Network

Rebecca Frauzem, Kasper Fjellerup, Rafiqul Gani

Thursday November 12, 2015

Department of Chemical and Biochemical Engineering, Technical University of Denmark

DTU Chemical Engineering
Department of Chemical and Biochemical Engineering

Overview

- Introduction
- Objective
- Methodology
- Evaluation Criteria
- Conceptual Case Study
- Conclusions

Introduction

9-10 billion people on earth

6-7 x global GDP growth

5-6 x production capacity for commodities (steel, chemicals, lumber, etc.)

3.5 x energy demand

7 x electricity demand

Siirola, J.J., 2012.

Objective

- Develop a systematic methodology for the sustainable design of captured CO_2 utilization processes
- Define criteria for the evaluation of alternative processing routes
- Develop methods and tools for the implementation of the methodology
- Determine feasibility, opportunities and sustainability of captured CO_2 utilization processes through a conceptual case study

Methodology Overview

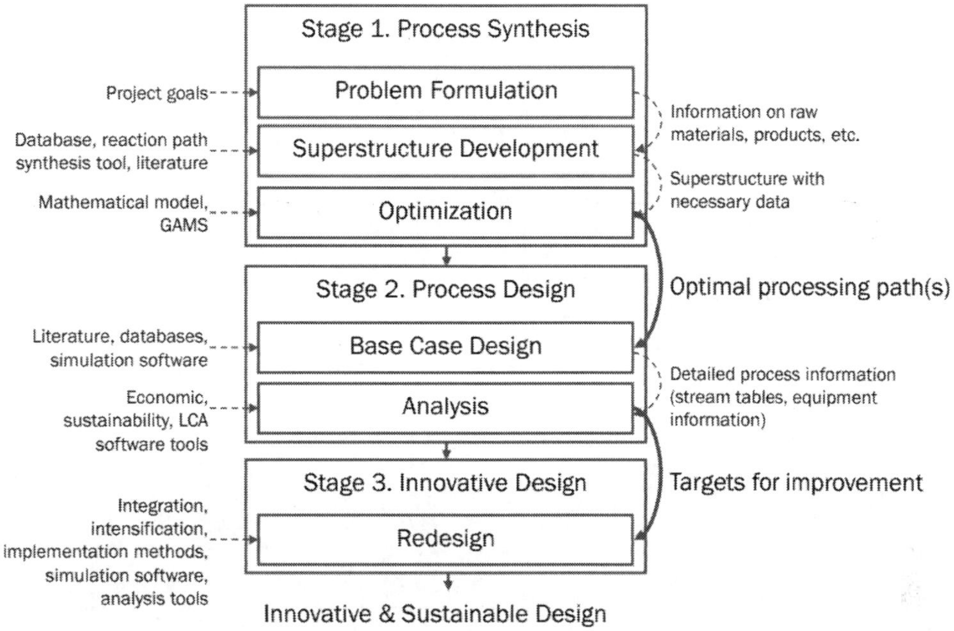

Stage 1 – Process Synthesis

- Focus on 1st stage
- Objective: <u>obtain processing path(s)</u>

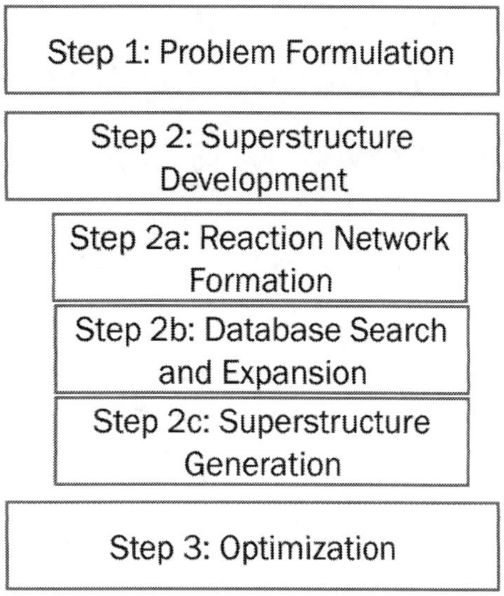

Step 2a: Reaction Network Generation

- Complete network of reactions
- Reaction path synthesis (RPS) performed via adopted CAMD method

Target Generation
- Problem statement OR with computer-aided molecular design (CAMD) generate reaction products

Reaction Path Synthesis (RPS)
- Use software tool to generate reactions

Thermodynamic Screening
- Screen for thermodynamic feasibility with Gibbs Free Energy

Network Formation
- Link all the reactions to form a network of utilization reactions

Cignitti, S., 2014

Step 2a: Reaction Network Generation (2)

- For a selection of C, H, O containing compounds, performed for 140 targets
- Have over 50 reactions

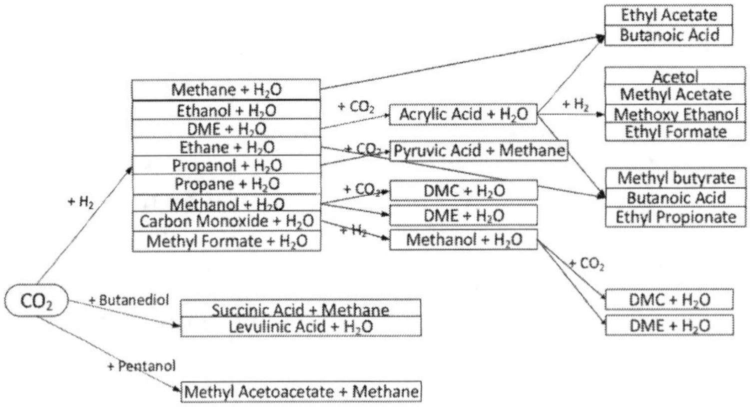

Cignitti, S., 2014

Step 2b: Database Search and Expansion

- Database implemented easy extraction and addition of data
- Include kinetic & equilibrium data, operating conditions

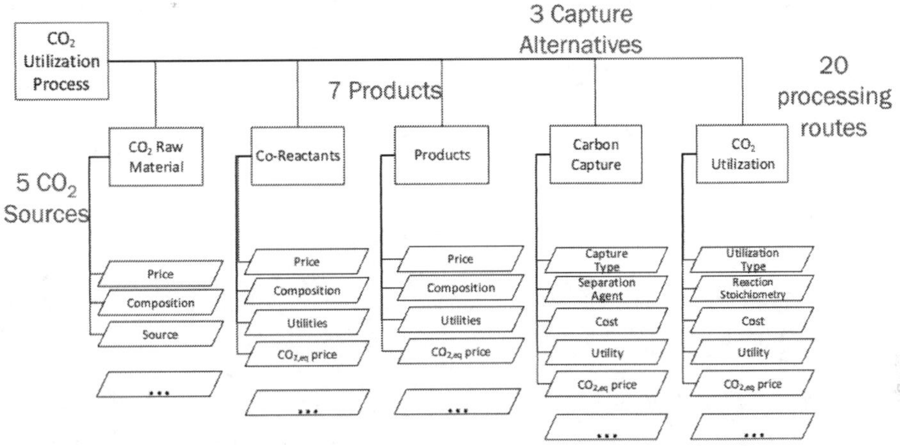

Step 2c: Superstructure Generation

- Reaction network needs to be formulated within the mathematical superstructure model

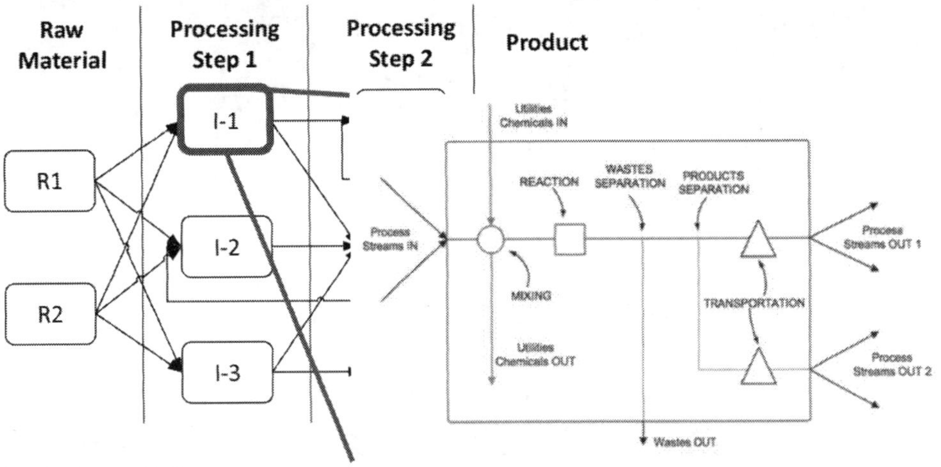

Quaglia, A. et al., 2014.

Step 2c: Superstructure Generation

- Mathematical formulation of MI(N)LP

$$\max f(x,y)$$

s.t. $h(x,y) = 0$, $g(x,y) \geq 0$ — g, h from the process model and process specification

Where the variables are continuous and discrete

$x \in X$ — x is cont. operating variables,

$x^{LO} \leq x \leq x^{UP}$ — bounded by a feasible region

$y \in \{0; 1\}^n$ — y are decision variables

Quaglia, A. et al., 2014.

Step 2c: Optimization

- For small problems, logic and heuristics used
  - Result: small selection of alternatives
- For large problems, mathematical programming necessary
  - Superstructure model
  - Data
  - Result: selection of the best processing path(s) for the objective function
- When data is not available, Stage 2 (process design) used to obtain information

Superstructure Example

- An example: DMC synthesis
  - 9 raw materials (including co-reactants)
  - 1 product
  - 10 processing steps
  - 4 routes
  - Up to 4 alternatives within the steps for each route
  - Objective function: minimize cost to maximize profit

Superstructure Example

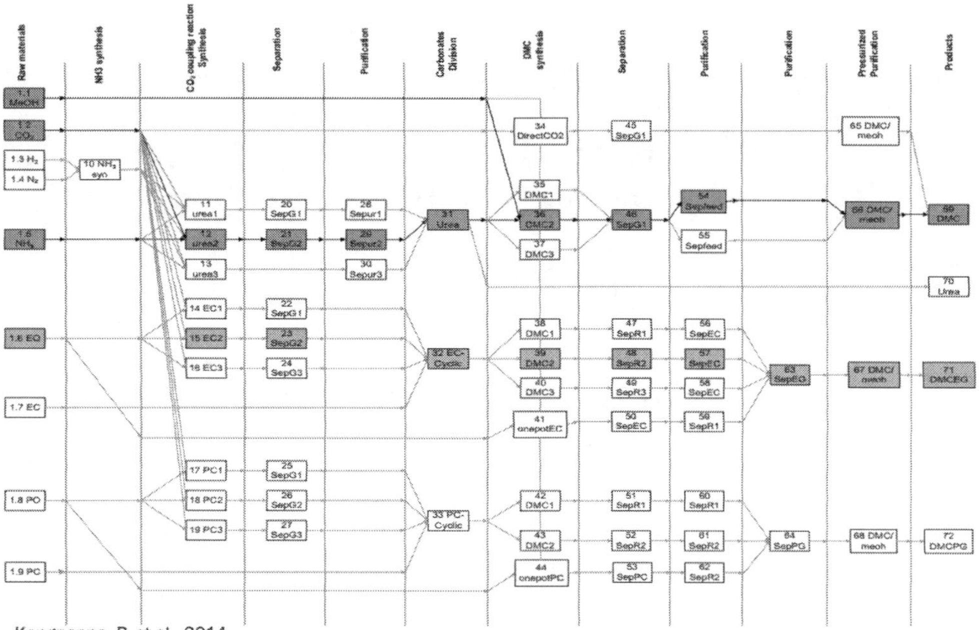

Kongpanna, P. et al., 2014.

Evaluation Criteria

- Need criteria to evaluate in Stage 2 & 3
- Primary: net CO_2 emission
- Additional evironmental and economic:
  - Economic: CAPEX, OPEX, payback period, return on investment (ROI)
  - Environmental: utility requirements, cooling water requirement, waste generated, global warming potential (GWP)

Net CO_2 emission $(t_{CO_2}/t_{Product})$

$$= \sum_{streams} CO_{2,\,out} + \sum_{utility} CO_{2,\,generated} - \sum_{streams} CO_{2,\,in}$$

12/11/2015 R. Frauzem: A Methodology for a Sustainable CO_2 Capture and Utilization Network 15

Conceptual Case Study

- Problem: sustainable CO2 utilization process to methanol
  - Methanol versatile and increasing in demand
- First: carbon capture cost (economic and environmental)
- Second: follow 3-stages for utilization process
- Result: innovative solutions by integrating and intensifying processes

12/11/2015 R. Frauzem: A Methodology for a Sustainable CO_2 Capture and Utilization Network 16

Carbon Capture

- For stationary emissions, carbon capture technology:
  - Pre-combustion, Post-combustion (for flue gases), Oxy-fuel
- To target emissions, look at post-combustion options
  - Mebranes (Metal organic framework (MOF), zeolites)
  - Absorption (MEA, DEA, etc.)
  - Adsorption
  - Other alternatives (bio-based, etc.)
- Absorption is industrial standard (especially for power-plant flue gas)
  - Disadvantage: energy intensive process (solvent recovery)

Carbon Capture (2)

- MEA absorption process for coal-fired power plant emissions (base case)
- To address energy demand in reboiler, manipulate reboiler duty by changing the purity requirement

- Decreasing to 97.2% CO_2
  - 10x reduction in utility cost
  - 1/2 the investment cost
- Now a net negative CO_2
  - -0.26 kg CO_2 eq./kWh

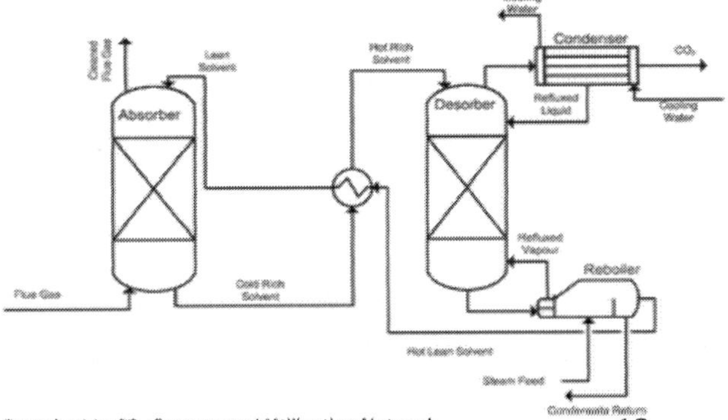

MacDowell, N. et al., 2010

Stage 1 – Methanol Processing Path(s)

DTU

- Follow Stage 1
- Using Reaction Path Synthesis, a network was formed
- Using database formulated a superstructure
- Use logic-based optimization
  - 2 alternatives: direct synthesis and bi-reforming

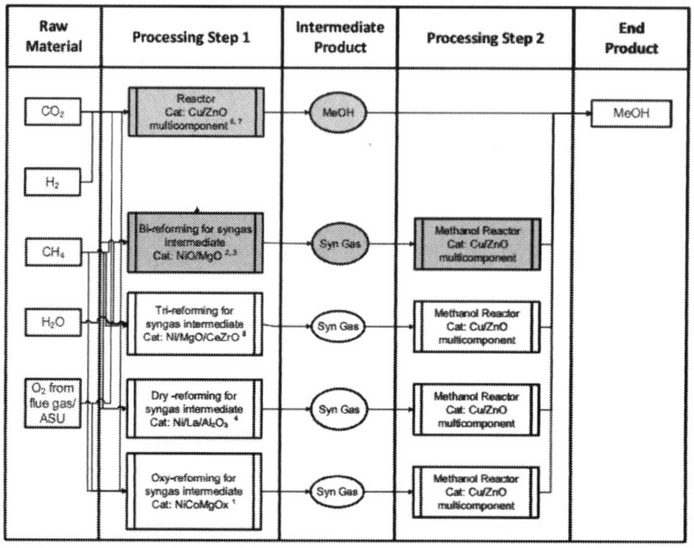

Stage 2 – Design and Analysis of Alternatives

DTU

- Design processes using simulations of existing processes
- Negative net CO_2 emission (-0.71 $t_{CO2,eq}/t_{MeOH}$) for direct hydrogenation of methanol
- Economically feasible (TCI = \$126 million)

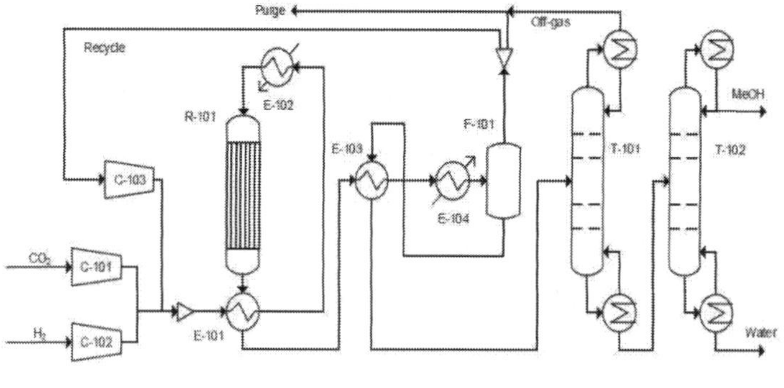

Stage 3 – Integration of Capture and Conversion

DTU

- Linking the processes, the overall sustainability improved
- Verified (economic feasibility) through rigorous simulation
- Net CO_2 emissions = $-0.45 t_{CO2,eq}/t_{MeOH}$
- Without integration, net CO_2 emissions = $2.08 t_{CO2,eq}/t_{MeOH}$

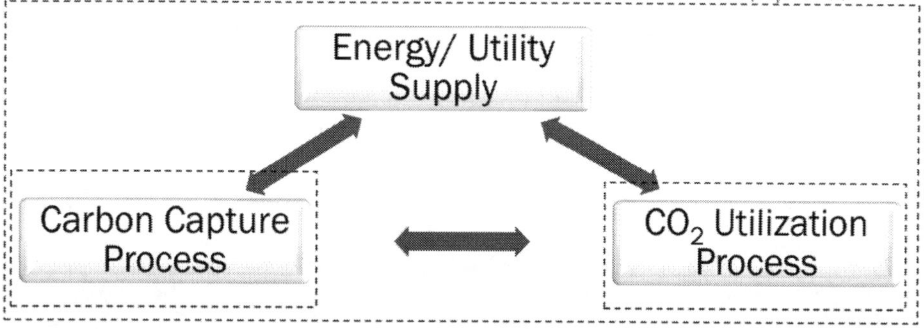

- New integrated (combined) capture-conversion routes can be generated!

Conclusions

DTU

- A methodology for the design of sustainable CO_2 capture-utilization processes has been developed
- A database has been created to enable the easy extraction and addition of data needed by the method
- A network of feasible CO_2 utilization reactions has been generated through an adopted CAMD algorithm
- Net CO_2 emission defined and used as an evaluation criteria to screen alternatives
- A conceptual case study developed as a motivational example
  - Economically feasible and sustainable to produce methanol from captured CO_2

Thanks for listening!

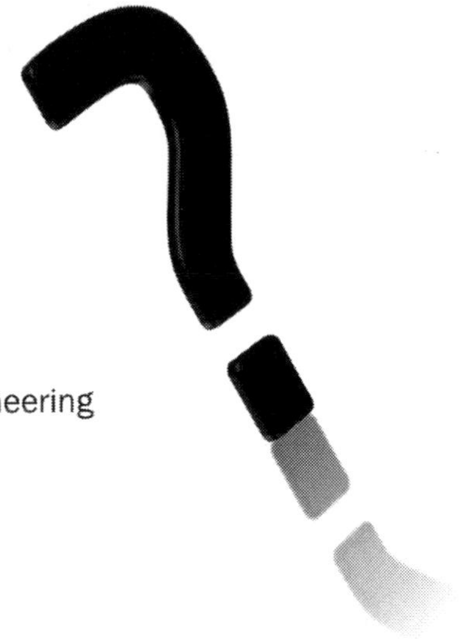

Rebecca Frauzem
Department of Chemical and Biochemical Engineering
Technical University of Denmark
Søltofts Plads Building 229
2800 Kgs. Lyngby
Denmark
rebfra@kt.dtu.dk

References

1. Siirola, J.J., 2012. Thoughts on Sustainability, Energy, Carbon Management, and Shale Gas. Seminar.
2. Cignitti, S., 2014. *Computer-Aided Reaction Path Synthesis*, Technical University of Denmark.
3. Quaglia, A. et al., 2014. Systematic network synthesis and design: Problem formulation, superstructure generation, data management and solution. *Computers & Chemical Engineering*, 72, pp.68–86. Available at: http://linkinghub.elsevier.com/retrieve/pii/S0098135414000787.
4. Quaglia, A. et al., 2012. Integrated business and engineering framework for synthesis and design of enterprise-wide processing networks. *Computers & Chemical Engineering*, 38, pp.213–223. Available at: http://linkinghub.elsevier.com/retrieve/pii/S0098135411003486 [Accessed October 28, 2014].
5. Kongpanna, P. et al., 2014. Techno-economic evaluation of different CO_2-based processes for dimethyl carbonate production. *Chemical Engineering Research and Design*, 93, pp.496–510.
6. Methanol Institute, 2015. The Methanol Industry. Available at: http://www.methanol.org/Methanol-Basics/The-Methanol-Industry.aspx [Accessed August 20, 2015].
7. MacDowell, N. et al., 2010. An overview of CO_2 capture technologies. Energy & Environmental Science, 3(11), pp.1645–1669.
8. Toyir, J., Miloua, R., Elkadri, N.E., Nawdali, M., Toufik, H., Miloua, F., Saito, M., 2009. Sustainable process for the production of methanol from CO2 and H2 using Cu/ZnO-based multicomponent catalyst. Phys. Procedia 2, 1075–1079. doi:10.1016/j.phpro.2009.11.065
9. Ushikoshi, K., Mori, K., Kubota, T., Watanabe, T., Saito, M., 2000. Methanol synthesis from CO2 and H2 in a bench-scale test plant. Appl. Organomet. Chem. 14, 819–825. doi:10.1002/1099-0739(200012)14:12<819::AID-AOC90>3.0.CO;2-#
10. Van-Dal, É.S., Bouallou, C., 2013. Design and simulation of a methanol production plant from CO2 hydrogenation. J. Clean. Prod. 57, 38–45. doi:10.1016/j.jclepro.2013.06.008

423479 Design and Techno-Economic Analysis of a Carbon Capture and Storage By Mineralization (CCSM) Process Using NaOH As an Alkaline Feedstock

Abstract:

As CO°©$_2$ capture technologies are approaching the commercialization stage, storage of the large amounts of CO_2 expected from the capture has emerged as a major problem. The conventional large scale storage method of injecting CO_2 into strata or ocean bears some critical problems such as adverse effects on the environment and potential leakage to atmosphere. For that reasons, the approach of *carbon capture and storage by mineralization (CCSM)* is getting much attention as it converts CO_2 into chemically more stable mineral carbonates, e.g., calcium carbonate ($CaCO_3$), magnesium carbonate ($MgCO_3$), which are also valuable industrial raw materials. In this work, a CCSM process using sodium hydroxide (NaOH) as an alkaline feedstock is designed by integrating the well-known chlor-alkali process with a CO_2 absorption process. In the designed process, NaOH solution made from electro-chemical reactions within the chlor-alkali process is transported to the CO_2 absorption process for capture and conversion of CO_2 (in flue gas emitted from a pulverized coal power plant) into sodium bicarbonate ($NaHCO_3$) and sodium carbonate (Na_2CO_3). The designed process should overcome the insufficient alkalinity supply and slow carbonation rate problems which are the main challenges in the existing CCSM processes. In particular, the designed process is suitable for our native country of Korea in consideration of the low electricity price and high accessibility to brine. First, the designed process is implemented in the commercial process simulator of AspenPlus^{TM}. Second, a techno-economic analysis of the designed process is carried out for a comparison with the standard carbon capture and storage (CCS) approach using the MEA absorption and injection into ocean, which is most close to commercial operation. Economics of the designed process are analyzed in terms of cost of CO_2 avoidance (\$/ton CO_2). Finally, a sensitivity analysis is performed to find influential parameters of the designed process with respect to overall economic performance.

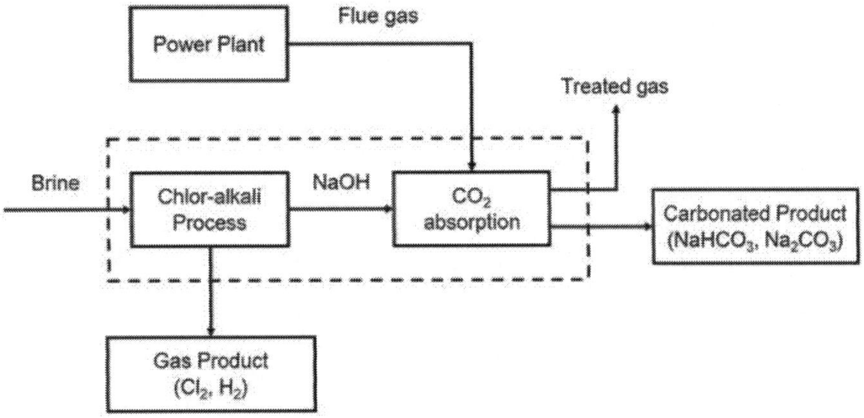

Figure 1 Schematic diagram of the designed CCSM process

References

[1] R. R. Chandran and D. T. Chin, *Electrochimica Acta, 1986, 31, 39.*

[2] G. Heddle, H. Herzog, and M. Klett, "The economics of CO2 storage," *Massachusetts Institute of Technology, Laboratory for Energy and the Environment,* 2003.

425714 Carbon Dioxide Utilization in the Polyurethane Supply Chain: An Environmental Perspective

Abstract:

Utilization of CO2 intends to close the carbon cycle in order to reduce both greenhouse gas emissions (GHG) and depletion of fossil resources. However, while intuition may suggest environmental benefits of CO2 utilization, these benefits cannot be taken for granted and an environmental assessment is required. Environmental benefits have already been identified for utilization of CO2 in polyols for polyurethane production [1]. However, the potential to utilize CO2 for polyurethanes goes beyond direct CO2 utilization in polyols: CO2 can also be used indirectly for basic chemicals in the polyurethane supply chain.

In this work, we therefore systematically explore the environmental potential of all direct and indirect CO2 utilization options for flexible and rigid PUR foams. The analysis is based on an LCA-based PUR supply chain optimization model using linear programming to identify PUR production with minimal environmental impacts.

The presented analysis shows that it is not environmentally optimal to utilize as much CO2 as possible in the PUR supply chain. We also demonstrate that the full CO2 utilization potential for environmental impact reduction can be exploited if hydrogen production becomes more environmentally friendly. The systematic analysis framework presented thus identifies routes towards more sustainable polyurethanes.

References:

[1] N. von der Assen and A. Bardow, *Green Chem.*, 2014, **16**, 3272–3280.

425943 Systems Design and Analysis of Direct Air Capture (DAC) of CO_2 Via Temperature Vacuum Swing Adsorption

Abstract:

CO_2 concentrations in the atmosphere have risen from a pre-industrial level of 270 ppm to around 400 ppm today. This rise has led to increased trapping of heat by the atmosphere and a potential for catastrophic climate change. To reduce the rate of CO_2 concentration increase, post combustion technologies have been proposed which are focused on removing CO_2 from point sources. However, this constraints the location of the CO_2 capture plant, and cannot mitigate other CO_2 emission sources, such as transportation, and cannot reduce the atmospheric concentration, only slow its increase. On the other hand Direct Air Capture (DAC) aims at removing CO_2 directly from air increasing the flexibility of deployment and enabling CO_2 concentration reduction in the future.

Our work proposes capturing CO_2 from air via temperature vacuum swing adsorption (TVSA) process using solid adsorbents. The solid sorbent is an amine impregnated MOF material (MIL-PEI) grown as films. Monolith contactors are used to support the films due to the low gas pressure drop that can be achieved during the adsorption phase of the cycle. Steam is used as a stripping agent during the desorption step. The amines in MIL-PEI are prone to oxidation/leaching during the desorption step if operated at high temperature. Vacuum swing is integrated with temperature swing process to minimize this risk. This will require less heating of the MOF surface since vacuum facilitates lowering the partial pressure of CO_2 over the MOF surface. The captured CO_2 will be utilized as part of the carbon source to develop a novel route for synthesis of 2,5 furandicarboxylic acid (FDCA) based on enzymatic carboxylation of furfural at near ambient conditions.

This talk will address the cyclic simulation of the DAC process and its overall economics. A partial differential algebraic model has been developed to solve coupled heat and mass transfer profiles of the three component system, viz, CO_2, the inert gas and the water. The experimental data from isotherm measurements of MIL-PEI-50 were used to estimate the parameters for CO_2 adsorption using the temperature dependent Langmuir isotherm model. Furthermore, the rate of adsorption of CO_2 is approximated using the linear driving force model (LDF). During the desorption step the steam is assumed to condense on the surface of the MOF but not to penetrate the MOF pores. The simulations for the TVSA model are performed using gPROMS, which is a commercial dynamic process modeling and optimization software. The governing equations are solved using method of lines with a finite difference based discretization for the axial and radial domains. The energy required for CO_2 removal is compared with the theoretical minimum energy required for ideal gas unmixing. Parasitic losses during the desorption step have been identified. They are mainly the heat of enthalpy to raise the sorbent and the monolith wall temperature during the desorption step. Energy analysis has been performed suggesting areas of improvement for reduction in these parasitic losses. It is shown that the energy requirements of the TVSA process will decrease as the sorbent capacity increases. It will also bring down the parasitic loss percentage. Optimization studies will further improve the process.

Comparison of Various Technological Options in a Smr-Hydrogen Plant in Refineries for Supplying CO2 Feedstock to Combined Reforming and Dry Reforming of Methane Based CO2 Conversion Processes

Roh, K., Korea Advanced Institute of Science and Technology (KAIST)
Lim, H.,
Ga, S.,
Lee, J. H., Korea Advanced Institute of Science and Technology (KAIST)
Originally presented on: 11/12/2015 13:54:00 - 14:15:00

Combined reforming (CR) and dry reforming (DR) of methane are two of the most promising CO_2 conversion reactions for utilizing concentrated CO_2 as a feedstock to produce syngas, which is used in various applications like liquid fuel synthesis.[1,2] In order to make these conversion processes more economically attractive, it is important to procure CO_2 feed as cheaply as possible. Steam methane reforming (SMR) based hydrogen pl ant in typical refineries is a suitable CO_2 emission source to target for this because there are process streams containing CO_2 at high partial pressure, giving potentials for an acceptably low CO_2 avoidance cost (\$/ton of CO_2).[3] The aim of this study is to compare various processing options for removing CO_2 in a SMR-H_2 plant and supplying it to the two CO_2 conversion processes, employing the combined reforming and dry reforming of methane, respectively. As the technological candidates, three CO_2 capture technologies applied to the H_2 purification unit upstream are considered: physical absorption (1-stage Selexol), chemical absorption (activated MDEA), and adsorption (Vacuum Swing Adsorption). In addition, the applicability of directly feeding the a CO_2 containing output stream of the H_2 purification into the CO_2 conversion processes is examined, an option that does not require any CO_2 capture facility. Methanol and long-chain hydrocarbons are selected as final products of the CR based and the DR based CO_2 conversion processes respectively because they guarantee the suitable syngas conditions for synthesizing the targeted products. Mass and energy balance information of each process is obtained by a commercial process simulator Aspen plus¢ç. To analyze the economic feasibility of each technological option, operating and capital costs are evaluated.

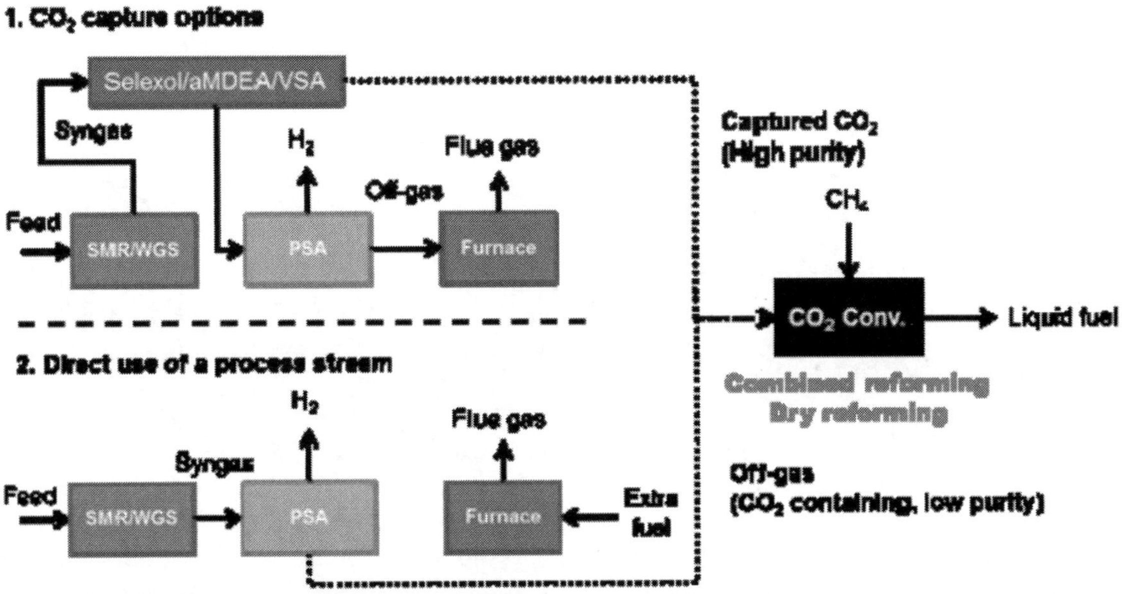

Figure 1. Four different strategies (three CO_2 capture options & process stream direct use) for integrating the H_2 plant with CO_2 conversion processes

Reference

[1] Olah, G. A.; Goeppert, A.; Prakash, G. S., Beyond oil and gas: the methanol economy. John Wiley & Sons: 2009.

[2] Gadalla, A. M.; Bower, B., The role of catalyst support on the activity of nickel for reforming methane with CO_2. Chemical Engineering Science 1988, 43 (11), 3049-3062

[3] Soltani, R.; Rosen, M. A.; Dincer, I., Assessment of CO_2 capture options from various points in steam methane reforming for hydrogen production. International Journal of Hydrogen Energy 2014, 39 (35), 20266-20275

430459 Optimal Design of Solvent-Based Post-Combustion CO2 Capture Plants Using Phase-Change Solvents

Abstract:

Solvent-based CO_2 capture using chemical absorption is a well-established and mature technology with broad applications in the industrial sector. However, as the cost introduced for solvent regeneration is rather high, resent research efforts focus on the identification of alternative solvents as well as flowsheet modifications that can reduce the energetic demands of the overall process without compromising the efficiency of the capture plant. The use of new phase-change fluids as alternative CO_2 capture solvents appears appealing and with a high potential for energy reduction [1]. Biphasic solvents include lipophilic amines which exhibit a liquid–liquid phase separation upon heating. As a result two liquid phases emerge including a concentrated organic phase containing the largest part of the amine solvent as well as the chemically bound CO_2 and an aqueous phase comprised mainly of water which can be easily separated. The non-thermal extraction of the aqueous phase prior to desorption using decanter drums enables stripping of the rich solution at much lower temperature than the 120 ^{o}C used in conventional regeneration. The low regeneration temperature of often less than 90 ^{o}C, together with the high cyclic CO_2 loading capacity considerably reduce the energetic desorption requirements [2]. Furthermore, stripper operation is enabled at elevated pressure, considerably increasing the desorption efficiency. However, as the number of phases may vary within the process due to thermodynamic conditions, tracking of the phase boundary is essential to determine the existence of the second liquid phase. This is challenging because the existence of the third phase causes discontinuities due to the alternating number of phases and the subsequent change in the model structure.

To address the above challenges we combine a phase boundary tracking method with the approximating properties of orthogonal collocation on finite elements (OCFE) model [3, 4] formulation to develop an accurate and reliable process model for three-phase reactive separation columns. Phase boundary tracking is greatly facilitated by the OCFE formulation as it is approached with the help of an efficient phase stability algorithm at the finite element breakpoints. The adaptive placement of the element breakpoint on the phase transition boundary enables the use of suitable modules that account explicitly for all phases that appear at certain regions within the column. Therefore, the associated mathematical discontinuity is properly handled and the model accuracy is greatly improved.

The energetic gains obtained from the use of a phase change solvent are further enhanced by the consideration of systematic structural and operating modifications imposed on a reference absorption/desorption flowsheet. Such modifications are realized with the help of a generalized process design framework [5]. These modifications include a plethora of choices such as a number of unconventional stream topologies, use of intercooler circuits, cascades of desorption columns as well as multi-pressure operation in the regeneration step. The employed module-based framework encompasses "building blocks" emulating the operation of basic processes, including chemical reactions, mass and heat exchange, as well as stream mixing and splitting. The location of equipment such as side-coolers or liquid-liquid separators as well as the existence and location of potential side feed/draw streams is implicitly determined through design optimization in the sense of minimizing the total process cost.

The proposed developments are illustrated considering an aqueous solution of hexylamine as the phase change solvent [2] in a case study addressing the effective treatment of an industry-derived CO_2-containing flue gas stream. A number of unconventional process flowsheets aiming towards the maximization of the driving forces within the separation columns is employed and used in design optimization studies. Vapor-liquid-liquid data are obtained for the hexylamine-water-CO_2 mixtures using the statistical associating fluid theory for potentials of variable range (SAFT-VR) [6].

Acknowledgements

The authors would like to thank Prof. Claire Adjiman, Prof. Amparo Galindo, Prof. George Jackson, and Dr. Alexandros Chremos, of Imperial College London for providing the SAFT-VR thermodynamic models. Funding from the European Commission under grant FP7-ENERGY-2011-1-282789-CAPSOL is gratefully acknowledged.

Cited References

[1] Svendsen H.F., Hessen E.T., Mejdell T. (2011). Carbon dioxide capture by absorption, challenges and possibilitites. Chem. Eng. J., 171, 718-724.

[2] Zhang J., Qiao Y., Agar D.W. (2012). Intensification of low temperature thermomorphic biphasic amine solvent regeneration for CO2 capture. Chem. Eng. Res. Des., 90, 743–749.

[3] Swartz C.L.E., Stewart W.E. (1987). Finite-element steady state simulation of multiphase distillation. AIChE J., 33, 1977.

[4] Damartzis T., Seferlis P., (2010). Optimal Design of Staged Three-Phase Reactive Distillation Columns Using Nonequilibrium and Orthogonal Collocation Models. Ind. Eng. Chem. Res., 49(7), 3275–3285.

[5] Damartzis T., Papadopoulos, A.I., Seferlis P. (2014). Optimum synthesis of solvent-based post-combustion CO_2 capture flowsheets through a generalized modeling framework. Clean. Techn. Environ. Policy, 16(7), 1363-1380.

[6] Mac Dowell N., Pereira F.E., Llovell F., Blas F.J., Adjiman C.S., Jackson G., Galindo A., (2011). Transferable SAFT-VR models for the calculation of the fluid phase equilibria in reactive mixtures of carbon dioxide, water, and n-alkylamines in the context of carbon capture, J. Phys. Chem. B, 115, 8155-8168.

433754 Dynamic Model Development and Validation of a MEA-Based CO2 Capture System

Abstract:

The U.S. Department of Energy's Carbon Capture Simulation Initiative (CCSI) is focused on developing computational tools and models for accelerating the development and commercialization of post-combustion CO_2 capture technologies. The process modeling team of CCSI is developing high-fidelity process models for a monoethanolamine (MEA) solvent system, which has high potential for commercial applications, for rapidly designing and scaling up such processes while also improving their operability. Previous works of the authors have determined key thermodynamic and transport models (density, viscosity, surface tension, heat capacity and VLE models) and are applied in the current investigation. One of the focuses of this work is to develop an accurate mass transfer model. In the existing literature, typically the mass transfer models and their parameters such as the liquid and gas-side mass transfer coefficients, diffusivity, and interfacial area are regressed using the data obtained from different experimental set-ups and scales, often in a sequential and sub-optimal way, leading to increased parametric uncertainties. For minimizing these uncertainties, a simultaneous regression of mass transfer coefficients, diffusivity and interfacial area was performed by using FOQUS, a tool developed in CCSI, leveraging the operating data from the commercial packings and wetted wall columns.

In the open literature, there is a clear lack of in-depth dynamic validation studies. In this work, the rigorous dynamic models developed in both Aspen Plus Dynamics® and gPROMS® platforms are validated using dynamic data collected from the US DOE's National Carbon Capture Center (NCCC) in Wilsonville, Alabama. The dynamic test runs were conducted by ensuring persistent of excitation to give key insights into model-form and parametric discrepancies of the process. The data include solvent composition and loading data that were collected every 5 minutes and analyzed in the NCCC laboratory. Due to the noisy and erroneous data and missing measurements of a number of key variables, a dynamic data reconciliation problem is solved. The transient behavior of the process and model validation provide strong insights that will be discussed in this presentation.

428845 Vector-Based Sustainability Analysis: A Fundamental Study on Sustainable Development

Abstract:

From the sustainability science point of view, sustainable development is a three-dimensional (3D), discretized state-transition problem, where the dimension is determined by the core concept of sustainability, which is triple-bottom-line based, and the state-transition reflects a fundamental feature of sustainability development. These require continuous assessment, prediction, decision making, and action taking. In most cases, information for sustainability assessment and decision making is uncertain due to data impreciseness and incompleteness. This makes solution identification and analysis for sustainable development very challenging.

In this paper, we formulate sustainability analysis problems as a general vector analysis problem. Mathematically, we introduce a sustainability vector, which is characterized by its magnitude and direction in a multi-dimensional space. To quantify the balance of triple-bottom-line-based development, we introduce a development angle. We also describe a space partition method by which a number of zones in development space can be defined and featured. As a sustainability improvement process involves a series of sustainability state transitions, we apply a vector calculus technique, in a discrete time-space domain, for decision analysis. We then introduce an algorithm to streamline the tasks in using this vector-based sustainability analysis method. A case study on sustainable biodiesel manufacturing is presented to demonstrate the key features and applicability of the proposed methodology.

428030 Statistical Algorithm for Sustainability Measurement and Decision Making

Abstract:

Statistical Algorithm for Sustainability Measurement and Decision Making

Rajib Mukherjee[1], Debalina Sengupta[2], Subhas K Sikdar[3]

1. Vishwamitra Research Institute, Crystal Lake, IL
2. Texas A&M University, College Station, TX
3. NRMRL, US EPA, Cincinnati, OH

Abstract

Sustainability of process, product or supply chain is generally assessed with metrics or indicators that covers all three areas of sustainability: environmental, economic and societal. These indicators are not represented in any absolute scale. In fact there is no such thing as absolute sustainability. In order to assess sustainability of product, process or supply chain, we have to use a relative scale. To compare two or multiple competing systems in a relative scale, the systems may be assumed as points in a vector space and they are compared with reference to each other. In this case, each system will act as a vector and the indicators will act as the components of the vector. In case of multiple systems to be compared, it is convenient to create an artificial reference state and compare the competing systems with reference to the hypothetical state. The hypothetical reference state may be created in various ways. In some case, the reference state can be created with the minimum value of each and every components of the vector. In some other cases, it could be the average of the components. The origin of the vector space where all components are zero can also be used as a reference state. All these methods of finding the reference state has their own merits and demerits in evaluating the relative sustainability of the competing systems. Once a suitable reference state is identified for the competing systems, the actual comparison can be done using an aggregate method where all the indicators are aggregated into a single index. Several alternative aggregate indices can be derived statistically from the indicator data. In the simplest form, the aggregation can be performed using the geometric mean of the indicators. Other than normal geometric mean, aggregate indices based on Euclidean distance, Mahalonabis distance, Canberra metrics, vector norm have also been proposed by various researchers with varied applications. Aggregation of the indicators in any of the methods requires the vectors to be normalized. This is to retain the variability of the component of the vectors but reduce the variability among the components. Normalization can be performed in various ways. Each method of normalization has its own significance in the evaluation of the aggregate.

In this work we will show different ways of reference selection, normalization and aggregation of sustainability indices and make relative assessment of different methods with examples. We will present what can we learn about a system from an aggregate reading that we cannot learn by comparing individual indicator values. We will also present statistical methods of identifying which indicators are necessary and sufficient for sustainability evaluation of specific system. This in turn will also determine which indicators are insignificant or redundant. Furthermore, we will also show the sensitivity of the system sustainability to changes in any one of the indicators and rank the indicators in order of their contribution to sustainability of the system. With our method of sustainability analysis with aggregated index and important indicator selection and their sensitivity, it would be easier to determine which indicators require attention in order to improve the product or process from sustainability viewpoint.

429925 Multi-Objective Optimization Combined with Input-Output and Eco-Cost Assessment for Decarbonizing the European Economy

Abstract:

Multi-objective optimization combined with input-output and eco-cost assessment for decarbonizing the European Economy

D. Cortés-Borda [a], L. Jimenez-Esteller [a], Guillén-Gosálbez [a,b],

a.Departament d'Enginyeria Química, Universitat Rovira i Virgili, Av. Països Catalans, 26, Tarragona E-43007, Spain.

b.Centre for Process Integration, School of Chemical Engineering and Analytical Science, The University of Manchester, Manchester M13 9PL, UK

The world is presently facing environmental and economic challenges derived from the excessive consumption of non-renewable sources. The accelerated increase of fossil-derived greenhouse gas (GHG) emissions has brought global warming to alarming levels. As a result, worldwide governments have set up international agreements in which the participants are committed to reduce their GHG quota by increasing the share of renewable energy sources in their energy mixes (among other alternatives).

So far, the global share of renewable energies (including wind, geothermal, solar, biomass and waste energies) represents only 1.8% of the world's energy consumption [1]. Further increases in the global renewable energy share is challenging due to the low competitiveness of such technologies. Hence, GHG saving strategies based on the deployment of renewable energies might result unsuccessful without environmental policies that place the renewable technologies in the same economic threshold than fossil fuels.

Since the late 90's, governments of the most developed countries have included taxation mechanisms in their policy agendas to overcome the competitiveness barriers that environmentally conscious alternatives face [2]. Defining environmental taxes on GHG intensive products re-directs the economy towards GHG saving alternatives. That is, the end users would prefer less GHG-intensive products (after an increase of GHG taxes); and consequently the producers would need to improve their environmental efficiency in order to remain competitive. Furthermore, environmental taxation leads to a "double dividend", since revenues from taxes could be re-invested to compensate the environmental damage [3] or to prevent future emissions by investing in cleaner production technologies [4-5]. Nowadays, the "eco-cost" is well known as the cost of preventing a given environmental burden [6]. Numerous studies have assessed the eco-cost in different industries/sectors [8-16], however, so far there are no studies integrating the eco-cost assessment of industries as a whole economy.

The environmentally extended input-output models (EEIO) [17] provide a comprehensive description of the inter-industry economic transactions, which allows linking the total economy (as a whole) with its corresponding environmental burdens and associated eco-costs. Furthermore, the flexibility of EEIO models allows allocating the environmental responsibilities both to consumers or producers, depending on whether the assessment is consumption-based or production-based.

Combining EEIO with multi-objective optimization algorithms is an appealing methodology to identify solutions that are economically and environmentally optimal in an economy. On this basis, the aim of this work is to minimize the eco-costs derived from the GHG emissions of the European Economy; while simultaneously maximizing the demand satisfaction of the final customers.

To this end, we formulate the EEIO model as a multi-objective optimization linear programming (LP) problem with two objective functions (the eco-cost of greening the European economy; and the total demand satisfaction); subject to a system of linear equations representing the macroeconomic transactions in an economy; and the environmental extension equations. The LP model leaves the final demand of each sector of the European Economy as a free variable that can range within realistic upper and lower bounds, in order to find the optimal solutions. To investigate the existence of tradeoffs between the two competitive objectives, we apply the epsilon-constraint method that generates a set of Pareto optimal solutions [18]; each of which presents a unique combination of demand satisfaction and eco-cost savings.

We applied our method to the 25-country European Union's Economy (EU-25) (The European Union according to the 2004 enlargement). The macroeconomic and environmental data was retrieved from the E3IOT database [19-20], which contains highly disaggregated input-output tables (i.e. 487 industrial sectors and household activities).

A preliminary analysis was first performed in order to compare the eco-cost of each sector with its corresponding total economic output. Hence, by comparing each sector's direct and indirect associated eco-costs (production-based and consumption-based, respectively) with their own economic output, we find that there are 16 sectors (from the production perspective) and 13 sectors (from the consumption perspective) that generate more eco-costs than the economic output they generate. Among these sectors, we find the electric service (utilities), petroleum refining, eating and drinking places, meat animals, poultry slaughtering and processing, fluid milk, to name some.

The optimization results provide insight into which sectors should be regulated first in order to reduce the GHG emissions and the associated eco-cost. Hence, the European Commission should firstly regulate the sectors identified by our model (by either investing in improving the efficiency or establishing stronger environmental regulations on their final demand) in order to move towards a more sustainable economy.

References

[1] N. Lior, "Sustainable energy development: The present (2009) situation and possible paths to the future," *Energy*, vol. 35, no. 10, pp. 3976–3994, Oct. 2010.

[2] P. Ekins, "European environmental taxes and charges: recent experience, issues and trends," *Ecol. Econ.*, vol. 31, no. 1, pp. 39–62, Oct. 1999.

[3] C. N. Eng-ct, R. Dones, T. Heck, C. Bauer, S. Hirschberg, P. Scherrer, I. Psi, P. Bickel, P. Preiss, U. Stuttgart, L. I. Panis, and I. De Vlieger, "ExternE-Pol Externalities of Energy : Extension of Accounting Framework and Policy Applications New energy technologies," no. x, pp. 1–76, 2005.

[4] J. G. Vogtlander and A. Bijma, "The 'Virtual Pollution Prevention Costs '99': A single LCA-based indicator for emissions," *Int. J. Life Cycle Assess.*, vol. 5, no. 2, pp. 113–124, 2000.

[5] J. G. Vogtländer, A. Bijma, and H. C. Brezet, "Communicating the eco-efficiency of products and services by means of the eco-costs/value model," *J. Clean. Prod.*, vol. 10, no. 1, pp. 57–67, Feb. 2002.

[6] J. G. D. U. of T. Vogtländer, "Ecocost value." [Online]. Available: www.ecocostsvalue.com.

[7] J. G. Vogtländer, E. Lindeijer, J.-P. M. Witte, and C. Hendriks, "Characterizing the change of land-use based on flora: application for EIA and LCA," *J. Clean. Prod.*, vol. 12, no. 1, pp. 47–57, Feb. 2004.

[8] J. Vogtländer, P. van der Lugt, and H. Brezet, "The sustainability of bamboo products for local and Western European applications. LCAs and land-use," *J. Clean. Prod.*, vol. 18, no. 13, pp. 1260–1269, Sep. 2010.

[9] M. A. Morales-Mora, E. Rosa-Dominguez, N. Suppen-Reynaga, and S. A. Martinez-Delgadillo, "Environmental and eco-costs life cycle assessment of an acrylonitrile process by capacity enlargement in Mexico," *Process Saf. Environ. Prot.*, vol. 90, no. 1, pp. 27–37, Jan. 2012.

[10] L. Čuček, R. Drobež, B. Pahor, and Z. Kravanja, "Sustainable synthesis of biogas processes using a novel concept of eco-profit," *Comput. Chem. Eng.*, vol. 42, pp. 87–100, Jul. 2012.

[11] E. Igos, E. Benetto, I. Baudin, L. Tiruta-Barna, Y. Mery, and D. Arbault, "Cost-performance indicator for comparative environmental assessment of water treatment plants.," *Sci. Total Environ.*, vol. 443, pp. 367–74, Jan. 2013.

[12] A. Mestre and J. Vogtlander, "Eco-efficient value creation of cork products: an LCA-based method for design intervention," *J. Clean. Prod.*, vol. 57, pp. 101–114, Oct. 2013.

[13] M. A. M. Mora, E. R. Dominguez, A. A. Ibarra, N. S. Reynaga, and S. A. M. Delgadillo, "A methodological improvement for assessing petrochemical projects through life cycle assessment and eco-costs," *Int. J. Life Cycle Assess.*, vol. 19, no. 3, pp. 517–531, Oct. 2013.

[14] N. M. van der Velden, M. K. Patel, and J. G. Vogtländer, "LCA benchmarking study on textiles made of cotton, polyester, nylon, acryl, or elastane," *Int. J. Life Cycle Assess.*, vol. 19, no. 2, pp. 331–356, Sep. 2013.

[15] R. Rivas-Hermann, J. Köhler, and A. E. Scheepens, "Innovation in product and services in the shipping retrofit industry: a case study of ballast water treatment systems," *J. Clean. Prod.*, Jun. 2014.

[16] S. Ferreira, M. Cabral, N. F. da Cruz, and R. C. Marques, "Economic and environmental impacts of the recycling system in Portugal," *J. Clean. Prod.*, vol. 79, pp. 219–230, Sep. 2014.

[17] W. Leontief, "Environmental repercussions and the economic structure : an input-output approach" vol. 52, no. 3, pp. 262–271, 1970.

[18] Ehrgott, M. Multicriteria Optimization, second ed. Springer, Berlin. 2005.

[19] G. Huppes, A. de Koning, S. Suh, R. Heijungs, L. van Oers, P. Nielsen and J.B. Guinée. Environmental impacts of consumption in the european union: High-resolution input-output tables with detailed environmental extensions. Journal of Industrial Ecology 10(3): 129-146. 2006

[20] R. Heijungs, A. de Koning, S. Suh and G. Huppes. Toward an information tool for integrated product policy: Requirements for data and computation. Journal of Industrial Ecology 10(3): 147-158. 2006.

433661 Assessing Sustainability By Life Cycle Assessment Versus Techno-Ecological Synergy

Abstract:

Life cycle assessment (LCA) has become the most common approach for assessing and designing sustainable systems. Over the last two decades, this method has been standardized and is a convenient approach to quantify environmental impact of processes through the life cycle with the availability of inventories and ease of software use. The primary benefit of including life cycle considerations in engineering decisions is that it reduces the chance of unintended harm by shifting a problem outside the narrow boundary of traditional engineering. Thus, LCA considers the impact of emissions and resource use of activities that contribute to the selected product or process from "cradle to grave" [1].

However, using LCA to measure the environmental impact of systems and develop sustainable solutions suffers from several shortcomings. For example, LCA encourages decisions that do "less bad" instead of "more good" in the sense that since this approach only estimates the impacts of systems relative to other similar systems, processes that have the least impact are deemed sustainable. Furthermore, since LCA is a macro-level analysis, it can be difficult to make improvements for specific processes, particularly if most of the impact lies outside the boundary of the plant scale, along the supply chain or use phase of a product. LCA methods also ignore essential interactions with *nature* that provide ecosystem goods and services that are essential for sustaining human activities and well-being. Although efforts are underway to incorporate into LCA the impacts on ecosystem services and biodiversity due to land use and land use change in a spatially explicit manner [2],[3] currently the role played by ecosystems in supporting engineering activities and the ecological limits imposed by these systems have been ignored. Eco-LCA [4], [5] a tool that explicitly accounts for the ecosystem service flow across the life cycle of products bridges the gap between conventional LCA by extending the boundary of analysis to include ecological systems but does not capture ecological carrying capacity. The approach of cradle-to-cradle (C_2C) design also attempts to overcome some shortcomings of LCA but this approach ignores life cycle aspects and the role of ecosystems.

The recently developed framework of Techno-Ecological Synergy (TES) attempts to overcome many of these shortcomings of LCA by explicitly capturing synergies between systems in technological and ecological spheres. Systems at multiple scales ranging from individual processes, supply chains and life cycles can be evaluated along with their relevant supporting ecological systems. Resource inputs to and emissions from technological systems create a demand for ecosystem services, while ecological systems have the capacity to sequester most of these pollutants thus supplying relevant ecosystem services. Absolute sustainability indices defined by Bakshi et.al 2015 [6] based on the overshoot in demand for ecosystem services over its supply can be used to assess the level of sustainability of systems. This metric system is multi-scale in nature, ranging from local to global. For a process to be locally-sustainable, emissions within a local boundary should not exceed the amount that could be sequestered by the surrounding ecosystem. A necessary condition to be satisfied for reaching the goal of sustainability is to maintain the demand for ecosystem services within the supply at the largest scale applicable to the ecological system.

This talk will provide practical insight into the pros and cons of using the TES framework to assess the sustainability of a system. Application of the TES framework to determine the level of sustainability of a system compared to that of the LCA approach will be demonstrated by a book manufacturing system - a traditional paper book vs. a book designed by the C_2C approach. The latter uses synthetic polypropylene sheets while the former uses pulp-based paper sheets. In the recycling stage, for traditional book, ink and paper cannot be separated easily, leading to down-cycling rendering both materials with lower qualities. On the contrary, synthetic paper in C_2C books can be "up-cycled" and reused since the ink can be washed off easily. Designers have established that by closing this material loop for synthetic paper, the C_2C book is expected to perform environmentally better.

In this case study, LCA and TES are applied to the production of both type of books to compare their environmental performances. From the LCA study, it can be inferred that C_2C book have a larger environmental impact than the traditional book. However, analyzing the production process using LCA technique does not provide an insight into the absolute level of sustainability of these systems. For instance, even though traditional book production has comparatively less environmental impact, the resource input and pollutant emission might still be well-above ecological carrying capacity. In contrast, TES can provide more insights for absolute sustainability at selected technological scale and corresponding ecological scale. Taking carbon sequestration service as an example, at local scale, it can be inferred that the pulp-based paper production for traditional book is sustainable because the demand for carbon sequestration service is less than the ability of surrounding ecosystem to provide the service. Similarly, at local scale, the synthetic paper production for C_2C book is unsustainable because the local demand for the service overshoots the supply from surrounding ecosystem. At life cycle scale, the demand for carbon sequestration service imposed by CO_2 emissions due to a traditional book exceeds the amount that can be sequestered, mainly because of the inclusion of emission-intensive processes, such as electricity generation and book glue production. From this multi-scale analysis, we can identify processes with higher overshooting and make technological improvements accordingly. Moreover, by using the TES approach to design for sustainability, we not only have the option to reduce the resource inputs and emissions as with the LCA approach, but also the option of increasing the supply of ecosystem services that can be achieved by proper landscape planning. This solution might be potentially more cost-effective and environmental-friendly than its technological counterpart.

Even though TES approach can provide us with many useful insights, the practical feasibility of this approach is limited by the availability of detailed local production information and proper ecological models. The former is necessary to quantify the demand for ecosystem services within the local ecological boundary and the latter is crucial for better quantifying the supply of ecosystem services at different scales. With the development of geographic information system and remote sensing technology, more accurate quantification of ecosystem service supply and demand at selected scale can be achieved. TES approach can then provide absolute sustainability metrics at different scales and for various ecosystem services, which can facilitate decision-making at all relevant scales.

[1] Vink, E. T., Rabago, K. R., Glassner, D. A., & Gruber, P. R. (2003). Applications of life cycle assessment to NatureWorks™ polylactide (PLA) production. Polymer Degradation and Stability, 80(3), 403-419.

[2] Geyer, R., Stoms, D. M., Lindner, J. P., Davis, F. W., & Wittstock, B. (2010). Coupling GIS and LCA for biodiversity assessments of land use. The International Journal of Life Cycle Assessment, 15(5), 454-467.

[3] Koellner, T., & Geyer, R. (2013). Global land use impact assessment on biodiversity and ecosystem services in LCA. The International Journal of Life Cycle Assessment, 18(6), 1185-1187.

[4] Zhang, Y., Singh, S., & Bakshi, B. R. (2010). Accounting for ecosystem services in life cycle assessment, Part I: a critical review. Environmental Science & Technology, 44(7), 2232-2242.

[5] Zhang, Y., Baral, A., & Bakshi, B. R. (2010). Accounting for ecosystem services in life cycle assessment, part II: toward an ecologically based LCA. Environmental Science & Technology, 44(7), 2624-2631.

[6] Bakshi, B. R., Ziv, G., & Lepech, M. D. (2015). Techno-Ecological Synergy: A Framework for Sustainable Engineering. Environmental Science & Technology, 49(3), 1752-1760.

[7] Braungart, M., McDonough, W., & Bollinger, A. (2007). Cradle-to-cradle design: creating healthy emissions - a strategy for eco-effective product and system design. Journal of Cleaner Production, 15(13-14), 1337-1348.

420408 Life Cycle Analysis of Three Methods to Treat Primary Clarifier Effluent: Aerated Activated Sludge Bed, Hrad Combined to a Trickling Filter and Trickling Filter

Abstract:

This talk will address techniques used to evaluate the value of two emerging wastewater treatment technologies proposed as sustainable alternatives to aerated activated sludge beds in waste water treatment plants. Results from a case study executed at demonstration scale in a wastewater treatment plant and in collaboration with industrial professionals will be presented. In particular, the outcomes from several assessments including energy balances and Life Cycle Analyses will be presented and evaluated for their impact relative to tradtional metrics used by industry experts to reach decision points on the inclusion or exclusion of new and emerging technology aimed at improving sustainability.

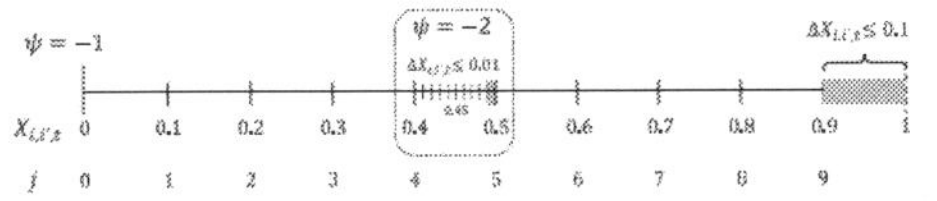

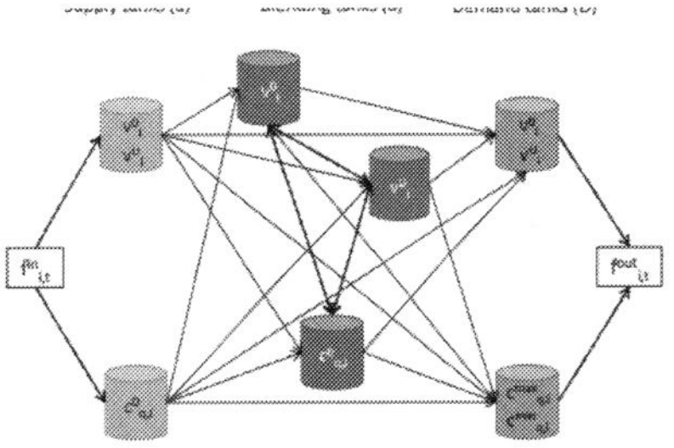

A More Efficient Formulation for the Multiperiod Blending Problem

Pedro M. Castro (pmcastro@fc.ul.pt)
http://www.researcherid.com/rid/C-3642-2008

Associate Researcher
CMAFCIO
Faculdade de Ciências
Universidade de Lisboa

IF INVESTIGADOR FCT

Ciências ULisboa

LISBOA UNIVERSIDADE DE LISBOA

Motivation

- Pooling problem of great importance in petroleum refineries (Haverly 78)
  - Given a set of liquid fuels, find most profitable blends that meet technical constraints
  - Pools are blending tanks, content to be determined by optimization
  - **Generalized pooling** (Meyer & Floudas 06)
- Challenging mathematical problem
  - Non-convex bilinear terms
    - Need for global optimization algorithms
  - Few formulations have been proposed
    - Selection of model variables, leading to relaxations of varying strength
      - p- (Haverly 78), q- (Ben-Tal et al. 94), pq- (Tawarmalani & Sahinidis 02), tp- and stp- (Alfaki & Haugland 13)
- Multiperiod pooling problem incorporates scheduling aspect
  - Supply and demand vary with time
  - MIQCP (MINLP), binary variables appear linearly in the constraints
    - Kolodziej et al. 13, Lotero et al. 15

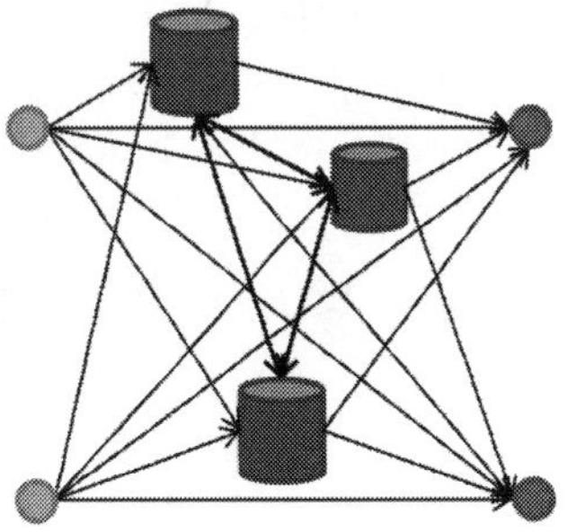

Problem statement

- Supply tanks (S)
  - Fuels with constant composition $c_{q,i}^{0}$
  - Variable input amount $f_{i,t}^{in}$ (kg)
- Blending tanks (B)
  - Variable composition (mass%)
  - Either flow in or out
- Demand tanks (D)
  - Composition bounds $c_{q,i}^{min}$, $c_{q,i}^{max}$
  - Time-dependent output $f_{i,t}^{out}$
  - Valuable products p_i (\$/kg)
- Other important parameters
  - Transfer costs $cs_{i,i'}^{fx}$ (\$), $cs_{i,i'}^{vr}$ (\$/kg)
  - Bounds on transfers $f_{i,i'}^{min}$, $f_{i,i'}^{max}$ (kg)
  - Initial and maximum inventory v_i^0, v_i^{max} (kg)

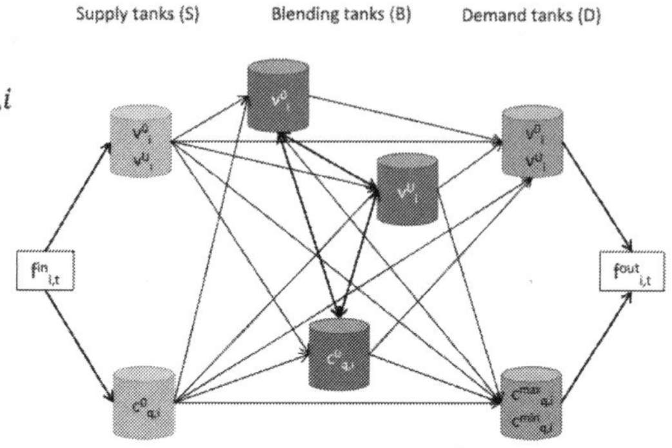

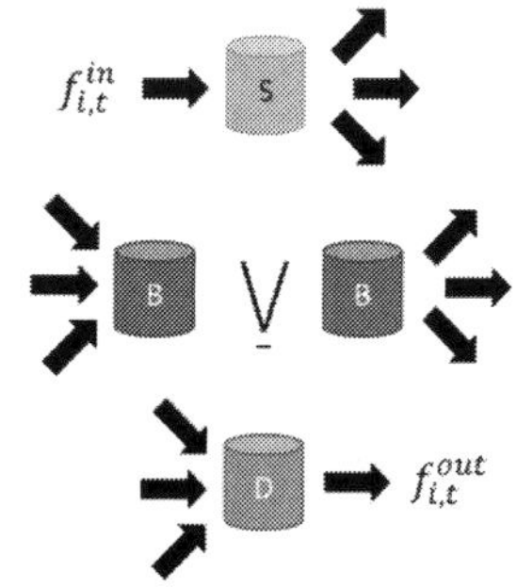

Total flows $F_{i,i',t}$ & compositions $C_{q,i,t}$

- Most intuitive formulation (Kolodziej et al. 13)
  - Shared with p- & most process network models
- Key constraints
  - Multiperiod mass balance
    - $V_{i,t} = v_i^0\big|_{t=1} + V_{i,t-1}\big|_{t>1} + f_{i,t}^{in} + \sum_{i' \in I_i} F_{i',i,t} - f_{i,t}^{out} - \sum_{i':i \in I_{i'}} F_{i,i',t} \ \forall i,t$
  - Mass balance for quality q (source of bilinear terms)
    - $C_{q,i,t}V_{i,t} = (c_{q,i}^0 v_i^0)\big|_{t=1} + (C_{q,i,t-1}V_{i,t-1})\big|_{t>1} + \sum_{i' \in I_i \cap S} c_{q,i'}^0 F_{i',i,t} +$ $\sum_{i' \in I_i \cap B}[(c_{q,i'}^0\big|_{t=1} + C_{q,i',t-1}\big|_{t>1})F_{i',i,t}] - \sum_{i':i \in I_{i'}}\left[\left(c_{q,i}^0\big|_{t=1} + C_{q,i,t-1}\big|_{t>1}\right)F_{i,i',t}\right] \ \forall q, i \in B, t$
  - Link to demand tank i' only if compositions are within bounds (big-M)
    - $\begin{bmatrix} Y_{i,i',t} \\ c_{q,i'}^{min} \le C_{q,i,t-1} \le c_{q,i'}^{max} \end{bmatrix} \underline{\vee} \begin{bmatrix} \neg Y_{i,i',t} \\ C_{q,i,t-1} \ge 0 \end{bmatrix} \ \forall q, i' \in D, i \in I_{i'} \cap B, t > 1$
  - Enforcing bounds on mass transferred
    - $f_{i,i'}^{min} Y_{i,i',t} \le F_{i,i',t} \le f_{i,i'}^{max} Y_{i,i',t} \ \forall i', i \in I_{i'}, t$
  - No simultaneous flow in and out of blending tanks
    - $Y_{i',i,t} + Y_{i,i'',t} \le 1 \ \forall i'', i \in I_{i''} \cap B, i' \in I_i, t$
  - Objective function
    - $\max \sum_{i'} \sum_{i \in I_{i'}} \sum_t [(p_i + p_{i'} - cs_{i,i'}^{vr})F_{i,i',t} - cs_{i,i'}^{fx} Y_{i,i',t}]$

Individual flows $F_{l,i,i',t}$ & split fractions $X_{i,i',t}$

- New approach for multiperiod pooling problem
  - Define liquid fuels l, discrete entities that keep their identity
    - Idea implicit in continuous-time formulations for crude oil scheduling
      - Priority slots of Mouret et al. 09, single time grid RTN of Castro & Grossmann 14
    - Identical to source based model in Lotero et al. 15 (uses more constraints but is proven stronger than compositions model)
  - Split fraction variables $X_{i,i',t} \in [0,1]$ hold fraction of contents in tank i leaving for i' during slot t
    - In steady-state problems sum of split fractions over outlet streams adds to 1
      - tp-formulation, Galan & Grossmann 98 (process networks)

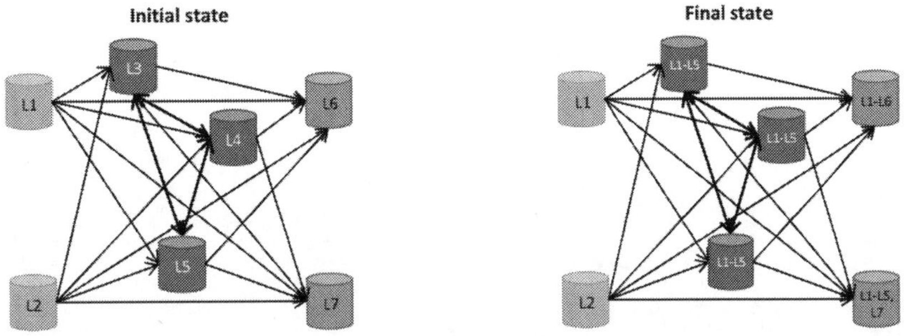

Individual flows & split fractions model

- Most variables gain liquid fuel index l
  - Mass balance
    - $V_{l,i,t} = v^0_{l,i}\big|_{t=1} + V_{l,i,t-1}\big|_{t>1} + f^{in}_{l,i,t} + \sum_{i' \in I_i: l \in L_{i'}} F_{l,i',i,t} - F^{out}_{l,i,t}\big|_{i\in D} - \sum_{i':i \in I_{i'}} F_{l,i,i',t}$ $\forall i, l \in L_i, t$
  - Link to demand tank i' only if compositions are within bounds (big-M)
    - $c^{min}_{q,i'} \sum_{l\in L_i} V_{l,i,t-1} - M(1 - Y_{i,i',t}) \le \sum_{l \in L_i} c_{q,l} V_{l,i,t-1} \le c^{max}_{q,i'} \sum_{l\in L_i} V_{l,i,t-1} + M(1 - Y_{i,i',t})$ $\forall q, i' \in D, i \in I_{i'} \cap B, t > 1$
  - Definition of split fraction (source of bilinear terms)
    - $F_{l,i,i',t} = \boxed{X_{i,i',t} V_{l,i,t-1}}$ $\forall i', i \in I_{i'} \cap B, l \in L_i, t > 1$
  - Other constraints
    - $f^{out}_{i,t} = \sum_{l \in L_i} F^{out}_{l,i,t}$ $\forall i \in D, t$
    - $f^{min}_{i,i'} Y_{i,i',t} \le \sum_{l \in L_i} F_{l,i,i',t} \le f^{max}_{i,i'} Y_{i,i',t}$ $\forall i', i \in I_{i'}, t$
    - $\sum_{l\in L_i} V_{l,i,t} \le v^{max}_i$ $\forall i, t$
    - $\max \sum_{i'} \sum_{i \in I_{i'}} \sum_t [(p_i + p_{i'} - cs^{vr}_{i,i'}) \sum_{l \in L_i} F_{l,i,i',t} - cs^{fx}_{i,i'} Y_{i,i',t}]$

Relaxation of bilinear term $X_{i,i',t}V_{l,i,t-1}$

- Standard McCormick (76)
  - $W_{l,i,i',t} = X_{i,i',t}V_{l,i,t-1}$; add four sets of constraints
- Multiparametric Disaggregation (Castro and co-workers 13, 15)
  - Normalized variable $X_{i,i',t}$ chosen for discretization
  - Binary variables $Z_{i,i',t,j,k}$ select digit j for position k
    - $X_{i,i',t} = \sum_{k=\psi}^{-1}\sum_{j=1}^{9} j \cdot 10^k \cdot Z_{i,i',t,j,k} + \Delta X_{i,i',t}$
    - $W_{l,i,i',t} = \sum_{k=\psi}^{-1}\sum_{j=1}^{9} j \cdot 10^k \cdot \hat{V}_{l,i,i',t,j,k} + \Delta W_{l,i,i',t}$
    - $V_{l,i,t-1} = \sum_{j=0}^{9} \hat{V}_{l,i,i',t,j,k}$
    - $v_{l,i,t-1}^{L} Z_{i,i',t,j,k} \le \hat{V}_{l,i,i',t,j,k} \le v_{l,i,t-1}^{U} Z_{i,i',t,j,k}$
    - $10^{\psi}(V_{l,i,t-1} - v_{l,i,t-1}^{U}) + v_{l,i,t-1}^{U}\Delta X_{i,i',t} \le \Delta W_{l,i,i',t} \le 10^{\psi}(V_{l,i,t-1} - v_{l,i,t-1}^{L}) + v_{l,i,t-1}^{L}\Delta X_{i,i',t}$
    - $v_{l,i,t-1}^{L}\Delta X_{i,i',t} \le \Delta W_{l,i,i',t} \le v_{l,i,t-1}^{U}\Delta X_{i,i',t}$
    - $\sum_{j=0}^{9} Z_{i,i',t,j,k} = 1$

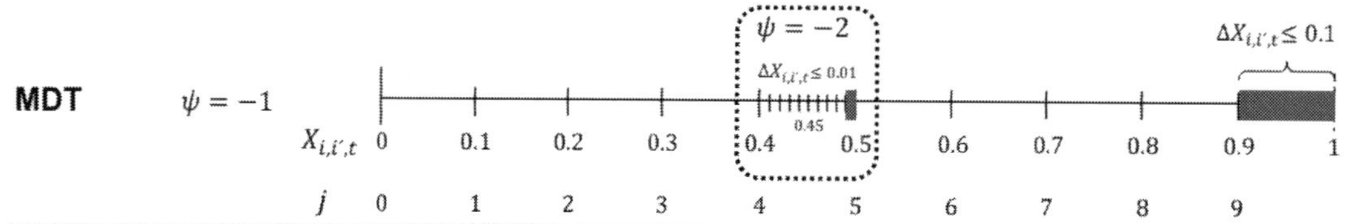

MILP-NLP global optimization algorithm

- Takes advantage of properties:
  - MILP relaxation from MDT equivalent to MINLP as $\psi \to -\infty$
  - Solution from MILP good starting point for NLP (Jia et al. 03)
    - Approach used for McCormick relaxation
- Overview (maximization problem, tolerance= ε)
  1. $\psi = -1$
  2. Solve MILP to get UB and values $y_{i,i',t}$ of binary variables $Y_{i,i',t}$
  3. Add constraint $Y_{i,i',t} = y_{i,i',t}$ reducing original MINLP to an NLP
  4. From MILP starting point, solve NLP to get feasible solution and LB
     - If infeasible, LB=$-\infty$
  5. If $\frac{UB-LB}{UB} \le \varepsilon$, STOP, solution is global optimum;
     else, $\psi = \psi - 1$, return to 2.

Why do we need global solvers?

- Difficulty already in finding feasible solutions
  - McCormick relaxation rather loose, MILP-NLP procedure fails

Formulation	Total flows and compositions			Individual flows and split fractions		
Problem	TCPUs	LB [$]	UB [$]	UB [$]	LB [$]	TCPUs
029	0.45	13359.4	13359.4	13359.4	13359.4	0.58
146	0.89	Infeasible	46917.0	45896.5	Infeasible	1.33
480	1.25	Infeasible	9705.4	9443.2	8671.6	1.80
531	0.83	Infeasible	20796.1	21380.2	Infeasible	0.72
718	0.45	7393.6	7393.6	7393.6	7393.6	0.47
721	0.60	Infeasible	14247.6	13526.8	13526.8	0.93
852	0.81	Infeasible	54458.7	53962.7	53962.7	1.33

New formulation not always tighter

- DICOPT also not good enough (assumes rMINLP is convex)

		Total flows and compositions					Individual flows and split fractions				
Problem	Binary variables	Total variables	Equations	Bilinear terms	CPUs	LB [$]	LB [$]	CPUs	Total variables	Equations	Bilinear terms
029	36	103	202	64	3.1	Inf.	Inf.	3.13	219	294	64
146	87	223	617	256	5.1	Inf.	43325.7	2.41	689	965	480
480	124	313	879	376	8.86	Inf.	Inf.	8.7	941	1383	720
531	104	273	732	358	5.91	Inf.	Inf.	9.24	878	1218	684
718	87	223	603	244	1.15	5625.6	7393.6	1.15	672	939	456
721	87	223	623	256	5.21	Inf.	13495.6	0.63	689	971	480
852	120	305	859	376	10.3	Inf.	53962.7	1.71	933	1363	720

Global solvers comparison

- Problems from new formulation easier to tackle!
- BARON 14.0 best overall performer
  - Can solve all 7×2 problems in <3600 CPUs
  - Major improvement over version 10.2
- GloMIQO 2.3 more likely to be fastest solver
  - Whenever UB from McCormick relaxation is optimal solution
- MDT algorithm can be at least one order of magnitude faster than both BARON and GloMIQO
  - Very efficient if no need to go into deep accuracy levels (low ψ values)

	CPUs						Gap [%], if higher than 0.0001%		
	Total flows and compositions			Individual flows and split fractions			Total flows and compositions		Individual flows and split fractions
Problem	New	BARON	GloMIQO	New	BARON	GloMIQO	New	GloMIQO	New
029	0.71	3.13	0.23	0.76	0.33	0.23	-	-	-
146	3600	1227	3600	3600	91.3	488	0.0007	0.7436	0.0003
480	3600	302	3600	32.0	453	3363	0.0452	0.1386	-
531	41.6	97.4	3600	3.48	43.3	252	-	0.0332	-
718	0.85	3.56	0.22	0.77	3.97	0.32	-	-	-
721	460	265	243	2.12	21.4	1.82	-	-	-
852	3.00	231	600	4.36	134	3.91	-	-	-

One of the constraints is too restricting!

- Every input stream to a demand tank must be within its composition bounds?
  - Avoids additional nonlinear constraints in original formulation
    - Artificial constraint to keep complexity manageable?
- Real constraint is for the blend to be within bounds!
  - Eliminates big-M constraints from the model
    - $c_{q,i}^{min} \sum_{i' \in I_i} \sum_{l \in L_{i'}} F_{l,i',i,t} \leq \sum_{i' \in I_i} \sum_{l \in L_{i'}} c_{q,l} F_{l,i',i,t} \leq c_{q,i}^{max} \sum_{i' \in I_i} \sum_{l \in L_{i'}} F_{l,i',i,t}$
  - Problems become easier to solve
  - Major improvements in profit $

		CPUs		
Problem	Improvement [%]	New algorithm	BARON	GloMIQO
029	0	0.57	0.33	0.29
146	5.72	468	72.6	159
480	41.6	2.68	94.7	2.75
531	12.2	2.69	75.1	1.43
718	0	0.62	4.30	0.48
721	19.1	1.35	28.7	0.65
852	2.45	50.3	1002	368

Optimal solution for problem 480

- Takes full advantage of network's operational flexibility
  - Worth to consider multiperiod problem over steady state

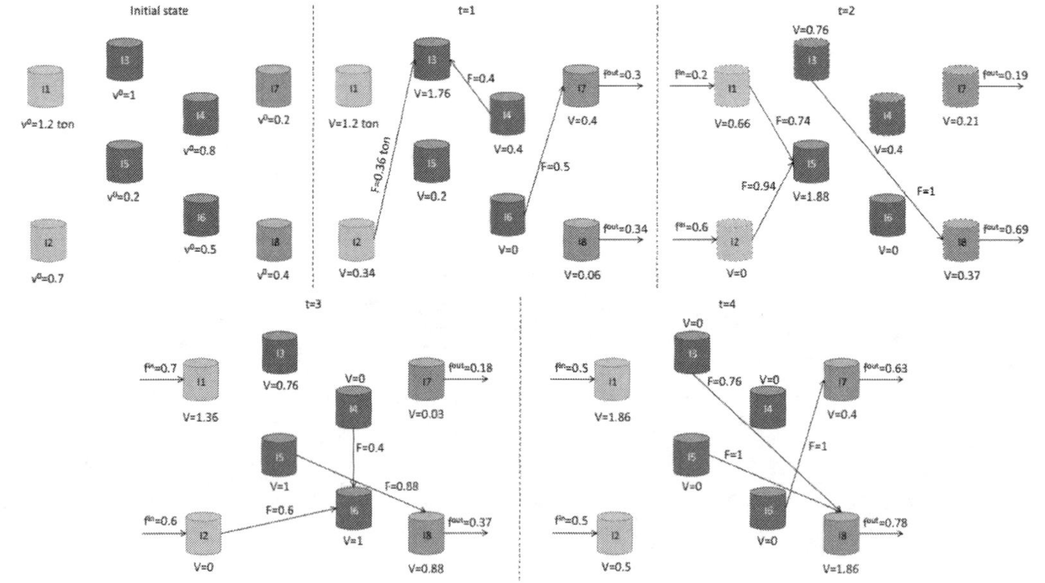

Optimum with new constraint (P480)

- Tanks I1-I2 can now send fuel directly to I7-I8 (if joined by others)
  - They actually bypass blending tanks, leading to 41.6% higher profit

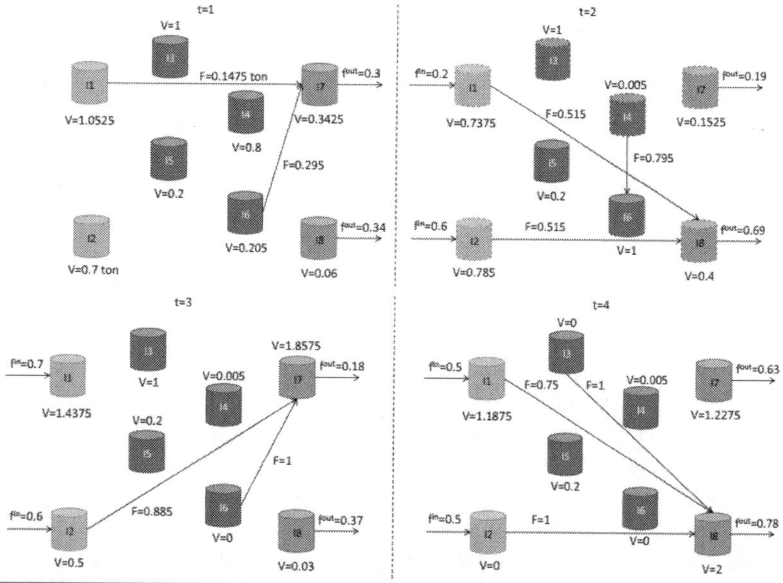

Conclusions

- New MINLP model for multiperiod pooling problem
  - Each blend initially present in a tank treated as distinct entity
    - Liquid fuel
  - Larger in size but typically tighter
    - Different sets of complicating constraints
  - Improved performance for all 5 optimization algorithms
- Critical analysis of important model constraint
  - Better to enforce inlet blend within composition bounds than every individual input stream (to a demand tank)
  - Big-M constraints no longer needed $\Rightarrow$ Improved performance
- Further details
  - AIChE J. 61 (2015) 3728-3738
- Acknowledgments:
  - Fundação para a Ciência e Tecnologia, Investigador FCT 2013 program

421769 A Multi-Time-Scale Approach for the Integrated Maintenance and Production Scheduling of Multipurpose Process Plants

Abstract:

Abstract

In this work we go beyond the traditional concept of maintenance, which focuses on performing disruptive activities based on short-term needs. In maintenance scheduling it is useful to consider all technical, administrative and supervision actions that intend to retain the productivity. This includes for instance cleaning operations, asset inspection, component lubrication or part-substitutions involving only a partial plant shutdown.

In the scheduling literature, the interaction of maintenance and production scheduling decision processes has already been widely explored. Dedopoulos at al. (1995) considers enforcing a set of maintenance tasks to be performed within the scheduling horizon through penalties. Hazaras et al. (2012) accounts for different types of maintenance: fixed time maintenance intervals as well as a set of maintenance with flexible start time to be determined by the optimization. Castro et al. (2014) discussed the optimal maintenance scheduling problem for a gas engine power plant where the production is driven by electricity sales assuming seasonal variations in electricity pricing. Biondi et al. (2015) presents an MILP model for short-term scheduling of production and maintenance in multipurpose process plants allowing to effectively "close the loop" between the two decision processes.

Here we take a step forward assuming that the dynamics of the degradation process leading to a maintenance need is not always fast enough to be captured within the production scheduling horizon. This is supported by the fact that maintenance tasks in real industrial processes are often less frequent. Nonetheless, it is crucial to take them into account while generating a production schedule as their impact on plant operation can be disruptive. In order to accomplish this objective, we introduce a multi-time-scale approach capable of dealing with the short and medium-term decisions needed to satisfy the production requirements, as well as, with the medium-long term ones dealing with maintenance.

The scheduling horizon is divided into two parts; short term and long term using a discrete time-representation with different granularity (e.g. one hour and one day) to solve an integrated optimization problem. Both the short-term and the long-term models are based on the STN process representation by Shah et al. (1993).

For the short-term model, we considered the model in Biondi et al. (2015) with an enhanced formulation dealing with the asset residual life update. Each asset of the plant is assumed to be able to work in different operation modes impacting the production performance (e.g. speed, throughput…) and degradation speed. Furthermore, each asset is supposed to have a predefined residual lifetime that can be restored through maintenance tasks.

For the medium-long term model, the aggregated STN representation by Maravelias et al. (2004) is extended to account for the residual life update and maintenance requirements. Instead of the proposed single relaxed MILP model for the whole planning horizon, an aggregated model is written for each time slot, allowing not only to get a solution that is closer to feasibility, but also to optimally plan maintenance requirement whenever needed.

The proposed approach is tested on different STN process representations and is able to deal with small and medium problem sizes. Maintenance tasks are planned on plant assets whenever needed, trying to influence the production as little as possible. Similarly, the production is scheduled on the plant in order to satisfy the production requests and to reduce the total asset degradation and maintenance costs.

Acknowledgements

The Marie Curie FP7-ITN research project "ENERGY-SMARTOPS", Contract No: PITN-GA-2010-264940 is acknowledged for financial support.

References

Biondi, M., Sand, G. , Harjunkoski, I. (2015), "Tighter integration of maintenance and production in short-term scheduling of multipurpose process plants", Computer Aided Chemical Engineering, Vol 37, 1937-1942

Castro, P. M., Grossmann, I. E., Veldhuizen, P. and Esplin, D. (2014), "Optimal maintenance scheduling of a gas engine power plant using generalized disjunctive programming". AIChE J., Vol 60, 2083–2097

Dedopoulos, I. T., Shah, N. (1995), "Optimal Short-Term Scheduling of Maintenance and Production for Multipurpose Plants", Industrial Engineering and Chemistry Research, 34 (1),192–201

Hazaras, M. J., Swartz, C. L. E., Marlin, T.E. (2012), "Flexible maintenance within a continuous-time state-task network framework", Computers & Chemical Engineering, Vol 46, 167–177

Maravelias, C. T., Grossmann, I. E. (2004), "A hybrid MILP/CP decomposition approach for the continuous time scheduling of multipurpose batch plants", Computers & Chemical Engineering, Volume 28, 1921–1949

Shah, N. , Pantelides, C. C., Sargent, R.W.H. (1993), "A general algorithm for short-term scheduling of batch operations—II. Computational issues", Computers & Chemical Engineering, Volume 17, Issue 2, 229–244

427175 Adjustable Robust Optimization for Scheduling Multipurpose Batch Plants Under Uncertainty

Abstract:

In this talk we develop a novel methodological framework to take into account parameter uncertainty during the scheduling optimization of multipurpose batch plants. Parameter uncertainty is typical in such a context, where parameters such as processing times, product demands, prices and availability of resources (e.g, equipment, personnel, utilities, storage, raw materials) may not be known with certainty at the time of decision-making. Deterministic optimization approaches, which operate on nominal (expected) values of uncertain parameters and which do not explicitly take into account parameter variation, may lead to suboptimal or infeasible solutions depending on the actual realization of these parameters. One of the main frameworks for mitigating the risk associated with parameter uncertainty is Robust Optimization (RO) [1,2], which focuses on generating solutions that retain their feasibility as long as the uncertain parameters assume values (i.e., "realize") from a pre-specified uncertainty set.

The RO approach has been applied on multipurpose batch plant scheduling in a number of recent papers [3-6] with very promising results, demonstrating the suitability of the framework to address parameter uncertainty in this context. In these previous studies, the RO framework was applied in the traditional, "static" manner, which handles all of the decisions as "here-and-now"; that is, as decisions that must be made before the beginning of the time horizon when the model optimization is conducted. As a consequence of this limitation of the static RO approach, one must commit to a solution that may prove suboptimal if a favorable realization of the parameters occurs. This limitation, however, is not motivated by actual practice, where in many cases decisions can be made "last-minute" or changed within the course of the scheduling horizon. Since decisions can in reality be delayed for some later time, when valuable information from the progress of the execution of the schedule will have become available ("wait-and-see"), efficiencies can be exploited and less conservative schedules obtained.

In this work, we extend the static approach and apply the framework of Adjustable Robust Optimization (ARO) [7] in the context of batch plant scheduling. Instead of generating a single "here-and-now" decision, the ARO framework generates a family of solutions (a.k.a., a policy) which adapts to the actual realization of the uncertainty as the latter is gradually revealed during the scheduling horizon. We thus demonstrate how real-time information can be exploited without having to resort to—typically expensive and possibly suboptimal—reoptimization processes (see, e.g., "reactive scheduling") and without compromising our chosen level of insurance against risk. We also present the concept of decision-dependent uncertainty sets in order to realistically model a typical scheduling environment where assumptions about parameter correlations can only be postulated for scheduling tasks that will in fact occur (as opposed to tasks that were defined for modeling convenience but never chosen).

A comprehensive computational study is presented across literature benchmark examples from which the performance and the value of the proposed framework is assessed. Our results show that the ARO framework can obtain solutions that are less conservative than the solutions obtained by static RO and that closely approach the theoretical limit of "worst-case-deterministic" solutions. In addition to the prevailing solution method that requires a model reformulation, we explore also solution strategies based on a cutting-plane method [8] in order to further enhance computational efficiency.

[1]. Ben-Tal, A., Nemirovski, A. (1999). Robust solutions of uncertain linear programs. *Operations research letters*, 25(1), 1-13.

[2]. Bertsimas, D., Sim, M. (2004). The price of robustness. *Operations research*, 52(1), 35-53.

[3]. Vin, J. P., Ierapetritou, M. G. (2001). Robust short-term scheduling of multiproduct batch plants under demand uncertainty. *Industrial & engineering chemistry research*, 40(21), 4543-4554.

[4]. Lin, X., Janak, S. L., Floudas, C. A. (2004). A new robust optimization approach for scheduling under uncertainty: I. Bounded uncertainty. *Computers & chemical engineering*, 28(6), 1069-1085.

[5]. Janak, S. L., Lin, X., Floudas, C. A. (2007). A new robust optimization approach for scheduling under uncertainty: II. Uncertainty with known probability distribution. *Computers & chemical engineering*, 31(3), 171-195.

[6]. Li, Z., Ierapetritou, M. G. (2008). Robust optimization for process scheduling under uncertainty. *Industrial & Engineering Chemistry Research*, 47(12), 4148-4157.

[7]. Ben-Tal, A., Goryashko, A., Guslitzer, E., Nemirovski, A. (2004). Adjustable robust solutions of uncertain linear programs. *Mathematical Programming*, 99(2), 351-376.

[8]. Mutapcic, A., Boyd, S. (2009). Cutting-set methods for robust convex optimization with pessimizing oracles. *Optimization Methods & Software*, 24(3), 381-406.

431340 Robust Optimization for Production - Distribution Coordination of Industrial Gases Supply-Chains Under Uncertainty

Abstract:

An industrial gases supply-chain coordinates production and distribution decisions at multiple plants or depots while satisfying several customer demands. Plants operate air separation units, which feature high electricity consumption to produce gaseous and liquid products. Detailed distribution is required to coordinate pipeline demands for gaseous products and truck availabilities for liquid products. Multiple depots are considered as the source of the trucks, then trucks visit the plants, and finally the trucks distribute the product to the customers. In order to ensure customer storage replenishments, the product can also be purchased form alternative sources. Additionally, detailed inventory management has been developed by considering the storage tanks and the consumption profile for each day at the customer location. The replenishment is then coordinated with the production and storage levels at the production plants. At the supply-chain level, production and distribution decisions must be coordinated to fulfill several customer demands, local demands, and guarantee a safety stock in the inventory levels while minimizing the overall cost. Fixed customer demands based on forecast data or consumption profiles are usually considered, but, it is often the case that these parameters are unknown or characterized using several scenarios, upper and lower levels, probability functions, etc.

When large scale problems are studied and recourse actions are expensive or unrealistic, the robust optimization approach becomes an important tool to obtain feasible solutions for all the uncertainty sets under consideration[1]. This work further elaborates on a mixed-integer linear programming (MILP) formulation to minimize the overall cost of production and distribution of industrial gases supply-chains.[2]Given the uncertain demands we propose to characterize a closed box demand uncertainty set reformulating: i) a robust counterpart to analyze the worst case scenario; ii) the robust counterpart considering budgets of uncertainty, in which budgets of uncertainty have been proposed to compute the tradeoff between robustness and expenses of the solutions.

This work proposes two reformulations of the deterministic MILP production-distribution coordination model to consider demand uncertainty for large-scale industrial gases problems. In the first approach the deterministic model has been reformulated in to a robust counterpart to consider the worst case approach. The second approach obtains the robust counterpart considering budgets of uncertainty that constrain the number of realizations of the worst case scenario[3]. Budgets of uncertainty reduce the robustness, but they yield more realistic solutions and provide flexibility to the decision makers. The main objective is to minimize the total cost of production and distribution of liquid products under demand uncertainty.

[1] Bertsimas D., Thiele A. (2006). A Robust Optimization Approach to Inventory Theory. Operations Research. (54), pp. 150-168.

[2] Marchetti, P. A., Gupta, V., Grossmann, I. E., Cook, L., Valton, P. M., Singh, T., Li, T., and André, J. (2014). Simultaneous Production and Distribution of Industrial Gas Supply-Chains. Computer and Chemical Engineering. (69), pp. 39-58.

[3] Zhang, Q., Grossmann, IE., Heuberger, C.F., Sundaramoorthy, A., Pinto J.M. (2015). Air separation with cryogenic energy storage: Optimal scheduling considering electric energy and reserve markets. (61), pp. 1547-1558.

435888 Optimal Procurement Contract Selection with Price Optimization Under Uncertainty for Process Networks

Abstract:

Manufacturing enterprises deal with uncertainty from both internal and external sources. Internally, production variability due to unplanned events may prevent the company to achieve its demand-driven production targets. Externally, fluctuations in supply and demand as well as market economic conditions pose challenges to efficient operation of the supply chain. One way to reduce the level of uncertainty on both the supply and the customer sides, and that is typically used by companies, is by making contractual agreements (Tsay, Nahmias, & Agrawal, 1999). In the context of this paper, a contract is a binding agreement in which the seller provides the specified product and the buyer pays for it under specific terms and conditions.

A different approach to managing uncertainty is pricing analytics (Phillips, 2005; Bodea & Ferguson, 2014), also known as price optimization. In formulating such a problem, selling prices become decision variables, and the demand of a product is modeled as a function of its price. Nonetheless, this still typically does not completely eliminate uncertainty, since stochastic environmental, economic, and market conditions remain uncontrollable by the manufacturing company.

In this paper, we combine the aforementioned uncertainty management strategies in a two-stage stochastic optimization model for multi-period, multi-site tactical production planning. Uncertain parameters include raw material and finished product price and availability/demand. We propose a manufacturer-centric approach in which the purchase contract structures are set by suppliers, and it is the manufacturer's decision to select which contract, if any, to sign. In addition, the manufacturer sets the selling price of its main products that may be used to design sales contracts with its customers. However, it is the customer's decision to select the sales contracts, if any, designed by the manufacturer. The contract design problem is not addressed in this paper, but is discussed as a future work. We demonstrate how different pricing models, linear and nonlinear, can be used within the proposed optimization framework. Throughout this paper, we refer to price as *unit* price (e.g., \$/kg, \$/t etc.).

More specifically, we use three quantity-based contract models (Park et al., 2006): discount after a certain purchased amount, bulk discount, and fixed duration contracts. With regards to pricing models, we argue that general regression models can be used to describe the relationship between selling price, demand, and possibly other predictors, such as economic indicators. For illustration purposes, we consider three demand-response models (i.e., selling price as a function of demand) that are typically encountered in the literature: linear, constant-elasticity, and logit. Uncertainty is considered and modeled in both supply (e.g., raw material spot market price) and demand (random nature of the residuals of the regression models). The proposed method is illustrated with two numerical examples of chemical process networks.

References

Bodea, T., and Ferguson, M. 2014. Segmentation, Revenue Management and Pricing Analytics. Routledge – Taylor & Francis. New York, NY. USA.

Park, M.; Park, S.; Mele, F. D.; and Grossmann, I. E. 2006. Modeling of Purchase and Sales Contracts in Supply Chain Optimization. Industrial & Engineering Chemistry Research. 45(14):5013–5026.

Phillips, R. L. 2005. Pricing and Revenue Optimization. Stanford Business Books – Stanford University Press. Palo Alto, CA. USA.

Tsay, A. A.; Nahmias, S.; and Agrawal, N. 1999. Modeling Supply Chain Contracts: A Review. In Tayur, S.; Ganeshan, R.; and Magazine, M., eds., Quantitative Models for Supply Chain Management, volume 17 of International Series in Operations Research & Management Science. Springer US. 299–336.

431883 Changeover Formulations for Discrete-Time Mixed-Integer Programming Scheduling Models

Abstract:

Scheduling problems can be addressed using a wide range of methods, including heuristics and metaheuristics, special-purpose scheduling algorithms, constraint programming, and mixed-integer programming (MIP). Mixed-integer programming methods can be used to model a range of problems under a number of different types of processing constraints, and multiple objective functions; while other methods are limited in the types of problems they can solve effectively. One limitation of MIP methods, especially discrete-time, however is the solution of problems with sequence-dependent changeover times, a feature that can have a significant effect on the scheduling of chemical processes. Accordingly, the goal of this work is to develop new methods for problems with sequence-dependent changeover times. We choose to study discrete-time models because they can be readily extended to account for a number of features (e.g., linear modeling of inventory and utility costs, intermediate release and due times, etc.).

We propose five new formulations to model sequence-dependent changeovers, which can be added to any discrete-time scheduling model to address problems with different objective functions and in various production environments. The new formulations are different from each other in terms of the number of tasks that are included in each constraint, and the number of time points that each type of constraint is written for. We also propose a valid inequality for makespan minimization problems, which significantly improves the computational performance. The new formulations are compared with literature models [1, 2, 3] in terms of tightness and computational performance.

Regarding the tightness of the models, we compare the relative strength of the formulations. Specifically, we develop nine propositions regarding the strength of pairs of formulations. When neither formulation is tighter, we find points that satisfy either one but not the other. It turns out that some of the new formulations are strictly tighter than the previously proposed ones. Furthermore, we prove that one of the proposed formulations is actually facet-defining for the problem that includes unit assignment constraints and changeover constraints; e.g., for a max profit problem where a task can be executed multiple times.

In terms of computation, we study different problems (single unit, parallel units, or parallel units with unequal capacities) and different objective functions (makespan, tardiness or cost minimization), using instances of various difficulty levels. Interestingly, we find that tighter formulations do not always lead to faster solution times. Finally, we show that some of the new formulations perform better than the previously proposed ones.

[1] Kondili, E.; Pantelides, C. C.; Sargent, R. W. H. A General Algorithm for Short-Term Scheduling of Batch-Operations .1. Milp Formulation. Comput Chem Eng 1993, 17 (2), 211-227.
[2] Shah, N.; Pantelides, C. C.; Sargent, R. W. H. A General Algorithm for Short-Term Scheduling of Batch-Operations .2. Computational Issues. Comput Chem Eng 1993, 17 (2), 229-244.
[3] Wolsey, L. A. MIP modelling of changeovers in production planning and scheduling problems. Eur J Oper Res 1997, 99 (1), 154-165.

431264 A Hybrid Scheduling Formulation for Integrated Scheduling and Process Control Using Discrete-Time Dynamic Models

Abstract:

Production scheduling and process control are historically separate layers of decision making within the scope of the operation of chemical processes. However, improvements in process economics provide a strong incentive for integrating control and dynamic considerations within the production scheduling framework [1, 2]. The integration of scheduling and control is a challenging task owing to the different time scales involved in making the respective decisions, and the corresponding need to balance long-term prediction with real-time execution [3].

In our previous work [4], we proposed using a (continuous-time), low-order representation of the input-output closed-loop dynamics of the process, which we referred to as a "scale-bridging model (SBM)," to incorporate dynamic information in the scheduling formulation. In this paper, we present an extension of our results to discrete-time SBMs. We focus on time-series modeling of the closed-loop process dynamics, focusing in particular on the use of autoregressive models with exogenous inputs (ARX models) to capture the relevant process components. Based on these discrete-time scale bridging models (DSBMs) [5], we develop a hybrid discrete- and continuous-time formulation of the integrated scheduling and control problem for continuous processes.

The DSBM is included in the scheduling calculation as a constraint, thereby providing information on the role that transitions and dynamics play in the process economics. Within this framework, we introduce a novel approach for using the DSBMs to determine transition times between product grades, using a reverse integration of the output error (RIE) of the system. The resulting integrated scheduling and control framework is computationally efficient owing to the fact that the DSBM and RIE calculation are linear. We also demonstrate that the proposed data-driven framework is widely applicable in industry, and show that DSBMs can easily be derived from operating data that are routinely collected from the operation of multi-product processes.
We present a multi-product CSTR case study to demonstrate the effectiveness of our approach. For comparison purposes, we show that the optimal schedule determined using the DSBM and RIE matches the results obtained when using a rigorous process model, while exhibiting superior computational performance.

[1] M. Baldea and I. Harjunkoski, "A systematic review of the integration of production scheduling and process control," Comput. Chem. Eng., 71:377–390, 2014.

[2] I. Grossmann. Enterprise-wide optimization: A new frontier in process systems engineering. AIChE J., 51:1846–1857, 2005.

[3] C.T. Maravelias and C. Sung. Integration of production planning and scheduling: Overview, challenges and opportunities. Comput. Chem. Eng., 33(12):1919-1930, 2009.

[4] J. Du, J. Park, I. Harjunkoski, and M. Baldea, "A Time-Scale Bridging Approach for Integrating Production Scheduling and Process Control", Comp. Chem. Eng http://dx.doi.org/10.1016/j.compchemeng.2015.04.026.

[5] C. Touretzky, I. Harjunkoski, and M. Baldea, "Autoregressive Reduced Order Models for Integrated Scheduling and Control of Process Systems", Submitted to the 2015 Conference on Decision and Control.

422854 Preprocessing and Tightening Methods for Time-Indexed Mixed-Integer Programming Chemical Production Scheduling Models

Abstract:

Over the last three decades, several research efforts have been made in order to incorporate different features that appear in scheduling problems. To this end, many Mixed-Integer Programming (MIP) models have been proposed to account for storage, utilities, limited resources, changeovers, maintenance, rescheduling, among other features, in addition to the basic assignment, sizing and timing of products in processing units. However, the main drawback of most of these models is that they can only handle small- to mid-scale instances. The challenge of solving large-scale industrial instances using MIP models has remained for the most part intractable.

Recently, Maravelias and coworkers have concentrated their efforts in enhancing the solution process for existing time-indexed, discrete and continuous, MIP scheduling models for problems in network production environments. They have shown that reformulations[1,2] and tightening methods[3,4] are particularly effective in reducing computational time and improving optimality gap of existing models, even when large-scale instances are considered[5].

In this work[6], we propose a series of preprocessing algorithms for the generation of strong valid inequalities (i.e. tightening) for MIP scheduling models, based on time- and inventory-related data. In particular, we use constraint propagation techniques to calculate different parameters that are then used to bound the number of times a subset of tasks can be executed in a feasible solution. We also extend some of the propagation ideas to generate three classes of new tightening constraints.

The first algorithm we propose calculates the effective *time window* available for a given task to be processed in a particular unit. To this end we divide the algorithm into two steps: (1) calculation of the *earliest start time* (EST) of a task in a given unit, using not only time-related information but also inventory data; and (2) calculation of the *shortest tail*(ST) to the end of the horizon of every task in its compatible units. The time window is then calculated by subtracting the EST and ST from the time horizon.

Next, a forward-propagation algorithm calculates the maximum feasible amount that every task can produce within its time window and the correspondent maximum amounts of each material that can be produced within the given horizon. The propagation reveals if a task is limited either by inventory (i.e. they have excess time in their window, but no additional input materials can be produced) or time (i.e. there is enough material to continue a task execution, but the time window is exceeded).

A new algorithm is then introduced to identify subsets of dependent tasks, whose number of batches are interrelated and their production is limited by one another. We define three main types of dependent groups: (I) tasks that share a processing unit, (II) tasks that share an input material, and (III) tasks that consume material produced by the same upstream unit. The first two types define explicit limitations due to the time window for the shared unit and the cumulative production of the common input material respectively. The third type defines an implicit relation between tasks based on the effect of limited material produced by dependent tasks in the upstream unit.

Finally, a fourth algorithm is used to determine the valid inequalities associated with the three types of dependent groups, based on the calculation of the convex hull for the feasible region defined by the number of batches of dependent tasks. These inequalities limit the number of times a subset of dependent tasks can be executed in a feasible solution.

The proposed methods result in tightening constraints expressed in terms of assignment binary variables (X_{ijt}=1 if task i is assigned to start on unit j at time point t) which are present in all time-indexed MIP models, therefore they are applicable to all time-indexed models accounting for a wide range of processing features.

We also propose an extension of the ideas introduced by the time window algorithm to the case where variable EST and ST are considered. Specifically, we develop tightening constraints expressed in terms of these new variables for (1) subsequent tasks, (2) tasks consuming a common material, and (3) tasks sharing the same unit. Nonetheless, we show that the constraints derived from the sequence of preprocessing algorithms produce the best results for both discrete- and continuous-time models.

We finally present an extensive computational study that includes 27 instances of three different networks commonly studied in the literature. Our methods are applied to one discrete-time (Shah et al.[7]) and two continuous-time (Sundaramoorthy and Karimi[8], Gimenez et al.[9]). We show that significant reductions in computational time and optimality gap are achieved; in many cases, up to two orders of magnitude decrease in computational time is achieved when these constraints are applied to the discrete-time model. We also apply these methods to solve to optimality a large-scale industrial instance that could not be solved by the original discrete-time formulation.

References

1. Velez, S.; Maravelias, C.T. Reformulations and Branching Methods for Mixed-Integer Programming Chemical Production Scheduling Models. *Ind. Eng. Chem. Res.* **2013**, 52: 3832-3841.
2. Merchan, A.F.; Maravelias, C.T. Reformulations of Mixed-Integer Programming Continuous-Time Models for Chemical Production Scheduling. *Ind. Eng. Chem. Res.* **2014**, 53: 10155-10165
3. Velez, S.; Sundaramoorthy, A.; Maravelias, C.T., Valid Inequalities Based on Demand Propagation for Chemical Production Scheduling MIP Models. *AIChE J.* **2013**, 59: 872-887.
4. Merchan, A.F.; Velez, S.; Maravelias, C.T. Tightening Methods for Continuous-Time Mixed-Integer Programming Models for Chemical Production Scheduling. *AIChE J.* **2013**, 59: 4461-4467
5. Velez, S.; Merchan, A.F.; and Maravelias, C.T. On the Solution of Large-Scale MIP Scheduling Models. *Chem. Eng. Sci.* **2015**, Submitted.
6. Merchan, A.F.; Maravelias, C.T. Preprocessing and Tightening Methods for Time-Indexed MIP Chemical Production Scheduling Models. *Comp. Chem. Eng.* **2015**, Submitted.
7. Shah, N.; Pantelides, C.C.; Sargent, R.W.H.. A General Algorithm for Short-Term Scheduling of Batch-Operations .2. Computational Issues. Comp. Chem. Eng. **1993**, 17: 229-244
8. Sundaramoorthy, A.; Karimi, I.A. A simpler better slot-based continuous-time formulation for short-term scheduling in multipurpose batch plants. Comp. Chem. Eng. **2005**, 60: 2679-2702
9. Gimenez, D.M.; Henning, G.P.; Maravelias, C.T. A novel network-based continuous-time representation for process scheduling: Part I. Main concepts and mathematical formulation. Comp. Chem. Eng. **2009**, 33: 1511-1528

410982 Using Prior Knowledge for Multivariable Model Identification for Integral Controllability

Thursday, November 12, 2015: 12:30 PM
Salon E (Salt Lake Marriott Downtown at City Creek)

Shyam Panjwani, University of Houston, HOUSTON, TX and Michael Nikolaou, Chemical & Biomolecular Engineering, University of Houston, Houston, TX

Integral controllability (IC) is necessary and sufficient to achieve robustness of decoupling multivariable control with integral action (Garcia and Morari, 1985). IC refers to the fact that the steady-state gain matrix model and actual plant must satisfy the eigenvalue inequality

$$-\mathrm{Re}\left[\lambda\left(\mathbf{G}\hat{\mathbf{G}}^{-1}\right)\right] < 0 \tag{1}$$

for all plants belonging to an uncertainty set U. The design of experiments for identification of the model, satisfying integral controllability is a challenge as the identified model and actual plant must satisfy a cumbersome eigenvalue-based inequality (eqn.). To alleviate this problem, Darby and Nikolaou (2009) proposed a sufficient condition for satisfaction of eqn. for all

$$J \,\square\, \sum_{k=1}^{n} c \frac{\|\hat{\mathbf{u}}_k\|_1}{\hat{\sigma}_k} \sqrt{\hat{\mathbf{v}}_k{}^T (\mathbf{M}^T\mathbf{M})^{-1} \hat{\mathbf{v}}_k} < 1 \tag{2}$$

Where

$$\hat{\mathbf{G}} = \sum_{k=1}^{n} \hat{\sigma}_k \hat{\mathbf{u}}_k \hat{\mathbf{v}}_k^T \tag{3}$$

and the set U is based on ellipsoidal uncertainty, resulting from standard least-squares estimation:

$$U = \left\{ \mathbf{G} = \begin{bmatrix} \mathbf{g}_1^T \\ \vdots \\ \mathbf{g}_n^T \end{bmatrix} \in \Re^{n \times n} \,\middle|\, \left(\mathbf{g}_i^T - \hat{\mathbf{g}}_i^T\right) \mathbf{M}^T \mathbf{M} \left(\mathbf{g}_i - \hat{\mathbf{g}}_i\right) \le c^2 \right\} \tag{4}$$

Equations and can be used to design experiments for an integral controllable model.

While model identification relies on data used to identify the model, it is always beneficial to use a priori knowledge while building such a model. It may be difficult to capitalize on this simple realization in practice, particularly when experiments must be designed for identification of a system that is partially known and whose model must satisfy integral controllability. The objective of this study is to propose an adaptive design of experiments which uses prior knowledge about the system, to identify integral controllable model. Prior knowledge may come in a variety of forms. The present study focuses on prior knowledge expressed in terms of linear equality constraints on rows of the steady-state gain matrix as

$$\mathbf{H}_i \mathbf{g}_i = \mathbf{0} \tag{5}$$

Using eqn. , a new sufficient condition for satisfaction of eqn. was derived as

$$J \square \sum_{k=1}^{n} c \frac{\|\hat{\mathbf{u}}_k\|_1}{\hat{\sigma}_k} \sqrt{\hat{\mathbf{v}}_k^T [(\mathbf{M}^T\mathbf{M})^{-1} - (\mathbf{M}^T\mathbf{M})^{-1}\mathbf{H}_k^T \{\mathbf{H}_k (\mathbf{M}^T\mathbf{M})^{-1}\mathbf{H}_k^T\}^{-1}\mathbf{H}_k (\mathbf{M}^T\mathbf{M})^{-1}]\hat{\mathbf{v}}_k} < 1 \tag{6}$$

The adaptive design strategy proposed by (Darby, 2008a) was modified using the above condition (eqn.). In summary, a numerical solution of the problem ***minimize*** *J(Q)* was developed, where the decision variable $\mathbf{Q}$ is a Cholesky

factor of the input covariance matrix

$\mathbf{C}_m \approx \frac{1}{t-1}\mathbf{M}^T\mathbf{M}$.

The proposed modified adaptive design method was tested on a 5x5 fluidized catalytic cracking (FCC) plant (Darby, 2008b). The results obtained from this modified adaptive experiment design (using prior knowledge) were compared with the experiment design strategy proposed by (Darby, 2008a) .Figure 1 shows that the proposed adaptive design produces inputs that generate data for identification of an IC-compliant model much faster (in an order-of-magnitude shorter time) than a comparable adaptive experiment design without use of prior knowledge.

Additional results and case studies as well as future directions will be included in the presentation.

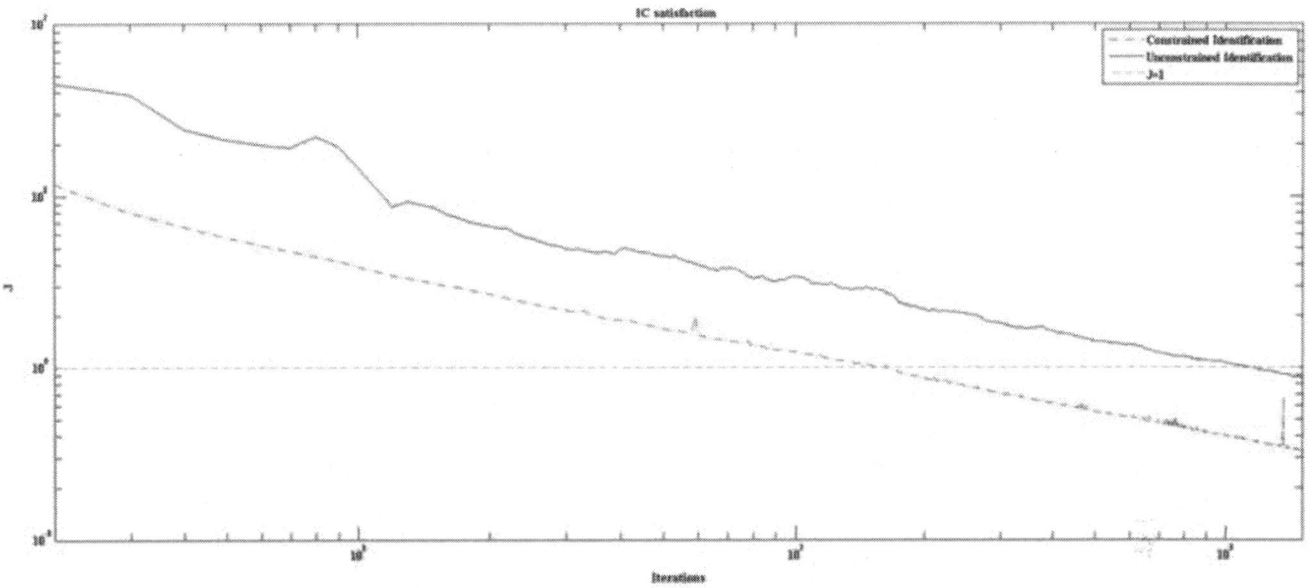

Figure 1: Satisfaction of Sufficient condition (eqn. & eqn.) in adaptive design References

Darby, M. L. (2008a). Studies of online optimization methods for experimental test design and state estimation. Chemical and Biomolecular Engineering. Houston, University of Houston. **PhD:** 152-153.

Darby, M. L. (2008b). "Studies of online optimization methods for experimental test design and state estimation. ."

Darby, M. L. and M. Nikolaou (2009). "Multivariable system identification for integral controllability." Automatica **45**(10): 2194-2204.

Garcia, C. E. and M. Morari (1985). "Internal model control: 2. design procedure for multivariate systems." Ind. Eng. Chem. Process Des. Dev. **24**: 472-484.

418360 Generalization of a Parameter Set Selection Procedure for Nonlinear Systems

Abstract:

Generalization of a Parameter Set Selection Procedure for Nonlinear Systems

Daniel Howsmon, Wei Dai, and Juergen Hahn

Systems of ordinary differential equations (ODEs) or differential algebraic equations (DAEs) are often used to describe dynamic systems, with applications in biochemical reaction networks [1], quantitative systems pharmacology [2], pharmaceutical processes [3], fermentation processes [4], and ecological systems [5] to name a few. The accuracy of these models depends on both the model structure and parameters. Whereas the structure is often chosen to best reflect the fundamental physics, chemistry, and/or biology of the system under study, the model parameters are sourced from available literature or estimated from experimental data.

To avoid over-fitting of the model to data, it is common that only a subset of the model parameters are estimated from experimental data and the remaining parameters are set to their nominal values. Parameter set selection procedures using collinearity index methods [5], genetic algorithms [6], column-pivoting methods [7], Gram-Schmidt orthogonalization methods [8], and clustering methods [9] have been proposed. All of these methods rely on local sensitivity analysis to describe the relationship between the model parameters and model output. Therefore, selecting uncertain parameters for estimation is dependent on their uncertain nominal values. This may result in suboptimal parameter sets, especially as the confidence in nominal parameter values decreases [10]. Global sensitivity analysis can overcome some of the challenges associated with local sensitivity analysis by simultaneously varying model parameters over the range of interest, directly incorporating the uncertainty in parameters. However, results from the Morris method [11], sampling-based methods [12], variance-based methods [13], and other global sensitivity analyses are difficult to interpret for the parameter set selection problem since they do not use the concept of sensitivity vectors.

To overcome the shortcomings of both local and global sensitivity analyses for parameter set selection, a procedure that uses local sensitivity analysis and dynamic optimization to simultaneously vary a large number of parameters was previously developed [14]. Briefly, a sensitivity cone containing all sensitivity vectors associated with a given parameter evaluated at different combinations of parameter values is constructed. The sensitivity cone is completely characterized by the nominal sensitivity vector and the angle between the cone surface and the nominal sensitivity vector. For a given parameter p_{i1}, this angle $_{i1}$ can be found from the following dynamic optimization problem:

$$\begin{aligned}
\min_p \quad & \cos^2(\phi_{i_1}) = \frac{\left[s_{i_1j}(p)\cdot s_{i_1j}(p_0)\right]^2}{\left\|s_{i_1j}(p)\right\|^2 \cdot \left\|s_{i_1j}(p_0)\right\|^2} \\
\text{s.t.} \quad & \frac{dx_k}{dt} = f_k(x,u,p), \qquad x_k(t_0) = x_k^0, \quad k = \{1,\dots,n\} \\
& \frac{ds_{i_1k}}{dt} = \sum_{r=1}^{n}\left(\frac{\partial f_k}{\partial x_r}\cdot s_{i_1r}\right) + \frac{\partial f_k}{\partial p_{i_1}}, \qquad s_{i_1k}(t_0) = 0, \quad k = \{1,\dots,n\} \\
& s_{i_1j}(p) = \left[s_{i_1j}(t_1)\dots s_{i_1j}(t_{end})\right]^T, \qquad p_{min} \le p \le p_{max}
\end{aligned}$$

Here, x R^n is the state vector, u R^m is the input vector, p R^ℓ is the parameter vector, and s_{ik} is the sensitivity of state k with respect to parameter i. The ODEs are discretized using collocation on finite elements to create a nonlinear programming problem. After the ℓ optimization problems are solved for a given uncertainty level, the angle between two sensitivity cones is determined as

$$\hat{\theta}_{i_1,i_2} = \max\{\theta^0_{i_1,i_2} - \phi_{i_1} - \phi_{i_2}, 0\}$$

where $\theta^0_{i_1,i_2}$ is the angle between the nominal sensitivity vectors for parameters i_1 and i_2. The cosine similarity scores

$$d_{i_1,i_2} = 1 - \cos(\hat{\theta}_{i_1,i_2})$$

are then used to cluster the parameters via a hierarchical clustering technique. However, clustering based upon such sensitivity cones may be too restrictive in cases of high nonlinearity or asymmetric uncertainty resulting in a very conservative estimate of the number of parameters to be estimated.

To overcome this limitation, a procedure that directly computes the minimum angle between groups of sensitivity vectors is developed. For each set of parameters $\{p_{i1}, p_{i2}\}$ this minimum angle can be directly found from the following dynamic optimization problem:

$$\begin{aligned}
\max_p \quad & \cos^2(\theta_{i_1,i_2}) = \frac{\left[s_{i_1j}(p)\cdot s_{i_2j}(p)\right]^2}{\left\|s_{i_1j}(p)\right\|^2 \cdot \left\|s_{i_2j}(p)\right\|^2} \\
\text{s.t.} \quad & \frac{dx_k}{dt} = f_k(x,u,p), \qquad x_k(t_0) = x_k^0, \quad k = \{1,\dots,n\} \\
& \frac{ds_{i_1k}}{dt} = \sum_{r=1}^{n}\left(\frac{\partial f_k}{\partial x_r}\cdot s_{i_1r}\right) + \frac{\partial f_k}{\partial p_{i_1}}, \qquad s_{i_1k}(t_0) = 0, \quad k = \{1,\dots,n\} \\
& \frac{ds_{i_2k}}{dt} = \sum_{r=1}^{n}\left(\frac{\partial f_k}{\partial x_r}\cdot s_{i_2r}\right) + \frac{\partial f_k}{\partial p_{i_2}}, \qquad s_{i_2k}(t_0) = 0, \quad k = \{1,\dots,n\} \\
& s_{i,j}(p) = \left[s_{i_1j}(t_1)\dots s_{i_1j}(t_{end})\right]^T, \qquad p_{min} \le p \le p_{max}
\end{aligned}$$

The angles computed from the solution to these optimization problems are then directly used in hierarchical clustering. This procedure is more flexible and can more accurately determine the minimum distance between groups of sensitivity vectors resulting in a more realistic number and selection of parameters for estimation. The only drawback is that the procedure is more computationally expensive as the sensitivity cone approach requires solutions to ℓ optimization problems and the presented technique requires solutions to $\ell(\ell - 1)/2$ optimization problems with 1.5x more optimization variables. However, this downside is not significant for the number of parameters that are typically found in problems under investigation. The work includes a comparison with the parameter cone approach, highlighting scenarios where the increased computational cost could be justified.

References

[1] S. Maiti, W. Dai, R. C. Alaniz, J. Hahn, and A. Jayaraman, "Mathematical Modeling of Pro- and Anti-Inflammatory Signaling in Macrophages," *Processes*, vol. 3, no. 1, pp. 1–18, Dec. 2014.

[2] O. Demin, T. Karelina, D. Svetlichniy, E. Metelkin, G. Speshilov, O. D. Jr, D. Fairman, P. H. van der Graaf, and B. M. Agoram, "Systems Pharmacology Models Can Be Used to Understand Complex Pharmacokinetic-Pharmacodynamic Behavior: An Example Using 5-Lipoxygenase Inhibitors," *CPT Pharmacomet. Syst. Pharmacol.*, vol. 2, no. 9, p. e74, Sep. 2013.

[3] K. V. Gernaey and R. Gani, "A model-based systems approach to pharmaceutical product-process design and analysis," *Chem. Eng. Sci.*, vol. 65, no. 21, pp. 5757–5769, Nov. 2010.

[4] W. Dai, D. P. Word, and J. Hahn, "Modeling and dynamic optimization of fuel-grade ethanol fermentation using fed-batch process," *Control Eng. Pract.*, vol. 22, pp. 231–241, Jan. 2014.

[5] R. Brun, P. Reichert, and H. R. Künsch, "Practical identifiability analysis of large environmental simulation models," *Water Resour. Res.*, vol. 37, no. 4, pp. 1015–1030, Apr. 2001.

[6] Y. Chu and J. Hahn, "Integrating parameter selection with experimental design under uncertainty for nonlinear dynamic systems," *AIChE J.*, vol. 54, no. 9, pp. 2310–2320, Sep. 2008.

[7] M. Velez-Reyes and G. C. Verghese, "Subset selection in identification, and application to speed and parameter estimation for induction machines," in , *Proceedings of the 4th IEEE Conference on Control Applications, 1995*, 1995, pp. 991–997.

[8] B. F. Lund and B. A. Foss, "Parameter ranking by orthogonalization—Applied to nonlinear mechanistic models," *Automatica*, vol. 44, no. 1, pp. 278–281, Jan. 2008.

[9] Y. Chu and J. Hahn, "Parameter Set Selection via Clustering of Parameters into Pairwise Indistinguishable Groups of Parameters," *Ind. Eng. Chem. Res.*, vol. 48, no. 13, pp. 6000–6009, Jul. 2009.

[10] Y. Chu and J. Hahn, "Quantitative Optimal Experimental Design Using Global Sensitivity Analysis via Quasi-Linearization," *Ind. Eng. Chem. Res.*, vol. 49, no. 17, pp. 7782–7794, Sep. 2010.

[11] M. D. Morris, "Factorial Sampling Plans for Preliminary Computational Experiments," *Technometrics*, vol. 33, no. 2, pp. 161–174, May 1991.

[12] G. M. (Univ of V. Hornberger and R. C. Spear, "Approach to the Preliminary Analysis of Environmental Systems," *J Env. Manage*, vol. 12, no. 1, pp. 7–18, 1981.

[13] R. W. Atherton, R. B. Schainker, and E. R. Ducot, "On the statistical sensitivity analysis of models for chemical kinetics," *AIChE J.*, vol. 21, no. 3, pp. 441–448, May 1975.

[14] W. Dai, L. Bansal, J. Hahn, and D. Word, "Parameter set selection for dynamic systems under uncertainty via dynamic optimization and hierarchical clustering," *AIChE J.*, vol. 60, no. 1, pp. 181–192, Jan. 2014.

420834 A Robust Data-Driven Reduced-Order Model for Real-Time Optimization of Steam-Methane Reformers Under Distributed Sensing

Abstract:

Large quantities of hydrogen are consumed in refineries and for production of important chemicals such as ammonia and methanol. Declining crude-oil quality and increased fertilizer demands, among others, have led to further increase in hydrogen demand. A significant portion (~80 %) of industrial hydrogen consumption is met via natural-gas steam methane reforming. This process takes place in a large scale, high-temperature, and highly energy-intensive unit called a steam methane reformer (SMR), where endothermic reforming reactions are carried out in hundreds of catalyst-filled tubes placed in a gas-fired furnace. A typical modern hydrogen production plant consumes a substantial amount (10^5 GJ) of natural gas per day. The overall productivity (energy consumed per unit H_2 produced) of the plant is strongly dependent on how efficiently the SMR is operated, which further depends on the spatial temperature distribution inside the furnace, where a more uniform distribution paves the way for reduced plant-wide energy use.

Rigorous SMR models, such as computational fluid dynamics (CFD) models, can provide accurate temperature distribution information. However, such models are not suitable for real-time, on-line use. In our previous work (Kumar et al., 2015), we demonstrated the feasibility of using data-driven empirical models to optimize fuel distribution to the furnace burners, and achieve a uniform furnace temperature profile. A reduced-order linear model was used to modulate the fuel distribution among the burners in a representative smaller-scale SMR system. It was shown that a reduced-order empirical model with much lower computational requirements, when developed using sufficiently rich data, can be a viable substitute to the detailed modeling of the complex thermal and flow interactions in the furnace.

In this work, the reduced-order modeling approach is extended to the full-scale SMR system. Specifically, we propose a novel data-driven reduced-order model, termed as egg-crate SMR (EC-SMR) model, for the SMR furnace. The EC-SMR model predicts the variations in the furnace temperature profile upon adjustments in the openings of the fuel-stream valves. The resultant redistribution of fuel among the burners upon valve adjustments is explicitly considered. The localized impact due to change in fuel-throughput of an individual burner is modeled as exponentially decaying function and the overall perturbation in the temperature distribution is approximated as a combined effect of local impacts from all the burners. We show that the model is robust to noise in data and overcomes several other shortcomings of previously published models. In particular, the fuel-redistribution based formulation leads to spatially 'smooth' predictions, unlike the PLS model predictions, of temperature variations, and allows for easy incorporation of nonlinear interactions, if desired. Being computationally parsimonious, the model is amenable for online optimization of the furnace temperature distribution. Comparisons of model predictions with actual plant measurements (obtained from an array of infrared cameras) show the excellent performance of the EC-SMR model in capturing the redistribution effects, which is crucial for elimination of temperature variability in the furnace at the level of valve manipulations. Subsequently, a case study is presented, whereby the EC-SMR model is employed in an optimization framework with the goal of minimizing temperature variability within the furnace.

References:

Kumar, A., Baldea, M., Edgar, T. F., & Ezekoye, O. A. (2015). Smart Manufacturing Approach for Efficient Operation of Industrial Steam-Methane Reformers. *Industrial & Engineering Chemistry Research, 54* (16), 4360-4370.

421231 On Identification of Well-Conditioned Nonlinear Systems: Application to Economic Model Predictive Control of Nonlinear Processes

Abstract:

Despite the increasing availability of reasonably accurate first-principles models for broad classes of chemical processes, there are still a large number of processes that involve physico-chemical phenomena that remain poorly characterized, and thus, they cannot be accurately modeled by first-principles models. The implication of this is that recent developments in model predictive control, and specifically in the profit-optimizing strategy of economic model predictive control (EMPC) (e.g., [1]-[3]), may not be readily applied to important chemical processes. One method for overcoming this issue is through the use of linear empirical models (e.g., [4]), which are widely used in industry and have been addressed in the context of EMPC in [5]. However, to more greatly optimize the economic objective function used in EMPC, it is desirable to take process nonlinearities into account in the process model. Nonlinear process models typically have no analytic solution, however, so they can only be solved within an EMPC optimization problem by using numerical integration. Nonlinear empirical models that are identified by the polynomial nonlinear state-space (PNLSS) ([6]) process identification methodology are not guaranteed to produce an empirical model that is well-conditioned such that it can be used for model predictive control.

This work focuses on economic model predictive control (EMPC) of nonlinear processes using well-conditioned polynomial nonlinear state-space models, which has not been addressed in the literature previously. Specifically, we initially address the development of a nonlinear system identification technique for a broad class of nonlinear processes which leads to the construction of polynomial nonlinear state-space dynamic models which are well-conditioned over a broad region of process operation in the sense that they can be correctly integrated in real-time using explicit numerical integration methods via time steps that are significantly larger than the ones required by nonlinear state-space models identified via existing techniques. Working within the framework of polynomial nonlinear state-space models, additional constraints are imposed in the identification procedure to ensure well-conditioning of the identified nonlinear dynamic models. This development is key because it enables the design of Lyapunov-based EMPC (LEMPC) ([2]) systems for nonlinear processes using the well-conditioned nonlinear models that can readily be implemented in real-time owing to the reduced computational burden required to compute the control actions within the process sampling time. Sufficient conditions are derived for closed-loop stability under the LEMPC scheme based on the well-conditioned nonlinear models when these models provide a sufficient degree of accuracy in the region where time-varying economically optimal operation is considered. Finally, both the system identification and LEMPC design techniques are applied to a classical chemical process example and significant advantages in terms of computation time reduction in LEMPC calculations are demonstrated.

[1] Angeli D, Amrit R, Rawlings JB. On average performance and stability of economic model predictive control. IEEE Transactions on Automatic Control. 2012;57:1615-1626.
[2] Heidarinejad M, Liu J, Christofides PD. Economic model predictive control of nonlinear process systems using Lyapunov techniques. AIChE Journal. 2012;58:855-870.
[3] Ellis M, Durand H, Christofides PD. A tutorial review of economic model predictive control methods. Journal of Process Control. 2014;24:1156-1178.
[4] Favoreel W, De Moor B, Van Overschee P. Subspace state space system identification for industrial processes. Journal of Process Control. 2000;10,149-155.
[5] Alanqar A, Ellis M, Christofides PD. Economic model predictive control of nonlinear process systems using empirical models. AIChE Journal. 2015;61:816-830.
[6] Paduart J, Lauwers L, Swevers J, Smolders K, Schoukens J, Pintelon R. Identification of nonlinear systems using polynomial nonlinear state space models. Automatica. 2010;46:647-656.

409898 Validation and Sensitivity Analysis of a Continuous Lumping Model for Hydrocracking on a Zeolite Catalyst

Abstract:

Validation and Sensitivity Analysis of a Continuous Lumping model for Hydrocracking on a Zeolite Catalyst

Per Julian Becker[1], Benoit Celse[1], Victor Costa[1], and Denis Guillaume[1]*

1. IFP Energies Nouvelles, Rond-Point de l'Echangeur de Solaize, 69360 Solaize, France

**E-mail: benoit.celse@ifpen.fr*

HIGHLIGHTS

- Validation of a Continuous Lumping Model.
- Parameter fitting on 57 experimental runs.
- Good prediction of conversion (370+ cut) and yield structure.
- Improved performance compared to correlative simulator.
- Sensitivity study of operating conditions and feed characteristics.

INTRODUCTION

Catalytic hydrocracking is used to convert heavy Vacuum Gas Oil (VGO) residue to more valuable middle distillate (150-370°C) and/or naphta (>150°C) cuts. A robust kinetic model allows the optimal process design and operating conditions to be chosen to maximize the desired cuts and product characteristics. Hydrocracking of VGO residue performed in a two-step process: 1) a hydrotreatment step in reactor R1, which serves mainly to remove nitrogen and sulfur from the feed; 2) a hydrocracking step in the second reactor (R2), which performs the main hydrocracking on a zeolite catalyst.

A continuous lumping model for R2 is proposed and validated in this work. This approach groups the individual components of the hydrocarbon mixture according to a continuous distribution with normalized boiling point $C(\theta,t)$. The cracking kinetics are governed by a continuous reactivity function $f(\theta)$, and a yield distribution $g(\theta,\theta^*)$. The inhibition of the zeolite catalyst due to residual organic nitrogen and ppNH3 are taken into account by I_{HDN}.

MATERIALS & METHODS

A total of 12 model parameters are adjusted using 57 experimental runs in a pilot plant, with different Iranian/Arabian feeds. About half the feeds were used in a two-step process (R1+R2) where gasses (H_2S, NH_3, and hydrocarbon C1-C4) are carried over from the first reactor. The remainder of the tests used pre-treated feeds, where the $ppNH_3$ was simulated by adding an equivalent amount of aniline additive. Feed characteristics (SIMDIS, sulfur/nitrogen content, aromatic carbon content etc.) for the feed of the second reactor were known. Temperature was varied between 360 – 400°C, and liquid hourly space velocities (lhsv), which corresponds to the inverse of contact time, was varied between 1.0 – 3.5 h^{-1}. Experiments were performed at 100 and 140 bar, with a $ppNH_3$ between 0.1 – 1.5 bar and a H_2/HC of 600 – 1400 l/l. An additional set of 15 experimental runs was used to validate the model. These tests used similar feeds and operating conditions as the ones used in the calibration database. Parameter identification was performed on a supercomputer, using a set of 96 randomized initial parameters, in order to ensure convergence of the optimizer to a global, rather than a local, minimum.

RESULTS & DISCUSSIONS

The comparison with a correlative model (with more than 30 parameters) for the experimental validation database (15 points) on conversion and yield was carried out. It shows that the continuous lumping model provides better simulation results. The parity graphs show that the validation points are well predicted, even for relatively high values of lhsv (>3 h^{-1}).

A sensitivity study allows to further validate the model. The influence of variations of the most important operating conditions on the model predictions for the conversion X_{370+} is quantified. The operating conditions studied are : 1) lhsv (liquid hour space velocity), 2) H2/HC, 3) the nitrogen content of the feed of R1 (NR1), 4) the residual nitrogen content of the feed of R2 (NR2). The shifts in ΔT (to obtain the right conversion), shown below, are within the ranges expected from experimental observations. More global sensitivity analyses using Sobol index may be carried out.

REFERENCES

[1] PJ Becker, B Celse, D Guillaume, H Dulot, V Costa, Hydrotreatment modeling for a variety of VGO feedstocks: a Continuous Lumping Approach, Fuel 139 (2015) 133-143

430607 A Meal Detection Algorithm Based on Continuous Glucose Measurements for Use in Artificial Pancreas Systems

Abstract:

A Meal Detection Algorithm Based on Continuous Glucose Measurements for use in Artificial Pancreas Systems

Sediqeh Samadi 1, Kamuran Turksoy 2, Jianyuan Feng1,Iman Hajizadeh1, Ali Cinar1, 2

1Department of Chemical and Biological Engineering, Illinois Institute of Technology, 10 West 33rd St., Chicago, IL 60616

2Department of Biomedical Engineering, Illinois Institute of Technology, 3255 S. Dearborn St., Chicago, IL 60616

Analysis of the process data and identification of temporal features caused by various underlying changes in process conditions are essential tasks in engineering problems such as detection, diagnosis and control problems. Historical process data presents information about the effect of the major process changes on observed variables. Modeling the effects of changes either quantitatively or qualitatively is useful for process simulation as well as real time change detection and identification. Once the changes are identified, the controller can make the appropriate remedial actions. In this work, trend analysis and qualitative model development are used to detect the effects of meals for use in artificial pancreas systems (AP). AP systems offer an important alternative for automating the regulation of blood glucose concentration (BGC) of patients with Type 1 Diabetes (T1D). But AP systems may infuse too much insulin that can cause hypoglycemia. Accurate closed-loop control is essential for developing AP systems that adjusts insulin infusion rates of insulin pumps and a multivariable AP developed at IIT was successful in reducing the number of hypoglycemic episodes [1,2]. Various known disturbances such as meals affect the BGC. Traditionally an insulin bolus is computed manually to balance the effects of a meal on BGC and administrated. However, the elimination of manual inputs to AP systems would be very appealing to patients. In AP systems, one of the biggest challenges is automatically handling a meal's effect that causes deviation of BGC from control target range. An AP control system equipped with a meal detection module can provide accurate predictions of change in BGC for use in control decisions. The current work focuses on the development of a meal detection algorithm that can be incorporated to AP control systems. In the meal prediction system designed, frequently measured subcutaneous glucose concentration from a Continuous Glucose Monitoring (CGM) system provides data both to the controller and the meal detection module. The proposed meal detection algorithm has two main steps. In the first step, the meal's effect on BGC variation is modeled by some qualitative variables. To do that qualitative trend analysis (QTA) method transforms glucose concentration time series to a sequence of non-overlapped segments. To each segment a member of a small set of qualitative variables is assigned. Each qualitative variable has a different combination of first and second derivative signs. The sequence of segments labeled with such variables called qualitative representation of CGM signal gives valuable information about glucose variation. A single qualitative variable and also a sequence of them may indicate the signs of meal's effect on glucose change and they are the essence of proposed detection algorithm. In the second step, a multi-stage clustering method separates the false detections from the true ones. The method is tested on real patients' data. The meals are detected before a remarkable increase in glucose concentration takes place.

The average increase in BGC between the beginning of the meal (known in the experiments conducted) and the time it is detected varies within the range of 10-30 mg/dl for different patients. The largest detection change was observed to be 50 mg/dl, which belongs to a patient whose change in blood glucose is rapid. Less than 10% of test meals (snacks) are not detected, and almost all are snacks. The number of false positive detection alarms is in range of 1-2 for different patients within around 50 hours experimental data. The impact of the proposed automated meal detection technique will be illustrated by including a module in our AP and comparing the characteristics of BGC profiles under closed-loop control with and without the automated meal detection module.

[1] Turksoy K, Bayrak E.S, Quinn L, Littlejohn E, Cinar A. Multivariable Adaptive Closed-Loop Control of an Artificial Pancreas Without Meal and Activity Announcement. Diabetes Technology & Therapeutics 2013. 15(5), 386-400.

[2] Turksoy K, Quinn L, Littlejohn E, Cinar A. An Integrated Multivariable Artificial pancreas Control System. Journal of Diabetes Science and Technology 2014. 8(3), 498–507.

426483 A Novel Insulin Delivery Approach Capable of Substantially Tight Variation of Blood Glucose Level

Abstract:

The potential for successful automatic insulin delivery has entered a new era due to recent technological advancements of insulin pumps and blood glucose sensors. However, for full automation and control twenty four hours a day/seven days a week (24/7) the control algorithm must be capable of tight control for major disturbances such as meals, activity and stress. Theoretically, the superiority of feedforward control (FFC) over all other control systems is that corrective action can be taken proactively to completely eliminate the effects of measured disturbances on the control variable. Thus, FFC has the potential to completely cancel the effect of any disturbance that it models. More specifically, in this FFC approach the objective is the development of an automatic insulin delivery system (AIDS) that nullifies the effects of eating, activity and stress. Accomplishing this goal will substantially reduce the variation in BGC.

While FFC has great potential, it is quite difficult to accomplish. From a search in the literature, FFC has seen very limited application in real processes in general and our work appears to be the only AIDS approach. Successful FFC starts with the ability to obtain an accurate causative model for inputs only. This is quite difficult to achieve for real data, especially in the use of free-living data, because of high cross correlation of inputs and changing levels of unmeasured disturbances. Our work takes a unique approach in the use of transfer function dynamic structures with physically interpretable parameters that are highly nonlinear in the models. Through published work were have demonstrated the superiority of our modeling approach causatively over all other methods theoretically as well as experimentally. In addition, our approach is quite efficient in capturing the variation in the response due to the measured inputs with large variations in unmeasured inputs, even those highly correlated with measured inputs that are modeled.

Even with an accurate FFC model, an effective FFC algorithm is needed for its implementation into a control system. Traditional FFC decomposes each input into a separate FFC and adds the output from each controller to obtain controller output or in this case the insulin rate to compensate for the model input changes. To obtain this value the model must be linearized and the final equations can be a highly complex high order differential equation. This work will present a simple control law that does not decompose or linearize the model. Thus, it keeps all the properties of the FFC model in its determination of FFC output.

In this work, 11 cases of two weeks of free living data collection on subjects with type 1 diabetes is modeled. The study was done in two parts. The first part takes a Wiener modeling approach which allows each input to have separate dynamic behavior. However, while the Wiener network (WN) is an excellent choice for the input transfer function model, in the response space, it is limited in representing the interaction of insulin and glucose in the blood. This drawback limits the ability to develop an accurate fit for insulin infusion rate, which is critical to controller performance. However, to model blood insulin and glucose interaction, one needs BIC at the sampling rate of BGC (i.e., every five minutes with the glucose sensor used in our studies). There is no such sensor in existence currently. To circumvent this need, we developed a semi-coupled network (SCN) for BIC and BGC that was inspired by the compartment modeling work of researchers in the literature. This led to the development of an unmeasured "pseudo" BIC variable. To our knowledge, this is the first use of such a variable in this context of subject-specific clinical modeling of BGC consisting only of inputs in free-living data collection. This novel change made a substantial advancement in obtaining a better phenomenological model and in providing a dynamic relationship between insulin infusion rate, consumed nutrients, and BGC.

This work consists of 13 input variables of three (3) food nutrients, seven (7) activity variables including one to measure stress, 24-hour clock time for circadian rhythm, and two (2) for insulin infusion. The activity variables were collected using the SenseWear® Pro3 Body Monitoring System (BodyMedia Inc., Pittsburgh, Pennsylvania), which is worn on the triceps of the subject's arm.

For a FFC model, the premier measure of performance is the correlation of the measured BGC and the fitted BGC (r_{fit}). The results of splitting the data sets into 1 week of training, 4 days of validation, and 3 days of testing are reported here. The average r_{fit} values of the 11 cases for the WN for training, validation and testing were 0.58, 0.59, and 0.59, respectively. The average r_{fit} values of the 11 cases for the \SCN for training, validation and testing were 0.64, 0.63, and 0.61, respectively. The closeness of the values for each of the three sets is an indication that the models did not significantly over-fit the data on unmeasured disturbances. Given that the values for each approach are close, this is an indication that both methods explained about the same amount of variation. While the SCN values are slightly higher they are not substantially higher, which seems to suggest that about all the variations due to the measured inputs were captured by both models. Thus, given this result, the superiority of the SCN is not revealed by r_{fit} but in its ability to capture or reveal true static and dynamic behavior.

Our static evaluation consisted of examining the steady state relationships of carbohydrates (x_C), insulin infusion rate (x_I) and BGC level (G). This evaluation consisted of using the SCN results of one subject and plotting x_I versus x_C with lines of constant G. First, the numbers for these variables were in the ranges used by the subject. Secondly, for a constant value of x_C (e.g., 60 gms), as G decreased (e.g. from 300 to 100 mg/dL), x_I increased indicating that it would take more insulin to maintain a lower BGC, which makes phenomenological sense.

The dynamic evaluation consisted of setting x_C to zero or setting x_I to zero for a selected subject's model. With x_C set to zero, the WN BGC shifted down slightly but the SCN BGC dropped to zero, as it should. Conversely, with x_I set to zero, the WN BGC shifted up slightly but the SCN BGC continued to rise over time, as it should. Thus, these static and dynamic evaluations support the SCN over the WN on grounds of phenomenological soundness.

An r_{fit} value around 0.6 may not seem very high, but this value does not need to be extremely high to make a significant difference in tightening glucose control. This amount may represent nearly all the variation in the response that is due to the set of inputs models. In this case the model did an exceptional job in capturing all the variation that could be captured and whatever this level is, the FFC has the potential to nullify all the measured inputs.

To understand the potential impact of a particular FFC model in this work we use its fit to illustrate how the variation would be reduced under the proposed FFC approach. More specifically, the variability is substantially reduced, with the standard deviation dropping from 73.9 to 8.7 mg/dL, a decrease of 88.3%. This improvement results in a level of variation similar to a person without diabetes. Thus, this AIDS has the potential to control BGC as well as the body does for people without diabetes.

Key words: Artificial Pancreas, Insulin Delivery System

420778 Sensor Error Detection and Functional Redundancy for Continuous Glucose Monitoring

Abstract:

Continuous glucose monitors (CGM) can help diabetes patients monitor their blood glucose concentration (BGC) with higher frequency and less manual effort compared to finger sticks. It receives information from a subcutaneous glucose sensor wirelessly. The sensor and CGM are also critical components of an artificial pancreas (AP) which relies on real time glucose concentration information as an input to calculate appropriate insulin infusion rates for patients with type 1 diabetes (T1D). The accuracy of glucose concentration provided by CGM become critical for glucose level management with AP systems. For most of available CGM sensors in the market, the glucose reading is calculated using the current signal is generated by the multilayer reaction in interstitial fluid. This current signal, considered as raw signal, need to be calculated by a predesigned filter to convert it to glucose concentration. However, the current signal may disturbed by local glucose fluctuations, the filter may not be accurate enough to convert the row signal to glucose concentration and the sensor signal may affected by interference from other devices or noise when communicating the data to the AP . As a result, outliers, missing data and large signal noise will make the CGM signal unreliable and cause the AP control systems to suggest incorrect insulin infusion flow rates that will drive patients to hypoglycemia or hyperglycemia.

In order to detect errors in CGM sensor signal, and replace detected erroneous signal with model based signal, two techniques, robust Kalman filter (RKF) and locally weighted partial least squares (LW-PLS), are introduced to build a hybrid CGM signal fault detection and redundancy analysis system. This innovative hybrid system leverages the advantages of automatic measurement error elimination with RKF and database supported prediction with LW-PLS. The system is designed with varying thresholds to distinguish between signal faults and real dynamic changes so that the number of erroneous fault detections can be reduced. This system is tested with clinical data from T1D patients. Data from 10 clinical experiments were analyzed, each experiment lasting for 56 hours. More than 500 CGM sensor errors were added to raw CGM data for detection and analysis. The results indicate that the proposed system can successfully detect most of the erroneous CGM signals and substitute with values computed by functional redundancy generated with the hybrid system.

431574 Error Detection for Chemical Plant Automation Logic Using Model Checking and Supervisory Control Theory

Abstract:

Analyzing the automation logic in a modern chemical plant requires systematic tools, due to the complexity of the discrete logic itself, as well as its interaction with the continuous process dynamics. The interaction between the discrete dynamics of the automation system and the continuous dynamics of the chemical process results in a hybrid system. Such systems are notoriously difficult to analyze due to the rapid explosion of the number of discrete states and the need to integrate differential equations to track the evolution of the continuous variables between discrete transitions. The number of discrete states in a typical industrial system may be 10^{100} or more.

The purpose of this work is to examine the application of a recently-developed method for detecting errors in the discrete logic of hybrid systems. We propose to apply supervisory control theory to reduce the error-detection problem for hybrid systems to a tractable discrete problem. The class of errors that this approach can detect includes the reachability of forbidden/undesirable discrete states, or situations in which a particular piece of equipment is prevented from ever switching on.

In the current paper we formulate the error detection problem as a series of computation tree logic (CTL) specifications of the form $AG(p_1 \wedge EF(p_2))$, which must be satisfied for the system to be free of errors. This type of specification mirrors the classic controllability and non-blocking requirement of supervisory control theory, which was developed by Ramadge and Wonham for discrete event systems. Symbolic model checking (which has previously been applied to verify discrete logic in chemical plant automation systems) is used at the core of our error-detection algorithm, along with the aforementioned supervisory control theory, to check if the specification can be satisfied. The combination of these techniques is novel in that it expands the class of logical errors which can be detected in hybrid systems without explicitly evaluating the continuous dynamics.

The method we develop in this research relies on conservative approximation of the system dynamics to ensure that an error is reported only if it is guaranteed to violate one of the specifications. We provide both process-independent tests, which should apply to any control system, and process-dependent tests that address specific aspects of a particular plant's operation. All of the specifications presented in this work are of the form mentioned previously, and therefore are compatible with the methods we develop. A sample problem, motivated by an industrial collaboration with The Dow Chemical Company, is included to demonstrate the application of the presented methods.

429704 Quality-Relevant Fault Detection and Identification for Batch Processes Based on Stochastic Programming

Abstract:

Batch processes play an important role in modern industry, producing high-value products such as food, plastics, pharmaceuticals, and semiconductors [1]. Since the products of batch processes often require extremely high quality, it is necessary to develop quality-relevant monitoring methods for process safety and quality enhancement [2]. Over the past years, various methods have been proposed to solve the problem using multivariate statistical monitoring methods, such as multiway partial least squares [3] and other output related methods [4-6]. However, most of these methods are based on the assumption that batch-to-batch variations are not obvious, which means that quality-relevant variables in each historical batch are identical and follow a similar trajectory. With the increased complexity and integrated nature of modern industrial processes, the quality variables of different batches may behave rather differently, primarily due to the stochastic nature of process parameters and other process uncertainties. Based on existing methods, the historical quality-relevant variables are calculated directly on the basis of historical process variables and mechanistic models with fixed coefficients. Thus, the quality-relevant data collected during actual operation may not quite agree with the historical monitoring model. When faced with the problem of online batch process monitoring, the stochastic features of different batch runs need to be taken into consideration.

To address this challenge, we propose a batch process monitoring method based on stochastic programming, which can make the monitoring model more robust and reliable. Based on historical process data, quality-relevant variables are generated with stochastic model coefficients and other model uncertainties, such as varying set-points and sensor noise. Then the stochastic programming model is introduced to classify historical batches into different scenarios by solving a constrained optimization problem [7,8]. According to this model, all batches are considered normal, and similar batches are integrated into the same scenario. A discrete probability density function is obtained to reveal the probability of each scenario. Then a mean trajectory is extracted from the quality-relevant variables of different scenarios and the residuals of each scenario can be obtained at the same time. For monitoring purposes, the control limit of the historical data is developed by the stochastic residuals. Once the residuals of online data are obtained, it can be compared with the control limit to detect quality-relevant faults. Following detection, this method can achieve online model updating by integrating a new normal batch into historical scenarios. Meanwhile, the information of faulty batches is extracted as well, including the abnormal residuals and the exact detection time to help fault classification and identification. The proposed approach is illustrated using a simulated penicillin fed-batch fermentation process.

References:

[1] F. Shen and Z. Song, "Multivariate trajectory-based local monitoring method for multiphase batch processes, *Ind. Eng. Chem. Res.*, **54**, 1313-1325, 2015.

[2] W. Sun, Y. Meng, A. Palazoglu, J. Zhao, H. Zhang and J. Zhang "A method for multiphase batch process monitoring based on auto phase identification," *J. Process Control*, **21**, 627-638, 2011.

[3] P. Nomikos and J.F. MacGregor, "Multi-way partial least squares in monitoring batch processes," Chemom. Intell. Lab. Syst., **30**, 97−108, 1995.

[4] Z. Ge, Z. Song and F. Gao, "Review of recent research on data-based process monitoring," *Ind. Eng. Chem. Res.*, **52**, 3543-3562, 2013.

[5] N. Lu and F. Gao, "Stage-based process analysis and quality prediction for batch processes," *Ind. Eng. Chem. Res.*, **44**, 3547-3555, 2005.

[6] C. Zhao, F. Wang, Z. Mao, N. Lu, and M. Jia, "Improved batch process monitoring and quality prediction based on multiphase statistical analysis," *Ind. Eng. Chem. Res., 47,* 835-849, 2008.

[7] J.R. Birge and F. Louveaux. *Introduction to stochastic programming*. Springer-Verlag, New York, 2011.

[8] C. Liu, Y. Fan and F. Ordonez, " A two-stage stochastic programming model for transportation network protection," *Comput. Oper. Res.*, **36**, 1582-1590,2009.

432192 Development of a Novel Ekf-GA Approach for Distributed Sensor Placement – Application to WGSR in an IGCC Plant

Abstract:

The efficiency of a process is highly affected by disturbances and faults, which in extreme cases may lead to unsafe conditions or to process shutdown and hence need to be detected early for preventive actions or maintenance planning. However, disturbances and faults may not be available for measurement directly and therefore need to be estimated using available measurements and the underlying model of the process. The number, type and location of measurements (termed as measurement model) will influence the accuracy of the estimate and hence careful consideration needs to be done for selection of measurements. In this work, we will describe an approach for synthesizing an optimal measurement model and its application to the water gas shift reactor (WGSR) as part of an integrated gasification combined cycle (IGCC) plant. The WGSR is one of the important units in an IGCC plant for satisfying power production and carbon capture goals. A first-principles model of the WGSR has been developed in our previous work where the reactor is described by a system of non-linear differential and algebraic equations (DAE). The design of the measurement model is done by using the Genetic algorithm (GA) and extended Kalman filter (EKF).

An enhancement to the existing EKF that not only handles the uncertainties in the differential equations, but also considers the presence of uncertainties in algebraic equations while satisfying other state constraints will be described. The GA is used to search the large number of possible combinations of sensors while the EKF is used to estimate the values of the desired variables for each set of sensors that is considered by the GA. However, this approach is cumbersome due to the vast search space and computational burden for nonlinear state estimation. To address the latter issue, reduction in computational burden is studied by developing simplified WGSR models using scaling analysis. For reducing the computational burden further, in-situ adaptive tabulation (ISAT), a storage and retrieval technique, is used. The reduction in computational burden and the accuracy of the resulting measurement model are compared against the use of the original model.

431009 Built-in-Tests for Thermal Fluid Systems of Aerospace Applications

Abstract:

The reliability, cost, safety and environmental impact of complex and uncertain engineering systems that are prone to faults, drives fault detection and isolation (FDI) methodologies to the forefront. More specifically, the efficiency and accuracy of these FDI methods are increasingly important among competitive industries, such as aerospace, where system downtime can have a significant impact on profit. Advancement of FDI methods can increase system reliability, availability, and safety through more accurate fault diagnosis. Model-based FDI methods entail advantages over hardware redundant or data-driven approaches in their absence of need for additional hardware, as well as the increased accuracy over wide operating regions, due to their use of first principles and/or empirical correlations.[1,2] Recent advances in model-based FDI methods are applicable to nonlinear systems of thermal fluid dynamic processes, such as aircraft environmental control systems (ECS). For instance, looking at the example of an aircraft ECS, significant fouling can occur over time within its primary heat exchanger that can lead to performance degradation and abrupt downstream faults. Current monitoring techniques, like Kalman filtering or observer-based methods, which utilize statistical inferences, such as cumulative sum tests, are incapable of accurately detecting the incipient phenomena of fouling over short periods of time due to the small deviations in the system performance during that time span.[3] As a result, the aerospace industry typically uses a shotgun approach in which the heat exchanger is removed for maintenance at arbitrarily chosen time intervals.

In this presentation we propose a method for initiated built-in-tests (iBIT) for aircraft ECS heat exchanger fouling detection, in which the extractable test information is maximized on the basis of a system model. Heat exchanger fouling detection, in terms of quantification of its severity, is critical for aircraft maintenance scheduling and safe operation. We focus on methods for offline fouling detection during aircraft ground handling, where the allowable variability range of admissible inputs is wider. We explore methods of optimal experimental design to estimate heat exchanger inputs and input trajectories that maximize the identifiability of fouling. Fouling metrics, such as thermal fouling resistance, are treated as parameters and analyzed along with a combination of uncertain system inputs, such as inlet temperatures, moisture content, and mass flows. The iBIT design vector, *i.e.* the system admissible inputs at which fouling is to be estimated, is manipulated to optimize experimental information and expressed via the Fisher Information matrix. Nominal and optimal iBITs are compared with respect to their capability to identify fouling, through parameter estimation, using the corresponding heat exchanger measurements. It is shown that the proposed iBIT methodology for heat exchanger fouling identification allows for accurate and precise estimation of the heat exchanger fouling, when this would have been infeasible with current conventional methods and without the addition of extra measurement devices or other BIT equipment. We close by generalizing the iBIT formulation for continuous non-linear systems with uncertainty.

Acknowledgment

This work was sponsored by the UTC Institute for Advanced Systems Engineering (UTC-IASE) of the University of Connecticut and the United Technologies Corporation. Any opinions expressed herein are those of the authors and do not represent those of the sponsor. Help and guidance by Modelon and Modelon-AB are gratefully acknowledged.

References

1. Hwang, I., Kim, S., Kim, Y. & Seah, C. E. A survey of fault detection, isolation, and reconfiguration methods. *IEEE Trans. Control Syst. Technol.* **18,** 636–653 (2010).

2. Simani, S. Model-based fault diagnosis in dynamic systems using identification techniques. (Universita degli Studi di MODENA e REGGIO EMILIA, 2000). doi:10.1007/978-1-4471-3829-7

3. Isermann, R. Model-based fault-detection and diagnosis - Status and applications. *Annu. Rev. Control* **29,** 71–85 (2005).

431564 A Novel Plant-Model Mismatch Estimation Method for Model Predictive Control

Abstract:

Model predictive control (MPC) has become the de-facto standard for multivariable control in the process industries. The widespread adoption of MPC systems can be attributed to their ability to deal with complex process interactions as well as with constraints imposed on the process states, and input and output variables. MPC performance is strongly dependent on the accuracy with which process behavior is predicted, which, in turn, is vitally dependent on the quality of the process model that is employed in the controller. Inherent process changes (induced, e.g., by fouling, corrosion, catalyst deactivation) lead to plant-model mismatch and can compromise the operation of the MPC system. In principle, these shortcomings could be addressed by performing regular model updates. This is, however, a costly undertaking and is generally only performed when closed-loop performance has degraded beyond the point where serious economic penalties are incurred in plant operations.

Current MPC performance assessment techniques typically adopt a two-step procedure. In the first step, the control performance degradation caused by the mismatch is identified and quantified. In the second step, the plant-model mismatch is estimated based on the degradation information. The first step has spurred several research efforts, with approaches based on multivariate statistical process control (MSPC) or controller performance benchmarking concepts being proposed. MSPC uses multivariate statistical tools, such as principal component analysis (PCA) and partial least squares (PLS) to identify abnormal operation [1]. Controller benchmarking defines a controller performance index using a theoretical or empirical metric [2, 3]. The second, mismatch estimation step, has received comparatively little attention. For instance, identifying the variable that contributes most to the degradation of controller performance using a control chart [4] has been proposed.

In this work, we introduce a novel plant-model estimation approach for MPC using the autocovariance matrices of process closed-loop data. We focus on estimation in the presence of moderate mismatch, in the sense that we seek early detection of plant-mode discrepancies based on operating data collected from seemingly "normal" operating conditions, where process economic performance is not sensibly degraded. We make the assumption that in this case the active set of the MPC does not change during operation. On this basis, we derive explicit expressions of the autocovariance matrices of the process inputs and outputs as functions of the magnitude of the plant-model mismatch. Subsequently, we solve an inverse problem to obtain an estimate of the plant-model mismatch from closed-loop input and output data. The proposed approach is illustrated in with several case studies.

[1] A. AlGhazzawi and B. Lennox. Model predictive control monitoring using multivariate statistics. Journal of Process Control, 19:314-327, 2009.
[2] T.J. Harris. Assessment of closed loop performance. Can. J. Chem. Eng., 67:856-861, 1989.
[3] J Yu and S.J. Qin. Statistical MIMO controller performance monitoring. part I: Data-driven covariance benchmark. Journal of Process Control, 18:277-296, 2008.
[4] J Yu and S.J. Qin. Statistical MIMO controller performance monitoring. part II: Performance diagnosis. Journal of Process Control, 18:297-319, 2008.

418567 Visualization and Data-Driven Monitoring of Batch Processes

Abstract:

Visualization and Data-Driven Monitoring of Batch Processes

Ray Wang, Michael Baldea, and Thomas F. Edgar

McKetta Department of Chemical Engineering

The University of Texas at Austin, 1 University Station C0400, Austin, TX 78712

Mark Nixon, Willy Wojsznis, Ricardo Dunia

Emerson Process Management, Austin, TX

Email: mbaldea@che.utexas.edu

The operation of batch processes differs significantly from that of continuous processes. In continuous processes, one or more steady states are well-defined and are typically maintained. On the other hand, in batch processes, the final product and its desired quality parameters are well-known, but are reached via an inherently transient path. This feature poses specific challenges in the monitoring of batch processes, and has led, e.g., to batch-specific modifications of projection-based data-driven monitoring and fault detection tools such as dynamic principal component analysis (DPCA).

In this paper, we present a novel, geometric, method for data-driven monitoring of batch processes. Our work relies on an extension of the time-explicit Kiviat (radial) plot visualization and fault detection framework that we have previously developed for continuous processes [Wang et al., AIChE J., submitted]. In this framework, each sample of a multivariate time series data set collected from process operations is represented in a radial plot. The radial plots are stacked vertically in the order in which the samples were acquired, resulting in a time-explicit representation of the multivariate time series. The geometric properties of this representation allow for establishing an elliptic confidence region for the centroids of the polygons corresponding to each data sample, which in turn can be used to define the normal steady-state and faulty operating states of the corresponding process.

In order to extend these ideas to batch processes, we redefine confidence ellipses: rather than computing a single confidence region using multiple samples collected from steady-state operation, we calculate separate confidence ellipses at every sample point, based on data acquired from multiple batches. This allows us to establish a "funnel" for batch trajectories, with the expectation that the trajectories of successful batches would evolve within the funnel, while faulty batches exit the funnel at some time point before completion. We study the effect of batch alignment on our proposed framework, and discuss its online implementation. Finally, we present an application to an industrial case study, along with a comparison of the fault detection performance of our methodology with that of other methods for batch process monitoring available in the literature.

434866 Performance Monitoring of Control Room Operators through Eye Gaze Analysis

Abstract:

Abstract

Safety in chemical process plant has become primary focus after the accidents like three mile island(1979), Esso's Longford refinery (1998),BP's Texas City refinery (2005) etc. to count a few. The investigation reports reveal the control room operator error as one of the leading factor for such incidents. Studies from Mannan (2004) suggest that almost 70 % of the critical incidents occurred just because of human error in plant operation. Analysis in high risk industries indicate that human errors typically originate from failure of proper situation assessment (SA) during abnormal plant conditions (Endsley 1988). It has been shown that analysis of cognitive behavior of humans can reveal their level of situation awareness, vital information that could help prevent human errors.

Studies on eye mind hypothesis suggest that gaze movement and cognitive behavior are highly correlated. Therefore, visual attention from eye gaze analysis can be used to get insight about level of situation awareness. For instance, Kilingaru (2013) used eye gaze analysis to quantify the expertise level of pilot. Based on the dwell time, duration of fixation of eye on Areas-of-Interest (AOIs), they mapped the cognitive state of the operator into three categories attention focusing, attention blurring, and misplaced attention. Attention focusing is characterized by a continuous dwell on a specific instrument over a period of time with a few glances at other instruments. On the other hand, short dwell time with many saccades between instruments indicates attention blurring. A short dwell on key instruments but with extended dwell outside the instrument panel is an indication of misplaced attention. This work demonstrated that information about pilots' cognitive states could be obtained by continuous monitoring of the fixation data. Eye tracking has been used in several disciplines to understand situation awareness, expertise level, learning ability and mental state of the human subjects. However, to our knowledge, these opportunities are yet to be explored in the chemical process industries.

In this study, we used a simulated ethanol plant in a MATLAB environment with an Human Machine Interface (HMI) which closely mimics a typical DCS interface. The experiment is interactive and the participant can use the HMI to get process information and manipulate the control valves. Information about the real-time values of the eleven measured variables, the list of standing alarms, and trend of any one variable (selected by the participant) can be seen in the snapshot of the HMI shown in Figure 1. The main objective of this work is to understand the cognitive behavior of the operator during disturbances. To accomplish this, the participants (students with control theory background) were given ten separate tasks. During the task, the participants take various actions to assess the situation and eliminate the disturbance by searching for appropriate information and perform corrective actions (opening / closing valves). These actions taken by the participant using the HMI including mouse movements, mouse clicks and slider movements are recorded during the experiment. TX300 eye tracker is used to record the eye gaze data of the participants. We used this data to calculate dwell time on various AOI to analyze the cognitive behavior of operator abnormal plant situation.

From eye gaze data analysis, we observed that although 91% of the tasks could be successfully controlled by the operators, patterns in the eye gaze behavior indicated that not all successful operations were similar; nearly 12% of the successful operators closely resembled operators who had failed to control the process. They were characterized by relatively lower attention on the primary variables (variables that are directly affected by the disturbance) which indicates a lower understanding of the underlying process dynamics. Dwell time analysis revealed that these participants had relatively lower dwell on the primary trend compared to other participants who had successfully completed the task. Our research also confirmed quantitatively that the use of trend information is important especially in the orientation phase of situation assessment which is in-line with the studies reported by Yin (2012). These insights can be used during operator training to evaluate operators' expertise in a deeper cognitive level and can also be used to develop an online measure to track the operator's situation awareness especially during disturbances and abnormal situations.

References

Mannan, Sam, ed. Lees' Loss prevention in the process industries: Hazard identification, assessment and control. Butterworth-Heinemann, 2004.

Gupta, J. P. "The Bhopal gas tragedy: could it have happened in a developed country?." *Journal of Loss Prevention in the process Industries* 15.1 (2002): 1-4.

Endsley, Mica R. "Design and evaluation for situation awareness enhancement." *Proceedings of the Human Factors and Ergonomics Society Annual Meeting*. Vol. 32. No. 2. SAGE Publications, 1988.

Hermens, Frouke, Rhona Flin, and Irfan Ahmed. "Eye movements in surgery: A literature review." *Journal of Eye Movement Research* 6.4 (2013): 1-11.

Kilingaru, K., Tweedale, J.W., Thatcher, S. and Jain, L.C. (2013). "Monitoring pilot "Situation Awareness"." *Journal of Intelligent and Fuzzy Systems,*24(3), 457-466.

Yin, S., 2012. Proactive monitoring in process contol using predictive trend displays. Ph.D. thesis, Nanyang Technological University, Singapore.

419472 Modelling Template for the Development of the Process Flowsheet

Abstract:

Models are playing important roles in design and analysis of chemicals/bio-chemicals based products and the processes that manufacture them. Model-based methods and tools have the potential to decrease the number of experiments, which can be expensive and time consuming, and point to candidates, where the experimental effort could be focused.

In this contribution a general modelling framework for systematic model building through modelling templates, which supports the reuse of existing models via its tools integration and model import and export capabilities, is presented.

Modelling templates in connection to other modelling tools within the modelling framework are forming a user-friendly system, which will make the model development process easier and faster and provide the way for unified and consistent model documentation. The modeller can use the template for their specific problem or to extend and/or adopt a model. This is based on the idea of model reuse, which emphasizes the use of a model not only for one specific application but also for future applications involving different needs and levels of detail to match different purposes. As the result the model developer can generate and test models systematically, efficiently and reliably. In this way, development of products and processes can be faster, cheaper and very efficient.

The developed modelling framework involves three main parts: 1) a modelling tool, that includes algorithms for model generation; 2) a template library, which provides building blocks for the templates (generic models previously developed); 3) computer aided methods and tools, that include procedures to perform model translation, model analysis, model verification/validation, model solution and model documentation.

In this work, the integrated use of all three parts in the modelling framework is highlighted through a development of a simple flowsheet model. Initially the model equations are obtained from the tool for model generation and then transferred to model analysis tool. Further, based on this model, a modelling template is created, which is used later for the modelling of Williams-Otto plant. The main goal is a demonstration of the template reuse for the system with similar repetitive objects.

422651 A Systematic Approach of Using Material Properties Data for Pharmaceutical Process Simulation

Abstract:

A Systematic Approach of Using Material Properties Data for Pharmaceutical Process Simulation

Jun Zhang[1], Frances Pereira[2], Ravendra Singh[1], Sean Bermingham[2], Fernando Muzzio[1], Rohit Ramachandran[1], Marianthi Ierapetritou[1]

[1]C-SOPS, Chemical and Biochemical Engineering, Rutgers University, Piscataway, NJ, USA

[2]Process Systems Enterprise Inc, London, UK

Pharmaceutical industry faces a "data rich" world since data are continuously generated from the substantial experiments employed in the process development activities. The data generation speed was anticipated as double every month, and it is still being promoted by the development of rapid experimental techniques. Benefits of harnessing this accumulated data include: narrowing down the design space of pharmaceutical process to be explored, reducing the amount of experiments to be implemented, and generating a clear landscape about target product and process performance [1]. However, less attention has been given on data utilization as indicated by the survey of pharmaceutical industry that pointed to less than 10% of data use. The reason of inefficient data utilization is that the data lack systematic organization and there is very little work on information extraction from such data sets. To address this issue, a systematic approach is developed which consists of material properties data representation, search function and multivariate data analysis.

The data representation can systematically represent material properties data, which is generated by different characterization devices, as a set of specifications; and these specifications would be organized as a XML file, which is an extensible markup language defined by W3C. Each data point of material properties is represented as a XML file with unique filename as the ID, and in each XML file, the measurements of material properties are organized as a hierarchical structure where each node represents a specification, i.e. a specific measurement's name with associated value.

The search function is developed to allow user to retrieve desired material properties data, which consists of user interface, ontology base and comparison algorithm. User interface allows user to define the specifications of data to be retrieved, e.g. API name, as well as the numerical criterion to be used for data searching, e.g. ±10% of specified viscosity. Ontology base consists of a set of ontologies that describe the relationship of terminologies referred in the materials properties data, e.g. Paracetamol is a type of API. The ontology establishes logic links among the data that allow more related data can be explored, which expands the search space that could help user to further understand the data. Comparison algorithm compares user's specifications with each XML file using numerical criterion and ontologies to return the data that is relevant with user's specifications.

Based on the returned data, partial lease square regression (PLS) algorithm is used to correlate the material properties information and process parameters with process output in order to generate a predictive model. Such predictive model ensures consistency of information of each unit's input and output, and it would greatly facilitate process simulation, especially the whole process consisting of different unit models that require inconsistent inputs and outputs.

As a case study, a specific blending process consisting of feeding, co-milling and blending, which has inconsistent model inputs and outputs, is used. gPROMs [2] is selected as the simulation platform to demonstrate how this framework can be implemented and how the data can be fully and flexibly used for process simulation study.

References:

[1] Zhang J, Hunter A, Zhou Y. A logic-reasoning based system to harness bioprocess experimental data and knowledge for design. Biochemical Engineering Journal, 2013, 74: 127-135.

[2] Boukouvala F, Niotis V, Ramachandran R, Muzzio F, Ierapetritou M, An integrated approach for dynamic flowsheet modeling and sensitivity analysis of a continuous tablet manufacturing process, Computers & Chemical Engineering, 2012, 42: 30-47.

413654 Unleashing the Potential of PAT with Real Time Knowledge Management and Quality Based Control

Abstract:

Unleashing the Potential of PAT with Real Time Knowledge Management and Quality Based Control

Martin Gadsby: Director of Optimal Industrial Automation Ltd Bristol, UK

After many years of being a technology used solely by researchers and industrial evangelists, PAT is finally making a very real and exponentially growing impact on the manufacture of pharmaceuticals as more and more companies realise that you truly can have improved product quality whilst at the same time achieving a reduction in cost and manufacturing time. Furthermore, the holistic approach embraced by PAT is having a growing impact on manufacturing in the food and chemical industries to name just two, where the perhaps occasional use of unconnected and 'uncontrolled' spectral instruments have been used for many years.

The application of PAT is a team effort where multiple skills, disciplines and technologies are required, and a technical keystone to a cost effective and regulatory compliant deployment is the central 'hub in the wheel', the PAT Knowledge Manager such as synTQ. This software package is used to tie together and coordinate in real time multiple multivariate and univariate data sources and consumers and facilitate the generation of quality predictions. Furthermore, a key point of PAT is process understanding, and you need a system that can turn data into information and then in turn to knowledge, this being disseminated to the users in a fashion that is digestible and usable. Process control is required – both feed forward and feedback, and you need a system that is regulatory compliant and is a fundamental tool in the continuous improvement programme that is such an essential part of maximising the PAT benefits.
Finally, you need a system that is scalable and usable from the laboratory to full GMP manufacturing, and can be used on discrete, batch and continuous processes.

This presentation sets out to describe in more detail some of the technical data management challenges that will be encountered on a PAT deployment journey, and possible technical and product solutions that are available for you to add to your PAT 'armoury'.

435029 Attention Aware Systems in Process Control Rooms through Real-Time Pupillometry

Abstract:

Abstract

The role of control room operator's in many complex, highly-automated plants is predominately on process monitoring and supervision. The performance of the plant operator under critical situation is significantly affected by poor HMI (Human Machine Interface), high alarm rates, fatigue, stress, high workload and harsh environments. Lately, there are significant developments in HMI design leading to a user friendly display in the control room; however, less attention is paid towards understanding the cognitive behavior and attention level of the plant operators which could be used in development of HMI. The cognitive abilities of the human are limited and therefore, it is necessary to understand the mental state of the operator. Negligence to realize the operator capability and attention could result in a workload bombardment and probably lead to accidents like Esso Longford & Texas City refinery explosion.

In the recent past, there has been an increased emphasis in the development of attention aware HMI systems; those that that are capable of adapting to and supporting human attentional processes, especially in situations involving multi-tasking and in highly dynamic environments (Tsianos et al. (2009), Weibel et al. (2012)). Studies have shown that metrics computed eye tracking data can be used as a method of identifying users' actual behavior in a hypermedia setting. Researcher are currently utilizing point of eye gaze from eye trackers as an alternate HMI (Human Machine Interface) paradigm with mouse movements and selections made based on user's eye movements (Roda and Thomas (2006)). In the field of avionics, attempts have been made to design HMI that are capable of assessing the pilot's focus of attention and make predictions about it in the near future and also determine when their attention needs to be redirected for efficient operation (Weibel et al. (2012)). However, to the best of our knowledge, these opportunities are yet to be explored in the chemical process industries. In this work, we make a novel attempt to utilize eye tracker data in understanding the attention level of control room operators, a vital information in development of attention aware HMI system.

Task evolved pupillometry response (TEPR), analysis of changes in pupil diameter during execution of tasks, is one of the commonly used approaches to analyze the attention level and mental state of a human under critical situations (Christine et al. (2011)). Pupillometry has been widely used to evaluate attention level (Laeng et al. (2012)), decision making process (Marshall 2000) and stress levels (Pedrotti et al. 2014). Experiments have revealed that during decision making, pupil dilation increases before making a decision. More recently, Jiang et al. (2013) reported that when participants performed a simulated surgical task with varying difficulty levels and different demands of mental workload, the pupil size increased significantly during the harder task and decreased rapidly while performing the easier task; changes in pupil size can therefore be used to measure the mental workload. TEPR has also been used to as an online measure to track the attention level of drivers (Di Stasi et al. (2010)).

In this study, we attempt to develop an online measure to identify the attention level of control room operator during plant abnormalities. To accomplish this, students having control theory background have performed the role of an operator in controlling the simulated ethanol plant under abnormalities. The experiment is interactive and the participant can use the HMI to get process information and manipulate the control valves. During the task, the participants take various actions to assess the situation and eliminate the disturbance by searching for appropriate information and perform corrective actions (opening / closing valves). These actions taken by the participant using the HMI including mouse movements, mouse clicks and slider movements are recorded during the experiment. TX300 eye tracker is used to record the eye gaze data of the participants. Analysis of data from eye tracker revealed that the majority of the participants had their pupil dilated after the occurrence of alarm, revealing the high cognitive work load due to plant abnormality. Further, clearance of last alarm (end of the task) results in decrease of cognitive demand which is reflected by contraction (decrease) of the pupil size for majority of the participants. These preliminary results indicate that analysis of pupil size variations along with key events during the task could help in development of an online measure for identifying the attention level of control room operators.

Reference

Tsianos, N., Germanakos, P., Lekkas, Z., Mourlas, C., & Samaras, G. (2009, July). Eye-tracking users' behavior in relation to cognitive style within an e-learning environment. In Advanced Learning Technologies, 2009. ICALT 2009. Ninth IEEE International Conference on (pp. 329-333). IEEE.

Weibel, N., Fouse, A., Emmenegger, C., Kimmich, S., & Hutchins, E. (2012, March). Let's look at the cockpit: exploring mobile eye-tracking for observational research on the flight deck. In Proceedings of the Symposium on Eye Tracking Research and Applications (pp. 107-114). ACM.

Roda, C., and Thomas, J. 2006. Attention aware systems: Theories, applications, and research agenda. Computers in Human Behavior 22, 4, 557–587

Laeng, Bruno, Sylvain Sirois, and Gustaf Gredebäck. "Pupillometry A Window to the Preconscious?." *Perspectives on psychological science* 7.1 (2012): 18-27.

Marshall, Sandra P. "Method and apparatus for eye tracking and monitoring pupil dilation to evaluate cognitive activity." U.S. Patent No. 6,090,051. 18 Jul. 2000.

Pedrotti, M., Mirzaei, M. A., Tedesco, A., Chardonnet, J. R., Mérienne, F., Benedetto, S., & Baccino, T. (2014). Automatic stress classification with pupil diameter analysis. *International Journal of Human-Computer Interaction*, *30*(3), 220-236.

Jiang, X., Zheng, B., Tien, G., Atkins, M. S., 2013. Pupil response to precision in surgical task execution. In: MMVR.

pp. 210–214.

Di Stasi, L. L., Marchitto, M., Antolí, A., Baccino, T., & Cañas, J. J. (2010). Approximation of on-line mental

workload index in ATC simulated multitasks.Journal of Air Transport Management, 16(6), 330-333.

421740 Organization of Mongodb Database in Modena Project

Abstract:

MoDeNa aims at developing, demonstrating and assessing an easy-to-use multi-scale software-modelling framework application under an open-source licensing scheme that delivers models with feasible computational loads for process and product design of complex materials. Four scales are linked together by this interconnected modelling-software framework namely the nano-, micro-, meso-, and macroscale. The orchestrator enables the linking of all scales which is a necessary condition to obtain an integral approach (MoDeNa).

The orchestrator employs document-oriented NoSQL datastore as a back-end repository. The datastore handles a great variety of data types including execution state, inputs and outputs. NoSQL is chosen primarily due to its flexibility. Unlike other relational datastores, it does not require a predefined data schema between all these different types of data. As the data grow in size and the data structure evolve with new data, NoSQL can easily adapt to these changes instead of modifying a complex relational data schema (Gunter et. al., 2012).

Datastore choice is MongoDB, a schemaless document store database developed in an open-source project (Shermin, 2013). It has several advantages among other document-oriented datastores, namely powerful and simple query language, ease of administration, and good performance on read-heavy workloads where most of the data can fit into memory (Gunter et. al., 2012). Its relative weakness of huge datasets and write-heavy workloads can be a trade-off for MoDeNa project.

This presentation focuses on the organization of the datastore. In MongoDB, there are collections which hold data of similar type, called documents. Documents can also store documents in embedded form which provide ease in read operations. In MoDeNa project, the input data is in JSON format, and holds models and initial data points which are inserted by a user. Calculated data points are stored along with the input data. The execution state data is generated by the software and stores the state and intermediate results for all tasks in the system. The input data and execution state data are stored in separate datastores. The data structure will be illustrated further in the presentation.

References

1. Shermin, M. (2013). An Access Control Model for NoSQL Databases.

2. Gunter, D., Cholia, S., Jain, A., Kocher, M., Persson, K., Ramakrishnan, L., ... & Ceder, G. (2012, November). Community Accessible Datastore of High-Throughput Calculations: Experiences from the Materials Project. In *High Performance Computing, Networking, Storage and Analysis (SCC), 2012 SC Companion:* (pp. 1244-1251). IEEE.

3. MoDeNa project website. Retrieved May 11, 2015 from http://www.modenaproject.eu/

429052 Process Safety - a Systems Perspective

Abstract:

A modern process plant has several layers of automated protection systems that work independently but in concert to ensure the economical operation of the plant within its safety, environmental, and product quality limits. These systems are typically control systems, alarm systems, and emergency shutdown (ESD) systems. Control systems including conventional and model-based controllers [1] are used to operate process plants properly and produce high-quality products consistently, alarm systems are used to alert operators to plant variables that are about to exceed or have already exceeded design limits, and ESD systems are used to halt plants that are in danger of exceeding plant limits. As safety systems are the last line of defense to prevent shut-downs and accidents, they must be independent of control systems. While control systems occasionally fail, safety systems must be highly reliable, and must be able to override control systems whenever necessary.

Alarm systems are designed such that important (primary) variables are paired with actionable safety systems [2]. As a result of the digital computer revolution, nowadays alarms are easy and inexpensive to configure and deploy; modern field alarm systems often have hundreds or thousands of configured alarms, and therefore, during plant operation (even normal operation), operators can receive a huge number of alarms [3]. Despite many studies on alarm system design and management [4, 5, 6], there are still many critical challenges that need to be addressed to further improve existing traditional alarm systems. For example, existing alarm systems are generally configured only for individual measured process variables. While this simplicity is a strength of existing alarm systems, it is also their weakness as they are unable to consider interactions among variables, leading to an excessive number of false alarms related to the interacting variables. Furthermore, they can alert plant personnel only to the present operation hazards that can be identified from process measurements and have already impacted processing plants.

The focus of this work is on the upper two layers of the automated protection systems; i.e., alarm systems and ESD systems. We will survey the present state of these systems and will put these systems in perspective from a systems engineering point of view. We will classify the existing systems into reactive and proactive systems. It will be shown (a) that existing safety systems are mostly reactive and (b) how these systems can benefit from having proactive capabilities. For example, a proactive model-based alarm system [7] can systematically address many weaknesses of the existing alarm systems. First, it systematically accounts for plant *nonlinearities* and *interactions* among plant variables. Second, it can provide alarms tied to unmeasured state variables. Third, it can identify *current* and *future* hazards in plant operation. These are capabilities that current alarm systems do not have. Also, leading indicators [8, 9] can alert operators that an accident or a plant trip may occur before it actually does, giving operators time to make corrective actions to dampen the effect of the upcoming abnormal event before it occurs, or before it propagates to an emergency shutdown or worse. Therefore, it is desirable to program automated safety systems to respond to leading indicators, thus having an immediate automated response to specific indicators.

References

[1] Leveson, N.G., and Stephanopoulos, G., "A System-Theoretic, Control-Inspired View and Approach to Process Safety", *AIChE J.*, 60(1), 2-14, 2014

[2] Modarres, M., and Cadman, T. "A Method of Alarm System Analysis for Process Plants", *Comput. Chem. Eng.*, 10(6), 557-565, 1986.

[3] Hollifield, B. R., and Habibi, E., *Alarm Management: A Comprehensive Guide: Practical and Proven Methods to Optimize the Performance of Alarm Management Systems*. ISA, 2011

[4] Wei, L., Guo, W., Wen, F., Ledwich, G., Liao, Z., and Xin, J., "An Online Intelligent Alarm-processing System for Digital Substations", *IEEE Trans. Power Deliv.*, 26(3), 1615-1624, 2011.

[5] Zhao, J., and Zhu, J., "Data Filtering Based Alarm Processing Strategy for Repeating Alarms", *J. Tsinghua Univ. Sci. Tech.*, 52(3), 277-281, 2012.

[6] Schleburg, M., Christiansen, L., Thornhill, N. F., and Fay, A., "A Combined Analysis of Plant Connectivity and Alarm Logs to Reduce the Number of Alerts in an Automation System", *J. Proc. Cont.*, 23(6), 839-851, 2013.

[7] Mohseni Ahooyi, T., Arbogast, J., Seider, W.D., Oktem, U.G., Soroush, M., "A Method of Proactive Model-based (Soft) Alarm System Design", *AIChE Annual Meeting*, 2015.

[8] Pariyani, A., Seider, W. D., Oktem, U. G., and Soroush, M. "Dynamic Risk Analysis Using Alarm Databases to Improve Process Safety and Product Quality: Part II—Bayesian Analysis", *AIChE J.*, 58, 3, 826-841, 2012.

[9] Siu, N., "Risk Assessment for Dynamic Systems: an Overview", *Reliab. Eng. Sys. Safety*, 43(1), 43-73, 1994.

431176 Prediction of Mineral Scales in Oil & Gas Production Systems Using a Machine Learning Approach

Abstract:

Precipitation of mineral scales is one of the most sedulous and expensive operational problems in Oil & Gas Industry, because mineral scales may lead to severe permeability loss, reduces flow areas of productions and injections wells and breakdown of production facilities equipment generating a hydrocarbon production drop and high maintenance costs. In this context, the first practical step is to obtain an accurate estimation using a reliable method able to identify which mineral scales and maximum amount are going to precipitate under different reservoir operating conditions (temperature, pressure and composition). It is also known that thermodynamic ion interaction models and equations of state, used to determine behaviour of different ions and gas compounds leading to mineral scale formations, are limited by the ranges of temperature and pressure on which they can be used, making the calculation very complex and allows high error values. As a response, in recent years, many works have been focused on the development of a simple and reliable method that allows the prediction, with high accuracy, of the inorganic scales behaviour under wide ranges of operating conditions, preventing production and economical issues.

In this study a reliable model based on machine learning techniques (Least Squares Support Vector Machines and Coupled Simulated Annealing) was developed aiming to predict the type and the amount of mineral scales precipitated and compared with thermodynamic ion interaction models (Debye-Hückel, Pitzer, Extended UNIQUAC and e-NRTL), equation of state (Soave-Redlich-Kwong) and finally with commercial softwares (ScaleChem and DownHole SAT) predictions. Based on the results shown, the model proposed is adequate for mineral scales prediction with higher accuracy than thermodynamic models, equations of state and softwares prediction under a wide range of operating conditions, resulting in the reliable and correct prediction of mineral scales precipitation in Oil & Gas processes.

15AIChE
Annual Meeting, Salt Lake City, UT

QbD PROCESS TECHNOLOGIES

informetric SYSTEMS

RUTGERS
THE STATE UNIVERSITY OF NEW JERSEY

Applying Batch Data Principles to Continuous Manufacturing for the Purposes of Data Management, Batch Reporting, Analytics and Traceability

Paul Brodbeck
Chief Technologist
QbD Process Technologies, Inc.

Bob Engel
Vice President
Informetric Systems Inc.

Ravendra Singh
Assistant Research Professor
Chemical and Biochemical Engineering Department,
Rutgers University, NJ, USA

Background

- Many pharmaceutical processes that have been implemented using traditional batch techniques are now evolving towards continuous manufacturing in order to improve manufacturing efficiency and product uniformity
- There is intense research and development associated with applying continuous process technology to drug manufacturing that was previously batch oriented
- Since there is not discrete separation of materials at various stages of the process, it can be difficult or impossible to link incoming raw and intermediate materials to final product
- The regulatory and traceability considerations are particularly critical in this area
- Residence Time Distribution (RTD) models represent a possible solution for product traceability that is required with respect to regulatory and safety guidelines, as well as Current Good Manufacturing Practice (cGMP)
- RTD models enable batch reporting, analysis and material traceability of continuous processes

Continuous Plant for OSD

Rutgers
Engineering Research Center
for Structured Organic Particulate Systems (ERC-SOPS)

Singh, R., Boukouvala, F., Jayjock, E., Ramachandran, R. Ierapetritou, M., Muzzio, F. (2012). Flexible Multipurpose Continuous Processing. PharmPro Magazine, 28 June, 2012.

Experimental Setup and Instrumentation

- NIR sensor has been integrated with process for real time monitoring and feedback control
- Chemometric tools have been used for real time NIR spectrum analysis in order to acquire a concentration input
- OPC communication protocol has been used to communicate the data across different software tools
- Control loops have been implemented in distributed process control system
- Controller outputs have been sent to the actuators through fieldbus devices

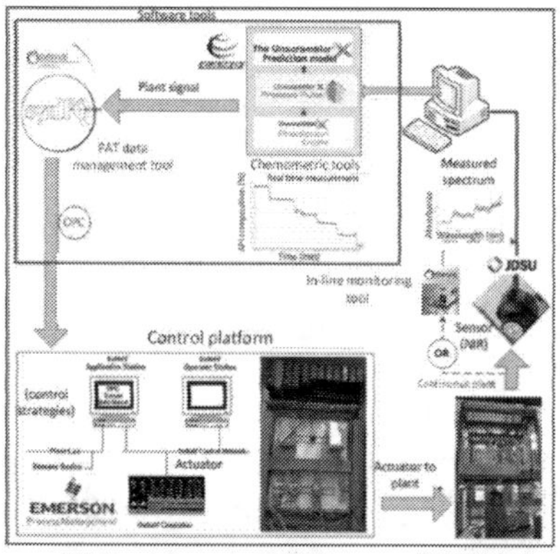

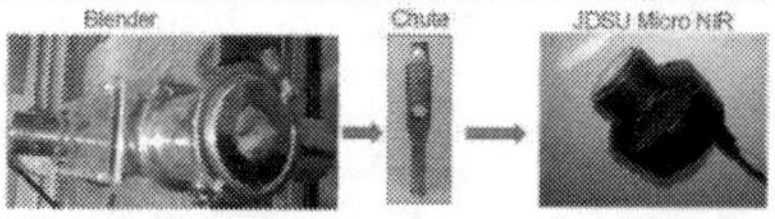

Singh, R., Sen, M., Ierapetritou, M., Ramachandran, R. (2015). Journal of Pharmaceutical Innovation, 10 (3), 233-253.

Residence Time Distribution (RTD)

- Models relationship between input and output concentrations
- Input pulse dissipates through unit operation as function of time
- Output concentration is characterized by Mean Residence Time (MRT) and distribution
- Maximum output concentration amplitude represents maximum possible response to input disturbance

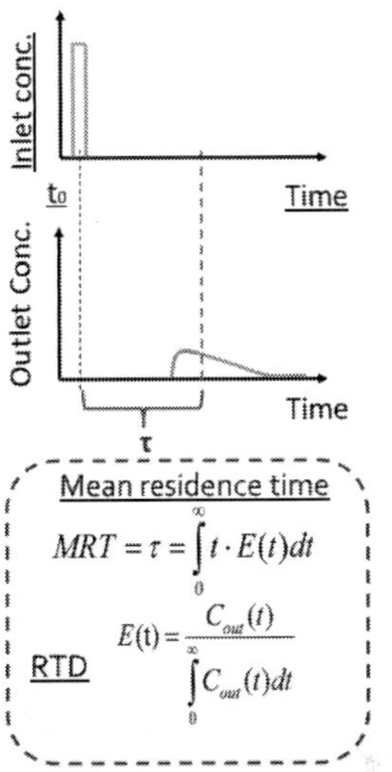

 15AIChE RUTGERS

QbD Process Technologies
ContinuousPlant™ Software Suite

informetric SYSTEMS QbD PROCESS TECHNOLOGIES

Continuous Plant / RTD Models

Full line RTD can be achieved by integrating the individual unit RTDs

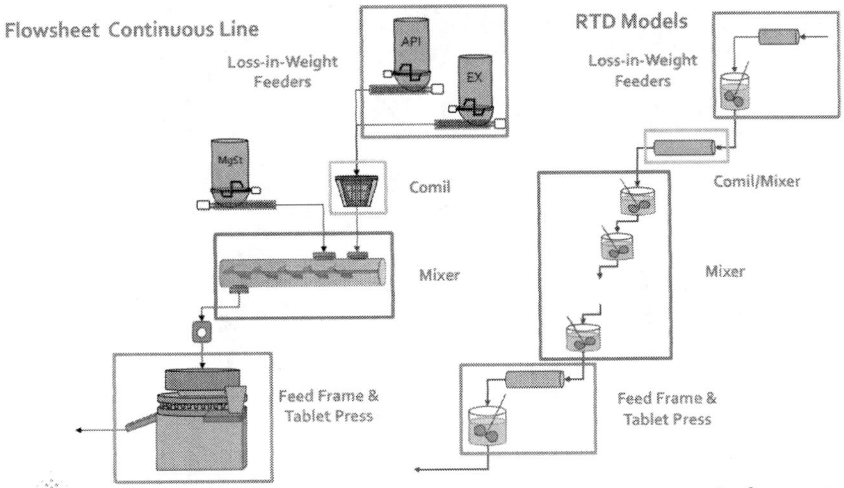

 15AIChE RUTGERS

QbD Process Technologies
ContinuousPlant™ Software Suite

informetric SYSTEMS QbD PROCESS TECHNOLOGIES

Residence Time Distribution (RTD)

Direct Compression RSD Formulas

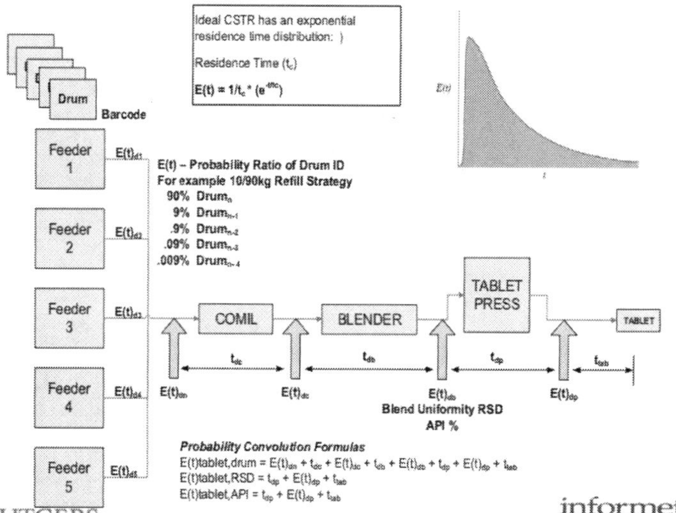

ContinuousPlant™ Software Suite

15AIChE Annual Meeting, Salt Lake City, UT — RUTGERS THE STATE UNIVERSITY OF NEW JERSEY — informetric SYSTEMS — QbD PROCESS TECHNOLOGIES

RTD Modeling for Material Tracing

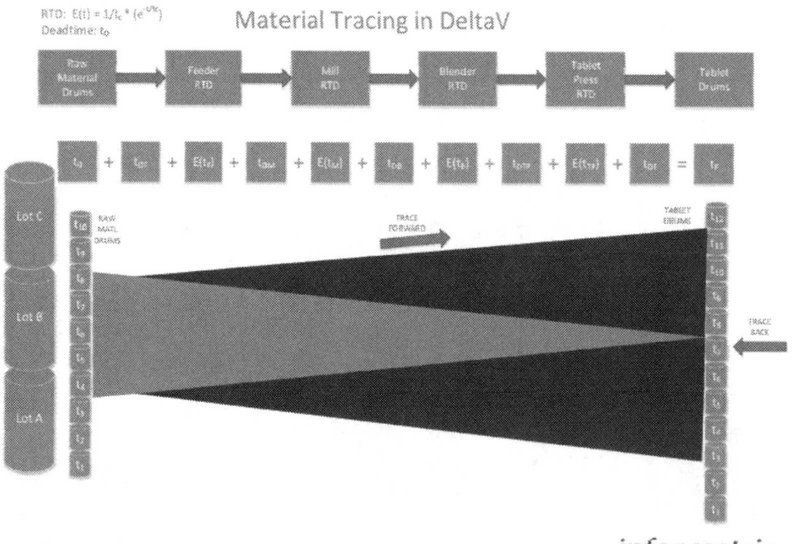

QbD Process Technologies
ContinuousPlant™ Software Suite

15AIChE Annual Meeting, Salt Lake City, UT — RUTGERS THE STATE UNIVERSITY OF NEW JERSEY — informetric SYSTEMS — QbD PROCESS TECHNOLOGIES

Reporting Architecture

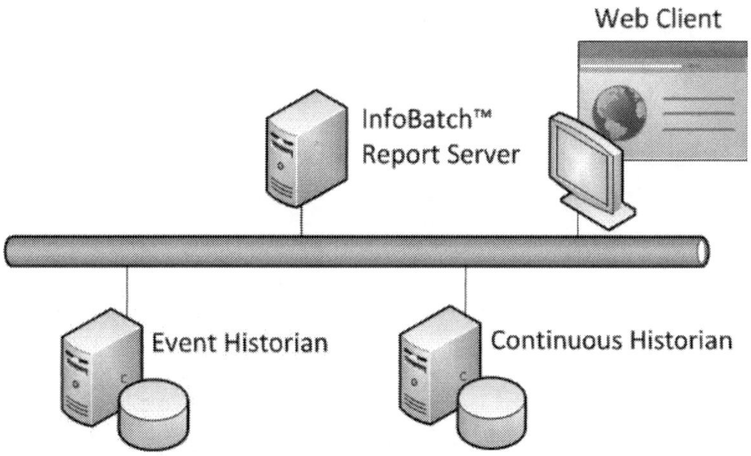

Reporting Stages

- Data Collection
  - Report server collects raw data from disparate data sources according to specified context model
- Aggregation
  - Report server aligns data
  - Report server performs intermediate calculations
- Rendering
  - Report server renders formatted report

Contextualization

- Traditional Batch
  - S88 Procedural Model (recipe driven)

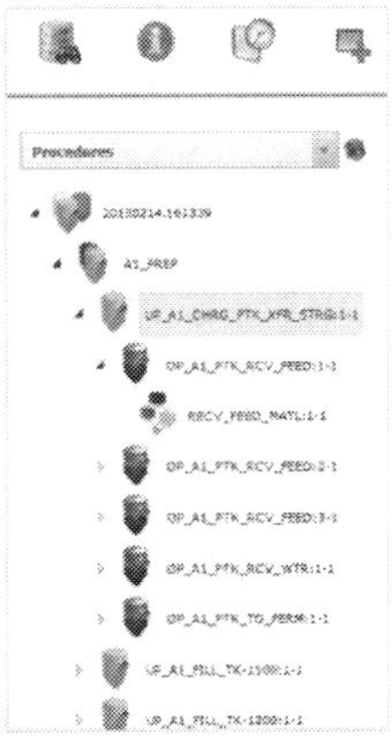

- Continuous
  - RTD Model Driven

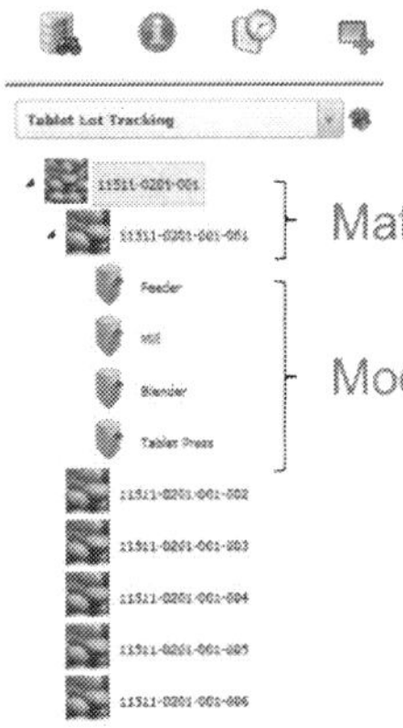

Material Tracing

Model based

15AIChE Annual Meeting, Salt Lake City, UT | RUTGERS THE STATE UNIVERSITY OF NEW JERSEY

QbD Process Technologies
ContinuousPlant™ Software Suite

informetric SYSTEMS | QbD PROCESS TECHNOLOGIES

Example Report – Tablet API Tracing

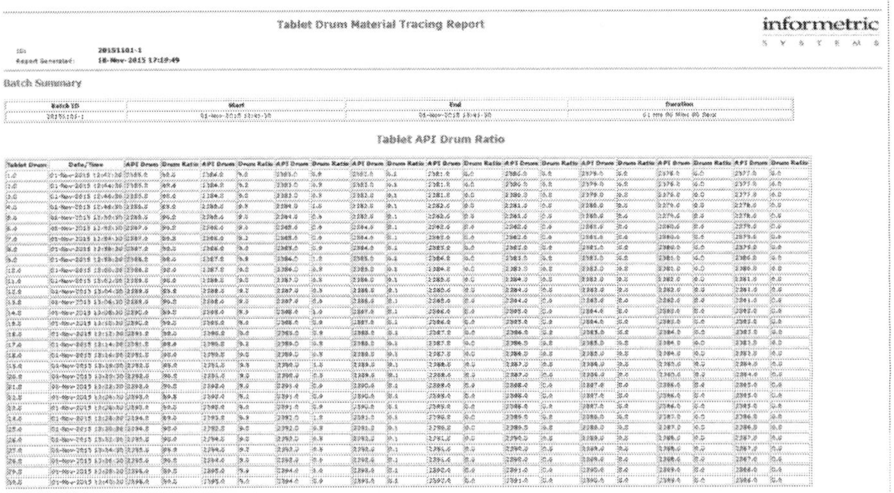

15AIChE Annual Meeting, Salt Lake City, UT | RUTGERS THE STATE UNIVERSITY OF NEW JERSEY

QbD Process Technologies
ContinuousPlant™ Software Suite

informetric SYSTEMS | QbD PROCESS TECHNOLOGIES

Example Report – Tablet API Tracing

Tablet Drum	Date/Time	Chart	API Drum	Drum Ratio	API Drum	Drum Ratio	API Drum	Drum Ratio	API Drum	Drum Ratio
1.0	01-Nov-2015 12:42:30	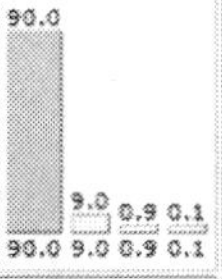 90.0 9.0 0.9 0.1	2385.0	90.0	2384.0	9.0	2383.0	0.9	2382.0	0.1
2.0	01-Nov-2015 12:44:30	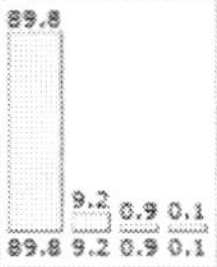 89.8 9.2 0.9 0.1	2385.0	89.8	2384.0	9.2	2383.0	0.9	2382.0	0.1
3.0	01-Nov-2015 12:46:30	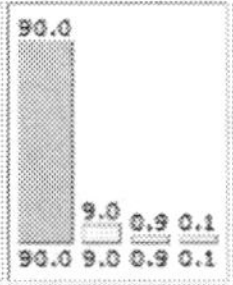 90.0 9.0 0.9 0.1	2385.0	90.0	2384.0	9.0	2383.0	0.9	2382.0	0.1

Summary

- The evolution of traditionally batch oriented processes to continuous manufacturing introduces challenges related to product traceability
- Residence Time Distribution (RTD) models enable the application of traditional batch reporting and analysis techniques to continuous manufacturing processes
- Contact Information:
  - Paul Brodbeck, QbD Process Technologies, Inc., paul.brodbeck@qbdprocess.com
  - Bob Engel, Informetric Systems Inc., bengel@informetric.com
  - Ravendra Singh, Rutgers University, ravendra.singh@rutgers.edu
  - References:

- "Modeling of Residence Time Distribution in Continuous Solid Oral Dose Pharmaceutical Manufacturing Processes". AIChE Annual Meeting 2014. Escotet-Espinoza, M.S., Rogers, A., Muzzio, F. J., and Ierapetritou, M. Department of Chemical and Biochemical Engineering, Rutgers University, Piscataway, NJ.
- Singh, R., Boukouvala, F., Jayjock, E., Ramachandran, R. Ierapetritou, M., Muzzio, F. (2012). Flexible Multipurpose Continuous Processing. PharmPro Magazine, 28 June, 2012.
- Singh, R., Sen, M., Ierapetritou, M., Ramachandran, R. (2015). Integrated moving horizon based dynamic real time optimization and hybrid MPC-PID control of a direct compaction continuous tablet manufacturing processJournal of Pharmaceutical Innovation, 10 (3), 233-253.

420816 Closed-Loop Average Economic Performance Under Real-Time Economic Model Predictive Control

Abstract:

Economic model predictive control (EMPC), an optimal control-based feedback control methodology, unites process control and economic process optimization into an integrated framework (e.g., [1]-[5]). The main objective of applying EMPC to control a chemical process over tracking model predictive control (TMPC) is to achieve improved closed-loop economic performance. It is well-known that closed-loop performance under EMPC without specific terminal conditions is not ensured in general. To address this problem, terminal conditions may be used. In [1], the asymptotic average economic performance under EMPC formulated with an equality terminal constraint was shown to be at least as good as the economic performance at the economically optimal steady-state. Other EMPC system have been proposed utilizing an equality terminal constraint or terminal region constraint which are designed on the basis of the optimal steady-state (e.g., [2]). Nevertheless, when terminal constraints are imposed in the EMPC optimization problem, recursive feasibility for systems with plant-model mismatch may not be ensured. On the other hand, when no terminal conditions are imposed in the EMPC problem, a sufficiently long prediction horizon is typically required for some guarantees on closed-loop performance under EMPC [3]. The use of a long prediction horizon may make EMPC computationally intractable for use in real-time operation. Moreover, an important issue from a practical stand-point for EMPC is real-time computation time. When the computation time is significant, closed-loop operation and/or performance degradation may occur.

In this work, a Lyapunov-based EMPC (LEMPC) scheme for nonlinear process systems is designed leveraging an auxiliary stabilizing control law and addressing real-time computation delay through an event-triggered implementation strategy. In particular, an auxiliary control law (e.g., TMPC or Lyapunov-based control law) is designed to stabilize a process at the economically optimal steady-state. The process model with the auxiliary control law are used to predict the solution of the system under the auxiliary control law over a finite-time horizon. Motivated by the desire to ensure economic performance improvement under EMPC compared to under the auxiliary control law, the computed solution is utilized in the LEMPC scheme as a terminal constraint. This type of terminal constraint is similar to that proposed in [4] for use in EMPC. To account for real-time computational time, the real-time implementation strategy for LEMPC proposed in [5] is extended to the proposed LEMPC with the terminal constraint designed on the basis of the solution of the system under the auxiliary control law. Recursive feasibility of the proposed LEMPC scheme, closed-loop stability, and finite-time and asymptotic average economic closed-loop performance are analyzed. The scheme is applied to a chemical process example to demonstrate the theoretical properties of the proposed LEMPC scheme.

[1] Angeli D, Amrit R, Rawlings JB. On average performance and stability of economic model predictive control. IEEE Transactions on Automatic Control. 2012;57:1615-1626.
[2] Amrit R, Rawlings JB, Angeli D. Economic optimization using model predictive control with a terminal cost. Annual Reviews in Control. 2011;35:178-186.
[3] Grune L. Economic receding horizon control without terminal constraints. Automatica. 2013;49:725--734.
[4] Ellis M, Christofides PD. On finite-time and infinite-time cost improvement of economic model predictive control for nonlinear systems. Automatica. 2014;50:2561--2569.
[5] Ellis M, Christofides PD. Real-time economic model predictive control of nonlinear process systems. AIChE Journal. 2015;61:555--571.

424544 Computationally Efficient Economic Model Predictive Control: Stability, Performance, Convergence

Abstract:

Traditionally, in the operation of chemical processes, economic optimization and process control are carried out in two different layers in a hierarchical system. Economic optimization is performed in the upper real-time optimization (RTO) layer based on steady-state process models according to certain economic metrics reflecting operation profits or costs. The outputs of the RTO layer are optimal operating points of the process which are sent to a lower layer -- the advanced process control layer. Model predictive control (MPC) is often used in the advanced process control layer. This hierarchical approach has been successful and has been widely used. However, in many processes (for example, the oilsand separation process), the dynamic performance is crucial and the optimization-control separation is either inefficient or inappropriate. Economic model predictive control (EMPC) removes the separation between optimization and control and addresses both in one single layer in the framework of model predictive control.

Motivated by the lack of an easy-to-implement approach for the construction of terminal cost for EMPC, in our previous work [1], we developed an approach to construct the terminal cost for EMPC and proposed two EMPC algorithms for both infinite-time and finite-time operations. In the approach for the terminal cost construction, an auxiliary nonlinear controller which renders the desired optimal steady state asymptotically stable is taken advantage of. The idea is to take the auxiliary nonlinear controller as a benchmark control system and to design EMPC algorithms that gives better economic performance than the auxiliary nonlinear controller. Two EMPC algorithms for infinite-time and finite-time operations were designed. The two EMPC algorithms give provable (practically) improved economic performance. Because of the terminal cost, there is no requirement on the length of the prediction horizon of the EMPC which means that the developed EMPC algorithms could be very computationally efficient.

In this work, we focus on the rigorous analysis of the stability, performance and convergence properties of the computationally efficient EMPC. We show that for strictly dissipative systems, the proposed economic MPC scheme achieves practical stability and near optimal asymptotic performance. Moreover, we show that exponential stability of the optimal steady state can be achieved if the auxiliary controller is exponentially stabilizing in a strong sense. This condition can be relaxed if the storage function satisfies further assumptions. Achievable performance under exponential stability is also characterized. These results will be demonstrated via numerical examples.

[1] Liu, S.; Zhang, J. & Liu, J. Economic MPC with Terminal Cost and Application to an Oilsand Primary Separation Vessel, Chemical Engineering Science, 2015.

435191 Application of Approximate Infinite Horizon Economic MPC to Electric Power Networks with Energy Storage

Abstract:

The method of Economic Model Predictive Control (EMPC) is conceptually attractive in that it replaces the traditional quadratic objective function of MPC with an economic based objective function. However, many have indicated concerns about the trade-off between stability and economic / computational performance of EMPC. In a previous effort the notion of approximate infinite-horizon EMPC (AIH-EMPC) was introduced and shown to improve all of the above properties when applied to an exceedingly simple reaction system. In the current effort, the proposed AIH-EMPC method is applied to an electric power network system. In addition to being a much more complicated system (in terms of size as well as the inclusion of integer variables to address unit commitment aspect of grid operation), this example possess an economic based quadratic objective function. (This objective is distinct from that of a regulation type quadratic objective function.) The resulting controller is shown to possess much improved characteristics with respect to computational effort while sacrificing very little in terms of economic performance.

429856 Integrated Real-Time Optimization and Control of Multiunit Batch Processes

Abstract:

Integrated real-time optimization and control of multiunit batch processes

Francesco Rossi,[a,b,*] *Daniel Casas Orozco,*[c] *Gintaras Reklaitis,*[a] *Flavio Manenti*[b]

[a] *Purdue University, School of Chemical Engineering, Forney Hall of Chemical Engineering, 480 Stadium Mall Drive, West Lafayette, IN 47907-2100, United States*

[b] *Politecnico di Milano, Dipartimento Chimica, Materiali e Ingegneria Chimica "Giulio Natta", Piazza Leonardo da Vinci 32, Milano 20123, Italy*

[c] *Universidad de Antioquia UdeA, Engineering Faculty, Chemical Engineering Department, Environmental Catalysis Research Group, Calle 70 No. 52-21, Medellín, Colombia*

frossi@purdue.edu

Despite the historical tendency for conversion of batch operations to their continuous processing equivalents, batch manufacturing remains in widespread use in a number of CPI sectors, such as pharmaceuticals and fine chemicals, fermentations and bio-chemical products, and special polymers. Indeed batch processing remains important on the industrial landscape and is likely to remain significant for the foreseeable future. Many processes operating in the batch mode produce high added-value products, are characterized by intrinsic non-ideal thermodynamic behavior and exhibit strongly non-linear dynamics, features which provide challenges for design and in operation. To address these challenges, there continues to be an ongoing body of research seeking safe and economical ways to design, optimize and control batch operations.

Several papers in the literature describe how to (efficiently) apply model-based online optimization and control strategies to batch units, but typically the focus is on a single process unit: a single reactor (Smets et al., 2004), distillation column (Greaves et al., 2003), crystallizer (Acevedo et al., 2015) or more specialized, unconventional equipment. There is limited literature on approaches to simultaneously consider multiple units or even an entire batch plant. For this reason, the authors of this contribution propose the development of integrated online optimization and optimal control methods for treating an entire integrated batch production line.

Specifically, we propose to: (I) model the entire batch process as a whole, also including design choices as independent variables (activation/deactivation of entire units, unit feeds & draws, unit to unit internal material/thermal streams, etc.); (II) specify a suitable performance function for the entire plant (either economic or performance based); and (III) apply an online optimization & control algorithm to the framework described in (I) and (II).

Depending on the perspective taken, the methodology proposed can be considered either as an online batch process design method or as an online model-based optimal control/optimization strategy. From a design perspective, the approach dynamically and optimally allows: (I) deciding when to stop/activate a unit and how long to keep it inactive; (II) choosing how to manage interconnections among units; (III) selecting optimal values for product draws and feeds (for fed-batch units only); (IV) determining the batch cycles in which the multiunit batch process will operate. From an operational perspective the online optimization/optimal control approach should promote: (I) the effective compensation of disturbances at the entire plant scale; (II) the selection of the optimal production levels based on the market demand; etc. All of these aspects need to be optimized together, accounting for their interactions. Therefore, the proposed integrated real-time optimal design and optimization/optimal control strategy seeks to guarantee improved performance compared to the conventional sequential approach, which combines offline plant optimal design with individual units online control/management developed separately.

The methodology discussed is validated and tested on a novel batch process for the production of β-pinene (Bauer et al., 1988), which includes a (catalytic) batch reaction followed by filtration and distillation steps. It is applied to all the batch plant units simultaneously and its performance is compared to the conventional sequential (offline optimal design & online optimization/optimal control) approach. Based on the cases presented, lessons will be drawn regarding the situations in which the simultaneous and sequential approaches have advantages.

References:

Acevedo, D., T. Yanssen, and Z.K. Nagy. "Multiobjective Optimization of an Unseeded Batch Cooling Crystallizer for Shape and Size Manipulation." Industrial & Engineering Chemistry Research 54 (2015): 2156-2166.

Greaves, M.A., I.M. Mujtaba, M. Barolo, A. Trotta, and M.A. Hussain. "Neural-network approach to dynamic optimization of batch distillation-Application to a middle-vessel column." Chemical Engineering Research and Design 81 (2003): 393-401.

Smets, I.Y., J.E. Claes, E.J. November, G.P. Bastin, and J.F. Van Impe. "Optimal adaptive control of (bio) chemical reactors: past, present and future." Journal of process control 14 (2004): 795-805.

Bauer, K., D. Garbe, and H. Surburg. "Ullmann's encyclopedia of industrial chemistry." Ullmann's Encyclopedia of Industrial Chemistry 11 (1988).

426881 A New Framework for Dynamic Parameter Estimation and Optimization of Batch Distillation Columns

Abstract:

Approximately 40,000 distillation columns are used in different industries in the US that account for 6% of the yearly US energy consumption [1]. Continuous distillation columns are generally used in large scale production facilities; optimization of these large-scale units has been considered widely in the literature. Smaller-use items are often processed in batch distillation columns. Optimization of batch distillation columns has gained less attention in the literature, mainly because of the transient nature of both the columns themselves and the market for smaller-use items [2]. Recently, more research studies have considered optimization of the batch distillation columns [3]–[6]. The first step in optimization of batch distillation columns is to develop a model. Several research studies have considered model development for batch distillation columns. The models studied in the literature are generally categorized as rigorous [7]–[9] or shortcut methods [10]–[12]. Rigorous models generally include detailed descriptions of mass transfer, energy transfer and thermodynamics, and are therefore able to account for system non-idealities. While these methods are more accurate, they have high computational costs. In shortcut models, several simplifying assumptions have been considered that decrease the computational time. However, lack of experimental results makes it difficult to determine the accuracy of these models, restricting their use to ideal systems. In this study, a third approach is used to develop a model that incorporates some of the non-idealities of the process not seen in the typical shortcut methods without increasing the computational cost. In this approach, a model is developed by fitting parameters to simplified forms of shortcut models. The use of fitting parameters incorporates system non-idealities into the model but requires experimental data to determine parameter values. Dynamic parameter estimation is used to reduce the experimental load. The accuracy of the parameters is then determined by implementing nonlinear statistics calculations as well as a sensitivity analysis. These calculations help analyze the number of parameters that can be estimated from the experimental data and return an acceptable range for those parameters. The framework used in this study is developed in a general approach that is applicable to other batch processes. The framework includes the use of $\ell 1$-norm dynamic parameter estimation, nonlinear statistics, parameter sensitivity analysis, and dynamic optimization. These techniques have not been applied together to batch distillation columns before in a holistic approach to dynamic optimization. The results obtained from dynamic optimization are also validated in this study.

The batch column used to collect data is a 38 tray, 2 inch, vacuum-jacketed and silvered Oldershaw column. A constant heat input is supplied from a reboiler heater. The reboiler is initially charged with 1.5 L of a 50/50 wt% mixture of methanol and ethanol for each run, with the goal being a product of 99 wt% methanol. The manipulated variable is the reflux ratio. A doublet test is implemented to collect data for model validation. Initially four parameters are assumed for fitting data to the model. These parameters are heater efficiency, vaporization efficiency, condenser molar holdup, and tray molar holdup. The maximum error observed between the model prediction and experimental data for the instantaneous distillate is 10%. For all but four measurements the model prediction is within 2%. The error between the prediction and measurement for the cumulative number of methanol is negligible.

Although the fit between the model and experimental data is excellent, a glance at the confidence interval reveals a wide region for all parameters except the heater efficiency. This might be because of the over-parametrization of the problem. A sensitivity analysis is therefore used to determine the number of the parameters that best represent the model. The singular value decomposition of the sensitivity matrix is implemented and it is found that there is one principle parameter (heater efficiency) that can be used to match the production data (number of moles of methanol). The singular value analysis also reveals that only one other parameter is required to estimate the composition of methanol. In this study, vaporization efficiency is selected as the second parameter and the other two parameters (condenser and tray holdups) are assumed fixed for further analysis. Implementing the confidence interval calculations for the model with two parameters shows a narrow confidence interval that confirms the results obtained from the sensitivity analysis. The model prediction with two parameters also match the experimental results.

The validated model is then used to maximize the amount of methanol production for a 90 minute batch time. The non-optimized base case production over a 90 minute run is 9.5 moles of 99.2 mol% methanol at a constant reflux ratio of 4. To optimize methanol production, reflux ratio is varied every 5 minutes by the optimizer. Optimization results show that the reflux ratio starts low, then increases in a nominally linear pattern. This is done to take advantage of the high methanol concentration at the top of the column after the startup period. The model prediction for the optimization case is 9.75 moles of 99.0 mol% methanol. Implementing the optimized reflux ratio in the batch column resulted in 10.8 moles of 99.8 mol% methanol obtained in the experiment. The difference between model prediction and experiment is 10% and 0.8% for overall production and product composition, respectively. The agreement between model and experiment is excellent and reflects the work done to validate the model. Methanol production using the optimized reflux ratio represents a 14% increase in column production over the base case with a reflux ratio of 4. The increase in the productivity of the column highlights the potential benefits in applying the proposed framework in larger batch distillation columns as well as other batch processes.

[1] A. Lucia and B. R. McCallum, "Energy targeting and minimum energy distillation column sequences," *Comput. Chem. Eng.*, vol. 34, no. 6, pp. 931–942, 2010.

[2] S. Ulas, U. M. Diwekar, and M. A. Stadtherr, "Uncertainties in parameter estimation and optimal control in batch distillation," *Comput. Chem. Eng.*, vol. 29, no. 8, pp. 1805–1814, 2005.

[3] Y.-S. Kim, "Optimal Control of Time-Dependent Processes (Design, Batch Distillation, Catalyst Deactivation)," 1985.

[4] U. Diwekar, *Batch Distillation: Simulation, Optimal Design, and Control*. CRC Press, 1995.

[5] B. T. Safrit and A. W. Westerberg, "Improved Operational Policies for Batch Extractive Distillation Columns," *Ind. Eng. Chem. Res.*, vol. 36, no. 2, pp. 436–443, 1997.

[6] K. H. Low and E. Sørensen, "Simultaneous optimal configuration, design and operation of batch distillation," *AIChE J.*, vol. 51, no. 6, pp. 1700–1713, 2005.

[7] M. Barolo and G. B. Guarise, "Batch Distillation of Multicomponent Systems with Constant Relative

Volatilities," *Chem. Eng. Res. Des.*, vol. 74, no. 8, pp. 863–871, 1996.

[8] Y. H. Kim, "Optimal design and operation of a multi-product batch distillation column using dynamic model," *Chem. Eng. Process. Process Intensif.*, vol. 38, no. 1, pp. 61–72, 1999.

[9] I. M. Mujtaba and M. A. Hussain, "Optimal operation of dynamic processes under process-model mismatches: Application to batch distillation," *Comput. Chem. Eng.*, vol. 22, Supple, pp. S621 – S624, 1998.

[10] M. Barolo and F. Botteon, "Calculation of parametric sensitivity in binary batch distillation," *Chem. Eng. Sci.*, vol. 53, no. 10, pp. 1819–1834, 1998.

[11] A. O. Converse and G. D. Gross, "Optimal Distillate-Rate Policy in Batch Distillation," *Ind. Eng. Chem. Fundam.*, vol. 2, no. 3, pp. 217–221, 1963.

[12] M. M. Lopes and T. W. Song, "Batch distillation: Better at constant or variable reflux?," *Chem. Eng. Process. Process Intensif.*, vol. 49, no. 12, pp. 1298–1304, 2010.

430115 Moving Horizon Based Real Time Optimization and Advanced Hybrid Model Predictive Control of Continuous Pharmaceutical Manufacturing Process

Abstract:

The pharmaceutical company is still largely a batch process industry. The strict regularity requirements from the regulatory agencies and different level of complexities involved with pharmaceutical processing has prevented the transition towards continuous manufacturing. However, Quality by Design (QbD), a novel initiative proposed by regulatory agencies, is a big leap in making this transition successful. QbD involves the use of process analytical technology (PAT) tools to monitor the process with a real time feedback system to provide a control strategy suitable for continuous manufacturing. Traditional control strategies such as proportional-integral-derivative (PID) and model predictive control (MPC) have been widely used in chemical and petrochemical industries in the past decades. However, pharmaceutical industries have been slow in adapting this technology in the traditional quality-by-testing (QbT) approach largely due to strict regulatory requirements.

In this work, a moving horizon based real time optimization (MH-RTO) technique has been integrated with a hybrid model predictive control (MPC) system. The integrated system has been applied in a continuous downstream API separation and purification process as well as tablet manufacturing process. The API separation and purification process consist of four continuous downstream unit operations (Crystallization, filtration, drying and mixing) while the continuous direct compaction tablet manufacturing process consist of four unit operations (feeders, co-mill, blender, tablet press). In the proposed approach, the integrated MH-RTO provides the optimal operational set points for the total production rate in real time. The MH-RTO takes into consideration the capital and operating cost, the market fluctuations, the product inventory, the product quality assurance strategy, the regulatory constraints, and the product rejections. An advanced hybrid model predictive control system then ensures that required production rate with desired quality is met with minimum resources and time. A robust optimization strategy and an efficient control system have been integrated to achieve the maximum profit. The MH-RTO integrated with a hybrid control strategy ensures the maximum possible profit irrespective of the market demand fluctuations. The basic advantage of the MH-RTO framework is that it takes into consideration the future demand and thus can lead to increased profit compared to a standard Real Time Optimization approach.

The objective of this presentation is two-fold. First to highlight the moving horizon based real time optimization technique and MPC-PID hybrid control architecture, and second to demonstrate its applications through two case studies: downstream API separation and purification process and tablet manufacturing process.

Imperial College London

Presentation ID# 409130

Whole-system optimisation of integrated wind-electricity-hydrogen networks for decarbonising the domestic transport sector in Great Britain

Dr. Sheila Samsatli

2015 AIChE Annual Meeting

Salt Lake City, UT

8-13 November 2015

wholeSEM

Centre for Process Systems Engineering Imperial College, London

EPSRC Engineering and Physical Sciences Research Council

Imperial College London

Outline

1. Introduction
2. Integrated wind-electricity-hydrogen networks model
3. Case studies
4. Conclusions

Imperial College London

Meeting domestic transport demand using only on-shore wind

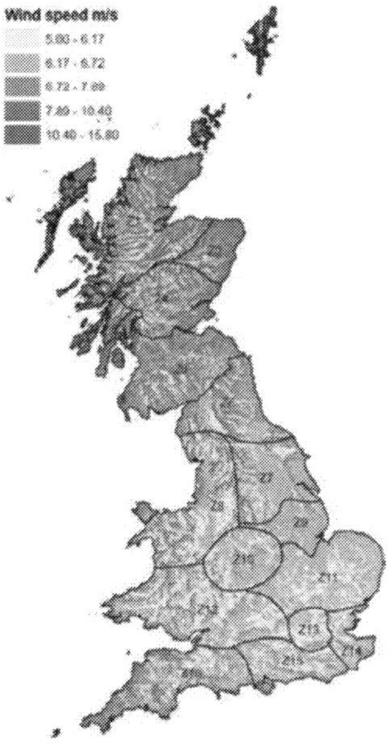

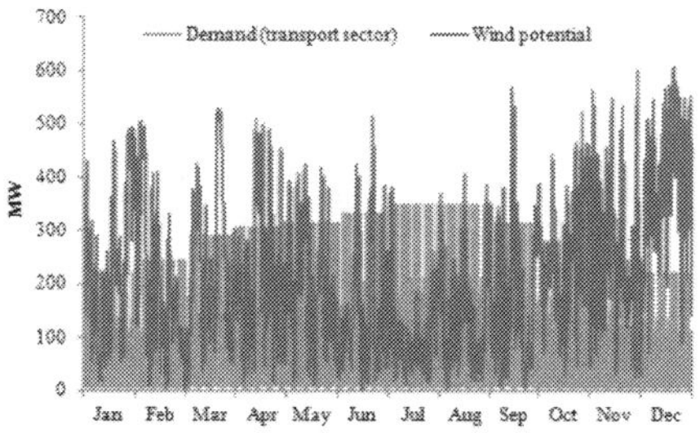

How many wind turbines and where?
Electricity or hydrogen?
What will the network look like?
Storage? What and where?
Technologies (fuel cells, electrolysers)?

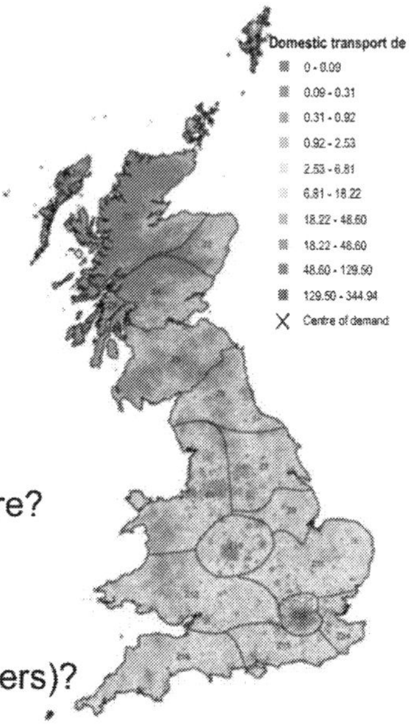

Wind availability

Demand

Imperial College London

Integrated wind-hydrogen-electricity networks model

- Modelled in STeMES*, a general model of energy systems
- Mixed Integer Linear Programming (MILP) model
  - Determines the design and operation of the network
- Spatio-temporal model
  - Spatially-distributed (demands, wind availability, location of technologies)
  - Dynamic (demands, wind availability, storage, operation of the network depends on time)
- Network of conversion, storage and transport technologies
- Implemented in AIMMS and solved with CPLEX solver

*S. Samsatli and N. Samsatli (2015). A general spatio-temporal model of energy systems with a detailed account of transport and storage. Computers and Chemical Engineering 80, 155-176, 0098-1354.

Imperial College London

Temporal representation

- Long-term strategic decisions
- Short-term operational issues (intermittency, dynamics of energy storage)

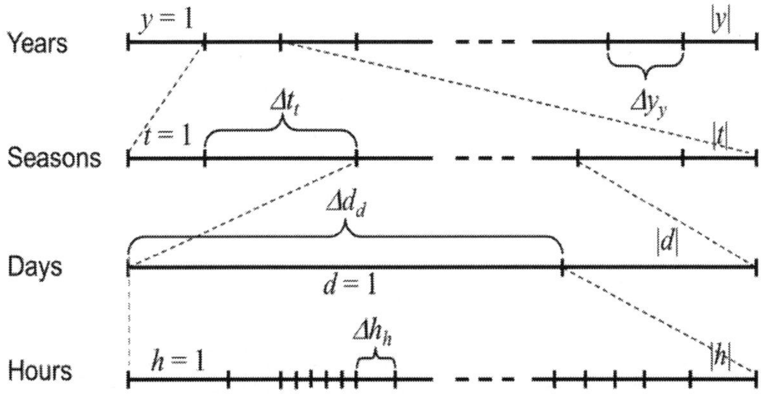

Without storage – very easy!
With storage – extra variables for initial inventories; extra constraints to link inventories within and between time levels

Imperial College London

Spatial representation

- GB divided into 16 transmission zones based on the National Grid SYS 17 study zones
- Each zone may contain a number of technologies for generation, conversion, storage and transport
- Transmission lines (e.g. hydrogen pipelines or electricity cables) may connect each zone to its neighbours

Imperial College
London

Spatio-temporal wind availability

Time-series wind data for each zone were obtained from **Virtual Wind Farm Model** of Iain Staffell

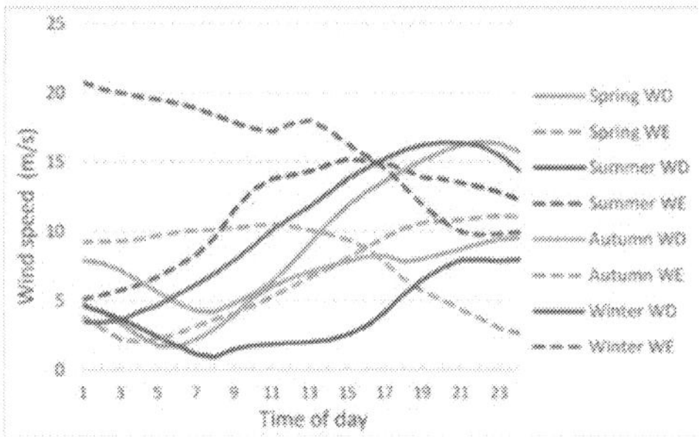

Temporal distribution of wind speed in zone 13 (similar graphs were derived for the other zones)

Imperial College
London

Spatio-temporal demand data

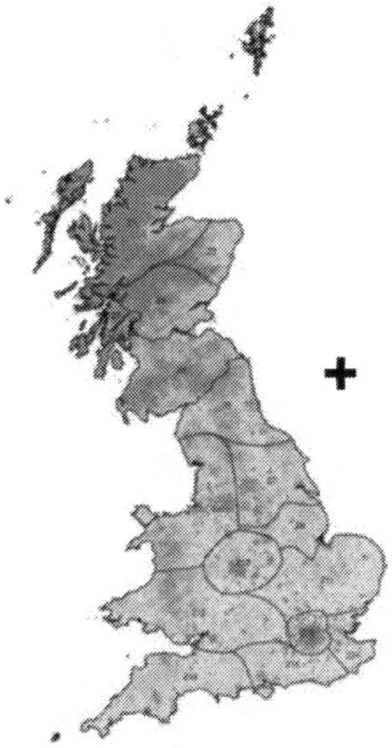

Hydrogen demand at 1km

+ Department of Transport statistics for vehicular usage at different times =

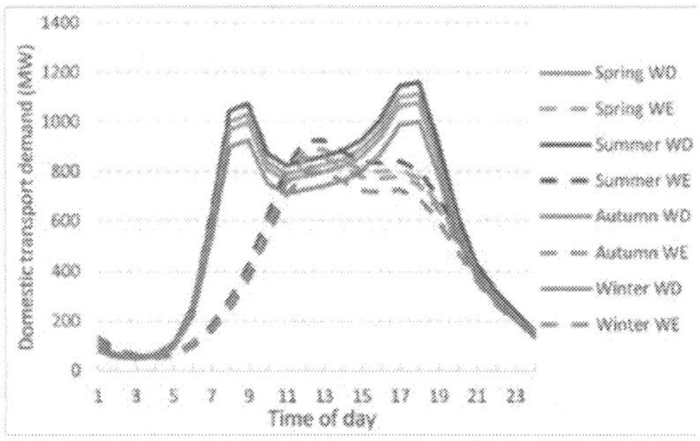

Temporal distribution of hydrogen demands in Z13 (similar graphs were derived for the other zones)

Imperial College
London

Wind turbine siting constraints

Criteria used to determine the total land area in each zone suitable for siting wind turbines

1. Annual wind speed of at least 5m/s at 45m above ground level
2. Slope of less than 15%
3. Access: a minimum distance of 500m from minor road network
4. Connectivity to National Grid: at least 200m but not more than 1500m from major road network
5. Not in SSSI (Sites of Special Scientific Interest)
6. Population impacts: at least 500m from DLUA (developed land used area
7. Water pollution: at least 200m from river
8. Wildlife and interference: at least 250m from woodland
9. Safety: at least 5km from airports
10. Not occupied by existing wind turbines including spacing between turbines of 5 rotor diameters

Imperial College
London

Land footprint constraint

- Total available area for wind turbines
  - Intersection of the 10 constraints
- Total available area in each zone defines the land footprint constraints in the model

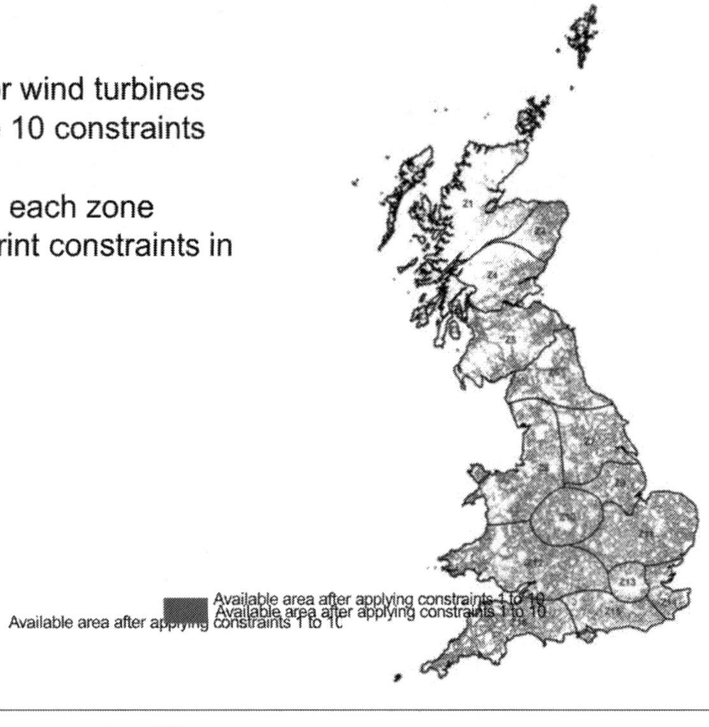

Imperial College
London

Resource-Technology Network

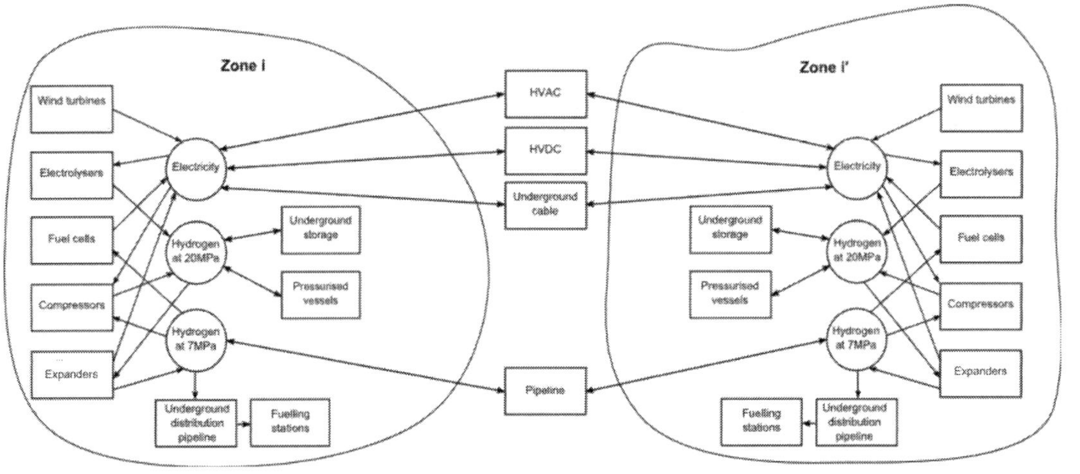

Network of conversion, storage and transmission technologies connecting 2 zones

Imperial College
London

Production technology

Wind turbines

- Considered standard on-shore wind turbines at commercial scale (i.e. rotor diameter of 100m)
- Assumed a minimum spacing of 5 rotor diameters between turbines
- Existing wind turbines can be used if it's cost-effective to do so

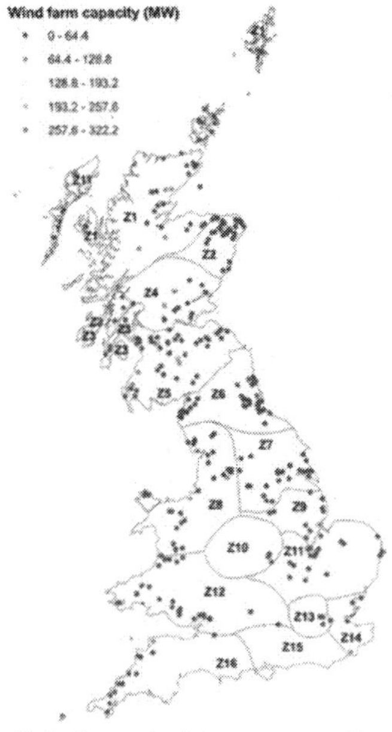

Existing wind farm capacity

Imperial College London

Conversion technologies

Electrolysers
- Max production rate of 69.38MW
- High pressure electrolysis at 20MPa

Fuel cells
- 41.63MW solid oxide fuel cell

Compressors
- Reciprocating compressors at different sizes

Expanders
- Reciprocating expanders at different sizes

Imperial College London

Storage technologies

Characterised by their:
- Maximum available capacity
- Injectability – maximum rate that H_2 gas can be injected into storage
- Deliverability – maximum rate that H_2 gas can be withdrawn from storage

Cushion gas - the volume of gas required to be kept in a facility in order to maintain operating pressure and cannot be recovered until the facility stops operation

Overground storage
- Compressed gas storage tanks at 20MPa
- Considered 3 sizes
  - Small - Max capacity: 0.36GWh; Injectability/deliverability: 15.13MW
  - Medium - Max capacity: 3.63GWh; Injectability/deliverability: 151.25MW
  - Large - Max capacity: 36.3GWh; Injectability/deliverability: 1512.50MW

Imperial College London

Underground storage

- Salt caverns, depleted oil/gas fields and aquifers
- Four underground storage considered in the model:

  Aldborough (Z7)
  - Salt cavern
  - Max capacity: 3.3TWh; Injectability: 9.17GW; Deliverability: 10.82GW

  Humbly Grove (Z15)
  - Depleted oil/gas field
  - Max capacity: 3.05TWh; Injectability: 3.79GW; Deliverability: 3.29GW

  Rough (Z7)
  - Depleted oil/gas field
  - Max capacity: 34TWh; Injectability: 10GW; Deliverability: 18.96GW

  Warmingham (Z8)
  - Salt cavern
  - Max capacity: 1.08TWh; Injectability: 4.88GW; Deliverability: 1.21GW

Imperial College London

Transmission technologies

Hydrogen pipeline
- Data obtained from simulation using the pipeline model in gCCS
- Diameter: 100cm, max inlet pressure of 7.1MPa

Electricity cables
- HVAC OHL, single circuit, 400kV, 1500MVA
- HVAC OHL, double circuit, 400kV, 2x1500MVA
- HVAC underground XLPE cable, single circuit, 400kV, 1000MVA
- HVAC underground XLPE cable, double circuit, 400kV, 2x1000MVA
- HVDC OHL, bipolar, $\pm$ 400kV, 1500MW
- HVDC underground cable pair, $\pm$ 350kV, 1100MW

Imperial College London

Centres of demand

- Transmission assumed to be between the centres of demand of each zone
- Obtained from the demand density at 1km, $D(x, y)$

$$x_z = \frac{\iint_{S_z} xD(x,y)\,dx\,dy}{\iint_{S_z} D(x,y)\,dx\,dy}$$

$$y_z = \frac{\iint_{S_z} yD(x,y)\,dx\,dy}{\iint_{S_z} D(x,y)\,dx\,dy}$$

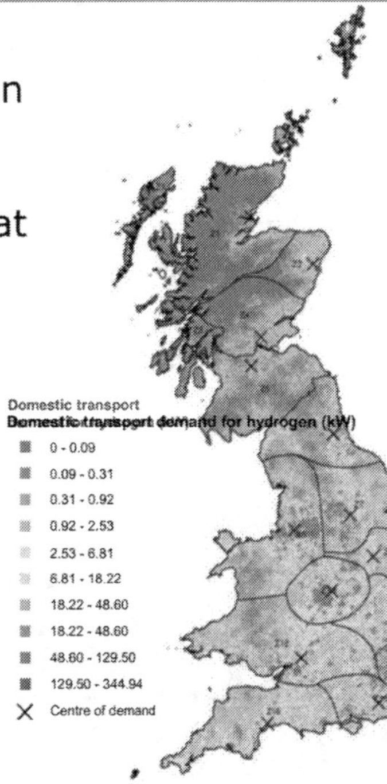

Imperial College London

Length of the distribution network

- Previous methods used a grid of square cells and assumed that demands are uniformly distributed.
- A better approximation uses the demand density, $D(x, y)$:

$$L_z^{network} = \iint_{S_z} \frac{D(x,y)}{C} \sqrt{(x - x_z)^2 + (y - y_z)^2}\,dx\,dy$$

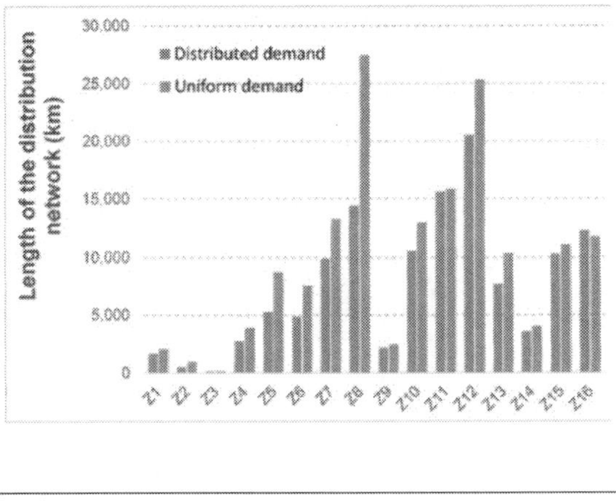

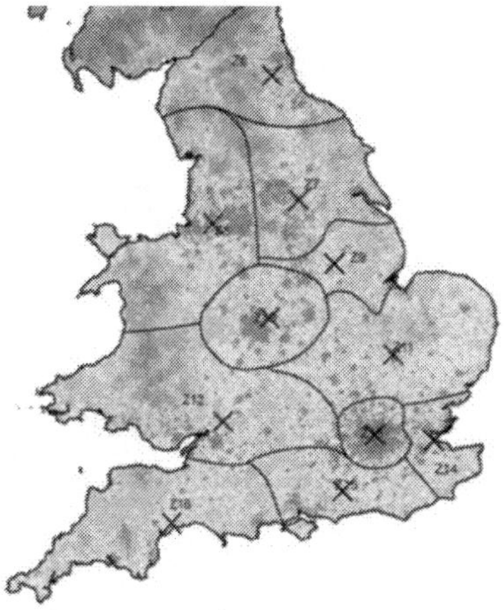

Imperial College London

Number of fuelling stations

- Number of fuelling stations:

$$N_z^{stations} = \left\lceil \frac{1}{C} \iint_{S_z} D(x,y)\, dx\, dy \right\rceil$$

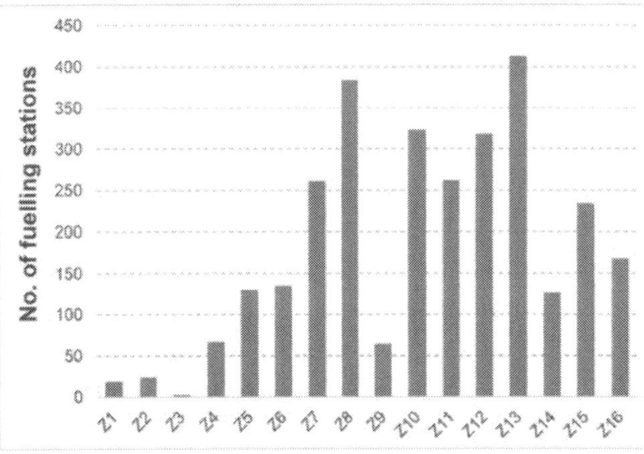

Total distribution cost ≈ £bn17/yr
Assuming:

- Fuelling station capacity of 1,500kg/day
- Distribution pipeline diameter of 20cm
- Capital charge factor of 3

Imperial College London

Production/conversion-storage-transmission network problem

Given:

- The hydrogen demand and availability of wind power
- Characteristics of each technology (e.g. unit costs: CAPEX, O&M; efficiency)

Determine:

- The optimal number, size and location of wind turbines, electrolysers, fuel cells, compressors, expanders and hydrogen storage
- Whether to transmit the energy as electricity or hydrogen or both
- The structure of the transmission network
- The hourly operation of each network component

Subject to:

- Max available land area for the technologies (only land footprints of wind turbines are considered in this study)
- Satisfying all of the demands

Objective:

- Minimise total system cost

Base case: choose the best from all of the possible technologies

(existing wind turbines not included)

Breakdown of costs [£M/yr]

(excluding distribution costs)

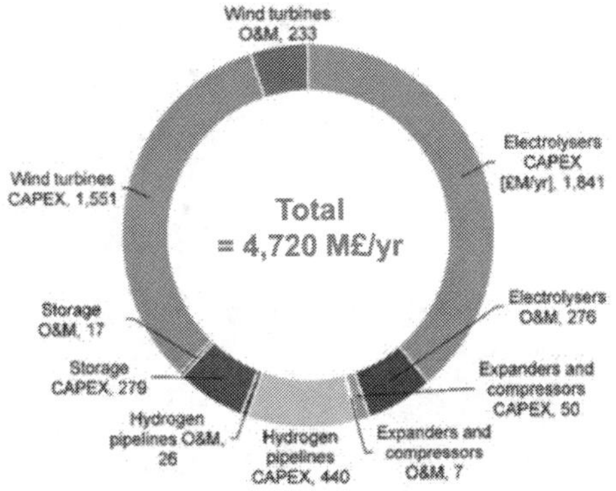

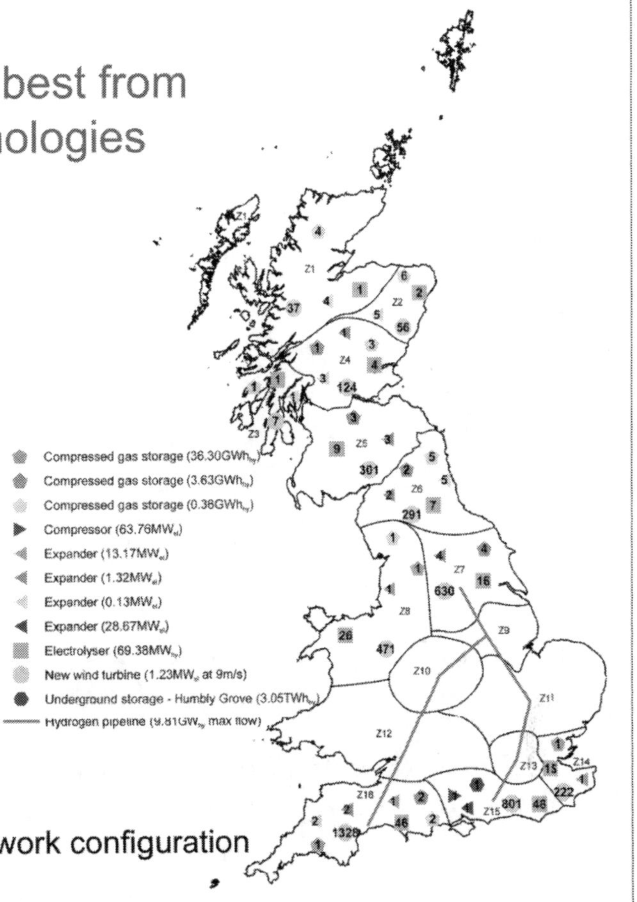

Least-cost network configuration

Base case (cont...)

Snapshot of the operation of the transmission network (pipeline) during weekdays in summer

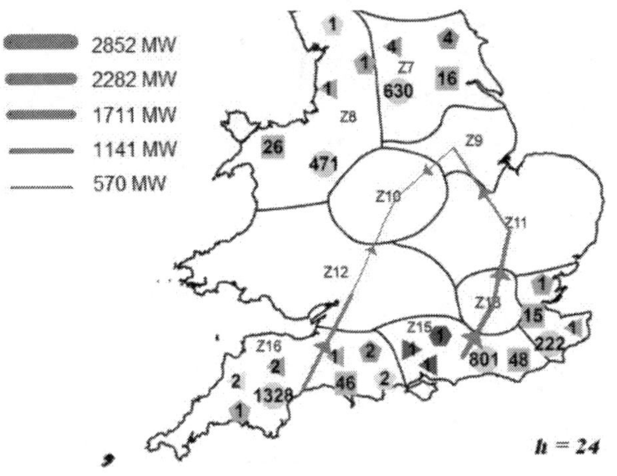

View in slide show mode for the animation

Base case (cont...)

Inventory of hydrogen in the Humbly Grove underground storage (Z15)

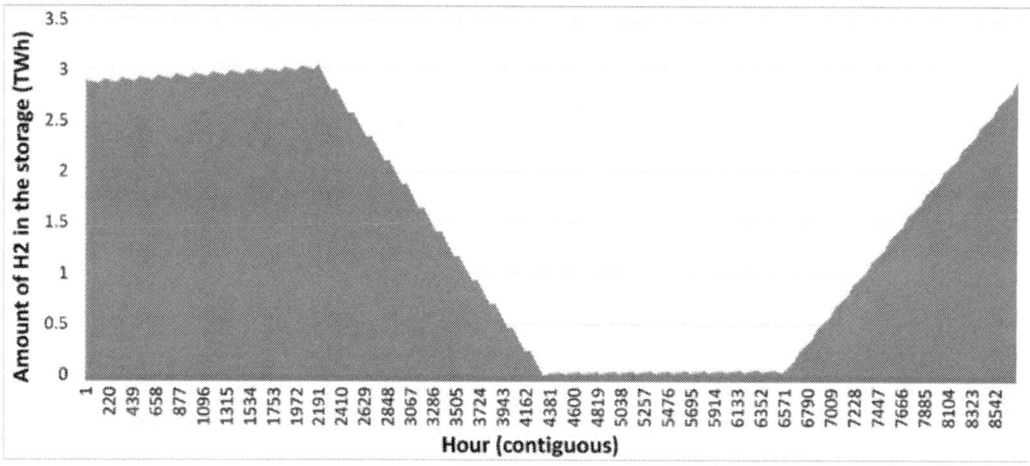

The underground storage is being used effectively as a seasonal storage.

Base case (cont...)

Inventory of hydrogen in overground storage (tanks) at different zones

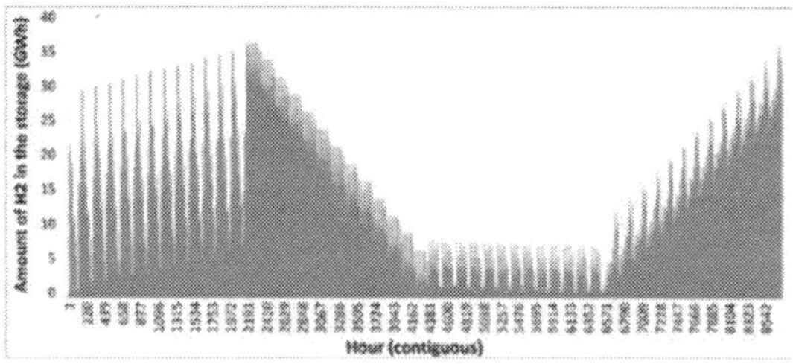

36.3GWh storage in zone 16

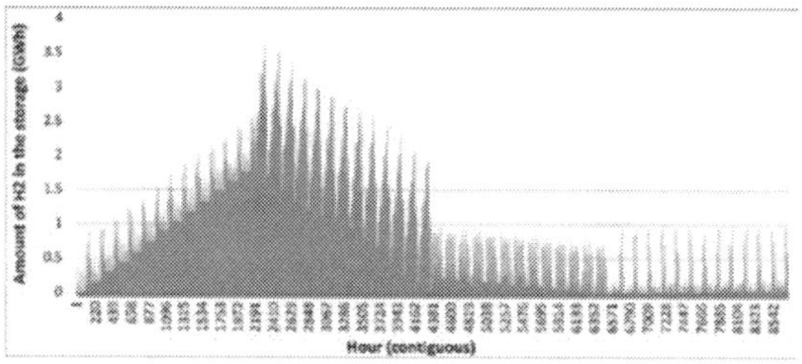

3.63GWh storage in zone 4

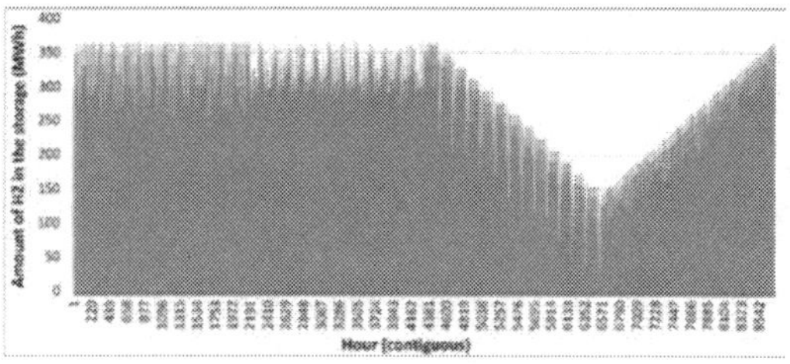

363MWh storage in zone 8

The tank storage are being used as seasonal storage as well as for hourly balancing

Case 2: The value of existing wind turbines

Breakdown of costs [£M/yr]
(excluding distribution costs)

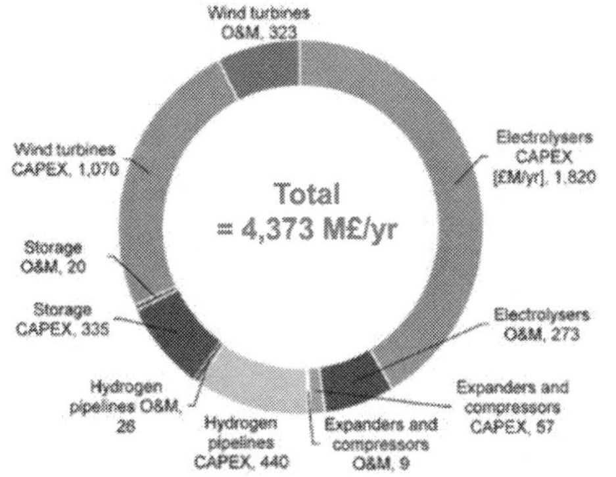

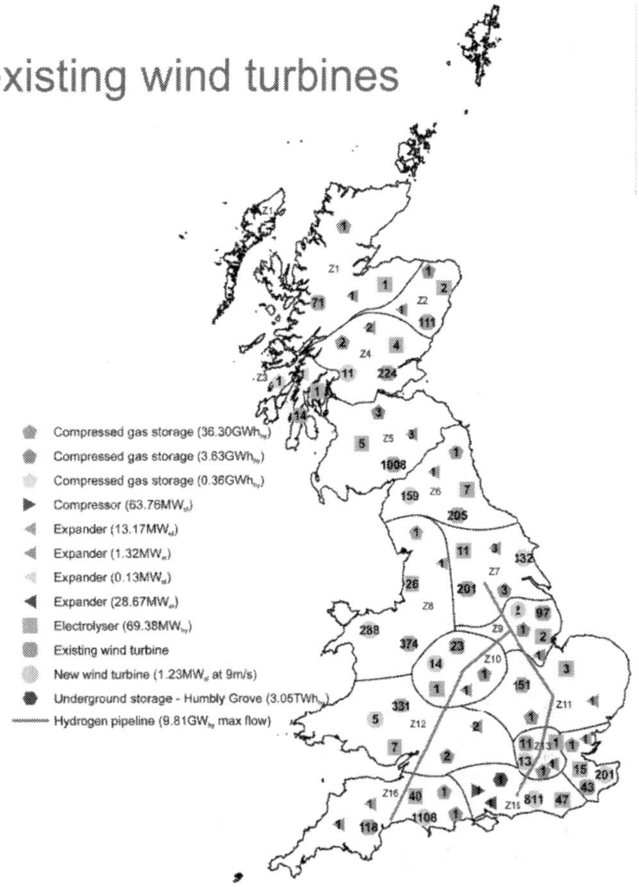

Comparison with base case

	% difference
Wind turbines	-22%
Conversion (Electrolysers, expanders and compressors)	-1%
Storage	20%
Transmission	0%
Overall network	**-7%**

Case 3: The value of underground storage

Breakdown of costs [£M/yr]
(excluding distribution costs)

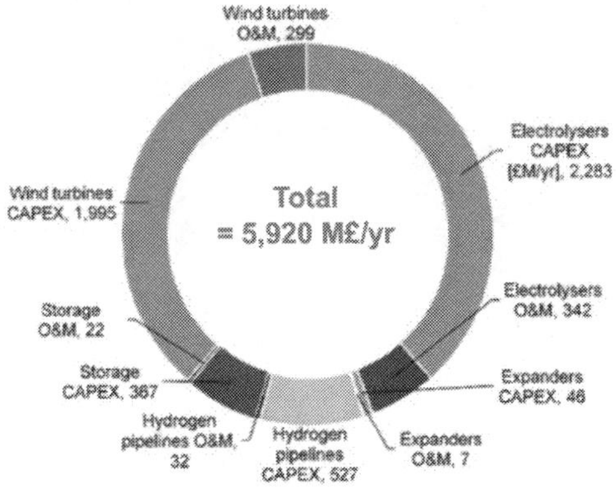

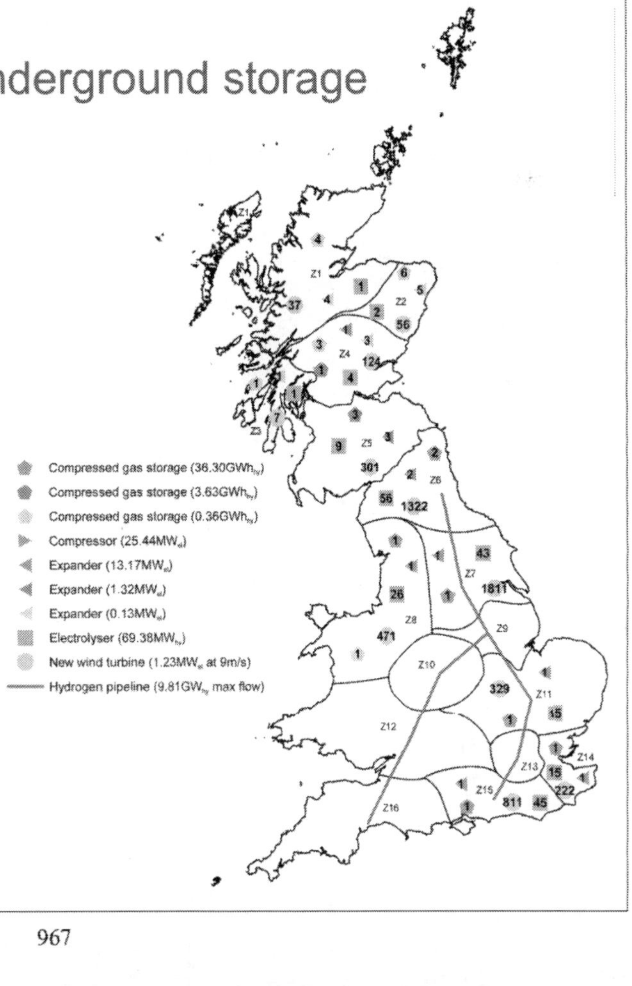

Comparison with base case

	% difference
Wind turbines	29%
Conversion (Electrolysers, expanders and compressors)	23%
Storage	32%
Transmission	20%
Overall network	**25%**

Case 4: The value of pipeline transmission

Breakdown of costs [£M/yr]
(excluding distribution costs)

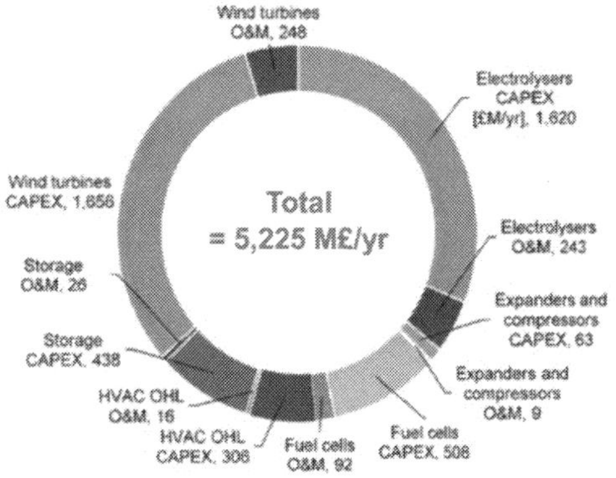

Comparison with base case

	% difference
Wind turbines	7%
Conversion (Electrolysers, expanders and compressors)	17%
Storage	57%
Transmission	-31%
Overall network	**11%**

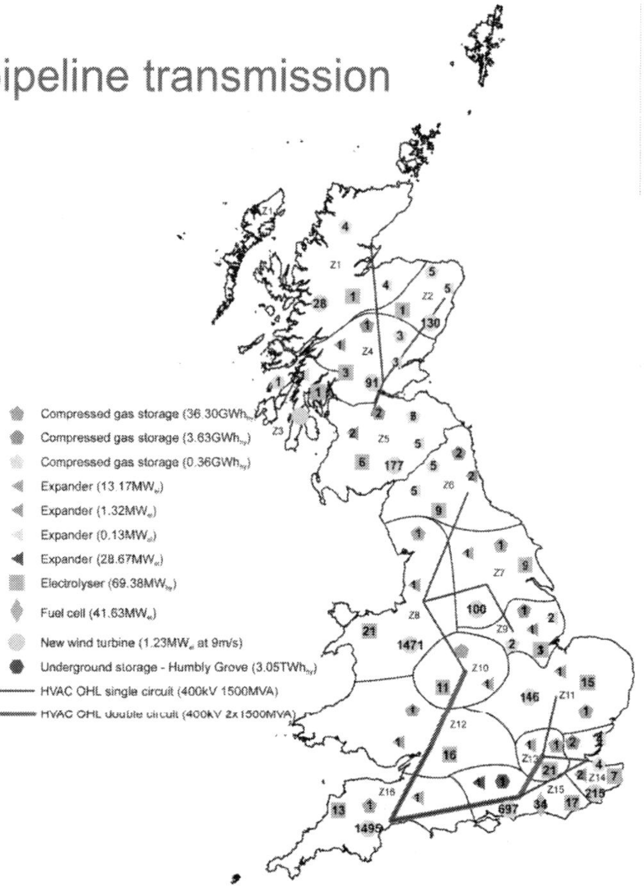

Case 5: The cost of underground electricity cables

Breakdown of costs [£M/yr]
(excluding distribution costs)

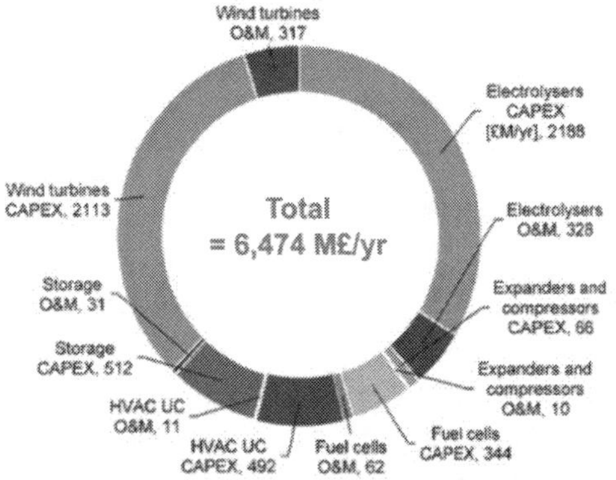

Comparison with base case

	% difference
Wind turbines	36%
Conversion (Electrolysers, expanders and compressors)	38%
Storage	83%
Transmission	8%
Overall network	**37%**

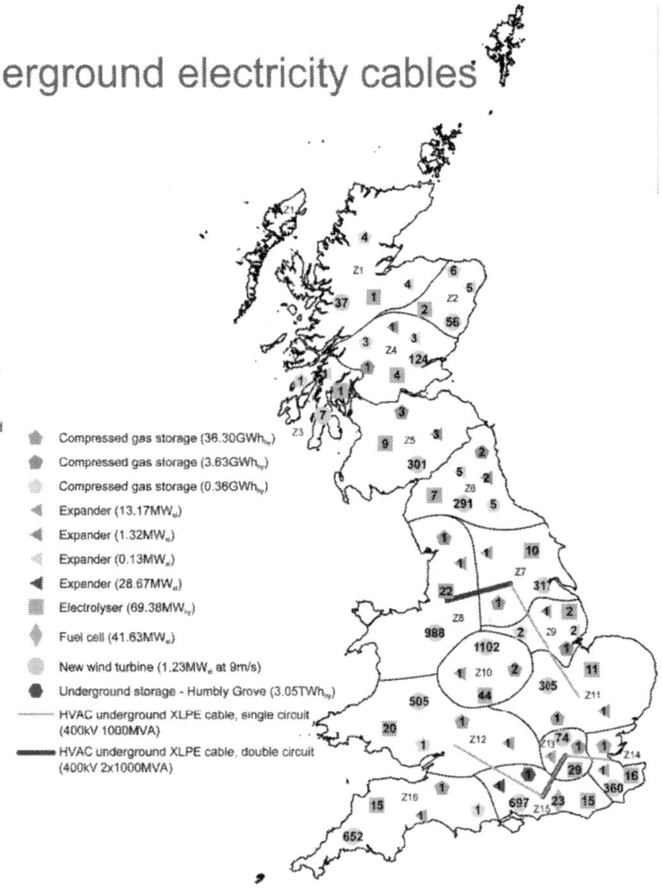

Imperial College
London

Conclusions

- All of the domestic transport demand can be met by wind
- Optimal solution (without using existing turbines)
  - Build a hydrogen pipeline network in the south of England and South Wales
  - Use the Humbly Grove underground storage
  - No wind turbines or technologies in the Midlands and Greater London
  - Northern Wales, England and Scotland self sufficient
  - Humbly Grove is used for seasonal storage
  - Storage tanks elsewhere used for hourly balancing as well as seasonal storage
- Results may change with the inclusion of more technologies, e.g. batteries, electric vehicles

S. Samsatli, I. Staffell, N. Samsatli (2015). Optimal design and operation of integrated wind-hydrogen-electricity networks for decarbonising the domestic transport sector in Great Britain. Int. J. of Hydrogen Energy (revised manuscript submitted).

Imperial College
London

Future work

- Electric vehicles
  - Determine optimal penetration of electric and FC vehicles
  - Allow electricity storage in some electric vehicles
- Pipeline storage
- Integrate natural gas networks
  - H_2 injection into gas pipelines
- Include demands from other sectors

Imperial College
London

Acknowledgements

- Dr. Iain Staffell for the wind data
- EPSRC funding through the wholeSEM consortium (EP/K039326/1)

418104 Accurate Simulation of Natural Gas Liquefaction Processes

Abstract:

We present a new compact formulation and accurate solution strategy for modeling and simulation of natural gas liquefaction processes. The critical process units in liquefaction processes are cryogenic multistream heat exchangers, which are notoriously difficult to simulate and design rigorously due to their complex heat transfer behavior, made all the more difficult by the presence of phase changes and non-ideal thermodynamics. We extend a newly developed model for multistream heat exchangers[1] to incorporate automatic detection and handling of phase changes in which the fluids are described by cubic equations of state. We formulate and solve our model as a small system of nonsmooth equations, rather than using the smoothing approximations or large optimization-based disjunctive programs which are found in the literature for handing these phenomena[2]. Additionally, we employ an extension of automatic differentiation to automatically calculate computationally relevant generalized derivatives[3] and a linear programming based Newton method[4]to guarantee that a precise, physically realizable result is found with a local quadratic convergence rate. A case study of the PRICO process is presented to demonstrate the model and solution algorithm.

---

[1] H. A. J. Watson, K. A. Khan, and P. I. Barton. "Multistream heat exchanger modeling and design." Submitted.

[2] R. S. Kamath, L. T. Biegler, and I. E. Grossmann, "Modeling multistream heat exchangers with and without phase changes for simultaneous optimization and heat integration," *AIChE Journal*, vol. 58, no. 1, pp. 190–204, 2012.

[3] K. A. Khan and P. I. Barton, "A vector forward mode of automatic differentiation for generalized derivative evaluation." In Press. *Optimization Methods & Software.*

[4] F. Facchinei, A. Fischer, and M. Herrich, "An LP-Newton method: nonsmooth equations, KKT systems, and nonisolated solutions," *Mathematical Programming*, vol. 146, pp. 1–36, 2014.

420114 Integrated Biomass and Natural Gas Refineries for the Co-Production of Liquid Fuels, Olefins, and Aromatics: Optimization Under Uncertainty

Abstract:

Biomass and natural gas are attractive feedstocks for the synthesis of alternative refineries towards the production of fuels and chemicals [1-3]. Biomass can provide environmental benefits through reduced greenhouse gas (GHG) emissions, whereas inexpensive natural gas can improve process economics. Recent work [1-3] suggested that these refineries could be economically competitive with chemicals production significantly increasing the plant profitability. However, volatility of the crude oil prices can shift the process economics significantly. Crude oil prices are around $50/bbl in 2015 compared to $100/bbl in 2014 [4]. In tandem, refinery gasoline prices fell to as low as $1.36/gal in 2015 from as high as $2.89/gal in 2014 [4]. Similarly, petrochemicals price index fell to as low as $850/MT in 2015 compared to a value of $1450/MT in 2014 [5].

A refinery design that is highly profitable at 2014 prices may become infeasible to operate at 2015 levels. Therefore, uncertainties in feedstock and product prices need to be systematically addressed. To this motive, we have incorporated a robust optimization framework [6-8] into the process synthesis superstructure developed for the production of liquid fuels and chemicals [1-3]. The process superstructure incorporates a hybrid feedstock for the coproduction of liquid fuels, olefins, and aromatics with multiple conversion pathways, and upgrading technologies. The large-scale non-convex non-linear optimization model (MINLP) for the process superstructure is improved with the robust counterpart constraint for each uncertain price parameter.

A set of case studies is investigated to evaluate different uncertainty sets, robust counterpart formulations, and probability of constraint violation. Each case study is solved to global optimality within a branch-and-bound framework to maximize the profit. Important topological trade-offs (which process to utilize, which products to produce) triggered by the robust optimization framework will be presented. The strength of robust optimization framework will be demonstrated via a comparison with deterministic case study results.

References:

[1]: Baliban, R. C.; Elia, J. A.; Floudas, C. A.; Biomass and Natural Gas to Liquid Transportation Fuels: Process Synthesis, Global Optimization, and Topology Analysis, 2013, Industrial and Engineering Chemistry Research, 52(9), 3381-3406.

[2]: Onel, O.; Niziolek, A. M.; Elia, J. A.; Baliban, R. C.; Floudas, C. A.; Biomass and Natural Gas to Liquid Transportation Fuels and Olefins (BGTL+C2_C4): Process Synthesis and Global Optimization, 2015, Industrial and Engineering Chemistry Research, 54, 359-385.

[3]: Niziolek, A. M.; Onel, O.; Elia, J. A.; Baliban, R. C.; Floudas, C. A.; Coproduction of liquid transportation fuels and C6_C8 aromatics from biomass and natural gas, 2015, AIChE Journal, 61(3), 831-856.

[4]: Energy Information Administration, This Week in Petroleum (May 6, 2015), 2015.

[5]: Platts Global Petrochemicals Index, 2015

[6]: Li, Z.; Ding, R.; Floudas, C. A.; A Comparative Theoretical and Computational Study on Robust Counterpart Optimization: I. Robust Linear Optimization and Robust Mixed Integer Linear Optimization, 2011, Industrial and Engineering Chemistry Research, 50, 10567-10603.

[7]: Li, Z.; Tang, Q.; Floudas, C. A.; A Comparative Theoretical and Computational Study on Robust Counterpart Optimization: II. Probabilistic Guarantees on Constraint Satisfaction, 2012, Industrial and Engineering Chemistry Research, 51(19), 6769-6788.

[8]: Li, Z.; Floudas, C. A.; A Comparative Theoretical and Computational Study on Robust Counterpart Optimization: III. Improving the Quality of Robust Solutions, 2014, Industrial and Engineering Chemistry Research, 53(33), 13112-13124.

421748 Mixed-Integer Programming Models for Long-Term, Quality-Sensitive Shale Gas Development

Abstract:

Mixed-Integer Programming Models for Long-Term, Quality-Sensitive Shale Gas Development

Markus G. Drouven[1] and Ignacio E. Grossmann[2]

Department of Chemical Engineering

Carnegie Mellon University

Pittsburgh, PA 15213

[1]mdrouven@cmu.edu, [2]grossmann@cmu.edu

Abstract

The production of shale gas from unconventional resource plays is transforming the energy landscape in the United States. Advances in production technologies, notably the dual application of horizontal drilling and hydraulic fracturing, allow the extraction of vast deposits of trapped natural gas that, until recently, were uneconomic to produce. The Energy Information Administration predicts that shale gas will account for 50% of total U.S. natural gas production by 2040 [1]. Natural gas demand is also expected to increase in the electric power and nearly all other industrial sectors. The future development of shale gas resources requires an extensive expansion of the existing gas production, transmission, and processing infrastructure. Virtually all stages of the shale gas supply chain need to be expanded and upgraded to match the ever growing natural gas supply and demand [3]. Since the necessary capital investments for drilling rigs, pipelines, boosting stations and midstream processing facilities are substantial, the long-term planning of upstream production and natural gas transmission is a key challenge.

The problem addressed in this work can be stated as follows. Within a potential shale gas development area an upstream operator has identified a set of candidate well pads from which shale gas may or may not be extracted. To extract the gas the operator can develop, i.e., drill and fracture a limited number of wells at every candidate pad. Ultimately, the operator wishes to sell extracted gas at a set of downstream delivery nodes which are typically located along interstate transmission pipelines. For this purpose a gathering system superstructure has been identified. This superstructure specifies all feasible, alternative options for laying out gathering pipelines to connect candidate well pads with the given set of delivery nodes. In addition, the superstructure indicates candidate locations for compressor stations as well as the location of existing processing plants.

The long-term shale gas development problem involves planning, design and strategic decisions. In terms of planning decisions the operator needs to decide: a) where, when and how many wells to drill at every candidate well pad, b) whether selected wells should be shut-in and, if so, for how long, and c) how to allocate drilling rigs over time. The design decisions involve: a) where to lay out gathering pipelines, b) what size pipelines to install, c) where to construct compressor stations, and d) how much compression power to provide. Finally, we consider strategic decisions that include: a) the selection of preferred downstream delivery nodes, b) the arrangement of delivery agreements, and c) the procurement of take-away capacity. The upstream operator's objective is to determine the optimal development strategy by making the right planning, design and strategic decisions such that the net present value is maximized.

The shale gas development problem requires operators to size equipment that is part of the shale gas gathering system such as pipelines and compressors. We take advantage of the discrete nature of the respective design variables, i.e., standardized pipeline diameters and compressor sizes, and systematically derive disjunctive models based on Generalized Disjunctive Programming (GDP) that yield tight continuous relaxations. Disjunctive models are usually transformed into mixed-integer constraints using either a Big-M (BM) or a Hull-Reformulation (HR) [3]. The continuous relaxation of the (HR) is at least as tight as and generally tighter than the (BM), but it requires disaggregated variables and constraints which increase its size. Similar to Castro & Grossmann [4], we show that in this particular case the proposed equipment sizing models can be transformed by means of a *compact* Hull Reformulation. This compact reformulation does not require the introduction of disaggregated variables or constraints and, hence, yields the best possible reformulation of the disjunctive models.

In general, the composition of the extracted shale gas can vary significantly within a shale play. In the Marcellus play, for instance, the gas quality ranges from "wet gas", which contains heavier hydrocarbons such as ethane, butane or propane, to "dry gas", which is almost exclusively methane, i.e., nearly pipeline-quality gas. In practice, a cluster of shale wells will feed different gas qualities into a gathering system at varying production rates over time. Hence, the composition of the gas delivered to midstream processors or downstream transmission lines varies temporally. These variations can create major challenges since upstream operators have to satisfy strict gas quality specifications at delivery nodes. Hence, it is critical to consider composition variations in shale gas development areas. In light of this reality we emphasize that the long-term shale gas development problem is highly *quality-sensitive*, i.e., the quality of the gas extracted within a particular development area determines decisively which development strategies are profitable and which ones may not even be feasible. In this work we address the quality-sensitive shale gas development problem with multiple delivery nodes, i.e., mixing and splitting within the gathering system is explicitly permitted. This problem corresponds to a network flow pooling problem that we model as a nonconvex MINLP. The non-convexities arise from bilinear terms in the flow balances around splitting nodes. We present a solution strategy that relies on an MILP approximation coupled with a restricted MINLP that yields near-global solutions in a reasonable period of time.

The proposed optimization model is applied to a set of real-world case studies based on historic development data that is provided by one of the largest upstream operators in the Marcellus Shale region. The results demonstrate clearly that previous, uncoordinated development strategies led to over-sized gathering systems that were heavily under-utilized at times. Moreover, the case studies allow us to actually quantify the economic value of advanced, computational tools in this domain.

References

[1] U.S. Energy Information Administration (EIA). Annual Energy Outlook with Projections to 2040. April 2013.

[2] Goellner, J. F. Expanding the Shale Gas Infrastructure. *AIChE CEP* August 2012, 49-52.

[3] Grossmann, I.E.; Trespalacios, F. Systematic Modeling of Discrete-Continuous Optimization Models through Generalized Disjunctive Programming. *AIChE J.* 2013. *59* (9). 3276-3295.

[4] Castro, P. M.; Grossmann, I.E. Generalized Disjunctive Programming as a Systematic Modeling Framework to Derive Scheduling Formulations. *Ind. Eng. Chem. Res.* 2012, *51*, 5781-5792.

[5] Cafaro, D. C.; Grossmann, I. E. Strategic Planning, Design, and Development of the Shale Gas Supply Chain Network. *AIChE J.* 2014. doi: 10.1002/aic.14405.

422011 Optimal Design and Synthesis of Shale Gas Processing and Ngls Recovery Process

Abstract:

The past decades have witnessed the fast development of the global shale gas industry. It is predicted that shale gas production will be substantially expanded in the future, resulting in a need for additional facilities to absorb the burgeoning supplies of shale gas [1]. As midstream infrastructure in the shale gas value chain, shale gas processing plants separate valuable constituents, such as natural gas liquids (NGL), from raw shale gas feedstocks, and remove the undesired constituents to meet product specifications in the downstream distribution network. Multiple technology alternatives have been developed for various separation operations in a shale gas processing plant, but the optimal extent of constituent separation and removal, as well as the best technology selection and operation strategy has not been determined for the entire shale gas processing and NGLs recovery. Moreover, sustainability has emerged as a critical consideration in the design and synthesis of energy systems [2]. Therefore, it is worth research effort to develop sustainable shale gas processing plants that demonstrate the best process economics while minimizing the overall environmental impacts.

In this work, we propose a comprehensive superstructure of shale gas processing and NGLs recovery for the production of sale gas and NGLs from raw shale gas. The superstructure is divided into four major sections, namely acid gas removal, dehydration, NGLs recovery, and nitrogen rejection [3, 4]. Received raw shale gas is compressed before sent to the acid gas removal processes to separate carbon dioxide and hydrogen sulfide. The purification alternatives in this section are based on the same chemical absorption technology, but equipped with various absorbents. Later in the second section, the purified gas from acid gas removal undergoes dehydration, which can be achieved by three alternative processes using absorption or condensation technologies. Next, the dry sweet gas go through the third section to recover NGLs via a distillation column and further fractionate major constituents of NGLs through a distillation train. The distillation column for NGLs recovery, or the demethanizer, is operated under relatively low temperature, and requires a complex utility system. Accordingly, we consider several NGLs recovery process alternatives, which differ with each other in terms of the configuration of the heat exchanger network. In the last section, extra nitrogen is split from the gas product before the final sale gas can be sold to the downstream customers. In addition to the mentioned technologies and processes, sulfur recovery is also integrated into the superstructure to enhance the overall economic performance [5].

Based on the proposed superstructure, we develop a multi-objective mixed integer nonlinear programming (MINLP) model to determine the most cost-effective and environmentally sustainable design of a shale gas processing plant. This model includes four types of constraints. Mass balance constraints accounts for the mass relationships in each equipment units using continuous design variables and integer variables for technology selection [6, 7]. Energy balance constraints describe the energy conservation of all equipment units [8]. Following the life cycle optimization framework [9], we conduct a techno-economic analysis in the economic evaluation constraints and minimize the annual total cost per unit amount of sale gas produced as one of the objective functions to determine the economic viability of the optimal process design. The remaining constraints evaluate the environmental impact of the sale gas product based on a cradle-to-gate life cycle analysis [10]. Accordingly, the system boundary covers the emissions embedded in raw shale gas acquisition, transportation, and sale gas production [11]. Both direct greenhouse gas emissions in the shale gas processing plant and indirect greenhouse gas emissions in the background processes are considered and aggregated into a carbon footprint indicator [12]. The other objective function, therefore, is to minimize the carbon footprint indicator per unit amount of sale gas produced. The resulting multi-objective MINLP problem is solved using a global optimization framework [13, 14]. The optimal results are plotted in a Pareto-optimal curve which reveals the tradeoffs between the competing objectives.

References

[1] J. F. Goellner. (2012) Expanding the shale gas infrastructure. *Chemical Engineering Progress*. 49-59.

[2] B. R. Bakshi and J. Fiksel, "The quest for sustainability: Challenges for process systems engineering," *AIChE Journal,* vol. 49, pp. 1350-1358, Jun 2003.

[3] C. He and F. Q. You, "Shale Gas Processing Integrated with Ethylene Production: Novel Process Designs, Exergy Analysis, and Techno-Economic Analysis," *Industrial & Engineering Chemistry Research,* vol. 53, pp. 11442-11459, Jul 16 2014.

[4] C. He and F. You, "Toward more cost-effective and greener chemicals production from shale gas by integrating with bioethanol dehydration: Novel process design and simulation-based optimization," *AIChE Journal,* vol. 61, pp. 1209-1232, 2015.

[5] C. He, F. Q. You, and X. Feng, "A Novel Hybrid Feedstock to Liquids and Electricity Process: Process Modeling and Exergoeconomic Life Cycle Optimization," *AIChE Journal,* vol. 60, pp. 3739-3753, Nov 2014.

[6] B. Wang, B. H. Gebreslassie, and F. Q. You, "Sustainable design and synthesis of hydrocarbon biorefinery via gasification pathway: Integrated life cycle assessment and technoeconomic analysis with multiobjective superstructure optimization," *Computers & Chemical Engineering,* vol. 52, pp. 55-76, May 10 2013.

[7] D. J. Garcia and F. You, "Multiobjective optimization of product and process networks: General modeling framework, efficient global optimization algorithm, and case studies on bioconversion," *AIChE Journal,* vol. 61, pp. 530-554, 2015.

[8] B. H. Gebreslassie, M. Slivinsky, B. L. Wang, and F. Q. You, "Life cycle optimization for sustainable design and operations of hydrocarbon biorefinery via fast pyrolysis, hydrotreating and hydrocracking," *Computers & Chemical Engineering,* vol. 50, pp. 71-91, Mar 5 2013.

[9] F. Q. You and B. Wang, "Life Cycle Optimization of Biomass-to-Liquid Supply Chains with Distributed-Centralized Processing Networks," *Industrial & Engineering Chemistry Research,* vol. 50, pp. 10102-10127, Sep 7 2011.

[10] B. H. Gebreslassie, R. Waymire, and F. You, "Sustainable design and synthesis of algae-based biorefinery for simultaneous hydrocarbon biofuel production and carbon sequestration," *AIChE Journal,* vol. 59, pp. 1599-1621, 2013.

[11] Q. Zhang, J. Gong, M. Skwarczek, D. J. Yue, and F. Q. You, "Sustainable Process Design and Synthesis of Hydrocarbon Biorefinery through Fast Pyrolysis and Hydroprocessing," *AIChE Journal,* vol. 60, pp. 980-994, Mar 2014.

[12] J. Gong and F. Q. You, "Optimal Design and Synthesis of Algal Biorefinery Processes for Biological Carbon Sequestration and Utilization with Zero Direct Greenhouse Gas Emissions: MINLP Model and Global Optimization Algorithm," *Industrial & Engineering Chemistry Research,* vol. 53, pp. 1563-1579, Jan 29 2014.

[13] J. Gong and F. Q. You, "Global Optimization for Sustainable Design and Synthesis of Algae Processing Network for CO2 Mitigation and Biofuel Production Using Life Cycle Optimization," *AIChE Journal,* vol. 60, pp. 3195-3210, Sep 2014.

[14] J. Gong and F. You, "Value-Added Chemicals from Microalgae: Greener, More Economical, or Both?," *ACS Sustainable Chemistry & Engineering,* vol. 3, pp. 82-96, 2015.

424630 Design and Economical Evaluation of Polygen Process to Co-Produce Synthetic Natural Gas (SNG) and Ammonia

Abstract:

Design and Economical Evaluation of POLYGEN process to co-produce Synthetic Natural Gas (SNG) and Ammonia

Bor-Yih Yu and I-Lung Chien

Department of Chemical Engineering, National Taiwan University, Taipei 106, Taiwan

The steady state design and economical evaluation of poly-generation (POLYGEN) process to co-produce synthetic natural gas (SNG) and ammonia is studied in this work. POLYGEN has been a widely studied topic recently, in which several different chemicals could be produced parallel at the same time. One of the two products, SNG, holds very similar composition and heat value to typical natural gas, and can be used as a replacement in industrial and home usages. Another product, Ammonia, is one of the most important inorganic chemical over the world, which could be used as the precursor of various kinds of chemicals, as fertilizers, or as cleaning agent. In POLYGEN process, the relative production rates for different chemicals could be adjusted based on different market demands, daily usages, and also changing political strategies.

The items covered by this work are illustrated in the following. In coal-based POLYGEN processes, syngas is first generated from coal gasification, and the raw syngas is then passes the treating stage. In syngas treating section, the sulfur contents in raw syngas stream from gasification section is first removed. After that, the syngas is then split into two streams. One of them passes through water-gas-shift-reaction (WGSR) and CO_2 removal section, and the other bypasses. After almost completely reaction, the one that passes WGSR and CO_2 removal has high purity in H_2. To meet the required H_2/CO ratio for the feed into SNG plant, part of the syngas after treatment is mixed with the one that bypasses. After syngas treatment and composition adjustment, SNG and ammonia are produced individually by methanation and ammonia synthesis reaction, respectively. For ammonia synthesis, a novel Ru/C catalyst, which has higher activity than the traditional Fe-based catalyst at lower pressure, is used to catalyze the reaction. Besides, some waste heat could be recovered by generating steam, and the steam is used to drive turbine to produce electricity for usage inside the plant-wide process.

In our previous study (Yu and Chien, 2015)[1], we illustrated that the SNG production price is lower than liquefied natural gas (LNG) importation price in Taiwan. The SNG production price is about 10.563 (USD/MMBTU) in a SNG-only plant. By POLYGEN process to co-produce SNG and ammonia, the SNG production cost could become even lower. If 20% of syngas is used to produce ammonia, the SNG production price will drop to 9.518 (USD/MMBTU), and if 40% for ammonia production, the SNG production price will further drop to 7.156 (USD/MMBTU). Thus, although the POLYGEN process leads to increasing total capital investment, it still has positive influences from the economic aspects. Besides, the flexibility of shifting the production rate of SNG or ammonia gives the odds for adapting the changes in market demand.

References:

1. Yu, B. Y; Chien, I. L. Design and Economic Evaluation of a Coal-to-Synthetic Natural Gas Process. Ind. Chem. Res. Res. **2015**, 54, 2339-2352.

431902 Application of Exergy Efficiencies in Complicated Cryogenic Processes

Abstract:

Application of Exergy Efficiencies in Complicated Cryogenic Processes

Donghoi Kim[1], Truls Gundersen[1]

[1]Department of Energy and Process Engineering, Norwegian University of Science and Technology (NTNU), Kolbjoern Hejes vei 1B, NO-7491, Trondheim, Norway

donghoi.kim@ntnu.no

Our world today relies heavily on low temperature processes for energy, food, medical, and even space science. To improve such systems, an energy balance has been used as a traditional tool for performance measurement. However, the method based on the first law of thermodynamics does not consider the quality of energy. Unlike the traditional method, exergy analysis is a relatively new way of evaluating processes, which applies both the first and second law of thermodynamics. Exergy is the maximum available work acquired by bringing a system to equilibrium with its environment. Exergy analysis allows identifying where exergy is destroyed in a process, in other words, the location of entropy generation. This gives guidelines to improve efficiency by highlighting units having the largest exergy destruction. In cryogenic processes, refrigeration is produced by consuming compression power, and power is pure exergy. Thus, it is evident to use exergy analysis for low temperature processes.
Such exergy has a distinctive behaviour between above and below ambient temperature when it comes to heat transfer. Above ambient, the maximum work (exergy) acquired is always less than the heat transferred, but asymptotically approaching the value of heat with increasing temperature. In contrast, below ambient temperature, exergy of heat is increasing exponentially with decreasing temperature, so the exergy value can be equal and even larger than the amount of heat. Thus, attention has to be given when using exergy as a performance parameter due to this discontinuity in exergy of heat. In addition, the temperature and pressure based physical exergy affect each other when a process stream experiences changes in its temperature or pressure, which requires caution when applying exergy analysis. As physical exergy relates to ambient conditions, it is also important to know whether a process is operated above, across, or below ambient.

In exergy analysis, there have been various definitions suggested for exergy efficiency. However, the existing definitions of exergy efficiency do not fully handle the above-mentioned issues. Most of these efficiencies are defined and formulated for above ambient processes. Thus, a new exergy efficiency referred to as the Exergetic Transfer Effectiveness (ETE) has been developed in our research group by carefully defining exergy sources and sinks [1]. Depending on the operating temperature level of a system, this efficiency thoroughly identifies the components of the exergy sources and sinks by decomposing the physical exergy into temperature and pressure based exergy. As a result, the ETE can be used for any processes operating above, across, and below ambient.

The main objective of this paper is to compare the ETE with other exergy efficiency definitions for some selected natural gas liquefaction processes such as dual mixed refrigerant (DMR) processes. The calculation of both system and equipment efficiencies are performed in order to comprehend the main sources of exergy destruction. In addition, the extension of the ETE for process units having a change in chemical exergy is presented. A discussion is also made on the practicality of the new exergy efficiency for design and optimization of cryogenic processes.

References

[1] Marmolejo-Correa, D. and T. Gundersen, "Low Temperature Process Design: Challenges and Approaches for using Exergy Efficiencies", *Computer Aided Chemical Engineering*, vol.29, pp. 1909-1913, 2011.

413802 Hybrid Models of Distillation Units for Planning, Scheduling, and Real Time Optimization of Refinery Operations

Abstract:

Distillation models presently used in real time optimization are rigorous tray to tray models. For instance, a crude distillation unit models for real time optimization predicts accurately product distillation curves and other bulk properties (e.g. sulfur, density, etc.) by using approx. 10k to 20k of nonlinear equation. The model is capable of predictions which are within 1% of the actual equipment performance. On the other hand, models used for planning and scheduling are much simpler (e.g. swing cut model) but they provide a significantly lower accuracy. Discrepancies between these models make it difficult to reconcile the results between the planning and scheduling objectives and the objectives for RTO.

Our goal is to develop accurate and yet small in size, easy to converge distillation models which include mass and energy balances, as well as prediction of composition related properties of the product streams. Such models will enable determination of optimal crude selection and associated optimal operating conditions in production planning. In addition, these models are a step towards optimizing operating conditions while scheduling the operations. Being able to use the same model for RTO will ensure that the operating targets are consistent across planning, scheduling, and RTO.

In this work, we present methodology for development of hybrid models of distillation units. The goal is to achieve accuracy on par with detailed tray to tray models and to have much smaller number of equations. In addition, the model needs to be linear or nearly linear to make it easy to include it in the overall plant model for production planning and for scheduling.

Crude unit model (about 200 mostly linear equations) removes commonly made assumption that the front end/back end of the adjacent product from the CDU are equidistant from the crude distillation curve. Model performance is demonstrated on a crude unit consisting of a preflash tower, atmospheric tower, and of vacuum tower. Model predictions are within 1% to 2% of the rigorous tray to tray model which has been used as a surrogate for the plant.

We also present models of several additional types of distillation towers (about 30 to 40 equations per tower) and illustrate their accuracy by comparing them with rigorous simulations.

This work provides a basis for consistent determination of optimal operating conditions in production planning, production scheduling, and real time optimization.

(745b) Robust Refinery Planning Under Uncertainty

Colvin, M., Aspen Technology, Inc.
Apap, R. M., Carnegie Mellon University
Varvarezos, D., Aspen Technology, Inc.

Originally presented on: 11/12/2015 15:36:00 - 15:57:00

Refinery planning is the problem of determining what crudes to purchase and how they will be processed in order to meet all demand requirements and specifications while optimizing operations. Decisions on which crudes to run are typically made in two stages, contract and spot, with contract decisions having to be made in advance of operations and before uncertainty in spot crude or product prices can be resolved. Crude purchased on contract tends to represent the majority of crude processed at a refinery. This forms a basic hedging strategy that ensures: (i) feed availability, (ii) cost assurance, and (iii) potential price discount. This of course limits the potential profit upside and it also depends on the location. Coastal refineries with more optionality tend to have a larger portion of their feedstock purchased from the spot market.

Refinery planning is modeled as a mixed integer, non-linear optimization problem that can contain tens to hundreds of thousands of equations and variables. Due to the size of the models, making optimal contract purchase decisions using stochastic programming techniques can quickly become computationally infeasible. In this paper, we explore ways to make the two stage crude purchase decision process under uncertainty computationally attractive. In particular, we propose a specialized partition strategy that categorizes, samples and optimizes over multiple uncertain parameters such as benchmark crudes, crack spreads, and key equipment availability, while maintaining a computationally tractable size problem. The algorithm occurs in two stages, deciding the crudes to purchase on contract in the first stage and deciding on spot crude purchases in the second stage. At each stage of the optimization a curve demonstrating the probability of outcomes is generated for the user to better understand the risk of a decision set and to potentially identify alternate solutions. Finally, we apply this approach to a particular refinery plan where we demonstrate the benefits of the proposed strategy.

416657 Refinery Planning Under Uncertainty Integrated with a Nonlinear Crude Distillation Unit Model

Abstract:

In the refining industry, the profit made is mainly dependent on the crude oil procurements and plant operations. In order to enhance the profitability of a refinery, an optimization formulation subject to the refinery model is employed to plan the optimal crude purchases and decide the operations according to that procurement. However, two major challenges in optimizing this system are that the crude qualities are not known accurately at the time of purchase and the LP- based refinery model is relatively too simple to reflect the true behavior of the refining process. Hence, the objective of this research is to provide a systematic and efficient approach to maximize refining profitability in the presence of uncertainties using more realistic refinery model.

Towards this end, a nonlinear model using Geddes fractional index (FI) [1,2] is applied to describe the behavior of the crude distillation unit (CDU) for yield prediction, and integrated with the entire plant-wide model [3] with pooling processes. To incorporate uncertainties in the crude oil properties, the optimization scheme is modeled as a scenario-based two-stage stochastic program. In stage I, the best crude oil combination is selected among a set of candidates and their purchase amounts are determined to maximize the expected profit across all scenarios. In stage II, the uncertainties are realized and the optimal operations, such as the cut point temperatures for the CDU and flow rates in the pooling processes are determined according to the real qualities of the crudes, associated with each scenario, such that all important quality specifications are satisfied.

The proposed scheme takes into account uncertainties in the crude quality, nonlinearity from the pooling and units, and logical decisions, resulting in a large-scale mixed-integer non-linear programming (MINLP) formulation. The global optimization of such problems is extremely difficult because of the existence of many nonlinear, non-convex terms. We first show that our well-developed feasibility and optimality-based domain reduction techniques can substantially reduce the solution time of the state-of-the-art software, such as ANTIGONE [4], to obtain an ε-optimal solution for the nominal scenario. Then the non-convex generalized Benders decomposition (NGBD) methodology [5, 6] is further enhanced by the new adaptive piecewise convex relaxation [7, 8] to solve the stochastic program more efficiently. The results of the stochastic scheme are compared with the deterministic approach to demonstrate the benefits of expected profit maximization and variance reduction. Even though the complexity of the new optimization model is substantially increased due to the multi-scenario and non-convex formulation, the proposed enhanced NGBD approach is still able to find and verify a high quality solution with a small relative gap, whereas the traditional NGBD and other commercial software cannot find a comparable solution in several days.

Reference

[1] R. L. A. Geddes, A general index of fractional distillation power for hydrocarbon mixtures. AIChE Journal, vol. 4, 1958, 389-392.

[2] A.M. Alattas, I. E. Grossmann and I. Palou-Rivera, Integration of nonlinear crude distillation unit model in refinery planning optimization, Industrial and Engineering Chemistry Research, vol. 50, 2011, 6860-6870.

[3] J. P. Favennec, Refinery operation and management, Editions TECHNIP, Paris, 2001.

[4] R. Misener and C. A. Floudas. ANTIGONE: Algorithms for coNTinuous / Integer Global Optimization of Nonlinear Equations, Journal of Global Optimization, vol. 59, 2014, pp. 503-526.

[5] X. Li, A. Tomasgard and P. I. Barton, Decomposition strategy for the stochastic pooling problem, Journal of Global Optimization, vol. 54, 2012, pp. 765-790.

[6] X. Li, A. Tomasgard and P. I. Barton, Nonconvex generalized Benders decomposition for stochastic separable mixed-integer nonlinear programs, Journal of Optimization Theory and Applications, vol. 151, 2011, pp. 425-454.

[7] X. Li, Y. Chen, P. I. Barton, Nonconvex generalized Benders decomposition with piecewise convex relaxations for global optimization of integrated process design and operation problems, Industrial and Engineering Chemistry Research, vol. 51, 2012, 7287-7299.

[8] Y. Chen, Optimal design and operation of energy polygeneration systems, Ph.D. Thesis, 2013.

420139 Integrated Gasoline Blending and Delivery Operations: Short-Term Scheduling Via a Novel Unit-Specific Event-Based Continuous-Time Formulation and Global Optimization Approach

Abstract:

Oil refinery operations involve three main segments, namely crude oil storage and processing, intermediate processing, and product blending and distribution [1-3]. While the first involves upstream operations from ship arrivals to crude distillation, the last involves downstream operations from the blending of hydrocarbon fractions to the storage and delivery of final products. Gasoline is one of the most profitable products of an oil refinery and can account for as much as 60-70% of total profit [4-6]. An oil refinery typically blends several gasoline cuts or fractions from various processes to meet its customer orders of varying specifications. The large numbers of orders, delivery dates, blenders, blend components, tanks, quality specifications, etc. make this problem of blending and scheduling highly complex and nonlinear. A heuristic treatment of the nonlinear blending and complex combinatorics can lead to inferior schedules and costly quality give-aways. Thus, scheduling using advanced concept and techniques of mixed-integer optimization are imperative for avoiding ship demurrage, improving order delivery and customer satisfaction, minimizing quality give-aways, reducing transitions and slop generation, exploiting low-quality cuts, and reducing inventory costs. The blending recipe and scheduling decisions make this problem a non-convex mixed-integer nonlinear optimization model (MINLP).

Early work [7-8] on this problem focused mainly on optimal blending of various intermediate fractions and some additives to meet product quality specifications. They did not provide an integrated solution that considers resources allocation and temporal decisions as well [5]. Several efforts [1,4,6,9-10] have been made to address the problem of scheduling of gasoline blending operations. Most of them have considered only select aspects of the full gasoline-blending problem. An integrated treatment of recipe, blending, scheduling, and distribution is missing. Recently, Li et al. [5] and Li and Karimi [11] developed single-grid and multi-grid slot-based continuous-time formulations for the simultaneous treatment of recipe, blending, and scheduling. They incorporated many problem features such as non-identical parallel blenders, minimum run length and amount in a blend run, constant blending rates, changeover, one blender charging at most one tank at a time, multi-purpose product tanks, piecewise constant profiles for blend component qualities and feed rates, etc. They developed a schedule adjustment procedure to solve successive MILPs instead of the non-convex MINLP. Castillo and Mahalec [12-13] addressed the same problem as that of Li et al. [5] and Li and Karimi [11] and developed a discrete-time formulation. They also imposed additional operational features including minimum blend run thresholds. To solve the large-scale optimization model, they proposed the inventory pinch based solution algorithm which decomposed the entire problem into three levels. As a result, no global optimization approach has been presented to solve this overall problem.

In the presentation, we first develop a new unit-specific event-based continuous-time formulation for the integrated treatment of recipe, blending, and scheduling of gasoline blending and delivery operations. We include all operational features of Li et al. [5], Li and Karimi [11], and Castillo and Mahalec [12-13] such as non-identical parallel blenders, minimum blend length and amount in a blend run, constant blending rate in a blend run, product independent, production dependent, and product sequence dependent transition times, one blender charging at most one tank at a time, multi-purpose product tanks, piecewise constant profiles for blend component qualities and feed rates, etc. We use linear blending indices to develop appropriate linear blending correlations for properties. The non-convexities mainly arise from forcing constant blending rates during blend runs. To solve this model to e- (e.g., 1%) global optimality, we propose a hybrid global optimization approach in which we first allow blending rates to vary during blending runs, then we follow the adjustment procedure from Li et al. [5] and Li and Karimi [11] to obtain feasible solutions where constant blending rates are enforced. These feasible solutions are used as starting points to initialize commercial solvers (e.g., CONOPT [14]) to generate local optimal solutions. After that, we fix the constant blending rates to construct MILP problems and develop an iterative MILP and NLP procedure to improve solution quality until some termination criteria are met. Finally, we use the global optimization solver ANTIGONE[16] to further improve the solution quality if the e- global optimality is not obtained after the iterative MILP and NLP procedure. We solve 14 industrial-scale examples from Li and Karimi [11] to illustrate the effectiveness of our proposed formulation and hybrid global optimization approach. The computational results demonstrate that our proposed formulation does improve the MILP relaxation of Li and Karimi [11]. Furthermore, all 14 examples are solved to be 1% global optimality within reasonable computational effort.

Keywords: gasoline; blending; scheduling; non-convex; mixed integer nonlinear programming (MINLP); global optimization

References

[1] Pinto, J. M.; Joly, M.; Moro, L. F. L. Planning and Scheduling Models for Refinery Operations. *Comput Chem Eng.* **2000**, *24*, 2259-2276.

[2] Li, J.; Karimi, I. A.; Srinivasan, R. Robust Scheduling of Crude Oil Operations under Demand and Ship Arrival Uncertainty, Presented at the AIChE Annual Meeting, San Francisco, CA, Nov. 12-17, 2006.

[3] Shah, N. K.; Ierapetritou, M. G. Short-Term Scheduling of a Large-Scale Oil-Refinery Operations: Incorporating Logistics Details. *AIChE Journal* 2011, 57, 1570-1584.

[4] Jia, Z. Y.; Ierapetritou, M. Mixed-Integer Linear Programming Model for Gasoline Blending and Distribution Scheduling. *Industrial and Engineering Chemistry Research* **2003**, *42*, 825-835.

[5] Li, J.; Karimi, I. A.; Srinivasan, R. Recipe Determination and Scheduling of Gasoline Blending Operations. *AIChE Journal* **2010**, *56*, 441-465.

[6] Mendez, C. A.; Grossmann, I. E.; Harjunkoski, I.; Kabore, P. A. Simultaneous Optimization Approach for Off-Line Blending and Scheduling of Oil-Refinery Operations. *Computers and Chemical Engineering* **2006**, *30*, 614-634.

[7] Dewitt, C. W; Lasdon, L. S.; Waren, A. D.; Brenner, D. A.; Melhem, S. A. OMEGA: An Improved Gasoline

Blending System for Texaco. *Interfaces* **1989**, *19*, 85-101.

[8] Rigby, B.; Lasdon, L. S.; Waren, A. D. The Evolution of Texaco's Blending Systems: From OMEGA to Starblend. *Interfaces* **1995**, 25, 64-83.

[9] Glismann, K.; Gruhn, G.; Short-Term Scheduling and Recipe Optimization of Blending Processes. *Computers and Chemical Engineering* **2001**, *25*, 627-634.

[10] Joly, M; Pinto, J. M. Mixed-Integer Programming Techniques for the Scheduling of Fuel Oil and Asphalt Production. *Trans IChemE. Part A*. **2003**, *81*, 427-447.

[11] Li, J.; Karimi, I. A. Scheduling Gasoline Blending Operations from Recipe Determination to Shipping using Unit Slots. *Ind. Eng. Chem. Res.* **2011**, *50*, 9156-9174.

[12] Castillo, P. A. C.; Mahalec, V. Inventory Pinch Based, Multiscale Models for Integrated Planning and Scheduling-Part I: Gasoline Blend Planning. *AIChE Journal* **2014**, *60*, 2158-2178.

[13] Castillo, P. A. C.; Mahalec, V. Inventory Pinch Based, Multiscale Models for Integrated Planning and Scheduling-Part II: Gasoline Blend Scheduling. *AIChE Journal* **2014,** *60*, 2475-2497.

[14] Drud, A. A GRG Code for Large Sparse Dynamic Nonlinear Optimization Problems, *Mathematical Programming* **1985**, *31*, 153–191.

[15] Misener, R.; Floudas, C. A. ANTIGONE: Algorithms for Continuous/ Integer Global Optimization of Nonlinear Equations. *Journal of Global Optimization* **2014**, *59*, 503-526.

429759 Bitumen Upgrading and Synthetic Crude Oil Blending Optimization

Abstract:

Bitumen upgrading is an important procedure of oil sands production. Its main objective is to upgrade the extra heavy bitumen into light synthetic crude oil (SCO) that can be processed in refineries. Bitumen can also be blended with SCO or condensate to reduce the viscosity for pipeline transportation. The SCO and bitumen products have to satisfy certain specifications (e.g., Sulphur concentration, density, and viscosity) so as to be processed in downstream refineries. Simultaneously producing various SCO and bitumen products and taking full advantages of available resources to increase the economic benefits are the main motivations for the bitumen upgrading and synthetic crude oil blending optimization problem.

There are a number of publications related to the blending optimization problem in literature. To name a few, the gasoline blending and distribution scheduling problem was studied in [1]; the work of [2] addressed the simultaneous optimization of the off-line blending and the short-term scheduling problem in oil-refinery applications; the scheduling of crude oil operations under demand uncertainty is studied in [3]; in the work of [4], the crude oil scheduling problem was modeled as a control problem and the crude oil blending and inventory management was tackled using the receding horizon strategy. In general, mathematical modeling of the blending problem lead to challenging nonlinear optimization problem. Therefore, creative methods are needed to circumvent the difficulties for the optimization.

Although much work has been done in the area of blending optimization, they are mainly limited to the crude oil or product blending in the conventional refineries. This paper presents a novel method which addresses the optimization of blending in oil sands upgrading plant. While process unit model (e.g., Distillation unit, Coker, Hydrotreater) and blending relations for various properties [5] are the major parts of the mathematical model, the objective function is based on profits calculated from product sales minus raw bitumen cost, hydrogen costs and unit operation cost.

Presence of nonconvex terms (i.e., bilinear and trilinear terms) causes both local and global optimization solvers to fail. To address this issue, a multistep optimization strategy is proposed to tackle the optimization problem. First, a set of mass ratio before each blender can be found through a small nonlinear optimization problem in which product's qualities play the major role. By knowing the optimal mass ratio of each blender which can provide desired product specifications, blending rules will be redundant hereafter and the problem will be reduced into a linear optimization problem. The simplified case is easy to solve and the optimality of its results can be guaranteed. In the next step, the obtained results are applied as an initial guess of the original nonlinear nonconvex blending optimization problem to improve the current optimal point using either local or and global optimization solver in GAMS. The method is illustrated through a case study of bitumen upgrading process with Coker unit and the required data is adapted from simulation studies of bitumen upgrading process [6].

References

[1] Jia, Z., M. Ierapetritou. Mixed-integer linear programming model for gasoline blending and distribution scheduling. *Industrial & Engineering Chemistry Research* 42 (2003): 825-835.

[2] Méndez, C. A., Grossmann, I. E., Harjunkoski, I., & Kaboré, P. A simultaneous optimization approach for off-line blending and scheduling of oil-refinery operations. *Computers & chemical engineering*, *30*(2006), 614-634.

[3] J. Li, R. Misener, and C. A. Floudas. Scheduling of Crude Oil Operations Under Demand Uncertainty: A Robust Optimization Framework Coupled with Global Optimization. *AIChE Journal*, 58 (2012) 2373–2396.

[4] A. A. Hamisu, Y. Cao, and A. Kokossis, Scheduling Crude Oil Blending and Tanks Inventory Control under CDU Demand Uncertainty: A Receding Horizon Approach. Proceedings of the 20th International Conference on Automation & Computing, Cranfield University, Bedfordshire, UK, 12-13 September 2014.

[5] M. A. Fahim, T. A. Al-Sahhaf, and A. Elkilani. Fundamentals of Petroleum Refining. Elsevier Science & Tech, 2009.

[6] A. Alvarez-Majmutov and J. Chen. Simulation of bitumen upgrading processes. *Digital Refining*, 2 PTQ Q2 2013.

430374 Modelling Liquid Drug Product Manufacturing: Method and Case Study

Abstract:

Modelling drug product manufacturing processes both for optimizing production schedules as well as for simulating process enhancements is an increasingly important topic in times of rising economic pressure on drug manufacturers. In real-life, production schedules resulting from the execution of drug production manufacturing models can only be consulted reasonably, if a consolidation of all relevant production constraints is achieved in one model. Such constraints include demand, shelf-life, due date, workforce availability as well as utilities and equipment limited resources. On top of these constraints, production schedules are often rescheduled due to product reprioritizations, raw material scarcities and holding time limitations, e.g., sterile holding times. Hence, optimal scheduling on the shop floor level is a very complex task, which can be a cornerstone for increasing production efficiency.

In this work, the aforementioned challenges have been solved and a slot-based formulation is presented that was inspired by, and was applied to a case study of liquid drug product production line. While considering demand, shelf-life, due date, product-dependent processing rates and sterile holding times, the formulation allows finding the optimal batch order with respect to makespan minimization. This goal is achieved by the introduction of two operation modes, similar to the concept described by Moniz et al. (2014). The model includes batches produced individually, and the more time-efficient variant, where multiple batches are produced within a campaign, i.e., cleaning and sterilization activities are reduced. A second key feature of this model is the complexity reduction of modelling the filling processes for liquid products. Such fillings are performed in several hundred working steps; however, grouping the filling tasks into process elements that need to be performed without major interruptions, e.g., weekends, allowed an efficient modelling with only six process elements.

As an industrial case study, the proposed model was applied for scheduling a filling line for liquid drug products of Roche in Switzerland. Results were used to validate the model and optimize a three-months production schedule; including orders of various biopharmaceutical drug products and respecting phases of technical services as well as media-fills. Production plans resulting from the application of the scheduling model withstood an expert evaluation on executability with real life constraints, which provided the successful validation of the scheduling algorithm. The makespan for batches as scheduled in the sampled production schedule by company experts and the schedule provided by the presented algorithm differ by about 2 % of the overall planning horizon duration. Such a difference of few percentages on the line entails important benefits to the processing plant.

References: Moniz, Samuel; Barbosa-Póvoa, Ana Paula; de Sousa, Jorge Pinho (2014): Simultaneous regular and non-regular production scheduling of multipurpose batch plants: A real chemical–pharmaceutical case study. In Computers & Chemical Engineering 67, pp. 83–102. DOI: 10.1016/j.compchemeng.2014.03.017.

431990 Flexible Turnaround Planning in Integrated Chemical Site Networks

Abstract:

An integrated chemical site is a network of plants interacting with each other. Each plant fulfills one or more roles in the network producing raw materials, intermediates, and/or producing to external markets. The plants or sites are typically grouped based on specific business envelopes, but may interface across multiple businesses, envelopes or even other sites. Though each plant or envelope runs its own preferred routine and preventive maintenance over short time horizons, planning and coordinating large-scale preventive maintenance tasks across the entire integration network has significant potential benefits.

Amaran et al. [1] demonstrated long-term (1-5 years) turnaround planning for continuous processes by modeling a rolling horizon maintenance scheme as a mixed integer linear programming (MILP) problem. The benefits include long term productivity improvements across the network, lower peak manpower utilization, novel schedules that consider practical limitations like contract-based manpower, availability and other seasonal constraints. Amaran et al. [2] captured uncertainties in turnaround duration and its effect on manpower allocation, production planning and downstream inventory build-up prior to turnarounds. This was demonstrated in a medium-term (6-12 months) planning framework with a combination of robust optimization and multi-stage stochastic programming. This helped to refine grouped turnaround schedules from the long-term planning model.

This paper addresses a methodology of how to make adjustments to an existing turnaround plan in order to respond to unexpected changes in the system over a short and medium-term horizon, such as rapidly changing market conditions, manpower availability, variability in production rates over time, and unplanned outages. In practice, plants are constantly monitored as the processes and markets change with time. The states of the systems may diverge more than that anticipated at the time of the original turnaround planning. For example, a plant may maintain high catalyst activity longer than usual and the rescheduling of an upcoming turnaround may have the highest economic benefit for that individual plant, but other plants in the network may actually be impacted by a delayed turnaround. In another scenario, the demand for products and market prices for raw materials could change sufficiently to prompt ramp-up in production. The converse market conditions would prompt a ramp-down in production and also change turnaround economics. Therefore, the capability to evaluate flexibility in the turnaround schedule is needed.

However, re-planning the turnaround schedule in response to endogenous and exogenous factors brings about risks. Mathematical programming techniques naturally lend themselves to understanding the trade-off of these risks with both impacts on site integration and the possibility of replenishing production through turnaround investments. In this work, we rely primarily on a decision-rule based approach to incorporate flexibility in rescheduling a turnaround. At each intermediate time period, a decision as to whether to (i) continue operation, or (ii) perform a "pit-stop" turnaround, or (iii) perform a full turnaround is sought based on the associated risk of loss. Each task has its own associated costs, benefits, resources and probabilities of occurrence.

We propose a flexible turnaround framework that assesses the sensitivity of turnaround schedules, justifies and supplements site-wide approximations in long-term and medium-term models, and offers a holistic framework for maintenance and turnaround planning in integrated chemical sites. The methodology is demonstrated through a case study on a large example site network.

References:

1. Amaran, S., N. V. Sahinidis, B. Sharda, M. Morrison, S. J. Bury, S. Miller and J. M. Wassick, Long-term turnaround planning for integrated chemical sites, Computers and Chemical Engineering, 72, 145-158, 2015.
2. Amaran, S., T. Zhang, N. V. Sahinidis, B. Sharda and S. J. Bury, Medium-term maintenance turnaround planning under uncertainty for integrated chemical sites, Computers and Chemical Engineering, submitted, 2015.

AUTHOR INDEX

AUTHOR INDEX

AUTHOR INDEX

AUTHOR INDEX

AUTHOR INDEX

AUTHOR INDEX

AUTHOR INDEX

AUTHOR INDEX